Handbook of Food Preservation

FOOD SCIENCE AND TECHNOLOGY

A Series of Monographs, Textbooks, and Reference Books

1. Flavor Research: Principles and Techniques, *R. Teranishi, I. Hornstein, P. Issenberg, and E. L. Wick*
2. Principles of Enzymology for the Food Sciences, *John R. Whitaker*
3. Low-Temperature Preservation of Foods and Living Matter, *Owen R. Fennema, William D. Powrie, and Elmer H. Marth*
4. Principles of Food Science
   Part I: Food Chemistry, *edited by Owen R. Fennema*
   Part II: Physical Methods of Food Preservation, *Marcus Karel, Owen R. Fennema, and Daryl B. Lund*
5. Food Emulsions, *edited by Stig E. Friberg*
6. Nutritional and Safety Aspects of Food Processing, *edited by Steven R. Tannenbaum*
7. Flavor Research: Recent Advances, *edited by R. Teranishi, Robert A. Flath, and Hiroshi Sugisawa*
8. Computer-Aided Techniques in Food Technology, *edited by Israel Saguy*
9. Handbook of Tropical Foods, *edited by Harvey T. Chan*
10. Antimicrobials in Foods, *edited by Alfred Larry Branen and P. Michael Davidson*
11. Food Constituents and Food Residues: Their Chromatographic Determination, *edited by James F. Lawrence*
12. Aspartame: Physiology and Biochemistry, *edited by Lewis D. Stegink and L. J. Filer, Jr.*
13. Handbook of Vitamins: Nutritional, Biochemical, and Clinical Aspects, *edited by Lawrence J. Machlin*
14. Starch Conversion Technology, *edited by G. M. A. van Beynum and J. A. Roels*
15. Food Chemistry: Second Edition, Revised and Expanded, *edited by Owen R. Fennema*
16. Sensory Evaluation of Food: Statistical Methods and Procedures, *Michael O'Mahony*
17. Alternative Sweeteners, *edited by Lyn O'Brien Nabors and Robert C. Gelardi*
18. Citrus Fruits and Their Products: Analysis and Technology, *S. V. Ting and Russell L. Rouseff*

19. Engineering Properties of Foods, *edited by M. A. Rao and S. S. H. Rizvi*
20. Umami: A Basic Taste, *edited by Yojiro Kawamura and Morley R. Kare*
21. Food Biotechnology, *edited by Dietrich Knorr*
22. Food Texture: Instrumental and Sensory Measurement, *edited by Howard R. Moskowitz*
23. Seafoods and Fish Oils in Human Health and Disease, *John E. Kinsella*
24. Postharvest Physiology of Vegetables, *edited by J. Weichmann*
25. Handbook of Dietary Fiber: An Applied Approach, *Mark L. Dreher*
26. Food Toxicology, Parts A and B, *Jose M. Concon*
27. Modern Carbohydrate Chemistry, *Roger W. Binkley*
28. Trace Minerals in Foods, *edited by Kenneth T. Smith*
29. Protein Quality and the Effects of Processing, *edited by R. Dixon Phillips and John W. Finley*
30. Adulteration of Fruit Juice Beverages, *edited by Steven Nagy, John A. Attaway, and Martha E. Rhodes*
31. Foodborne Bacterial Pathogens, *edited by Michael P. Doyle*
32. Legumes: Chemistry, Technology, and Human Nutrition, *edited by Ruth H. Matthews*
33. Industrialization of Indigenous Fermented Foods, *edited by Keith H. Steinkraus*
34. International Food Regulation Handbook: Policy • Science • Law, *edited by Roger D. Middlekauff and Philippe Shubik*
35. Food Additives, *edited by A. Larry Branen, P. Michael Davidson, and Seppo Salminen*
36. Safety of Irradiated Foods, *J. F. Diehl*
37. Omega-3 Fatty Acids in Health and Disease, *edited by Robert S. Lees and Marcus Karel*
38. Food Emulsions: Second Edition, Revised and Expanded, *edited by Kåre Larsson and Stig E. Friberg*
39. Seafood: Effects of Technology on Nutrition, *George M. Pigott and Barbee W. Tucker*
40. Handbook of Vitamins: Second Edition, Revised and Expanded, *edited by Lawrence J. Machlin*
41. Handbook of Cereal Science and Technology, *Klaus J. Lorenz and Karel Kulp*
42. Food Processing Operations and Scale-Up, *Kenneth J. Valentas, Leon Levine, and J. Peter Clark*
43. Fish Quality Control by Computer Vision, *edited by L. F. Pau and R. Olafsson*
44. Volatile Compounds in Foods and Beverages, *edited by Henk Maarse*
45. Instrumental Methods for Quality Assurance in Foods, *edited by Daniel Y. C. Fung and Richard F. Matthews*
46. *Listeria*, Listeriosis, and Food Safety, *Elliot T. Ryser and Elmer H. Marth*
47. Acesulfame-K, *edited by D. G. Mayer and F. H. Kemper*
48. Alternative Sweeteners: Second Edition, Revised and Expanded, *edited by Lyn O'Brien Nabors and Robert C. Gelardi*
49. Food Extrusion Science and Technology, *edited by Jozef L. Kokini, Chi-Tang Ho, and Mukund V. Karwe*
50. Surimi Technology, *edited by Tyre C. Lanier and Chong M. Lee*
51. Handbook of Food Engineering, *edited by Dennis R. Heldman and Daryl B. Lund*
52. Food Analysis by HPLC, *edited by Leo M. L. Nollet*
53. Fatty Acids in Foods and Their Health Implications, *edited by Ching Kuang Chow*
54. *Clostridium botulinum*: Ecology and Control in Foods, *edited by Andreas H. W. Hauschild and Karen L. Dodds*

55. Cereals in Breadmaking: A Molecular Colloidal Approach, *Ann- Charlotte Eliasson and Kåre Larsson*
56. Low-Calorie Foods Handbook, *edited by Aaron M. Altschul*
57. Antimicrobials in Foods: Second Edition, Revised and Expanded, *edited by P. Michael Davidson and Alfred Larry Branen*
58. Lactic Acid Bacteria, *edited by Seppo Salminen and Atte von Wright*
59. Rice Science and Technology, *edited by Wayne E. Marshall and James I. Wadsworth*
60. Food Biosensor Analysis, *edited by Gabriele Wagner and George G. Guilbault*
61. Principles of Enzymology for the Food Sciences: Second Edition, *John R. Whitaker*
62. Carbohydrate Polyesters as Fat Substitutes, *edited by Casimir C. Akoh and Barry G. Swanson*
63. Engineering Properties of Foods: Second Edition, Revised and Expanded, *edited by M. A. Rao and S. S. H. Rizvi*
64. Handbook of Brewing, *edited by William A. Hardwick*
65. Analyzing Food for Nutrition Labeling and Hazardous Contaminants, *edited by Ike J. Jeon and William G. Ikins*
66. Ingredient Interactions: Effects on Food Quality, *edited by Anilkumar G. Gaonkar*
67. Food Polysaccharides and Their Applications, *edited by Alistair M. Stephen*
68. Safety of Irradiated Foods: Second Edition, Revised and Expanded, *J. F. Diehl*
69. Nutrition Labeling Handbook, *edited by Ralph Shapiro*
70. Handbook of Fruit Science and Technology: Production, Composition, Storage, and Processing, *edited by D. K. Salunkhe and S. S. Kadam*
71. Food Antioxidants: Technological, Toxicological, and Health Perspectives, *edited by D. L. Madhavi, S. S. Deshpande, and D. K. Salunkhe*
72. Freezing Effects on Food Quality, *edited by Lester E. Jeremiah*
73. Handbook of Indigenous Fermented Foods: Second Edition, Revised and Expanded, *edited by Keith H. Steinkraus*
74. Carbohydrates in Food, *edited by Ann-Charlotte Eliasson*
75. Baked Goods Freshness: Technology, Evaluation, and Inhibition of Staling, *edited by Ronald E. Hebeda and Henry F. Zobel*
76. Food Chemistry: Third Edition, *edited by Owen R. Fennema*
77. Handbook of Food Analysis: Volumes 1 and 2, *edited by Leo M. L. Nollet*
78. Computerized Control Systems in the Food Industry, *edited by Gauri S. Mittal*
79. Techniques for Analyzing Food Aroma, *edited by Ray Marsili*
80. Food Proteins and Their Applications, *edited by Srinivasan Damodaran and Alain Paraf*
81. Food Emulsions: Third Edition, Revised and Expanded, *edited by Stig E. Friberg and Kåre Larsson*
82. Nonthermal Preservation of Foods, *Gustavo V. Barbosa-Cánovas, Usha R. Pothakamury, Enrique Palou, and Barry G. Swanson*
83. Milk and Dairy Product Technology, *Edgar Spreer*
84. Applied Dairy Microbiology, *edited by Elmer H. Marth and James L. Steele*
85. Lactic Acid Bacteria: Microbiology and Functional Aspects, Second Edition, Revised and Expanded, *edited by Seppo Salminen and Atte von Wright*
86. Handbook of Vegetable Science and Technology: Production, Composition, Storage, and Processing, *edited by D. K. Salunkhe and S. S. Kadam*
87. Polysaccharide Association Structures in Food, *edited by Reginald H. Walter*
88. Food Lipids: Chemistry, Nutrition, and Biotechnology, *edited by Casimir C. Akoh and David B. Min*
89. Spice Science and Technology, *Kenji Hirasa and Mitsuo Takemasa*

90. Dairy Technology: Principles of Milk Properties and Processes, *P. Walstra, T. J. Geurts, A. Noomen, A. Jellema, and M. A. J. S. van Boekel*
91. Coloring of Food, Drugs, and Cosmetics, *Gisbert Otterstätter*
92. *Listeria*, Listeriosis, and Food Safety: Second Edition, Revised and Expanded, *edited by Elliot T. Ryser and Elmer H. Marth*
93. Complex Carbohydrates in Foods, *edited by Susan Sungsoo Cho, Leon Prosky, and Mark Dreher*
94. Handbook of Food Preservation, *edited by M. Shafiur Rahman*

Additional Volumes in Preparation

Food Safety: Science, International Regulation, and Control, *edited by C. A. van der Heijden, Sanford Miller, Maged Younes, and Lawrence Fishbein*

Fatty Acids and Their Health Implications: Second Edition, Revised and Expanded, *edited by Ching Kuang Chow*

Handbook of Food Preservation

edited by

M. Shafiur Rahman

Horticulture and Food Research Institute of New Zealand
Auckland, New Zealand

MARCEL DEKKER, INC. NEW YORK • BASEL

ISBN: 0-8247-0209-3

This book is printed on acid-free paper.

Headquarters
Marcel Dekker, Inc.
270 Madison Avenue, New York, NY 10016
tel: 212-696-9000; fax: 212-685-4540

Eastern Hemisphere Distribution
Marcel Dekker AG
Hutgasse 4, Postfach 812, CH-4001 Basel, Switzerland
tel: 44-61-261-8482; fax: 44-61-261-8896

World Wide Web
http://www.dekker.com

The publisher offers discounts on this book when ordered in bulk quantities. For more information, write to Special Sales/Professional Marketing at the address above.

Current printing (last digit):
10 9 8 7 6 5 4 3 2 1

PRINTED IN THE UNITED STATES OF AMERICA

Preface

Food preservation is an action or method of maintaining foods at a desired level of properties or nature for their maximum benefits. In general, each step of handling, processing, and storage affects the characteristics of food, which may be desirable or undesirable. Thus, understanding the effects of each preservation method on foods is critical in food processing. This book is the first definitive source for most of the methods of food preservation.

The processing of food is not as simple or straightforward as in the past. It is now moving from an art to a highly interdisciplinary science. A number of new preservation techniques are being developed to satisfy current demands of economic preservation and consumer satisfaction in nutritional and sensory aspects, convenience, absence of preservatives, low demand of energy, and environmental safety. Better understanding and manipulation of these conventional and sophisticated preservation methods could help to develop high-quality, safe products by better control of the processes and efficient selection of ingredients. Food processing needs to use preservation techniques ranging from simple to sophisticated; thus any food process must acquire the requisite knowledge about the methods and technology and then apply the skill to acquire and implement the technology. Keeping this in mind, this volume has been developed to contain fundamental and practical aspects of most of the preservation methods of food important to practicing industrial and academic food scientists, technologist, and engineers. Innovative technology in preservation is being developed in the food industry that can extend shelf life, minimize risk, or improve functional, sensory, and nutritional properties. The large and ever-increasing number of food products and new preservation techniques available today creates a great demand for a handbook of food preservation methods. This book emphasizes practical, cost-effective, and safe strategies for implementing preservation techniques and dissects the exact mode or mechanisms involved in each preservation method by highlighting the effect of preservation methods on food properties.

This book is divided into four parts. Part 1, "Preservation of Fresh Food Products," encompasses the overview of food preservation and postharvest handling of foods. This part stands alone for those who want a basic background in postharvest technology for foods of plant and animal origin. It also gives valuable background information on the classification of foods, causes of food deterioration, and classification of food preservation methods with the mode of their action.

Part 2, "Conventional Food Preservation Methods," presents comprehensive details on glass transition, water activity, drying, concentration, freezing, irradiation, modified atmosphere, hurdle technology, and use of natural preservatives, antioxidants, pH, and nitrites. Each chapter covers the mode of preservation actions and their applications in food products.

Part 3, "Potential Food Preservation Methods," details new and innovative preservation techniques, such as pulsed electric fields, ohmic heating, high-pressure treatment, edible coating, encapsulation, light, and sound.

Part 4, "Enhancing Food Preservation by Indirect Approach," describes areas that indirectly help food preservation by improving quality and safety. These areas are packaging, hazard analysis and critical control point (HACCP) techniques, and managing profit and quality. Packaging as an integral part of the processing and preservation of foods is discussed in detail.

This is the first book published that combines all methods of preservation in this major field of science and technology. It is an invaluable resource for practicing and research food technologists, engineers, and scientists and a valuable text for upper-level undergraduate and graduate students in food, agriculture/biological science, and engineering.

Writing a book is an endless process, so the editor would appreciate receiving new information and comments to assist in future compilations. I am confident that this book will prove to be interesting, informative, and enlightening to readers.

I would like to thank Almighty Allah for giving me life and the opportunity to gain knowledge to write this book. I would like to express deep appreciation to the numerous academics, scientists, engineers, and technologists who were generous with their comments on the idea of developing this book. Special thanks to all contributors for their efforts in preparing the manuscripts and for their cooperation and patience in meeting publishing deadlines. I wish to express my gratitude to HortResearch for giving me the opportunity and facilities to initiate such an exciting project. I am also grateful to Drs. Conrad Perera and John Shaw for their support. Appreciation is extended to everyone at Marcel Dekker, Inc., for continued assistance and successful development of this book, especially Ms. Maria Allegra and Mr. Rod Learmonth. I wish to express my gratitute to the HortResearch Library staff who assisted me patiently with online literature searches and interlibrary loans.

I sincerely acknowledge the sacrifices made by my parents, Asadullah Mondal and Saleha Khatun, during my early education. Appreciation is due all my teachers, especially Mr. Habibur Rahman, Professors Nooruddin Ahmed, Iqbal Mahmud, Khatiqur Rahman and Jasimuz Zaman, and Dr. Shahab Uddin for their encouragement and help in all aspects to pursue higher education and research. I wish to thank my relatives and friends, especially Professor Md. Mohar Ali Bepari and Dr. Md. Moazzem Hossain, for their continued inspiration. I am grateful to my wife, Shilpi, for her patience and support during this work and for so much love and understanding, and to my little children, Radhin and Deya, for allowing me to work at home

M. Shafiur Rahman

Contents

Contributors

Elizabeth A. Baldwin Citrus and Subtropical Products Laboratory, Agricultural Research Service, U. S. Department of Agriculture, Winter Haven, Florida

Gustavo V. Barbosa-Cánovas Biological Systems Engineering, Washington State University, Pullman, Washington

Titus De Silva Montana Wines Ltd., Auckland, New Zealand

Robert H. Driscoll Department of Food Science and Technology, The University of New South Wales, New South Wales, Australia

T. V. Gamage Auckland, New Zealand

M. Marcela Góngora-Nieto Biological Systems Engineering, Washington State University, Pullman, Washington

Leon G. M. Gorris Agrotechnological Research Institute, Wageningen, The Netherlands

Theodore P. Labuza Department of Food Science and Nutrition, The University of Minnesota, St. Paul, Minnesota

Gerard La Rooy Business Process Improvement, Auckland, New Zealand

Lothar Leistner International Food Consultant, Kulmbach, Germany

Aurelio López-Malo* Biological Systems Engineering, Washington State University, Pullman, Washington

Enrique Palou* Biological Systems Engineering, Washington State University, Pullman, Washington

Janet L. Paterson Department of Food Science and Technology, The University of New South Wales, New South Wales, Australia

Ronald B. Pegg Department of Biochemistry, Memorial University of Newfoundland, and PA Pure Additions, Inc., St. John's, Newfoundland, Canada

Herman W. Peppelenbos Agrotechnological Research Institute, Wageningen, The Netherlands

Anne Perera Food and Nutrition Consultancy Service, Auckland, New Zealand

**Current affiliation*: Departmento de Ingeniería, Química y Alimentos, Universidad de las Américas-Peubla, Cholula, Mexico

Conrad O. Perera Horticulture and Food Research Institute of New Zealand, Auckland, New Zealand

Jan Pokorný Department of Food Science, Prague Institute of Chemical Technology, Prague, Czechia

M. Shafiur Rahman Horticulture and Food Research Institute of New Zealand, Auckland, New Zealand

M. N. Ramesh Food Engineering Department, Central Food Technological Research Institute, Mysore, India

M. A. Rao Department of Food Science and Technology, Cornell University–Geneva, Geneva, New York

Fereidoon Shahidi Department of Biochemistry, Memorial University of Newfoundland, and PA Pure Additions, Inc., St. John's, Newfoundland, Canada

Eddy J. Smid Agrotechnological Research Institute, Wageningen, The Netherlands

Barry G. Swanson Department of Food and Science and Human Nutrition, Washington State University, Pullman, Washington

Humberto Vega-Mercado Washington State University, Pullman, Washington

Alfredo A. Vitali Fruthotec, Ital, Campinas, Brazil

1

Purpose of Food Preservation and Processing

M. SHAFIUR RAHMAN
Horticulture and Food Research Institute of New Zealand, Auckland, New Zealand

I. WHAT ARE FOODS?

Foods are materials, in a raw, processed, or formulated form, which are consumed orally by humans or animals for growth, health, and satisfaction or pleasure. Unlike drugs, there are generally no limitations on the amount of foods that may be consumed [6]. Chemically, foods are mainly composed of water, lipids, fats, and carbohydrates with small proportions of minerals and organic compounds. Minerals in the form of salts and organic substances are present in foods as vitamins, emulsifiers, acids, antioxidants, pigments, polyphenols, or flavors [9].

II. TYPES OF FOODS

Raw foods generally originate from two major sources: plant and animal. Borgstrom [2] classified foods as summarized in Table 1. The classification of foods by other professionals may differ from that in Table 1. Some foods holds an intermediate position, such as honey, being a plant product but collected and processed by animals, i.e., bees [2]. The different classes of foods are defined below:

Perishable foods: foods that spoil quickly, within 1 or 2 days (e.g., meat and fish).
Nonperishable foods: foods in which spoilage is relatively slow (e.g., fruits and vegetables).
Harvested foods: foods removed from the medium of immediate growth (plant, soil, or water).
Raw foods: foods in the earliest or primary state, after harvesting or slaughter, not having been subjected to any treatment apart from cleaning and size grading [1].
Fresh foods: harvested items retaining their complete structural integrity or state as a whole without perceptible evidence of physical, chemical, or microbiological change. The main steps in preparing fresh foods are washing, curing, cooling, some postharvest treatments, and storage in cold rooms or modified atmosphere.

TABLE 1 Classification of Common Raw Food Sources

I. Plant Origin
A. Grains or cereals: wheat, corn, sorghum barley, oats, rye millet, rice, buckwheat
B. Pulses: red kidney bean, lima bean, navy bean, pea, lentil, broad bean, cowpea, chickpea, vetch (fitches)
C. Fruits
a. Tropical fruits: banana, plantain, pineapple, papaya, guava, mango, passionfruit, breadfruit, avocado, zapote, cherimoya, naranjilla, Surinam cherry
b. Subtropical fruits
  i. Citrus fruits: orange, lemon, tangerine, grapefruit, citron, lime, kumquat
  ii. Others: figs, pomegranate, olives, persimmon, (cactus figs), peijabe
c. Deciduous fruits: pome (seed) fruits, apple, grapes, pear, quince
c. Stone fruits: peach, cherry, plum, apricot
d. Berries: strawberries, raspberries, blackberries, loganberries, boysenberries, cloudberries, cranberries, lingonberries (whortleberries) elderberries, black currants, red currants, gooseberries, rose hips
D. Melons and squashes: cantaloupe, honeydew, watermelon, squashes
E. Vegetables
a. Leafy vegetables: cabbage, Brussels sprouts, spinach, celery, artichoke, leeks, lettuce, endive, bamboo shoots, hearts of palm, herbs
b. Root vegetables: carrot, radish, parsnip, turnip, rutabaga, salsify
c. Seeds: green peas, green beans, lima beans, okra
d. Others: cauliflower, broccoli, cucumbers, onions, garlic, tomatoes
F. Tuber products: white potatoes, sweet potatoes (yams), taro cassava (maniok), jerusalem artichoke (topinambur), true yams (Dioscorea spp.) earth almonds
G. Nuts: almond, beech, Brazil nut, breadnut, butternut, cashew, chestnut, filbert, peanut (groundnut), pecan, pinole, pistachio, walnut
H. Fungi
a. Fat type: baker's yeast, brewers' yeast, food yeast
b. Protein type: champignon, truffles, morels, cantharels
I. Honey (nectar)

II. Animal Origin
A. Milk: cow, buffalo, goat, sheep, horse, reinder, caribou, camel, yak, elephant
B. Meat and entrails
a. Domesticated: beef, veal, pork, mutton, goat, reindeer, camel, llama, alpaca, rabbit, horse, yak, eland, saiga
b. Game and venison: deer, moose, elk, caribou, hare, hippopotamus, monkey, polar bear
C. Poultry
a. Domesticated fowl: chicken, turkey, duck, goose, pigeon, (squab), pheasant, guinea hen
b. Wild fowl: grouse, patridge, quail, pheasant, capercaillie, ptarmigan
c. Miscellaneous: swallow's nest
D. Eggs: chicken, duck, seagull
E. Fish (flesh, roe, livers)
a. Salt water
  i. Wild: herring, sardine, tuna, mackerel, salmon, cod, haddock, pollock, whiting, hake, redfish (ocean perch), plaice, sole, flatfish, dolphinfish, swordfish, shark, dogfish
  ii. Cultivated: milkfish
b. Fresh water
  i. Wild: pike, pike perch, whitefish, perch, trout, salmon, sturgeon
  ii. Cultivated: eel, carp, tench, catfish, buffalo fish, tilapia
F. Shellfish
a. Crustacean: crabs, lobsters, crayfish, crawfish, shrimp, prawns
b. Molluscan
  i. Pelecypods oyster, clam, blue mussels, scallop, squid, octopus
  ii. Gastropods: snails (escargots)
G. Miscellaneous: frogs' legs, rattlesnakes, turtles, sea cucumbers

(*continued*)

Table 1 Continued

J. Manna: ash tree, oak, tamarisk, alhagi
K. Sugars: sugar cane, sugar beet, maple syrup, palm sugar (date)
L. Oilseeds: soybean, olive, cottonseed, peanut, sunflower, palm kernels, coconut (copra) rapeseed, sesame
M. Seaweeds: laver, noro (*Porphyra* spp.), kombu (*Laminaria* spp.), wakame (*Undaria pinnatifida*)
N. Beverage ingredients: coffee, tea, cocoa, yerba mate, mint, fenugreek, tilia
O. Spices

Source: Ref. 2.

Minimally processed foods: foods having the quality of fresh foods without conventional food-preservation techniques (e.g., freezing, canning, drying) and are convenient to prepare, serve, and consume. The main steps are trimming, shelling, cutting, slicing, dicing, and storage in packaging at low temperatures to prolong quality.

Preserved foods: products changed little during manufacture, in which the main preservation methods do not change the individuality of the foods (e.g., canned, frozen, and dehydrated foods) [2].

Manufactured foods: products in which the raw substance loses its individuality one or more basic methods of preservation (e.g., sausages, cured meats, jams and marmalades, wines) [2].

Formulated Foods: products prepared completely based on mixing and processing of individual ingredients to result in a shelf-stable products (e.g., cakes, biscuits, breads, ice cream).

Primary derivatives or pure products: components obtained from the raw product through purification (e.g., starch from corn or potato, sugar from beets, fats and oils from seeds and fish).

Secondary derivatives: products derived by further steps from basic components, for example, fat hardened by hydrogenation.

Synthetic foods: products made through microbial or chemical synthesis, for example, vitamins, nutrients, etc.

Functional foods: can have a positive impact on an individual's health, physical well-being, slowing the aging process, and mental health.

Medical foods: can prevent and treat diseases.

III. WHAT IS FOOD PRESERVATION?

Food preservation involves action taken to maintain foods with the desired properties or nature for as long as possible. It lies at the heart of food science and technology, and it is the main purpose of food processing.

IV. WHAT TO PRESERVE?

It is important to know the properties one wants to preserve. One property may be important for one product, but detrimental for others. For example, collapse and pore formation occur during the drying of foods. This can be desirable or undesirable, depending on the desired quality of the dried product; for example, crust formation is desirable for long bowl life in the case of breakfast cereal ingredients, and quick rehydration is necessary (i.e., no crust) for instant soup ingredients. Structural collapse in dried foods releases encapsulated aroma and initiates lipid oxidation in the protective amorphous matrices formed during drying, thus structural collapse should be avoided.

V. WHY PRESERVE?

The processing and storage of all foods is vital for the continuous supply of foods during seasons and off-seasons. One very important consideration that differentiates the agricultural process from most other industrial processes is their seasonal nature. The main reasons for processing foods are:

To overcome mistakes in agriculture
To produce value-added products
To provide variation in diet

The agricultural industry produces raw food materials in different sectors. Mistakes in agricultural production can be overcome by avoiding wrong areas, times, and amounts of food raw materials as well as by increasing storage life using simple methods of preservation.

Value-added food products can give better-quality foods in terms of improved nutritional, functional, and sensory properties. Consumer demand for more healthful and convenient foods also affects food preservation and processing. For example, there is at present a decline in the consumption of animal and animal products and a growth in cereal and cereal-based products as well as fruit and vegetable products. Coupled with this there is an interest in more varieties of food and more exotic foods and beverages.

Food habits have been a very important aspect of human society since its origin. Food habits depend on socioeconomic and cultural factors. What one eats, why it is eaten, and how it is eaten were once thought to be solely a gastronomic issue. In fact, food is always more that just a source of nourishment. People eat not only to meet their physiological needs, but also in response to social needs and pressures. Basic appetite requires one to secure an adequate diet that will provide the energy and various nutrients necessary for metabolic functioning. Our nutritional needs are well understood. However, there are also psychological desires that are satisfied by the act of eating food that is pleasant to the senses. Changing trends and lifestyles invariably change food consumption patterns and behavior, which in turn is translated into diversified food habits. These changing trends in food consumption and lifestyle have further solidified food practices and patterns [3].

Eating should be pleasurable to the consumer, and not be boring. People like to eat wide varieties of foods having different tastes and flavors. Variation in the diet is necessary in underdeveloped countries in order to reduce reliance on a specific type of grain (e.g., rice or wheat). The consumption load on rice or wheat can be reduced by replacing them in the diet with, for example, potato. The gradual evolution of foods—particularly ethnic foods—is very much alive today, as ingredients are added or subtracted to create new foods. One example of this continuing progress is *fusion* foods, which combine the flavors of different regions and cultures [8].

VI. MAJOR PROCESSES OF FOOD DETERIORATION

During storage and distribution, foods are exposed to a wide range of environmental conditions. Environmental factors such as pressure, temperature, humidity, oxygen, and light can trigger several reaction mechanisms that may lead to food degradation. As a consequence of these mechanisms, foods may be altered to such an extent that they are either rejected by or harmful to the consumer [11].

Food handling and mechanical, physical, chemical, and microbial effects are the leading causes of food deterioration and spoilage. Damage is caused by the mishandling of foods during harvesting, processing, and distribution, which will lead to a reduced shelf life of foods. Bruising of fruits and

Table 2 Sources of Contamination

Prior to processing	1. Soils, seawater, and fresh water
	2. Raw materials (field and orchard)
	3. Harvest, storage
During and after processing	1. Processing equipment
	2. Added ingredients
	3. Water used in processing
	4. Processed food (packaging, storage, distribution)

Source: Refs. 2, 7.

vegetables during harvesting and postharvest handling leads to the development of rot. Crushing of dried snack foods during distribution seriously affects their quality. Tuberous and leafy vegetables lose water when kept in atmospheres with low humidity and subsequently wilt. Dried foods kept in high humidity may pick up moisture and become soggy [11]. Condensation of moisture on foods or a damp atmosphere favors microbial growth, occasionally promotes insect development, and may indirectly lead to deterioration resulting in destructive self-heating [2].

Mechanical damage is conducive to spoilage. Bruises and wounds are such defects, and they frequently cause further chemical and microbial deterioration. Peels, skins, and shells constitute natural protection against this kind of spoilage [2].

In the case of frozen foods, fluctuating temperatures are often destructive, for example, fluctuating temperatures cause recrystallization of ice cream, leading to an undesirable sandy texture. Freezer burn is a major quality defect in frozen foods that is caused by the exposure of frozen foods to fluctuating temperatures. These large fluctuations may cause a phase change by thawing or refreezing foods. Similarly, phase changes involving melting and solidifying of fats are detrimental to the quality of candies and other lipid-containing confectionery items. Shriveling occurs due to the loss of water from harvested fruit and vegetables.

The four sources of microbial contaminants are soil, water, air, and animals (insects, rodents, and humans). The sources of contaminants are shown in Table 2. The living organisms shown in Table 3 spoil food after harvesting during handling, processing, and storage. Spoilage microorganisms emanating from several major sources are shown in Table 3 [2].

Each microorganism has (1) an optimum temperature at which it grows best, (2) a minimum temperature, at which growth no longer takes place, and (3) a maximum temperature, above which all development is suppressed. Bacteria that grow particularly well at low temperatures are called

Table 3 Organisms that Spoil Foods

1. Microorganisms
   a. Fungi: mold and yeast
   b. Bacteria
   c. Phages
   d. Protozoa
2. Insects and mites
   a. Directly by eating (infestation)
   b. Indirectly by spreading diseases (fruitfly, housefly)
3. Rodents
   a. Directly by consuming food
   b. Indirectly by spreading disease

Source: Ref. 2.

Table 4 Storage Life of Some Fresh Foods at Normal Atmospheric Conditions

Food	Terminology	Normal storage life
Meats, fish, and milk	Perishable	1–2 days
Fruits and vegetables	Semiperishable	1–2 weeks
Root crops	Semiperishable	3–4 weeks
Grains, pulses, seeds, nuts	Nonperishable	12 months

psychrophilic (*cryophilic*) or low-temperature organisms. Bacteria with an optimum temperature of 20–45°C are *mesophilic*, and those with an optimum temperature above 45°C are *thermophilic* [2].

Microbial growth in foods results in food spoilage with the development of undesirable sensory characteristics, and in certain cases the food may become unsafe for consumption. Microorganisms have the ability to multiply at high rates when favorable conditions are present. Prior to harvest, fruits and vegetables generally have good defense mechanisms against microbial attack, however, after separation from the plant, they can easily succumb to microbial proliferation. Similarly, meat upon slaughter is unable to resist rapidly growing microbes [11]. The pathogenicity of certain microorganisms is a major safety concern in the processing and handling of foods in that they produce chemicals in foods that are toxic to humans. Their growth on foods may also result in undesirable appearance and off-flavors. Microbial or chemical contaminants are also of concern in food deterioration. Chemicals from packaging materials may also be a source of food contamination.

Several chemical changes occur during the processing and storage of foods. These changes may cause food to deteriorate by reducing its sensory and nutritional quality. Many enzymic reactions change the quality of foods. For example, fruits when cut tend to brown rapidly at room temperature due to the reaction of phenolase with cell constituents released in the presence of oxygen. Enzymes such as lipoxygenase, if not denatured during the blanching process, can influence food quality even

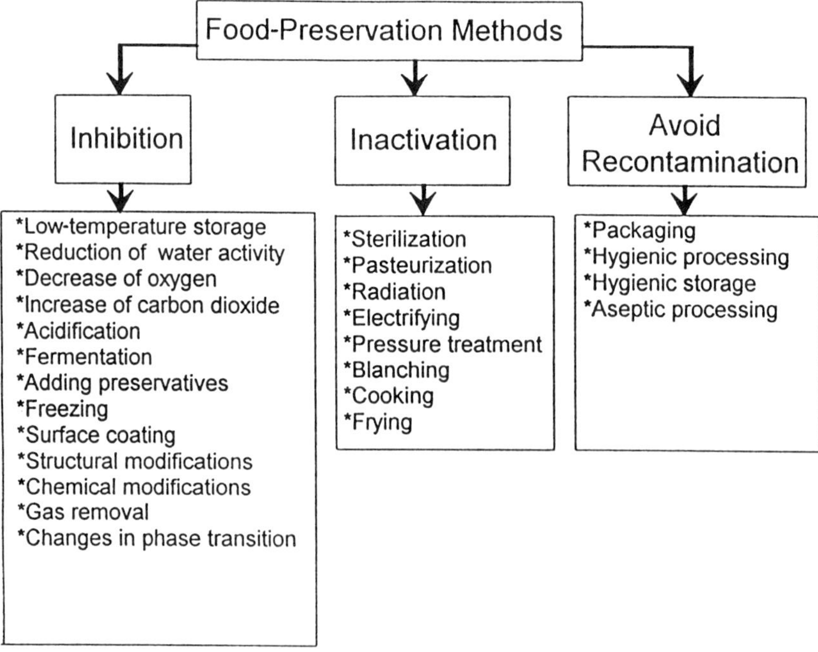

Figure 1 Major food preservation techniques. (From Ref. 4 and 5.)

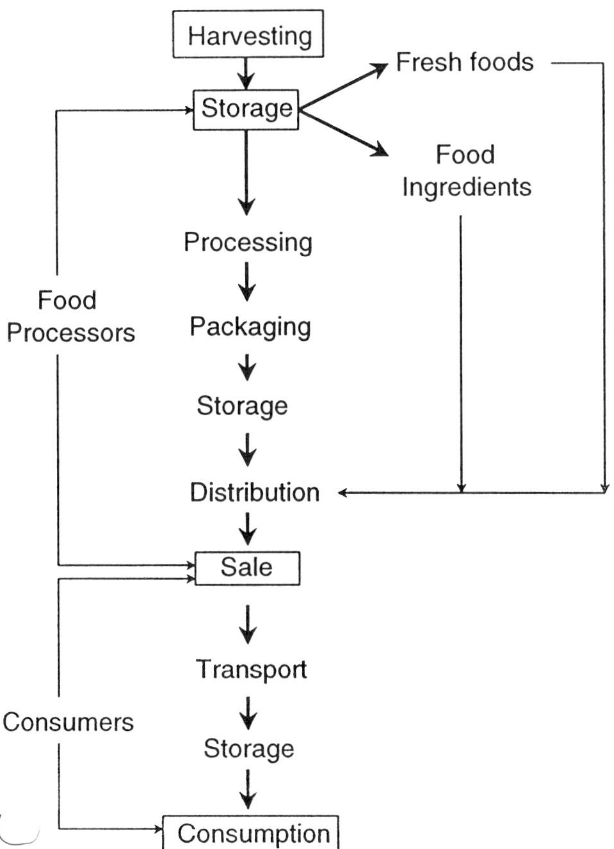

FIGURE 2 Various stages of food production, manufacture, storage, distribution, and sale.

at subfreezing temperatures. In addition to temperature, other environmental factors such as oxygen, water, and pH induce deleterious changes in foods that are catalyzed by enzymes [11].

The presence of unsaturated fatty acids in foods is a prime reason for the development of rancidity during storage as long as oxygen is available. While development of off-flavors is markedly noticeable in rancid foods, the generation of free radicals during the autocatalytic process leads to other undesirable reactions, for example, loss of vitamins, alteration of color, and degradation of proteins. The presence of oxygen in the immediate vicinity of food leads to increased rates of oxidation. Similarly, water plays an important role; lipid oxidation occurs at high rates at very low water activities.

Some chemical reactions are induced by light, such as a loss of vitamins and browning of meats. Nonenzymic browning is a major cause of quality change and degradation of the nutritional content of many foods. This type of browning reaction occurs due to the interaction between reducing sugars and amino acids, resulting in the loss of protein solubility, darkening of lightly colored dried products, and development of bitter flavors. Environmental factors such as temperature, water activity, and pH have an influence on nonenzymic browning [11].

VII. HOW TO PRESERVE?

Foods are perishable or deteriorative by nature. The storage life of fresh foods at normal atmospheric conditions is presented in Table 4. Some foods spoil readily, and others keep for longer but limited

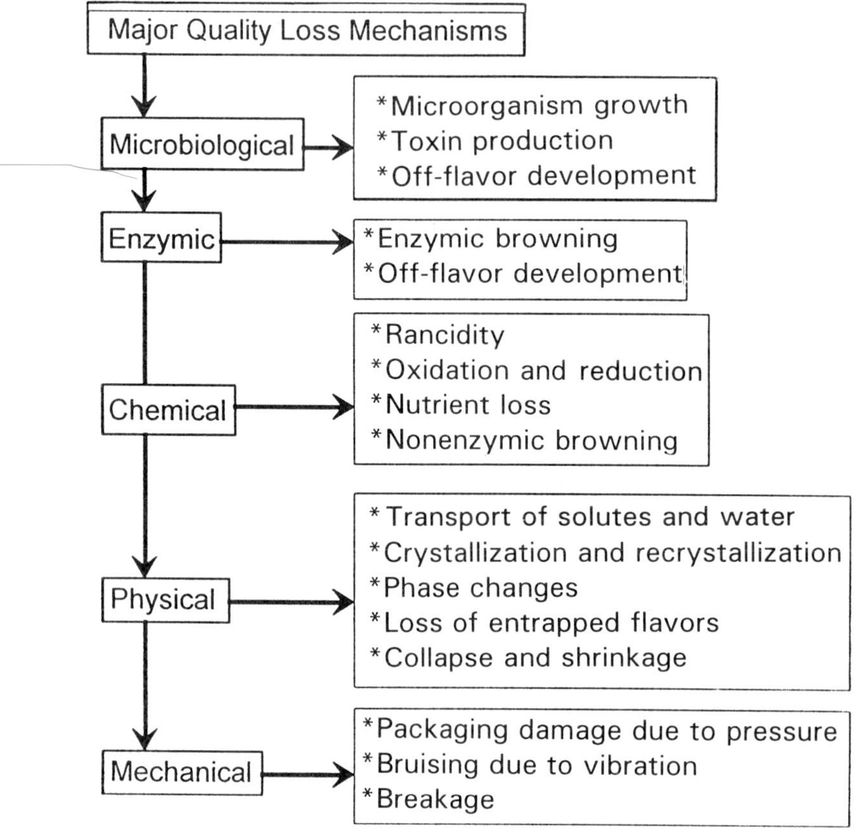

Figure 3 Quality-loss mechanisms. (From Ref. 4.)

periods [2]. Numerous changes take place in foods during processing and storage. In many cases deterioration is more insidious due to biochemical processes and may not immediately become visible [2]. It is well known that conditions in which foods are processed and stored may adversely influence the quality attributes in foods.

Based on the mode of action, major food-preservation techniques can be categorized as: (1) slowing down or inhibiting chemical deterioration and microbial growth, (2) directly inactivating bacteria, yeasts, molds, or enzymes, and (3) avoiding recontamination before and after processing [4,5]. A number of techniques or methods from the above categories are shown in Figure 1.

VIII. HOW LONG TO PRESERVE?

After storage for a certain period, one or more quality attributes of a food may reach an undesirable state. At that time, the food is considered unsuitable for consumption and is said to have reached the end of its shelf life. In studying the shelf life of foods, it is important to measure the rate of change of a given quality attribute [11]. The product quality can be defined using many factors, including appearance, yield, eating characteristics, and microbial characteristics, but ultimately the final use must provide a pleasurable experience for the consumer [10]. Loss of quality is very dependent on food type and composition, formulation (for manufactured foods), packaging, and storage conditions [4]. The various stages of food production, manufacture, storage, distribution, and sale are shown in Figure 2. Quality loss can be minimized at any stage and thus quality depends on the overall control

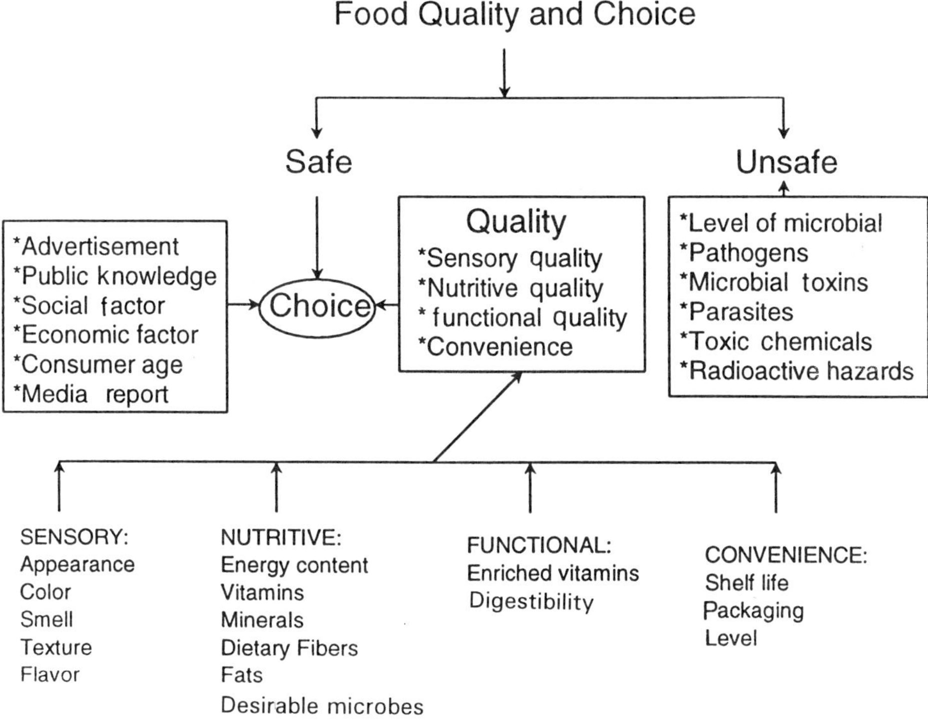

FIGURE 4 Factors affecting food quality, safety, and choice.

of the processing chain. The major quality-loss mechanisms and consequences are shown in Figures 3 and 4. When preservation fails, the consequences range broadly from extremely hazardous to color loss [4]. Thus, the length of preservation depends on the above factors. In many cases, very prolonged storage or shelf life is not needed, which simplifies both the transport and marketing of the foodstuff.

IX. FOR WHOM TO PRESERVE?

The preference for foods depends on age, sex, and cultural background. Nutritional requirements and food restrictions apply to different population groups. Food poisoning can be fatal, especially in infants, pregnant women, the elderly, and those with depressed immune systems. The legal aspects of food preservation are different for human foods and animal foods. Thus, it is necessary to consider the group for whom the products are being manufactured.

REFERENCES

1. M. Barnett and J. R. Blanchfield, What does it mean?, *Food Sci. Technol. 9*:93 (1995).
2. G. Borgstrom, *Principles of Food Science*, The Macmillan Co., London, 1968.
3. R. F. Florentino, Food habits and their role in food technology, *Development of Food Science and Technology in South East Asia* (O. B. Liang, A. Buchanan, and D. Fardiaz, eds.), IPB Press, Bogor, 1993, p. 13.
4. G. W. Gould, Introduction, *Mechanisms of Action of Food Preservation Procedures* (G. W. Gould, ed.), Elsevier Applied Science, London, 1989, p. 1.
5. G. W. Gould, Overview, *New Methods of Food Preservation* (G. W. Gould, ed.), Blackie Academic and Professional, Glasgow, 1995.

6. G. James, The therapeutic goods authority (TGA) perspective on functional foods, Foods of the Future Conference September 18–19, 1995, Sydney.
7. S. E. Martin and Z. J. Ordal, Sources of food spoilage microorganisms, *The Safety of Foods* (H. D. Graham, ed.), Avi Publishing, Westport, CT, 1980, p. 54.
8. D. Pszczola, Discovering new ethnic foods, *Food Technol. 50*:32 (1996).
9. A. Raemy and P. Lambelet, Thermal behaviour of foods, *Thermochim. Acta 193*:417 (1991).
10. J. G. Sebranek, Poultry and poultry products, *Freezing Effects on Food Quality* (L. E. Jeremiah, ed.), Marcel Dekker, New York, 1996, p. 85.
11. R. P. Singh, Scientific principles of shelf life evaluation, *Shelf Life Evaluation of Foods* (C. M. D. Man and A. A. Jones, ed.), Blackie Academic and Professional, Glasgow, 1994, pp. 3–24.

2

Postharvest Handling of Foods of Plant Origin

T. V. GAMAGE
Auckland, New Zealand

M. SHAFIUR RAHMAN
Horticulture and Food Research Institute of New Zealand, Auckland, New Zealand

I. HARVESTING FACTORS AND QUALITY

The ultimate quality of fresh or processed foods of plant origin depends on: (a) harvesting factors, (b) harvesting methods, and (c) postharvest handling procedures [39]. Quality attributes of food and processed products usually differ in many ways. Consumers judge fresh products by a different set of criteria than they do processed foods.

A. Preharvest Factors

Preharvest factors affect the nutritive value, postharvest yield, and quality. These factors include genetic factors, climatic factors, biotic, edaphic, chemical, and other factors, and the combined effects of these factors [39,84].

1. Genetic Factors

Cultivar and rootstock selection are important due to differenes in raw product composition, durability, and response to processing. For example, the type of rootstock used influences the quality of orange quality [84]. In many cases, fruit cultivars grown for fresh market sale are not the optimal cultivars for processing, such as canning, drying, and freezing [39]. New varieties have been developed to achieve higher yield, disease resistance, improved organoleptic properties, improved nutritional value, and reduced undesired toxic compounds [84].

2. Climatic Factors

The growing region and environmental or climatic conditions specific to each region significantly affect the quality of plant produce. These conditions include temperature, humidity, light, wind, soil texture, elevation, and rainfall [39,84]. Light intensity significantly affects vitamin concentration, and temperature influences the transpiration rate, affecting mineral uptake and metabolism [39]. The duration, intensity, and quality of light affect the quality at harvest. In tomatoes, leaf shading of fruits produces a deeper red color during the ripening period than in the case of tomatoes exposed to field light [22]. Moreover, tomatoes grown in full sunlight contain more sugar and dry matter than those grown in the shade [100]. In the case of citrus, fruits exposed to the sun were lighter in weight, with thinner rind, higher in solids, and lower in acids and juice than those that were shaded or grown inside a canopy. The differences in day length and light quality affect the product physiology (e.g., onion varieties developed for short-day climates will not produce large bulbs) [97]. In purple cabbage or eggplants, the formation of anthocyanin pigments is controlled by short-wavelength light in the blue and violent regions [68]. Salunkhe et al. [84] mentioned that in general, the effects of light on the synthesis of any nutrient in a plant depends on the plant species, the amount and duration of light, temperature, and soil fertility. Thiamine synthesis in plants is stimulated by light and generally occurs in the leaves and increases in concentration until the plant is mature. Turnips harvested in the morning contain more riboflavin than those harvested at other times of the day [84]. Fruits produced in regions having a cool climate usually are more acidic than those grown in warmer regions [90].

3. Environmental Factors and Cultural Practices

Environmental and cultural factors include soil type, soil nutrient and water supply, pruning, thinning, pest control or chemical spray, and density of planting [39,68,84]. Fertilizer addition may significantly affect the mineral content of fruit, while other cultural practices such as pruning and thinning may influence the nutritional composition by changing fruit crop load and size [39]. The closer the planting, the less sweet the fruits will be. Among leafy vegetables, leaves are larger and thinner under conditions of low light intensity [68]. Many physiological disorders may also be traced to the nutrient status of the soil, and respiration of harvested fruits is also affected by fertilization [68]. Potatoes grown in sandy, gravelly, or light loamy soils, low moisture conditions, or low-fertility soils have consistently higher dry matter content than those grown in peat or low-moisture soils. A high nitrogen:potassium ratio increases the tendency of the potato to darken after cooking. Phosphorus deficiency in soil also results in the darkening of potatoes. Pineapple plants receiving undue amounts of nitrogen produce tart, white, opaque fruits with poor flavor characteristics [84]. Pesticide spray residues may give rise to off-flavors in fresh and processed products. Excessive use of pesticides may even produce harmful metabolites and toxicity, which are not necessarily destroyed during processing [84].

B. Maturity at Harvest

Maturity at harvest is one of the primary factors affecting fruit, vegetable, and other crop compositions, quality, and storage life. Achieving optimum harvest maturity is vital to ensure maximum postharvest life of the produce. Although most fruits reach peak eating quality when harvested fully ripe, they are usually picked mature, but not ripe, to decrease mechanical injury during postharvest handling. Immature fruits are more subject to shriveling and mechanical damage and are of inferior quality when ripened. Overripe fruits are likely to become soft, mealy, and tasteless soon after harvest. Fruits picked either too early or too late in the season are more susceptible to physiological disorders and have a shorter storage life than those picked at midseason [39]. Both immature and overripe harvesting can cause extensive produce loss, thus maturity indices are important criteria for harvesting. The optimum maturity of produce for fresh consumption and processing is determined according to the purpose for which it is used. The exact degree of maturity considered best for canning may not be best for dehydration, freezing, or making jams or preserves [84]. For example, fully ripened fruits should be used for drying or preserve making to achieve the best flavor, but for fresh marketing they would be too susceptible to damage. For fresh market, fruits must be harvested at prime maturity.

Different indices are used to identify maturity, which vary among types and cultivars of the produce. The common methods used for maturity indices use (a) visual means (size and shape, overall color, skin color, flesh color, presence of dried outer mature leaves, drying of plant body, elapsed days from full bloom to harvest, mean heat units during development, development of an abscission layer, surface morphology and structure, and fullness of fruit), (b) physical means (ease of separation or abscission, flesh firmness, tenderness, specific gravity, or density), (c) chemical analysis (soluble solids, starch, acidity, sugar:acid ratio, juice content, oil content, tannin content), (d) computation (days from bloom in relation to date of bloom), and (e) physiological indices (respiration and internal ethylene concentration). It is beneficial to combine various indices of maturity to assess most accurately the stage at which the product may be harvested [39,41,70,83] In addition to harvesting indices, picking time and weather conditions also affect quality. Thus, the harvesting of fruits and vegetables during or immediately after rain should be avoided, and harvest is preferably carried out during the cooler part of the day (usually early morning) to avoid shriveling and wilting [77].

C. Harvesting Methods

Harvesting provides the first opportunity for mechanical damage of plant produce, therefore, the choice of harvesting method should allow for maintenance of quality [39]. Harvesting can be done by manually or mechanically. The advantages of manual harvesting are (a) accurate selection and grading according to maturity, (b) minimal damage to the product, (c) minimal capital investment, (d) mechanical devices can be used as aids. The disadvantages are that manual harvesting management of a labor force, and (b) is relatively slow. The advantages of mechanical harvesting are that it is fast and requires little labor management. The disadvantages are that fruits and vegetables may experience damage during handling, such as skin abrasion and tissue bruising and trained personnel and special field layout and cropping patterns are necessary. The use of improper machinery and equipment during mechanical harvesting may cause serious product losses [83]. When knives are used to harvest leafy vegetables or bananas, care must be taken to avoid mechanical injury. Pickers can be trained in methods of identifying produce that is ready for harvest [41].

The harvesting system used and its management have a direct effect on the incidence and severity of mechanical injuries. Management of the harvesting operation, whether manual or mechanical, can have a major impact on the quality of the harvested produce. Thus, management procedures should include the following for best results: (a) determining optimum time of harvest regarding fruit maturity and climatic conditions, (b) training and supervising of workers, and (3) effective quality control procedures [39].

Vegetables other than root crops should not be placed directly on the soil, nor should they be exposed to sunlight, heat, or rain. Such exposure can lead to a high internal temperature, which can be detrimental [79]. Simple shade or grass cover can provide protection for the harvested products until they are removed from the field. However, some root crops are benefited by brief exposure to the sun in order to dry off the surface or facilitate removal of adhering soil. [83].

II. POSTHARVEST PHYSIOLOGY

A. Classification of Foods of Plant Origin

Foods of plant origin can be classified based on different criteria. Plant parts used as food are broadly divided into three classes: (a) seeds and pods, (b) bulbs, roots, tubers, and corms (potatoes, sweet potatoes, yams, taro, cassava, beets, carrots, parsnips, radishes, turnips, salsify, onions, and garlic), and (c) flowers (cauliflower and broccoli), fruits (tomatoes, melons, eggplant, okra, peppers, beans, and peas), buds, stems, and leaves (cabbage, kohlrabi, collards, asparagus, rhubarb, lettuce, celery, and all leafy green) [83].

Fruits are commonly classified according to where they are grown: temperate zone, subtropical, and tropical. Kader and Barrett [39] list the following examples:

Temperate zone fruits: pome fruits (apple, pear, quince), stone fruits (apricot, cherry, nectarine, peach, plum), small fruits and berries (grape, strawberry, raspberry, blueberry, blackberry, cranberry)

Subtropical fruits: citrus fruits (grapefruit, lemon, lime, orange, pummelo, tangerine, and mandarin), noncitrus fruits (avocado, cherimoya, fig, kiwi, olive, pomegranate)

Tropical fruits: major tropical fruits (banana, mango, papaya, pineapple), minor tropical fruits (carambola, cashew apple, durian, guava, longan, lychee, mangosteen, passion fruit, rambutan, sapota, tamarind)

B. Postharvest Physiology

The postharvest period begins with the separation of food items from the medium of immediate growth or production and ends when the food enters the process of preparation for final consumption or further preservation [37]. When plant materials are attached to the parent plant (before harvest), losses due to respiration and transpiration are replaced by photosynthates and nutrients supplied from the plant [26]. The plant material continues to respire during postharvest period.

Fruits and vegetables pass through five distinct phases: (a) development (morphological and chemical completion of tissue), (b) young or premature (developmental period before the onset of maturation), (c) mature (completion or fullness of growth and edible quality; most of the maturation processes must be completed while the produce is still attached to the plant), (d) ripe (maximum aesthetic and edible quality), and (e) senescent (produce is worthless and unedible). The duration and rate of these stages vary according to the kind and variety of the product [70,83,84].

The large amounts of starch, acid, and phenolics present in the flesh of young or premature fruits make them inedible. At the young stage the flesh contains little sugar, i.e., sucrose, glucose, or fructose. The flesh cells enlarge considerably and sugar content increases, while starch, acid, and phenolic contents decrease as the fruits approach maturity. In addition, certain volatile compounds develop, giving the fruit its characteristic aroma and flavor. Chlorophyll degradation (loss of green color) and synthesis of carotenoids (yellow and orange colors) and anthocyanins (red and blue colors) take place in both the skin and the flesh with fruit ripening. All fruits soften as they ripen due to changes in cell wall composition and structure [39]. The details of compositional changes of fruits during maturation are given by Kader and Barrett [39]. The important enzymes found in foods of plant origins are given in Table 1.

1. Ripening

The ripening state is important since produce is edible at this stage. The major changes to occur during fruit ripening are (a) degradative, such as chlorophyll breakdown, starch hydrolysis, and cell wall degradation, and (b) constructive, such as formation of carotenoids and anthocyanins, synthesis of aroma volatiles, and ethylene formation [8]. The changes during ripening are structural or physical, chemical and nutritional, and biochemical or enzymic [84].

A wide range of structural changes occurs in fruits during ripening. These are [67,84]:

1. Changes occur in cell wall thickness, cell wall adhesion
2. Plasmalemma becomes permeable
3. Amount of intercellular spaces contribute to softening (prime indication of ripening)
4. Changes occur in plastids
5. Chloroplasts change, e.g., transform into chromoplasts
6. Changes occur in color and texture
7. Epicuticular wax forms a visible and distinctive structure
8. Cuticle becomes thicker
9. Epidermal hairs decrease or disappear
10. Endocarp is usually lignified

The details of compositional and morphological changes during ripening and senescence can be found in Mattoo et al. [55], Baker [4], Salunkhe et al. [84], and Kader and Barrett [39]. Chemical changes occur in carbohydrates, organic acids, amino acids and proteins, lipids, pigments, pectic substances, volatile components, enzymes.

Table 1 Enzymes Important to Plant Foods

Enzyme	Action	Result
Polyphenoloxidase	Catalyzes oxidation of phenolics	Formation of brown polymers
Polygalacturonase	Catalyzes hydrolysis of glycosidic bonds between adjacent polygalacturonic acid residues in pectin	Tissue softening
Pectinesterase	Catalyzes deesterification of galacturonans in pectin	Tissue firming
Lipoxygenase	Catalyzes oxidation of lipids	Production of off-odor and off-flavor
Ascorbic acid oxidase	Catalyzes oxidation of ascorbic acid	Loss of nutritional quality
Chlorophyllase	Catalyzes removal of phytol ring from chlorophyll	Loss of green color
Amylolytic enzymes	Hydrolysis of amylose and amylopectin	Breaking glucosidic linkages
Cellulases and hemicellulases	Hydrolysis of cell wall components	Breaking of glucosidic and xylosidic linkages
Proteases	Hydrolysis of proteins	Breaking of peptide bonds
Lipases and esterases	Hydrolysis of lipids and phospholipids	Breaking of ester bonds
Phytase	Action on phytate	Liberates phosphate groups
Glucose oxidase	Oxidation of glucose	Production of hydrogen peroxide

Source: Refs. 39, 76.

With advances in maturity, the pulp weight increases, with a gradual decrease in peel weight in the case of banana. With maturation, a gradual increase in acidity of the pulp is observed. At the initial stages of fruit growth, the concentration of total sugars, including reducing and nonreducing, is low. As maturity progresses, total sugars rise sharply with the appearance of glucose and fructose.

Each type of fruit has a unique assemblage of volatile compounds, and an increase in aroma volatiles contributes to the flavor, taste, and smell of the fruits. These volatiles are present in extremely small quantities (< 100 μg/g fresh weight). The total amount of carbon involved in the synthesis of volatiles is less than 1% of that expelled as carbon dioxide [39]. Volatiles are formed in climacteric fruits by ethylene, although ethylene itself does not have strong aroma and does not contribute to typical fruit aroma [39]. The main volatile compounds are divided into acids, alcohols, esters, carbonyls, aldehydes, ketones (low molecular weight compounds), and hydrocarbons In both climacteric and nonclimacteric fruit the most important aroma volatile compounds that increase during ripening are the esters. Pesis [72] listed different volatile components developed during the ripening process of fruits. The characteristic or optimum flavor developed at a specific stage of the ripening process. Feijoa is a very aromatic fruit that has its best aroma and flavor after natural abscission, but during storage it loses its special flavor [86]. Tomatoes change from green to red during ripening, and most of the aroma volatiles have a peak in concentration during the pink-to-red stage. In mango, the characteristic flavor appears to develop only after the half-ripe stage (climacteric stage), and the extent of flavor generation depends on the temperature conditions during storage [27]. There is also a dramatic increase in aroma volatiles in nonclimacteric fruits. Esters, acetals, alcohols, and aldehydes are formed in strawberry [92]. In pineapple there is a dramatic increase in ester production during ripening [28], while in citrus there are no dramatic changes in volatile production during storage. In grapefruit and pomelo, the changes in volatiles are very minor during storage, except for the important increase in nootkatone, which contributes significantly to the grapefruit flavor

Nonethylene, nonrespiratory organic volatiles may also have physiological and/or quality effects on fresh produce. Toivonen [94] reviewed these aspects in detail. These volatiles include terpenes, carboxylic acids, alcohols, aldehydes, sulfur compounds, ammonia, and jasmonates. The factors influencing accumulation of nonethylene volatiles are stress and injury, disease, varietal influence, and atmospheric composition [94]. Toivonen [94] states that potassium permanganate, extensively used for the removal of ethylene, cannot be used to remove these volatiles. Toivonen proposed that selective molecular sieves (synthetic zeolites) and boric acid may have potential in this regard. Removal of some volatiles, such as C_6-aldehydes, thiocyanates, and sulfur compounds, may lead to enhanced bacterial growth.

2. Transpiration or Water Loss

Transpiration is a process of water loss from fresh produce, resulting in a loss of salable weight, appearance (wilting and shriveling), textural quality (softening, flaccidity, limpness, crispness, and juiciness), and nutritional quality [39]. It is generally held that in fresh fruits and vegetables the signs of shrivel become objectionable when weight loss reaches about 5% of the harvested weight [69,83]. Most fruits and vegetables with 5–10% loss in moisture content are visibly shriveled as a result of cellular plasmolysis [84].

The transpiration rate depends on (a) the structure and condition of the surface, (b) the suface-to-volume ratio of produce, (c) the relative humidity and temperature of storage, (d) air movement or circulation, (e) atmospheric pressure, (f) the rate of cooling after harvest. The transpiration rate during refrigerated storage condition can be grouped into three ranges [96]: high (500–850 mg/kg hr mmHg; e.g., carrots and parsnips), intermediate (100–250; e.g., cabbage and rutabagas), and low (10–80; e.g., potatoes and onions).

Nature and Structure of the Surface

The outer protective covering (dermal system) of produce governs the regulation of water loss. The main sites of transpiration in plants are the hydathodes, stomata, epidermal cells, lenticels, trichomes (hairs), and cuticle. The surface structure consists of characteristics like the number of stomates in the epidermis, type of surface, tissues underlying the skin, and the structure and thickness of wax coating on the surface (cuticle). The cuticle is composed of surface waxes, cutin embedded in the wax, and carbohydrate polymers. The thickness, structure, and chemical composition of the cuticle vary greatly among products and among development stages of a given fruit or vegetable.

Stem scars are an important pathway for water loss in the tomato, whereas lenticles provide the pathway in potatoes. A hydrophobic waxy coating consisting of a complex and well-ordered structure of overlapping platelets provides greater protection against water loss than a thick, structureless coating. The epidermal cells underneath the waxy layer are also structured compactly with minimal space between adjacent cells. In leafy vegetables, small openings located at intervals in the epidermis called stomates allow water loss and gas exchange. Transpiration is faster in plants with a greater number of hairs, and very fine hairs may not modify the transpiring potential of a plant surface. Water loss can also be reduced by the application of waxes and other water-resistant skin coatings to the surface of the produce or by providing appropriate packaging such as plastic bags or film wraps [67,83,84,99].

Mechanical Damage

The surface cracks that may develop during growth or result from the mechanical handling of produce, such asa abrasions, impact bruises, vibration bruises, scratches, surface cuts, punctures, and other injuries that either remove or weaken the protective outer layers, also cause water loss. At early stages of maturity, some fruits and vegetables have the ability to repair and seal off the damaged area. The capacity of wound healing diminishes in most cases as the plant organs mature. However, some tuber and root crops retain their ability to seal off wound areas even after harvest. Moreover, mechanical damage causes a release of active enzymes, exposes internal tissue to environmental contamination, increases respiration, promotes chemical and enzymatic reactions (i.e., browning), allows the spread of harmful microorganisms, and induces an overall quality decline. Resistance to cracking in tomatoes is correlated with flatness of the epidermal cells [26,67,81,83,99].

A bruise is a simple and irreversible phenomenon, but defining all the ways in which bruising can be caused is quite complex. Impact bruising results when the dynamic failure stress or strain level for the produce tissue is exceeded. The factors affecting impact are drop height, velocity just before impact, mass of each produce item, radius of curvature of the impacting surfaces at the point of contact, stiffness of tissue beneath each surface, area of contact for each surface, and failure stress or strain level for each tissue [2].

Surface Area and Volume Ratio

The higher the ratio of surface area to unit volume, the greater is the loss of water by evaporation. Thus, at the same conditions a leaf will lose water and weight much faster than a fruit, and a small fruit or root or tuber will lose weight faster than a larger one [83].

Temperature, Humidity, and Air Movement (Environmental Factors)

Temperature, relative humidity, and air movement also affect water loss. In general, higher surface temperature increases the rate of transpiration, and higher relative humidity of storage atmosphere reduces it. Thus, it is common to lower the temperature or raise the relative humidity during storage. One practical way of minimizing transpiration is to cool the produce quickly via hydrocooling using antifungal chemicals, which will both cool the fruits and control the adhering fungal growth [84]. Surface condensation should be avoided to protect against mold growth. Air movement and atmospheric pressure also influence water loss [83]. Air circulation or velocity increases evaporation of moisture from the surface. Water evaporates more rapidly at lower atmospheric pressures [84]. Moreover, the types of packing and loading density and depth also affect water loss [48]. Thus, it is common to use low-temperature, high-humidity conditions to control transpiration.

3. Respiration

Respiration is an indicator of metabolic activity of living produce. All living organisms convert matter into energy through a fundamental process of life called respiration. In plants, respiration primarily involves the enzymatic oxidation of sugars to carbon dioxide and water, accompanied by a release of energy. Changes in proteins, lipids, and organic acids also take place through respiration [83]. The rate of deterioration (degree of perishability) of fruits is generally proportional to their respiration rate [39]. There are three phases of respiration: (a) breaking down of polysaccharides into simple sugars, (b) oxidation of sugars to pyruvic acid, and (c) aerobic transformation of pyruvate and other organic acids into carbon dioxide, water, and energy [73].

The rate of respiration is often a good indication of the storage life of a crop: the higher the rate, the shorter the life; the lower the rate, the longer the life [81]. The climacteric peak can be prolonged by reducing the rate of respiration. The average value of respiratory intensity was measured in 18 vegetable crops at 0°C; their intensity increased 1.7 times at 5°C, 2.8 times at 10°C, 4.5 times at 15°C, and 6.0 times at 20°C [16]. Burzo [16] classified vegetables into four groups and Kader and Barrett [39] into five, as shown in Table 2.

The respiration rate depends on both internal and external factors. The internal factors include (a) quantity of substrate (predominantly sugars), (b) size, shape, cell morphology, and maturity, (c) structure of the peel, (d) volume of the intercellular spaces, and (e) chemical composition of the tissue, which affects the solubility of oxygen and carbon dioxide. The external factors include (a) temperature, (b) availability of ethylene, oxygen, carbon dioxide, and growth regulators (a 3–5% reduction of oxygen concentration would not have an adverse effect on produce, but a comparable increase in carbon dioxide could suffocate and ruin certain fruits and vegetables; ethylene production rates of fresh fruits are reduced by storage at low temperature and by reduced oxygen [<8%] and/or elevated carbon dioxide above 1%), (c) fruit injury, disease, incidence, and water stress, and (d) removal of heat from respiration [39,73,83,84]. A classification of fruits and vegetables according to ethylene production rates can be found in Table 3.

Ethylene affects the physiological processes of plants. As a plant hormone, ethylene regulates many aspects of growth development and senescence and is physiologically active in trace amounts (<0.1 ppm). It is a natural product of plant metabolism and is produced by all tissues of higher plants and by some microorganisms [39]. Fruits can be divided into two groups based on their respiration pattern: nonclimacteric and climacteric. Examples of nonclimacteric and climacteric fruits and vegetables are given in Table 4.

Nonclimacteric Fruits

Nonclimacteric fruits ripen on the tree, and are not capable of continuing their ripening process once removed from the plant. These produce very small quantities of ethylene and do not respond to ethylene treatment, except in terms of degreening in citrus fruits and pineapples by degradation of chlorophyll [39].

Climacteric Fruits

Climacteric fruits can be harvested mature and ripened off the plant. These produce much larger quantities of ethylene in association with their ripening, and exposure to ethylene treatment will result in faster and more uniform ripening [39]. The respiration rate is minimal at maturity and remains rather constant, even after harvest. The rate will rise abruptly to the climacteric peak only when ripening is about to take place, after which it will slowly decline [73].

4. Undesirable Life Processes

Sprouting

Sprouting in vegetables, such as onion, ginger, garlic, and potatoes, can be a serious cause of quality loss. This process is related to dormancy and rest. Dormancy is a condition of quiescence due to internal or external factors, and rest is a phenomenon in which sprouting does not occur in spite of a favorable environment (e.g., potatoes that have no period of rest) [69,83].

Rooting

Rooting is initiated by high humidity, especially in roots and tubers. It may result in a rapid decay, shriveling, and exhausting of food reserves [69,83].

Table 2 Classification of Vegetables Based on Respiration Rate

Class		Respiration rate (mg/kg hr)		Example
		10°C	20°C	
I	Very low	<10	<40	Nuts, dates, dried fruits
II	Low	10	40	Potatoes, onions, cucumbers, apple, pear, kiwi, pomegranate, Chinese date (jujube)
III	Moderate	10–20	40–80	Peppers, carrots, tomatoes, eggplant, citrus fruits, banana, cherry, nectarine, peach, plum, avocado (unripe), persimmon
IV	High	20–40	80–120	Peas, radish, apricot, fig (fresh), avocado (ripe), cherimoya, papaya
V	Very high	>40	>120	Mushrooms, green onions, cauliflower, dill, parsley, melon, okra, strawberry, blackberry, raspberry

Source: Refs. 16, 39.

Table 3 Classification of Fruits and Vegetables According to Ethylene Production Rates

Class		Rate (ml C_2H_4/kg hr)	Commodities
I	Very low	<0.1	Artichoke, asparagus, cauliflower, cherry, strawberry, pomegranate, leafy vegetables, potato
II	Low	0.1–1.0	Blueberry, cranberry, cucumber, eggplant, okra, olive, pepper, persimmon, pineapple, pumpkin, raspberry, tamarillo, watermelon
III	Moderate	1.0–10.0	Banana, fig, guava, melon, honeydew, mango, plantain, tomato
IV	High	10.0–100.0	Apple, apricot, avocado, cantoupe, feijoa, kiwi, nectarine, papaya, peach, pear, plum
V	Very high	>100.0	Cherimoya, apple, passionfruit, sapota

Source: Ref. 40.

TABLE 4 Examples of Nonclimacteric and Climacteric Fruits and Vegetables

Class	Type	Example
I	Nonclimacteric	Berries (blackberry, cherry, cranberry, raspberry, strawberry), grape, citrus (grapefruit, lemon, lime, orange, mandarin, orange, and tangerine), pineapple, pomegranate, lychee, tamarillo, loquat, cucumber, fig, melon
II	Climacteric	Apple, pear, quince, persimmon, apricot, nectarine, peach, plum, kiwi, avocado, banana, plantain, mango, papaya, cherimoya, sapodilla, guava, passion fruit, pawpaw, tomato

Source: Refs. 39, 84.

Seed Germination

Seed germination of mature fruits is favored during storage of, for example, chayote, tomatoes, papaya, and pod-bearing vegetables. This process is detrimental to quality in some cases [69].

Greening

The exposure of potatoes to light during storage may produce green tissues containing solanine, which has toxic properties [83]. Thus, light intensity should be minimized.

Toughening

Green beans and sweet corn may toughen due to the development of spongy tissues when storage is unduly prolonged [83].

Elongation of Existing Structure

Asparagus, carrot, beet, and kohlrabi may show signs of elongation [69].

Tropic Response

The response to gravity and light may cause a bending of tissues. Products with uneven shapes are difficult to package, often adversely affecting their salability [83].

5. Physiological Disorders

Plants require a balance mineral intake from soil for proper development. Physiological disorders due to nutritional dificiencies may affect the skin, flesh or core region of the produce. Factors involved include maturity at harvest, cultural practices, climate during growing season, produce size, harvesting, and handling practices [84]. Deficiency of certain minerals (calcium, boron, potassium, copper, iron, cobalt) can result in certain disorders; for example, calcium deficiency causes bitter pit in apples [39].

6. Postharvest Diseases or Infections

Postharvest diseases are initiated in produce in the following ways: (a) at an early stage of development when attached to the plant, (b) by direct penetration of certain fungi or bacteria through the intact cuticle or through wounds and/or natural openings in the surface, and (c) through injuries in cut stems or damage to the surface. While most microorganisms can invade only damaged produce, a few are able to penetrate the skin of healthy tissue. Initially, only one or a few pathogens may invade and break down the tissues, followed by a broad-spectrum attack of several weak pathogens, resulting in complete loss of the commodity due to the magnified damage. Fungicides and bactericides are used to protect produce from fungi and bacteria. Postharvest diseases can be controlled b prevention of infection, eradication of incipient infections, and retarding the progress of pathogen in the produce [83,84].

The low pH of many fruits is a major factor influencing the composition of their microflora. In general, most yeasts and molds grow well under acid conditions, and thus fungi are often the predominant microorganisms in fruit products [90]. At least 215 species of yeast and fungi are important in foods [20], and it is estimated that 32 genera are associated with fruits and fruit products [90]. Only a few species of yeast are pathogenic for humans and other animals [65], and none of the pathogenic species are common contaminants of fruits or fruit products [90].

Mold-infected raw fruit may become soft after processing because pectinases were not inactivated by thermal treatment [64]. More than 100 different toxic compounds are produced by some 200 mold species [46]. For example, aflatoxins are primarily found in cereals and corn, cottonseed, and peanuts; patulin appears to be most common mycotoxin in processed fruits [90]. In general, bacteria that cause disease in humans are not associated with fruit products, and many do not tolerate the low pH of fruits. There are, however, a few exceptions [90].

7. Biochemical Changes

Chemical and enzymatic changes may cause tissue softening, off-flavors, pigment loss and off-colors, and an overall decline in nutritional value and taste [26].

III. POSTHARVEST HANDLING AND TREATMENT

A. Postharvest Factors Affecting Quality

Postharvest losses can occur when produce is handled in the orchard, during transportation, throughout the handling system (sorting, sizing, ripening, cooling, and storage), and between harvesting and consumption or processing [39]. Postharvest factors are, therefore, environmental atmosphere, handling methods, postharvest treatments, and time period between harvesting and consumption.

1. Environmental Atmosphere

Humidity

Water loss after harvest leads to rapid shriveling, wilting, and loss of crispness, and plant tissues become touch or mushy and eventually inedible. A direct loss in marketing is a result of the reduction of salable weight [83]. Relative humidity influences water loss, decay development, incidence of some physiological disorders, and uniformity of fruit ripening. Condensation of moisture on the surface of the produce (sweating) over long periods of time is probably more important than relative humidity in enhancing decay. The optimal or proper relative humidity is 0.85–0.90 for most fruits and 0.90–0.98 for most vegetables except for dry onions and pumpkins (0.70–0.75). Some roots require almost 1.0 relative humidity [26,39]. High relative humidity (RH) limits the weight loss of plant materials. The advantage of a high RH with respect to weight loss is offset by stimulation of mold growth on the surface.

Atmospheric Gas Composition

Components of the atmosphere, such as oxygen, carbon dioxide, and ethylene, can greatly affect respiration rate and storage life. In addition, some natural volatile components of the produce as well as added ones also affect the growth of microorganisms in produce. Reduction of oxygen and elevation of carbon dioxide may be intentional (modified or controlled-atmosphere storage) or unintentional. Such changes can promote or retard deterioration. The magnitude of such changes depends upon the commodity, cultivar, physiological age, oxygen and carbon dioxide levels, temperature, and duration of storage [39]. Storage in hypobaric or subatmospheric conditions can reduce the partial pressure of ethylene and oxygen, thus slowing the ripening and senescent process of fruits and vegetables [83].

Ethylene is a ripening hormone produced by almost all higher plants. It is of great importance in the postharvest physiology of perishable fruits and vegetables. Ethylene can cause serious disorders in leafy vegetables and flowers at very low concentrations. For example, head lettuce exposed to minute amounts of ethylene, 1 ppm or less, shows reddish-brown or olive spots within a few days. This disorder is called russet spotting. Thus, products sensitive to ethylene should never be held in the same storage room with products like apples and cantaloupes that emit great amounts of ethylene [83]. The ripening of tomatoes can be retarded by high carbon dioxide levels and low oxygen levels [7,33].

Ethylene is the atmosphere can be reduced by adequate ventilation or by forcing air through filters made of activated charcoal (brominated) or alkaline $KMnO_4$ [83].

Temperature

Temperature during storage: Temperature management is the most important tool for the extension of shelf life and maintenance of the quality of plant produce. Low-temperature storage can most effectively extend the shelf life of plant products. Storage at a few degrees above the freezing point is desirable to avoid freezing injury. In terms of optimum storage temperature, fresh fruits and vegetables can be broadly classified into four groups, as proposed by Floros [26] (Table 5). Storage temperatures of fruits and vegetables for short periods are given in Table 6. If the storage period is 5 days or less, relative humidity is maintained between 0.85 and 0.95, and the ethylene level is kept below 1 ppm by ventilating or using a scrubber [41]. The classification of fresh horticultural crops based on their relative perishability is given in Table 7.

Chilling injury: Chilling injury is a physical disorder induced by low, nonfreezing temperatures that occurs in certain susceptible plants or produce [84]. The consequences are (a) change in internal color (spots or areas may become soft and tan, brown, or black), (b) external discoloration, (c) pitting of surface or skin (early symptom of chilling injury), (d) uneven ripening, (e) off-flavor development, (f) accelerated incidence of surface molds and decay, (g) loss of nutritive value (e.g., β-carotene, provitamin A, and ascorbic acid) in some products, and (h) loss of ripening and increase in susceptibility to fungal spoilage.

The symptoms of chilling injury may not be evident while the produce is held at chilling temperatures, but they become apparent after transfer to a higher temperature. Fruits and vegetables that are native to subtropical or tropical areas are not adaptable to low temperatures, either in the field or in storage. The extent of chilling injury depends on four factors: (a) temperature and relative humidity (more injury at lower humidity), (b) duration of exposure to a given temperature, (c) low temperature sensitivity of the products, and (d) ethylene content. A fluctuation in storage conditions should be avoided [21,26,39,83,98]. The susceptibility of various fruits and vegetables to chilling injury is presented in Table 3.8.

The fatty acids play an important role in chilling injury, and injury to plant cell membranes is due to the change in the fluidity of membrane lipids. At higher temperatures, the membrane lipids are fluid or mobile with low viscosity, and below a critical temperature these lipids enter a gel-like state and become immobile. This affects the properties of the membranes, particularly the activities of the membrane-associated enzymes involved in energy production and protein synthesis and membrane permeability. Plants of tropical origin tend to have more highly saturated fats than plants of temperate regions, thus are more susceptible to chilling injury. Since membranes are composed of an ordered arrangement of lipid and protein molecules, changes in lipids may result in a change in membrane permeability [71,84].

Chilling injury can be controlled by temperature preconditioning, regulating humidity, maintaining a controlled atmosphere, and breeding of resistant varieties. The effect of the gradual lowering of temperature prior to storage is apparently related to the type of postharvest metabolism involved; this reduces chilling injury in some fruits, such as banana and avocado. Low humidity or increased transpiration aggravates the symptoms of chilling injury. Modification of the storage atmosphere also affects pitting. A concentration of 7% oxygen was optimal in preventing chilling injury. An atmosphere of pure oxygen produced severe pitting of limes, but one without oxygen resulted in even more [71]. Thus, the optimum concentration should be maintained during storage.

High-temperature treatment: Postharvest heat treatment also has a positive effect on fruit quality during storage [52]. Heating of apples to 38°C for 3 or 4 days before storage suppressed soft-

Table 5 Optimum Storage Temperature for Fresh Fruits and Vegetables

Class	Storage type	Optimum temperature (°C)	Commodity
I	Cold storage	0–5	Apples, apricots, artichokes, asparagus, beans (lima) beets, broccoli, Brussels sprouts, cabbage, cantaloupes, carrots, cauliflower, celery, cherries, collards, corn, dates, figs, grapes, kiwi, lettuce, mushrooms, nectarines, onions (green), oranges, peaches (ripe), pears[a] (ripe), peas, plums, radishes, rhubarb, spinach, strawberries, turnips
II	Cold storage	5–10	Avocados (ripe), beans (snap), blueberries, cranberries, cucumbers, eggplant, melons (ripe), okra, peppers, pineapple (ripe), squash (summer), tangerines
III	Slightly cool	10–18	Bananas, coconuts, grapefruit, limes, lemons, mangoes, melons (unripe), nuts, papayas, pears[a] (unripe), pumpkins, squash (winter), sweet potatoes, tomatoes
IV	Room temperature	18–25	Avocados (unripe), nectarines (unripe), onions (dry), peaches (unripe), potatoes, watermelons

[a]The optimum storage temperature for pears is 3–7°C (ripe) and 16–20°C (unripe).
Source: Ref. 26.

Table 6 Storage Temperatures for Short Periods

Class	Storage type	Temperature (°C)	Commodities
I	Low	0–2	Vegetables: anise, artichoke, argula*, asparagus*, bean sprouts, beet, Belgian endive*, bok choy, broccoli*, broccoflower*, Brussels sprouts*, cabbage*, cantaloupe, carrot*, cauliflower*, celeriac, celery*, chard*, chicory*, collard*, cut vegetables, daikon*, endive*, escarole*, garlic, green onion*, herbs (not basil), horseradish, Jerusalem artichoke, kale, kohlrabi, leek*, lettuce*, mint, mushroom, mustard green*, parsley*, parsnip, raddichio, radish, rutabaga, rhubarb, salsify, shallot, spinach*, snow pea*, sweet corn, sweet pea*, Swiss chard, turnip, turnip greens*, waterchestnut, watercress*; Fruits: apple, apricot, avocado (ripe), blackberry, blueberry, cherry, currant, cut fruits, date, fig, gooseberry, grape, kiwi*, nectarine, peach, pear (Asian and European), persimmon*, plum, prune, quince, raspberry, strawberry
II	Medium	7–10	Vegetables and melons: basil, beans (snap etc.), cactus leaves, cucumber*, eggplant*, Juan Canary melon, kiwano, okra*, pepper (bell and chili), squash (summer, soft rind*), tomatillo, watermelon*; Fruits: avocado (unripe), cactus pear (tuna), carambola, chayote, cranberry, feijoa, guava, kumquat, longan, lychee, mandarin, olive, orange, passionfruit, pepino, pineapple, pomegranate, tamarillo, tangelo, tangerine
II	Moderate	16–18	Vegetables and melons: casaba melon, casava, Crenshaw melon, dry onions, ginger, honeydew melon, honeydew melon, jicama, potato, Persian melon, pumpkin, squash (winter and hard rind), sweet potato*, taro, tomato (ripe mature green), yam*; Fruits: atemoya, banana, breadfruit, cherimoya, coconut, graprfruit*, lemon*, lime*, mango, mangosteen, papaya, plantain, pummelo, rambutan, sapote, soursop

*Products marked with an asterisk are sensitive to ethylene damage
Note: Relative humidity = 0.85–0.95 and ethylene < 1 ppm.
Source: Ref. 93.

TABLE 7 Classification of Fresh Horticultural Crops According to Relative Perishability and Potential Storage Life in Air at Near Optimum Temperature and Relative Humidity

Class	Perishability	Potential storage life (weeks)	Commodity
I	Very high	<2	Apricot, blackberry, blueberry, cherry, fig, raspberry, strawberry, asparagus, bean sprouts, broccoli, cauliflower, green onion, leaf lettuce, mushroom, muskmelon, pea, spinach, sweet corn, tomato (ripe), most cut flowers and foliage, minimally processed fruits and vegetables
II	High	2–4	Avocado, banana, grape, guava, loquat, mandarin, mango, melons (honeydew, Crenshaw, Persian), nectarine, papaya, peach, plum, artichoke, green beans, Brussels sprouts, cabbage, celery, eggplant, head lettuce, okra, pepper, summer squash, tomato (partially ripe)
III	Moderate	4–8	Apple and pear (some cultivars), grape (sulfer dioxide treated), orange, grapefruit, lime, kiwi, persimmon, pomegranate, table beet, carrot, radish, potato (immature)
IV	Low	8–16	Apple and pear (some cultivars), lemon, potato (mature), dry onion, garlic, pumpkin, winter squash, sweet potato, taro, yam, bulbs and other propagules of ornamental plants
V	Very low	>16	Tree nuts, dried fruits and vegetables

Source: Ref. 38.

ening [44,75] and decreased storage disorders such as superficial scald and bitter pit [14,53]. Prestorage heat plus calcium dip produced a synergistic benefit greater than either alone in maintaining fruit firmness [50] and decreasing storage disorders [42]. The synergistic effect of a heat treatment combined with a calcium dip was obtained only if the dip was conducted after heat treatment [45]. Lurie et al. [52] found that heating of apples caused formation of areas of amorphous wax and fewer surface cracks than unheated apples. A continuous wax layer on heated fruit decreased the movement of external calcium into the apple.

Low temperature inhibits the ripening process, but it may cause chilling injury [19]. Thus, a compromise temperature should be selected to avoid both chilling injury and ripening. The compromise temperature for most varieties of tomatoes is 12°C, but since ripening is not prevented at this temperature, most tomatoes can be stored at this temperature for only 2 weeks [34]. Temperatures above 35°C were found to inhibit fruit ripening [9,43]. However, storage of fruit at elevated temperatures may not feasible due to excessive weight loss and loss of normally ripening ability after extended periods [23,95]. Holding tomatoes for various times at 38°C before storage of up to 4 weeks at 2°C prevented chilling injury, while holding tomatoes at 20°C before storage did not affect chilling injury development. Temperatures higher than 38°C were not generally as effective as 38°C, and 42 or 46°C caused heat damage [51]. Typical heat treatments in hot water and hot forced air treatments for pathogen of fruits and vegetables are shown in Table 9.

Heating the surface of fruits to a few degrees below the injury threshold has been shown to eradicate or delay the development of incipient infections by pathogenic fungi [84]. Heat treatments are used for quarantine security against insects found on fresh fruits. Heat treatment can be done by using hot air, vapor heat, and hot water. Heat treatments have the advantages of low cost, relatively simple application equipment, and no chemical residues left on the treated commodity [84]. Vapor heat treatment is a relatively expensive operation in terms of the initial investment in equipment and operating cost. It requires an air-tight and moisture-proof treating room equipped with automatic temperature and humidity controls and a boiler for steam generation [1]. In general, both hot water and vapor heat treatment caused excessive damage and cause peel injury to grapefruit. Hot water produced more decay than forced hot air [57]. The hot water–treated grapefruit had a 5-fold higher incidence of aging than that treated with vapor heat, and using gibberellic acid before heat treatments reduced decay [59]. Surface injuries are generally favorable sites for infection by postharvest decay organisms. Brown et al. [13] found that heat treatments promoted lignin formation and accelerated peel injury healing, which also inhibits fruit rot in grapefruit.

Low-temperature treatment: Heather et al. [32] found that cold disinfestation treatment efficacy can meet international market requirements. They also found that in mandarins and tangelos disinfested by cold treatment at 1°C for 16 days, no fruit flies survived.

Compatibility in Mixed Loads

Often it is necessary to transport or store several types of produce at once. The problems created by shipping incompatible commodities together may be quite severe. The factors for determining the compatibility of products are (a) temperature, (b) relative humidity, (c) modification of the atmosphere, (d) release of volatile components harmful to other products, (e) products that produce or absorb objectionable odors or flavors, and (f) difficulties in loading shipping containers or storage facilities of different sizes and shapes [18,84]. Compatibility of fruits and vegetables for storage are given in Table 10.

2. Postharvest Handling Methods

Postharvest handling systems involve the channels through which harvested produce reaches the processing facility or consumer [39].Various postharvest handling systems are used for plant materials. These may vary with the type of product, end use of the product, and affordable or acceptable level of technology [58]. The efficient management of production, handling, storage, pricing, and marketing is linked with an integral production and distribution system [83]. Handling methods should be chosen such that they maintain quality and avoid undesirable decay.

Table 8 Chilling Injury Temperature of Fruits and Vegetables Stored at Moderately Low but Nonfreezing Temperatures

Class	Commodity	T_{inj}[a]	Injury when stored between 0°C and safe temperature[b]
A (0–5°C)	Apple (some cultivars)	2–3	Internal browning, brown core, soggy breakdown, soft scald
	Asparagus	0–2	Dull, gray-green, limp tips
	Avocados	4.5–13	Grayish-brown discoloration of flesh
	Beans (lima)	1–4.5	Rusty brown specks, spots, or areas
	Cranberries	2	Rubbery texture, red flesh
	Guavas	4.5	Pulp injury, decay
	Melons (cantaloupes)	2–5	Pitting, surface decay
	Melons (watemelons)	4.5	Pitting, objectionable flavor
	Oranges (California)	3	Pitting, brown stain
	Oranges (Arizona)	3	Pitting, brown stain
	Pomegranates	4.5	Pitting, external and internal browning
	Potatoes	3	Mahogany browning, sweetening
	Tamarillos	3–4	Surface pitting, discoloration
B (6–10°C)	Beans (snap)	7	Pitting and russeting
	Cucumbers	7	Pitting, water-soaked spots, decay
	Eggplants	7	Surface scald, alternaria rot, blackening of seeds
	Limes	7–9	Pitting, turning tan with time
	Melons (honeydew)	7–10	Reddish-tan discoloration, pitting, surface decay, failure to ripen

	Melons (Casaba)	7–10	Pitting, surface decay, failure to ripen
	Melons (Crenshaw)	7–10	Pitting, surface decay, failure to ripen
	Melons (Persian)	7–10	Pitting, surface decay, failure to ripen
	Okra	7	Discoloration, water-soaked areas, pitting, decay
	Olives (fresh)	7	Internal browning
	Papayas	7	Pitting, failure to ripen, off-flavor, decay
	Peppers (sweet)	7	Sheet pitting, alternaria rot on pods and calyxes, darkening of seed
	Pineapples	7–10	Dull green when ripened
	Pumpkins (hardshell)	10	Decay, especially alternaria rot
	Pumpkins (suashes)	10	Decay, especially alternaria rot
	Tomatoes (ripe)	7–10	Watersoaking and softening decay
C (11–20°C)	Bananas (green or ripe)	11.5–13	Dull color when ripened
	Grapefruit	10	Scald, pitting, watery breakdown
	Jicama	13–18	Surface decay, discoloration
	Lemons	11–13	Pitting, membranous staining, red blotch
	Mangoes	10–13	Grayish scaldlike discoloration of skin, uneven ripening
	Sweet potatoes	13	Decay, pitting, internal discoloration, hard core when cooked
	Tomatoes	13	Poor color when ripe, alternaria rot

[a]Approximate lowest safe temperature (°C).
[b]Symptoms often apparent only after removal to warm temperatures, as in marketing.
Source: Ref. 31.

Table 9 Typical Heat Treatments for Pathogen Disinfection of Fruits and Vegetables

Commodity	Temperature (°C)	Relative humidity	Time (min)	Possible injuries
A Hot water				
Apple	45		10	Reduced storage life
Grapefruit	48		3	—
Green beans	52		0.5	—
Lemon	52		5–10	—
Mango	52		5	No stem rot control
Melon	57–63		0.5	—
Orange	53		5	Poor degreening
Papaya	48		20	—
Peach	52		2.5	Mortile skin
Pepper (bell)	53		1.5	Slight spotting
B Hot forced air				
Apple	45	1.00	15	Deterioration
Melon	30–60	Low	35	Marked breakdown
Peach	54	0.80	15	—
Strawberry	43	0.98	30	—

Source: Ref. 3.

B. Postharvest Treatments

Wills et al. [99] found that the effectiveness of postharvest physical and chemical treatments depends on (a) the ability of the physical or chemical treatment (or agent) to reach the pathogen, (b) the level and sensitivity of the infection, (c) the sensitivity of the vegetables produced, and (d) the time of infection and extent of development of microbial decay. Physical and chemical postharvest treatments are discussed below.

1. Physical Treatments

Curing

Curing is a postharvest healing process of the outer tissues via development of a wound periderm, which acts as an effective barrier against further infection and water loss [12]. It is usually accomplished by holding the produce at high temperature and high relative humidity for several days while harvesting wounds heal and a new and protective layer of cells is formed [41]. Thus, the purposes of curing are (a) to cure and heal wounds of tubers and bulbs that occurred during harvesting, (b) to strengthen the skin, (c) to dry superficial leaves, such as on onion bulbs, which protect against microbial infection during subsequent storage and distribution, (d) to develop skin color in case of onion, and (e) to reduce water loss during postharvest in plant products, such as potatoes, sweet potatoes, cassavas, yams, onions, and garlic [83].

Curing is generally carried out at the farm level using high temperature and humidity at little or no cost [83]. While curing can be initially costly, the extension of storage life makes the practice economically worthwhile [41]. If local weather conditions permit, crops can be undercut in the field, windrowed, and left to dry for 5–10 days. The dried tops of the plants can be arranged to cover and shade the bulbs during the curing process, protecting the produce from excess heat and sunburn [41]. The best curing conditions for certain crops are given in Table 11. One day or less at 35–45°C and 0.60–0.75 relative humidity is recommended if forced heated air is used for curing onions and other bulbs [41].

Precooling

Prompt precooling conserves the weight of the product and reduces the rate of metabolic activity. Thus, rapid cooling produces acceptable appearance, texture, flavor, and nutritive value in strawberries [61]. Precooling is a means of removing field heat, and its aims are to quickly slow down respiration, minimize microbial growth, reduce transpiration rate, and ease the load on cooling system [84]. Common methods of precooling are room cooling, forced air cooling, package icing, hydrocooling, and vacuum cooling. Forced air cooling achieves very rapid cooling by placing product in a cold room. Package ice cooling is accomplished by filling containers with prescribed amounts of crushed ice or a liquid water–ice slurry combination. Hydrocooling is a rapid way to cool large batches of product by spraying or flooding the commodity with near-freezing water. The near-freezing water cools the product about 15 times faster than air, allowing for greater harvesting and marketing flexibility. The above methods of precooling may cause chilling injury if the temperature is below optimum. Highly perishable fruits, such as strawberries, bush berries, and apricots, should be cooled to near 0°C within 6 hours of harvest. Other fruits should be cooled to their optimum temperature within 12 hours of harvest [39].

Cleaning, Disinfection, and Rinsing

Additional steps, such as washing with disinfectants, and applications of waxes or other protective agents are also used in plant products. The main goals of cleaning are to (a) eliminate contamination, such as insects, surface dirt, and soil particles, (b) separate pesticides, fertilizers, and residues of other field-applied chemicals from the product, (c) reduce the microbial load of the incoming material, and (d) enhance the appearance of the product. Washing may have beneficial or detrimental effects. Some commodities, such as strawberries, mushrooms, and cherries, are not generally washed since it would encourage microbial growth. Washing cucumbers reduces the storage life to half that of unwashed

Table 10 Compatiblity Groups for Storage of Fruits and Vegetables

Group	Temperature (°C)	RH	Commodity
1	0–2	0.90–0.95	Apples, apricots, beets (topped), berries, (except cranberries), cashew apple, cherry (Barbados), cherries, coconuts, figs (not with apples), grapes (without sulfur dioxide), horseradish, kohlrabi, leeks, longan, loquat, lychee, mushrooms, nectarines, oranges[a], parsnips, peaches, pears, pears (Asian), persimmons, plums, pomegranates, prunes, quinces, radishes, rutabagas, turnips
2	0.2	0.95–1.00	Amaranth[b], anise[b], artichokes[b], asparagus, bean sprouts, beets[b], Belgian endive, berries (except cranberries), bok choy, broccoli[b], Brussels sprouts[b], cabbage[b], carrots[b], cauliflower, celeriac[b], celery[b], cherries, corn (sweet)[b], daikon[b], endive[b], escarole[b], grapes (without sulfur dioxide), horseradish, Jerusalem artichoke, kiwi, kohlrabi[b], leafy greens, leeks[b] (not with figs or grapes), lettuce, lo bok, mushrooms, onions[b] (green, not with figs, grapes, mushrooms, rhubarb, or corn), parsley[b], parsnips[b], peas[b], pomegranate, raddichio, radishes[b], rhubarb, rutabagas[b], salsify, scorzonera, snow peas, spinach, turnips[b], waterchestnut, watercress[b]
3	0.2	0.65–0.75	Garlic, onions (dry)

4	4.5	0.90–0.95	Cactus leaves, cactus pears, caimito, cantaloupes[b], clementine, cranberries, lemons[a], lychees, kumquat, mandarin[a], oranges (California and Arizona), pepino, tamarillo, tangelos[a], tangerines[a], ugli fruit[a], yucca root
5	10	0.85–0.90	Beans, calamondin, chayote, cucumber, eggplant, haricot vert, kiwano, malanga, okra, olive, peppers, potatoes, pummelo, squash (summer, soft shell), tamarind, taro root
6	13–15	0.85–0.90	Atemoya, avocados, babaco, bananas, bitter melon, black sapote, boniato, breadfruit, canistel, carambola, cherimoya, coconuts, feijoa, ginger root, granadilla, graprfruit, guava, jaboticaba, jackfruit, langsat, lemons[a], limes[a], mamey, mangoes, mangosteen, melons (except cantaloupes), papaya, passionfruit, pineapple, plantain, potatoes (new), pumpkin, rambutan, santol, soursop, sugar apple, squash (winter, hard shell), tomatillos, tomatoes (ripe)
7	18–21	0.85–0.90	Jicama, pears (for ripening), sweet-potatoes[c], tomatoes (mature green), watermelon[c], white sapote, yams[c]

Group 1: Many products in this group produce ethylene.
Group 2: Many products in this group are sensitive to ethylene.
Group 3: Moisture will damage these products.
Group 4 and 5: Many of these products are sensitive to ethylene and sensitive to chilling injury.
[a]Citrus treated with biphenyl may give odors to other products.
[b]Can be top iced.
[c]Separate from pears and tomatoes due to ethylene sensitivity.
Source: Ref. 56.

Table 11 Optimum Conditions for Curing of Different Crops

Commodity	Temperature (°C)	Relative humidity	Days required	Ref.
Cassava	30–40	0.90–0.95	2.5	41
Cassava	25–40	0.80–0.85	7–14	78
Potato	15–20	0.90–0.95	5–10	41
Sweet potato	30–32	0.85–0.95	4–7	41
Sweet potao	29–32	0.80–0.90	4–7	78
Sweet potato	30–33	0.85–0.95	5–7	78
Yams	32–40	0.90–1.00	1–4	41

fruit. Prolonged washing may produce a water-soaked appearance, and moisture penetration may aid in pathogen access through the wounds. The natural waxy layer on the skin may be partly removed by washing [1]. The effectiveness of the washing operation depends on the characteristics of the water, e.g., acidity, hardness, mineral content, temperature, and microbial count. It is also depends on the amount of water used, the force applied, and whether brushing or rubbing is practiced. In some cases, soft nylon brushes are used for scrubbing. Dry brushing with or without air blast is performed to remove loose scales, soil, or dust in some cases, e.g., onions, garlic, potatoes, sweet potatoes, cantaloupes, and melons [84].

Several cleaning and disinfecting compounds are mentioned in Table 12. Chlorine solutions are used to clean raw products and food-processing equipment. The recommended levels of chlorine in solution are 1–3 ppm for rinsing and 50 ppm for sanitizing [26,49]. Generally hypochlorous acid (added to chlorine gas or hypochlorite salt) is used to sanitize the water for wash. The effectiveness of hypochlorous acid depends on the pH and temperature. It is necessary to maintain the effective concentration of the acid in the wash water, especially when the water is recycled. In commercial operations 50–100 ppm chlorine at pH 7.5–8.5 is frequently employed when the washing water carries a substantial amount of soil and organic matter. Use of chlorinated water at 10–200 mg/kg rapidly kills the vegetative cells of yeast and bacteria. It is also reported that postharvest washing is not an effective method for removing fungi from infected tissues. Sulfamic acid and other amines are added to water to form *N*-chloramines, which tend to stabilize the concentration of active chlorine. Sodium *o*-phenylphenate (SOPP) is also occasionally used to reduce the number of pathogenic microorganisms in produce treatment water. SOPP is noncorrosive, it improves the stability of solution, and it is compatible with chemicals that react with chlorine. 2-Aminobutane (phosphate) is added to the SOPP solution in some cases [84].

Table 12 Cleaning and Disinfecting Compounds Used in Fruits and Vegetable Processing

Cleaning compounds	Alkaline salts, alkyl sulfates, ethylene oxide with alkyl phenols, soaps, strong acids (e.g., nitric, hydrochloric), strong alkalines (e.g., sodium hydroxide), weak acids (e.g., gluconic, citric)
Disinfecting compounds	Caustic soda (sodium hydroxide), chlorine and its derivatives, formaldehyde, hydrogen peroxide, idophores, ozone, peracids (e.g., peracetic acid), silver ions, steam, strong acids (e.g., nitric, phosphoric, sulfuric), sulfites and sulfur dioxide, quaternary ammonium products, trichloroiodine

Source: Ref. 26.

If fruit is excessively dirty, a detergent may be used prior to using the sanitizing agent. The final rinse should be with fresh and clean water. Removal of excess surface water may be necessary following washing. This can be done using blotting rollers or air flow over the produce [39]. If this is not done, increased relative humidity around the produce, such as stone fruits and potatoes, can cause infection and subsequent decay [84].

Waxing

The purpose of waxing or applying another edible coating is to (a) reduce the rates of respiration and transpiration, (b) enhance product gloss, (c) generate modified atmosphere, (d) protect from fungus, (e) protect from insect or mechanical injury, and (f) cure tiny injuries and scratches on the surface [1,84]. Wax coating is not always desirable. The coating may modify the inside atmosphere, leading, for example, to low oxygen levels that could encourage fermentation.

2. Chemical Treatments

Chemical treatments can be carried out in the atmospheric gas or while dipping produce in a water solution.

Fumigation

Fumigation of grapes with sulfur dioxide has been standard practice for controlling decay since the 1930s. Salunkhe et al. [84] described how the usual practice is to fumigate grapes as soon as possible after harvest with 1% sulfur dioxide (v/v) for 20 minuted to sterilize the surface of the berries and any injuries made during harvest. This initial treatment is followed by a periodic fumigation with a 0.25% sulfur dioxide at 7- to 10-day intervals during storage [60]. In some cases the quality, e.g., color and texture, of fruits are also improved by sulfur dioxide treatment. The major disadvantages of sulfur dioxide are that it is corrosive to metal surfaces in the storage room and treatment chamber and it bleaches the point of attachment of the stem [84].

Nitrogen trichloride (NCl_3) fumigation treatment has been used extensively to control sporulation and spread of pathogenic fungi during storage. In recent years this practice has declined because of corrosion problems. NCl_3 hydrolysis in a moist environment leads to HOCl, which is probably responsible for decay control and for corrosion [84]. The biphenyl is impregnated into fruit wraps oron to the paper sheets placed at the bottom and the top of the fruit container to inhibit fungi. It sublimes into the atmosphere surrounding the fruit and inhibits the development of decay. The main problem with using biphenyl is that it leaves a residue on the surface which gives off a slight hydrocarbon odor [84].

Fumigants such as ethylene dibromide and methyl bromide have been traditionally used as quarantine treatment for horticultural crops. Objections are rising to the use of these chemicals due to health risks and environmental pollution. Ethylene dibromide has been banned; methyl bromide is still used, but it has also been targeted by authorities [101].

The potential alternatives to chemical fumigation are (a) low- and high-temperatures treatments, (b) atmospheres with very low oxygen and/or very high carbon dioxide, and (c) atmospheres with natural insecticidal volatiles. Low temperatures may cause chilling injury in some cases, whereas high temperatures cause heat injury and a short postharvest life. Atmospheres with very low oxygen ($\leq$0.5%) and/or very high carbon dioxide ($\geq$50%) are insecticidal. However, not all fresh fruits and vegetables can tolerate such extreme atmospheres. The use of these atmospheres is only feasible if they cause no detrimental effects to the produce. The advantages of using insecticidal atmospheres include: (a) they leave no toxic chemicals on the produce, (b) they are environmentally safe, and (c) they are competitive in cost with chemical fumigants. The disadvantages are that it takes longer to kill insects with insecticidal atmospheres than with fumigants, and these extreme atmospheres may cause anaerobiosis and fermentation in fresh horticultural crops [101]. Yahia [101] mentioned that during long-term storage insecticidal atmospheres can be used for grains, nuts, and raisins, since no detrimental effects have been reported. In certain fresh fruits or vegetables, such as mango, papaya, and avocado, they can also be applied. They also found that the potential injury caused by high carbon dioxide to insects is more pronounced than that caused by low oxygen.

Dipping Treatments

Cherries washed with a 3% acetic acid solution were found to be very satisfactory without a wetting agent. The wetting agents reduced the requirement of acid concentration and also improved the quality.

Sprays and dips of calcium chloride delay softening and senescence of fruits by cross-linking between polygalacturonide chains and calcium in cell walls, resulting in an extension of shelf life. The formation of pectic acids from pectates may cause softening of fruits due to a separation of calcium from the pectic acid, ultimately resulting in changes in the ultrastructure of the cell wall [84].

Low calcium levels in apple fruit flesh are strongly related to high levels of storage disorders, such as bitter pit [25]. Soil fertilization with calcium salts will not necessarily ensure higher levels in fruits, thus increasing calcium levels are generally achieved by orchard sprays or by postharvest dips [54]. Calcium can improve texture by reacting with pectic acid to form calcium pectate. Calcium and other divalent ions are useful in controlling respiration rate, texture or firmness loss, storage disorders, ethylene production, and microbial decay [25,26].

Ryu and Holt [82] showed that surface disinfection of apples with an aqueous solution of cinnamon oil, containing cinnamaldehyde as the principal component, significantly reduced spoilage by molds on most cultivars tested. The addition of Tween 80 (0.05%) and ethanol (3%) may dissolve a wax layer, increasing the susceptibility of the cinnamon oil–treated apples towards fungal spoilage. Gibberellic acid applied preharvest to grapefruit retards the yellowing of the peel and maintains increased peel resistance to ovipositioning by the fruit fly [29].

Tomatoes are particularly vulnerable to the microbial spoilage at calyces and wound sites on the surface. Calyces are usually the first sites on tomatoes on which visible growth of fungi appears. Compared to the fruit surface, the calyx carries the major part of the microbial load, consisting of epiphytic bacteria and molds. Microorganisms associated with the wax layer probably have less chance to proliferate because of the limited availability of substrates necessary for growth in this microenvironment [88]. The major pathogens affecting the postharvest life of tomato fruit are *Alternaria alternate, Botrytis cinerea,* and *Rhizopus stolonifer* [91]. These fungi do not grow when water activity drops below 0.88, 0.93, and 0.93, respectively [47].

Disinfection of tomatoes with sodium hypochlorite before packaging greatly reduced subsequent microbial spoilage [7]. However, several countries have abandoned the use of hypochlorite for the disinfection of foods, and natural plant-derived compounds with antimicrobial activity and low mammalian toxicity could be good alternatives [88].

The highly fungicidal activity and low mammalian toxicity of natural cinnamaldehyde make this compound a possible candidate for the applications as a surface disinfectant for foods [36]. *Trans*-cinnamaldehyde is generally used to flavor foods and is effective when applied in aqueous solution [87]. The application of this compound is less effective via gas phase, since it is readily oxidized to cinnamic acid when exposed to air [88].

Smid et al. [88] showed that the surface treatment of tomatoes with *trans*-cinnamaldehyde significantly reduced the number of potential spoilage bacteria and fungi. This reduction in microbial load postponed the appearance of visible fungal spoilage of the tomatoes, and thus prolonged the shelf life of fruit subsequent storage under modified atmosphere. A part of the surface-associated cinnamaldehyde evaporated or oxidized to less active cinnamic acid. Another portion probably remain accumulated in wax layer of the tomato epidermis due to the hydrophobic nature of cinnamaldehyde. Consequently epiphytic microorganisms were exposed to the antimicrobial compound for a longer period, prolonging storage life [88]. At high doses such plant source compounds may affect odors, thus plant compounds should be selected for both efficacy and minimal interference with the natural odor of the product. The selection of chemical treatment depends on the purpose. It is necessary to consider toxicity, residue in the product, and legal aspects before applying any chemical treatment. A list of chemicals used for dipping treatments are compiled in Tables 13 and 14.

Cleanliness and hygienic conditions can also reduce produce loss. Disinfection of storerooms by spraying with 5% lysol or 2% formalin before storing produce is recommended. Painting the walls with antifungal chemicals and fumigation with paraformaldehyde are also effective [70].

Table 13 Chemicals Used in Dipping Treatments

Main purpose	Chemicals
Regulation of ripening and senescence	Benzylaminopruine (BA), benzyladenine, chlorophenoxyacetic acid (CIPA), cycocel (CCC), cytokinin, dichlorophenoxy acetic acid, gibberellic acid (GA), kinetin, β-napthoxyacetic acid, cycloheximide, actinomycin-D, vitamin K, maleic acid, ethylene oxide, sodium dehydroacetic acid, calcium carbide, 2,4-dichlorophenoxy acetic acid, 2,4,5-trichlorophenoxyacetic acid, trichlorophenoxypropionic acid, indol-acetic acid (IAA), naphthalene-acetic acid (NAA), auxin ethylene
Growth retardant	Maleic hydrazide (MH), *p*-quinone, semi-carbazide, 2-thiouracil, phenylhydrazine, isonicotinic acid hydrazide, isopropyl-*N*-chlorophyl carbamate (CIPC), tetrachloronitrobenzene (TCNB), pentachloronitrobenzene (PENB)
Antimicrobial agent	Dehydroacetic acid, methyl formate, methyl bromide, benomyl, thiabendazole, aminobutane, kocide 101, captan, orthophenylphenate (SOPP), thiabendazole, sodium carbonate, borax, sodium hydroxide, hexamine, 2,6-dichloro-4-nitroaniline (DCNA)
Ethylene absorbent	Potassium permanganate
Nonethylene and nonrespiratory volatile absorbents	Selective molecular sieves (synthetic zeolites), boric acid

Source: Refs, 78, 84, 94.

Table 14 Chemicals Used for Ripening and Senescence

Chemicals	Effects	Commodities
Cytokinin	Delays chlorophyll degradation and senescence	Leafy vegetables, spinach, pepper, bean, cucumber
Benzyladenine	Delays chlorophyll degradation and senescence	Cherry
Benzylaminopurine	Delays chlorophyll degradation and senescence, reduce protein loss	Sweet cherry, cauliflower, endive, parsley, snap beans, lettuce, radish, onions, cabbage, Brussels sprouts, sprouting broccoli, mustard greens, radish top, celery, asparagus
Kinetin	Delays chlorophyll degradation and senescence	Leafy vegetables, spinach, pepper, bean, cucumber, cherry
Gibberellin	Retards maturation, ripening, senescence, retards chlorophyll disappearance, increases peel firmness, delays accumulation of carotenoids	Tomato, banana, kiwi, citrus (orange, grapefruit)
Maleic hydrazide[a]	Inhibition of sprouting	Onion, sugar beet, turnip, carrot, potato
Maleic hydrazide	Delays ripening	Mango, tomato, sapota fruit
Alar (cold)	Enhances firmness	Fruits
Alar (hot)	Increases the permeability of external tissue and activates the carotene-synthesizing enzymes	
Alar	Delays deterioration and discoloration	Mushroom
Alar	Preserves chlorophyll	Leaves of beans
Alar	Inhibits synthesis of solanine	Potato
Cycocel	Retards senescence and inhibits deterioration	Vegetables
CIPC	Controls sprouting	Potato

[a]Analog of maleic hydrazide are *p*-quinone, semi-carbazide, 2-thiouracil, phenylhydrazine, and isonicotinic acid hydrazide.
Source: Ref. 84.

IV. PACKAGING, STORAGE, AND DISTRIBUTION

Postharvest handling systems involve the channels through which harvested fruits reach the processing facility or consumer [39]. The quality of fresh produce is affected markedly by the length of time between picking and storage [84]. Packing, transportation, storage, and retail distribution are also important for postharvest handling of plant materials. Physical and mechanical injuries can occur during the handling, grading, and packaging operations before and after shipment [83]. To avoid excessive physical damage, proper precautions are necessary during harvesting, loading, unloading, transportation, sorting, packaging, and storage [26]. The total system approach follows a batch of produce from the field to the retail market, and identification of the critical points of quality loss need to be identified [35].

A. Dumping and Packaging

Dumping may be accomplished by manual or automatic dumping in water to reduce mechanical damage, and field heat [84]. Dumping of produce can be in water, in flotation tanks, or as dry dumping. The produce should be able to withstand wetting, and if dry dumping systems are used, they should be well padded to reduce impact bruising. A bin cover may also be used to allow inverting the bin and to regulate the flow of produce out of the bin [39]. Sorting, sizing, and grading also enhance the quality and provide better control of processing or storage conditions. These can be done based on color, size, maturity, etc.

Using properly designed containers for transport and marketing can maintain a products freshness, succulence, and quality. Incorrectly designed containers can cause significant mechanical damage and bruising [83]. When large-sized products, such as watermelons, muskmelons, pumpkins, yams, and cabbages, are transported in bulk using trucks, products should be carefully stacked and adequately covered to protect them from the environment [83].

A packaging unit is used to facilitate moving produce from one location to another. It provides protection. Stacking should be possible without collapse or pressure damage to the produce. The packaging contributes greatly to the efficient marketing of fresh fruits and vegetables. The benefits of packaging are [30]: (a) it serves as an efficient handling unit, (b) it provides convenient warehouse or home storage unit, (c) it protects quality and reduces waste by avoiding mechanical damage, reducing moisture loss, providing beneficial modified atmosphere, providing clean or sanitary produce, preventing pilferage, (d) it provides service and sales motivation, (e) it reduces the cost of transport and marketing, and (f) it facilitates the use of new modes of transportation.

The proper packaging can protect fresh produce from environmental factors, such as moisture and light. Its major purpose is to avoid bruising and abrasion. The container must be strong enough to withstand stacking and the impact of loading and unloading without bruising or scarring tender produce. Thus, containers may require the use of liners, pads, trays, or tissue wraps to prevent damage from contact with rough surfaces or adjacent produce.

Short-term or long-term storage of fresh produce may be needed before processing to regulate product flow and extend the processing season [39].

1. Box Storage

Produce can be packed in a box (wooden or paper) with absorbent lining or padding materials. The main limitations of the box storage method are the cost of the box and the extra labor required for packing [78]. Lug boxes and pallet boxes are also used. Pallet boxes are used for bulk handling, which saves loading and unloading time and reduces manual labor [84].

2. Bag Storage

Plastic films, mesh, or net or plastic-lined paper may also be used to pack fresh produce and the selection depends on the cost and the type of produce.

3. Individual Seal Packaging

Individual seal packaging creates a water-saturated atmosphere around fruit, which reduces water loss and shrinkage. It is also referred as individual shrink wrapping, shrink wrapping, and unipackaging. The advantages of individual seal packaging are: (a) it is an alternative to expensive traditional refrigeration and sophisticated controlled atmosphere storage, (b) it doubles and sometimes triples the shelf life of fruits and vegetables as measured by appearance, firmness, shrinkage, weight loss, and other keeping qualitites, (c) it also delays physiological deterioration better than cooling to optimal temperature alone, (d) it reduces chilling injury in some cases, such as in citrus, (e) it can be combined with cooling, and (f) it delays the changes related to senescence, such as deterioration of cell membrane integrity and softening. The limitations are that off-flavors are caused by poor gas exchange and decay and spoilage are enhanced due to the phytotoxic microatmosphere (low oxygen, excessive carbon dioxide and ethylene) that may develop around the enclosed fruit. Moreover, film that contains fungicides can be used to reduce toxic residue in products, and films containing ethylene-absorbing substances can also be used. Perforated film is used to allow higher gas exchange rates and avoid accumulation of ethylene in the enclosed microatmosphere [26,35].

Sealing of stem end of a fruit or vegetable with molten paraffin or some other coating can also increase its shelf life by controlling respiration. The major potential sides of infection in harvested mangoes are the stem end and natural openings like lenticels and stomata. In mango, spoilage due to stem end rot and anthracnose limits its storage potential [89].

4. Padding Materials

The placement of produce in containers during packing must be done in a way to avoid cuts, punctures, pressure bruises, abrasion, or friction [84]. The placement or immobilization should also be maintained to reduce damage. This can be done using various types of trays or by certain volume fill techniques, such as padding or cushioning [84]. Natural padding materials include leaves, straw, grass, and coconut husks.

Mechanical damage occurs in the postharvest handling system primarily in two ways: impact forces and compressive forces. The compressive forces act on the product when it is handled in bulk and normally involves static loads (in bins, stacks) or dynamic compressive loads (bin handling and transport). Excessive impact occurs during harvesting, grading, handling, and transportation, and excessive compression loads occur during bulk handling and package handling. In a study of a tomato-handling system, Campbell et al. [17] found that up to 40% of the crop sustained mechanical damage. They also found that one third of the damage occurred at harvest and was related to bin design and filling techniques. The bins were found to be too deep and overfilled, both of which resulted in compression damage to the fruit. In the case of apples, most bruising occurred due to excessive compressive forces. This was exacerbated by transportation on short-tyned forks, on the rear of the trailers, and on heavy-duty leaf-sprung trucks traveling on unsealed roads [15]. Bollen et al. [11] found that rejection level bruising occurred at locations near the bin edge and center bottom and was highly dependent on bin condition and/or transport duration. The fruits in the middle of the bin experience bruising to a nonreject level in both transport and bin condition regimes. Bruising was considerable at the top of the bin near the sides and was likely to be due to impact damage since bin condition and transport duration have little influence on damage levels.

Compression damage is the primary cause of damage to fruit while it is handled in bulk. The force on the product is transferred from the vehicle transporting the bin to the produce. The energy is dissipated through movement of the product and absorption by the produce. The severity of the levels of bruising resulting from bulk handling has been reported in various studies. The forces vary considerably within bins according to the load paths due to produce stacking patterns. This loading pattern is also influenced by the bin design and the transport method [66].

Padding materials are the principle method of reducing damaging impacts in harvesting and postharvest handling systems. Padding materials should have the following properties: (a) ability to absorb the impact energy without damaging the produce, (b) not impart a high rebound energy to the produce, (c) durability by internal structure fatigue (impact energy absorption ability) and surface wear (needs lower thickness), (d) at least 60% of the impact energy absorbed so that produce rebound is minimized, (e) cushion cleanup, sanitation, and compatibility with water, fungicides, waxes, and cleaning solutions must be excellent, and (f) cushioning physical properties (thickness, stiffness). In terms of severity of impact handling systems may be classified into two broad categories: low-energy systems (less than 0.9 J), such as for apples and stonefruit, and high-energy systems (around 1.5 J), such as for potatoes. The materials that can be used are PVC, polyethylene, neoprene, polyurethane, wool carpet, polypropylene, poron, or no bruze. The materials are usually made with a porous internal structure and specific surface characteristics [2,10].

Fresh produce is mechanically or manually handled several times or packing lines before it arrives at the point of consumer purchase. Packing line equipment and other harvesting and postharvest handling equipment are traditionally designed and installed using many transfers from one operation to the next. During this handling the produce hits (impacts) hard surfaces or other produce. Cushioning and velocity control devices can be chosen that will avoid bruising in handling systems. A cushioning material must provide effective energy absorption and dissipation and not create the critical stress/strain level in the produce tissue that will initiate bruising. Bruising can be caused by intermittent shocks, compressive forces, or prolonged low-level vibrations occurring during transportation of produce from the orchard or field to the packing house, and from the packing house to the retail store [2]. The packing line should be designed to have the appropriate drop height or roll velocity. If hard surfaces on the equipment are adequately cushioned and the roll velocity of each item is controlled to a low enough level, impact bruises can be avoided.

B. Storage

In processing, storage plays a significant role. Salunkhe et al. [84] mentioned the following purposes: (a) to extend the availability of fresh produce in the market, (b) to ensure continuous supply of quality raw materials to the processing line, (c) to extend the length of the processing season, (d) to hold raw materials obtained during favorable price situations, (e) to condition certain commodities, such as potatoes, onions, and garlics, and (f) to ripen certain fruits, such as mangoes and bananas [80].

1. Natural Storage

Vegetables such as potato, yam, sweet potato, garlic, and ginger are kept in situ for several months after they attain maturity. They are then removed when desired. It is easy and economical because it does not involve any extra expenditure and fabrication for storing. The vegetables are removed before the rainy season to prevent rotting and sprouting [70].

2. Field Storage

Underground storage in pits and trenches by mounting soil on the surface is not suitable for short-term storage. Hay or straw and then soil is used to protect the surface from water leakage and freezing. Pits are used for storing beet, potato, carrot, turnip, cabbage, parsnip, etc. The disadvantages include expensive labor, variable climatic conditions, and adverse weather conditions, such as cold and wet [70,83].

3. Ventilated Storage

Underground rooms or aboveground warehouses are commonly used for storage. In both cases, design and type of construction depends on the environment and the nature of the product to be stored. For example, a storage structure for sweet potatoes differs from that for onions [83]. It can be designed for cold or hot, moist or ventilated storage.

4. Controlled-Atmosphere Storage

In this method, produce is placed in an air-tight room and the gas composition of the room atmosphere is maintained at the desired level for best quality.

REFERENCES

1. E. K. Akamine, H. Kitagawa, H. Subramanyam, and P. G. Long, Packinghouse operations, *Postharvest Physiology, Handling, and Utilization of Tropical and Subtropical Fruits and Vegetables* (E. B. Pantastico, ed.), AVI Publishing, Westport, CT, 1975, p. 267.
2. P. R. Armstrong, G. K. Brown, and E. J. Timm, *Harvest and Postharvest Technologies for Fresh Fruits and Vegetables* (L. Kushwaha, R. Serwatowski, and R. Brook, eds.), American Society of Agricultural Engineers, St. Joseph, MI, 1995, p. 183.
3. R. Barkai and D. J. Phillips, Postharvest heat treatments of fresh fruits and vegetables for decay control, *Plant Dis.* (*Nov.*):1085 (1991).
4. J. E. Baker, Morphological changes during maturation and senescence, *Postharvest Physiology, Handling, and Utilization of Tropical and Subtropical Fruits and Vegetables* (E. B. Pantastico, ed.), AVI Publishing, Westport, CT, 1975, p. 128.
5. E. A. Baldwin, M. O. Nisperos-Carriedo, and M. G. Moshonas, Quantitative analysis of flavour and other volatiles and for certain constituents of two tomato cultivars during ripening, *J. Am. Soc. Hortic. Sci. 116*:265 (1991).
6. R. Barki-Golan and D. J. Phillips, Postharvest heat treatments of fresh fruits and vegetables for decay control, *Plant Dis.* (*Nov.*):1085 (1991).
7. R. Bhowmik and J. C. Pan, Shelf life of mature green tomatoes stored in controlled atmosphere and high humidity, *J. Food Sci. 57*:948 (1992).
8. J. B. Biale and R. E. Young, Respiration and ripening in fruits-retrospect and prospect, *Recent Advances in the Biochemistry of Fruits and Vegetables* (J. Friend and M. J. C. Rhodes, eds.), Academic Press, London, 1981, p. 1.
9. M. S. Biggs, W. R. Woodson, and A. K. Handa, Biochemical basis of high-temperature inhibition of ethylene biosynthesis in ripening tomato fruits, *Physiol. Plant. 72*:572 (1988).
10. A F. Bollen and B. T. D. Rue, Padding materials for handling horticultural products: Development of an evaluation procedure, *Harvest and Postharvest Technologies for Fresh Fruits and Vegetables* (L. Kushwaha, R. Serwatowski, and R. Brook, eds.), American Society of Agricultural Engineers, St. Joseph, MI, 1995, p. 129.
11. A. F. Bollen, I. M. Woodhead, and B. T. D. Rue, Compression forces and damage in the postharvest handling system, *Harvest and Postharvest Technologies for Fresh Fruits and Vegetables* (L. Kushwaha, R. Serwatowski, and R. Brook, eds.), American Society of Agricultural Engineers, Michigan, 1995, p. 168.
12. R. H. Booth, Postharvest deterioration of tropical root crops: losses and their control, *Trop. Sci. 16*:49 (1974).
13. G. E. Brown, M. A. Ismail, and C. R. Barmore, Lignification of injuries to citrus fruit and susceptibility to green mold, *Proc. Fla. State Hort. Soc. 91*:124 (1978).
14. D. Burmeister, Prediction and study of bitter pit in apples using Mg^+ induced bitter pitlike symptoms, Ph.D. thesis, Michigan State University, East Lansing, MI.
15. C. L. Burton, G. K. Brown, N. L. S. Pason, and E. J. Timm, Apple bruising related to picking and hauling impacts, Paper No. 89-6049, Summer Meeting, Quebec, Canada.
16. I. Burzo, Influence of temperature level on respiratory intensity in the main vegetable varieties, Proc. Symp. on Postharvest Handling of Vegetables, *Acta Hortic. 116*:61 (1980).
17. D. T. Campbell, S. E. Prussia and R. L. Shewfelt, Evaluation postharvest injury to fresh market tomatoes, *J. Food Destrib. Res.* (*Sept.*)*:16* (1986).
18. W. Chace and E. B. Pantastico, Principles of transport and commercial transport, *Postharvest Physiology, Handling, and Utilization of Tropical and Subtropical Fruits and Vegetables* (E. B. Pantastico, ed.), AVI Publishing, Westport, CT, 1975, p. 444.
19. H. M. Couey, Chilling injury of crops of tropical and subtropical origin, *HortSci. 17*:162 (1982).
20. T. Deak and L. R. Beuchat, Identification of foodborne yeasts, *J. Food Prot. 50*:243 (1987).
21. C. Dennis, *Postharvest Physiology of Vegetables* (J. Weichmann, ed.), Marcel Dekker, New York, 1987.
22. E. L Denisen, Tomato color as influenced by variety and environment, *Proc. Am. Soc. Hort. Sci. 51*:349 (1948).

23. I. Eaks, Ripening, respiration, and ethylene production of 'Haas' avocado fruits at 30 to 40°C, *J. Am. Soc. Hort. Sci. 103*:576 (1978).
24. E. A. Estes, Feasibility and affordability considerations in precooling fruits and vegetables, *Harvest and Postharvest Technologies for Fresh Fruits and Vegetables* (L. Kushwaha, R. Serwatowski, and R. Brook, eds.), American Society of Agricultural Engineers, St. Joseph, MI, 1995, p. 390.
25. I. B. Ferguson, Calcium in plant senescence and fruit ripening, *Plant Cell Environ. 7*:477 (1984).
26. J. D. Floros, The shelf life of fruits and vegetables, *Shelf Life Studies of Foods and Beverages: Chemical, Biological, Physical, and Nutritional Aspects* (G. Charalambous, ed.), Elsevier Science Publishers B. V., Amsterdam, 1993, p. 195.
27. A. S. Gholap, C. Bandyopadhyay, and G. B. Nadkarni, Aroma development in mango fruit, *J. Food Biochem. 10*:217 (1986).
28. W. A. Gortner, G. G. Dull, and B. H. Kraus, Fruit development, maturation, ripening, and senescence: biochemical basis for horticultural terminology, *Hortic. Sci. 2*:141 (1967).
29. P. D. Greany, R. E. McDonald, W. J. Schroeder, P. E. Shaw, M. Aluja, and A. Malavasi, Use of gibberellic acid to reduce citrus fruit susceptibility to fruit flies, *Bioregulators for Crop Protection and Pest Control* (P. A. Hedin, ed.), American Chemical Society, Washington, DC, 1995, p. 40.
30. R. E. Hardenburg, Principles of packaging, *Postharvest Physiology, Handling, and Utilization of Tropical and Subtropical Fruits and Vegetables* (E. B. Pantastico, ed.), AVI Publishing, Westport, CT, 1975, p. 283.
31. R. E. Harderburg, A. E. Watada, and C. Y. Wang, *The Commercial Storage of Fruits, Vegetables, and Florist and Nursery Stocks,* USDA, Agricultural Handbook No. 66, 1986.
32. N. W. Heather, L. Whitfort, R. L. McLauchlan, and R. Kopittke, Cold disinfestation of Australian mandarins against Queensland fruit fly *(Diptera: tephritidae)*, *Postharv. Biol. Technol. 8*:307 (1996).
33. M. Herregods, Storage of tomatoes, *Acta Hortic. 20*:137 (1971).
34. G. E. Hobson, The short term storage of tomato fruit, *J. Hort. Sci. 56*:363 (1981).
35. J. J. Jen, Postharvest handling and processing of selected fruits and vegetables, *Trends in Food Processing II* (A. H. Ghee, N. Lodge, and O. K. Lian, eds.), Singapore Institute of Food Science and Technology, Singapore, 1989, p. 261.
36. P. M. Jenner, E. C. Hagan, J. M. Taylor, E. L. Cook, and O. G. Fitzhugh, Food flavourings and compounds of related structure. I. Acute oral toxicity, *Food Cosmet. Toxicol. 2*:327 (1964).
37. A. Kader, Postharvest biology and biotechnology: an overview, *Postharvest Technology of Horticultural Crops* (A. A. Kader, ed.), University of California, Publication No. 3311, Davis, CA, 1992, p. 15.
38. A. A. Kader, Postharvest handling, *The Biology of Horticulture—An Introductory Textbook* (J. E. Preece and P. E. Read, eds.), John Wiley & Sons, New York, 1993, p. 353.
39. A A. Kader and D. M. Barrett, Classification, composition of fruits, and postharvest maintenance of quality, *Processing Fruits: Science and Technology*, Vol. 1, *Biology, Principles, and Applications* (L. P. Somogyi, H. S. Ramaswamy, and Y. H. Hui, eds.), Technomic Publishing Company, Lancaster, PA, 1996, p. 1.
40. A. A. Kader, R. F. Kasmire, F. G. Mitchell, M. S. Reid, N. F. Sommer, and J. F. Thompson, *Postharvest Technology of Horticultural Crops*, University of California, Davis, 1985.
41. L. Kitinoja, and A. A. Kader, *Small-scale Postharvest Handling Practices: a Manual for Horticultural Crops*, 3rd ed., University of California, Davis, CA, 1995.
42. J. D. Klein and S. Lurie, Time, temperature, and calcium interact in scald reduction and firmness retention in heated apples, *HortSci. 29*:194 (1994).
43. J. D. Klein and S. Lurie, Heat treatments for improved postharvest quality of horticultural crops, *HortTechnol. 2*:316 (1992).
44. J. D. Klein and S. Lurie, Prestorage heat treatment as a means of improving poststorage quality of apples, *J. Am. Soc. Hortic. Sci. 103*:584 (1990).
45. J. D. Klein, S. Lurie, and R. Ben Arie, Quality and cell wall components of 'Anna' and 'Granny Smith' apples treated with heat, calcium, and ethylene, *J. Am. Soc. Hortic. Sci. 115*:954 (1990).

46. P. Krogh, The role of mycotoxins in disease of animals and man, *J. Appl. Bacteriol. Symp. Suppl. 18*:99S (1989).
47. L. Lacey, Pre- and postharvest ecology of fungi causing spoilage of foods and other stored products, *J. Appl. Bacteriol. Symp. Suppl.* 11S (1989).
48. C. P. Lentz, L. van den Berg, and R. S. McCullough, Study of factors affecting temperature, relative humidity and moisture loss in fresh fruit and vegetable storages, *J. Inst. Can. Technol. Aliment. 4*:146 (1971).
49. B. S. Luh and J. G. Woodroof, *Commercial Vegetables Processing*, 2nd ed. AVI/Van Nostrand Reinhold, New York, 1989.
50. S. Lurie and J. D. Klein, Calcium and heat treatments to improve storability of 'Anna' apples, *HortSci. 27*:36 (1992).
51. S. Lurie and A. Sabehat, Prestorage temperature manipulations to reduce chilling injury in tomatoes, *Postharv. Biol. Technol. 11*:57 (1997).
52. S. Lurie, E. Fallik, and J. D. Klein, The effect of heat treatment on apple epicuticular wax and calcium uptake, *Postharv. Biol. Technol. 8*:271 (1996).
53. S. Lurie, J. D. Klein, and R. Ben Arie, Postharvest heat treatment as a possible means of reducing superficial scald of apples, *J. Hortic. Sci. 65*:503 (1992).
54. J. L. Mason, Increasing calcium content of calcium sensitive tissues, *Commun. Soil. Sci. Plant Anal. 10*:349 (1979).
55. A. K. Mattoo, T. Murata, E. B. Pantastico, K. Chachin, K. Ogata, and C. T. Phan, Chemical changes during ripening and senescence, *Postharvest Physiology, Handling, and Utilization of Tropical and Subtropical Fruits and Vegetables* (E. B. Pantastico, ed.), AVI Publishing, Westport, CT, 1975, p. 103.
56. B. M. McGregor, *Tropical Products Transport Handbook*, USDA Office of Transportation, Agricultural Handbook 668, 1989.
57. R. G. McGuire, Market quality of grapefruit after heat quarantine treatments, *HortSic. 26*:1393 (1991).
58. A. R. Miller, Physiology, biochemistry and detection of bruising (mechanical stress) in fruits and vegetables, *Postharv. News Information* 2:53N (1992).
59. W. R. Miller and R. E. McDonald, Comparative responses of reharvest GA-treated grapefruit to vapor heat and hot water treatment, *HortSci. 32*:275 (1997).
60. K. E. Nelson and H. B. Richardson, Storage temperature and sulfur dioxide treatment in relation to decay and bleaching of stored table grapes, *Phytopathology 57*:950 (1967).
61. M. C. N. Nunes, A. M. M. B. Morais, J. K. Brecht, and S. A. Sargent, Quality of strawberries after storage is reduced by a short delay to cooling, *Harvest and Postharvest Technologies for Fresh Fruits and Vegetables* (L. Kushwaha, R. Serwatowski, and R. Brook, eds.), American Society of Agricultural Engineers, St. Joseph, MI, 1995, p. 15.
62. H. E. Nursten, Volatile compounds: the aroma of fruits, *The Biochemistry of Fruits and their Products* (A. C. Hulme, ed.), Academic Press, New York, 1970, p. 239.
63. M. O'Brien, Bulk handling methods, *Postharvest Physiology, Handling, and Utilization of Tropical and Subtropical Fruits and Vegetables* (E. B. Pantastico, ed.), AVI Publishing, Westport, CT, 1975, p. 246.
64. J. M. Ogawa, J. Rumsey, B. T. Manji, G. Tate, J. Toyoda, E. Bose, and L. Dugger, Implications and chemical testing of two rhizopus fungi in softening of canned apricots, *Calif. Agric. 28*:6 (1974).
65. H. J. Pfaff, M. W. Miller, and E. M. Mrak, *The Life of Yeasts*, 2nd ed., Harvard University Press, Cambridge, 1978.
66. D. W. Pang, F. Bollen, A. McDougall, and B. D. Rue, Simulation of bulk apple handling to determine bruising levels, *Harvest and Postharvest Technologies for Fresh Fruits and Vegetables* (L. Kushwaha, R. Serwatowski, and R. Brook, eds.), American Society of Agricultural Engineers, St. Joseph, MI, 1995, p. 152.
67. E. B. Pantastico, General introduction: Structure of fruits and vegetables, *Postharvest Physiol-*

ogy, Handling, and Utilization of Tropical and Subtropical Fruits and Vegetables* (E. B. Pantastico, ed.), AVI Publishing, Westport, CT, 1975, p. 1.

68. E. B. Pantastico, Preharvest factors affecting quality and physiology after harvest, *Postharvest Physiology, Handling, and Utilization of Tropical and Subtropical Fruits and Vegetables* (E. B. Pantastico, ed.), AVI Publishing, Westport, CT, 1975, p. 25.
69. E. B. Pantastico, H. Subramanyam, M. B. Bhatti, N. Ali, and E. K. Akamine, Preharvest physiology: Harvest indices, *Postharvest Physiology, Handling, and Utilization of Tropical and Subtropical Fruits and Vegetables* (E. B. Pantastico, ed.), AVI Publishing, Westport, CT, 1975, p. 56.
70. E. B. Pantastico, T. K. Chattopadhyay, and H. Subramanyam, Storage and commercial storage operations, *Postharvest Physiology, Handling, and Utilization of Tropical and Subtropical Fruits and Vegetables* (E. B. Pantastico, ed.), AVI Publishing, Westport, CT, 1975, p. 314.
71. E. B. Pantastico, A. K. Mattoo, T. Murata, and K. Ogata, Physiological disorders and diseases, *Postharvest Physiology, Handling, and Utilization of Tropical and Subtropical Fruits and Vegetables* (E. B. Pantastico, ed.), AVI Publishing, Westport, CT, 1975, p. 339.
72. E. Pesis, Introduction of fruit aroma and quality by post-harvest application of natural metabolites or anaerobic conditions: 1. Biosynthesis and degradation of aroma volatiles during post-harvest life, *Modern Methods of Plant Analysis 18*:19 (1996).
73. C. T. Phan, E. B. Pantastico, K. Ogata, K. Chachin, Respiration and respiratory climacteric, *Postharvest Physiology, Handling, and Utilization of Tropical and Subtropical Fruits and Vegetables* (E. B. Pantastico, ed.), AVI Publishing, Westport, CT, 1975, p. 86.
74. J. Pino, R. Torricella, and F. Orsi, Correlation between sensory and gas chromatographic measurements on grapefruit juice volatiles, *Nahrung 30*:783 (1986).
75. S. W. Porritt and P. D. Lidster, The effect of prestorage heating on ripening and senescence of apples during storage, *J. Am. Soc. Hortic. Sci. 103*:584 (1978).
76. K. Poutanen, Enzymes: an important tool in the improvement of quality of cereal foods, *Trends Food Sci. Technol. 8*:300 (1997).
77. F. J. Proctor, J. P. Goodliffe, and D. G. Coursey, Postharvest losses of vegetables and their control in the tropics, *Vegetable Productivity* (C. R. W. Spalding, ed.), McMillan, London, 1981.
78. V. Ravi, J. Aked, and C. Balagopalan, Review on tropical root and tuber crops. I. Storage methods and quality changes, *Crit. Rev. Food Sci. Nutri. 36*:661 (1996).
79. J. E. Richard, O. J. Burden, and D. G. Coursey, Studies on the insulation of tropical horticultural produce, *Acta Hortic. 84*:115 (1978).
80. R. Rodriguez, B. L. Raina, E. B. Pantastico, and M. B. Bhatti, Quality of raw materials for processing, *Postharvest Physiology, Handling, and Utilization of Tropical and Subtropical Fruits and Vegetables* (E. B. Pantastico, ed.), AVI Publishing, Westport, CT, 1975, p. 467.
81. A. L. Ryall, and W. J. Lipton, *Handling, Transportation and Storage of Fruits and Vegetables,* Vol. 1, Vegetables and Melon AVI Publishing, Westport, CT, 1972.
82. D J. Ryu and D. L. Holt, Growth inhibition of *Penicillium expansum* by several commonly used food ingredients, *J. Food Protect. 56*:862 (1993).
83. D. K. Salunkhe and B. B. Desai, *Postharvest Biotechnology of Vegetables,* CRC Press, Boca Raton, FL, 1984.
84. D. K. Salunkhe, H. R. Bolin, and N. R. Reddy, *Storage, Processing, and Nutritional Quality of Fruits and Vegetables,* Vol. 1: *Fresh Fruits and Vegetables*, CRC Press, Boca Raton, 1991.
85. M. Sawamura, T. Tsuji, and S. Kuhwahara, Changes in the volatile constituents of Pummelo during storage, *Agric. Biol. Chem. 53*:243 (1989).
86. G. J. Shaw, P. J. Ellingham, and E. J. Birch, Volatile constituents of feijoa headspace analysis of intact fruit, *J. Sci. Food Agric. 34*:743 (1983).
87. E. J. Smid, Y. De Witte, and L. G. M. Gorris, Secondary plant metabolites as control agents of postharvest *Penicillium* rot on tulip bulbs, *Postharv. Biol. Technol. 6*:303 (1995).
88. E. J. Smid, L. Hendriks, H. A. M. Boerrigter, and L. G. M. Gorris, Surface disinfection of tomatoes using the natural plant compound *trans*-cinnamaldehyde, *Postharv. Biol. Technol. 9*:343 (1996)

89. D. H. Spalding, Resistance of mango pathogens to fungicides used to control postharvest diseases, *Plant Dis. 66*:1185 (1982).
90. D. F. Splittstoesser, Microbiology of fruit products, *Processing Fruits: Science and Technology*, Vol. 1, *Biology, Principles, and Applications* (L. P. Somogyi, H. S. Ramaswamy, and Y. H. Hui, eds.), Technomic Publishing Company, Lancaster, PA, 1996, p. 261.
91. N. F. Sommer, R. J. Forlage, and D. C. Edwards, Postharvest diseases of selected commodities, *Postharvest Technology of Horticultural Crops* (A. A. Kader, ed.), University of California, Division of Agriculture and Natural Resources, Davis, CA, 1992, p. 117.
92. R. Teranishi, J. W. Corse, W. H. McFadden, D. R. Black, and A. I. Morgan, Volatiles from strawberries. I. Mass spectral identification of more volatile components, *J. Food Sci. 28*:478 (1963).
93. J. F. Thompson and A. A. Kader, *Postharvest Outreach Program*, University of California, Davis, CA, 1995.
94. P. M. A. Toivonen, Non-ethylene, non-respiratory volatiles in harvested fruits and vegetables: their occurrence, biological activity and control, *Postharv. Biol. Technol. 12*:109 (1997).
95. M. Tsuji, H. Harakawa, and Y. Komiyama, Changes in shelf life and quality of plum fruit during storage at high temperature, *J. Jpn. Soc. Hortic. Sci. 52*:469 (1984).
96. L. Van den Berg, C. P. Lentz, Moisture loss of vegetables under refrigerated storae conditions, *Can. Inst. Food Technol. 4*:143 (1971).
97. D. H. Wallace, Genetics, environment and plant resources, *Vegetable Training Manual* (R. L. Villareal, and D. H. Wallace, eds.), Coll. Agric. Coll. Laguna, 1969, p. 80.
98. R. B. H. Wills and S. L. Gibbons, Use of very low ethylene levels to extend the postharvest life of Hass vocado fruit, *Int. J. Food Properties 1*:71 (1998).
99. R. H. H. Wills, T. H. Lee, D. Graham, W. B. McGlasson, and E. G. Hall, *Postharvest: An Introduction to the Physiology and Handling of Fruit and Vegetables*, New South Wales University Press, Sydney, 1981.
100. G. W. Winsor, Some factors affecting the quality and composition of tomatoes, *Acta Hortic. 93*:35 (1979).
101. E. M. Yahia, Insecticidal atmospheres for tropical fruits, *Harvest and Postharvest Technologies for Fresh Fruits and Vegetables* (L. Kushwaha, R. Serwatowski, and R. Brook, eds.), American Society of Agricultural Engineers, St. Joseph, MI, 1995, p. 282.

3

Postharvest Handling of Foods of Animal Origin

M. Shafiur Rahman
Horticulture and Food Research Institute of New Zealand, Auckland, New Zealand

I. POSTSLAUGHTER HANDLING OF MEAT

A. Causes of Meat Deterioration

Meat is the edible flesh of any of a number of species of animal or bird, both wild and domesticated. The soft and flabby condition of the muscle and the oily condition of fat immediately following slaughter change. The muscle becomes stiff and hard, and the fat becomes firm. The meat again becomes soft after hanging or conditioning, but does not return to its previous flabby, condition. This may be due to the setting of fat and postmortem changes. It is generally accepted that spoilage occurs when bacterial numbers reach 10^7 /cm^2 on the meat surface, but there remains some disagreement on these numbers relative to quality acceptance. The cause of meat spoilage is not mere numbers of microflora but the metabolites produced using meat ingredients as substrate [6]. Bailey et al. [6] identified 179 volatile compounds from meat during refrigerated storage of samples inoculated with bacteria previously considered as contaminants of meat.

Temperature is the most important extrinsic factor affecting the growth of microorganisms on meat. These organisms are usually classified on the basis of their temperature growth range as thermophiles, mesophiles, and psychrophiles, according to their ability to grow at high (45–65°C), medium (25–40°C), and low (10–30°C) temperatures ranges respectively [6].

Rigor mortis is the change that takes place in muscle after death. These changes include (a) hardening, stiffening, and shortening of the muscle, (b) loss of transparency, (c) loss of elasticity, (d) joints becoming stiff and immovable. These changes generally appear about 10 hours after death, are pronounced at about 24 hours, and then gradually pass off. They start at the head and then extend to the limbs and the rest of the body. The development of rigor mortis is influenced by (a) the atmospheric temperature—a high temperature hastens its onset and a low temperature retards it—(b) the state of the animal before death, (c) any reduction in muscle glycogen during life.

The most immediate change caused by bleeding or death is the loss of oxygen carried by circulating blood. One of the results is that adenosine triphosphate (ATP) is no longer produced and the ATP already present in the muscles is gradually reduced. As it disappears, the two major muscle proteins, myosin and actin, combine to form actomyosin, which causes the muscles to contract and to become extensible and firm (hardening). Also because of the loss of ATP, inorganic phosphate is formed, which stimulates the breakdown of glycogen to lactic acid. The low pH caused by this acid renders the muscle proteins more vulnerable to attack by muscle enzymes, which are held in check during life. The breakdown of those proteins produces a very suitable medium for the growth of bacteria.

The ultimate pH of meat is governed by the animal's preslaughter reserves of muscle glycogen. After death the muscle breaks down glycogen via the anaerobic glycolytic pathway to produce lactic acid, thus lowering muscle pH. If the animal's glycogen reserves were depleted preslaughter, for example, by stress or exercise, insufficient lactic acid is produced to lower the pH of the muscle to its normal value, around 5.6 [27]. The elevated pH affects meat quality, such as appearance, water-holding capacity, tenderness, and cooked flavor [13]. In the extreme case meat is transformed to have a dark, firm, and dry appearance [13]. At pH values above 5.8, the keeping quality of fresh chilled meat is adversely affected because of bacterial growth due to the lower contents of glucose, lactic acid, and pH [39]. Preslaughter injections of adrenaline were used to control the ultimate pH level [13].

As the body temperature falls, fat solidifies. The fat ultimately oxidizes, i.e., acquires a rancid off-flavor owing to the fact that there is no longer blood to produce antioxidants. One of the most important attributes of meats is color. If the color is not protected during its shelf life, a brown color in fresh meat and grey in cured meat will develop as a result of oxidation. The basic pigment of raw meat is myoglobin. The pigment deteriorates to brown due to the formation of metamyoglobin via oxidation of myoglobin when meat is exposed to low levels of oxygen and enzymes are active. At this stage meat is said to have lost its bloom. This is the reason why vacuum-packed raw beef is purple. The bright red color is due to the oxygenation of myoglobin to oxymyoglobin at high levels of oxygen, e.g., greater than 20% [32]. Factors affecting the shelf life of meat and meat products are summarized in Table 1.

The effect of temperature on the rate of decrease in pH reflects its effect on the rate of lactic acid production. Cassens and Newbold [18] studied the changes in pH and in the levels of some phosphate fractions with time in beef muscle stored at 1, 15, and 37°C soon after slaughter. An effect of temperature on the metabolism of an organic phosphate fraction stable to both acid and alkali was noted. At 15°C the decrease in pH of muscle, creatine phosphate, and other acid-labile phosphate occurred more slowly than at 37°. Lesiak et al. [73] studied the effects of postmortem temperature (0, 12, and 30°C) and time on the water-holding capacity of hot-boned turkey breast and thigh muscle. They found that higher temperature and longer storage time induced greater drip losses in breast. Longer storage time induced greater drip losses, but the least drip loss occurred at 12°C in thigh muscle. Higher temperature increased the supernatant weight in breast but decreased it in thigh.

Table 1 Factors Affecting Shelf Life

Intrinsic Factors
Type of animal, e.g., porcine, bovine
Breed and feed regime
Age of animal at time of slaughter
Initial microflora
Chemical properties, e.g., peroxide value, pH, acidity, redox potential
Availability of oxygen
Extrinsic Factors
Processing conditions and control
Hygiene, standard of personnel, and equipment cleaning
Quality-management system, such as HACCP procedures
Temperature control
Packaging system: materials equipment, gases
Storage types

Source: Ref. 32.

A variety of flavors influence meat flavor. In general, some of the factors that influence flavor are the animal's age, breed, sex, the type of diet and nutrients given the animal, processing operations, and cooking methods. In addition to desirable flavor, off-flavors are also developed. Lipid oxidation appears to be the focal point for dealing with the origin of particular off-flavors described as oxidative rancity, grass-fed beef flavor, and warmed-over flavor [3].

B. Postslaughter Handling of Meat

1. Slaughter Conditions

The preslaughter conditions or factors affecting meat quality are (a) organs like liver and kidney, which may contain microorganisms (muscle of healty animals is generally sterile), (b) stress on the animal during transport, (c) feed availability in the period preceding slaughter, and (d) genetic and physiological traits [122].

The rate of microbial spoilage of meat varies widely, depending on initial microbial contamination, temperature, pH, presence of oxygen, presence of nutrients, and presence of inhibitory substances including carbon dioxide. The most common forms of spoilage of raw meat take place slowly, and considerable growth of microorganisms can occur without detracting from the eating quality of the meat. Spoilage, when it becomes apparent, takes the form of souring or slime production on the surface, which are readily recognized by the typical consumer before the meat becomes unsuitable for consumption. Less common spoilage takes the form of putrefaction, producing offensive odors and flavors associated with the breakdown of nitrogen-containing substances [48]. The most important factors in handling fresh meat are speed of handling, control of temperature, and good hygiene conditions. Slaughter practices must minimize both physical and microbiological contamination of carcasses. Smulders [122] found that integrated hygiene control (i.e., Hazard Analysis Critical Control Point) along the meat-production line can be the processor's most effective approach to increasing the storage life of these products.

Meats are subject to contamination from a variety of sources within and outside the animal. Such contamination can occur during slaughter and processing. The level of microbial contamination of a freshly dressed carcass and the composition of the flora depend on the technical structure of the slaughterhouse and the hygienic conditions during the slaughter dressing procedures [117]. Ismail et al. [59] identified a total of 34 fungal genera, represented by 6 species and one variety, from air, water, walls, and floors of slaughterhouses. They also mentioned that surroundings may constitute significant sources of mold contamination for beef carcasses. Ishikawa [58] mentioned four major factors: material, machine, method, and workers.

Material: The Animal

The principal sources of microflora are (a) normal microflora of the skin, such as staphylococci, micrococci, pseudomonads, yeast and molds, (b) microorganisms of fecal and soil origin, and (c) microorganisms present in the intestines or on the hide or fleece of the live animal. The extent of these sources depends on [122] the stress of moving animals, mixing animals at market, insufficient disinfection of transport vehicles, and cross-infection during extended transport time. Showering upon arrival at the slaughterhouse reduces the contamination in many cases, however, washing poultry is generally considered impractical as it is extremely difficult to get the birds dry again [122].

Machine: Equipment and Utensils

Equipment, utensils, and slaughter facilities (design, layout, interior finishes, movement of materials, and operators) should be properly designed, and proper cleaning and disinfection procedures should be in place. The floor and walls of slaughter facilities should be smooth to allow proper cleaning and disinfection. The cleanliness of the slaughter process and chill floor of the slaughterhouse can be used to improve microbial quality of meat [62]. Captive bolts and sticking knives have been shown to be a source of internal contamination of organs as well as muscle [122]. Knives, steels, and aprons of slaughter personnel who handle beef carcasses before dehiding may be an important source of contamination; polishing equipment is probably the most important factor in contaminating pig carcasses [44,125]. Other cross-contamination sources include protective gloves often used in the dressing and boning line and surfaces of cutting tables for further breakdown of carcasses into subprimals [124,128,135]. Procedures must be in operation specifying cleaning methods, schedules, and frequency for both the structure of and the equipment in the factory. The cleanliness of cutting boards and the sterilization of boning knives should be ensured by dipping in hot water (82°C). Cutting boards should not be used for more than 4 hours without cleaning or turning [32].

Method: Slaughter and Fresh Meat Processing

Scalding water contaminated with dirt and feces is a major source of bacterial contamination. Water can enter the lungs and stick wounds [122]. Immersion of carcasses in scalding water of about 60°C will remove hairs from the pig's epidermis and may reduce bacterial contamination [125]. Scalding carcasses by steam has been proposed as an alternative to avoid recontamination [43]. The most hazardous steps in the beef dressing line are removal of the horns and the freeing of the skin around the lower parts of the legs and the sternal region [129]. In fresh meat the size reduction of meat is an important factor, for example, mincing or dicing, with a consequent increase in surface area plus the opportunity for microbiological contamination [32].

Workers: The Slaughter Personnel

The most important sources of bacteria from humans include the oral and nasal cavity, digestive tract, and the skin. These can be reduced by (a) washing, regular cleaning, and disinfection, (b) improving nonhygienic working conditions, (c) low initial bacterial loads, (d) hiring only skilled and careful slaughter personnel, and (e) management commitment to the control of quality and safety [122].

2. Chilling

Chilled storage at temperatures less than 10°C is the simplest procedure for storing meat products. For long storage the temperature should be as near to the freezing point as practicable (i.e., –1.5°C) and relative humidity should be controlled within 0.85–0.95 to prevent drying or condensation on the surface. The main spoilage organisms during aerobic low-temperature storage are *Pseudomonas* species, whereas under vacuum storage the major spoilage organisms are *Lactobacillus* species and yeasts [32].

3. Electrical Stimulation of Meat

Electrical stimulation is the process of passing an electric current through a body or carcass of freshly slaughtered animals. In electrical stimulation of meat a voltage gradient of 5–10 V/cm is applied as AC pulses through electrodes fixed at opposite ends of the long axis of muscle [72]. Electrical stimulation applied with low voltage as soon as possible after slaughter gives uniform effects throughout muscles. When stimulation is delayed 30–45 minutes, high voltage must be used [102]. Based on the time to reach pH 6.0, Carse [17] showed that this time decreased markedly as the voltage increased up to 250 V. In all frequencies used in their experiment, an increase in stimulation voltage from 50 to 320 V increased the pH more than 60% for a 30-second stimulation period [20]. The increase of voltage also affects the tenderness of cooked legs and loins from lambs [19]. It was suggested that the resulting increase in tenderness was mainly due to the prevention of cold shortening [12,34,132].

The effect of electrical stimulation on tenderness is highly dependent on subsequent cooling rate [82]. Very slow cooling sometimes accelerates the already high rate of pH fall to such an extend that the tissue is significantly toughened [102]. The use of low-voltage (40 V) electrical stimulation and normal chilling rates could provide maximum economy in handling of grade beef carcasses [91]. Polidori et al. [102] stated that most authors agree that low voltage is more practical than high voltage. For safety reasons a low-voltage system is more attractive for application under commercial conditions. In general, the lower the voltage, the less the danger to the operator and the less stringent the requirements imposed by electricity authorities for guarding of carcass and equipment [102].

The choice of an effective stimulation period for beef carcasses or side also depends on the extent of postmortem delays. For delays up to 20 minutes a 90-second period is recommended, but with longer delays, up to 60 minutes postmortem, the period is 120 seconds [87]. The higher initial pH requires longer time to reach pH 6.0.

The frequency of the applied voltage is an important determining factor. pH values were higher at 40 and 75% when frequencies changed from 10 and 20 pulses/sec to 50 and 100 pulses/sec [20]. The frequency optimum of around 9–16 pulses/sec seems to be hold for most muscles of sheep and beef carcasses [102]. Longer stimulation periods of 120 seconds at 9–16 pulses/sec gave the highest pH in the case of beef. The advantages of lower stimulation frequencies is the lower energy input, which reduces heating at the electrode contacts and in the musculature. For example, at 14.28 pulses/sec the energy input is only one-seventh that at 100 pulses/sec [20].

A decrease in bacterial count was observed when inoculated meat samples were stimulated in sterile petri dishes with 21 mA current at 60 Hz and 2 seconds duration shocks for 4 minutes [92]. They also found a reduction in the thermal resistance of bacteria after electrical stimulation.

Palaniappan et al. [98] reviewed hypotheses on the mechanisms by which electrical stimulation of meat can affect microorganisms. These are:

1. A reduction in the muscle pH value retards the microbial growth [69].
2. Electrical stimulation impairs the metabolism of bacterial cells [106].
3. Electrical stimulation initiates the release of some proteolytic enzymes from the meat tissue, which causes the bacterial death [31].
4. Changes in meat oxidation-reduction potential or generation of free radicals causes bacterial reduction [89].

Electrical stimulation soon after slaughter is followed by a rapid fall in pH while muscle temperature is still high [81]. The biochemical changes include increased glycolysis as a result of accelerated adenosine triphosphatase and phosphorylase activities [67].

Electrical stimulation of prerigor muscle has received considerable attention as a means to increase tenderness or alleviate the effects of cold shortening [102]. Polidori et al. [102] reviewed the effects of the use of electrical stimulation of carcasses of animals meat on sensorial characteristics. The reported data can be summarized as follows:

1. Muscle temperature can fall and pH is reduced greatly.
2. Electrical stimulation hastened rigor changes by preventing the carcasses from cold shortening, one of the major cause of meat toughness, due to tearing and stretching or due to the release of lysosomal enzymes.
3. Electrical stimulation also results in improvement in meat color, increased marbling, and enhancement of flavor. Electrical stimulation accelerates glycolysis by reducing the concentration of adenosine triphosphate (ATP) and other high-energy phosphates and by rapid fall in pH and parallel increase in muscle lactate and fall in glycogen [102].
4. Electrical stimulation improves the flavor development [121]. It has been postulated that the accelerated aging process increased tenderness and produced chemical compounds that may be responsible for the aged meat flavor [102]. Aging of carcasses after slaughter for 8–14 days at 0–4°C, to improve their tenderness, has been practiced for many years and still remains an important procedure for producing tender meat [68]. Calkins et al. [16] found that differences in the concentration of creatine phosphate, adenine nucleotides, and their derivatives caused differences in beef steaks.
5. The major improvement in tenderness of electrically stimulated meat was originally due to prevention of cold shortening [17]. Other mechanisms have been reviewed by Polidori et al. [102].
6. Heat ring development can be significantly reduced or alleviated by the use of electrical stimulation [112]. Heat ring develops due to the differential chill rate within the

ribeye muscle, which causes differences in color and rigor development from the exterior to the interior portions of the muscle.

Safety has been of utmost importance during the implementation of electrical stimulation in New Zealand, Australia, the United States, and Europe. In some instances safety concerns have effectively prevented commercial adoptation of the process [102].

4. Decontamination of Meat and Poultry Carcasses

Many decontamination treatments are used in the meat industry. These can be grouped as chemical, physical, and combinations of the two [10].

Hot Water Ringing

Washing and trimming are used to decontaminate carcasses. Barkate et al. [7] used hot water washing at 95°C to increase the surface temperature of beef carcasses to 82°C for 10 seconds which resulted in reductions in bacterial number. Smith [120] reported that more than 3-log reductions were possible in the case of *Escherichia coli, Salmonella, Aeromonas hydrophila, Yersinia enterocolitica, Pseudomonas fragi,* and *Listeria monocytogenes* from the surface of beef tissue by applying 80°C water for 10 or 20 seconds. Similarly, spray washing with water at 83.5°C for 10 or 20 seconds resulted in 2.2-log and 3.0-log reductions of bacterial counts, respectively [24]. A reduction of 95% was possible in total aerobic plate counts on beef sides by treatment with spray washing at 75°C and 300 kPa [104].

The spray washing was more effective in reducing bacterial counts and visible fecal contamination when pressure and temperature were increased [24,41,65]. Kelly et al. [65] reported that the bacterial numbers on lamb carcasses decreased from 0.05 to 1.0 log/cm^2 when the temperature of the spray washing was increased from 57 to 80°C.

Gorman et al. [41] reported that the temperature of water in spray washing can be reduced using a combination of knife trimming and spray washing. They reported that using hot water to achieve 74°C at the surface in beef resulted in a 3-log CFU/cm^2 reduction, while a combination of knife trimming and spray washing with warm water less than 35°C resulted in reductions of 1.4–2.3 log CFU/cm^2. Reagan et al. [107] also reported that spray washing was as effective as trimming in decontamination of beef carcasses. Delmore et al. [26] also indicated that the decontamination of beef carcasses could be achieved by knife trimming followed by spray washing or by spray washing followed by hot water rinsing. However, trimming had some disadvantages: (a) removal of bacteria by hot water was more consistent than that by knife trimming [41,107], (b) trimming can be highly variable and its efficacy can be affected by the skill and/or diligence of the individual, (c) recontamination or cross-contamination may also occur [37].

Chlorinated water is also used for washing. The Europian Union does not allow the addition of chlorine to process water during poultry and meat processing [10]. Chlorine levels do not normally exceed 50 ppm, which results in a reduction in microbial load only by 1-log cycle [60]. Application of 200 mg/liter (20 ppm) chlorine appears to reduce bacteria substantially on poultry, pork, and beef. In some cases more than 200 mg/liter is required, for example, in the case of low initial counts on beef carcasses, and in the case of poultry carcasses a level of 300–400 ppm is required to effectively eradicate *Salmonella* [133]. The effectiveness of chlorine for bacterial reduction can be improved by combining it with an organic acid such as acetic acid or by raising the temperature of the solution [30].

Surface Treatment by Organic Acids

Surface treatment with organic acids is a more realistic option to eliminate pathogens without adversely affecting quality. Smulders [122] reviewed acid treatment in detail. The antimicrobial effects of organic acids are well documented. Many fermented foods are preserved by the formation of lactic acid. Moreover, acids also contribute to the flavor in addition to keeping quality.

The antimicrobial action of acids depends on three factors: pH, extent of dissociation, and specific effect related to acid molecules [57]. Effective growth inhibition by an acid only occurs when an appropriate amount of the undissociated molecule is present [57]. This amount may be obtained either by applying more acid or by lowering the pH. Most organic acids are therefore effective only at low pH, i.e., 5.5. However, in some cases pH > 6 may also be effective for some acids [122]. Ingram et al. [57] mentioned that the different antimicrobial activities (specific effect) of various organic acids is related to the potency to penetrate a cell, the part of the cell that is attacked, and the chemical nature of that attack. For example, heteropolar molecules are surface active, and thus affect the bacterial cell surface and its permeability, and the moderate lipophilic part penetrates the cell membranes. The acid that penetrates affects the cell biology.

Efficacy of meat decontamination by acids: The factors influencing the efficacy of acid treatments are [122] (a) the nature of the meat surface and initial level of contamination, (b) initial bacterial load, (c) types of acids used, (d) concentration and temperature of the acids, (e) types of microorganisms present on the surface, and (f) waiting time after slaughter and duration of acid treatment.

The fat content of meat affects the efficiency of its decontamination [28,29,45,61]. It is easier to reduce microorganisms on fat than on lean beef. The main reason for this is that the buffering capacity of lean tissue is many times greater than that of fat tissue, which allows a much faster initial drop in surface pH [122].

Prasai et al. [105] suggested that acids may not be able to properly contact firmly attached bacteria in the crevices of pork skin. A high degree of contamination of the meat surface with organic material and increasing levels of organic material in scalding tanks reduces the efficiency of acid treatments and requires higher acid concentrations [122]. Initial levels of bacterial contamination on meat significantly affect the treatment and the pattern varies between bacterial species [122].

Mountney and O'Malley [88], after considering 11 acids, recommended acetic, adipic, and succinic acids rather than hydrochloric, lactic, or sorbic acid as being most effective in extending the shelf life of poultry. Chung and Goepfert [21] recommended acetic and propionic acid as the most effective agents for *Salmonella* after examining 13 acids. Some authors also recommended lactic acid. Mixtures of acids have been used for their synergistic effects on microorganisms, but synergistic effects were not always observed. Smulders [122] concluded that both acetic and lactic acid seem to produce the desired microbial reduction and that a mixture of the two may enhance the effectiveness.

The concentration of acid used is usually 1–4% and the temperature 15–55°C. In general, higher acid concentrations and temperatures produce the best antimicrobial effects [122]. Different types of microorganisms react differently to acids, for example, yeasts and molds are more acid tolerant than bacteria. Smulders [122] reviewed different aspects of acid sensitivity of various pathogenic and spoilage microorganisms.

Bacteria are attached to the meat surface by (a) retention (i.e., retained in a liquid surface film), (b) entrapment (i.e., by the specific microtopography of the surface), (c) adsorption (i.e., by short-range attractive forces in a solid-liquid interface), and (d) adhesion (i.e., by intimate contact with polymer bridging or fimbriae and holdfasts) [122]. Thus, the meat surface should be treated quickly after dressing or boning before contaminating organisms have colonized on the surface. Attachment is generally considered to be a two-step process: reversible association with a surface followed by irreversible adherence [83]. Kim et al. [66] described a washing procedure applied to remove reversible attached bacteria associated with the surface water film. Bacterial cell attachment to surfaces is a complex interaction between cells, surfaces, and liquid phases [36,84]. Piette and Idziak [100,101] found that composition and ionic strength of the liquid phase were more important than other factors in some meat-spoilage bacteria. Adhesion of bacteria to surfaces could involve specific interaction between complementary surface structures of microorganisms and physicochemical characteristics, such as charges or surface free energies [15]. Selective blocking of carboxyl or amino groups at the bacterial cell surface by chemical means [101] and coating the cell with antiserum [66] did not reduce attachment to meat. Hydrophobicity and surface charge of bacterial cells are generally involved in attachment. Kim et al. [66] found that live and gamma-irradiated *Salmonella typhimurium* cells had similar cell surface charges, whereas cells grown in a chemically defined medium had different surface charges. In both cases cells had similar surface hydrophobicity. However, differences in cell surface charge did not affect attachment rate, thus irreversible attachment of *Salmonella* to poultry skin was nonspecific. A sufficient contact time should be allowed to achieve desired acid-microbe interaction [122]. The modes of application, such as spraying, immersion, and electrostatic dispersion, also have an impact on the efficiency of the process.

Effect of acid treatment on quality: Acid treatment may also affect the sensory and functional properties of meat, such as color, flavor, odor, and drip loss. In general, treatment with lactic, acetic, and citric acids at low concentrations does not affect color adversely. Higher concentrations resulted in bleaching of lean tissue and caused off-color of fat tissue [122]. For example, in veal carcasses the use of lactic acid is limited to 1.25% but can be 2% if all fat is trimmed off [123]. When many blood spots are present on the meat surface, the coagulation of blood may cause rusty brown-black spots. This is more evident at higher acid concentration [122]. Smulders [122] concluded that decontamination with acetic or lactic acid at effective concentrations of 1–2% scarcely impairs the sensory quality of meat surfaces. The sensory scores are more readily affected by acetic acid than lactic acid, therefore mixtures may alleviate color problems. Smulders also mentioned that the slight visible discoloration forms immediately after application of acid at low concentrations, and usually disappears upon diffusion of the acid. Acetic, propionic, lactic and formic acids were considered acceptable from a toxicological perspective [35].

Phosphates, Hydrogen Peroxide, and Ozone

Inorganic phosphates, hyrogen peroxide, and ozone are also used for meat carcass decontamination. Trisodium phosphate treatment is officially accepted and implemented in the poultry slaughter process [10]. It does not cause undesirable sensory effects that would be detectable by the consumer [52]. A 10–12% alkaline solution could be used [10].

The formation of radicals by hydrogen peroxide damages nucleic acids, proteins, and lipids, providing a bactericidal-bacteriostatic effect [64]. Hydrogen peroxide as a poultry carcass decontaminant was shown to be minimally effective at 0.5%(v/v) in water. At this level a temporary bleaching and bloating of the carcasses and excessive foaming of chiller water is observed [10]. Bolder [10] concluded that the application of hydrogen peroxide for decontamination seems to be an effective and safe method to control the spread of pathogens.

Ozonated water is also used for washing of carcasses [107]. Sheldon and Brown [115] demonstrated that ozonated water produced neither visual defects to the carcasses, nor sensory off-flavors, but bacterial count reduction was poor, i.e., less than 1-log cycle and with no increase in shelf life. Gorman et al. [41] mentioned that spraying beef carcasses with water followed by a spray treatment with ozonated water resulted in effective bacteriological sanitation.

Chlorine and Chlorine Dioxide

Aqueous chlorine is widely used in food processing to control microbial growth. Its bactericidal activity decreases in alkaline conditions and/or at high levels of organic matter. Furthermore, potentially toxic/mutagenic reaction products, including trihalomethanes, are formed during chlorine treatment of food components [75,140].

Chlorine dioxide has received much attention due to its advantages over aqueous chlorine. It is seven times more potent than aqueous chlorine in killing bacteria in poultry-processing chill water [74], its bactericidal activity is not affected by alkaline conditions and/or the presence of high levels of organic matter [23,143], and it is less reactive than aqueous chlorine in interacting with organic compounds, such as unsaturated fatty acids, their methyl easters [141], and tryptophan and their derivatives [114]. The Food and Drug Administration approved a 3 ppm residue of chlorine dioxide for controlling microbial populations in poultry-processing water [75].

Antibiotics

It is possible to improve hygiene levels by means of antibiotics [8]. The most frequently used and least costly of such antibiotics are chlorotetracycline and oxytetracycline. It was concluded that the most effective applications involved combinations of antibiotics [70]. In many cases application of antibiotics to meat faces criticism and protest. Many do not agree with antibiotic application to meat as a legal procedure because antibiotics may cause toxic and allergic reactions in consumers due to bacterial resistance and cumulative effects in the human body. Thus, a strict control on applications should be maintained [8].

II. POSTHARVEST HANDLING OF FISH AND SEAFOOD

A. Fish and Seafood Spoilage

During life, fish are protected by a skin that secretes antimicrobial compounds, such as lysozyme, and by antibodies in the blood. Lean fish contain 20% protein, less than 5% lipid, with little carbohydrate, whereas fatty fish contains 10–30% lipid. The pH of the flesh is near neutral. The flesh is highly buffered due to the presence of phosphates and creatine in the muscle and has a low oxidation-reduction potential [38].

Harvesting results in the death of fish with the following consequences: (a) cessation of the supply of energy, (b) cell membranes are no longer energized, and molecules and ions can freely diffuse, (c) antimicrobials are no longer produced or distributed, (d) microflora penetrates the skin from outside surface and flesh from the intestines and gills [38]. During this period the contractile mechanism can still operate, permitting the muscle to contract and relax [103]. The muscle consumes ATP in order to remain relaxed, thus when ATP is no longer sufficient the muscle will contract—this is rigor mortis. All muscle contracts in rigor mortis, either because ATP is exhausted or because the pH has fall sufficiently. Suzuki [130] showed linewidths of NMR spectra of water in flat fish in prerigor, rigor, and postrigor stages. The linewidth in rigor stage was broader than in the prerigor and postrigor stages, but in the postrigor stage it became narrower than in the prerigor stage. The same has also been observed in sea bass, but only in the case of starved carp where rigor mortis was obscure and the change in the width of lines was not clear [130]. Blanshard and Derbyshire [9] investigated the state of water in muscles as a function of time after death by spin-lattic relaxation. They found that during rigor there was a net transfer of water from the free phase of the region giving rise to a more rapidly relaxing signal.

Marine fish contain trimethylamine oxide (TMAO) as an osmoregulator. Some bacteria or endogenous enzymes can reduce TMAO to trimethylamine (TMA, which has the odor of stale fish) and formaldehyde. The bacteria can obtain energy from this reaction and use it as electron acceptor in the absence of oxygen. Fresh water fish containing TMAO usually have a longer shelf life, even if the level of TMAO is low. Thus, the level of TMAO-reducing bacteria is important in order to preserve marine fish. There is little fermentation activity in the microflora due to the low carbohydrate content of most fish [38]. The initial quality of seafood is related to the species, the growing area conditions, the fishing or harvesting techniques, the seasonal biological changes in muscle and other organs, and the postharvest storage and processing conditions [14]. The ultimate quality is linked to the biochemical changes in the major constituents and microbial load and type in the fish or seafood and water from which they are harvested. The organoleptic changes are caused by bacteria. The active bacteria are psychrotrophic or cold loving and well adapted to growth under chill conditions [38].

For nondestructive testing, gill odor has been considered the most reliable and reproducible characteristic, but cooked flavor without added condiments is the most precise [38]. The most distinct stages of spoilage are shown in Table 2. The concentration of TMA is used as an index of spoilage either alone or as a component of the total volatiles containing ammonia and other amines. An alternative chemical index is hypoxanthine, which is derived from the breakdown of adenosine triphosphate. After the death of a fish, ATP in the meat decomposes to uric acid by a series of catabolic enzymes [108]. Hypoxanthine is an intermediate of these reactions and accumulates as storage time increases. Another more complex measurement is the potassium value, which is based on ATP, adenosine diphosphate (ADP), adenosine monophosphate (AMP), and hypoxanthine [38]. Spooilage results in changes in the electric properties of the skin and tissue, and changes in the dielectric properties of whole fish and of skin-on portions can be measured in a nondestructive manner [38].

TABLE 2 Characteristics of Phase Changes During Spoilage of Cod Fish Stored in Ice

Phase	Storage (days)	Characteristic changes
I	0–6	No marked sign of spoilage
II	7–10	No odor production
III	11–14	Production of some sour, slightly sweet to fruity odors
IV	>14	Production of hydrogen sulfite and other sulfide compounds, fecal, and strong ammonia odors

Source: Ref. 116.

Postharvest fish muscle quality can be improved by and a delay in onset of rigor mortis is associated with low stress or rapid killing methods [63]. Jerrett et al. [63] mentioned that a combination of behavioral conditioning, conservative handling practices, and chemical anaesthesia can be used to minimize the extent of premortem exercise and thereby provide rested fish. They also mentioned that it was important to reduce preharvest exercise in the production of high-quality fish muscle.

The amount of drip that is lost from muscle depends largely upon changes in the water-holding capacity of the muscle protein after death. Once water-holding capacity falls below the water content of the muscle, excess water will be lost, either immediately or later during processing or cooking [103]. Drip loss is influenced by a combination of time and temperature of the immediate postmortem period [103].

1. Bacterial Spoilage

The deterioration of fresh fish is primarily due to bacterial action. It is recognized that only some of the bacteria present is responsible for producing the off-odors, off-flavors, appearance, and textural changes that constitute spoiled fish [38]. The composition of the microflora is dependent on different factors, such as environment and season. Spoilage bacteria and mechanical handling of fish affect the degradation of nucleotides in fish. In cod fillets the presence of spoilage bacteria increased the rate of degradation of inosine to hypoxanthine. This indicates that bacterial enzymes play an active role in contributing to the degradative process. Moreover, handling of and mechanical damage to muscle tissue may increase degradation of inosine monophosphate and inosine by making their substrates more accessible due to enzyme decompartmentalization [22]. The synthesis of proteases depends on the nutrient source, and secretion of proteases increases with the decrease in available nutrients, reaching a maximum concentration by late log or early stationary phase of the bacterial growth. The level of protease secretion is low during the initial stage of bacterial spoilage, i.e., 10^6 CFU/g of flesh. Above 10^6 CFU/g there is high protease secretion [22].

Bacterial spoilage is evident in fish even at 0 to –4°C, but spoilage can be prevented below –10°C [116]. For cod stored in ice, there is a 2- to 3-day lag period with a logarithmic increase by day 10 in the bacterial flora, generally with counts up to $10/^8cm^2$ skin or 10^8/g muscle. *Pseudomonas* species dominate up to 90% by the day 12. They produce the spoilage odors, such as ammonia, and volatile sulfur compounds, such as mercaptans and hydrogen sulfide [42]

The main hazards of fish and seafood are pathogenic bacteria (aquatic environment, from humans or animals, biogenic amine producers, spoilage bacteria), parasites, biotoxins, and viruses [56]. All aquatic environments can harbor spores of *Clostridium botulinum*, which can contaminate fish both in marine and fresh water environments. *Vibrio parahaemolyticus* is the lead cause of food poisoning in Japan, where much fish is eaten raw. It does not grow below 10°C and dies out at chill temperature and is heat labile. As it is halophilic, it survives salting and thus smoking. Control is therefore achieved by proper chill storage. There has been a global increase in the presence of algal toxins in shellfish. More toxins are being identified. The major concern in recent years to the fish-processing industry has been *Listeria monocytogenes*. This organism can grow at chill temperatures and is not inhibited by the levels of salt in smoked products [38]. The bacterial flora of water reflects the flora of fish harvested from the area. The flora of cold water fish are predominately gram-negative, while flora in tropical fish are predominately mesophilic gram-positive microorganisms [22]. The dominant microflora of cold water fish species are *Pseudomonas, Alteromonas, Moraxella, Acinetobacter, Vibrio, Flavobacterium,* and *Cytophaga* [77]. Gram-positive organisms found on warm water fish species are predominately *Micrococcus* and *Bacillus* [22]. Psychrotrophic bacteria are found in almost all types of refrigerated and frozen foods. The outgrowth of pseudomonands in spoiling fish is due to their efficient use of free amino acids (especially methionine and cysteine) and peptides in nonprotein nitrogen fraction during the early stage of spoilage, and their secretion of potent proteases, thus promoting proteolysis after low molecular weight components have been exhausted. The primary use of all nitrogenous compounds is through oxidative deamination, which results in the accumulation of ammonia, volatile fatty acids, and sulfur-containing compounds [77]. Volatile sulfur-containing compounds are believed to produce spoilage odors. These compounds have extremely low thresholds, and their origin is usually traced to nonprotein nitrogenous compounds in the skin and flesh.

Some volatile compounds that have been isolated from spoiling fish muscle are ethyl mercaptan, methyl mercaptan, dimethyl sulfide, dimethyl disulfide, hydrogen sulfide, acetaldehyde, propionaldehye, diacetyl, ethanol, methanol, acetone, acetoin, butanal, methyl butanal, and ethanal [77]. Selected volatile compounds in spoilage fish and their probable sources are compiled in Table 3.

2. Color Changes

Fresh fish has a translucent appearance due to the even scattering of incident light. With an increase in spoilage, there is a gradual disintegration of myofibrils, resulting in their wider and more random intracellular distribution. The fish surface then appears opaque because the incident light is unevenly scattered [22,47].

The color of fish flesh changes during low-temperature and freezing storage. The flesh becomes yellow due to oxidation of carotenoid pigments and lipids in the tissues. Other factors that may result in yellowing are lipid oxidation and reactions with carbonyl-amines. The red color changes in whole orange roughy during ice storage are shown in Table 4 [22].

Factors affecting pigment loss are as follows: (a) myetoperoxidase from fish leukocytes cause rapid discoloration of β-carotene in the presence of hydrogen peroxide and iodide or bromide ions due to the breakage of double bonds, (b) free radical addition, free radical abstraction, or singlet oxygen addition to the double bonds of β-carotene, (c) temperature and concentration of storage oxygen, (d) progressive increase in pH from 7 to 8 during storage; potassium β-oxyacrolein, the enolic salt of malondialdehyde, formed self-condensation reactions and polymerized at pH 7–8 to form fluorescing compounds [22]. The appearance of surface slime on whole orange roughy stored in ice is shown in Table 5 as a function of storage days.

3. Textural Changes

The texture of fresh fish changes from "firm" and "moist" to "mushy" and "runny" during storage. These changes in texture are due to tissue softening as a result of myofibrillar disintegration and connective tissue weakening. During storage the spoiling storage intracellular and extracellular pro-

Table 3 Selected Volatile Compounds and Their Probable Source in Spoiling Fish

Compound	Probable source
Hydrogen sulfide	Cysteine
Dimethyl sulfide	Methionine
Methyl mercaptan	Methionine
Acidic, propionic, butyric, and hexanoic acid esters	Glycine, leucine, serine
Trimethylamine	Trimethylamine oxide
Dimethylamine	Triethylamine oxide
Ammonia	Urea, various amino acids

Source: Ref. 77.

TABLE 4 Color Changes Appearance of Gills in Whole Orange Roughy Stored in Ice

Storage (days)	Color changes	Appearance of gills
0	Red/orange	Dark red
4	Red/organe	Dark red
6	Orange on fins, head, and tail	Dark red
9	Slightly blotched, body faded	Dark red, slightly milky slime
11	Blue steel-gray with tinges of orange	Brown/red or bleached, sticky, creamy, slime
13	Bleached pale grey or blue tail with head pale orange	Brown/red or bleached with brown slime on gills
16	Washed-out grey/blue with pale head	Brown or bleached with brown slime

Source: Ref. 113.

teases degrade myofibrillar proteins [22]. The changes in texture of whole orange roughy stored in ice for 16 days are shown in Table 6. The sensory changes of white fish are presented in Table 7. Quality loss of stored fish muscle may take the form of development of unpleasant odors and flavors, excessively soft texture, loss of liquid-holding capacity, and development of a dry or tough texture upon cooking [96]. Postmortem tenderization of fish muscle is one of the major problems related to fish freshness and its quality [99]. The causes of postmortem tenderization postulated so far are as follows [1]: (a) weakening of rigor between myosin and actin as seen by the decrease of Mg^{+}-ATPase activity, (b) breaking down of Z-disc structure of myofibrils, and (c) degradation of titin [99]. Ando et al. [1] also suggested that postmortem tenderization of fish muscle is closely related to the gradual disintegration of the extracellular matrix structure after death. The weakening of the Z-line depends on a proteolytic mechanism, which removes α-actinin from this structure by action of a calcium-dependent neutral proteinase: calpain [99]. In addition, other myofibrillar proteins such as titin and nebulin are also implicated in postmortem weakening of fish [34] and meat muscle [2]. Papa et al. [99] studied α-actinin release and its degradation from myofibril Z-line in postmortem white dorsal muscle from bass and sea trout stored at 4° and 10°C. Using α-actinin–specific antibodies, they showed that this protein is rapidly released within the first 24 hours for the two species, and reaches a plateau within 4 days. The release and proteolysis of α-actinin are time- and temperature-dependent processes that take place at early stages of fish storage. The proteolysis of α-actinin seems to be dependent on fish species.

In fish intramuscular connective tissue, type I and V collagens are presented. Sato et al. [109,110] demonstrated that type V collagen was solubilized specifically in softened rainbow trout muscle. Similarly, type V collagen became solubilized in softened sardine muscle after 1 day of chilled

TABLE 5 Appearance of Surface Slime on Whole Orange Roughy Stored in Ice

Storage (days)	Color and consistency of surface slime
0	No slime
4	Clear slime
6	Clear or slightly cloudy slime
9	Clear or slightly cloudy slime
11	Clear or slightly cloudy slime
13	Brown slime on body
16	Thick yellow slime

Source: Ref. 113.

Table 6 Changes in Whole Orange Roughy Stored in Ice

Storage (days)	Odor changes in gills	Textural changes in flesh
0	Slightly seaweedy	Firm and resilient
4	Slightly seaweedy	Firm and resilient
6	Mild and sweet	Firm and resilient
9	Fish meal, sweet, salty, briney	Firm and resilient
11	Sweet, salty, mussel, soapy, oily	Retains finger indentation, no gaping, slightly soft
13	Metallic, stale, seaweedy, and slightly rotting odors	Retains finger indentation, soft
16	Strong rotting and putrid	Retains finger indentation, no gaping, soft

Source: Ref. 113.

storage, whereas tiger puffer muscle did not show significant softening, changes in structure of connective tissues, or biochemical properties of collagens [111]. This was due to the presence of more type I collagen in tiger puffed muscle than in sardine, carp, and mackerel [111]. Thus, degradation of type V collagen caused disintegration of the thin collagen fibrils in pericellular connective tissue, weakening pericellular connective tissue, and resulting in postharvest softening. Similarly, Tachiban et al. [131] reported that the degradation of Z-discs of ordinary muscle was faster in cultured red sea bream than in the wild counterpart.

The liquid-holding capacity of muscle is highly influenced by fibril swelling/contraction and the distribution of fluid between intra- and extracellular locations [94]. Moreover, changes in muscle structure are an important factor. Changes in muscle structure are strongly influenced by temperature, ionic strength, and chemical composition due to season, and maturation of muscle and pH. Ofstad et al. [96] concluded that the liquid-holding capacity of raw fish seems to be dependent on two main factors: genetic differences in the muscle protein, and the postmortem muscle pH and subsequent time-dependent muscle degradation. They found that salmon muscle possessed much better liquid-holding properties than the cod muscle, as did wild cod better than fed cod regardless of the storage time. The myofibrils of the salmon muscle were denser, and intra- and extracellular spaces were filled by fat and a granulated amorphous material. The denaturation characteristics of myosin, actin, and a sarcoplasmic protein differed between salmon and cod, indicating the stability of the myosin-actomysin complex. Postmortem degradation of the endomysial layer and the sarcolemma may further facilitate the release of liquid. Thus, the release was related to species-specific structural features and better stability of the muscle proteins. The severe liquid loss of fed cod was due to a low pH–induced denaturation and shrinkage of the myofibrils [96].

Holes and slits appear between the myotomes (muscle segments) because of breakage of the minute tubes of connective tissue that issue from the myocommata (connective tissue sheets) and run between and around the muscle cells. This phenomenon is known as gaping in the musculature of fish [79]. Love and Haq [79] found that a low pH leads to much gaping and vice versa. At a given temperature, the gaping increased with time, and subsequent freezing increased it further [79].

4. Lipid Oxidation

Lipid oxidation is a major cause of quality deterioration in fish and seafood due to lipid content and the extend of polyunsaturation [22,54]. Catalysts for lipid oxidation are molecular and singlet oxygen, metals such as iron and copper, and enzymes such as lipoxygenase [14]. The postmortem changes in fish muscle related to lipid oxidation are (a) decrease in ATP, (b) increase in ATP breakdown products, e.g., hypoxanthine, (c) changes in xanthine dehydrogenase to xanthine oxidase, (d) loss of reducing compounds, e.g., ascorbate, glutathione, (e) increase in content of low molecular weight

Table 7 Sensory Changes in White Fish

		Raw		Cooked	
Quality	Score	Gill odor	General appearance	Odor or flavor	Texture
Fresh	10	Seaweed, sharp metallic	Convex eyes, shiny red blood, flesh translucent	Slight sweet/meaty	Dry
Spoiling	7	Bland, loss of tanginess	Eyes flat, dull skin, no translucence	Bland/Neutral	Firm/Succulent
Stale	5	Milky, mousy, yeasty	Sunken eyes, browning of gills, opaque mucus, waxy flesh	Some amines	Softening
Putrid	3	Acetic acid, old boots odor sour sink	Eyes very sunken, yellow flesh, brown blood, yellow bacterial slime	Bitter, sulfites	Mushy

Source: Ref. 38.

transition metals, (f) conversion of heme (Fe II) pigments to oxidized form (Fe III), (g) loss of structural integrity of membranes, (h) loss of antioxidants in membranes, e.g., tocopherols, (i) inability of muscle cell to maintain calcium gradients [54]. These changes make the tissue more susceptible to oxidation, especially through changes in membranes. In addition, size reduction, such as fillets and minces, can enhance lipid oxidation by exposing more lipid to oxygen. The destruction of antioxidants such as tocopherols by heat, enzymes, and salts also plays an important role in lipid oxidation [14]. The glutathione peroxidase activity is located in various fish muscles and presumably protects muscle from oxidative deterioration of lipid during storage and processing [139]. Nakano et al. [90] reported that glutathione peroxidase increased significantly during storage, which suggested that the increase in enzyme activity could protect fish muscles from oxidative deterioration during storage and processing. Watanabe et al. [139] reported that total glutathione peroxidase activity in Japanese jack mackerel and skipjack tuna fish muscles decreased gradually during stroage at 4°C.

Lipid oxidation can be minimized by (a) reducing oxygen access (vacuum packaging or edible coating), (b) maintaining natural antioxidants or adding antioxidants, (c) minimizing increases or prooxidants (e.g., iron), (d) maintaining low temperature, (e) minimizing salt constituents, and (f) removing unstable lipids (e.g., subcutaneous fat) and dark muscle, which contains more fat [14].

5. Changes in Protein

Proteins in fish and seafood are subjected to significant changes during postharvest storage, processing, and both ice and uniced storage. The proteinases are responsible for changes, such as hydrolysis, which can result from animal or from spoilage microflora especially during the later stages of spoilage [47]. The types of proteases and their inhibitors are given in Table 8. The solubility of proteins during washing increased when fish were held for a longer time and/or at higher temperatures. This is because the degradation of myosin heavy chain and actin increased rapidly at longer storage times and/or elevated temperatures, resulting in a higher loss of total protein during washing. Lin and Park [76] studied the effect of postharvest storage temperatures and times on proteolysis of Pacific whiting. Myosin heavy chain degraded rapidly during postharvest storage at low temperatures (0–5°C), and greater degradation occurred at elevated temperatures. Actin degradation was similar to that of myosin heavy chain, but to a lesser degree. Degradation of both was highly correlated to protein solubility. Low temperatures reduced, but did not completely inhibit proteolysis [76]. It is well established that the myofibrillar proteins differ in stability depending on the habitat temperature of the species [50]. During iced and frozen storage, the thermal characteristics of myosin subunits deteriorate faster in cold water than in warm water fish [25,53].

Table 8 Types of Proteases and Their Inhibitors

Protease	Inhibitor
Serine protease	Phenylmethyl sulfonyl fluoride
	Dispropyl fluorophosphate
Thiol protease	Heavy metal iodoacetamide
	N-ethyl maleimide
	Anipain
	Leupetin
Metalloprotease	EDTA
	O-Phenanthroline
	8-Hydroxyquinoline
Acid protease	Pepstain
	Diazoacetyl norleucine methyl ester
	Epoxy (*p*-nitrophenoxy) propane

Source: Ref. 138.

Large amounts of glycogen in mammalian muscle can result in low final pH. The higher the postmortem temperature, the quicker is the onset of rigor, the shorter its duration, and the more severe the contraction. However, if meat is chilled below about 13°C soon after death, severe contraction takes place almost immediately, resulting in permanent toughening of the meat. This is called cold shortening [103].

Venugopal [138] studied the sites of attack by proteases on proteins and found that the points of cleavage for carboxypeptidase, aminopeptidase, endoprotease and proteinase were carboxyl-terminal, amino-terminal, and internal peptide bonds, respectively.

B. Postharvest Treatments of Fish and Seafood

Warm and cold water fish do not always react in the same way to handling and may often need different treatments to achieve the desired results [103]. Unlike meat, fish is hunted traditionally, thus giving less scope for manipulating the immediate pre- and postmortem conditions. In the case of cold water fish, higher temperatures accelerate the onset and resolution of rigor as in meat, but cold shortening is not caused. This may be because these fish function normally at temperatures below 5°C. Another difference is that fish never seem to attain such low pH values as are sometimes encountered in meat.

1. Washing

Fish may be gutted, i.e., their intestines removed. This is essential for some species, as otherwise the digestive enzymes and the bacteria in the gut would soon attack the flesh. In other species this may not be necessary due to the anatomical location of the gut cavity relative to the edible parts [38]. It is customary to wash fish to remove blood and any remnants of guts. However, washing can remove some of the natural antimicrobial secretions and may not be advantageous to the storage life [38].

The limited efficacy of heat treatments directly applied to whole fish by immersion (hot wash) is probably due to slow heat penetration and/or changes in microflora interactions [136]. Vaz-Pires et al. [137] isolated 80 bacterial isolates from sea scad to study their heat stability. They found *Shewanella colwellinia* to be more heat resistant. The main target site of damage of the heat treatment tested in their experients was the cell wall.

2. Chilling or Icing

Most fish are chilled with ice soon after capture unless they are meant only for local markets. Ice is a very good coolant. It is also cheap, and it must be made from pure, unpolluted and bacteria-free water. Chilling reduces the temperature of fish to 0°C [38]. The shelf life of raw fish as a function of temperature is shown in Table 9.

Table 9 Approximate Shelf Life of Raw Fish as a Function of Temperature

Fish	Temperature (°C)	Shelf life (days)
Cod	0	16
	5	7
	10	4
	16	1
Herring	0	10
	5	4
Salmon	0	2
	10	5
Plaice	0	18
	10	8

Source: Ref. 38.

In fish during ice storage the penetration of microorganisms occurred primarily from the intestines, the skin flora being found in the flesh only during the later stages of spoilage (8–17 days, depending on species) [51,55]. It was also observed that the quality and keeping time of the trout was reduced when the fish was exposed to physical stress. The infection level in the fish muscle increased with increasing physical stress and was higher for feeding than for starving trout [51]. When gutting the fish it was observed that the intestines appeared pale and bloodless in the treated samples. This could be the result of a stress condition, which increased the production of adrenalin, thus forcing the blood into the muscles and gills. There was good correlation between the log count and the organoleptic score when the bacteria count was higher than 100/g. Herborg and Villadsen [51] mentioned that the lack of correlation at lower bacterial counts may be expected, since autolytic spoilage processes will be more active in the earlier storage period and will be the predominant factor influencing organoleptic assessment.

Gibson [38] mentioned that the importance of rapid chilling cannot be overstressed. Chilling reduces the rates of chemical and biochemical change as well as microbial growth. The temperature at which fish live is relatively low, thus chilling of fish does not have as great an effect as the chilling of meat from warm-blooded animals [38].

Hanpongkittikun et al. [49] identified *Staphylococcus aureus, Salmonella,* and *Vibrio parahaemolyticus* from shrimp samples during ice storage. Samples were taken from controlled harvesting and open market. Controlled harvested shrimp had a shelf life of 8 days, samples from open market 4 days.

Chilling to 0°C in ice or refrigerated areas is an essential requirement for quality retention for most fish and seafood. Storage at –2 to –4°C is better than at 0°C or 2–5°C from a microbial point of view, but not necessarily from a quality point of view due to toughening and autolytic changes such as lipid hydrolysis. Shelf life may be extended further by use of sorbate or sulfite [14]. Good hygienic practices are also needed to reduce the possibility of contamination of products. This is achievable on land but is more difficult on board fishing vessels [38].

3. Chemical Treatment

Chlorine and Chlorine Dioxide

Lin et al. [75] found that a commercial chlorine dioxide killed *Escherichia coli, Listeria monocytogenes*, and its streptomycin-resistant strain at 15, 10, and 7.5 ppm. They also found that aqueous chloride dioxide was more effective than aqueous chlorine in killing *L. monocytogenes* on fish cubes and in washed-off solutions. Fish cubes treated with aqueous chlorine showed no visual change in color. The treated solutions became lightly milky. A light brown color occurred on fish cubes treated with aqueous chloride dioxide at 400 ppm. The treated solutions had a light pink (40 and 100 ppm) to light yellow (200 and 400 ppm) color, with some turbidity. The fish cubes treated with aqueous chlorine or chlorine dioxide contained no detectable chlorine residues, but commercially chlorine dioxide solution showed chlorite and some free and combined chlorine, especially at 200 and 400 ppm.

Hydrogen Peroxide

The shelf life of fish can be increased by dipping in hydrogen peroxide solution. Hydrogen peroxide acted as a preservative as well as a bleaching agent, thus yielding a higher quality product with an extended shelf life. A major point of concern had been that hydrogen peroxide treatments would lead to excessive oxidative rancidity, thereby causing a marked decrease in the overall quality of product [118]. Sims et al. [118] indicated that hydrogen peroxide was entirely dissipated in the flesh within 0.5 hour when raw fillets of herring were immersed in dip solutions containing up to 600 ppm hydrogen peroxide. The color observations carried out on samples from the treatments indicated that the fillets treated with hydrogen peroxide were considerably whiter than those of untreated fillets.

Lactic Acid Bacteria

The addition of living cultures of lactic acid bacteria is used in fish to control pathogen growth. *Listeria monocytogenes* is difficult to control in lightly salted (<6% NaCl in aqueous phase), pH (above

5), and storage temperature around 5°C. Wessels and Huss [142] mentioned that *L. monocytogenes* in lightly preserved fish products can be controlled using food-grade lactic acid bacteria. The effect was due not to lactic acid inhibition, but to the production of the natural preservative nisin by the lactic acid bacteria. Sodium chloride up to 4% allowed for efficient growth and nisin production, while 5% sodium chloride resulted in very slow growth and no detectable nisin.

Enzyme Inhibitors

The tenderization or flesh softening of seafood has generally been attributed to the activity of endogenous muscle proteases in the postmortem animal. The undesirable postmortem activities of these enzymes have been controlled by low-temperature and chemical treatments [4]. The use of plasma glycoprotein and α_2-macroglobulin (the active component in egg white plasma hydrolysates) can be used to inhibit several endogenous proteases [78]. α_2-Macroglobulin noncompetitively inhibited the proteases, in decreasing order: cathepsin D > trypsin > chymotrypsin > collagenase. The inhibitor's activity depended on the size of the substrate molecule, the size of the enzyme, and the relative specificity of the enzyme. Thus, it could control undesirable proteolytic activities. In intact fish and other muscle foods, it may be restricted to tissue penetration due to membrane barriers and the relatively large molecular size of the inhibitor. Proteases in surimi and other minced muscle foods should be readily inhibited by it. The activity did not seem to be adversely affected by low temperatures (4–7°C) [4].

Prawns develop blackspot (melanosis) in chilled and frozen storage. Sulfite is a reducing agent used to prevent this discoloration. It reverse the formation of colored compounds (quinones) and in addition acts as a competitive inhibitor of polyphenol oxidase, the enzyme that causes the production of the pigment melanin [80,145]. Recently it was found that 4-hexylresorcinol binds irreversibly to polyphenol oxidase, inhibiting its action [85]. The likelihood of adverse reactions in humans from the low levels found in prawns is slight [119]. Sensory panel evaluation found no effect on the taste, texture, visual appearance, and development of normal colors after cooking by treating prawns with 4-hexylresorcinol. Slattery et al. [119] used Everfresh® (which contains 4-hexylresorcinol) to inhibit polyphenol oxidase in trawled and farmed prawns. In comparison with sodium metabisulfite treatment, Everfresh® (0.2% 4-hexylresorcinol) provided greater protection against blackspot, particularly on the body of the prawn during storage on ice, in refrigerated sea water, and in ice after frozen storage. When Everfresh® was used according to the manufacturer's recommendations, residues of 4-hexylresorcinol were less than 2 μg/g in prawn flesh.

III. POSTHARVEST HANDLING OF MILK

Milk is a polyphasic normal secretion of the mammalian glands, containing 3.9% fat, 3.3% protein, 5.0% lactose, and 0.7% minerals [144]. Milk proteins include α-, β-, and κ-caseins, β-lactoglobulin, α-lactalbumin, serum albumin, lactotransferrin, immunoglobulins, and β_2-microglobulin [33]. The term milk is understood as reffering to cow's milk unless other species are mentioned specifically. The acidity of normal fresh milk varies from 0.10 to 0.26% lactic acid with a pH of approximately 6.5. The addition of acids results in the formation of a precipitate, which appears as a soft, white, jellylike mass, known as curd, with more or less separation of nearly clear fluid or whey. It is important that the milk used for processing have acceptable flavor characteristics. Various weed, feed, and cowy flavors can be transmitted to milk by the cow's respiratory or digestive system. These are considered normal and acceptable up to certain level, and excessive amounts can cause off-flavors that are difficult to remove by processing. Salty flavor can arise from cows in late lactation and infected with mastitis. Milk diluted with water can taste flat and be lacking in typical flavor [71].

Undesirable changes in raw milk are initiated by microbiological growth and metabolism or by chemical or enzymic reactions. Temperature is critical for dairy food quality and shelf life. Cold temperatures are used to minimize microbial growth in raw milk until it can be processed and to extend the shelf life of nonsterile dairy foods. The microorganisms in raw milk just prior to pasteurization may include heat-susceptible pathogens as well as spoilage types [71]. Psychrotrophs became an escalating problem for the dairy industry during the introduction of refrigerated storage of raw milk. Psychrotrophic microorganisms capable of growing in milk at temperatures close to 0°C are represented by both gram-negative and gram-positive bacteria. For example, the gram-negative bacteria are *Pseudomonas, Achromobacter, Serratia, Alcaligenes, Chromobacterium*, and *Flavobacterium*; and gram-positive bacteria are *Bacillus, Clostridium, Corynebacterium, Streptococcus, Lactobacillus* and *Microbacterium* [127]. In aerated milk at 4°C, many strains of *Pseudomonas* spp. can produce sufficient proteinases to hydrolyze all of the available casein into soluble peptides [86,126]. The enzyme activity from psychrotrophs stimulates the growth of starter lactic acid bacteria in milk [127]. The quality of milk may be affected by heat-resistant enzymes secreted by psychrotrophs in raw milk before heat treatment or other enzymes and metabolites that are produced by microflora during cold storage. The shelf life of dairy products is given in Table 10.

The spoilage of milk and dairy products is characterized by taste and odor changes, such as sour, putrid, bitter, malty, fruity, rancid, and unclean. The type of spoilage may also cause undesirable body, texture, and functional changes [71]. In milk about 40% of the milk solids is lactose, a major substrate for microbial fermentation in milk. Microorganisms use one of the two following methods to start fermentation: by the lactase enzyme (β-D-galactosidase) or by hydrolyzing the phosphorylated lactose by β-D-phosphogalactoside galactohydrolase. Microorganisms containing the lactase enzyme include *Escherichia coli, Streptococcus thermophilus, Streptococcus lactis, Lactobacillus bulgaricus, Lactobacillus plantarum*, and *Bacillus subtilis*. Lactic acid bacteria converts lactose to lactic acid and other byproducts. Milk with a detectable acid/sour flavor is considered unacceptable commercially. Cold storage temperatures and sanitary storage and processing conditions for raw milk and cream can prevent the high acid/sour flavors to develop [71].

A malty flavor or odor can occur in milk if *Streptococcus lactis* var. *maltigenes* grows and metabolizes amino acids in milk to aldehyde and alcohols. The fruity flavors in dairy foods can be caused by the metabolic activity of lactic acid and psychrotrophic bacteria with the formation of esters. Flavor defects in milk described as putrid, bitter, and unclean may be caused by the growth and metabolism of psychrotrophic bacteria. The lipase enzyme is often active at low temperatures, causing lipolyzed flavor. Gram-negative psychrotrophic bacteria are actively lipolytic [71].

Low-temperature storage can reduce the frequency of raw milk collection from dairy farms to just two or three times a week, and further storage of milk in the dairy plant over weekends [127]. Bulk milk tanks usually ranging from 0.8 to 19 m^3 are used to receive, cool, and hold the milk. As the cows are mechanically milked, the milk flows through sanitary pipelines to an insulated stainless steel bulk tank. An electric agitator stirs the milk, and mechanical refrigeration begins to cool it even during milking, from 32.2 to 10°C within the first hour, and from 10 to 4.4°C within the next hour. Some large dairy farms may use a plate or tubular heat exchanger to rapidly cool the milk. The temperature of the blended milk must be below the 7.2°C during the second and subsequent milkings [5].

Since the milk is picked up from the farm tank daily or every other day, cooled milk may be stored in an insulated silo tank. Milk in the farm tank is pumped into a stainless steel tank on a truck for delivery to the plant or receiving station. The tanks are well insulated and the temperature rise should not be more than 1.1 K in 18 hours when testing the tank full of water and the average gradient between the water and the atmosphere surrounding the tank is 16.7 K [5].

The most common grades of raw milk are Grade A and Manufacturing Grade. The dairy farmer must meet state and federal standards to produce Grade A milk. In addition to the state requirements, a few municipal governments also have raw milk regulations. The dairy farmer must have healthy

Table 10 Shelf Life of Dairy Products

Product	Temperature (°C)	Shelf life (days)
Market milk	<4	12–14
Cottage cheese	2–4	25–30
Yogurt, sour cream, and dairy dip	<4	30–60
Cured cheese	<4	Several months

Source: Ref. 71.

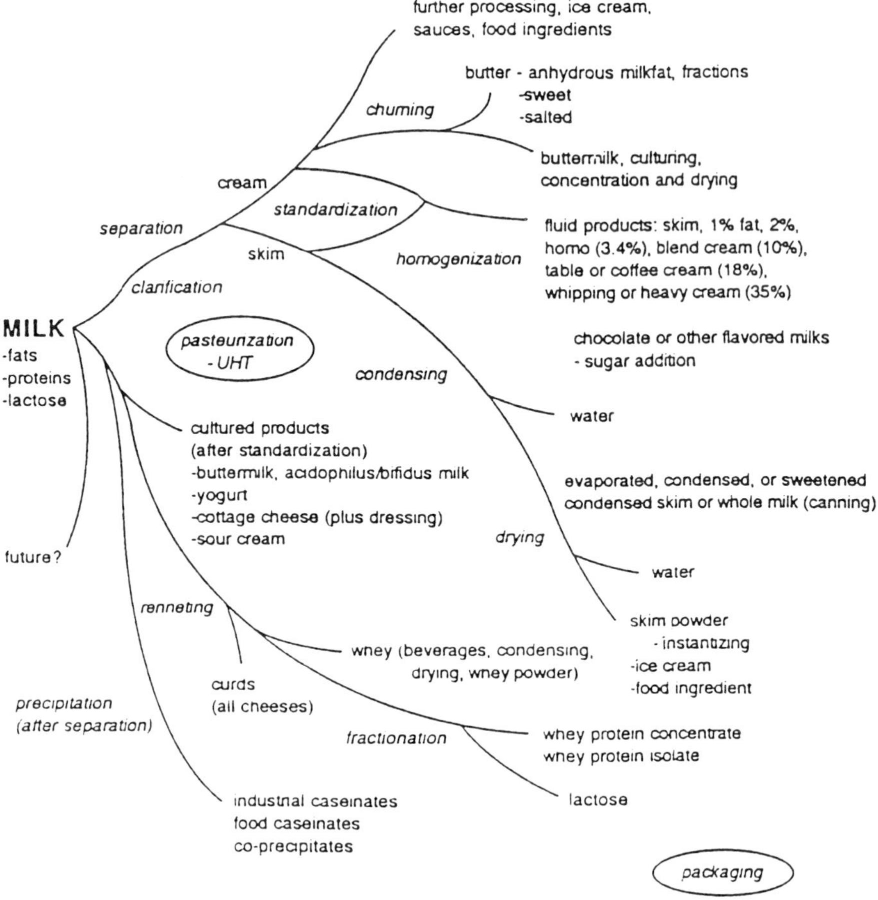

FIGURE 1 The family of dairy products manufactured from milk. (From Ref. 40.)

cows, have adequate facilities (barn, milkhouse, and equipments), maintain satisfactory sanitation of these facilities, and have milk with a bacteria count less than 1×10^5 per ml for individual producers. Commingled raw milk cannot have more than 3×10^5 counts per ml. The milk should not contain pesticides, antibiotics, or sanitizers, show no positive results of drug residue or objectionable flavors and odors [5].

Most dairy processing plants either receive raw milk in bulk from a producer or arrange for pickup directly from the dairy farms. Storage tanks, from 4 to 230 m^3 made of stainless steel lining and well insulated, may be required for nonprocessing days and emergencies. An average 18-hour temperature change of no more than 1.6 K should be in the tank filled with water, and gradient from the surrounding air is 16.7 K. For horizontal storage tanks, the allowable temperature change under the same conditions is 1.1 K. The tank may need cooling depending on the initial milk temperature and holding time. A plate heat exchanger may be connected or the tank surface around the lining may be cooled by passing a refrigerant or by circulation of chilled water or glycol solution. Agitation is essential to maintain uniform milk fat distribution. Milk held in large tanks, such as the silo type is continuously agitated with a slow-speed propeller driven by a gearhead electric motor or with filtered compressed air [5].

A number of milk products are available from which water has been partially removed. In the manufacture of evaporated milk, slightly more than half of the water is removed by heating the milk at 48.9–60°C in vacuum pans. After evaporation the milk is homogenized, cooled, placed in cans, and sterilized at 115.6°C for 15 minutes. Most evaporated milk is fortified with vitamin D [11]. The family of dairy products manufactured from milk is shown in Figure 1.

REFERENCES

1. M. Ando, H. Toyohara, Y. Shimizu, and M. Sakagushi, Post mortem tenderisation of rainbow trout (*Oncorhyncus mykiss*) muscle caused by gradual disintegration of the extracellular matrix structure, *J. Sci. Food Agric. 55*:589 (1991).
2. T. J. Anderson and F. C. Parrish, Post mortem degradation of titin and nebulin of beef steaks varying in tenderness, *J. Food Sci. 54*:748 (1989).
3. A. J. S. Angelo and A. M. Spanier, Lipid oxidation in meat: mechanism and control, *Shelf Life Studies of Foods and Beverages* (G. Charalambous, ed.), Elsevier Science Publisher B. V., Amsterdam, 1993, p. 35.
4. I. N. A. Ashie and B. K. Simpson, α-Macroglobulin inhibition of endogenous proteases in fish muscle, *J. Food Sci. 61*:357 (1996).
5. ASHRAE Handbook, *Refrigeration Systems and Applications*, American Society of Heating, Refrigerating, and Air-Conditioning Engineers, Inc., Atlanta, 1994.
6. M. E. Bailey, K. O. Intarapichet, K. O. Gerhardt, R. A. Gutheil, and L. A. Noland, Bacterial shelf-life of meat and volatile compounds produced by selected meat spoilage organisms, *Shelf Life Studies of Foods and Beverages* (G. Charalambous, ed.), Elsevier Science Publishers B. V., Amsterdam, 1993, p. 63.
7. M. L. Barkate, G. R. Acuff, L. M. Lucia, and D. S. Hale, Hot water decontamination of beef carcasses for reduction of initial bacterial numbers, *Meat Sci. 35*:397 (1993).
8. M. S. Basol and U. Gogus, Methods of antibiotic applications as related to microbial quality of lamb by PCA and bioluminescence, *J. Food Sci. 61*:348 (1996).
9. J. M. V. Blanshard and W. Derbyshire, *Water Relations of Food* (R. B. Duckworth, ed.), Academic Press, London, 1974, p. 568.
10. N. M. Bolder, Decontamination of meat and poultry carcasses, *Trends Food Sci. Technol. 8*:221 (1997).
11. G. Borgstrom, *Principles of Food Science*, The Macmillan Company, London, 1968.
12. P. E. Bouton, A. L. Ford, P. V. Harris, and F. D. Shaw, Electrical stimulation of beef sides, *Meat Sci. 4*:145 (1980).
13. T. J. Braggins, Effect of stress-related changes in sheepmeat ultimate pH on cooked odor and flavor, *J. Agric. Food Chem. 44*:2352 (1996).
14. K. A. Buckle, Postharvest technology of seafood products, *Postharvest Technology for Agricultural Products in Vietnam* (B. Champ, and E. Highley, eds.), ACIAR Proceedings No. 60, Australian Centre for International Agricultural Research, Canberra,1995, p. 128.
15. H. J. Busscher, A. H. Weerkamp, H. C. Van der Mei, A. W. J. Van Pelt, H. P. De Jong, and J. Arends, Measurements of the surface free energy of bacterial cell surfaces and its relevance for adhesion, *Appl. Environ. Microbiol. 48*:980 (1984).
16. C. R Calkins, T. R. Dutson, G. C. Smith, and Z. L. Carpenter, Concentration of creatine phosphate, adenine nucleotides, and their derivatives in electrically stimulated and nonstimulated beef muscle, *J. Food Sci. 47*:1350 (1982).
17. W. A. Carse, Meat quality and the acceleration of post-mortem glycolysis by electrical stimulation, *J. Food Technol. 8*:163 (1973).
18. R. G. Cassens and R. P. Newbold, Effects of temperature on the post-mortem metabolism in beef muscle, *J. Sci. Food Agric. 17*:254 (1966).
19. B. B. Chrystall, Electrical stimulation, refrigeration and subsequent meat quality, Paper E7.3, Proceedings 24th Eur. Meat Res. Workers Conference, Kulmbach, Germany, 1978,
20. B. B. Chrystall and C. E. Devine, Electrical stimulation, muscle tension, and glycolysis in bovine sternomandibularis, *Meat Sci. 2*:49 (1978).
21. K. C. Chung and J. M. Goepfert, Growth of *Salmonella* at low pH, *J. Food Sci. 35*:326 (1970).

22. J. Colby, L. G. Enriquez-Ibarra, and G. J. Flick, Shelf life of fish and shelfish, *Shelf Life Studies of Foods and Beverages* (G. Charalambous, ed.), Elsevier Science Publishers B. V., Amsterdam, 1993, p. 85.
23. R. N. Costilow, M. A. Uebersax, and P. J. Ward, Use of chlorine dioxide for controlling microorganisms during the handling and storage of fresh cucumbers, *J. Food Sci. 49*:396 (1984).
24. K. R. Davey and M. G. Smith, A laboratory evaluation of a novel hot water cabinet for the decontamination of sides of beef, *Int. J. Food Sci. Technol. 24*:305 (1989).
25. J. R. Davies, D. A. Ledward, R. G. Bardsley, and G. Poulter, Species dependence of fish myosin stability to heat and frozen storage, *Int. J. Food Sci. Technol. 29*:287 (1994).
26. L. R. G. Delmore, J. N. Sofos, J. O. Reagan, and G. C. Smith, Hot-water rinsing and trimming/washing of beef carcasses to reduce physical and microbiological contamination, *J. Food Sci. 62*:373 (1997).
27. C. E. Devine and B. B. Chrystall, High ulimate pH in sheep, *Dark-Cutting in Cattle and Sheep* (S. U. Fabiansson, W. R. Shorthose, R. D. Warner, eds.), Proceedings of an Australian Workshop, Australian Meat and Live-stock Research and Development, Sydney, 1989, p. 55.
28. J. S. Dickson, Control of *Salmonella typhimurium*, *Listeria monocytogenes*, and *Escherichia coli* 0157:H7 on spray chilling system, *J. Food Sci. 56*:191 (1991).
29. J. S. Dickson, Acetic acid action on beef tissue surfaces contaminated with *Salmonella typhimurium*, *J. Food Sci. 57*:297 (1992).
30. J. S. Dickson and M. E. Anderson, Microbiological decontamination of food animal carcasses by washing and sanitizing systems: a review, *J. Food Prot. 55*:133 (1992).
31. T. R. Dutson, G. C. Smith, and Z. L. Carpenter, Lysosomal enzyme distribution in electrical stimulated ovine muscles, *J. Food Sci. 45*:1097 (1980).
32. R. C. Eburne and G. Prentice, Modified-atmosphere-packed ready-to-cook and ready-to-eat meat products, *Shelf Life Evaluation of Foods* (C. M. D. Man and A. A. Jones, eds.), Blackie Academic & Professional, London, 1994, p. 156.
33. W. N. Eigel, J. E. Butler, C. A. Ernstrom, H. M. Farrll, V. R. Harwalkar, R. Jennes, and R. McL Whitney, Nomenclature of proteins of cow's milk: fifth revision, *J. Dairy Sci. 67*:1599 (1984).
34. G. Eikelenboom, F. J. M. Smulders, and H. Ruderus, The effect of high and low voltage electrical stimulation on beef quality, *Meat Sci. 15*:247 (1985).
35. P. S. Elias, *Elimination of Pathogenic Organisms from Meat and Poultry* (F. J. M. Smulders, ed.), Elsevier, Amsterdam, 1987, p. 345.
36. M. Fletcher and G. I. Loeb, Influence of substratum characteristics on the attachment of a marine pseudomonad to solid surfaces, *Appl. Environ. Microbiol. 37*:67 (1979).
37. FSIS-USDA, Immediate actions: Cattle clean meat program, FSIS Correlation Packet, Interim guidelines for inspectors, Food Safety and Inspection Service, United States Department of Agriculture, Washington, DC, 1993.
38. D. M. Gibson, Preservation technology and shelf life of fish and fish products, *Shelf Life Evaluation of Foods* (C. M. D. Man and A. A. Jones, eds), Blackie Academic & Professional, London, 1994, p. 72.
39. C. O. Gill and K. G. Newton, Current topics in veterinary medicine and animal science, *Microbiology of DFD Beef* (D. E. Hood, and P. V. Tarrant, ed.), Martinus Nijhoff Publishers, Hague, 1981, p. 305.
40. H. D. Goff and M. E. Sahagian, Freezing of dairy products, *Freezing Effects on Food Quality* (L. E. Jeremiah, ed.), Marcel Dekker, New York, 1996, p. 299.
41. B. M. Gorman, J. N. Sofos, J. B. Morgan, G. R. Schmidt, and G. C. Smith, Evaluation of hand-trimming, various sanitizing agents and hot water spray-washing as decontamination interventions for beef brisket adipose tissue, *J. Food Prot. 58*:899 (1995).
42. L. Gram, G. Trolle, and H. H. Huss, Detection of specific spoilage bacteria from fish stored at low (0°C) and high (20°C) temperatures, *Int. J. Food Microbiol. 4*:65 (1987).
43. H. Gratz, *Schlachten und Vermarkten 76*:111 (1976).

44. F. H. Grau, *Elimination of Pathogenic Organisms from Meat and Poultry* (F. J. M. Smulders, ed.), Elsevier, Amsterdam, 1987, p. 221.
45. G. G. Greer, *Meat Focus Int. May*:207 (1993).
46. N. F. Haard, Enzymes from marine organisms as food processing aids, *J. Aquatic Food Prod. Technol. 1*:17 (1992).
47. N. F. Haard, *Advvances in Seafood Chemistry: Composition and Quality* (G. J. Flick, and R. E. Martin, eds.), Technomic Publishing, PA, 1992, p. 1992.
48. R. S. Hannan, Properties of meat, Lecture note, Leeds University, Leeds, 1985.
49. A. Hanpongkittikun, S. Siripongvutikorn, D. L. Cohen, Black tiger (*Penaeus monodon*) quality changes during iced storage, *ASEAN Food J. 10*:125 (1995).
50. R. Hastings, G. W. Rodger, R. Park, A. D. Matthe, and E. M. Anderson, Differential scanning calorimetry of fish muscle: the effect of processing and species variation, *J. Food Sci. 50*:503 (1985).
51. L. Herborg and A. Villadsen, Bacterial infection/invasion in fish flesh, *J. Food Technol. 10*:507 (1975).
52. R. Hollender, F. G. Bender, R. K. Jenkins, and C. L. Black, Research note: consumer evaluation of chicken treated with trisodium phosphate application during processing, *Poultry Sci. 72*:755 (1993).
53. B. K. Howell, A. D. Matthews, A. P. Donnelly, Thermal stability of fish myofibrils: a differential scanning calorimetric study, *Int. J. Food Sci. Technol. 26*:283 (1991).
54. H. O. Hultin, Oxidation of lipids in seafoods, *Seafoods: Chemistry, Processing Technology and Quality* (F. Shahidi, and J. R. Botta, ed.), Blackie Academic, London, 1994, p. 49.
55. H. H. Huss, D. Dalsgaard, L. Hansen, H. Ladefoged, A. Petersen, and L. Zittan, The influence of hygiene in catch handling on the storage life of iced cod and plaice, *J. Food Technol. 9*:213 (1974).
56. H. H. Huss, Development and use of the HACCP concept in fish processing, *Int. J. Food Microbiol. 15*:33 (1992).
57. M. Ingram, F. J. H. Ottoway, and J. B. M. Coppock, *Chem. Ind. 42*:1154 (1956).
58. K. Ishikawa, *Guide to Quality Control*, Asian Productivity Organisation, Tokyo, 1976, p. 1.
59. M. A. Ismail, A. H. A. Elala, A. Nassar, and D. G. Michail, Fungal contamination of beef carcasses and the environment in a slaughterhouse, *Food Microbiol. 12*:441 (1995).
60. O. J. James, R. L. Brewer, J. C. Prucha, W. O. Williams, and D. R. Parham, Effects of chlorination of chill water on bacteriologic profile of raw chicken carcasses and giblets, *J. Am. Vet. Med. Assoc. 200*:60 (1992).
61. B. Jarvis, and C. S. Burke, *Inhibition and Inactivation of Vegetative Microbes* (F. A. Skinner, and Hugo, W. B. eds.), Academic Press, New York, 1977, p. 345.
62. K. W. F. Jericho, G. C. Kozub, J. A. Bradley, V. P. J. Gannon, E. J. Glosteyn-Thomas, M. Gierus, B. J. Nishiyama, R. K. King, E. E Tanaka, S. D'Souza, and J. M. Dixon-MacDougall, Microbiological verification of the control of the processes of dressing, cooling and processing of beef carcasses at a high line-speed abattoir, *Food Microbiol. 13*:291 (1996).
63. A. R. Jerrett, J. Stevens, and A. J. Holland, Tensile properties of white muscle in rested and exhausted chinook salmon (*Oncorhynchus tshawytscha*), *J. Food Sci. 61*:527 (1996).
64. B. J. Juven and M. D. Pierson, Antibacterial effects of hydrogen peroxide and methods for its detection and quantitation, *J. Food Prot. 59*:1233 (1996).
65. C. A. Kelly, J. F. Dempster, and A. J. McLoughlin, The effect of temperature, pressure and chlorine concentration of spray washing water on numbers of bacteria on lamb carcasses, *J. Appl. Bact. 51*:415 (1981).
66. K. Y. Kim, H. S. Lillard, J. F. Frank, and S. E. Craven, Attachment of *Salmonella typhimurium* to poultry skin as related to cell viability, *J. Food Sci. 61*:439 (1996).
67. A. C. Kondos and D. G. Taylor, Effect of electrical stimulation and temperature on biochemical changes in beef muscle, *Meat Sci. 19*:207 (1987).
68. M. Koohmaraie, J. E. Schollmeyer, and T. R. Dutson, Effect of low-calcium-requiring calcium

activated factor on myofibrils under varying pH and temperature conditions, *J. Food Sci. 51*:28 (1986).
69. A. W. Kotula, Microbiology of hot boned and electrostimulated meat, *J. Food Prot. 44*:545 (1981).
70. W. E. Kramlich, A. M. Pearson, and F. W. Tauber, *Processed Meat*, 3rd edition, Avi, Westport, CT, 1980, p. 320.
71. W. S. LaGrange, and E. G. Hammond, The shelf life of dairy products, *Shelf Life Studies of Foods and Beverages* (G. Charalambous, ed.), Elsevier Science Publishers B.V., Amsterdam, 1993, p. 1.
72. R. A. Lawrie, *Meat Science*, 4th ed., Pergamon Press, New York, 1985.
73. M. T. Lesiak, D. G. Olson, C. A. Lesiak, and D. U. Ahn, Effects of postmortem temperature and time on the water holding capacity of hot-boned turkey breast and thigh muscle, *Meat Sci. 43*:51 (1996).
74. H. S. Lillard, Levels of chlorine and chlorine dioxide of equivalent bactericidal effect in poultry processing water, *J. Food Sci. 44*:1594 (1979).
75. W. Lin, T. Huang, J. A. Cornell, C. Lin, C. Wei, Bactericidal activity of aqueous chlorine and chlorine dioxide solutions in a fish model system, *J. Food Sci. 61*:1030 (1996).
76. T. M. Lin and J. W. Park, Protein solubility in Pacific whiting affected by proteolysis during storage, *J. Food Sci. 61*:536 (1996).
77. J. Liston, *Chemistry and Biochemistry of Marine Food Products* (R. E. Martin, G. J. Flick, and D. R. Ward, eds.), AVI Publishing, Westport, CT, 1982, p. 27.
78. M. A. Lorier and B. L. Aitken, Method for treating fish with α_2-macroglobulin, U.S. patent 15,013,568 (1991).
79. R. M. Love and M. A. Haq, The connective tissues of fish. III. The effect of pH on gaping in cod entering rigor mortis at different temperatures, *J. Food Technol. 5*:241 (1970).
80. C. F. Madero and G. Finne, Properties of phenoloxidase from gulf shrimp, Proc. 7th Ann. Trop. and Subtrop. Fish. Technol. Conf. Americans, 1982, p. 328.
81. B. B. Marsh, Effects of early post-mortem muscle pH and temperature on meat tenderness, *Proc. Ann. Reciprocal Meat Conf. 36*:131 (1983).
82. B. B. Marsh, T. P. Ringkob, R. L. Russel, D. R. Swartz, and L. A. Pagel, Effects of early post-mortem glycolytic rate on beef tenderness, *Meat Sci. 21*:241 (1987).
83. K. C. Marshall, R. Stout, and R. Mitchell, Mechanism of the initial events in the adsorption of marine bacteria to surfaces, *J. Gen. Microbiol. 68*:337 (1971).
84. S. McEldowney, and M. Fletcher, Effect of growth conditions and surface characteristics of aquatic bacteria on their attachment to solid surfaces, *J. Gen. Microbiol. 132*:513 (1986).
85. A. J. McEvily, R. Iyengar, and S. Otwell, Sulfite alternative prevents shrimp melanosis, *Food Technol. 45*:80 (1991).
86. R. C. McKellar, *Enzymes of Psychrotrophs in Raw Food*, CRC Press, Boca Raton, FL, 1989.
87. MIRINZ, Specification for accelerated conditioning of beef, Meat Ind. Res. Inst., Hamilton, New Zealand, RM 73, 1978.
88. G. J. Mountney and J. O'Malley, *Poultry Sci. 44*:582 (1965).
89. B. Mrigadat, G. C. Smith, T. R. Dutson, L. C. Hall, M. O. Hanna, and C. Vaderzant, Bacteriology of electrically stimulated and unstimulated carcasses, *J. Food Prot. 43*:686 (1980).
90. T. Nakano, M. Sato, and M. Takeuchi, Glutathione peroxidase of fish, *J. Food Sci. 57*:1116 (1992).
91. A. Y. M. Nour, L. A. Gomide, E. W. Mills, R. P. Lemenager, and M. D. Judge, Influence of production and post-mortem technologies on composition and palability of USDA select grade beef, *J. Anim. Sci. 72*:1224 (1994).
92. H.W. Ockerman and J. Szczawinski, The effect of electrical stimulation on the microflora of meat, *J. Food Sci. 48*:1004 (1983).
93. H.W. Ockerman and J. Szczawinski, Combined effects of electrical stimulation and methods of meat preservation upon the survival of bacteria, *J. Food Process. Preserv. 8*:47 (1984).

94. G. Offer and J. Trinick, On the mechanism of water holding in meat: the swelling and shrinkage of myofibrils, *Meat Sci. 8*:245 (1983).
95. R. Ofstad, B. Egelandsdal, S. Kidman, R. Myklebust, R. L. Olsen, and A. Hermansson, Liquid loss as effected by post mortem ultrastructural changes in fish muscle: cod (*Gadus morhua* L) and salmon (*Salmo salar*), *J. Sci. Food Agric. 71*:301 (1996).
96. R. Ofstad, S. Kidman, and A. Hermansson, Ultramicroscopical structures and liquid loss in heated cod (*Gadus morhua* L) and salmon (*Salmo salar*) muscle, *J. Sci. Food Agric. 72*:337 (1996).
97. S. W. Otwell and M. Marshall, Screening alternatives to sulphiting agents to control shrimp melanosis, Proc. 11th Ann. Trop. and Subtrop. Fish. Technol. Conf. Americans, 1986, p. 35.
98. S. Palaniappan, S. K. Sastry, and E. R. Richter, Effects of electricity on microorganisms: a review, *J. Food Proc. Pres. 14*:393 (1990).
99. I. Papa, C. Alvarez, V. Verrez-Bagnis, J. Fleurence, and Y. Benyamin, Post mortem release of fish white muscle α-actinin as a marker of disorganisation, *J. Sci. Food Agric. 72*:63 (1996).
100. J. P. Piette and E. S. Idziak, Adhesion of meat spoilage bacteria to fat and tendon slices and to glass, *Biofouling 5*:3 (1991).
101. J. P. Piette and E. S. Idziak, A model study of factors involved in adhesion of *Pseudomonas fluorescens* to meat, *Appl. Environ. Microbiol. 58*:2783 (1992).
102. P. Polidori, R. G. Kauffman, and F. Valfre, The effects of electrical stimulation on meat quality, *Ital. J. Food Sci. 8*:183 (1996).
103. R. G. Poulter, R. W. H. Parry, and E. C. Blake, Recent advances on the effect of immediate post mortem handling on the quality and yield of tropical fish, *Food Science and Technology in Industrial Development* (S. Maneepun, P. Varangoon, and B. Phithakpol, eds.), Institute of Food Research and Product Development, Bangkok, 1988, p. 1001.
104. V. H. Powell and B. P. Cain, A hot water decontamination system for beef sides, *CSIRO Food Res. 47*:79 (1987).
105. R. K. Prasai, G. R. Acuff, L. M. Lucia, J. B. Morgan, S. G. May, and J. W. Savell, Microbiological effects of acid decontamination of pork carcasses at various location in processing, *Meat Sci. 32*:413 (1992).
106. M. Raccach, and R. L. Henrickson, The bacteriological quality of meat from electrically stimulated carcasses, *Proc. 26th Eur. Meeting Meat Res. Workers 2*:70 (1980).
107. J. O. Reagan, G. R. Acuff, D. R. Buege, M. J. Buyck, J. S. Dickson, C. L. Kastner, J. L. Marsden, J. B. Morgan, R. Nickelson II, G. C. Smith, and J. N. Sofos, Trimming and washing of beef carcasses as a method of improving the microbiological quality of meat, *J. Food Prot. 59*:751 (1996).
108. T. Saito, K. Arai, and M. Matsuyoshi, A new method for estimating the freshness of fish, *Bull. Jpn. Soc. Sci. Fish. 24*:749 (1959).
109. K. Sato, C. Ohashi, K. Ohtsuki, and M. Kawabata, Type V collagen in trout (*Salmo gairdneri*) muscle and its solubility change during chilled storage of muscle, *J. Agric. Food Chem. 39*:1222 (1991).
110. K. Sato, A. Koike, R. Yoshinaka, M. Sato, and Y. Shimizu, Postmortem changes in type I and V collagens in myocommatal and endomysial fractions of rainbow trout (*Oncorhynchus mykiss*) muscle, *J. Aquat. Prod. Technol. 3*:5 (1994).
111. K. Sato, M.. Ando, S. Kubota, K. Origasa, H. Kawase, H. Toyohara, M. Sakaguchi, T. Nakagawa, Y. Makinodan, K. Ohtsuki, and M. Kawabata, Involvement of type V collagen in softening of fish muscle during short-term chilled storage, *J. Agric. Food Chem. 45*:343 (1997).
112. J. W. Savell and G. C. Smith, Electrical stimulation effects on meat tenderness. muscle structure and he quality indicating characteristics of meat, Proceedings Annual Meeting Res. Dev. Assoc. Mil. Food Packag. Syst., New York, 1979, p. 1.
113. D. N. Scott, G. C. Fletcher, M. G. Hogg, and J. M. Ryder, Comparison of whole with headed and gutted orange roughy stored in ice: sensory, microbiology and chemical assessment, *J. Food Sci. 51*:79 (1986).

114. A. C. Sen, J. Owusu-Yaw, W. B. Wheeler, and C. I. Wei, Reactions of aqueous chlorine and chlorine dioxide with tryptophan, N-methyl tryptophane, and 3-indolelactic acid: kinetic and mutagenicity studies, *J. Food Sci. 54*:1057 (1989).
115. B. W. Sheldon and A. C. Brown, Efficacy of ozone as a disinfectant for poultry carcasses and chill water, *J. Food Sci. 51*:305 (1986).
116. J. M. Shewan, *J. Appl. Bact. 34*:299 (1970).
117. M. Sierra, E. Gonzalez-Fandos, M. Garcia-Lopez, M. C. G. Fernandez, and B. Moreno, Contamination of lamb carcasses at the abbatoir. Microflora of freshly dressed lamb carcasses: indicators and spoilage organisms, *Arch. Lebenmittelhyg. 46*:135 (1995).
118. G. G. Sims, G. E. Cosham, and W. E. Anderson, Hydrogen peroxide bleaching of marrinated herring, *J. Food Technol. 10*:497 (1975).
119. S. L. Slattery, D. J. Williams, and A. Cusack, A sulphite-free treatment inhibits blackspot formation in prawns, *Food Australia 47*:509 (1995).
120. M. G. Smith, Destruction of bacteria on fresh meat by hot water, *Expidemiol. Infect. 109*:491 (1992).
121. G. C. Smith, Effects of electrical stimulation on meat quality, color, grade, heat ring and palatability, *Advances in Meat Research* (D. H. Pearson, and T. R. Dutson, eds.), AVI Publishing Co., Westport, CT, 1985, p. 121.
122. F. J. M. Smulders, Preservation by microbial decontamination; the surface treatment of meats by organic acids, *New Methods of Food Preservation* (Gould, G. W. ed.), Blackie Academic and Professional, Glasgow, 1995, p. 253.
123. F. J. M. Smulders and C. H. J. Woolthuis, *J. Food Prot. 48*:838 (1985).
124. F. J. M. Smulders and G. Eikelenboom, *Accelerated Processing of Meat* (A. Romita, C. Valin, and A. A. Taylor, eds.), Elsevier Applied Science, London, 1987, p, 79.
125. J. M. A. Snijders, Hygiene Bij Het Slachten Van Varkens, Ph.D. thesis, Utrecht University, 1976.
126. T. Sorhaug and L. Stepaniak, Microbial enzymes in the spoilage of milk and dairy products, *Food Enzymology* (P. F. Fox, ed.), Elsevier, 1991, p. 169.
127. T. Sorhaug and L. Stepaniak, Psychrotrophs and their enzymes in milk and dairy products: quality aspects, *Trends Food Sci. Technol. 8*:35 (1997).
128. M. E. Stiles and L. K. Ng, Enterobacteriaceae associated with meats and meat handling, *Appl. Environ. Microbiol. 41*:867 (1981).
129. F. Stolle, *J. Appl. Bacteriol. 50*:235 (1981).
130. T. Suzuki, State of water in sea food, *Water Activity: Influences on Food Quality* (L. B. Rockland, and G. F. Stewart, eds.), Academic Press, 1981, p. 743.
131. K. Tachibana, T. Misima, and M. Tsuchimoto, Changes of ultrastructure and cytochemical Mg^{2+} ATPase activity in ordinary muscle of cultured and wild red sea bream during storage in ice, *Nippon Suisan Gakkaishi 59*:721 (1993).
132. D. G. Taylor and A. R. Marshall, Low voltage electrical stimulation on beef carcasses, *J. Food Sci. 45*:144 (1980).
133. J. S. Teotia and B. F. Miller, Destruction of Salmonellae on poultry meat with lysozyme, EDTA, X-ray, microwave and chlorine, *Poultry Sci. 54*:1388 (1975).
134. H. Tsuchiya, S. Kita, and N. Seki, Post mortem changes in α-actinin and connectin in carp and rainbow trout muscle, *Nippon Suisan Gakkaishi, 58*:793 (1992).
135. E. G. M. Van Klink and F. J. M. Smulders, *J. Food Prot. 9*:660 (1989).
136. P. Vaz-Pires, C. Capell, and R. Kirby, Low-level heat-treatment to extend shelf-life of fresh fish, *Int. J. Food Sci. Technol. 29*:405 (1994).
137. P. Vaz-Pires, P. Gibbs, and R. M. Kirby, Effect of heat on microorganisms isolated from freshly caught chilled scad (*Trachurus trachurus*), *Int. J. Food Sci. Technol. 31*:277 (1996).
138. V. Venugopal, Extracellular proteases of contaminant bacteria in fish spoilage: a review, *J. Food Protect. 53*:341 (1990).
139. F. Watanabe, M. Goto, K. Abe, and Y. Nakano, Glutathione peroxidase activity during storage of fish muscle, *J. Food Sci. 61*:734 (1996).

140. C. I. Wei, D. L. Cook, and J. R. Kirk, Use of chlorine compounds in the food industry, *Food Technol. 39*:107 (1985).
141. C. I. Wei, A. C. Sen, M. F. Fukayama, H. A. Ghanbari, W. B. Wheeler, and J. R. Kirk, Reactions involving HOCl or ClO_2 with fatty acids under aqueous conditions and mutagenicity of reaction products, *Can. Inst. Food Sci. Tech. Technol. 20*:19 (1987).
142. S. Wessels and H. H. Huss, Suitability of *Lactococcus lactic* subsp. *lactis* ATCC 11454 as a protective culture for lightly preserved fish products, *Food Microbiol. 13*:323 (1996).
143. G. C. White, *Handbook of Chlorination for Potable Water, Wastewater, Cooling Water, Industrial Processes, and Swimming Pools*, Van Nostrand Reinhold, New York, 1972.
144. D. W. S. Wong, W. M. Camirand, and A. E. Pavlath, Structures and functionalities of milk proteins, *Crit. Rev. Food Sci. Nutr. 36*:807 (1996).
145. J. Zawistowski, C. G. Biliaderis, and N. A. Michael, Polyphenol oxidase, *Oxidative Enzymes in Foods* (D. S. Robinson, and N. A. M. Eskin, eds.), Elsevier, New York, 1991, p. 217.

4

Glass Transition and Other Structural Changes in Foods

M. Shafiur Rahman
Horticulture and Food Research Institute of New Zealand, Auckland, New Zealand

Many physical modifications made in ingredients or foods during processing and formulation can affect the stability of the food during storage. Such modifications can also improve the sensory, nutritional, and functional properties of foods. Changes experienced by foods during processing include glass formation, crystallization, caking, stickiness, oxidation, gelatinization, and collapse. Through precise knowledge and understanding of such modifications, one can develop safe, high-quality foods for consumption.

I. GLASS TRANSITION IN FOODS

A. Glass Transition

Phase transitions in foods can be divided into two groups: first-order and second-order. At first-order transition temperature, the physical state of a material changes isothermally from one state to another (e.g., solid to liquid, liquid to gas) by release or absorption of latent heat (e.g., melting, crystallization, condensation, evaporation). Second-order transition occurs (e.g., amorphous state to glass state) without release or absorption of latent heat [41].

Glass transition is a second-order time-temperature-moisture–dependent transition, which is generally characterized by a discontinuity in physical, mechanical, electrical, thermal, and other properties of a material. Food materials are in an amorphous state below the glass transition temperature and are rigid and brittle. Glasses are not crystalline with a regular structure, but retain the disorder of the liquid state. In kinetic terms, glass temperature is defined as the temperature at which the viscosity of a material reaches 10^{13}–10^{14} Pa s and the molecular diffusion rate is many years [54]. Molecular mobility increases 100-fold above glass transition [2]. Above glass transition, the viscosity of the matrix may decrease to a point where it becomes too low to support its own weight, resulting in flow or deformation of the matrix [50].

B. Why Glass Transition?

It has been mentioned in the literature that foods can be considered very stable at the glassy state, since below glass temperature compounds involved in deterioration reactions take many months or even years to diffuse over molecular distances and approach each other to react. The hypothesis has recently been stated that this transition greatly influences food stability, as the water in the concentrated phase becomes kinetically immobilized and therefore does not support or participate in reactions. Formation of a glassy state results in a significant arrest of translational molecular motion, and chemical reactions become very slow.

C. Factors Affecting Glass Transition

1. Equilibrium and Nonequilibrium

Complex foods exist in states of either unstable nonequilibrium or metastable equilibrium, but never in true thermodynamic equilibrium [13]. Fennema [13] defined the terminology as follows.

Thermodynamic Equilibrium

Any food consisting of only one phase requires minimization of free energy to attain thermodynamic equilibrium. For foods containing two or more phases, thermodynamic equilibrium requires that the chemical potential be equal, in every part of the system, for each substance present. Chemical potential determines whether a substance will undergo a chemical reaction or diffuse from one part of a system to another. An equilibrium state can be attained through many possible paths, i.e., the same properties must be obtainable at a given temperature regardless of whether the temperature is approached by cooling or warming.

Metastable Equilibrium

Metastable equilibrium refers to a state of pseudo-equilibrium, or apparent equilibrium, which is stable over practical time periods but is not the most stable state possible. A metastable state can exist (i.e., conversion to a more stable equilibrium state will not occur) when the activation energy for conversion to a more stable equilibrium state is so high that the rate of conversion is of no practical importance.

Nonequilibrium

Nonequilibrium refers to a state that is inherently unstable, i.e., change to a more stable state is likely to occur at a rate of practical importance. The exact rate at which destabilization occurs depends on the particular system and the conditions to which it is exposed.

2. Cooling and Metastability

Typical Cooling Curves

Typical cooling curves during freezing are shown in Figure 1 [41]. The abrupt rise in temperature due to the liberation of the heat of fusion after initial supercooling represents the onset of ice crystallization. Cooling below the initial freezing point of the sample without formation of ices is defined as supercooling [13]. Pure water can be undercooled by several degrees before the nucleation phenomenon begins. Once the critical mass of nuclei is reached, the system nucleates at point a in Figure 1 and releases its latent heat faster than heat is being removed from the system [41]. The temperature then increases instantly to the initial freezing temperature at point b.

In aqueous solutions point a is not as low as pure water, since the added solute will promote heterogeneous nucleation, thereby accelerating the nucleation process. Solute greatly decreases the amount of supercooling for two reasons: faster nucleation and lowered freezing point. In very concentrated solutions, it is sometimes even difficult to induce supercooling [16].

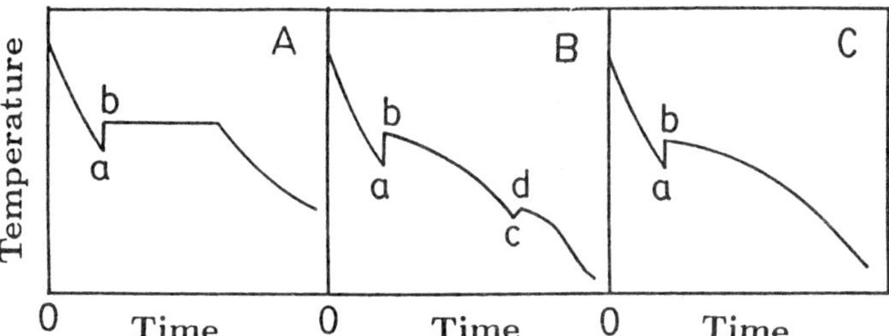

Figure 1 Typical cooling curves: (A) water, (B) solution, (C) food; (a) ice crystallization, (b) equilibrium freezing point, (c) initiation of solute crystal, (d) eutectic point. (From Ref. 41.)

After crystallization is completed, the temperature drops as the sensible heat is released in case of water (Fig. 1A). In solutions, supersaturation continues due to the freezing of water, and solute crystals may form by releasing latent heat of solute crystallization, causing a slight jump in temperature from c to d (Fig. 1B). These points are known as eutectic points.

In solutions with multiple solutes or foods it is difficult to determine the eutectic points. Many different eutectic points might be expected, but each plateau would be quite short if small quantities of solutes were involved (Fig. 1C). If a material is heated from a frozen or glass state, then the onset of ice melting is called melting point of ice. The freezing point or melting point is considered as an equilibrium process, i.e., neither cooling or heating nor rate affects the phase transition point of ice crystallization.

Cooling Rate and Thermodynamic Equilibrium

Fennema's [13] schematic depiction of a binary system is used to simplify the presentation. The basic format of the figures is shown in Figure 2, a plot of sample temperature versus equilibrium state. The columns represent different rates of cooling: equilibrium cooling, moderate cooling, and rapid cooling. The left column (equilibrium cooling) represents cooling at an exceedingly slow rate, the middle column (moderate cooling) represents cooling at a moderate rate consistent with commercial prac-

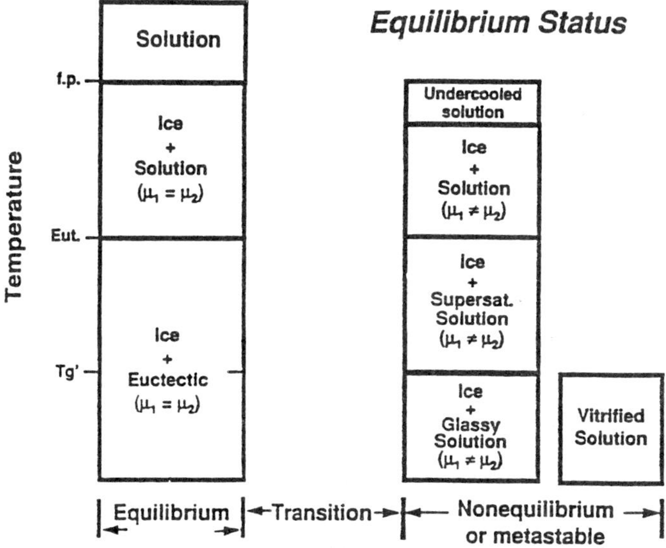

Figure 2 Equilibrium status based on rate of cooling. (From Ref. 13.)

tice, and the right column (very rapid cooling) represents cooling at a rate that is exceedingly rapid. Cooling will follow a downward path in the columns.

Equilibrium cooling: Cooling in first column in Figure 2 is in accord with thermodynamic equilibrium, which is not possible under practical circumstances, nor it is desirable. Thermodynamic freezing is, however, worthy of consideration for conceptual reasons. The path begins at point S, which represents an aqueous solution containing one solute at thermodynamic equilibrium (Fig. 3). Cooling must occur at an exceedingly slow rate to preserve equilibrium conditions [62], which will eventually bring the solution to its initial freezing point T. At this point, an ice crystal must be added to avoid undercooling and subsequent nucleation of ice, both of which are nonequilibrium events. With the ice crystal in place, further cooling will result in the formation of additional pure ice crystals and a decline in the freezing point of the unfrozen phase. Eventually the temperature decreases to the saturation or eutectic point U. Again at this point a small crystal of solute must be added to avoid supersaturation and subsequent nucleation of solute, both of which are nonequilibrium events. Fennema [13] noted that solute crystallization is mandatory to sustain thermodynamic equilibrium, and it is common even in the presence of seed crystals of solute that the solute will not crystallize at a subeutectic temperature. However, it is assumed here to form solute crystallization. Further cooling results in crystallization of ice and solute in constant proportion, leaving the unfrozen phase unchanged in composition and freezing point. This dual crystallization process continues at constant temperature until crystallization of water and solute is as complete as possible. Further cooling will simply lower the sample temperature with no further change in physical state.

Moderate cooling: The metastable or nonequilibrium pathways are quite different from the equilibrium cooling. Both paths start in the solution at point A, and cooling brings the sample to its initial freezing point B. At this point further cooling at a moderate rate, consistent with commercial practices, results in undercooling to point C. This nonequilibrium supercooling eventually results in nucleation, release of latent heat of crystallization, and, if cooling is relatively slow, a reversal in temperature almost to the initial freezing point B. As ice forms with further cooling, the freezing point of the solution phase declines. With further cooling, more of the original water converts to ice and the solute eventually attaints its saturation concentration D (eutectic point). Further cooling typically does not result in nucleation of solute crystals, rather the solution becomes increasingly supersaturated with solute, and this condition is normally metastable. Continued cooling to point E will cause the supersaturated unfrozen phase to convert to a metastable, amorphous solid (a glass) with a very

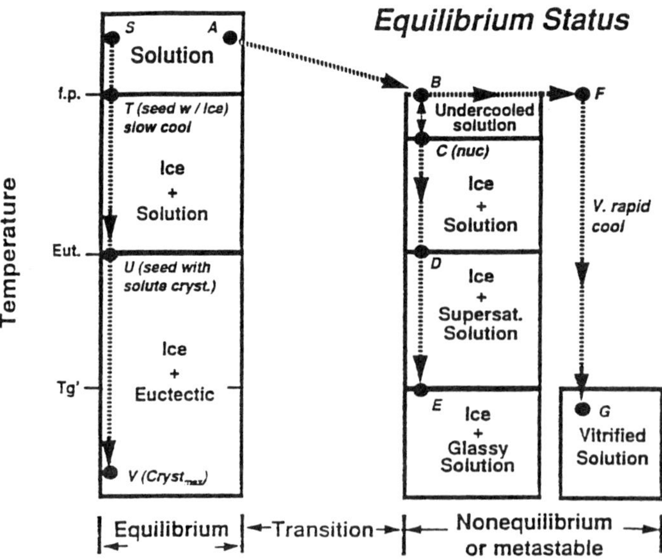

FIGURE 3 Equilibrium status and path of cooling. (From Ref. 13.)

high viscosity (about 10^{12} Pa s). This temperature is the glass transition temperature (T_g), which is usually determined by the composition of the sample and the rate of cooling. If cooling has been slow by commercial standards, the unfrozen solution can be maximally freeze-concentrated, and T_g, under this circumstance, will assume quasi-invarient value known as T'_g, which is dependent only on the solute composition of the sample. In practice, maximum freeze concentration is usually not obtained, and the observed T_g differs from T'_g. A number of numerical data are compiled by Rahman [41].

Very fast cooling: A third possible cooling pathway (Fig. 3) involves very rapid removal of heat from very small samples. This path has no commercial significance for foods. Thus, the ABFG path results in vitrification of the entire sample. The glass temperature is not an equilibrium process, and the cooling rate also affects the glass temperature. With very slow cooling during freezing there is a possibility of heterogeneous nucleation of ice and less possibility of ice formation. In the extreme case (slowest cooling), the material may transform from liquid to glass at the melting point without further ice formation.

3. Phase Diagram

The relation between freezing point and vapor pressure for pure water and simple aqueous solution are shown in Figure 4. The freezing point of water occurs at point A (P^v_w, θ_w). The freezing point of simple solution occurs at B (P^v_s, θ_s). $\theta_w - \theta_s$ is the freezing point depression. Point B also represents an equilibrium conditions between pure ice and aqueous solution.

A phase is defined as a homogeneous portion of a material or system having physically distinct bounding surface. Foods may be single or multiphase and can be characterized based on the phases present. A water solute phase diagram is shown in Figure 4. Curve AB represents equilibria between the solution and solid (ice) formed and can be called the freezing curve. Any point on curve AB is the freezing point of the solution. Line BC of Figure 5 represents equilibria between the solution and solute hydrate. The solution at any point on line BC is said to be saturated, and line BC is referred to as a solubility curve. Line AB has a negative slope showing the expected decrease in freezing point with increasing solute concentration. The positive slope of the line BC indicates the increase in solubility with increasing temperature. The freezing point and solubility curves intersect at the eutectic (or cryohydric) point B, where ice and solute crystallize simultaneously. The eutectic temperature, represented by line EBF of Figure 5, is the highest temperature at which maximum solidification of the system can be achieved.

4. State Diagram

In complex systems, food freeze concentration progresses and various solutes eventually reach their eutectic temperatures. However, solute crystallization at the eutectic point is unlikely due to the very

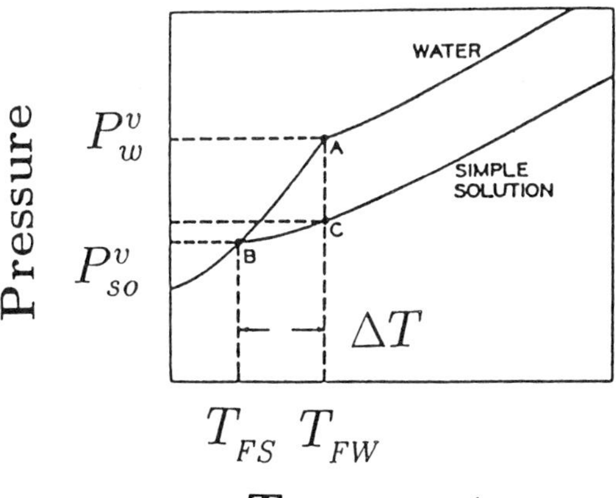

FIGURE 4 Vapor pressure versus temperature showing freezing point. (From Ref. 14.)

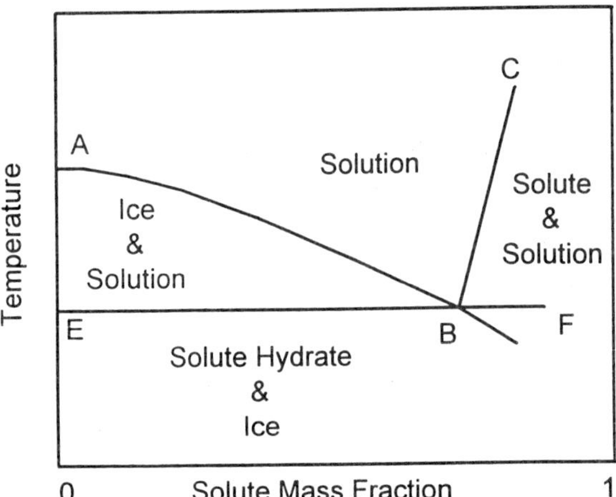

Figure 5 Phase diagram at constant pressure. (AB) Freezing curve, (BC) solubility line, (EBF) eutectic line.

low temperatures, extremely high viscosities, and resulting low diffusion rates and limited solute mobility [54]. Thus, the concentration of the solute exceeds the saturation point, resulting in a supersaturated state. In the state diagram in Figure 6, the freezing line (AB) and solubility line (BC) are shown in relation to the glass transition line (DEF). The glass forms at a characteristic glass transition temperature (point E) lower than the eutectic temperature (point B), and the water content at point E is the unfreezeable water. The glass transition temperature decreases from the T_{gs} of pure amorphous material to a theoretical T_{gw} of pure water. T'_g and X'_s are two parameters which reflect the

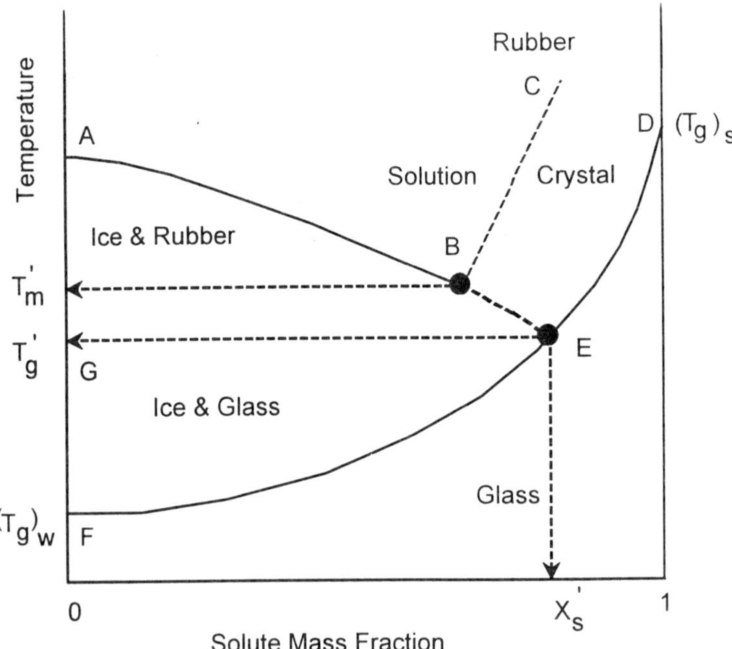

Figure 6 State diagram (AB) freezing curve, (BC) solubility curve, (B) eutectic point, (DEF) glass transition line, (E) glass transition temperature at maximul freeze concentration.

physical state of the noncrystallizing solute. T'_g is the maximally freeze-concentrated glass transition temperature, corresponding to the extended intersection of the equilibrium liquidous curve with the glass curve in the state diagram. X'_s is the composition of solute at T'_g. T'_m is the end point of freezing curve at B.

The formation of the amorphous state and its relation to equilibrium conditions are shown in Figure 7. The amorphous states (glass or rubber) are nonequilibrium states with time-dependent properties, and changes between equilibrium states and the glassy state always occur through the rubbery state [47].

5. Factors Affecting Glass Transition

The glass transition can be achieved by [13] (a) intrinsic conditions, or the alternation of product composition (water content, pH, oxygen tension, appropriate choice of chemical additives, e.g., emulsifiers, surface active agents, hydrocolloids, cryoprotectants, etc.), and (b) extrinsic conditions, or the control of process and storage conditions (temperature, composition of the atmosphere, protection from light, contamination, and mechanical damage). The glass transition temperature in food can be increased by adding more solutes of higher molecular weight.

D. Applications of Glass Transition

The metastable equilibrium and nonequilibrium states in complex foods are important in creating or retaining certain desirable properties. Thus, it is a major goal of the food scientist/technologist to maximize the number of desirable attributes existing in metastable equilibrium and to achieve acceptable stability of these desirable attributes [13]. The physicochemical processes governed by the glass temperature of foods are given in Table 1.

1. Microbial Stability

The microbial stability of food has long been estimated by its water activity: the lower the water activity, the more microbiologically stable the food. One defect in this concept is that microbial stability is affected by the nature and type of the solute at a given water activity. Another weakness is

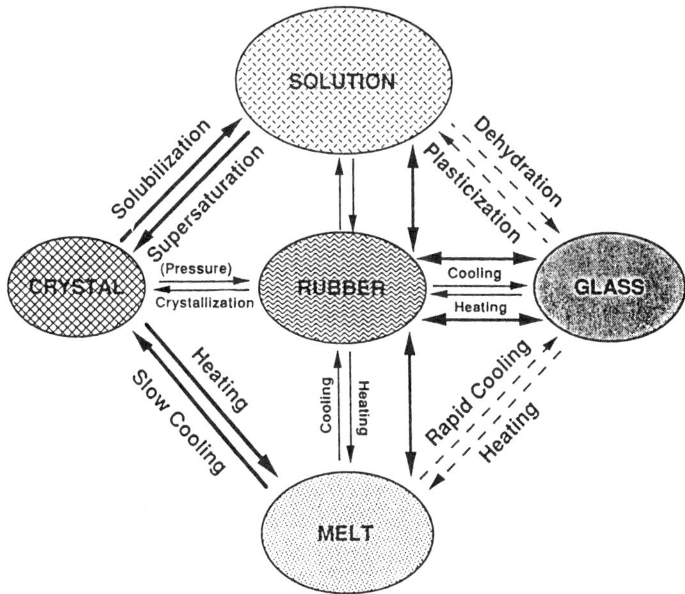

FIGURE 7 The physical state of material. (From Ref. 47.)

Table 1 Physicochemical Processes Governed by Glass Transition Temperature of Foods and Food Ingredients

Physicochemical properties
1. Agglomeration, lumping, and caking
2. Aroma retention
3. Collapse and stickiness
4. Crystallization or graining
5. Diffusivity
6. Enzyme activity
7. Microorganism inactivation
8. Nonenzymic brownig
9. Oxidation
10. Plating
11. Recrystallization
12. Sandiness or graininess
13. Structural collapse

that water activity is defined at equilibrium, whereas food at low and intermediate moisture are not in a state of equilibrium. In the dynamic state water may be migrating from one component of food to another. This nonequilibrium state is difficult to predict by equilibrium state defined by water activity.

Due to the above factors, Slade and Levine [56] and Franks [15] maintained that water activity can serve as a useful, but not sole indicator of microbial safety. Slade and Levine's [56[hypothesis was that *water dynamics* or glass-rubber transition may be applied instead of water activity to predict the microbial stability of concentrated and intermediate-moisture foods. However, Chirife and Buera [10] maintained that glass-rubber transition would not be useful in predicting with confidence the microbial stability of foods. They analyzed data from the literature and concluded that water activity and glass transition are two different entities. The mobility factors (i.e., glass transition) in addition to water activity are not useful for a better definition of microbial stability of foods. Water activity is a solvent property, and glass is a property related to the structure of food. Thus, both properties are needed for understanding food-water relationships at different conditions [10,58,60].

Sapru and Labuza [49] studied the inactivation of bacterial spores and their glass transition temperature. Spores at glass transition have high heat resistance, and above glass they are easy to inactivate. At a given temperature, the inactivation rate decreases with the increase of glass transition temperatures of spores.

It is now common in the food-processing and preservation industry that the BET or GAB monolayer value used is the critical water content, at which dehydrated foods are more stable. The water activity at the monolayer water content is also called the *critical water activity*. Similarly, Roos [44] defined the critical water content using the glass behavior of foods where the glass transition temperature is 25°C. The critical moisture content and critical water activity from glass transition increased linearly with the increase in molecular weight. Thus, higher molecular weight dry solutes can be stored at higher moisture content without deterioration.

2. Diffusion Process

The physical state of a food is a factor affecting the rates of chemical reactions. Slade and Levine [54] and Karel and Saguy [24] have proposed that glass transition affects diffusion-controlled chemical reactions through the decrease of diffusion coefficient. The decrease in diffusivity is due to the changes in viscosity and mobility. Parker and Ring [37] studied the water diffusivity in maltose-water mixtures above and below the glass transition. Point when they plotted diffusivity and viscosity as a function of T_g/T, the slopes for diffusivity and viscosity with temperature were not parallel, but instead were extremely divergent when $T_g/T > 0.8$. This evidence indicated that the decrease of diffusivity near or below glass is not solely due to the decrease of viscosity.

3. Oxidation

Shimada et al. [53] (methyl linoleate in lactose-gelatin matrix) and Labrousse et al. [26] showed that oxidation of unsaturated lipids entrapped in sugar-based matrices is affected by physical changes such as collapse or crystallization occurring above glass transition. The encapsulated oil was released as a consequence of the crystallization of amorphous lactose. The released oil underwent rapid oxidation, while encapsulated oil remained unoxidized.

4. Nonenzymic Browning

Karmas et al. [25], Karel et al. [23], and Buera and Karel [7] indicated that phase transitions with physical aspects of the matrix are factors affecting the rates of nonenzymatic browning reactions. Nonenzymic browning below glass transition was very slow. The systems used were vegetables, dairy products, and model food systems with amino acids and sugars in a PVP matrix. In this case both moisture and glass transition affect the reaction rate [7]. This is due to the changes in diffusion coefficient below glass transition when nonenzymatic reactions take place in the diffusion limited region. The Williams-Landel-Ferry (WLF) equation was valid to predict the reaction rate constant as a function of moisture and temperature above glass [7].

No general rule is observed as to whether water activity or glassy state of the system, as dictated by glass transition temperature, affects the rates of chemical reactions in reduced-moisture solid food systems. Bell [2] studied the kinetics of nonenzymatic browning pigment formation in a model PVP (different molecular weight) matrix. The browning rates of matrices having different glass temperature, but constant water activity, were significantly different except when all were in the glassy state. As the system changed from a glassy state to a rubbery state, the rate of browning increased sevenfold. The rate of browning also increased as water activity increased from 0.33 to 0.54, but then appeared to plateau with further increases in water activity. In addition, the concentration of reactants in the aqueous microenvironment had a significant impact on the rate of brown pigment formation.

O'Brien [34] studied the rate of nonenzymic browning in freeze-dried model systems containing lysine with glucose or sucrose or trehalose at pH 2.5 and a water activity of 0.33. The temperatures were 40, 60, and 90°C. All systems were in a rubbery state at 90°C, whereas at 40 and 60°C trehalose was in a mixed amorphous glass-crystalline system and glucose and sucrose were in a rubbery state. The rate of nonenzymatic browning in the trehalose system was much lower than that in the sucrose or glucose system depending on temperature. The rate constant was in the order: glucose → sucrose → trehalose. The presence of crystalline material in the trehalose system at 40 and 60°C may have influenced the overall rate of hydrolysis, stabilizing the system. At 90°C, all systems were in the rubbery state and there were substantial differences between the stability of sucrose and trehalose. Thus, glass transition is not only the controlling rate of browning, but mainly the rate-limiting sucrose hydrolysis step since glucose had a higher rate than sucrose. The effect of glass transition was much less effect than previously reported.

Bell and Hageman [3] studied the kinetics of aspartame degradation in the PVP model system at constant temperature (25°C) and pH (7.0) as a function of water activity and glass transition independently. Degradation reaction rates at constant water activity, but different glass transition temperature, were not significantly different, and rates at a similar distance from T_g, but different water activities, were significantly different. Thus, the rate of aspartame degradation was significantly influenced by the water activity, while the effect of the glass transition temperature on the reaction was negligible.

5. Denaturation of Protein

The properties and functionality of the protein depend on whether it exists in the native or the denatured state; maintaining protein structure and functionality is important in food science. The temperature of denaturation decreased with increasing moisture content to some plateau, where further increases in moisture no longer influenced the denaturation temperature [5]. Bell and Hageman [5] studied the denaturation temperature of globular proteins (β-lactoglobulin, ovalbumin, and ribonuclease) in a dry state as a function of glass transition temperatures of polyhydroxy compounds (water, glycerol, sorbitol, sucrose, and trehalose). The component was in the 25–33% w/w. The thermal stability of the protein correlated with the glass transition temperatures of the polyydroxy component; the lower the T_g of the component, the greater was the degree of protein destabilization. They hypothesized that in the dry state, the additives were acting as plasticizers, enhancing the mobility and thus the unfolding of the globular proteins.

6. Sensory Properties

The effect of molecular weight on glass temperature of starch hydrolysis products (SHP) is shown in Figure 8. The plateau region in Figure 8 indicates the useful range of gelation, encapsulation, cryostabilization, thermochemical stabilization, and facilitating of drying process. The lower end corresponds to the area of sweetness, browning reactions, and cryoprotection. The intermediate region at the upper end of the steeply rising portion represents the area of antistaling ingredients. The map can be used to choose individual SHPs or mixtures of SHPs and other carbohydrate to achieve the desired complex functional behavior for specific product applications [27]. For example, the synthesis of SHPs capable of gelation from solution should be designed to yield materials of DE ≤ 6 and $T'_g \geq -8°C$. Similarly, potato starch maltodextrins of 5–6 DE (25% w/w) produced thermoreversible, fat-mimetic gels, and tapioca SHPs of DE ≤ 5 also form fat-mimetic gels from solution to develop fat-replaceable ingredients [6,42,43].

Levine and Slade [27] developed a linear relation between dextrose equivalent (DE) and glass transition temperature of maximally freeze-concentrated solution of commercial SHPs. Thus, the correlation can be used to approximately calculate DE for dextrin and maltodextrin of unknown SHP, which can make unnecessary more tedious and time-consuming classical experimental methods for DE determination [33].

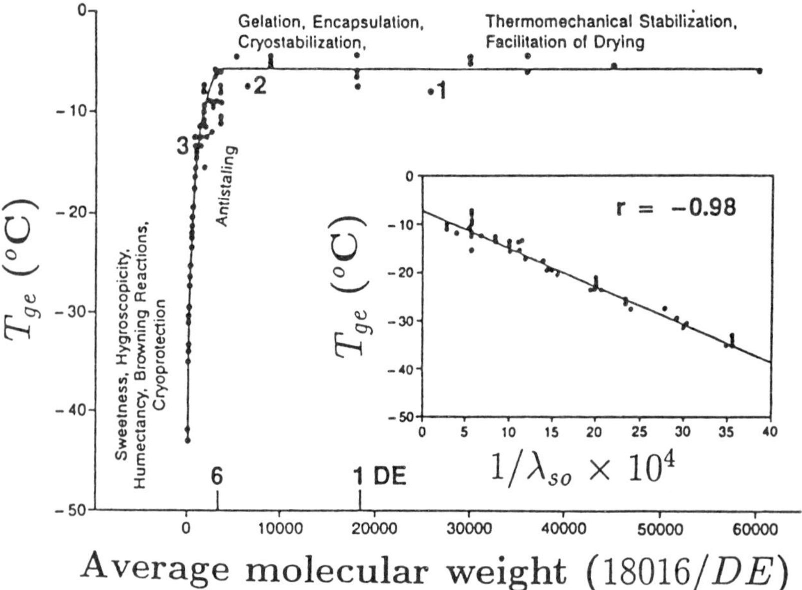

FIGURE 8 Effect of molecular weight on glass temperature of starch hydrolysis product (SHP). (From Ref. 27.)

7. Hydrolysis

The effect of glass transition on different chemical reactions is not as clear as in cases where physical changes occur. This is due to the multiple roles of water in foods, such as plasticizer, reactant or product of chemical reactions, and pH [8]. One of the chemical reactions proposed to occur only in the rubbery phase (i.e., above glass transition) is sucrose inversion in acid-containing amorphous powders [28]. Buera et al. [8] investigated the effect of glass transition on the rate of acid-catalyzed sucrose hydrolysis in an amorphous polymeric matrix of polyvinylpyrrolidone (PVP). No direct relationship was found between sucrose hydrolysis in a PVP matrix and T_g or $(T - T_g)$. Glass transition is not a key factor determining the rate of sucrose hydrolysis. The major effect on the rate of hydrolysis was related to changes in pH, which is moisture dependent. Knowledge of the actual pH of a system and the possible changes that may occur during concentration or drying is necessary for better understanding of chemical changes in low- and intermediate-moisture foods.

Sucrose hydrolysis is an acid-containing (low pH = 3.1) amorphous starch powder (native or pregelatinized) occurred to a significant extent in the glassy state [50]. The mobility effects do not control the extent of reaction. Hydrolysis (31–85% remaining sucrose) was observed at different water content and temperature below glass [50]. Little reaction occurred at moisture contents below the so-called BET monolayer. Temperature was a critical factor controlling sucrose inversion.

8. Enzyme Inactivation

Schebor et al. [51] studied the stabilization of the enzyme invertase (β-fructofuranosidase) by its incorporation in aqueous model systems of trehalose, maltodextrin, and polyvinylpyrrolidone followed by freeze-drying and desiccation to zero moisture content. When the systems were heated at 90°C for thermal inactivation of invertase, the enzyme was protected by maltodextrin and polyvinylpyrrolidone but not significantly protected by trehalose, although all systems were in the glassy state. Cardona et al. [9] observed significant inactivation of invertase when maltodextrin and polyvinylpyrrolidone kept well below their glass transition, but the enzyme was fairly stable in rubbery trehalose systems. At moisture contents that allowed trehalose crystallization, rapid thermal inactivation of invertase was observed. The invertase inactivation in heated systems of reduced moisture could not be predicted on the basis of glass transition, and this was particularly true for trehalose, for which it was evident that the glassy state was not the main stabilizing factor.

II. GELATINIZATION AND RETROGRADATION

A. Starch Gelatinization Mechanism

Starch granules are insoluble in cold water due to closer, more orderly packing of the starch granules at the surface of a granule than its interior. If a suspension of starch in water is heated, diffusion of water into granules causes swelling (i.e., increasing the granule volume) and disrupts hydrogen bondings, and subsequently water molecules become attached to the hydroxyl groups in starch. The linear amylose diffuses out of the swollen granule during and after gelatinization and makes up the continuous gel phase outside the granules. This phase transformation is termed as gelatinization. As long as water is abundant, gelatinization will occur at a fixed temperature range, normally 60–70°C [20].

Water in the amorphous parts of starch acts as a plasticizer by decreasing the glass transition temperature. Gelatinization temperatures are mainly influeced by moisture, solutes (ionic and non-ionic), and processing pretreatment conditions. However, gelatinization is a melting process caused by heat in the presence of water.

Olkku and Rha [29] summarized the steps of gelatinization as follows: (a) granules hydrate and swell to several times their original size, (b) granules lose their birefringence, (c) clarity of mixture increases, (d) rapid increase in consistency occurs and reaches a maximum, (e) linear molecules dissolve and diffuse from ruptured granules, (f) uniformly dispersed matrix forms a gel or paste-like mass.

B. Importance of Gelatinization

1. Food Product Quality Determination

Gelatinization of starch plays an important role in determining the structural and textural properties of many foods. The proportion of raw and gelatinized starch in ready-to-serve starchy products may be critical in determining acceptability. The texture of many foods, such as breakfast cereals, beverages, rice, noodles, pasta, and dried soups, depends on the fraction of gelatinized starch in the product [18]. Gelatinization of starch is important in processes such as baking of bread, gelling of pie fillings, formation of pasta products, and thickening of sauces to produce a desirable texture or consistency [35].

Parboiling of rice is an ancient process in which paddy rice soaked in warm water for several hours, steaming the soaked rice—usually under pressure to gelatinize the rice in the endosperm—followed by drying and milling. This facilitates removal of the hulls and results in less damage to the rice. Parboiled rice has an advantage in that the grains show little or no tendency to clump or become gluey when prepared for eating, and when cooked it is firmer and less cohesive. Gelatinization temperature is important in rice and rice products in terms of consumers acceptance. For this reason, many consumers prefer this type of rice.

The intensity of heat treatment in the gelatinization process also affects the amount of soluble starch in the milled rice, which is probably related to the degree of gelatinization. The presence of an insoluble amylose complex appears to be responsible for the above characteristics of parboiled rice [40].

The stickiness of rice can be manipulated by controlling the gelatinization process. Low molecular weight amylose in mashed potatoes give a gluey, sticky, or gummy texture.

2. Nutritional Quality Determination

The nutritional value of milled parboiled rice is higher than that of white rice because gelatinized rice is easy to digest.

C. Retrogradation

The term retrogradation is used to describe changes that occur as a result of the cooling and storage of starch products. It refers to the association and crystallization of starch in water causing effects, such as precipitation, gelation, and changes in consistency and opacity. The most common negative effect of retrogradation in foods is deceased storage stability. Amylopectin starch or modified starch—which attract small side chains that will act as a steric hindrance to recrystallization—is usually used to avoid retrogradation [20].

The change in texture and flavor of starch-based products during storage is commonly called staling. The product becomes dry and hard, often caused by starch retrogradation. Davidou et al. [12] studied the staling of white bread at ambient temperature. An increase in rigidity was observed, which was attributed both to starch retrogradation and to changes in the organization of the amorphous part of crumb. Complexes between the native lipids of flour and amylose were formed during the first 2 days of storage. They appeared to reduce the maximum level of starch retrogradation. The glass transition temperature of crumb was not significantly modified by these structural changes.

III. CRYSTALLIZATION AND RECRYSTALLIZATION

A. Crystallization

Crystallization is the conversion process of a substance from an amorphous solid, liquid, or gaseous state to the crystalline state. The formation of ice crystal by freezing is the most common example of crystallization in food processing. Other amorphous food components such as sucrose, fructose, and salts also undergo a crystallization process during processing, preservation, and storage of food products. Similarly, crystallization also occurs in lipids. Crystals have a more orderly structure than amorphous materials. Thus, controlling the crystallization process is important for stable and acceptable formulated food products as well as in separation processes, such as freeze concentration and pure solute recovery.

A liquid phase must be supersaturated for crystals to be formed or existing crystals to grow. Supersaturation can be achieved by cooling and/or evaporation (cooling crystallization, evaporating crystallization, and vacuum crystallization), drawning out or a reaction. Sometimes a third substance, known as a displacement agent, is added to a solution. The agent reduces the solubility of the dissolved substance and thus leads to supersaturation. This process is known as drawning out crystallization. When two or more solutes react together to change the solubility and then precipitate out, it is called precipitation crystallization. The above classification has been done based on the mechanism of crystallization. In a crystallization process, there may be formation of crystals from dissolved solute or solvent. Thus, solute crystallization and/or solvent crystallization may occur based on the component being crystallized. In solvent crystallization, kinetics is often controlled by mass transfer, whereas in many cases heat transfer is the limiting factor in melt crystallization [32]. Time, temperature, and saturation affect the solute crystallization process. The rate of cooling during freezing affects the ice crystal size in foods.

1. Crystallization Dynamics

Supersaturation is the main driving force for solute crystallization in food products containing a variety of constituents, such as protein, carbohydrate, salt, and gum. Supersaturation is typically expressed as the ratio of supersaturated syrup concentration to concentration at solubility [19].

Other factors, such as agitation, impurities, and activity of foreign nucleating agents, also enhance or retard crystal formation by influencing metastable limits [19]. Below glass transition the crystallization process is retarded for months to years due to the decrease of mobility of solute. A DSC thermogram of the crystallization of sucrose is shown in Figure 9. The crystallization process is also time, temperature, and moisture dependent (Fig. 10). The dynamics of nucleation are also important to the shelf life of sugar-based products, where the onset of nucleation may occur and reduce the product quality. The stability of many milk products depends on the crystallization of amorphous lactose during storage. Palmer et al. [36] studied the crystallization of amorphous sucrose at 24°C and 0.30–0.325 relative humidity. They observed that the surface of the particles crystallized first, folowed by a rapid crystallization of the remaining amorphous sucrose inside each particle. The second step proceeds at a fast rate due to a local increase in relative humidity, because the crystalline crust imposes a sufficient barrier to diffuse water released by crystallization. Makower and Dye [30] observed that a critical relative humidity is needed to initiate crystallization. They observed 0.12 for sucrose and 0.05 for glucose below which no crystallization occurred in 3 years.

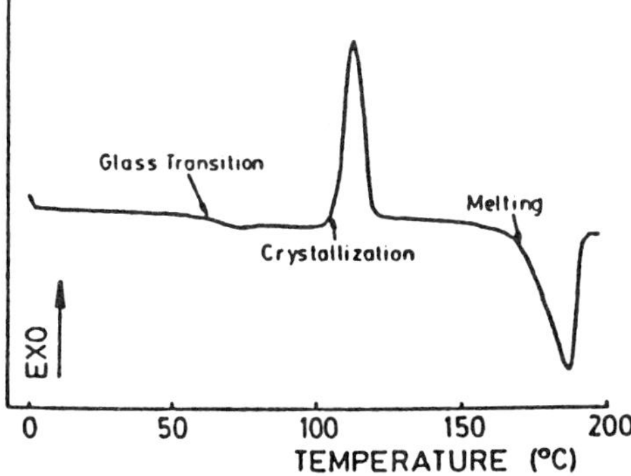

FIGURE 9 Typical phase transitions of amorphous sucrose (glass transition: 62°C, instant crystallization at 103°C, melting point: 185°C). Crystallization may occur at any temperature between T_{gs} and T_c depending on holding time. (From Ref. 45.)

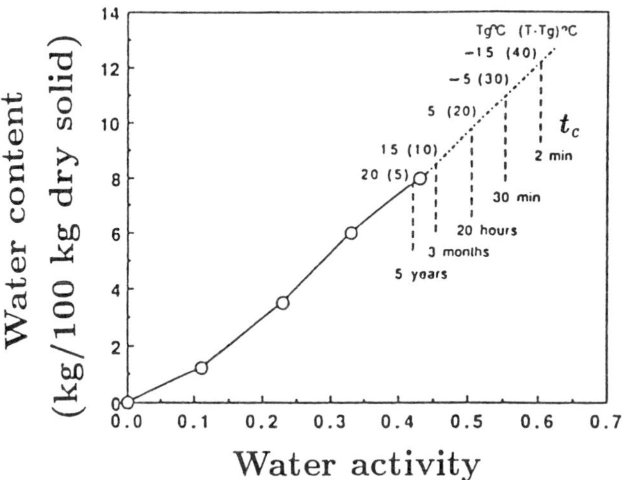

FIGURE 10 Effect of temperature dependence of crystallization time on the sorption isotherm of amorphous lactose at 25°C. (From Ref. 46.)

2. Crystallization Kinetics

Crystallization is a three-step mechanism: (a) nucleation (homogeneous or heterogeneous formation of nuclei), (b) propagation (growth of crystals from nuclei by intermolecular association, and (c) maturation (crystal perfection by annealing of metastable microcrystallites and/or continuation of slow growth) [19,55]. The crystallization kinetics of partially crystalline polymers in terms of crystallization rate as a function of temperature are shown in Figure 11. The crystallization temperature depends on the temperature difference between glass and melting temperature. In food systems heterogeneous nucleation occurs at active sites in the foods, such as on insoluble solids, and the container surface. The rate of nucleation (number of nuclei/m^3 s) in food systems mainly depends on supersaturation, temperature, energy input via agitation, and level of additives in formulation. The growth rate of crystal depends on supersaturation, temperature, solution viscosity, agitation rate, pH, number density of crystals, and level of additives [19]. The growth may be diffusion or surface incorporation controlled. Roos and Karel [46] studied the isothermal crystallization time (t_c) of amorphous lactose at varying temperatures and moisture contents. Isothermal crystallization time was determined at the time at the peak of the exotherm. Roos and Karel [46] found from crystallization dynamics that crystallization occurs at a constant value for $T_c - T_g$, so they recommended the WLF equation be used to determine the time of crystallization as:

$$\log\left(\frac{t_c}{t_g}\right) = \frac{-17.44(T - T_g)}{51.6 + (T - T_g)} \tag{1}$$

The experimental data and WLF equation for lactose are shown in Figure 12, indicating how useful the WLF equation can be in predicting crystallization time.

B. Recrystallization

Recrystallization is the process by which crystals grow as small ones disappear or reduce in number and size. This process can lead to lower-quality food products. The overall effect is to increase the mean crystal size and broaden the cryatal size distribution. Small crystals have a high surface area–to–volume ratio and consequently a high excess surface energy. The recrystallization rate is influenced

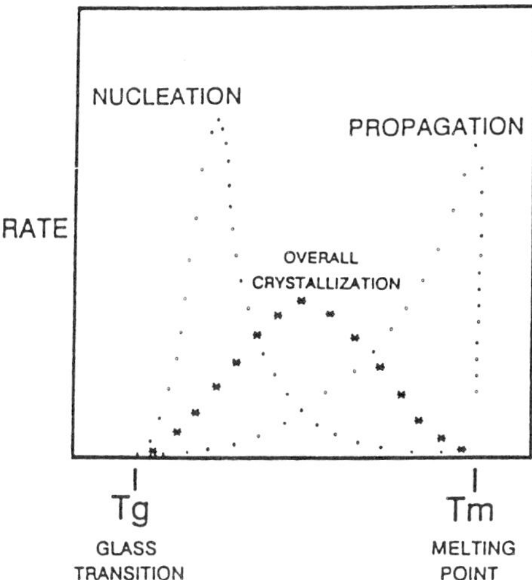

FIGURE 11 Crystallization kinetics as a function of temperature. (From Ref. 28.)

by temperature, solute concentration, and viscosity of liquid phase. The mechanisms involved in recrystallization in frozen foods are migration and accretion [57].

IV. CAKING, AGGLOMERATION, STICKINESS, AND COLLAPSE

A. Caking

Aguilera et al. [1] stated that collapse, stickiness, agglomeration, and caking are related phenomena. Caking is a phenomenon by which a free-flowing powder is first transformed into lumps, then into an agglomerated solid, and ultimately into a sticky material, resulting in a loss or gain of function-

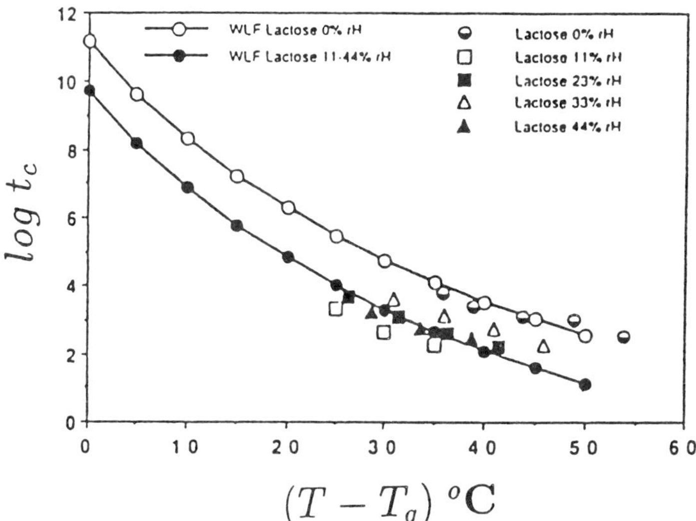

FIGURE 12 Crystallization time as a function of $T - T_g$ at different moisture content. (From Ref. 46.)

ality and affecting the quality. Caking may also occur as a result of recrystallization, either after fat melting or after solubilization at crystal surfaces [1]. The process may also be initiated by surface wetting due to moisture or cooling, by electrostatic attraction between particles [39], or by sticking through a mechanism of viscous flow driven by surface energy.

Aguilera et al. [1] indicated that a strict definition of caking is difficult to formulate, because changes in a particulate system depend on time, temperature, moisture, and position within the powder and involve many different stages. These stages are (a) free flowing, (b) bridging, (c) agglomeration, (d) compaction, and (e) liquefaction [1].

Bridging, the initial stage in caking, occurs as a result of surface deformation and sticking at contact points between particles, without a measurable decrease in system porosity. At this stage, small interparticle bridges may even disintegrate under mild shaking [1], and a mass of powder resists movement and is no longer free flowing [38].

Agglomeration, a later stages, involves an irreversible consolidation of bridges, but the high porosity of the particulate system is maintained, resulting in particle clumps with structural integrity [1]. This is the stage of enhancing or controlling interparticle adhesion [17].

Compaction, an even more advanced stage of caking, is associated with a pronounced loss of system integrity as a result of thickening of interparticle bridges owing to flow, reduction of interparticle spaces, and deformation of particle clumps under pressure [1]. The created agglomerates stabilize in some cases under the action of external forces [17].

In the final stage of caking, interparticle bridges disappear as a result of sample liquefaction and extensive flow owing to the high moisture content. This stage usually involves the solubilization of low molecular weight fractions and the presence of hygroscopic behavior [1].

The caking tendency of powders may be evaluated and predicted, and the means of preventing this phenomenon are important to obtain high-quality stable products. Mannheim et al. [31] states the need for a good understanding of the interparticle forces acting within a powder and the factors affect its flow properties under adverse conditions and prolonged storage. The factors affecting flow properties include particle size and shape, the moisture content of the powder, pressure applied, temperature, and chemical composition. Many methods are available for evaluating the cohesiveness of powders and the strength of agglomerates [31]. The methods available for flow conditioning and inhibition of caking are [31] (a) drying to low moisture content, (b) handling the powder at low-humidity atmospheres and packaging in high-barrier packages, (c) storage at low temperatures, (d) storage in desiccation, (e) agglomeration, and (f) addition of anticaking agents.

The different stages of caking are commonly encountered in powders and related substances. Sticking properties can also be used to advantage, such as in the creation of agglomerated products, particularly instant coffee and milk [61]. In the agglomeration of powders, particles are forced to interact and form granules with improved shape, appearance, and handling properties, including better flow characteristics, improved packing density, dust-free operation, and faster dissolution in liquids [52]. The agglomeration process should be controlled to obtain the desired powder properties. Temperature and humidity are the important factors in controlling stickiness, therefore these factors should be considered during processing and packaging.

Aguilera et al. [1] plotted glass transition temperature, sticky point temperature, and collapse temperature of dehydrated fruit juices in the same graph as a function of moisture content. Although some data scattering is evident, the glass transition temperature for the mixtures of water and the main nonaqueous component of a sample appear to predict reasonably well the phenomena of collapse and stickiness as a function of temperature. Thus, the glass-rubber concept can also be used to explain these processes. Caking can begin to occur rapidly at temperature above a glass of 20–30°C [11], and stickiness occurs generally at 10–20°C above the glass temperature [48].

B. Collapse

In freeze-drying, water is removed by sublimation of ice and forming a cake. Void spaces are formed from spaces previously occupied by ice crystals formed during freezing. When freeze-dried cake is heated, at a certain temperature a change in structure generally occurs known as collapse [21].

The temperature at which food material loses its structure and volume during storage and processing is called the collapse temperature. The transition at which dry food powder transforms to a viscous state is termed the sticky point. At this stage a sharp increase in the force required to stir a powder is observed, which is indicative of a more advanced stage of caking [1]. These phenomena depend on time, temperature, and moisture content, and also affect the sensory behavior and processing of many food materials. The loss of initial structure often results in the loss of desirable product qualities, although in some cases controlled manipulation of these changes is used to produce improved products [59]. In freeze-drying, collapse of capillaries in the dry layer results in puffing, poor aroma retention, poor rehydration characteristics, and loss of desirable structure. Collapse may cause adequate mobility to produce rearrangements of internal structure. Various important quality defects result from collapse [22]. In dried powders microscopic flow causes stickiness, leading to caking, agglomeration, and other visual defects in food products.

Collapse of the amorphous structure of dried food components results in disruption of internal pockets of molecular or larger size in which volatiles such as flavor compounds are entrapped until released by dissolving the entrapping matrix [22]. As some collapse and flow occurred, there was a decrease in the projected area up to the point where the complete liquefication of the flake resulted in the final equilibrium shape. When the fractional collapse was plotted against the loss of entrapped flavors occurring because of exposure to the increased temperature (or increased moisture) causing the collapse, excellent correlations were obtained [22].

Collapse also allows penetration of gases and vapors into foods. Droplets of oxidizable lipids incorporated in an emulsion containing carbohydrates or polymers may be freeze-dried and then washed with a nonpolar solvent such as hexane. All of the lipids are then entrapped and shielded from atmospheric oxygen. Subsequent exposure to water vapor, which plasticizes the structure, allows penetration of oxygen and consequent deterioration of lipids [22].

REFERENCES

1. J. M. Aguilera, J. M. del Valle, and M. Karel, Caking phenomena in amorphous food powders, *Trends Food Sci. Technol. 6*:149 (1995).
2. L. N. Bell, Kinetics of non-enzymatic browning in amorphous solid systems: distinguishing the effects of water activity and the glass transition, *Food Res. Int. 28*:591 (1996).
3. L. N. Bell and M. J. Hageman, Differentiating between the effects of water activity and glass transition dependent mobility on a solid state chemical reaction: aspartame degradation, *J. Agric. Food Chem. 42*:2398 (1994).
4. L. N. Bell and M. J. Hageman, A model system for differentiating between water activity and glass transition effects on solid state chemical reactions, *J. Food Quality 18*:141 (1995).
5. L. N. Bell and M. J. Hageman, Glass transition explanation for the effect of polyhydroxy compounds on protein denaturation in dehydrated solids, *J. Food Sci. 61*:372 (1996).
6. E. E. Braudo, E. M. Belavtseva, F. F. Titova, I. G. Plashchina, V. L. Krylov, V. B. Tolstoguzov, F. R. Schierbaum and M. Richter, Struktur und Eigenschaften von Maltodextrin-hydrogelen. *Starch 31*:188 (1979).
7. M. P. Buera and M. Karel, Application of the WLF equation to describe the combined effects of moisture and temperature on non-enzymatic browning rates in food systems, *J. Food Process. Preserv. 17*:31 (1993).
8. M. D. P. Buera, J. Chirife, and M. Karel, A study of acid-catalyzed sucrose hydrolysis in an amorphous polymeric matrix at reduced moisture contents, *Food Res. Int. 28*:359 (1995).
9. S. Cardona, C. Schebor, M. P. Buera, M. Karel, and J. Chirife, Thermal stability of invertase in reduced-moisture amorphous matrices in relation to glassy state and trehalose crystallization, *J. Food Sci. 62*:105 (1997).
10. J. Chirife and M. D. P. Buera, Water activity, glass transition and microbial stability in concentrated/semimoist food systems, *J. Food Sci. 59*:921 (1994).
11. L. Chuy and T. P. Labuza, Caking and stickiness of dairy-based food powders as related to glass transition, *J. Food Sci. 5*:43 (1994).

12. S. Davidou, M. Le Meste, E. Debever, and D. Bekaert, A contribution to the study of staling of white bread: effect of water and hydrocolloid, *Food Hydrocolloids 10*:375 (1996).
13. O. Fennema, Metastable and nonequilibrium states in frozen food and their stabilization, *Food Preservation by Moisture Control. Fundamentals and Applications* (G. V. Barbosa-Canovas and J. Welti-Chanes, eds.), Technomic Publishing Co., Lancaster, PA, 1995, p. 243.
14. O. R. Fennema, W. D. Powrie, and E. H. Marth, *Low-Temperature Preservation of Foods and Living Matter*, Marcel Dekker, New York, 1973.
15. F. Franks, Water activity: a credible measure of food safety and quality? *Trends Food Sci. Technol. 2*:68 (1991).
16. F. Franks, Complex aqueous systems at subzero temperatures, *Properties of Water in Foods* (D. Simatos and J. L. Multon, eds.), Martinus Nijhoff Publishers, Dordrecht, 1985.
17. B. A. Graham, Recent advantages in agglomeration during spray drying, *Food Aust. 49*:184 (1997).
18. H. S. Guraya and R. T. Toledo, Determining gelatinized starch in a dry starchy product, *J. Food Sci. 58*:888 (1993).
19. R. W. Hartel, Controlling sugar crystallization in food products, *Food Technol. 47*:99 (1993).
20. A. Hermansson and K. Svegmark, Developments in understanding of starch functionality, *Trends Food Sci. Technol. 7*:345 (1996).
21. M. Karel, Role of water activity, *Food Properties and Computer-Aided Engineering of Food Processing Systems* (R. P. Singh and A. G. Medina, eds.), Kluwer Academic Publishers, New York, 1989, p. 135.
22. M. Karel, The significance of moisture to food quality, *Developments in Food Science 2* (Chiba et al., eds.), Kodansha Ltd., Tokyo, 1979, p. 347.
23. M. Karel, M. P. Buera, and Y. Roos, Effects of glass transitions on processing and storage, *The Science and Technology of the Glassy State in Foods* (J. M. V. Blanshard and P. J. Lillford, eds.), Nottingham University Press, Nottingham, 1993, p. 13.
24. M. Karel and I. Saguy, Effects of water on diffusion in food systems, *Water Relationships in Foods* (H. Levine and L. Slade, eds.), Plenum Press, New York, 1991, p. 157.
25. R. Karmas, M. P. Buera, and M. Karel, Effect of glass transition on rates of nonenzymatic browning in food systems, *J. Agric. Food Chem. 40*:873 (1992).
26. S. Labrousse, Y. Roos, and M. Karel, Collapse and crystallization in amorphous matrices with encapsulated compounds, *Sci. Aliments 12*:757 (1992).
27. H. Levine and L. Slade, A polymer physico-chemical approach to the study of commercial starch hydrolysis products (SHPs), *Carbohydrate Polym. 6*:213 (1986).
28. H. Levine and L. Slade, Interpreting the behavior of low moisture foods, *Water and Food Quality* (T. M. Hardman, ed.), Elsevier Applied Science, New York, 1989, p. 71.
29. D. Lund, Influence of time, temperature, moisture, ingredients and processing conditions on starch gelatinization, *Crit. Rev. Food Sci. Technol. 20*:249 (1984).
30. B. Makower and W. B. Dye, Equilibrium moisture content and crystallization of amorphous sucrose and glucose, *J. Agric. Food Chem. 4*:72 (1956).
31. C. H. Mannheim, J. X. Liu, and S. G. Gilbert, Control of water in foods during storage, *J. Food Eng. 22*:509 (1994).
32. A. Mersmann, Fundamentals of crystallization, *Crystallization Handbook* (A. Mersmann, ed.), Marcel Dekker, New York, 1994.
33. D. G. Murray and L. R. Luft, Low-DE corn starch hydrolysates, *Food Technol. 27*:32 (1973).
34. J. O'Brien, Stability of trehalose, sucrose and glucose to nonenzymic browning in model systems, *J. Food Sci. 61*:679 (1996).
35. J. Olkku and C. Rha, Gelatinization of starch and wheat flour starch—a review, *Food Chem. 3*:293 (1978).
36. K. J. Palmer, W. B. Dye, and D. Black, X-ray diffractometer and microscopic investigation of crystallization of amorphous sucrose, *J. Agric. Food Chem. 4*:77 (1956).
37. R. Parker and S. G. Ring, Diffusion in maltose-water mixtures at temperatures close to the glass transition, *Carbohydrate Res. 273*:147 (1995).
38. H. Pasley and P. Haloulos, Stickiness—a comparison of test methods and characterisation param-

eters, *IDS Drying '94*, Proceedings of the 9th International Drying Symposium, Gold Coast, 1-4 Aug., 1994.
39. M. Peleg, *Physical Properties of Foods* (M. Peleg and E. Bagley, eds.), Avi Publisher, Westport, CT, 1983, p. 293.
40. R. J. Priestley, Studies on parboiled rice: Part 1—Comparison of the characteristics of raw and parboiled rice, *Food Chem. 1*:5 (1976).
41. M. S. Rahman, *Food Properties Handbook*, CRC Press, Inc., New York, 1995.
42. M. Richter, F. Schierbaum, S. Augustat, and K. D. Knoch, U.S. patent 3,3962,454, 8 June (1976).
43. M. Richter, F. Schierbaum, S. Augustat, and K. D. Knoch, U.S. patent 3,986,890, 19 October (1976).
44. Y. H. Roos, Water activity and physical state effects on amorphous food stability. *J. Food Eng. 16*:433 (1993).
45. Y. H. Roos, Phase transitions and transformation in food system, *Handbook of Food Engineering* (R. Heldman and D. B. Lund, eds.), Marcel Dekker, New York, 1992.
46. Y. Roos and M. Karel, Differential scanning calorimetry study of phase transitions affecting the quality of dehydrated materials, *Biotechnol. Prog. 6*:159 (1990).
47. Y. Roos and M. Karel, Applying state diagrams to food processing and development, *Food Technol. 45*:66 (1991).
48. Y. Roos, *Phase Transition of Foods*, Academic Press, 1995.
49. V. Sapru and T. P. Labuza, Glassy state in bacterial spores predicted by polymer glass-transition theory, *J. Food Sci. 58*:445 (1993).
50. C. Schebor, M. D. P. Buera, J. Chirife, and M. Karel, Sucrose hydrolysis in glassy starch matrix, *Food Sci. Technol. 28*:245 (1995).
51. C. Schebor, M. P. Buera, and J. Chirife, Glassy state in relation to the thermal inactivation of enzyme invertase in amorphous dried matrices of trehalose, maltodextrin and PVP, *J. Food Eng. 30*:269 (1996).
52. H. Schubert, *Int. Chem. Eng. 21*:363 (1981).
53. Y. Shimada, Y. Roos, and M. Karel, Oxidation of methyl linoleate encapsulated in amorphous lactose-based food models, *J. Agric. Food Chem. 39*:637 (1991)
54. L. Slade and H. Levine, Beyond water activity: recent advances based on an alternative approach to the assessment of food quality and safety, *Crit. Rev. Food Sci. Technol. 30*:115 (1991).
55. L. Slade and H. Levine, A food polymer science approach to structure-property relationships in aqueous food systems: non-equilibrium behavior of carbohydrate water systems, *Water Relationships in Food* (H. Levine and L. Slade, eds.), Plenum Press, New York, 1991.
56. L. Slade and H. Levine, Structural stability of intermediate moisture foods-a new understanding, *Food Structure—Its Creation and Evaluation*, (J. R. Mitchell and J. M. V. Blanshard, eds.), Butterworths, London, 1987, p. 115
57. R. L. Sutton, I. D. Evans, and J. F. Crilly, Modeling ice crystal coarsening in concentrated disperse food systems, *J. Food Sci. 59*:1227 (1994).
58. J. R. Taylor, Glass-state molecular mobility, *Food Ind. South Africa 48*:29 (1995).
59. S. Tsourouflis, J. M. Flink, and M. Karel, Loss of structure in freeze-dried carbohydrates solutions. Effect of temperature, moisture content and composition, *J. Sci. Food Agric. 27*:509 (1976).
60. C. Van den Berg, Food-water relations: progress and integration, comments and thoughts, *Water Relations in Foods* (H. Levine and L. Slade, eds.), Plenum Press, New York, 1991, p. 21.
61. D. A. Wallack and C. J. King, Sticking and agglomeration of hygroscopic, amorphous carbohydrate and food powders, *Biotechnol. Prog. 4*:31 (1988).
62. J. Zhao and M. R. Notis, Phase transition kinetics and the assessment of equilibrium and metastable states, *J. Phase Equil. 14*:303 (1993).

5

Food Preservation by Heat Treatment

M. N. Ramesh
Central Food Technological Research Institute, Mysore, India

I. INTRODUCTION

Of the many techniques used to preserve food, only a few rely on killing the relevant infecting microorganisms. Thermal inactivation is still the most widely used process in this category. Other processes rely on physical exclusion of microorganisms after filtration or sterilization, e.g., superclean or aseptic systems. Most procedures, however, merely rely on slowing of growth of the food-spoiling organisms.

The advantages of using heat for food preservation are as follows: (a) heat is economical, (b) heat is safe and chemical-free foods can be produced, (c) the product becomes more tender and pliable with the desired cooked flavor and taste, (d) the majority of the spoilage microorganism are heat labile and can be killed, and (e) thermally processed foods, when packed in sterile containers, have a very long shelf life. The disadvantages of using heat are as follows: (a) overcooking may lead to textural disintegration and undesired cooked flavor, and (b) nutritional deterioration results from high-temperature processing.

II. PASTEURIZATION

A. Introduction

Pasteurization is a process of heat treatment used to inactivate enzymes and to kill relatively heat-sensitive microorganisms that cause spoilage with minimal changes in food properties (e.g., sensory and nutritional). It is also defined as "mild heat treatment" for avoiding microbial and enzymatic spoilage. It is used to extend the shelf life of food at low temperatures (usually 4°C) for several days (e.g., milk) or for several months (e.g., bottled fruit). Heating liquid foods to 100°C is employed to destroy heat-labile spoilage organisms, such as non–spore-forming bacteria, yeast, and molds.

B. Purpose of Pasteurization

The primary objective of pasteurization is to free the food of any microorganisms that might cause deterioration or endanger the consumer's health. The severity of the heat treatment and the resulting extension of shelf life are determined mostly by the pH of the food. In low-acid foods (pH > 4.5), the main purpose is the destruction of pathogenic bacteria, whereas below pH 4.5, destruction of spoilage microorganisms or enzyme inactivation is usually more important. Table 1 shows different pasteurization conditions for foods.

Pasteurization does not aim at killing spore-bearing organisms, such as the thermophilic *Bacillus subtilis*, but these organisms and most other spore-bearing bacteria cannot grow in acidic fruit juices, and consequently their presence is of no practical significance. Pasteurization of carbonated juices need only be conducted at such a temperature and for such a time that yeasts and molds are destroyed. Yeast is killed by heating at 60–65°C and the resistance mold spore negative in most cases a temperature of 80 C for 20 minutes. But molds require oxygen for growth and for this reason heavily carbonated juice can be pasteurized safely at 65°C which destroy yeast cells, most still (non-carbonated) juices must be pasteurized at 80°C. Juices of high acidity may be pasteurized at lower temperatures (60–65°C).

Processing containers of food that have a naturally low pH (e.g., fruit pieces) or in which the pH is artificially lowered (e.g., pickles) is similar to canning. In acid products like tomatoes, mangoes, bananas, etc. (pH 4.0–4.4), yeast, molds, and bacteria (both thermophilic and mesophilic) can grow. The main risk of spoilage is from spore-forming species other than *Clostridium botulinum*, especially *B. coagulans* among the aerobes and *Clostridium pasteurianum* and *Clostridium thermosaccharolyticum* among the anaerobes. In high-acid foods (pH < 3.9) like pineapple juice, spoilage is generally caused by non–spore-forming bacteria (*Lactobacillus* and *Leuconostoc*), yeast, or molds [1].

Fruits with a pH lower than 4.5 contains enzyme systems such as catalase, peroxidase, polyphenol oxidase, pectin esterase, etc., in addition to spoilage organisms. Unless inactivated, these enzymes are likely to cause undesirable changes in the canned products. Some of these enzymes, particularly peroxides, have higher heat resistance than the spoilage organisms and have been used in evolving the thermal processing of canned fruits.

C. Types of Pasteurization

There are several types of pasteurization:

1. In-package pasteurization: Inside packages, heating to the level of sterility is not required. A gradual change in temperature is preferred in some containers.
2. Pasteurization prior to packaging: Preheating is good for foods that are sensitive to high temperature gradients.
3. Batch pasteurization: This is also called the low temperature, short time process. Here fluid foods like milk are held in a tank where they are heated to 62.8°C for 30 minutes. A batch pasteurizer consists of a steam-jacketed kettle or a tank equipped with steam coils in which the juice or milk is heated to the desired temperature [4].
4. Continuous pasteurization: This is also called the high temperature, short time process. Foods like milk is subjected to 71.7°C for about 15 seconds or more by flowing through different heat exchangers. In continuous pasteurization generally plate heat exchanger, tubular heat exchanger, scraped surface heat exchanger are used depending on the viscosity of the fluid food material. The heating medium is usually steam or water.

D. Achieving Desired Pasteurization

Broadly, pasteurization can be achieved by a combination of time and temperature such as (a) heating foods to a relatively low temperature and maintaining for a longer time, e.g., holding pasteurization and pasteurization by overflow method, or (b) heating foods to a high temperature and holding it for a short time only.

Pasteurization can be performed in two ways: (a) by first filling sterile containers with the product and then pasteurizing or (b) by pasteurizing the product first and then filling sterile containers.

Table 1 Purpose of Pasteurization of Different Foods

Food	Main purpose	Subsidiary purpose	Minimum processing conditions
pH < 4.5			
Fruit juice	Enzyme inactivation (pectinesterase and polygalacturonase)	Destruction of spoilage microorganisms (yeast, fungi)	65°C for 30 min; 77°C for 1 min; 88°C for 15 sec
Beer	Destruction of spoilage microorganisms (wild yeasts, *Lactobacillus* species) and residual yeast (*Saccharomyces* species)	—	65–68°C for 20 min (in bottle) 72–75°C for 1–4 min at 900–1000 kPa (in can)
pH > 4.5			
Milk	Destruction of pathogens (Brucella microabortis, Mycobacterium tuberculosis)	Destruction of spoilage organisms and enzymes (*Coxiella burnetti*)	63°C for 30 min; 71.5°C for 15 sec
Liquid egg	Destruction of pathogens (*Salmonella seftenburg*)	Destruction of spoilage microorganisms	64.4°C for 2.5 min 60°C for 3.5 min
Ice cream	Destruction of pathogens	Destruction of spoilage microorganisms	65°C for 30 min 71°C for 10 min 80°C for 15 sec

E. Pasteurization Equipment

1. Pasteurization of Packaged Foods

In packaged foods like beer and fruit juices, in-container processing is applied. When the container is glass, generally hot water processing is used to reduce any damage due to thermal shock. After processing, the container is cooled to 40°C, which also facilitates evaporation of the surface water. This minimizes external corrosion of metal containers or caps and accelerates the setting of adhesives used in labels.

Water Bath Pasteurization

For acidic food products that can be adequately pasteurized at temperatures of 100°C or below, a water bath is one of the simplest methods of heating for pasteurization. The water bath may be either a rectangular steel tank or a vertical retort. The product is packed in retort crates or in racks and immersed in the bath for pasteurization. Cooling may be carried out in the same tank used for heating, or the containers may be moved from the heating tank to a cooling tank. Heating and cooling also may be carried out in steps. Essentially the same procedure is followed for processing of meats, pickles, applesauce, and other acidic food products [2].

The continuous water bath is an improvement over batch operation and is used by both pickle processors and fruit canners for pasteurization where high production rates are required. A conveyor belt moves through the tank at a selected speed to provide adequate time in the bath to accomplish pasteurization. The tank is usually divided into sections, each of which is heated and controlled separately.

In continuous water bath pasteurizers, the jars and cans must proceed down an incline into the tank and up an incline when they come out of the tank. Since there is considerable hazard in conveying glass containers up or down an incline, plants that pasteurize glass-packed products usually use water spray or steam pasteurizers [2].

Continous Steam or Water Spray Pasteurization

The continuous water spray pasteurizer is extensively used for pasteurizing beer and acidic food products. In this type of unit, bottles or cans are conveyed through the pasteurizer either by a walking beam or by a continuous belt conveyor. It is common practice to have as many as six different temperature zones or sections throughout the pasteurizer to obtain maximum efficiency. These are first preheat, second preheat, pasteurizing zone, precool, cooling, and final cooling zone. Water spray units are designed so the water in the first preheat zone drains off the jars in the precool zone, and the water that is sprayed in the precool zone is used in the first preheat zone. In this way a considerable amount of heat is recovered and reused and a reduced amount of cooling water is required. Cooling water is also recirculated [2].

Glass containers should not be subjected to excessive thermal shock; when heating products in glass containers it is recommended that the thermal shock temperature difference be kept below 21°C and under no condition to exceed 38°C. When cooling a hot product in a glass container, temperatures are more critical; 10°C is a desirable maximum and under no conditions should the temperature change exceed 21°C.

Several sections are necessary in both continuous steam and water spray pasteurizers to heat glass containers efficiently; however, metal containers may be pasteurized in the same equipment. Through the use of sectionalized equipment, it is possible to have high-temperature heating and low-temperature cooling of glass containers with a minimum amount of thermal shock breakage. The water spray–type unit has been very successful in the pasteurization of beer and similar products where the operation proceeds under ideal conditions.

The steam pasteurizer is simply a tunnel open at both ends with a conveyer along the bottom. Cloth baffles are hung between each section, but these are not adequate to hold the steam in the pasteurizer against strong air currents. The rate of heat transfer from the steam-air mixture to the food container is not constant in the steam pasteurizer, but varies with steam temperature and steam velocity.

Tunnel Pasteurization

Hot water sprays are used to heat containers as they pass through the different heating zones of the tunnel and provide an incremental rise in temperature until pasteurization is achieved. Cold water sprays then cool the containers as they continue through the tunnel. Steam tunnels have the disadvantage of faster heating, giving shorter residence time, and smaller equipment. Savings in energy and water are achieved from heat recovery from the hot product and recirculating water. Temperatures in the heating zone are gradually increased by reducing the amount of air in the steam-air mixtures, and cooling takes place using water sprays or by immersion in a water bath.

2. Pasteurization of Unpackaged Liquids

Long-Hold or Vat Pasteurizing

Vat or tank-type heat exchangers are used for the long-hold method of pasteurization. Here the raw product is pumped into the vat, heated to the pasteurizing temperature, held for the required time, and pumped from the vat through cooling equipment. With most vat pasteurizers circulation of the heating medium can be started as soon as the filling of the vat is begun. In this way some heating of the product takes place during filling so that the heating time can be shortened. With some designs of pasteurizing vats, cold water can be circulated over the outside of the inner liner as soon as the holding period is completed, thus doing part of the cooling in the vat [3].

It is considered good practice with all heat-exchange equipment for dairy products to use a heating medium (hot water or steam vapor) only a few degrees warmer than the milk, resulting in less accumulation of milkstone on heating surfaces and less danger of injury to cream line or flavor.

Advantages: Vat pasteurizers are well suited for small plants and for low-volume products in larger operations. They can handle a variety of products with a wide range of physical characteristics. They are especially well adapted to the processing of cultured products such as buttermilk and sour cream, which, in addition to being pasteurized and cooled, require mixing for the incorporation of starter, several hours of quiescent holding for incubation, agitation for breaking the curd, and final cooling in the tank.

Disadvantages: There are several disadvantages to consider. Vat pasteurization is normally a batch operation and is inherently slow, although the flow can be made continuous by the use of three or more vats (depending upon the holding, heating, filling, and emptying times). The operation may even be made automatic by use of complex and expensive controls. In the great majority of batch operations, manual controls are used, and constant attention must be given by the operator to prevent overheating, overholding, and burning. Another disadvantage is that regenerative heating is not possible in the vat, so both heating and cooling is relatively expensive.

Heat Exchanger Pasteurizer

Small-scale batch pasteurization is carried out in open boiling pans or in scrapped surface heat exchangers. Generally less viscous liquids are pasteurized by plate heat exchanger. Some products, such as fruit juices and wines, require deaeration before pasteurization to prevent oxidative changes during storage. This can be achieved by spraying liquids into the vacuum chamber after which dissolved air is removed.

The plate heat exchanger consists of a series of thin vertical stainless steel plates. The plates form parallel channels held tightly together in a metal frame and separated by rubber gaskets to produce a watertight seal. The plates are corrugated to induce turbulence for a high heat transfer rate. A flow diagram is shown in Figure 1.

The advantages of heat exchangers over in-bottle processing include (a) more uniform heat treatment, (b) simpler equipment and lower maintenance costs, (c) reduced space requirements and labor costs, (d) greater flexibility for different products, and (e) greater control over pasteurization conditions.

A number of systems for pasteurizing milk have been used commercially. The first were batch systems employing holding tanks. Milk was heated in a jacketed tank to a temperature of 63°C and held for 30 minutes. This type of system is now rarely found, but it can be suitable for small opera-

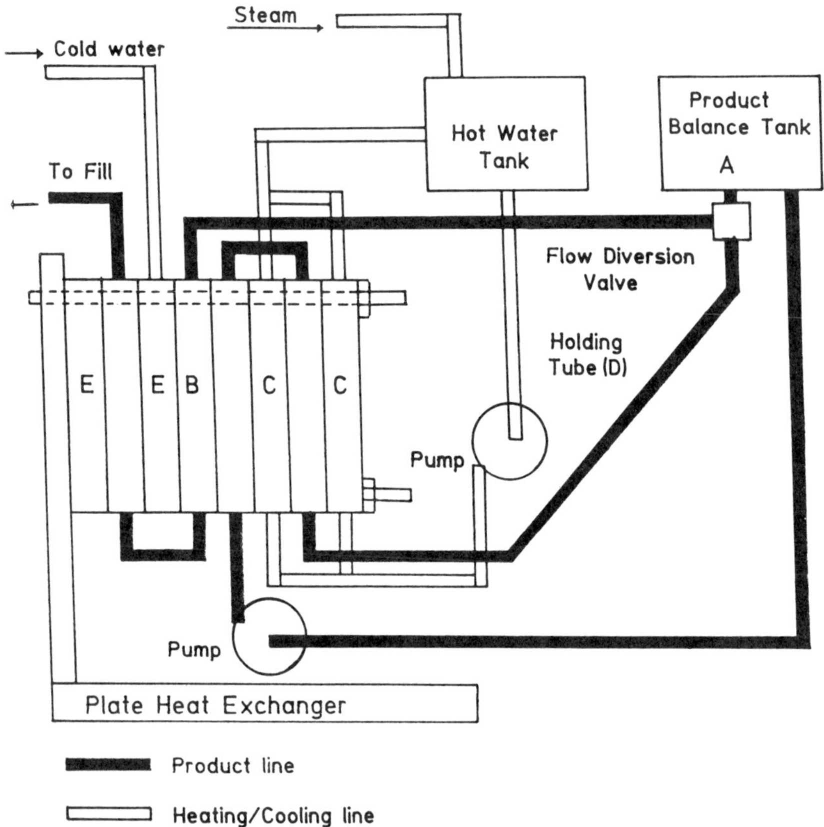

FIGURE 1 Plate heat exchanger pasteurizer.

tions. Improvements on the batch system came with the advent of the continuous-holding or retarding systems. Holding times and temperatures are the same; however, the tanks automatically fill, hold, and empty in a timed cycle. The system of choice in most modern dairies is now the high-temperature, short-time (HTST) process.

The heat exchanger has the following advantages over the batch and continuous-holding systems: (a) lower initial cost due to elimination of holding tanks, (b) less labor required as the system incorporates mechanized circulation cleaning, (c) saves space (about 10,000 liters/hr can be pasteurized in 4.5 m^2), (d) increased flexibility (capacity of the plant and processing rate are easily controlled), (e) ease of recording and safeguarding pasteurization temperature requirements (milk can be readily diverted if it does not reach minimum safe pasteurized temperatures), and (f) lower operating costs (plant can be almost entirely automatically controlled).

The capacity of the equipment varies according to the size and number of plates—up to 80,000 liters/hr. Other types of heat exchangers are also used for pasteurization. In particular, the concentric tube heat exchanger is suitable for more viscous food and is used with dairy products, mayonnaise, tomato ketchup, and baby food. It consists of a number of concentric stainless steel coils, each made from double- or triple-walled tube. Food passes through the tube, and heating and cooling water is recirculated through the tube walls. Liquid food is passed from one coil to the next for heating and cooling, and the heat is regenerated to reduce energy costs. Pasteurized food is immediately deposited into cartons or bottles. Care with cleaning and hygiene is therefore necessary.

High-Temperature, Short-Time Pasteurizers

HTST pasteurizers are continuous-flow systems using tubular, plate, swept surface, direct steam, in conjunction with a timing pump, a holder, and controls for temperature and flow rate. The great majority of HTST pasteurizers use plate-type heat exchangers with sections for regenerative heating and cooling. This is also referred to as a flash pasteurizer [3].

Continuous pasteurizes assure that all of the product of an entire run receives uniform treatment. HTST pasteurizers employing regenerative heating are much more economical to operate than batch pasteurizers. In the application of controls, the general requirements for flow rate, temperature, and pressure must be considered, for these are the factors that govern proper operation and public health safety.

Flow rate through a continuous pasteurizer is regulated by the metering or timing pump. A positive displacement pump of the rotary or piston type is used almost exclusively for milk and milk products. Often variable speed drives are employed so that the flow rate can be changed when desired. A continuous pasteurizer must include synchronization of holding time and flow rate.

Controlling temperature includes maintaining a uniform product temperature at some set value at or above the legal minimum and diverting the flow, directing it back through the system if, at the end of the holder, it is below the legal minimum temperature. Usually, a safety thermal limit–recording controller is used, which keeps continuous record of the temperature.

The pressure is especially important in two areas of a continuous pasteurizer—the regenerator and the flow-diversion valve. Where product-to-product regeneration is used, it is necessary, for public health reasons, to maintain at least 7 kPa more pressure on the pasteurized side than on the raw side so that any leakage through the heat exchanger can be identified, thus eliminating the possibility for contaminating the pasteurized product.

In order to prevent mixing of air into the product and inefficiency of the pump due to air leakage into the system, the entire system is operated at a positive pressure (above atmospheric pressure). A centrifugal booster pump is employed between the product storage tank and the regenerator. This will ensure that the pasteurized product is always under higher pressure than the raw product.

To facilitate this, it is necessary to (a) size the booster pump correctly to deliver the rated capacity at a predetermined pressure, and (b) equip the booster pump with a pressure-actuated switch located at the outlet of the pasteurized regenerator set so that the pump can run only when the pressure is at least 7 kPa greater than that on the raw-product side. If the cooler section does not produce enough back-pressure on the pasteurized regenerator to satisfy the minimum 7 kPa difference, it may be necessary to install a restrictor in the line.

The other pressure requirement is needed on the diverted milk line, since it may affect the holding time during diversion. If, upon testing, the holding time during diverted flow is shorter than that of forward flow, a restricting orifice should be placed in the diverted line.

Flash pasteurization: The process of heating fruit juices for only a short time at a temperature higher than the pasteurization temperature of the juice is called flash pasteurization. In this method the juice is heated rapidly for about 1 minute to a temperature about 5°C higher than the pasteurization temperature, filled into airtight containers under steam to sterilize the steam, and then cooled. This process can also be used for orange juice, apple juice, grape juice, etc.

The advantages of this process are that it (a) minimizes flavor loss, (b) aids in the retention of vitamins, (c) effects economy in time and space, (d) helps to keep the juice uniformly cloudy, (e) heats juice uniformly, reducing cooked taste to a minimum, and (f) effects beneficial enzyme inactivation in addition to destruction of viable microorganisms.

Deaerator and flash pasteurizer: Freshly extracted and screened juices contain an appreciable quantity of oxygen, which should be removed before packing. The special equipment used for this purpose is called a deaerator. The deaerated juice is then heated in flash-pasteurization equipment. Commercial deaerators and flash pasteurizers differ greatly in design, construction, and capacity. The deaerator and flash pasteurizer has been used successfully for fruit juices like tomato and pineapple and orange.

Ultra-High-Temperature Pasteurizers

The equipment for ultra-high-temperature (UHT) pasteurizers is much the same as for HTST units. The controls are similar, but the operating temperature points are higher. The holder is, of course, much smaller for minimum pasteurizing time. Generally, with a holding time in the order of 3 seconds, it is impossible to determine the holding time accurately by tests like those used with HTST pasteurizers, and calculated holding times are preferred [3].

Where ultra-high-temperature treatment is desired because of its greater bacterial destruction or its beneficial effects on body and texture in ice cream and where confusion exists regarding requirements for UHT pasteurization, a UHT treatment may be given following regular pasteurization. This may be accomplished with a direct-steam heater installed downstream from the flow diversion valve or with steam-vacuum flavor-treating equipment.

The Vacreator

The vacreator is a special type of pasteurizing apparatus used particularly in the butter industry. The product is fed into a steam-heated chamber where it is flashed at a temperature of 90–96°C under 13.5–30 kPa of vacuum. It then passes into a second chamber of higher vacuum (50–67 kPa) and is reduced to a temperature of 72–82°C. From there it passes into a third high-vacuum chamber (91–95 kPa) and is reduced to a temperature of 38–46°C. This process is claimed to be very effective for pasteurization and at the same time to remove undesirable odors and flavors. It also employs a continuous-type machine and is especially adapted for use on high-viscosity material, although it will also operate on plain fluid milk [3].

F. Quality of Pasteurized Foods

1. Color, Flavor, and Aroma

The main cause of color deterioration in fruit juices is enzymatic browning by polyphenoloxidase. This is promoted by the presence of oxygen, and fruit juices are therefore routinely deaerated prior to pasteurization. The difference between the color of raw and pasteurized milk is due to homogenization; pasteurization has no measurable effect. Other pigments in plant and animal products are also unaffected by pasteurization [1].

Even a small loss of volatile aroma compounds during pasteurization of juices causes a reduction in quality and may also unmask other "cooked" flavors. Volatile recovery may be used to produce high-quality juices, but it is not routine. Loss of volatiles from raw milk removes a haylike aroma and produces a blander product.

2. Nutrients

In fruit juices, losses of vitamin C and carotene are minimized by deaeration. Changes to milk are confined to a 5% loss of serum proteins and small changes in the vitamin content (Table 2). Most pasteurized food products have a low pH because either the natural pH of the system is low or the product has been fermented to produce an acidic environment. Since most heat-labile nutrients are relatively stable in acid conditions, nutrient losses in such products are relatively minor. Although

Table 2 Vitamin Losses During Pasteurization of Milk

	Method of pasteurization	
Vitamin	HTST (%)	Holder (%)
Thiamine	10	10
Vitamin C	10	20

Source: Ref. 1.

thermal losses during pasteurization may be small, oxidative losses can be high. Thus pasteurization of liquid foods such as fruit juices, beer, and wine is generally accomplished in indirect heat exchangers like plate or double tube rather than open film-type pasteurizers. Often fluids are deaerated prior to pasteurization. The most important nonacid liquid food is milk. The effect of pasteurization on the nutrients in milk has received considerable attention [5].

G. Packaging of Pasteurized Foods

Both bottles and cartons take into account the properties of milk and provide packaging acceptable to consumers worldwide. Glass bottles have the advantage of being easily cleaned, transparent, and rigid, but the great disadvantages of high weight and fragility.

Increasingly, milk is also packaged in gable top (PurePak, Elopak) or other cartons (TetraBrik). Even though the equipment may be expensive to install, the advantages include a lower price per unit of milk and a lower risk of contamination from the air during filling [6]. Smaller quantities have been packaged in plastic pouches. A cylindrical milk carton with a reclosable pouring lid has been introduced [7].

While suitable for sterilized milk, glass bottles are a problem for "long-life" milk. The question of container is therefore of vital interest to the dairy industry, as about 50% of all milk produced is sold in liquid form [7]. Sunlight can destroy riboflavin and vitamin C in milk, producing a taint by the oxidation of fat [7] and protein. This led to the use of brown glass bottles, which hold back the light rays responsible. However, taint is very rare and brown bottles are not very attractive; it has also been found that milk becomes sour faster in brown than in colorless bottles [8].

1. Returnable Bottles

For economic reasons, the use of the returnable glass bottles has continued over many years. The glass bottle will take a long time to disappear because of its economic advantage, the traditions of the industry, and the attitude of the consumer. Other factors such as transport costs have led to the use of nonreturnable packaging, although more recently "green" considerations have led to the reintroduction of returnable bottles [7,9].

The advantages of nonreturnable containers are (a) elimination of returned empties, (b) elimination of collecting, sorting, and washing, (c) elimination of foreign object problem, (d) elimination of glass fragment problem, and (e) reduction in transport costs. The disadvantages are (a) possible increases in costs of packages, (b) lack of consumer acceptance, (c) delivery problems resulting in lower total sales, (d) hygiene problems, and (e) environmental considerations.

Plastic containers and plastic-coated cartons are nearly sterile by virtue of their method of manufacture. No sterilizing process is necessary for pasteurized milk, but for the aseptic filling of milk, sterilization is essential. So far, TetraBrik has proved most effective. Containers for this purpose must be sterilized immediately before filling.

2. Glass Bottles

The traditional glass bottles used for fruit juices and juice beverages provide many advantages. Glass is not susceptible to mold growth and is impermeable to odorous vapors and liquids. Hot-filling and in-bottle pasteurization are generally employed for pure fruit juices or products that do not contain preservatives. Any microbiological contamination on the inner surfaces of the bottle and the closure is destroyed by the hot liquid, and adequate sterility is obtained without heating the container [7].

Glass bottles can also be covered with a polystyrene shield, which enables them to be reduced in weight without risking breakage. Sleeves give protection and graphics can be added easily. Some bottles are shrink-wrapped with plastic sleeves.

3. The PET Bottle and Other Plastic Containers

PET bottles are displacing those made from PVC for products such as edible oils and mineral waters as well as glass bottles for carbonated products. Improvements in processing technology have resulted in the appearance of stretch-brown PVC bottles.

Other forms of plastic container have also been used [10,11], e.g., the Plastocan, a coextruded plastic container with conventional aluminum easy-open can ends, and the Rigello container, a multilayered polypropylene foil extrusion with a spherical bottom and tear-off cap assembled in a paperboard cylinder. New combinations of materials in can form are also being developed. Coca Cola patented a PET/aluminum can with easy-open top [12]. High-barrier plastics cans that will be recyclable are under investigation. Orange juice has also been packed in clear polypropylene bottles, which provide good oxygen and moisture barrier properties [13]. Tamper-evident pull-tab closures are used on this container. Paperboard basket carriers, plastic clips (on bottle necks), and shrink films are used to provide multipacks holding three, four, or six units.

4. Cans

Fruit juices and fruit juice concentrates are frequently distributed in cans [14]. The most common of these are standard tin-plate containers, but specially lacquered and coated cans are also used, especially for high-acid products. Cans are usually hot-filled, but sometimes are aseptically filled. Cold-filling after pasteurization is occasionally employed, but refrigerated or frozen storage is then advisable. Products preserved with benzoic acid can also be filled cold after pasteurizing, but sulfated products are incompatible with cans. The juice tends to deteriorate in the cans due to corrosion and an increasing amount of tin and iron in the product.

In the normal hot canning process, the juice is first deaerated to improve its flavor stability and then pasteurized to destroy microorganisms and inactivate enzymes. After hot-filling into the cans, the lids are applied and sealed immediately before cooling, which forms a slight vacuum in the headspace as the liquid contracts. This is desirable as the presence of oxygen encourages corrosion (cold-filling operations usually involve undercover gassing, in which the headspace is replaced by carbon dioxide immediately before sealing the lid).

Carbonated beverages are susceptible to metal pick-up and are therefore packaged in lacquered two-piece aluminum cans or three-piece tin-plate with side seams having a special tab design to withstand the internal pressure. Warming the filled cans immediately before packaging is important, otherwise the cans when filled with cold carbonated liquid attract a layer of condensation from the atmosphere and may corrode on the outside. Frozen orange juice concentrate has been distributed in composite paperboard or plastic cartons of approximately 170 ml capacity [10]. There are many pack variations, including cartons with tear-off ends or left unpasteurized to provide maximum freshness of flavor. Spoilage may result if these are not frozen.

Beverage cans are also sold in multipacks of four, six, or more [15]. The most common form of overwrap, which assists in handling and distribution, is a plastic ring carrier, which slips underneath the rim of the can and grips tightly throughout distribution. Paperboard multipacks are also popular, as are shrink wraps.

5. Cartons

Pasteurized fruit juice and soft drinks can be packaged very successfully in polyethylene-coated cartons or in plastic containers [15]. These products have a limited shelf life when stored in a refrigerator. Materials selected must not absorb flavor components from the juice. In addition, acid diffusion into the plastic material can delaminate the package. Polyethylene is the most common surface-contact material and is regarded as chemically stable to most food products. Packaging materials must also provide the best possible barrier to light, as light affects the color and nutritive value of fruit juices.

Aseptic filling of pasteurized fruit juices and other drinks into a TetraPaks and other systems (e.g., Combibloc, PurePak, Elopak) has also become popular, giving the product an extended shelf life. Such products have advantages over hot-filled products or nonaseptically packaged products, which need a chilled distribution chain [18].

H. Energy Aspects of Pasteurization

Indirect heating through a heat exchanger involves much more energy than direct heating with infusion heating. To reduce the energy input and hence improve the efficiency of the system, a regenerative method is adopted. For example, the total energy difference between the 90 and 80% regeneration capacity of a system processing milk is about 62,131 kcal/hr. This shows that even a 10% increase in regenerative efficiency can save considerable energy [16]. Regeneration of heating and cooling streams is now an accepted energy-conservation technique.

The heat energy consumption for pasteurization will be about 30 MJ per 1000 liters of milk, and correspondingly the cooling energy is about 4 kWh per 1000 liters. Normally, pasteurizers will have the facility of regeneration, with an efficiency of 75–92%. Efficiency can be increased by increasing the number of plates in the regeneration section. But this increases the pumping pressure and hence the electric energy. Hence, the maximum regeneration is about 90%. Table 3 gives the steam and electrical requirements of a large pasteurization unit [17,18]. Increasing regeneration decreases the steam and refrigeration requirement but increases the electricity repaired for pumping. These energy requirements have also been converted into their primary fuel equivalents. This facilitates examining processes on an uniform energy basis. The processing, storage, and distribution of pasteurized milk requires an estimated 2200 kJ/kg of milk [19].

III. BULK PACKAGING

A. Introduction

The thermal-processing operation requires the heating of food products. For a low-acid food product (pH > 4.5), this process requires heating the product to temperatures above 100°C, usually in the range 115–130°C, for a time sufficient to achieve a 12 log reduction of the spores of *C. botulinum* as defined in Department of Health Code of Practice No. 10. Current practices are, however, to move to even higher temperatures and consequently shorter process times, maximizing the organoleptic and nutrient retention within the product. The time-temperature procedure required to render a product commercially sterile must be carefully determined using established procedures.

Canned foods might be described as full-moisture, ambient-temperature stable food products regardless of the package form or the preservation process employed. Stability is defined as the absence of microbiological safety problems, but not necessarily retention of all desirable attributes of the food under all storage conditions.

B. Processing Equipment

The food-processing industry produces a wide range of products in a variety of containers necessitating the need for an equally wide range of processing techniques and hence retort designs and operating procedures. Retorting systems can be classified in several ways.

1. Classification of Retorting Systems

Container-Processing Method

In batch systems, the retort is filled with product, closed, and then put through a processing cycle. In continuous retorting systems, containers are continuously fed into and out of the retort. Batch re-

TABLE 3 Steam and Electrical Requirements of a Pasteurizer

Regeneration (%)	80	85	90	95
Steam (kg/liter)	0.012	0.01	0.005	0.003
Power (kWh/1000 liters)				
Pumping	0.66	0.88	1.54	2.86
Refrigeration	3.52	2.64	1.76	0.88
Cooling				
Primary energy (MJ/liter)				
Steam	0.072	0.053	0.034	0.021
Electricity	0.059	0.050	0.047	0.053

Source: Refs. 17, 18.

torts are available in a number of configurations for various applications including static, rotary, steam-heated, and water-heated with or without air overpressure. The air overpressure is necessary to maintain the integrity of the containers during retort operating cycles for glass and flexible containers.

Heating Medium

The following media can be used in the retort:

1. Saturated steam: This condenses on the outside of the container and latent heat is transferred to the food. Air trapped inside the retort should be evacuated to avoid insulating boundary film around the cans. This causes underprocessing by preventing steam from condensation. The procedure of removing air from the retort is known as venting.
2. Hot water: Hot water or overpressure air is mainly used to avoid thermal shock in the case of glass and to avoid surface burning or overheating near container walls in the case of thin flexible packaging materials.
3. Flames: High-temperature radiation by flames (1770°C) leads to short processing time with improved food quality. An approximate 20% energy savings is possible compared to conventional canning.

2. Processing Equipment

Batch/Still Report (Horizontal and Vertical)

Batch steam retorts: These are usually arranged either vertically (Fig. 2) or horizontally (Fig. 3) and are used for canned products, which are placed into baskets immediately after seaming and are then placed into the body of the retort shell. The retort is made out of a metal shell pressure vessel which is fitted with inlets for steam (A), water (B), and air (E) and has outlet ports for venting (D) air during retort come-up and for draining (C) the retort at the end of the cycle. A pocket for instruments, thermometer, temperature-recording probe, and pressure gauge is located on the side of the vessel. To ensure adequate steam movement around the instruments, in particular the temperature sensors, the instrument pocket is fitted with a constant steam bleed (D). On vertical retorts the lid is hinged at the top and secured to the shell during processing by several bolts. In horizontal steam retorts the door is usually on the end of the machine, which can swing open. A safety valve and pressure-relief valve (F) are also provided for safety of the equipment [20].

The operating cycle for this type of retort involves bringing the retort up to a temperature of around 100°C and then allowing steam to pass through the vessel to the atmosphere long enough so that all air in the retort and between the cans is removed (venting) before the retort is finally brought up to the operating pressure and processing temperature. At the end of the processing time, the steam is turned off and a mixture of cooling water and air is introduced into the retort to cool the cans. The purpose of the air is to maintain the pressure in the retort following the condensation of the residual steam after the initial introduction of cooling water. If this pressure is not maintained, the containers may deform due to pressure imbalances between the internal pressure in the cans and the retort. As the temperature drops, the pressure in the retort may be controlled and gradually reduced until atmospheric pressure is reached and water can be allowed to flow through the retort, cooling the cans to a temperature of about 40°C before they are removed from the retort. Cans are removed from the retort at this temperature since this allows the surface of the cans to dry rapidly by evaporation, reducing the risk of leaker spoilage [21]. The water is preferably sprayed, or alternatively the retort may simply be filled and allowed to stand long enough for the cans to cool to the unloading temperature.

Both of these systems are static in operation. For other types of product it is possible to assist the rate of heat penetration by agitating the cans in the steam environment by rotation either about the horizontal axis in a horizontal retort or in the vertical plane in a vertical retort.

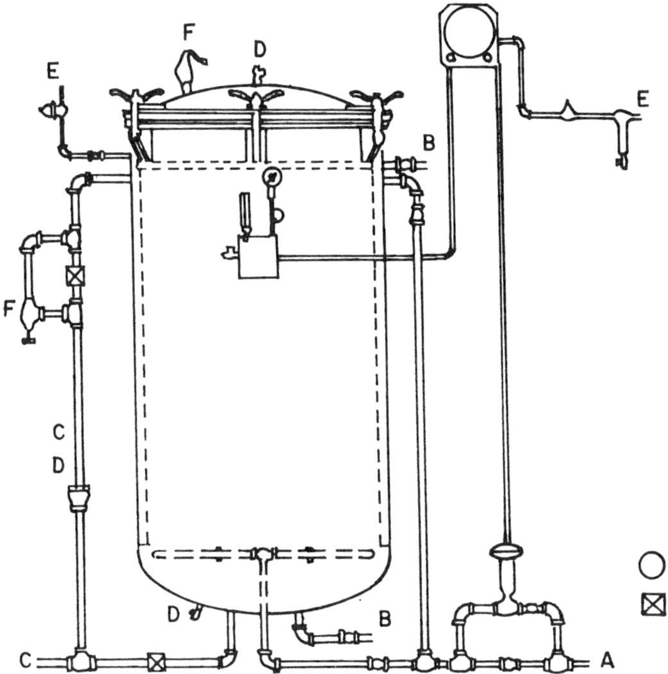

FIGURE 2 Vertical batch retort. (From Ref. 7.)

Steam/air retort systems: The use of glass and plastic containers has increased the use of alternative retorting systems. With these types of containers it is usually not sufficient to rely on the strength of the containers alone to counteract the build-up of internal pressure during heating, but a constant overpressure of air is required to ensure the integrity of the package during heating. Thus, the heating medium used in this type of retort is often a mixture of steam and air in proportions designed to provide the necessary steam temperature and air overpressure to maintain package integrity. In order to ensure adequate mixing of the steam and air, these retorts are fitted with a fan sys-

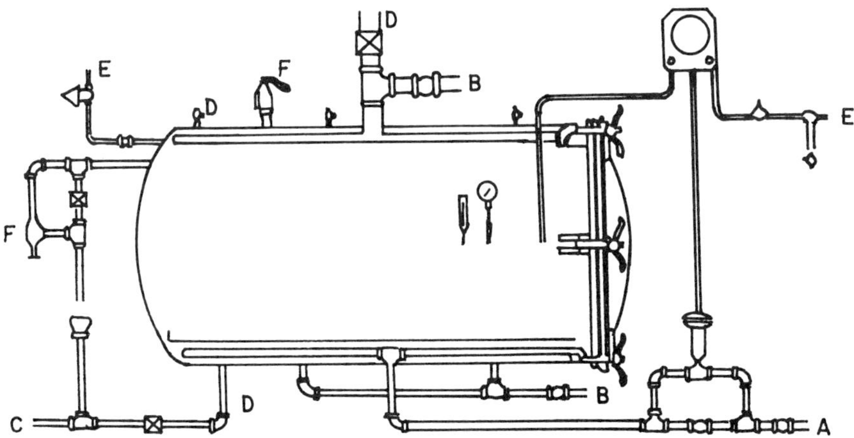

FIGURE 3 Horizontal batch retort. (From Ref. 7.)

tem to disperse the steam and air, eliminating the possibility of development of cold spots in processing chamber [20].

Controlling this type of retort system can be difficult, particularly ensuring an adequately uniform temperature distribution in the retort environment when the steam is mixed with cold compressed air. Here, unlike in the case of saturated steam retorts, the presence of air must not permit a reduction in the partial pressure of the steam and hence retort temperature, but only provide the overpressure needed to ensure package integrity. However, the steam and air must be mixed so that pockets of cold steam/air mix do not form in the retort and lead to inadequate processing of the cans.

Water-Processing Retorts

This system is mainly used for the processing of glass jars. Raining water techniques (Fig. 4) require the use of either an external steam injection system or heat exchanger system outside the direct environment of the retort. In the latter case the cold water feeding the system is combined with the recycled heating medium and raised to the temperature required in the retort before being admitted to the sterilization chamber through a spray arrangement. The containers will be arranged to allow good contact between the hot water heating medium and the product either using spacer bars or distribution plates. It is imperative that a good distribution of the water occurs, as otherwise stratification may occur and certain containers will receive inadequate heat process. Control of the temperature in this system is difficult, but the safest practice is to base the thermal process received by the product on the outlet temperature of the retort, i.e., the temperature measured in the return line to the heat exchanger [20].

The velocity of water in these retorts when passing over the package is of vital importance as this will influence the rate of heat transfer to the product due to its effect on the heat transfer coefficient. This is different from saturated steam retort processes, where the heat transfer coefficient can be considered infinite [22].

Horizontal Circulating Water Retorts

Horizontal water retorts can be with or without reels for end-over-end basket rotation. The process vessel, or autoclave, is conventionally connected to a separate pressurized water supply tank from which heated water, preheated to a temperature of 5–10°C above the intended process temperature, can be rapidly pumped to the process vessel to reduce come-up time (Fig. 5) [23,24].

The initial step of the process involves preheating the process water in the storage drum to the programmed temperature. After the four cars filled with cans are loaded into the working retort and the door is closed, the process water is transferred to the working retort. The programmed temperature, pressure, and rotation are maintained. After sterilization, the process water is pumped back to the storage drum for reheating before the next batch. Cold water in the working retort is used to cool the product.

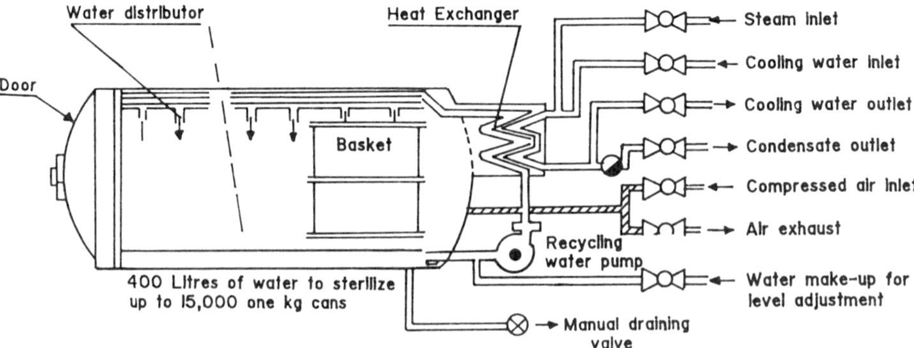

FIGURE 4 Water-processing retorts. (From Ref. 20.)

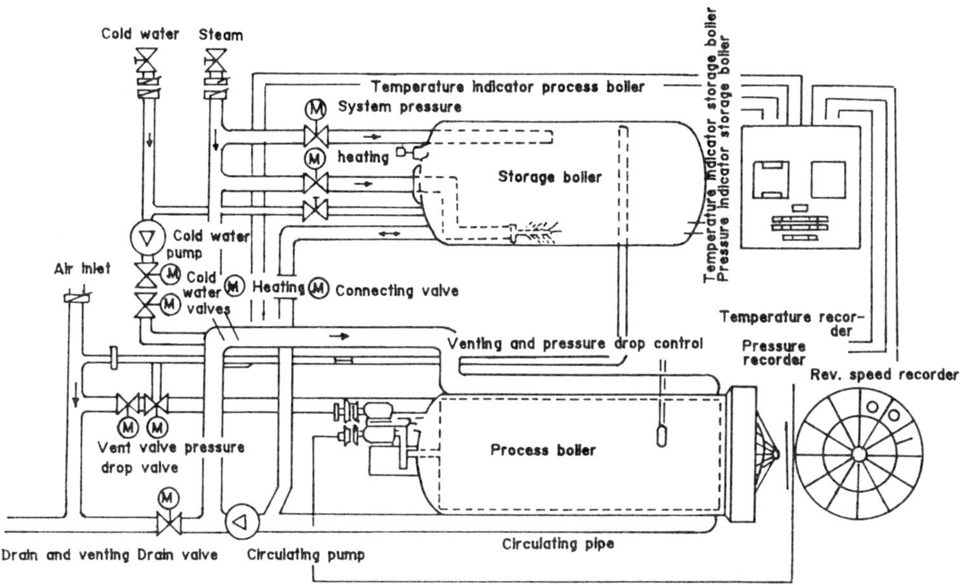

FIGURE 5 Horizontal circulating water retort. (From Ref. 23.)

The temperature/pressure differential between the heated water reservoir and the processing vessel (together with a compressed air override, if needed) provides the necessary overpressure required for sterilizing pouches, semi-rigid trays, buckle-prone cans having large areas of flat surface (half steam table trays, meat cans), and glass jars. Overpressure during heating can contribute to heat penetration efficiency in cans with large surfaces relative to product depth (and with conventional headspace volumes) by compressing the cover against the product surface. For certain fluid products, agitation may enhance heat penetration and thereby permit a reduced process.

Agitation may reduce the heating time but very definitely not the lethality required. It is conventional, in fact, to assign a higher F value as the basis of processing conditions in agitating cooks on the legitimate grounds that the peripheral product does not receive the high lethality, relative to the cold spot, during agitation-induced convection that characterizes straight conduction heating retort processes. Even when agitation is not a part of the process in a reel-equipped circulated water retort, it is recommended that the baskets be turned at minimum rpm (usually 6–8) to assist in uniform temperature distribution around the containers.

Alternatively, heat distribution is also accomplished by circulating water using a centrifugal pump, which continuously draws water through multiple holes spaced along the bottom of the process vessel and reintroduces it, after passage through a steam injector, at a high flow rate through a similar bank of horizontal holes in the top. Since heat distribution is dependent on a combination of forced water circulation throughout the baskets and mechanical movement of the containers through the water, dividers between the layers of containers must provide a minimum resistance to flow.

Crateless Retort Systems

The vertical retort has grown larger, lost its crates, and become automated. These new retorts have been recognized as a universal symbol of low-acid food processing. They are usually 2.5 m high and 2 m in diameter with four to five times greater capacity than the conventional three-basket vertical retort. The crateless retort is filled with water, which acts as a cushion for cans which are filled from an automatic conveyor. Steam is admitted through the top opening and this forces the water out of the retort through the bottom opening. The hot water can be recycled in another retort or in the next cycle. After processing, cooling water is let in through the bottom and is discharged through overflow. After the cooling cycle, the retort is drained off and the bottom door is partially opened. The cans then fall onto a shaker screen and are conveyed by belt to the unscrambler [23,25].

The advantages of this system include (a) labor savings, (b) steam savings, (c) no crates necessary, and (d) flexibility of process and can size. The only disadvantage is the denting of cans, which can be suitably reduced with proper attention and use of cushioning systems.

Continuous Rotary Pressure Sterilizer

Continuous pressure cookers can mechanize the basket entry and exit. Continuous agitation adds another dimension: forced convection within the can in order to improve heat transfer. By increasing agitation, cooking time may be reduced. The cooker consists of at least one cooker shell, a cooler shell, and a positive feeding device. Filled cans from the closing machine enter the line by means of a positive feed device, which times the cans in synchronization with the rest of the line. From the feeding device, the can is transferred to a rotary valve on the cooker, which is designed to prevent the escape of steam from the cooker shell. The cans are then ejected from the reel to a rotary transfer valve and into the next shell for additional cooking and cooling depending upon particular requirements [23–27].

The cooler shell is approximately two-thirds full of water to provide flood cooling of the cans as they progress through the shell. Water enters at the discharge end and exits at the feed end of the shell for a counterflow cooling effect. The reel in the cooler has a series of baffles that seals off the central portion of the reel to confine the movement of cooling water to that area occupied by the cans. The combination of the reel baffles and counterflow movement of water ensures efficient usage of cooling water and the controlled uniform cooling of cans.

The can agitation, which permits short-time, high-temperature cooking, and rapid, efficient cooling, is provided by the spiral-reel mechanism. Each revolution of the reel produces a three-phase cycle.

Fixed Reel Travel Phase: This phase takes place in the upper place of the rotation cycle for a distance of approximately 200 degrees around the periphery of the shell in which the cans are carried about on a central axis.

Free Rotation Phase: This phase occurs in the lower portion of the shell, where the cans roll on the spiral tees for a distance of approximately 100 degrees. In this area, agitation of the product in the can takes place due to a free rolling action of the cans about their own axis.

Transitional Phase: This phase takes place on both sides of the shell's periphery, wherein free rolling agitation commences, after fixed reel travel, and where it stops after passing through the free rotation phase.

Cooling techniques employed for pressure-cooked products are also important. In the cooking operation, the pressure developed inside the can is counterbalanced by the steam pressure in the cooking vessel. When the cans are transferred from the cooking to the cooling operation, adequate external pressure must be maintained in order to prevent the can from bursting and/or buckling under certain conditions. Three methods are used for cooling: pressure cooling, open (atmospheric) cooling, and split cooling

The advantages of the system are (a) labor savings, (b) improved product quality, (c) savings in floor space, (d) flexibility, (e) reduced can damage, (f) reduced processing times, (g) higher sterilization temperatures and capacity, and (h) less steam and water usage.

Continuous atmospheric cookers are the standard for high-acid foods, like tomatoes and fruits. The operation is followed by continuous atmospheric cooling.

Hydrolock Continuous Cooker/Cooler

The Hydrolock is a continuous, agitating cooker/cooler for high-speed short-time sterilization of a wide variety of container sizes and shapes. The system is applicable to the processing of cans, glass

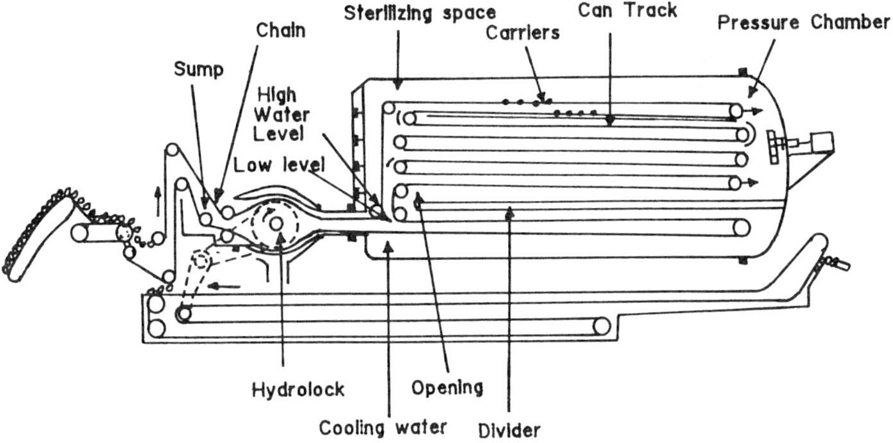

FIGURE 6 Hydrolock continuous cooker. (From Ref. 29.)

jars, semi-rigid plastic and metal containers, and retortable pouches. It is also capable of processing plastic and metal containers with heat-sealed closures [23,24,26–31].

The basic parts of the Hydrolock system (Fig. 6) are the water lock, cooker/cooler, chain carrier system, cooling system, and water-circulating system. Containers enter and travel through the process between two parallel conveyor chains. These chains enter and leave through water into a rotating pressure lock, sealed partly by water and partly by mechanical means. This facilitates preheating of incoming and precooling of outgoing containers. After loading through the lock, the containers are continuously conveyed through the steam chamber and finally into precooling water through which the conveyor passes. Containers exit through the same rotating pressure lock through which they entered and pass along a cooler conveyor.

The Hydrolock is equipped to provide overhead pressure during the cooling cycle to retain the container integrity. Final product cooling is completed in two passes of atmospheric cooling below the pressure vessel. Cans roll in shallow water in a stainless steel "pan" being pushed by stainless steel rods attached at their ends to roller chains. Any heating medium can be used with the system-saturated steam, water, or a steam-air mixture. When an overriding air pressure is required, as with glass containers, aluminum cans, plastic containers, or flexible pouches, air is mixed with the steam by means of one or more turbo fans, which produce a homogeneous mixture of the two gases.

Hydrostatic Pressure Sterilizer

This sterilization method is more commonly known as "hydrostatic sterilization" because the steam pressure in these units is maintained by water pressure. Hydrostatic cookers are continuous pressure cookers in which the operating pressure is maintained by water pressure. Hydrostatic cookers (Fig. 7) have two components: water chambers and steam chambers. The temperature of the steam in the steam chamber is controlled by pressure produced by the water legs and can be regulated by moving the level of water in the leg [23,24,28,32].

Containers are conveyed into the cooker through a water leg at 80°C. This is the down-traveling water leg, and the container temperature begins to increase. As the containers move down this leg, they encounters progressively hotter water. In the lower part of this leg the water temperature reaches about 100°C, and then, near the water seal area next to the steam chamber, the water temperature is about 107°C. In the steam chamber the can is exposed to a temperature of 115–130°C, the steam temperature being set to suit the product undergoing sterilization. Upon leaving the steam chamber, the can again goes through a water seal into water at a temperature of about 107°C, where the cooling cycle commences under pressure.

The advantages of this system include (a) savings in floor space, (b) large reduction in steam (50%) and water (70%) costs because of regenerative heating and cooling, compared to batch steam

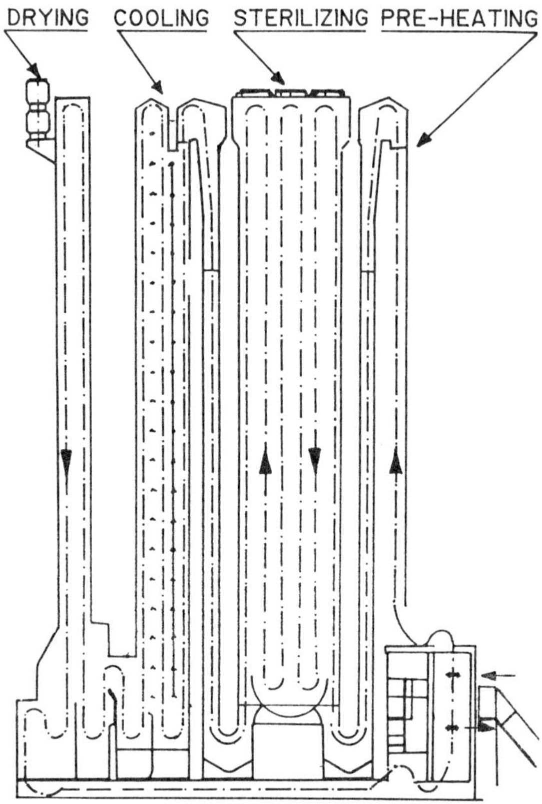

FIGURE 7 Hydrostatic pressure sterilizer (internal details). (From Ref. 23.)

retorts, (c) high capacity operation, (d) capability of processing all sizes of cans, glass containers, and retort pouches, (e) constant temperature operation, (f) less cooling water needed, (g) containers subjected to minimum thermal shock, (h) containers are handled gently because of low chain speed, (i) low labor requirements, and (j) the brines in vegetables such as peas and beans are clear because of absence of agitation.

Hydrostatic cookers are, on the other hand, expensive, large, require heavy structural steel to house, and the installation costs are high. Other disadvantages of hydrostatic cookers are the impossibility of vigorously agitating the can and the very large capital investment required. Comparing hydrostatic sterilizers with continuous rotary cookers, the use of hydrostatic cookers is justified when the following conditions are encountered: (a) the unit is operated year round, preferably on a two-shift basis, (b) the product is sensitive to high processing temperatures, and (c) the filler speed is at least 400 cans per minute.

A direct comparison is very difficult because of the many factors involved. Continuous pressure cookers have a lower installed cost than hydrostatic cookers, and they require less maintenance because they are simpler machines with less moving parts.

Hydrostatic Helix

The hydrostatic cooker has been reduced to a far more compact size. It has no mechanical valves or locks and thus can be a truly continuous motion retort. The helical pump or hydrostatic helix consists of a rotating coiled tube in which each turn of the coil is charged at the intake partly with liquid and partly with air. The coil rotates about a horizontal axis. With no pressure at the discharge, the rotating coil may meter liquid at a rate proportional to its rotational speed. With a discharge backpressure, the liquid in each coil turns forms a series of additive hydrostatic legs. The hydrostatic head developed is a function of the number of turns of the helix and the diameter. When the coil is rotated, liquid can enter the coil by gravity flow for one-half turn only, when the first turn (acting as a manometer) is in the upright position. As the coil continues to turn through the next half turn, only air can enter because the manometer is inverted. Thus, equal volumes of liquid and gas are alternately introduced into the helix in a repetitive cycle. The helical pump thus operates with many short columns of gas in contrast to commercial hydrostatic retorts that employ a few, tall, unbroken legs. There have been no commercial applications of this scheme in the food industry [33].

Continuous Pallet Sterilizer

Hydrostatic sterilizers, because of their size and the necessary complexity of their water recirculation systems, have become so expensive to construct and erect that the market is now limited to a few of the world's largest food and pharmaceutical companies [23]. The continuous pallet sterilizer is basically a continuous vertical retort through which cans are transported on pallets. The feed and discharge of the pallets is effected, without pressure loss, through venting locks. Each filled, unprocessed pallet load is conveyed by a rack and pinion arrangement into the lock. After the outside pressure door of the infeed lock is closed, steam is introduced, first at atmospheric pressure to purge air from the pallet and chamber, and thereafter under pressure to equilibrate the lock with the retort.

After the venting-equilibration cycle, the pallet is moved forward until it is at the base of the retort. The pallets slowly ride upward on their four railroad-like wheels. The processed pallets leave the top of the retort through a "let-down" lock. The flexibility of a retort, in terms of container type, is shared by the continuous pallet retort by virtue of the capacity of a round vessel to withstand much higher pressures than a rectangular hydrostatic sterilizer tower. Hot water sprays, overpressured with air, and steam with superimposed air pressure are, in addition to pure steam, possible sterilizing media. The potential thus exists for continuous processing of pouches, semirigid aluminum containers, institutional half steam table trays, and glass jars.

Flash 18 Process

This process, designed for solid foods, is so named because it operates at 18 psi (124 kPa) overpressure and includes rapid heating in the operation. The outstanding characteristics of this continuous canning process are the filling of cans in a pressurized room under 18 pounds of air pressure at a temperature of 124°C, the closing of the cans under the same conditions, and the elimination of retorting. Under normal atmospheric pressure, it is not possible to fill foods at a temperature above 100°C. By raising the air pressure in the filling and can-sealing chamber, the boiling temperature is raised and it is possible to fill cans at the sterilization temperature [24,28].

Food products may be prepared in a conventional mixer or cooker outside the pressure chamber. After preparation, the food is pumped through a high-pressure pump into a steam injector, which raises the product temperature to 125–130°C and holds it for 30–90 seconds. The food emerges from the steam injector in a deaerator in the pressurized chamber where the added steam and entrained air are flashed off. The food at 128°C is now filled into nonsterile cans under flowing steam. At this high temperature the product and the containers are sterilized. Cans are cooled under pressure and conveyed out of the pressure chamber where final cooling takes place.

The main advantages of this process are the continuous heating of the food without disintegration of the solid components; the brighter color and improved flavor; enhanced consistency and texture of foods; the elimination of "cooked" flavor from canned meats and vegetables; one-shot filling into cans rather than metering-in first the solid components and then the brine or sauce; and the lack of requirements for presterilization of the cans.

As for disadvantages, investment in equipment for this process is very high, and there is the perpetual problem of finding workers willing and able to work under the unusual pressurized conditions. The operators must enter and exit through locks for compression and decompression. Employees work a 4-hour shift including the time in the locks. In addition, there is a license fee for the right to use the process.

Flame Cookers

Infrared radiation as an indirect heat source has been developed into a flame cooker. Flame cookers attempt to increase the temperature differential between the heating source and the food product in order to increase the rate of heat penetration. By increasing the rate of agitation, probability of burn-on is markedly reduced [24,25,29,33].

Gas burner operation at 1100°C provides the heat source to impart the a high-temperature, short-time effect. The cans are placed in very close proximity (just a few millimeters away) to the burners and are kept in constant rotation, with a temperature differential within can and contents not exceeding 1°C. Thus, even fully lithographed cans may be heated without damage. There is, however, no way to impart counterpressure, so cans must be rigid enough to withstand internal steam pressure. With low-viscosity products, extremely high rates of temperature increase (e.g., 0.5°C/2 sec) of contents is possible. The unit depends on continuous axial rotation (about 120 rpm) to move the cans along the burners and to obtain internal turbulence. Some units have a steam preheat section.

A flame sterilizer is shown in Figure 8 [34]. The Steriflamme units consists basically of three sections. The first is a steam preheater, where the cans are heated to a temperature of approximately 100°C. In the second section the cans roll through a series of open flames at 1100°C produced by especially constructed gas burners. The rolling motion of the cans increases the rate of heat transfer into the whole mass of the food. Next the cans pass through an intermittently heated burner holding section for about 4.5 minutes. Spray cooling follows the heating cycle. The total elapsed time in the cooker is generally less than that required for batch retorting.

The advantages of this system are (a) continuous motion, (b) high capacity, (c) high fuel efficiency, and (d) low cost, since retorts and retort baskets need not be heated.

Fluidized Bed Sterilization

The fluidized bed retort is a cooker in which sand or ceramic pellets are used as the heat transfer medium. The medium is kept hot and fluid by a flame underneath and an air stream. The particles behave much as boiling liquid. Cans move through the bed meeting the same resistance they would if the medium were a thick liquid and receiving some abrasive effect from the particles [29].

Advantages of this system include (a) close temperature control, (b) high and controllable temperature differential, (c) no pressure chamber required, (d) continuous process, (e) several sizes may be sterilized simultaneously, and (f) large equipment. Disadvantages include (a) the possibility of burning and discoloration of the can surface, (b) damage to can seals due to abrasion, and (c) small can sizes.

Hot Sterilization

Hot air is employed as the heating medium in a Swedish-designed sterilizer described by Ecklund and his associates. Hot air with very high velocity (approximately 610 m/min) is employed to decrease

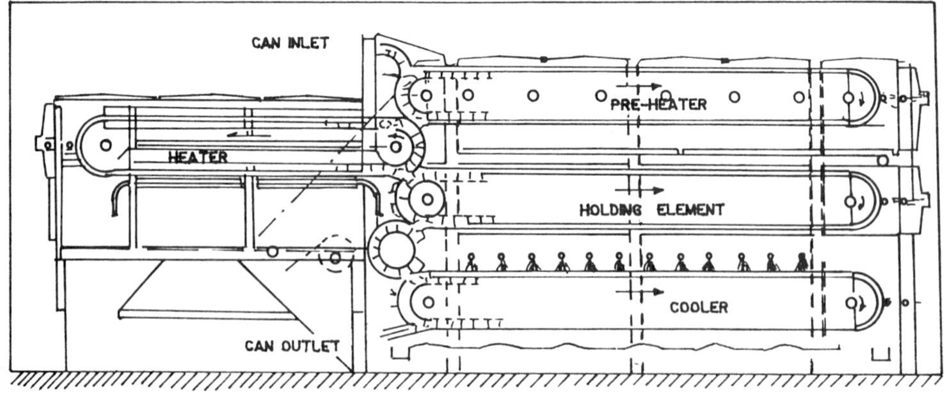

FIGURE 8 Flame sterilizer (longitudinal section). (From Ref. 34.)

the thickness of the nonturbulent air layer adjacent to the can surface. High-velocity air in excess of 150°C also creates a large temperature differential between the surface and the contents. Cans are axially rolled through to create forced convection within the can contents, thus reducing the possibility of burning or overcooking [25].

C. Pretreatment

1. Receiving Raw Products and Packaging Materials

Incoming raw materials, ingredients, and packaging components should be inspected upon receipt to ensure that they are suitable for processing in terms of variety, quality, and combination. Raw materials should be received in an area separate from the processing areas. Prior to being placed in inventory, ingredients susceptible to microbiological contamination that would render them unsuitable for processing should be examined for microbiological conditions for use in processing. Products should be held prior to processing in such a manner as to minimize the growth of microorganisms. All of the above are recommended procedures to be followed in receiving raw materials and packages for both low-acid and high-acid foods. For acidified and low-acid foods, they are mandatory as part of the FDA Good Manufacturing Practices regulations.

2. Separation of the Edible Portion

Many products as received at the factory require some special treatment to separate the edible portion. Peas must be removed from the shells (vined), fruit must be peeled and cored, corn must be husked, fish must be cleaned, etc. For each such operation special machinery has been developed so that a large amount can be handled quickly.

3. Washing

Immediately upon receipt at the cannery, raw food materials are thoroughly washed. The objectives of washing are to separate soil and foreign materials and to reduce the load of spoilage bacteria naturally present in foods. Reduction of the number of bacteria present increases the effectiveness of the sterilization process. Washing also improves the quality and appearance of foods.

Washing is done by equipment in which the products are subject to high-pressure water sprays or strong-flowing streams of water while passing along a moving belt or while being tumbled on agitating or revolving screens. Sometimes a flotation type of washer is also used to remove chaff or other extraneous material. Washing is sometimes preceded by mechanical removal of soil and other fine materials that adhere to the food. Stones and other heavy objects are also separated by density difference.

In some canning procedures, operations whose functions are not primarily to clean the raw material may exert a cleaning effect. Thus, blanching or scalding serves the additional purpose of cleaning the food. The same applies to the water spray used to cool foods after blanching.

4. Size Grading

Many products are graded using a series of screens having holes of different diameters. For products like green beans a series of parallel bars with varying distances between them are used. Some machines revolve and others vibrate. Each product undergoes certain size-grading procedures that have been proven to give most satisfactory results. Larger round units may be put in a long narrow trough in the bottom of which are moving rollers, ropes, or cables close together at one end and further apart at the other. The small units pass through first and the larger ones toward the open end.

5. Inspecting

All products must be inspected before they go into the cans. Modern machines are very efficient, but personal inspection is still needed for proper assurance of freedom from imperfections. This is usually done by passing the washed produce over an inspection belt. This should be designed so that the inspectors can sit comfortably close to and facing the belt. Both hands should be free to use easily.

Special attention should be given to proper lighting. The light source should be ample but diffused rather than glaring and should be so placed and shaded so that it does not shine in the workers' eyes.

6. Blanching

Blanching is an operation in which a raw food material is immersed in water at 88–99°C or exposed to live steam. Water temperature must be well controlled at a desirable level. The blanching operation varies according to the maturity and type of vegetable used. Blanching of a given vegetable or fruit is performed for the following reasons:

1. To inhibit enzymatic action. Natural product enzymes are inactivated and thus oxidative and other types of chemical reactions are also inhibited, which contributes to higher product quality and nutritive value by avoiding undesirable changes in natural color and flavor, as well as reduction in the content of certain vitamins.
2. To expel respiratory gases. Raw fruit and vegetables contain intracellular gases similar in composition to air but somewhat higher in oxygen and carbon dioxide. The release of gases prevents strain on the can seams during heat processing and favors development of a higher vacuum in the finished product. Another desirable effect is a reduction of internal can corrosion by reducing the oxygen content of headspace gases. Headspace oxygen acts as a depolarizer in electrochemical corrosion reactions, thus increasing the rate of corrosion.
3. To reduce initial microbial load in the products.
4. To soften food. A softer product is easier to fill in container, and higher drained weights are obtained.
5. To facilitate preliminary operations. Peeling, dicing, cutting, and other preparatory steps are accomplished more easily and efficiently.
6. To set the natural color of certain products under optimum conditions.
7. To remove raw flavors from food.
8. As an additional cleaning measure.

Blanching is usually accompanied in equipment especially designed for individual products. The equipment must be designed to subject the raw materials to a particular temperature range for an optimum period of time. The shortest blanching time that accomplishes the desired objectives usually yields the best product. Many vegetables and some fruits are blanched.

Another blanching is an important step in canning, the following points should also be considered:

1. It may change texture, color, and flavor of the food because of heating process.
2. It increases the loss of soluble solids, especially in case of water blanching
3. It may change the chemical and physical state of nutrients and vitamins.
4. It has adverse environmental impacts such as greater water and energy use, and the problem of effluent disposal.

7. Exhausting

The exhausting of containers for the removal of air should be controlled so as to meet the conditions for which the process was designed. Vacuum in canned foods may be obtained by preheating foods prior to closing. In producing vacuum by this means, the product may be heated prior to filling, or it may be heated both before and after filling. Heat in this case is employed to expand the product, to expand and drive out the occluded and dissolved gases in the product, and to reduce the air in the headspace before closure. The length of heating and the final temperature attained before closure have a very important relationship to the ultimate vacuum in the can.

Heating may be accomplished by passing the filled can through a steam or hot water exhaust box. It is common to refer to exhaust box treatment as "thermal exhaust" and to preheating before filling as "hot fill." Exhaust boxes are generally best adapted for canned foods that can be readily heated, such as brine- and syrup-packed fruits and vegetables. The major disadvantages of exhaust boxes are their bulkiness and their large steam requirements.

In mechanical vacuum closure by high-vacuum closing machines, the filled cans while cold or at a rather low temperature are passed into a clincher, which loosely clinches the covers without forming an air-tight seal. The cans are then transferred through a suitable valve into a vacuum chamber, subjected to vacuum for an instant while in the vacuum chamber, sealed, and then ejected through another valve. Vacuums drawn on the machine while the cans are in the vacuum chamber may vary over a wide range, depending on the desired final vacuum and on the temperature of the liquid contents. This method of exhausting air from canned foods subjects the contents to a vacuum for a rather short time before closure. Therefore, the air is withdrawn mainly from the headspace and only partially from the product itself, and proper adjustment of the headspace is necessary for proper performance.

D. Quality of Packaged Foods

1. Foods of Plant Origin

The purpose of heat sterilization is to extend the shelf life of foods while minimizing the changes in nutritive value and eating quality. Differences between the heating characteristics of microorganisms, enzymes, and sensory or nutritional components of foods are exploited to optimize processes for the retention of nutritional and sensory qualities. This is achieved in practice by a reduction in size or cross-sectional area of containers, by agitation during processing, or by aseptic processing.

The extent of thermal processing a food receives depends upon the composition and physical characteristics of the product and is the result of a combination of time and temperature. Physicochemical changes occurring during processing and storage are the factors that determine the product quality in terms of its sensory properties and nutrient content. Reactions take place during both the process itself and subsequent storage. Generally the changes that occur during storage are slow, particularly when compared with those occurring in an equivalent unprocessed material, and it is on this basis that heat preservation is effective in providing materials outside their normal seasons and in a convenient prepared, often formulated, form ready for consumption or reheating and then consumption.

The physical and chemical reactions that occur during processing, desirable or undesirable, are often more significant and certainly occur much more rapidly than those that occur during storage. As previously noted, the degree of heat processing varies according to the product. In turn the changes that occur during processing are influenced by the time and temperature of the process and the composition and properties of the food material [35] and its environment [36].

Sensory Quality

The heat process itself has a major effect upon the quality of a food product and is responsible for a range of changes. Starch gelatinization and structural protein denaturation have a direct influence on the texture of a food. Heat-induced reactions such as the Maillard reaction affect the color and flavor as well as the nutritional status of the food [37,38]. One of the most significant reactions, however, is oxidation, which can occur during the process and throughout subsequent storage. Flavor [39], color [40], and occasionally textural changes [41,42] have all been shown to be related to oxidation, although in the majority of cases the exact mechanisms remain unelucidated. Before any oxidative event can take place, contact with molecular oxygen must have occurred at some point in the history of the food, even as part of the biochemistry of the food components or ingredients as living organisms. In general, contact that occurs before the heat process is less important than that during or after processing, since it is the manipulative and thermal procedures of food production that have the greatest effect on tissue damage and the resultant mixing of cell contents from different materials.

Sensory evaluation can be used to measure either the effects of processing conditions on foods or their acceptability (like-dislike). The latter tests are carried out by untrained people, the former by trained assessors, usually working under controlled conditions of lighting, temperature, sample size, etc., and using well-established techniques [43].

The actual test method used depends on the purpose of the investigation, e.g., a difference test would be used if the purpose was to find out whether a process had changed a product, a description test to find out how the process had affected it. For all tests, those carrying out the evaluations should have been trained in the assessment method.

Appearance (e.g., color, shape, surface texture, and translucency) is assessed visually under standard lighting conditions and against a constant background that will not affect the color (e.g., gray). Flavor is made up of taste (sweet, salt, bitter, acid) and odor (volatile compounds), and is usually evaluated by mouth at the temperature at which the food (or drink) is normally consumed. A product can be tested for total flavor (as in a difference test), as a specific flavor (e.g., a taint), or broken down into the individual flavor attributes in quantitative descriptive analysis (stone and sides), the latter two procedures require specific training. Texture can be evaluated visually (e.g., viscosity), by feel (e.g., softness of fruit), but more generally by mouth when the food is manipulated by tongue, cheeks, and teeth. Similar techniques are used for flavor analysis, of which the most detailed is the texture profiling method [44].

Texture

Two types of tissue damage occur during the heating of plant material: destruction or damage to the semipermeable cell membranes, and disruption of the intercellular structures with resultant cell separation [45]. The effects of these types of tissue damage are a loss in cell turgor and cellular adhesion, which give rise to loss of crispness and softening of the heat-processed product.

Other major changes in the texture of heated foods arise from the denaturation of proteins. Even on relatively mild heating, conformational change affecting the tertiary structure of proteins can be observed [46]. Denaturation of the proteins may follow. The hydrogen bonds, maintaining the secondary and higher structure of the protein, rupture and predominantly random coil configuration occurs [47]. This can lead to considerable changes in the chemical and physical properties of proteins due to losses in solubility, elasticity, and flexibility [46,48]. This mechanism also causes enzyme inactivation and breakdown of proteinaceous toxins and antinutrients. The resultant turbidity leads to either a precipitate or gel, which will greatly alter their water-holding capacity and can also lead to increased thermal stability [49].

Starches are the basic reserve carbohydrate form in plants and are widely used in processed foods as thickeners. Starch gelatinization components at a range of temperatures, which corresponds to the solvation of the macromolecules and is dependent upon the type of starch present. The difference in behavior can be partly explained by the relative properties of two components, amylose and amylopectin. Amylose gives an opaque solution, which sets to a firm gel when cooled, whereas amylopectin gives a viscous translucent paste, which retains its fluidity on cooling. In order for gelation to occur, starches must be exposed to both heat and water, but even within a single product gelatinization is uneven.

In fruits and vegetables, softening is caused by hydrolysis of pectic materials, gelatinization of starches, and partial solubilization of hemicelluloses, combined with a loss of cell turgor. Calcium salts may be added to blanching water or to brine or syrup to form insoluble calcium pectate and thus to increase the firmness of the canned product. Different salts are needed for different types of fruit (e.g., calcium hydroxide for cherries, calcium chloride for tomatoes, and calcium lactate for apples) owing to differences in the proportion of demethylated pectin in each product.

Color

The color of a food product is determined by the state and stability of any natural or added pigments and the development of any coloration during processing and storage. Natural pigments are generally unstable compounds which are broken down on heating but whose stability is dependent upon many factors. In fruits and vegetables, chlorophyll is converted to pheophytin, carotenoids are isomerized from 5,6-epoxides to less intensely colored 5,8-epoxides, and anthocyanins are degraded to brown pigments. Anthocyanins are fairly heat-stable compounds but take part in a wide range of reactions, e.g., with ascorbic acid, sugar breakdown products such as hydroxymethyl furfural, and other reactive phenolics, which bring about their breakdown [50]. Factors that accelerate degradation include high levels of oxygen in the product and storage temperature.

Conversely, anthocyanins can be undesirable in a product and can be produced on thermal treatment of lencoanthocyanidin [51,52]. They give rise to defects such as very dark broad beans and red gooseberries. Other problems can occur with anthocyanin pigments due to the formation of metal complexes, for example, the bluing of red fruits and the pinking of pears when exposed to tin [53,54]. The flavonoid rutin, present in asparagus, can also form a complex with iron causing dark discoloration in lacquered cans where iron dissolution can occur [40] and in which the colorless tin complex is not formed.

Carotenoids are mostly fat soluble and are responsible for yellow, orange, and red coloration. They are unsaturated compounds and are therefore susceptible to oxidation giving rise to off-flavor and bleaching. In addition, two types of isomerization can occur, namely, *cis-trans* isomerization and epoxide isomerization, which can give rise to lightening of color. The storage temperature is considered to have a greater effect on isomerization than on the heat process itself.

The two major groups of porphyrin-based pigments are chlorophyll and the heme compounds, both of which are very sensitive to heat. On processing, chlorophyll is converted to pheophytin with an associated loss of green color [40]. Several approaches have been taken to try to reduce the color loss, such as adjusting the pH [55,56] and the use of HTST treatments. In the latter case, although improvements were observed after processing, these were lost during storage [57]. Betalins are water-soluble pigments which are susceptible to oxidation and loss of red color. Browning of heat-preserved beetroot products is an example where residual oxygen in the product or headspace causes the appearance of a chocolate brown color.

In addition to the breakdown of pigments, oxidation and the Maillard reaction can produce colors during processing and storage. Heat processing itself in the presence of oxygen has a major effect on end product quality, demonstrated by a comparison of products packed in plain tin-plate cans with the identical material processed in lacquered cans or glass jars. In the plain tin-plate container, dissolution of the tin during processing removes a major proportion of oxygen from the pack and little is available to react with the food. Some products such as pale fruits, tomatoes and tomato formulations, mushrooms, and milk products are particularly susceptible to such heat-induced oxidative changes. It has been demonstrated that a brownish color develops in beans dipped in tomato sauce packed in different container types [58]. Rose and Blundstone [58] showed that the major source of color change in beans in tomato sauce was due to the formation of melanoidins from Maillard reaction products.

Ascorbic acid is often utilized as an antioxidant and can be effective in improving color in certain products, e.g., mushrooms. It can, however, be degraded to produce reactive compounds, which can then further react to form brown pigments.

Flavor

Generally, heat preservation does not significantly alter the basic flavors of sweetness, bitterness, acid, or salt. In fruits and vegetables, changes are due to complex reactions that involve the degradation, recombination, and volatilization of aldehydes, ketones, sugars, lactones, amino acids, and organic acids. Major changes can, however, occur in the volatile flavor components. One of the most important sources of volatiles is lipid oxidation or oxidative rancidity. Lipid oxidation can be brought about during both processing and storage where oxygen is available and is a particular problem in fatty foods and some vegetables, notably cereals, legumes, and pulses. The proposed chemical mechanisms of oxidation are well documented [39,59] and are outside the scope of this book. Three stages are involved: initiation, propagation, in which highly reactive hydroperoxides are formed, and termination. The initiation stage takes place in the presence of catalysts, such as metal ions or metalloproteins, but can also be brought about by heat or light. The reaction does, however, have a low activation energy (4–5 kcal/mole). The hydroperoxides formed take part in secondary reactions to give rise to a range of volatiles, including aldehydes, ketones, and alcohols and it is these that produce typical rancid or stale off-flavors.

Volatile flavor compounds are also produced via the Maillard reaction. Since the first scheme for this reaction was proposed by Hodge [60], a great deal of research has been undertaken. The reaction occurs during heating and extended storage, is influenced by water activity, with an optimum flavor generation at intermediate values of around 30% water [61], and is accelerated by high pH and buffers such as phosphates and citrates [62]. The first stage of the reaction is fairly well defined and involves the condensation between carbonyl groups of the reducing carbohydrates and the free amino acids or protein and rearrangement to produce Amadoryi compounds. This leads to a loss of protein nutritional quality, as described later, but does not affect the sensory properties significantly [63]. The second stage is very complex and gives rise to numerous products, many volatiles, and is responsible for many characteristic flavors and off-flavors in food materials.

Loss of volatile constituents can also present problems in heat-preserved foods. A breakdown of essential oils in citrus products can result from oxidation. There is also a need for research into the loss or absorption of volatiles through modern packaging.

Nutrients

Both physical and chemical reactions occur in heat-preserved foods, which influence nutritive value (Table 4). Physical factors such as the loss of soluble nutrients, or leaching, can be significant for products in which a carrying liquid is discarded before consumption. Chemical reactions include heat damage to labile nutrients such as vitamins. One of the most fundamental changes that can occur in a heat-preserved product is the movement of water and solids within the food material during processing, storage, and reheating. In a formulated product or a product in which the entire pack contents are consumed, such changes can be largely disregarded, from the nutritional point of view, in that they do not alter the total amount of the nutrients consumed. Products packed in a liquor that is discarded before consumption, however, often exhibit dilution, dehydration, or loss of total solid materials from the edible portion. In such cases the interpretation of any apparent changes in composition needs to be made with these considerations in mind. Sterilized soy products may show an

Table 4 Effect of Heat Processing on Major Nutritional Components

Nutrient	Effect
Dry matter	Loss of total solids into canning liquor
	Dilution
	Dehydration
Protein	Enzymic inactivation
	Loss of certain essential amino acids
	Loss of digestibility
	Improved digestibility
Carbohydrate	Starch gelatinization and increased digestibility
	No apparent change in content of carbohydrate
Dietary fiber	Generally no loss of physiological value
Lipids	Conversion of *cis* fatty acids to *trans* by oxidation
	Loss of essential fatty acid activity
Water-soluble vitamins	Large losses of vitamins C and B_1 due to leaching and heat degradation
	Increased bioavailability of biotin and niacin due to enzyme inactivation
Fat-soluble vitamins	Mainly heat stable
	Losses due to oxidation of lipids
Minerals	Losses due to leaching
	Possible increase in sodium and calcium levels by uptake from canning liquor

increase in nutritional value owing to an unidentified factor that decreases the stability of the trypsin inhibitor in soybeans.

Proteins: Heat preservation can lead to both desirable and undesirable changes in the nutritive quality of proteins. They are susceptible not only to heat but also to oxidation, alkaline environment, and reaction with other food constituents such as reducing sugars and lipid oxidative products. The total amount of crude protein generally appears relatively unchanged due to heat processing [64,65] but can suffer from leaching into the liquid component of some products [66]. The crude protein levels, however, appear to be stable during subsequent storage of canned vegetables [64,65]. The changes that do occur are associated with tertiary structure and functionality, as discussed earlier, and chemical changes related to digestibility and amino acid availability.

Canning of potatoes also leads to losses of amino acids, although this has been shown to vary depending on the specific gravity of the potato [67]. Lysine is again particularly vulnerable with a reduction in availability of about 40%. Some of the losses found in canned potatoes, however, may be due to the leaching of the protein into the brine [66], although the major cause of loss of amino acids during heat preservation is the Maillard reaction. Soybeans and many other legumes also undergo improved protein digestibility and bioavailability, especially of the sulfur-containing amino acids, on heating due to inactivation of trypsin inhibitors and unfolding of the major seed globulins.

Vitamins: The effect of heat preservation on vitamins is generally detrimental, although mild heating conditions can have beneficial effects on the bioavailability of certain vitamins, particularly biotin and niacin. This is due to the inactivation of enzymes and binding agents [68]. The stability of vitamins varies under different conditions, with vitamin C and thiamine being more susceptible to degradation through heating. Fat-soluble vitamins are more stable, although they can be degraded by oxidation, especially when heated. Losses of water-soluble vitamins during processing can be considerably higher.

Vitamin C is the most labile of the vitamins and can be lost during storage of the fresh material, food preparation, washing, and blanching as well as by degradation on heating and leaching into a carrying liquor during the process. Studies on garden peas and carrots have shown that as much vitamin C can be lost during storage of the fresh produce for 7 days prior to cooking as is lost by canning. Much of the vitamin C lost during canning is leached into the canning liquor. Thiamine is the most heat sensitive of the B vitamins, especially under alklaine conditions, and it is also susceptible to leaching during any washing or blanching stages. Thiamine less labile than vitamin C and retention of 60–90% is usual in canning [69].

Folic acid and pyridoxine are also susceptible to degradation by heating and, in the case of folic acid, also by oxidation. Canning of potatoes can lead to vitamin losses of up to 30% [70]. Riboflavin and niacin are both relatively stable on heat preservation, although riboflavin is very sensitive to light and will undergo degradation in the presence of both heat and light together [71]. Heat-preserved foods often require less cooking than do fresh foods, and the differences in vitamin content between fresh and processed foods at the point of consumption can often be negligible. In canned fruits and vegetables, significant losses of all water-soluble vitamins may occur, particularly ascorbic acid.

Minerals: Minerals are generally stable to most of the conditions encountered during heat preservation, i.e., heat, air, oxygen, acid, or alkaline. Losses of minerals, however, can occur during processing, especially of vegetables, due to leaching into canning liquor. Conversely, certain minerals (e.g., sodium and calcium) can be taken up by the food from the cooking or canning liquids. Comparisons between fresh and canned vegetables have shown higher ash content in canned products in all cases. This is due to the uptake of sodium and to lesser extent of calcium from the brine. Between 15 and 50% of potassium can be lost primarily by leaching on the canning of vegetables. Slight leaching of zinc and negligible changes in iron content also occur during processing. Heating has been seen to increase the bioavailability of iron in spinach; the presence of fructose also leads to increased iron bioavailability [72]

Carbohydrates: Carbohydrates are less susceptible than most other food compounds to chemical changes during heat preservation. The levels of total and available carbohydrates in vegetables have been found to be very stable on canning and subsequent storage of the canned vegetables. However, there are some effects of heat on various carbohydrates. The effect of sugar on protein and iron bioavailability and the relationship between starch, texture, and palatability have been discussed earlier. Gelatinization of the starch also increases the digestibility of foods. A good example of this [70,73] is the potato, which in its raw state is largely indigestible. The exact effect of heat preservation on various types and constituents of dietary fiber has not been fully investigated. Cellulose, the main constituent of dietary fiber, hemicelluloses, and pectins are together responsible for the structure and texture of plant foods [74,75] and can be disrupted by heating, which leads to a softening of the food and increased palatability, generally without any loss in the physiological value of the dietary fiber. Overheating can lead to breakdown of cells enabling water-soluble nutrients (e.g., certain minerals, vitamins, and pectins) to be leached out. Although dietary fiber is considered to be largely unaffected by heat processing, the exact relationship between time/temperature conditions, dietary fiber breakdown, and the extent of nutrient loss due to fiber breakdown requires further study.

Lipids: Lipids, especially unsaturated lipids, are prone to oxidation when heated in the presence of air or oxygen, resulting in losses in nutritional value of the food product. Although the major effect of lipid oxidation is on the flavors of foods, oxidation can lead to a conversion of the natural *cis* fatty acids to *trans* fatty acids [72]. The digestion and absorption of *trans* fatty acids is comparable to that of the *cis* fatty acids, and their nutritional value as an energy source is not affected. However, *trans* fatty acids do not generally possess essential fatty acid activity, i.e., as precursors of prostaglandins, thromboxanes. This activity is dependent on a *cis*-9, *cis*-12 methylene interrupted double-bond system, but if sufficient linoleic acid is consumed the *trans* fatty acids do not appear to inhibit essential fatty acid metabolism [76,77]. Lipid oxidation has also been implicated in the loss of protein quality and can inhibit the activity of the fat-soluble vitamins A, D, and E as well as of vitamin C and folate. The oxidation of fats in processed foods, however, can be controlled by the exclusion or minimization of oxygen and the use of antioxidants. The effects of heat preservation on the nutritional value of fats can generally be considered as negligible.

2. Foods of Animal Origin

Color

The time-temperature combinations used in canning have a substantial effect on most naturally occurring pigments in meat foods. The red oxymyoglobin pigment is converted to brown metmyoglobin, and purplish myoglobin is converted to red-brown myohemichromogen. Maillard browning and caramelization also contribute to the color of sterilized meats. However, this is an acceptable change in cooked meats. Sodium nitrite and sodium nitrate are added to some meat products to reduce the risk of growth of *C. botulinum*. The resulting red-pink coloration is due to nitric oxide myoglobin and metmyoglobin nitrite. Loss of color is often corrected using permitted synthetic colors.

Flavor and Aroma

Complex changes also occur in canned meats (e.g., pyrolysis, deamination and decarboxylation of amino acids, degradation, Maillard reactions, caramelization of carbohydrates to furfural and hydroxymethylfurfural, and oxidation and decarboxylation of lipids). Interactions between these components produce more than 600 flavor compounds in 10 chemical classes [78,79]. Other volatile have been identified as having a significant effect on the flavor of foods, perhaps one of the most dramatic of which is the development of "catty taint." This is an extremely unpleasant and potent odor produced by the reaction of unsaturated ketones with natural sulfur-containing components of the food [78,79]. Heating is essential in the formation of the taint, and incidents have been widespread due to the diverse availability of the unsaturated ketones. Examples include processed meat products using meat from a cold store painted with a material containing mesityl oxide as a solvent contaminant [80], canned ox tongues that had been hung on hooks coated with a protective oil [81], and pork packed in cans with a side seam lacquer that had been dissolved in an impure solvent [81,82].

Texture

Changes in texture are caused in canned meats by coagulation and a loss of water-holding capacity of proteins, which produces shrinkage and stiffening of muscle tissues. Softening is caused by hydrolysis of collagen, solubilization of the resulting gelatin, and melting and dispersion of fats through the product. Polyphosphates are added to some products to bind water. This increases the tenderness of the product and reduces shrinkage. Small changes in the viscosity of milk are caused by modification of κ-casein, leading to an increased sensitivity to calcium precipitation and coagulation.

Nutrients

Canning causes the hydrolysis of carbohydrates and lipids, but these nutrients remain available and the nutritive value of the food is not affected. Proteins are coagulated and, in canned meats, losses of amino acids are 10–20%. Reductions in lysine content are proportional to the severity of heating but rarely exceed 25%. The loss of tryptophan and to a lesser extent, methionine reduces the biological value of the proteins by 6–9%. Vitamin losses are mostly confined to thiamine (50–75%) and pantothenic acid (20–35%). However, there are large variations owing to differences in the types of food, the presence of residual oxygen in the container, and methods of preparation (peeling and slicing) or blanching. In some foods, vitamins are transferred into the brine or syrup, which is also consumed. There is thus a smaller nutritional loss. Heat sterilization of meat leads to a reduction in digestibility of the meat proteins and damages amino acids, especially the essential sulfur-containing amino acids and lysine, with 10–15% losses in beef [83].

3. Conclusion

When considering the effect of heat preservation on the quality of foods, two important points should be considered:

1. Many of the changes that occur, of a sensory or a nutritional nature, do so during the thermal process and are not restricted to heat-preserved foods. In many instances the process replaces conventional cooking prior to consumption. Reheating the heat-preserved food is a relatively mild treatment, which does not significantly affect the quality.
2. Heat-preserved foods make available to the consumer a wider choice of sensory experience and nutritional requirements without constraint of seasonality and the burden of preparation.

E. Types of Food Packaging

1. Introduction

A hermetically sealed container means a container designed and intended to be secure against the entry of microorganisms and to maintain the commercial sterility of its contents after processing. The container is an essential factor in the preservation of foods by canning. After canned foods are sterilized, it is the container that protects the canned food from spoilage by recontamination with microorganisms. It is most important for the success of the canning operation to use good-quality, reliable containers and properly adjusted closing machines. Thus, the seams and closures produced will be within the guidelines necessary to prevent access of microorganisms into the container during the cooling operation and during the shelf life of the product.

2. Tin-Plate Cans

Today the types of cans available include the following [29]: (a) tin-plate body and ends, (b) tin-plate body and one end, aluminum convenience end, (c) three-piece aluminum can (rare, but available and used, with adhesive side seam, for alcoholic cocktails), (d) tin-free steel with tin end, tin-free steel end, aluminum end, or a combination, (e) tin-free steel body, (f) adhesive-joined side seam, (g) welded side seam, (h) draw and iron two-piece aluminum can, (i) conventional top chime, (j) neck-in top flange so that chime is flush with body, (k) draw and iron two-piece steel cans (not commercially available except in small sizes for aerosol cans).

Tin-Free Steel (TFS)

Tin-free steel is steel that has been rendered electrochemically passive by means other than tinning. Steel rusts in the presence of oxygen and dissolves in food to impart undesirable colors and flavors. By placing an inert layer on the surface of the same low-carbon steel used for electrolytic tin-plating, it is at least partially protected against oxidation and dissolution. The chrome layer acts a corrosion barrier. The oxide layer reduces chrome surface oxidation and covers discontinuities in the surface. The inorganic chrome/chrome oxide coatings, of course, must be protected with organic coating systems similar to those used for tin-plate steel but improved to protect the much thinner and therefore more sensitive base coatings [29,84].

Two-Piece Cans

All of the major and secondary firms manufacturing can-making and can-handling equipment are producing a two-piece draw and iron can. To date, the only significant commercial can is a very small 28–57 g (1–2 oz) unit. It is evident that a two-piece steel can would eliminate the long seam and one double seam and thus preclude two sources of potential leakage. The amount of metal used would be reduced to below that used for a three-piece can. Steel is a cheaper metal than aluminum, but it is far more difficult to form. A two-piece steel can could offer the advantages of a two-piece aluminum can, but at a lower price [29].

Draw and iron: In this technique, the can is made in two steps. A disc is blanked from sheet stock and drawn into a cup. The can is lengthened or deepened by ironing the sides in a stretching operation that reduces wall thickness to about one third of its original gauge. The drawn and ironed can is then washed clean of lubricant, lithographed, interior spray-coated, necked-in, and flanged. A two-piece cup-like can ready to receive contents and a double seam end is the resulting product [85–87].

Draw and redraw: This is essentially an impact extrusion process in which a punch applies pressure to an aluminum slug causing the metal to flow around the punch and form a deep, seamless cup. Metal required for the walls is taken from the bottom, so deep draws are not possible even when the second or redrawing step is performed. The result is not necessarily a smooth wall cup as obtained from the draw-and-iron process, but rather a cup-like container, which can have a slightly flared-out base containing the remainder of the slug aluminum. The open end may then be flanged with or without the neck feature [85].

Three-Piece Cans

Three piece "sanitary cans," consisting of a can body and two end pieces, are used to seal heat-sterilized foods hermetically as well as for other food products like powders, syrups, and cooking oils. Presently, the three-piece cans are being widely used, and other cans like two-piece cans, aluminum cans, and other flexible containers are slowly replacing them [29].

American Can MiraSeam: Although tin-free steel has been in existence for some time its fabrication poses a real problem. Tin-plate may be formed into can bodies using mechanical locks and solder at rates of up to 500 per minute. In this process, the thin layer of thermoplastic nylon is laid along one edge of the tin-free steel blank. The two edges of the blank are heated to about 260°C, bumped together, and cooled [84].

Conoweld: The other major commercial process for fabricating tin-free steel cans is forge-welding, also called the Conoweld process. In forge-welding, the body blank coating is stripped away and the blank is formed into a cylinder with a 1.6-mm overlap held together by four spot welds. The cylinder is then subjected to high-intensity welding heat along the entire seam by passing it between two rotating copper electrode wheels. A solid continuous strip of metal is formed along the edge [84,88].

Soldered side seam: The starting material for this type of container is usually in coil form, which is cut into sheets and fed to a slitter to produce individual can-body blanks. The blanks are then notched in a bodymaker (i.e., removal of metal at juncture or crossover of side seam and end), rolled and hooks are formed on both edges. Flux, as well as serrations and vents, promotes wetting and flow of solder into the seam [29]. After the body cylinder has been formed from the blank, it passes automatically over a burner followed by a solder roll, which introduces molten solder into the side seam folds. Excess solder is removed by wiping. Finally, a stripe of lacquer may be applied to the interior and exterior surfaces of the soldered side seam for product compatibility. Subsequent steps include flanging and beading the bodies, double-seaming one end into the bodies, and testing for integrity.

Cemented side seam: This type of container has a lap seam with a polyamide adhesive. It is the preferred technology for beer and soft drink containers, and it has been used successfully for processed luncheon meat cans [29]. Advantages of cemented side seams include high-speed production, use of economical tin-free steel (TFS), ability to run interchangeably with three-piece cans, compatibility with customer filling equipment, abuse resistance, and ability to accept wrap-around lithography.

Welded side seam: This type of container is fabricated by applying electric resistance heating to weld the container. This special technique maintains the heat effect of the weld to as narrow a strip of the can as possible. In this process the can body is then formed into a cylinder and the diameter is fixed by four tack welds spaced about 40 mm apart along the length of the can. The can body is then continuously seam-welded by passing a high current through the overlap can edges as it passes between two rotating copper electrode wheels [29].

Half-Size Steam Table Trays

The half-size steam table tray is a low-profile, two-piece, retortable institutional-size container that is formed from chromed steel plate, preenameled with modified vinyl inside and out. It has a capacity of approximately 3 kg [No. 10 can] (603 × 700). The lid is double seamed like a standard three-piece cylindrical can. Especially modified double-seaming machines are used [86]. Advantages of the steam table tray over No. 10 cans are said to include reduction in sterilization time by as much as 60% because heat reaches the center of the product in the container and achieves sterility much faster; packing of canned foods previously available only as frozen; tray serves the purposes of shipping container, heating container, and serving container; cleaning and washing of pan at the food service outlet eliminated; and lead side seams eliminated. Advantages over frozen foods include shelf-stability, no energy usage during transportation and storage, and no need to defrost. Disadvantages of the tray are slower line speeds and considerably more labor-intensive tray handling than round, cylindrical cans.

Aluminum Cans

Greater attention is being given to aluminum for fabricating cans and other containers for processed foods. Its use to date has been mainly in applications where there is some inherent advantage over the tin-plate such as lower shipping expense, freedom from food and can black sulfide discoloration or rust, easier puncture opening, and where special easy-opening features are desirable [29,86]. Steel cans are so well established in the canning industry that exceptionally good reasons are required before a change of material is contemplated. The present and future use of aluminum for cans, particularly in the large market for processed food use, is to a great extent dependent on the price at which it may be sold to the users, relative to that of an equivalent steel can.

On the other hand, while aluminum might appear to be more costly per unit area than steel, it would be wrong to draw the conclusion that aluminum cans are in no position to compete with tin-plate cans. The materials are sufficiently different in their properties for steel to do well under conditions that are not conducive to the use of aluminum, and for aluminum to replace tin-plate in some instances. When all applicable factors are considered, there are instances where aluminum cans offer advantages of product quality, economy, and national strategic value for the canning of certain food products. The use of easy-open lids is also a significant factor with strong appeal. Aluminum cans do not rust. Their appearance, always bright, can be an important sales feature. An important advantage of aluminum cans is that they are lead-free, but aluminum cans dent easily, abrade, and do not operate interchangeably with steel cans because of weight and because of the tendency for aluminum to bind on steel railings.

Features of aluminum cans include the following:

Interior coatings are required to obtain acceptable shelf life.
At pH 5.0 or above, simple enameling is satisfactory: within 2–3 years at room temperature, however, hydrogen springers usually occur.
Acid products like fruit require better coatings and even then allow shelf life of only 12–18 months.
No significant differences exist in color and flavor of foods in aluminum or tin-plated cans. No blackening occurs with sulfur-bearing foods; aluminum may tend to bleach highly pigmented foods [89].

Aluminum and tin-plate perform equally well with low-acid foods; tin-plate was generally superior for acid foods. Coating was, of course, required for the aluminum. The drawn aluminum can, with a diameter about twice its height, is gaining increasing commercial importance, being used for cheese snacks, dips, and now meat salads. These shallow-drawn cans are made from precoated-H34 or H35 temper aluminum. Draw-and-redraw cans are used commercially for meats, pet foods, processed meats, milk puddings, carageenan gels, fruits in syrups, etc. Impact-extruded aluminum cans have been used commercially for beer by Coors for its 200-g cans (distributed under refrigeration).

Aluminum will be employed increasingly for food canning because of its ease of forming into cans more readily used by consumers. Improvements in steel cans were stimulated by the introduction of aluminum. The compatibility of aluminum with almost all foods has been proven, provided, of course, appropriate linings are properly applied. The two-piece can is possible to make in many different sizes and shapes to fit product and market situations. The canner now has a far wider choice of shapes, sizes, forms, decoration, openings, etc.

Collapsible Tubes

Aluminum may also be used in the form of collapsible tubes for packaging processed food products. Sterilized foods packaged in collapsible tubes for the feeding of astronauts and high-altitude aviators have been developed. The aluminum tube fitted with a hollow-handled plastic spoon, which can be attached to the neck of the tube, should make a desirable and convenient package for feeding infants or bedridden patients [29].

Composite Cans

Another development is the foil/fiber can, more commonly called the composite can. Used at first for refrigerated biscuit dough, this material is now being used for frozen concentrated orange juice. The composite spiral can of fiber/polyethylene/aluminum foil has the major share of the juice and juice drink frozen concentrate market. Composite cans have been successfully employed for shortening and, with polyvinylidene chloride coating, for vacuum-packing of roasted and ground coffee. There has been considerable publicity on the use of composites for beers, hot fills, pasteurized, and even retorted foods [29]. The three major areas of interest appear to be materials to impart strength and protection, improved interior and exterior coatings, and better gasketing materials to effect better seals.

Can opening: For luncheon meat a 340-g (12 oz) cemented lap seam with dimensions the same as the traditional key opening tin-plate can has been developed. The can body is made from coated aluminum, bonded on the long seam with the thermoplastic adhesive. The aluminum has been scored so that by pulling a ring tab, the consumer can break the aluminum and then tear the entire body away from the ends, thus exposing the contents [29,90].

Convenience tops: The advantage of easy-open ends is obvious; no mechanical opener is required. The integral rivet aluminum easy-open end is the major system being used because aluminum may be scored and torn with relatively low force. A substantial fraction, however, employs an adhered that scored steel easy-open closures are now commercially feasible. The high tensile strength of steel has made it difficult to tear and thus, until recently, unsuitable for convenience opening. Aluminum was, and still is, the basic material employed for easy-opening closures [29,90]. Convenience closures must machine satisfactorily on canners' equipment, be economical, be easy opening without danger to the consumer, have no adverse effect on the product, either by itself or in combination with other packaging materials, and open only by consumer action.

Lacquers

Metal containers have an integral and sometimes an external protective coating. These coatings are generally applied as liquids consisting of a dispersion or a solution of resins or polymers in a solvent. Such lacquers may be applied before or after fabrication of the container. These protective coatings are necessary to (a) protect the metal from the acidity of the contents, (b) avoid contamination of product by metal ions from package, and (c) act as a barrier to external corrosion/aberration.

Four basic coating processes are in common use [91]:

1. Roller coating in sheet form: This process is essential for sheet-coating of tin-plate for welding cans since the process allows for stenciling, i.e., the provision of plain margins, which permits resistance welding of the side seam. Plate for sheet-fed drawn-and-redrawn (DRD) lines is also prepared in this way.
2. Coil coating: This process is applied where very large volumes of material of essentially the same specification are required. Capital cost is high, as is the cost of stoppages to effect either substrate or coating changes. Typical applications are TFS for DRD cans, although its use for food can ends is anticipated. In addition to cost considerations, lacquering in the coil permits the use of reverse roller coating, which produces a more uniform film ideal for drawn can manufacture. Stenciling necessary for welding cannot be achieved in the coil process.
3. Spraying: This process is used for made-up containers drawn and wall-ironed (DWI) and as one means of applying side-stripe protection on welded cans.
4. Electrocoating: This process is an alternative for applying protective films to finished can bodies, such as DWI and DRD cans made of tin-plate or aluminum.

The advantages are (a) lower materials usage and thinner, more uniform lacquer films, (b) lower solvent emissions to the atmosphere, and (c) higher standards of process monitoring and quality control.

3. Glass Containers

Glass containers are ideally suited for the packaging of many foods and are widely used for the canning of fruits, vegetable, and juices. Meat and poultry products may also be packed in glass containers. When combined with the appropriate closure, the glass jar or bottle provides an inert, hermetic, durable packaging medium for a wide variety of foods. The transparency of glass makes it the ideal choice for many products displayed for the consumer on the retail shelf. In addition, the resealability and storage characteristics of glass containers give them added consumer appeal [86].

Hermetic closures are required for any food products that are subject to microbiological spoilage such as baby foods, prepared infant formulas, fruits, vegetables, meat products, juices, jams, jellies, preserves, tomato sauce, chicken products, processed pet foods, pasteurized pickles, and many others where preservation using heat is required. Functionally speaking, hermetic closures consist of a metal shell made of either tin-plate, tin-free steel, or aluminum, an inside coating, and an impervious sealing material. This material may be rubber, either natural or synthetic, a plastisol or sheet polyvinyl chloride material, or other suitable plastics. Hermetic seals may be applied to packages in many different ways under many different conditions. They can be applied with a wide variety of capping equipment. Hermetic seals can be achieved by being pushed on, crimped on, rolled on, screwed on, or turned on as in the case of lug caps. They may be crowns, side seal closures, rolled-on closures, screw caps, or lug caps.

4. Retortable Pouches

Retortable flexible containers are laminate structures that are thermally processed like a can, are shelf-stable, and have the convenience of frozen boil-in-the-bag products. The materials for flexible containers must provide superior barrier properties for a long shelf life, seal integrity, toughness, and puncture resistance and must also withstand the rigors of thermal processing. Retortable flexible containers may be retort pouches or semi-rigid containers [86].

Products Packed in Retortable Flexible Containers

The following types of foods may be packed in a retort pouch: meats, sauces with or without particulate, soups, fruits and vegetables, specialty items like potato salad, bakery products, and pet foods. Some of the entrees packed this way are meatballs and gravy or tomato sauce, chicken a-la-king, chicken stew, beef stew, ravioli, spaghetti and meat sauce, Hungarian goulash, beef stroganoff, barbecue chicken, and sukiyaki. Generally, any product currently packaged in cans or glass can be packaged in flexible containers. Some products that now appear only in frozen form can be transferred to semi-rigid container where the product can be layered.

Structure of Flexible Containers

The structure of the retortable pouch in general use today is a three-ply laminate composed of 12 μm polyester film, adhesive laminated to 9–18 μm aluminum foil, which is either laminated to a 76 μm polypropylene film or extrusion coated with 76 μm polypropylene resin. The polyester film is used for high temperature resistance, toughness, and printability.

F. Energy Aspects of Packaging

Energy analysis of the food sterilization unit is useful in two respects: first, it provides quantitative information on energy requirements of use in designing the energy-generating and delivery system, and second, it evaluates the modes of energy loss. Information obtained from the energy analysis can be used for quantifying energy-conservation practices [92]. Energy required for manufacturing, transporting, and processing were estimated for two alternative systems (canning line and retort pouch line), each capable of producing 43.3 metric tons of processed spinach per 8-hour shift. The following conclusions can be drawn:

1. Container manufacturing requires more than 80% of the energy required for each system.
2. A pouch-processing line will have much higher electrical requirements than a comparable canning line. However, the costs associated with electrical use are small compared to total costs.
3. The difference in processing energy requirements for food production in retort pouches and cans is small and will not significantly affect the decision to adopt retort pouch or can processing technology.
4. The total energy requirement for a retort pouch packaging system is significantly less than that for a can packaging system.
5. Container and energy costs for a retort pouch packaging system are significantly lower than those for a comparable complete can packaging system.
6. A comprehensive economic analysis must be conducted before a decision to adopt retort pouch processing technology can be made.

A dominant factor influencing total energy use in the canning industry is the heat requirement of food sterilization. Unger [93] observed that continuous cookers now in use in canneries are typically more energy-efficient than batch processing in retorts. Energy consumption rates for various sterilizing equipment were recently compared by Ferrua and Col [94]. They investigated the energy requirements of a rotary pressure retort, a rotary atmospheric retort, and a flame sterilizer, and estimated the overall heating efficiency to be 47.7, 31.2, and 27.5%, respectively. Casimir [95] presented comparative costs of the heat required for sterilization of canned products by different equipment (Table 5). There is some lack of agreement among various investigators on the energy requirements of different equip-

TABLE 5 Energy Costs Required for Thermal Processing of Canned Foods by Different Equipment

Processing equipment	Comparative costs of heat		
Static retort	100[a]	100[b]	100[c]
Continuous rotary atmospheric retort	—	—	64
Continuous rotary pressure retort	—	—	46
Hydrostatic retort	20	56	—
Fluidized-bed retort	—	38	—
Microwave retort	1230	—	—
Flame sterilization	56	—	88

[a]Values in this column from Ref. 95.
[b]Values in this column from Ref. 97.
[c]Values in this column from Ref. 94.

ment, probably as a consequence of different experimental procedures and possible neglect of important sources of energy loss.

Sampson [96] conducted a thermal energy analysis on a stationary retort. He reported that only 16.7% of the steam supplied to the stationary tort was used in heating the cans and contents. The remainder was lost as follows: 36.4% passed out of vents, 16.4% was used to heat the retort and crates, 11.2% was used to heat the condensate in the bottom of the retort, and 19.3% was lost through radiation. The study indicates significant loss of steam during venting. Steam requirements in a cannery were discussed in considerable detail by Lopez [24]. Data from different canneries showed steam consumption to be quite consistent for the retorting operations, averaging 3 kg/min of steam per 24 No. 2 cans. During venting, the peak of steam consumption may vary between 1135 and 2720 kg/hr for a standard three- to four-crate retort, depending upon the size of the steam inlet line. The peak demand drops off to an operating demand of 45–68 kg/hr after the vent valve is closed and the retort reaches operating temperature. Lopez [24] emphasized that estimation of the total steam consumption requirements of a cannery requires knowledge of the steam demands, both peak and operating, for each individual piece of equipment.

A novel fluidized-bed retort described by Jowitt and Thorne [97] and Thorne [98] involves heating and cooling of cans in a fluidized bed of sand or other granular material of high density. Fuel savings can be significant with a fluidized-bed retort. Since the heating medium (usually heated air) does not go through a phase change, recycling of the heating medium would improve the energy efficiency of the equipment.

IV. ASEPTIC PACKAGING

A. Introduction

The development of HTST processing methods for sterilizing in a continuous flow has brought about the need for aseptic packaging. It is only though the use of aseptic packaging that the benefits of HTST treatment can be fully realized. It is the packaging that is usually the major cause of contamination of the product [86]. Aseptic packaging exhibits the greatest quality improvement over conventional canning when viscous low acid products are processed. Many products can be commercially sterilized prior to packaging by continuous processes so that their organoleptic and nutritional quality is not significantly affected. Products such as puddings, sauces, dips, and pastes are currently aseptically processed. In the techniques applied to aseptic packaging, continuous heat exchangers can be designed so that any temperature profile may be applied.

B. Sterilization Systems

1. Typical Aseptic Systems

The key components of aseptic systems—timing or metering pump, product heater, holding tube, cooler and back pressure valve are characteristic of practically all aseptic processing systems. The type of aseptic processing equipment selected is basically dependent on the pH, the viscosity or consistency of the product and on whether it contains particulate and their size [86].

Factors that affect scheduled aseptic processes for optimum quality and safety are [86]:

1. Targeted sterilization value (Fo value)
2. Heat penetration rates (viscosity, particles)
3. Maximum temperature product can withstand
4. Allowable heating and cooling times in lethal temperature range not detrimental to product
5. Time-temperature relationship and lethality in holding tube
6. Lethality during heating and cooling
7. Holding tube size (length and diameter)
8. Product vapor pressure and required back-pressure
9. Product formulation and characteristics (pH)

The sterilization value required for commercial sterilization of low-acid foods and the effect of heat on the destruction rate of microbes (z value) associated with the product and potential spoilage microorganisms should be determined. The literature values for the same product or process could be used. Products that are especially heat labile must receive a sterilization process to achieve minimum quality loss and adequate commercial sterility.

2. Validation of Aseptic Processing and Packaging

The first step is to review the applicable regulations. Products with a pH greater than 4.6 and those with pH less than 4.6 fall into different regulatory framework. If pH is achieved through the addition of acid, then the product should be regulated [99]. After reviewing the applicable regulations, the next step is the prepurchase or preinstallation review of processing, filling, and packaging equipment. The processor should collaborate closely with the equipment supplier and the process authority to review the overall aseptic design to assure that the system can deliver a commercially sterile product and that the system can be cleaned after production and sterility maintained. The review should also include the methodology of packaging material sterilization. The instrumentation and control systems of the equipment should also be discussed in the review. Particular attention should be paid to the location of sensors adequate to document proper equipment sterilization and maintenance of the product and equipment sterility.

Once purchase decisions are made and the equipment is being installed, the location and orientation of sensors, valves, etc. should be checked. Instruments and sensors should be calibrated for accuracy before start-up. After installation is completed and the system is in operation, testing should be performed to assure that the processing, filling, and packaging equipment is functioning properly before starting the commercial production. This testing should include the preproduction sterilization cycle for the product sterilizer, piping, and all product contact surfaces within the aseptic zone of the packaging equipment. Also, the location and adequacy of temperature-measuring devices should be evaluated with reference to measured temperatures or biological destruction, and recommendations made if changes are required.

Once the sterilization process for product and equipment have been established, inoculated packs of the product should be processed for confirmation of the system operation. If everything works properly at least four commercial production runs of uninoculated product should be packed and incubated, followed by 100% examinations for evidence of spoilage.

C. Processing Equipment

1. Processing Systems

Infusion Sterilization

Steam injection sterilization: This is the most rapid method of heating the product, facilitating the attainment of sterilization temperature within seconds. Combined with the rapid method of cooling by injection of the hot product into a vacuum chamber and evaporation of an equivalent amount of water, a very high-quality product is obtained. This method is usually combined with heating and cooling in heat exchangers to the low temperature range (80°C) [100].

Liquid infusion into steam: This system involves infusion of a thin film of liquid into a steam atmosphere, facilitating rapid heating. Cooling is also achieved by infusion of the liquid into a vacuum chamber.

The advantages of this system are (a) it is a versatile processing method designed primarily to heat and cool fluid foods in a matter of seconds, (b) it produces the fastest heating methods, which minimizes flavor changes and product damage normally associated with high processing temperatures (especially important for low-acid products that require sterilizing up to 149°C), (c) it may be prepiped packaged systems or field-assembled components arranged to meet specific plant space requirements, (d) acquisition costs are low when high flow rates are being processed, and (e) there are few moving parts and service costs are low. The disadvantages are: (a) the method is suitable for particle-free liquids only, and (b) heat recovery is reduced to about 50%.

Tubular Aseptic Sterilizer

Tubular aseptic sterilizing is an indirect heating/cooling method that uses stainless steel coiled or straight tubular heat exchangers. The tubing diameter is relatively small compared to product flow. As a result, extremely high flow velocities within the tubing maximizes turbulcncc. High turbulence induces rapid heat transfer [29,86,100]. The tubing diameter is suited to the product flow and viscosity. Tubes are fabricated into coils or bundles and placed, along with special media baffles, into stainless steel jackets. Hot water, steam, or cold water pass through these jackets to heat or cool the product flowing within the tubes. A series of horizontal tubular heat exchangers and a vertical holder tube hold the product at the required sterilizing temperature for the required time. For low-acid products, this is generally in the area of 149°C with a holding time of 2–4 seconds. High-acid products would normally be heated to around 93°C and held for approximately 30 seconds. From the holding tube, the product flows to another series of vertical tubular heat exchangers for cooling.

These systems have the following advantages: (a) they provide high heat-transfer rates and a scrubbing action that reduces "burn-off" or fouling in the tubes, resulting in a very short processing time, helping to preserve the natural flavor of the product, (b) they have considerable flexibility in the range of products they can handle and the temperature range at which a specific product is processed, (c) they are completely self-contained, requiring only product and utility hook-ups to be made during installation, (d) there are no gaskets to replace on the high-temperature side, (e) most are available with regeneration as an option. Regeneration may run as high as 85% depending on flow rates, product characteristics, and regeneration option used.

Swept Surface Sterilizer

This type of heat exchanger is similar to a tube heat exchanger but is provided with a central rotating shaft carrying a scarping device for the heated surfaces. This prevents burning and fouling of foods at the surface and also provides a mixing action. The system is used when a viscous material or one containing small, discrete particles is to be processed. Swept surface sterilizing is an indirect heating/cooling method. With the continual removal of product from the cylinder wall, the product film is reduced to an absolute minimum, permitting long processing runs without product build-up on the heat exchanger wall [86,100,101].

Its advantages are: (a) it is capable of processing heat-sensitive products, (b) it is versatile for aseptic processing, (c) products can be processed over a broad temperature range and viscosity with or without particulate, (d) the various horizontal and vertical configurations allow this form of heat exchanger to be adapted to specific systems or plant requirements, and (e) it may be used in series with other types of heat exchangers for products such as starches that might increase in viscosity due to processing.

Plate Sterilizer

Plate heat exchangers (described in Sec. II) can also be used for aseptic processing.

2. System Comparison

The above systems can be compared as follows:

1. Infusion into steam: Fastest heat exchange system; handles heat-sensitive products best; can remove volatile flavor compounds.
2. Tubular heat exchanger: Easier to maintain; faster heat exchange than systems mentioned below.
3. Swept surface heat exchanger: Capable of handling viscous or solid-containing product.
4. Plate heat exchanger: The least expensive; has high throughput characterized by low pressure operation.

Certain products like milk or ice cream mix are highly susceptible to flavor changes when subjected to prolonged high temperatures and should be processed using a system that produces rapid uniform heating and cooling. The swept surface heat exchanger represents a gradual heating and cooling profile beneficial for heavy viscous products and products with particulate. Rapid heating and cooling is not as critical here, as slower heating allows penetration of particulate and produces uniform flavor.

The plate heat exchanger represents an even more gradual increase and decrease in product temperature. Juices and juice products are not as heat sensitive as dairy products and can be processed on a plate heat exchange system providing the solids or pulp content are not sufficient to cause fouling. Normally, these products need only be heated to 93°C for 30 seconds to produce adequate sterilization. Specific products and systems will dictate their own profile. The comparison shown here illustrates the difference in the four methods of heating/cooling and aids in selecting the proper aseptic processing system for the product.

3. Packaging Systems

Aseptic packaging refers to the filling of a cold commercially sterile product under sterile conditions into a presterilized container and closure under sterile conditions to form a seal that effectively excludes microorganisms. Aseptic literally means the exclusion of microorganisms from the environment. Aseptic processing is really a method of packaging because foods are not sterilized or cooked or otherwise altered by aseptic methods. Rather, they are handled or moved by aseptic methods to assure that they retain the microbiological quality with which they started. In general, aseptic packaging is coupled with HTST or UHT methods of food sterilization, and the two processes are joined in a complete system to product what are referred to in the trade as aseptically processed foods. However, of all the equipment that has been or is being proposed, none is an actual integrated system. The processor must purchase the sterilizer and the aseptic packager as separate units and then tie them together.

Aseptic filling and packaging systems can be classified into categories based on the type of packaging material and the method of forming the container (Table 6) [102].

Table 6 Classification of Aseptic Filling and Packaging Systems

Category	Examples of systems
I. Metal and rigid containers sterilized by heat	
A. Steam/metal containers	Dole canning systems Drum fillers, e.g., Scholle, FranRica
B. Hot air/composite can	Dole hot air system
II. Webfed paperboard sterilized by hydrogen peroxide	TetraPak (BrikPak) International Paper
III. Preformed paperboard containers	Combibloc LiquiPak
IV. Preformed, rigid/plastic containers	Metalbox Freshfill Gasti Crosscheck
V. Thermoform-fill-seal	Benco Asepack Bosch Servac Conoffast Thermoforming USA
VI. Flexible plastic containers	
A. Bag-in-box type	Scholle LiquiBox
B. Pouches	Asepak Prepac Prodo Pak Inpaco
C. Blow molded	Bottlepack Serac ALP

Source: Ref. 102.

Form/Fill/Seal Machines

Form/fill/seal machine for pouches: Figure 9 shows the principle of operation of an aseptic vertical form/fill/seal machine for three-sided sealed pouches [103]. The packaging material from a reel, usually a complex multilayer material, is sterilized by hydrogen peroxide in a heated bath, which is the siphon lock to a sterile chamber with a slight overpressure of sterile air. In this chamber the film is dried, folded over a shoulder to form a tube, and sealed at the long seam. Then the tube, which is closed at the bottom of the cross seal, may be drawn to the nonsterile exterior of the chamber through a tightly fitting flexible lock. Sterile filling inside the chamber is performed using a sine filler. In the tube above, the contents are protected by a neutral atmosphere of sterile nitrogen, which maintains a very low oxygen concentration in the headspace of the packs. Grippers spread the sealing zones and vertically reciprocating sealing bars with cutting knives outside the sterile cabinet are transporting down, sealing, and cutting off the pouches. The pack output is 15–35 pouches per minute, depending on size. Products that are, at present, filled by these machines include various tomato products, sauces such as cheese sauce, and pizza sauce with particulate. Meal constituents and curries could also be filled [100]. The filling system has CIP and SIP characteristics. Presterilization of the filling system with pressurized steam and of the sterile chamber of the machine by condensed hydrogen peroxide vapors, as well as heated air, is performed automatically.

Thermoform/fill/seal machine for cups and trays: Both film for the cups and trays and the lid are drawn from rolls and transported into the totally closed sterile cabinet through a heated hydro-

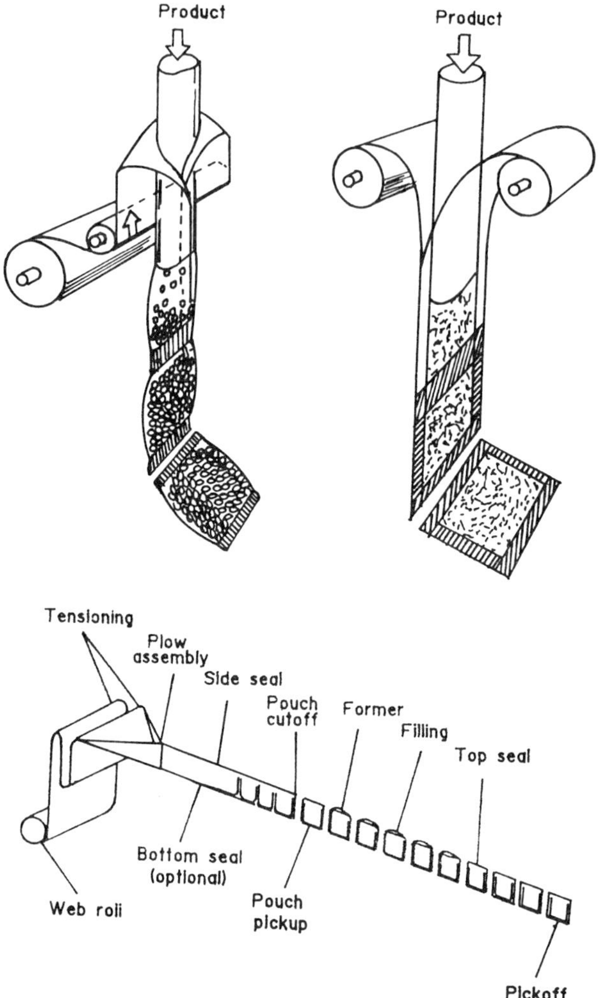

FIGURE 9 Form/fill/seal machines. (From Ref. 103.)

gen peroxide bath. The lower film is heated locally, thermoformed with plug assistance by pressurized sterile air, and the formed packs are then filled. Filling is performed by a special piston filler with reciprocating valves having cutting edges. This filler is able to deposit mixtures with particulate some millimeters in size. The shafts of the sliding valves and pistons penetrate the vessel for the product that has to be filled. At the mechanical drive external nonsterile air is separated from the sterile air above the product by rolling diaphragms. After filling, the lid is applied to the filled web and sealed at the rims. Headspace gas flushing may be performed. Now the webs may be transported to the nonsterile exterior through a contour lock without risk of infection, where the final sealing of the packs, notching, and cutting is carried out [104].

Filling and closing line for bottles and jars: The containers, which are precleaned and heated by a special rinser, enter the sterilizing machine in one lane; they are sterilized in several lanes, upside down, by treating inside and outside with hydrogen peroxide vapors and then drying with sterile air. The containers are inverted and transported intermittently to a piston filler. In the next stage the containers are closed with metal caps, which were sterilized with pressurized steam when entering the machine. For liquid products, magnetic-inductive metering devices are used for filling. For plastic bottles heat-seal foil closures are applied [104].

The TetraPak System

The principle of this system is to take the packaging material directly from the reel and to form it continuously into a tube. The tube is sealed below before filling and above after filling. The main advantage of this pack is that there is no headspace in the finished pack. The precreased web of packaging material unwinds over rollers, which soften the transverse crease. The web is immersed in a hydrogen peroxide bath for sterilization. The packaging material is a lamination made of pare, foil, and polyethylene. This combination provides a light and gas barrier, strength, and heat-sealing properties [29,86,105].

Conoffast System (Continental Can Co.)

This system employs prefabricated and sealed pouches that may be internally sterilized by one of four methods. Empty pouches are fed through an ultraviolet chamber to minimize external pouch contamination and then filled using hypodermic needles in a superheated steam atmosphere. The needle is withdrawn and the puncture area is heat-sealed. Ionizing radiation has been suggested to sterilize the pouch material surfaces. Steam pressure sterilization restricts the packaging materials to those that can resist the temperatures. To sterilize the area in which filling and sealing is performed and to sterilize the needle, superheated steam at 145°C can be used [29,86,105].

The Conoffast system also utilizes the basic package forming-filling-sealing system. Sterilization of the inside of the package in this system is based on the high temperatures generated within the thermoplastic resins during the extrusion process used to produce the multilaminated packaging material. During the aseptic packaging operation, the top plastic sheet is delaminated, exposing a sterile surface of the packaging material to the food in a sterile environment within the Conoffast unit. Before filling, the packaging material is thermoformed into cup-shaped containers. Sterile food product is filled into the package under sterile conditions. Then the packages are sealed with packaging material from which one layer has been delaminated, again exposing the food to contact with a sterile surface. Sterility during the forming, filling, and sealing operations is maintained by performing these operations in a sterile environment under positive pressure.

The Combibloc System

In this system premade carton blanks are used, which are die-cut, creased, side-seamed, and printed at the factory origin. This facilitates a more perfect flame-welded seam, thus ensuring good integrity of the seal, and the packaging machinery can handle different sizes with a simple height adjustment. A carton blank is drawn from the magazine by suction pads and placed on a mandrel. The sealing surface at the bottom of the carton is softened by hot air. The bottom is folded, pressed, and sealed against the end face of the mandrel. The top is prefolded and then passed on to the aseptic zone, where it is sterilized by hydrogen peroxide spray. After filling, the package top is folded and sealed by ultrasonic welding [105].

The International Paper System

In this system the packaging material is taken from a reel. From the reel the web passes through a series of scoring rollers into a hydrogen peroxide bath for sterilization. The horizontal seals are made by alternating jaws and an induction heater. Individual packages move to the final folding and sealing station for sealing of top and bottom flaps [105].

The Gasti System

This system operates with preformed cups made of plastic or aluminum. This facilitates greater flexibility and allows container quality to be approved in advance. In operation, preformed cups are dispensed from a magazine and are sterilized using hydrogen peroxide vapor. The cups are then filled with product in the sterile section and are sealed with presterilized lids [105].

The Liqui-Pak System

This system uses a combination of two sterilizing media—hydrogen peroxide and ultraviolet light—to obtain aseptic packages. This approach has a synergistic effect, which results in more effective bactericidal action than high concentrations of hydrogen peroxide and ultraviolet light used individually. The cartons travel through the sterile area against a flow of filtered air. The filler is unique, with a plastic bellows mechanism for sterile dispensing of the product. After filling, the gable-carton is heat-sealed in a conventional manner [105].

The Metal Box "FreshFill" System

This system uses preformed cups sterilized by hydrogen peroxide. The product is filled by a multihead filler, with the filling chamber isolated from other machines areas by sterile air overpressure generated by an ultrafilter. The filled cups are sealed using a presterilized foil material and stamped out in the conventional manner. Before start-up, the filling chamber, fillers, and supply line are sterilized with steam at 130°C for 20 minutes [105].

Avoset System

This system packs fluid dairy products in glass and cans, and then in aerosol containers. The system sterilizes containers and product separately and brings them together in a sterile environment. The entire equipment is placed in a controlled environment, and all critical elements are sterilized. An operator is present to monitor the equipment. Although sterility is assured, distribution under refrigeration to retard biochemical changes is generally recommended [29,105,106].

Manton-Gaulin (Pet Inc.)

This system is basically a glove box. Polyethylene bags are sterilized in the glove box with an ethylene oxide mixture and is heated to 49°C for 6 hours. After sterilization, the gas is replaced by sterile air. Sterile mix is aseptically pumped to the filling nozzle within the glove box. Using glove box techniques, the operator fills one bag at a time through the rigid spout. The spout is heat-seal closed with a laminated foil material. The filled bag leaves the glove box through a chlorinated water trap and is dried outside of the aseptic filling area [29,105].

Scholle containers with a semi-automatic aseptic system for filling 6-gallon polyethylene bags have been used. Sterile conditions are maintained by pressurized sterilized air and a continuous spray-mist of chlorine solution over a hinged-cap fitment on the bag during filling.

D. Quality of Aseptically Processed Foods

The basic consideration for HTST sterilization is that for each 10°C (18°F) increment bacteria is reduced by a factor of 10, while the rate of destruction of nutrients and of other chemical reactions affecting product color and flavor increases by a factor of approximately 3. This is called the z value. Therefore, the higher the sterilization temperature, the larger the difference between destruction rates. This is the concept that explains one of the most important advantages of aseptic processing and packaging systems. It is the reason why organoleptic and nutritive value changes in products processed by aseptic processing systems are less pronounced than in products processed by other systems, such as conventional canning [107].

Chemical and flavor changes during high-temperature heating are particularly severe in low-acid foods, which require more severe heat treatment to be sterile. At higher processing temperatures, bacterial spore destruction is much faster than the destruction of food constituents. Foods processed by aseptic processing are better in color and higher in thiamine than those made by conventional packaging processes. Aseptic processing is used commercially in the dairy industry and for fruit juices and purees, pea soups, sauces, tomato paste, etc.

The effect of HTST sterilization on the nutritive value of meat products has been a subject of great interest. Protein quality may be degraded by the destruction of one or more essential amino acids, formation of inter- and intramolecular bonds resistant to digestive enzymes, and alteration in the rate at which the various amino acids are released from protein, resulting in a mixture of amino acids that may be less efficient for metabolism and assimilation. The essential amino acids tryptophan, methionine, and lysine are also destroyed by HTST sterilization. Destruction of methionine and accumulation of sulfur dioxide are the main chemical indices of the effect of cooking meat proteins at 120–150°C. Protein denaturation is not the only change that occurs during heat processing. Hydrolysis of proteins and polypeptides also takes place. Collagen, elastin, and reticulin compose the connective tissue of meat and are insoluble in water and salt solutions. Collagen is transformed by heating into soluble gelatin. In this conversion, some of the cross-links are broken, resulting in shortening and disorganization of the protein chains. The conversion also accounts for higher protein solubility in processed meat products. Aseptic strained meat has been shown to be more nutritious and higher in thiamine retention.

Aseptically processed meat and vegetable products lose thiamine and pyridoxine, but other vitamins are largely unaffected. There are negligible vitamin losses in aseptically processed milk, and lipids, carbohydrates, and minerals are virtually unaffected. Riboflavin, thiamine, pantothenic acid, biotin, nicotinic acid, and vitamins B_6 and B_{12} are unaffected. Nutrient losses also occur during periods of prolonged storage, and these should also be considered when assessing the importance of sterilized foods in the diet.

The use of HTST processes is particularly adaptable to aseptic processing. The destruction of nutrients during the thermal processing is dependent on (a) time-temperature treatment used as the basis of the process and (b) the rate of heat transfer into the product. In an aseptic processing system, as the processing temperature is about 150°C for a very short period, nutrient retention is greatly enhanced. Ammerman [108] has reported the effects of heat treatments of equal microbial lethality on selected food constituents including nutrients, color, proteins, and flavor compounds. He has illustrated that retention of vitamin C in tomato juice improves during HTST processing. For natural products containing enzymes, the limitation of the benefits of HTST processing occurs when the basis of the process shifts from microbes to enzymes at a temperature of 132–143°C for shorter periods.

In an evaluation of HTST aseptic processing [109], it was found that thiamine retention was significantly greater in HTST products than in conventionally canned and retorted products. For pyridoxine, the benefit of HTST was not as evident, as destruction of pyridoxine is not as temperature dependent as that of thiamine. HTST aseptic canning also results in significant improvements on organoleptic qualities [110]. Most reports on the effect of thermal processing on nutrients only contain information on the content of a specific nutrient after the thermal process and give the percentage retention or loss of that nutrient. As there are numerous processing methods and time-temperature possibilities for accomplishing commercial sterilization, it is incorrect to assume that the nutrient losses reported in the literature represent the average or norm for the industry. Hence, data on such nutrient losses are of limited value but can be used as guidelines for selecting an optimum process schedule [110]. Nutrient losses range from 0 to 91%, depending on the nutrient and product. These losses represent the sum of the losses during the entire processing period, including blanching.

In some studies [111–113], the sterilization temperature of foods has been optimized to maximize retention of nutrients or minimize production of an undesirable product. In these studies, microbial inactivation in the heaters was included in the determination of the required sterilizing value.

E. Packaging Aseptically Processed Foods

1. Types of Packages

Packages used for aseptic packaging may be divided into three classes: rigid, semi-rigid, and flexible [86].

1. Rigid means that the shape or contours of a filled and sealed container are neither affected by the enclosed product nor deformed by an external mechanical pressure

of up to 70 kPa. Rigid containers include metal cans, glass containers, and rigid cups. Among them are the cans used in the Dole canning system, the metal drums used by the Fran-Rica aseptic drum-filling systems, the glass containers used by the Dole and other systems for aseptic packaging of fruit juices in consumer sizes, and prefabricated and form-fill-seal cups.

2. Semi-rigid means that the contours of a filled, sealed container are not affected by the enclosed product under normal atmospheric temperature and pressure, but can be deformed by an external pressure of less than 70 kPa. Among the semi-rigid containers are those made of laminations of paper/plastic film/aluminum foil mainly used by the TetraPak, Ex-Cell-O, and PurePak processes, the thermally formed packages used in the International Paper Company system and in the Conoffast system of the Continental Can Company, and aseptic plastic bottles.
3. Flexible means that the shape and contours of a filled, sealed container are affected by the enclosed product. Flexible packages are bags or pouches with or without exterior support. Among these are aseptic pouches in consumer sizes, the bag-in-box system, and the bag in metal drum or wooden bin.

2. Basic Characteristics of the Aseptic Packaging Area

The packaging area or "clean" filling room should also receive a fair amount of attention. The critical factors are as follows [114]:

Floors, walls, and ceilings sealed with crevice-free resin coating are preferred over the usual tiled construction.
Sewers should be outside the filling room.
Floors should be dry.
All air should be well filtered, and the room should be operated under a positive pressure.
Packaging operations should be isolated from the interdepartmental traffic pattern. Only essential personnel should be in the filling room.
Employees should wear clean uniforms and single-service hats.
Package storage should be protected from gross external contamination.
Light fixtures should be flush-mounted to prevent dust and dirt contamination or accumulation.
Wash basins should have germicidal soap and foot-operated fixtures.
No visitors should be allowed in the filling room. Observation windows or a balcony should be installed instead.

The aseptic packaging unit must fill a container under conditions that do not allow recontamination of the product or the container. Modern aseptic packaging systems rely more heavily on shrouded packaging units than on the plant environment. A realistic definition of aseptic packaging used in industry [114] states that "commercial aseptic packaging is the filling of a substantially sterile product into a substantially sterile container, while maintaining a bacteriological quality of the packaged product to be treated as a sterile commodity prior to its delivery to the ultimate consumer."

Other important considerations in the use of a sterile filling and packaging unit are the speed of packaging and the maintenance required for the machine.

3. Intrinsic Properties of Packaging Materials

There are a set of strict requirements for aseptic packaging systems [114–116]. The packaging materials should

Have very low permeability to water, gases (oxygen), volatile compounds, and light.
Be inert, i.e., not impart flavors or taints to the product.
Be capable of being hermetically sealed by suitable means or by closure conforming to the requirements already given.
Be sterilizable.
Be durable.
Provide an effective barrier against microbial contamination.
Be appealing to consumer.
Be resistant to the absorption of moisture.
Not contain substances that are toxic or harmful to health.
Be unaffected by any chemical and/or heat treatment that may be applied preparatory to filling.
Not be susceptible to deteriorative changes when stored for prolonged periods of time under the different climatic conditions likely to be encountered; they should remain durable for a period of time compatible with the anticipated shelf life of the product.
Conform to specifications.
Provide for a minimum waste of usable material(s) during preparation and filling of the containers.
Be easily disposable, thereby constituting minimal possible ecological problems.
Be light in weight but sufficiently robust to withstand shock or rough treatment likely to occur during handling and transportation.
Be relatively inexpensive.
Be marked or printed in such a manner that they conform to legal requirements and are descriptive of the product and the conditions that should apply to storage, expected shelf life, etc.

4. Sterilization of Packaging Materials

Package sterilization has been accomplished by using a number of methods and combinations [117]. These methods are listed in Table 7 [118].

Superheated Steam Systems

In this system, sterilization of the container and its closure is accomplished by the application of heat using superheated steam. The advantage of this system is that it can achieve high temperatures at the atmospheric pressure, but microorganisms are more resistant to superheated steam than saturated steam.

Dry Hot Air Systems

Hot air sterilization has advantages and disadvantages similar to superheated steam. There are currently no units of this type utilized in the production of low-acid foods, but the equipment has been used for the production of juice and beverages.

Hydrogen Peroxide Systems

A number of systems utilize hydrogen peroxide in combination with heat and/or other adjuncts. In one such system, the packaging material is not metal and it comes in rolls rather than in preformed containers. The system also utilizes a different sterilizing medium. The rolls are continuously fed into a vertical machine, which sterilizes, forms, fills, and seals the package. Sterilization is accomplished with a combination of hydrogen peroxide and heat.

A second system is similar to the one above. The main difference is how the heat is applied to the package surface. This system provides the heat necessary for sterilization by means of a heated stainless steel drum. Contact with the drum heats the peroxide and effects sterilization.

TABLE 7 Methods for Sterilizing Aseptic Packages

Method	Application	Advantages/Disadvantages
Superheated steam	Metal containers	High temperature at atmospheric pressure; microorganisms are more resistant than in saturated steam
Dry hot air	Metal or composite juice and beverages containers	High temperature at atmospheric pressure; microorganisms are more resistant than in saturated steam
Hot hydrogen peroxide	Plastic containers, laminated foil	Fast and efficient method
Hydrogen peroxide/UV combination	Plastic containers (preformed cartons)	UV increases effectiveness of hydrogen peroxide
Ethylene oxide	Glass and plastic containers	Cannot be used where chlorides are present or where residuals would remain
Heat from coextrusion process	Plastic containers	No chemicals used
Radiation	Heat-sensitive plastic containers	Can be used to sterilize heat-sensitive packaging materials; expensive; problems with location of radiation source

A third system also uses packaging material that comes in rolls. The rolls are continuously fed into the machine, which forms, fills, and seals the package. Sterilization is accomplished with the combination of hydrogen peroxide and heat. The packaging material travels through a bath of hot hydrogen peroxide, which softens the material for forming. Cups are then formed, filled, and sealed with lids that have also traveled through a hydrogen peroxide bath.

A fourth system utilizes preformed cups to which a lid foil is heat-sealed after filling. The cups are fed into the machine where they are sterilized by a spray of peroxide followed by heating. The lid material is sterilized by being passed through a peroxide bath.

Another system, which can utilize preformed cartons, sprays the inside of the carton with low concentrations of hydrogen peroxide. This sprayed carton then passes under a UV light source, which acts synergistically with hydrogen peroxide in destroying microorganisms. Results of tests using suspensions of microorganisms have shown this combination to be very effective.

Heat of Extrusion System

This is a form-fill seal packaging system, but it differs from the systems discussed earlier. This system relies on the temperature reached by thermoplastic resins, during the co-extrusion process used to produce multilayer packaging material, to produce a sterile product surface. During production, the multilayer package material is fed into the machine where it is delaminated under sterile conditions. This removes a layer of material and exposes the sterile product contact surface. The container material is then thermoformed into cups. The lid material, which is also delaminated, is then sealed onto the cup after filling.

5. Aseptic Packaging

In general, aseptic packaging is coupled with HTST or UHT methods of food sterilization, and the two processes are joined in a complete system to produce aseptically processed foods. However, of all the equipment that has been or is being proposed, none is an actual integrated system. The processor must purchase the sterilizer and the aseptic packager as separate units and then tie them together.

Drums

The drum is placed under the filling chamber and then raised up to seal against a gasket. Saturated steam is introduced into the drum, and the interior of the drum is pressurized with steam. After a total of 2.5-minute cycle, a filling tube lowers into the drum and dispenses sterile product to fill the drum. The filling tube retracts, and a sealing head with magnetically attached lid swings over the drum. Steam pressure is employed to apply pressure to crimping jaws to fix the gasketed lid into place. Alternatively, an empty drum and cover are placed in a filling retort, and both are steam-sterilized under pressure. Sterile product is then filled into the drum, and the lid is placed. The drum is removed, and another cycle is initiated. The capacity of a system with two filling retorts is 24 drums per hour [29,119]. An advantage claimed for the drum system is lowered shipping weight because one large reusable container is used rather than 75 No.10 cans plus 12.5 corrugated cases. Product recovery is higher because there is far less surface area to drain. Further, the labor cost of emptying one 208-liter container is significantly less than the cost of opening 75 cans.

Tanks

Rather than pulping, finishing, and concentrating into paste for bulk handling, tomatoes are simply chopped, sterilized, and filled aseptically into 100-gallon tanks. For example, tomatoes are washed, chopped into chunks, heated in tubular heaters, cooled or reheated, deaerated, and subsequently filled into tanks under a nitrogen blanket. Storage tanks are galvanized steel, lined with a baked-on epoxy coating. Tanks are chemically sterilized before filling [29]. The basic objective is to provide tomato processors with a large source of raw material that can be converted into many different products over a long period. Among the products that can be made from chopped tomatoes are paste, puree, sauce, catsup, pizza sauce, juice, and chili sauce.

Glass Containers

Aseptic packaging in glass has not been broadly successful on a commercial basis. Juice is heated to and held 9 seconds at 93°C and cooled to 1°C in a heat exchanger. One- and 2-liter bottles are cleaned by rinsing with water. After washing, bottles are discharged into a closed area blanketed with 99.9% sterile, dehumidified air. The filler is sterilized with boiling water prior to operation. Closures are steam-sterilized and taken into the clean room. Sterile juice is filled into the sterilized jars and capped in the clean area. The product, known as aseptic cold-pack juice, is reported to have long shelf life under 10°C [29]. At room temperature storage, hot-filled (conventional) juice has an acceptable shelf life of one month as compared to 3 months for cold-filled juice. The basic problem with glass is that the maximum temperature differential glass containers can withstand is approximately 15°C. Thermal shock between inner and outer surfaces, for example, could lead to unequal thermal expansion sufficient to crack the glass.

Plastic Containers

As long as the container can exclude microorganisms and prevent passage of gas, the container material need not be rigid. Metal in heavy gauges and glass are both fabricated into rigid containers. No rigid plastics are employed commercially for sterile packaging. All plastic materials are partially permeable to moisture and gas. Because of implied low-strength characteristics, food products that have been sterilized are packaged in semi-rigid materials. Paperboard with appropriate coatings is also utilized to form semi-rigid packages [29].

Two basic types of semi-rigid systems are in commercial use: (a) thermoform, fill and seal, and (b) preformed cups, fill, seal [29].

Gable-top paperboard cartons: In this system the blanks in flat form in stacks are packed by the converter into corrugated shipping cases sealed with paper tape to restrict air flow. Blanks are sterilized by exposure to ethylene oxide gas. In the ethylene oxide treatment, a vacuum is drawn on the case and the vacuum is displaced with ethylene oxide gas either alone or in mixture with another active or inert gas. Treatment temperature is 38–65°C, and the time period is on the order of several hours. The high temperature stimulates the activation of microbiological spores to the vegetative form where they are more vulnerable to the gas mixture [120–122]. In this system, the sterilized blanks are "aseptically" removed from their carrier in a clean room–type area adjacent to the aseptic packaging equipment. Portions of the equipment in contact with product-contact surfaces of the blank are made of stainless steel for easy sanitation. A sterile blank is picked off and placed on a mandrel where bottom flaps are heated and sealed. Open-top cartons are then removed from the forming mandrel, and they are resterilized with a finely atomized mist of hydrogen peroxide injected directly into the package.

Saf-Pak: In this system packages employed are in the form of shallow cups, either aluminum or plastic. The cups are nested, sealed in polyethylene bags, and sterilized using ethylene oxide methods. After sterilization, the nested cups are transported to the aseptic packaging area, where they are aseptically removed from their polyethylene shroud and placed in the shrouded clean environment. Sterilized food is filled through the equipment into cups that have been denested into holders. After filling, still under the shroud, the cups are sealed. Products that undergo Saf-Pak processing have a guaranteed shelf life of 4–6 months without refrigeration [29,105,106].

Flexible Packages

The two TetraPak AB systems are frequently viewed as flexible packaging systems. Technically, neither is a flexible packaging system because both make use of paperboard rather than a true flexible material, but this would be of academic concern in any event. The polyethylene pouching material is sterilized using a hydrogen peroxide bath. Thimmonier vertical form, fill, and seal milk pouching equipment has been converted for aseptic packaging [123].

Reclosable Aseptic Packaging System

Reclosing is useful for products such as milk and wine. Shaking is important for products such as pulpy fruit juices. The system today consists of a closure with two parts. The visible part is a rectangular polypropylene chassis with a cap that snaps open and shut on a hinge. The device is glued to the top corner surface of the pack with hot-melt adhesive [124].

Underneath the cap is the second part of the system—an aluminum pull tab that covers the pouring hole and provides good tamper evidence. The pull tab, which is wielded to the inner liner of the package, is easy to peel off. It reveals a long, pear-shaped hole that has excellent pouring characteristics. The reclosing system is extremely sanitary, unlike "punch"-style cap devices that require users to poke a finger into the product in order to open the package.

The advantages of this system are as follows:

1. A high-quality long-life product is achieved by the aseptic filling process. Ease of stackability, space savings in terms of transportation (both filled and unfilled cartons), low weight, cost savings in carton material production, and high level of environmental friendliness are the result.
2. The reclosable system increases the intrinsic value of the carton. The reclosable, opening and pouring device provides the consumer with easy opening, spill-free pouring, and reliable reclosure as well as hygienic handling.

3. After reclosure the product is protected and can be shaken as necessary. The product quality remains unimpaired, protected from foreign flavors. The carton with the system fitted is ideal for refrigerator storage. The new carton image is completed by the slim, elegant shape suggestive of high quality.

6. Aseptic Canning System (Dole Process)

Aseptic canning systems have been developed to minimize quality changes that may occur in slow-heating foods processed in conventional retorting systems. Heat exchangers are used in aseptic canning systems to rapidly sterilize and cool food products before they are filled aseptically into sterile cans. Cream sauces, soups, and products containing rice, cheese, or high tomato content are particularly improved by the HTST processing used in the aseptic canning techniques. Both low-acid and high-acid products may be processed and packaged aseptically [86].

The standard Dole equipment with fully instrumented controls consists of five components [105]: (a) a tunnel heated by superheated steam, in which empty cans are sterilized at a temperature of 220°C for 40 seconds (this temperature is critical, as above 235°C the solder binding will be lost, (b) a cooling section in which the sterilized containers are cooled by spraying sterile water on the outside of the container, (c) a cover sterilizer, (d) a filling section, and (e) a presterilized and aseptic can-seaming section. The system is fully instrumented with controls.

In the heat-exchange systems used in conjunction with the Dole system, the liquidor semi-liquid food product is pumped continuously under pressure through the heating section of the sterilizer, in which it is quickly brought up to sterilization temperature (135–150°C), then through a holding section for the determined length of time to ensure commercial sterilization, and finally through a cooling section to the Dole aseptic canning system. The heat-exchange system, in which the product is heated and held under pressure for the time necessary to complete sterilization, constitutes in effect a continuous-flow pressure cooker. Four general types of heat-exchange equipment—steam injection, swept surface or scraper, tubular, and plate type—are used.

The Dole aseptic canning system is being used in an increasingly wide range of commercial operations. Dairy products successfully packed with the Dole system include whole milk, evaporated and concentrated milks, flavored milks and other dairy drinks, milk-based baby food formula, cream, butter, and some types of cheese. Whole milk and concentrated milks processed by aseptic fill methods are capable of being stored for extended periods of time and shipped without seriously affecting flavor or nutritive value. The delicate flavor structure of milk products is preserved better by the Dole system than by conventional canning methods, which subject the products in the container to an extended heating period [86].

With the substantial number of Dole installations in the field and the wide application of the system, an increasing number of food products are being commercially packed successfully Heat-sensitive products such as white sauce and hollandaise sauce, a broad range of soups, tomato paste, pear, peach, and banana purees, baby foods, and nonmilk formulas, meat purees, puddings, and pet food, and many other standard and specialty items—all are packed advantageously by the Dole aseptic canning system. Practically any product capable of being pumped through heat-exchange equipment will experience quality improvement not afforded by conventional methods.

F. Energy Aspects of Aseptic Processing

Studies of the energy aspects of aseptic processing have been widely carried out on milk, and most of the data available are on milk products. The energy requirements of aseptic processing units containing steam injection without regeneration or tube heat exchangers as applied to the processing of milk have been evaluated. For steam infusion the system consumed about 1000 kJ/kg of milk [125]. For tube heat exchangers the system consumed about 400 kJ/kg of milk [126]. The total energy requirement in the HTST system for milk processing was about 225 kJ/kg of milk. The break-up of the energy was 98 kJ/kg of milk thermal, 30 kJ/kg electrical, and 97 kJ/kg refrigeration. The steam infusion system requires about 360 kJ/kg of milk as it is a direct contact heating system [19]

With aseptic packaging processed milk cooling below 29°C is not required. This fact saves more energy, using about 100 kJ/kg of milk. Also, aseptic processing of products does not require post-processing refrigeration, resulting in further savings of 900 kJ/kg of product [19]. The most widely used heat exchangers in aseptic processing of foods are the plate and tubular type. If direct heating and cooling of product to sterilizing temperatures is used, the energy consumed is considerable. Therefore, use of heat regeneration is important [127].

The plate heat exchanger holds a distinct edge in this regard; regeneration efficiencies of 90% (direct) or 85% (indirect) have been used. Tubular units can achieve regeneration by indirect methods, whereby a secondary water flow exchanges heat from the preheater tubes to the cooler tubes. Unfortunately, indirect regeneration requires four times the surface as does direct for the same heat recovery; as a result regeneration efficiencies over 70% are rare.

Steam injection/infusion systems are not efficient: for example, with milk duties regeneration accounts for only slightly more than 50% of the total heat input. Regeneration with swept surface units, while not impossible, is not used due to capital cost considerations.

V. ULTRA-HEAT TREATMENT

A. Introduction

The HTST process is defined as sterilization by heat for times ranging from a few seconds to 6 minutes. The term HTST is used to describe any short-time heat process for which the exact conditions of time and temperature is determined by the nature of the product [128]. The UHT process employs in-line sterilization heat treatments within the temperature range of 130–150°C, with holding times of 2 seconds. More generally, 2–8 seconds are employed. The term UHT is used in a more restricted sense, i.e., when referring to dairy products processed at temperatures in the range of 130–150°C.

B. UHT Processing Systems

UHT plants can be divided into two main categories of heat treatment: direct and indirect heating. With indirect systems, a physical conducting barrier (heat exchanger), usually made of stainless steel, separates the product and the heating medium (hot water or steam). The geometry of the heat exchanger may be of the tubular, plate, or scraped-surface type. Burton [129] has classified the systems used for milk and milk products on the basis of the type of heat-exchange equipment used. The same equipment is used for nondairy products, but other equipment has been developed for specific purposes, e.g., for processing viscous culinary products or products containing particulate. Moreover, systems are very often designed specifically for certain conditions and/or products. The classification given in Table 8 is based on that from Burton [129] for dairy products, but it has been modified and extended to give broader coverage.

1. Indirect Systems

APV Paracel

The Paracel pasteurizer is a tubular heat exchanger specifically designed to handle fruit juices containing cellular material. It can also form the basis of a HTST installation in which small particles are to be processed with minimal damage. It is modular in design, giving good flexibility in use, thus enabling a wide range of products to be processed. Recirculating water gives up to 90% heat regeneration when used as a pasteurizer for juice, thus reducing steam and cooling water requirements to low levels [129].

Cherry Burrell Spiratherm

The Spiratherm uses a tubular heat exchanger (spiratherm), which has a series of spiral tubes in cylindrical heating chambers, for heating above 70°C and for cooling the product. The heating medium can be either hot water or steam. If water is used, the Spiratherm cylinder is fitted with baffles to direct

Table 8 High-Temperature Processing Plants

Type	Manufacturer	Country of origin
Indirect heating		
Tubular	APV (Paracel)	Britain
	Cherry Burrell (Spiratherm)	United States
	Nuova Frau (Sterflux)	Italy
	Stork (Sterideal)	Holland
Plate	Ahlborn	West Germany
	Alfa-Laval (Steritherm)	Sweden
	APV (Ultramatic)	Britain
	Sordi	Italy
Direct heating		
Steam-into-product (injection)	Alfa-Laval (VTIS)	Sweden
	APV (Uperiser)	Britain
	Cherry Burrell	United States
	Rossi & Catelli	Italy
	Stork (Steritwin)	Holland
Product-into-steam (infusion)	Crepaco (Ultratherm)	United States
	Laguilharre	France
	Thermovac	France
	Nuova Frau Dasi	Italy/United States
	Pasilac (Palarisator)	Denmark
Other types		
Electric resistance heating	Elecster	Finland
Friction	Atad	France
Particulate systems	APV (Jupiter)	Britain
	Steriglen System	Australia

Source: Ref. 129.

water flow and to increase the heat-transfer rates. If the heating medium is steam, product flow velocities of 2.4–6.7 m/sec are used to give good heat transfer and to reduce deposit formation. Where lower heat-transfer rates can suffice (lower temperature heating sections), flow velocities of 2.4–3.5 m/sec are adequate. Spiratherms are suitable for a wide range of food products, namely, dairy creams and 3:1 concentrated milk, fruit juices, and beverages, baby formulas, ketchup, and sauces. Product tubes within the Spiratherm will withstand 1035 kPa pressure. The Spiratherm heat exchanger provides regeneration on an indirect basis up to 70% [105,130].

Nuova Frau Sterflux

This system uses concentric tube heat exchangers. The product is taken from a constant level reservoir by means of a centrifugal pump to the first heat exchanger and is heated to 75°C by the hot ongoing fluid. It is then degassed and pumped to the second-stage regenerative heater, and the temperature is raised to 121°C. At this temperature the product is homogenized and passed to the main heating section, which is serviced by saturated steam. The pressure generated by the homogenizer is sufficiently high to ensure that the sterile circuit is always at a higher pressure than the incoming liquid. The sterilizing temperature is 136–138°C. The sterile product is cooled, first by the incoming liquid, as part of the heat–recovery process, and then by cold water. Product leaves the plant at about 20°C [129].

The Sterideal System

This system is based on concentric tubes heat exchangers. A high degree of natural turbulence and high-pressure drop give good heat transfer. The high-pressure drop can be tolerated because no gaskets are used; however, manual cleaning is difficult. Pumping pressure is supplied by the homogenizer [131–134]. Product is pumped by a centrifugal pump from the main storage tank to a five-cylinder positive displacement pump, which is the pumping section of the homogenizer. The first tubular regenerator preheats the product to 65°C before it passes through the first homogenizing valve at 2070 kPa pressure. The second regenerator heats the product to 120°C. The product is then passed to the tubular heating section, where it is heated indirectly to 135–150°C. A steam inlet valve maintains the required temperature. Product-holding sections are optional. Sterilized product passes through the cooling section of the regenerator and then through the second homogenizing valve and the regenerator at 5170 kPa pressure. Product leaves the regenerator at about 30°C and then passes to mains and chilled water coolers before passing to the packaging machine. The plant automatically goes to its rinse, cleaning, and sterilizing cycle if proper sterilization temperatures are not maintained.

The Ahlborn Process

This plant uses plate-type exchangers and includes two deaeration stages. Liquid is pumped from the header tank to the first plate heat exchanger and is heated to 70°C by regenerative action against the hot sterile product. From here, two parallel homogenizers pump the product to the second heater, and the product temperature is raised to 100°C before it enters a holding section. The purpose of this is to stabilize the product against deposit formation. The third heat exchanger uses the hot sterile product as the heat-exchange medium to raise its temperature to 130°C [135,136]. The final heat exchanger uses steam as the heating medium to raise the product temperature to 140°C. After a short hold, the product is cooled in the sections of the heat exchanger where a high degree of regeneration is achieved. A heat exchanger is used to heat water for presterilization of the plant. It is also used to cool the sterilizing water circulating in the plant prior to commencing production.

Alfa-Laval Steritherm

This plant uses plate heat exchangers with snap-in gaskets specially developed for UHT processing. The product enters through a balance tank and is pumped to the regenerative section of the heat exchanger, in which the incoming milk is heated to 66°C by the hot sterilized product. The warm product passes to a nonaseptic homogenizer, in which it is subjected to a two-stage homogenization process. After homogenization, the product is heated to the sterilization temperature of 137°C by hot water circulating in the secondary circuits of the heat exchanger. The heated product passes through the holding tube and is retained for 4 seconds. After the holding tube it passes to a cooling section and is cooled to about 76°C by means of water in the water circuit. It is further cooled to 20°C in the regenerative section of heat exchanger by the incoming product [129,137,138].

Sterilization of the plant is achieved with hot water at 137°C circulating for 30 minutes. At the end of this period the outlet section is gradually cooled and the plant is ready for production. Sterile product forces water from the system until undiluted product is available at the filler or aseptic tank. The reverse operation, is possible if supply of product ceases or there is a prolonged stoppage. Cleaning is carried out by a programmed CIP method using caustic and acid solutions. Cleaning is automatically controlled as is the plant operation, the degree of sophistication depending on requirements.

APV Ultramatic

This system is based on the use of Paraflow plate heat exchanger. A number of options are available depending on the requirements, e.g., length of run, degree of regeneration, automation, and type of product [129,139,140]. The development of new gasket material that could withstand higher pressures allowed for processing temperature of 135–138°C. With a temperature of 135, a holding time of 2 seconds is required to sterilize the product. Sterilization of the plant is achieved by circulating water at 130°C through the product circuit for 30 minutes. The balance tank is by passed once the plant is purged of air and operating temperatures have been reached, thus conserving energy.

The raw product is fed from the balance tank by pump through the regenerative section of the heat exchanger and the temperature is raised to 85°C. The preheated product then passes through an homogenizer back into the heating section of the Paraflow, and it is heated to approximately 138°C, achieved by the use of hot water heated by steam in a service heater. Sterilizing temperature is maintained for 5 seconds in a holding tube, after which the product is cooled to 100°C by the water circulating in the secondary circuit. Final cooling to 20°C is achieved by regeneration against incoming raw product.

Sordi Steriplak S2

The Sordi sterilizer uses plate-type heat exchangers but has only one deaeration stage. Product from the header tank at 5°C is pumped into the first regeneration section and is heated to about 60°C by the outgoing product. At this temperature it is homogenized and fed to the second regenerative section and its temperature is raised to 92°C. The product then passes to the stabilizing section and is held at this temperature for 20–40 seconds in order to reduce deposit formation later [129,136].

It then passes to the third-regeneration section where the temperature is raised to about 120°C corresponding to 85% heat regeneration. The second centrifugal pump feeds the product into the final sterilization section of the plate heat exchanger, which uses superheated water as the heating medium. The hot water is generated by heating with steam at approximately 820 kPa pressure, and its temperature is automatically adjusted so that the temperature of the product coming from the sterilizer is 138°C. It is held at this temperature for 4 seconds before cooling to 113°C in the third-regeneration section. The product passes to the degasser in order to remove undesirable volatiles. It is extracted by a third centrifugal pump and directed in counterflow through the second- and first-regeneration sections. The final cooling to 20°C is brought about by indirect cooling with tap or refrigerated water. If the temperature of sterilization falls below a preset value, the product is diverted from the packaging machines and recycled to the balance tank, with simultaneous acoustic and visual warning.

2. Direct UHT Process

Vacutherm Instant Sterilizer (VTIS, Alfa-Laval)

This sterilizer is a steam-into-product heating system. It utilizes a flow desertion valve and an expansion cooling vessel in the diverted product line [130,143,144]. The general design of the plant is in four sections: preheating and cooling, homogenization, sterilization, and cleaning in place (CIP) treatment [129,141]. The product is pumped from the storage tank into the float hopper, from which it is pumped by the centrifugal pump to the plate heat exchanger, where its temperature is raised by means of hot water to 75°C. The water temperature is maintained by injecting steam into the water in the injection nozzle. A positive displacement pump delivers the product further into the steam injection nozzle, and the product is heated to the sterilization temperature of 140°C in a fraction of a second and is maintained for 3–4 seconds by passing through the holding pipe and return valve and the vacuum vessel.

In the vacuum vessel, the product is cooled very rapidly to 76°C. An aseptic centrifugal pump draws the product from the vacuum vessel and delivers it to the homogenizer. From the latter, the product flows to the aseptic section in the plate heat exchanger, where it is cooled to 20°C by the same water that was used earlier for heating the incoming product. Thus cooled and sterilized product can go directly to an aseptic packaging machine.

APV Uperisation

Uperiser direct heating systems are available in two models. One utilizes a tubular preheater, the other a plate-preheater. The plate-type system allows a larger processing capacity. The tubular preheater uses low-pressure steam, while the plate-type utilizes vacuum steam. Product cooling is by regeneration with the plate Uperiser, while the tubular system uses a separate aseptic cooler. It is a steam-into-product system, using an injection head of unique design which imparts spiral flow and turbulence to the product [129,142].

Raw product is fed from the balance tank by the feed pump through preheaters and its temperature raised. By action of pump and valve, the liquid flow rate and pressure is regulated for injection of steam in the Uperising unit to raise the temperature to 144°C and held for 5 seconds before the product is cooled by flashing in the expansion vessel where it is held under reduced pressure. This action removes undesirable volatile and cools the product to 80°C without altering its composition. The product that collects in the level reservoir is removed by the pump and fed to the homogenizer, which in turn pumps it through the two-stage plate cooler to the filler. Part of the vapor liberated in the expansion vessel is used to heat the product in the first preheater; the remainder is liquefied in the condenser and extracted by the condensate pump. Product is pumped aseptically from the expansion chamber to a homogenizer and then to the sterile cooler (tubular, plate) before aseptic filling.

Cherry Burell Aro-Vac

This system uses a steam-into-product injection process. The incoming product is heated by regeneration to 70°C and then passed, by a timing pump, to the steam heater, which raises the temperature to 145°C. After being subjected to a short hold of 3 seconds, the product is expanded in the Aro-Vac chamber so that the temperature falls to 70°C. The product is extracted from the chamber by an aseptic pump and homogenizer. Since the flash-cooling system is under negative pressure, special precautions are taken to steam-seal potential areas of leakage. Cooling is carried out by indirect means with water at 15°C and 1°C [133,143].

Rossi and Catelli

In this system, product is fed via a centrifugal pump to a tubular preheater, where the condensing vapor from the expansion vessel heats the incoming liquid to 80°C. The product is then heated by direct steam injection to raise the temperature to 142°C, and after a 4-second hold is flash-cooled to 83.5°C in an expansion vessel. From this vessel it is fed to an aseptic homogenizer and then cooled by water in shell and tube heat exchangers. For products containing volatile constituents, an aroma recovery unit is attached [129].

Stork Steritwin

This is a combination unit comprising tubular heaters and a second heater, which is either a direct steam injection unit or a triple tube indirect unit. The first-regeneration unit heats the incoming liquid to 65°C prior to homogenization. From the homogenizer the product is passed to the second regenerator and the temperature is raised to between 100 and 120°C. After final heating to 138°C, the product is flash-cooled in an expansion vessel. The sterile product is extracted from vessel by means of a Mono pump and regeneratively cooled by the incoming raw product. Final cooling is by cold water [129].

Crepaco/APV Ultratherm

This system is based on product-into-steam or infusion type. The infusion heater uses inverse distributor cups, which cause the product to flow in free-falling, umbrella-shaped films, which give good steam contact [144]. A centrifugal booster pump sends the product from a balance tank through plate heat exchanger, the first stage of which is a regeneration section. The second part is a heating section serviced by recirculating hot water. A product flow control pump is interposed between the two. From the preheater the product passes to the steam infusion vessel and the temperature is raised to the required sterilizing temperature. The hot product is removed from the infuser by a positive timing pump into a continuous, upward-sloping holding tube maintained under pressure by a back-pressure valve. It then discharges into the expansion chamber for cooling.

Laguilharre

Product is pumped by a high-pressure pump through two preheaters. In the first, product is preheated by vapor from the expansion cooling vessel, while the second preheater is heated with steam from the heating vessel. Temperature of the product after the preheaters is 75°C. The product is then sprayed into the heating chamber where it is heated to the steam temperature as it falls to the bottom of the chamber. The steam bleed to the second preheater is removed from the top of the heating chamber. The bleed also allows oxygen released from the product to escape from the chamber which could otherwise interfere with the action of the steam in the heating chamber [143,145]. The controller regulates the steam pressure and maintains the processing temperature at 140°C. The pressure forces the product to the expansion chamber, where it is flash-cooled to 75°C. The vapor is used in the first preheater, which acts as a vacuum producer. Water vapor and noncondensable gases from preheater are removed by a vacuum pump. Product is pumped aseptically from the expansion chamber and cooled in tubular coolers to 5°C with cold water.

Thermovac

Thermovac uses plate heat exchangers. Two plate heat exchangers preheat the product to 75°C. The second preheater is heated by low-pressure steam to give the correct product temperature for mixing with high-pressure steam in the heating vessel. The preheater is heated indirectly by the vapors from the expansion chamber, which is an ejection condensor that condenses the vapors from the expansion chamber and maintains the proper vacuum for flash-cooling. The condensing water is circulated through the first preheater to give regeneration and further condensing of extracted vapors. A vacuum pump removes noncondensable gases from the expansion chamber [145]. The product is mixed with steam in the heating vessel and then flash-cooled in the expansion chamber to 75°C. Two plate heaters are used for further cooling of the product before aseptic packaging.

The Da-Si Sterilizer

The main feature of this process is a novel heat-exchanger design. The product comes from preheaters at 75°C via a tube and passes through a bank of control valves that control flow. Gauges measure pressure as the product flows through individual feed lines to 100-mm screens, which shape it into continuous, laminar free-falling films only 0.5 mm thick. Superheated steam at 448 kPa enters the chamber through a series of inlet nozzles. The product is heated to 145°C by saturated steam. Moisture added is removed later in the Da-Si process [146]. Crusting is prevented in the sterilizing chamber by maintaining the Teflon-coated cone-like bottom at minimum product temperature. Input/output flow of product is maintained by a low-level liquid seal and a liquid level control interconnected with the main feed line. Sterile product leaves the exchanger at the bottom. The Da-Si sterilizer has no moving parts and need not be disassembled since steam injection makes the system self-cleaning. CIP spray balls are also integrated into the chamber for use after each run.

Pasilac Palarisator

The product is pumped from the balanced storage tank to the first tubular preheater. Vapors from the expansion cooling chamber pass through the ejector condensor and are used in the first preheater. A second tubular preheater using low-pressure steam raise the product temperature to 75°C. Product then passes through a three-way diversion valve into the heating vessel. Steam pressure in the heating vessel is maintained to give a product temperature of 145°C. The product then flows to the expansion chamber. Air released from the product in the heating chamber is extracted and send to the condensor and is removed by a vacuum pump. An aseptic pump passes the product through a small tubular water cooler to an aseptic homogenizer. The product is then pumped aseptically through water coolers to a sterile holding tank [143,145].

3. Other Types of System

Elecster

This unit is suitable for the sterilization of milk. An electric current is used to raise the temperature to 140°C by conductive heating. To reduce burning, the milk is forced through the pipe coils under pressure, giving good turbulence. Regenerative cooling from the incoming milk is used. Investment costs are about half that of more conventional UHT plants, and no refrigeration is required [129,147].

Atad

This equipment uses the friction generated by a disc rotating at 5000 rpm between two fixed discs spaced 0.3 mm apart. The product, after regenerative preheating, is passed in at the center and leaves at the circumference. Friction heats the liquid rapidly to a sterilizing temperature of 140°C. Cooling is carried out in a conventional heat exchanger [105]. As with other equipment of the very rapid heating type, enzyme survival may be a problem. The rotating discs give rise to homogenization of fat emulsions, and it is obvious that the clearances do not permit handling of products with suspended matter of any appreciable size. Friction sterilization has been used on various dairy products, meat extracts, dessert puddings, and fruit beverages.

Jupiter System

The central figure of the Jupiter process is the double cone aseptic processing vessel (DCAPV). The basis of the Jupiter system consists of the separate processing of solid and liquid phases to ensure that each phase receives the optimum heat treatment. In particular, overcooking of the liquid phase is avoided [148,149]. The DCAPV is a rotating jacketed vessel, supported on bearings and with inlet and outlet trunnions through which the product and service flows. Within the inlet trunnion, a stationary core contains an inlet pipe for the injection of steam, air, and liquids, a movable drain and vent system, and a thermometer dipole. The dipole continually monitors the steam and solid bed temperatures during processing. The draining system acts as a vent during sterilization and can be lowered into the product to remove any excess condensate and cooling liquid before sauce addition. The outlet trunnion houses the product pipe, which runs from the outlet valve on the vessel via a rotary joint to the filler reservoir [148].

Those parts of the plant that must remain sterile, i.e., the liquid plant and associated equipment, the filler and its reservoir, are isolated from the double cone by sterile barriers. The double cone is rendered unsterile during the loading operation but is resterilized along with all associated pipe work during the processing operation. A special high-lift Zephyr valve has been developed to enable adequate clearance for large particles. Large hollow trunnions are designed and built to service the double cone. All sequences are automated and controlled by microprocessors [148].

After loading, the solids are heated by direct steam in the vessel headspace by indirect jacket heating while the vessel is rotated between 2 and 20 rpm. Any gas evolved during the process is vented via the drain and vent system. During the heating cycle, preheated liquid (water or stock) may be injected into the vessel to ensure optimum cooking conditions and in some instances to protect fragile products from damage. Meanwhile, sterilized product from the previous batch feeds from the reservoir to the aseptic fillers. When the required heat treatment has been achieved, the product is cooled by replacing steam with cold water in the jacket and by admitting sterile air through the vessel and pipe work. The sterile air is maintained at a small positive pressure during the remainder of the process cycle. At the end of the cooling cycle, any excess liquid or condensate is removed from the stationary vessel via the drain system lowered into the product bed. The final blended product is transferred to the filler reservoir under aseptic conditions. The vessel and transfer pipe work are then rinsed, ready for the next batch.

Steriglen

The principle of the Steriglen process is to heat the product under pressure, allow the temperature to equilibrate, and then cool in a continuous unit, using vibrators and the partial fluidizing action of steam or cold nitrogen gas in the heating and cooling legs, respectively, to give product flow [129]. The solids are fed to the heating section, which consists of a vibrating perforated plate through which superheated steam is injected. The product moves through the heater on a bed of steam as a 25- to 45-mm-deep fluidized bed. The superheated steam, under pressure at a temperature of 155°C, exhausts its superheat and part of its latent heat to the product. The product reaches a predetermined mass average temperature and then drops through a header and decompression valve to the first of three equilibration zones, where the maximum temperature is reduced to the mass average temperature (155–133°C). The pressure drops in the header from 380 to 300 kPa, which is maintained in the first equilibration zone. The second zone is held at a pressure of 200 kPa, and the temperature of the product remains at the sterilizing temperature of 133°C. The product then passes through another valve to the first flash-off stage, where the pressure is 140 kPa and the product temperature falls to 125°C. Finally, the product drops through another header and valve to the cooler unit, where the temperature is reduced by a combination of evaporative cooling and the action of bold nitrogen passing through the perforated carrier plate.

C. UHT Processing of Milk and Milk Products

The International Dairy Federation has suggested that UHT milk should be defined as a milk that has been subjected to a continuous-flow heating process at a high temperature for a short time and has afterwards been aseptically packaged. The heat treatment is to be at least 135°C for one or more seconds [155]. The processed product is stored and distributed in a cooled state, at 5°C, during its short 10–20 days of usable life. Consequently, refrigeration equipment is essential for commercial utilization of the processed milk. Production of milk by an alternate route whereby the product becomes "sterilized" rather than pasteurized has a number of advantages. The UHT treatment allows sterilization, resulting in a marketable product [150].

D. UHT Processing of Juice Products

The biggest in road into the aseptic packaging market has been achieved in the area of juice and juice drinks. These systems are much simpler than those required for milk products, as the problems of very high temperature and fouling are not encountered. It is, however, necessary that juices be deaerated prior to aseptic processing in order to reduce the effect on the juice of the sterilizing temperature and also to extend the shelf life of the package by minimizing its oxygen content at the time of closure [127]. Deaeration must, however, be done in such a way that during the removal of oxygen as little of the aroma as possible is removed. It is preferable that the removal of air be carried out at room temperature or lower with a mild vacuum: the design of the deaerator must be able to deaerate properly under these conditions. The need to retain aroma precludes the use of steam injection/infusion plants.

While juice does not have the fouling tendencies of milk, it is considerably more heat sensitive: Hence, small-diameter tube or plate heat exchanger systems are very common. High thermal efficiency without undue processing time is achieved with the plate heat exchanger, as it is able to use direct regeneration. There is a growing interest in the packaging of juice products containing a high level of large particulate (nonhomogenized) pulp. For such products, the tubular system using indirect regeneration, although more expensive than the plate heat exchanger, is preferred.

E. Quality of UHT Processed Products

1. Milk and Milk Products

Safety and consistency are the two absolute components of quality. Often in an attempt to differentiate or create uniqueness for a product, a particular area of consistency is promoted as most desirable and identified as a quality factor. Adherence to recommended operating procedures for any process or equipment will ensure freedom from pathogenic organisms and toxic substances. Since this is the first criterion for any acceptable food-processing system, the design aspects have thoroughly incorporated the necessary features to ensure product safety. Two important considerations for consistency of product parameters in milk subjected to UHT treatment are (a) UHT milk having similar characteristics of the current 14-day shelf life coded pasteurized product, and (b) the changes that take place in UHT during its extended (90–180 day) shelf life [151,152]. Although the factors influencing the two criteria overlap, the process itself is the major determining factor for the first consideration, while the character of the raw milk is of greater importance for the second. The very high temperatures used in the UHT process can cause immediate chemical changes in some food components.

UHT treatment of milk will cause denaturation of at least some of the heat-sensitive whey proteins. Proteins are, to a certain extent, denatured during UHT processing [153]. This is particularly the case with whey proteins. The direct heating process causes less denaturation (60–70%), while the indirect process denatures about 75–80% of proteins. β-Lactoglobulin is more affected than is α-lactalbumin. However, this denaturation does not necessarily affect the nutritive value of the processed milk.

Whey protein denaturation was identified as one of the primary reasons for reduced clottability of UHT-treated milk by rennet [154]. The same phenomenon, together with the need for post-sterilization homogenization, may also be involved in the detrimental effects of UHT processing of whipping cream on its whippability [155,156]. These are rather major changes of product functionality, which must be overcome by changes in the process design and or specific product reformulation suitable for UHT processing. Chemical reactions between the major components of the system may also occur as a result of the high temperature used. According to Hostettler [157], formation of a stable complex between milk fat and milk protein is induced by homogenization and appears to be accentuated by subsequent UHT treatment of milk. This fat-protein complex may have been the cause of the insoluble fat clumps observed in UHT-processed, aseptically bottled milk after 102 days of storage [158]. A covalent complexing between casein and lactose upon UHT heating has been reported [159]. Some of these chemical reactions involving protein could be significant from the nutritional standpoint, especially in formulated foods with high concentrations of these reactions.

Other chemical changes caused by heat treatment may affect the market quality of the final products. Nonenzymatic browning (Maillard reaction) has been a common defect of in-bottle sterilized milk concentrates. Modern UHT methods, particularly direct heating, have minimized this problem in fluid milk, as shown by differences in the hydroxymethylfurfural contents of UHT and in-bottle sterilized milk [160]. Other color changes in UHT processing of nondairy foods may be associated with heat degradation of pigments, particularly in fruit and vegetable products. As with microorganisms and enzymes, environmental factors such as pH, water activity, presence or absence of oxygen, as well as varietal differences, will influence these orientation values substantially. A major quality attribute affected by high-heat processing is flavor. In UHT milk and other dairy products, interesting and important flavor-related chemical changes occur. It appears that the effect of the UHT processing per se on final product flavors is slight, especially when the freshly produced milk is stored for several days to decrease the intensity of the cooked flavor present immediately after the processing [161]. Loss of nutritive value may also be caused by a UHT process due to changes in amino acid availability in heated proteins. In milk, some whey protein denaturation occurs even with the most advanced direct heating systems.

In the case of milk, flavors developed during heating have been the primary factor responsible for poor acceptance of the aseptically packaged product [162]. Although the use of the DASI system [163] for heating yields a product with better flavor compared to those heated using other systems, there is still enough flavor difference between sterilized milk and pasteurized refrigerated milk to affect acceptability. A quality factor in heated milk that can be easily measured is the browning caused by the Maillard reaction. Oxygen in milk may have an effect on flavor. Both direct and indirect methods produce a similar quality product except that the direct method produces a processed milk with lower oxygen levels. Dissolved oxygen promotes the oxidation of ascorbic acid to dehydroascorbic acid. It was also found that the stability of folic acid is increased when oxygen is removed from milk. The rate of destruction of both ascorbic and folic acids seems to depend upon the dissolved oxygen concentration in milk.

Dissolved oxygen also has an effect on milk flavor. It has been suggested that the initial oxygen content does not really matter to the final flavor unless the product is used in the first 5 days after processing. On the other hand, the presence of oxygen, as already mentioned, may be required to react with the sulfhydryl (SH) groups to avoid the "cabbage" taste that is associated with UHT milk. A rapid decrease in SH content is observed within the first 10 days after processing. This suggests a correlation between oxygen content and decrease in SH groups.

The specific cooked flavor characteristic can be minimized by selection of the direct heating processes discussed earlier. The improved flavor is a result of the reduced contact with burned-on deposits associated with the indirect method. Flash-cooling of the direct process is also effective in reducing the oxygen content of the milk, this being the most important criterion for flavor improvement. Zadow [164], working with direct processing, found that oxygen level significantly affected the keeping qualities of UHT milk. A residual level of 3.5 ppm was felt to be optimum. Products with higher levels developed oxidized flavors upon storage at room temperature for 90 days; products with lower levels exhibited poor flavor performance with a dominant cooked flavor for the first 2 weeks.

2. Food Products

The effect of very high temperatures on flow properties of some formulated foods should be considered. Viscous fluids containing heat-coagulable proteins and/or easily gelatinizable starches could exhibit considerable thickening upon processing, especially with indirect systems. For example, a severe burn-on of puddings and custards containing carrageenan, xanthan gum, and locust bean gum occurred in an indirect scraped-surface heater [165]. This problem was eliminated by using a direct steam infusion unit. The more rapid heating not only eliminates the burn-on problem, but also allows the custards to be aseptically canned in a liquid state and left to slowly gel during a 2-week storage period. In general, due to the limited amount of heat applied as compared to canning, the flavor of many UHT products will resemble that of the fresh rather than canned counterparts [152].

3. Nutritional Quality

Many effects of any sterilization process on nutritional quality are likely to be found in the inactivation of heat-labile vitamins. In this regard, UHT processing is much less damaging than ordinary canning. It is generally accepted that no major losses of any of the heat-labile vitamins occur as a result of the UHT heating per se. Most of the D values are much too high for the vitamins to be substantially affected by the treatment, especially since the z values are also large. Vitamin losses and other undesirable nutritional changes in UHT products may occur during prolonged storage, often aggravated by insufficient precautions against exclusion of oxygen from the packages. No reports of clearly identifiable losses of nutritional quality have been found for other foods processed by the UHT technique. No significant changes have been found in the nutritional quality of lipids, minerals, and carbohydrates. Concerning vitamins, the stable water- and fat-soluble vitamins are little affected by either direct or indirect UHT processing [151]. The fat-soluble vitamins A,D,E, and β-carotene and the water-soluble vitamins of the B complex, thiamine, riboflavin, pantothenic acid, biotin, nicotinic acid, B_6 and B_{12}, are little affected.

From the point of view of spore destruction UHT treatment can be unnecessarily severe. Burton [166] investigated the destruction of spores in capillary tubes and UHT treatment by direct heating system and concluded that data obtained in the laboratory should not be extrapolated to UHT plants because under certain conditions spores contained in capillary tubes are found to be more heat resistant than spores subjected to plant sterilization. Reduced time or temperature would have a significant impact on the flavor character, denaturation, and deposition of the milk proteins. The high-heat treatment of the UHT process alters organoleptic parameters other than flavor.

In another research study [151] it was found that milk subjected to UHT processing was too creamy, rich, or thick compared to regular HTST processed 2% fat milk. They reported that it was more like homogenized milk, and they had become used to the less rich, more watery consistency of 2% fat milk. It is interesting to note that when 2% fat milk was introduced as a new product into the marketplace, it was necessary to add up to 3% milk solids to satisfy the consumer's demand for the equivalent smooth rich body of the standard 3.5% fat milk in the marketplace. Perhaps the acceptance of two changes to a consistency standard in one generation is too much to expect. Sensory evaluation of UHT milk based on either difference or preference testing indicates that the product is inferior to regularly pasteurized milk [167]. Research characterizes this difference as a combination of cooked flavor and heavier mouthfeel. Any possible reduction in the severity of the heat treatment necessary to ensure an absence of spores in the finished product would most likely have favorable benefits.

F. Packaging of UHT-Processed Foods

Packaging is very critical for maintaining the long shelf life of the UHT products as well as for marketing the product. Packaging and handling of the sterilized product is roughly 30% more than the pasteurized milk.. In general, there are four basic categories of containers used for packaging UHT sterile product [116]. These are:

1. Glass bottles or jars sealed with plastic-lined "crown" or twist-off closures
2. Heat-sealed paperboard laminates in carton form
3. Heat-sealed sachets or pouches of pliable plastic materials; heat-sealed semirigid plastic tubes or bottles
4. Suitable sealed tinned steel drums or hermetically sealed tinned steel cans; hermetically sealed aluminum cans

1. Glass Containers

The principles of using dry heat to sterilize the container are followed by a prolonged cooling period to prevent thermal shock breakage on filling. In this system, infrared heat from electric elements is used to sterilize the bottles, which are rotated as they pass the heaters in order to give a uniform temperature rise. Alternatively, dry heat or chemical sterilization can be used. Saturated steam can act as a clean and efficient sterilizing agent when used at 150°C for a few seconds. Bottles are filled in the same chamber after sterilization [115,151]. Glass bottles have been used for pasteurized milk for many years, but UHT milk packaging needs more rigorous controls, and perhaps with further developments bottles may well be used for such a process.

The advantages of this type of container are (a) glass is inert and impermeable to gases, vapors, and liquids, (b) glass containers have high resistance to compression or internal pressure, (c) bottles can be filled at a very fast rate (generally more than 4000 per hour), (d) glass can be recycled and does not present many environmental problems, (e) glass can be relatively inexpensive, and (f) bottle returns lower the expenses. The disadvantages are (a) glass containers are usually very heavy and add to handling expenses, (b) the containers are very fragile, (c) for UHT packaging, colored or coated glass is necessary to keep out light, (d) problems of cleaning bottles contaminated with hydrocarbons persist, and (e) glass bottles may not be considered "fashionable."

2. Heat-Sealed Paperboard Laminates

There are at least five different types of machines in commercial use for the packaging of UHT products in paper cartons [115,116,168].

Tetrahedral Type (TetraPak)

Two types of laminates are used [169]. They are, from outside to inside, (a) waxed paper/polyethylene lined and (b) are waxed paper/polyethylene/aluminum foil/polyethylene. The first type is gas and vapor permeable and intended for a proper shelf life of 1–2 weeks. The second type is gas and vapor impermeable and intended for a proper shelf life of at least 3 months. As the laminated paper leaves the roll, a reinforcing polyethylene strip is thermally welded along the edge. The strip then passes through a bath of hydrogen peroxide solution containing a small amount of wetting agent. The strip is formed into a vertical tube, and the reinforced longitudinal seam is thermally welded. Product enters the tube through a filling pipe and the tube is sealed transversally by two jaws at right angles to each other. The packages are then shaped and formed and packed into bigger cartons or crates.

Advantages of these cartons include (a) low levels of spoilage, (b) very little or no air is present in the filled cartons, and (c) they need less laminate material per volume. Disadvantages include (a) the shape presents problems of handling and therefore public resistance, (b) they cannot be stacked easily, (c) the largest size of filling is only 1 liter, and (d) they are prone to mishandling by school children.

The TetraPak is being gradually replaced by the more acceptable TetraBrik cartons.

TetraBrik (Rectangular Types)

Two TetraBrik designs are available, each of which produces cartons of fixed dimensions with the height of the carton being determined by the desired capacity. The two types of carton material available [170] are (from the outside to inside):

1. Duplex consisting of polyethylene/Kraft composite paper (external face white, internal face brown)/polyethylene-lined
2. Aluminum duplex laminate consisting of polyethylene/Kraft composite paper (external face white, internal face brown)/polyethylene/aluminum foil/polyethylene lined

Type 1 is intended for products requiring a shelf life of 4–5 weeks. Type 2 is intended for products requiring a shelf life of 4–5 months. The basic forming and filling scheme is similar to that for tetrahedral cartons.

Advantages of these containers include (a) the carton shape provides easy stacking and handling, (b) absence of air helps maintain the quality of milk for a long time, and (c) air "knives" are more efficient at removing traces of hydrogen peroxide. The disadvantages include (a) scissors are needed to cut open the carton, (b) sealing "through" product may add to undesirable flavor, (c) laminate material with aluminum foil is not recyclable and may cause environmental problems, and (d) since there is no air in the carton, care has to be taken not to spill milk while pouring it out.

Zupak Cartons

Two types of Zupak machines are available. One is the type that handles cartons with capacities of 1 or 1.5 liters. The other type handles 0.5- or 0.25-liter cartons. A machine can be changed to process either of the two sizes in less than 10 minutes. The laminate used is very similar to the one used for TetraBrik cartons, and the filling method is also very similar. However, the sterilization of the laminate is achieved by cold hydrogen peroxide followed by evaporation with hot sterile air [115,116].

Selfpack Cartons

In this type of carton 1.0, 0.5, 0.25, and 0.2 liter can be packaged from a separate machine. The laminate used is similar to the one used for TetraBrik, and the filling method is also very similar. The laminate is sterilized with cold hydrogen peroxide, followed by evaporation in a current of sterilized air at 180°C [151].

PurePak Cartons

In these containers a five-ply laminate carton material is used for the preformed blanks consisting of polyethylene/paper/polyethylene-2/aluminum foil/polyethylene. The cartons used are delivered as preformed blanks, which in sealed paperboard boxes are sterilized by ethylene oxide gas. These blanks are then stacked in the magazine feed of the machine. The blanks are first automatically shaped to form a "sleeve," after which the base is heat-sealed. After filling, the carton top is folded to a gabled form and heat-sealed, away from the milk [171].

Advantages of these containers include (a) heat sealing of carton is not "though" but "on top" of product, (b) the gabled top provides better pourability and can be tucked back in for storing unused product in the refrigerator, (c) presence of air in the carton avoids spillage while pouring. The disadvantages include (a) the filling rate of the machine is quite slow, (b) separate gassing of preformed cartons adds to total expense, (c) ecological problems may result due to aluminum foil, (d) the gabled top, if not pressed, presents stacking problems, and (e) air in container may affect quality of long-life products.

The Combibloc Filler System

The Combibloc system is similar to the others described above. It represents an intermediate between TetraBrik and PurePak systems. Brick-shaped cartons are used, but these are preformed as the PurePak system. The outstanding features of the system are the speed of packaing—5100 packages per hour—and the ability to alternate between 1- and 0.5-liter cartons in a matter of minutes [151].

3. Plastic Containers

While for carton packaging gas and light imperbeaility can be achieved quite easily by adding a layer of aluminum foil to the laminate, most plastics have a problem of permeability to both gases and light. This problem can, however, be overcome by pigmenting the plastic film (to avoid light) and using composite layers (to overcome gas and vapor permeability). The composite layers provide protection comparable to an aluminum foil sandwich. Such films are available and are being used on the Pre-Pak machines [169].

These composites are obtained by coextrusion of different plastics such as (a) coextrusion of black and white polyethylene with polyvinylidene chloride (PVDC), (b) coextrusion of black and white polyethylene with polypropylene, and (c) coextrusion of black and white polyethylene with PVDC and aluminum foil.

There are basically three groups of plastic materials which are preferably used for packaing mass-produced dairy products:

1. Polystyrene, which can be "standard" type or "impact-resistance" type depending upon the demands on the package. Polystyrene is the most widely used material for making cups for yogurt, cream, cottage cheese, etc. and has good thermoforming properties.
2. Polyethylene of high and low density is used for bottles and pouches.
3. Polyvinyl chloride (PVC) is also used for cups for butter, cheese, etc.

The plastic containers that have been developed for UHT milk and dairy products are of three main types: (a) films or laminates in sachet or pouch forms, (b) bottle or tube-type semirigid blow-minded containers, (c) semirigid containers of the form/fill or preformed (thermoformed) type.

Sachets or Pouches

The sachet machine is placed in a sterile environment which is under positive pressure. The laminate from the roll is first immersed for 15 seconds in a bath of ethanol. Excess ethanol is squeezed off between rollers and the laminate is then subjected to UV radiation. After another exposure to UV, the sachets are filled, sealed, and guillotined. The filled containers are passed through a bath of 1% formalin solution, which serves as an exit airlock from the enclosed area. Recovery and reuse of the ethanol used as initial sterilant is an attractive feature of this system [151].

Advantages of this system are that (a) sachets are very light and easy to handle, (b) the filling rates can be quite high, and (c) ethanol recycling lowers the expense of operation. The disadvantages include (a) scissors are needed to cut the sachets, and (b) potential ecological problems exist due to the presence of aluminum.

Bottles or Tubes

Granulated polyethylene is fed to the machine under pressure and a hot melt emerges in a tube form. This is then blow-molded into a bottle form, filled in situ, heat-sealed, and discharged. The bottle pack is fully aseptic. Single or multicavity molds (one to six) can be fitted into the machine. The speed of output is reported to be 1000 liters per cavity per hour. Sizes of bottles range from 280 ml to 14.5 liters [116].

Form/Fill or Thermoformed Packages

These containers are made from a mixture of plastics (e.g., polystyrene/PC/polystyrene), which is heated to over 140°C and then formed and pressed into rectangular or round shapes by means of plug-type formers. Multiplug formers and multihead fillers are generally used, after which the containers are sealed with plastic-coated aluminum foil. Hydrogen peroxide is used as the sterilant, and output rates are in the region of 500 0.5-liter containers per hour [172].

The advantages of using plastics materials for UHT product are (a) the packs are light and hence easily transported, (b) output rates can be very high, (c) raw material is usually granulated, resulting in ease of operation, (d) several different varieties of plastic are available, (e) many plastics are re-usable, and (f) composite plastics have a very low gas transmission rate. The disadvantages are (a) increasing the hydrocarbon cost increases the plastics cost, (b) some plastics have been known to react with product, and (c) the problem of disposal of the plastic bottles.

4. Metal Cans

Metal cans were the first containers to be aseptically filled. The cans are sterilized in a tunnel at atmospheric pressure by steam at about 200°C, superheated with gas flames. After this, the cans are precooled by steam condensate before being filled and finally seamed under aseptic conditions. Size ranges from 130 ml to 22.7 liters are available. Output rates are about 5000 liters/hr. In spite of their advantages, metal cans have been phasing out slowly from a number of food industries. The reasons are chiefly environmental and safety related [151].

G. Energy Aspects of UHT-Processing Systems

In a typical direct-heating system there would be no change in the circuit if the injection head was replaced with an infusion lead. The circuit contains three milk pumps and the high-pressure pump of the homogenizer. Often a fourth pump is required to deal with vapors from the expansion chamber. This might represent 110 kW for the pumps and homogenizer at a capacity of 8000 liters/hr. Consequently, the direct system requires more power and pumps in order to operate, resulting in higher capital costs [151]. An indirect plate-type UHT system requires a single centrifugal pump and a homogenizer with an installed capacity of 90 kW. A tubular indirect system requires only a single high-pressure pump to provide both homogenizing pressure and pressure drop through the heat exchanger.

Economic utilization of heat in the direct system is poor. Because of expansion cooling of the milk at 85°C, regenerative heating is not possible beyond about 80°C. If the incoming milk temperature is 5°C with a final processing temperature of 145°C, the maximum amount of regeneration possible is 50%. Regeneration can be increased by raising the temperature at which direct heating is applied to more than 85°C. However, the benefits of rapid heating in the absence of heat exchangers are progressively lost. With indirect systems regeneration levels of greater than 80% can be obtained. Another loss in thermal efficiency for the direct system lies in the high quality of steam required to mix with the milk. A separate boiler would be needed for the sterilizer alone; such a small unit might incur energy costs higher than the costs of using steam from a main boiler.

One disadvantage of indirect systems is that deposits accumulate on their heat exchangers, which interfere with heat transfer. This means the heating medium temperature must rise during a process run in order to maintain sterility. Deposits also restrict milk flow through the heat exchanger, which requires higher pump pressures for flow maintenance. With plate exchangers, there is a limit to the permissible pressure increase, which is set by the bursting strength of the gaskets. Plate exchangers will operate between 1 and 6 hours depending on operating conditions and type of product. The need for regular cleaning reduces the efficiency of indirect plants, especially plate-type plants. Deposit formation occurs in direct plants, but to a much smaller extent. Deposit usually forms at the restriction valve between the expansion chamber and holding tube. However, the deposit build-up is so slow that runs of 20 hours are quite common. The specific energy consumption per unit of milk produced is lower for the direct than for the indirect process.

UHT processing of food products requires a greater quantity of thermal and electrical energy in the processing plant than conventional pasteurization or HTST processing [173–175]. Also, UHT-processed food products do not require any refrigeration, saving an estimated 900 kJ/kg of product [176]. Part of the additional energy consumed in UHT processing can be potentially recovered and reused [177]. In one of the case studies involving a steam-infusion UHT system at the University of Maryland dairy plant, the UHT system consumed between 475 and 570 kJ/kg of milk. The sterilizer utilized about 73% of the total energy required by the UHT system. The thermal energy was 430–524 kJ/kg and the electrical energy was 46 kJ/kg of milk. Thus, the sterilizer was the process consuming the greatest amount of energy with a thermal efficiency of 98% [176]. In another study involving a free-falling film steam-infusion UHT sterilization process with regeneration, the energy [176] requirement ranged from 573 to 667 kJ/kg. The energy consumption of the steam-infusion UHT sterilization system is linear with the sterilization temperature.

VI. COOKING

A. Introduction

Some foods are consumed raw and others are cooked. There are several reasons for this. Cooking improves the flavor of the food, for instance; the flavor of uncooked flour or sour apples is not very pleasant, but when the flavor has been converted into bread and the apples stewed with sugar, their flavor is much improved. On the other hand, fresh strawberries are not cooked as this would spoil the delicious flavor of the raw fruit [178]. Cooking may also improve the attractiveness of food. Eating a raw chop is not much relished, but after cooking, it has an appetizing appearance and a good smell. Even more important, cooking may make a food more digestible. It would be difficult to eat the flesh of a raw chop (or uncooked flour) even if you wanted to, but after cooking, it is much more tender and so easier to chew and digest. Finally, cooking may improve the keeping quality of a food and make it safe. For example, milk may be boiled to delay the souring process and kill bacteria. The preservation of food by heat treatment is quite distinct from cooking.

Cooking is only one part of food preparation. Apart from the actual cooking process, ingredients may have to be blended together and they may need special preparation by soaking, sieving, or chopping. Seasoning, spices, herbs, and sauces may be used to improve the flavor, color and garnishes may be added to improve attractiveness, and texture may be improved by grinding, mashing or mixing.

B. Cooking Methods

Cooked food is food that has been changed in various ways by heat treatment. The heat may be applied in a number of ways; it may be dry or moist, it may be applied by means of fat or by infrared radiation [178] (Table 9).

1. Dry-Heat Methods

When food is cooked in an oven it is said to be baked. Baking is a rather slow method of cooking, but it has the advantage that large quantities of food can be cooked evenly. Sometimes the food to be cooked is put into the oven in a container containing a little fat; food cooked in this way is said to be roasted. Meat and potatoes are the foods most often cooked by roasting. Cooking temperatures used in an oven vary from below 100°C (very slow) to about 260°C (very hot). Broiling is another method of applying dry heat. The food to be broiled is placed beneath a red-hot source of heat, usually a glowing metal grid. Radiant heat is directed onto the surface of the food which is rapidly heated. Broiling heat is applied to the top surface of the food, and the food should be turned from time to time. Infrared grilling makes use of heat rays which have longer wavelengths than visible light. Some of the radiation used in normal grilling is of this kind, but in infrared cookery the proportion of infrared radiation is much increased, and this reduces cooking time to such an extent that a steak, for example, may be cooked in a minute.

2. Moist-Heat Methods

Although cooking with water involves using low temperatures, it is a relatively quick method of cooking because water has a great capacity for holding heat and for transferring this heat rapidly to food by means of convection. In moist heat cooking food is heated by either water or steam. Boiling uses boiling water; simmering uses water near, but below the boiling point and is similar to both stewing (for meat and juice) and poaching (for fish). Boil-in-the-bag cooking uses boiling water indirectly, but because the food is sealed in the bag this method prevents loss of flavor and soluble nutrients into the cooking water. In steaming, steam is used directly to heat the food or indirectly to heat the container. Although steaming is slower than boiling, cooking may be sped up by the use of a pressure cooker, in which steam is produced at higher than normal pressure. The increase in pressure raises the temperature at which water boils, so the cooking temperature is increased and the cooking time is reduced.

Table 9 Summary of Cooking Methods

Method of heating	Method of cooking	Description	Foods
Dry heat	Baking	Cooking carried out in an oven	Potatoes, fish, cakes
	Roasting	Baking with the addition of fat	Large joints of tender meat, potatoes
	Grilling	Using direct radiant heat	Small cuts of tender meat, e.g., chops and steak; fish
Moist heat	Boiling	Using boiling water	Eggs, fish, large joints of tough meat, vegetables,
	Stewing and poaching	Using hot water below its boiling point	Meat in stews and hot pots, fruit, fish, eggs
	Steaming	Using steam from boiling water	Fish, vegetables, sweet puddings
	Pressure cooking	Using water boiling above its normal boiling point	Meats, vegetables (used in general for foods which take a long time to cook by normal boiling or steaming)
Fat	Frying	Using hot fat	Meat or sausages, potatoes, eggs, fish
Infrared	Equal to rapid grilling	Using infrared radiation	Small, tender cuts of meat
Microwave	Equal to rapid grilling	Using microwaves	All foods

Source: Ref. 178.

In essence a pressure cooker is a pot with a well-fitting lid arranged so that steam can be safely generated under pressure. The pot and lid lock together by means of a groove to make the cooker pressure-tight. The food to be cooked and the required amount of water are put into the pot, which is then closed. When the closed pot is heated, air is driven out through the air vent until the cooker is full of steam. In pressure cookers with a pressure indicator, the vent then closes and pressure builds up to the value required. Slow heating only is then needed to maintain this pressure, which is shown by the pressure indicator. Should the pressure rise too much, steam automatically escapes through the air vent. The fusible plug is a second safety device; this will melt if the cooker overheats or boils dry.

3. Frying

In frying, food is cooked in hot fat. Fat has a much higher boiling point than water and can be heated almost to its boiling point without smoking. Frying is a quick method of cooking because of the high temperature used. In shallow frying, a shallow pan is used and enough fat is added to cover the bottom of the pan. Although such a method is quick, heating of the food is uneven and it should be turned from time to time. Lard, drippings, and vegetable oils (e.g., olive oil, corn oil, and cottonseed oil, often blended together) are best for shallow frying. In deep frying, a deep pan and plenty of fat are used, so that when the food is added it is completely covered by the fat, which should be very hot. Temperatures of between 150 and 200°C are usually used, and the temperature of the fat may be checked with a thermometer. Such a method is quick and the food is cooked evenly on all sides. Refined vegetable oils or cooking fats, which are made by hardening a blend of vegetable, animal, and marine oils, are best for deep frying.

4. Microwave Cooking

In ordinary cooking heat is applied to the outside of food and it gradually penetrates to the inside. In microwave cooking, the heat is generated within the food. In a microwave oven microwaves penetrate the food and are converted into heat within the food. Thus the whole food heats up very quickly. Microwaves can only penetrate food to a depth of 3–5 cm; thus, small pieces of food are cooked very quickly. Larger pieces of food are cooked more slowly, however, because where the microwaves cannot penetrate the food is heated by conduction.

The great advantage of microwave cooking is its speed. For example, a fish fillet is cooked in only 30 seconds, a chop in one minute, a chicken in 2 minutes, and a baked potato in 4 minutes. For this reason microwave ovens can be especially useful in places, such as canteens, snack bars, and hospitals, where food often has to be kept hot for long periods before it is eaten. Microwave ovens also have some disadvantages. For instance, cooking times must be carefully controlled. In addition, food does not turn brown or develop crispness in a microwave. Crispy bacon or conventional-looking brown crusty chops need to be finished off under the broiler.

5. Slow Cooking

The use of microwave ovens and pressure cookers is intended to speed up cooking, but recently cookers designed to slow down the cooking process have been introduced. Slow cookers are electrically heated and made of a material with good insulating properties such as earthenware so that heat transfer to the food is slow and a steady temperature is maintained during cooking. Slow cookers work on low power (1 kW) so that the cooking temperature remains below 100°C. The result is that food is cooked at a low, even temperature over a long period—usually 4–6 hours. Slow cookers are much more economical to operate than conventional ovens. Slow cooking is ideal for cooking casseroles, stews, and cheaper, tougher cuts of meat so that they becomes tender, facilitating reduction in weight loss by evaporation. It also prevents loss of juices because the slow cooker is sealed and no moisture escapes.

C. Application of Heat

The three common systems of heat application are [179]:

1. Indirect heating by combustion gases conducted through flues, radiators, or past surfaces of the baking chamber such as the underside and backside of the chamber
2. Semi-direct heating in which part of the combustion gases are forced into baking chamber to create pronounced convection currents
3. Direct heating using electricity or gas with ribbon-type burners

A fourth, indirect method that utilizes high-pressure steam tubes is used to a very limited extent. The indirect firing system utilizes isolated combustion chambers from which the hot combustion gases are circulated by either suction or pressure through a bank of radiator tubes and are either returned to the combustion chamber or vented to an exhaust flue. The radiator tubes, which transverse the baking chamber, give up their heat by radiation and convection to the baking chamber. Depending upon the size of the oven, several combustion chambers or unit heaters may be used, each equipped with a burner, circulating fans, ducts, radiator tubes, and temperature controllers. The radiator tubes can be so arranged that radiated heat is applied to both the top and bottom of the pans. Because of the inherent limitation of the radiating heating surface that can be designed into an oven, the heating efficiency of indirect fired ovens is generally less than that of direct-fired ovens.

The semi-direct firing system resembles the indirect system in using a separate combustion chamber and relying on radiator tubes to carry the hot combustion gases to the baking chamber. Here, however, the radiators are provided with either thin slots or small holes so that the hot gases are forced into the oven chamber, thereby creating extensive convection chambers, which are further augmented by a forced draft. In this system, the baking effect is produced by convectional and radiational heat, and in this way advantage is taken of the special benefits offered by both methods of heat transfer.

In direct gas-fired ovens, ribbon burners are placed directly in the baking chamber, crosswise to the travel of the oven trays or conveyors. Originally, an air-gas mixture was supplied to the burners by an aspirator in which the combustion air was drawn into the mixing unit by the vacuum created when the gas was forced through the gas orifice under pressure. Each burner has its own gas control valve which must be manipulated individually for heat control. The aspirator system has a number of disadvantages, principal among which is its tendency to flame-out under certain operating conditions, thereby creating hazardous oven conditions.

A safer and more efficient firing method has been devised in which gas and air are mixed prior to being fed through a common header to the individual ribbon burners. In this system, filtered air is supplied to an air-gas mixing unit. One such unit is provided for each oven zone consisting of as many as 15 or more ribbon burners. In this way, all burners in one zone can be throttled uniformly by means of a single zone control valve. The unit mixer in which the air and gas are mixed is under pressure for each individual ribbon burner. With this system, the gas and air headers for each top and bottom heat zone are provided, respectively, with a zero gas regulator and an air valve controlled by an automatic temperature controller. The gas regulator and air valve operate together to uniformly supply the correct amounts of gas and air to the unit mixer in the zone. For balancing the heat distribution across the width of the oven, flame distributor–type burners are used, which provide a three-zone control of the burner flame across the oven.

The direct gas-fired system is the simplest and most efficient method of oven heating. By proper design and distribution of ribbon burners, uniform gas consumption can be obtained throughout the oven. The velocity of the combustion gases coming out of the burner ports, together with the natural convection currents created by the generated heat, supply sufficient turbulence to produce fairly uniform temperatures throughout the chamber as long as the size, number, and location of the burners are adequate and the control equipment is correctly engineered and installed. In some direct-fired ovens forced convection is applied to obtain a more efficient utilization of the heat energy by recirculating the chamber atmosphere past the ribbon burners rather than venting it to outside.

In the steam-tube system, oven heating is performed by a series of high-strength, hermetically sealed tubes which are partially filled with water or a heat-stable liquid possessing a high boiling point. The tubes are installed in the oven so that a short portion of each tube protrudes into the combustion chamber, where direct heat is applied to cause the water to vaporize into high-pressure steam. The balance of the tube extends into the baking chamber. The tubes are slightly inclined toward the fire box so that steam condensate returns to the heated ends to be revaporized. Because saturated steam at atmospheric pressure has a temperature of 100°C, and because baking temperatures are normally within the range of 182–232°C, it is evident that considerable pressure must be developed in the steam tubes to attain baking temperatures.

Heat transfer in this system is principally by radiation and is, hence, governed by the physical law that states that the heat rays radiated from any one source vary inversely to the square of their distance. Therefore, the steam tubes must attain much higher temperatures than the actual baking temperature. The magnitude of the pressure developed within the tubes is indicated by the fact that a temperature of 335°C requires a steam pressure of 13800 kPa. The use of steam tubes as a method of oven heating has been largely discontinued as it is relatively inefficient and lacks the necessary flexibility of heat control.

D. Cooking Equipment

The objectives of the cooking process are to reduce the moisture content of the product mixture, to melt, to solubilize, to caramelize, if necessary, and to invert. There are many types of cooking equipment, which may be classified in several ways for the convenience of discussion. According to the working mode, they are batch cookers, semicontinuous cookers, and continuous cookers. According to the method of heating, they come as direct-fired cookers or steam-jacketed cookers. According to the working pressure, they are atmospheric cookers, pressure cookers (steam-injection cookers), and vacuum cookers.

1. Batch Cookers

Steam Kettles

Food preparation of liquids and pumpable foods (including particles) is carried out in kettles or in heat exchangers. Kettles are used in batch operations. Heat is supplied to the foods through the kettle wall from condensing steam in the steam jacket covering most of the kettle wall. The jacket is equipped with an air-venting valve and a condensate-removal system with a steam trap. The food is transported to and from the wall into the bulk of the food by a flow created by the density differences between hot and cold product or by mechanical agitation [180]. The free convection flow develops easily when the heat supply is sufficient for a rapid steam pressure build-up in the jacket. In the flow pattern, only a thin layer of the fluid close to the wall will be involved in the heat transfer. The heated fluid will rise to the top of the volume and spread over the top surface. The major part of the volume will slowly fall towards the bottom, where it will contact the heat-transfer surface again.

Models of steam kettles differ in relation to (a) depth, which may be deep or shallow, (b) steam-jacketing, which may be full or 2/3, (c) mounting, which may be on legs, a pedestal, or wall-mounted, (d) type, such as tilting or stationary, and (e) source of steam, which may be direct or self-generated. The materials commonly used are aluminum and stainless steel; the finish may be dull or polished. All steam equipment should have safety valves and pressure gauges. Those having self-generated steam should have an automatic low-water cutout and a thermostatically controlled cutout heat. Tilting kettles should have a secure device for stopping them at any desired degree of tilt.

Steam Cookers

Steam cookers vary in (a) number of compartments (single or in stacks of two or three) and size, (b) source of steam, (c) type of base, and (d) design. Steamers should have heavy-duty gaskets. Cooking containers must be suitable for the material to be cooked, should minimize handling, and should permit suitable load size for workers to lift. The use of serving pans may often minimize transfer.

Vacuum Cookers

This system is used for the production of hard candy, jelly, gum candies, and low boiled sweets. The main components of the vacuum cooker consist of a cooking chamber, a kettle under it with a vacuum facility, a stirrer inside the cooking chamber with a variable-speed drive, a vacuum pump with spray condensor and control system. The finished product is discharged into the kettle and hence is called drawoff kettle [181]. The ingredients are fed into the cooking chamber by dosing or manually. The cooking temperature and time are controlled automatically. The product after cooling is discharged into the draw-off kettle through suction created by vacuum pump. During this process, the cooked mass will be evaporated under vacuum pump. The draw off kettle is lowered after reversing the vacuum to unload the product.

Rotary Cereal Cooker

This is a single rotating pressure cooker, controlled either automatically or manually and mounted on a horizontal axis. The cooker consists of a stainless steel barrier with slide valve or door for charging and discharging; steam is injected at 200 kPa working pressure through a flexible hose and rotary union [182]. The pressure cooker consists of an all-welded stainless steel double-conical barrel with annular rings at each end to which are bolted end plates. The barrel has a centrally positioned opening fitted with door for charging and discharging. Steam injection is through a 50-mm-diameter stainless steel braided flexible hose and rotary union mounted at the end of each stub shaft. Steam passes through the center of each shaft into a stainless steel steam chest, which is integral with the shaft and end plate. Steam injection and exhaust is controlled through a system of stainless steel ball valves. Pressure in the cooker is indicated at each end by a pressure gauge.

2. Continuous Cookers

Pressure Cookers/Blanchers

The Balfour continuous pressure cooker: The Balfour continuous pressure cooker can handle particulate foodstuffs ranging from cereals and diced vegetables to regular whole beet roots and potatoes, including diced meat. It is designed for the continuous feeding of process materials into and removal from the cooking vessel, which is utilizing steam under pressure. The retention time of the process material can be strictly controlled and varied according to requirements [182]. The main cooking section consists of a plain helical screw conveyor operating within a circular body—the body being designed to operate at the maximum steam pressure. Pressure lock systems are fitted at both the inlet and outlet ends. The size of the cooker is directly proportional to the required cooking time and inversely proportional to the total process time [183].

The Turboflo blancher/cooker: The Turboflo hydrostatic blancher/cooker uses a unique steam injection and energy-circulation system for blanching and cooking potatoes, vegetables, fruits, meats, and poultry. It combines the benefits of energy efficiency and improved product quality and saves processing time and also requires less floor space. The cooker achieves process time savings of up to 30%. The fully enclosed and insulated chamber prevents steam from escaping and saves energy cost. Products blanched in this cooker retain more nutrients and have both better color and better taste and thus customer appeal. The energy-circulation method features a fan-driven steam path, which penetrates the product mass evenly to assure thorough blanching/cooking of product. The modular section design provides maximum flexibility for food processors [184].

Steam Cookers

A continuous steam cooker has been developed at CFTRI. The equipment is a continuous conveyor with a facility for open steaming into the chamber. A water inlet is provided through a flow meter to add a measured quantity of water during processing. The chamber is steam-jacketed for additional heating. A variable-speed drive is provided to vary the residence time of cooking. A rotary valve is fixed at the inlet end to control the material feed rate. The conveyor speed and the rotary valve speed are matched with sprocket and chain drive [185]. A stationary water-draining device having an SS trough and SS and nylon sieving screens is installed at the discharge of the machine, which greatly helps in the quick separation of water from the cooked product.

The product to be cooked is carefully washed and water is fed into the cooker at a measured rate (flow rate depending upon the output and variety). Steam is let into the system by opening the steam valves at controlled pressures. Steam supply is maintained at a constant rate for both the spreader and the jacket. The product is continuously fed from the hopper through the rotary valve at the required rate. At the end of the set residence time uniformly cooked product is discharged through the outlet chute. The product is collected from the mesh and is ready for consumption.

Microfilm Cookers

Microfilm cookers are available as atmospheric or vacuum cookers. They are suitable for all types of high boiling and can also be used for low boiling. They are very efficient cookers using the scraped-film principle, which results in extremely rapid cooking. This reduces process inversion to a minimum. The scraped film also means that confectionery containing milk products can be handled on this machine without the risk of burning or the need for frequent cleaning [182]. The principal uses of this machine are for feeding depositing plant or continuous cooling bands. In the case where a depositor is to be fed, a vacuum cooker should be chosen. The plant starts with a free-standing stainless steel reservoir for dissolved syrup. This is fitted with inlet filters. The syrup is pumped from the holding tank into the cooker by an infinitely variable syrup feed pump, which has a manually adjusted control. The preheater, rotor, syrup feed pump, discharge pump, vacuum pump, metering pumps, electrical control panel, and steam controls are all mounted on a column framework.

The rotor is a scraped film evaporator made up of a bronze steam-jacketed tube with a high-speed rotor fitted inside the center. The rotor has hinged blades, which wipe the inner surface of the tube. The sugar is spread in a very thin film and is moved through the cooking tube by a combination of gravity and the design of the hinged blades. The cooked sugar is discharged from the microfilm cooker in one of two ways, either by gravity from the base of the rotor or by a discharge pump and delivery pipe. Gravity discharge can only be used in the case of atmospheric cooking. Where vacuum cooking is used a discharge pump is necessary to withdraw the sugar from the vacuum. A discharge pump can also be used with an atmospheric cooker.

E. Effects of Cooking on Nutrients

During cooking great changes take place in the nature of food. Different foods behave in different ways when cooked. These effects apply to all foods [178].

1. Fats

When fats are heated they melt, and if they contain water, it is driven off as water vapor. At 100°C fats containing water appear to boil; this is caused by the water being given off as steam. Fats are stable to heat and can be heated almost to their boiling point before they start to break down. It is because of this fact—and also because they have high boiling points—that fats are used for cooking. When fats are heated too much, they break down, producing an unpleasant-smelling smoke. Fat on the outside of meat and in bacon darkens in color on strong heating, and if the temperature is too high some breakdown and charring may occur.

2. Carbohydrates

When exposed to dry heat, carbohydrates are broken down and darken in color. For example, sucrose browns on caramelization and finally chars and becomes black, while starch is broken down into more easily digested dextrin and also darkens and eventually chars. Many foods that contain both sugars and protein turn golden brown and change flavor on heating. These changes occur in the toasting of bread and the baking of bread, cakes and biscuits, and contribute to the pleasant flavor and attractive color of these products.

When a mixture of starch and water is heated, the starch granules absorb water and swell and gelatinize, forming a thick white paste. This is why starchy material (e.g., flour) is used to thicken sauces. On cooling, the paste sets and forms a gel. Uncooked starchy foods are difficult to digest because the digestive juices cannot penetrate into the starch grains. Cooking causes the starch granules to swell and gelatinize.

The polysaccharides starch, cellulose, and pectin are important constituents of fruit and vegetables. On cooking, insoluble cellulose changes little, except to soften, whereas starch softens as it gelatinizes and pectin becomes more soluble and dissolves somewhat, allowing cells to separate and making the fruit and vegetable easier to eat. Fruit with a high pectin content, such as apples, becomes soft and pulpy on cooking.

3. Proteins

Proteins undergo great changes when they are heated. Many proteins coagulate when heated; for example, egg white coagulates when it is heated above 60°C. As proteins coagulate they become solid. For example, when milk is heated a sking form because some of the proteins have coagulated. Cheese is another important protein food, and when it is heated it softens and on further heating some of the proteins coagulate and the cheese becomes stringy and tough. Not all proteins coagulate on heating, which is important when considering how to cook protein foods. Collagen and elastin, for example, are two important insoluble proteins in meat, and because they are not soluble they are not easily digested. Their presence in meat makes it tough, and as the cheaper cuts of meat usually contain more collagen and elastin than more expensive ones, they are usually tougher. Tough meat must be cooked in a way that will make it tender. If such meat is cooked at high temperatures for long periods it remains tough or may even become tougher! Tough meat needs to be cooked slowly using low temperatures; both dry heat and moist heat methods may be used.

Tough meat is often cooked slowly using moist heat (e.g., stewing). This converts the tough collagen into gelatin. Gelatin is a soluble protein and so is easily digested. Slow cooking using dry heat is also effective in converting collagen into gelatin. Elastin softens on cooking, but not to the same extent as collagen.

4. Mineral Elements

Heat does not affect mineral salts found in food because they are stable substances that do not break down at the temperatures used in cooking. Moist-heat methods of cooking, such as stewing and boiling, cause loss of salts, which are soluble in water. Boiled fish, for example, is rather tasteless because of the considerable loss of mineral salts that occurs during cooking. However, the salts are present in the water in which the fish has been boiled, and this liquid or stock can be used for making a tasty sauce to eat with the fish.

5. Vitamins

Dry-heat cooking methods destroy those vitamins which are unstable to heat. Vitamin C is destroyed at quite low temperatures, and so all methods of cooking cause some loss of this vitamin. To make the loss as small as possible, foods containing vitamin C should be cooked for as short a time as possible and should be eaten as soon as they are cooked. Two of the B vitamins, thiamine and riboflavin, are unstable at high temperatures. Riboflavin is the more stable of the two, and little is lost except at high cooking temperatures, such as those used in rapid grilling. Thiamine is largely destroyed at high temperatures, such as are used in grilling and roasting. Cooking with moist heat causes loss of water-soluble vitamins, as well as those that are destroyed at low temperatures. Vitamin C is both soluble in water and unstable to heat, and therefore some loss during cooking cannot be avoided. Vitamin C is also destroyed by oxygen present in air and dissolved in cooking water. The rate of destruction is hastened by enzymes present in the plant or fruit. These enzymes are set free by crushing or chopping.

The B vitamins are soluble in water in varying degrees, thiamine being the most soluble. A considerable proportion of the thiamine in foods may be lost during cooking, especially if they are boiled in alkaline solutions. For this reason alkaline substance, such as sodium bicarbonate, are not added to green vegetables to prevent loss of green color during cooking. The amounts of the other B vitamins lost during cooking are small and not important. Vitamin A and D are insoluble in water and stable except at high temperatures. There is therefore little, if any, loss of these vitamins during cooking.

REFERENCES

1. P. Fellows, Processing by application of heat, *Food Processing Technology: Principles & Practice* (P. Fellows, ed.), Ellis Horwood, England, 1988, p. 221.
2. I. J. Pflug and W. B. Esselen Heat sterilization, *Fundamentals of Food Canning Technology* (J. M. Jackson and B. M. Shinn, eds.), Van Nostrand Reinhold, New York, 1979, p. 71.
3. A. W. Farrall, Pasteurizing equipment, *Engineering for Diary Food Products* (A. W. Farrall, ed.), John Wiley & Sons, New York, 1980, p. 363.
4. N. I. Barclay, J. D. Potter, and A. L. Wiggins, Batch pasteurization of liquid whole egg, *J Food Technol. 19*:605 (1984).
5. J. E. Ford, J. W. G. Porter, S. Y. Thompson, J. Tootmill, and J. Edwards-Webb, Effects of UHT processing and subsequesnt storage on the vitamin content of milk, *J. Dairy Res. 36*:447 (1969).
6. K. J. Burgess, Dairy products, *Food Industries Manual*, 2nd ed. (M. E. Ranken ed.), Blackie, Glasgow, 1988.
7. F. A. Paine and H. Y. Paine, Fresh and chilled foods, *A Hand Book of Food Packaging*, 2nd ed. (F. A. Paine and H. Y. Paine ed.) Chapman & Hall, New York, 1982, p. 224.
8. G. Stehle, Trends in packaging techniques for milk products and fruit, *Neue Verpack 41*(10):56 (1988).
9. Hassen, A dressing down of the bottle freaks, *North Eur. Food Dairy J. 55*:81 (1989).
10. A. J. Iversen, Cartons for liquids, *Modern Processing, Packaging and Distribution Systems for Food* (F. A. Paine, ed.), Blackie, Glasgow, 1987, p. 86.
11. L. Karjalainen, Packaging of carbonated beverages, *Modern Processing, Packaging and Distribution Systems for Food* (F. A. Paine ed.), Blackie, Glasgow, 1987.
12. K. Kimura and G. Mitsu, Discussion on development and innovation of food packaging, *Food Policy* (*Jpn.*) *3*:65 (1989).
13. F. A. Paine and H. Y. Paine, Juices, soft drinks and alcoholic beverages, *A Hand Book of Food Packaging*, 2nd ed. (F. A. Paine and H. Y. Paine ed.) Chapman & Hall, New York, 1982, p. 339.
14. J. A. G. Rees and J. Bettison, Heat preservation, *Food Industries Manual* (M. D. Ranken, ed.), Blackie, Glasgow, 1988, p. 477.
15. J. V. Bousom, Carriers; beverage, *The Wiley Encyclopedia of Packaging Technology* (M. Bakker, ed.), Wiley, New York, 1986, p. 129.
16. S. Mardon, Energy saving in European dairy industry, *Diary Ind. Int. 47*(6):9 (1982)
17. R. L. Upadhya, Recent developments in energy conservation in dairy industry, *Indian Dairyman 39*(1):1 (1987).
18. P. S. Harris, Energy conservation in dairy industry, *J. Soc. Diary Technol. 3*(3):133 (1978).
19. D. I. Chandarana, B. C. Frey, L. E. Stewart, and J. F. Mattick, UHT milk processing-effect on process energy requirements. *J Food Sci. 49*:977 (1984).
20. P. S. Richardson and J. D. Selman, Heat processing equipment, *Processing and Packaging of Heat Preserved Foods* (J. A. G. Rees and J. Bettisson, eds.) Chapman & Hall, New York, 1990, p. 50.
21. R. H. Thorpe, D. Atherton, and D. S. Steele, Technical Manual 2, Campden Food Preservation Research Association, Chipping Campden, Glos, UK, 1975.
22. G. Tucker, and P. Clark, Computer modeling for the control of sterilization processes, Technical Memorandum 529, Campden Food and Drink Research Association, Chipping Campden, Glos, UK, 1989.
23. W. E. Perkins, Discontinuous and continuous retorts, *Introduction to the Fundamentals of Thermal Processing* (Sleeth, ed), IFT, Chicago, Nov 15–17, 1979, p. 82.
24. A. Lopez, *A Complete Course in Canning*, Vol. 1 (A. Lopez ed.) The Canning Trade Inc., 1987, p. 182.
25. D. J. Casimir, New equipment for the thermal processing of canned foods, *Food Technol. N.Z. 4*;290 (1969).
26. B. L. Manfre, Criteria for the selection of heat processing equipment, Part I. Equipment, *Canner/Packer 138*(10):21 (1969).

27. B. L. Manfre, Criteria for the selection of heat processing equipment, Part II. Economics, *Canner/Packer 138*(12):21 (1969).
28. R. D. Blanchett, F. D. Hickey, and J. Reimers, Controlled agitation retorting cuts heating time, retains high quality of cream-style corn, *Food Proc. 30*(1):16 (1969).
29. A. L. Brody, Food canning in rigid flexible packages, *CRC Food Sci. Nutr. 2*:187 (1971).
30. P. L. Goldfarb, Pouch for low-acid foods, *Modern Pack. 43*(12):70 (1970).
31. F. K. Lawler, New sterilizer made in France, *Food Eng. 39*(7):73 (1967).
32. W. Faasen and C. Hoogzand, Aspects of Technology, engineering and design of hydrostatic high-speed processing of foods in glass and metal containers, XXII Intern. Congress of Food Science and Technology, Warsaw, 22–27 August, 1966.
33. D. R. Farkas, M. E. Lazar, and W. C. A. Rockwell, A rotating hydrostatic helix for transferring solid-liquid mixtures under pressure or vacuum, *Food Technol. 23*:180 (1969).
34. M. Beauvois, G. Thomas, and H. Cheftel, A new method for heat-processing canned foods, *Food Technol. 15*(4):5 (1961).
35. T. Niederauer, The influence of technological processes on the nutrient value of foods, *Riechstoffe, Aromen, Kosmetica 29*(6):118 (1979)
36. O. Fennema, Chemical changes in food during processing: an overview, *Chemical Changes in Food During Processing* (T. Richardson and W. J. Finley, eds.) AVI Publishers, Westport, CT, 1985, p. 1.
37. J. Mauran, Effects of processing on proteins and food processing and nutrition: an overview, Proceedings of the XIII International Congress of Nutrition, 1985, p. 762.
38. R. F. Hurrell and K. J. Carpenter, Maillard reaction in food, *Physical, Chemical and Biological Changes in Food Caused by Thermal Processing* (T. Hoyem and O. Kvale, eds.) Applied Science, London, 1977, p. 168.
39. R. J. Hamilton, The chemistry of rancidity in foods, *Rancidity in Foods* (Allen and R. J. Hamilton, eds.), Elsevier Applied Science, 1989, p. 1.
40. M. L. Woolfe, Pigments, *Effects of Heating on Foodstuffs* (R. J. Priestly, ed.) Applied Science. London, 1979, p. 77.
41. R. L. Ory, A. J. St. Angelo, Y. Y. Gwo, G. J. Flick, Jr., and R. R. Mod, Oxidation induced changes in food, *Chemical Changes in Food During Processing* (T. Richardson and J. W. Finley, eds.), AVI Publishing Co., Westport, CT, 1985, p. 205.
42. S. A. Matz, Effects of non-enzymatic chemical changes, *Changes in Food Texture*, AVI Publishing Co., Westport, CT, 1962, p. 262.
43. Methods for Sensory Analysis of Food Part I, Introduction and General Guide to Methodology, British Standard Institution, London, 1980.
44. A. S. Szczeskiak, M. A. Brandt, and J. H. Friedman, Development of standard rating scales for mechanical parameters of texture and correlation between the objective and sensory methods of texture evaluation, *J. Food Sci. 28*(4):397 (1963).
45. S. A. Matz, Blanching, cooking & canning, *Food Texture*, AVI Publishing Co., Westport, CT, 1962, p. 177.
46. J. W. Finley, Environmental effects on protein quality, *Chemical Changes in Food During Processing* (T. Richardson and J. E. Finley, eds) AVI, Westport, CT, 1985, p. 443.
47. D. A. Ledward, Proteins, *Effects of Heating on Foodstuffs* (J. R. Priestley, ed.), Applied Science, London, 1979, p. 1.
48. A. E. Bender, Effects of processes, *Food Processing and Nutrition*. Academic Press, London, 1978.
49. J. E. Kinsella, Relationship between structure and functional properties of food proteins, *Food Proteins* (P. P. Fox and J. J. Carden, eds.), Applied Science, London, 1982, p. 51.
50. K. Simpson, Chemical changes in natural food pigments. *Chemical Changes in Food During Processing* (T. Richardson and J. W. Finley, eds.) AVI Publishing Co., Westport, CT, 1985, p. 409.
51. J. B. Adams and H. A. W. Blundstone, *The Biochemistry of Fruits and Their Products*, Vol. 2 (A. C. Hulme, ed.), Academic Press, London, 1971, p. 513.

52. J. B. Adams, and M. M. Ongley, Changes in the polyphenols of red fruits during heat processing, Technical Bulletin No. 23, Campden Food and Drink Research Association, 1972.
53. B. V. Chandler and K. M. Clegg, Pink discoloration in canned pears I: role of tin in pigment formation, *J. Sci. Food Agric. 21*:315 (1970).
54. C. T. Timberlake and P. Bridie, Anthocyanins in developments, *Food Colours* (J. Wolford, ed.), Applied Science, 1980, p. 115.
55. G. J. Malecki, British patent no. 772, 062 (1957).
56. G. J. Malecki, British patent no. 915, 429 (1963).
57. F. M. Clydesdale, Chlorophyllase activity in green vegetables with reference to pigment stability in thermal processing, Ph.D. thesis, Univ. of Massachusettes, Amherst, 1966.
58. D. J. Rose and H. A. W. Blundstone, The reproduction of the effects of plain tinplate in other forms of containers, Technical Memo No. 522, Campden Food and Drink Research Association, 1989.
59. F. D. Gunstone, Chemical properties, *The Lipids Handbook* (F. D. Gunstone, J. L. Harwood, and F. B. Padley, eds.), Chapman and Hall, London, 1986, p. 449.
60. J. E. Hodge, Chemistry of browning reactions, *J. Agric. Food Chem. 1*:928 (1953).
61. M. L. Wolfram and C. S. Rooney, Chemical interactions of amino compounds and sugars VIII. Influence of water *J. Am. Chem. Soc. 75*:5435 (1953).
62. J. Saunders and F. Jervis, The role of buffer salts in nonenzymatic browning. *J. Sci. Food Agric. 17*:245 (1966).
63. R. F. Hurrell and K. J. Carpenter, Mechanisms of heat damage in proteins. 4—The reactive lysine content of heat damaged material as measured in different ways. *Fr. J. Nutr. 32*:589 (1974).
64. M. N. Hall, M. C. Edwards, M. C. Murphy, and R. J. Pither, A comparison of the composition of canned, frozen and fesh garden peas as consumed, Technical Memo No. 553, Campden Food and Drink Research Association, 1989.
65. M. N. Hall, M. C. Edwards, M. C. Murphy, and R. J. Pither, A comparison of the composition of canned, frozen and fresh carrots as consumed, Technical Memo No. 553, Campden Food and Drink Research Association, 1989.
66. R. N. Choudhuin Roy, A. A. Joseph, V. A. Daniel, M. Narayana Rao, M. Swaminathan, A. Srinivasan, and V. Subramanyan, Effect of cooking, frying, baking and canning on the nutritive value of potato, *Food Sci. (Mysore) 12*(9):253 (1963).
67. A. S. Jaswal, Effects of various processing methods on free and bound amino acid content of potatoes. *Am. Pot. J. 50*:86 (1973).
68. A. E. Bender, Effects of food processing on vitamins, *Food Processing and Nutrition*. Academic Press, London, 1985, p. 786.
69. A. Benterud, Vitamin losses during thermal processing, *Physical, Chemical and Biological Changes in Food Caused by Thermal Processing* (T. Hoyem and O. Kvale, eds.) Applied Science, London, 1977, p. 199.
70. J. Woolfe, Processing, *The Potato in the Human Diet* (J. Woolfe, ed.), Int. Potato Centre and Cambridge University Press, London, 1987, p. 139.
71. R. J. Priestley, Vitamins, *Effects of Heating on Foodstuff* (R. J. Priestley ed.), Applied Science, London, 1979, p. 121.
72. O. Fennema, Food processing and nutrition: an overview, Proceedings of the XIII International Congress of Nutrition, 1985, p. 762.
73. G. G. Birch, Chemical, physical and biological changes in carbohydates induced by thermal processing. *Physical, Chemical and Biological Changes in Food Caused by Thermal Processing* (T. Hoyem and O. Kvale, eds.), Applied Science, London, 1977, p. 152.
74. C. T. Greenwood and D. N. Mann, Carbohydrates, *Effects of Heating on Foodstuffs* (R. J. Priestly, ed.) Applied Science, London, 1979, p. 35.
75. R. Cottrell, *Nutrition in Catering*, Parthenon Publishing Group, 1987, p 1.
76. J. L. Harwood, A. Cryer, and M. I. Gurr, Medical and agricultural aspects of lipids, *The Lipid Handbook* (F. D. Gunstone, J. L. Harwood, and F. B. Padley, eds.), Chapman and Hall, London, 1986, p. 527.

77. Trans Fatty Acids, BNF Task Force report.
78. F. Aylward, G. Coleman, and D. R. Haisman, Catty odours in food: the reaction between mesityl oxide and sulphur compounds in foodstuffs, *Chem. Ind. 5*:1563 (1967).
79. F. Aylward, G. Coleman, and D. R. Haisman, Catty taints in foodstuffs, Technical Memo No. 71, Campden Food and Drink Research Association, 1967.
80. R. L. S. Patterson, Catty odours in food: their production in meat stores from mesityl oxide in paint solvents, *Chem. Ind. 6*:584 (1968).
81. Catty taints in foods. *Food Trade Rev. 39*(9):47 (1969).
82. N. Goldenberg and J. R. Mathesan, Off-flavours in foods: a summary of experience 1948-74. *Chem. Ind. 13*:551 (1975).
83. K. Czerenski and K. Jarsabek, Changes in the biological value during thermal processing of meat under different conditions, measured by means of the fluorodinitrobenzol determination of available lysine, *Przenyol Spozywezy 18*:714 (1964).
84. Cannon and Howard, The tin free steel revolution, *Packaging Institute Professional,* Spring (1969)
85. P. C. Althen, The aluminum can, *Food Technol. 19*:764 (1965).
86. A. Lopez, *A Complete Course in Canning*, Vol. 2 (A. Lopez ed.) The Canning Trade Inc. Publ. 1987, p. 62, 111.
87. P. C. Althen and Phillip, Status of the aluminum can, *Tech. Quart. 2*:20 (1964).
88. W. T. Chiappe, Forge-welded seams, *Modern Packaging 43*(3):82 (1970).
89. A. Lopez and M. Jiminez, Canning fruit and vegetable products in aluminum containers, *Food Technol., 23*:1200 (1969).
90. H. A. Eden, Evolution of easy-open aluminum cans for beer cans, *Tech. Quart. 4*(2):168 (1968).
91. T. A. Turner, Packaging of heat preserved foods in metal containers, *Processing and Packaging of Heat Preserved Foods*, (J. A. G. Rees and J. Bettisson, eds.) Chapman & Hall Publ. 1990, p. 12.
92. J. F. Steffe, J. R. Williams, M. S. Chinnan, and J. R. Black, Energy requirements and cost of retort pouch vs can packaging systems, *Food Technol. 34*(9):39 (1980).
93. S. G. Unger, Energy utilization in the leading energy-consuming and processing industries. *Food Technol. 29*(12):33 (1975).
94. J. P. Ferrua and M. H. Col, Energy consumption rates for steam equipment. *Canner/Packer 144*(1):44 (1975).
95. D. J. Casimir, Flame sterilization, *CSIRO Food Res. Quart.*:34 (1975).
96. D. F. Sampson, Some aspects of the technology of processing, *Sterilization of Canned Foods*, American Can Co., Maryland, 1953, p. 205.
97. R. Jowitt and S. N. Thorne, Evaluates variables in fluid retorting, *Food Eng. 43*(11):60 (1971).
98. S. N. Thorne, Heat processing of canned foods in fluidized beds, *Food Technol. Aust. 24*(3):132 (1972).
99. D. T. Bernard, A. Gavin III, V. N. Scott, B. D. Shafer, K. E. Stevenson, J. A. Unverferth, and D. I. Chandrana, Validation of aseptic processing and packaging, *Food Technol. 44*:119 (1990).
100. N. Buchner, Aseptic processing aand packaging of food particulates, *Aseptic Processing and Packaging of Particulate Food* (E. M. A. Willhoft, ed.), Blackie Academic and Professional, London, 1993, p. 5.
101. V. R. Carlson, *Aseptic Processing*, Cherry-Burrell Corp., Chicago, 1969.
102. K. E. Stevenson, Procedings of National Food Processors Association Conference—Capitalizing on Aseptic II, Food Processors Institute, Washington DC, 1985, p. 59.
103. N. N. Potter and J. H. Hotchkirs, Form fill sealing machines, *Food Science*, 5th ed. (H. H. Potter and J. H. Hotchkirs, eds.), Chapman & Hall, 1995, p. 487.
104. R. B. Davis and D. T. Maunder, New system for aseptic pouch packing, *Modern Packaging 40*(10):157 (1967).
105. A. C. Hersom, Aseptic processing and packaging of food, *Food Rev. Int. 1*(3):223 (1986).
106. E. Glaser, Recent directions for aseptically packaged fluids, *Food Prod. Dev. 2*(3):60 (1969).
107. B. S. Luh, C. G. Gonzalez-Acuna, S. Leonard, and M. Simone, Aseptic canning of foods—V, Chemical and flavour changes in strained beef, *Food Technol. 18*:212 (1964).

108. G. R. Ammerman, The effect of equal lethal heat treatment at various time and temperatureon selected food components. Ph.D. thesis, Purdue Univ., W. Lafayette, IN, 1957.
109. R. Johnson, M.S. thesis, University of Georgia, Athens, 1973.
110. G. J. Everson, J. Chang, S. Leonard, B. S. Luh, and M. Simone, Aseptic canning of foods II & III: thiamine and pyridoxine retention as influenced by processing methods, storage time, temperature and type of container, *Food Technol. 18*:84 (1964).
111. D. Lund, Effects of heat processing on nutrients, *Nutritional Evaluation of Food Processing*, 3rd ed., Van Nostrand Reinhold Co., New York, 1988, p. 341.
112. B. Hallstrom and P Dejmek, Optimization and comparitive evaluation of UHT plants, *Milchwissenschaft 32*(6):447 (1977).
113. H. A. Lehniger and W. A. Beverloo, Physical operations, *Food Process Engineering* (H. A. Lehniger and W. A. Beverloo, eds.), D. Reidel Publishing Co., Dordrecht, 1975, p. 324.
114. R. E. Lisiecki, Aseptic packaging, PurePak Laboratories Publication, Ex-cell-O Corp., Detroit, 1968.
115. H. Burton, Aseptic packaging, Proceedings of Ultra High Temperature Processing of Dairy Products, Agricultural Institute, Republic of Ireland, Society of Dairy Technology, London, 1970, p. 17.
116. T. R. Ashton, Aseptic packaging of UHT milk, in Monograph on UHT Milk, Intern. Dairy Federation, Brussels, 1972, p. 100.
117. K. A. Ito and K. E. Stevenson, Sterilization of packaging materials using aseptic systems, *Food Technol. 38*:60 (1984).
118. C. P. Collier and C. T. Townsend, The resistance of bacterial spores to superheated steam, *Food Technol. 10*:92 (1956).
119. C. C. Witmer, A. Malvick, B. Snyder, and K. Robe, Simplest aseptic drum filler-drum is pressure chamber, *Food Proc. 31*(4):44 (1970).
120. F. M. Joffe and C. O. Ball, Kinetics and energetics of thermal inactivation and the regeneration rates of a peroxidase system, *J. Food Sci. 27*:587 (1962).
121. B. S. Luh, C. G. Gonzalez-Acuna, S. Leonard, and M. Simone, Aseptic canning of foods, VI. Hematin and volatile sulfur compounds in strained beef, *Food Technol. 18*:216 (1964).
122. A. M. Swanson and M. E. Seehafer, Recent research developments in processing sterilized milk concentrates, Bulletin No. 6, Department of Agricultural Economics, University of Illinois, 1965.
123. H. Wainess, Long-life dairy products, Dairy & Ice Cream Field, 24 Nov., 1970.
124. N. Hawker, Reclosable packaging system, *Food Technol. Eur. 1*(3):62 (1994).
125. R B. Biziak, Energy use in UHT sterile milk processing. M.S. thesis, North Carolina State Univ., Raleigh, 1981.
126. R. B. Biziak, K. R. Swartzel, and V. A. Jones, Energy evaluation of an UHT shell and tube processing system, *J. Food Sci. 47*(6):1875 (1982).
127. D. F. Dinnage, Aseptic processing of liquid food, Proceedings of National Food Processors' Association Conference, Oct 11–12, Washington, 1983, p. 32.
128. C. O. Ball and F. C. W. Olson, *Sterilization in Food Technology: Theory and Practice, Calculations*, McGraw-Hill, New York, 1957.
129. H. Burton, New monograph on UHT milk, International Dairy Federation Document 133, 1981, p. 80.
130. V. R. Carlson, Proceedings International Conference on UHT Processing and Aseptic Packaging of Milk and Milk Products, November 27–29, 1979, sponsored by Department of Food Science, North Carolina State University, 1980, p. 185.
131. H. Burton, Components achieve flexibility in sterile milk processing, *Am. Dairy Rev. 12*:28 (1967).
132. H. Burton, The Stork Sterideal process for sterile milk production, *Am. Dairy Rev. 8*:38 (1967).
133. H. Burton, UHT processing plants for the production of sterile milk—III: The Stork Sterideal Process, *Dairy Ind. 30*:700 (1965).
134. K. Rozema, Proceedings International Conference on UHT Processing and Aseptic Packaging of Milk and Milk Products, November 27–29, 1979, sponsored by Department of Food Science, North Carolina State University, 1980, p. 213.

135. H. G. Kessler and F. P. Horak, Testing and technical apprisal of the GEA Ahlborn UHT installation type IHS, *Dtsch. Milchwirtsch. 32*(37):1378 (1981).
136. H. Burton, Ahlborn and Sordi UHT plants for processing sterile milk, *Am. Dairy Rev. 12*:28 (1967).
137. K. Bake, Alfa laval steritherm; a new generation of UHT plants, *Molk. Techn. 43*:26 (1979).
138. K. Nilsson, Proceedings International Conference on UHT Processing and Aseptic Packaging of Milk and Milk Products, November 27–29, 1979, sponsored by Department of Food Science, North Carolina State University, 1980, p. 201.
139. H. Burton, The AVP Ultramatic process for sterile milk production, *Am. Dairy Rev. 7*:46 (1967).
140. J. D. Ridgway, Commercial operation of UHT plant utilizing an aseptic carton filler, *Am. Dairy Rev. 7*:52 (1967).
141. H. Burton, Use of Alfa-laval equipment for the production of sterile milk, *Am. Dairy Rev. 9*:32 (1967).
142. H. Burton, The uperisation process for the production of sterile milk, *Am. Dairy Rev. 9*:38 (1967).
143. H. Burton, UHT processing plants for the production of sterile milk—V: Laguilharre, *Diary Ind. 30*:872 (1965).
144. J. Fowler, Proceedings International Conference on UHT Processing and Aseptic Packaging of Milk and Milk Products, November 27–29, 1979, sponsored by Department of Food Science, North Carolina State University, 1980, p. 194.
145. H. Burton, Processing sterile milk with plants by Laguilharre, Palarisator and Thermovac, *Am. Dairy Rev. 10*:40 (1967).
146. L. E. Slater, Falling film sterilizer comes of age, *Food Eng. 12*:88 (1973).
147. J. Barabas, Evaluation of suitability of the Elecster sterilization equipment, *Prum. Potravin. 30*(5):582 (1979).
148. T. E. Szemblenski, Equipment and systems used in aseptically processed food products containing discreate particulate matter, Procedings of National Food Processors' Association Conference, Oct 11–12, Washington, 1993, p. 32.
149. A C. Hersom and D. T. Shore, Aseptic processing of foods comprising sauce and solids, *Food Technol. 35*(5):53 (1981).
150. W. F. Hall and G. M. Trout, *Milk Pasteurization*, AVI Publishing, Westport, CT, 1968
151. N. Kosaric, B. Kitchen, C. J. Panchal, J. D. Sheppd, K. Kennedy, and A. Sargant, UHT milk-production, quality and economics. *CRC Crit. Rev. Food Sci. Nutr. 34*(9):49 (1981).
152. P. Jelen, Review of basic technical principles and current resaerch in UHT processing of foods, *Can. Inst. Food Sci. Technol. J. 16*(3):159 (1982).
153. J. W. G. Porter and S. Y. Thomson, The nutritive value of UHT milk, in Monograph on UHT Milk, International Dairy Foundation, Brussels, 1972, p. 56.
154. J. Pien, Chemical and physico-chemical aspect and laboratory control, IDF seminar on UHT milk, Malmo, Sweden, May 24–27, 1971.
155. H. Hanni, E. Fluckiger, and H. Eger, Comparative study of composition and physical properties of pasteurized and UHT whipping cream, *Schweiz. Milchztg. 106* (1980).
156. C. H. Foster and P. Jelen, Improving the whipping properties of UHT heavy cream, 24th Annual Conference, CIFST, Winnipeg, 1981.
157. IDF, New monograph on UHT milk, International Dairy Federation Bulletin, Document 133, Brussels, 1981.
158. P. Jelen, Effects of light and storage temperature upon the properties of UHT sterilized bottled milk, *Prum. Pot. 27*(10):294 (1972).
159. L. G. Turner, H. E. Swaisgood, and A. P. Hansen, Interaction of lactose and proteins of skin milk during UHT processing, *Dairy Sci. 61*:384 (1978).
160. J. Mottar and M. Naudts, Quality of UHT milk compared to pasteurized and in-bottle sterilized milk, *Le Lait 59*:476 (1979).
161. R. S. Mehta, Milk processed at ultra-high temperatures-a review, *J. Food Prot. 43*:212 (1980).
162. OTA (Office of Technology Assessment, U.S. Congress), Emerging Food Marketing Technolo-

gies, a Preliminary Analysis, U.S. Government Printing Office, Washington, DC, 1978, Stock No. 052-003-00612-0.
163. M. F. Dios, Proceedings Aseptipak 84, Schotland Business Research Inc., Princeton, NJ, 1984, p. 205.
164. J. G. Zadow, The influence of oxygen on flavour of UHT whole milk, XVII Int. Dairy Congress, Melbourne, Australia, 1970, p. 182.
165. S. Ranieri, Aseptic canning process keeps shelf-stable custard line, *Food Prod. Dev. 13*(11):32 (1979).
166. H. Burton, An introduction to ultra-high-temperature processing and plant, *J. Soc. Dairy Technol. 30*(3):135 (1977).
167. Report on UHT Milk, Ontario Milk Marketing Board Task Force, Torono, 1976.
168. W. H. Clifford and S W. Gyeszly, Food packaging, *Food Engineering System,* Vol. 1, Operations (A. W. Farrall, ed.), AVI Publishing, Westport, CT, 1976, p. 431.
169. C. R. Oswin, *Plastic Films and Packaging*, Applied Science, London, 1975, p. 15.
170. Tetrapak Company Publication No. 45, AB TetraPak, Lund, Sweden, 1976.
171. Purepack (XLO), Instruction manual and parts catalogue for NLL, EX-Cell-O Corp. Detroit, 1976.
172. S. E Farnham, *A Guide to Thermoformed Plastic Packaging*, Caners Books, Boston, 1972.
173. M. S. Okos and C. C. Chen, Energy consideration for aseptic processing systems, Proceedings of a Conference Sponsored by Food Science Institute, Purdue University, West Lafayette., IN, 1978.
174. D. I. Chandarana, Energy aspects of a steam infusion milk sterilization process, M.S. thesis, University of Maryland, College Park, MD, 1982.
175. D. I. Chandarana, B. C. Fery, L. E. Stewart, and J. F. Mattick, Effect of temperature and holding time on enrgy required to process UHT milk, ASAE paper No. 82-6019, ASAE St. Joseph, MI, 1982.
176. B. C. Frey, D. I. Chandarana, L. E. Stewart, and J. F. Mattick, Unit process energy requirements for steam infusion UHT milk processing, *Trans ASAE*:1956 (1984).
177. B. C. Frey, D. I. Chandarana, L. E. Stewart, and A. N. Manor, Review of heat recovery methods for food processing plants, ASAE paper No. 79-3510, ASAE, St. Joseph, MI, 1979.
178. A. Cameron, Cooking I: methods and effects on nutrients, *The Science of Food and Cooking*, 2nd ed. (A. Cameron, ed.) Edward Arnold Publishers, London, 1978, p. 187.
179. E. J. Pyer, Oven equipment, *Baking Science and Technology*, Vol. 2 (E. J. Pyer, ed.) Siebel Publishing Co., Chicago, 1973, p. 1128.
180. Ohlsson, Boiling in kettles and scraped surface heatexchangers, Proceedings of IUFoST, Sweden, June 8–11, 1968, p. 71.
181. L. M. Cheng, Cooking machines *Food Machinery for the Production of Cereal Foods, Snack Foods and Confectionary* (L. M. Cheng ed.) Wood Head Publication, New York, 1992, p. 224.
182. Baker Perkins Catalogue on Microfilm Cookers and Cereal Cookers, Baker Perkins, Westfield Road/ Peterborough, England.
183. Continuous pressure cooker, *Food Manf. 50*(4):68 (1975).
184. Turboflow blancher/cooker, *Food Technol. N.Z. 10*(4):37 (1975).
185. M. N. Ramesh and P. N. Srinivasa Rao, Development and performance evaluation of a continuous rice cooker, *J. Food Eng. 27*:377 (1996).

6

Drying and Food Preservation

M. SHAFIUR RAHMAN AND CONRAD O. PERERA
Horticulture and Food Research Institute of New Zealand, Auckland, New Zealand

I. DRYING BASICS

A. What Is Drying?

Dehydration is a time-honored method of preserving food—one of the most important for the food-processing industry. The sun-drying of fish and meat was practiced as long ago as 2000 B.C., and dried vegetables have been sold for about a century and dried soups for much longer [69]. Whereas drying in earlier times was done in the sun, now many types of sophisticated equipment and methods are being used to dehydrate foods. During the past few decades, considerable efforts have been made to understand some of the chemical and biochemical changes that occur during dehydration and to develop methods for preventing undesirable quality losses.

Drying is a method of water removal to form final products as solids, while concentration means the removal of water while retaining the liquid conditions. While one of the main reasons fermentation is used in food processing is to develop desired flavor or aroma, loss of flavor or aroma compounds is the main problem of drying in terms of quality.

The costs of processing, packaging, transportation, and storage are less for dried products than for canned and frozen foods. The main purposes of drying are to increase shelf life, reduce packaging and storage costs, lower shipping weights, improve sensory attributes, encapsulate flavors, and preserve nutritional value in some cases [4]. Three factors should be considered before selecting a drying process: the quality of the products in relation to requirement, the economics of the process, and the environmental impact on the process.

Water may also be removed from solids by means of mechanical force. The products can be pressed or compressed by centrifugal force and pressure. This process, termed mechanical dewatering, is used as a preprocessing step before thermal drying in some cases. Dewatering causes a loss of soluble solids with the separated water, thus may result in a loss of quality and nutritive value.

B. Classification of Drying

Drying processes can be broadly classified, based on the water-removing method applied, as (a) thermal drying, (b) osmotic dehydration, and (c) mechanical dewatering. In thermal drying a gaseous or void medium is used to remove water from the material, thus thermal drying can be divided into three types: (a) air drying, (b) low air environment drying, and (c) modified atmosphere drying. In osmotic dehydration solvent or solution is applied to remove water, whereas in mechanical dewatering physical force is used to remove water. In mechanical dewatering centrifugal force or pressure is applied to materials with a physical barrier (i.e., membrane) to keep liquid and solid phases separated. Consideration should be given to many factors before selecting a drying process. These factors are (a) the type of product to be dried, (b) properties of the finished product desired, (c) the product's susceptibility to heat, (d) pretreatments required, (e) capital and processing cost, and (f) environmental factors. Cohen and Yang [43] concluded that there is no one best technique for all products.

C. Biochemical Aspects

Heat, solvent, or mechanical force is usually used in the drying processes. Thus, the effects of these factors on microorganisms and enzymes in foods should also be a major concern during drying, although the main purpose of drying is preservation. Drying reduces the water activity, thus preserving foods by avoiding microbial growth and deteriorative chemical reactions. (More details of water activity and food preservation are available in Chapter 4.) The effects of heat on microorganisms and enzymes is also important in the drying of foods. Thus, care should be given to maximize all types of biological inactivation for food preservation. On the other hand, care must be given to reduce the damage during drying of bacterial cultures and bioactive materials. Thus, detrimental effects of drying may be desirable or undesirable, depending on the purpose of the drying process.

Inactivation of enzymes during spray-drying is typically due to high temperatures. For bacteria, inactivation during drying typically results from both high temperatures and dehydration [122]. Lethal thermal injury to the pathogen *Coxiella burnetii* coincides with alkaline phosphatase inactivation during the pasteurization of milk [61].

Fu et al. [72] studied the lethal thermal injury to the bacteria *Lactococcus lactis* during the spray-drying of cell paste in condensed skim milk at pH 6.3. They found that as the outlet air temperature increased from 77 to 120°C, the percent survival decreased from 59 to 1.1%. A period of 12–16 hours was required before the cells produced significant amounts of lactic acid during growth in reconstitution of dried culture. Thermal inactivation could be minimized by using small drops, low outlet air temperatures, and modified spray-dryer designs. In contrast to alkaline phosphate inactivation during pasteurization of milk, bacterial injury generally results from thermal and dehydration effects. Freeze-drying produces less damage to microflora than other conventional drying methods. Rybka and Kailasapathy [190] found that only freeze-dried *Lactobacillus acidophilus* met suggested minimum levels. *Bifidobacterium* levels were 10 times lower than the minimum required in yogurts. This was due to metabolic injury by long-time drying and storage conditions.

Biological products are thermosensitive and xerosensitive. In drying such materials, such as vitamins, antibiotics, enzymes, yeasts, molds, it is very important which of their properties are most valuable for further utilization of such products. In drying of microbial cultures in the food industry, a maximum possible number of live microorganisms should be preserved [215]. Starter cultures of dried lactic acid bacteria are used in the production of a number of food and feed products. Many compounds have been tested to improve the survival of lactic acid bacteria during freeze-drying. These are polyols, polysaccharides, disaccharides, amino acids, proteins, vitamins, and various salts [36]. Couture et al. [46] mentioned that several additives improve the survival of culture in milk or milk-based formulas. Polyethylene glycol, dextran, and bovine albumin increased survival during freeze-drying of milk [68]. Some freeze-dried cultures are more stable during storage if they are microencapsulated or immobilized in polymers [37,102,104]. Champagne et al. [35] studied the effect of addition of gelatin, xanthan gum, or maltodextrins on survival during freeze-drying and stability during storage of four lactic acid bacteria in a milk-sucrose medium. They found that none of the polymers generated marked improvement of the milk-sucrose medium in terms of survival of bacteria during freeze-drying. Gelatin improved the storage stability of few types of freeze-dried cultures, but certain additives had a detrimental effect on some cultures.

Linders et al. [123] found that thermal and dehydration inactivation occurred during drying of *Lactobacillus plantarum*. Thermal inactivation can be minimized by optimizing process conditions. They found that maltose and sucrose decreased the dehydration inactivation of *L. plantarum* by fluidized bed drying, whereas lactose had little effect. The order of effectiveness of the disaccharides in the protection of *L. plantarum* against drying corresponds with the number of equatorial –OH groups available. The decrease of dehydration inactivation at a certain water content cannot be explained by the change in water activity caused by this addition. The results indicated that specific interactions between components are important.

A water-replacement hypothesis has been proposed in the literature to explain the protective effect of sugars on cell membrane [42,48]. According to this hypothesis, sugars can replace the water around the polar membrane phospholipids and proteins, thus preventing the membrane from disruption during drying. The number of equatorial –OH groups of mono- and disaccharides is related to their ability to stabilize the membrane [48,101].

The introduction of sugars and polyhydric alcohols into the solvent medium has been found to stabilize biological macromolecules in solution, mainly preventing the loss of enzymic activity, inhibiting irreversible aggregation, or increasing the thermal denaturation temperature [50]. Ascorbate oxidase has potential clinical and food use. The activity of ascorbate oxidase purified from green zucchini squash decreased on freeze-drying [50]. D'Andrea et al. [50] found that the enzyme activity of rehydrated samples was strictly correlated with the addition of lyoprotectants, freezing temperature, enzyme concentration, and storage time. The addition of lyoprotectants such as sucrose, mannose, glucose, or polyvinylpyrrolidone improved enzyme stabilization. A notable loss of enzyme activity was observed when freezing was performed by immersion in liquid nitrogen in comparison to samples frozen at –20 or –75°C, thus short freezing at very low temperature was detrimental to stabilization.

Cassava contains a high concentration of cyanogenic glycosides, which may give rise to hydrocyanic acid by enzymatic hydrolysis. Monroy-Rivera et al. [146] investigated cassava detoxification by air-drying. More than 70% of free and bound cyanide can be eliminated using drying at 60°C (air speed: 1.5 m/sec, humidity: 0.003 kg water/kg dry air).

Martin and Stott [133] observed respiration of sultana grapes during drying. Respiration is slowed by drying but not necessarily abolished, and rehydration did not restore lost respiration rates [85].

D. Water in Foods

The concept of bound and free water has been developed from drying principles. The product containing no water is termed as *bone-dry*. Water in foods exists in different forms or states. Water in foods having properties different from those of pure water can be defined as bound water. In the literature different forms of bound water are defined [176], for example, unfreezable, immobile, monolayer, and nonsolvent water. Rahman [176] states that the fraction of bound water depends on the definition and measurement techniques. The state of water in foods is important for the drying process as well as the stability of foods during storage. The binding energy of different states of bound water affects the drying process, since it requires more energy to remove bound water than free water.

E. Endpoint of Drying

Drying is a term used for the reduction in moisture content from an initial value to some acceptable final value. Equilibrium in drying system is the ultimate endpoint for the process. Water activity (isotherm) is commonly used to estimate the equilibrium point in the cases of thermal and osmotic drying processes. In mechanical dewatering, the magnitude of the applied force and rheological properties of the foods affect the equilibrium point.

Generally meat, fish, and dairy products are dehydrated to a moisture content of 3% or less; vegetable products usually to 5%; and cereal products frequently to as much as 12% [192]. A maximum moisture level is usually established for each dried product separately based on desired acceptable quality after drying and during storage. Different attributes of quality can be targeted, thus endpoint should be determined from all aspects, such as safety first and then consumer acceptance.

F. Heating Methods and Heat Transfer

Heating of air by electric heater or flue gas is the conventional heating method used for drying. In this case, heat transfer from the gas to the product occurs mainly through convection. Heating method is another important aspect of drying in terms of quality as well as energy cost.

Microwave and dielectric heating use the electromagnetic wavelength spectrum as a form of energy, which interacts with the materials, thus generating heat and increasing the drying rate dramatically. Dielectric drying uses frequencies in the range of 1–100 MHz, whereas microwave drying uses frequencies in the range of 300–300,000 MHz [43]. The major drawback of microwave heating is the nonuniformity of the energy within the chamber, which can be partially offset by the use of waveguides and a rotating tray [43].

El Abd et al. [59], who studied drying of clover seeds using infrared and electrically heated drying, found a saving in power consumption in the case of infrared heating.

II. THERMAL DRYING

A. Air-Drying

1. Convection Air-Drying

This is the simplest drying technique, taking place in an enclosed and heated chamber. The drying medium, hot air, is allowed to pass over the product, which has been placed in open trays. Convection drying is often a continuous process and is most often used for products that are relatively low in value [43]. Air-drying is usually accomplished by passing air at regulated temperature and humidity over or through the food in a dryer. Different types of dryers are now available.

Mechanism of Drying

State diagrams show the physical state of a material as a function of the concentration and temperature. A state diagram used to indicate the temperature-concentration dependency during drying is shown in Figure 1. This generic diagram could be used to understand different quality-modifying events, such as collapse, crystallization, and glass-rubber transformation.

The mechanisms of moisture transfer depend mainly on the types or physical state of food materials and drying process. The food materials can be classified as (a) homogeneous gels, (b) porous materials with interconnecting pores or capillaries, and (c) materials having an outer skin that

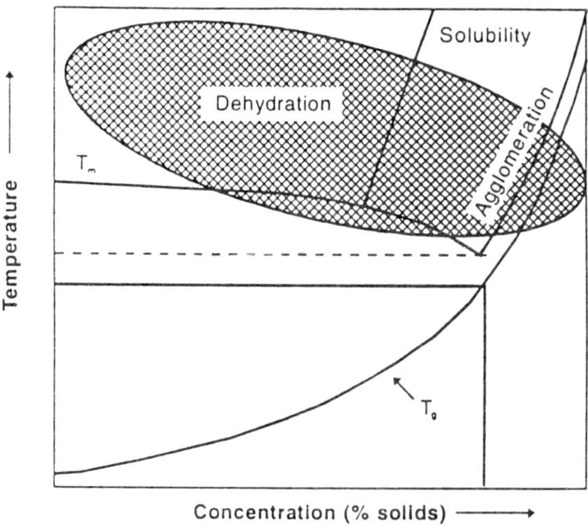

FIGURE 1 Location of dehydration in a state diagram. (From Ref. 155.)

is the main barrier to moisture flow [47]. The types or structure of foods always play an important role in drying process.

Drying is simultaneously a heat and a mass transfer process. Hence, there are two resistances: heat transfer and mass transfer. During the *constant rate period*, it is considered that there exists a thin film of water on the slice and there is no internal or external mass transfer resistance. Hence, the drying is controlled by external heat transfer. In the *falling rate period*, the drying is controlled by internal mass transfer resistance. The absence of a constant rate period indicates that the drying is controlled from the beginning by internal mass transfer resistance. The moisture content at which the drying period changes from a constant to a falling rate can be considered the critical moisture content. The critical moisture content depends on the characteristics of the foods and the drying conditions. The critical moisture contents varied from 0.78 to 0.83 (kg/kg, wet basis) for vegetables and 0.85 to 0.89 (kg/kg, wet basis) for fruits [193]. At high moisture contents liquid flow due to capillary forces dominates. At decreasing moisture content the amount of liquid in the pores also decreases and a gas phase is built up, causing a decrease in liquid permeability. Gradually the mass transfer is taken over by vapor diffusion in a porous structure. At the saturation point there is no longer liquid available in the pores and mass transfer is taken over completely by vapor diffusion [45].

The moisture is transferred from the solid materials by diffusion or capillary mechanisms. In a diffusion mechanism, the driving force is concentration gradient. Water diffusion can be in the form of liquid or vapor. In the case of liquid diffusion osmotic pressure could be the driving force for water movement. In the capillary mechanism the moisture moves due to surface tension force and does not conform to the laws of diffusion. A porous material contains a complicated network of interconnecting pores and channels, and at the surface mouths of pores of various sizes exist. As water is removed, a meniscus is formed across each pore, which sets up capillary forces by the interfacial tension between the water and the solid. Capillary forces act in a direction perpendicular to the surface of the solid. It has been suggested that a combined mechanism of capillary forces and vapor diffusion is responsible for moisture movement in drying of potato [78,79]. The drying experiments of Saravacos and Charm [193] with surface-active agents failed to show any important of capillary forces during the dehydration of potatoes and other vegetables. Surfactants are known to reduce the surface tension of water, thus increasing the capillary forces in porous materials. Thus, capillary flow is not significant in the vegetables studied by Saravacos and Charm [193]. Waananen and Okos [230] showed that during drying of pasta at temperature close to the boiling point, liquid flow dominates moisture transport at high moistures and vapor flow is significant only at low moistures. Achanta and Okos [4] reviewed the shrinkage of different biopolymers and concluded that shrinkage on drying is equal to the volume of moisture leaving, thus it is conceptually difficult to justify that capillary flow is important during the drying of high-moisture biopolymers.

The strength of capillary forces at a given point in a pore depends on the curvature of the meniscus, which is a function of the pore cross section. Small pores develop greater capillary forces than large ones, thus large pores tend to empty their water content first [138]. In large pores, the capillary forces are small. The force of gravity is then large in comparison with the capillary forces, and there is a directional effect due to gravity [138].

Drying Curve

There are three major stages of drying (Figs. 2 and 3):

1. Transient early stage, during which the product is heating up (transient period)
2. Constant or first period, in which moisture is comparatively easy to remove (constant rate period)
3. Falling or second period, in which moisture is bound or held within the solid matrix (falling rate period)

Typical drying rate curves are shown in Figures 2 and 3. The moisture content at which the change from the first to the second period occurs is known as the critical moisture content.

Typically two falling rate periods are observed for both hygroscopic and nonhygroscopic solids [221]. The first falling rate period is postulated to depend on both internal and external mass trans-

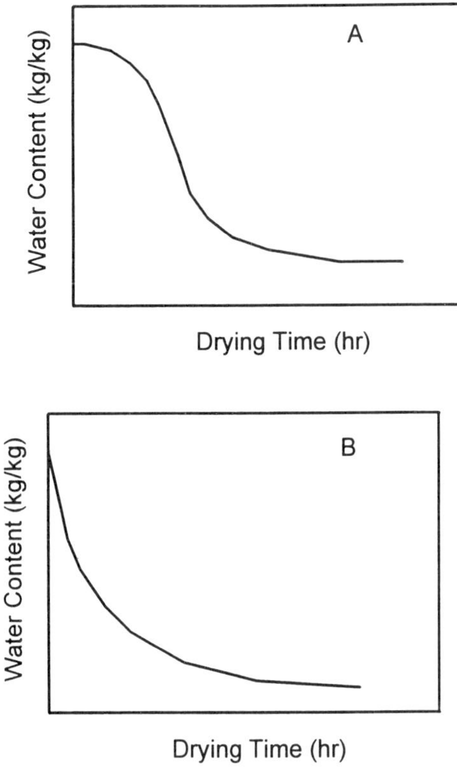

Figure 2 Typical drying curves (water content vs. drying time): (A) with a lag period, (B) without a lag period.

fer rates, while the second period, during which drying is much slower, is postulated to depend entirely on internal mass transfer resistance only. The slower rate may be due to the solid-water interaction or glass-rubber transition [4].

The drying behaviors of food materials depend on the porosity, homogenity, and hygroscopic properties. The immediate entrance into the falling rate is characteristic of hygroscopic food materials. Lee et al. [117] studied the effect of sodium sulfate on the surface evaporation of a porous medium during the constant rate period. The drop of the drying rate was significant due to the decrease of surface vapor pressure and the change of liquid surface curvature due to meniscus effects by surface tension.

Energy Aspect of Air-Drying

Drying is one of the most energy-intensive processes in the food industry. Apart from the rise of energy costs, legislation on pollution and sustainable and environmentally friendly technologies have created greater demand for energy-efficient drying processes in the food industry. Thus, novel thinking in the technology of drying methods and dryer design is evident. The food industry could save much money by avoiding costly energy waste. Improving energy efficiency by only 1% could result in an as much as 10% increase in profits [16]. Conducting an energy survey is the traditional way to approach the problem. The energy survey analyzes the energy defect level at each stage of processing and strategies for their remedy [16].

Energy losses in air-drying: Heat losses during drying can be grouped as: heat loss with the exhaust air, heat loss with the product, radiation heat loss from the dryer, heat loss due to leakage of

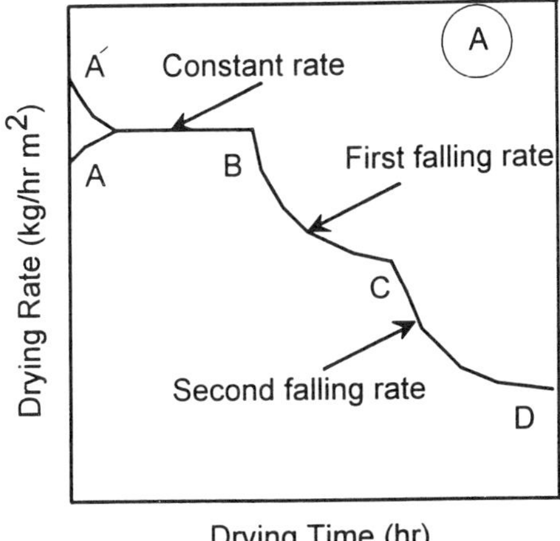

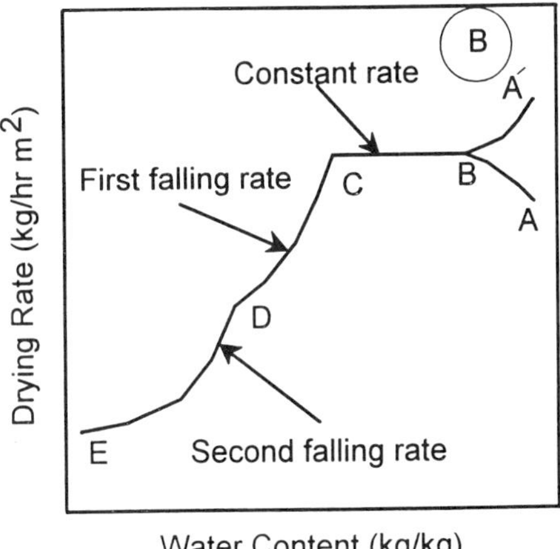

FIGURE 3 Typical drying rate curves: (A) drying rate versus drying time, (B) drying rate versus water content.

air from the dryer, and heat loss due to overdrying of products. Table 1 shows the possible energy savings for walnut dehydration.

Recirculating exhaust air in grain dryers is popular because of its energy conservation and effect on grain quality. The higher-humidity air damages grains to a lesser extent than low-humidity air.

TABLE 1 Energy Savings for Walnut Dehydration

Method	Possible savings (%)
Preventing overdrying	25–33
Recirculation of drying air	25
Reducing airflow rate	≤25
Improved burner design and operation	≤10
Insulation of drying	3–4

Source: Ref. 216.

Grains are severely damaged by high drying temperatures [210]. Thus, by changing the dryer design energy losses can be avoided while achieving higher product quality.

New concepts in dryer design for energy efficiency: Energy can be saved by (a) reducing drying time or increasing throughput (better control), (b) avoiding heat losses, and (c) heat recovery from exhaust gas and dried product.

The potential for energy conservation by design and changes in drying operation is significant. Strumillo and Lopez-Cacicedo [216] reviewed most of the methods of energy recovery from exhaust air, including heat exchangers (pipe and plate types), thermal wheel, heat pipe installation, and run-around coil. These methods recover mainly sensible heat from the exhaust, while most of the heat is lost with the latent heat of water vapor in the exhaust air. This latent heat can be recovered by condensing out water using a refrigeration system. However, a refrigeration system will consume extra power before further use. Among these methods the heat pump dryer (using a dehumidifier) has a high potential for use in food industry (discussed later).

2. Solar or Open Air-Drying

In solar drying, radiation energy from the sun is used. Solar drying is a nonpolluting process and uses renewable energy. Moreover, it is an abundant energy source that cannot be monopolized [93]. Solar drying has several drawbacks, which limit its use in large-scale production. These are (a) the need for large areas of space and for high labor inputs, (b) the difficulty in controlling the rate of drying, and (c) inset infestation and microbial contamination [43,93]. More details of solar drying construction and operation are given by Imre [93].

3. In-Store Drying

In-store drying can also be called low-temperature in-bin drying. It may be used where grain remains in store until milled or sold The main advantages of two-stage drying are [147] reduced energy requirements, increased drying system capacity, and improved product quality. However, weather conditions in tropical climates are less favorable for in-store drying due to high ambient temperatures and relative humidities. Two-stage drying can produce good quality by preventing discoloration of high-moisture grain and reducing cracking of skin dry kernels [212]. Jebson and He [97] developed high-quality dehydrated garlic powder by a two-stage drying technique, in which most of the garlic was dried as large pieces at high temperature, but a smaller portion was dried as thin slices at lower temperature.

4. Explosive Puff-Drying

Explosive puff-drying uses a combination of high temperature and high pressure and a sudden release of the pressure (explosion) to flush superheated water out of a product. This method gives the product good rehydratability. However, the high heat can degrade the food quality, and the explosion puffing may compromise the product integrity [43].

5. Spray-Drying

Spray-drying is used to remove water from a free-flowing liquid mixture, thus transforming it into a powder product. The fluid that is to be dried is first atomized by pumping it through either a nozzle or a rotary atomizer, thus forming small droplets with large surface areas. The droplets immediately come into contact with a hot drying gas, usually air. The liquid is very rapidly evaporated, thus minimizing contact time and heat damage. Disadvantages include (a) the size of the equipment required to achieve drying is very large, and (b) very oil materials might require special preparation to remove excessive levels of fat before atomization [43]. Ultrasonication in the chamber can be used instead of complex atomization to produce small-diameter droplets in spray-drying.

Drying in a spray-dryer typically occurs in four successive stages [106]:

1. The feed is transformed into droplets by an atomizer.
2. The droplets, once formed, lose water by diffusion while maintaining an approximately spherical shape.

3. Once the temperature of a droplet approaches or exceeds the boiling point of water, morphological changes occur, typically originated by nucleation of one or more bubbles formed from dissolved air. These bubbles can repeatedly grow and break through the surface, causing formation of blowholes and/or ramiform protrusions, followed by collapse and repetition of this cycle many times.
4. Once the particle is dry enough, the final shape sets into place, and any final evaporation of water occurs.

6. Fluidized Bed Drying

This technique involves movement of particulate matter in an upward-flowing gas stream, usually hot air. Fluidization mobilizes the solid particulates, thus creating intimate contact between the hot carrier gas and the solids. Drying occurs by convection. The hot gas is introduced into the bottom of a preloaded cylindrical bed and exits at the top. In some cases, a vibratory mechanism is used to increase the contact of the product with the hot gas. Fluidized bed drying is usually carried out as a batch process and requires relatively small, uniform, and discrete particles that can be readily fluidized [43]. The main advantages of fluidized bed drying are uniform temperature and high drying rates, thus less thermal damage. A rotating chamber is also used with fluidized bed, thus increasing centrifugal force to further increase the drying rate and mixing.

A suspension of photosynthetic bacteria in a highly heat-sensitive and thermolabile biomaterial must be dried carefully to ensure the required product quality. Expensive freeze-drying and high vacuum-drying are commonly used for dehydration of these products. Pan et al. [159] found that the vibrating fluid bed technique can be successfully used if the material is absorbed on a bed of solid carriers. The use of solid carrier (sea sand or wheat bran), porous or nonporous, prevents the biomaterial from deterioration due to thermal shock. Wheat bran is preferable because of its large sorption capacity and the possibility of its use as a product component.

7. Spouted Bed Drying

In a spouted bed dryer, the heated gas enters the chamber at the center of a conical base as a jet. The particles are rapidly dispersed in the gas, and the drying occurs in an operation similar to flash-drying. This work very well with larger pieces that cannot be dried in a fluidized bed dryer [43].

8. Ball Drying

In this method the material to be dried is added to the top of the drying chamber through a screw conveyor. The material within the drying chamber comes into direct contact with heated balls made from ceramic or other heat-conductive material. Drying occurs primarily by conduction. Hot drying air is passed through the bottom side of the chamber. When the product arrives at the bottom of the chamber, it is separated from the balls and collected [43].

9. Rotary Drum Drying

Rotary drum dryers are cylindrical shells 1–5 m in diameter, 10–40 m in length, and rotating at 1–8 rpm with a circumferential speed of approximately 0.2–0.4 m/sec. These conditions depend on the product types used for drying. They are designed to operate at a nearly horizontal position, inclined only by 2–6° to maintain the axial advance of solids, which are fed from the upper end of the dryer body [166].

10. Drum Drying

This technique removes water from a slurry, paste, or fluid that has been placed on the surface of a heated drum. The dryer may comprise either a single or a double drum. Drum drying is typically a continuous operation, and care must be taken to ensure that the product that is to be dried adheres well to the drying surface; in some cases, it may be necessary to modify the liquid product by additives to change its surface tension or viscosity [43].

11. Microwave Drying

Microwave drying is rapid, more uniform in the case of liquids, and more energy efficient than conventional hot air-drying [51]. Applying microwave energy under vacuum affords the advantages of both vacuum-drying and microwave drying, providing improved energy efficiency and product quality. The energy can be applied by pulsed or continuous mode. Pulsed microwave drying is more efficient than continuous drying. In both cases, drying efficiency improves when lower pressure is applied [235].

Microwave drying may also result in a poor-quality product if not properly applied [99]. Surface scorching is one of the most common problems [83]. Yongsawatdigul and Gunasekaran [236] found that microwave vacuum-drying of cranberries resulted in a better quality (increased redness and softer texture) than conventional hot air-drying. The microwave operating conditions have an effect on the quality of the dried cranberries. The storage stability of the product dried by the microwave vacuum method was comparable to that of conventionally dried cranberries.

Prabhanjan et al. [171] studied conventional and microwave-assisted convective air-dried carrots. Microwave drying resulted in a substantial decrease (25–90%) in drying time. Rehydration properties were improved by drying at higher power levels for carrots [171] and apples [98,157]. However, for carrots low power levels gave good color after rehydration. Yongsawatdigul and Gunasekaran [237] studied microwave-assisted vacuum-drying of cranberries. They found that pulsed application of microwave energy was more efficient than continuous application. In both cases, drying efficiency improved when lower pressure (5.33 kPa) was applied. Shorter power-on time and longer power-off time provided more favorable drying efficiency in the pulsed mode. Power-on time of 30 seconds and power-off time of 150 seconds was the most suitable setting for maximum drying efficiency for carrots. The moisture-extraction rates of the continuous and pulsed modes varied from 0.717 to 1.003 and from 0.800 to 1.434 kg water/kW-hr, respectively [237].

B. Low Air Environment Drying

1. Smoking

The use of wood smoke to preserve foods is nearly as old as open air-drying. Although not primarily used to reduce the moisture content of food, the heat associated with the generation of smoke also gives a drying effect. Smoking has been mainly used with meat and fish. Smoking not only imparts desirable flavor and color to some foods, some of the compounds formed during smoking have a preservative effect (bactericidal and antioxidant) [43].

Smoking is a slow process, and it is not easy to control. Wood smoke is extremely complex—more than 400 volatiles have been identified [144]. Wood smoke contains nitrogen oxides, polycyclic aromatic hydrocarbons, phenolic compounds, furans, carbonylic compounds, aliphatic carboxylic acids, tar compounds, carbohydrates, pyrocatechol, pyrogallols, organic acids, bases, as well as carcinogenic compounds like 3,4-benzpyrene. Nitrogen oxides are responsible for the characteristic color of smoked food, whereas polycyclic aromatic hydrocarbon components and phenolic compounds contribute to its unique taste. These three chemicals are also the most controversial from a health perspective [144]. Smoke solutions are available, either condensed products from the dry distillation of wood or synthetically prepared mixtures of phenols. Synthetic smokes produce results close to actual smoke curing, and harmful components can be eliminated from synthetic smokes [29]. The odor, composition of flavor compounds, and antimicrobial activity of the smoke is highly dependent on the nature of the wood. Some studies found that beech and oak woods product the wood smoke with the best sensory properties [82]. Herbs, spices, or pine cones may be added to product unique, aromatic smoke flavors [89].

Color development in smoked fish is a complex process. The Maillard reaction with glycolic aldehyde and methyglyoxal in the dispensing phase of smoke plays the dominant role [219]. Hot-smoked fish produced by exposing the fish to sawdust smoke had good flavor but poor color [2]. Several types of synthetic colors have been used to color kippers in England [188]. Paprika has also been used as a seasoning and to impart color to smoked fish [2]. Abu-Bakar et al. [2] used caramel to improve the color of hot-smoked spanish mackerel (*Scomberomorus* spp.), chub mackerel (*Rastrelliger kanagurta*), kurau (*Polynemus* spp.), and skinless squid (*Loligo* spp.) mantle. Spanish mackerel, chub mackerel, and squid immersed in brine containing 0.4, 2.0, and 0.6% caramel (w/v), respectively, for 30 minutes at 25°C produced the most acceptable color. Smoked products with golden-yellow to light brown coloring were preferred by the panelists.

The use of wood smoke in preventing lipid oxidation in meat and fish products has been investigated [13,116]. Polyphenols derived from the smoke acted as antioxidants. Woolfe [234] found that smoke-drying initiated lipid oxidation in herring (*Sardinella aurita*) as evidenced from peroxide values. The site of initiation was bound lipids in contact with the proteins, and final moisture content was the predominating factor affecting rate of oxidation. Sheehan et al. [207] found that fat levels in raw salmon affect texture, oiliness, and color of smoked salmon during storage. McIlveen and Vallely [143,144] believe that good-quality smoked processed cheese may be possible.

2. Vacuum-Drying

Vacuum-drying of food involves subjecting it to a low pressure and a heating source. The vacuum allows the water to vaporize at a lower temperature than at atmospheric conditions, thus foods can be dried without exposure to high temperature. In addition, the absence of air during drying diminishes oxidation reactions. Thus, color, texture, and flavor of dried products are improved. Narasimham and John [153] studied the product quality of diced breadfruit both vacuum-dried and freeze-dried. The product quality parameters (shrinkage, rehydration, taste, texture, and appearance) were equivalent to those in conventionally freeze-dried product. In controlled low-temperature vacuum-drying, the product was preheated in a through-flow drier at 60°C for 1 hour and dried at 0.5 mmHg pressure. The product temperature during drying decreased from 50 to –5°C, thus the cost-intensive freezing process at extremely low pressure could be avoided.

3. Freeze-Drying

In freeze-drying the material that has been frozen is subject to a pressure below the triple point (0°C) and heated to cause ice sublimation to vapor. A schematic diagram of the different states of water with triple point is shown in Figure 4. This is used to retain vitamins and heat-sensitive components for a high-quality dried food product.

Freeze-drying utilizes a high vacuum to remove water from a solid phase (ice) as a vapor phase without going through a liquid phase. Since the material remains frozen and drying takes place at low

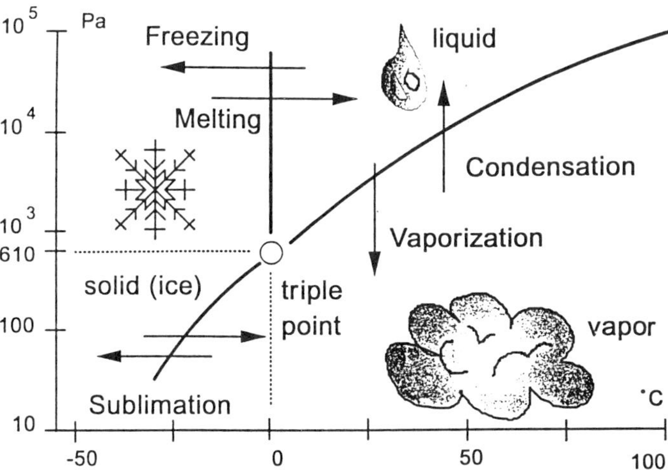

Figure 4 Schematic diagram of the different states of water showing triple point. (From Ref. 155.)

temperature, no heat damage occurs. In addition, there is little or no loss in qualities of the product, and because the removal of ice crystals leaves a porous honeycomb-type structure, the product tends to rehydrate rapidly. However, freeze-drying is a slow and expensive process. The long processing time requires additional energy to run the compressor and refrigeration units, which makes the process very expensive for commercial use. Thus, it is most suitable for drying of high-value products [43].

4. Heat Pump Drying

Principles

The heat pump dryer is a further extension of a conventional convection air-dryer with an in-built refrigeration system. The heat pump works in the same way as a refrigerator, with an evaporator that absorbs heat and a condenser that emits heat, effectively pumping heat from a low-temperature evaporator to a higher-temperature condenser. The dry heated air is supplied continuously to the product to pick up moisture. The humid air passes through the evaporator of the heat pump where it condenses, thus giving up its latent heat of vaporization to the refrigerant in the evaporator. This heat is used to reheat the cool dry air passing over the hot condenser of the heat pump. Thus, in a heat pump the latent heat recovered in the process is released at the condenser of the refrigeration circuit and used to reheat the air within the dryer [135,164]. A schematic diagram of the operation of a typical heat pump dryer is shown in Figure 5.

Potential Advantages for Food Industry

The use of the heat pump dryer offers several advantages over conventional hot air dryers for the drying of food products, including higher energy efficiency, better product quality, the ability to operate independently of outside ambient weather conditions, and zero environmental impact. In

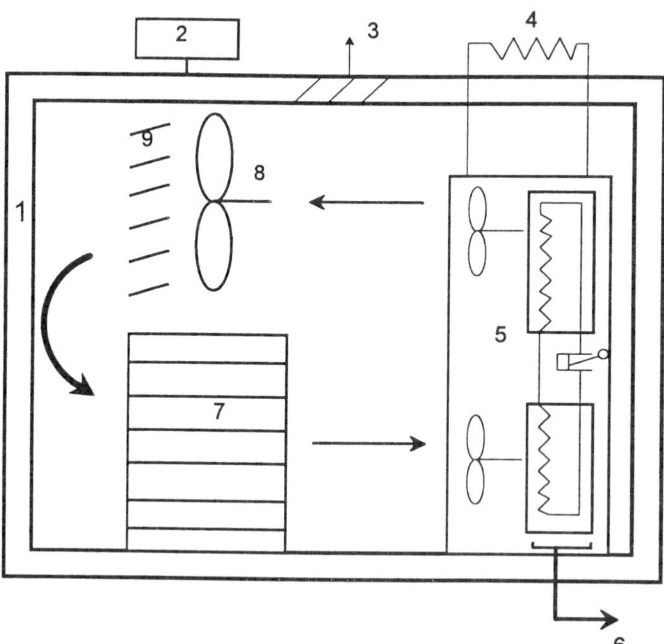

FIGURE 5 A schematic diagram of the operation of a typical heat pump dryer. 1, Vapor-sealed and insulated structure; 2, humidifier; 3, overheat vent; 4, external condenser; 5, heat pump dehumidifer; 6, condensate; 7, product tray; 8, primary air circulation fan; 9, air distributor. (From Ref. 164.)

addition, the condensate can be recovered and disposed of in an appropriate manner, and there is also the potential to recover valuable volatiles from the condensate [164].

Quality improvement: Potential improvements in the quality of dried products by optimum drying conditions is a major advantage of using heat pump drying of foods. Usually dried products have low aroma volatile content, suffer a loss of heat-labile vitamins, and have high incidence of color degradation. Ginger dried in a heat pump dryer was found to retain over 26% of gingerol, the principal volatile flavor component responsible for its pungency, compared to only about 20% in rotary dried commercial samples [135]. The higher volatile retention in heat pump–dried samples may be due to reduced degradation of gingerol at the lower drying temperature used compared with the commercial dryer temperatures. The loss of volatiles varies with concentration, with the greatest loss occurring during the early stages of drying when the initial concentration of the volatile components is low [196]. Since heat pump drying is conducted in a sealed chamber, any compound that volatilizes will remain within it, and the partial pressure for that compound will gradually build up within the chamber, retarding further volatilization from the product [164].

Development of a brown center sometimes occurs in macadamia nuts if high-moisture nuts are dried at elevated temperatures [173]. Van Blarcom and Mason [227] found that heat pump drying of macadamia nuts did not result in the above defect, even when they were dried at 50°C. Mason [134] studied the heat pump drying of macadamia kernels and herbs with temperature and relative humidity ranges of 30–50°C and 0.10–0.50, respectively. Freshly harvested macadamia nuts can be dried rapidly up to a moisture content of 0.015 with no loss in quality, and there was no significant difference of quality when dried under the conditions mentioned above. This may be due to the faster drying rates associated with the heat pump drying process [164]. The losses in color, flavor, and nutritive value associated with dried products are attributed to nonenzymatic browning. It is recognized that the rate of reaction for nonenzymatic browning in dried products is highest at moisture levels that are commonly attained towards the end of the drying cycle, when the drying rate is low and the product temperature approaches that of the drying medium. However, the lower drying temperatures used throughout the drying cycle in heat pump dryers reduce the extent of nonenzymatic browning reactions.

The color and aroma of herbs (e.g., parsley, rosemary, and sweet fennel) can be improved when compared with the commercial products. The sensory values were nearly doubled in case of heat pump–dried herbs compared to commercially dried products. There was no significant difference in the quality of herbs dried below moisture content 0.04 for the experimental drying temperatures (40 and 50°C) and relative humidity (0.30 and 0.40).

A wide range of quality characteristics can be obtained by running the dryer within wide ranges of temperature and relative humidity. The use of modified atmospheres for drying of sensitive materials such as food products is another important potential aspect of heat pump drying technology. During drying oxygen-sensitive materials such as flavor compounds and fatty acids can undergo oxidation, giving rise to poor flavor, color, and rehydration properties. Using modified atmospheres to replace air will permit new dry products to be developed without oxidative reactions occurring [164].

Energy efficiency: Removal of water in its liquid state rather than the vapor state allows the latent heat of vaporization to be captured, and only a small amount of sensible heat is lost with condensate. The energy spent in drying is usually expressed as the "specific moisture-extraction rate" or SMER. The SMER for a well-designed heat pump dryer lies between 1 and 4 kg/kWh, whereas the SMER of single-pass hot dryer is only 0.95 kg/kWh [164]. A general comparison of heat pumps with vacuum- and hot air-drying is presented in Table 2.

Process efficiency: Heat pump drying has the ability to operate at set conditions independently of outside ambient weather conditions. In addition, it is environmentally friendly, i.e., no gases or fumes are given off to the atmosphere. The condensate can be recovered and disposed of in an appropriate manner, and there is also the potential to recover valuable volatiles from the condensate [164]. Since drying takes place in a closed system, a low air-leakage rate gives negligible heat loss.

Strommen [214] studied the drying of stockfish (unsalted) and klipfish (salted) in a heat pump dryer and found that the drying time was lowered by a factor of 4 with a high quality level. The

TABLE 2 General Comparison of Heat Pump Dryer with Vacuum and Hot Air-Drying

Parameter	Hot air-drying	Vacuum drying	Heat pump drying
SMER (kg water/kWh)	0.12–1.28	0.72–1.2	1.0–4.0
Drying efficiency (%)	35–40	≤70	95
Operating temperature range (°C)	40–90	30–60	10–65
Operating % RH range	Variable	Low	10–65
Capital cost	Low	High	Moderate
Running cost	High	Very high	Low

RH: Relative humidity; SMER: specific moisture-extraction rate.
Source: Ref. 164.

klipfish was dried from 0.55 to 0.45 water content (wet basis) and the stockfish was dried from 0.80 to 0.20 water content. The inlet temperature and relative humidity in the tunnel were 20–25°C and 0.73, respectively. For heavily salted fish about 27°C, burn spots were found. The above authors found total energy consumption for oil burner air flow, automatic control of humidity at the exhaust, and heat pump to be 875, 479, and 125 kWh/ton, respectively.

Increasing the humidity in the drying air slows down the drying process but improves the energy efficiency [27]. In general, heat pump drying efficiency and capacity are dependent on temperature and humidity. In general, the SMER increases with an increase in humidity in the dryer [164]. In a conventional air dryer at low temperature (10–30°C) it is not possible to run the drying operation due to high ambient relative humidity (0.70–0.90), but heat pump drying can be performed at these low temperature since the relative humidity can be lowered to 0.10.

The thermal insulation and gas-tightness of the seals of the chamber structure is important in achieving high energy efficiency for the heat pump. In addition to the electrical energy required to drive the compressor, energy is also required to preheat the product and chamber structure, to drive the fan for primary air flow over the product that is to be dried, and to replace any heat loss through conduction and air leakages. The motors driving the fan and the compressor can be located within the chamber so that the residual heat produced by them is absorbed within the drying chamber instead of being lost to the atmosphere [164].

Progress and Applications

There are a number of technological problems to be overcome before the process can be applied to the food industry.

Capital cost: The capital cost of a heat pump dryer is higher than for a conventional hot-air dryer due to additional refrigeration system requirement. However, its cost should be much less than that of vacuum or freeze-drying.

Limited drying temperature: While low-temperature drying is a potential advantage, too low a temperature will limit the drying rate, which has implications for throughput. Also, slower drying rates at low temperature may give rise to potential microbial growth problems [164].

Process control and design: Like vacuum or freeze dryers, heat pump dryers are more amenable to batch drying because the drying takes place in a hermetically sealed container. The construction of continuous drying may involve high engineering modeling and design costs. Therefore, benefits need to be evaluated on the basis of cost rather than energy efficiency alone [164].

Microbiological safety: Most of the vegetative cells of microorganisms will be destroyed by normal hot air drying at 60–80°C with only few exceptions (e.g., heat-resistant bacteria, yeast, and molds) [70]. Although there are some concerns about the potential for growth of microorganisms at the temperatures used in heat pump dryers, in practice there have been no reports of increased numbers of microorganisms in heat pump dried foods compared to those dried by conventional means [27]. Serious microbiological problems may arise if the dryer is designed poorly. The problems of microbial growth should be the focus of further research.

5. Superheated Steam-Drying

Superheated steam is used as a drying medium. The main advantages of this type of drying are that it can provide an oxygen-free medium for drying and process steam available in the industry can be used without any capital cost. An oxygen-free medium has the potential to give high-quality food products, however, negligible information is available in the literature.

III. OSMOTIC DEHYDRATION

A. Osmotic Process

Osmotic dehydration of foods has potential advantages for the food-processing industries. This dehydration process generally will not allow a product of low moisture content to be considered shelf-stable. Consequently, an osmotically treated product should be further processed (generally by air-, freeze-, or vacuum-drying methods) to obtain a shelf-stable product, or the process could be used as a pretreatment for canning, freezing, and minimal processing.

Osmotic dehydration is the process o water removal by immersion of a water-containing cellular solid in a concentrated aqueous solution (syrup or brine). The driving force for water removal is the chemical potential between the solution and the intracellular fluid. If the membrane is perfectly semipermeable, solute is unable to diffuse through the membrane into the cells. However, it is difficult to obtain a perfect semipermeable membrane in food systems due to their complex internal structure, and there is always some solute diffusion into the food and leaching out of the food's own solute. Thus, mass transport in osmotic dehydration is actually a combination of simultaneous water and solute transfer processes (Fig. 6). The main steps in the preparation of osmotically dehydrated products are shown in Figure 7.

The osmotic dehydration process can be characterized by equilibrium and dynamic periods [177]. In the dynamic period the mass transfer rates are increased or decreased until equilibrium is reached. Equilibrium is the end of osmotic process, i.e., the net rate of mass transport is zero. The removal of water is mainly by diffusion and capillary flow, whereas solute uptake or leaching is only by diffusion. The fundamental knowledge for prediction of mass transport is still a grey area, although

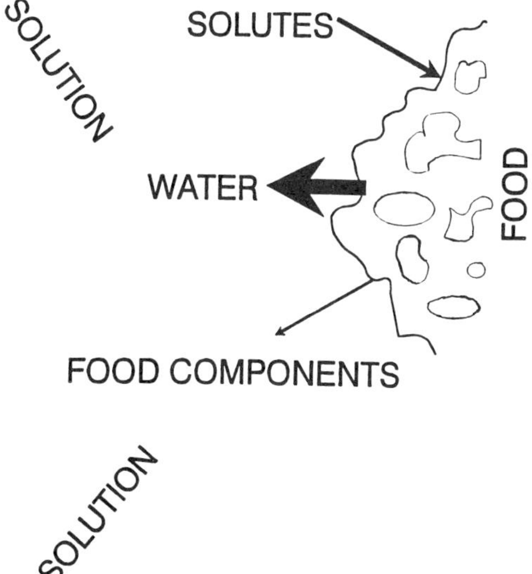

FIGURE 6 Water and solute transfer osmotic process.

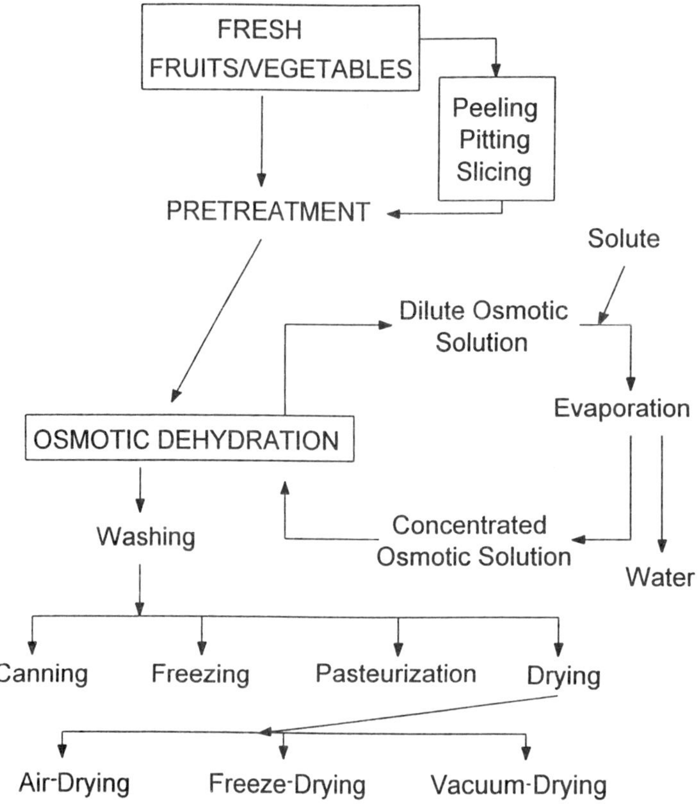

FIGURE 7 Major steps in the processing of osmotic dehydrated products.

considerable efforts have been made over the past decade to improve the understanding of mass transfer in osmotic dehydration [181].

B. Potential Advantages for Industry

The use of the osmotic dehydration process in the food industry has several advantages:

1. Quality improvement in terms of color, flavor, and texture
2. Energy efficiency
3. Packaging and distribution cost reduction
4. Chemical treatment not required
5. Product stability and retention of nutrients during storage

1. Quality Improvement

It is well established that osmotic dehydration improves product quality in terms of colour, flavor, and texture. Rahman [177,181,220] reviewed the merits of osmotic dehydration for product quality improvement and process efficiency. However, the literature does not offer a fundamental understanding of the mechanisms of flavor entrapment in the food matrix, color retention, physics of textural improvement. The phenomena that may be involved in aroma retention are the adsorption of volatiles onto the infused solute matrix, physico-chemical interactions between volatiles and other substances, and micro-regional encapsulation, in which volatile compounds are immobilized in "cages" formed in association with dissolved solids [39,40,65–67,211,229].

Osmotic dehydration treatment of apples prior to freeze-dehydration prevented cellular damage during the freezing step and had little adverse effect on the tissue of freeze-dehydrated apple slices [118].

2. Energy Efficiency

Osmotic dehydration is a less energy-intensive process than air- or vacuum-drying because it can be conducted at low temperatures. Lenart and Lewicki [120] found that energy consumption is osmotic dehydration at 40°C with syrup reconcentration by evaporation was at least two times lower than in convection air-drying at 70°C. In the frozen food industry, high energy levels are used for freezing because of the large quantity of water present in fresh foods. A significant proportion of this energy could be saved if plant materials were concentrated prior to freezing [90]. A reduction in the moisture content of food can reduce refrigeration load during freezing. This is typified by salting, which is one of the oldest methods for preserving fish and vegetables. The high level of solute in osmotically treated products decreases water activity and preserves them, and thus an energy-intensive drying process can be avoided. Following osmotic treatment, the resultant syrup can be used in the juice or beverage industries as a byproduct, improving process economy only in the case of osmotic drying without salt.

3. Packaging and Distribution Cost

Partially concentrating fruits and vegetables prior to freezing saves packaging and distribution costs [22]. The product quality is comparable to that of conventionally processed products. The process is referred to as "dehydrofreezing."

4. Chemical Treatment

Canning of apples is not practiced commercially due to some inherent problems associated with the high gas volume in apple tissue, the difficulty of its removal during exhausting, its low drained weight, and its mushy texture [206]. There have been a few attempts at canning apple slices using calcium chloride, a firming agent, to improve the texture [49]. However, using osmotically treated apple pieces in the canning process resulted in a firmer texture and improved quality of the product [206]. This process is known as "osmo-canning."

Chemical treatment to reduce enzymic browning can be avoided by using an osmotic agent [169]. Sugar has two effects: (a) effective inhibition of polyphenol oxidase, the enzyme that catalyzes oxidative browning of many cut fruits, and (b) prevention of the loss of volatile flavor during further air- or vacuum-drying [232]. However, if the final product after air-drying contains 10–20% moisture, enzymic and nonenzymic browning will cause slow deterioration of color and flavor [167]. Therefore, Ponting [167] suggested adding a blanching step after the osmotic process and using sulfur dioxide during or after the osmotic step if the final moisture content of the product is more than 20%.

5. Product Stability During Storage

The product obtained by osmotic process is more stable than untreated food materials during storage due to low water activity by solute gain and water loss. At low water activity, deteriorative chemical reactions and growth and toxin production by microorganisms in the food are low.

In canning high-moisture fresh fruit and vegetable, water can flow from the product to the syrup brine, causing dilution. This can be avoided using the "osmo-canning process" to improve the product stability [206]. Similarly, the use of osmo-dehydrofrozen apricot and peach cubes in yogurt can improve consistency and reduce whey separation [77].

Solute exerts a germicidal effect. Salt reduces the solubility of oxygen in the substrate, thereby restricting the growth of aerobes.

C. Factors Affecting the Osmotic Dehydration Process

1. Type of Osmotic Agent

The most commonly used osmotic agents are sucrose for fruit and sodium chloride for vegetables, fish, and meat. Other osmotic agents include glucose, fructose, lactose, dextrose, maltose, polysaccharide, maltodextrin, corn starch syrup, and combinations of these osmotic agents. A number of researchers have investigated the use of binary mixtures of solutes with sucrose as a means of reducing solute cost and improving the effectiveness of osmosis [87,94]. The ultimate choice of blend will depend on many factors, such as solute cost, organoleptic compatibility with the end product, and additional preservation action by the solute.

2. Concentration of Osmotic Solution

Both the water loss to equilibrium level and drying rate increase with an increase in osmotic syrup concentration, since the water activity (i.e., mass transfer driving force) of syrup decreases with an increase in solute concentration in the syrup [21,22,60,119,121,129,132,178]. A dense solute barrier layer at the surface of the product is formed with an increase in syrup concentration, thus enhancing the dewatering effect and reducing the loss of nutrients during the process [198,199]. A similar solute barrier is also formed in the case of syrups with higher molecular weight solutes at even low concentration.

3. Temperature of Osmotic Solution

The rate of osmosis is markedly affected by temperature [18]. This is the most important parameter affecting the kinetics of water loss and solute gain. Water loss increases with increase of temperature, whereas solid gain is less affected by temperature. In the case of high temperature, solute cannot diffuse as easily as water through the cell membrane, and thus the approach to osmotic equilibrium is achieved primarily by flow of water from the cell [178]. This type of equilibrium results in a lower solute gain by the product.

4. Properties of Solute Used in Osmosis

The osmotic process is affected by the physicochemical properties of the solutes employed. These differences arise mainly from differences in molecular weight, ionic state, and solubility of solute in water. According to the principle of osmosis, the rate of water loss from the fruit to the syrup having large molecular weight solute is lower than that of syrup having small molecular weight solute when both syrups are at the same mass concentration. This is due to low vapor pressure of the syrup having low molecular weight solute. However, contrary to the above physical chemistry principle, for osmotic syrups with equal concentrations at the early stage, those with high molecular weight solutes will have a greater rate of water removal and lower solute transfer (due to low solute penetration), compared with those of low molecular weight. This was demonstrated using a model agar gel [180].

The pH of the syrup can also affect the osmotic process. Acidification increases the rate of water removal by changes in tissue properties and consequential changes in the texture of fruits and vegetables [150]. Water removal was maximal at pH 3 for apple rings using corn syrup [44]. In more acidic solution (i.e., at pH 2) the apple ring became very soft. However, firmness was maintained at pH values between 3 and 6. The softening may be due to hydrolysis and depolymerization of the pectin.

5. Agitation of Osmotic Solution

Osmotic dehydration can be enhanced by agitation or circulation of the syrup around the sample [44,87,119]. However, the improvement is so small that in some cases it might be more economical to use no agitation when consideration is taken of equipment needs and breaking of fruits [169]. The effect of agitation on the osmotic dehydration of kiwi slices was found to depend on the syrup-to-fruit mass ratio and syrup concentration [163].

6. Geometry of the Material

Osmotic concentration behavior will depend on the geometry of sample pieces, due to the variation of surface area per unit volume (or mass) and diffusion length of water and solutes involved in mass transport. Contreras and Smyrl [44] found that mass loss was about 1.3 times higher when apple slice thickness decreased from 10 to 5 mm. Lerici et al. [121] found that solute gain increased as the ratio of surface area to minimum diffusion length increased, while water loss increased to a maximum (depending on the shape) and then decreased. This decrease in water loss was probably due to a reduction of diffusion caused by high solid gain at the surface and consequent formation of a solute layer. At the same operating conditions, fresh fruit having different sizes and shapes can result in final products with very different characteristics [121].

7. Osmotic Solution and Food Mass Ratio

Both solid gain and water loss increase with the increase of syrup-to-food mass ratio [64,169]. Uddin and Islam [225] studied the effect of syrup-to-fruit slice mass ratio on osmotic treatment of pineapple at 21°C. They observed that weight loss increased until the syrup-to-fruit ratio was 4:1, but by increasing the ratio up to 6:1 no further gain was observed. Thus, they defined the optimum ratio for pineapple as 4:1. The rate of solid gain and water loss increased significantly when the syrup-to-fruit mass ratio was increased from 1:1 to 6:1 in osmotic drying of potato in 61% sucrose syrup [119]. At equilibrium solute content in potato was the same at syrup:fruit ratios of 1:1 and 10:1. Thus, syrup-to-fruit mass ratio has negligible effect on potato composition at equilibrium.

8. Physicochemical Properties of Food Materials

The chemical composition (protein, carbohydrate, fat, salt etc.), physical structure (porosity, arrangement of the cells, fiber orientation and skin), and pretreatments may affect the kinetics of osmosis in food. A steam-blanching step for 4 minutes before osmosis gave lower water loss and higher solid gain when applied to fresh potato slices [94]. The loss of membrane integrity due to heating was the cause of the poor osmotic concentration behavior [94]. Freezing raw fruit disrupts cells and results in poor osmosis of thawed fruit [167]. Saurel et al. [199] studied osmosis of frozen apple without thawing in ethylene glycol and polyethylene glycol at 30–70°C. They observed similar results as with fresh apple.

9. Operating Pressure

Vacuum osmotic dehydration results in a change of behavior of mass transfer in fruit-sugar solution systems [62,63,163,208]. Vacuum treatments intensify the capillary flow and increase water transfer, but have no influence on solute uptake [62]. The total water transfer results from a combination of traditional diffusion and capillary flow and is affected by the porosity or void fraction of the fruit [63,208].

D. Problems Associated with Osmotic Dehydration in the Food Industry

A number of technological problems must be overcome before the process can be applied to the food industry.

1. Product Sensory Quality

The main disadvantage of the osmotic process is that it may increase saltiness or sweetness, or decrease the acidity of the product, which is not desirable in some cases. This can be avoided by controlling the solute diffusion and optimizing the process to improve the sensory assessment of the product. Edible semipermeable membrane coatings can also be used to reduce solute uptake and increase water loss [34].

2. Syrup Management

Microbial validation of the process for long-time operation and reuse of the syrup by recycling are important factors for industrial applications [181]. The management of syrup is the major challenge to make the process industrially viable. These include syrup composition and concentration, syrup recycling, solute addition, reuse of the syrup, and waste disposal. The cost of the syrup is a key factor in the success of the process. The compositional changes related to leaching from the fruit or vegetable may influence the product quality (color, acids, sugar, minerals, vitamins, etc.). Microbial contamination can increase with the number of times that the syrup is recycled. The control of solute composition in recycling for single-solute syrup is easier than mixed-solute syrups. During the recycling process, the diluted syrup can be reconcentrated by evaporation and/or reverse osmosis. The syrup-to-fruit mass ratio should be kept as low as possible to reduce the production costs.

3. Process Control and Design

Inadequate information about experiments presented in the literature and the limited data available have precluded effective design and control of this process by the food industry. Further studies are necessary to have a clear understanding of the variation of equilibrium and rate constants with process variables and characteristics of the food materials. Most of the osmosis studies have been concerned with the qualitative prediction of the processing factors, but more quantative prediction is necessary for industrial use in process design and control. On-line measurements of syrup properties can provide continuous control of the process.

Fruits and vegetables tend to float on the concentration syrup due to the higher density of the syrup. Moreover, the viscosity of the syrup exerts considerable mass transfer resistance and causes difficulty in agitation, and the syrup tends to adhere to the surface of the food material.

Breakage of food material pieces may occur by flow of syrup in case of continuous process or by mechanical agitation in case of batch process. The increase of slat concentration in fish during osmosis slowed the subsequent air-drying process [96,185].

Equilibrium is the endpoint of osmosis, but for practical purposes a number of other factors should be considered to ensure the quality of the final product. These include damage to the cells and development of off-flavor due to longer processing time and reuse of the syrup [177]. Finally, adequate packaging systems should be used to make sure that consumers are getting good-quality products.

These problems should be addressed in further research work on osmotic dehydration of foods.

IV. PRETREATMENTS IN DRYING

Pretreatment is common in most drying processes in order to improve product quality or process efficiency. The main drying pretreatments are blanching, dipping, and sulfiting. In recent years an improvement in quality retention of dried products by altering processing strategy and/or pretreatment has gained much attention.

A. Blanching

1. Purpose of Blanching

Enzymes are responsible for off-flavor development, discoloration or browning, deterioration of nutritional quality, and textural changes in food materials. Thus, the main purpose of blanching is to inactivate the naturally occuring enzymes present in foods.

2. Advantages of Blanching

The blanching step is used in the freezing or drying process to (a) inactivate the naturally occuring enzymes, (b) remove gases from vegetable surfaces and from intercellular spaces, (c) reduce the initial microorganism load (e.g., the microbial count is reduced by a factor of 2000 for peas and over 40,000 for potatoes) [69], (d) to clean raw food materials initially, (e) to facilitate preliminary operations such as peeling and dicing, and (f) to improve color, texture, and flavor under optimum conditions.

3. Disadvantages of Blanching

The disadvantages of blanching are that (a) it may change texture, color, and flavor because of the heating process, (b) it increases the loss of soluble solids, especially in the case of water blanching, (c) it may change the chemical and physical state of nutrients and vitamins, and (d) it has adverse environmental impacts, such as large water and energy use and problems of effluent disposal.

4. Effects of Blanching on Dried Food Products

Blanching is a process of preheating the product by immersing in water or steam. In water blanching a wide variation of temperature may be possible, but it can cause a loss of soluable nutrients. Water-soluble vitamins may be lost due to leaching, thermal destruction, or enzymic oxidation during water blanching. Alzamora et al. [7] studied the water-soluble vitamin losses during blanching of peas and found that (a) blanching of peas at higher temperatures can reduce leaching losses, (b) leaching losses can be reduced more in static water blanching than in agitated blanching, (c) small leaching losses of vitamins occur during water cooling of blanched peas, and (d) the larger the peas, the lower the leaching losses. The loss of mannitol during blanching of mushrooms was studied by Biekman et al. [20]. The decrease of the mushroom mass and the increase of the mannitol concentration in blanching fluid was measured as a function of blanching time and temperature (50–100°C). An initial lag phase was observed in the experiments.

Blanching is also thought to accelerate the drying process by cracking the skin or outer surface of food materials. Abdelhaq and Labuza [1] studied the level of sulfiting of apricots from 0 to 2000 ppm and air-drying at 50–80°C. The quality of dried apricots was judged by extent of browning development and hardness determination. Sulfiting and subsequent air-drying using 800–1000 ppm sulfur dioxide at any temperature in the range 50–80°C was found to be best. Sulfiting was the major factor in controlling dry apricot quality and would be hard to reduce. Blanching reduced the drying time of apricots by 10–20%.

Blanching results in some degree of chlorophyll degradation with the subsequent formation of pheophytin. The extent of chlorophyll conversion is related to the degree of blanching. Peroxidase activity is widely used as an index of blanching because peroxidase is the most heat-stable enzyme found in vegetables.

In unblanched dehydrated spinach, destruction of chlorophyll is positively correlated with moisture content and was affected very little by oxygen content of the storage atmosphere, whereas destruction of carotene was greatly dependent on oxygen in the storage atmosphere and independent of moisture content of dried spinage [55]. Maharaj and Sankat [130] studied the effect of blanching on the quality of dehydrated dasheen leaves. Blanching in water or in alkali (0.06% magnesium carbonate) (at 100°C for 10 sec) followed by sucrose infusion prior to air-drying at 60°C resulted in better products than steam-blanched (at 96°C for 6 min) or unblanched leaves. Texture, rehydration, flavor, color, and aroma were comparable with soups made from freshly harvested ingredients. Green color was preserved as reflected by chlorophyll and pheophytin contents as well as hue angle measurements. This may be due to the prevention of pheophytin formation by addition of alkalizing agent during blanching. Usiak et al. [226] studied the effect of blanching on the rheological properties of applesauce. Consistometer values and yield stress increased as blanching temperature increased from 35 to 59°C and decreased from 71 to 83°C. The consistency index and serum viscosity were almost unchanged by blanching temperature, but both decreased with increasing storage time of fresh fruit. Blanching apples at 59–71°C for about 20 minutes before cooking and pulping into applesauce increased sauce thickness considerably. At water activity below 0.6, red delicious apple slices showed increased hardness in blanched, sulfited and untreated control slices as measured by the force required to shear the slices. As their water activity was lowered toward 0.29, blanched slices became about 2.5 times harder than the sulfited or untreated samples. Brown color development in the slices were maximal at water activity near 0.7 and less on either side of this water activity [19].

Disruptions by blanching increased exposure of macromolecular complexes to the aqueous phase, suggesting enhanced intraparticulate water binding as the mechanism for increased juice cloud stability [142]. Formation of this gel-like network by a blanching pretreatment could result in retention of water by dried fruit, giving softer fruit pieces. The plasticizing effects of water retained by the expanded macromolecular complex in fruit blanched predrying may result in a longer shelf life of cereal and fruit mixtures [19].

One minute in steam and boiling water proved sufficient to inactivate the enzyme peroxidase of cherry. Such high temperatures, however, resulted in a mushy fruit having a dull, reddish-brown cast. Longer periods of time only increased the mushiness and discoloration [124]. Alderman and Newcombe [6] found 3 and 4 minutes of blanching at 57.2°C to be most satisfactory for desirable skin color, flesh color, and general appearance. The dehydrated cherry blanched at the temperature and time mentioned above gave negative results when tested for peroxidase. Higher temperature tended to destroy the red color of skin and soften the flesh to transform mushy texture and loss of water-soluble pigments of the cherry. This low temperature had no effect on the skin waxy layer.

Water blanching time (1–30 min) and temperature (50–100°C) had significant effects on the crispness of banana chips fried in soybean oil at 190°C, and a choice of specific levels of these factors can increase crispness. Blanching whole green bananas for 22 minutes at 69°C before peeling and frying was predicted to produce the crispiest chips, almost three times higher than for chips made from unblanched bananas [95].

Mohamed and Hussein [145] mentioned that low-temperature, long-time (LTLT) blanching (70°C for 20 min) together with calcium treatment can be used to improve the texture of rehydrated dried carrots when compared to high-temperature, short-time (HTST) blanching (100°C for 3 min). LTLT blanching allowed pectin methylesterase to deesterify pectin, which can then react with calcium to form salt bridges. LTLT blanching with dipping solution containing 2% glycerol and 1% calcium chloride with other components (sodium metabisulfite, L-cysteine-HCl, *N*-acetyl-L-cysteine) also affected the properties. L-Cysteine-HCl (0.3%) was found to be most effective in preventing ascorbic acid loss and obtaining a product with the highest rehydration ability, compared to pretreatments with 0.3% *N*-acetyl-L-cysteine and 0.1% sodium metabisulfite. On the other hand, 0.1% sodium metabisulfite was most effective in preserving the carotenoid content of dried carrots. Ascorbic acid and rehydration ability were more adversely affected by long drying time than high drying temperature, while carotenoids were more sensitive to high drying temperature than drying time. At 60°C drying temperature was good for ascorbic acid retention and rehydration ability, while 40°C drying temperature was good for carotenoid content and color of dried carrots [145].

Quintero-Ramos et al. [174] used a double blanch consisting of a LTLT blanch followed by a STHT blanch for carrots. They used whole carrots at varying temperatures (50–65°C) and five time periods (15–90 min), then blanched again for 6 minutes at 100°C. The high-temperature blanch was designed to inactivate enzymes that have the potential to generate off-flavors during storage in dry conditions. They found that this method before dehydration gave significant increases in firmness when the blanch temperature was 60 or 65°C and the blanching time 45 minutes or longer. Blanching at 70 or 55°C or lower gave a less firm product in the same blanch time. The double-blanch process had little effect on rehydration ratios.

Understanding the changes in flavor before, during, and after blanching could aid processors in limiting major flavor changes and determining the quality of products before further drying. The quality of carrot flavor and texture was affected by initial treatment of water blanching, resulting in a low concentration of important flavor compounds and lower intensity rating of sensory attributes. Shamaila et al. [204] analyzed flavor volatiles by dynamic headspace analysis of blanched carrots in boiling water for 0–300 seconds. Most terpenoid volatiles (sabinene, β-pinene, β-myrcene, limonene, *trans*-caryophyllene, α-humulene, β-bisabolene, and α-farnesene) decrease by at least 50% within 60 seconds of blanching. Quality attributes of color, texture, raw carrot aroma, sweetness, flavor, and overall impression decreased with blanching time, while cooked carrot aroma increased. Blanching time correlated with flavor and sensory attributes [204]. Limiting blanching time, rapid cooling, alternative blanching methods, and/or combinations may result in processed carrots with better flavor. Processing methods that include steam and/or microwave blanching and nonthermal enzyme inactivation such as with high-pressure processing may have fewer detrimental effects on flavor and maintain a firmer texture. The rapid loss of aroma compounds during water blanching may contribute to the lack of desirable residual raw carrot aroma in a processed carrot product flavor [204].

Blanching may affect the rehydration of fruits and vegetables, although drying conditions is the most important factor affecting rehydration [85].

Optimum conditions of blanching (time and temperature) should be used to achieve the desired quality of dried products. Thus, adequate measures should be taken to avoid the adverse effects of blanching.

B. Sulfiting

Sulfur dioxide preserves the texture, flavor, vitamins content, and color that make food attractive to the consumer. Sulfur dioxide treatment is used widely in the food industry to reduce fruit darkening rate during drying and storage. The sulfur dioxide taken up by the foods displaces air from the tissue in plant materials, softens cell walls so that drying occurs more easily, destroys enzymes that cause darkening of cut surfaces, shows fungicidal and insecticidal properties, and enhances the bright attractive color of dried fruits. It also inhibits darkening of dried fruit during storage and during distribution to the consumer and preserves ascorbic acid and carotene [140,213].

Sulfiting treatment can be done by burning sulfur or soaking in sulfite solution. The smoke of sulfur dioxide can be produced by burning sulfur with oxygen in the air and then circulating to the smoking chamber. Potential advantages in using a bisulfite solution are [213] (a) decreased air pollution, (b) better control of the sulfuring process, (c) greatly shortened sulfuring time, and (d) decreased desorption losses during drying.

Permitted levels of sulfur dioxide and other additives (solutes) in dried foods varies country to country. According to IFT [91] food legislation, fruits can contain the highest levels of sulfur dioxide of all food products The allowed limit is 2000 mg SO_2/kg dried fruit.

At present sodium metabisulfite is the main compound used in the preservation of food products to generate sulfur dioxide and its corresponding anions. In solution, bisulfite generates SO_2 (aq), HSO_3^-, and SO_3^{2-}. The dominant species, for a pH between 2 and 7, is the hydrogen sulfite (HSO_3^-) ion. At higher pH values, SO_3^{2-} is dominant. However, for pH values lower than 1.86, SO_2 (aq) is the most abundant species [187].

The chemical reactions of sulfur dioxide when it is added to fruits and other food products are complex [25]. When sulfur dioxide is absorbed into fruit, or a product of similar pH, it is converted mainly to bisulfite ion. The bisulfite ion can remain free and available to retard the formation of Maillard-type compounds, and it can also reversibly bind to certain compounds, such as the carbonyl groups of aldehydes [75]. The highest portion of bisulfite ion is available in solution at pH 3.5 [213]. This bound sulfite is considered to have no retarding effect on product deterioration [31], thus it is important to know the factors that influence binding [25]. Bolin and Jackson [25] found that about 30–40% of sulfur dioxide was in the bound form in dried apricots. One factor that influences sulfur dioxide binding is the availability of oxygen to the system. With oxygen present, some of the sulfur dioxide present can be oxidized irreversibly to form the sulfate. By removing these molecules of free sulfur dioxide from the system, the equilibrium is shifted and more bound sulfur dioxide is released. The rate of free sulfur dioxide oxidation varies during drying, depending on the drying procedure used, with a 50% greater rate occurring in sun-dried fruit than in dehydrated fruit.

Bolin and Jackson [25] studied the effect of pH on sulfur dioxide binding in a glucose-bisulfite model solution. There was an increase in percent binding as the solution pH value increased, with a maximum being reached at about pH 4, followed by a reduction in binding. In model solutions the binding increased from 48 to 57% when pH changed from 2.8 to 3.9. In dried apples, the binding at these same pH levels reacted differently, dropping from about 69 to 51%. This difference of the higher percentage of binding occurring in the lower-pH fruit could be caused by the multiplicity of different binding components other than glucose in fruit. For example, two strong binding compounds, acetaldehyde and pyruvic acid, behave very differently [31]. In acetaldehyde, as the pH is raised from 3 to 4, the equilibrium constant of the bisulfite compound decreases. The reverse occurs with pyruvic acid. Moreover, pectins binds strongly and sucrose exhibits no binding, while mannose binds three times more strongly than glucose [92]. Paterson et al. [160] found that the potato starch intrinsic viscosity is reduced after pasting in sulfite at 95°C, indicating that some polysaccharide degradation occurred. It is suggested that this is the reason for the increased release of polysaccharide form the starch granule when pasted in the presence of sulfite. It was also found that the mechanism was oxidative-reductive depolymerization of the starch polysaccharides.

During air-drying of apricots a phenomenon known as white center may form in the pit cavity. This could be caused by desiccation of cells in the pit cavity due to general heat stress experienced by the tree and water loss from the fruits. Consumers may mistake this for some kind of mold. The defect may be overcome in practice by quickly dipping cut fruits in a sulfite solution before drying [140].

Trongpanich [222] found that sulfiting and blanching affected (dip in 1% sodium metabisulfite for 20 min) the texture and rehydration rate of dried bamboo shoots by increasing elasticity and decreasing the rate of rehydration. Preblanching and sulfiting had a greater effect than sulfiting alone. Microphotographs of a cross section through the phloem of the rehydrated samples showed wrinkles on the cell well. The most wrinkling was observed in the preblanched and sulfited sample. Transmission electron micrographs also showed changes in the cell wall constituents due to the pretreatments, and degradation gradually increased with storage time. Abdelhaq and Labuza [1] found that as the level of added sulfite increased, the hardness of unblanched apricots dried to 28% moisture decreased. Scanning electron micrographs of sulfited apricot and peach tissues showed that cell walls had ruptured and collapsed, resulting in a smaller cell volume than in untreated fruit. Also a decrease in penetration force was required for the sulfited fruit [172].

Chan and Cavaletto [38] found that papaya puree with a high concentration of sulfur dioxide (1105 mg SO_2/kg) showed slightly lower rates of drying than papaya puree without sulfite or with a low concentration of sulfite (552 mg SO_2/kg). Moyls [152] showed that finished apple leather, which had no added sulfite, had a larger void space and rigid structure, while added sulfite reduced the rigidity of the structural components and porosity of the product. Thus, a nonsulfited product will dry rapidly like a foam mat, while the sulfited product with the cell wall collapsed or partially collapsed will dry more slowly. Mahmutoglu et al. [131] found that increasing initial sulfer dioxide concentrations and solar drying increased drying rates of apricots. Both during drying and storage, the initial level of sulfur dioxide was seen to decline considerably.

A number of factors affect sulfur dioxide uptake by fruits and vegetables, including concentration and temperature of dipping solution, time of dipping, geometry and conditions of samples (i.e., peeled or unpeeled, whole or sliced), and agitation of solution [186]. The absorption of sulfur into apricot halves mainly occurred on the cutting surface, and no absorption through the apricot skin was observed [213].

C. Dipping Pretreatments

Dipping treatment with chemicals is also used in addition to blanching or sulfiting. The dipping treatment is a process of immersion of foods in a solution containing additives. The concentration level is usually below 5% and dipping time is usually below 5 minutes, whereas osmotic dehydration is usually carried out at higher concentrations and for long processing times. The main purpose of the dipping treatment is to improve drying characteristics and quality. Chemicals used for dipping treatment are summarized in Table 3.

1. Enhanced Drying Rate

Certain surfactants appear to enhance the rate of rehydration of dried tissue [85] (see Table 3). Among these compounds, methyl and ethyl oleate worked the best [24,165]. Increased drying rates for cherries, blueberries, and prunes were observed after dipping treatment with ethyl oleate [168]. Methyl oleate has realized the greatest usage because of economics and its higher taste threshold [81]. A carbonate-oleate combination was superior to their use alone in accelerating drying rate. Eissen et al. [58] found that the drying rate of grapes increased approximately 30% after dipping in a mixture of

Table 3 Chemicals Used for Dipping Treatment

Type	
Chemicals	
Esters	Methyl oleate, ethyl oleate, butyl oleate
Salts	Potassium carbonate, sodium carbonate, sodium chloride, potassium sorbate, sodium polymetaphosphate
Organic acids	Oleic acid, steric acid, caprillic acid, tartaric acid, oleanolic acid
Oils	Olive oil
Alkali	Sodium hydroxide
Wetting agents	Pectin, tween, nacconol
Others	Sugar, liquid pectin
Surfactants[a]	
Nonionic	Monoglycerides, diglycerides, alkylated aryl polyester alcohol, polyoxyethylene sorbitan monostearate, sorbitan monostearate, D-sorbitol, polyoxyethylene
Anionic	Sodium oleate, stearic acid, sorbitan heptadecanyl sulfate
Cationic	Dimethyl-benzyl-octyl ammonium chloride

[a]From Ref. 194.

0.4% olive oil plus 7% potassium carbonate, and better results were also obtained using 2% ethyl oleate plus 2.5% potassium carbonate. Riva and Peri [184] concluded that dipping in an ethyl oleate alkaline solution (2% ethyl oleate plus 2.5% potassium carbonate) resulted in a 25–33% increase in drying rate in comparison to untreated grapes. Bolin and Stafford [23] studied the effect of fatty acid esters and carbonates on enhancing of air-drying rates of grapes. Methyl oleate (2%) and potassium carbonate (2%) (1:1) produced a greater increase in drying rate. A synergistic effect results from the combinated use of alkali carbonates and methyl oleate for the drying of grapes. When excess carbonate was used, drying was further accelerated. Sodium carbonate was less effective than potassium carbonate, however, the cost of sodium salt is about a fifth that of potassium.

Weitz et al. [231] studied the surface dipping treatment of prunes for solar drying. Dipping prunes in 4% methyl oleate was the most effective treatment because of the faster drying rate attained and the better appearance of the final product. An interesting alternative is the employment of olive oil–potassium carbonate emulsions, as they are less expensive and can be easily obtained. There is a 7:1 cost advantage between equal amounts of a 2% methyl oleate solution and a 2% olive oil plus 2% potassium carbonate solution.

Microwave drying of grapes pretreated with 2% ethyl oleate in 0.5% sodium hydroxide solution resulted in comparatively good-quality raisins with reduced drying times. Pretreatment with 3% ethyl oleate in 0.5% sodium hydroxide solution led to a similar product without any major advantage over the former. Grapes treated with 3% ethyl oleate in 2.5% potassium carbonate solution took longer to dry. Sodium hydroxide treatment resulted in inferior quality in terms of color and appearance [223].

A substantial increase of the drying rate of corn by the addition of ethyl oleate to the dipping solution was observed [217]. Raouzeos and Saravacos [182] found that a shorter drying time and good quality of solar-dried product was achieved by pretreatment of grapes in a hot (80°C) solution of 0.5% sodium hydroxide and 2.0% ethyl oleate. They stated that a potassium carbonate solution might be preferable because of safer handling.

Dipping sweet cherries in 0.5–2% solution of boiling sodium carbonate for 5 to 20 seconds followed by a rinse in cool water checked the skins of fruits sufficiently to prevent case hardening during drying. The formation of fine checks on the surface increased the rate of drying. It is normal commercial practice in Australia to spray fresh plums with a near-boiling dilute 0.1% sodium hydroxide solution before they enter the drying tunnel. This has the effect of checking or cracking of the wax layer on the skin of fruit, facilitating escape of moisture from the flesh and consequently accelerating drying [139]. Bayindirli [14] found that treatment with 9% sodium hydroxide at 90°C for 60 seconds was sufficient to peel tomatoes.

Esters affect the waxy surface of fruits by altering the physical arrangement of the surface wax platelets, thus allowing moisture to more readily evaporate from the fruit. This was confirmed using electron microscopy for grapes [170] and sweet cherries [86]. Grncarevic et al. [80] concluded that dipping increased the hydrophilic groups on the wax surface by reversible attachment of long-chain fatty acids and their esters. The increase in hydrophilic groups on the normally lipophilic wax surface would form a sequence of attachment sites to facilitate the transfer of water through the crystalline wax layer. The addition of potassium carbonate is necessary, possibly acting by saponification of fatty acids such as oleic, steric, and oleanolic acids, which are known constituents of grape wax. The conversion of wax from the hydrophobic to the hydrophilic condition is regarded as an important step, and it appears that potassium carbonate influences this reaction [33]. Riva and Peri [184] found that sodium hydroxide dipping had largely superficial effects on grapes due to the solubilization of the waxy material and to intensive mechanical damage of the skin, thus accelerating only the first stage of drying. On the other hand, dipping in an ethyl oleate alkaline solution causes a decrease in the resistance of both the skin tissues and internal diffusion, resulting in a drying time one third to one quarter shorter than that of untreated grapes.

Saravacos and Charm [194], who used surfactants (0.1–0.2%) as a pretreatment for porous fruits such as apples and peaches, found that the drying rate increased during the early drying period (constant rate period). There was no significant effect on the drying rates of potatoes, carrots, onions, garlic, or seedless grapes. The surfactants had no effect on the falling rate period in fruits or vegetables. The equilibrium moisture content and rehydration were not affected by surfactant treatment. The shrinkage properties of potatoes and apples during dehydration were unaffected by the surfactants [194].

Saravacos et al. [196] studied the effect of ethyl oleate on drying rates of starches and seedless grapes in a pilot plant air-dryer. Incorporation of ethyl oleate in unheated pastes of high-amylose starch significantly increased the drying rate during the initial stage of drying, but had little effect on the drying rates for granular high-amylopectin starch pastes and gelatinized starch. Ethyl oleate may act as a wetting agent, increasing the evaporation rate of water from porous surfaces in the early stage of drying. Unheated high-amylose starches are highly porous, whereas granular high-amylopectin starch pastes and maize starches have low porocity. Ethyl oleate, acting as surfactant, may have increased the spreading of free water within the sample (wetting), resulting in an increased drying rate. In grapes, ethyl oleate evidently acts on the grape skin by dissolving the waxy components, which offer high resistance to moisture transfer. Surface-active agents were found to have little effect on the drying rates of grapes or other fruits with waxy skin [194]. Thus, the action of ethyl oleate can be explained by its dissolving action on the waxy components and the cell walls of the grapes and by its wetting (surface-active) effect on the resulting porous structure of the grape skin.

The use of wetting agents definitely decreased the time required to dehydrate cherries, while the addition of pectin or sugar seemed to have the opposite effect of slowing down the drying process [6].

2. Quality Improvement

Potassium carbonate treatment of alfalfa prior to drying improved color compared to nontreated sample [161]. This may be due to the alkalinity of potassium carbonate. Radler [175] studied the prevention of browning during drying by dipping treatment of emulsion oil (2% ethyl oleate plus 2.5% potassium carbonate in water). The enzyme polyphenol oxidase is generally regarded as being responsible for the browning of plant foods. Radler observed that the enzyme was not affected directly by the presence of active compounds of dipping oil, although the drying rate was increased significantly. The activity of the polyphenol oxidase was greatly reduced with increasing sugar concentrations. Thus, rapidly drying prevented browning of pretreated samples. The dipping treatment did not affect the respiration of berries [175]. Dried squid color showed significant improvement by soaking in 0.6% potassium sorbate plus 4.0% sodium polymetaphosphate and adjusting the pH of solution to 5.9 by acetic acid [74].

D. Freezing Pretreatment

Kompany et al. [107] studied the effect of freezing pretreatment during air-, vacuum-, and freeze-drying of fruits and vegetables. They found that the rehydration rate of dried products increased to a level comparable to that for freeze-dried products. It was also noticed that the longer the duration of freezing, the better the kinetics of rehydration of the dried product. This was due to the formation of large ice crystals by slow freezing. Drying rate was not dependent on the pressure in the drying chamber, ranging between 20 and 50 mmHg, and above 50 mmHg the drying rate decreases rapidly with pressure.

E. Cooking Pretreatment

Cooking at different pressure levels before drying can destroy microorganisms and affect the physicochemical properties of dried products. Arganosa and Ockerman [9] developed a stable dried beef product (8–12% water content) by nitrate and nitrite curing using pressure cooking and accelerated forced-air oven drying at 79°C. The dried beef was acceptable up to at least 6 weeks of storage at 32–33°C considering sensory properties, residual nitrite, TBA values, and microbial counts. Del-Valle and Marco [52] also developed a simple process based on blanching and precooking followed by dehydration for the production of quick-cooking (4–5 min) beans. The organoleptic score of the product is comparable to that for freshly cooked beans, and the process sharply decreased butterflying (a quality defect) during drying. Bean flavor development in cooking increased with increasing water: bean ratio, cooking pressure, and time. Air-drying temperature (80 and 100°C) and velocity had no effect on quality. Sultana seedless grapes dipped in alkali and pretreated in a microwave oven and dried in a solar dryer showed a reduction in moisture content of 10–20%. The microwave-treated grapes dried nearly two times faster than controls in a convection dryer. Blanching in boiling water had the same effect on the drying rate as microwaves [109].

V. QUALITY FACTORS OF DRIED FOODS

Multiplication of microorganisms should not occur in properly processed dehydrated foods, but they are not immune to other types of food spoilage. If dried foods are safe in terms of pathogenic microbial count and toxic or chemical compounds, then acceptance depends on the flavor or aroma, color, appetizing appearance, texture, taste, and nutritional value of the product.

Dehydrated foods containing fats are prone to develop rancidity after a period, particularly if the water content is reduced too much. From nonfatty vegetables, such as cabbage, as much water as possible should be removed, because this helps to conserve ascorbic acid. The storage life of dehydrated food is much increased, and the loss of vitamin A and ascorbic acid much decreased, in the absence of oxygen. By completely filling the container with compressed dehydrated food, the amount of oxygen can be reduced to a minimum. Replacement of air in the container with nitrogen is far preferable: most dehydrated foods can be stored for years or more in sealed tins in which the air has been replaced by nitrogen. Lack of initial microbial and pesticide contamination is also important to dried food quality.

A. Selection of Variety

Suitable varieties of produce with desired maturity should be used to achieve a dried product with the best quality. Pea variety and maturity play an important role in controlling the quality of dehydrated peas [32,151]. Lynch and Mitchell [127] observed that maturity was the most important single factor that determined the quality of dehydrated green peas. The relationship of cultivar and coating treatment to the quality of precooked dehydrated pinto beans was investigated by Su and Chang [218]. Blanching (100°C, 15 min), soaking (82.2°C, 1 hr), cooking (100°C, 20 min), coating (70°C, 5 min), and dehydration (65°C, 5–6 hr) were used. Twelve types of biopolymers including modified starches, dextrins, maltodextrins, and alginate, and Fiesta, Othello, and Topaz varieties were used as coating materials. Eight biopolymers effectively reduced splitting of beans without detrimental effects on rehydration, firmness, or color. Cultivars had a strong effect on wholeness of bean products, and Fiesta cultivar was the best for dehydration. Residual trypsin inhibitor activity of precooked pinto beans was 5%, which is good for the product. Peas of different varieties and effects of pretreatment were studied by Beerh and Kurien [17], who found early Waranasi sweet and Duke albany types to be the best for drying.

B. Shrinkage or Collapse

The texture, density, wettability, rehydration capacity, and mechanical properties of dehydrated foods depend directly on drying-process conditions. For example, freeze-drying typically results in biological products with porous crust and superior rehydration capacity, whereas hot air-drying results in a dense product with an impermeable crust. Depending on the end use, these properties may be desirable or undesirable. If a long bowl life is required for a cereal product, a crusty product that prevents moisture reabsorption may be preferred. If a product with good rehydration capacity is required, a slow drying process, such as freeze-drying, may be preferred [3]. Dehydrated vegetables in instant noodles are a good example of this type of product. Thus, a fundamental understanding of the phenomena that control the above-mentioned properties is necessary to achieve the desired quality. Some of these phenomena are discussed in the next section.

Shrinkage is the change in volume during processing due to, for example, moisture loss during drying, ice formation during freezing, and formation of pore by puffing. Shrinkage or collapse during drying plays a significant role in determining the rate of drying and the quality of the dried product. Shrinkage occurs because biopolymers cannot support their weight in the absence of water [4].

Shrinkage is an important phenomenon impacting dried food product quality by reducing product wettability, changing product texture, and decreasing product absorbency. It is a result of capillary collapse, a normal occurrence as fluid is evaporated or removed from a capillary. Shrinkage prevents moisture replacement by air.

The apparent shrinkage during processing can be defined as the ratio of the apparent volume at a given moisture content and the initial apparent volume of the materials before processing. Two types of shrinkage are usually observed in food materials: isotropic and anisotopic. Isotropic shrinkage can be described as the uniform shrinkage of all geometric dimensions of a material. Anisotropic shrinkage can be described as the nonuniform shrinkage of different dimensions of a material.

In fish and seafood, shrinkage in the direction parallel to muscle fibers is significantly different to that perpendicular to the fibers during air-drying [11,179], whereas most fruits and vegetables shrink isotropically [179].

1. Mechanism and Control of Shrinkage or Collapse

Some raw materials may end up as completely different products, depending on the type of drying process applied. For example, drying at 5°C results in a porous product with good rehydration capacity, whereas drying at 80°C results in a dense product with poor rehydration capacity [100]. Products baked at a low temperature have a different texture than those baked at a higher temperature. The low-temperature product has a crumbly texture, whereas that baked at high temperatures has a crispy crust. It is this interrelationship between the drying-process conditions and the final product quality that is not completely understood [3]. Genskow [76] and Achanta and Okos [3] mentioned several mechanisms that affect the degree of collapse, and understanding these mechanisms would aid in achieving desired shrinkage or collapse in the products.

A product's bulk density can be varied significantly by changing the drying temperature [76,136]. If spray-drying temperature is maintained significantly higher than the boiling point of the solvent, the product forms a skin, and internal solvent vaporization causes formation of puffed product granules with a low bulk density. A relatively low drying temperature may result in a high-density product if slightly above the material glass transition temperature [4].

The following physical mechanisms play an important role in the control of shrinkage or collapse:

Surface tension: considers collapse in terms of the capillary suction created by a receding liquid meniscus
Plasticization: considers collapse in terms of the plasticizing effect solvents have on various polymer solutes
Electrical charge effects: considers collapse in terms of van der Waals electrostatic forces
Mechanism of moisture transport effects
Gravitational effects

2. Plasticization

Biopolymers are by nature viscoelastic, thus shrinkage or collapse on removal of moisture occurs at a finite rate. The rate at which shrinkage occurs is related to the viscoelastic properties of the matrix moisture. The higher the viscosity of the mixtures, the lower the rate of shrinkage and vice versa. Shrinkage and glass transition are interrelated, and significant shrinkage can be noticed during drying only if the temperature of drying is higher than the T_g of the material at that particular moisture content [4].

The extent of shrinkage and the density of the final product typically depend on the mobility of moisture in the material and the rate at which the material is dried. This could be clearly understood by comparing freeze-drying and hot air-drying of vegetables. In freeze-drying, since the temperature of drying is above the T_g, the material is in the glassy state, hence shrinkage is negligible. As a result the final product is very porous. In hot air-drying, since the temperature of drying is above the T_g, the material is in the rubbery state and substantial shrinkage occurs. This is the reason why the product of hot air-drying is dense and shriveled. Karathanos et al. [100] mentioned that maintaining the material at its initial moisture content and air-drying below the T_g is enough to reduce shrinkage. However, this is not as efficient as freeze-drying to reduce shrinkage, since air-drying at that temperature is very slow. Thus, residence time can be long enough to result significant shrinkage in the case of air-drying in a frozen state [4].

C. Case Hardening or Crust Formation

Since outer layers necessarily lose moisture before interior portions, the concentration of moisture in these layers is less than in the interior, and the surface layers shrink against an unyielding, constant-volume core. This surface shrinkage causes checking, cracking, and warping. This type of shrinkage causes moisture gradient and resistance near the surface.

In extreme cases, shrinkage and drop in diffusivity may combine to give a skin practically impervious to moisture, which encloses the bulk of the material so that interior moisture cannot be removed. This is called *case hardening* [138]. In food processing, case hardening is also commonly known as crust formation. Warping, checking, cracking, and case hardening can be minimized by reducing the rate of drying, thereby flattening the concentration gradients in the solid and protecting the entire piece against shrinkage. The rate of drying is controlled most readily by controlling the humidity of the drying air.

Crust (or shell) formation may be either desirable or undesirable in food products. If a dried product is to be rehydrated before use, crust is undesirable, but for breakfast cereals where long bowl life is desirable, crust is necessary. In microencapsulation of flavors, rapid crust formation is required to prevent flavor losses. The thickness of the crust that forms is a function of the drying conditions. The faster the drying rate, the thinner the crust. Slow drying above glass will result in a product with thick crust [4]. During rapid drying, surface moisture drops to a low value so that the surface, which was initially rubbery, undergoes a rapid transition to the glassy state. If the internal moisture is not high enough to replace the moisture lost from the surface, a dense crust is formed. Since the curst is in a glassy state, slow surface drying rate prevents further shrinkage. Achanta and Okos [4] noted that crust formation may be inhibited by (a) allowing the drying rate to be slow enough that moisture loss from the product surface is replenished by moisture from the inside, (b) drying below the glassy state, and (c) reducing the time scale of internal diffusion by reducing characteristic length, since smaller particles give lower internal resistance. Crust formation is also important in explosion puffing. In this case the high-moisture product is exposed to rapid drying conditions such as high temperature and vacuum. This results in rapid surface drying and crust formation. The impermeable crust coupled with the extreme drying conditions results in rapid moisture vaporization and causes large internal pressures to build up, resulting in product expansion/puffing. During the expansion stage, stress build-up in the glassy surface may cause the surface to crack, allowing vapor to escape. Cracking of the crust may result in the loss of some of its impermeability, and the degree of product expansion may decrease. Achanta and Okos [4] mentioned that experimental studies of the dynamics of crust formation are absent in literature.

D. Stress Development and Cracking or Breakage

Checking and breakage of dried foods has two undesirable consequences: loss of valuable product and loss of consumer satisfaction [4]. Cracking is detrimental to grain quality since affected kernels are more susceptible to mold attack during storage and pathogenic invasion after seeding. Cracked grains are also of lower organoleptic quality, which limits their use in direct food preparation. The destructive effect of cracks is of prime importance for seed-quality grains because of reduced germination [110]. A biopolymer during drying shrinks at the surface, causing tensile stress at the surface and compressive stress on the inside. When the tensile stress on the material surface exceeds the material breaking strength, the material cracks. Figure 8 illustrates this concept.

Internal cracking in the starchy endosperm of a grain is induced by mechanical stress due to the high humidity gradient inside the kernel and/or to thermal stress. The fissure is a large internal fracture usually found to be perpendicular to the long axis of grain [205]. Fissures in rice grains result in kernal breakage, particularly during processing. Very rapid drying causes case hardening on the surface of the grain, dry sealing the moisture within the inner layers, causing tensile stress at the grain surface and compressive stress at the core. When the moisture gradient decreases after drying, moisture from the central portion of the grain diffuses to the surface, causing it to expand while the internal portions contract due to moisture loss. The result is compression at the surface and tension in the central portions of the grain. When the compressive stress at the grain surface exceeds the tensile strength of its interior, the kernel pulls apart, resulting in fissures [54]. Most fissuring occurs within 48 hours after drying, but additional fissures develop at a low rate for another 72 hours thereafter [111,112]. Drying rate is a significant factor contributing to the development of fissures in rice, which is detrimental to rice quality [12,112,114]. High or fast drying rate is the main cause of breakage. If the drying rates of paddy are high, in particular for air temperatures above 45°C, internal cracking can occur. Arora et al. [10] showed that a temperature difference larger than 43°C between air-drying and rice kernels may result in serious cracking, and drying temperature should be below 53°C. The humidity difference between the grain and air also influences the process of cracking [113]. De Peuty et al. [54] found that breakage can be decreased at higher drying temperatures by increasing air humidity. Bonazzi et al. [26] found no loss of quality due to thermal gradients if paddy is exposed to temperatures between 30 and 90°C without drying. On the other hand, drying at ambient humidity in the same air temperature as above leads to a rapid increase in the percentage of broken kernels. This proportion increases with the drying rate. Thus, humidity of drying is the major factor for keeping quality. At the beginning of a drying process the effect due to the moisture gradient appears to be dominant. The reason is that a difference in moisture content of 0.01 kg/kg (macaroni—radius:length ratio of 1.7) developed a maximum tangential stress of 12.4 MPa as compared to one of 0.131 kPa associated with a temperature gradient of 11 K [56].

Sarker et al. [197] found that high initial moisture contents of rice grains and high drying air temperatures caused the most fissured grains. HTST drying caused grains to fissure soon after drying. The transient moisture gradient is highest perpendicular to the long axis of the grain and located near the grain middle. The steepest gradient is near the surface of the endosperm, and gradients after short drying periods produce fissures more quickly than gradients after long drying periods.

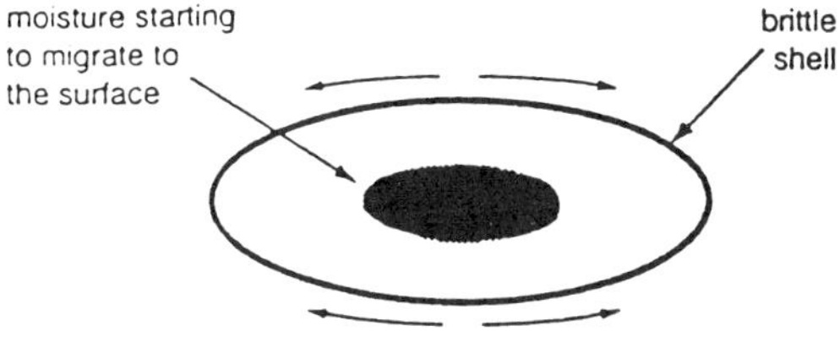

(a) grain immediately after drying showing brittle surface under tension

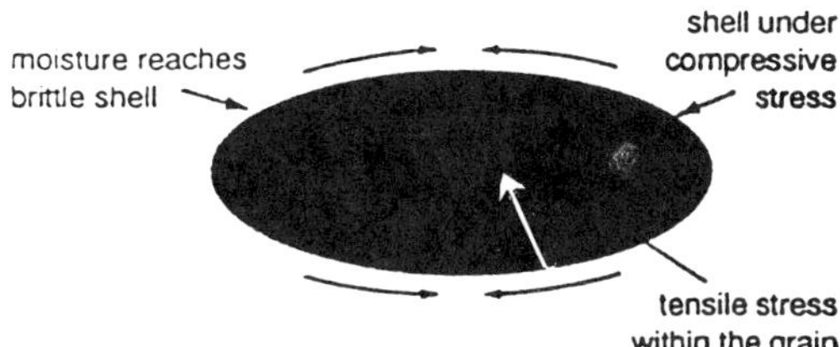

(b) brittle shell swells slower than layers beneath as it absorbs more moisture

(c) grain fissures to release stress

FIGURE 8 Concept of rice fissuring due to moisture migration within the grain. (From Ref. 30.)

In microwave drying, stress cracking can be even more pronounced due to superposition of the pressure gradient that may build up within the material under certain drying conditions [224]. Adu eta al. [5] successfully applied the theory of viscoelasticity in explaining stress development in soybeans during pulsed microwave drying. Kudra et al. [110] studied the stress cracking of wheat by soft x-ray photography during microwave drying. They found that wheat variety Panda is more vulnerable than the variety Kadett. The higher level of initial damage before harvesting should promote further destruction during more intense drying. There are definite effects from both the inlet air temperature and the microwave power density on stress cracks.

E. Volatile Development or Retention

In addition to the physical changes, drying generates flavor or releases flavor from the foods. Drying of fresh bell peppers greatly changed the composition of volatiles: most volatiles evaporated and new volatile odor compounds were formed by chemical reactions [126,228]. Such changes in volatiles might affect the aroma of fresh peppers after drying. The Spanish *Capsicum* spp. scored relatively high for off-aroma, rubbery, and sharp, while the Hungarian samples generally scored higher for flavor similar to tomato, hay, and sweet aroma [233]. Luning et al. [125] studied the aroma of fresh and not air-dried bell peppers. They found that hot air-drying decreased levels of the odor compounds 3-hexenal, 2-heptanone, 2-hexenal, 2-hexenal, hexanol, 3-hexanol, 2-hexenol, and linalool. Drying increased the level of 2-methylpropanal and 2- and 3-methylbutanal. The increase of volatile compounds, with cacao, spicy, sweaty, rancid and sweet, caramel-like odor notes during drying, may be correlated with some typical aroma attributes of dried bell peppers like savory, rancid/sweaty, sweet/stickly, and hay-like aroma.

Beasley and Dickens [15] recognized that off-flavors were produced when drying air temperatures were above 35°C. They observed that the amount of off-flavor detected appeared to be a function of drying air temperature and peanut moisture content, and off-flavor was more likely to occur in immature peanuts than in mature peanuts. Off-flavors resulting from high-temperature drying can be passed on to peanut butter and roasted peanuts. Osborn et al. [158] measured the concentration of acetaldehyde, ethanol, and ethyl acetate in mature (black class) and immature (orange class) peanuts during drying at 35 and 40°C. The concentration of these compounds within the kernel increased during drying, and the highest concentration was observed in immature kernels dried at 40°C. The factors that determined the production rates of these volatiles were drying air temperature, peanut maturity, and moisture content. They also suggested that acetaldehyde and ethyl acetate may be better indicators of off-flavor.

High-temperature drying cycles are widely used for industrial pasta making for several reasons: reduction of processing time, increase in plant output, and significant improvement of both cooking quality and hygienic standards of the dried pasta. However, high-temperature cycles promote extensive heat damage in dried pasta, mainly related to off-color and off-flavor, loss of protein, and formation of unnatural new compounds [183]. Resmini et al. [183] concluded that in order to keep the advantages of high temperatures in pasta drying without promoting extensive Maillard reaction compounds and related undesirable effects, the whole pasta-making process must be modified from a more biotechnological point.

Retention or loss of volatile compounds during spray-drying can be vital for product quality. A substantial volatile loss occurred during the first three stages of spray drying, and there should be zero or very little loss of volatiles during the fourth stage due to selective diffusion [106]. Losses can occur during atomization, from undisturbed drops, and as a result of morphological development. Several factors affect volatile retention, including control of atomizer pressure or rotation speed, choice of spray angle, configuration of air input, alteration of air temperature profile, feed concentration, presence of an oil phase and/or suspended solids, foaming of the feed, feed composition, and stream blanketing of the atomizer [106]. Senoussi et al. [202] studied the operating conditions for maximal retention of sensitive tracers (tyrosinase, vanillin, diacetyl, ascorbic acid, thiamine hydrochloride) for three different spray dryers. Feed solution concentration and drying air temperature, together with the type of dryer and method of atomization, are important parameters to be considered. Support materials (maltodextrin, sucrose, skim milk, and whole milk) also had a significant effect on retention. The quality is well retained in maltodextrin and skim milk, and retention is impaired by decreasing maltrodextrin solid content from 50 to 40% or replacing 20% maltodextrin with sucrose. The retention of ethanol in a single droplet of model food with aqueous maltodextrin on spray drying was also studied by Furuta et al. [73]. They found that retention increased with increasing initial concentration of maltodextrin, increasing air temperature and velocity and decreasing humidity.

The losses of volatile components during spray drying are governed by a ternary diffusion process within the drops. The loss rates are greatest in the immediate vicinity of the atomizer, before sufficient drying of the drop surfaces occurs for a selective diffusion mechanism to become dominant. In the selective diffusion mechanism, the surface water content is reduced sufficiently so that the diffusion coefficients of volatile substances become substantially lower than that of water [103,105]. Losses of volatile acetates in sucrose solutions have been monitored as a function of location in the region near the atomizer [103]. Foaming and incorporation of an emulsified oil phase or surfactants all have substantial effects on volatile loss [105]. The surfactants were effective in reducing volatile loss from sucrose solution by virtue of suppression of droplet circulation and oscillation [71].

Kompany and Rene [108] studied the effect of freezing methods on the retention of aroma (1-octene-3-ol, 3-octanone, 3-octanol, benzaldehyde, and benzyl alcohol) during freeze-drying of mushrooms. They found that slow freezing results in higher retention of 1-octene-3-ol in the freeze-dried mushroom.

F. Nonequilibrium or Glass Transition

The glass transition theory explaining quality changes during drying is reviewed by Achanta and Okos [4].

G. Caking and Stickiness

Caking and stickiness of powders, desirable or undesirable, occur in dried products. Caking is desirable for tablet formation and undesirable when a dry free-flowing material is required. To reduce caking during drying, a logical option is to dry rapidly so that moisture content drops to a level where caking is inhibited. This rapid drying will form a crust, which may be undesirable, thus product optimization or solutes in product formulation may be considered.

Tendencies to form surface folds on particles during spray drying are governed by the viscosity of the concentrated solution. Stickiness and agglomeration tendencies also depend upon the viscosity of the concentrated solution, surface tension, particle size, and exposure time [105]. For viscosities below the critical value, stickiness usually occurs. The predicted critical viscosity was within the range of 10^9–10^{11} mPas. For amorphous particles with concentrate viscosities greater than this value, there will be insufficient flow to cause stickiness during short (1–10 sec) contacts [189,200]. Downton et al. [53] postulated a mechanism of sticking and agglomeration through viscous flow, driven by surface tension and forming bridges between particles. The size of a "bridge" needed to cause sticking of particles was estimated as 0.1–1.0% of the particle diameter on the basis of the strength of "bridges" between particles [189,200].

H. Functional Properties

Protein solubility of corn decreases with an increase in air-drying temperature [128,141,162]. Saline-soluble protein denaturation (property of the germ) begins at more than 75°C and leads to complete denaturation at or above 100°C [149]. Mourad et al. [149] found that drying methods (thin layer, slow drying and fluidized bed, fast drying) did not considerably influence protein denaturation, which indicates that this denaturation is a purely thermal process, and thus drying rate had a negligible effect [149].

The influence of air-drying temperature on grain millability is characterized by an increase of starch fraction in the fine fiber. A reduction in the starch yield is directly attributed to the thermal denaturation of protease in the endosperm of the grain [57]. Degradation of wet milling quality (property of endosperm) was observed at temperatures higher than 95°C. Grains with low protein solubility do not necessarily indicate poor millability, because the former measures a property of the germ, whereas the latter is an endosperm property [149].

In spray-drying, powder from rotary atomizers normally has a higher bulk density than powder from pressure nozzle atomizers. Thus, an atomization system accompanying operating conditions is necessary for desired bulk density. Concentration of feed to higher solids normally increased bulk density. The characteristics of the feed are also important. Slurry feeds dry to higher bulk density powders than do solutions. Preheating to reduce viscosity and improve the homogeneity of the atomized spray can also increase bulk density. An increase in inlet air temperature influences the bulk density of many products by decreasing its value and cooling, and transporting powder in a conveying system increase bulk density [137].

Sadowska et al. [191] studied the air-drying conditions (initial moisture, 10–22%; temperature, 60–120°C) and processability of dried rapeseed. The rheological properties by the oil-point test were strongly correlated with initial moisture content before drying and air-drying temperature when they dried to a final moisture content of 6.5%.

I. Color Retention or Development

High temperature and long drying time degrade a product's original color. Color in foods can be preserved by minimal heat exposure or applying HTST or short time with pH adjustment [28,84]. At lower water activities chlorophyll is bound in noninteractive components, or water may not be available for the reaction to form pheophytins from chlorophyll [201]. Dutton et al. [55] observed that pheophytin content of the dried spinach decreased as the moisture level was lowered for preservation of green color. Another cause of color degradation may be due to enzymatic browning causing rapid darkening, mainly of the leafy portions. The formation of dark pigments via enzymatic browning is initiated by the enzyme polyphenol oxidase. Another reason for discoloration is photo-oxidation of pigments, caused by light in combination with oxygen leading to severe discoloration. This phenomenon was reported to occur in meat by Anderson et al. [8].

Patil et al. [161] found that the color of dried alfalfa stems was better when they were dried in smaller chop lengths. The increase in drying air temperature had a beneficial effect on the color of chopped alfalfa dried up to 175°C.

Nketsia-Tabiri and Sefa-Dedeh [156] studied the quality of air-dried salted tilapia as a function of salting time (0–24 hr), air-drying time (6–20 hr), and temperature (40–60°C). At each salting time the mean hardness score was dependent on the drying temperature and time, while the color and acceptability scores were influenced by the drying temperature. The highest quality level can be achieved at any temperature by manipulating salting and drying time.

J. Rehydration of Dried Products

The removal of water from plant cells changes the cells' physical properties, which affects the rehydration of dried products. These effects include (a) loss of osmotic pressure, (b) change of permeability in the protoplasmic membrane, (c) crystallization of polysaccharide gels in the cell wall, (d) coagulation of protoplasmic proteins, (e) formation of pores, and (f) change of pH. The factors affecting the rehydration of dried products can be categorized as intrinsic and extrinsic.

1. Intrinsic

Dehydration increases the crystallization of polysaccharide gels by bridging reactive polymers groups closer together. The free hydroxyl groups of polysaccharides in fresh vegetables and fruits have a secondary valence, which is almost completely fulfilled by water. These hydroxyl groups lose their noncovalently bound water upon dehydration. The shrinkage of the plant cells enables adjacent polysaccharides molecules to be drawn together and thus fulfil the hydroxyl groups valence [115].

Drying causes skins to toughen and makes it difficult for the water to penetrate into dried foods. Predehydration treatments are also usually designed to improve rehydration properties.

Pricking prior to dehydration helped in quick dehydration and better rehydration of green peas. However, the extent of advantage gained due to pricking depended on the pea variety and also varied with pea maturity [203].

2. Extrinsic Factors

Neubert et al. [154] studied the effects of the pH of the rehydration medium on the rehydration ability of dehydrated celery. They tested the pH range 2.6–11.5 and concluded that pH had no appreciable effect on the rehydration of celery that had had no predrying treatment. Horn and Sterling [88] found dried carrots to be rehydrated to a maximum extent at pH values 2 and 12. The greater hydration at low pH may be as a result of the smaller number of hydrogen ions present to catalyze hydrolysis. Similarly, carrots rehydrated to the greatest extent at pH 12 possibly because the concentration of the hydroxyl anions was great enough at this pH to cause a substantially greater degree of oxidation of the hemicellulose and pectinates in the cell walls of carrots.

The effect of anions on the inhibition of crystallization (retrogradation) of starch followed in the order of: $SCN^- \rightarrow (PO_4^{3-}, CO_3^{2-}) \rightarrow I^- \rightarrow NO_3^- \rightarrow Br^- \rightarrow Cl^- \rightarrow C_2H_3O_2^- \rightarrow F^- \rightarrow SO_4^{2-}$ [148]. Ciacco and Fernandes [41] showed that rate of retrogradation of wheat starch gels followed the lyotropic series. Sieh and Sterling [209] reported that neutral polysaccharides, e.g., glycogen, obtained an electric charge in a salt solution. The order of the anions of a sodium salt that gave glycogen an increasing electrophoretic mobility is: $SO_4^{2-} \rightarrow$ tartrate$^- \rightarrow Cl^- \rightarrow Br^- \rightarrow C_2H_3OO_2^- \rightarrow HCO_3^- \rightarrow$

$H_2PO_4^- \rightarrow F^- \rightarrow NO_3^- \rightarrow HPO_4^{2-} \rightarrow I^- \rightarrow SCN^- \rightarrow CO_3^{2-} \rightarrow PO_4^{3-} \rightarrow OH^-$. Glycogen is distilled water and in sodium citrate had no electrophoretic mobility. The above order of anions was approximately the same order as the lyotropics series for starch precipitation [148].

The effect of different anions and of solution pH on the degree of rehydration (change in volume) of dehydrated carrots was measured by Horn and Sterling [88]. Carrots had significantly greater rehydrated volume when rehydrated in distilled water than in any salt solution. The order was: $H_2O \rightarrow SCN^- \rightarrow$ NO\eq $\rightarrow Cl^- \rightarrow$ SO\eq $\rightarrow$ citrate. This effect may be due to a steric hindrance to the penetration of the anions into the plant cells.

Color of cherries rehydrated in a sugar solution was superior to that of cherries rehydrated in water. This indicates that the sugar solution caused less pigment loss [6].

REFERENCES

1. E. H. Abdelhaq and T. P. Labuza, Air drying characteristics of apricots, *J. Food Sci. 52*:342 (1987).
2. A. Abu-Bakar, M. Y. Abdullah, and K. Azam, The effects of caramel on the quality of smoked fish, *ASEAN Food J. 9*:116 (1994).
3. S. Achanta and M. R. Okos, Impact of drying on biological product quality, *Food Preservation by Moisture Control, Fundamentals and Applications* (G. V. Barbosa-Canovas and J. Welti-Chanes, eds.), Technomic Publishing, Lancaster, PA, 1995, p. 637.
4. S. Achanta and M. R. Okos, Predicting the quality of dehydrated foods and biopolymers—research needs and opportunities, *Drying Technol. 14*:1329 (1996).
5. B. Adu, L. Otten, and R. B. Brown, Preventing stress cracking during microwave drying: viscoelastic approach, ASAE Meeting, St. Joseph, MI, Paper No. 92-6007.
6. D. C. Alderman and B. Newcombe, Dehydration of montmorency cherries, *Mich. Quart. Bull. 28*:97 (1945).
7. S. M. Alzamora, G. Hough, and J. Chirife, Mathematical prediction of leaching losses of water soluble vitamins during blanching of peas, *J. Food Technol. 20*:251 (1985).
8. H. J. Anderson, G. Bertelsen, A. Ohlen, and L. H. Skibsted, Modified packaging as protection against photodegradation of the color of pasteurized sliced, *Meat Sci. 28*:77 (1990).
9. F. C. Arganosa and H. W. Ockerman, The influence of curing ingredients, packaging method and storage on the biochemical and sensory qualities and acceptability of a dried beef product, *J. Food Proc. Pres. 12*:45 (1987).
10. V. K. Arora, S. M. Henderson, and T. H. Burkhardt, Rice drying cracking versus thermal, and mechanical properties, *Trans. ASAE 16*:320 (1973).
11. M. Balaban and G. M. Pigott, Shrinkage in fish muscle during drying, *J. Food Sci. 51*:510 (1986).
12. T. Ban, Rice cracking in high rate drying, *Jpn. Agric. Res. Quart. 6*:113 (1971).
13. A. Banks, *J. Sci. Food Agric. 3*:250 (1952).
14. L. Bayindirli, Mathematical analysis of lye peeling of tomatoes, *J. Food Eng. 23*:225 (1994).
15. E. O. Beasley and J. W. Dickens, Engineering research in peanut curing, Technical Bulletin No. 155, North Carolina Agricultural Experiment Station, Raleigh.
16. M. Beedie, Energy savings—a question of quality, *South Afr. J. Food Sci. Technol. 48*(3):14, 16 (1995).
17. O. P. Beerh and S. Kurien, Influence of variety, maturity and some pre-drying treatments on the quality of dried peas, *Indian Food Packer 30*:27 (1976).
18. C. I. Beristain, E. Azuara, R. Cortes, and H. S. Garcia, Mass transfer during osmotic dehydration of pineapple rings, *Int. J. Food Sci. Technol. 25*:575 (1990).
19. T. Beveridge and S. E. Weintraub, Effect of blanching pretreatment on color and texture of apple slices are various water activity, *Food Res. Int. 28*:83 (1995).
20. E. S. A. Biekman, H. I. Kroese-Hoedeman, and E. P. H. M. Schijvens, Loss of solutes during blanching of mushrooms (*Agaricus bisporus*) as a result of shrinkage and extraction, *J. Food Eng. 28*:139 (1996).

21. R. N. Biswal and M. Le Maguer, Mass transfer in plant materials in contact with aqueous solutions of ethanol and sodium chloride: equilibrium data, *J. Food Proc. Eng. 11*:159 (1989).
22. R. N. Biswal, K. Bozorgmehr, F. D. Tompkins, and X. Liu, Osmotic concentration of green beans prior to freezing, *J. Food Sci. 56*:1008 (1991).
23. H. R. Bolin and A. E. Stafford, Fatty acid esters and carbonates in grape drying, *J. Food Sci. 45*:754 (1980).
24. H. R. Bolin, V. Petrucci, and G. Fuller, Characteristics of mechanically harvested raisins produced by dehydration and by field drying, *J. Food Sci. 40*:1036 (1975).
25. H. R. Bolin and R. Jackson, Factors affecting sulfur dioxide binding in dried apples and apricots, *J. Food Proc. Pres. 9*:25 (1985).
26. C. Bonazzi, F. Courtois, C. Geneste, B. Pons, M. C. Lahon, and J. J. Bimbenet, Experimental study on the quality of rough rice related to drying conditions, Drying '94, Proceedings of the 9th International Drying Symposium, Gold Coast, 1994, p. 1031.
27. P. Britnell, S. Birchall, S. Fitz-Payne, G. Young, R. Mason, and A. Wood, The application of heat pump dryers in the Australian food industries, *Drying '94* (V. Rudolph, R. B. Keey, and A. S. Mujumdar, eds.), University of Queensland, Brisbane, 1994, p. 897.
28. K. A. Buckle and R. A. Edwards, Chlorophyll color and pH changes in HTST processed green pea puree, *J. Food Technol. 5*:173 (1970).
29. Preservation using chemicals, *Short Course Manual on Food Processing and Preservation*, Bangladesh University of Engineering and Technology, Dhaka, 1983, p. 1.
30. M. C. Bulaong, Modelling the head rice yield of high moisture grains after high temperature drying, M.Sc. thesis, University of New South Wales, Sydney, 1994.
31. L. F. Burroughs and A. H. Sparks, Sulphite-binding power of wines and ciders, III. Determination of carbonyl compounds in a wine and calculation of its sulphite-binding power, *J. Sci. Food Agric. 24*:207 (1973).
32. J. S. Caldwell, C. W. Culpepper, B. D. Ezell, M. S. Wilcox, and M. C. Hutchins, The dehydration of peas, *The Canner 103*:13 (1946).
33. T. C. Chambers and J. V. Possingham, Studies of the fine structure of the wax layer of sultana grapes, *Aust. J. Biol. Sci. 16*:818 (1963).
34. W. M. Camirand, R. R. Forrey, K. Popper, F. P. Boyle, and W. L. Stanley, Dehydration of membrane-coated foods by osmosis, *J. Sci. Food Agric. 19*:472 (1968).
35. C. P. Champagne, F. Mondou, Y. Raymond, and D. Roy, Effect of polymers and storage temperature on the stability of freeze-dried lactic acid bacteria, *Food Res. Int. 29*:555 (1996).
36. C. P. Champagne, N. Gardner, E. Brochu, and Y. Beaulieu, The freeze-drying of lactic acid bacteria. A review, *Can. Inst. Food Sci. Technol. J. 24*:118 (1991).
37. C. P. Champagne, N. Morin, R. Couture, C. Ganon, P. Jelen, and C. Lacroix, The potential of immobilized cell technology to produce freeze dried, phage-protected cultures of *Lactococcus lactis*, *Food Res. Int. 25*:419 (1992).
38. H. T. Chan and C. G. Cavaletto, Dehydration and storage stability of papaya leather, *J. Food Sci. 43*:1723 (1978).
39. J. Chirife and M. Karel, Volatile retention during freeze drying of aqueous suspensions of cellulose and starch, *J. Agr. Food Chem. 21*:936 (1973).
40. J. Chirife, M. Karel, and J. Flink, Studies on mechanisms of retention of volatile in freeze-dried food models: the system PVP-n-propanol, *J. Food Sci. 38*:671 (1973).
41. C. F. Ciacco and J. L. A. Fernandes, Effects of various ions on the kinetics of retrogradation of concentrated wheat starch gels, *Starch 31*:51 (1979).
42. S. Clegg, The physical properties and metabolic status of artemia cysts at low water contents: the "water replacement hypothesis," *Membranes, Metabolism and Dry Organisms* (A. C. Leopold, ed.), Comstock Publishing Associates, London, 1986, p. 374.
43. J. S. Cohen and C. S. Yang, Progress in food dehydration, *Trends Food Sci. Technol. 6*:20 (1995).
44. J. E. Contreras and T. G. Smyrl, An evaluation of osmotic concentration of apple rings using corn solids solutions, *Can. Inst. Food Technol. J. 14*:310 (1981).

45. W. J. Coumans and W. M. A. Kruf, Transport parameters and shrinkage in paper drying, Drying '94, Proceedings of the 9th Internal Drying Symposium, Gold Coast, p. 1205.
46. R. Couture, D. Gagne, and C. P. Champagne, Effect dedivers additifs sur la survie a la lyophilisation de *Lactococcus lactis*, *J. Inst. Can. Sci. Technol. Alim. 5*:224 (1991).
47. J. Crank, Some mathematical diffusion studies relevant to dehydration, *Fundamental Aspects of Dehydration of Foofstuffs*, Macmillan, London, 1958, p. 37.
48. L. M. Crowe, R. Mouradian, J. H. Crowe, S. A. Jackson, and C. Womersley, Effects of carbohydrates on membrane stability at low water activities, *Biochim. Biophys. Acta 769*:141 (1984).
49. R. L. Dang, R. P. Singh, A. K. Bhatia, and S. K. Verma, Studies on kashmir apples-canning as rings, *Indian Food Packer. 30*:9 (1976).
50. G. D'Andrea, M. L. Salucci, and L. Avigliano, Effect of lyoprotectants on ascrobate oxidase activity after freeze-drying and storage, *Proc. Biochem. 31*:173 (1996).
51. R. V. Decareau, *Microwave in the Food Processing Industry*, Academic Press, New York, 1985.
52. F. R. Del-Valle and E. Marco, Production of quality quick-cooking beans by a cooking/dehydration process, *J. Food Proc. Pres. 12*:83 (1988).
53. G. E. Downton, J. L. Flores-Luna, and C. J. King, Mechanism of stickiness in hygroscopic, amorphous powders, *Ind. Eng. Chem. Fundam. 21*:447 (1982).
54. M. A. Du-Peuty, A. Themelin, J. F. Cruz, G. Arnaud, and J. P. Fohr, Improvement of paddy quality by optimising of drying conditions, Drying '94, Proceedings of the 9th International Drying Symposium, Gold Coast, p. 929.
55. H. J. Dutton, G. F. Bailey, and E. Kohake, Dehydrated spinach, *Ind. Eng. Chem. 35*:1173 (1943).
56. P. L. Earle and N. H. ceaglske, Factors causing the checking of macaroni, *Cereal Chem. 26*:267 (1949).
57. S. R. Eckhoff and C. C. Tso, Starch recovery from steeped corn grist as affected by drying temperature and added commercial protease, *Cereal Chem. 68*:319 (1991).
58. W. Eissen, W. Muhlbauer, and H. D. Kurtzbach, Solar drying of grapes, *Drying Technol. 3*:63 (1985).
59. M. Z. El-Abd, M. N. Naim, and Y. A. El Tawil, Drying of clover seeds using infrared and electrically heated dryers, Proceedings of the Third International Drying Symposium, Vol. 2 (J. C. Ashworth, ed.), Drying Research Ltd., Wolverhampton, 1982, p. 407.
60. D. F. Farkas and M. E. Lazar, Osmotic dehydration of apple pieces: effect of temperature and syrup concentration on rates, *Food Technol. 23*:688 (1969).
61. P. J. Fellows, *Food Processing Technology: Principles and Practices*, Ellis Horwood, New York, 1992, p. 505.
62. P. Fito, Modelling of vacuum osmotic dehydration of food, *J. Food Eng. 22*:313 (1994).
63. P. Fito and R. Pastor, Non-diffusional mechanisms occurring during vacuum osmotic dehydration, *J. Food Eng. 21*:513 (1994).
64. J. M. Flink, Dehydrated carrot slices: influence of osmotic concentration on drying behaviour and product quality, *Food Process Engineering* (P. Linko, Y. Malkki, J. Olkku, and J. Larinkari, eds.), Applied Science Publishers, London, 1979, p. 412.
65. J. Flink and M. Karel, Retention of organic volatiles in freeze-dried solutions of carbohydrates, *J. Agr. Food Chem. 18*:295 (1970).
66. J. Flink and M. Karel, Effects of process variables on retention of volatiles in freeze-drying, *J. Food Sci. 35*:444 (1970).
67. J. Flink and T. P. Labuza, Retention of 2-propanol at low concentration by freeze drying carbohydrate solutions, *J. Food Sci. 37*:617 (1972).
68. G. Font de Valdez, G. S. De Giori, A. P. De Ruiz Holgado, and G. Oliver, Comparative study of the efficiency of some additives in protecting lactic acid bacteria against freeze-drying, *Cryobiology 22*:574 (1983).
69. B. A. Fox and A. G. Cameron, *Food Science a Chemical Approach*, Hodder and Stoughton, London, 1982.
70. W. C. Frazier and D. C. Westhoff, *Food Microbiology*, 3rd ed., McGraw-Hill, New York, 1978.

71. D. D. Frey and C. J. King, The effects of surfactants on mass transfer during spray drying of aqueous sucrose solutions, AIChE Meeting, Denver, CO, August 1983.
72. W. Fu, S. Suen and M. R. Etzel, Inury to *Lactis* subsp. *lactis C2* during spray drying, Drying '94, Proceedings of the 9th International Drying Symposium, Gold Coast, 1994, p. 785.
73. T. Furuta, S. Tsujimoto, M. Okazaki, and R. Toei, Retention of volatile component in a single droplet during drying, *Engineering and Food. Vol. 2: Processing and Applications* (B. M. McKenna, ed.), Elsevier Applied Science, Essex, 1984, p. 33.
74. W. Garnjanagoonchorn and S. Lertsupakul, Quality development of dried squid, *Food Science and Technology in Industrial Development* (S. Maneepun, P. Varangoon, and B. Phithakpol, eds.), Institute of Food Research and Product Development, Bangkok, 1988, p. 196.
75. H. Gehman and E. M. Osman, The chemistry of the sugar-sulfite reactions and its relationship to food problems, *Adv. Food Res. 5*:53 (1954).
76. L. R. Genskow, Considerations in drying consumer products, *Drying '89* (A. S. Mujumdar and M. Roques, eds.), Hemisphere Publishing, New York, 1990.
77. R. Giangiacomo, D. Torreggiani, M. L. Erba, and G. Messina, Use of osmodehydrofrozen fruit cubes in yogurt, *Ital. J. Food Sci. 6*:345 (1994).
78. P. Gorling, Drying behavior of vegetable substances, *V.D.I. Forsch. Gebiete Ingenieurw 22*:5 (1956).
79. P. Gorling, Physical phenomena during the drying of foodstuffs, *Fund. Asp. Dehydrated Foods*, Macmillan, London, 1958, p. 42.
80. M. Grncarevic, F. Radler, and J. V. Possingham, The dipping effect causing increased drying of grapes demonstrated with an artificial cuticle, *Am. J. Enol. Vitic. 19*:27 (1968).
81. D. G. Guadagni and A. E. Stafford, Factors affecting the threshold of methyl and ethyl oleate emulsions in raisin and raisin paste, *J. Food Sci. 44*:782 (1979).
82. M. D. Guillen and M. L. Ibargoitia, Volatile components of aqueous liquid smokes from *Vitis vinifera* L shoots and *Fagus sylvatica* L wood, *J. Sci. Food Agric. 72*:104 (1996).
83. S. Gunasekaran, Grain drying using continuous and pulsed microwave energy, *Drying Technol. 8*:1039 (1990).
84. S. M. Gupte and F. J. Francis, Effect of pH adjustment and high temperature short time processing in color and pigment retention in spinach puree, *Food Technol.*:141 (1964).
85. G. J. Haas, H. E. Prescott, and C. J. Cante, On rehydration and respiration of dry and partially dried vegetables, *J. Food Sci. 39*:681 (1974).
86. W. O. Harrington, C. H. Hills, S. B. Jones, A. E. Stafford, and B. R. Tennes, Ethyl oleate sprays to reduce cracking of sweet cherries, *HortScience 13*:279 (1978).
87. J. Hawkes and J. M. Flink, Osmotic concentration offruit slices prior to freeze dehydration, *J. Food Proc. Pres. 2*:265 (1978).
88. G. R. Horn and C. Sterling, Studies on the rehydration of carrots, *J. Sci. Food Agric. 33*:1035 (1982).
89. S. Hughes, Back to the good old days, *Food Proc. 55*:27 (1993).
90. C. C. Huxsoll, Reducing the refrigeration load by partial concentration of foods prior to freezing, *Food Technol. 35*:98 (1982).
91. Expert Panel on Food Safety and Nutrition. Quality of fruits and vegetables, scientific status summary. *Food Technol.* (6):99–106 (1990).
92. M. Ingram and K. Vas, combination of sulphur dioxide with concentrated orange juice. I—Equilibrium states, *J. Sci. Food Agric. 1*:21 (1950).
93. L. L. Imre, Solar drying, *Handbook of Industrial Drying* (A. S. Mujumdar, ed.), Marcel Dekker, New York, 1987, p. 357.
94. M. N. Islam and J. N. Flink, Dehydration of potato II. Osmotic concentration and its effect on air drying behavior, *J. Food Technol. 17*:387 (1982).
95. J. C. Jackson, M. C. Bourne, and J. Barnard, Optimization of blanching for crispness of banana chips using response surface methodology, *J. Food Sci. 61*:165 (1996).
96. A. C. Jason, A study of evaporation and diffusion processes in the drying of fish muscle, *Fun-*

damental Aspects of the Dehydration of Foodstuffs, Society of Chemical Industry, London, 1958, p. 103.

97. R. S. Jebson and Y. He, Studies on the drying behaviour of garlic and a novel technology to produce high quality garlic at a low cost, *Trans. IChemE. 72(C)*:73 (1994).
98. M. R. Jeppson, Consider microwaves, *Food Eng. 11*:49 (1964).
99. P. G. Jolly, Temperature controlled combined microwave convection drying, *J. Microwave Power 21*:65 (1986).
100. V. Karathanos, S. Anglea, and M. Karel, Collapse of structure drying of celery, *Drying Technol. 11*:1005 (1993).
101. H. Kawai, M. Sakurai, Y. Inoue, R. Chujo, and S. Kobayashi, Hydration of oligosaccharides: anomalous hydration ability of trehalose, *Cryobiology 29*:599 (1992).
102. L. Kearney, M. Upton, and A. McLaughlin, Enhancing the viability of *Lactobacillus plantarum* inoculum by immobilizing cells in calcium alginate beads incorporating cryoprotectants, *Appl. Environ. Microbiol. 56*:3112 (1990).
103. T. G. Kieckbusch and C. J. King, Volatiles loss during atomization in spray drying, *AIChE J. 26*:718 (1980).
104. H. S. Kim, B. J. Kamara, I. C. Good, and G. L. Enders, Method for the preparation of stability microencapsulated lactic acid bacteria, *J. Ind. Microbiol. 3*:253 (1988).
105. C. J. King, Review paper: transport process affecting food quality in spray drying, *Engineering and Food., Vol. 2: Processing and Applications* (B. M. McKenna, ed.), Elsevier Applied Science, Essex, 1984, p. 559.
106. C. J. King, Spray drying: retention of volatile compounds revisited, Drying '94, Proceedings of the 9th International Drying Symposium, Gold Coast, 1994, p. 15.
107. E. Kompany, K. Allaf, J. M. Bouvier, P. Guigon, and A. Maureaux, A new drying method of fruits and vegetables—quality improvement of the final product, *Drying 91* (A. S. Mujumdar and I. Filkova, eds.), Elsevier Science Publishers, Amsterdam, 1991, p. 499.
108. E. Kompany and F. Rene, Effect of freezing conditions on aroma retention in frozen and freeze dried mushrooms (*Agaricus bisporus*), *J. Food Sci. Technol. 32*:278 (1995).
109. A. E. Kostaropoulos and G. D. Saravacos, Microwave pre-treatment for sun-dried raisins, *J. Food Sci. 60*:344 (1995).
110. T. Kudra, J. Niewczas, B. Szot, and G. S. V. Raghavan, Stress cracking in high-intensity drying: identification and quantification, Drying '94, Proceedings of the 9th International Drying Symposium, Gold Coast, 1994, p. 809.
111. O. R. Kunze and M. S. U. Choudhury, Moisture adsorption related to the tensile strength of rice, *Cereal Chem. 49*:684 (1972).
112. O. R. Kunze, Fissuring of the rice grain after heated air drying, *Trans. ASAE 22*:1197 (1979).
113. O. R. Kunze and C. W. Hall, Relative humidity changes that cause brown rice to crack, *Trans. ASAE 9*:396 (1965).
114. O. R. Kunze, Moisture adsorption in cereal grain technology—a review with emphasis on rice, *Appl. Eng. Agric. 7*:717 (1991).
115. J. Kuprianoff, Bound water in food, *Fundamental Aspects of the Dehydration of Foodstuff* , Society of Chemical Industry, London, 1958, p. 14.
116. C. H. Lea, *J. Soc. Chem. Ind. 52*:57T (1933).
117. M. Lee, L. Chou, and J. Huang, The effect of salt on the surface evaporation of a porous medium, Drying '94, Proceedings of the 9th International Drying Symposium, Gold Coast, 1994, p. 223.
118. C. Y. Lee, D. K. Salunkhe, and F. S. Nury, Some chemical and histological changes in dehydrated apple, *J. Sci. Food Agric. 18*:89 (1967).
119. A. Lenart and J. M. Flink, Osmotic concentration of potato. I. Criteria for the end-point of the osmosis process, *J. Food Technol. 19*:45 (1984).
120. A. Lenart and P. P. Lewicki, Energy consumption during osmotic and convective drying of plant tissue, *Acta Aliment. Polon. 14*:65 (1988).

121. C. R. Lerici, G. Pinnavaia, M. D. Rosa, and L. Bartolucci, Osmotic dehydration of fruit: influence of osmotic agents on drying behavior and product quality, *J. Food Sci. 50*:1217 (1985).
122. L. C. Lievense, M. A. M. Verbeek, T. Taekema, G. Meerdink, and K. van't Riet, Modelling the inactivation of *Lactobacillus plantarum* during a drying process, *Chem. Eng. Sci. 47*:87 (1992).
123. L. J. M. Linders, G. I. W. De Jong, G. Meerdink, and K. van't Riet, The effect of disaccharide addition on the dehydration inactivation of *Lactobacillus plantarum* during drying and the importance of water activity, Drying '94, Proceedings of the 9th International Drying Symposium (V. Rudalph, R. B. Keey, and A. S. Mujumdar, eds.), Gold Coast, 1994, p. 945.
124. E. H. Lucas and D. L. Bailey, A simple rapid quantitative method of assaying peroxidase activity in dehydrated vegetables and fruits, *Mich. Agric. Exp. Sta. Quart. Bul. 26*: (1944).
125. P. A. Luning, D. Yuksel, R. V. D. V. De Vries, and J. P. Roozen, Aroma changes in fresh bell peppers (*Capsicum annuum*) after hot-air drying, *J. Food Sci. 60*:1269 (1995).
126. P. A. Luning, R. Van der Vuurst de Vries, D. Yuksel, T. Ebbenhorst-Seller, H. J. Wichers, and J. P. Roozen, Combined instrumental and sensory evaluation of flavor of fresh bell peppers (Capsicum annuum) harvested at three maturation stages, *J. Agric. Food Chem. 42*:2855 (1994).
127. L. J. Lynch and R. S. Mitchell, When to harvest canning peas, Aust. CSIRO Cir. No. 5-P, 1955.
128. M. M. Mac Masters, F. R. Earle, H. H. Hall, J. H. Ramser, and G. H. Dungan, Studies the effect of drying conditions up on the composition and suitability for wet milling of artificially dried corn, *Cereal Chem. 31*:451 (1954).
129. T. R. A. Magee, A. A. Hassaballah, and W. R. Murphy, Internal mass transfer during osmotic dehydration of apple slices in sugar solutions, *Ir. J. Food Sci. Technol. 7*:147 (1983).
130. V. Maharaj and C. K. Sankat, Quality changes in dehydrated dasheen leaves: effects of blanching pre-treatments and drying conditions, *Food Res. Int. 29*:563 (1996).
131. T. Mahmutoglu, Y. B. Saygi, M. Borcakli, and G. Ozay, Effects of pretreatmentdrying method combinations on the drying rates, quality and storage stability of apricots, *Food Sci. Technol. 29*:418 (1996).
132. M. Marcotte and M. Le Maguer, Repartition of water in plant tissues subjected to osmotic process, *J. Food Proc. Eng. 13*:297 (1991).
133. R. J. L. Martin and G. L. Stott, The physical factors involved in the drying of sultana grapes, *Aust. J. Agric. Res. 8*:444 (1957).
134. R. L. Mason, Application of heat pumps to drying food products, *Food Aust. 41*:1070 (1989).
135. R. L. Mason, P. M. Britnell, G. S. Young, S. Birchall, S. Fitz-Payne, and B. J. Hesse, Development and application of heat pump dryers to the Australian food industry, *Food Aust. 46*:319 (1994).
136. K. Masters, *Spray Drying Handbook*, Halsted Press, New York, 1985.
137. K. Masters, Impact of spray dryer design on powder properties, drying, *Drying 91* (A. S. Mujumdar and I. Filkova, eds.), Elsevier Science Publishers, Amsterdam, 1991, p. 56.
138. W. L. McCabe and J. C. Smith, *Unit Operations of Chemical Engineering*, 3rd ed., McGraw-Hill Book Company, New York, 1976.
139. D. McBean, M. W. Miller, J. I. Pitt, and A. A. Johnson, Prune drying in Australia—a reappraisal of methods, *CSIRO Food Pres. Quart. 26*:2 (1966).
140. D. McBean, Drying and processing tree fruits, Division of Food Research Circular No. 10, Commonwealth Scientific and Industrial Research Organization, Australia, 1976.
141. T. A. McGuire and F. R. Earle, Changes in the solubility of corn protein resulting from artificial drying of high moisture corn, *Cereal Chem. 15*:179 (1958).
142. D. L. McKenzie and T. Beveridge, The effect of storage, processing and enzyme treatment on the microstructure of cloudy spartan apple juice particulate, *Food Microstructure 7*:195 (1988).
143. H. McIlveen and C. Vallely, The development and acceptability of a smoked processed cheese, *Br. Food J. 98*:17 (1996).
144. H. McIlveen and C. Vallely, Something's smoking in the development kitchen, *Nutr. Food Sci. 6*:34 (1996).
145. S. Mohamed and R. Hussein, Effect of low temperature blanching, cysteine-HCl, N-acetyl-L-

cysteine, Na metabisulphite and drying temperatures on the firmness and nutrient content of carrots, *J. Food Proc. Pres. 18*:343 (1994).

146. J. A. Monroy-Rivera, A. Lebert, C. Marty, J. Muchnik, and J. J. Bimbenet, Simulation of cyanoglucosidic compounds elimination in cassava during drying, *Drying 91* (A. S. Mujumdar and I. Filkova, eds.), Elsevier Science Publishers, Amsterdam, 1991, p. 463.
147. V. R. Morey, R. J. Gustafson, and H. A. Cloud, Combination high-temperature, ambient-air drying, *Trans. ASAE 24*:509 (1981).
148. M. K. S. Morsi and C. Sterling, Crystallization in starch: the role of ions in the lyotropic series, *J. Polym. Sci. Part A 1*:3547 (1963).
149. M. Mourad, M. Hemati, and C. Laguerie, Evaluation of the technical quality of corn kernels submitted to various techniques of drying at different operating conditions, Drying '94, Proceedings of the 9th International Drying Symposium, Gold Coast, 1994, p. 1023.
150. J. H. Moy, N. B. H. Lau, and A. M. Dollar, Effects of sucrose and acids on osmovac-dehydration of tropical fruits, *J. Food Proc. Pres. 2*:131 (1978).
151. J. C. Moyer, C. J. Tressler, W. T. Tapley, K. A. Wheeler, F. A. Lee, W. I. Zimnerman, F. W. Tanner, R. E. Hilt, Z. I. Kertesz, and D. K. Tressler, Varietal adaptation of New York vegetables to dehydration, *The Canner 97*:15 (1943).
152. A. L. Moyls, Drying of apple purees, *J. Food Sci. 46*:939 (1981).
153. P. Narasimham and P. J. John, Controlled low temperature vacuum dehydration as an alternative to freeze-drying of diced breadfruit, *J. Food Sci. Technol. 32*:305 (1995).
154. A. M. Neubert, C. W. Wilson, and W. H. Miller, Studies on celery rehydration, *Food Technol. 22*:1296 (1968).
155. H. H. Nijhuis, E. Tirringa, H. Luyten, F. Rene, P. Jones, T. Funebo, and T. Ohlsson, Research needs and opportunities in the dry conservation of fruits and vegetables, *Drying Technol. 14*:1429 (1996).
156. J. Nketsia-Tabiri and S. Sefa-Dedeh, Optimization of process conditions and quality of salted dried tilapia (*Oreochromis niloticus*) using response surface methodology, *J. Sci. Food Agric. 69*:117 (1995).
157. F. S. Nury and D. K. Salunkhe, Effects of microwave dehydration on components of apples, U.S. Agric. Res. Serv., ARS, 1968, p. 74.
158. G. S. Osborn, J. H. Young, and J. A. Singleton, Measuring the kinetics of acetaldehyde, ethanol, and ethyl acetate within peanut kernels during high temperature drying, *Trans. ASAE 39(3)*:1039 (1996).
159. Y. K. Pan, J. Z. Pang, Z. Y. Li, A. S. Mujumdar, and T. Kudra, Drying of heat-sensitive bioproducts on solid carriers in vibrated fluid bed, Drying '94, Proceedings of the 9th International Drying Symposium, Gold Coast, 1994, p. 819.
160. L. Paterson, J. R. Mitchell, S. E. Hill, and J. M. V. Blanshard, Evidence for sulfite induced oxidative depolymerisation of starch polysaccharides, *Carboyhydrate Res. 292*:143 (1996).
161. R. T. Patil, S. Sokhansanj, and T. Larsen, Effect of drying of alfalfa chop on color preservation, Drying '94, Proceedings of the 9th International Drying Symposium, Gold Coast, 1994, p. 1037.
162. J. W. Paulis and J. S. Wall, Fractionation and characterisation of alcohol-soluble reduced corn endosperm glutelin proteins, *Cereal Chem. 56*:1223 (1977).
163. C. O. Perera, A report on drying of kiwifruit (confidential), Horticulture and Food Research Institute of New Zealand Ltd., Auckland, 1990.
164. C. O. Perera and M. S. Rahman, Heat pump drying, *Trends Food Sci. Technol. 8*(3):75 (1997).
165. V. Petrucci, N. Canata, H. R. Bolin, G. fuller, and A. E. Stafford, Use of oleic acid derivatives to accelerate drying of Thompson seedless grapes, *J. Am. Oil Chem. 51*:77 (1974).
166. B. E. Platin, A. Erden, and O. L. Gulder, Modelling and design of rotary dryers, *Proceedings of the Third International Drying Symposium*, Vol. 2 (J. C. Ashworth, ed.), Drying Research Ltd., Wolverhampton, 1982, p. 466.
167. J. D. Ponting, Osmotic dehydration of fruits- recent modifications and applications, *Proc. Biochem. 8*:18 (1973).

168. J. D. Ponting and D. M. McBean, Temperature and dipping treatment effects on drying rates and drying times of grapes, prunes and other waxy fruits, *Food Technol. 24*:85 (1970).
169. J. D. Ponting, G. G. Watters, R. R. Forrey, R. Jackson, and W. L. Stanley, Osmotic dehydration of fruits, *Food Technol. 20*:125 (1966).
170. J. V. Possingham, T. C. Chambers, F. Radler, and M. Grncarevic, Cuticular transpiration and wax structure and composition of leaves and fruit of *Vitis vinifera*, *Aust. J. Biol. Sci. 20*:1149 (1967).
171. D. G. Prabhanjan, H. S. Ramaswamy, and G. S. V. Raghavan, Microwave-assisted convective air drying of thin layer carrots, *J. Food Eng. 25*:283 (1995).
172. G. Prestamo and C. Fuster, Proc. of the IUFost Int. Symp. on Chemical Changes During Food Processing 1, 1984, p. 269.
173. K. Prichavudhi and H. Y. Yamamoto, Effect of drying temperature on chemical composition and quality of macadamia nuts, *Food Technol. 19*:1153 (1965).
174. A. Quintero-Ramos, M. C. Bourne, and A. Anzaldua-Morales, Texture and rehydration of dehydrated carrots as affected by low temperature blanching, *J. Food Sci. 57*:1127 (1992).
175. F. Radler, The prevention of browning during drying by the cold dipping treatment of sultana grapes, *J. Sci. Food Agric. 15*:864 (1964).
176. M. S. Rahman, *Food Properties Handbook*, CRC Press, Boca Raton, FL, 1995.
177. M. S. Rahman, Osmotic dehydration kinetics of foods, *Indian Food Indus. 15*:20 (1992).
178. M. S. Rahman and J. Lamb, Osmotic dehydration of pineapple, *J. Food Sci. Technol. 27*:150 (1990).
179. M. S. Rahman and P. L. Potluri, Shrinkage and density of squid flesh during air drying, *J. Food Eng. 12*:133 (1990).
180. A. Raoult-Wack, S. Guilbert, and M. Le Maguer, Simultaneous water and solute transport in shrinking media—Part 1. Application to dewatering and impregnation soaking process analysis (osmotic dehydration), *Drying Technol. 9*:589 (1991).
181. A. L. Raoult-Wack, Recent advances in the osmotic dehydration of foods, *Trends Food Sci. Technol. 5*:255 (1994).
182. G. S. Raouzeos and G. D. Saravacos, Solar drying of raisins, *Drying Technol. 4*:633 (1986).
183. P. Resmini, M. A. Pagani, and L. Pellegrino, Effect of semolina quality and processing conditions on nonenzymatic browning in dried pasta, *Food Aust. 48*:362 (1996).
184. N. Riva and C. Peri, Etude du sechage des raisins. 1. Effet de traitements de modification de la surface sur la cinetique du sechage, *Sci. Aliments 3*:527 (1983).
185. S. F. Roberts, Methods of fish salting, *Cured Fish Production in the Tropics*, University of Philippines in the Visayas, Diliman, Quezon City, 1986.
186. L. R. Ross and R. H. Treadway, Factors affecting the sulfur dioxide uptake in sulfited pre-peeled potatoes, *Am. Potato J. 38*:9 (1961).
187. C. Rossello, J. Canellas, I. Santiesteban, and A. Mulet, Simulation of the absorption process of sulphur dioxide in apricots, *Food Sci. Technol. 26*:322 (1993).
188. A. Ruiter, Colour of smoke foods, *Food Technol. 33*:54 (1979).
189. H. Rumpf, Die Wissenschaft des Agglomerierens, *Chem. Ing. Technol. 46*:1 (1974).
190. S. Rybka and K. Kailasapathy, The survival of culture bacteria in fresh and freeze-dried AB yoghurts, *Aust. J. Dairt Technol. 50*:51 (1995).
191. J. Sadowska, J. Fornal, A. Ostaszyk, and B. Szmatowicz, Drying conditions and processability of dried rapeseed, *J. Sci. Food Agric. 72*:257 (1996).
192. H. Salwin, Defining minimum moisture contents for dehydrated foods, *Food Technol. 13*:594 (1960).
193. G. D. Saravacos and S. E. Charm, A study of the mechanism of fruit and vegetable dehydration, *Food Technol. 1*:78 (1962).
194. G. D. Saravacos and S. E. Charm, Effect of surface-active agents on the dehydration of fruits and vegetables, *Food Technol. 1*:91 (1962).
195. G. D. Saravacos, E. Tsami, and D. Marinos-Kouris, Effect of water activity on the volatile com-

ponents of dried fruits, *Mechanisms of Action of Food Preservation Procedures* (G. W. Gould, ed.), Elsevier Applied Science, London, 1988, p. 347.

196. G. D. Saravacos, S. N. Marousis, and G. S. Raouzeos, Effect of ethyl oleate on the rate of air-drying of foods, *J. Food Eng. 7*:263 (1988).
197. N. N. Sarker, O. R. Kunze, and T. Strouboulis, Transient moisture gradients in rough rice mapped with finite element model and related to fissures after heated air drying, *Trans. ASAE 39*:625 (1996).
198. R. Saurel, A. Raoult-Wack, G. Rios, and S. Guilbert, Mass transfer phenomena during osmotic dehydration of apple. I. Fresh plant tissue, *Int. J. Food Sci. Technol. 29*:531 (1994).
199. R. Saurel, A. Raoult-Wack, G. Rios, and S. Guilbert, Mass transfer phenomena during osmotic dehydration of apple. II. Frozen plant tissue, *Int. J. Food Sci. Technol. 29*:543 (1994).
200. H. Schubert, Kapillardruck und Zugfestigkeit von feuchten Haufwerken aus kornigen Stoffen, *Chem. Ing. Technol. 45*:396 (1973).
201. S. J. Schwartz and T. V. Lorenzo, Chlorophylls in foods, *Crit. Rev. Food Sci. Nutri. 29*:1 (1990).
202. A. Senoussi, E. Dumoulin, A. Lebert, and Z. Berk, Spray drying of food liquids: behaviour of sensitive tracers in polysaccharides and milk, Drying '94, Proceedings of the 9th International Drying Symposium, Gold Coast, p. 905.
203. W. H. Shah and N. A. Sufi, Effect of maturity and pricking on dehydration and rehydration characteristics of various varieties of green peas, *Pakistan J. Sci. Res. 31*:174 (1979).
204. M. Shamaila, T. Durance, and B. Girard, Water blanching effects on headspace volatiles and sensory attributes of carrots, *J. Food Sci. 61*:1191 (1996).
205. A. D. Sharma and O. R. Kunze, Post-drying fissure developments in rough rice, *Trans. ASAE 25*:465 (1982).
206. R. C. Sharma, V. K. Joshi, S. K. Chauhan, S. K. Chopra, and B. B. Lal, Application of osmosis-osmo-canning of apple rings, *J. Food Sci. Technol. 28*:86 (1991).
207. E. M. Sheehan, T. P. O'Connor, P. J. A. Sheehy, D. J. Buckley, and R. FitzGerald, Effect of dietary fat intake on the quality of raw and smoked salmon, *Irish J. Agric. Food Res. 35*:37 (1996).
208. X. Q. Shi and P. F. Maupoey, Mass transfer in vacuum osmotic dehydration of fruits: a mathematical model approach, *Food Sci. Technol. 27*:67 (1994).
209. J. B. Sieh and C. Sterling, charging of glycogen in salt solutions, *Biochim. Biophys. Acta 184*:281 (1969).
210. S. Sokhansanj, Quality of food grains in recirculating hot-air dryers, *Proceedings of the Third International Drying Symposium*, Vol. 2 (J. C. Ashworth, ed.) Drying Research Ltd., Wolverhampton, 1982, p. 253.
211. J. Solms, F. Osman-Ismail, and M. Beyeler, The retention of volatiles with food components, *Can. Inst. Food Sci. Technol. J. 6*:10 (1973).
212. G. S. Srzednicki and R. H. Driscoll, In-store drying and quality maintenance in grain, *Postharvest Technology for Agricultural Products in Vietnam* (B. R. Champ and E. Highley, eds.), Australian Centre for International Agricultural Research, Canberra, 1995, p. 42.
213. A. E. Stafford, H. R. Bolin, and B. E. Mackey, Absorption of aqueous bisulfite by apricots, *J. Food Sci. 37*:941 (1972).
214. I. Strommen, New equipment in fish drying, Proceedings of the Third International Drying Symposium, Vol. 1 (J. C. Ashworth, ed.), Drying Research Ltd., Wolverhampton, 1982, p. 295.
215. C. Strumillo, A. S. Markowski, and J. Adamiec, Selected aspects of drying of biotechnological products, *Drying 91* (A. S. Mujumdar and I. Filkova, eds.), Elsevier Science Publishers, Amsterdam, 1991, p. 36.
216. C. Strumillo and C. Lopez-Cacicedo, Energy aspects of drying, *Handbook of Industrial Drying* (A. S. Mujumdar ed.), Marcel Dekker, New York, 1987, p. 823.
217. J. A. Suarez, M. Loncin, and J. Chirife, A preliminary study on the effect of ethyl oleate dipping treatment on drying rate of grain corn, *J. Food Sci. 49*:236 (1984).
218. H. L. Su and K. C. Chang, Dehydrated precooked Pinto bean quality as affected by cultivar and coating biopolymers, *J. Food Sci. 60*:1330 (1995).

219. D. G. Tilgner, Z. Sirkoski, H. Urbanowicz, and Z. Nowak, Smoke curing with the dispersing phase of various curing smokes, *Technol. Mesa. Spec. Edn. 62*: (1962).
220. D. Torreggiani, Osmotic dehydration in fruit and vegetable processing, *Food Res. Int. 26*:59 (1993).
221. R. E. Treybal, *Mass Transfer Operations*, 3rd ed. McGraw-Hill, Singapore, 1981.
222. K. Trongpanich, Effect of pretreatments on texture of dehydrated bamboo shoot, *Development of Food Science and Technology in South East Asia* (O. B. Liang, A. Buchanan, and D. Fardiaz, eds.) IPB Press, Bolgor, 1993, p. 485.
223. T. N. Tulasidas, G. S. V. Raghavan, and E. R. Norris, Effects of dipping and washing pre-treatments on microwave drying of grapes, *J. Food Proc. Eng. 19*:15 (1996).
224. I. W. Turner and P. G. Jolly, Combined microwave and convective drying of a porous material, *Drying Technol. 9*:1209 (1991).
225. M. B. Uddin and N. Islam, Development of shelf-stable pineapple products by different methods of drying, *J. Inst. Engrs. Bangladesh 13*:5 (1985).
226. A. M. G. Usiak, M. C. Bourne, and M. A. Rao, Blanch temperature/time effects on rheological properties of applesauce, *J. Food Sci. 60*:1289 (1995).
227. A. Van Blarcom and R. L. Mason, Low humidity drying of macadamia nuts, Proceedings of the Fourth Australian Conference on Tree and Nut Crops, Lismore, NSW, 1988, p. 239.
228. S. M. Van-Ruth and J. P. Roozen, Gas chromatography/sniffing port analysis and sensory evaluation of commercially dried peppers (*Capsicum annuum*) after rehydration, *Food Chem. 51*:165 (1994).
229. A. Voilley and D. Simatos, Retention of aroma during freeze- and air-drying, *Food Process Engineering* (P. Linko, Y. Malkki, J. Olkku, and J. Larinkari, eds.), Applied Science Publishers, London, 1979, p. 371.
230. K. M. Waananen and M. R. Okos, Analysis of bulk flow transfer, *Technol. Today*, 1991, p. 289.
231. D. A. Weitz, M. A. Lara, and R. D. Piacentini, Dipping treatment effects on simulated prune solar drying, *Can. Inst. Food Sci. Technol. J. 22*:133 (1989).
232. A. G. Wientjes, The influence of sugar concentrations on the vapor pressure of food odor volatiles in aqueous solutions, *J. Food Sci. 33*:1 (1968).
233. C. Wilkins, Paprika chemistry and its relationship to spice quality, *Spices, Herbs and Edible Fungi* (G. Charalambous, ed.), Elsevier Science B.V., Amsterdam, p. 381.
234. M. L. Woolfe, The effect of smoking and drying on the lipids of West African herring (*Sardinella* spp.), *J. Food Technol. 10*:515 (1975).
235. J. Yongsawatdigul and S. Gunasekaran, Pulsed microwave vacuum drying of cranberries. Part I. Energy use and efficiency, *J. Food Proc. Pres. 20*:121 (1996).
236. J. Yongsawatdigul and S. Gunasekaran, Microwave-vacuum drying of cranberries: Part II. Quality evaluation, *J. Food Proc. Pres. 20*:145 (1996).
237. J. Yongsawatdigul and S. Gunasekaran, Microwave-vacuum drying of cranberries: Part I. Energy use and efficiency, *J. Food Proc. Pres. 20*:121 (1996).

7

Fruit Juice Concentration and Preservation

M. A. RAO
Cornell University—Geneva, Geneva, New York

ALFREDO A. VITALI
Fruthotec, Ital, Campinas, Brazil

I. INTRODUCTION

Fruit juices and other fluid foods are concentrated in order to reduce their volume and mass, which in turn results in reduced costs of packaging, storage, and transportation. Juice concentration also results in lower water activity, which aids in shelf-life extension. Concentrated fruit juices also have aided the development of many new food products and the economic utilization of perishable crops during peak harvest periods.

Frozen concentrated orange juice (FCOJ) is by far the dominant juice in consumption and trade, with Brazil and Florida being the major players. For example, from July 1996 to June 1997, Brazil exported about 1.037 million metric tons of 65 °Brix FCOJ [100]. The projected production of organic juice in Florida for the 1997–1998 season is about 1.137 million metric tons of 65 °Brix orange juice [100]. Brazil also exports pineapple juice and other types of tropical fruit juices and pulp. The per capita consumption of apples and the import of apple juice into the United States has also increased [17].

In this chapter, the methods for concentration of fruit juice will be discussed with respect to the underlying principles, as well as the advantages and the disadvantages of each method. Reviews dealing with the technological aspects of concentration of fruit juices have been presented by Heid and Casten [35], Ramteke et al. [65], Nagy et al. [59], and Somogyi et al. [83]. Also very useful are reviews dealing with concentration as an unit operation, such as a discussion of evaporators for fluid foods [55], energy considerations in the concentration of fluid foods [79], the engineering principles and economics of freeze concentration [80,95], and the engineering principles of aroma recovery [13]. Rao [66] reviewed concentration of apple juice, and some of the sections in this chapter are based on that work.

Prior to 1940, concentrated fruit juices were largely used as ingredients in soft drinks so that moderate flavor changes were acceptable. The development of equipment and methods for concentration of fruit juices accelerated when demand arose for concentrates that could be reconstituted with water to yield juices that resembled the starting product [35]. Another use for concentrated fruit juices and purees is as fillings in bakery products. In the present review, an attempt have been made to draw upon previous reviews and published research articles in order to cover the technological, engineering, and the economic considerations of fruit juice concentration. Prior to discussion of the principles and the commercial application of the concentration methods, with particular reference to the concentration of fruit juice, the physicochemical properties of fruit juices will be discussed.

Because FCOJ is the most important of all fruit juices, its production is illustrated in Figure 1. The processes for the production of concentrated apple and grape juices are similar and simpler because there are no suspended solids; the dissolved pectins are removed enzymatically from the single strength juice, filtered, and concentrated up to about 72 °Brix.

A. Physical, Thermal, and Rheological Properties of Fruit Juices

Physical, thermal, and rheological properties of fruit juices are essential for designing handling and processing equipment and unit operations. The physical properties of foods have been reviewed by Okos [61] and Rahman [64], and useful data on fruit juices can be found in those works. The principal components of fruit juices are sugars, but the quality of fruit juices is determined to a large extent by their color and flavor and the sugar:acid ratio. Table 1 contains the sugar and acid content of a few juices reported in the literature.

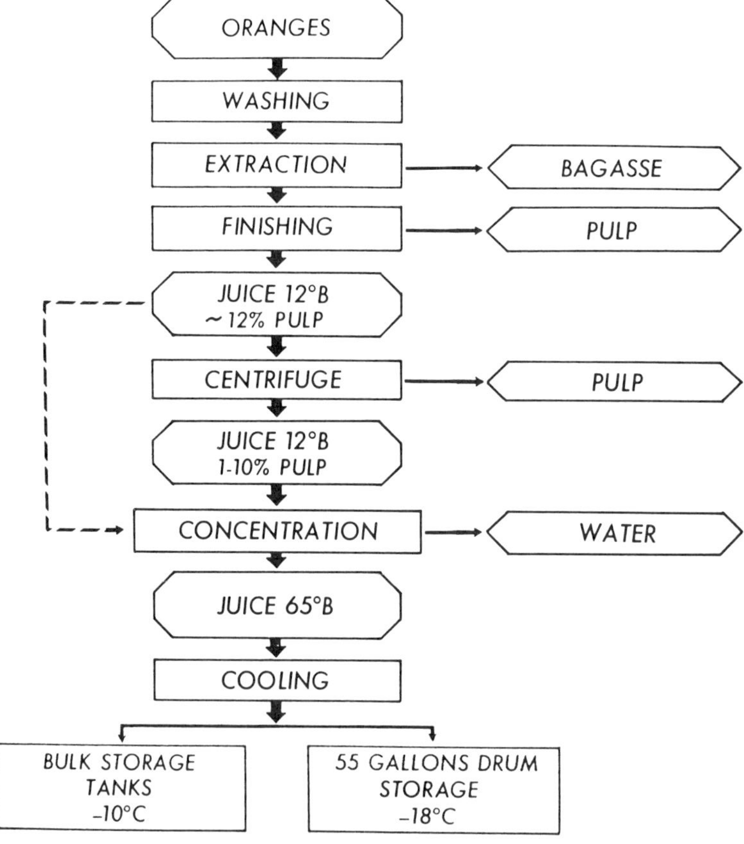

FIGURE 1 Schematic diagram of steps in production of frozen concentrated orange juice (FCOJ). (From Ref. 101.)

Table 1 Typical Sugar and Acid Content of Different Fruit Juices

<table>
<tr><th rowspan="3">Fruit juice</th><th colspan="4">Sugar content (% by mass)</th><th rowspan="2">Acid</th><th rowspan="3">Ref.</th></tr>
<tr><th colspan="2">Reducing sugars</th><th rowspan="2">Sucrose</th><th rowspan="2">Total sugars</th></tr>
<tr><th>Fructose</th><th>Glucose</th><th>(% by weight)</th></tr>
<tr><td>Orange, Valencia</td><td>15.80</td><td>12.50</td><td>33.78</td><td>62.08</td><td>4.60</td><td>25</td></tr>
<tr><td>Apple</td><td colspan="2">7.95</td><td>2.95</td><td>10.90</td><td>0.02</td><td>36</td></tr>
<tr><td>Grape</td><td colspan="2">15.40</td><td>1.25</td><td>16.65</td><td>1.20</td><td>36</td></tr>
<tr><td>Grapefruit</td><td colspan="2">4.25</td><td>2.15</td><td>6.40</td><td>1.15</td><td>36</td></tr>
<tr><td>Pineapple</td><td colspan="2">4.50</td><td>8.95</td><td>13.45</td><td>0.80</td><td>36</td></tr>
<tr><td>Orange, Pera</td><td colspan="2">28.60</td><td>30.80</td><td>59.40</td><td>4.66</td><td>101</td></tr>
<tr><td>Orange, Naval</td><td colspan="2">29.00</td><td>30.70</td><td>59.70</td><td>3.96</td><td>101</td></tr>
</table>

Chen [25] suggested that the density of sucrose solutions can also be used to estimate the density of fruit juices. Noting that weights determined for density calculations usually are not corrected for buoyancy due to air, for the apparent density of sucrose solutions in g/ml at 20°C, Chen [25] proposed the equation:

$$\rho = \sum_{n=0}^{5} b_n(°\text{Brix})^n \tag{1}$$

where n = 0, 1, . . . , 5, and $b_0 = 0.997174$, $b_1 = 3.857739 \times 10^{-3}$, $b_2 = 1.279276 \times 10^{-5}$, $b_3 = 6.191578 \times 10^{-8}$, $b_4 = -1.7774448 \times 10^{-10}$, and $b_5 = -4.199709 \times 10^{-13}$. The accuracy of the above equation was estimated to be $\pm 1 \times 10^{-5}$ g/ml. The density of fruit juices decreases slightly with increase in temperature, and accurate density values should be used when dealing with large volumes. Bayindirli [10,11] presented equations for calculating the densities of apple and grape juices as a function of °Brix and temperature. For apple juice, in the concentration range 14–39 °Brix and temperature range 20–80°C, the equation for density is:

$$\rho = 0.83 + 0.35 \exp[(0.01)(°\text{Brix})] - 0.000564\ T \tag{2}$$

where T is the temperature in K. For example, the density of 30 °Brix apple juice at 20°C calculated from the above equation is 1.137 g/ml. For density of grape juice in the concentration range: 19–35 °Brix and temperature 20–80°C, the equation is slightly different:

$$\rho = 0.74 + 0.43 \exp[(0.01)(°\text{Brix})] - 0.000555\ T \tag{3}$$

In both Eqs. (2) and (3), the exponential dependence on concentration and a linear dependence on temperature are noteworthy. For example, the density of 30 °Brix apple juice at 20°C (293.1 K) calculated from the above equation is 1.14 g/ml and that of grape juice is 1.16 g/ml.

For specific heat of fruit juices other than orange juice, c_p(kJ kg^{-1} K^{-1}), Chen [25] suggested an equation derived from data on sucrose solutions:

$$c_p = 4.197[1 - x_s\{0.57 - 0.0018(T - 20)\}] \tag{4}$$

where x_s is the weight fraction of solids in the juice and T is the temperature in °C. For the specific heat of orange juice, Chen [25] suggested a slightly different equation, which is valid for the concentration range 0–70 °Brix and temperature range 0–100°C:

$$c_p = 4.197[1 - x_s\{0.64 - 0.0018(T - 20)\}] \tag{5}$$

The thermal conductivity of fruit juices, k(W m^{-1} K^{-1}), can be calculated from an equation presented by Chen [25] that was credited to Riedel:

$$k = (0.565 + 0.0018T - 0.000006T^2)(1 - 0.54x_s) \tag{6}$$

where x_s is the mass fraction of solids in the juice and t is the temperature in °C.

B. Rheological Behavior of Fruit Juices

The viscosity or the rheological behavior of a fruit juice is important because it increases from a few mPa·s for the single strength juice to several mPa·s for concentrated juices [69,102,103]. By definition, rheology is the study of deformation and flow of matter by force. Therefore, rheological properties of fluids, such as concentrated orange juice, are based on flow response of foods when subjected to either a shearing stress or a shearing rate at a fixed temperature (Fig. 2). Extensive discussion on rheological behavior of fluid foods can be found elsewhere [67], and only a brief description is given here.

We note that depectinized and filtered apple and grape juices are Newtonian fluids, i.e., their viscosity at a specific concentration is dependent only on temperature. Fluids such as water and sugar solutions at a fixed temperature exhibit flow behavior independent of applied shear and are called

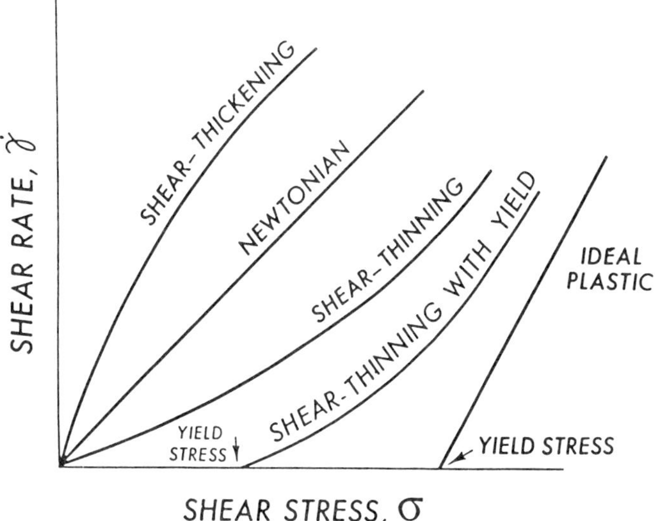

FIGURE 2 Illustration of flow behavior of fluid foods. Apple and grape juices are Newtonian fluids, but FCOJ is a shear-thinning fluid.

Newtonian foods. They can be described by Eq. (7) in terms of the shear stress (σ) and the shear rate ($\dot{\gamma}$).

$$\sigma = \eta\dot{\gamma} \tag{7}$$

Because all depectinized and filtered juices are essentially sugar solutions, they can be expected to behave as Newtonian fluids. In contrast, concentrated orange juice and undepectinized apple juice are shear-thinning (pseudoplastic) non-Newtonian fluids [75,102,103] so that their apparent viscosity depends on the effective shear rate in the evaporator. The nonNewtonian nature arises from the presence of dissolved pectins, which are high molecular weight compounds, and the insoluble orange pulp. However, serum from FCOJ is a nearly Newtonian fluid [102].

In general, viscosity of fruit juices decreases with increase in temperature. The effect of temperature on the viscosity of fruit juices can be described by the Arrhenius relationship:

$$\eta = \eta_\infty \exp(E_a/RT) \tag{8}$$

In Eq. (8), η is viscosity (Pa s), η_∞ is a constant (Pa s), E_a is the activation energy (J/mol), R is the gas constant (value), and T is the absolute temperature (K). The magnitude of activation energy indicates the influence of temperature on the viscosity of a liquid. It depends on the composition of the liquid and on the range of temperatures employed for its determination. For example, activation energy for depectinized apple juice is higher than that for the undepectinized juice. Also, its magnitude increases with increase in sugar concentration. A practical application of this behavior in the evaporation of fruit juices is that it would be preferable to subject the most concentrated juices to high temperatures so that they can flow relatively easily. Figure 3 illustrates typical magnitudes of activation energy for depectinized apple and grape juices as a function of sugar concentration.

Because depectinized and filtered apple and grape juices are Newtonian fluids, equations were derived by Bayindirli [10,11] that can be used to estimate their viscosities as a function of concentration and temperature. The general equation for viscosity of both apple and grape juices is of the form:

$$\frac{\eta}{\eta_{wa}} = \exp\left\{\frac{A(^\circ\text{Brix})}{[100 - B(^\circ\text{Brix})]}\right\} \tag{9}$$

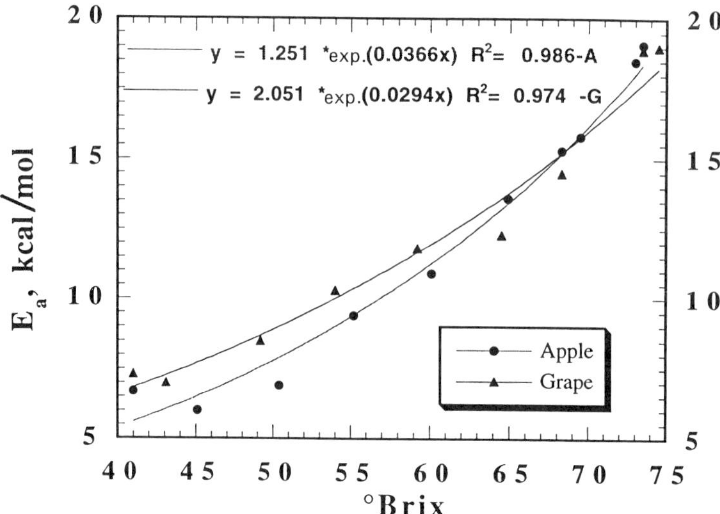

FIGURE 3 Flow activation energy of apple and grape juices as a function of sugar content (°Brix). (Data from Ref. 69.)

where η is the viscosity (mPa·s) of depectinized and filtered apple juice, η_{wa} = viscosity of water at the same temperature (mPa·s), and A and B are constants. For apple juice in the concentration range 14–39 °Brix and temperature 293–353 K (20–80°C), the constants A and B are:

$$A = -0.24 + (917.92/T)$$

$$B = 2.03 - 0.00267\ T \tag{10}$$

For grape juice in the concentration range 19–35 °Brix and temperature 293–353 K (20–80°C), the constants A and B are:

$$A = -3.79 + (1821.45/T)$$

$$B = 0.86 + 0.000441\ T \tag{11}$$

In both Eqs. (10) and (11), the temperature T is in K. For the purpose of illustration, from Eqs. (9), (10), and (11), the viscosity of 30 °Brix apple juice at 20°C was estimated to be 4.0 mPa·s and that of grape juice 2.8 mPa·s, respectively.

In engineering operations, such as pumping and mixing, on non-Newtonian fluid foods, the power law model has been used extensively:

$$\sigma = K\dot{\gamma}^n \tag{12}$$

The Arrhenius relationship for the consistency index is:

$$K = K_\infty \exp(E_{ak}/RT) \tag{13}$$

where K_∞ is the frequency factor, E_{aK} is the activation energy (J/mol), R is the gas constant (value), and T is temperature (K).

The effect of concentration (c) of soluble solids (°Brix) and insoluble solids (pulp) on either apparent viscosity or the consistency index of the power law model of FCOJ can be described by exponential relationships [102,103]. Equations (14) and (15) are applicable to the consistency index (K) of the power law model [Eq. (12)]. In the case of FCOJ, it should be noted that insoluble solids are expressed in terms of pulp content determined on a 12 °Brix sample by centrifugation for 10 min at 360 × *g*.

$$K = K^c \exp(B_K^c \ ^\circ\text{Brix}) \tag{14}$$

$$K = K^P \exp(B_K^P \ \text{Pulp}) \tag{15}$$

where K^c, K^P, B_K^c, and B_K^P are constants. The combined effects of temperature and concentration can be described by combining the above models [Eqs. (13)–(15)].

It is clear from the above discussion that the rheological properties of FCOJ will depend on the °Brix, pulp content, size and shape of the pulp particles, pectin content of FCOJ sample, and the temperature at which the data was obtained. Carter and Buslig [18] studied particle size distribution in commercial FCOJ samples. Mizrahi and coworkers conducted systematic studies on FCOJ, which they described as a physicochemical approach [52–54]. They used a modified Casson equation to describe the flow behavior of FCOJ.

Because the power law model [Eq. (12)] is used in the determination of pumping and mixing power requirements, literature values of the power law parameters of several FCOJ samples [28,102,103] are given in Table 2. The values in Table 2 reflect the influence of fruit varieties, °Brix, pulp content, and temperature. In particular, the strong influence of temperature on the consistency index of the power law model should be noted. The constants in Eqs. (13)–(15) that describe the effects of temperature, °Brix, and pulp content on the consistency index of the power law model are summarized in Tables 3, 4, and 5, respectively [102,103]. These data should be used, albeit with caution, in reaching ballpark values of the power law parameters of FCOJ from temperature, °Brix, and pulp content data.

II. COLLIGATIVE PROPERTIES OF FRUIT JUICES

Colligative properties depend on the number but not on the nature of molecules in a system, and they play an important role in a number of separation processes and water activity. The colligative properties of ideal solutions are independent of the chemical nature of the solute. Typical colligative properties include boiling point elevation, freezing point depression, and osmotic pressure. The thermodynamic basis of colligative properties is discussed here in brief.

A. Chemical Potential and Activity Coefficient

The Gibbs free energy, G, for a closed system is defined as:

$$G = H - TS = E + PV - TS \tag{16}$$

where H is enthalpy, S is entropy, E is internal energy, T is absolute temperature, P is pressure, and V is volume. In a multicomponent system in which chemical species may be gained or lost, G is a function of T, P, and the number of moles of a species i, n_i. The partial molar Gibbs free energy $\overline{G}_i$ of a species i, or the chemical potential (μ_i), is defined as:

$$\mu = \overline{G}_i = \left(\frac{\partial G}{\partial n_i}\right)_{t,p,n_j} \tag{17}$$

Therefore, chemical potential is the change in free energy of a system due to an infinitesimal change in number of moles of a constituent i when temperature, pressure, and mole quantities of other constituents are held constant. It indicates the driving force in the transfer of a component; the transfer occurs from a system with high μ_i to a system with a lower μ_i. At equilibrium, μ_i is equal for all systems and states. Specifically, when a solution and its vapor are in equilibrium, the chemical potential in the liquid (L) and the vapor (V) phases are equal:

$$\mu_{iL} = \mu_{iV} \tag{18}$$

The chemical potential of a gas can be expressed in terms of measurable functions and in terms of its value at its standard state, μ_i^o:

Table 2 Power Law Parameters (K and n) of Concentrated Orange Juice Samples

COJ	(°Brix)	Method	Temp (°C)	K ($Pa \cdot s^n$)	n	Ref.
Hamlin early	64.2	conc cylinder	25.0	4.1	0.59	25,28
9.8% pulp		Haake RV12	15.0	6.0	0.60	
			0.0	9.2	0.68	
			−10.0	14.3	0.71	
Hamlin late	63.3	conc cylinder	25.0	1.9	0.73	25,28
10.3% pulp		Haake RV12	15.0	8.1	0.56	
			0.0	12.8	0.62	
			−10.0	13.9	0.71	
Pineapple early	63.3	conc cylinder	25.0	2.6	0.64	25,28
9.8% pulp		Haake RV12	15.0	5.9	0.59	
			0.0	8.9	0.68	
			−10.0	12.2	0.71	
Pineapple late	65.8	conc cylinder	25.0	8.6	0.53	25,28
11% pulp		Haake RV12	15.0	13.4	0.54	
			0.0	18.6	0.64	
			−10.0	36.4	0.63	
Valencia early	68.8	conc cylinder	25.0	5.1	0.54	25,28
8% pulp		Haake RV12	15.0	6.7	0.61	
			0.0	14.0	0.62	
			−10.0	27.2	0.61	
Valencia late	66.8	conc cylinder	25.0	8.4	0.54	25,28
9.8% pulp		Haake RV12	15.0	11.8	0.57	
			0.0	18.6	0.64	
			−10.0	41.4	0.63	

Navel 11.1% pulp	65.1	conc cylinder Haake RV2	29.50	0.90	0.74	102,103
			19.90	1.60	0.72	
			10.10	2.70	0.73	
			−0.7	5.90	0.71	
			−5.0	7.90	0.72	
			−9.3	10.80	0.74	
			−14.1	14.60	0.76	
			−18.5	29.20	0.71	
Pera 3.4% pulp	64.9	conc cylinder Haake RV2	30.0	0.43	0.82	102,103
			20.0	0.67	0.83	
			10.0	1.58	0.79	
			0.0	3.24	0.79	
			−5.0	4.9	0.78	
			−10.0	7.11	0.79	
			−14.0	10.57	0.81	
			−18.0	18.32	0.80	
Pera 5.7% pulp	65	conc cylinder Haake RV2	30.0	0.68	0.80	102,103
			20.0	1.25	0.77	
			10.0	2.06	0.78	
			0.0	4.47	0.76	
			−5.0	6.49	0.77	
			−10.0	8.8	0.79	
			−14.0	15.59	0.76	
			−18.0	24.45	0.76	
Pera 8.6% pulp	65.3	conc cylinder Haake RV2	30.0	1.44	0.73	102,103
			20.0	1.89	0.76	
			10.0	3.85	0.73	
			0.0	7.32	0.74	
			−5.0	11.14	0.73	
			−10.0	15.77	0.74	
			−14.0	25.34	0.73	
			−18.0	42.79	0.70	

(*continued*)

Table 2 Continued

COJ	(°Brix)	Method	Temp (°C)	K ($Pa \cdot s^n$)	n	Ref.
Pera	64.8	conc cylinder	30.0	2.1	0.70	102,103
11.1% pulp		Haake RV2	20.0	2.92	0.72	
			10.0	5.61	0.70	
			0.0	11.45	0.69	
			–5.0	16	0.69	
			–10.0	23.91	0.69	
			–14.0	34.62	0.70	
			–18.0	58.89	0.67	
Valencia	65.3	conc cylinder	29.6	2.51	0.68	102,103
21.2% pulp		Haake RV2	19.7	6.09	0.61	
			10.2	8.25	0.65	
			–0.4	18.46	0.63	
			–5.1	24.49	0.63	
			–9.7	40.58	0.60	
			–14.2	59.69	0.61	
			–18.3	109.9	0.55	

TABLE 3 Effect of Temperature on Rheological Properties of Concentrated Orange Juice[a]

Sample	K_∞ (Pa·s^n)	K_{ak} (kcal/gmol)	Temp. range (°C)	Ref.
Pera orange juice, 65°Brix,				
0% Pulp	7.672E-11	12.9	30–(−18)	102,103
3.4%	9.881E-10	11.9		
5.7%	3.130E-08	10.3		
8.6%	1.742E-08	10.8		
11.1%	4.195E-08	10.6		
Pera orange juice, 5.7% Pulp,				
65°Brix	3.13E-08	10.3	30–(−18)	
62°Brix	1.39E-07	9.7		
58°Brix	1.31E-07	8.8		
56°Brix	8.95E-08	8.9		
52°Brix	6.62E-08	8.5		
Pera orange juice, 4.6% Pulp				
65°Brix	1.86E-08	10.6	30–(−18)	
62°Brix	3.77E-08	9.6		
58°Brix	2.30E-07	8.6		
56°Brix	1.09E-07	8.8		
52°Brix	1.66E-07	8.3		

[a]Effect of temperature model based on consistency index, $K = K_\infty \exp(E_{ak}/RT)$, where T is temperature (K), K_∞ is collision factor or value at infinite temperature, R is gas constant, and E_{ak} is activation energy based on consistency index.

$$\mu_i = \mu_i^o + RT\ln P_i \tag{19}$$

A new function, fugacity, is introduced such that for real gases at all pressures:

$$\mu_i = \mu_i^o + RT\ln f_i \tag{20}$$

It is also specified that:

$$\lim_{P \to 0} \frac{f_i}{P} = 1 \tag{21}$$

TABLE 4 Effect of Soluble Solids (°Brix) on Power Law Consistency Index (K) at Several Temperatures

Temperature (°C)	K^c (Pa·s^n)	B_K^c ([°Brix]$^{-1}$)
−10	2.02×10^{-4}	0.164
−5	2.14×10^{-4}	0.183
0	4.42×10^{-5}	0.179
10	6.64×10^{-5}	0.161
20	5.01×10^{-5}	0.158
30	4.01×10^{-5}	0.153

Source: Ref. 103.

Table 5 Effect of Pulp Content on Power Law Consistency Index (K) at Several Temperatures

Temperature (°C)	K^P (Pa·s^n)	B_K^P ([pulp, %]$^{-1}$)
−18	10.28	0.159
−10	3.77	0.165
0	1.63	0.181
10	0.69	0.200
20	0.34	0.206
30	0.20	0.225

Source: Ref. 102.

For a pure material in a solution, we can define the fugacity coefficient as:

$$\varphi_i = \frac{f_i}{P} \tag{22}$$

By definition, the activity coefficient of a constituent i is:

$$\gamma_i = \frac{f_i^*}{x_i f_i^o} \tag{23}$$

where, f_i^* is the fugacity of component i in solution and f_i^o is the fugacity at a standard state; for a pure component, the right-hand side reduces to the activity, a_i:

$$a_i = \frac{f_i}{f_i^o} \tag{24}$$

B. Water Activity

From the above, the water activity, a_w, at a constant temperature can be shown to be:

$$a_w = \left(\frac{f_w}{f_w^o}\right)_T \tag{25}$$

Assuming the vapor pressure correction factor for the solution and water to be the same, the ratio f_w:f_w^o may be replaced by P_w:P_w^o, so that:

$$a_w = \left(\frac{P_w}{P_w^o}\right)_T \tag{26}$$

where P_w is the vapor pressure of water in the food at a temperature T and P_o^w is the vapor pressure of pure water at the same temperature T.

Because a_w is a very important, if not the most important, factor affecting the shelf life of intermediate-moisture foods, numerous studies have been conducted of its measurement, prediction, and role in preservation of intermediate-moisture foods ($0.65 < a_w < 0.90$). In addition to conservation, deterioration reactions can be influenced by a_w. For example, Toribio et al. [97] found that nonenzymatic browning in stored concentrated apple juice samples (65–90.5 °Brix) was found to occur between water activities 0.53 and 0.55.

Leung [42] reviewed the thermodynamic basis of water activity and typical magnitudes of water activity of foods. In an ideal solution, from Raoult's law, a_w is proportional to the mole fraction of water. However, fruit juices are not ideal solutions. Therefore, a number of equations have been developed to estimate a_w, and a few that are pertinent to fruit juices will be discussed here. For experimental determination of a_w, a number of techniques have been developed [40,42,64,92], including change in electrical conductivity of immobilized salt solution, change in electrical capacitance of thin polymer films, dewpoint by chilled mirror technique, change in longitudinal dimension of water-sorbing fiber, partial water vapor pressure by manometric system, and relative weight of moisture sorbed by anhydrous hydrophilic solid, e.g., microcrystalline cellulose. It appears that electric hydrometers, which measure change in electrical conductivity, have become popular for measurement of a_w because of the ease with which they can be used and their reproducibility, but the sensors are affected by polyols and volatile amines.

For determination of a_w, fruit juices may be considered to consist mainly of different sugars, and the presence of small amounts of compounds that impart characteristic color and flavor may be neglected. Typical proportion of sugars in several juices is shown in Table 6. Norrish presented an equation for water activity of a solution, a_{ws}, when a single solute is present:

$$a_{w,s} = X_1 \exp(-K_s X_s^2) \tag{27}$$

where X_1 is the mole fraction of water, X_s is the mole fraction of the solute, and K_s is a constant. For multiple nonelectrolyte solutes, Ferro Fontán et al. [32] modified Norrish's equation [60] for a_w to obtain:

$$(a_w)_M = X_1 \exp(-K_M X_s^2) \tag{28}$$

where K_M is given by the equation:

$$K_M = \sum_s K_s x_s (\overline{M}/M_s) \tag{29}$$

where x_s is the weight ratio of solute s, so that $\Sigma_s x_s = 1$, M_s is the molecular weight of a solute, and $\overline{M}$ is an average molecular weight defined as:

$$\overline{M} = \left(\sum_s \frac{x_s}{M_s} \right)^{-1} \tag{30}$$

Values of K_M and $\overline{M}$ calculated by Ferro Fontán et al. [32] are given in Table 6. Estimated a_w values of lemon, orange, apple, and grape juices at 25°C are shown in Figure 4. Because the K_M and $\overline{M}$

Table 6 Proportion of Sugars in Different Fruit Juices[a] Used for Calculation of Parameters K_M and $\overline{M}$

	Proportions of major sugars (w/w, %)				
Fruit juice	Fructose	Glucose	Sucrose	K_M	$\overline{M}$
Apple	62.4	14.8	22.7	2.81	202
Grape	35.8	64.2	0.0	2.25	180
Lemonade	6.4	6.4	87.1	5.55	307.2
Orange	21.2	23.9	54.8	3.90	243.6
Pineapple	33.6	52.2	14.1	2.59	193.2
Prune	24.8	75.0	0.1	2.25	180

[a]From Ref. 106.
Source: Ref. 32.

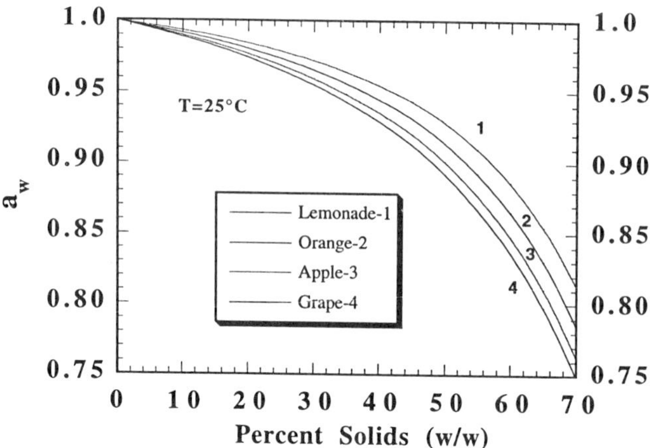

FIGURE 4 Estimated a_w values at 25°C of lemon, orange, apple, and grape juice. (Adapted from Ref. 32.)

of prune juice and grape juice are equal, the a_w versus solids curve of grape juice also represents that of prune juice.

Values of $a_w$an be estimated using other methods, such as freezing point depression due to the addition of a solute. The partial derivative of μ_i with respect to pressure can be written as:

$$\left(\frac{\partial \mu_i}{\partial P}\right)_T = RT\ d(\ln P_i) \tag{31}$$

The above equation after integration and replacing pressure with fugacity becomes:

$$\mu_i - \mu_i^o = RT \ln a_i \tag{32}$$

Because at equilibrium the chemical potential of the liquid (μ_L) and solid (μ_S) phases are equal, from Eq. (16) an equation for freezing point depression can be derived:

$$-\ln X_i = \frac{\Delta H_f(T_o - T)}{RT\ T_o} \approx \frac{\Delta H_f \Delta T_f}{RT_o^2} \tag{33}$$

where X_l is the mole fraction of the solvent, T is the freezing point of the solution, T_o is the freezing point of the solvent, ΔT_f is the freezing point depression, ΔH_f is the molar latent heat of fusion, and R is the gas constant. Recalling Raoult's law, Eq. (33) can be written as:

$$-\ln a_w = \frac{\Delta H_f(T_o - T)}{RT\ T_o} \approx \frac{\Delta H_f \Delta T_f}{RTT_o} \tag{34}$$

where $T_o = 273.15$ K is the freezing point of water, $\Delta T_f = (T_o - T)$ and ΔH_f is the molar latent heat of fusion of ice over the range T_o and T. Chen [23] examined means to estimate bound water and effective molecular weights from experimental ΔT_f data and corrected a_w for deviation from ideality. The Raoult's law for water activity:

$$a_w^o = \frac{1 - x_s}{1 - x_s + x_s(M_w/M_s)} \tag{35}$$

The modified Raoult's law for bound water is [79]:

$$a_w = \frac{1 - x_s - bx_s}{1 - x_s - bx_s + x_s(M_w/M_s)} \tag{36}$$

where M_w is the molecular weight of water, M_s is the molecular weight of a solute, and b is the amount of bound water per unit weight of solid (kg bound water/kg of solid). From the above equation, as expression for ΔT_f can be obtained:

$$\Delta t_f = \frac{1856}{M_s} \ln\left[\frac{1 - x_s - bx_s}{1 - x_s - bx_s + x_s(M_w/M_s)}\right] \tag{37}$$

If the molecular weight of the solute is not known, b can be evaluated from freezing point depression data at two different concentrations:

$$b = \frac{x_{s1}\Delta T_{f2} - x_{s2}\Delta T_{f1}}{x_{s1}x_{s2}(\Delta T_{f2} - \Delta T_{f1})} - 1 \tag{38}$$

The procedure for estimating a_w consists of (a) determination of the bound water per unit weight of solid, b, (b) calculation of effective molecular weight $\overline{M}$ of the mixture of solutes [e.g., eq. (30)] and using it in place of M_s, and (c) calculation of a_w from Eq. (36). In Figure 5, a_w values of orange juice estimated using models of Ferro Fontán et al. and Chen, as well as experimental data given in Chen, are all found to be in good agreement with each other; the experimental values plotted are means of lower and higher values.

Relationships between a_w and ΔT_f have been derived that allow estimation of a_w from ΔT_f or vice versa. For example, Ferro Fontán [32] presented the equation:

$$-\ln a_w = 0.0096934\ \Delta T_f + 4.761 \times 10^{-6}\ (\Delta T_f)^2 \tag{39}$$

For $a_w > 0.88$, Schwartzberg [80] presented a simpler equation that predicted values of a_w within 2%:

$$-\ln a_w = 0.009687\ \Delta T_f \tag{40}$$

For example, for a concentrated fruit juice sample with a freezing point depression of 10°C, values of a_w estimated from Eqs. (39) and (40) were 0.907 and 0.908, respectively.

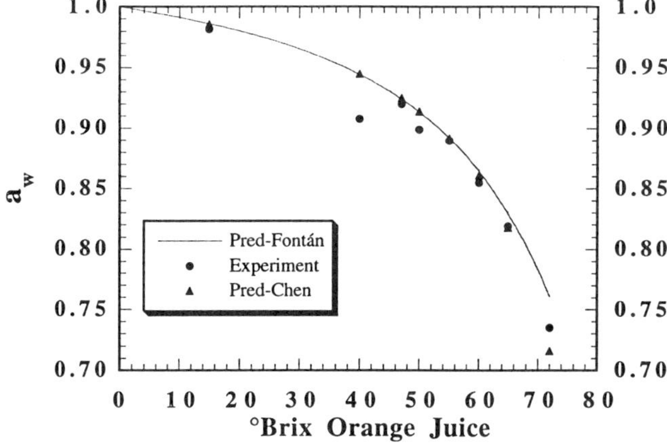

FIGURE 5 Estimated a_w values at 25°C of orange juice. (Adapted from Refs. 23,32.)

C. Rate of Reaction

In foods, most chemical and physical changes can be described by either zero-order or first-order kinetics. For a zero-order reaction:

$$-\frac{dA}{dt} = K_0 \tag{41}$$

After integration and denoting the initial composition as A_o:

$$A = A_o - K_o t \tag{42}$$

where K_o is the zero-order reaction rate constant. For a first-order reaction:

$$-\frac{dA}{dt} = K_1 A \tag{43}$$

where K_1 is the first-order reaction rate constant. After integration, one can obtain:

$$A = A_o \exp(-K_1 t) \tag{44}$$

The D-value has been used extensively to characterize microbial death kinetics. The D-value is the time in minutes for a one log cycle change of a quantity, such as microbial population, at a constant temperature. The D-value and K_1 are related by the equation:

$$K_1 = \frac{2.303}{D} \tag{45}$$

The notation D_T is used to denote the D-value at a temperature T.

In addition to describing the change in a quantity as a function of time at a fixed temperature, one must be able to describe the effect of temperature on the quantity. The reaction rate constant can be related to temperature according to the Arrhenius relationship. For example, for a first-order reaction rate constant:

$$K_1 = K_r\left[\exp(-E_a/R)\left(\frac{1}{T} - \frac{1}{T_r}\right)\right] \tag{46}$$

where K_1 is the first-order reaction rate constant, K_r is the reaction rate constant at a reference temperature, E_a is the activation energy (J/mol), T is the temperature of the system (K), and T_r is the reference temperature (K). It should be noted that whereas the activation energy of chemical reactions is negative, because viscosity decreases with increase in temperature it is positive for viscosity changes.

The z-value is used in microbiology to characterize the influence of temperature on the D-value, and it is the temperature change that results in a 10-fold change (one log cycle) in the D-value. The z-value and E_a are related by the equation:

$$E_a = \frac{2.303RTT_r}{z} \tag{47}$$

In addition to the z-value and E_a, the Q_{10} value is used as a measure of the temperature coefficient of chemical and biological reactions and is defined as the change in the reaction rate constant for a change of 10°C:

$$Q_{10} = \frac{K_0 \text{ at } (T+10)°C}{K_0 \text{ at } T°C} \text{ or } \frac{K_1 \text{ at } (T+10)°C}{K_1 \text{ at } T°C} \tag{48}$$

For many chemical reactions, the Q_{10} about 2.5–4. The relationship between Q_{10} and z-value is:

$$z = \frac{18}{\log Q_{10}} \tag{49}$$

Nonenzymatic browning is of concern in many fruit juices. Therefore, the kinetics of nonenzymatic browning in concentrated apple juice during storage [96] and of grapefruit juice during thermal and concentration processes [74] have been studied. The kinetics of microorganism death is very important both from a public health and an economic loss due to spoilage point of view.

D. Freezing

Since water is often the major constituent in foods, many of the property changes during freezing follow that of water. Thus, it is instructive to study how water properties change during freezing. Often an apparent specific heat is defined as:

$$(c_p)_{apparent} = \frac{\Delta H}{\Delta T} \tag{50}$$

which includes the effect of latent heat and is used to describe phase-change processes. In the above equation, ΔH is change in enthaly due to change in temperature, ΔT [68].

III. MICROBIOLOGICAL CONSIDERATIONS

Because of their toxin-forming ability in low-acid foods, spores of *Clostridium botulinum* have received considerable attention; their commonly accepted D-value and z-value at 121.1°C (250°F) are 0.21 minute and 10°C, respectively. However, in this chapter the emphasis will be on microorganisms that can affect fruit juices.

With the discovery that *Escherichia coli* O157:H7 in unpasteurized commercial apple juice [7] was the cause of a fatality, pasteurization of all high-acid fruit juices is very important, if not essential. There was a major outbreak of *Escherichia coli*–related illness (45 cases) between September and November 1996 in the United States and Canada [7], but a few cases were reported as early as 1991. The source of *E. coli* has been attributed to contact of the apples with animal (farm or wild) feces prior to processing. The basic requirements for production of juice safe for consumption still are that the fruit used should be of edible quality and not rotten, free from fecal contamination, and properly flumed/washed (brush-washed, if necessary). McLellan and Splittstoesser [47] suggested that heating apple juice for only 0.1 minute at 71.1°C eliminates *E. coli* contamination. Kozempel et al. [39] estimated that pasteurization of apple juice on a commercial scale was estimated to cost less than 2 cents per liter.

Although not of public health concern, spoilage of fruit juice by bacilli was first reported by Cerny et al. [19]. They reported having detected an odor of "disinfectant" or "detergent" in fruit juice that could lead to economic loss. Splittstoesser et al. [88], working with isolates of *Bacillus* species from commercial pasteurized apple juice and hot-filled apple cranberry beverage, observed growth at pH 3.5–4.0 in apple juice and white grape juice. They reported D-values of 56–57 minutes at 85°C, 16–23 minutes at 90°C, and 2.4–2.8 minutes at 95°C, and a z-value between 7.2 and 7.7 °C. For 16 °Brix Concord juice, D_{90} values were 11–14 minutes with a z-value of 6.2–6.9°C [90]. A significant increase in D-value with increase in soluble solids was noted: for example, the D_{90} in 65 °Brix Concord juice exceeded 120 minutes. Uboldi Eiroa et al. [99] worked with *Alicyclobacillus acidoterrestris* from orange juice and reported D-values between 60.8 and 94.5 minutes at 85°C, 10 and 20.6 minutes at 90°C, and 2.5 and 8.7 minutes at 95°C. The z-value obtained was between 7.2 and 11.3°C. In general, acid-tolerant microorganisms possess little heat resistance, as seen from the D-values at relatively low temperatures given in Table 7.

TABLE 7 Heat Resistance of Various Acid-Tolerant Microorganisms

Microorganism	Medium	D-value (min)	Ref.
Molds			
Eurotium herbariorum ascospores	Grape juice	$D_{70} = 2.5$	89
Neosartorya fischeri ascospores	Apple juice	$D_{85} = 18$	89
Yeasts			
Saccaromyces cerevisiae			
ascospores	Apple juice	$D_{55} = 106$	89
vegetative cells	Apple juice	$D_{55} = 0.90$	
vegetative cells	Orange juice	$D_{55} = 1.04$	81
Bacteria			
Escherichia coli O157:H7	Apple juice	$D_{52} = 12$	89
Streptococcus lactis	Orange juice	$D_{55} = 1.43$	81

IV. CONCENTRATION OF FRUIT JUICES

Concentration of fruit juices and other fluid foods involves the removal of water. There are several methods for water removal from liquid foods, and the principal ones that are applicable to fruit juices juice are (a) by means of evaporation, (b) by reverse osmosis, and (c) by freezing.

A. Concentration By Evaporation

Concentration of a fruit juice can be achieved by evaporating water, the major component, from the single-strength fruit juice. Evaporation of water is also accompanied by the boiling off and loss of flavor compounds, so that it has become accepted commercial practice to strip the aroma compounds prior to concentration. Walker [105] reviewed the developed of equipment and methods for the recovery of volatile flavor compounds in fruit juices. The principles of recovery of volatiles that contribute to the aroma of fruit juices have been well discussed by Bomben et al. [13], and they will be discussed only in brief here.

1. Principles of Recovery of Volatile Compounds

In a typical aroma-recovery process, the volatile aroma compounds are removed from the juice and concentrated to an aroma-rich solution. Typically, the volatile compounds are removed from the juice by partial evaporation of the juice, and they are concentrated by rectification to 1/100 to 1/150 of the original volume of food. Figure 6 is a schematic diagram of an aroma-recovery process based on partial juice evaporation [66]. Another way of removing aroma compounds is by steam-stripping them from the juice in a countercurrent vapor-liquid contacting device such as a sieve plate column [76]. However, the method of partial evaporation has found wide commercial use because evaporation is a common method for water removal in commercial operations.

Bomben et al. [13] pointed out that aroma compounds found in foods have higher molecular weights than water and that thermodynamic quantities such as the activity coefficient and the vapor pressure of the compound determine the volatility of a compound, which is a measure of the ease with which a compound escapes from a food.

Moyer and Saravacos [58] discussed the technical aspects of fruit juice aroma recovery. They suggested that the design of aroma-recovery equipment must be based on a few key components and that both low- and high-boiling compounds must be considered. In this respect, it should be noted that knowledge of aroma compounds in fruit juices is incomplete in many cases. Roger and Turkot [72] discussed the designing of distillation equipment for fruit aromas with particular emphasis on Concord grape juice. Vapor-liquid equilibrium data must be available for each aroma compound, typically at the low concentrations at which some of the key compounds are present in juices. For example, in the design of a distillation column for Concord grape juice, the composition range of methyl an-

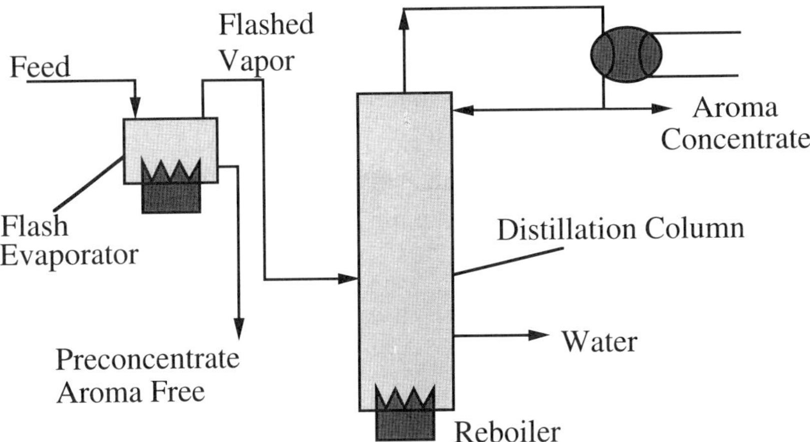

FIGURE 6 Schematic diagram of aroma recovery prior to concentration by evaporation. (Adapted from Ref. 66.)

thranilate in the liquid and vapor as mole fraction was 10^{-7}–10^{-4}. Recent improvements in aroma rectification and recovery include cooling aroma concentrates to below 0°C and the use of secondary condensers for handling components difficult to condense [4].

In many food solutions, volatile flavor compounds are present in very low concentrations and solute-solute interactions may be neglected; this implies that the activity coefficient is constant at its value at infinite dilution (γ^{∞}). From reliable values of γ^{∞}, it is possible to predict the vapor-liquid equilibrium over the entire range of composition [78]. For the design of evaporation systems, data are needed on activity coefficients of aroma components in the juices or solutions to be concentrated, whereas for aroma distillation columns, coefficients in aqueous solutions are needed. Nevertheless, some systems may have concentrations of one substance above the infinite dilution range. For example, values of activity coefficient in solutions with high levels of ethanol would be desirable for processing partially fermented juices or in the production of wines with low alcoholic content. Sancho et al. [73] determined infinite dilution activity coefficients of several fruit juice aroma compounds by the inert gas-stripping method [41] and compared them with other published data (Table 8).

2. Evaporators for Fruit Juice Concentration

Systematic efforts to concentrate fruit juices at low temperatures and pressures in order to obtain high-quality concentrated juices began in the twentieth century. The equipment developed between 1899 and 1928 included vacuum pans with steam jackets, steam coils, and calandria type heaters as well as continuous rising film side-tube heaters with vapor separators. Much of the development work was done with orange juice, and the significant advance in the production of frozen concentrated orange juice took place at the U.S. Department of Agriculture, Winter Haven, Florida, Research Laboratory. The process involved (a) concentrating orange juice to about 60% solids at temperatures below 27.7°C, (b) adding sufficient single-strength juice to reduce the solids to 42% solids, and (c) freezing and storing the concentrate in air-tight containers under vacuum [35]. The historical development of evaporators for producing frozen concentrated orange juice has been described in detail by Berry and Veldhuis [12].

The techniques developed for concentrated orange juice have had a major influence on the concentration of other fruit juices. For example, from 1946 to 1963 vacuum concentration was a major method of fruit juice concentration. Heid and Casten [35] pointed out that fruit juice concentrators may be classified in a number of ways: (a) single pass or re-recirculating, (b) single or multiple stage, (c) single or multiple effect, and (d) having a surface- or mixing-type condenser. The word "stage" is used to indicate the flow of the product (or juice) through the evaporator. Therefore, the first stage

Table 8 Values of Infinite Dilution Activity Coefficient, γ^{∞}, of Aroma Compounds in Water at 25°C

Compound	Experimental results	Literature data	Percent difference
Ethyl acetate	68.3 ± 4.1	64[a]	−6
Butyl acetate	1178 ± 30	1300[a]	10
Pentyl acetate	4874 ± 167	[b]	—
Isopentyl acetate	3974 ± 156	[b]	—
Hexyl acetate	21031 ± 494	23000[a]	9
Ethyl butyrate	1062 ± 74	840[c]	−21
Hexanal	1213 ± 7	1050[a]	−13
trans-2-Hexenal	581 ± 44	[b]	—
Benzaldehyde	1001 ± 44	[b]	—
Hexanol	963 ± 5	720[c]	−25
Ethanol	3.49 ± 0.31	3.55[d]	2
Isobutanol	65.4 ± 0.5	[b]	—
2-Methyl butanol	200 ± 3	[b]	—
3-Methyl butanol	206 ± 5	[b]	—
Acetone	7.0 ± 0.5	7.7 ± 0.3[e]	10

[a]From Ref. 20.
[b]No experimental values at 25°C were found in the literature.
[c]From Ref. 43.
[d]From Ref. 84.
[e]From Ref. 104.
Source: Ref. 73.

receives the single-strength juice, and the finished product leaves the last stage. The term "effect" is used to indicate the flow of vapor (or steam) through the evaporator. Therefore, the first effect of an evaporator receives steam from a boiler. The second effect receives the vapors "boiled off" from the first effect and so on down the line [70]. The juice stages in an evaporator do not necessarily coincide with the effects, i.e., the feed juice may be introduced into any effect. Because the volume of juice decreases with concentration, as a general rule, in an evaporator the number of stages are either equal to or more than the number of effects [22,33]. Figure 7 is a schematic diagram of a five-effect, seven-stage falling film tubular evaporator that illustrates the concepts of effects and stages.

Because the evaporation of water requires a large amount of energy, about 2.33 MJ/kg (960 Btu/lb) of water at atmospheric pressure, techniques have been developed to utilize the vapor generated as the energy source for evaporating more water. These evaporators are known as multiple-effect evaporators, and good general descriptions of such systems employed in the chemical industry can be found in Standiford [91] and McCabe and Smith [45].

3. High-Temperature, Short-Time Evaporators

It should be noted that modern evaporators for fruit juice concentration are high-temperature, short-time (HTST) evaporators that minimize heat-induced changes in the juices, which can be classified into plate and tubular types. In the former, the juice is concentrated on the walls of rectangular channels formed by adjacent plates, while in the latter, the juice is concentrated on the internal surfaces of vertical tubes. Whereas Figure 7 is an illustration of a tubular evaporator, Figure 8 is a line diagram of a plate evaporator. The increased use of HTST evaporators for concentrating fruit juices can be credited to the development of temperature-accelerated short-time (TASTE) evaporators.

Multiple-effect evaporator systems can be operated either in forward flow or in reverse flow. In the former, the single-strength juice is fed in to the effect in which fresh steam is used, and in the latter, fresh steam is introduced in the stage from which concentrated juice is removed. Most juice

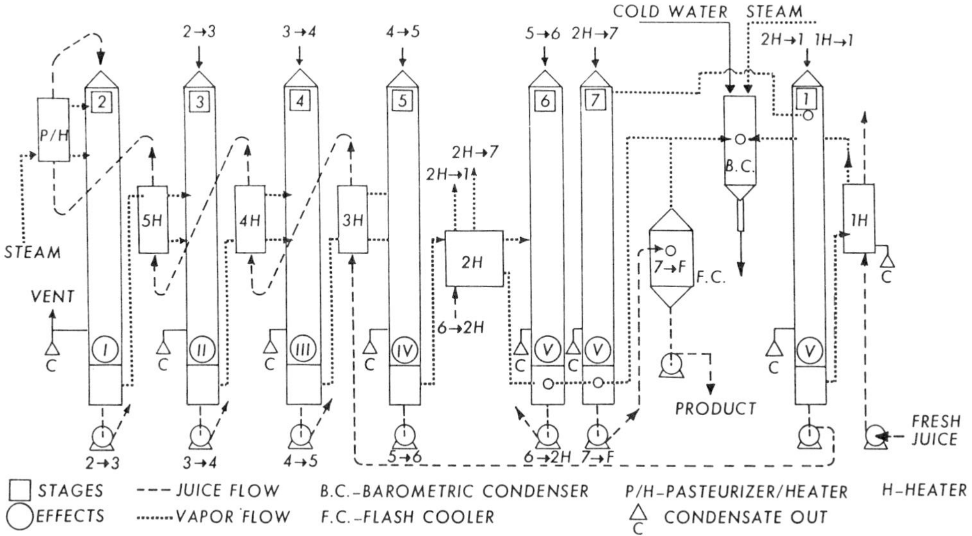

FIGURE 7 Schematic diagram of a five-effect seven-stage TASTE evaporator. (From Ref. 33.)

evaporators are operated in the forward flow mode, i.e., the first stage is also the first effect [70]. Schwartzberg [79] reported that most food-product evaporators in the United States have two to four effects and that in Europe systems containing up to nine effects have been employed; further, most liquid food evaporators are operated in the forward feed mode. Modern fruit juice evaporators are designed for flexibility and for efficient operation so that, as indicated in Figure 7, a number of juice flow options are possible.

4. Energy Consumption in Evaporation

To a good approximation, it can be assumed that the steam consumption in a multiple-effect evaporator with N effects is given by the relationship:

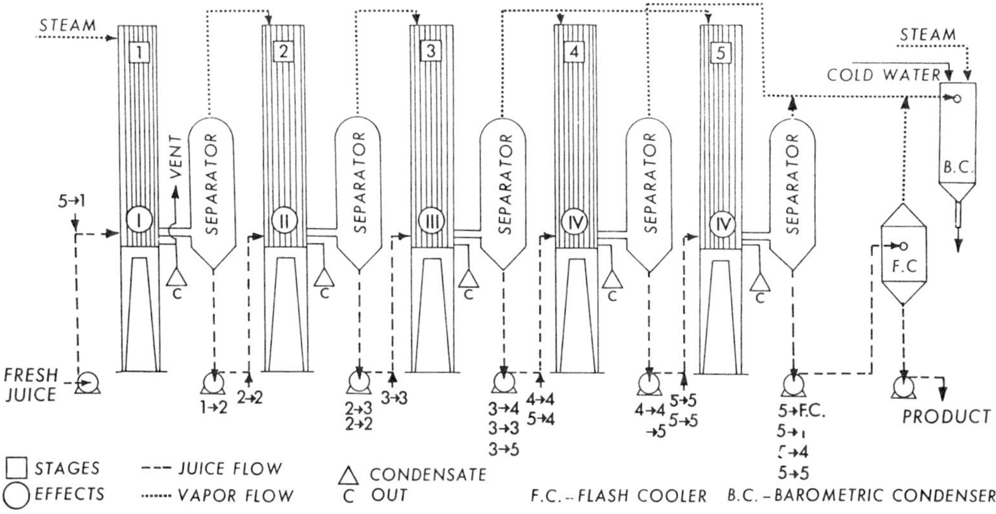

FIGURE 8 Schematic diagram of a four-effect five-stage plate evaporator. (From Ref. 33.)

$$\text{Steam consumption (kg)} = \frac{\text{kg of steam to evaporate one kg of water}}{N} \tag{51}$$

This is based on the assumption that the water removed from each effect is the same and is equal to 1/N of the total water removed. However, in feed forward evaporators the quantity of water in each successive effect increases because of flashing when partially concentrated juice passes from an effect with a higher boiling temperature to one with a lower boiling temperature. The opposite effect takes place in reverse flow evaporators in that the amount of water removed from each effect decreases as the juice is transferred from one effect to another [79].

Relatively few energy-consumption studies have been conducted on the performance of evaporators for the concentration of fruit juices in commercial plants. The reported studies dealt with the evaporation of orange juice [22,33], and because of the significant amount of orange juice evaporated, they will be discussed here. One convenient way of expressing the performance of an evaporator is in terms of steam economy or steam efficiency, defined as:

$$\text{Steam economy} = \frac{\text{kg water evaporated}}{\text{kg steam}} \tag{52}$$

The steam economy of multiple-effect evaporator systems can be expressed as CN, where C is a constant that is less than one and N is the number of effects of the system. The steam efficiencies of two tubular and two plate evaporators in a concentrated orange juice plant were found to be 0.85N and 0.82N, respectively [33]. These data reflect the performance of the evaporators under industrial operating conditions.

5. Heat-Transfer Considerations in Evaporation

In an evaporator, the rate of evaporation of water depends on the rate of heat transfer between the heating medium, generally steam, and the juice. The rate of heat transfer, q (W), can be expressed by the relationship:

$$q = U\,A\,\Delta T \tag{53}$$

In Eq. (53), A is the surface area of heat transfer (m^2), ΔT is the overall temperature difference (°C or K), and U is the overall heat-transfer coefficient ($W\ m^{-2}\ K^{-1}$). In a typical system (Fig. 9) for

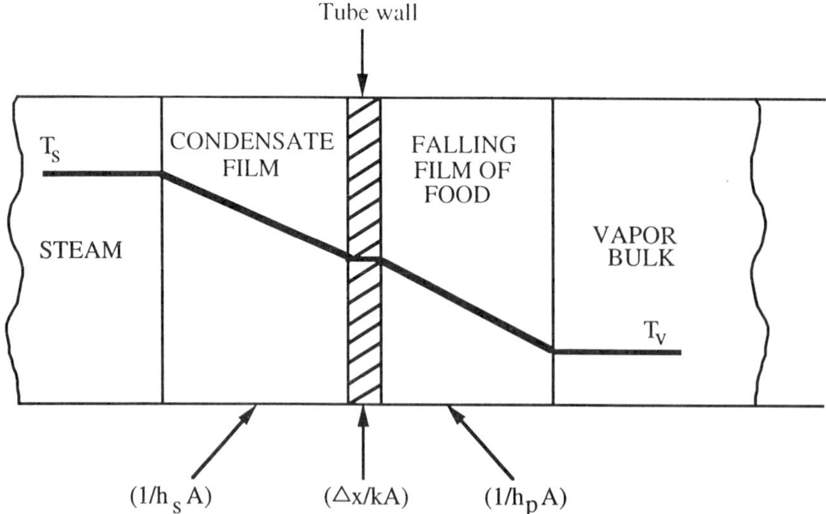

FIGURE 9 Illustration of heat transfer across a tube wall from steam to a film of food.

evaporating juice, the overall resistance to heat transfer, 1/UA, can be broken down into the resistance on the side of the heating medium ($1/h_sA$), that due to the wall of the metal surface ($\Delta X/kA$), and that on the side of the product ($1/h_pA$):

$$\frac{1}{UA} = (1/h_sA) + (\Delta X/kA) + (a/h_pA) \tag{54}$$

In Eq. (54), the heat-transfer resistance due to the wall of the metal surface will be low because of the low magnitude of the wall thickness and the high magnitude of the thermal conductivity (k) of stainless steel, the metal commonly employed in food-processing equipment. The resistance ($1h_sA$) on the steam side also will be low because of the high magnitude of the steam-side heat-transfer coefficient (h_s); typical magnitudes of h_s range between about 11,400 and 28,400 W m^{-2} K^{-1} for condensing steam [44].

The resistance to heat transfer on the product side ($1/h_pA$) and, in particular, the product side heat-transfer coefficient (h_p) depends on the effective velocity of the falling liquid film and its physical and thermal properties such as the density, viscosity, and thermal conductivity. In addition to the factors dependent on the juice, the extent of fouling on the heat-transfer surface plays a major role in the rate of heat transfer. Because fouling reduces heat-transfer rates, cleaning of the evaporator surfaces is scheduled at regular intervals for their efficient operation.

From the above discussion, it is clear that in a multiple effect evaporator system the magnitude of the overall heat-transfer coefficient varies with the stage and effect of the system. Magnitudes of overall heat-transfer coefficient for a four-effect, six-stage evaporator-concentrating orange juice presented by Chen [21] are given in Table 9. The table also contains the product temperature and the concentration range in each stage of the evaporator. From the data in the table it is clear that the magnitude of the overall heat-transfer coefficient depends on the concentration of the juice.

For apple juice, data obtained on a pilot-scale evaporator [77] indicate that the overall heat-transfer coefficient depends on the soluble solids (sugars and pectic substances). The overall heat-transfer coefficients for depectinized and filtered apple juice at 54.4°C ranged from 1987 W m^{-2} K^{-1} at 15 °Brix to 1192 W m^{-2} K^{-1} at 65 °Brix. The overall heat-transfer coefficients of cloudy apple juice were significantly lower than those of clear juice, and the 60 °Brix concentrate formed a solid gel upon cooling; the overall heat-transfer coefficients at 54.4°C ranged from 1420 W m^{-2} K^{-1} at 15 °Brix to 795 W m^{-2} K^{-1} at 55 °Brix.

Moresi [57] presented a comprehensive discussion of fluid flow and heat transfer in falling film evaporators as well as the design and costs of their components. It was shown that for the concentration of 40,000 kg/hr of apple juice from 11 to 72 °Brix in a multiple-effect falling film evaporator system, the optimum number was five effects for either forward- or backward-flow systems.

B. Juice Concentration by Reverse Osmosis

The impetus for the development of reverse-osmosis (RO) membranes was to find an effective method for desalination of sea water. In the early 1950s, Professor Samuel Yuster of the University of California at Los Angeles determined, using the Gibbs adsorption equation, that a hydrophilic surface in contact with brine will have a thin layer of about 0.3–0.4 nm of pure water adjacent to the surface. However, efforts to skim the pure water failed, and efforts were shifted to other means of removal of water. In 1958, Sourirajan discovered the semi-permeability of a cellulose acetate membrane. During the summer of 1958, Drs. Loeb and Sourirajan teamed up with the aim of increasing the permeation rate of water and produced an asymmetrical RO membrane. The asymmetry and the development of methods to produce a thin film over a much thicker layer of highly porous support material led to the commercial success of cellulose acetate membranes for RO applications. Cheryan [26] reviewed concentration of liquid foods by reverse osmosis, including methods of manufacture of RO membranes.

Table 9 Heat Transfer Coefficient and Operating Conditions of a Four-Effect, Six-Stage Orange Juice Evaporator

	Effect	Stage	Product temp. (°C)	Juice concentration (°Brix)		Overall heat transfer coefficient ($W\ m^{-2}\ K^{-1}$)
				In	Out	
Evaporator	1	1	81.9	12.0	14.7	1930
	2	2	67.8	14.7	19.5	1817
	3	3	53.6	19.5	29.8	1420
	4	4	37.8	29.8	44.8	710
	4	5	37.8	44.8	55.6	539
	4	6	37.8	55.6	65.0	426
Preheater	1–4	1–6	21.1–96.1	12.0	12.0	3407

Source: Ref. 21.

1. Osmotic Pressure

When a solution and a pure solvent are separated by a membrane permeable to the solvent (A), water in the case of fruit juices, and not to the solute (B), the solvent will move toward the solution. At equilibrium the chemical potential of the solvent, μ_A, on both sides of the membrane is the same, but an osmotic pressure (Π) would develop. Due to the presence of the solute, μ_A will be lowered by $\Delta\mu_A$, which is counteracted due to the imposed osmotic pressure:

$$\Delta\mu_A = -RT \ln a_A = \Pi\overline{V}_A \tag{55}$$

where $\overline{V}_A$ is the average molar volume of the solvent, which is assumed to be independent of pressure. For an ideal dilute solution, the above equation can be written as:

$$\Pi\overline{V}_A = -RT \ln X_A = -RT \ln X_B \tag{56}$$

For a dilute ideal solution, the osmotic pressure can be calculated from:

$$\Pi = \frac{n_B \; RT}{V_A} = C_B \; RT \tag{57}$$

where n_B is the number of moles of solute and C_B is the molar concentration of the solute. The above equation is known as Van Hoff's law of osmotic pressure [42]. For nonideal solutions, an osmotic coefficient (ϕ) is defined:

$$\phi = \frac{-m_A \ln a_A}{v \; m_B} \tag{58}$$

where v is the number of moles of ions formed from one mole of electrolyte, a_A is the activity of the solvent (water in fruit juices), and m_A and m_B are the molal concentrations of solvent and solute, respectively. The equation for osmotic pressure of non ideal solutions is:

$$\Pi = \frac{RT \, \phi \, v \, m_B}{m_A \; V_A} \tag{59}$$

Osmotic pressure is the measure of the resistance to flow of water through the membrane exerted by the solute. It is proportional to solute concentration and varies with the activity and the molecular weight of the solute. Reverse osmosis occurs when an applied pressure is exerted on the more concentrated solution in excess of the differential pressure. This causes the water to flow from the more concentrated solution to the less concentrated solution. Therefore, before flow of water can occur from the more concentrated side to the less concentrated side, the applied pressure must exceed the differential osmotic pressure across the membrane.

2. Estimation of Osmotic Pressure of Sucrose Solutions and Fruit Juices

In Table 10, osmotic pressure data of sucrose solutions of different concentrations as well as estimated values based on Eqs. (55) and (56) [56] are given. In fruit juices, the major contributors to osmotic pressure are sugars (hexoses and disaccharides) and organic acids [49]. Osmotic pressures of fruit juices are about 1.3 MPa, while those of concentrated fruit juices can be over 10 MPa.

It would be desirable to be able to estimate the osmotic pressure of fruit juices from their composition. Thijssen [94] presented an empirical equation for estimating the osmotic pressure of fruit juices:

$$\text{Osmotic pressure (MPa)} = \frac{13.375c}{(1 - c)} \tag{60}$$

TABLE 10 Osmotic Pressure of Sucrose Solutions at 20°C

Concentration (mol/liter)	Osmotic pressure (kPa)		
	Experimental	Calculated [Eq. (55)]	Calculated [Eq. (56)]
0.098	262	247	239
0.192	513	553	469
0.282	771	792	689
0.370	1027	1035	902
0.453	1292	1279	1104
0.533	1560	1520	1297
0.610	1837	1763	1490
0.685	2119	2003	1672
0.757	2403	2244	1844
0.825	2700	2480	2006

Source: Ref. 56.

In Eq. 60, c is the mass fraction of dissolved solids. Matsuura et al. [46] also presented an equation for estimating the osmotic pressure of fruit juices:

$$\text{Osmotic pressure (MPa)} = \frac{(24.546x_c)}{(1 - 3.94x_c)} \tag{61}$$

where x_c refers to the carbon mass fraction in the fruit juice.

3. Crossflow Membrane Operation

Crossflow operation of RO systems has been developed so that concentration of fluids can be conducted at high rates. The major difference between crossflow operation and the perpendicular flow in a typical mechanical filtration is that in the former the feed solution flows across and parallel to the membrane and not directly through it. In perpendicular flow filtration, there are only the influent and effluent streams. A portion of the feed, usually the solvent (water), is forced through the membrane as permeate, while unpermeated solutes and solids in solution are carried off with the remainder of the feed as concentrate. The crossflow and continuous sweeping action due to the flow minimizes blockage of the membrane's pores and buildup of particles, resulting in longer membrane use prior to cleaning. A comprehensive review of crossflow membrane technology was presented by Paulson et al.

4. Flux Equation

The volumetric flux ($m^3\ s^{-1}\ m^{-2}$) of permeate can be related to the operating pressure, the osmotic pressure, and the membrane characteristics by the relationship [86]:

$$N = A\{P - \Pi(X_{A2}) + \Pi(X_{A3})\} \tag{62}$$

where N is the solvent (water) flux through the membrane ($m^3\ s^{-1}\ m^{-2}$), A is the permeability constant of the membrane, P is the operating pressure, Π is the osmotic pressure, X_{A2} is the concentration of the boundary solution on the high-pressure side, and X_{A3} is the concentration of permeate. A number of factors affected the flux of permeate through a RO membrane:

1. *Effect of pH*. The pH of the fluid to be concentrated affects the level of hydrolysis of the membrane. The recommended pH range for cellulose acetate membranes is 2–8 and 4.5–12.5 for polyamide membranes [5]. Foods that are basic reduce the life of membranes, and the rate of hydrolysis increases exponentially with increasing pH [62].

2. *Effect of Temperature.* Temperature plays an important role as it affects the permeation rate and the chemical and the physical degradation of RO membranes. The viscosity of water decreases with increase in temperature so that at high temperatures the water flux rates will increase. Temperature also affects the solubility of precipitation of various solutes, and this affects the permeation rate of solutes. High operating temperatures lead to hydrolytic stability of the membrane and shortens its life. For example, cellulose acetate membranes distort appreciably at 65.5°C.
3. *Concentration Polarization.* Concentration polarization occurs when the diffusion rate of a solute from near the membrane surface back to the bulk solution is less than the rate of solute transport from the bulk solution to the membrane surface. The net result is that a solute-rich boundary layer is formed next to the membrane, which impedes mass transfer of permeate and increases the differential osmotic pressure and, consequently, the permeate flux is reduced.

Because diffusion back to the bulk solution will be slow when large organic molecules are present, severe concentration polarization can occur in foods that contain large organic molecules. Further, the large organic molecules could polymerize into larger insoluble molecules increasing further concentration polymerization. Other large molecules such as metal hydroxides tend to be closely pressed together to form a dynamic membrane [5].

5. Membrane Construction and Configuration

Reverse-osmosis membranes have a thin top layer of smaller pores with the bottom supporting layer having large pores. For example, spiral wound cellulose acetate membranes made by Osmonics Inc. (Minnetonka, MN) have a top layer with pores in the 0.5-nm range. Therefore, molecules much larger than the pores or greater than a molecular weight of about 200 daltons are separated by a simple sieve process. The total membrane thickness was 0.08–0.13 mm, with the top layer being only 1% of the total thickness. For the purpose of comparison, a sugar molecule is approximately 1 nm in diameter, and whey protein is approximately 200 nm in diameter. The support layer is used to transfer water after it has passed the top layer. If there is excessive pressure, the support will compact, which in turn leads to reduced porosity and concomitant decrease in permeate flux.

Because RO requires higher operating pressures than ultrafiltration (UF), there are more restrictive design considerations [62]. The configurations commonly employed for RO applications in the food industry have been described in detail by Paulson et al. [62] and are summarized here.

Tubuar Element

In this configuration, which is relatively uncommon, the feed and concentrate streams flow through the inside of the tube, usually about 1.3 cm in diameter. The advantages of this configuration are the ease in cleaning the element and the applicability of simple fluid dynamics for predicting the performance. The major disadvantages of this configuration are the relatively small membrane area per unit volume of the membrane (low packing density), high pumping energy requirements, and high cost.

Hollow-Fiber Element

In this configuration for RO, the fibers are very small (42 μm id, 85 μm od) and the pressure flow is on the outside of the fibers. The small size allows for very high packing densities that are 10–20 times that of spiral wound membranes. The major disadvantages of this configuration are the low flux rates and susceptibility to plugging and fiber breakage.

Plate and Frame Element

In this configuration, flat sheets of membranes are placed between plates which form flow channels 0.5–1.0 mm in width. The membrane and its supports are sandwiched together in large numbers to form modules with membrane areas of about 26.94 m^2 (290 ft^2), similar to plate and frame presses. The advantages of this design are that only the membranes need to be replaced and, often, they can be cleaned and reused. The disadvantages are high capital costs, the large number of pressure seals to prevent leakage, and high labor costs.

Spiral Wound Element

In this configuration, a membrane sheet is wound around a central permeate collection tube resulting in high packing density, which in turn depends on the spacer material used. The flow channel height in this configuration ranges from 0.64 to 0.86 mm. The advantages of this configuration include ease with which the elements can be replaced and, because the membrane is always in compression, the high transmembrane operating pressures that can be employed. One disadvantage of this configuration, because of the typical mesh style spacer material, is the possibility of channel blockage with process streams that contain high suspended solids or fibrous materials.

6. Mechanism of Separation in RO Membranes

RO membranes allow permeation of some solutes in fruit juices but repel others. A number of mechanisms have been proposed to explain the selectivity of RO membranes to solutes in aqueous systems. Perhaps the most consistent with all known experimental facts is the "preferential sorption-capillary (PSC) flow" [87]. Cheryan [24] pointed out that the basic equations of PSC flow model are somewhat similar to those of the solution-diffusion model with the modification that (a) the concentration at the membrane surface on the high-pressure side is related to the bulk concentration by the film theory, and (b) equations for the pure water and the solute permeability coefficients in terms of the operating variables are required. Alvarez et al. [3] reported good agreement between experimentally determined values of permeate flux of apple juice and those predicted using the solution-diffusion model combined with the film model.

According to the PSC flow mechanism, RO occurs due to the preferential sorption of constituents of a fluid mixture and their permeation through the porous membrane. For RO to take place, it is essential that a membrane has the right chemical nature (polar and nonpolar effects) and that its pores are of the appropriate size and number (steric effect). The polar parameter indicates the extent of a solute's acidity (proton-donating characteristic) or basicity (proton-accepting characteristic). The nonpolar parameter quantifies the extent of hydrophobic interactions between the nonpolar part of the solute and the membrane. These two factors influence the composition of the solution at the interfacial layer near the membrane surface.

A concentration gradient thus exists with the preferentially sorbed interfacial layer rich in one of the constituents of the bulk solution. Application of pressure forces the interfacial layer through the membrane capillaries, resulting in a permeate solution that is different in composition from the bulk solution. The capillary or pore refers to any void space between the high- and low-pressure sides of the membrane. Either a solute or the solvent can be preferentially sorbed at the membrane/solution interface, depending on the chemical interactions between the membrane and the solution. Therefore, solute-solvent-membrane material interactions govern the extent of preferential sorption.

7. Criteria for Separation of Components of Fruit Juices

The physicochemical criteria for the separation of fruit juice components using cellulose acetate membranes were discussed by Matsuura et al. [46]. For the purpose of the present discussion, the major water-soluble components in fruit juices are considered to be sugars 10–20%, inorganics and flavor compounds 1%, and very small quantities of volatile flavor compounds, vitamins, and proteins.

The PSC flow mechanism is applicable for the removal of water and various solutes from fruit juices. The polar effect of the undissociated alcohols, aldehydes, ketones, and esters can be described by the Taft number [86].

$$\sigma^* = (1/2.48)[\log(k/k')_B - \log(k/k')_A] \tag{63}$$

where σ^* is the Taft number, k and k′ are rate constants for hydrolyses of RCOOR′ and CH COOR′, respectively, and B and A refer to alkaline and acid hydrolyses carried out for the same R′ under identical experimental conditions. The Taft number is a measure of the electron-withdrawing power of the substituent group in a polar molecule. Generally a more negative Taft number results in an increase in the basicity of the molecule and hence higher solute rejection by the cellulose acetate membrane and retention in the concentrate.

The separation of acids is dependent on the degree of dissociation and the pKa. For dissociated acids that exist as ions, electrostatic repulsion by the negatively charged cellulose acetate membrane keeps these ions from permeating through. At high pKa, the acidity (hydrogen bonding ability) of the molecule is low enough to result in preferential sorption for water over solutes. An increase in the number of polar functional groups in the acid results in higher preferential sorption for water over the acid. Solute rejection by the membrane decreases in the order citric (monohydroxy-tri-carboxylic acid) > tartaric (dihydroxy-dicarboxylic acid) > malic (monohydroxy-dicarboxylic) > malic (monohydroxy-dicarboxylic) > lactic (monohydroxy-monocarboxylic) > acetic, propionic, butyric, valerlic, benzoic (monocarboxylic acids).

In the separation of sugars, their very high negative Taft number leads to a better rejection by the membrane than all other fruit juice compounds. A summary of the Taft number criterion governing the relative separation of alcohols, aldehydes, ketones, esters, undissociated acids and sugars is shown in Table 11. The Taft number for a solute corresponds to a critical pore diameter (equal to twice the thickness of the preferentially sorbed interfacial pure water layer) on the membrane surface. There is no necessary correlation between the critical pore diameter and the size of solute and solvent molecules. The critical pore diameter could be larger than either solute or solvent, and solute rejection could still take place. The critical pore diameter could also be different for different membrane material-solution systems [85].

A more negative Taft number corresponds to a larger critical pore diameter on the membrane surface. If the pore diameter on the membrane surface is fixed and corresponds to the critical pore diameter for a solute's Taft number, all other solutes with higher Taft numbers will permeate through. For example, if the pore size of the membrane is the critical pore diameter corresponding to the Taft number for glucose, then glucose, fructose, and sucrose will be retained and all other solutes with Taft number greater than that of glucose will permeate through the membrane.

8. Flux Rates of Apple Juice

Flux rates of apple juice were determined using two spiral wound cellulose acetate (CA) membranes with 90% and 99% salt rejection ratings (CA-90 and CA-99) and an experimental polyamide 99% salt rejection (PA 99) (Osmonics Inc., Minnetonka, MN) membrane Figure 10 is a schematic diagram of the experimental set-up that was used. Both CA membranes (97% and 99% salt rejection) had a 1.394 m^2 (15 $feet^2$) membrane area, while the polyamide (99% salt rejection) had a 0.929 m^2 (10 $feet^2$) membrane area. The regression equations and the corresponding R are listed in Table 12.

The operating conditions, such as temperature and pressure, during the tests were close enough for the different membranes so that their performance could be compared. As expected, the CA-97 had a higher flux rate over the entire concentration range because it had a lower salt rejection with presumably larger average pore size. The PA-99 membrane also had a higher flux rate that the CA-

Table 11 Relationship Between Relative Magnitudes of Taft Number and Solute Rejection by Membranes

Relative Taft number	Result
σ^* solute < σ^* water	Water preferentially sorbed at the membrane/ solution interface
σ^* solute = σ^* water	No solute rejection by membrane; solute permeates through
σ^* solute > σ^* water	Solute preferentially sorbed at the membrane/ solution interface; solute rejection or permeation dependent on the operating conditions

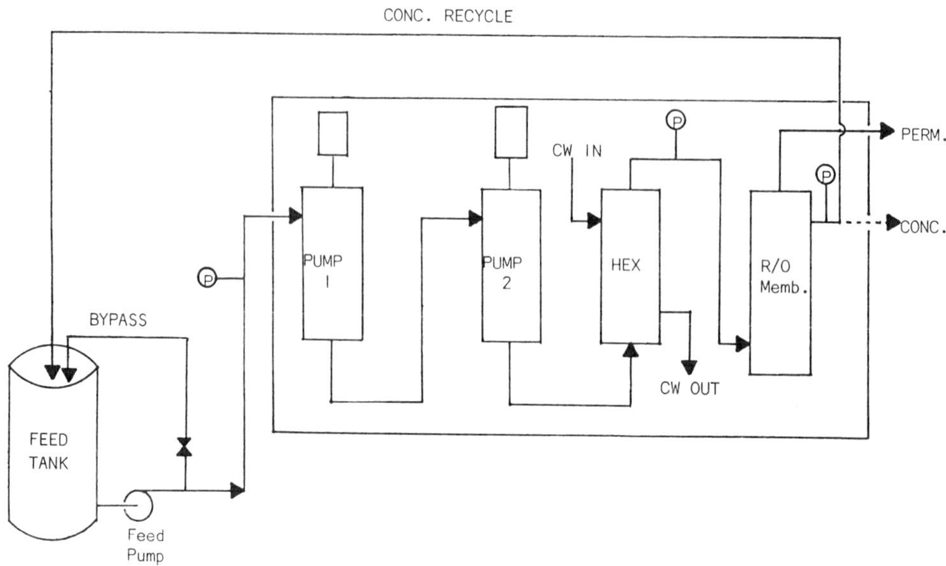

Figure 10 Schematic diagram of a reverse-osmosis concentration unit for apple juice. (From Ref. 27.)

99 over the entire concentration range. Merson et al. [50] also reported higher permeate flux rates for a poly(ester/amide) membrane compared to that of a cellulose acetate membrane. As expected, the flux rate decreased with time due possibly to an increase in the osmotic pressure at the high-pressure side of the membrane, membrane pore blocking by solutes or other macromolecular material and membrane compaction. Practically, this means that membranes will need to be cleaned or replaced at specified intervals to maintain an economically justified higher flux rates.

9. Retention of Aroma Volatiles

A 1-μl apple calibration mixture consisting of 14 odor-active compounds that have been identified to have important apple aroma [30] and an internal standard of dodecane were used to identify the compounds retained. Not all compounds in the calibration mix were detected. Most of the butyrates were not present because the apples used were subjected to long controlled-atmosphere storage, which

Table 12 Apple Juice Concentration by Reverse Osmosis: Regression Equations and R for CA-97, CA-99, and PA-99 Membranes

Membrane	Regression Eq.	R(%)
a) Permeate Flux (y) versus °Brix Concentrate (x)		
CA-97	y = 17.6 – 0.726x	97.1
CA-99	y = 13.0 – 0.648x	97.1
PA-99	y = 16.7 – 0.67 x	96.7
b) Permeate Flux (y) versus Hours of Operation (x)		
CA-97	y = 8.25 – 0.919x	98.2
CA-99	y = 4.57 – 0.319x	96.6
PA-99	y = 7.27 – 0.621x	98.6

Source: Ref. 27.

TABLE 13 Apple Juice Aroma Compounds Retained in Reverse-Osmosis Concentration Using a Cellulose Acetate and a Polyamide Membrane and Their Relative Ranking

Retention Ranking	CA-99	PA-99
Highest	*trans*-2-Hexenal	Ethyl-2-methyl-butyrate
	n-Hexanal	*trans*-2-Hexenal
	Isopentyl acetate	Isopentyl acetate
		n-Hexanal
	Butyl acetate	Butyl acetate
		Ethyl butyrate
Lowest	Hexyl acetate	Hexyl acetate

Source: Ref. 27.

was reported to be unfavorable to butyrate retention in the apple [1]. Compounds that were identified are summarized in Table 13 for the CA-99 and PA-99 membranes. Additional details regarding retention of apple aroma compounds by the CA-99 and the PA-99 membranes can be found in Chua et al. [27].

10. RO Application to Citrus and Other Juices

Braddock et al. [16] have studied parameters affecting RO concentration of ethanol, acetaldehyde, and other aroma compounds in essences from commercial orange juice evaporators. Kane et al. [37] found that in RO concentration of lemon juice aroma, the recovery of total volatiles was 79.9% and the loss of volatile compounds was attributed to adsorption in the polymeric matrix of the membrane; some loss was attributed to cleaning.

RO can be used in conjunction with evaporation to obtain high-quality concentrated juices. For example, Duxbury [31] reported a processing technique that maximized the natural quality of pure apple and pear juice concentrates by minimizing product exposure to vacuum evaporation. Key elements were the use of a preconcentration RO process that removes 50–75% of moisture normally removed in vacuum evaporation, followed by a microfiltration process (low-temperature pasteurization and increased juice clarity). The filtered preconcentrated product is then further concentrated to 70 °Brix in a vacuum evaporator. Cross [29] described a patented process in which UF and RO were used to concentrate orange juice to about 42 °Brix, a concentration much higher than those reported with CA and polyamide membranes. For concentration of fruit juices containing suspended solids, such as tomato juice [48] and orange juice [29], the suspended solids are removed, typically by either centrifugation or ultrafiltration [29,38], and the clear serum is concentrated by reverse osmosis. Clarified serum was concentrated to 60 °Brix [29] and by using two different membranes, one capable of withstanding elevated pressures, concentrated products with improved flavor could be produced. The much higher soluble solids concentrations possible using newly developed membranes should lead to greater use of RO.

C. Freeze Concentration

It has been known for a long time that aqueous solutions could be concentrated by freezing part of the water out as ice and mechanically separating the ice by draining, pressing, or centrifuging. Earlier, Tressler [36] reviewed the historical development of concentration by freezing as well as its advantages and disadvantages, and Thijssen [93,94] reviewed the technical and economic aspects of concentration by freezing. Schwartzberg [80] presented an extensive discussion of the relevant topics of freeze concentration: thermodynamics, types of freezers, nucleation and ice crystal growth (ripening), and washing of the solutes in wash columns.

Concentration by freezing has several advantages over concentration by evaporation, the predominant method of concentration of fruit juices. In terms of energy consumption, freezing requires

335 kJ/kg (144 Btu/lb) of water frozen out as compared to about 2.23 MJ/kg (960 Btu/lb) of water evaporated in a single-effect evaporator. Even in multiple-effect evaporators it is difficult to achieve energy consumption of the order of 326 kJ (140 Btu/lb) of water.

Another concern is the heat damage, C, at a constant temperature in evaporation and other thermal processes that can be assumed to be directly proportional to the residence time, t_r, or the juice molecules:

$$C = kt_r \tag{66}$$

The effect of temperature on k can be expressed by the Arrhenius relationship:

$$k = k_T \exp(-E_{ac}/RT) \tag{67}$$

where k_T is a constant, E_{ac} is activation energy for the chemical change, T is the absolute temperature (K), and R is the gas constant. For many thermally sensitive reactions E_{ac} can be assumed to be between 42 and 120 J/mol so that at a constant residence time in a process an increase of about 10°C results in an increase in thermal decomposition by a factor of 1.7–5.0 [94]. Because of low operating temperatures, a major advantage of concentration by freezing is the minimal loss of quality factors such as color and volatile aroma constituents. Currently the expression freeze concentration is used for concentration by freezing and removing water, and it will be used in the rest of this chapter.

Braddock and Marcy [15] concentrated pasteurized Valencia and Temple orange juices to 45 °Brix. Except for considerable pulp reduction of feedstream juices, there were few differences from normal citrus juice–recovery procedures for freeze concentration. Since the product retained most of the aroma constituents of fresh juice, careful handling and high-quality fee juice prior to freeze concentration was much more important than for evaporation. Braddock and March [14] extracted pineapple juice and adjusted by finishing and centrifugation to two pulp levels: 12% and 2% by volume. They determined physical, chemical, and sensory differences in heat-stabilized single-strength juices, evaporator-concentrated, and freeze-concentrated juices and found no significant differences between samples for degree Brix, % acid, total hexose, vitamin C, browning index, color, or viscosity. Pulp content was lower in both freeze- and evaporator-concentrated samples. Statistically significant ($p < 0.001$) flavor differences were detected between freeze- and evaporator-concentrated juices. Flavor of reconstituted freeze-concentrated juice was comparable to single-strength juice and preferable to evaporator concentrate [14].

Disadvantages of freeze concentration include loss of the liquid food with ice and high operating and capital costs of freeze-concentration equipment. The level of concentration that can be achieved by freeze concentration is limited due to the formation of eutectic mixtures by water with other components of liquid foods. In the case of fruit juices, the eutectic point occurs at about –23.3°C. In addition, the high viscosity at low temperatures and high concentrations is another hurdle in handling the concentrated product.

1. Principles of Freeze Concentration

Freezing points for several fruit juices and wine as a function of concentration of solids are presented in Figure 11 [80]. For example, by freezing apple juice that normally contains about 10% dissolved solids to –11°C, its concentration can be increased to 50% solids at equilibrium by removal of water as ice crystals. Freezing point depression for juices containing monosaccharides can be calculated from Riedel's empirical equation:

$$\Delta T_f = 10x_s + 50x_s^3 \tag{68}$$

Based on Eq. (36), Schwartzberg [80] derived the equation:

$$T = \frac{-103.23(18.02/M_s)x_s}{[1 - (1 + b)x_s]} \tag{69}$$

Further, when the freezing point was calculated using $E = (18.02/M_s) = 0.0988$ and $b = 0.0524$, on average the values of T were within 2% of those from Eq. (68) (Table 14). Orange juice that contains

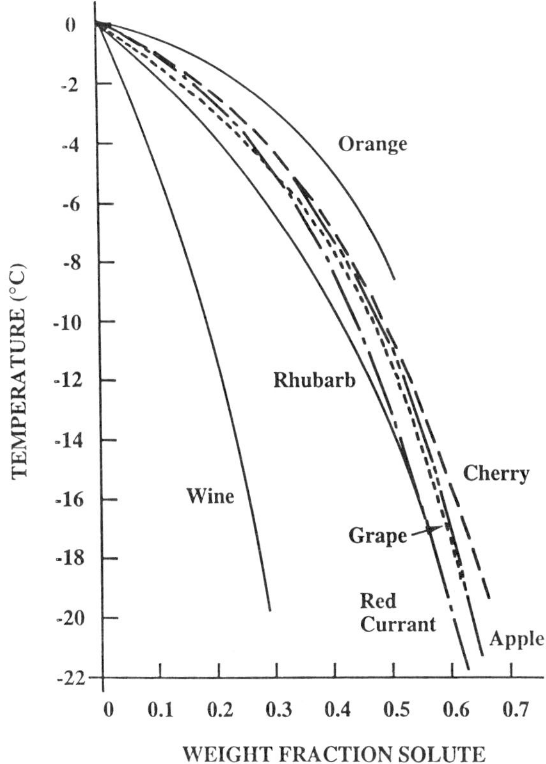

FIGURE 11 Freezing points of fruit juices and wine as a function of sugar concentration. (From Ref. 80.)

a high proportion of sucrose has freezing points slightly higher than those estimated using Eqs. (68) and (69).

Figure 12 is an illustration of a basic freeze-concentration process. The crystallizer in which the crystal growth takes place is usually a stirred vessel in which the driving force for crystallization

TABLE 14 Freezing Point of Fruit Juices as a Function of Solids Fraction According to Riedel and Schwartzberg Models

Mass fraction solids	Freezing point (°C)	
	Riedel	Schwartzberg
0.10	−1.05	−1.14
0.15	−1.67	−1.82
0.20	−2.40	−2.58
0.25	−3.28	−3.46
0.30	−4.35	−4.47
0.35	−5.64	−5.65
0.40	−7.20	−7.05
0.45	−9.06	−8.72
0.50	−11.25	−10.76
0.55	−13.82	−13.32
0.60	−16.80	−16.60
0.70	−24.15	−27.11
0.75	−28.59	−36.30

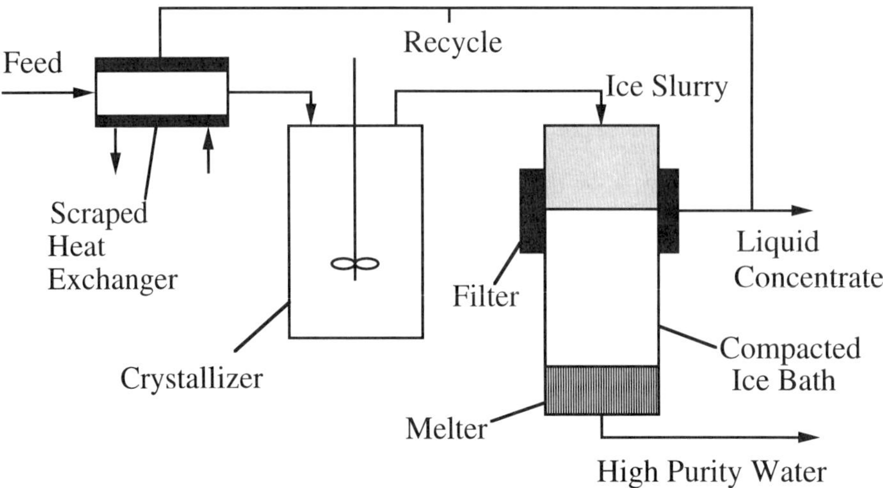

FIGURE 12 Schematic diagram of a freeze-concentration process. (From Ref. 66.)

is provided either by a supercooled crystal-free feed stream or by direct or indirect cooling of the ice suspension in the crystallizer. The crystals formed are of high quality even though small imperfections can be found in the lattice [93]. Therefore, theoretically, freeze concentration is the most selective dewatering process. Hartel [34] classified nucleation mechanisms into primary and secondary.

In practice, deviations from the ideal dewatering process occur because of the mother liquor adhering to the ice crystals. Consequently, the selectivity of concentration is totally determined by the quality of the ice-liquid food separation. In orange juice, removal of pulp and heat stabilization prior to freeze concentration resulted in lower losses of solids in the wash column and a higher quality product, respectively [15]. The separation is performed in a centrifuge, a press, a wash column, or a combination of all these devices.

For efficient separation of ice from the mother liquor in basket centrifuges and wash columns, the specific surface area of the crystals must be as small as possible. In presses, the separation is insensitive to the ice crystal shape and size. However, the capacity of all the separation devices increases with an increase in the size and uniformity of the crystals. Therefore, the design and operation of crystallizers are aimed at the formation of spherical crystals of uniform size [93].

The residence time required to obtain a fixed mean crystal diameter is about inversely proportional to the supercooling of the liquid in the crystallizer. The number of crystals per unit mass is directly proportional to the residence time and the nucleation rate of the crystals. The latter is a second-order process with respect to the number of crystals in the crystallizer and probably of a higher order with respect to supercooling. The use of a high cooling rate results in strong supercooling near the heat-removing interphase and consequently to high nucleation rates and to small crystals.

2. Crystal Growth

The factors affecting growth of ice crystals have been discussed in detail by Thijssen [83], and they will be described only briefly here. In continuously operating crystallizers, the average ice crystal size is a complex function of the mean residence time of the liquid in the crystallizer, the manner in which the mother liquor is cooled, the cooling rate, the concentration of dissolved solids, and the degree of turbulence in the suspension. The growth rate of the crystals is determined by the heat and mass transfer in the liquid phase and by the liquid-solid interphase. Thijssen [93] showed that the crystal growth velocity, V, is related to operating parameters and physical properties by the expression:

$$V = \frac{\Delta c_w}{\rho_w - c_w}\left[\frac{2D}{d} + 0.47\left(\frac{\varepsilon}{\nu^3}\right)^{0.22} Sc^{1/3} D\, d^{-0.12}\right] \tag{70}$$

where Δc_w is the difference in water concentration between that in the bulk liquid and that at the crystal surface, ρ_w is the density of water, c_w is the water concentration in the bulk liquid, D is the diffusion coefficient of water in the liquid phase, d is the diameter of ice crystal, ε is the power consumption of the agitator per unit mass of suspension, ν is the kinematic viscosity (ratio of viscosity and density) of the liquid phase, and Sc, the Schmidt number, is the ratio of the kinematic viscosity and the diffusivity. Schwartzberg [80] presented a more detailed examination of ice crystal growth. The presence of dissolved polymers, e.g., pectins, affects ice crystallization [82] so that their removal by either enzymes or ultrafiltration results in a more efficient freeze-concentration process. Another important factor is that ice crystal growth decreases with increase in concentration of soluble solids; for example, ripening rates were four times less at 42% sucrose concentration than at 10%.
From Eq. (70), it can be deduced that the rate of crystal growth increases with the stirring power, strongly decreases with liquid viscosity, and is quite insensitive to crystal size. With respect to viscosity of the fruit juice, Rao et al. [69] reported that it increases exponentially with concentration. Further, the Arrhenius activation energy, E_a, in Eq. (13) is high at low temperatures, and it increases with concentration of dissolved solids, so that the rate of ice crystal growth in fruit juices at low temperatures will be low and will decrease as the concentration of dissolved solids increases in the crystallizer.

3. The Crystallizer

Crystallizers can be classified as either direct-cooled or indirect-cooled, depending on the manner in which the heat of crystallization is removed. In the former, the heat of crystallization is removed by either vacuum evaporation of a part of water or by evaporation of a nonmiscible refrigerant in contact with the aqueous solution. In the food industry, only indirect-cooled crystallizers have been employed [93]. There are three main types of indirect-cooled crystallizers.

In the first type, heat removal and crystallization take place separately. The feed stream is supercooled by about 4–6°C in a heat exchange, ensuring that no crystals are formed and fed to a stirred crystallizer. With this type, it is claimed that local supercoolings can be avoided and that spherical crystals with a mean diameter greater than 1 mm are obtained [93].

In the second type, the feed stream is nearly completely frozen by bringing the liquid into contact with a cold wall and the solid slabs are mechanically removed and fed to an ice-liquid separator. The main advantage of this type is that even very dilute solutions can be concentrated in one step to 40% or more. However, separation of the phases is difficult because of the small ice crystals.

The last type is capable of producing pumpable ice slurries and most freeze-concentration crystallizers are of this type. The nucleation and crystal growth occurs at a cold wall, and the ice crystals are removed continuously by scraper blades and brought into suspension.

4. Ice-Liquid Separators

Presses, centrifuges, and wash columns can be employed to separate ice crystals from concentrated juice. Among presses, both hydraulic piston and screw type presses can be employed [93]. A major concern is the loss of concentrate occluded to the ice, and this loss decreases with increase in cake pressure. Nevertheless, the loss of solids at maximum pressure amounts to 9% for a 10% solution that is concentrated to 40% in one step.

Among centrifuges, both batch and continuous types can be used. The loss of solids with ice is a function of the centrifugal force, residence time of the ice in the centrifuge, and amount of wash water. The residual amount of liquid food in ice is determined by the equilibrium between centrifugal and capillary forces acting on the liquid present in the interstices between the crystals. Because capillary forces are approximately proportional to the specific surface area of the crystals, the residual liquid will be minimal for uniform, spherical, and large crystals.

Wash columns perform two distinct operations. First, the concentrate is separated from ice crystals by draining and filtration or light pressing. Second, the compact ice mass is washed with water in a countercurrent fashion. Wash columns that were operated under normal atmospheric pressure for desalination of sea water are not suitable for the highly viscous liquid food concentrates. Therefore, wash columns operating under large pressure gradients are used for liquid foods.

Figure 13 is a schematic diagram of a Phillips wash column [93]. A slurry containing 30–40% crystals originating in a crystallizer is fed continuously under pressure to the wash column. In the filtration section, the majority of the concentrate is removed through a wall filter. The remaining crystal mass flows to the end of the column where it is completely melted. Most of the melt is removed as nearly pure water (solids content less than 0.2%), and about 5% is forced to flow countercurrently to the impure crystals as reflux liquid. The energy for crystal movement and washing (reflux driving force) is supplied by a pulse unit in the nearly pure water product stream. The capacity of the wash column increases with increase in permeability of the ice bed. In turn, the permeability of the ice bed increases strongly with increase in mean crystal size and with uniform crystal size distribution.

5. Energy Consumption and Economics

Compared to evaporation and reverse osmosis, liquid foods of exceptional quality can be produced by freeze concentration because the process temperature is so low that chemical and biochemical reactions are of little importance. It appears that freeze concentration would be especially suitable for undepectinized filtered fruit juices. However, due to the high viscosities at high solids concentration and low temperatures, concentration rates would decrease as concentration progresses. In addition, care must be taken to eliminate the loss of dissolved solids with the ice, and the economics of capital and operating costs play an important role. Schwartzberg [80] estimated the energy consumption for concentration of a 12% soluble solids juice fed at 20°C to 50% in a four-stage freeze concentration system to be 360 kJ (100 kWh) per 1000 kg of water removed. However, energy consumption data of industrial juice freeze-concentration units operating under realistic conditions are not available. Renshaw et al. [71] presented a comprehensive analysis of costs for concentration and drying of foods that pointed out the important role of economy of scale, especially in freeze concentration. For example, increasing the water-removal capacity from 450 to 4500 kg/hr lowered the estimated

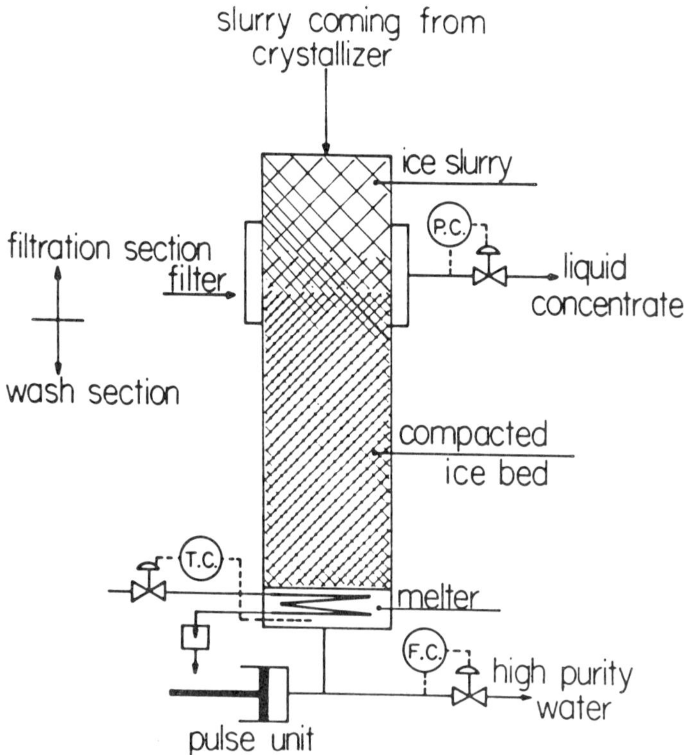

FIGURE 13 Illustration of a Phillips wash column. (From Ref. 93.)

cost by a factor of about four (from 25 to 6 cents/kg/hr), and further increase in the capacity to 45000 kg/hr lowered the estimated cost by another 50% (from 6 to 4 cents/kg/hr).

V. NEW JUICE-PROCESSING TECHNOLOGIES

A number of relatively new thermal and nonthermal preservation technologies are being tested in academic and industrial laboratories. For fruit juices, these include ultrafiltration that serves the purpose of both moderate concentration and removal of large particles and molecules, and high-energy light pulses that result in temperatures up to 2000°C on the surface of a food. In addition, high-voltage pulsed electric field (PEF) and high-pressure (HP) technology have been the subjects of recent research.

High-voltage PEFs rupture the cellular walls of microorganisms. The feasibility of high-voltage PEF treatment as a food-pasteurization technique was studied in a static treatment unit in which 25-ml batches of foods were processed with electric field strengths up to 40 kV/cm [107]. While maintaining fluid food temperature at <25°C, *E. coli* ATCC 11229 and *Saccharomyces cerevisiae* suspensions in test media simulated milk ultrafiltrate and apple juice were treated with pulsed electric fields. A three- to four-decimal reduction in microbial viability was observed. Bai-Lin et al. [8,9] found that the principal parameters affecting PEF treatment of apple juice to inactivate *S. cerevisiae* were electric field intensity, treatment time, and number of pulses. PEF proved cost-effective for nonthermal apple juice pasteurization, using HTST pulses requiring less than 10% of the electrical energy necessary for HTST pasteurization. Application of 10 2.5-μs pulses of 35 kV/cm or 2 2.5-μs pulses of 50 kV/cm achieved 6D inactivation of *S. cerevisiae* inoculated at 2×10^6 colony-forming units (cfu)/ml. The former treatment resulted in greater than 3-week shelf life extension to juices stored at 4 or 25 °C. Although the studies on *E. coli* were conducted with ATCC 11229 [63] rather than *E. coli* O157:H7, the PEF technique shows considerable promise for pasteurization of apple and other fruit juices.

High-pressure technology, of the order of 4,000–10,000 atmospheres, has been used in the food industry for about 15 years (mainly in Japan) with the manufacture of HP-processed jams, fruit juices and fruit desserts (i.e., high-acid foods). It is believed that high pressures destroy protein structure, thus damaging the vital systems of microorganisms. HP processing successfully inactivates microorganisms but is less effective at inactivating bacterial spores and some enzymes. Processing at moderate pressures (200-499 MPa) can be combined with pasteurization temperature (60–90°C) for effective preservation of food. HP processing can be performed in-container without affecting barrier and mechanical properties of packaging materials [51].

Aleman et al. investigated pulsed and static ultra-high-pressure (UHP) treatments as preservation methods for fruit juice. Specifically, sinusoidal and step-pressure pulses were compared with static pressure treatments for inactivation of *S. cerevisiae* 2407-1a in unsweetened pineapple juice. Inactivation was not observed after 40–4000 fast sinusoidal pulses (10 Hz) at 4–400 seconds in the 235-270 MPa range. Static 270 MPa treatments at 40 and 400 seconds resulted in 0.7 and 5.1 decimal reductions, respectively. Slower 0–270 MPa step pulses at 0.1 (10 pulses), 1 (100 pulses), and 2 (200 pulses) Hz with 100 seconds total on-pressure time resulted in 3.3, 3.5, and 3.3 decimal reductions, respectively. A comparable static pressure treatment resulted in 2.5 decimal reductions. Changing the on-pressure/off-pressure time ratio (3:1, 3:3, and 1:3) demonstrated that treatments with longer on-pressure times, e.g., 0.6 second on-pressure and 0.2 second off-pressure, were significantly more effective in inactivating *S. cerevisiae*, resulting in an approximately 4 decimal reduction in 100 seconds [2].

REFERENCES

1. T. E. Acree, J. Barnard, and D. G. Cunningham, A procedure for the sensory analysis of gas chromatographic effluents, *Food Chem. 14*:273 (1984).
2. G. D. Aleman, E. Y. Ting, S. C. Mordre, A. C. O. Hawes, M. Walker, D. F. Farkas, and J.

A. Torres, Pulsed ultra high pressure treatments for pasteurization of pineapple juice, *J. Food Sci. 61*:388 (1996).

3. V. Alvarez, S. Alvarez, F. A. Riera, and R. Alvarez, Permeate flux prediction in apple juice concentration by reverse osmosis, *J. Membrane Sci. 127*:25 (1997).
4. Essence recovery keys quality in fruit pulp concentration, *Food Eng. Int.* (Sept):69 (1976).
5. Engineering memo #18, Osmonics, Inc., Minnetonka, MN, 1977.
6. Reverse osmosis membranes cleared for bulk liquid food, *Food Chem. News 26*(42):41 (1984).
7. Outbreak of Escherichia coli O157:H7 infections associated with drinking unpasteurized commercial apple juice—British Columbia, California, Colorado, and Washington, *MMWR 45*(44):975 (1996).
8. Q. Bai-Lin, C. Fu-Jang, G. V. Barbosa-Canovas, and B. G. Swanson, Nonthermal inactivation of Saccharomyces cerevisiae in apple juice using pulsed electric fields, *Lebensm. Wiss. Technol. 28*:564 (1995).
9. Q. Bai-Lin, U. R. Pothakamury, H. Vega, O. Martin, G. V. Barbosa-Canovas, and B. G. Swanson, Food pasteurization using high-intensity pulsed electric fields, *Food Technol. 49*(12):55 (1995).
10. L. Bayindirli, Mathematical analysis of variation of density and viscosity of apple juice with temperature and concentration, *J. Food Proc. Pres. 16*:23 (1992).
11. L. Bayindirli, Density and viscosity of grape juice as a function of concentration and temperature, *J. Food Proc. Pres. 17*:147 (1993).
12. R. E. Berry and M. K. Veldhuis, Process of oranges, grape fruit and tangerines, *Citrus Science and Technology* (S. Nagy, P. E. Shaw, and M. K. Veldhuis, eds.), The AVI Publishing Co., Westport, CT, 1977.
13. J. L. Bomben, S. Bruin, H. A. C. Thijssen, and R. L. Merson, Aroma recovery and retention in concentration and drying of foods, *Advances in Food Research*, Vol. 20 (C. O. Chichester, E. M. Mrak, and G. F. Stewart, eds.), Academic Press, New York, 1973, p. 1.
14. R. J. Braddock, and J. E. Marcy, Freeze concentration of pineapple juice, *J. Food Sci. 50*:1636 (1985).
15. R. J. Braddock and J. E. Marcy, Quality of freeze concentrated orange juice, *J. Food Sci. 52*:159 (1987).
16. R. J. Braddock, G. D. Sadler, and C. S. Chen, Reverse osmosis concentration of aqueous-phase citrus juice essence, *J. Food Sci 56*:1027 (1991).
17. M. G. Brown, R. L. Kilmer, and K. Bedigian, Overview and trends in the fruit juice processing industry, *Fruit Juice Processing Technology*, (S. Nagy, C. S. Chen, and P. E. Shaw, eds.), Agscience, Inc., Auburndale, FL, 1993, p. 1.
18. R. D. Carter, and B. S. Buslig, Viscosity and particle size distribution in commercial Florida frozen concentrated orange juice, Proceedings of the Florida State Horticulture Society, 1977, p. 130.
19. G. Cerny, W. Hennlich, and K. Poralla, Spoilage of fruit juice by bacilli: isolation and characterisation of the spoiling microorganism, *Z. Lebens. Unters. Forsch. 179*:224 1985.
20. S. K. Chandrasekaran, and C. J. King, Retention of volatile flavor components during drying of fruit juices, *Chem. Eng. Symp. Ser. 67*(108):122 (1971).
21. C. S. Chen, Evaporator technology for citrus, Paper presented at the 1981 Summer Meeting American Society of Agricultural Engineers, June 21–24, Orlando, FL.
22. C. S. Chen, Citrus evaporator technology, *Trans. Am. Soc. Agric. Engrs. 25*:1457 (1982).
23. C. S. Chen, Calculation of water activity and activity coefficient of sugar solutions and some liquid foods, *Lebensm. Wissens. Technol. 20*:64 (1987).
24. C. S. Chen, Physicochemical principles for the concentration and freezing of fruit juices, *Fruit Juice Processing Technology* (S. Nagy, C. S. Chen, and P. E. Shaw, eds.), Agscience, Inc., Auburndale, FL, 1993, p. 23.
25. C. S. Chen, Physical and rheological properties of fruit juices, *Fruit Juice Processing Technology* (S. Nagy, C. S. Chen, and P. E. Shaw, eds.), Agscience, Inc., Auburndale, FL, 1993, p. 56.
26. M. Cheryan, Concentration of liquid foods by reverse osmosis, *Handbook of Food Engineering* (D. R. Heldman and D. B. Lund, eds.), Marcel Dekker, New York, 1992, p. 393.

27. H. Chua, M. A. Rao, T. E. Acree, and D. G. Cunningham, Concentration of apple juice by reverse osmosis, *J. Food Proc. Eng. 9*:231 (1988).
28. P. G. Crandall, C. S. Chen, and R. D. Carter, Models for predicting viscosity of orange juice concentrate, *Food Technol. 36*(5):245 (1982).
29. S. Cross, Membrane concentration of orange juice, *Proc. Florida State-Horti. Soc. 102*:146 (1989).
30. D. G. Cunningham, T. E. Acree, J. Barnard, R. M. Butts, and P. A. Braell, Charm analysis of apple volatiles, *Food Chem. 19*:137 (1986).
31. D. D. Duxbury, Apple, pear juice concentrates 100% pure, *Food Proc. 53*(4):78, 80 (1992).
32. C. Ferro Fontán, J. Chirife, and R. Boquet, Water activity in multicomponent nonelectrolyte solutions, *J. Food Technol. 18*:533 (1981).
33. J. Gasparino Filho, A. A. Vitali, F. C. P. Viegas, and M. A. Rao, Energy consumption in a concentrated orange juice plant, *J. Food Proc. Eng. 7*:77 (1984).
34. R. W. Hartel, Evaporation and freeze concentration, *Handbook of Food Engineering* (D. R. Heldman and D. B. Lund, eds.), Marcel Dekker, New York, 1992, p. 341.
35. J. L. Heid and J. W. Casten, Vacuum concentration of fruit and vegetable juices, *Fruit and Vegetable Juice Processing Technology* (D. K. Tressler and M. A. Joslyn, eds.). AVI Publishing Co., Westport, CT, 1961, p.
36. D. K. Tressler, Preservation by freezing, *Fruit and Vegetable Juice Processing Technology* (D. K. Tressler and M. A. Joslyn, eds.), AVI Publishing Co., Westport, CT, 1961, p. 164.
37. L. Dane, R. J. Braddock, C. A. Sims, and R. F. Matthews, Lemon juice aroma concentration by reverse osmosis, *J. Food Sci. 60*:190 (1995).
38. S. S. Koeseoglu, J. T. Lawhon, and E. W. Lusas, Use of membranes in citrus juice processing, *Food Technol. 44*(12):90 (1990).
39. M. Kozempel, A. McAloon, and W. Yee, The cost of pasteurizing apple cider, *Food Technol. 52*(1):50 (1998).
40. T. P. Labuza, K. Acott, S. R. Tatini, R. Y. Lee, J. Flink, and W. McCall, Water activity determination: a collaborative study of different methods, *J. Food Sci. 41*:910 (1976).
41. J. Leroi, J. Masson, H. Renon, J. Fabries, and H. Sannier, Accurate measurement of activity coefficients at infinite dilution by inert gas stripping and gas chromatography, *Ind. Eng. Chem. Proc. Des. Dev. 16*:139 (1977).
42. H. K. Leung, Water activity and other colligative properties of foods, *Physical and Chemical Properties of Food* (M. R. Okos, ed.), American Society of Agricultural Engineers, St. Joseph, MI, 1986, p. 138.
43. D. Marinos-Kouris, and G. D. Saravacos, Volatility of organic compounds in aqueous sucrose solutions, paper No F4.27, presented at 5th International Congress of Chemical Engineering (CHISA' 75), Prague, August 25–29, 1975.
44. W. E. McAdams, *Heat Transmission*, McGraw-Hill, New York, 1954.
45. J. L. McCabe, and J. M. Smith, *Unit Operations in Chemical Engineering*, 3d ed., McGraw-Hill, New York, 1976.
46. T. Matsuura, A. G. Baxter, and S. Sourirajan, Concentration of fruit juices by reverse osmosis using porous cellulose acetate membranes, *Acta Aliment. 2*:109 (1973).
47. M. R. McLellan, and D. F. Splittstoesser, Reducing risk of E. coli in apple cider, *Food Technol. 50*(12):174 (1996).
48. C. A. Merlo, W. W. Rose, L. D. Pedersen, E. M. White, and J. A. Nicholson, Hyperfiltration of tomato juice: pilot scale high temperature testing, *J. Food Sci. 51*:403 (1986).
49. R. L. Merson and A. I. Morgan, Juice concentration by reverse osmosis, *Food Technol. 22*:631 (1968).
50. R. L. Merson, G. Paredes, and D. B. Hosaka, Concentrating fruit juices by reverse osmosis, *Ultrafiltration Membranes and Applications* (A. R. Cooper, ed., Plenum Publishing Corp., 1980, p. 405.
51. B. Mertens, Developments in high pressure food processing. I, *Int. Zeitschrift Lebensm. Tech. Mark. Verpack. Anal 44*(3):100 (1993).

52. S. Mizrahi, A review of the physicochemical approach to the analysis of the structural viscosity of fluid food products, *J. Text. Stud. 10*:67–82 (1979).
53. S. Mizrahi and Berk, Z. Flow behavior of concentrated orange juice, *J. Text. Stud. 1*:342 (1970).
54. S. Mizrahi, and Firstenberg, R., Effect of orange juice composition on flow behaviour of six-fold concentrate, *J. Text. Stud. 6*:523 (1975).
55. J. G. Moore, and W. E. Hesler, Evaporation of heat sensitive materials, *Chem. Eng. Prog. 59*(2):87 (1963).
56. W. J. Moore, *Physical Chemistry*, 4th ed., Prentice-Hall, Englewood Cliffs, NJ, 1972.
57. M. Moresi, Design and optimisation of falling-film evaporators, *Developments in Food Preservation - 3* (S. Thorne, ed.), Elsevier, New York, 1985.
58. J. C. Moyer and G. D. Saravacos, Scientific and technical aspects of fruit juice aroma recovery, Presented at the 7th International Fruit Juice Congress, Cannes, France, September 5, 1968.
59. S. Nagy, C.S. Chen, and P. E. Shaw, *Fruit Juice Processing Technology*, Agscience, Inc., Auburndale, FL, 1993.
60. R. S. Norrish, An equation for the activity coefficients and equilibrium relative humidities of water in confectionery syrups, *J. Food Technol. 1*:25 (1966).
61. M. R. Okos, *Chemical and Physical Properties of Food*, American Society of Agricultural Engineers, St. Joseph, MI, 1986.
62. D. J. Paulson, R. L. Wilson, and D. D. Spatz, Crossflow membrane technology and its applications, *Food Technol. 38*(12):77 (1984).
63. U. R. Pothakamury, G. V. Barbosa-Canovas, and B. G. Swanson, Electron microscopic studies of *E. coli* and *S. aureus* subjected to high voltage pulsed electric fields, IFT Annual Meeting, 1995, p. 218.
64. S. Rahman, *Food Properties Handbook*, CRC Press, Boca Raton, FL, 1995.
65. R. S. Ramteke, N. I. Singh, M. N. Rekha, and W. E. Eipeson, Methods for concentration of fruit juices: a critical evaluation, *J. Food-Sci. Technol.* (India). *30*:391 (1993).
66. M. A. Rao, Concentration of apple juice, *Processed Apple Products* (D. L. Downing, ed.), Van Nostrand Reinhold, New York, 1989, p. 137.
67. M. A. Rao, Rheological properties of fluid foods, *Engineering Properties of Foods*, 2d ed. (M. A. Rao and S. S. H. Rizvi, eds.), Marcel Dekker, New York, 1995, p. 1.
68. M. A. Rao, and A. K. Datta, Food process engineering: thermodynamic and transport properties, *Encyclopedia of Agricultural Science*, Vol. 2 (C. E. Arntzen, ed.), Academic Press, New York, 1994, p. 381.
69. M. A. Rao, H. J. Cooley, and A. A. Vitali, Flow properties of concentrated juices at low temperatures, *Food Technol. 38*(3):113 (1984).
70. H. Rebeck, Economics in evaporation, Proceedings of 16th Annual Short Course for the Food Industry, University of Florida, Gainesville, FL, 1976.
71. T. A. Renshaw, S. F. Sapakie, and M. C. Hanson, Concentration economics in the food industry, *Chem. Eng. Progr. 78*(5):33 (1982).
72. N. F. Roger and V. A. Turkot, Designing distillation equipment for volatile fruit aromas. *Food Technol. 19*(1):69 (1965).
73. M. F. Sancho, M. A. Rao, and D. L. Downing, Infinite dilution activity coefficients of apple juice aroma compounds, *J. Food Eng. 34*:145 (1997).
74. I. Saguy, I. J. Kopelman, and S. Mizrahi, Extent of nonenzymatic browning in grapefruit juice during thermal and concentration processes: kinetics and prediction, *J. Food Proc. Pres. 2*:175 (1978).
75. G. D. Saravacos, Effect of temperature on viscosity of fruit juices and purees, *J. Food Sci. 35*:122 (1970).
76. G. D. Saravacos, J. C. Moyer, and G. D. Wooster, Stripping of high-boiling aroma compounds from aqueous solutions, Research Circular 21, New York State Agricultural Experiment Station, Geneva, NY, 1969.
77. G. D. Saravacos, J. C. Moyer, and G. D. Wooster, Concentration of liquid foods in a pilot-scale falling film evaporator, Food Sciences Bulletin, No. 4, New York State Agricultural Experiment Station, Geneva, NY, 1970.

78. L. B. Schreiber, and C. A. Eckert, The use of infinite dilution activity coefficients with Wilson's equation, *Ind. Eng. Chem. Proc. Des. Dev. 10*:572 (1971).
79. H. G. Schwartzberg, Energy requirements for liquid food concentration, *Food Technol. 31*(3):67 (1977).
80. H. G. Schwartzberg, Food freeze concentration, *Biotechnology and Food Process Engineering* (H. G. Schwartzberg and M. A. Rao, eds.), Marcel Dekker, New York, 1990, p. 127.
81. R. Shomer, U. Cogan, and C. H. Mannheim, Thermal death parameters of orange juice and effect of minimal heat treatment and carbon dioxide on shelf-life, *J. Food Proc. Pres. 18*:305 (1994).
82. C. E. Smith, and H. G. Schwartzberg, Ice crystal size changes during ripening in freeze concentration, *Biotech. Prog. 1*:111 (1985).
83. L. P. Somogyi, D. M. Barrett, and Y. H. Hui, *Processing Fruits: Science and Technology*, Vol. 2, *Major Processed Products*, Technomic Publishing, Lancaster, PA, 1996.
84. F. Sorrentino, A. Voilley, and D. Richon, Activity coefficients of aroma compounds in model food systems, *AIChE J. 32*:1988 (1986).
85. S. Sourirajan, Reverse osmosis—a general separation technique, *Reverse Osmosis and Synthetic Membranes* (S. Sourirajan, ed.). National Research Council of Canada, Ottawa, Canada, 1977, p. 1.
86. S. Sourirajan and T. Matsuura, Physicochemical criteria for reverse osmosis separations, *Reverse Osmosis and Synthetic Membranes* (S. Sourirajan, ed.). National Research Council of Canada, Ottawa, Canada, 1977, p. 5.
87. S. Sourirajan and T. Matsuura, *Reverse Osmosis/Ultrafiltration Process Principles*, National Research Council of Canada, Ottawa, Canada, 1985, p. 5.
88. D. F. Splittstoesser, J. J. Churey, and C. Y. Lee, Growth characteristics of aciduric sporeforming bacilli isolated from fruit juices, *J. Food Prot. 57*:1080 (1994).
89. D. F. Splittstoesser, Microbiology of fruit products, *Processing Fruits: Science and Technology*, Vol. 1, *Biology, Principles, and Applications* (L. P. Somogyi, H. S. Ramaswamy, and Y. H. Hui, eds.), Technomic Publishing, Lancaster, PA, 1996, p. 261.
90. D. F. Splittstoesser, C. Y. Lee, and J. J. Churey, Control of *Alicyclobacillus* in the juice industry, IFT Annual Meeting Abstracts, 1997, p. 111.
91. F.D. Standiford, Evaporation, *Encyclopedia of Chemical Technology*, 2nd ed., Vol. 8, Interscience, New York, 1962, p. 559.
92. L. Stoloff, Calibration of water activity measuring instruments and devices: collaborative study, *J. Assoc. Off. Anal. Chem. 61*:1166 (1978).
93. H. A. C. Thijssen, Freeze concentration of food liquids, *Food Manuf. 44*(7):49 (1969).
94. H. A. C. Thijssen, Concentration processes for liquid foods containing volatile flavors and aromas, *J. Food Technol. 5*:211 (1970).
95. H. A. C. Thijssen, Freeze concentration, *Advances in Proconcentration and Dehydration of Foods* (A. Spicer, ed.), John Wiley & Sons, New York, 1974,
96. J. L. Toribio and J. E. Lozano, Nonenzymatic browning in apple juice concentrate during storage, *J. Food Sci. 49*:889 (1984).
97. J. L. Toribio, R. V. Nunes, and J. E. Lozano, Influence of water activity on the nonenzymatic browning of apple juice concentrate during storage, *J. Food Sci. 49*:1630 (1984).
98. J. A. Troller, The water relations of food-borne bacterial pathogens. A review, *J. Milk Food Technol. 36*:276 (1973).
99. M. N. Uboldi Eiroa, V. C. A. Junqueira, and F. B. Schmidt, Alicyclobacillus in orange juice: occurrence and heat resistance of spores. *J. Food Prot.* (Submitted).
100. University of Florida, Florida Citrus Economic Research Service, Gainesville, FL, March 1998.
101. A. A. Vitali, Rheological behavior of frozen concentrated orange juice (in Portuguese), Ph.D. thesis, Escola Politecnica, University of São Paulo, São Paulo, Brazil, 1983.
102. A. A. Vitali and M. A. Rao, Flow properties of low-pulp concentrated orange juice: serum viscosity and effect of pulp content, *J. Food Sci. 49*:876 (1984).
103. A. A. Vitali and M. A. Rao, Flow properties of low-pulp concentrated orange juice: effect of temperature and concentration, *J. Food Sci. 49*:882 (1984).

104. A. Voilley, Activity coefficients of aroma compounds and water activity in model systems, *J. Food Proc. Eng. 8*:159 (1986).
105. L. H. Walker, Volatile flavor recovery, *Fruit and Vegetable Processing Technology* (D. K. Tressler and M. A. Joslyn, eds.), AVI Publishing Co., Westport, CT, 1961.
106. B. K. Watt and A. L. Merrill, *Composition of Foods*, Agricultural Handbook No. 8, USDA, ARS, Washington, DC, 1963.
107. Q. Zhang, G. A. Monsalve, G. V. Barbosa-Canovas, and B. G. Swanson, Inactivation of *E. coli* and *S. cerevisiae* by pulsed electric fields under controlled temperature conditions, *Trans. ASAE 37*:581 (1994).

8

Food Preservation by Freezing

M. Shafiur Rahman
Horticulture and Food Research Institute of New Zealand, Auckland, New Zealand

I. THE FREEZING PROCESS IN BIOLOGICAL MATERIALS

Freezing provides a significantly extended shelf life and has been successfully employed for the long-term preservation of many foods. Freezing changes the physical state of a substance by changing water into ice when energy is removed in the form of cooling below freezing temperature. Usually the temperature is further reduced to storage level (e.g., –18°C). The freezing process can be clearly shown using freezing or cooling curves and phase diagrams. The terminology of the freezing process (e.g., precooling, supercooling, freezing, tempering, eutectic, ice nucleation, and glass transition) are discussed in Chapter 5.

II. MODE OF PRESERVING ACTION

The freezing of foods slows down, but does not stop, the physical and biochemical reactions that govern their deterioration [54]. There is a slow progressive change in organoleptic quality during storage, which does not become objectionable for some time [136]. The loss of quality of frozen foods due to storage depends primarily on storage temperature and the length of time in storage. Microbial growth is completely stopped below –18°C, and both enzymic and nonenzymic changes continue at much slower rates during frozen storage [109].

The freezing process reduces the random motion and rearrangement of molecules in the matrix [67,87,109]. Freezing involves the use of low temperatures and reactions take place at slower rates as temperature is reduced. The presence of ice and an increase in solute concentration can have significant effects on the reactions and state of the matrix [129]. The final influence of temperature on chemical reactions due to freezing may be [122] (a) normal stability (a temperature decrease results in a slower reaction rate, thus better stability when foods are stored), (b) neutral stability (the temperature has no influence on the reaction rate), or (c) reversed stability (a temperature decrease results in an increased reaction rate). Regardless of the type of aqueous system, concentration during freezing causes the unfrozen portion to undergo marked changes in such physical properties as ionic strength, pH, viscosity, water activity, surface and interfacial tension, and oxidation-reduction potential. It is important to note that oxygen is almost totally expelled from ice crystals as they are formed [122]. Reid [129] reviewed three types of cell damage due to freezing: osmotic damage, solute-induced damage, and structural damage.

In slow cooling, ice forms slowly in the external cells. If there is sufficient time, water from the cells migrates out by osmotic pressure. This results in cell shrinkage and can damage the membranes. This water does not return to the cells on thawing due to the damage of cell wall, and the consequence is drip loss [129].

The concentration of the solute increases as freezing progresses. Thus, high solute concentrations of the unfrozen matrix, in particular high salt, can cause damage to many polymeric cell components and may kill the cell [106]. This concentration effect is present whether freezing is fast or slow. Cryoprotectants, such as sugars, are usually added to the aqueous phase to reduce salt-induced damage [129].

In addition to the concentration effect, the formation of ice within the cell may cause damage to the delicate organelle and membrane structure of the cell. As a consequence, enzyme systems may be released, leading to a variety of effects, including off-flavor production. This can be prevented by blanching, a prefreezing heat treatment that denatures enzymes [129].

Reid [129] maintains that it is possible to design and control the freezing process through an awareness of the mechanisms of damage. Freezing preservation is far from perfect, and awareness of this fact is needed if techniques are to be developed to overcome known shortcomings and to assure that this method remains competitive with the other major methods of food preservation [44]. A strategic quality approach (quality enhancement) may provide a higher success rate for new frozen food products [137]. Product quality improvement and energy reduction or increased process efficiency are major issues related to the freezing process.

III. QUALITY OF FROZEN FOODS

A. Freezing Rate and Quality

Controlling the freezing process, including careful prefreezing preparation and postfreezing storage of the product, is an important aspect of achieving a high-quality product [54]. An important factor in the quality of frozen foods is the freezing rate. Generally, fast freezing produces better quality frozen food than slow freezing, although the reason for this is not as well understood as is sometimes stated. The rate of freezing of plant tissue is important because it determines the size of the ice crystals, cell dehydration, and damage to the cell walls.

In the case of animal tissue, the concentration of salt within the cells is higher than that in the extracellular region. Consequently, freezing will start outside the cells due to the freezing point depression induced by the solute concentration in the cells. As soon as ice appears, the solute concentration rises. This is a process of freeze concentration. At some point, osmotic pressure difference will cause water to flow from within the semipermeable cells to the extracellular region in order to balance the chemical potentials. This dehydration of the cell is accompanied by a shrinkage of the cell, which not normally lethal. The freezing rate affects this process because rapid freezing results in less cell dehydration (since water has less time to diffuse out of the cell), less breakage of cell walls, and less textural damage. The more rapid the crystallization, the smaller the ice crystals and the less damage is caused by the process of freeze concentration. Consequently, a reverse situation holds for thawing. Slow heating allows an equilibrium to be reached as the melted water diffuses back through the cell wall.

In the case of plant tissue, there is evidence that large ice crystals can cause mechanical damage to cell walls in addition to cell dehydration. In agarose gels, large ice crystals (100–300 μm) with increasing interstitial spaces grow under slow freezing conditions at –25°C, while small ice crystals (1–2 μm) form during rapid freezing in liquid nitrogen [10]. Bevilacqua et al. [14] measured the diameter of the intracellular dendrites and extracellular ice crystals for meat frozen under simulated conditions similar to industrial freezing. They correlate the ice crystal diameter with the characteristic freezing time. De Kock et al. [30] studied the effect of freezing rate (cryogenic, fast; mechanical, slow) on the quality of cellular and noncellular precooked starchy foods. Quality was determined immediately after freezing, as well as after frozen storage, by chemical, physical, microscopical, and sensory methods. The rapid freezing of cellular starchy food resulted in a better quality product than slow freezing immediately after freezing. Rapid freezing of noncellular starchy food, however, produced a product that was only slightly better in quality than the slowly frozen product. After storage, the quality of the rapidly frozen cellular product was still better in quality than the slowly frozen product, although the difference was smaller. The slight advantage gained by rapid freezing of the noncellular product was lost during storage [30].

Symons [143] expressed that undue emphasis on the importance of freezing speed is sometimes misleading. Unless freezing is excessively slow—days or weeks rather than hours—most products are comparatively insensitive to the speed of freezing. In any case an increase in volume of around 10% is associated with freezing most foods. Broadly speaking, faster is marginally better than slower in most products. This is particularly true for fruit and vegetable products, but less so for animal tissue [143]. Moreover, the initial advantage obtained by fast freezing is lost during storage due to recrystallization [30].

Although fast freezing has advantages, some products will crack or even shatter if exposed directly to extremely low-temperature liquid for a long period of time. Recently Hung and Kim [67] reviewed the fundamental aspects of freeze cracking and strategies for its prevention. The mechanisms to explain freeze cracking vary. The proposed mechanisms are:

1. Volume expansion: The volume expansion due to the formation of ice and the amount of empty space in a microstructure are the primary factors affecting the degree of mechanical damage to cells during freezing. In addition, differences in moisture content, composition, or amount of unfreezable water may cause different degrees of cracking [45].
2. Contraction and expansion: Cracking may also occur to relieve the product of internal stress from nonuniform contraction during rapid cooling [157], [127]. On the other hand, both contraction and expansion may cause freeze cracking [135].
3. Internal stress: Fast freezing will cause crust formation at the surface, which serves as a shell and prevents further volume expansion when the internal portion of the unfrozen material undergoes freezing. This process then contributes to internal stress buildup later in the freezing process. The freezing crack will occur if the internal stress exceeds the strength of the exterior frozen material during processing [82]. The distribution of the stress is the controlling factor, and it is governed by absorbing (dissipating) the stress into the structure or reflecting the stress to cause buildup of internal stress [83].

Miles and Morley [107] studied the internal pressures and tensions in meat during freezing, frozen storage, and thawing. A maximum stress of almost 60 bars is possible. They found that during freezing internal compression developed at a rate that increased as freezing progressed, and most of the pressure was developed after the center had started to freeze. Generally the circumferential tension in the outer surface of the muscle reached a breaking point and a shallow crack formed along the length of the muscle or the surface yielded, causing a bulge [107]. Kim and Hung [83] found that size, moisture content, density, modulus of elasticity, Poisson's ratio, and porosity all had significant influence on freeze cracking. However, no single property completely explained the development of freeze cracking [83].

Excessive freezing speeds can ruin a product. The build-up of internal pressure during very rapid freezing shatters the already frozen external layers and produces very small crystals, leading to scattering of incident light [143]. The current practice of quick freezing is generally chosen in order to save processing time (cost) and factory space. Moreover, rapidly chilled muscles become tough on freezing and thawing, a phenomenon known as cold shortening (not a problem in poultry) [143].

In foods containing microbial cultures, it is important to maintain their activity. Rapid freezing cause detrimental effects on the yeast activity of frozen dough. This may be due to the formation of intracellular ice crystals invariably lethal to yeast cell membranes [105]. Yeast activity decreased significantly when the rate of cooling was increased from 0.98 to 1.57°C per minute [111].

B. Microbial Aspects

The detrimental effects of freezing on microorganisms may be desirable or undesirable, depending on the type of food product. In frozen foods without any added beneficial cultures, microbial growth or spoilage is not desirable, whereas care must be given to reduce the damage in cells during freezing of foods containing beneficial microbial cultures.

The maximum recommended storage temperature at which microbiological spoilage ceases lies between –9 and –12°C. Although microbiological spoilage can be avoided at these temperatures, the enzymes present in the product will still play a part in spoilage. Hence hygienic conditions or heat processed (blanched or cooked) will enjoy a longer shelf life [143].

Freezing causes the apparent death of 10–60% of the viable microbe population and gradually increases during frozen storage [109]. Microorganisms differ considerably in their sensitivity to freezing, thus the main concern is organisms that are likely to survive the freezing treatment and grow when the product is thawed. There is considerable variation in the ability of bacteria to resist damage by freezing [2]. In general gram-negative bacteria are less resistant to freezing death than are gram-positive bacteria. Nonsporulating rods and spherical bacteria are the most resistant, while bacterial spores, such as *Clostridium* and *Bacillus*, remain unaffected by freezing. Bacteria in the stationary phase are more resistant than those in the log phase [109,119,136]. Genera commonly encountered in frozen food include *Pseudomonas, Achromobacter, Flavobacter, Micrococcus, Lactobacillus, Corynebacterium*-like catalase-positive rods, enterococci, *Streptococcus lactis, S. lactis*-like streptococci, *Aerococcus*, and *Pediococcus* [109,142].

Gianfranceschi and Aureli [55] studied the survival of *Listeria monocytogenes* during freezing (–50°C) and frozen storage (–18°C) when inoculated into chicken breast, ground beef, spinach, mozzarella cheese, and codfish. They observed only a slight decrease in the viable population ranging from 0.1 to 1.6 log after 57 minutes at –50°C. *Listeria monocytogenes* cells were more resistant to death and injury when they were inoculated in ground beef and chicken breast, whereas they were less resistant in fish. A further reduction in viability of survival cells (up to 1.0 log) was detected after 240–300 days of storage at –18°C [55].

The effects of freezing on several foodborne pathogens were reviewed by El-Kest and Marth [41]. The mode of damage to the bacteria cells were reviewed by Archer et al. [2]. The principal site of damage to bacterial cells during freezing has been shown to be the membrane. Very rapid cooling of cells from room temperature to –150°C resulted in more lipid crystallization before any rearrangement of intramembrane particles could occur. This leads to damage being limited mainly to the area around the cytoplasmic membrane. In slowly frozen samples phase separation of the outer and cytoplasmic membranes was induced, causing the outer membrane to be split off by extracellular ice crystal formation. This damage could be reduced by addition of a cryoprotectant, which modifies ice crystal formation. Damage to membranes leads to the leakage of internal cell materials, such as potassium ions, β-galactosidase, low molecular solutes, amino acids, RNA, and single- and double-strand DNA. The release of these substances has been correlated negatively with cell viability [2].

Another type of damage is osmotic dehydration of the cell caused by extracellular ice formation and the resulting increase in extracellular solute concentration. This process causes the intracellular macromolecules to move close to the membranes. The resulting repulsive forces give rise to large anisotropic stresses in the membranes, resulting in deformation, phase separation, and formation of a nonlamellar phase. Moreover, addition of salt and lowered pH also play a role in the complex nature of freeze injury and cell survival [2,156].

In fermented foods, such as yogurt, frozen storage should increase the viability of beneficial cultures incorporated for their potential health benefits and control of other spoiling microorganisms. The potential beneficial roles of bifidobacteria in the human intestine reported include antagonistic effects on enteropathogenic bacteria, breakdown of carcinogenic *N*-nitrosamines, and suppression of liver tumorigenesis [78]. Thus, bifidobacteria is incorporated in dairy products. Kebary [78] studied the visibility of *Bifidobacterium bifidum* in the fermented dairy product zabady. The numbers of bifidobacteria surviving after 5 weeks of frozen storage (–25°C) of zabady was higher ($>10^7$) than the minimum level necessary to achieve the beneficial effect of bifidobacteria (10^5–10^6/ml). The total bacterial count decreased as the amount of added bifidobacteria increased. This could be due to the effect of antimicrobial substances produced by bifidobacteria [78]. In eight strains of *Lactobacillus acidophilus*, higher rates and greater activity were always obtained by storing cultures at –80°C, but most strains stored at –30°C also survived well. The viability of frozen cultures was affected more by storage temperature than by cooling rate [47].

The plate counts decreased less than 1 log cycle and fermentation activity was 40–70% when cultures of *Lactobacillus delbruchii* subsp. *bulgaricus* were stored at –80°C for 1 year. However, fermentation activity was less than 10% when cultures were stored for 1 year at –30°C [46]. The fermentation activity of *Streptococcus salivarius* subsp. *thermophilus* was similarly reduced to 10–60% after 1 year of storage at –30°C. Protective solutes can be used to improve survival.

Yeast cells in bakery products do not withstand freezing well. This can be partially compensated for by increasing the amount of yeast used in the formulation or adding improved yeast strains having a better survival rate in freezing [143]. The freeze-thaw–tolerant yeast should have a high trehalose content in addition to reduced activity [110]. Trehalose has been reported to perform a cryoprotectant function in the yeast cell [113]. Although the amount of yeast in the formulation could be increased, much higher levels of yeast have a detrimental effect on aroma and flavor of the baked product [97]. When dough pieces were made up and frozen immediately after mixing, yeast activity remained stable after prolonged storage periods. When fermentation was allowed to proceed after mixing and before freezing, the yeast became less tolerant of freezing temperature and its activity declined. This may be due to a change in yeast cell membrane sensitivity [71].

C. Physical Changes and Quality

1. Free and Bound Water

Different types of water are present in frozen foods. These can be broadly categorized as free and unfreezable water. Bound water does not freeze even at very low temperatures. A major cause of product degradation is the amount of unfrozen water present in the frozen matrix. Unfrozen water is known to be reactive, particularly during frozen storage, rendering the product susceptible to deteriorative and enzymatic reactions and limiting its frozen shelf life. Thus, the concept of glassy state is being applied to frozen food stability, since molecular mobility reduces the reaction rates of the unfrozen water matrix and other components [54]. (More details are given in Chapter 5.) Generally unfreezable water molecules in aqueous solution are immobilized translationally or rotationally by solutes [108]. The amount of unfreezable water can be experimentally measured for different types of foods.

2. Recrystallization

Ice recrystallization during frozen storage influences the product quality in different ways. Recrystallization of solutes and ice in frozen foods is also important to quality and shelf life. A polymer is least prone to crystallization at a temperature below glass. In general the rate of recrystallization is highest at the midpoint between the melting and glass transition temperatures.

Fluctuations in product temperature of 2–3°C, as are likely to be found in bulk cold stores kept at –18°C or colder, are unlikely to cause perceptible damage even over long periods [143]. On the other hand, frequent large fluctuations during retail display and during the carry-home period cause ice crystals to ripen or grow, coalesce, and move to the product surface. This leads ultimately to a freeze-dried product if packaging is permeable to moisture, allowing the sublimed or evaporated water vapor to escape. The loss of moisture results in toughening of animal tissue and greater exposure to any oxygen present.

3. Retrogradation

Quality loss by staling and starch retrogradation occurs most rapidly at chill temperatures in backed goods. After baking, starch from the loaf progressively crystallizes and loses moisture. Until a critical point of moisture loss is reached, freshness can be restored by heating and reabsorbing starch crystals. A tight wrap helps to keep the moisture content high for a certain amount of time. The complete crystallization of starch poduces the crumbly texture of stale bread. Slow freezing is to be avoided in order to reduce the time spent at chill temperatures. Amylase is a useful antidote to bread staling. In general, moisture migration during frozen storage is the principal cause of staling [3,143].

4. Protein Denaturation

Some protein denaturation and solubility changes are known to occur as a result of freezing, but the practical significance of these changes is not clear [136]. Fish muscle has a unique arrangement of muscle fibers. It is divided into a number of segments called myotomes, which are separated from one another by a thin sheath of connective tissue called the myocomma or myoseptum [133]. Quality loss during cold storage of meat is characterized by increasing loss of water-holding capacity, a decrease in protein extractability, a decrease in sulfhydryl groups, and a slight loss of ATPase activity [130,136]. In frozen meat water losses are related to the denaturation of myofibrillar protein [43]. The water-holding capacity of the meat and biochemical properties of actomyosin, such as enzymic activity, viscosity, and surface hydrophobicity, are affected by freezing. The expressible moisture in adductor muscles increased during freezing and frozen storage. These changes were accompanied by actomyosin denaturation. The myosin and paramyosin of the actomyosin complex were most affected [115]. An amino group from some essential amino acids, such as lysine, can react with the carbonyl group of reducing sugars during processing or storage [56]. Peptides and amino acids are also increased in the drip fluid, as are nucleic acids, indicating protein changes and structural cellular damage [136]. During freezing water molecules freeze out and migrate to form ice crystals, resulting in the disruption of the organized H-bonding system that stabilizes the protein structure, and hydrophobic as well as the hydrophilic regions of protein molecules become exposed to a new environment. This may allow the formation of intermolecular cross-linkages either within a protein molecule or between two adjacent molecules [133].

5. Freezer Burn

Moisture loss by evaporation from the surface of a product leads to freezer burn, an unsightly, white color that can be mistaken for mold but is resolved on rehydration during cooking unless it is severe [143]. It is usually in the form of patches of light-colored tissues, produced by evaporation of water, which leave air pockets between meat fibers [102]. Dehydration of product or freezer burn may occur when freezing an unpackaged product in blast freezers unless the velocity of air is kept to about 2.5 m/sec and the period of exposure to air is controlled. This dehydration can be controlled by humidification, lowering of storage temperatures, or better packaging [3].

A single package of spinach and cauliflower experienced a 1.5-fold increase in loss of moisture per 2.8°C rise in temperature between –17.8 and –6.7°C [35]. The loss of moisture occurs faster when held in a temperature cycle than when held at constant temperature during frozen storage [121].

6. Glass Formation

The glass transition (described in Chapter 5) has a dramatic effect on frozen food quality. The product is most stable below T'_g, and moisture has little influence on T'_g. The presence of low molecular weight solutes lowers the T'_g, and high molecular weight solutes exert little effect. This means that with increasing maturity many vegetables display a decrease in sugars and an increase in starch, thus increasing the T'_g and the stability of frozen foods.

7. Functional Properties

The rheological properties of fresh and frozen thawed okra dispersions were significantly different when measured at 20–80°C. The dispersions were pseudoplastic with both consistency coefficient and flow behavior index influenced by temperature. The consistency coefficient was higher for unfrozen sample than frozen thawed sample when measured at 20 and 50°C, but the reverse was observed at 80°C. The flow behavior indices were not different at any temperature between 20 and 80°C [114].

Texture is important in frozen leafy vegetables [51]. After freezing-thawing, firmness decreased, rupture strain increased, and consequently crispness decreased [50,51]. The rate of freezing was critical to tissue damage. The optimum freezing rate for carrots was established as –5°C/min using a programmable freezer and based on texture and histological structure [49,52]. Chinese cabbage leaves crack when frozen rapidly with liquid nitrogen. The optimum freezing rate for Chinese cabbage was –2°C/min considering tissue softening and drip loss. Freezing-thawing accelerated release of pectin, but the freezing rate did not greatly affect pectin release [51]. Fruits, such as strawberries, apples, peaches, and citrus, contain thin-walled cells with a large proportion of intracellular water, which can freeze, resulting in extensive cell rupture and radical alteration of the mechanical properties of the material [31]. Khan and Vincent [80] studied the mechanical damage induced by controlled freezing in apple and potato. As the tissue freezes, ice crystals form extra- or intracellularly, pushing the cells apart or rupturing cell walls and producing large voids within the tissue. Changes in the mechanical behavior (wedge penetration, tensile, and compression) of the material were directly related to the degree of cell damage by freezing.

Mashl et al. [103] studied the unidirectional freezing of a gelatinized corn starch–water mixture and found that at freezing velocities less than or equal to 7.5 μm/sec, starch granules were alternately pushed or entrapped by the advancing solid/liquid interface producing a segregated structure consisting of alternating high-starch and low-starch bands, thus segregation of the starch within the product occurred, which is detrimental to consistency, texture, and appearance. At a velocity of 10 μm/sec, the frozen product was homogeneous.

The development of rancid flavors and progressive toughening accompanied by the development of cold store flavor are the principal sensory changes in seafoods [143]. The firming of soft texture characteristic of young fish with the early onset of protein denaturation is preferred by most taste panelists [79]. Flavor change is probably more critical than texture since this can occur early [26]. The denaturation of myosin increased in a frozen solution. The rate of formation of insoluble, high molecular weight protein aggregates increased as the temperature decreased below the freezing point and reached a maximum near the eutectic point of the solution [19]. Because of the concentration effect, pH can also change during freezing. A decrease in pH to more favorable values for degradation result in faster protein denaturation [122]. Fish gradually loses its juiciness and succulence after freezing and subsequent frozen storage. In gadoids, the chemical breakdown of trimethylamine oxide (TMAO) to dimethylamine (DMA) and formaldehyde (FA) and the subsequent cross-linking of FA to muscle proteins produced textural breakdown and resulted in a cottony or spongy texture. In this case free water exists loosely like a sponge. When eaten, the fish muscle loses all its moisture during the first bite, and subsequent chewing results in a very dry and cottony texture [133]. In nongadoid species, such as crab, shrimp, and lobster, muscle fibers also tend to toughen and to become dry during freezing and storage.

Thawing and refreezing can lead to quality deterioration [68]. Dyer et al. [40] reported accelerated deterioration for refrozen cod fillets stored at –23°C. Changes in enzyme acivities of α-glucosidase and β-*N*-acetylglucosaminidase in rainbow trout on thawing, refreezing, and frozen storage were observed, but they did not relate to differences in sensory attributes [112]. Cowie and Little [27] reported no correlation between decreasing protein solubility and sensory attributes for cod stored at –29°C. Thawed and refrozen fish muscle displayed a faster decline in myofibril protein solubility than once-frozen fish and had reduced water-holding capacity, but analysis of proton spin-spin relaxation times indicated no changes in water location. The decline in protein solubility was not caused by complete protein unfolding. Long thawing times of 30 hours before refreezing and storage resulted in cooked fish having a gray appearance and stale flavor [68]. Whole fish when thawed exhibits less textural change than filleted fish due to the presence of the backbone, which provides as structural support for the flesh [133].

The functional properties of cheese, which also changed after freezing and thawing, include meltability, stretchability, elasticity, and free oil formation, cohesiveness, etc. Meltability is the capacity to flow together and form a uniform continuous melted mass. Stretchability is the tendency upon pulling to form fibrous strands that elongate without breaking. Elasticity is the capacity of the fibrous strands to resist permanent deformation. Free oil formation is the separation of liquid fat from the melted body into oil pockets. Viscosity is due to particle segregation or coagulation. Luck [100] concluded that frozen storage was suitable for cream cheese, unripened camembert, and brick cheese, but not for Gouda or Cheddar cheese. Mozzarella cheese, originated in Italy, is consumed worldwide largely due to the popularity of pizza and similar foods. Mozzarella differs from most cheeses in that it is often consumed in a melted state. Consequently, the functional properties such as meltability, stretchability, elasticity, and free oil formation are important to the quality of the product. Cervantes et al. [22] concluded that freezing (one-week storage) and thawing did not affect the quality of mozzarella cheese quality as assessed by compression, beam bending, and sensory evaluation. Dahlstrom [28] showed poor meltdown, acid flavor, fat leakage, free surface moisture, bleached discoloration, and poor cohesiveness immediately after thawing, but normal characteristics reappeared after the thawed cheese was aged for 1–3 weeks. Bertola et al. [12] studied the freezing rate and frozen storage (3 months at –20°C) of mozzarella cheese. The functional quality loss (meltability, apparent viscosity, free oil formation) can be avoided as long as the product was aged 14–21 days at 4°C before being consumed. Again freezing mozzarella cheese that had been ripened for about 14 days produced a product similar (hardness, adhesiveness, cohesiveness, springiness, and nonprotein nitrogen) to refrigerated cheese at the same stage of aging [13].

Gaping in fish fillet may be observed to worsen if fish are slowly frozen. Love and Haq [98] showed that the rate at which whole cod were frozen had little effect on the gaping of the fillets cut after thawing, although very slow freezing did cause a small increase [98].

D. Chemical Changes and Quality

1. Rancidity

Oxygen is the bugbear of almost all frozen foods, leading to oxidative rancidity (if any unsaturated lipids are present), loss of color, and development of off-flavors. Freezing results in a concentration of solutes, which catalyze the initiation of oxidative reactions and disrupt and dehydrate cell membranes, exposing membrane phospholipids to oxidation. Membrane phospholipids are highly unsaturated and have been demonstrated to be the initiation point of oxidation in muscle tissue [136].

The degradation of lipids in frozen peas during storage at –18°C led to flavor damage due to formation of hydroperoxides, thiobarbituric acid, and fatty acids, particularly in unblanched samples [65]. The lipid oxidation was mainly due to lipoxygenase and lipohydroperoxidase breakdown products [24]. The hydrolyzed and oxidized products of lipids affect the quality of frozen vegetables [131].

Lipid hydrolysis occurs in fish during storage, which may affect lipid oxidation. Free fatty acids are believed to be more readily oxidized than the equivalent esters when lipoxygenase is involved [61,118,134]. Lipid oxidation in mackerel minces occurred continually as long as the samples were exposed to air independent of hydrolytic activity, but was deactivated or retarded by cooking the sample or by lowering the storage temperature (–40°C). Lipid oxidation was observed not only in the free fatty acids, but also in the triacylglycerides and the phospholipids extracted from mackerel mince [70].

Poultry fat becomes rancid during very long storage periods or at extremely high storage temperatures. Rancidity in frozen whole poultry stored for 12 months is not a serious problem if the bird is packaged in essentially impermeable film and held at –18°C or below. Danger of rancidity is greatly increased when poultry is cut up before freezing and storage because of the increased surface exposure to atmospheric oxygen [3]. Antioxidants like BHA, BHT, or tocopherols and metal chelators such as pyrophosphates, tripolyphosphates, or hexametaphosphates are effective in reducing oxidation. The distribution of antioxidants in meat is difficult, thus including tocopherols in animal feed results in deposition of tocopherols in membrane locations. This is much more effective in preventing the initial step with phospholipids [136].

Malonaldehyde is one of the decomposition products of auto-oxidation of polyunsaturated lipid materials in food. Malonaldehyde is the main component in the TBA (2-thiobarbituric acid) value that is used to evaluate the degree of oxidation of lipids. Malonaldehyde reacts with myosin from trout. The rate of reaction of malonaldehyde with α-amino groups of myosin was greater at –20°C than at 0°C and almost equal to that at 20°C [122].

Oxidant level should be increased in frozen dough formulations, as oxidants increase dough strength. A higher shortening level is recommended for frozen dough production. Generally, shortening protects dough structure from damage due to ice crystallization [71].

2. Color, Flavor, and Vitamin Loss

Color Loss

The most important color changes in fruits and vegetables are related to three biochemical or physicochemical mechanisms [20]: (a) changes in the natural pigments of vegetable tissues (chlorophylls, anthocyanins, carotenoids), (b) development of enzymic browning, and (c) breakdown of cellular chloroplasts and chromoplasts. Pineapple for processing should be of optimum ripeness with yellow color, good aroma and flavor, and be free from blemishes, such as black heart, water blister, yeasty rot, or brown spot. For frozen pineapple slices, semi-translucent, highly colored slices are generally considered the most attractive and have the best flavor. Pineapple color is important because it is often the basis for judging product acceptability. The golden color of pineapple fruit is mainly due to carotenoids, which become more predominant with ripening as chlorophyll content decreases. Heat processing, freezing, and thawing lead to cell disintegration, pigment degradation, and isomerization of carotenoids [21,66,140]. Bartolome [8] evaluated the influence of freezing (cold room −18°C and air-blast freezer −50°C) and frozen storage (−18°C for 0–12 months) on the color and sensory quality of pineapple slices (Smooth Cayenne and Red Spanish cultivars). No differences were found in sensory analysis (color and appearance) between the cultivars, frozen at different rates, compared to fresh product, or after 1 year of frozen storage. However, both cultivars were suitable for freezing.

Color and flavor are important sensory attributes and vitamin content a functional attribute of frozen foods. The green color of vegetables is lost due to chlorophyll degradation during freezing and frozen storage resulting from the conversion of chlorophyll to pheophytins and/or to the destruction of both chlorophyll and pheophytins, giving a dull khaki color. During storage chlorophyll is converted to pheophytin with a loss of green color and vitamin C; these can be used as objective indicators of quality [109,143].

Chlorophyll was bleached during fat peroxidation, oxidation of glycolic acid by α-hydroxy acid dehydrogenase, and chlorophyllase, which hydrolyze the phytal ester group of chlorophylls and pheophytins [146]. Storage temperature and time, acidity, and blanching time affect the loss of chlorophyll in frozen vegetables. A 10-fold increase in the conversion rate occurred with an approximate 8.3°C increase in temperature. Blanching decreased the loss of chlorophyll during frozen storage [109]. Various inorganic salts, such as sodium chloride, potassium chloride, potassium sulfate, sodium sulfate, and sodium or ammonium bicarbonate, have been used to reduce chlorophyll loss [20].

The maximum stability of carotenes in frozen spinach was 2 years at −28.9, 1 year at −6.7, and 7 days at 4.4°C [36]. Carotene retention curves were sigmoidal with three regions: initiation, acceleration, and retardation. They were typical of autocatalytic reactions. Lipoxygenases were the major enzymes involved in carotene degradation [57]. Moharram and Rofael [109] reviewed carotene degradation in frozen vegetables.

In poultry, a light surface color for carcasses is considered important and is best achieved with rapid surface freezing, which generates a smooth chalky white surface. This is achieved by supercooling the product and forcing nucleation of a high number of small ice crystals. These crystals stay small because there is little water migration to already formed crystals during such a fast process. Numerous small ice crystals cause the surface to reflect light and appear white in color [136]. An alternative approach is to crust freeze the outer part of the carcass rapidly using liquid brine immersion, spray systems, or cryogenics like liquid nitrogen or carbon dioxide, and then to move the partially frozen bird to air blast or cold storage for the remainder of the process. A freezing front migration rate of 2–5 cm/hr is recommended to achieve fast freezing effects and 0.1–0.2 cm/hr for slow freezing [102,136].

Darkening of bones is a condition that occurs in immature chickens and has become more prevalent as broilers are marketed at younger ages. Darkening may arise during chilled storage or during the freezing and defrosting process. It occurs because some of the heme pigment normally contained in the interior of the bones of particularly young chickens leaches out through spongy areas and discolors the adjacent muscle tissues [3,136]. Leaching only occurs in carcasses from relatively young birds because the bones are not completely calcified and are more porous than in mature birds [136]. The development of dark bones was greatly reduced by a combination of freezing and storage at –35°C and immediate cooking after rapid thawing [18]. Aside from this combination, the freezing rate, time between slaughter and freezing, temperature and length of storage, and temperature fluctuations during storage have no marked influence in preventing this discoloration [3]. While eating qualities do not change, the appearance constitutes a negative factor in consumer acceptance [136].

Flavor and Aroma Loss

Freezing affects the flavor and aroma of frozen foods. For example, freezing of strawberries is usually associated with a reduction in aroma and the development of off-flavor. The decrease in aroma is due to a rapid decomposition and diffusion of esters [32,39], whereas the concentrations of franeol and mesifurane linked to strawberry flavor are not affected by freezing [39]. The off-flavor of frozen strawberries differs from that of frozen vegetables [77,104,148]. Off-flavor in frozen vegetables is usually due to insufficient blanching. Deng et al. [33] found that the development off-flavor in frozen thawed strawberries was due to the chemical production of H_2S rather than enzymatic activity. The identity of H_2S was verified both chemically and using gas chromatography–mass spectroscopy analyses. The olfactory properties by sensory analysis indicated the presence of sulfurous compounds. Usually H_2S is derived from the sulfur-containing amino acids cysteine or methionine during processing. Dent et al. [33] also showed that amino acid was not the main precursor of the off-flavor compounds, but that off-flavor development in frozen strawberries can be attributed to the breakdown of cells by freezing, thereby decreasing the pH in the cytosol, which in turn leads to the release of sulfide ion as H_2S. The duration of the production of H_2S was longer in strawberries at –40° and –80°C than at –20°C. This may be due to the low boiling point of H_2S (–59.0°C). Vigorous crushing of fresh strawberries also gave rise to the production of H_2S. Thus, structural damage is one of the important factors.

In fish and seafood, formaldehyde is formed during cold storage by enzymic decomposition of trimethylamine oxide (TMAO). It is a good objective criterion of time-temperature exposure in frozen gadoid species [16]. The formaldehyde reacts with proteins, thereby decreasing their solubility in salt and buffer solutions [128,143]. Santos-Yap [133] mentioned that changes in the flavor of fish and seafood generally occur in three distinct phases during frozen storage: (a) gradual loss of flavor due to loss of or decrease in concentration of some flavor compounds, (b) detection of a neutral, bland, or flat flavor, and (c) development of off-flavors due to the presence of compounds such as acids and carbonyl compounds that are products of lipid oxidation.

Vitamin Loss

Retention of nutritional components in foods is a concern when any type of preservation method is used, but freezing is probably the least destructive [136]. The destruction of vitamin C (ascorbic acid) occurs during freezing and frozen storage. This loss is influenced by blanching conditions, types of freezing, types of package, and time-temperature conditions [109]. The loss is mainly due to oxidation and/or the action of ascorbic acid oxidase [131]. A 10-fold increase in the rate of loss of ascorbic acid occurs per 8.9°C rise in storage temperature of frozen vegetables [11]. Generally frozen vegetables stored at –24°C displayed better ascorbic acid retention than those at –12 and –18°C, respectively [109]. Blanching improves ascorbic acid retention in vegetables. A combination of microwave energy and steam or water blanching yielded frozen products with better ascorbic acid retention than conventional procedures [37].

Vitamin B losses sometimes occur in frozen meat products. B vitamin losses may be significant in frozen poultry products, but most losses are the result of the subsequent thawing and cooking treatments rather than of the freezing process [136].

3. Release of Enzymes

The disruption of plant or animal tissues by freezing leads to the release of enzymes bound to the structures. Beef and pig skeletal muscle contain two isozymes of glutamic-oxalacetic transaminase: one in the mitochondria and the other in the sarcoplasm [85]. Hamm and Kormendy [62] found that freezing and thawing causes a remarkable increase of glutamic-oxalacetic transaminase activity in the muscle press juice. Fish contains malic enzymes in two forms: free and latent. The latter is solubilized by disruption of the tissue caused by freezing and thawing [59]. Barbagli and Crescenzi [5] found that the activity of cytochrome oxidase in extracts of tissues after freezing and thawing was increased by 15 times in chicken's liver, by 2.5 times in trout, and by 4 times in beef muscle compared to extracts of unfrozen samples. Thus, a method was developed to distinguish between fresh and frozen meat based on the enzymes released [5,58,62,63].

Around 0°C enzymic breakdown of protein becomes the principal cause of quality loss, and below –8°C microbiological spoilage ceases and protein denaturation coupled with oxidative rancidity in fatty species become the chief factors affecting quality [143].

4. Hydrolysis

Generally, starch in vegetables does not change significantly during frozen storage [86]. Rofael [131] observed no significant changes in starch of beans, peas, okra, or mallow during storage at –18°C for one year. The reducing sugars of these frozen vegetables were increased during storage due to the hydrolysis of both oligo- and polysaccharides of these products. Thus the amount of reducing sugars is a good indicator of storage life [109]. In melons, total cell wall polysaccharides decreased more during the first 5 months than during the second 5 months of frozen storage. This suggested that pectins and hemicellulose fractions were modified and solubilized by either mechanical or enzyme-catalyzed changes in cell wall polymers [139]. However, freezing preservation of pineapple slices led to minimal changes in soluble solids and sugar content (fructose, glucose, and sucrose), pH, titratable acidity, and nonvolatile organic acids (citric and malic acids) after a year of frozen storage at –18°C [7,8].

5. Acetaldehyde Formation

The formation of acetaldehyde in frozen vegetables increases during storage, thus is an indication of shelf life [131]. Acetaldehyde is a product of aerobic fermentation of pyruvate in plant tissues [74]. The amount of acetaldehyde formation depends on the pretreatment, such as blanching time, and storage period [23,109,131]. Chow and Watts [23] found that acetaldehyde increased when fresh vegetables were heated beyond the minimum required for enzyme inactivation. Dimethylamine content, formaldehyde content, and shear force measurement correlated very well with sensory texture score of frozen red hake [93].

The enzymatic breakdown of triethylamine oxide (TMAO) to dimethylamine (DMA) and formaldehyde affect textural changes in fish species during frozen storage. Further, formaldehyde's contribution to protein changes in muscle during frozen storage would clarify the toughening mechanism. Frozen storage and fluctuation in temperature affect both dimethylamine and formaldehyde formation in frozen fish [90].

E. Processing and Packaging Factors

1. Pretreatments for Freezing

It is important to realize that successful freezing will only retain the inherent quality present initially in a product and will not improve quality characteristics, thus quality level prior to freezing is a major consideration. The use of high-quality initial materials based on standards and grades is essential to high-quality frozen products. The level of intrinsic product quality, such as freshness, suitability of variety or genetics for freezing, soil nutrients for foods of plant origin, dietary factors for foods of animal origin, harvesting or slaughtering methods, and processing such as blanching, cooking, chilling, and addition of antioxidants also have profound effects. Microbiological quality prior to freezing remains a major determinant of postthaw quality. Although freezing can reduce some pathogens, there is also usually significant survival. Thus other methods must be used to ensure elimination of pathogenic organisms from frozen poultry [75,136]. The commonly used pretreatments are discussed in the following sections.

Blanching

Most vegetables and some fruits are blanched before freezing. Blanching destroys the semipermeability of cell membranes, destroys cell turgor, removes intercellular air, filling these spaces with water, and establishes a continuous liquid phase. As a result, ice crystallization can occur through the entire matrix of food without interruption during the freezing process. It also affects texture, color, flavor, and nutritional quality by inactivating enzymes. Cell turgor is an important component of the eating quality of many fruits. It is produced by the internal pressure of the cell contents. Reduced turgor is perceived as softness and lack of crispness and juiciness. When turgor is an important product characteristic, blanching and freezing may not be acceptable. If the product is cooked before consumption, retention of turgor through earlier processing is not necessary since thermal treatment is more severe than blanching or freezing [109,129]. Blanching has other advantages, such as destruction of microorganisms and wilting of leafy vegetables assisting in packaging [20].

Properly blanched vegetables have a long shelf life at frozen food temperatures, enabling them to be exported all over the world and to span the seasons [143]. Blanching of fruits may be detrimental in many cases, resulting in (a) rapid discoloration by enzymic browning [143], (b) loss of texture, (c) formation of cooked taste, (d) some loss of soluble solids, especially in water blanching, and (e) adverse environmental impact due to energy requirements and disposal of used water [20].

Blanching at 70–105°C is associated with the destruction of enzyme activity. Hot water blanching is usually carried out between 75 and 95°C for between 1 and 10 minutes, depending on the size of the individual vegetable pieces. High-pressure steam blanching is more energy efficient than water blanching. It is important that cooling be carried out shortly after blanching, especially for products to be frozen [20].

The enzymes involved in the production off-flavor are catalase, lipoxygenase, and peroxidase, and their heat stability varies with the types of vegetables and fruits. Peroxidase and catalase seem to be the more heat stable, thus could be used as an index of adequate blanching for vegetables. A 95% loss of enzyme activity following blanching is considered adequate. The quality of blanched frozen vegetables was improved if some peroxidase activity remained at the end of the blanching process. The activities of most enzymes are greatly dependent on pH of the tissue or the blanching water. Additives, such as citric acid, sodium chloride, and carbonates, can be used in water depending on the purpose [20,25,109]. Bottcher [17] reported that the highest-quality products were obtained when the following percentages of peroxidase activity remained: peas, 2–6.3%; green beans, 0.7–3.2%; cauliflower, 2.9–8.2%; and Brussels sprouts, 7.5–11.5%. It was concluded that the complete absence of peroxidase activity indicated overblanching [20].

Heat Treatments

Texture is an important quality attribute of frozen fruits and vegetables. Loss of tissue firmness, disruption of the cell membrane, and excessive softness are the major consequences to be avoided [125]. Low temperature and long-time pretreatment were useful in improving the texture of frozen vegetables by avoiding excessive softness. Carrots heated for 30 minutes at 60°C and frozen above –5°C/min (optimum rate) should escape both cell damage and excessive softening [49,144]. The deesterification of pectin by pectin methylesterase during preheating prevented transelimination of pectin. Fuchigami et al. [53] found that preheating followed by quick freezing was effective in improving excessive softness and cell damage. The optimum preheating occurred with 30 minutes at 60°C or 5 minutes at 70°C, and the optimum freezing was –5 to –50°C. Preheated carrots retained a firmer texture than those blanched in boiling water. After preheating, the amount of high methoxyl pectin decreased and low methoxyl pectin increased. Quick freezing resulted in better texture than slow freezing. Loss of texture was accompanied by release of pectin. Slow freezing accelerated release of pectin as compared to quick freezing. Preheated carrots were slower to release pectin. The degree of esterification of pectin substances in raw carrots decreased during preheating, freezing, and thawing. Cell damage in quick-frozen carrots was slight.

Product preparation prior to freezing may include cutting, deboning, slicing, and other operations to provide greater convenience. A greater variety of products cooked prior to freezing is becoming increasingly popular with consumers. These include breaded and fried portions, cured and smoked products, and items in marinades or broths [136]. The freezing rate of precooked chicken affects the quality of the product. Breaded precooked drumsticks frozen with liquid nitrogen are susceptible to cracking, separation of meat from the bone, and developing small areas of white freezer burn next to the surface [3]. Cooked products are likely to exhibit greater increases in lipid oxidation than raw products during storage. This is due to the oxidative change and higher TBA values making the product more susceptible to further oxidative changes during frozen storage. Antioxidants are very effective in cooked chicken during frozen storage [73,120,150].

Dipping Pretreatments

In many cases foods are dipped or soaked in different solutions before freezing and the type of solute used depends on the desired purpose. Apples are commonly treated by soaking slices in a 1% salt solution in order to remove intercellular air. Fruits are also dipped in ascorbic acid and sugar solutions to minimize browning or blanched for a short time to inactivate enzymes [129].

Paredi et al. [115] studied the effect of dipping in polyphosphates on the biochemical properties of adductor muscles during freezing and stored at –30°C. Immersion in polyphosphate solution was effective in reducing water loss in stored muscle. In addition it delayed the decrease in enzymatic activity (Mg^{2+}-ATPase) and provided some protection for the myosin light chains without affecting either the extractability or the viscosity of actomyosin from frozen stored muscles.

In many cases the frozen product is protected by a suitable glazing compound. A glaze acts as a protective coating against the two main causes of deterioration during storage: dehydration and oxidation. It protects against dehydration by preventing moisture from leaving the product and against oxidation by mechanically preventing air contact with the product. Oxidation can also be minimized if the glaze carries a suitable antioxidant [3]. For products intended for short-term storage, glazing can be practically utilized as a viable alternative to storage without a protective covering [133]. Moreover, glazing can be a cheaper alternative to expensive packaging systems for fish stored at –20°C [72].

The different glazes available include inorganic salt solutions of sodium acid phosphate, sodium carbonate, and calcium lactate, alginate solution, also known as "Protan" glaze, antioxidants, such as ascorbic and citric acids, glutamic acid, and monosodium glutamate, and other edible coatings, suh as corn syrup solids [153].

Bacterial Ice Nucleators or Antifreeze Proteins

The application of bacterial ice nucleators to the freezing of some model food systems and real foods, such as salmon, egg white protein, and cornstarch gels, elevates nucleation temperatures, reduces freezing times, and improves the quality, such as flavor and texture. These can also be used in freeze concentration of fresh foods for modification of their properties [149]. The use of bacterial ice nucleators is a unique application of biotechnology, as it directly improves freezing processes [94]. When bacterial cells were added to isotropic aqueous dispersions of hydrogels composed of proteins and polysaccharides, the bulk of the water was converted into directional ice crystals at subzero temperatures not lower than –5°C and resulted in the formation of anisotropically textured products [94]. Details of this topic are reviewed by Wolber [155] and Li and Lee [94]. Antifreeze proteins, found in polar fish and cold-tolerant insects and plants, can affect freezing in several different ways: (a) by lowering the freezing temperature, (b) by retarding recrystallization on frozen storage, and (c) by promoting ice nucleation causing supercooled solutions to freeze more rapidly [54]. Mizuno et al. [108] studied the effect of solutes on the antifreeze and immobilizing activities of water. The antifreeze activities of saccharides that consisted of glucose were higher than others, and in salts those that possessed a higher ionic charge had higher antifreeze activities. In water-soluble amino acids, a few amino acids that formed no eutectic mixture above –20°C had especially high antifreeze activities. The high antifreeze activity is caused by high immobilizing activity for water molecules, and the immobilizing mechanism varied with the type of solute [108]. The antifreeze proteins depress the freezing point by attaching to ice crystals and interfering with water molecules joining the ice lattice. Recent computer modeling suggests that at least for one antifreeze peptide, the molecules are arranged in an antiparralel fashion with cooperative side-to-side-binding [96]. There are two groups of antifreeze proteins: antifreeze glycoproteins and antifreeze proteins. The primary structure of the antifreeze glycoproteins is a repeating (Ala-Ala-Thr) sequence with a disaccharide attached to the threonine residue. The antifreeze proteins have various structures. Type I proteins have an α-helical structure, whereas Type II and III proteins have some unusual secondary structures. Synthetic antifreeze peptides may have also practical applications in foods [96].

Osmotic Concentration

Osmotic concentration of vegetables prior to freezing is a pretreatment that can improve end-product quality [20]. It is well established that osmotic dehydration improves the product quality in terms of color, flavor, and texture. The merits of osmotic dehydration for product-quality improvement and process deficiency have been reviewed [124,126,147]. The effects of sugar on the quality of frozen fruits are reviewed by Skrede [141]. However, in the literature there is not much fundamental information about the mechanisms of flavor entrapment in the food matrix, color retention, and physics of textural improvement.

In the frozen food industry, high energy levels are used for freezing because a large quantity of water is present in fresh foods. A significant proportion of this energy could be saved if plant materials were concentrated prior to freezing [69]. A reduction in the moisture content of food can reduce refrigeration load during freezing. Partially concentrating fruit and vegetables prior to freezing saves packaging and distribution costs [15]. The product quality is comparable with that of conventional products. The process is referred to as dehydro-freezing.

Cryoprotection

Meat and fish muscle is susceptible to freeze denaturation, which decreases gel-forming potential, water-holding capacity, and protein solubility. Cryoprotectants are generally added to protect fish myofibrillar proteins from freeze denaturation during frozen storage [117]. Polydextrose, sucrose, and sorbitol have been reported to protect against freeze denaturation of Alaskan pollack surimi [116]. These agents are low in cost, safe, and have good solubility and beneficial functional effects [101]. Sucrose is usually combined with sorbitol to reduce sweetness. The cryoprotective effect of sugar is enhanced by adding polyphosphate [91]. Polydextrose provided to be an effective cryoprotectant for both pre- and postrigor beef. Arakawa and Timasheff [4] found that cryoprotectants increase the surface tension of water as well as the binding energy, preventing withdrawal of water molecules from the protein and thus stabilizing the protein. Phosphates had no cryoprotective effect but did increase pH and enhance protein extractability, which may enhance gel-forming and water-holding properties [102].

Park et al. [117] found that cooked gel strength was unaffected by freezing of beef or pork surimi-like materials for 48 hours, and addition of cryoprotectants (3 or 6% sorbitol, 3% glycerol, or 3% sucrose) before freezing had no effect on gel-forming ability. The washed myofibrillar proteins from beef muscle were quite stable during freeze-thaw treatment up to 6% sodium chloride. No difference in gel-forming ability after freezing with or without added salt was found. Wierbicki et al. [154] also reported no detrimental effects on water-holding capacity due to salting of meat prior to freezing. The interaction between salt ions and muscle proteins occurs rapidly, compared to the normal process of shrinking or coagulation of muscle proteins [34]. Dondero et al. [38] studied the cryoprotective effects of 18, 20, 25, and 36 DE maltodextrins at 8% (w/w) in surimi from jack mackerel stored at –18°C for 27 weeks. They found that 20 and 25 DE maltodextrins as well as sucrose or sorbitol mixtures were most effective in stabilizing surimi proteins during frozen storage [38].

Poultry meat showed little deterioration upon freezing and isolated myofibrillar systems made by the surimi procedure are less stable [101]. Kijowski and Richardson [81] found that mechanically recovered meat from broilers had reduced functionality when no cryoprotectants were used. Sorbitol or sucrose showed some protection of gel-forming ability of frozen samples, and sorbitol or sucrose with tripolyphosphate gave stronger gels after freezing or freeze-drying than fresh samples. The combined presence of sorbitol, sucrose, and tripolyphosphate restored most functional properties of frozen or freeze-dried material to that of the fresh material. Most of the loss of functionality during freezing or freeze-drying was caused by loss of solubility of myosin and, to a lesser extent, actin. Freeze-drying had a greater effect when no additive or NaCl was present. The blast-frozen and freeze-dried samples with no cryoprotectants had a very coarse structure with no obvious fine network system. In the presence of sorbitol or sucrose there was a finer meshwork for freeze-dried material, which was finer still for frozen material. In the presence of sorbitol or sucrose with tripolyphosphate, the network was even finer but with less obvious spaces in the matrix for both freeze-dried and frozen material. These were observed by scanning electron microscope.

Whole egg and yolk products are fortified with salt or sugar before freezing to prevent coagulation during thawing. The selection of additive depends upon the finished product specifications. Salt (10%) is added to yolks used in mayonnaise and salad dressings, and sugar (10%) is added to yolks used in baking, ice cream, and confectionary. Egg whites are not fortified as they do not have gelation problems during defrosting [3]. Table 1 shows the effects of freezing on the functional properties of liquid egg products.

High molecular weight (HMW) polymer cryoprotectants have the following advantages over low molecular weight (LMW) cryoprotectants [84]:

1. HMW polymers do not generally penetrate the cell membrane and remain in the extracellular suspensions and/or outer surface of the cell.
2. HMW polymers does not produce a significant freezing point depression within the range of concentrations that can be applied in practice.
3. There is no binary eutectic, i.e., the hydrated polymer does not crystallize from aqueous solution as LMW agents do.
4. HMW polymers have the ability to keep a substantial portion of the solution from freezing.

Although the presence of HMW polymeric cryoprotectants is limited to the extracellular suspension medium, HMW additives affect the intracellular composition by the efflux of intracellular water due to chemical potential change across the membrane when extracellular ice is formed [84].

Irradiation

High-dose irradiation can produce changes in the chemical composition and taste of fish and seafood. A combination of irradiation and freezing can be used in foods. A combination treatment involving freezing in conjunction with irradiation has recently been proposed as a means of retarding spoilage. It has been reported that parts of Europe irradiate frozen seafoods from Asia to eliminate microbial pathogens such as *Salmonella* [132].

2. Storage and Display

Packaging, storage, and display also affect frozen food quality. Loss of quality in frozen foods is a gradual process, the changes being slow or very slow, cumulative, and irreversible [143]. Optimum quality requires care at every stage of processing, packaging, storage, and marketing sequence.

Storage temperature is important for frozen food. Symons [143] mentioned that the speed of freezing was not as important to product quality as the maintenance of adequately cold temperatures (–18°C or less) during distribution.

A package for frozen product should (a) be attractive and appeal to the consumer, (b) protect the product from external contamination during transport and handling and from permeable gases and moisture vapor transfer, (c) allow rapid, efficient freezing and ease of handling, and (4) be cost-effective. To provide the greatest protection, a package must be well evacuated of air (oxygen) using a vacuum or gas-flushed system and provide an adequate barrier to both oxygen and moisture [3,136].

Table 1 General Effects of Freezing Rate, Storage Time, Storage Temperature, Thawing Rate, and Additives on Functional Properties of Liquid Egg Products

	Affect on functional properties		
Factor	Egg albumen[a]	Egg yolk[b]	Whole egg[c]
Freezing rate	Slower rate causes: reduced viscosity and increased foam foam stability	Slow rate causes: increased viscosity and gelation	Same as liquid egg yolk but less severe
Storage time	Longer time causes: reduced viscosity and increased foam stability	Longer time causes: increased viscosity and gelation	Same as liquid egg yolk but less severe
Storage temperature	Lower temperature causes: reduced viscosity and increased foam stability	−18°C results in maximum increase in viscosity and gelatin	Same as liquid egg yolk but less severe
Thawing rate	Faster rate causes: some protein denaturation	Slower rate causes: increased viscosity and gelation	Same as liquid egg yolk but less severe
Additives	None normally needed	2% NaCl and 8% sucrose inhibits gelatin; 10% used commercially	None normally needed

[a]Freezing usually has only a slight effect on egg albumen properties.
[b]Freezing often has a drastic effect on egg yolk viscosity.
[c]Freezing has a greater effect on whole egg properties than egg albumen but less than the effect on egg yolk.
Source: Ref. 29.

Since cost is involved in vacuum or modified-atmosphere packaging, these should be used only when necessary for quality. For example, vacuum packaging need not be used if lipid oxidation is not the limiting factor affecting the shelf life of a product.

The shelf life of frozen foods kept in open display cabinets at −15°C packed in 23 different types of plastic, cardboard, and laminate was studied. It was found that aluminum foil–laminated and metallized packages gave the best results. This is due to low levels of oxygen permeability, water vapor transmission, and light transparency and less fluctuating temperatures [1].

Two terms used to describe the shelf life of frozen foods are practical storage life (PSL) and just noticeable difference (JND). Practical storage life is the level of quality expected for the product by the ultimate consumer. Just noticeable difference is usually determined by a trained taste panel, and then multiplied by an arbitrary figure, generally between 2 and 5, to arrive at a practical storage life [143]. In some sensitive products, such as peaches, cauliflower, and red pigmented fish, the PSL may be close to the JND [143]. Most frozen products enjoy a shelf life of many months of even years [143].

Quality losses from frozen food increase log-linearly with the storage temperature when greater than −18°C [89]. The rate of quality loss increases about 2–2.5 times for every 5°C increase over −18°C [56,87,88]. In poultry it has been suggested that shelf life is likely to change by a factor of 3.5 for each 10°C change (increase or decrease) [136]. In seafood kept at around 0°C, enzymic breakdown of protein becomes the main cause of quality loss; between −8°C microbiological spoilage ceases and protein denaturation coupled with oxidative rancidity in fatty species become the chief factors affecting quality [143].

Some types of foods, such as fish, pork, animal organs, fried chicken, and spinach, can be maintained in a high-quality state for only 3–7 months at −20°C, whereas other foods, such as beef, sugared fruits, many bakery products, and many vegetables, can be maintained in a high-quality state for more than 12 months at −20°C [44]. Fish stored at −29°C will have a shelf life of more than a year [3]. Practical storage life values determined by the International Institute of Refrigeration (IIF) are reported by Symons [143].

F. Cold Chain Tolerance and Quality

1. Temperature Cycling

The steps in the frozen food cold chain are freezing, transport by refrigerated vehicle or container, distribution store, retail display cabinet, the unrefrigerated period between retail outlet and home, and time in a home freezer before being consumed in the frozen state, thawed, or end-cooked. Temperature abuse at any of the above steps causes quality deterioration. Time-temperature indicators have been proposed to monitor the lack of adequately cold temperatures during the cold chain. Fluctuation in storage temperature may contribute to deterioration of frozen foods [109].

2. Time-Temperature Tolerance Indicators

The concept of time-temperature tolerance (TTT) to describe frozen food stability is important. Physicochemical, chemical, or biological reactions give an irreversible indication (usually visible) of the history of the product. These indicators are placed on the outside of the packages and combine the time and temperature conditions to which they have been exposed [20]. Temperature history indicators do not provide a precise record of temperature as it changes with time, as do time-temperature recorders or digital data-acquisition systems, but are less costly [151]. Indicators that respond continuously for all temperature conditions are said to be full-history indicators, whereas devices that respond only for the period of time during which a temperature threshold has been exceeded are called partial-history indicators [152]. More detailed review of time-temperature indicators is given by Taoukis et al. [145].

IV. FREEZING METHODS

The overall cost of freezing preservation is lower than that of canning and/or drying if the freezer can be kept full [64]. If the material enters the freezer at just above the freezing point, a more controlled crystallization occurs compared to material at ambient temperature. Different types of freezing systems are available for foods. No single freezing system can satisfy all freezing needs because of the wide variety of food products and process characteristics [67]. The selection criteria of a freezing method are the type of product, reliable and economic operation, easy cleanability and hygienic design, and desired product quality [3].

Although all commercial freezing processes are operated at atmospheric conditions, there are potential applications of high-pressure assisted freezing and thawing of foods. The pressure-induced freezing point and melting point depression enables the sample to be supercooled to low temperature (e.g., –22°C at 207.5 MPa), resulting in rapid and uniform nucleation and growth of ice crystals on release of pressure. Other results include increased thawing rates, the possibility of nonfrozen storage at subzero temperatures, and various high-density polymorphic forms of ice [76]. Details of the applications of this process are reviewed by Kalichevsky et al. [76].

A. Freezing by Contact with a Cooled Solid: Plate Freezing

In this method the product is sandwiched between metal plates and pressure is usually applied for good contact. Plate freezers are only suitable for regular-shaped materials or blocks. When the product has been frozen, hot liquid is circulated to break the ice seal and defrost. Spacers should be used between the plates during freezing to prevent crushing or bulging of the package.

B. Freezing by Contact with a Cooled Liquid: Immersion Freezing

In this method food is immersed in low-temperature brine to achieve fast temperature reduction through direct heat exchange [67]. The fluids usually used are salt solutions (sodium chloride), sugar solutions, glycol and glycerol solutions, and alcohol solutions. The solutes used must be safe to the product in terms of health, taste, color, and flavor, and the product must be denser than fluids. Dilution from the foods may change the concentration, thus it is necessary to control the concentration to maintain a constant bath temperature. In order to ensure that the food does not come into contact with liquid refrigerants, flexible membranes can be used to enclose the food completely while allowing rapid heat transfer [54]. The water loss and salt gain, respectively, were less than 2 and 1 g/100 g initial gelatin gel in immersion freezing with a sodium chloride solution. The salt penetration was hindered by formation of an ice barrier [99].

A mixture of glycerol and glycol is a liquid-liquid medium, thus can also be used since there is no eutectic point. As the temperature is lowered, a point is reached where ice crystals are formed as slush. The temperature at which slush ceases to flow is called the flow point. Methanol or ethanol can also be used. Although the methanol will be removed during cooking, it is poisonous whereas ethanol is safe. Alcohols also pose a fire hazard in processing plants.

C. Freezing by Contact with a Cooled Gas

1. Cabinet Freezing

In this method cold air is circulated in a cabinet where the product is placed in a tray. The moisture pick-up from the product may deposit on the cooling coils as frost, which acts as insulation.

2. Air-Blast Freezing

In this method the temperature of the food is reduced with cold air flowing at a relatively high speed. A cabinet freezer with air velocity at least 5 m/sec generates high heat-transfer rates [67]. Air velocities between 2.5 and 5 m/sec give the most economical freezing. Lower air velocities result in slow product freezing, and higher velocities increase unit freezing costs considerable [3]. This method can be further divided into tunnel freezing, belt freezing, and fluidized bed freezing, depending on how the air interacts with the product [67].

Fluidized Bed Freezing

A fluidized bed freezer consists of a bed with a perforated bottom through which refrigerated air is blown vertically upwards. The air velocity must be greater than the fluidization velocity. This freezing method is suitable for small particulate food bodies of fairly uniform size, e.g., peas, diced carrots, corns, and berry fruits. The high degree of fluidization improves the heat-transfer rate and results in good use of floor space.

Belt Freezing

The first mechanized air-blast freezers consisted of a wire mesh belt conveyor in a blast room for continuous product flow. Uniform product distribution over the entire belt is required to achieve uniform product contact and effective freezing. Controlled vertical air flow forces cold air up through the product layer, thereby creating good contact with the product particles and increasing the efficiency. The principal current design is the two-stage belt freezer. Temperatures used are usually –10 to –4°C in the precool section and –32 to –40°C in the freezing section [3].

Spiral Freezing

A spiral belt freezer consists of a long belt wrapped cylindrically in two tiers, thus requiring minimal floor space. The spiral freezer uses a conveyor belt that can be bent laterally. It is suitable for products with a long freezing time (generally 10 min to 3 hr) and for products that require gentle handling during freezing. It also requires a spatial air-distribution system [3].

Tunnel Freezing

In this process products are placed in trays or racks in a long tunnel and cool air is circulated over the product.

D. Cryogenic Freezing

In cryogenic freezing liquefied gases are placed in direct contact with the foods. Food is exposed to an atmosphere below –60°C through direct contact with liquid nitrogen or liquid carbon dioxide or their vapor [67]. This is very fast method of freezing, thus adequate control is necessary for achieving quality products. It also provides flexibility by being compatible with various types of food products and having a low capital cost [54]. The rapid formation of small ice crystals greatly reduces the damage caused by cell rupture, preserving color, texture, flavor, and nutritional value. The rapid freezing also reduces the evaporative weight loss from the products, provides high product throughput, and has low floor space requirements [54]. Thermal diffusivity of the food will, however, restrict the heat transfer of heat from the product to the freezing medium [54]. Cryogenic gases can also be advantageously applied to produce a hard, frozen crust on a soft product to allow for easier handling, packaging, or further processing [95]. The cryo-mechanical technique utilizes a cryogenic gas to create a frozen crust on a fluid product, after which the product may be conveyed to a conventional mechanical freezer. This combination process offers the advantages of both systems [54].

The advantages of liquid nitrogen are that it is colorless and odorless and is chemically inert and boils at –195.8°C [138]. It is usually used for high-value products due to the high capital cost for gas compression. The product can be exposed to a cryogenic medium in three ways: (a) the cryogenic liquid is directly sprayed on the product in a tunnel freezer, (b) the cryogenic liquid is vaporized and blown over the food in a spiral freezer or batch freezer, or (c) the product is immersed in the cryogenic liquid in an immersion freezer [60,138].

REFERENCES

1. P. Ahvenainen and Y. Malkki, Influence of packaging on the shelf life of frozen carrot, fish and

ice cream, *Thermal Processing and Quality of Foods* (P. Zeuthen, J. C. Cheftel, C. Eriksson, M. Jul, H. Leniger, P. Linko, G. Vos, and G. Varela, eds.), Elsevier Applied Science, London, 1984, p. 528.

2. G. P. Archer, C. J. Kennedy, and A. J. Wilson, Position paper: Towards predictive microbiology in frozen food systems- a framework for understanding microbial population dynamics in frozen structures and in freeze-thaw cycles, *Int. J. Food Sci. Technol. 30*:711 (1995).
3. ASHRAE Handbook, *Refrigeration Systems and Applications*, American Society of Heating, Refrigerating, and Air-Conditioning Engineers, Atlanta, GA, 1994.
4. T. Arakawa and S. N. Timasheff, Stabilization of protein structure by sugars, *Biochemistry 21*:6536 (1982).
5. C. Barbagli and G. S. Crescenzi, Influence of freezing and thawing on the release of cytochrome oxidase form chicken's liver and from beef and trout muscle, *J. Food Sci. 46*:491 (1981).
6. L. G. Bartolome and J. E. Hoff, Firming of potatoes: Biochemical effects of preheating, *J. Agric. Food Chem. 20*:266 (1972).
7. A. P. Bartolome, P. Ruperez, and C. Fuster, Non-volatile organic acids, pH and titratable acidity changes in pineapple fruit slices during frozen storage, *J. Sci. Food Agric. 70*:475 (1996).
8. A. P. Bartolome, P. Ruperez, and C. Fuster, Changes in soluble sugars of two pineapple fruit cultivars during frozen storage, *Food Chem. 56*:163 (1996).
9. A. P. Bartolome, P. Ruperez, and C. Fuster, Freezing rate and frozen storage effects on color and sensory characteristics of pineapple fruit slices, *J. Food Sci. 61*:154 (1996).
10. P. S. Belton and I. J. Colquhoun, Nuclear magnetic resonance spectroscopy in food research, *Spectroscopy 4*(9):22 (1989).
11. G. Bennett, J. F. Cone, M. L. Dodds, J. C. Garrey, N. N. Guerrant, J. G. Heck, J. F. Murphy, J. E. Nicholas, J. S. Perry, R. T. Piercf, and M. D. Show, *Some Factors Affecting the Quality of Frozen Foods*, Bull. of Penn. State Univ., State College, No. 580, 1954.
12. N. C. Bertola, A. N. Califano, A. E. Bevilacqua, and N. E. Zaritzky, Effect of freezing conditions on functional properties of low moisture mozzarella cheese, *J. Dairy Sci. 79*:185 (1996).
13. N. C. Bertola, A. N. Califano, A. E. Bevilacqua, and N. E. Zaritzky, Textural changes and proteolysis of low-moisture mozzarella cheese frozen under various conditions, *Food Sci. Technol. 29*:470 (1996).
14. A. Bevilacqua, N. E. Zaritzky, and A. Calvelo, Histological measurements of ice in frozen beef, *J. Food Technol. 14*:237 (1979).
15. R. N. Biswal, K. Bozorgmehr, F. D. Tompkins, and X. Liu, Osmotic concentration of green beans prior to freezing, *J. Food Sci. 56*:1008 (1991).
16. R. I. Boeri, M. E. Alamandos, A. S. Ciarlo, and D. H. Gianni, Formaldehyde instead of dimethylamine as a measure of total formaldehyde formed in frozen Argentine hake (*Merluccius hubbsi*), *Int. J. Food Sci. Technol. 28*:289 (1993).
17. H. Bottcher, Enzymic activity and quality of frozen vegetables. I. Remaining residual activity of peroxidase, *Nahrung 19*:173 (1975).
18. A. W. Brant and G. F. Stewart, Bone darkening in frozen poultry, *Food Technol. (4)*:168 (1950).
19. H. Buttkus, Accelerated denaturation of myosin in frozen solution, *J. Food Sci. 35*:559 (1970).
20. M. P. Cano, Vegetables, *Freezing Effects on Food Quality* (L. E. Jeremiah, ed.), Marcel Dekker, New York, 1996, p. 247.
21. M. P. Cano and M. A. Marin, Pigment composition and color of frozen and canned kiwi fruit slices, *J. Agric. Food Chem. 40*:2141 (1992).
22. M. A. Cervantes, D. B. Lund, and N. F. Olson, Effects of salt concentration and freezing on mozzarella cheese texture, *J. Dairy Sci. 66*:204 (1983).
23. L. Chow and B. M. Watts, Origin of off odors in frozen green beans, *Food Technol. 23*:113 (1969).
24. H. M. Churchill, A. O. Scott, and C. K. Erb, A study of the biochemical and chemical causes of quality changes in frozen vegetables, Campden Food and Drink Research Association, Technical Memorandum, No. 517, 1989.

25. J. K. Collins, C. L. Biles, E. V. Wann, P. Perkins-Veazie, and N. Maness, Flavour qualities of frozen sweetcorn are affected by genotype and blanching, *J. Sci. Food Agric. 72*:425 (1996).
26. J. J. Connell and P. Howgate, Consumer evaluation of fresh and frozen fish, *Inspection and Quality* (R. Kreuzer, ed.), Fishing News (Book), Surrey, 1971, p. 155.
27. W. P. Cowie and W. T. Little, The relationship between the toughness of cod stored at –29°C and its muscle protein solubility and pH, *J. Food Technol. 1*:335 (1966).
28. D. G. Dahlstrom, Frozen storage of low moisture part skim mozzarella cheese, *M. S. Thesis*, Univ. Wisconsin, Madison (1978).
29. P. L. Dawson, Effects of freezing, frozen storage, and thawing on eggs and egg products, *Freezing Effects on Food Quality* (L. E. Jeremiah, ed.), Marcel Dekker, New York, 1996, p. 337.
30. S. De Kock, A. Minnaar, D. Berry, and J. R. N. Taylor, The effect of freezing rate on the quality of cellular and non-cellular par-cooked starchy convenience foods, *Food Sci. Technol. 28*:87 (1995).
31. J. M. DeMan, P. W. Voisey, V. F. Rasper, and D. W. Stanley, *Rheology and Texture in Food Quality*, Van Nostrand Reinhold/Avi, New York, 1976.
32. H. Deng and Y. Ueda, Effects of freezing methods and storage temperature on flavor stability and ester contents of frozen strawberries, *J. Jpn. Soc. Hort. Sci. 62*:633 (1993).
33. H. Deng, Y. Ueda, K. Chachin, and H. Yamanaka, Off-flavor production in frozen strawberries, *Postharv. Biol. Technol. 9*:31 (1996).
34. F. E. Deatherage and R. Hamm, Influence of freezing and thawing on hydration and changes of the muscle proteins, *Food Res. 25*:623 (1960).
35. W. C. Dietrich, M. D. Nutting, R. L. Olson, F. E. Lindquist, M. M. Boggs, G. S. Bohart, H. I. Neumann, and H. J. Morris, Time-temperature tolerance of frozen foods. XVI. quality retention of frozen green snap beans in retail packages, *Food Technol. 13*:136 (1959).
36. W. C. Dietrich, M. M. Boggs, M. D. Nutting, and N. E. Weinstein, Time-temperature tolerance of frozen foods. XIII. quality changes in frozen spinach. *Food Technol. 14*:155 (1960).
37. W. C. Dietrich, C. C. Huxsoll, and D. G. Guadagni, Comparison of microwave, conventional and combination blanching of Brussels sprouts for frozen storage, *Food Technol. 24*:613 (1970).
38. M. Dondero, C. Sepulveda, and E. Curotto, Cryoprotective effect of maltodextrins on surimi from jack mackerel (*Trachurus murphyi*), *Food Sci. Technol. Int. 2*:151 (1996).
39. C. Douillard and E. Guichard, The aroma strawberry (*Fragaria ananassa*): characterisation of some cultivars and influence of freezing, *J. Sci. Food Agric. 50*:517 (1990).
40. W. J. Dyer, D. I. Fraser, D. G. Ellis, and D. R. Idler, Quality changes in stored refrozen cod fillets, *Supp. Bull. Inst. Inter. Froid Annexe 1*:515 (1962).
41. S. El-Kest and E. H. Marth, Freezing of *Listeria monocytogenes* and other microorganisms: a review, *J. Food Prot. 55*:639 (1992).
42. O. Fennema, Loss of vitamins in fresh and frozen foods, *Food Technol. 31*:32 (1977).
43. O. R. Fennema, Comparative water holding capacity of various muscle foods, *J. Muscle Food 1*:363 (1990).
44. O. Fennema, Frozen foods: Challenges for the future, *Food Aust. 45*:374 (1993).
45. O. R. Fennema and W. D. Powrie, Fundamentals of low temperature food preservation, *Adv. Food Res. 13*:220 (1964).
46. R. Foschino, C. Beretta, and G. Ottogalli, Study of optimal conditions in freezing and thawing for thermophilic lactic cultures, *Ind. Latte 28*:49 (1992).
47. R. Foschino, E. Fiori, and A. Galli, Survival and residual activity of *Lactobacillus acidophilus* frozen cultures under different conditions, *J. Dairy Res. 63*:295 (1996).
48. M. Fuchigami, Relationship between maceration and pectic change of Japanese radish roots during cooking, *J. Home Econ. Jpn. 37*:1029 (1986).
49. M. Fuchigami, N. Hyakumoto, K. Miyazaki, T. Nomura, and J. Sasaki, Texture and histological structure of carrots frozen at a programmed rate and thawed in an electrostatic field, *J. Food Sci. 59*:1162 (1994).
50. M. Fuchigami, N. Hyakumoto, and K. Miyazaki, Texture and pectic composition differences in raw, cooked, and frozen-thawed Chinese cabbages due to leaf position, *J. Food Sci. 60*:153 (1995).

51. M. Fuchigami, K. Miyazaki, N. Hyakumoto, T. Nomura, and J. Sasaki, Chinese cabbage midribs and leaves physical changes as related to freeze-processing, *J. Food Sci. 60*(6):1260 (1995).
52. M. Fuchigami, N. Hyakumoto, and K. Miyazaki, Programmed freezing affects texture, pectic composition and electron microscopic structures of carrots, *J. Food Sci. 60*:137 (1995).
53. M. Fuchigami, K. Miyazaki, and N. Hyakumoto, Frozen carrots texture and pectic components as affected by low-temperature-blanching and quick freezing, *J. Food Sci. 60*:132 (1995).
54. R. M. George, Freezing processes used in the food industry, *Trends Food Sci. Technol. 4*:134 (1993).
55. M. Gianfranceschi and P. Aureli, Freezing and frozen storage on the survival of *Listeria monocytogenes* in different foods, *Ital. J. Food Sci. 8*:303 (1996).
56. S. A. Goldblith, Food processing, nutrition and the feeding of man during the next 25 years, Symposium of World Food Supply and Refrigeration, Frigoscandia, Stockholm, 1975.
57. M. Goldman, B. Horev, and I. Saguy, Decolorization of beta-carotene in model systems simulating dehydrated foods. Mechanisms and kinetic principles, *J. Food Sci. 48*:751 (1983).
58. E. Gould, Observations on the behaviors of some endogenous enzyme systems in frozen-stored fish flesh, *The Technology of Fish Utilization* (R. Kreuzer, ed.), Fishing News (Books) Ltd., London, 1965, p. 126.
59. E. Gould, An objective test for determining whether "fresh" fish have been frozen and thawed, *Fish Inspection and Quality Control* (R. Kreuzer, ed.), Fishing News (Books) Ltd., London, 1971, p. 72.
60. R. Gupta, Use of liquid nitrogen to freeze in the freshness, *Seafood Export J. 24*:33 (1992).
61. R. J. Hamilton, The chemistry of rancidity in foods, *Rancidity in Foods* (J. C. Allen and R. J. Hamilton, eds.), Elsevier Applied Science, New York, 1989, p.1.
62. R. Hamm and L. Kormendy, Transaminase of skeletal muscle. 3. Influence of freezing and thawing on the subcellular distribution of glutamic-oxalacetic transaminase in bovine and porcine muscle, *J. Food Sci. 34*:452 (1969).
63. R. Hamm and D. Masi, Routinemethode zur Unterscheidung swischen frischer Leber und aufgetauter Gefrierleber, *Fleischwirtschaft 55*:242 (1975).
64 R. S. Harris and E. Kramer, *Nutrition Evaluation of Food Processing*, 2nd ed., Avi Publishing Co., Westport, CT, 1975.
65. H. M. Henderson, J. Kanhai, and N. A. M. Eskin, The enzymic release of fatty acids from phosphatidylcholine in green peas (pisum satiuum), *Food Chem. 13*:129 (1983).
66. A. S. Hodgson and L. R. Hodgson, Pineapple juice, *Fruit Juice Processing Technology* (S. Nagy, C. S. Chen, and P. E. Shaw, eds.), AgScience Inc., Auburndale, 1993, p. 378.
67. Y. C. Hung and N. K. Kim, Fundamental aspects of freeze-cracking, *Food Technol. 50*:59 (1996).
68. R. Hurling and H. Mcarthur, Thawing, refreezing and frozen storage effects on muscle functionality and sensory attributes of frozen cod (*Gadus morhua*), *J. Food Sci. 61*:1289 (1996).
69. C. C. Huxsoll, Reducing the refrigeration load by partial concentration of foods prior to freezing, *Food Technol. 35*:98 (1982).
70. K. T. Hwang and J. M. Regenstein, Lipid hydrolysis and oxidation of mackerel (*Scomber scombrus*) mince, *J. Aquatic Food Product Technol. 5*:17 (1996).
71. Y. Inoue and W. Bushuk, Effects of freezing, frozen storage, and thawing on dough and baked goods, *Freezing Effects on Food Quality* (L. E. Jeremiah, ed.), Marcel Dekker, New York, 1996, p. 367.
72. M. Jadhav and N. Magar, Preservation of fish by freezing and glazing. II. keeping quality of fish with particular reference to yellow discoloration and other organoleptic changes during prolonged storage, *Fish Technol. 7*:146 (1970).
73. P. P. Jantawat and L. E. Dawson, Stability of broiler pieces during frozen storage, *Poultry Sci. 56*:2026 (1977).
74. M. A. Joslyn, *Cryobiology*, 9th ed., Academic Press, London, 1966.
75. M. Jul, *The Quality of Frozen Foods*, Academic Press, London, 1984.

76. M. T. Kalichevsky, D. Knorr, and P. J. Lillford, Potential food applications of high-pressure effects on ice-water transitions, *Trends Food Sci. Technol. 6*:253 (1995).
77. K. Kaneko, K. Hashizume, Y. Ozawa, and R. Masuda, Change in quality of strawberries by freezing and freeze storage, *Rep. Natl. Food Res. Inst. 52*:18 (1988).
78. K. M. K. Kebary, Viability of *Bifidobacterium bifidum* and its effect on quality of frozen zabady, *Food Res. Int. 29*:431 (1996).
79. T. R. Kelly, Quality in frozen cod and limiting factors on its shelf life, *J. Food Technol. 4*:95 (1965).
80. A. A. Khan and J. F. V. Vincent, Mechanical damage induced by controlled freezing in apple and potato, *J. Text. Stud. 27*:143 (1996).
81. J. Kijowski and R. I. Richardson, The effect of cryoprotectants during freezing or freeze drying upon properties of washed mechanically recovered broiler meat, *Int. J. Food Sci. Technol. 31*:45 (1996).
82. N. K. Kim, Mathematical modeling of cryogenic food freezing, Ph. D. dissertation, Univ. of George, Athens, 1993.
83. N. K. Kim and Y. C. Hung, Freeze-cracking in foods as affected by physical properties, *J. Food Sci. 59*:669 (1994).
84. C. Korber, K. Wollhover, and M. W. Scheiwe, The freezing of biological cells in aqueous solutions containing a polymeric cryo-protectant, *Properties of Water in Foods* (D. Simatos and J. L. Multon, eds.), Martinus Nijhoff Publishers, Dordrecht, 1985, p. 511.
85. L. Kormendy, G. Gantner, and R. Hamm, Isozime der Glutamatoxalacetat Transaminase im Skeletmuskel von Schwein und Rind, *Biochem. Z. 342*:31 (1965).
86. Z. Kosmala, M. A. Urbaniak, and G. A. Rydz, Some chemical and sensory properties of green peas and French beans frozen without blanching and stored in a frozen state, The XVI International Congress of Refrigeration, 1983, p. 623.
87. A. Kramer, Effects of freezing and frozen storage on nutrient retention of fruits and vegetables, *Food Technol. 33*:58 (1979).
88. A. Kramer and J. W. Farquhr, Testing of time-temperature indicating and defrost devices, *Food Technol. 30*:50 (1976).
89. T. P. Labuza, Drying food: technology improves on the sun, *Food Technol. 30*:34 (1976).
90. E. L. Leblanc and R. J. Leblanc, Effect of frozen storage temperature on free and bound formaldehyde content of cod (*Gadus morhua*) fillets, *J. Food Proc. Pres. 12*:95 (1988).
91. C. M. Lee, Surimi process technology, *Food Technol. 38*:69 (1984).
92. C. Y. Lee, M. C. Bourne, and J. P. Van Buren, Effect of blanching treatment on the firmness of carrots, *J. Food Sci. 44*:615 (1979).
93. J. J. Licciardello, E. M. Ravesi, R. C. Lundstrom, K. A. Wilhelm, F. F. Correia, and M. G. Allsup, Time-temperature tolerance of frozen red hake, *J. Food Quality 5*:215 (1982).
94. J. Li and T. Lee, Bacterial ice nucleation and its potential application in the food industry, *Trends Food Sci. Technol. 6*:259 (1995).
95. G. Londahl and B. Karlsson, *Food Technol. Int. Eur.*, 90–91 (1991).
96. P. J. Lillford and C. B. Holt, Antifreeze proteins, *J. Food Eng. 22*:475 (1994).
97. K. Lorenz and W. C. Bechtel, Frozen dough variety breads: Effect of bromate level on white bread, *Bakers Dig. 39(4)*:53 (1965).
98. R. M. Love and M. A. Haq, The connective tissues of fish. IV. Gaping of cod muscle under various conditions of freezing, cold-storage and thawing, *J. Food Technol. 5*:249 (1970).
99. T. Lucas and A. L. Raoult-Wack, Immersion chilling and freezing: Phase change and mass transfer in model food, *J. Food Sci. 61*:127 (1996).
100. H. Luck, Preservation of cheese and perishable dairy products by freezing, *S. Afr. J. Dairy Technol. 9*:127 (1977).
101. G. A. MacDonald and T. Lanier, Carbohydrates as cryoprotectants for meats and surimi, *Food Technol. 45*:150 (1991).

102. R. W. Mandigo and W. N. Osburn, Cured and processed meats, *Freezing Effects on Food Quality* (L. E. Jeremiah, ed.), Marcel Dekker, New York, 1996, p. 135.
103. S. J. Mashl, R. A. Flores, and R. Trivedi, Dynamics of solidification in 2% corn starch-water mixtures: Effect of variations in freezing rate on product homogeneity, *J. Food Sci. 61*:760 (1996).
104. R. Masuda, K. Kaneko, K. Hashizume, Y. Ozawa, K. Iono, and I. Yamashita, Effects of freezing method and the following storage condition of strawberries on the quality of jam and juice drink, *Rep. Natl. Food Res. Inst. 52*:25 (1988).
105. P. Mazur, Physical and temporal factors involved in the death of yeast at sub-zero temperatures, *Biophys. J. 1*:247 (1963).
106. P. Mazur, The role of intracellular freezing in the death of cells cooled at supraoptimal rates, *Cryobiology 14*:251 (1977).
107. C. A. Miles and M. J. Morley, Measurement of internal pressures and tensions in meat during freezing, frozen storage and thawing, *J. Food Technol. 12*:387 (1977).
108. A. Mizuno, M. Mitsuiki, S. Toba, and M. Motoki, Antifreeze activities of various food components, *J. Agric. Food Chem. 45*:14 (1997).
109. Y. G. Moharram and S. D. Rofael, Shelf life of frozen vegetables, *Shelf Life Studies of Foods and Beverages* (G. Charalambous, ed.), Elsevier Science Publishers B.V., Amsterdam, 1993.
110. O. Neyreneuf and J. B. Van Der Plaat, Preparation of frozen French bread dough with improved stability, *Cereal Chem. 68*:60 (1991).
111. O. Neyreneuf and B. Delpuech, Freezing experiments on yeasted dough slabs. Effects of cryogenic temperatures on the baking performance, *Cereal Chem. 70*:109 (1993).
112. K. Nilsson and B. Ekstrand, Enzyme leakage in muscle tissue of rainbow trout (*Onchorynchus mykiss*) related to various thawing treatments, *Z. Lebs. Unters. Forsch. 198*:253 (1994).
113. Y. Oda, K. Uno, and S. Otha, Selection of yeasts for bread making by the frozen dough method, *Appl. Environ. Microbiol. 52*:941 (1986).
114. A. O. Olorunda and M. A. Tung, Rheology of fresh and frozen okra dispersions, *J. Food Technol. 12*:593 (1977).
115. M. E. Paredi, N. A. De Vido de Mattio, and M. Crupkin, Biochemical properties of actomosin and expressible moisture of frozen stored striated adductor muscles of *Aulacomya ater ater* (molina): effects of polyphosphates, *J. Agric. Food Chem. 44*:3108 (1996).
116. J. W. Park, T. C. Lanier, and D. P. Green, Cryoprotective effects of sugar, polyols and/or phosphates on Alaska pollack surimi, *J. Food Sci. 53*:1 (1988).
117. S. Park, M. S. Brewer, F. K. McKeith, P. J. Bechtel, and J. Novakofski, Salt, cryoprotectants and preheating temperature effects on surimi-like material from beef or pork, *J. Food Sci. 61*:790 (1996).
118. H. B. W. Patterson, Safeguarding quality and yield, *Handling and Storage of Oilseeds, Oils, Fats, and Meal*, Elsevier Applied Science, New York, 1989, p. 1.
119. A. C. Peterson, *Proceeding Low Temperature Microbiological Symposium*, Cambell Soup Co., Campden, NJ, 1961.
120. J. Pikul, D. E. Lesczynski, P. J. Bechtel, and F. A. Kummerow, Effects of frozen storage and cooking on lipid oxidation in chicken meat, *J. Food Sci. 49*:838 (1984).
121. F. Pizzocaro, E. Senesi, and R. Monteverdi, Stability of lipids in frozen cauliflower in relation to heat pretreatment, *Ind. Alimentari 25*:372 (1986).
122. K. Porsdal and F. Lindelov, Acceleration of chemical reactions due to freezing, *Water Activity: Influences on Food Quality* (L. B. Rockland, and G. F. Stewart, eds.), Academic Press, New York, 1981.
123. M. C. Quintero-Ramos, M. C. Bourne, and Anzaldua-Morales, Texture and rehydration of dehydrated carrots as affected by low temperature blanvhing, *J. Food Sci. 57*:1127 (1992).
124. M. S. Rahman and C. O. Perera, Osmotic dehydration: A pretreatment for fruit and vegetables to improve quality and process efficiency, *Food Technol. 25*:144 (1996).
125. A. R. Rahman, W. L. Henning, and D. E. Wescott, Histological and freeze drying and compression, *J. Food Sci. 36*:500 (1971).

126. A. L. Raoult-Wack, Recent advances in the osmotic dehydration of foods, *Trends Food Sci. Technol. 5*:255 (1994).
127. R. M. Reeve and M. S. Brown, Historical development of the green bean pod as related to culinary texture. 2. Structure and composition at edible maturity, *J. Food Sci. 33*:326 (1968).
128. H. Rehbein, Does formaldehyde from cross-links between myofibrillar proteins during frozen storage of fish muscle?, *Fish Inspection and Quality* (R. Kreuzer, ed.), Fishing News, Surrey, 1971.
129. D. S. Reid, Fruit freezing, *Processing Fruits: Science and Technology. Vol. 1., Biology, Principles and Applications* (L. P. Somogyi, H. S. Ramaswamy, and Y. H. Hui, eds.), Technomic Publishing, Lancaster, CA, 1996, p. 169.
130. D. S. Reid, N. F. Doong, A. Foin, and M. Snider, Studies on the frozen storage of fish, *Refrigeration Science and Technology. Storage Lives of Chilled and Frozen Fish Products*, International Institute of Refrigeration, Paris, 1985.
131. S. D. Rofael, Effect of storage time, food processing plants and marketing conditions on the quality of some frozen vegetables, Ph.D. thesis, Alex University Alexandria, Egypt, 1992.
132. B. Salvage, *Frozen Food Rep. 3*:28–33 (1992).
133. E. E. M. Santos-Yap, Fish and seafood, *Freezing Effects on Food Quality* (L. E. Jeremiah, ed.), Marcel Dekker, New York, 1996, p. 109.
134. S. Schwimmer, Enzyme involvement in off-flavor from the oxidation of unsaturated lipids in nondairy foods, *Source Book of Food Enzymology*, AVI Publishing, Westport, CT, 1981, p. 421.
135. A. Sebok, I. Csepregi, and G. Beke, Cracking of fruits and vegetables during freezing and the influence of precooling, Intl. Cong. of Refrigeration, Montreal, Canada, Aug. 10-17, 1991.
136. J. G. Sebranek, Poultry and poultry products, *Freezing Effects on Food Quality* (L. E. Jeremiah, ed.), Marcel Dekker, New York, 1996, p. 85.
137. R. L. Shewfelt, M. C. Erickson, Y. C. Hung, and T. M. M. Malundo, Applying quality concepts in frozen food development, *Food Technol. 51*:56 (1997).
138. Z. Sham and M. Marpaung, The application of liquid nitrogen in individual quick freezing and chilling, *Development of Food Science and Technology in South East Asia* (O. B. Liang, A. Buchanan, and D. Fardiaz, eds.), IPB Press, Bogor, 1993, p. 80.
139. V. Simandjuntak, D. M. Barrett, and R. E. Wrolstad, Cultivar and frozen storage effects on muskmelon (*Cucumis melo*) colour, texture and cell wall polysaccharide composition, *J. Sci. Food Agric. 71*:291 (1996).
140. K. L. Simpson, Chemical changes in natural food pigments, *Chemical Changes in Food During Processing* (T. Richardson and J. W. Finley, ed.), The AVI Publishing Co., Inc, Westport, CT, 1985, p. 409.
141. G. Skrede, Fruits, *Freezing Effects on Food Quality* (L. E. Jeremiah, ed.), Marcel Dekker, New York, 1996, p. 183.
142. D. F. Splittstoesser and W. P. Wettergreen, The significance of coliforms in frozen vegetables, *Food Technol. 18*:134 (1964).
143. H. Symons, Frozen foods, *Shelf Life Evaluation of Foods* (C. M. D. Man and A. A. Jones, eds.), Blackie Academic & Professional, London, 1994, p. 298.
144. S. Tamura, Effects of pretreatment on firmness, morphological structure and loss of a taste substance of frozen carrot, *J. ARAHE 1*:61 (1991).
145. P. S. Taoukis, B. Fu, and T. P. Labuza, Time-temperature indicators, *Food Technol. 44*:70 (1991).
146. N. E. Tolbert and R. H. Burris, Light activation of the plant enzyme which oxidizes glycolic acid, *J. Biol. Chem. 186*:791 (1950).
147. D. Torreggiani, Osmotic dehydration in fruit and vegetable processing, *Food Res. Int. 26*:59 (1993).
148. Y. Ueda and T. Iwata, Off-odor of strawberry by freezing, *J. Jap. Soc. Hort. Sci. 51*:219 (1982).
149. M. Watanabe and S. Arai, Bacterial ice-nucleation activity and its application to freeze concentration of fresh foods for modification of their properties, *J. Food Eng. 22*:453 (1994).
150. J. E. Webb, C. C. Brunson, and J. D. Yates, Effects of feeding antioxidants on rancidity development in pre-cooked, frozen broiler parts, *Poultry Sci. 51*:1601 (1972).

151. J. H. Wells and R. P. Singh, Response characteristics of full-history time-temperature indicators suitable for perishable food handling, *J. Food Proc. Pres. 12*:207 (1988).
152. J. H. Wells and R. P. Singh, Performance evaluation of time-temperature indicators for frozen food transport, *J. Food Sci. 50*(2):369 (1985).
153. F. Wheaton and A. Lawson, *Processing Aquatic Food Products*, John Wiley and Sons, New York, 1985.
154. E. Wierbicki, L. E. Kunkle, and F. E. Deatherage, Changes in the water holding capacity and cationic shifts during the heating and freezing and thawing of meat as revealed by a simple centrifugal method for measuring shrinkage, *Food Technol. 11*:69 (1957).
155. P. K. Wolber, Bacterial ice nucleation, *Adv. Microb. Physiol. 34*:203 (1993).
156. J. Wolfe, Z. J. Yan, and J. M. Pope, Hydration forces and membrane stresses—cryobiological implications and a new technique for measurement, *Biophys. Chem. 49*:51 (1994).
157. E. R. Wolford and M. S. Brown, Liquid-nitrogen freezing of green beans, *Food Technol. 19*:1133 (1965).

9

Natural Antimicrobials for Food Preservation

EDDY J. SMID AND LEON G. M. GORRIS
Agrotechnological Research Institute, Wageningen, The Netherlands

I. INTRODUCTION

The spoilage and poisoning of foods by microorganisms is a problem that is not yet under adequate control despite the range of robust preservation techniques available (e.g., freezing, sterilization, drying, preservatives). In fact, food manufacturers increasingly rely on more mild preservation techniques to comply with the consumer demand for foods with a more natural appearance and nutritious quality than can be achieved by the robust techniques. In addition, consumers increasingly refuse foods prepared with preservatives of chemical origin, which still is everyday practice to achieve sufficiently long shelf life for foods and a high degree of safety with respect to foodborne pathogenic microorganisms. To meet the consumer criteria, food manufacturers are searching for new, more natural alternatives that sufficiently assure the safety of their products in the retail chain. The search for natural alternatives to chemicals is a logical one, because nature has long been a very generous source of antimicrobial compounds, many of which play an important role in the natural defense or competition systems of living organisms (ranging from microorganisms to insects, animals, and plants). Many plants contain compounds that have some antimicrobial activity, collectively referred to here as *green chemicals*. Spices and herbs, for instance, are well known to inhibit bacteria, yeasts, and molds and have traditionally found wide use in food preservation as well as for medicinal purposes. The use of spices and herbs or their extracts is often less effective than the use of their active ingredients, for which a number of attractive applications have been identified, as will be discussed. With respect to natural antimicrobial activity associated with microorganisms, referred to as *biopreservatives*, a mainstream field of study currently is the use of lactic acid bacteria. These bacteria have a long and safe tradition in food fermentation, and many potential applications as food preservatives have been established. The bacteriocins produced by lactic acid bacteria are especially promising, which will be discussed in detail. The use of natural antimicrobials in practice is subject to legislatory requirements which can be quite different in the various parts of the world, and this needs to be considered when discussing new developments in this area of food preservation.

II. RATIONALE FOR THE USE OF NATURAL ANTIMICROBIAL COMPOUNDS

In many countries worldwide, there is a rapidly growing demand for environmentally friendly, safe preservatives to be used for mild food preservation. Traditional food-preservation techniques have undesirable effects on the appeal of fresh food products and artificial preservatives are increasingly being banned. As a consequence, a variety of fresh or minimally processed, highly perishable vegetables have emerged on the marketplace having undergone milder preservation techniques, such as a combination of refrigeration and modified-atmosphere packaging (see Chapter 17). Mild preservation techniques can control product spoilage caused by microorganisms to some extent, mainly because they are used in adherence with the "hurdle technology" (combined processes) concept [66], as discussed in Chapter 19. However, it is now becoming more evident that potential safety hazards may occur with the mild preservation systems due to the survival and growth of certain foodborne pathogens. Of special concern are cold-tolerant (psychotrophic) pathogens, like *Listeria monocytogenes, Yersinia enterocolitica*, and *Aeromonas hydrophila*, which may grow to concern levels during the long shelf life these perishable foods may have. Mesophyllic pathogens (i.e., *Salmonella* spp., *Staphylococcus aureus*, enteropathogenic *E. coli*, and *Bacillus cereus*) pose a health hazard when temperature abuse occurs. Thus, there is an urgent need for the introduction of additional safety factors with these mild preservation techniques.

Chemical preservatives, such as sorbate, benzoate, etc., have for long been used as reliable preservative factors to control a number of microbial hazards. However, such compounds do not satisfy the concept of "natural" and "healthy" food that consumers prefer and that the food industry, consequently, needs to manufacture. The negative reaction to chemical preservatives in our society is strongly increasing, despite the fact that such compounds are as yet indispensable in food processing. As a result, replacement of chemicals by more natural alternatives can only be relevant when necessary (i.e., when the chemical alternatives are no longer acceptable) and possible (i.e., when natural substitutes are indeed (eco-)toxicologically safe to use and effective in practice). The necessity is underlined by many in agro-industry, legislatures, and consumer organizations. The possibility is supported by many studies performed by academics and food industrialists. It is clear that natural alternatives are not always as potent as existing chemicals and that the clever use of combined processes may be a prerequisite for optimal functionality. Also, it is evident that even natural alternatives will have to pass legislatory scrutiny and that the label "natural" should not be confused with inherent safety.

Nature is well known to contain many different types of antimicrobial compounds, which play an important role in the natural defense or competition systems of all kinds of living organisms, ranging from microorganisms to insects, animals, and plants. In this chapter, natural antimicrobials from plants and microorganisms will be discussed only, since these may be the most feasible substitutes for chemical food preservatives considering practical, legislatory, and ethical aspects.

Regarding the development of natural antimicrobial compounds from plants (collectively called green chemicals) for food preservation, research is now focused on the potential use of phytoalexins, organic acids, and phenols. In addition, promising results have been obtained with essential oils from herbs and aromatic plants. Such essential oils consist of mixtures of esters, aldehydes, ketones, and terpenes with broad-spectrum antimicrobial activity. The toxicological basis of many herbs and spices as well as of their active components has been studied [59], and often they are known to be food-grade or even GRAS (generally recognized as safe).

With respect to the natural antimicrobial activity derived from microorganisms (referred to as biopreservatives), the most promising ongoing development in food preservation is the use of lactic acid bacteria (LAB). LAB are GRAS organisms and have a long and safe tradition in food-fermentation practices. Use of these organisms or of the antimicrobial compounds they produce has been successfully achieved in many different types of foods. Most prominently, bacteriocins produced by LAB have been under investigation worldwide for food-preservation purposes. Bacteriocins are proteins with a rather narrow antimicrobial spectrum, as compared to traditional preservatives. This apparent disadvantage is compensated for by the possibility to use these compounds for targeted control and by the fact that they are not persistent in the environment and are destroyed in the human stomach.

In the rest of this chapter, basic knowledge about the occurrence and antimicrobial properties of those natural antimicrobials of plant and microbial origin will be presented that is relevant and feasible in modern food preservation. In fact, a wealth of knowledge on the topic is available in scientific literature and elsewhere, but only a small sample will be discussed here to illustrate the ongoing quest for useful natural antimicrobials [43,80].

III. NATURAL ANTIMICROBIALS OF PLANT ORIGIN

Plants have for centuries been appreciated for their antimicrobial or medicinal activity. Certain of these plans would be suitable to cultivate instead of lower value crops, thus improving cultivation revenues, which are currently under economic pressure. In many instances, antimicrobials in plants (green chemicals) function in the resistance or defense systems against microbial diseases or pests. Often, they have a particular taste or smell, which has led to them being used in the perfume and fragrance industry. Herbs and spices have been used since ancient times not only as "tastemakers," but also as preservatives or antioxidants [13,24,71]. A wide selection of literature exists describing the favorable properties and identifying the active components of plants.

The majority of antimicrobial plant compounds are identified as secondary metabolites, mainly being of terpenoid or phenolic biosynthetic origin. The rest are hydrolytic enzymes (glucanases, chitinases) and proteins acting specifically on membranes of invading microorganisms with antimicrobial activity [17,105]. In general, no sharp chemical division can be made between constitutive and induced antimicrobials [44]. Based on the accumulating data on various plant compounds involved in disease resistance, Ingham [58] proposed categorizing the chemical defense systems of plants into preinfectional and postinfectional factors. Preinfectional factors are constitutive antibiotics, also called *prohibitins*, which are synthesized and stored in specialized tissues where they slow down or arrest in situ microbial growth instantly upon infection. Examples of prohibitins are essential oil components with antimicrobial activity. Preinfectional factors that require a postinfectional increase in concentration for an adequate effect are called *inhibitins*. In addition, two types of postinfectional factors can be distinguished: postinhibitins and phytoalexins. Compounds belonging to the first class are toxic metabolites formed after infection by hydrolysis or oxidation of preformed compounds. The second class includes antimicrobial compounds, which are synthesized upon invasion of the host plant [106]. In this chapter, a brief overview will be given of natural antimicrobial compounds in plants that belong to one of four categories: phytoalexins, phenols, organic acids, or essential oil components.

In general, herbs and spices and several of their antimicrobial constituents are GRAS, either because of their traditional use without any documented detrimental impact or because of dedicated toxicological studies. Their application in crop protection and food preservation should be facilitated by this feature, but to date plants still are a poorly exploited source of alternative antimicrobial agents. The enormous potential of plants as a source of antimicrobial compounds is well illustrated by a review of Wilkins and Board [112], who report over 1389 plants as potential green chemical sources, and more specifically by the identification of over 250 new antifungal metabolites in plants between 1982 and 1993 [44]. Obviously, not all of the potential plant sources would quality to be introduced in our agriculture practices, simply because they would only grow in specific environments. Also, whenever a plant is considered to be exploited as a green chemical source, a thorough evaluation will have to be carried out of its value with respect to the net economy of its cultivation and actual production of the green chemical (be it the whole crop, an extract, or a purified compound), the market value of this antimicrobial preparation, and the costs for going through legislative procedures. Many of the potential sources my not pass this evaluation.

A. Phytoalexins

Phytoalexins are defined as host-synthesized, low molecular weight, broad-spectrum antimicrobial compounds whose synthesis from distant precursors is induced in plants in response to microbial infection or treatment of plant tissues with a range of naturally occurring or synthetic, artificial compounds (biotic or abiotic elicitors) [32]. More than 200 different phytoalexins have by now been identified in more than 20 plant families. Phytoalexins are broad-spectrum antibiotics, generally active against phytopathogenic fungi. In contrast to the preformed prohibitins, disease resistance due to phytoalexins is a dynamic process, requiring de novo synthesis of secondary metabolites. Also, the enzymes responsible for synthesis of phytoalexins are themselves synthesized in response to exposure to microbes or other effective stimuli [26]. Elicitors, the compounds triggering the synthesis of phytoalexins, range in nature from bacterial proteins [7,51,109] to fungal fatty acids [77] to host plant–derived oligosaccharides [79].

The antimicrobial activity of phytoalexins is often directed against fungi [102], although activity has also been reported towards bacteria [68]. Gram-positive bacteria have been found to be more sensitive than Gram-negative bacteria. Isoflavonoids, characterized by a C_6-C_3-C_6 basic skeleton structure, are among the most important chemical classes of phytoalexins [133], and studies on their application outside the natural sources have been undertaken [110]. The leguminosa are known for production of isoflavonoid phytoalexins, for example, pisatin (Fig. 1a) from *Pisum sativum*, phaseollin from *Phaseolus vulgaris*, and glyceollin from *Glycine max* [96]. The production of isoflavonoid phytoalexins in plant cell and tissue cultures has aroused much interest [34,35], and this could be a method for larger-scale artificial production when a plant itself has no sound commercial potential.

For structurally related compounds in the group of isoflavonoid phytoalexins, it was found that an increase in lipophilicity correlates positively with increased antifungal activity [5]. Terpenoid phytoalexins such as rishitin (Fig. 1b) are mainly found in the family of *Solanaceae*, e.g., in potato tubers. The in vitro activity of rishitin against bacteria was found to be inhibited by low levels of the divalent cations Ca^{2+} and Mg^{2+} [69], indicating that these compounds act on the cytoplasmic membrane of the target microorganisms.

Another major group of phytoalexins, also referred to as disease- or pathogenesis-related proteins, comprises chitinases, thionins, zeamatins, thaumatins, etc. [16,17,100]. Some of these proteins are involved in the synthesis of other phytoalexins or phenolic compounds as constitutive or inducible enzymes. Others reportedly have a direct antimicrobial effect. Because they are proteins, they would be completely digested in the stomach and thus would have no impact on the health of a consumer. Chitinases target chitin, a major component of the cell wall of most phytopathogenic fungi and also of the skeletal structure of most invertebrates, e.g., insects and mites. Healthy plants normally contain low levels of chitinase, but their production is induced following pathogen attack. The induced chitinase accumulates either intracellularly or in the intercellular space, where their activity is required. Because vertebrates and higher plants do not contain chitin, no adverse impact is known, reinforcing the appeal of chitinases for fungal control. The use of chitinases as antifungal agents has been studied successfully in the laboratory for almost 10 years [78,88], but practical application has not yet been realized. The same holds true for thionins, a group of small polypeptides with antifungal and antibacterial activity that occur in cereal endosperm, e.g., in barley, oat, and maize [16]. A closely related compound is viscotoxin from mistletoe. Extracts from this plant have been used against a variety of diseases and are still part of many herbal remedies.

Although the use of phytoalexins in food preservation has been suggested in many reviews [38,43,56,72], there are still very few examples of the actual use of these compounds in food pres-

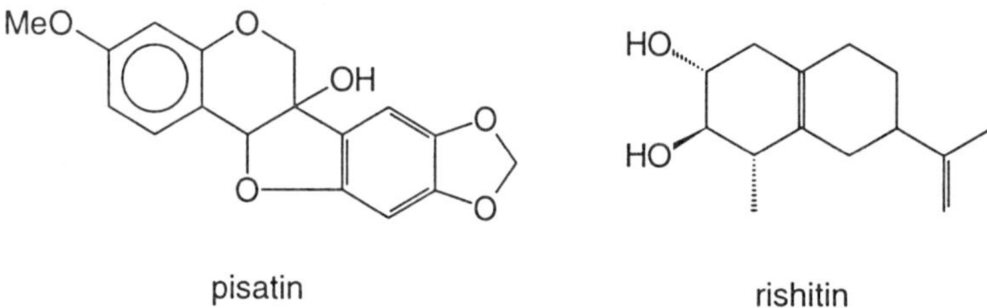

FIGURE 1 Phytoalexins from *Pisum sativum* (left) and *Solanum tuberosum* (right).

ervation. This is possibly due to the fact that phytoalexins in general show adequate antimicrobial effects at relatively high concentrations. In planta, this may not be a problem since these compounds accumulate locally to high concentrations, specifically in wounded plant tissue. The high concentrations necessary in food matrixes when applied from an exogenous source and their occasional cytotoxicity [8] hamper the application of these compounds as food-preservative agents. The development of analogs with higher specific activity and reduced toxicity could facilitate the application of these types of compounds.

B. Organic Acids

Citric, succinic, malic, and tartaric acids are commonly found in fruits (e.g., citrus, rhubarb, grapes, pineapples) and vegetables (e.g., broccoli, carrots). Through their use as acidulants or antioxidants in foods, benefit is taken as well from their antimicrobial properties. Lactic and propionic acid do not occur naturally in foods other than in trace amounts, although they are readily formed during natural fermentation. The antimicrobial activity of the various acids is extensively documented [60]. They target cell walls, cell membranes, metabolic enzymes, protein synthesis systems, and genetic material. Thus, they are active against a wide range of microorganisms. The organic acids contained in crops may well contribute to the natural crop resistance. Many organic acids or their derivatives are already applied as food preservatives.

C. Phenolic Compounds

At the beginning of the twentieth century, it was believed that plants contained compounds that were toxic towards invading fungi [107]. Initially, the abundant presence of phenolic compounds combined with their apparent in vitro activity towards many microorganisms was taken as an indication that these compounds could fulfill the primary role of the chemical defense system of plants. However, the role of plant phenolics in the chemical defense of plants against invading microorganisms is still unclear. Nevertheless, it has been appreciated that a vast range of phenolic compounds contribute to the defense mechanisms of plant tissues as well as to the sensory (taste, odor, appearance) and nutritional qualities of fresh or processed plants. Phenolics are characterized by an aromatic ring bearing one or, more frequently, several hydroxy substituents, including functional derivatives. Phenolic compounds usually occur conjugated, e.g., to sugars as β-D-glucopyranosides. The phenolic compounds are classified into three groups: simple phenols and phenolic acids (e.g., *p*-cresol, 3-ethylphenol, hydroquinone, protocatechuic, vanillic, gallic, syringic, ellagic acids), hydroxycinnamic acid derivatives (e.g., *p*-coumaric, caffeic, ferulic, sinapic acids), and flavonoids [72]. The latter group is the most important single group of phenolics in food, comprising catechins, proanthocyanins, anthocyanidins, and flavons, flavonols, and their glycosides. Finally tannins, a polymeric form of phenolics, are an important group of plant phenolics, unified by the common ability to precipitate protein from aqueous solution. The antimicrobial activities of the naturally occurring phenolics from olives, tea, and coffee have been studied in more detail than those from other sources, which may in part be due to the high value of the products being processed [72]. Phenolics from spices, such as gingeron, zingerone, and capsaicin, have been found to inhibit germination of bacterial spores. Native plant phenolics are important food-preservative factors and have, as a group, an impressive antimicrobial spectrum, although their deliberate use as food preservatives is rarely exploited.

D. Essential Oils and Their Components

Essential oils are mostly derived from spices and herbs but can also be isolated from fruits, roots, and stems of plants. Some oils and isolated plant compounds are used in food as flavoring agents. Derived from their functionality in plants, these compounds show a wide range of interesting biological activities [71]. Some compounds have been shown to attract flies for pollination, whereas others show a distinct insect-repelling activity. Others attract herbivores for seed distribution or show fungicidal or bactericidal activity to suppress infection by plant pathogenic microorganisms. Compounds that are approved for use in food and combine antimicrobial activity with low mammalian toxicity have great potential for application as natural food preservatives.

The antimicrobial activities of extracts obtained from spices, herbs, and other aromatic plants or parts thereof using organic solvents or steam distillation have been recognized for many years. Plants and plant extracts have been used since antiquity in folk medicine and food preservation, providing a range of compounds possessing pharmacological activities [28]. Most commonly, the active antimicrobial compounds are found in the essential oil fraction. With many herbs and spices, these compounds contribute to the characteristic aroma and flavor. Essential oils are mostly soluble in alcohol and to a limited extent in water. They consists of mixtures of esters, aldehydes, ketones, and terpenes [47]. Essential oil components with a wide spectrum of antimicrobial effect include thymol from thyme and oregano, cinnamaldehyde from cinnamon, and eugenol from clove.

The impact of essential oils on bacteria, especially on pathogens, has been extensively studied in the laboratory, and significant variations have been noted. For example, *Escherichia coli* was found to be more vulnerable than *Pseudomonas fluorescens* or *Serratia marcescens* to the essential oils of sage, rosemary, cumin, caraway, clove, and thyme [37], whereas *Salmonella typhimurium* was more sensitive to oregano and thyme oils than *Pseudomonas aeruginosa* [74]. Deans and Ritchie [27] studied the effect of 50 plant essential oils on 25 genera of bacteria and concluded that both Gram-positive and Gram-negative bacteria are susceptible, but the levels of impact were highly variable. Tassou and Nychas [101] have shown that the essential oil of *Pistacia lentiscus* var. *chia* (mastic gum) inhibits the growth of the food pathogen *Salmonella enteritidis* in skimmed milk. Mold growth on black table olives was found to be suppressed by methyl eugenol and the essential oil from *Echinophora sibthorpiana* [62]. The use of mustard oil in homogenized, canned beef was investigated by Drdak et al. [33]. A concentrations of 0.1% allylisothiocyanate the active antimicrobial compound in mustard oil, did not cause unacceptable sensory effects, allowed sufficient thermosterilization, and resulted in a microbially safe product. A recent detailed review by Nychas [72] summarizes findings that essential oil compounds from many different plant sources inhibit many foodborne pathogens (Table 1). *Staphylococcus aureus, Listeria monocytogenes, Aeromonas hydrophila, S. typhimurium*, and *Clostridium botulinum* are to some degree sensitive to extracts from linden flower, orange, lemon, grapefruit, mandarin, sage, rosemary, oregano, thyme, cinnamon, cumin, caraway, clove, thyme, allspice, mastic gum, and onion. However, most researchers inevitably came to conclude that the effectiveness of essential oils decreased when experiments were conducted in vivo. This could well be due to specific components of the food matrix, such as proteins and fats, which immobilize and inactivate the essential oil components.

The antifungal effects of essential oil components from several herbs, spices, and other plant materials have been investigated against important food spoilage or mycotoxigenic species of *Penicillium* and *Aspergillus*, but contradictory results were obtained [2,6,20,74,83]. While some found inhibitory activity, other researchers actually noted stimulating effects. Again, the food matrix may have had a decisive influence here, and it is recommended to standardize the experimental set-up accordingly.

Because essential oils contain a variety of compounds from different chemical classes, it is not possible to isolate a single mechanism by which these compounds act on microorganisms. An important common feature of essential oil components is their high degree of hydrophobicity. Therefore, these compounds partition preferentially into biological lipid bilayers as a function of their own lipophilicity and the fluidity of the membrane [73]. Accumulation of lipophilic compounds into biological membranes enhances their availability to the cell and therefore may inhibit cell vitality [89,91]. Despite the high degree of ordering of solutes in a lipid bilayer compared with bulk liquid phase [92], a good correlation between the partitioning coefficient of various lipophilic compounds in membrane/buffer and octanol/water two-phase systems has been observed [89,90]. Therefore the octanol/water partitioning coefficients, which are known for many different compounds present in essential oils, can be used to assess the potential antimicrobial effect of these compounds [89]. However, the presence of specific reactive groups in compounds, the variability in membrane composition, and the metabolic capacities of the target organisms make a reliable prediction of the toxicity of compounds based solely

TABLE 1 Antimicrobial Spectrum of Essential Oils from Herbs, Spices, and Plants

Acetobacter sp.	*Mycobacterium* sp.
A. calcoacetica	*M. phlei*
Aeromonas hydrophila	*Mucor* sp.
Alcaligenes sp.	*Neisseria* sp.
A. faecalis	*Neisseria sicca*
Arthrobacter sp.	*Pediococcus* sp.
Aspergillus niger	*Penicillium* sp.
A. flavus	*P. chrysogenum*
A. ochraceus	*P. citrinum*
A. parasiticus	*P. patulum*
Bacillus sp.	*P. roquefortii*
B. cereus	*Propionibacterium acnes*
B. subtilis	*Pityrosporum ovale*
Beneckea natriegens	*Proteus vulgaris*
Brevibacterium ammoniagenes	*Pseudomonas sp.*
B. linens	*P. aeroginosa*
Brochothrix thermosphacta	*P. clavigerum*
Campylobacter jejuni	*P. fluorescens*
Candida albicans	*P. fragi*
Citrobacter freundii	*Rhizopus sp.*
Clostridium botulinum	*Saccharomyces cerevisiae*
C. perfringens	*Salmonella sp.*
C. sporogenes	*S. enteriditis*
Corynebacterium sp.	*S. pullorum*
Edwardsiella sp.	*S. senftenberg*
Enterobacter aerogenes	*S. typhimurium*
Erwinia carotovora	*Sarcina marcencens*
Escherichia coli	*Staphylococcus aureau*
Flavobacterium suaveolens	*Trichophyton mentagrophytes*
Klebsiella pneumoniae	*Yersinia enterocolitica*
Lactobacillus sp.	*Vibrio parahaemolyticus*
L. minor	
L. plantarum	
Leuconostoc cremoris	
Listeria monocytogenes	
Micrococcus sp.	
M. luteus	
Moraxella sp.	

Source: Adapted from Ref. 72.

on their hydrophobicity difficult, if not impossible. This is exemplified by carvone and cinnamaldehyde, two compounds with comparable hydrophobicities but different antifungal mechanisms. Both compounds inhibit growth of *Penicillium hirsutum* when administered via the gas phase [93]. Full suppression of growth by carvone was observed only as long as the compound was present in the atmosphere. On the other hand, fungal growth inhibition by *trans*-cinnamaldehyde was found to be strictly irreversible. In conclusion, carvone acts as a fungistatic agent, whereas *trans*-cinnamaldehyde acts as a fungicide. The mechanism behind this difference in antifungal activity was investigated using *Saccharomyces cerevisiae* as a model organism [95]. Cinnamaldehyde was found to cause a (partial) collapse of the integrity of the cytoplasmic membrane, which leads to excessive leakage of metabolites and enzymes from the cell and finally loss of viability. In agreement with its fungistatic rather than fungicidal effect, loss of membrane integrity was not observed with carvone [95].

Considering any exploitation of essential oils, it should be stressed that large variations may occur in the yield of active compounds or total oil with the plant genotype and with different extraction methodologies, and also that variations are to be expected in the essential oil composition of the same species according to geographical location and environmental and agronomical conditions, as well as differences in essential oil content with diurnal rhythm. It is clear that essential oils or their active components are by no means a ready-to-use source from a production point of view, and many parameters need to be carefully standardized in detail in that respect.

E. Example of Application of Antimicrobials from Plants

Among the essential oil components, the volatile monoterpenes and aldehydes have attracted the recent interest of research and food industries because they can be used as food preservatives that leave a negligible amount of residues. Regarding application in practice, however, the volatile nature of the very potent compounds requires the development of suitable slow-release formulations or tailored packaging systems to maintain their functional activity for a sufficient time. For instance, with carvone, the prime monoterpene in essential oil of caraway (*Carum carvi* L.) seeds (Fig. 2a), a powerful antifungal effect has been found, which is already exploited for the protection of potato tubers under storage conditions [50]. However, carvone is gradually lost from the storage environment and has to be administered regularly.

Cinnamaldehyde, the major compound in cassia oil (Fig. 2b), shows potent antifungal activity against several food-associated fungi like *Penicillium* sp., *Fusarium* sp., and *Aspergillus* sp. [75]. Cinnamaldehyde also has been shown to possess antiaflatoxigenic properties [70]. When exposed to air, cinnamaldehyde is readily oxidized to cinnamic acid. Therefore, gas phase application of this compound is less effective [93]. However, its very potent fungicidal activity and low mammalian toxicity [59] make this natural compound an interesting candidate for application as a surface disinfectant for foods. An example of the use of cinnamaldehyde in food preservation is its potential use as a surface disinfectant for tomatoes [91]. Tomatoes are particularly vulnerable to microbial spoilage at calyces and wound sites on the fruit surface. The major pathogens affecting the postharvest life of tomato fruit are *Alternaria alternata, Botrytis cinerea*, and *Rhizopus stolonifer*. Calyces are usually the first part of the tomato on which fungi appears. It has been shown that disinfection of tomatoes with sodium hypochlorite before packaging greatly reduced subsequent microbial spoilage. However, several countries have abandoned the use of hypochlorite for disinfection of foods, and natural plant-derived compounds with sufficient antimicrobial activity and low mammalian toxicity such as cinnamaldehyde could be good alternatives. Smid et al. [91] investigated the reduction of spoilage-associated fungi and bacteria on whole tomatoes packaged under modified-atmosphere conditions. Tomatoes were treated for 30 minutes with a solution containing 13 mM cinnamaldehyde and stored at 18°C in sealed plastic bags. Under these conditions the development of the microbial population was recorded on treated and untreated (control) tomatoes. On day 4, visible fungal growth was

FIGURE 2 Structures of carvone (left) and *trans*-cinnamaldehyde (right), two secondary plant metabolites with antifungal activity.

observed on calyces of untreated fruits. *Penicillium* sp. was found to be the dominant fungal species on the calyx. The calyx of cinnamaldehyde-treated tomatoes remained free from visible fungal growth for at least 9 days. These observations agree with the microbial analysis of the tomatoes (Fig. 3). After 2 days of storage, pronounced growth of the bacterial population was observed on control tissues treated with 0.85% NaCl. After 4 days of storage, a significant increase in the size of the bacterial population was detected on untreated tomatoes. In contrast hardly any development of the bacterial population was detectable on cinnamaldehyde-treated tomatoes (Fig. 3a). As expected, visible fungal growth on calyces of both untreated and NaCl-treated tomatoes appearing at day 4 corresponded with a rapid increase in size of the fungal population. The size of the fungal population on cinnamaldehyde-treated tomatoes remained small under day 11 (Fig. 3b).

Fungicidal and bactericidal compounds from natural sources, such as cinnamaldehyde, may offer attractive possibilities for disinfection of fresh and minimally processed fruits and vegetables. A bottleneck for practical use of these compounds may be not the efficacy, but rather specific odors associated with such compounds at higher dosages. To overcome this problem, antifungal plant metabolites should be selected for both efficacy and minimal interference with the natural odor of the product.

IV. NATURAL ANTIMICROBIALS OF MICROBIAL ORIGIN

Microorganisms produce a wide range of components that influence the growth of other microorganisms present in their environment. Often these components increase the competitive edge of the producing organism and as such are an important feature of their survival and proliferation. Regarding food preservation, the most important single group of organisms to be considered as a source of biopreservatives are the lactic acid bacteria (LAB). For centuries these have been used in food fermentation to produce stable food products, including dairy (cheese), meat (sausages), and vegetable

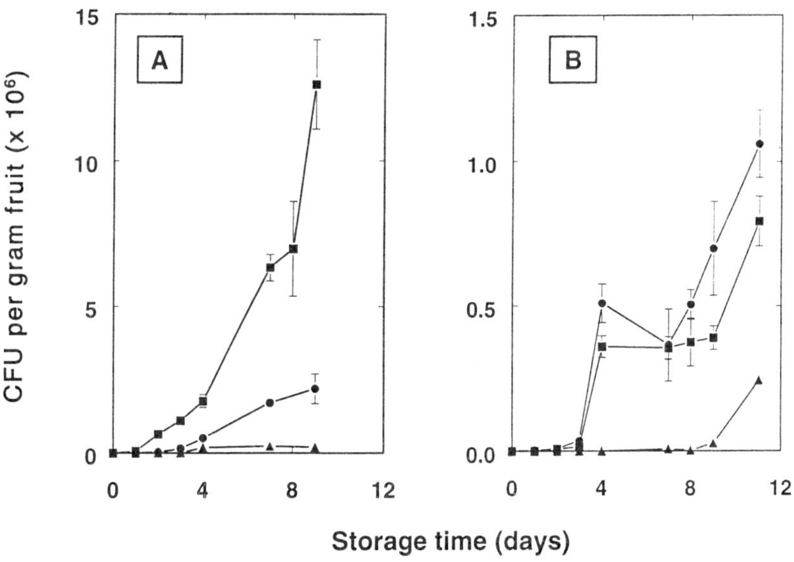

Figure 3 Development of the microbial spoilage organisms on tomatoes stored for 11 days at 18°C in sealed plastic bags. The microbial population on cinnamaldehyde-treated fruits (▲), NaCl-treated fruits (■), and untreated fruits (●) was monitored for 9 (bacteria; panel A) or 11 days (fungi; panel B). The data represent mean values of triplicate measurements, and each data point is calculated from a sample of five tomato fruits. Standard errors of the mean are indicated by error bars. (Modified from Ref. 94.)

(sauerkraut) products. The fact that fermented products, which naturally contain these microorganisms and the antimicrobials they may produce, have been consumed traditionally without a negative health effect, has given LAB GRAS status [42]. Lactic acid bacteria may produce both antimicrobial compounds with a relatively broad inhibition spectrum (i.e., organic acids and hydrogen peroxide) as well as compounds with a rather narrow antimicrobial spectrum (i.e., bacteriocins). The use of LAB as biopreservatives is possible via the application of the producing organism as a so-called "protective culture" to the food product and relying on its proliferation and consequent competition with the microorganisms to be suppressed. Alternatively, preparations of the active antimicrobial compounds may be utilized, with the advantage of an instant and more controllable effect. While the use of protective cultures in most countries needs only to be declared on the product, the use of antimicrobial metabolites such as the bacteriocins is subject to specific rules and regulations in food legislation.

Several recent reviews give more detail on bacteriocins and LAB in relation to their potential for food preservation [1,21,30,55,76,84,87,99].

A. Lactic Acid Bacteria as Protective Cultures

The use of lactic acid bacteria as starter cultures in the production of fermented meats, dairy products, and vegetables is one of the oldest food-processing practices utilized and meant to stabilize food products while obtaining specific, desired sensory and organoleptic properties. The success of the fermentation process depends on the competitiveness of starter cultures, and it is exactly for this reason that LAB have been so widely used. The many different antimicrobials they produce are able to counteract a wide range of competitors that would cause problems in the fermentation process. In recent years, some research has been developed into the use of LAB in food-processing applications where the outgrowth of specific problem microorganisms is to be controlled. In this case, the selected LAB are referred to as *protective cultures* and should affect pathogens or spoilage microorganisms without any negative impact on the sensory or organoleptic characteristics of the food product. Production of acids as the main antimicrobial agents is often detrimental to food quality and is not a suitable mechanism of action for protective cultures. LAB that produce a minimum amount of acids but expel bacteriocins in their environment do offer good options as protective cultures.

Whereas LAB as starter cultures have become widely used and accepted, their use as protective cultures is still under development. Exploitation depends partly on legislative hurdles that relate to the consideration that protective cultures that rely for their activity on bacteriocins are intended for use as preservative agents and function by use of compounds not yet generally recognized as safe in many countries, as discussed elsewhere in this chapter. This is in sharp contrast to the legal status of starter cultures, which are considered to be processing aids or ingredients and not preservatives, and for which the mode of action seems not to be a decisive issue from the legislatory point of view. Recently, it has been advocated to employ molecular biology tools to improve the performance of starter and protective cultures with regard to their production or preservation capacity [39]. Whereas, within the current food legislations, natural strains already have limited access to practice, it is expected that the use of genetically modified strains will not be approved of more easily.

The effort to develop protective cultures has been increasing over recent years but up to now has been confined to laboratory studies. Some review papers on the topic are available [40,55,56,63,81,87,108,114], although several are published in sources that may not be readily accessible. A number of interesting developments with respect to the use of LAB as protective cultures for several different food categories are discussed below.

1. Meat Products

In a screening exercise involving 221 strains of *Lactobacillus* species evaluated for their ability to inhibit the growth of microorganisms commonly occurring in meat products [85], a wide range of bacteria was found to be affected by individual strains: e.g., *Serratia marcescens* (by 47% of the strains), *Citrobacter freundii* (47%), *Proteus vulgaris* (67%), *S. typhimurium* (9%), and *Brochothrix thermosphacta* (87%). In most cases, the inhibitory activity of the protective culture was attributed to lactic acid formation, although 6 of the 221 LAB isolates (all isolates of *Lactobacillus sake*) formed a bacteriocin contributing to the inhibition of *L. monocytogenes*. Experiments performed by the same researchers on comminuted cured pork (German-type fresh Mettwurst) with pH 5.7 were aimed at control of *L. monocytogenes* and showed that a strain of *L. sake* producing a suitable antilisterial bacteriocin was able to reduce the growth potential of the pathogen by about 1 log cycle [86]. A mutant of *L. sake* that did not produce the bacteriocin did not affect the number of *Listeria* inoculated into this product. In another study using *L. sake* as the protective culture, the control of *L. monocytogenes* in vacuum-packaged sliced Brühwurst (cooked sausages) was emphasized [65]. Sliced sausage samples were inoculated with a mixture of four *L. monocytogenes* serovars, fortified with either one of two bacteriocin-producing strains of *L. sake*, isolate Lb706, which produced sakacin A, and isolate Lb674, which produced sakacin 674, or of a non–bacteriocin-producing strain of *L. sake* and stored for up to 28 days at 7°C. Whereas the non–bacteriocin-producing LAB reduced counts of *L. monocytogenes* but not to an acceptable extent, both bacteriocin-producing strains of *L. sake* were able to control growth of *L. monocytogenes* adequately at the high initial counts tested.

Using bacteriocin-producing and non–bacteriocin-producing strains of *Pediococcus acidilactici* for protection of turkey summer sausages against *L. monocytogenes*, Luchansky [67] found that the pathogen could be reduced by the bacteriocin-producer by 3.4 log cycles, but by only 0.9 log cycle when the non–bacteriocin-producing strain was used. In vacuum-packaged wiener and frankfurter sausages, proliferation of *L. monocytogenes* inoculated in the products was suppressed for over 60 days by addition of *P. acidilactici* JD1-23 at 10^7 CFU/g product, whereas the viable count of the pathogen increased from 10^4 to 10^6 in the control [12]. Degnan et al. [29] observed a clear antilisterial effect of yet another bacteriocin-producing strain of *P. acidilactici* in vacuum-packaged wieners stored at abuse temperature (25°C), where the addition of the protective culture resulted in a reduction of *L. monocytogenes* counts by 2.7 log cycles within 8 days while pathogen counts increased by 3.2 log cycles in sausages without added pediococci. In bacon, a pediocin-producing strain of *P. acidilactici* has been used in combination with reduced levels of nitrite to prevent toxin production by outgrowth of *C. botulinum* spores. Here, the protective culture would grow during conditions of temperature abuse, producing lactic acid and inhibitory pediocins. Strain *P. acidilactici* H, isolated form fermented sausage, exhibited a broader range of bactericidal activity than any other pediococcal bacteriocin due to the production of a bacteriocin termed pediocin AcH.

Pediocin producers have also been used as protective cultures relying on their lactic acid production, rather than on the production of a bacteriocin. Hutton et al. [57] used the "Wisconsin process" (a combination of lactic acid starter culture and sucrose) to prevent toxigenesis by *C. botulinum* in reduced nitrite bacon. In chicken salads these authors found that a combination of *P. acidilactici* and glucose prevented botulinum toxigenesis. When the chicken salad was temperature abused, the protective culture catabolized available glucose to lactic acid, which caused a decrease in the pH of the product. Pathogen challenge tests verified that the rate and extent of lactic acid accumulation in the chicken salad during temperature abuse was sufficient to preclude botulinum toxigenesis.

Kotzekidou and Bloukas [64] studied the effect of protective cultures on the shelf life of sliced vacuum-packed cooked ham. They found that cooked ham produced with *Lactobacillus alimentarius* and *Staphylococcus xylosus* as protective cultures was acceptable up to 28 days, while control ham has a shelf life of 21 days. The activity of the protective cultures was directed to micrococci, staphylococci, and *B. thermosphacta*. Meat salads with relatively high pH values (pH 6.0–6.5) were studied by Hennlich and Cerny [53] for potential application of LAB as protective cultures in limiting the hygienic risks caused by food salmonellae, staphylococci, or clostridiae. The risk of pathogen growth in these foods is most apparent under temperature-abuse conditions, and the research showed that distinct cultures of lactic acid bacteria are indeed able to decrease microbial risks due to foodborne pathogens at elevated temperature. Whereas they do not reduce spoilage by bacilli, yeasts or fungi, the protective cultures used could reduce the growth of pathogens and actually spoiled the food before the pathogens could grow to hazardous levels.

Andersen [4] recently reported on a commercial protective culture developed for fresh sausages (called "FloraCarn L2"), where contamination during or after processing is a possible hazard and the protective culture can be used as an additional safety and quality factor. FloraCarn L2 was tested in fresh British sausage mince and was shown to suppress the indigenous microflora and *B. thermosphacta*. In fresh coarse chopped sausages, the protective culture inhibited the possible development of indigenous coliform bacteria during storage.

Research on protective cultures has not always found potential positive applications for bacteriocin-producing LAB. Targeting at control of L. monocytogenes in meats during long-term storage, Buncic et al. [19] tested *L. sake* 265 (Lb 265) and *Lactobacillus casei* 52 (Lb 52) isolated from chilled meat products as protective cultures. Although both starter cultures produced bacteriocin at 4°C, they were not able to suppress growth of *L. monocytogenes* inoculated at 10^3 CFU/g on vacuum-packaged, raw beef (pH 5.3–5.4) during 23 days storage at 4°C when they were inoculated at the same low level. The protective cultures were equally ineffective when applied on vacuum-packaged emulsion-type sausages (pH 6.4) inoculated with *L. monocytogenes* and stored at 4°C for 23 days. Apparently, the amounts of bacteriocin produced in situ by the low initial numbers of protective cultures employed were not sufficient to inhibit or reduce *L. monocytogenes* on chilled meats to any significant extend, whereas higher initial numbers of lactic acid bacteria are not desirable in chilled meats for product quality reasons.

2. Fish and Seafood

Wessels and Huss [111] studied the use of protective cultures as inhibitors of *L. monocytogenes* in lightly preserved fish products. Co-culture of the pathogen with a nisin-producing strain of *Lactococcus lactis* subsp. *lactis* at 30°C resulted in a decline of the pathogen from 5×10^5 to <5 CFU/ml within 31 hours. However, when the protective culture was inoculated on slices of commercial cold smoked salmon stored at 10°C for 21 days, no net growth was detectable. Despite this lack of evidence for in situ proliferation of the protective culture, on cold smoked salmon slices co-inoculated with *L. monocytogenes* (10^4 CFU/g) and the protective culture (3×10^6 CFU/g), the population of the pathogen declined by a half log cycle during the first 15 days, then increased at a rate slightly lower than that of the control not inoculated with the lactococcus. Although a complete reduction of the pathogen was not achieved, the experiments proved the point that control of proliferation was feasible under practical conditions.

The use of bacteriocins from LAB for the preservation of brined shrimps, which are usually protected from microbial deterioration by addition of sorbic or benzoic acid, was tested by Einarsson and Lauzon [36]. Three different bacteriocins were evaluated (nisin Z, carnocin U149, and bavaricin A) for their biopreservative potency. With nisin Z, the most effective bacteriocin, a delay in bacterial growth was observed that resulted in an extension of the shelf life by 21 days (from 10 to 31 days). The strongest preservative effect was found with sodium benzoate and potassium sorbate, which completely inhibited microbial growth for 59 days when added to the brined shrimps at levels of 0.05–0.1% (w/w).

In a recent overview paper, Huss et al. [56] presented an update on biopreservation used with fish products as they discussed a range of relevant topics: biopreservation as a full or partial alternative to salt or chemical additives, protective cultures and their characteristics, selection of protective cultures, and limitations to the application of protective cultures.

3. Dairy Products

To control growth of clostridia in cheese spreads, which cause the so-called "late blowing" of the product (a combination of gas formation and butyric acid production), Zottola et al. [115] proposed to add nisin-producing lactococci. Many clostridia are sensitive to nisin, and the use of the protective culture resulted in a significant extension of the shelf life of the product. Specifically, spoilage by *Clostridium sporogenes* was reduced in the nisin-containing cheese spreads.

Contamination by *L. monocytogenes* can also cause problems in the production of cheeses, especially in products such as the Italian cheeses taleggio, gorgonzola, and mozzarella, in which the pH rises during ripening and maturation. Giraffa et al. [41] showed that *Enterococcus faecium* added during the manufacture of taleggio cheese releases a stable, antilisterial bacteriocin. An advantage of the rather narrow activity spectrum of bacteriocins from LAB is apparent from this study: the pathogen was suppressed by the added protective culture, whereas the activity of the thermophilic starter used in the cheese-making process was not affected.

Stecchini et al. [97] investigated the control of postprocess contamination of mozzarella cheeses by bacteriocin-producing strains of *Lactococcus lactis*. They observed that heat-treated cultures of such strains added to mozzarella cheese inoculated with *L. monocytogenes* and packaged in small bags resulted in a decrease in the initial counts of *Listeria*. *Listeria* counts remained significantly below those of the samples prepared without the addition of biopreservatives during a storage period of 2–3 weeks at 5°C.

Giraffa [40] presented a concise state-of-the-art review on the use of biological preservation with dairy products. In this overview, practical applications of protective cultures of LAB to increase the hygienic level of dairy products were reported as well as sensitivity of pathogens during the cheese-making process, survival of pathogens (i.e., *Listeria*) in cheese, cheese made with raw milk, and antimicrobial metabolites of LAB with emphasis on bacteriocins. The author concludes that biological preservation cannot replace good manufacturing practice (GMP) but offers an additional tool for improving the food quality.

4. Vegetable Products

Bacteriocin-producing LAB also show potential for the biopreservation of foods of plant origin, especially minimally processed foods such as prepackaged mixed salads and fermented vegetables. Vescovo et al. [140] observed a reduction of the high initial bacterial loads of ready-to-use mixed salads when bacteriocin-producing LAB were added to the salad mixtures. Furthermore, bacteriocin-producing starter cultures may be useful in the fermentation of sauerkraut [18,49] or olives to prevent the growth of spoilage organisms. In the fermentation of Spanish-style green olives, a bacteriocin-producing strain of *Lactobcillus plantarum* dominated the indigenous LAB without adversely affecting the organoleptic properties of the product [82]. In contrast, a non–bacteriocin-producing variant of this strain was outnumbered by the natural *Lactobacillus* population.

From studies of Cerny and Hennlich [22] on the use of LAB as protective cultures in potato salad to control food poisoning by salmonellae and toxin-producing staphylococci or clostridiae, several prospects became evident. In mayonnaise-based potato salads with pH values of 5.5–6.0 that were exposed to ambient temperatures for up to one week, the protective cultures greatly reduced the hygienic risks, although they did not increase the shelf life of those products.

Hennlich [54] reported on the selection and evaluation of LAB isolated from potato salads as protective cultures for chilled delicatessen salads, assuming that they were well adapted ecologically. Important criteria for selection were minimum growth temperature, rate of acidification at refrigeration temperature, and rapid growth and acid formation at abuse temperature (mimicking interruption of the cold chain). These criteria were adequately met by *Lactobacillus casei* ILV 110 and *L. plantarum* ILV 3. When used as protective cultures (10^4 CFU/g minimum), these strains inhibited the normal spoilage flora of delicatessen salads and also suppressed growth of *E. coli* and *C. sporogenes* inoculated into meat salads during storage at chill temperature. One of the isolates, *L. plantarum* ILV 3, was found to be suitable as a protective culture for weakly acidic delicatessen salads (pH 5.0–6.0) as well.

Cerny [23] studied the inhibitory effect of a range of lactic LAB (*Leuconostoc cremoris, L. lactis* var. *diacetylis*, *L. lactis* subsp. *lactis, L. lactis* subsp. *cremoris*, and *Lactobacillus casei*) was studied on the growth of several indicator microorganisms (*E. coli*, *Staphylococcus saprophyticus*, and *C. sporogenes*) in mayonnaise-based meat and potato salads (pH 5.5–6.5; prepared using pasteurized ingredients to eliminate endogenous LAB). It was found that addition of *L. cremoris* as a protective culture to potato salad completely controlled *E. coli* and *C. sporogenes* growth at room temp. *L. lactis* subsp. *lactis* (inoculation level 10^3–10^6 CFU/g) suppressed *E. coli* (10^2–10^4 CFU/g) in meat salad stored at room temperature. Importantly, it was concluded that best protective effects were observed when the ratio of *L. lactis* subsp. *lactis* to *E. coli* was greater than 10:1.

B. Bacteriocins Produced by Lactic Acid Bacteria

Bacteriocins are small proteins produced by many bacterial genera, including lactic acid bacteria. Most of the bacteriocins produced by LAB inhibit the growth of other lactic acid bacteria, but some are bactericidal to a number of food pathogens and food-spoilage bacteria. In all cases, these other bacteria are gram-positive. Thus, the bacteriocins or their producers can probably not be used as a general safety hurdle, but could still be used to form a specific hurdle to suppress the growth of notorious gram-positive pathogens such as *L. monocytogenes, C. botulinum*, and *B. cereus*.

Although many different bacteriocins have currently been identified and their potential use as food preservatives is apparent, the exploitation in current practice is limited to two bacteriocins: nisin and pediocin. The limited exploitation of bacteriocins is mainly due to the rather small bactericidal range of most bacteriocins, their low efficiency of production, their limited stability in the food matrix, and, overall, their disputed regulatory status. In fact, considering the limitations to practical application, only a few of the new bacteriocins have sufficiently favorable assets in comparison to nisin and pediocin that would warrant the effort of pursuing implementation in practice. Nevertheless, the increasing doubt about the safety of traditional chemical preservatives such as nitrite and propionate fuels the revival of interest seen in applied research today aimed at the introduction of natural preservative factors such as the bacteriocins.

Ever since the identification of the inhibitory activity of a strain of *Lactococcus lactis* subsp. *lactis* in 1928, LAB have been increasingly scrutinized by bacteriocin production. The inhibitory agent was later termed nisin, the first known and most extensively studied bacteriocin of LAB. Table 2 lists some of the potential applications of nisin. Today, more than 30 different bacteriocins produced by some 17 species of lactic acid bacteria have been identified, and much information has been obtained on the biochemistry and range of bactericidal activity. For food preservation, advantageous features of several bacteriocins are their relatively high heat resistance and inhibition of Gram-positive foodborne pathogens and spoilage organisms. Much attention has been given to the inhibition of *L. monocytogenes*. This cold-tolerant bacterium, which can result in a high mortality rate, occurs in many different foods, causing problems specifically in dairy (soft cheeses) and meat (pate, sausages) products. Also, the bactericidal impact of several bacteriocins on spore-forming bacteria, such as *Bacillus* and *Clostridium* species, has been subject of research for many decades and indicates the greater potential these bacteriocins could have in food preservation. In the following, a brief overview of the research on three interesting bacteriocins is presented with data taken from a variety of sources [25,30,31,48,55,61,76,85,99].

1. Nisin

Nisin is a protein consisting of 34 amino acids, which is stable to autoclaving and effectively inhibits growth of important Gram-positive foodborne pathogens like *L. monocytogenes* and *S. aureus*, and prevents outgrowth of spores of many species of *Clostridium* and *Bacillus*. It is especially active in acidic food matrixes. The bacteriocin is produced by some strains of *Lactococcus lactis* subsp. *lactis*, although different strains may produce structural variants deviating slightly in exact amino acid com-

TABLE 2 Foods and Beverages in Which the Bacteriocin Nisin Has Been Used

Food product	Function or use
Swiss-type cheese	Prevention of blowing faults caused by *Clostridia*
Milk	Extension of shelf life
Tomato juice	Allows lower heat-processing requirements
Canned foods	Control of flat sour caused by thermophilic spoilage bacteria
Sauerkraut	Optimizing starter function by improving competitiveness
Beer	Inhibition of spoilage by lactic acid bacteria
Wine	Control of spoilage by lactic acid bacteria

Source: Ref. 42.

position. Originally nisin was considered for use as an antibiotic, but because its range of inhibition is limited, it was not judged suitable for therapeutic use. However, nisin is completely degraded in the alimentary tract and it therefore can be used safely as a food additive. Its potential use as a food preservative was first demonstrated through the successful employment of nisin-producing cultures in the manufacture of Swiss-type cheeses. Due to their inhibition of gas-producing clostridia, blowing of the cheeses is prevented. Although vegetative cells of these organisms are killed or reduced in number by normal processing conditions, the heat-resistant spores require an excessive "botulinum cook" or the use of chemical additives to prevent their outgrowth. Nisin may be used as a natural additive to inhibit spore outgrowth or reduce their heat resistance.

Nisin has been used in conjunction with other preservative measures to enhance product safety or quality. In canned foods such as vegetables, soups, and puddings, nisin has been applied in conjunction with heating to successfully counteract heat-resistant spores of flat-sour thermophilic bacteria. Normal heating and nisin may be combined for milk production in countries where pasteurization, refrigeration, and transportation facilities are not adequate and where it is difficult to assure the supply of good quality milk to the public. When nisin is used with acetic, lactic, or citric acid, the effectiveness of blanching and pasteurization treatments may be better than with nisin or the organic acids alone. The use of nisin in combination with nitrite in meat products has been reported frequently. Although the combined application may allow for less nitrite to exert an identical degree of inhibition of clostridia compared to nitrite alone, the meat systems seem to influence the effectiveness of nisin strongly. Inhibition of *L. monocytogenes* in raw meat, for instance, may continue for 2 weeks at 5°C, but both the inhibitory effect and the nisin-related activity rapidly diminish at room temperature. Comparable findings hold for clostridia suppression in bacon and sausages. Conceivably, binding of nisin to meat particles and high salt concentration may reduce the amount of nisin in solution where it may be active.

2. Pediocin

Pediocins are bacteriocins produced by LAB of the genus *Pediococcus*. The first report on pediocin production dates back to 1975, when it was found that *Pediococcus pentosaceus* inhibited growth and acid production of *Lactobacillus plantarum*, and undesirable competitor in mixed-brine cucumber fermentation. The active agent, designated as pediocin A, inhibited a broad range of LAB as well as several clostridia, *S. aureus*, and *B. cereus*. The finding implied that pediocin production might be a favorable asset of starter cultures in the fermentation of sausages and vegetables, where reported staphylococci and naturally competing LAB are the major concern, respectively.

Several applications of pediocins have been assessed with regard to food safety. Pediocin PA-1, produced by a strain of *Pediococcus acidilactici*, has been shown to inhibit growth of *L. monocytogenes* inoculated into cottage cheese, half-and-half cream, and cheese sauce for 1 week at 4°C, whereas rapid growth to high cell densities was observed in the control samples (no bacteriocin added). The activity of pediocin PA-1 was not affected by fat or proteins present in the foods, while a synergistic action was noted between the effect of the bacteriocin and lactic acid. Extensive tests have shown that this pediocin is nontoxic, nonimmunogenic, and readily hydrolyzed by gastric enzymes. The potent antilisterial activity and the effectiveness of pediocin AcH and other pediocins as biopreservatives has by now been well established experimentally in beef wieners, semi-dry sausage, frankfurters, and fresh meat.

3. Sakacin

Sakacins, a group of bacteriocins produced by *Lactobacillus sake*, owe their discovery to the intensive search for natural antimicrobial compounds capable of increasing the shelf life of raw meat by inhibiting growth of meat-spoilage microorganisms and controlling *L. monocytogenes*. Several different antimicrobials are known to be produced by strains of *L. sake*, which normally reside on meat products. These strains are well adapted to the conditions in meats and conceivably are the best competitors in this food environment.

Lactocin S, produced by strain *L. sake* 45 isolated from naturally fermented sausage, is inhibitory against a range of LAB, including organisms from the same sausages. A similar bacteriocin is produced by a strain of *L. sake* isolated from Spanish dry sausages. However, the bactericidal range of this compound is much wider, including LAB and several Gram-positive foodborne pathogens, e.g., *L. monocytogenes, S. aureus, C. botulinum*, or *C. sporogenes*.

4. Other Bacteriocins and Combined Treatments

Most other bacteriocins identified are interesting mainly from a food-quality point of view, since their bactericidal activity is directed towards closely related LAB only. The impact of this in the quality of starter cultures has been mentioned before. Bacteriocin-producing strains of *Lactobacillus helveticus* (producing helveticins and lactocins), *Lactobacillus acidophilus* (lactacins, acidophilucin), and *Lactobacillus plantarum* (plantaricins, plantacin) have been most extensively studied in this respect. With regard to food safety, the reported inhibition of *C. botulinum* and even the gram-negative *A. hydrophila* by plantacin BN–producing strains of *L. plantarum* is very noteworthy.

Several members of the genus *Carnobacterium*, a group of LAB that have been found in large numbers in chilled meat products, have been found to produce bacteriocins (carnocins) or bacteriocin-like compounds in relatively high amounts at chill temperatures, which would give them a favorable competitive edge over psychrotrophic foodborne pathogens and spoilage organisms. Again, the bactericidal range is restricted mainly to the closely related LAB, but inhibition of *L. monocytogenes* and *A. hydrophila* has been reported too.

Although Gram-negative bacteria, yeasts, and molds are in general not sensitive to the action of bacteriocins from LAB, the presence of chelating agents, surfactants, or osmotic shock (high salt) may sensitive them. A combined preservation scheme would be advantageous here, as shown by Stevens et al. [98] for several combination treatments with nisin that inactivate *Salmonella* and other Gram-negative species. Other reports highlighted the specific benefits of combining nisin with EDTA [15], citrate [14], lysozyme and citric acid [3], pediocin [45], and siderophores [52] in improving the inhibitory activity of nisin towards Gram-negatives or even in extending its inhibitory spectrum to cover Gram-negatives.

C. Natural Occurrence of Bacteriocin Producers

Lactic acid bacteria naturally occur on many different foods (e.g., fruits and vegetables) or are used often in their production. Much research in the recent past has been devoted to tracing bacteriocin-producing LAB in fresh and fermented food products. Although the methods used were not always standardized, it is generally accepted that only a very small number of isolates obtained from food products are able to produce bacteriocins and that the spectrum of inhibition is highly variable. Vaughan et al. [103], for instance, evaluated LAB isolated from cheese, milk, meat, fruits, and vegetables for bacteriocin production targeted at a number of spoilage and pathogenic bacteria. Approximately 1000 isolates from each of the food categories were tested to inhibit *S. Listeria innocua*, and *Pseudomonas fragi*. LAB isolated from cheese, milk, and meat samples inhibited *L. innocua* rather than the other target strains. LAB isolated from vegetable material generally inhibited *S. aureus*. The majority of active strains was effective against only one of the indicators, but a few were inhibitory to two or three of the target microorganisms. In our own laboratory, only 9 out of 890 isolates taken from fresh and modified-atmosphere–stored vegetables were found to be bacteriocinogenic [9,10]. This indicates that indeed only a small part of the total population of LAB has this ability. Investigations into the capacity of different strains of *L. lactis* subsp. *lactis* present in different culture collections throughout the world to produce nisin have shown that there is drastic difference in this respect even between isolates of the same subspecies.

D. Application of Bacteriocins and Bacteriocin-Producing Cultures

Bacteriocins can be applied to food systems by three basic methods [42]. First, a pure culture of the viable bacteriocin-producing LAB can be applied, which offers an indirect way to incorporate bacteriocins in a food product. The success of this type of application depends on the ability of the bacteriocin-producing LAB to grow and produce the bacteriocin to the required extent in the food under the prevailing environmental conditions (temperature, pH, etc.). Second, a (semi-)purified preparation of the bacteriocin can be employed. In this way, the dosage of the bacteriocin can be most accurate and thus its effect most predictable. However, application of (semi-)purified preparations is limited by national regulations concerning food additives. Finally, a crude bacteriocin preparation is obtained by growing the bacteriocin-producing LAB on a complex, natural substrate (e.g., milk). This mode avoids the use of a purified compound while still being able to use a preparation of known and constant activity. This latter method is now employed for the industrial-scale production of nisin preparations. A nisin-producing LAB is grown in milk whey at optimal temperature. During the course of incubation, nisin is expelled into the substrate. At a sufficiently high level, the substrate is pasteurized, which kills the bacteria but does not affect the heat-stable nisin.

A suitable application system for natural antimicrobials of microbial origin was recently developed for the control of *L. monocytogenes* on minimally processed vegetables [9,10]. This study set out with the concept that a suitable protective culture, in order to grow well and produce sufficient amounts of the bacteriocin, should be well adapted to the ecosystem it is used in and therefore might best be obtained from the food product considered. LAB occur on most if not all minimally processed vegetables, although they generally account for only about 1% of the natural microflora. To identify bacteriocin-producing LAB, 890 LAB isolates were obtained from different fresh and modified-atmosphere–stored vegetables and screened for their ability to produce bacteriocins [9,10]. Only nine isolates were found that could adequately control *L. monocytogenes* on artificial growth media. Three isolates were found to have the required characteristics: one strain of *Enterococcus mundtii* and two strains of *Pediococcus parvulus*. Both types produced a bacteriocin that effectively controlled growth of *L. monocytogenes* in in vitro studies [9,10]. Both pediococci, however, only produced significant amounts of bacteriocin at temperatures over 15°C and were not really suited for any application at lower temperature. The bacteriocin produced by both strains was fully identified and characterized and appeared to be identical to pediocin PA-1, formerly only known to be produced by *Pediococcus acidilactici* [9]. *E. mundtii* produced significant amounts of a bacteriocin even at 4–10°C [11]. Thus, although it is not a LAB and does not have GRAS status, the organism is very suitable to test whether biopreservation can be the required safety hurdle for certain psychrotrophic pathogens. On laboratory media—sterile vegetable extract in agar—the application of the mundticin producer as a protective culture at 8°C was very promising (Fig. 4a). However, on fresh, nonsterile produce, no activity was found. Either the production of mundticin on produce at low temperature is not sufficient or the mundticin is inactivated after production (enzymatic inactivation, adsorption to produce). Since the application of partially purified bacteriocin was found to significantly delay the growth of *L. monocytogenes*, the inactivation may not be the most prominent problem. Although the direct application of *E. mundtii* on mungbean sprouts was not effective in reducing the initial viable count of *L. monocytogenes* or its growth potential, a decline of 2 log units in the initial numbers was achieved when the produce was dipped in a solution of mundticin prior to contamination with the pathogen (Fig. 4b). Identical results were obtained when the product was treated with a mundticin-containing alginate film. The increase of the viable count of the pathogen after 5 days may, again, be attributed to proteolytic degradation and growth of part of the *Listeria* population that was not affected by the intact mundticin. Noteworthy is that the counts of the pathogen did not exceed the initial inoculation level for approximately 8 days. Thus, use of food-approved bacteriocins in a dipping solution or as part of an edible coating may have good potential as a biopreservative treatment for minimally processed vegetables.

V. LEGISLATORY ASPECTS

Existing food legislation in most countries would not favor the use of natural compounds purified from their natural source, unless these compounds have genuinely acquired GRAS status. The purification

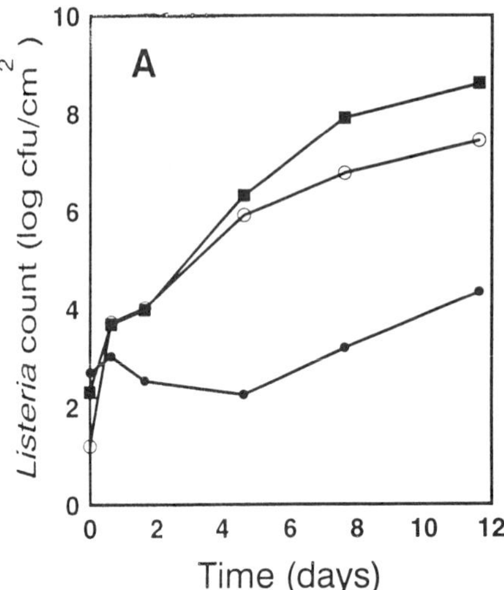

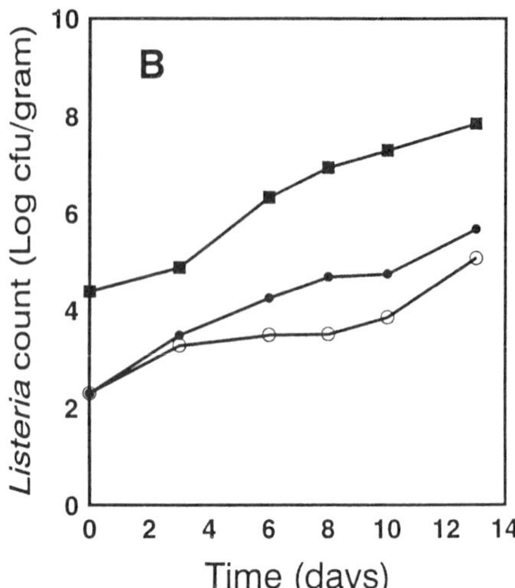

Figure 4 (A) Growth of *L. monocytogenes* (1:1 mix of strain LDCD681 and strain LDCD 1087) on vegetable agar medium in the absence of *E. mundtii* (■), co-cultured with the bacteriocinogenic *E. mundtii* (●), or co-cultured with the nonbacteriocinogenic *E. mundtti* DSM3848 (○), using initial *E. mundtti* levels of 10^6 cfu/cm^2. Incubations were performed at 8°C under a constant flow of 1.5% O_2 and 20% CO_2, balanced with N_2. (B) Growth of *L. monocytogenes* on mungbean spouts after treatment with purified bacteriocin of *E. mundtti*. The product was dipped in sterile water (■), dipped in sterile water containing 200 BU/ml of mundticin (○) or coated with an alginate film containing 200 BU/ml mundticin (●). Again, incubations were performed under 1.5% O_2/20% CO_2/78.5% N_2 at 8°C. (Modified from Ref. 10.)

process would bring green chemicals into the same category as synthetic chemical compounds, thus significantly lengthening the procedure for marketing approval and hampering economic implementation in practice. In fact, in most cases the legislative viewpoint on green chemicals or biopreservatives may be that they are new food additives or are applied for new purposes and consequently would require a nontoxicity record, despite their possible GRAS status. A more favorable form of application would thus be the inclusion of the spice or herb that contains the desired active ingredient or of the bacteriocin-producing strain in the food preparation because this still may be regarded as the most natural type of source.

The current regulatory status of bacteriocins and bacteriocin-producing organisms is a clear example of the current controversy between the use of the active compound or the natural source as a whole. In 1969, a joint FAO/WHO expert committee accepted nisin as a legal food additive, although it was not until 1988 that it was approved by the U.S. FDA for use in certain pasteurized cheese spreads. Presently nisin is permitted in at least 50 countries for the inhibition of clostridia in cheese and canned foods. None of the other bacteriocins known to date has a fully approved legal status as a food additive, although also the application of a pediocin-producing strain of *P. acidilactici* has been approved by the USDA for use in reduced nitrite bacon to aid in the prevention of botulinum toxin production by outgrowth of *C. botulinum* spores. The regulation of bacteriocin preparations from LAB stands in sharp contrast to the common use of these organisms as starter cultures. Moreover, LAB are commonly consumed in high numbers in fermented or cultured products and are often present as indigenous contaminants in many retail products. The general conception would be that the introduction of bacteriocins in foods at levels analogous to those capable of being produced by starter cultures should be as safe as the consumption of the cultured products themselves.

VI. FUTURE OUTLOOK

Food preservation by natural means has become a major challenge for food-manufacturing industries of all sizes and is dictated by the changes in consumer attitude in recent years towards chemical preservatives. All foods can be processed to extremes using physical methods that render them sterile and thus microbiologically safe. However, such foods would be unmarketable because consumers favor foods that are "natural" and "as-good-as-fresh" because they associate such products with a healthy diet. Current research trends in food microbiology and food technology focus on mild physical preservation techniques and the use of natural antimicrobial compounds.

Food preservatives of natural origin are generally considered as potential, safe sources of antimicrobials, but their effective use in practice has been established in only a few cases. Any antimicrobial extract or purified compound from a natural source will have to undergo tough toxicological scrutiny whenever its safe use is not guaranteed by well-documented data. Toxicological data for natural antimicrobials are often lacking and are as expensive to assemble as data for chemical compounds. The economy of changing from the range of still available synthetic chemicals to green chemicals will dictate whether commercialization is feasible at all. In many countries legislation has been passed to achieve significant reductions—in some cases even a total elimination—of chemical preservatives within the next decade. As may follow from the data briefly reviewed in this chapter, RTD institutions and food industries have identified a good number of possibilities for natural antimicrobials for future food preservation. However, successful marketing relies heavily on proper communication between industries, governments, and consumers. Negotiating marketing with legislatory bodies up to now has been mainly on a national level, which poses a major stumbling block to the introduction of natural alternatives. The cost for industries of obtaining legislatory approval for marketing are relatively high and to date have not specifically encouraged the search for natural alternatives. This increase in legislative pressure towards nonchemical strategies may favor their economic odds, but it could be even more helpful if the procedures of approval were accelerated on a worldwide scale in favor of natural antimicrobial compounds.

REFERENCES

1. T. Abee, L. Krockel, and C. Hill, Bacteriocins, modes of action and potentials in food preservation and control of food poisoning, *Int. J. Food Microbiol. 28*:169 (1995).
2. S. E. Aktug and M. Karapinar, Inhibition of foodborne pathogens by thymol, eugenol, menthol and anethole, *Int. J. Food Microbiol. 4*:161 (1987).
3. W. Anderson, Compositions having antibacterial properties and use of such compositions in suppressing growth of micro-organisms, European Patent 0466244A1 (1992).
4. I. Andersen, Bioprotective culture for fresh sausages, *Fleischwirtschaft 77*:635 (1997).
5. A. Arnoldi and L. Merlini, Lipophilicity-antifungal activity relationships for some isoflavonoid phytoalexins, *J. Agr. Food Chem. 38*:834 (1990).
6. M. A. Azzouz and L. B. Bullerman, Comparative antimycotic effects of selected herbs, spice plant components and commercial antifungal agents, *J. Food Prot. 45*:1298 (1982).
7. C. J. Baker, E. W. Orlandi, and N. M. Mock, Harpin, an elicitor of the hypersensitive response in tobacco caused by *Erwinia amylovora*, elicits active oxygen production in suspension cells, *Plant Physiol. 102*:1341 (1993).
8. J. G. Banks, R. G. Board, and N. H. C. Sparks, Natural antimicrobial systems and their potential in food preservation of the future, *Biotechnol. Appl. Biochem. 8*:103 (1986).
9. M. H. J. Bennik, E. J. Smid, and L. G. M. Gorris, Vegetable associated *Pediococcus parvulus* produces pediocin PA-1, *Appl. Environ. Microbiol. 63*:2074 (1997).
10. M. H. J. Bennik, W. Van Overbeek, E. J. Smid, and L. G. M. Gorris, Biopreservation for the control of *Listeria monocytogenes* on minimally processed, modified atmosphere stored vegetables. (submitted).
11. M. H. J. Bennik, A. Verheul, T. Abee, G. Naaktgeboren-Stoffels, L. G. M. Gorris, and E. J.

Smid, Interaction of nisin and pediocin PA-1 with closely related lactic acid bacteria that manifest over 100-fold differences in bacteriocin sensitivity, *Appl. Environ. Microbiol. 63*:3628 (1997).

12. E. D. Berry, R. W. Hutkins, and R. W. Mandigo, The use of a bacteriocin-producing *Pediococcus acidilactici* to control postprocessing *Listeria monocytogenes* contamination of frankfurters, *J. Food Prot. 54*:681 (1991).
13. L. R. Beuchat, Antimicrobial properties of spices and their essential oils, *Natural Antimicrobial Systems and Food Preservation* (V. M. Dillon and R. G. Board, eds.), CAB International, Wallingford, 1994, p. 167.
14. P. Blackburn, J. Plak, S. A. Gusik, and S. Rubino, Nisin composition for use as enhanced, broad range bactericides, U.S. patent WO 89/1239 (1989).
15. P. Blackburn, J. Polak, S. A. Gusik, and S. Rubino, Novel bacteriocin compositions for use as enhanced broad range bactericides and methods of prevention and treating microbial infection, U.S. Patent WO 90/09739 (1990).
16. H. Bohlmann and K. Apel, Thionins, *Ann. Rev. Plant Physiol. Plant Mol. Biol. 42*:227 (1991).
17. D. J. Bowles, Defense-related proteins in higher plants. *Ann. Rev. Biochem. 59*:873 (1990).
18. F. Breidt, K. A. Crowley, and H. P. Fleming, Controlling cabbage fermentations with nisin and nisin-resistant *Leuconostoc mesenteroides*, *Food Microbiol. 12*:109 (1995).
19. S. Buncic, S. M. Avery, and S. M. Moorhead, Insufficient antilisterial capacity of low inoculum *Lactobacillus* cultures on long term stored meats at 4°C, *Int. J. Food Microbiol. 34*:157 (1997).
20. D. R. L. Caccioni and M. Guizzardi, Inhibition of germination and growth of fruit and vegetable postharvest pathogenic fungi by essential oil components, *J. Essential Oil Res. 6*:173 (1994).
21. V. Carolissen-Mackay, G. Arendse, and J. W. Hastings, Purification of bacteriocins of lactic acid bacteria: problems and pointers, *Int. J. Food Microbiol. 34*:1 (1997).
22. G. Cerny and W. Hennlich. Minderung des Hygienerisikos bei Feinkostsalaten durch Schutzkulturen. Teil II: Kartoffelsalat, *Int. Z. Lebensm. Technol. Verfahrenstech. (ZFL) 42*:1-2, 12 (1991).
23. G. Cerny, Einsatz von Schutzkulturen zur Minderung des Hygienerisikos bei Lerbensmitteln, *Lebensmitteltechnik 23*:448, 450 (1991).
24. H. G. Cutler, Natural product flavour compounds as potential antimicrobials, insecticides, and medicinals, *Agro-Food-Industry-Hi-Tech. 6*:19 (1995).
25. M. A. Daeschel, Applications of bacteriocins in food systems, *Biotechnology and Food Safety* (D. D. Bills, and S. D. Kung, eds.), Butterworth-Heinemann, Boston, 1990, p. 91.
26. A. G. Darvill and P. Albersheim, Phytoalexins and their elicitors: a defence against microbial infection in plants, *Ann. Rev. Plant Physiol. 35*:243 (1984).
27. S. G. Deans and G. Ritchie, Antibacterial properties of plant essential oils, *Int. J. Food Microbiol. 5*:165 (1987).
28. S. G. Deans and K. P. Svoboda, Biotechnology and bioactivity of culinary and medicinal plants, *AgBiotechnol. News Inform. 2*:211 (1990).
29. A. J. Degnan, A. E. Yousef, and J. B. Luchansky, Use of *Pediococcus acidilactici* to control *Listeria monocytogenes* in temperature-abused vacuum-packaged wieners, *J. Food Prot. 55*:98 (1992).
30. J. Delvse-Broughton and M. J. Gasson, Nisin, *Natural Antimicrobial Systems and Food Preservation* (V. M. Dillon, and R. G. Board, eds.), CAB International, Wallingford, 1994, p. 99.
31. J. Delves-Broughton, P. Blackburn, R. J. Evans, and J. Hugenholtz, Applications of the bacteriocin nisin, *Antonie van Leeuwenhoek 69*:193 (1996).
32. R. A. Dixon, P. M. Dey, and C. J. Lamb, Phytoalexins: enzymology and molecular biology, *Adv. Enzymol. 55*:1 (1983).
33. M. Drdak, A. Rajniakova, and V. Buchtova. Mustard phytoncides utilization in the process of food preservation, Bioavailability '93, Nutritional, chemical and food processing implications of nutrient availability, Fed. Eur. Chem. Soc., 1993, p. 457.
34. H. Dornenburg and D. Knorr, Elicitation of chitinases and anthraquinones in *Morinda citrifolia* cell cultures, *Food Biotechnol. 8*:57 (1994).

35. H. Dornenburg and D. Knorr, Generation of colors and flavors in plant cell and tissue cultures, *Crit. Rev. Plant Sci. 15*(2):141 (1996).
36. H. Einarsson and H. Lauzon, Biopreservation of brined shrimp (*Pandalus boreatis*) by bacteriocins from lactic acid bacteria. *Appl. Environ. Microbiol. 61*:669 (1995).
37. R. S. Farag, Z. Y. Daw, F. M. Hewedi, and G. S. A. El-Baroty, Antimicrobial activity of some Egyptian spice essential oils, *J. Food Prot. 52*:665 (1989).
38. J. Giese, K. T. Chung, and C. A. Murdock, Natural systems for preventing contamination and growth of microorganisms in foods, *Food Technol. 10*:361 (1991).
39. R. Geisen and W. H. Holzapfel, Genetically modified starter and protective cultures. *Int. J. Food Microbiol. 30*:315 (1996).
40. G. Giraffa, Lactic and no-lactic acid bacteria as a tool for improving the safety of dairy products, *Ind. Aliment. 35*:244–248, 252 (1996).
41. G. Giraffa, N. Picchioni, E. Neviani, and D. Carminati, Production and stability of an *Enterococcus faecium* bacteriocin during taleggio cheesemaking and ripening, *Food Microbiol. 12*:301 (1995).
42. L. G. M. Gorris and M. H. J. Bennik, Bacteriocins for food preservation, *Int. Z. Lebensm. Technol. Verfahrenstech. (ZFL) 45*(11):65 (1994).
43. G. W. Gould, Industry perspectives on the use of natural antimicrobials and inhibitors for food applications. *J. Food Prot. (suppl.)*:82–86 (1996).
44. R. J. Grayer and J. B. Harborne, A survey of antifungal compounds from higher plant, 1982-1993, *Phytochemistry 37*:19 (1994).
45. M. B. Hanlin, N. Kachayanand, P. Ray, and B. Ray, Bacteriocins of lactic acid bacteria in combination have greater antibacterial activity, *J. Food Prot. 56*:252 (1993).
46. J. B. Harborne, The role of phytoalexins in natural plant resistance, *Natural Resistance of Plants to Pests* (M. B. Green and P. A. Hedin, eds.), American Chemical Society, Washington, DC, 1986, p. 22.
47. L. L. Hargreaves, B. Jarvis, A. P. Rawlinson, and J. M. Wood, The antimicrobial effects of spices, herbs and extracts from these and other food plants, The British Food Manufacturing Industries Research Association, Scientific and Technical Surveys no. 88, 1975.
48. L. J. Harris, H. P. Fleming, and T. R. Klaenhammer, Developments in nisin research, *Food Res. Int. 25*:57 (1992).
49. L. J. Harris, H. P. Fleming, and T. R. Klaenhammer, Novel paired starter culture system for sauerkraut, consisting of a nisin-resistant *Leuconostoc mesenteroides* strain and a nisin-producing *Lactococcus lactis* strain, *Appl. Environ. Microbiol. 58*:1484 (1992).
50. K. J. Hartmans, P. Diepenhorst, W. Bakker, and L. G. M. Gorris, The use of carvone in agriculture: sprout suppression of potatoes and antifungal activity against potato tuber and other plant diseases, *Ind. Crops Prod. 4*:3 (1995).
51. S. Y. He, H. C. Huang, and A. Collmer, *Pseudomonas syringae* pv *syringae* harpin (PSS)—a protein that is secreted via the HRP pathway and elicits the hypersensitive response in plants, *Cell 73*:1255 (1993).
52. I. M. Helander, A. Von Wright, and T. M. Matilla-Sandholm, Potential of lactic acid bacteria and novel antimicrobials against gram-negative bacteria, *Trends Food Sci. Technol. 8*:146 (1997).
53. W. Hennlich and G. Cerny, Minderung des Hygienerisikos bei Feinkostsalaten durch Schutzkulturen. Teil I: Fleischsalat, *Int. Z. Lebensm. Technol. Verfahrenstech. (ZFL) 41*:806 (1990).
54. W. Hennlich, Sicherer Hygieneschutz. Leistungsanforderungen an Schutzkulturen in Feinkostsalaten, *Lebensmitteltechnik 27*:51 (1995).
55. W. H. Holzapfel, R. Geisen, and U. Schillinger, Biological preservation of foods with reference to protective cultures, bacteriocins and food-grade enzymes, *Int. J. Food Microbiol. 24*:343 (1995).
56. H. H. Huss, V. F. Jeppesen, C. Johansen, and L. Gram, Biopreservation of fish products - a review of recent approaches and results, *J. Aquatic Food Prod. Technol. 4*:5, 37 (1995).
57. M. T. Hutton, P. A. Chehak, and J. H. Hanlin, Inhibition of botulinum toxin production by *Pediococcus acidilactici* in temperature abused refrigerated foods, *J. Food Safety 11*:255 (1991).

58. J. L. Ingham, Disease resistance in plants: the concept of pre-infectional and post-infectional resistance, *Phytopathol. Z. 78*:314 (1973).
59. P. M. Jenner, E. C. Hagan, J. M. Taylor, E. L. Cook, and O. G. Fitzhugh, Food flavourings and compounds of related structure. I. Acute oral toxicity. *Food Cosmet. Toxicol. 2*:327 (1964).
60. J. J. Kabara and T. Eklund, Organic acids and esters, *Food Preservatives* (N. J. Russel and G. W. Gould, eds.), Blackie Academic, London, 1991, p. 200.
61. W. J. Kim, Bacteriocins of lactic acid bacteria - their potentials as food biopreservative, *Food Rev. Int. 8*:299 (1993).
62. M. Kivanc and A. Akgul, Mould growth on black table olives and prevention by sorbic acid, methyl-eugenol and spice essential oil, *Nahrung 34*:369 (1990).
63. H. Knauf, Starterkulturen fuer die Herstellung von Rohwurst und Rohpoekelwaren: Potential, Auswahlkriterien und Beeinfluessungsmoeglichkeiten, *Fleischerei 46*:4, 6, 8 (1995).
64. P. Kotzekidou and J. G. Bloukas, Effect of protective cultures and packaging film permeability on shelf life of slice vacuum packed cook ham, *Meat Sci. 42*:333 (1996).
65. L. Kroeckel and U. Schmidt, Hemmung von *Listeria monocytogenes* in vakuum-verpacktem Bruehwurstaufschnitt durch bacteriocinogene Schutzkulturen, *Mitteilungsbl. Bundesanst. Fleischforsch. 33*:428 (1994).
66. L. Leistner and L. G. M. Gorris, Food preservation by hurdle technology, *Trends Food Sci. Technol. 6*(1):41 (1995).
67. J. Luchansky, Genomic analysis of *Pediococcus* starter cultures used to control *Listeria monocytogenes* in turkey summer sausage, *Appl. Environ. Microbiol. 58*:3053 (1992).
68. B. M. Lund and G. D. Lyon, Detection of inhibitors of *Erwinia carotovora* and *E. herbicola* on thin-layer chromatograms, *J. Chromatogr. 110*:193 (1975).
69. G. D. Lyon, Attenuation by divalent cations of the effect of the phytoalexin rishitin on *Erwinia carotovora* var. *atroseptica*, *J. Gen. Microbiol. 109*:5 (1978).
70. A. L. E. Mahmoud, Antifungal action and antiaflatoxigenic properties of some essential oil constituents, *Lett. Appl. Microbiol. 19*:110 (1994).
71. N. Nakatani, Antioxidative and antimicrobial constituents of herbs and spices, *Spices, Herbs and Edible Fungi* (G. Charalambous, ed.), Elsevier Science, Amsterdam, 1994, p. 251.
72. G. J. E. Nychas, Natural antimicrobials from plants, *New Methods of Food Preservation* (G. W. Gould, ed.), Blackie Academic, London, 1995, p. 58.
73. K. Oosterhaven, B. Poolman, and E. J. Smid, S-carvone as a natural potato sprout inhibiting, fungistatic and bacteriostatic compound, *Ind. Crops Prod. 4*:23 (1995).
74. N. Paster, B. J. Juven, and H. Harshemesh, Antimicrobial activity and inhibition of aflatoxin B1 formation by olive plant tissue constituents, *J. Appl. Bacteriol. 64*:293 (1988).
75. A. Pauli and K. Knobloch, Inhibitory effects of essential oil components on growth of food-contaminating fungi, *Z. Lebensm. Unters. Forsch. 184*:10 (1987).
76. B. Ray and M. A. Daeschel, Bacteriocins of starter culture bacteria, *Natural Antimicrobial Systems and Food Preservation* (V. M. Dillon, and R. G. Board, eds.), CAB International, Wallingford, 1994, p. 99.
77. K. E. Ricker and R. M. Bostock, Evidence for release of the elicitor arachidonic acid and its metabolites from sporangia of *Phytophthora infestans* during infection of potato, *Physiol. Mol. Plant Pathol. 41*:61 (1992).
78. W. K. Roberts and C. P. Selitrennikoff, Plant and bacterial chitinases differ in antifungal activity, *J. Gen. Microbiol. 134*:169 (1988).
79. A. Roco, P. Castaneda, and L. M. Perez, Oligosaccharides released by pectinase treatment of citrus limon seedlings are elicitors of the plant response, *Phytochemistry 33*:1301 (1993).
80. S. Roller, The quest for natural antimicrobials as novel means of food preservation, *Int. Biodeterior. Biodegrad. 36*:333 (1995).
81. M. Rozbeh, N. Kalchayanand, R. A. Field, M. C. Johnson, and B. Ray, The influence of biopreservatives on the bacterial level of refrigerated vacuum packaged beef, *J. Food Safety 13*:99 (1993).
82. J. L. Ruiz-Barba, D. P. Cathcart, P. L. Warner, and R. Jimenez-Diaz, Use of *Lactobacillus*

plantarum LPCO10, a bacteriocin producer, as a starter culture in spanish-style green olive fermentations, *Appl. Environ. Microbiol. 60*:2059 (1994).
83. J. Salmeron, R. Jordano, and R. Pozo, Antimycotic and antiaflatoxigenic activity of oregano (*Origanum vulgare* L.) and thyme (*Thymus vulgaris* L.), *J. Food Prot. 53*:697 (1990).
84. U. Schillinger, Bacteriocins of lactic acid bacteria, *Biotechnology and Food Safety* (D. D. Bills and S. D. Kung, eds.), Butterworth-Heinemann, Boston 1990, p. 54.
85. U. Schillinger and F. K. Luecke, Lactic acid bacteria as protective cultures in meat products, *Fleischwirtschaft 70*:1296 (1990).
86. U. Schillinger, M. Kaya, and F. K. Luecke, Behaviour of *Listeria monocytogenes* in meat and its control by a bacteriocin-producing strain of *Lactobacillus sake*, *Appl. Bacteriol. 70*:473 (1991).
87. U. Schillinger, R. Geisen, and W. H. Holzapfel, Potential of antagonistic microorganisms and bacteriocins for the biological preservation of foods, *Trends Food Sci. Technol. 7*:158 (1996).
88. A. Schlumbaum, F. Mauch, U. Vögeli, and T. Boller, Plant chitinases are potent inhibitors of fungal growth, *Nature 324*:365 (1986).
89. J. Sikkema, J. A. M. De Bont, and B. Poolman, Interactions of cyclic hydrocarbons with biological membranes, *J. Biol. Chem. 269*:8022 (1994).
90. J. Sikkema, J. A. M. De Bont, and B. Poolman, Mechanisms of membrane toxicity of hydrocarbons, *Microbiol. Rev. 59*:201 (1995).
91. J. Sikkema, B. Poolman, W. N. Konings, and J. A. M. De Bont, Effects of the membrane action of tetralin on the functional and structural properties of artificial and bacterial membranes, *J. Bacteriol. 174*:2986 (1992).
92. S. A. Simon, W. L. Stone, and P. B. Bennett, Can regular solution theory be applied to lipid bilayer membranes?, *Biochim. Biophys. Acta 550*:38 (1979).
93. E. J. Smid, Y. De Witte, and L. G. M. Gorris, Secondary plant metabolites as control agents of postharvest *Penicillium* rot on tulip bulbs, *Postharvest Biol. Technol. 6*:303 (1995).
94. E. J. Smid, L. Hendriks, H. A. M. Boerrigter, and L. G. M. Gorris, Surface disinfection of tomatoes using the natural plant compound *trans*-cinnamaldehyde, *Postharvest Biol. Technol. 9*:343 (1996).
95. E. J. Smid, J. P. G. Koeken, and L. G. M. Gorris, Fungicidal and fungistatic action of the secondary plant metabolites cinnamaldehyde and carvone, *Modern Fungicides and Antifungal Compounds* (H. Lyr. P. E. Russell, and H. D. Sisler, eds), Intercept Ltd., Andover, Hants, 1996, p. 173.
96. D. A. Smith and S. W. Banks, Biosynthesis, elicitation and biological activity of isoflavonoid phytoalexins, *Phytochemistry 25*:979 (1986).
97. M. L. Stecchini, V. Aquili, and I. Sarais, Behavior of *Listeria monocytogenes* in mozzarella cheese in presence of *Lactococcus lactis*, *Int. J. Food Micriobiol. 25*:301 (1995).
98. K. A. Stevens, B. W. Sheldon, N. A. Klapes, and T. R. Klaenhammer, Nisin treatment for inactivation of *Salmonella* species and other gram-negative bacteria, *Appl. Environ. Microbiol. 58*:3613 (1991).
99. M. E. Stiles, Biopreservation by lactic acid bacteria, *Antonie van Leeuwenhoek 70*:331 (1996).
100. A. Stintzi, T. Heitz, V. Prasad, S. Wiedemann-Merdinoglu, S. Kauffmann, P. Geoffroy, M. Legrand, and B. Fritig, Plant pathogenesis-related proteins and their role in defense against pathogens, *Biochimie 75*:687 (1993).
101. C. C. Tassou and G. J. E. Nychas, Antimicrobial activity of the essential oil of mastic gum (*Pistacia lentiscus* var. chia) on gram positive and gram negative bacteria in broth and in model food system, *Int. Biodeterior. Biodegrad. 36*:411 (1995).
102. H. D. Van Etten, Antifungal activity of pterocarpans and other selected isoflavonoids, *Phytochemistry 15*:655 (1976).
103. E. E. Vaughan, E. Caplice, R. Looney, N. O'Rourke, H. Coveney, C. Daly, and G. F. Fitzgerald, Isolation from food sources, of lactic acid bacteria that produced antimicrobials, *J. Appl. Bacteriol. 76*:118 (1994).
104. M. Vescovo, C. Orsi, G. Scolari, and S. Torriani, Inhibitory effect of selected lactic acid bacteria on microflora associated with ready-to-use vegetables, *Lett. Appl. Microbiol. 21*:121 (1995).

105. A. J. Vigers, W. K. Roberts, and C. P. Selitrennikoff, A new family of plant antifungal proteins, *Mol. Plant-Microbe Interact. 4*:315 (1991).
106. J. R. L. Walker, Antimicrobial compounds in food plants, *Natural Antimicrobial Systems and Food Preservation*, V. M. Dillon, and R. G. Board, eds.), CAB International, Wallingford, 1994, p. 181.
107. H. M. Ward, Recent researches on parasitism of fungi, *Ann. Bot. 19*:1 (1905).
108. H. Weber, Dry sausage manufacture. The importance of protective cultures and their metabolic products, *Fleischwirtschaft 74*:278 (1994).
109. Z. M. Wei, R. J. Laby, C. H. Zumoff, D. W. Bauer, S. Y. He, A. Collmer, and S. V. Beer, Harpin elicitor of the hypersensitive response produced by the plant pathogen, *Erwinia amylovora*, *Sci. 257*:85 (1992).
110. M. Weidenbörner and H. C. Jha, Antifungal activity of flavonoids and their mixtures against different fungi occurring on grain, *Pesticide Sci. 38*:347 (1993).
111. S. Wessels and H. H. Huss, Suitability of *Lactococcus lactis* subsp. *lactis* ATCC 11454 as a protective culture for lightly preserved fish products, *Food Microbiol. 13*:323 (1996).
112. K. M. Wilkins and R. G. Board, Natural antimicrobial systems, *Mechanisms of Action of Food Preservation Procedures* (G. W. Gould, ed.), Elsevier Applied Science, London, 1989, p. 285.
113. C. A. Williams and J. B. Harborne, Isoflavonoids, *Methods in Plant Biochemistry* Vol. I. *Plant Phenolics* (J. B. Harborne, ed.), Academic Press, London, 1989, p. 421.
114. L. Wimmer, Natuerliche Haltbarkeit, *Fleischerei 47*:12 (1996).
115. E. A. Zottola, T. L. Yezzi, D. B. Ajao, and R. F. Roberts, Utilization of cheddar cheese containing nisin as an antimicrobial agent in other foods, *Int. J. Food Microbiol. 24*:227 (1994).

10

Antioxidants in Food Preservation

JAN POKORNÝ
Prague Institute of Chemical Technology, Prague, Czechia

I. RANCIDITY OF FATS, OILS, AND FATTY FOODS

Rancidity is a objectionable defect in food quality. Fats, oils, or fatty foods are deemed rancid if a significant deterioration of the sensory quality is perceived (particularly aroma or flavor, but appearance and texture may also be affected, e.g., in fish). Rancidity includes several types of changes, but most often the degradation due to changes in lipid constituents is considered the main feature of rancidification.

A. Types of Rancidity

Several types of rancidity exist in fats, oils, and fatty foods, not always connected with oxidation reactions. The most important types are listed in Table 1.

Lipolytic rancidity is mainly due to lipases (triacylglycerol-acyl hydrolases, EC 3.1.1.3), which catalyze the cleavage of triacylglycerols (triglycerides) into free fatty acids and partial glycerol esters (monoacylglycerols or monoglycerides, diacylglycerols or diglycerides). In most fats and oils, the presence of free fatty acids is not perceptible by human senses, therefore not considered as flavor deterioration. Milk fats [1] are an exception, as they contain esters of butyric acid; free butyric acid, produced by lipolysis, imparts a typical disagreeable off-flavor, resembling rancid butter. In oils such as coconut oil or palm kernel oils, caproic, caprylic, and capric acids are released by lipolysis, which results in a soapy off-flavor. Such deterioration is frequently observed in stored food products containing coconut. The soapy flavor is due not to soaps, but to all derivatives (free fatty acids, soaps, or methyl esters) having a hydrocarbon chain of 6–10 carbon atoms.

Oxidative rancidity, the most important type of food rancidity, discussed in Sec. I.B in more detail. Chapter 11 will discuss oxidative rancidity exclusively.

Flavor reversion is a type of rancidity, typical for soybean oil, that is connected with minute absorption of oxygen. It appears during storage of fully refined, blended soybean oils and imparts a "beany" off-flavor to the product. It results from specific oxidation products, probably particular oxygen-substituted fatty acids (containing a furan group), present in soybean oil.

Ketonic rancidity, with its characteristic floral off-flavor, is often observed during storage of foods containing short- or medium-chain fatty acids (4–10 carbon atoms), such as those of milk fat or coconut fat [2]. It is caused by microbial degradation of medium-chain fatty acids into their re-

Table 1 Types of Rancidity Occurring in Fats, Oils, and Fatty Foods

Type of rancidity	Main substances producing rancidity	Type of chemical reaction	Materials subject to the type of rancidity
Lipolytic	Lower fatty acids, medium-chain fatty acids	Enzymic hydrolysis	Milk fat, palm-seed oils
Oxidative	Lower aldehydes and ketones	Autoxidation, enzymic oxidation	Polyunsaturated edible oils
Flavor reversion	Oxygen-substituted cleavage and rearrangement products	Oxidation, cleavage, and rearrangement	Soybean oil
Ketonic	2-Alkanones (methyl ketones)	β-Oxidation and enzymic decarboxylation	Milk fat, palm-seed oils

spective 2-alkanones [Fig. 1, Eq. (1)]. Methyl ketones (2-alkanones) impart floral, fruity, or blue cheese flavor notes to rancid foods.

B. Mechanism of Oxidative Rancidity

The sensory perception of rancidity is due to the presence of volatile substances possessing 3–12 carbon atoms; several classes of compounds are active in producing rancidity, such as aldehydes, ketones, alcohols, or even hydrocarbons. Generally, unsaturated derivatives are more active than the respective saturated derivatives. The mechanism of their production is not relevant, but usually oxidation products are more active than products formed as a result of other types of reaction [3].

Oxidative rancidity (also called autoxidation) is a free radical chain reaction consisting of three main phases: initiation, propagation, and termination [4,5]. During autoxidation, all unsaturated fatty acids bound in lipids are slowly oxidized. Polyunsaturated fatty acids are the least stable components, being easily attacked by air oxygen. At higher temperatures, saturated fatty acids are oxidized as well. The sensitivities of other substances containing a hydrocarbon chain (e.g., terpenes, such as carotenoids, sterols, and lower terpenes, present in essential oils) are often underestimated.

1. Initiation Reactions

During the oxidation of polyenoic (essential) fatty acids, the methylene group adjacent to two double bonds ($-CH=CH-CH_2-CH=CH-$), as is present in linoleic or linolenic acids, is the primary site of oxygen attack [5]. It is easily converted into the respective free radical—$R-H \Rightarrow R^* + H^*$. In monoenoic fatty acids, free radicals are formed by cleavage of a hydrogen atom on either side of the double bond. Most often, free radicals are formed by cleavage of a hydroperoxide molecule [Fig. 1, Eq. (5)]. Hydroperoxides are usually present in trace quantities in raw materials, where they are produced by singlet-oxygen or enzyme-catalyzed oxidation.

2. Propagation and Termination Reactions (Primary Reactions)

The unsaturated fatty acid radical easily absorbs a molecule of oxygen, forming a peroxy radical [Fig. 1, Eq. (2)]. The peroxy radical, being very reactive, abstracts a hydrogen atom from another molecule of a polyunsaturated fatty acid, forming a hydroperoxide and an alkyl free radical [Fig. 1, Eq. (3)]. The hydroperoxide molecule is easily cleaved with formation of a free radical [Fig. 1, Eq. (4)]. This

$$\text{R-CH}_2\text{-CH}_2\text{-COOH} \Rightarrow \underset{\text{OOH}}{\text{R-CH-CH}_2\text{-COOH}} \Rightarrow \text{R-CO-CH}_3 + \text{CO}_2 \quad [1]$$

$$\text{R*} + \text{O}_2 \Rightarrow \text{R-O-O*} \quad [2]$$

$$\text{R-O-O*} + \text{R-H} \Rightarrow \text{R-O-OH} + \text{R*} \quad [3]$$

$$\text{R-O-O-H} \Rightarrow \text{R-O-O*} + \text{H*} \text{ or } \text{R-O*} + \text{H-O*}$$

$$\text{or: 2 R-O-O-H} \Rightarrow \text{R-O-O*} + \text{R-O*} + \text{H}_2\text{O} \quad [4]$$

$$\text{-CH=CH-C*-CH=CH-} + \text{O}_2 \Rightarrow \underset{\text{OOH}}{\text{-CH}}\text{-CH=CH-CH-CH=CH- or} \quad [5]$$

$$\text{-CH=CH-CH=CH-}\underset{\text{OOH}}{\text{CH}}\text{-}$$

$$\text{2 R* or R* and R-O-O* or 2 R-O-O*} \Rightarrow \text{relatively stable products} \quad [6]$$

$$\text{R-O-O-H} + \text{Me}^{2+} \Rightarrow \text{R-O-O*} + \text{Me}^{+} + \text{H}^{+} \quad [7]$$

$$\text{R-O-O-H} + \text{Me}^{+} \Rightarrow \text{R-O*} + \text{H-O}^{-} + \text{Me}^{2+} \quad [8]$$

FIGURE 1 Mechanism of lipids rancidification.

reaction sequence can be repeated many times, therefore, it is called a chain reaction. During this reaction, the double-bond system of the original polyunsaturated fatty acid is usually isomerized into a conjugated dienoic system [Fig. 1, Eq. (5)]. The chain may be interrupted by the recombination of two free radicals [Fig. 1, Eq. (6)].

3. Metal Catalysis of Peroxidation in Polyunsaturated Lipids

The decomposition of hydroperoxides in autoxidizing lipids is catalyzed by transient-valency metal ions [6] [Fig. 1, Eq. (7)]. Two free radicals are produced from each hydroperoxide, which initiate further reaction chains. The lower-valency metal ion regenerates by reaction with another molecule of hydroperoxide [Fig. 1, Eq. (8)]. Therefore, minute traces of copper and iron and, to lesser degree, manganese and cobalt are important promotors of lipid oxidation.

4. Enzymic Lipid Oxidation

Various enzymes, mainly lipoxygenases (linoleate oxidoreductases, E. C. 1.13.11.12), catalyze the oxidation of linoleic, linolenic, and related essential fatty acids [7]. The products differ from those formed in the course of autoxidation by their stereoselectivity and positional selectivity. Several isoenzymes are usually present [8]. Lipoxygenases are mainly accompanied by the respective lyases, which cleave hydroperoxides into different low molecular weight compounds.

5. Secondary Reactions

Hydroperoxides, which are the most important primary reaction products, may decompose, even at room temperature, forming free radicals [Fig. 1, Eq. (4)], which initiate a further oxidative reaction chain. Therefore, it is advisable to add substances that will stabilize hydroperoxides to food material.

Hydroperoxides of polyunsaturated fatty acids have no particular odor or flavor, but they are easily cleaved at the adjacent double bond, resulting in the formation of various volatile low molecular weight (3–12 carbon atoms) aldehydes, hydrocarbons, alcohols, or ketones [9]. Only these secondary oxidation products impart the rancid off-flavor to oxidized fats and oils. They are easily oxidized in turn, giving rise to low molecular weight fatty acids and other tertiary reaction products. Lipid polymers, mainly dimers, are produced as well as cleavage products, but they do not substantially affect the sensory value.

C. Measurement of Oxidative Rancidity

Oxidative rancidity is very complex, therefore, several analyses of different reaction products are usually needed to identify and measure the degree of rancidity.

1. Sensory Methods

Rancidity is a phenomenon of perception, so that only psychometric methods are suitable for its measurement. Therefore, sensory analysis methods are fundamental for analysis [10]. The degree of rancidity is the final goal of analysis, but usually sensory profile methods are used to estimate all flavor descriptors responsible for the final perception of rancidity. Even with expert panels of assessors, the analysis should be repeated 10–20 times to obtain reliable results.

2. Chemical Methods

Lipid hydroperoxides, which are the primary reaction products, are easily determined iodometrically, but the peroxide value has no simple relation to the actual rancidity because hydroperoxides are odorless and tasteless [11]. The benzidine or (better) *p*-anisidine value measures the browning reactions of the respective aromatic amines with aldehydic oxidation products; they react with aldehydes or ketones, and the value depends on the degree of unsaturation as well as on other factors. The 2-thiobarbituric acid value, determined at 530 nm, is the measure of the condensation product with malonaldehyde or with a hydroperoxide produced by oxidation of 3-alkenals or 2,4-alkadienals. If the condensation products are measured at 450 nm, the absorbance is correlated with total aldehydic oxidation products, which correlate better with rancidity than the substances reactive with formation of red products [12]. Both of these methods are easy, rapid, and cheap, however, their specificity is questionable. Several other spectrophotometric methods were reported in the literature, mostly based on a reaction with aldehydes, e.g., the conversion of aldehydes into 2,4-dinitrophenylhydrazones and the determination of color intensity.

3. Chromatographic Methods

The best method for the measurement of rancidity is gas-liquid chromatographic (GLC) analysis of volatiles, isolated by the dynamic headspace analysis or other procedures [13]. The contents of propane, hexane, hexanal, 2,4-decadienals, and other products are correlated with the degree of rancidity. Nonvolatile products may be determined with reverse-phase high-performance liquid chromatography (HPLC) and polymers with high-performance size-exclusion chromatography (HPSEC).

D. Nutritional and Sensory Aspects of Oxidative Rancidity

1. Effects on Nutritional Value

The least resistant components of lipids are essential fatty acids, which not only lose their physiological activity by autoxidation, but they may turn into antinutritive agents when oxidized. Free radicals formed as intermediary products may initiate the development of cardiovascular diseases or cancer in vivo (see Sec. III.C). Lipid hydroperoxides decompose liposoluble vitamins, such as vitamin E, vitamin C, vitamin A or its provitamins – carotenes [14]. Lipid oxidation products possess direct toxicity as well, cyclic dimers and hydroperoxides of aldehydic oxidation products being particularly toxic. Hydroperoxides and ketones react with primary amine groups of proteins, decreasing thus the biological value [15]. Similarly, they react with amino acids and vitamins of the B group.

2. Effects on Sensory Value

As already stated, rancidity is a process resulting in changes of the sensory value of fats, oils, and other foods. Very minute rancidity is not objectionable to most consumers; on the contrary, it makes the flavor richer and thus more acceptable. Larger amounts of secondary oxidation products, however, cause negative consumer response that may lead to the rejection of rancid food. Virgin olive oil or fried foods are typical examples of products where minute amounts of oxidation are desirable but higher amounts are objectionable. Rancidity is a very complex phenomenon, and rancid foods may be described as, for example, old, warmed-over, cardboard, wet dog, or dumpy. Small amounts of oxidation products of polyunsaturated fatty acids impart fried flavor to foods [16]. Higher amounts of the same compounds produce the perception of burned flavor.

E. Reduction of Rancidity in Foods

Rancidity may be limited by decreasing the storage temperature, the access of oxygen, and the degree of unsaturation of the lipid fraction (see Sec. VI). When application of the above methods is not possible or is not satisfactory, the best way to control rancidity is with the addition of antioxidants (see below).

F. Sources of Food Oxidants

Fats, oils and related compounds, which turn rancid on oxidation, are mostly oxidized by air oxygen, which penetrates foods and is dissolved in both aqueous and lipid phases. The reaction may be catalyzed by enzymes, but the oxidant is still air oxygen. Other oxidants (Table 2) are of minor importance. In the presence of photosensibilizers (such as chlorophylls) and in the light, ordinary triplet oxygen is converted into single oxygen, which is 100–300 times more reactive [17]. A singlet oxygen molecule is added to a double bond of unsaturated lipids, and the intermediary unstable product is rapidly isomerized into a hydroperoxide. Hydrogen peroxide is easily cleaved, resulting in formation of free radicals, which readily oxidize unsaturated fatty acids.

TABLE 2 Oxidants in Food Products

Oxidant	Importance	Occurrence
Air (triplet) oxygen (autoxidation)	Most important in processed and stored foods	General
Enzymatically catalyzed oxidation	Stored raw materials	Oilseeds, nuts, cereals, legumes
Singlet oxygen	In the light in presence of photosensitizers	Edible oils, green foods
Ozone	Very low in foods	Essential oils
Quinones	In foods subject to enzymic browning	Fruits, vegetables, potatoes
Metals	Initiation of free radical oxidation	Meat, fruits
Superoxide anion	Mainly in in vivo systems	Meat
Hydrogen peroxide	In presence of ascorbic acid	Fruits, vegetables
Lipid hydroperoxides	In presence of polyunsaturated acids and carotenoids	Fruits, vegetables, fatty foods

III. REDUCTION AND CONTROL OF RANCIDITY IN FOODS

In the preindustrial period, food materials were used for immediate food production, and the products were usually consumed within a few hours or a day. The problem of rancidity was less important then than now, when food products are often stored for days or months before consumption. Therefore, processes deteriorating food during storage are substantially more important.

A. Why Antioxidants Are Necessary

The autoxidation of lipids is initiated by free radicals; hydroperoxides produced by autoxidation are decomposed, producing free radicals, which initiate further oxidation reactions. Therefore, autoxidation is called an autocatalytic reaction. In the beginning, the concentration of free radicals is very low, and oxidation is slow. Gradually, the concentration of hydroperoxides and other oxidation products increases, the concentration of free radicals formed during their decomposition increases as well, and thus the overall oxidation rate increases (Fig. 2). The storage of fat-containing food materials is limited by the period of slow oxidation.

The stage of very slow oxidation in the beginning of storage is called the induction period. The induction period may be prolonged by addition of antioxidants, which are not able to entirely eliminate the oxidation reactions even when they are active in prolonging the storage time.

B. Definitions of Antioxidants and Antioxidant Types

In a broad sense, antioxidants are all substances that can protect materials (not only foods) against autoxidation, irrespective of the mechanism of action. More exactly, such compounds should be called inhibitors of oxidation, and only substances that inhibit oxidation by reaction with free radicals should be called antioxidants [18]. Antioxidants may also inhibit the decomposition of lipid hydroperoxides,

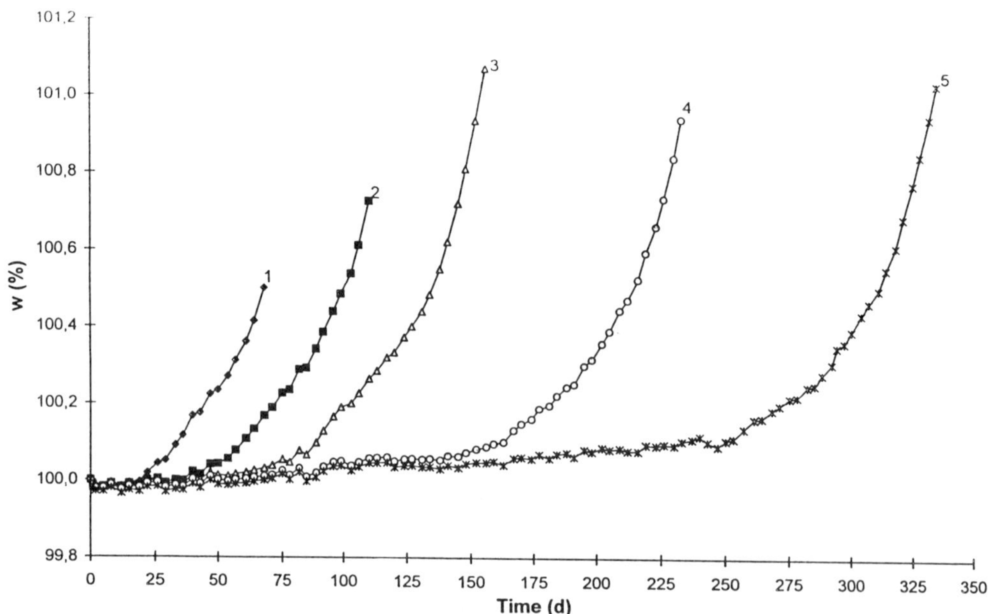

FIGURE 2 Course of oxidation reaction during storage of stabilized rapeseed oil under conditions of the Schaal oven test at 40°C. W = Weight increase; Time = storage time; 1 = control (without antioxidants), oil with hexane rosemary extract; 2 = 0.01% extract; 3 = 0.02% extract; 4 = 0.05% extract; 5 = 0.10% extract.

which would otherwise form free radicals [19]. The relative effect on oxidative chain breaking and on hydroperoxide stabilization may depend on the concentration. In this chapter, the broader sense is applied [20].

The most important types of antioxidants are summarized in Table 3. Phenolic antioxidants and their synergists are the most important representatives of these compounds in food applications.

C. Antioxidants in the Diet and Their Protective Effect Against Diseases

Until recently, humans consumed a diet low in polyunsaturated fats, while large amounts of various phenolic substances were regularly consumed. The free radical content of such a diet in vivo is low. In the last few decades, the consumption of polyunsaturated fats, especially edible oils, has increased several times, mainly as a prevention against some cardiovascular diseases. Polyunsaturated fatty acids are easily oxidized in vivo resulting in formation of free radicals [21]. Therefore, the application of antioxidants in the human diet for the suppression of free radicals is very important [22].

1. Cardiovascular Diseases

Numerous papers and reviews have been published on antioxidants and cardiovascular disease [21]. Free radicals produced by lipid oxidation damage the walls of blood vessels and some particles circulating in the blood. They are bound firmly into serum lipoproteins, forming physically and covalently bound products. Such modified lipoproteins are attacked by leukocytes as foreign bodies. They absorb those lipoproteins, forming sponge cells. These cells are deposited, preferentially in places on cell walls, damaged by free radicals. The atherosclerotic plaques produced in this way are the cause of atherosclerosis. Atherosclerosis is a very dangerous disease frequently found in developed societies with relatively large older populations. Therefore, the prevention of atherosclerosis has become very important.

The best way of preventing cardiovascular disease is to increase the concentration of antioxidants in the blood. This is most easily achieved by increasing the antioxidant intake in the diet. Free radicals present in blood are both lipophilic (lipid peroxy radicals) and hydrophilic (produced by decomposition of hydrogen peroxide, superoxide anion, or other active forms of oxygen), therefore, the best way to enrich foods is by adding both lipophilic (tocopherols, carotenes) and hydrophilic antioxidants (ascorbic acid). These three inhibitors may act as synergists [23]. Polyphenols in red wine probably have a favorable effect, similarly to polyphenols in fruits [24]. Some drugs also have antioxidant activity in vitro [25].

2. Cancer

Free radicals may attack nucleic acids, modifying their structure and changing their genetic code [26]. Therefore, they have mutagenic, teratogenic and cancerogenic activities. Lipid free radicals may convert other substances into potential carcinogens. The activity of various antioxidants (e.g., tocopherols, sulfur, and selenium compounds) is therefore important in cancer prevention.

Table 3 Types of Oxidation Inhibitors (Antioxidants)

Group of compounds	Mechanism of action	Example of inhibitor
Antioxidants	Reaction with free radicals	Propyl gallate, tocopherols
Hydroperoxide deactivators	Reaction with hydroperoxides	Cysteine, selenomethathione
Synergists	Regeneration of an antioxidant	Ascorbyl palmitate
Singlet oxygen quenchers	Transformation of singlet oxygen into triplet oxygen	Carotenes
Chelating agent	Binding heavy metals into inactive complexes	Polyphosphates, citric acid

3. Aging

Free radicals induce changes in polypeptide chains of proteins, resulting in cross-linking [27]. The formation of insoluble polymers is the main cause of aging. Therefore, the presence of antioxidants, which inhibit polymerization, in the body may delay the degenerative changes connected with the process of aging.

D. Synthetic Antioxidants

Natural antioxidants were originally used to protect foods against oxidation, but their activities were generally low and unreliable as they were mixtures of several compounds with different antioxidant properties as well as inactive impurities. The biological variability also contributed to the unreliable activity of natural antioxidants. They were soon replaced by pure synthetic compounds, which were cheaper and possessed reproducible activities.

1. Types of Phenolic and Heterocyclic Antioxidants

Many compounds are active as antioxidants, but only a few are used because of very strict safety regulations. Most of them are phenolic derivatives, usually substituted by more than one hydroxyl or methoxy group. Among heterocyclic compounds containing nitrogen, only ethoxyquin is used (2,6-dihydro-2,2,4-trimethylquinoline), but exclusively in feeds. Diludine (a substituted dihydropyridine derivative) is used for the stabilization of carotenes and other pharmaceutical preparations, but it is active in fats and oils, too [29].

Synthetic phenolic antioxidants are mostly *p*-substituted, while natural phenolic compounds are most *o*-substituted [30]. The *m*-substituted compounds are inactive. The *p*-substituted substances are preferred because of their lower toxicity. Structures of the most important compounds are given in Figure 3. Gallates are esters of gallic acid, which is a natural compound, but propyl, octyl, and dodecyl esters are not found in nature. Synthetic phenolic antioxidants are always substituted by alkyls to improve their solubility in fats and oils and to reduce their toxicity. More details can be found in the literature [31]. Mixtures of phenolic antioxidants often show synergistic activities, e.g., BHT and BHA [32]. In addition to their antioxidant activity, most phenolic antioxidants possess antimicrobial activity in foods [33].

It should be mentioned here that α-tocopherol, D-ascorbic acid, and other antioxidants may be synthesized, nevertheless, they are considered nature-identical compounds. Therefore, they will be discussed under natural antioxidants.

2. Mechanism of Action of Synthetic Antioxidants

Antioxidants react with peroxy radicals produced in oxidized lipids [Fig. 4, Eq. (1)], forming a hydroperoxide molecule and a free radical of the antioxidant [6]. These free radicals are relatively stable, so that they do not initiate a chain autoxidation reaction unless present in very large excess. They react [34] in a similar way with an alkoxy free radical formed during the decomposition of hydroperoxides [Fig. 4, Eq. (2)]. Free antioxidant radicals are deactivated by reaction with a lipidic peroxy or alkoxy radical [Fig. 4, Eqs. (3) and (4)] or with another antioxidant radical [Fig. 4, Eq. (5)]; dimers (and even trimers) are formed in this way, which have a modest antioxidant effect of their own. By the action of some synergists, such as ascorbic acid, the original antioxidant may be regenerated. Quinones are formed from phenolic antioxidants by reaction with two peroxy radicals. When antioxidants are present in excess, the reaction of antioxidant free radicals with oxygen may become important [Fig. 4, Eq. (6)], and even their reaction with polyunsaturated fatty acids [Fig. 4, Eq. (7)] has some impact on the course of oxidation. A large excess of antioxidants may thus result in conversion of antioxidant into prooxidant activity.

E. Natural Antioxidants

In the last 20 years, people have become afraid of technology and science, particularly of chemistry, and have turned to natural compounds in the diet [35]. Consumers prefer food products stabilized with

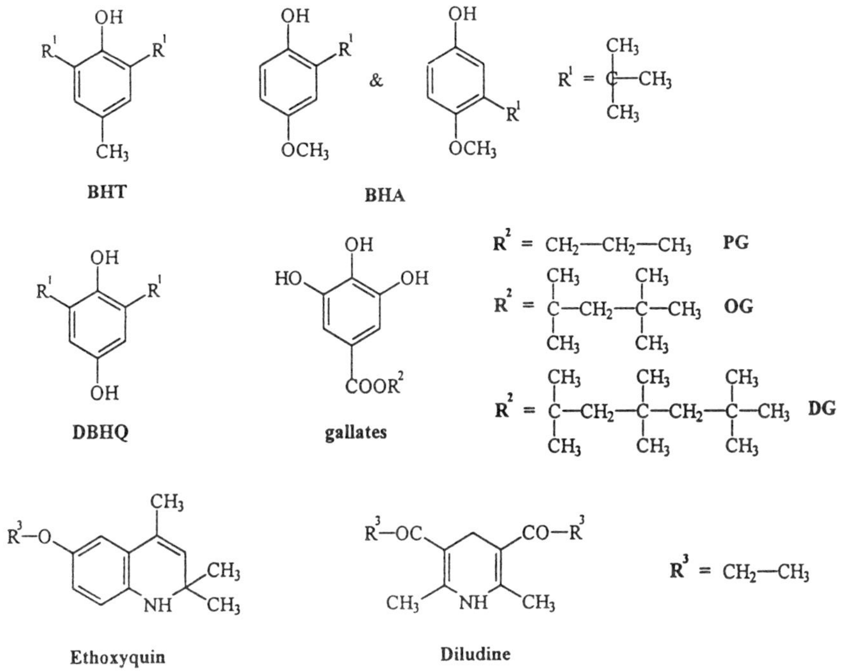

BHT BHA

$R^1 = C(CH_3)_3$

DBHQ gallates

$R^2 = CH_2—CH_2—CH_3$ PG

$R^2 = C(CH_3)_2—CH_2—C(CH_3)_2—CH_3$ OG

$R^2 = C(CH_3)_2—CH_2—C(CH_3)_2—CH_2—C(CH_3)_2—CH_3$ DG

Ethoxyquin Diludine

$R^3 = CH_2—CH_3$

FIGURE 3 Chemical structures of the most important synthetic antioxidants. BHT = Di-*tert*-butylhydroxytoluene; BHA = *tert*-butylhydroxyanisole; DBHQ = di-*tert*-butylhydroquinone; PG = propyl gallate; OG - octyl gallate; DG - *tert*-dodecyl gallate.

natural antioxidants [36]. Research and development laboratories have shown great interest in exploring natural antioxidants. Such research benefits the manufacturer, too, as natural compounds are subject to fewer legal regulations.

Almost all plants, microorganisms, fungi, and even animal tissues contain antioxidants of various types, which for various reasons (e.g., availability, food safety, economics) can be used as sources of antioxidants only in certain cases. For instance, of 147 plants tested, 107 extracts showed measurable antioxidant activities [37]; the majority of natural antioxidants are phenolic compounds. Natural phenolic antioxidants are usually accompanied by other inhibitors of oxidation, such as synergists, singlet oxygen quenchers, and chelating agents.

$$\text{A-H} + \text{R-O-O*} \Rightarrow \text{A*} + \text{R-O-OH} \quad (1)$$

$$\text{A-H} + \text{R-O*} \Rightarrow \text{A*} + \text{R-OH} \quad (2)$$

$$\text{A*} + \text{R-O-O*} \Rightarrow \text{A-O-OR} \quad (3)$$

$$\text{A*} + \text{R-O*} \Rightarrow \text{A-OR} \quad (4)$$

$$\text{A*} + \text{A*} \Rightarrow \text{A-A} \quad (5)$$

$$\text{A*} + O_2 \Rightarrow \text{A-O-O*} \quad (6)$$

$$\text{A*} + \text{R-H} \Rightarrow \text{A-H} + \text{R*} \quad (7)$$

FIGURE 4 Mechanism of reactions of antioxidants with free radicals. See text for details.

1. Types and Sources of Antioxidants

In higher plants, where phenolic compounds are very common, two series of compounds are of particular interest, namely, derivatives of the benzoic acid and cinnamic acid series [38]. The aromatic cycles substituted by two or three phenolic groups in the *ortho*-position are particularly important; some hydroxyl groups may be methoxylated. Gallic acid is a typical representative of the benzoic acid series, while caffeic acid is the most typical derivative of the cinnamic acid series. Catechins and flavones are more complicated compounds (see below), where the antioxidant activity is located in a pyrocatechol or pyrogallol radical bound in the molecule.

The best method of application of natural antioxidants is to use natural food components (e.g., cereals, nuts, fruits, vegetables) because they are regarded as safe and no special approval for their application is necessary. Another possibility is to use natural food ingredients, such as spices. Natural compounds derived from nonfood materials, such as gingko leaves [39], should be tested for toxicity before application. A natural antioxidant, nordihydroguaiaretic acid (NDGA), extracted from the creosote bush was originally used in food stabilization [40], especially of edible fats, but it is not permitted now because it has not passed more recent safety tests.

Tocopherols

Tocopherols are the most common antioxidants, as they are present, at least in traces, in nearly all food materials. They are derivatives of chroman with a diterpenic (phytol) side chain; the active configuration is the phenolic group in the benzene cycle, located *para* to the oxygen atom bound in the adjacent dihydropyrone cycle. There are four tocopherols, which differ in methyl substitution (Fig. 5). The most important antioxidant of this group is d-α-tocopherol, which has lower antioxidant activity in edible oils than other tocopherols, but is more easily absorbed in the intestines, therefore, it possesses an in vivo antioxidant activity and a vitamin E activity as well. The in vivo activity decreases in the order $\alpha > \beta > \gamma > \delta$ tocopherols [41], while in bulk fats and oils it decreases in the order $\delta > \beta,\ \gamma > \alpha$ tocopherols [42].

Tocopherols are present mostly in foods of plant origin. Their content in edible oils is particularly high (Table 4). Between 20 and 50% of tocopherols are lost during edible oil refining (especially during deodorization), but they are often replaced by addition of α-tocopherol acetate of a deodorization concentrate (collected during the deodorization step of oil refining) into refined oils. Tocopherol acetate is added instead of free tocopherol because it is more resistant against oxidation during storage. In animal foods, tocopherols are present only in negligible amounts.

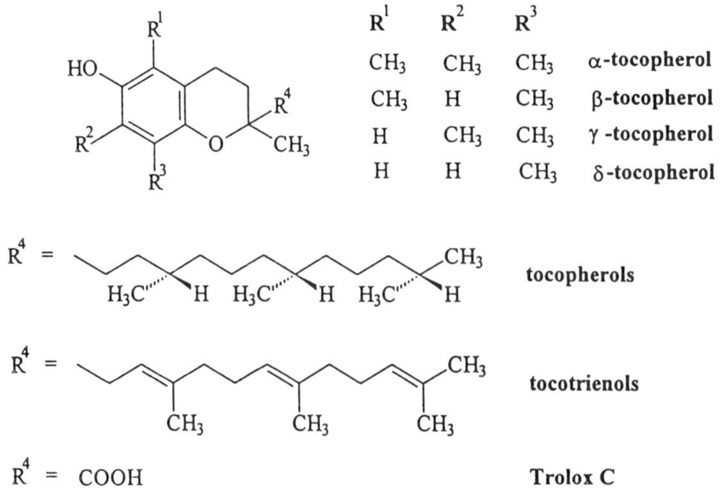

R^1	R^2	R^3	
CH_3	CH_3	CH_3	**α-tocopherol**
CH_3	H	CH_3	**β-tocopherol**
H	CH_3	CH_3	**γ-tocopherol**
H	H	CH_3	**δ-tocopherol**

R^4 = **tocopherols**

R^4 = **tocotrienols**

R^4 = COOH **Trolox C**

FIGURE 5 Chemical structures of tocopherols.

Table 4 Tocopherol Content in Vegetable Oils

Oil	α-Tocopherol (mg/kg)	β- + γ-Tocopherols (mg/kg)	δ-Tocopherol (mg/kg)	Total tocopherols (mg/kg)
Soybean	80–150	210–780	60–400	300–1400
Rapeseed	120–300	220–450	5–15	380–750
Sunflower	550–900	25–110	0–5	580–980
Corn germ	120–400	420–780	10–50	680–1250
Peanut	80–320	120–280	5–30	90–550
Olive	80–180	0–5	0	80–180
Cottonseed	550–700	200–280	0–5	800-980
Rice bran	500–650	230–340	0–5	750–950

In oilseeds, cereals, and other plant products, tocopherols are accompanied by dehydrotocopherols, tocodienols, and tocotrienols, which have one, two, and three double bonds in the side chain, respectively [43]. The last group is present in cereal flours and in palm fruits and palm oil. The related plastochromanol-8 has activity similar to α-tocopherol [44].

Tocopherols added to food products may be either synthetic products or natural concentrates, obtained most often from deodorization sludges from oil refining, from wheat or corn germs, or from other sources (even by synthesis). Natural tocopherols have optical activity, unlike synthetic products.

In foods, tocopherols act as relatively weak antioxidants; by reaction with free radicals, they are converted into quinones, spirodimers, and various other compounds [45], as well as in copolymers with oxidized lipids.

Antioxidants from Oilseeds

Several important oilseeds are sources of antioxidants other than tocopherols (Table 5). During the processing of oilseeds or oil-bearing fruits, antioxidants are partially extracted into crude oils.

The best known oxidation inhibitors are those present in olives, which are the fruits of *Olea europaea (oleracea)* L. Virgin, oil, produced by pressing olive fruits under low temperature, contains several antioxidants derived from hydroxytyrosol (Fig. 6), a derivative of tyrosine [46]. The antioxidants are present mostly as glycosides in the pericarp (i.e., are water soluble) and remain mainly in the residue after pressing (pomace). During ripening, storage, or pressing, glycosides may be hydrolyzed into the respective aglycones; for instance, oleoeuropein is present with its aglycone in virgin olive oils, but some antioxidants are partially reduced during pressing [47]. In the sterol fraction, olive oil contains Δ^5-avenasterol, which acts as an antipolymerization agent under frying conditions [46];

Table 5 Antioxidants from Oilseeds

Oilseed	Systematic name	Type of substance
All seeds	—	Tocopherols
Palm fruit	*Elaeis guineensis*	Tocotrienols
Olive fruit	*Olea europea*	Oleoeuropein aglycone
Sesame seed	*Sesamum indicum*	Sesamol
Cottonseed	*Gossypium hirsutum*	Gossypol
Rapeseed	*Brassica napus*	Sinapic acid
Flaxseed	*Linum usitatissiumum*	Lignans
Rice bran	*Oryza sativa*	Oryzanol

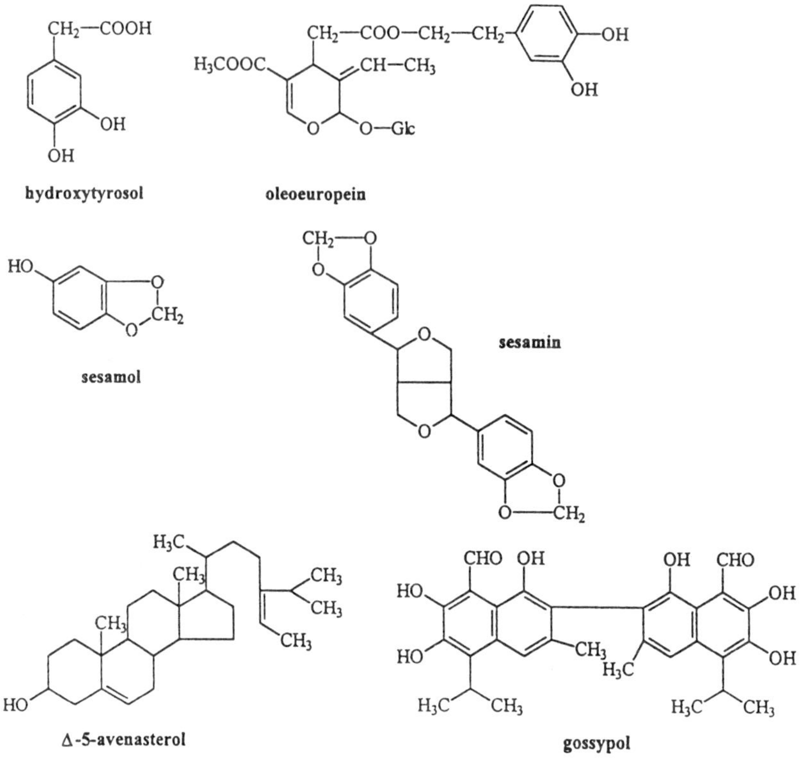

hydroxytyrosol

oleoeuropein

sesamol

sesamin

Δ-5-avenasterol

gossypol

Figure 6 Chemical structures of some natural antioxidants from seeds.

the antioxidant activity is due to the extraordinarily high stability of the respective free radical, formed by reaction of avenasterol with peroxy radicals.

Sesame seed from *Sesamum indicum* L. contains lignan analogs, which are, together with their decomposition products, present in crude sesame oil and extracted meal [48]. They are only partially eliminated during refining so that they contribute to the well-known stability of sesame oil. The most active compound in sesamol (Fig. 6), which is accompanied by several structurally related compounds [49]. The antioxidant activity is affected by roasting and steaming of seeds before extraction [50].

Sunflower seeds (*Helianthus annuus* L.) are rich in polyphenols, therefore, extracted meals are frequently discolored by interactions of oxidized polyphenols with protein. Seeds contain chlorogenic acid and related phenolic derivatives, which are only partially decomposed during processing. Polyphenoloxidases oxidize polyphenols into quinones, which react with amine groups, imparting brown discoloration to seed meals.

Peanuts (*Arachis hypogaea* L.) contain flavonoids, tannins, and various other phenolic compounds [51], concentrated in hulls [52]. Several papers report the preparation of hull extracts for the stabilization of foods.

Cottonseed (*Gossypium hirsutum* or *Gossypium barbadense*) at one time contained gossypol, a complicated polyphenolic compound with aldehydic groups possessing antioxidant properties. Because of the toxicity of gossypol (it is active as a male contraceptive agent), cottonseed-extracted meal could not be used as feed in large amounts. Therefore, modern cultivars (glandless cottonseed) of this plant are gossypol-free. Gossypol was a powerful antioxidant, but glandless cottonseed still contains flavonoid antioxidants, such as quercetin and tocopherols [53].

Soybeans also contain phenolic compounds with antioxidant activity, mostly flavones and isoflavones [54], which may stabilize lipids not only in beans, but also in soy products, such as tofu or tempeh [55]. Isoflavones possess a hormone-like activity, which should be accounted for when evaluating the possible inhibitory effect.

Rapeseed (*Brassica napus*) and the related turnip rapeseed (*Brassica rapa*) are relatively rich in phenolic compounds, among which sinapic acid prevails. It acts as a moderate antioxidant in polyunsaturated oils [56]. In rapeseed, sinapic acid is mostly bound to choline, forming sinapine, which is nearly inactive as an antioxidant [57]. Structurally related tannins are also present in rapeseed. Evening primrose (*Oenothera biennis*) contains high levels of phenolics, mainly derivatives of protocatechuic and gentisic acids [58].

Antioxidants from Cereals and Grain Legumes

Cereals, one of the most important components of the diet, contain several types of phenolic compounds, especially in bran. Oat is considered relatively efficient from this standpoint [59]; oat extracts were the first natural antioxidants studied in detail before the World War II. Extracts were patented and used in oils and other foods. Phenolic compounds present in oat seed are partially bound in lipids, and therefore they are liposoluble.

Rice bran is used for the production of oil, therefore, it is collected and extracted with solvents. During the process, some phenolic compounds, such as oryzanol [60], pass into crude rice oil. Rice oil is particularly rich in antioxidants (including tocopherols) and low in polyunsaturated fatty acids, therefore, it is very stable during storage, like olive oil.

Maillard products formed during the baking or roasting of food originate from sugars and amino acids; they have a stabilizing effect on lipids in foods, especially in baked products.

The content of phenolic substances in grain legumes is mostly low, but nevertheless, they might stabilize foods if added in substantial amounts as an ingredient. Phenolic substances (mainly flavanols) are concentrated in legume hulls [61]. These phenolic derivatives may be insoluble (tannins or lignins), but some derivatives are partially soluble in oil. This is why they sometimes become efficient in stabilizing the lipid fraction. Antioxidants from peas were isolated and their antioxidant activities reported [62]. Polar bean extracts showed high activities in emulsions [63].

Compounds from Fruits and Vegetables

The most important group of compounds active as antioxidants in fruits and vegetables consists of various flavones and related compounds. Some substances in this group act as cofactors of vitamin C, increasing its vitamin activity. The probable mechanism involves their antioxidant activity, which may protect vitamin C against destruction. In wine and deep-colored fruit juices, various anthocyanins are present, which are important antioxidants in the aqueous phase. Red wine has been recommended as a source of antioxidants to protect humans against the development of cardiovascular diseases, but the effect has not yet been sufficiently confirmed.

In addition to the above compounds, various terpenic derivatives could act as potent inhibitors of lipid oxidation. The effect of terpenes will be discussed in the next section, as terpenes are more important in spices.

In the category of fruits, the best investigated substances are those isolated from citrus fruits. As their activity is stronger in water emulsions because of the hydrophilic character of flavones and related compounds, ethanolic or aqueous extracts (which contain much more hydrophilic substances than hexane or ether extracts) seem to impart higher activities to emulsions or other polar food systems than hexane and ethyl acetate extracts.

Various nuts (e.g., macadamia) contain antioxidants; these substances are extracted into the respective oils by pressing. Inhibitors of oxidation have been detected in extracts from hop [64].

The most important extracts from fruits and vegetables are listed in Table 6. Most are various pyrocatechol derivatives. Onion and garlic contain efficient inhibitors; because of their pungent flavor, they are suitable only for meat products, snacks, or cheeses [65].

Potent antioxidants can be obtained from various nonfood plant products, such as gingko leaves, but the substances from such products (i.e., not used for food purposes) should be tested for safety before application.

Table 6 Antioxidants in Fruits and Vegetables

Species	Systematic name	Type of substance
Citrus	*Citrus* sp.	Flavonoids, carotenoids
Plums	*Prumus* sp.	Phenolics
Persimmon, kaki	*Diospyros kaki*	Procyanins, catechins
Red wine	*Vitis vinifera*	Anthocyanins
Pineapple	*Ananas comosus*	Flavanols
Onion	*Allium cepa*	Sulfur compounds
Garlic	*Allium sativum* ssp. *vulgare*	Sulfides, disulfides
Green pepper	*Capsicum annuum*	Flavonoids
Carrots	*Daucus carota*	Carotenoids, flavonoids
Betel	*Areca catechu*	Oleoresins, eugenol, hydroxychavicol
Legume	*Leguminosae*	Flavonoids
Green olives	*Olea europaea*	Anthocyanins
Mustard	*Sinapis alba*	Phenolics, isothiocyanates
Oak (wood)	*Quercus ilex*	Phenolic acids, lignins

Extracts from Herbs and Spices

Herbs are mostly leaves or stems from various plants used for the preparation of infusions, extracts, or soups. Many species of this group of food materials are active [37,66], mainly because of their content of phenolic compounds.

The most important representative of this group of substances are tea leaves, from both green and fermented tea (*Camellia sinensis* or *C. assamica* L.). Green tea contains a high percentage (about 20%) of catechins and related compounds. The mixture mainly consists of catechins, epicatechins, gallocatechins, and the respective gallates. Extracts from green tea are, therefore, very active—comparable to synthetic antioxidants—while extracts from fermented tea leaves are less active because a substantial part of the catechin has been converted into less active tea pigments [67,68]. During the fermentation of tea, catechins are partially oxidized into the respective quinones, which dimerize into tea pigments—theaflavins and thearubigins. Wastes left over after the preparation of tea infusions may be used for the extraction of antioxidants. The antioxidant activity in lard may be ranked in descending order as follows: epigallocatechin gallate > epigallocatechin > epicatechin gallate > epicatechin. Raw materials for the preparation of herbal teas are less efficient, but various herbs used in China, Japan, and other Far East countries contain efficient antioxidants, too. Algae contain brominated 3,4-dihydroxybenzene derivatives [70].

Spices are used for conditioning meals and bakery products. Several species possess antioxidant activities [71], some of which are listed in Table 7. Wastes left after the distillation of essential oils from spices could be used as raw material for the extraction of antioxidants.

The most active substances are produced from rosemary (*Rosmarinus officinalis* L.), which is the only commercially available antioxidant of this group, and sage (*Salvia officinalis* L.), which contains carnosic acid, carnosol, and other active substances (Fig. 7). Other active spices include thyme, juniper, oregano, gingseng, nutmeg, etc. In most cases, it is sufficient to add spices to the food products before or after preparation, but extracts may be added instead of spices. The addition of pure substances (isolated from spices) is not recommended, as the application of pure compounds would be subject to legal restrictions. Some spices alter the flavor of food products, but deodorized material (i.e., after removal of essential oil by steam distillation) could be used without affecting the flavor. Spices that have antioxidant activity often possess antibacterial activity, too. (Detailed information can be obtained from Ref. 71.)

TABLE 7 Antioxidants from Herbs and Spices

Species	Systematic name	Substance	Ref.
Rosemary	*Rosmarinus officinalis*	Carnosic acid, carnosol	76
			84
Sage	*Salvia officinalis*	Carnosic acid	26
Tanshen, danshen	*Salvia miltiorrhiza*	Carnosol	82,83
Thyme	*Thymus vulgaris*	Thymol, quinones	75
Satureja	*Satureia hortensis*	Flavonoids	75
Clove	*Eugenia caryophyllata*	Eugenol, gallates	89,129
Black pepper	*Piper nigrum*	Ferulic acid	90
Ginger	*Zingiber officinalis*	Cassumarin, ginerol	72
Juniper	*Juniperus communis*	Phenolics, resins	88
			90
Oregano	*Origanum vulgare*	*o*-Substituted phenolic acids	77,130,131
Fennel	*Foeniculum vulgare*	Dihydrocoumarins	92
			77
Curcuma	*Curcuma longa*	Curcumin	88
Spearmint	*Mentha piperita*	Flavonoids	71
Lavender	*Levandula angustifolia*	Flavonoids	88
Hop	*Humulus lupulus*	Flavonids, anthocyanins	132
Allspice	*Pimenta officinalis*	Flavonids	133

2. Mechanism of Action

The mechanism of action of natural phenolic antioxidants is essentially the same as that of synthetic antioxidants. The only difference is that they are usually present in mixtures with related compounds of varying activities and/or with synergists, such as phospholipids, amino acids, or terpenes. If they are added to food as unprocessed ingredients, the microstructure of the tissue can play a role.

F. Other Oxidation Inhibitors

In foods and feeds, phenolic antioxidants are rarely present without other types of inhibitors of oxidation reactions. Other inhibitors are added with extracts of natural antioxidants or may be even added to enhance the antioxidant activity of the system. Even synthetic antioxidants are often applied in mixtures with synergists and chelating agents.

1. Substances Decomposing Hydroperoxides in a Nonradical Way

Different substances may convert hydroperoxides into less reactive hydroxyl derivatives. Sulfur compounds are very active substances, which are mostly bound in proteins. Thiols, such as cysteine, are oxidized into disulfides [Fig. 8, Eq. (1)]. Cystine may be further oxidized by another molecule of a hydroperoxide into sulfinic acid [Fig. 8, Eq. (2)]. Similarly, sulfides, such as methionine, react with hydroperoxides, forming sulfoxides [Fig. 8, Eq. (3)]. The product may react with another molecule of a hydroperoxide, resulfing in the respective sulfone [Fig. 8, Eq. (4)]. Onion and garlic contain sulfur compounds possessing hydroperoxide-decomposing activity [93,94]. Selenium-containing amino acids (selenocysteine or selenomethionine), present in traces in natural proteins, react in a similar way. In the body, they help to retain vitamin E [95]. Selenium has more importance as part of an antioxidative enzyme, selenoglutathione oxidase, inactivating free radicals and other oxidants, particularly hydrogen peroxide.

Hydroperoxides may react with free amine groups of proteins [Fig. 8, Eq. (5)]; imines are formed from the intermediary product by subsequent dehydration. Various basic groups, such as

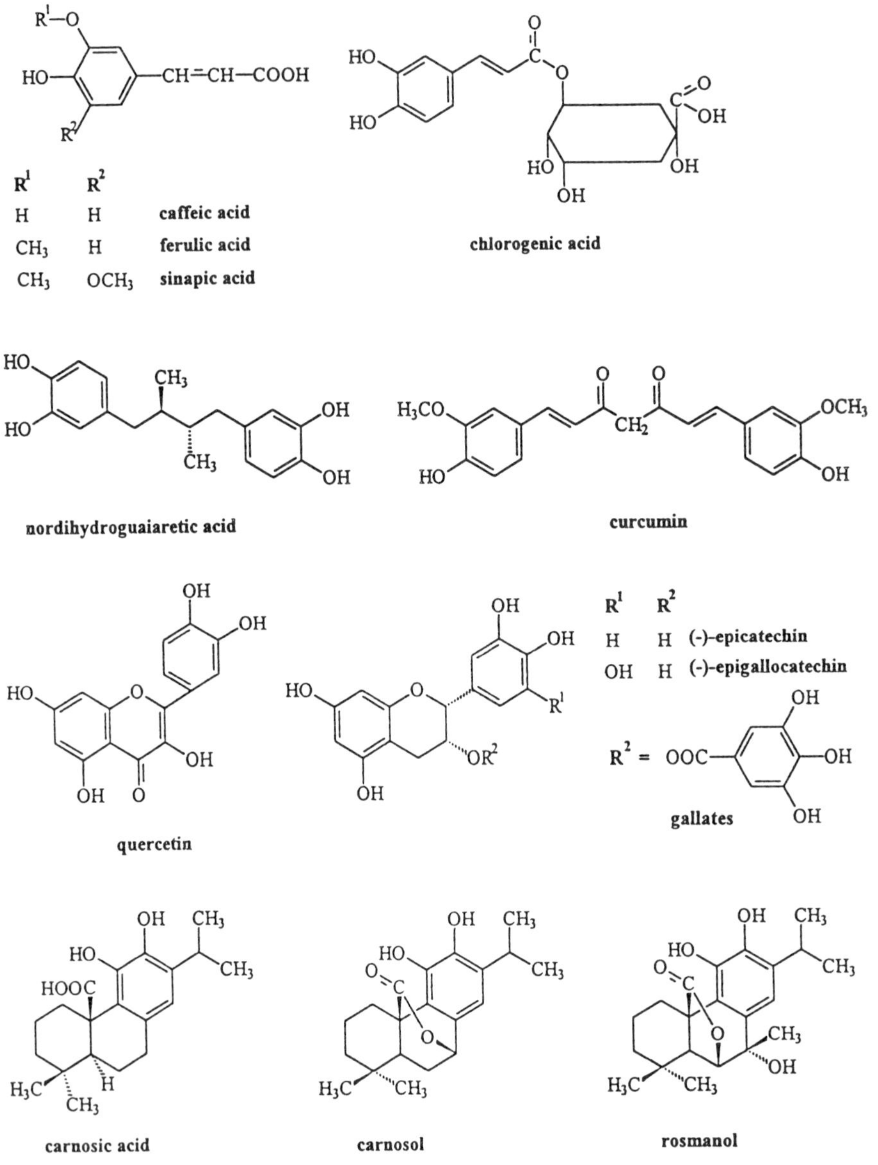

R^1	R^2	
H	H	caffeic acid
CH_3	H	ferulic acid
CH_3	OCH_3	sinapic acid

chlorogenic acid

nordihydroguaiaretic acid

curcumin

quercetin

R^1	R^2	
H	H	(-)-epicatechin
OH	H	(-)-epigallocatechin

R^2 = OOC– gallates

carnosic acid

carnosol

rosmanol

FIGURE 7 Chemical structures of some natural antioxidants from herbs and spices.

histamine or indole, can deactivate hydroperoxides following similar mechanisms. Carnosine, a histidine dipeptide, was found to be efficient in decomposing hydroperoxides [96]. Many amino compounds can deactivate free radicals, therefore, they are anticarcinogenic. Free amino acids are present in protein hydrolysates, which are thus active, e.g., in meat products [97].

2. Synergists

Synergists are substances that have no antioxidant activity of their own, but that can increase the activity of an antioxidant. The most frequently used synergists are polyvalent inorganic (phosphoric acid) or organic acids. Ascorbic acid and its isomer, erythorbic acid (earlier called isoascorbic acid), are the best known and most often used synergists. Citric acid is frequently used, followed by tartaric

$$2\,R^1\text{-SH} + R^2\text{-OOH} \Rightarrow R^1\text{-S-S-}R^1 + R^2\text{-OH} \quad (1)$$

$$R^1\text{-S-S-}R^1 + R^2\text{-OOH} \Rightarrow R^1\text{-S-SO-}R^1 + R^2\text{-OH} \quad (2)$$

$$R^3\text{-S-CH}_3 + R^2\text{-OOH} \Rightarrow R^3\text{-SO-CH}_3 + R^2\text{-OH} \quad (3)$$

$$R^3\text{-SO-CH}_3 + R^2\text{-OOH} \Rightarrow R^2\text{-SO}_3\text{-CH}_3 + R^2\text{-OH} \quad (4)$$

$$R^4\text{-NH}_2 + R^2\text{-OOH} \Rightarrow R^4\text{-N[OH]-}R^2 + H_2O \quad (5)$$

FIGURE 8 Deactivation of lipid hydroperoxides by sulfur and nitrogen compounds. See text for details.

acid, malic acid, and different amino acids. To increase the solubility in fats and oils, ascorbic acid is often esterified with higher fatty acids, such as palmitic acid, and citric acid is added as an ester. Ascorbyl palmitate is active as a synergist to tocopherols in vegetable oils and lard [98].

Phospholipids belong to this group of substances as well [99]. They enhance the activity of tocopherols [100]. Phosphatidylethanolamine acted as a synergist in fish oil [101]. It was reported as a more active synergist of flavonoids in lard than phosphatidylcholine [102].

The activity of synergists can be partially explained by their metal chelating activity, but also by their efficiency in converting free radicals into ions or by their effect to restitute original antioxidants from the respective free radicals. Most synergists act after several mechanisms.

Most natural products contain the above synergists in various amounts so that they can be used as sources of both phenolic antioxidants and of synergists, acting in the mixture after several mechanisms.

3. Singlet Oxygen Quenchers

As explained in Section II, singlet oxygen may be formed from ordinary triplet oxygen by action of light in the presence of photosensitizers. The singlet form of oxygen is very reactive, therefore, it is extremely important to deactivate it back to the triplet form very rapidly. Various substances are effective from this standpoint, but in food materials, carotenes and other carotenoids are the most active and the most frequently applied derivatives [103]. Carotenes are frequently present in plant materials; β-carotene (Fig. 9) is the most important representative of this group. Oxygen-substituted derivatives of carotenoids, called xanthophylls, have similar deactivating activity, even when they have no provitamin activity like that of carotenes. Carotenoids are efficient free radical scavengers as well. The structurally related vitamin A has a similar activity, but it is usually present in much smaller quantities in foodstuffs than are carotenoids.

4. Metal-Chelating Agents

Metal ions of transient valency, such as copper, iron, cobalt, chromium, or manganese, are very active prooxidants (see Sec. I). They may be active not only in the ionic form, but also in complexes. For instance, iron is an active prooxidant in heme derivatives. Therefore, substances that are able to bind these metals into inactive complexes can inhibit oxidation reactions of lipids.

Many metal-chelating substances are present in foods, especially in plant materials. Phytates (salts of phytic acid—*myo*-inositol hexaphosphate), phospholipids, and oxalates are the most common representatives of this group. Synergists, such as phosphoric acid, citric, tartaric, malic or ascorbic acids, also possess pronounced chelating activities. Polyphosphates are added to inactivate iron, especially in meat products [104]. Even some phenolic antioxidants may bind metals into complexes, e.g., quercetin. Silicone oil is often added to frying oils to inhibit degradation; the activity is, at least partially, attributed to prevention of dissolution of iron in oil during frying [105].

$S + h\nu \Rightarrow {}^1S^*$ excited singlet sensitizer

sensitizer $\Downarrow$ intersystem crossing mechanism

$S \Rightarrow {}^3S^*$ excited triplet sensitizer

$\Downarrow$

triplet oxygen ${}^3O_2 \Rightarrow {}^1O_2$ singlet oxygen + substrate $\Rightarrow$ oxidized substrate

$\Downarrow$ β-carotene or carotenoids

${}^3O_2$ triplet oxygen

Figure 9 Quenching of singlet oxygen produced during photooxidation.

The chelating activity depends on the pH value, water content, and other factors of the medium. Usually, metals are not completely inactivated by any agent; only their prooxidant activities decrease by chelation. We have shown in phosphatidic acids and in pheophytins that the chelating activity is often very high. Nevertheless, it should be kept in mind that even minute traces of free metals (not bound in complexes) are sufficient to efficiently promote autoxidation.

IV. APPLICATIONS OF ANTIOXIDANTS IN FOODS

Antioxidant activities depend not only on the antioxidant structure, but also on many other factors, such as the composition of the lipid fraction, availability of oxidants, presence of various other inhibitors or promotors of oxidation, presence of nonlipidic components, water, microstructure, temperature, etc. Therefore, the literature data should be always verified using a specific substrate and processing or storage conditions.

A. Stabilization of Fats and Oils

Animal fats, such as pork lard, beef tallow, or milk fat, contain very low amounts of natural antioxidants, therefore, their stability against oxidation is low in spite of their relatively low degree of unsaturation. Fortunately, both synthetic and natural antioxidants are very active in the stabilization of animal fats. Most often, mixtures of antioxidants and synergists are used for stabilization [106]. Lipid-soluble antioxidants show good results, but polar antioxidants could be used, too. When such fats, stabilized with polar antioxidants, are used for the preparation of food, e.g., for cooking or baking, the antioxidant activity is partially lost by extraction of antioxidants into the aqueous phase. Therefore, the carry-through effect (ability to preserve the antioxidant activity even in foods containing water) should be always tested.

Vegetable oils are rather difficult to stabilize because of their high content of polyunsaturated fatty acids. Their advantage is the presence of natural antioxidants in edible oils, mostly tocopherols. An addition of phenolic antioxidants usually shows limited efficiency, but the addition of synergists is helpful. Ascorbyl palmitate, phospholipids, or organic acids are useful as inhibitors of vegetable oils. Most polyunsaturated oils are treated with phosphoric, citric, or other polyvalent acids during processing (they are added in the stage of deodorization) so that some residual synergists are often present. Oils containing higher levels of natural antioxidants, such as olive oil, sesame oil, rice oil, etc., are sufficiently stable without additional stabilization. Further addition of tocopherols to most polyunsaturated vegetable oils is inefficient, as their natural content in oils is very close to the optimum [107].

Frying oils present a special case, as they are exposed to air oxygen at high temperatures (130–200 °C) during frying. Many phenolic antioxidants lose their antioxidant activity, observed at storage temperatures, under frying conditions to some extent. Common synthetic antioxidants are distilled off with water vapor during frying, therefore, it is better to use nonvolatile inhibitors. Carotenes and other hydrocarbons were found to be efficient in frying oils, as they form a monomolecular film on the surface, protecting against the access of oxygen. Frying oils are efficiently stabilized by the addition of minute amounts of silicone oils, which form a thin layer protecting against the access of oxygen. Frying oils with low polyunsaturated fatty acid content, such as low-linoleic sunflower oil or palmolein, are usually sufficiently stable under frying conditions that they do not require additional stabilization.

B. Applications of Antioxidants in Fat Emulsions

The activity of antioxidants in emulsions is very different from that measured in bulk fats and oils. Water-soluble antioxidants, such as propyl gallate or flavonoids, are extracted into the aqueous phase so that they lose their activity. Nonpolar synthetic antioxidants remain in the fat phase (sometimes in the form of less efficient micelles) or are accumulated at the interphase so that their activity is usually high. Polar antioxidants are more active in a nonpolar medium and nonpolar antioxidants in a polar medium, as was shown in case of Trolox versus α-tocopherol on the one hand and ascorbic acid versus ascorbyl palmitate on the other hand, when tested in bulk oil and in emulsion [108]. Some antioxidants of semipolar nature may accumulate on the water-oil interphase (Fig. 10) and prevent the diffusion of oxygen into the fat phase. The activity depends on the type of emulsion. Oil-in-water (O/W) emulsions are better protected against oxidation than water-in-oil (W/O) emulsions, as oil droplets are isolated from the access of oxygen by the aqueous phase and the interfacial layer. In the presence of tocopherols, phospholipids are active ingredients in emulsions [109]. When rosemary extracts are applied as antioxidants, carnosic acid and rosmarinic acid are more active than carnosol in oils, but carnosic acid and carnosol are equally active in emulsions, more so than rosmarinic acid [110].

C. Applications of Antioxidants in Foods

The efficiency of antioxidants in foods depends very much on the water content [111]. Generally, dry foods, such as dehydrated soups, dried milk [112], dried meat, etc., are very sensitive to oxidation, as air oxygen has free access to the film of lipids on nonlipidic particles. The stabilization is then less effective as the initiation rate of the autoxidation reaction is high and antioxidants are soon decomposed on storage or heating.

In water-containing foods, the lipid fraction is relatively stable, as it is usually protected by a layer of hydrated proteins or carbohydrates against the access of air oxygen. The addition of nonpolar antioxidants is effective, while polar antioxidants may lose their activity by passage into the aqueous phase, similarly as in fat emulsions.

Proteins and many other components of foods have protective action and act as synergists of other inhibitors, enhancing the effect of antioxidants [113]. Chelating agents are often present in foodstuffs as natural components, but heavy metals may be present as well, for instance, heme derivatives in animal products.

Because of the complexity of foods, it is necessary to test any addition of antioxidants to stabilize the particular material and to optimize the mixture of inhibitors.

D. Applications of Antioxidants in Packaging

Air oxygen penetrates the food material from outside, therefore, it is useful to protect food surfaces. Meat or fish products are often protected by packaging impregnated with antioxidants or by application of suitable antioxidants on the surface of materials, even when they are not packed.

Most often, foods are distributed packed; the packaging material is then of great importance [114]. If the material is permeable for oxygen, antioxidants may be added to the packaging to inhibit

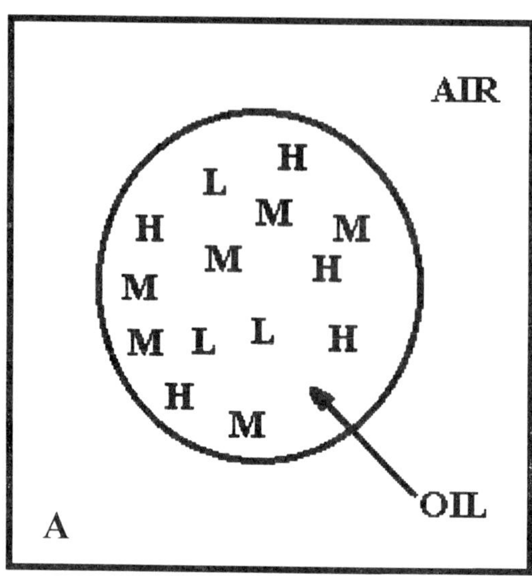

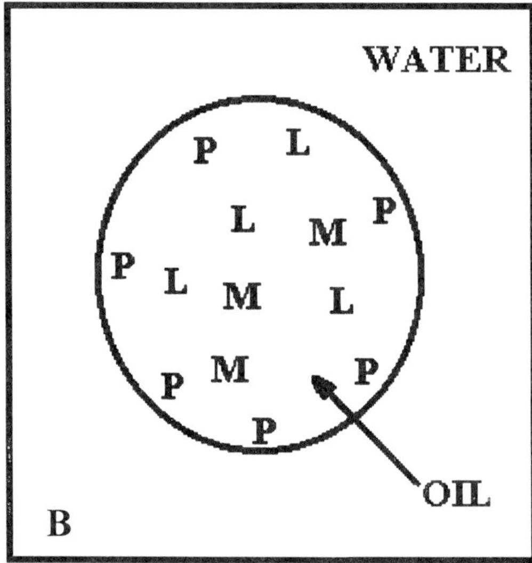

Figure 10 Distribution of lipophilic and hydrophilic antioxidants at the oil/water interphase and in oil. (A) System of bulk oil, in contact with air. (B) Emulsion of oil in water. L = Lipophilic antioxidants; H = hydrophilic antioxidants; M = micelles of lipophilic antioxidants; P = polarized lipophilic antioxidants on the interphase.

the diffusion of oxygen. These antioxidants may enter into the packed food [115], especially high-fat foods. Obviously, the only antioxidants that can be used in packaging are those that are allowed in foods, and they should be applied only in such amounts that the content in foods does not exceed the legal limits.

Packaging materials may be protected against oxidation by antioxidants, even when the producer does not intend to stabilize food in this way. The case is the same as above, but eventual extraction of antioxidants into food should be accounted for.

E. Legal Permitted Levels of Antioxidants

The use of synthetic antioxidants is restricted to a few compounds, which have passed very complicated and expensive tests to prove their safety [116–118]. Such tests include the generation tests and teratogenic and carcinogenic trials using at least three species of laboratory animals, among which at least one is nonrodent. Therefore, no new antioxidants (with a few exceptions) are being tested and permitted, and the addition of the antioxidants to food is usually restricted to 0.02% on a fat basis. Permitted antioxidants are different in different countries and subject to change; some examples are given in Table 8. For practical reasons, it is suitable to add mixtures of antioxidants, which usually have higher activities than single compounds and which guarantee that the limits for single compounds have not been exceeded.

Some natural materials are generally regarded as safe (GRAS); their use is not restricted by legislation (e.g., tocopherols, phospholipids, amino acids, etc.). Some common food ingredients, such as spices, are also not subject to restrictions. For pure compounds isolated from natural materials or extracts from nonedible plant materials, it is necessary to prove that there are no risks involved in their use.

V. ANALYSIS OF ANTIOXIDANTS

In the analysis of antioxidants, two aspects are important: determination of the content of antioxidants, and determination of the antioxidant activity.

A. Content of Antioxidants

The analysis of antioxidants consists of two steps [118]: isolation of antioxidants from the substrate and the purification of the extract, and quantification of antioxidants in the extract. Several methods have been standardized and can be found in the respective books of analytical standards [119–121].

1. Isolation of Antioxidants from Fats, Oils, and Foods

The analysis of antioxidants is very simple for pure preparations, but it is very difficult to prepare the sample for analysis of fats, oils, and other foods. Two difficulties are faced by the analyst: how to achieve quantitative yield during the extraction and purification of extracts, and how to prevent

TABLE 8 Antioxidants Used in Edible Fats and Oils

Antioxidant	Abbreviation	Applications
Propyl gallate	PG	Fats and oils
Dodecyl gallate	DG	Fats, emulsions
tert-Butylhydroxytoluene	BHT	Fats and oils, foods
tert-Butylhydroxyanisole	BHA	Fats and oils, foods
Di-*tert*-butylhydroquinone	DBHQ	Vegetable and fish oils, frying oils
Ascorbyl palmitate	AP	Vegetable oils, synergists
Tocopherols	To	General use in fats, oils
Citric acid, esters	CA	Synergists, edible oils
Thiodipropionic acid		Synergist
Lecithin		Synergist
Carotenes		Frying oils, edible oils, emulsified oils
Silicone oil		Frying oils

the oxidation of antioxidants during the operation. Usually, antioxidants, being semipolar substances, are extracted using semipolar solvents, and the removal of solvents is performed under nitrogen. The extract may be purified by solid phase extraction on cartridges or using traditional column chromatography. Sometimes it is possible to use a precolumn for the subsequent chromatographic separation.

Antioxidants naturally present or added to foods may be partially converted into degradation products by reaction with oxygen or free radicals. Therefore, the content found by the analysis needs not exactly correspond to the amount originally present or the amount added to foods.

2. Determination of Antioxidants After Isolation from the Substrate

Three important methods exist for the determination of antioxidants: spectrophotometric methods, chromatographic methods, and electrochemical methods (Table 9).

Spectrophotometric Methods

Spectrophotometric methods are very simple and do not require expensive equipment. Usually, the redox capacity of antioxidants is utilized and the sample is treated with a suitable redox indicator. The change of coloration is then recorded.

Another method is the reaction with ferric salts, with their reduction into ferrous salts. A compound reacting with either ferrous or the residual ferric salts is then added, and the coloration of the complex produce is measured [121]. Determination of tocopherols using the Emmerie-Engel method is a typical example.

Another very common method is the reaction of an antioxidant with the phosphomolybdate complex and the measurement of the coloration of molybden blue produced by reduction (proposed by Folin and Ciocalteu, which is unspecific, as it is applicable for all phenolic, not only *ortho-* or *para-*disubstituted compounds).

Spectrophotometric methods are not sufficiently selective, as all polyphenolic compounds usually react similarly and are determined. Therefore, they are suitable only in cases where a single antioxidant has been used or after previous separation or if the determination of total antioxidants is sufficient.

Chromatographic Methods

Antioxidants may be determined by gas-liquid chromatography [119], mainly after derivatization (silylation of phenolic hydroxyls). Standard methods for determination are available. Semipolar columns, such as columns packed with Carbowax 20 M, are satisfactory.

Analysis using high-performance liquid chromatography (HPLC) has been widely used in the last two decades [120]. Antioxidants are best determined with use of reversed-phase chromatography on octadecylated silica gel with ultraviolet detection. For the determination of very polar substances, such as propyl gallate or ascorbic acid, the direct phase would be acceptable.

Table 9 Methods for Analysis of Antioxidants

Method	Examples	Applications
Spectrophotometric	Redox titration with dichloro-phenolindophenol	Ascorbyl palmitate, propyl gallate, tocopherols
Chromatographic	Emmerie-Engel, Folin-Ciocalteu	Pyrocatechols, BHA, BHT
Electrochemical	GLC after derivatization, reverse-phase HPLC, TLC, polarography, voltammetry	All antioxidants, gallates, ascorbic acid, tocopherols

Thin-layer chromatography (TLC) was often used before the development of HPLC, and several standard methods based on TLC are still available [121]. TLC might be used [122] with sophisticated apparatus (such as high-performance TLC or TLC-FID).

Electrochemical Methods

Electrochemical methods are based on the determination of the amount of electric energy necessary to oxidize the substrate. The determination of all three tocopherol types using pulsed voltammetry is a typical example [123]. It is possible to combine electrochemical methods with chromatographic methods, for instance, to use HPLC equipped with an electrochemical detector.

B. Determination of Antioxidant Activity

During the storage or heating of fats, oils, or foods, the rate of oxidation reactions is rather low in the beginning, as nearly all free radicals produced in the system are inactivated by antioxidants. This initial period is called the induction periods. When antioxidants present in the substrate are exhausted, free radicals are no longer deactivated; they remain in the system and initiate oxidation chains and the reaction rate rapidly increases.

When antioxidants have been added to food substrates, the induction period becomes longer as antioxidants, present in higher amounts than in the original substrate, are exhausted after a longer time than in a nonstabilized substrate. The relative increase of the induction period due to the addition of antioxidants is called the protection factor. Protection factors are often high in case of low-unsaturated substrates, containing no natural antioxidants, such as pork lard (of the order of 20–70). In contrast, they are low (of the order of 2–4) in polyunsaturated substrates, such as edible oils, containing tocopherols or other natural antioxidants.

The best method for the determination of antioxidant activity would be storage (or frying or any other application), but the procedure would be very long and expensive. Usually, it is not possible to wait a long time to learn the results. Therefore, accelerated methods, using higher temperature or catalysts, are almost exclusively used [124].

The oldest and most precise method is the Schaal oven test, where the sample is stored at 40 or 60°C under free access of oxygen. Analysis lasts several days to several months, therefore, methods using higher temperatures are often used. The longest known method of this type is the AOM (active oxygen method) or (earlier) Swift procedure [120], where the sample is aerated at 97.7 (a boiling water bath) or 110°C, gases are collected, either monitored by sensory analysis or collected in water, and changes of the carbonyl content of the water phase or changes of the conductivity are recorded. The increase in conductivity is due to the presence of lower fatty acids, especially formic acid, which originate in later stages of oxidation by secondary reactions. The Rancimat method is based on the same principle, but the procedure is automatic [125]. The Rancimat method has been used a greater deal recently due to its simplicity, availability, and reproducibility. Oxidograph (ML Aarhus, Denmark) records changes in the amount of absorbed oxygen under very similar conditions.

The results of accelerated methods do not correspond exactly to the results obtained under storage conditions [126]. Frankel [127] raised objections against methods using high temperatures and recommended using tests at lower temperatures—in the range of 20–60°C.

The ASTM (American Society for Testing Materials) Oxygen Bomb and Oxipres (ML Aarhus) methods are based on the recording of pressure changes during oxidation at high oxygen pressure but without aeration, which is closer to actual conditions of food preparation and storage.

Various methods operating at lower temperatures have been developed; in these cases, the oxidation is catalyzed either by photosensitizers or by addition of metals. Their results may not correspond to those obtained during storage in the dark or without the presence of metals.

The end of the induction period may be measured in different ways, such as on the basis of oxygen absorption or decrease of oxygen pressure, weight increase [128], increase of peroxide value, determination of conjugated dienes, or increase of 2-thiobarbituric acid. As already mentioned, the AOM and Rancimat methods are based on the formation of lower fatty acids. The best method is the determination of changes in the sensory value, but this method requires a great deal of time and trained personnel. It could be replaced by GLC determination of volatiles, such as pentane, hexanal, or 2,4-decadienals, which are responsible for off-flavours. The HPLC determination of total oxidized products is not sufficiently sensitive for determination of the induction period. The method used for the determination of antioxidant activity should operate under conditions very similar to those of real application.

C. Alternatives Methods for Stable Foods Without Antioxidants

Theoretically, it is possible to prepare stable foods without the addition of antioxidants. The easiest way is to modify the recipe to include components rich in natural antioxidants and to exclude components rich in polyunsaturated lipids. Another method is to prevent the access of oxygen, e.g., by impermeable packaging, often combined with packaging in a vacuum or in an inert gas, such as nitrogen. Oxygen may be removed by addition of oxygen scavengers, such as a combination of D-glucose and glucose oxidase.

Storage at very low temperatures may be recommended, too. Nevertheless, there is a danger that when water freezes, all water present becomes frozen so that the food material becomes dry; under these conditions, the protective layer of hydrated proteins is damaged and the lipid fraction leaks from the natural emulsions or liposomes, so that lipids become exposed to air oxygen.

The rapid turnover of foods possible in computerized food store facilities effectively eliminates the necessity of longer storage with the addition of antioxidants.

VI. FUTURE TRENDS

In the near future, it is possible that the application of synthetic antioxidants will decrease still further and that of natural antioxidants will increase. In the more remote future, it is possible that consumers will become more accepting of safe synthetic antioxidants as a rational alternative to natural antioxidants. Eventually the use of antioxidants will be gradually eliminated.

REFERENCES

1. A. Scanlon, L. A. Sather, and E. A. Day, Contribution of free fatty acids to the flavour of rancid milk, *J. Dairy Sci. 48*:1582 (1965).
2. B. Kellard, D. M. Busfield, and J. L. Kinderlerer, Volatile off-flavour compounds in desiccated coconut, *J. Sci. Food Agr. 36*:415 (1985).
3. J. Pokorný, Fats, oils and other lipids, *Chemical Changes During Food Processing* (J. Davidek, J. Velíšek, and J. Pokorný, eds.), Elsevier, Amsterdam, 1990, p. 169.
4. N. A. Porter, S. E. Caldwell, and K. A. Mills, Mechanism of free radical oxidation of unsaturated lipids, *Lipids 30*:277 (1995).
5. H.W.-S. Chan, The mechanism of autoxidation, *Autoxidation of Unsaturated Lipids* (H.W.-S. Chan, ed.), Academic Press, London, 1987, p. 1.
6. J. Pokorný, Major factors affecting the autoxidation of lipids, *Autoxidation of Unsaturated Lipids* (H.W.-S. Chan, ed.), Academic Press, London, 1987, p. 141.
7. H. W. Gardner, Lipoxygenase as a versatile biocatalyst, *JAOCS 73*:1347 (1996).
8. G. Piazza, *Lipoxygenase and Lipoxygenase Pathway Enzymes*, AOCS Press, Champaign, IL, 1996.
9. W. Grosch, Reactions of hydroperoxides—products of low molecular weight, *Autoxidation of Unsaturated Lipids* (H.W.-S. Chan, ed.), Academic Press, London, 1987, p. 95.
10. K. Warner, Sensory evaluation of oils and fat-containing foods, *Methods to Assess Quality and Stability of Oils and Fat-Containing Foods* (K. Warner and N. A. M. Eskin, eds.), AOCS Press, Champaign, 1995, p. 49.
11. P. J. White, Conjugated diene, anisidine value, and carbonyl value analyses, *Methods to Assess*

Quality and Stability of Oils and Fat-Containing Foods (K. Warner and N. A. M. Eskin, eds.), AOCS Press, Champaign, 1996, p. 159.
12. R. Marcuse and J. Pokorný, Modified TBA test: higher correlation with sensory evaluation of oxidative rancidity by modified TBA test, *Fat Sci. Technol. 96*:185 (1994).
13. R. Przybylski and N. A. M. Eskin, Methods to measure volatile compounds and the flavor significance of volatile compounds, *Methods to Assess Quality and Stability of Oils and Fat-Containing Foods* (K. Warner and N. A. M Eskin, eds.), AOCS Press, Champaign, 1995, p. 107.
14. P. B. Addis, Occurrence of lipid oxidation products in foods, *Food Chem. Toxicol. 24*:1021 (1986).
15. H. W. Gardner, Lipid hydroperoxide reactivity with proteins and amino acids, *J. Agr. Food Chem. 27*:220 (1979).
16. J. Pokorný, Flavor chemistry of deep fat frying in oil, *Flavor Chemistry of Lipid Foods* (D. B. Min and T. H. Smouse, eds.), AOCS Press, Champaign, 1989, p. 113.
17. D. G. Bradley and D. B. Min, Singlet oxygen oxidation of foods, *Crit. Rev. Food Sci. 31*:211 (1992).
18. J. Pospíšil, Antioxidants and related stabilizers, *Oxidation Inhibition in Organic Materials* (J. Pospíšil and P. P. Klemchuk, eds.), CRC Press, Boca Raton, FL, 1989, p. 33.
19. A. Hopia, S.-W. Huang, and E. N. Frankel, Effect of α-tocopherol and Trolox on the decomposition of methyl pherol and Trolox on the decomposition of methyl linoleate hydroperoxides, *Lipids 31*:357 (1996).
20. M. T. Satue, S.-W. Huang, and E. N. Frankel, Effect of natural antioxidants in virgin olive oils on oxidative stability of refined, bleached and deodorized olive oil, *JAOCS 72*:1131 (1995).
21. B. Freis, *Natural Antioxidants in Human Health and Disease*, Academic Press, San Diego, 1994.
22. O. I. Aruoma, Assessment of potential prooxidant and antioxidant actions, *JAOCS 73*:1617 (1996).
23. E. Niki, N. Noguchi, H. Tsuchihashi, and N. Gotoh, Interaction among vitamin C, vitamin E and β-carotene, *Am. J. Clin. Nutr. 62*:1322S (1995).
24. E. N. Frankel, A. L. Waterhouse, and P. L. Teissedre, Principal phenolic phytochemicals in selected California wines and their antioxidant activity in inhibiting oxidation of human low-density lipoproteins, *J. Agr. Food Chem. 43*:890 (1995).
25. O. I. Aruoma, Characterization of drugs as antioxidant prophylactics, *Free Radical Biol. Med. 20*:675 (1996).
26. O. L. Aruoma, Assessment of potential prooxidant and antioxidant actions, *JAOCS 73*:1617 (1996).
27. R. J. O'Brien, Oxidation of lipids in biological membranes and intracellular consequences, *Autoxidation of Unsaturated Lipids* (H.W.-S. Chan, ed.), Academic Press, London, 1987, p. 233.
28. S. Therisson, F. Gunstone, and R. Hardy, The antioxidant properties of ethoxyquin and of some of its oxidation products in fish oil and meal, *JAOCS 69*:806 (1992).
29. L. Kouřimská, J. Pokorný, and G. Tirzitis, The antioxidant activity of 2,6-dimethyl-3,5-diethyoxycarbonyl-1,4-dihydropyridine in edible oils, *Nahrung 37*:91 (1993).
30. J. Pokorný, Stabilization of foods, *Oxidation Inhibition of Organic Materials*, Vol. I (J. Pospíšil and P. P. Klemchuk, eds.), CRC Press, Boca Raton, FL, 1989, p. 347.
31. J. F. Hudson, *Food Antioxidants*, Elsevier Applied Science, London, 1990.
32. K. Omura, Antioxidant synergism between butylated hydroxyanisole and butylated hydroxytoluene, *JAOCS 72*:1565 (1993).
33. M. Raccachi, The antimicrobial activity of phenolic antioxidants in foods, *J. Food Safety 6*:141 (1984).
34. K. Kikuyawa, A. Kunugi, and T. Kurechi, Chemistry and implications of degradation of phenolic antioxidants, *Food Antioxidants*, Elsevier, Barking, 1990, p. 65.
35. J. Pokorný, Natural antioxidants for food use, *Trends Food Sci. Technol. 2*:223 (1991).
36. J. Löliger, Natural antioxidants for the stabilization of foods, *Flavor Chemistry* (D. B. Min and T. Smouse, eds.), AOCS, Champaign, 1989, p. 302.
37. A. Kasuga, Y. Ayonagi, and T. Sugahara, Antioxidant activities of edible plants, *Nippon Shokuhin Kogyo Gakkaishi 35*:828 (1988).

38. K. Herrmann, Hydroxyzimtsäuren und Hydroxybenzoesäuren enthaltende Naturstoffe in Pflanzen, *Fortschr. Chem. Org. Naturst. 35*:73 (1978).
39. J.-G. Shen, B.-L. Zhao, M.-F. Li, Q. Wan, and W.-J. Xin, Inhibitory effects of *Ginkgo biloba* extract on oxygen free radicals, nitric oxide, and myocardial injury in isolated ischemic-reperfusion hearts, *Proc. Int. Symp. Natural Antioxidants*, AOCS Press, Champaign, 1995, p. 453.
40. E. P. Oliveto, Nordihydroguaiaretic acid, *Chem. Ind.*:677 (1972).
41. C. J. Dillard, V. C. Gavino, and A. L. Tappel, Relative antioxidant effectiveness of α-tocopherol and γ-tocopherol in iron loaded rat, *J. Nutr. 113*:2266 (1983).
42. F. V. Timmermann, Tocopherole-Antioxidative Wirkung bei Fetten und Ölen, *Fat Sci. Technol. 92*:201 (1990).
43. C. Mariani and G. Bellan, Sulla presenza di tocoferoli, diidrotocoferoli, tocodienoli, tocotrienoli negli oli vegetabili, *Riv. Ital. Sost. Grasse 73*:533 (1996).
44. N. Olejnik, M. Gogolewski, and M. Nogala-Ka^{3}ucka, Isolation and some properties of plastochromanols, *Nahrung 41*:109 (1997).
45. A. Kamal-Eldin and L.-A. Appelqvist, The chemistry and antioxidant properties of tocopherols and tocotrienole, *Lipids 31*:671 (1996).
46. D. Boskou, Olive oil composition, *Olive Oil Chemistry and Technology* (D. Boskou, ed.), AOCS Press, Champaign, 1996, p. 52.
47. D. M. Colguhoun, B. J. Hieks, and A. W. Reed, Phenolic content of olive oil is reduced in extraction and refining, *J. Clin. Nutr. 5*:105 (1996).
48. A. Kamal-Eldin and L.-A. Appelqvist, Variations in the composition of sterols, tocopherols and lignans in seed oils from four *Sesamum* species, *JAOCS 71*:149 (1994).
49. Y. Fukuda and M. Namiki, Recent studies on sesame seed and oil, *Nippon Shokuhin Kogyo Gakkaishi 35*:552 (1988).
50. F. Shahidi, R. Amarowicz, H. A. Abou-Gharbia, and A. A. Y. Shehata, Endogenous antioxidants and stability of sesame oil as affected by pressing and storage, *JAOCS 74*:143 (1997).
51. A. H. Y. Abdel, Flavonoids of cotton seeds and peanuts as antioxidants, *Riv. Ital Sost. Grasse 62*:147 (1985).
52. P. D. Duh, D.-B. Yeh, and G.-C. Yen, Extraction and identification of an antioxidative component from peanut hulls, *JAOCS 69*:814 (1992).
53. C. C. Whittern, E. E. Miller, and D. E. Pratt, Cottonseed flavonoids as lipid antioxidants, *J. Am. Oil Chem. Soc. 61*:1075 (1984).
54. D. E. Pratt, Natural antioxidants from soybeans and other oilseeds, *Autoxidation of Food and Biological Systems* (M. C. Simic and M. Karel, eds.), Plenum Press, New York, 1980, p. 283.
55. K. Murata, Antioxidative stability of tempeh, *J. Am. Oil Chem. Soc. 65*:799 (1988).
56. R. Zadernowski, H. Nowal-Polakowska, and B. Losow, Natural antioxidants in seeds of plant species, *Acta Acad. Sci. Techn. Olsten 27*:107 (1995).
57. R. Amarowicz, M. Karamac, J. P. D. Wanasundara, and F. Shahidi, Antioxidant activity of hydrophobic phenolic fractions of flaxseed, *Nahrung 41*:178 (1997).
58. R. Zadernowski, H. Nowak-Polakowska, and I. Konopka, Effect of heating on antioxidative activity of rapeseed and evening primrose extracts, *Pol. J. Food Nutr. Sci 5*:15 (1996).
59. K. J. Duve and P. J. White, Extraction and identification of antioxidants in oats, *J. Am. Oil Chem. Soc. 68*:365 (1991).
60. M. Diack and M. Saska, Separation of vitamin E and γ-oryzanols from rice bran by normal phase chromatography, *JAOCS 71*:1211 (1995).
61. A. Troszyñska, A. Bednarska, A. Latosz, and H. Koz^{3}owska, Polyphenolic compounds in the seed coat of legume seeds, *Pol. J. Food Nutr. Sci. 6*:37 (1997).
62. B. Grzeskowiak, Z. Pazola, and M. Goglewski, Yellow pea and its products as a source of natural antioxidants, *Acta Alim. Polon. 13*:122, 141 (1987).
63. T. Tsuda, T. Osawa, T. Nakayama, S. Kawakishi, and K. Ohshima, Antioxidant activity of pea bean extract, *JAOCS 70*:909 (1993).
64. M. Oyaizu, H. Ogihara, K. Sekemoto, and U. Naruse, Antioxidant activity of extracts from hop, *Yukagaku 42*:1003 (1993).

65. D. Jurdi-Haldeman, J. H. MacNeil, and D. M. Yared, Antioxidant activity of onion and garlic juices in stored cooked ground lamb, *J. Food Prot. 50*:411 (1987).
66. T. Hirosue, M. Matsuzawa, I. Irie, H. Kawai, and Y. Hosogai, On the antioxidative activities of herbs and spices, *Nippon Shokuhin Kogyo Gakkaishi 35*:631 (1988).
67. B. Xie, H. Shi, Q. Chen, and C. T. Ho, Antioxidant properties of fractions and polyphenol constituents from green oolong and black teas, *Life Sci. 17*:77 (1993).
68. Z. Y. Chen, P. T. Chan, H. M. Ma, K. P. Fung, and J. Wang, Antioxidative effect of ethanol tea extracts on oxidation of canola oil, *JAOCS 73*:375 (1996).
69. T. Matsuzaki and Y. Hara, Antioxidative activity of tea leaf catechins, *Nippon Shokuhin Kogyo Gakkaishi 59*:129 (1985).
70. K. Fujimoto, H. Omura, and T. Kaneda, Screening for antioxygenic compounds in marine algae and bromophenols as effective principles in a red alga, *Bull. Jpn. Soc. Sci. Fish. 51*:1139 (1985).
71. N. Naketani, *Antioxidant and Antimicrobial Constituents of Herbs and Spices*, Elsevier, Amsterdam, 1994.
72. I. D. Lee, Y. S. Kim, and C. R. Ashmore, Antioxidant property in ginger rhizome and its application to meat products, *J. Food Sci. 51*(11):20 (1986).
73. M. Takácsová, A. Príbela, and M. Faktorová, Study of antioxidant effects of thyme, sage, juniper and oregano, *Nahrung 39*:241 (1995).
74. H. Kikuzaki, and N. Nakatani, Structure of a new antioxidative phenolic acid from oregano, *Agr. Biol Chem. 53*:519 (1989).
75. N. V. Yanishlieva and E. M. Marinova, Antioxidant activity of selected species of the family *Lamiaceae* grown in Bulgaria, *Nahrung 39*:458 (1995).
76. M. E. Cuvelier, H. Richard, and C. Berset, Antioxidant activity and phenolic composition of pilot plant and commercial extracts of sage and rosemary, *JAOCS 73*:645 (1996).
77. S. Ivanov, A. Seher, and H. Schiller, Antioxidants in fatty oil of *Foeniculum vulgare* Miller, 2, *Fette Seifen Anstrichm. 81*:105 (1979).
78. R. S. Faraq, A. Z. M. A. Badei, and G. S. A. El-Baroty, Influence of thyme and clove essential oils in cottonseed oil oxidation, *J. Am. Oil Chem. Soc. 66*:800 (1989).
79. O. I. Aruoma, B. Halliwell, R. Aeschbach, and J. Löliger, Antioxidant and pro-oxidant properties of active rosemary constituents: carnosol and carnosic acid, *Xenobiotica 22*:257 (1992).
80. H. F. Liu, A. M. Booren, J. I. Gray, and R. L. Czechel, Antioxidant efficacy of oleoresin rosemary and sodium tripolyphosphate in restructured pork steaks, *J. Food Sci. 57*:803 (1992).
81. S. Chevolleau, J. F. Mallet, E. Ucciani, J. Gamisano, and M. Gruber, Antioxidant activity in leaves of some Mediterranean plants, *JAOCS 69*:1269 (1992).
82. M. H. Gordon and X. C. Weng, Antioxidant properties of extracts from tanshen, *Food Chem. 44*:119 (1992).
83. X. C. Weng and M. H. Gordon, Antioxidant activity of quinones extracted from tanshen (*Salvia orrhiza* Burge), *J. Agr. Food Chem. 40*:1331 (1992).
84. J. W. Wu, M.-H. Lee, C. T. Ho, and S. S. Chang, Elucidation of the chemical structure of natural antioxidants from rosemary, *J. Am. Oil Chem. Soc. 59*:339 (1982).
85. S. S. Chang, B. Matijasevic, O. A. L. Hsieh, and C. L. Huang, Natural antioxidants from rosemary and sage, *J. Food Sci. 42*:1102 (1978).
86. V. Holzmannová, Kyselina rosmarinová a její biologická aktivita, *Chem. Listy 90*:486 (1995).
87. S. L. Richheimer, M. W. Bernart, G. A. King, M. C. Kent, and D. T. Bailey, Antioxidant activity of lipid-soluble phenolic diterpenes from rosemary, *JAOCS 73*:507 (1996).
88. K. D. Economou, V. Oreopoulou, and C. D. Thomopoulos, Antioxidant activity of some plant extracts of the family *Labiateae*, *J. Am. Oil Chem. Soc. 68*:109 (1991).
89. R. E. Kramer, Antioxidants in clove, *J. Am. Oil Chem. Soc. 62*:111 (1985).
90. M. Takácsová and A. Príbela, Antioxidant effect of some phytoncides from spices, *Bull. Potrav. Výsk. 32*:67 (1993).
91. N. Noguchi, E. Komuro, E. Niki, and R. L. Wilson, Action of curcumin as an antioxidant against lipid peroxidation, *Yukagaku 43*:1045 (1994).

92. Y. H. Kim, H. K. Yoon, and K. S. Chang, The contents of phenolic compounds and antioxidant activities in various year stored red and white ginsengs, *Res. Rep. Agr. Sci. Technol. 11*:295 (1984).
93. S. Phelps and W. S. Harris, Garlic supplementation and lipoprotein oxidation susceptibility, *Lipids 28*:475 (1973).
94. M. Takácsová, M. Drdák, I. Šimek, and P. Szalai, Investigation of antioxidant effects of garlic extract in vegetable oil, *Potrav. Vědy 8*:141 (1990).
95. H.-S. Lee and A. S. Csallany, The influence of vitamin E and selenium on lipid peroxidation and aldehyde dehydrogenase activity in rat liver and tissue, *Lipids 29*:345 (1994).
96. C. K. Chow, V. L. Tatum, C. C. Yeh, N. H. Maynard, W. Ibrahim, G. Bruckner, G. A. Boissonneault, and C. B. Hong, Antioxidant function of carnosine, a natural histidine-containing dipeptide, Proceedings of International Symposium on Natural Antioxidants (L. Packer, ed.), AOCS Press, Champaign, 1996, p. 155.
97. W. Janitz, J. Korczak, and A. Ciepelewska, Ocena wlaœciwoœci przeciwutleniaj[1]cych i jeho pochodnych w miêsie odzyskanym mechaniczne z drobiu, *Roczn. Akad. Roln. Poznaniu 233*:37 (1992).
98. E. M. Marinova and N. V. Yanishlieva, On the activity of ascorbyl palmitate during the autoxidation of two types of lipid systems in the presence of α-tocopherol, *Fat Sci. Technol. 94*:448 (1992).
99. Z. Réblová and J. Pokorný, Effect of lecithin on the stabilization of foods, *Phospholipids: Characterization, Metabolism, and Novel Biological Applications* (G. Cevc and F. Paltauf, eds.), AOCS Press, Champaign, 1995, p. 378.
100. T. Segawa, S. Hara, and Y. Totani, Synergistic effect of phospholipids for tocopherol, *Yukagaku 43*:515 (1994).
101. Y. Totani, Antioxidant effect of nitrogen-containing phospholipids on autoxidation of polyenoic oil, *Yukagaku 46*:3 (1997).
102. B. J. F. Hudson and J. I. Lewis, Polyhydroxy flavonoid antioxidants for edible oils: phospholipids as synergists, *Food Chem. 10*:111 (1983).
103. S. H. Lee and D. B. Min, Effects, quenching mechanisms, and kinetics of carotenoids in chlorophyll-sensitized photooxidation of soybean oil, *J. Agr. Food Chem. 38*:1630 (1990).
104. C. Y. W. Ang and D. Hamm, Effect of salt and sodium tripolyphosphate on shear, thiobarbituric acid, sodium and phosphorus values of hot-stripped broiler breast meat, *Poultry Sci. 65*:1532 (1986).
105. H. Kusaka, K. Tokue, K. Morino, S. Ohta, and K. Yokomizo, Influence of silicone oil on the dissolution of iron into heated fats and oils, *Yukagaku 35*:1005 (1986).
106. V. Stankova and S. Ivanov, Studies on the efficient complex stabilization of lard, *Meat Ind. 17*:32 (1984).
107. W. Heimann and H. von Pezold, Über die prooxygene Wirkungen von Antioxygenen, *Fette Seifen Anstrichm. 59*:330 (1957).
108. E. N. Frankel, S.-W. Huang, J. Kanner, and J. B. German, Interfacial phenomena in the evaluation of antioxidants, *J. Agr. Food Chem. 42*:1054 (1994).
109. N. Watanaabe, S. Hara, and Y. Totani, Antioxidative effects of phospholipids in an emulsifying system, *Yukagaku 46*:21 (1997).
110. E. N. Frankel, S. W. Huang, R. Aeschbach, and E. Prior, Antioxidant activity of rosemary extract and its constituents in bulk oil and oil-in-water emulsions, *J. Agr. Food Chem. 44*:131 (1996).
111. J. Löliger, The use of antioxidants in foods, *Free Radicals and Food Additives* (O. I. Aruoma and B. Halliwell, eds.), Taylor and Francis, London, 1991, p. 121.
112. V. N. Wade, R. El-Tahriri, and R. J. M. Crawford, The autoxidative stability of anhydrous milk fat with and without antioxidants, *Milchwissenschaft 41*:479 (1986).
113. J. Pokorný, J. Davídek, V. Chocholatá, J. Pánek, H. Bulantová, W. Janitz, H. Valentová, and M. Vierecklová, Interactions of oxidizing ethyl linoleate with collagen, *Nahrung 34*:159 (1990).
114. T. P. Labuza and W. M. Breene, Application of active packaging for improvement of shelf life and nutritional quality of fresh and extended shelf-life foods, *J. Food Proc. Pres. 13*:1 (1989).

115. K. Terada and Y. Naito, Migration of BHT through airspace in food package system, *Packag. Technol. Sci. 2*:165 (1989).
116. R. Haigh, Safety and stressity of antioxidants: EEC approval, *Food Chem. Toxicol. 24*:1031 (1986).
117. D. L. Madhavi and D. K. Salunkhe, Toxicological aspects of food antioxidants, *Food Antioxidants* (D. L. Madhavi, S. S. Deshpande, and D. K. Salunkhe, eds.), Marcel Dekker, New York, 1995, p. 371.
118. K. Roberts and S. Dilli, Analytical chemistry of synthetic food antioxidants, *Analyst 112*:933 (1987).
119. *Official Methods of Analysis of AOAC INTERNATIONAL* 16th ed., 3rd rev., AOAC, Washington, DC, 1997.
120. *AOCS Official Methods and Recommended Practices*, AOCS Press, Champaign, 1993.
121. *IUPAC Standard Methods for the Analysis of Oils, Fats and Derivatives*, Blackwell, Oxford, 1987.
122. C. H. van Peteghem and D. A. Kekeyser, Systematic identification of antioxidants in lards, shortenings and vegetable oils by TLC, *J. Ass. Off. Anal. Chem. 64*:1331 (1981).
123. S. S. Atuma, Electrochemical determination of vitamin E in margarine, butter and palm oil, *J. Sci. Food Agr. 26*:393 (1975).
124. P. J. Wan, Accelerated stability methods, *Methods to Assess Quality and Stability of Oils and Fat-Containing Foods* (K. Warner and N. A. M. Eskin, eds.), AOCS Press, Champaign, 1994, p. 179.
125. W. J. Woestenberg and J. Zaalberg, Determination of the oxidative stability of edible oils—interlaboratory test with the automated Rancimat method, *Fette Seifen Anstrichm. 88*:53 (1986).
126. G. Reynhout, The effect of temperature on the induction time of stabilized oil, *J. Am. Oil Chem. Soc. 68*:483 (1991).
127. E. N. Frankel, In search of better methods to evaluate natural antioxidants and oxidative stability in food lipids, *Trends Food Sci. Technol. 4*:220 (1993).
128. J. A. García-Mesa, M. D. Luque de Castro, and M. Valcárcel, Factors affecting the gravimetric determination of the oxidative stability of oils, *JAOCS 70*:245 (1993).
129. R. S. Farag, A. Z. M. A. Badei, and G. S. A. El-Baroty, Influence of thyme and clove essential oils on cottonseed oil oxidation, *J. Am. Oil Chem. Soc. 66*:800 (1988).
130. C. Banias, V. Oreopoulou, and C. D. Thomopoulus, The effect of primary antioxidants and synergists on the activity of plant extracts in lard, *J. Am. Oil Chem. Soc. 69*:520 (1992).
131. S. A. Vekiari, V. Oreopoulou, C. Tzia, and C. D. Thomopoulos, Oregano flavonoids as lipid antioxidants, *J. Am. Oil Chem. Soc. 70*:483 (1993).
132. M. Oyaizu, H. Ogihara, K. Sekemoto, and U. Naruse, Antioxidative activity of extracts from hops, *Yukagaku 42*:1003 (1993).
133. N. Nakatani, Antioxidative compounds from spices, *Nippon Nogeikagaku Kaishi 62*:170 (1987).

11

Water Activity and Food Preservation

M. Shafiur Rahman
Horticulture and Food Research Institute of New Zealand, Auckland, New Zealand

Theodore P. Labuza
The University of Minnesota, St. Paul, Minnesota

I. BASICS OF WATER ACTIVITY

A. Terminology

1. Water Activity

Water is an important constituent of all foods. In the middle of the twentieth century scientists began to discover the existence of a relationship between the water contained in a food and its relative tendency to spoil. They also began to realize that the active water could be much more important to the stability of a food than the total amount of water present. Scott [150] clearly stated that the water activity of a medium correlated with the deterioration of food stability due to the growth of microorganisms. Thus, it is possible to develop generalized rules or limits for the stability of foods using water activity. This was the main reason why food scientists started to emphasis water activity rather than water content. Since then, the scientific community has explored the great significance of water activity in determining the physical characteristics, processes, shelf life, and sensory properties of foods. The water activity of fresh foods, as shown by Chirife and Fontan [37], is 0.970–0.996.

Water activity, a thermodynamic property, is defined as the ratio of the vapor pressure of water in a system to the vapor pressure of pure water at the same temperature, or the equilibrium relative humidity of the air surrounding the system at the same temperature.

A number of methods have been reported in the literature to measure or estimate the water activity of foods. Water activity–measurement methods include the following: colligative properties method, equilibrium sorption rate method, salt hygroscopicity method, hygrometric instrument method, isopiestic transfer method, and suction (matric) potential method. The accuracy of most methods lies in the range of 0.01–0.02 water activity units [140]. Details of the various measurement techniques are described by Labuza [102], Rizvi [140], and Rahman [134].

2. Sorption Isotherm

The moisture sorption isotherm is the dependence of moisture content on the water activity of one sample at a specified temperature. It is usually presented in graphical form or as an equation. Brunauer

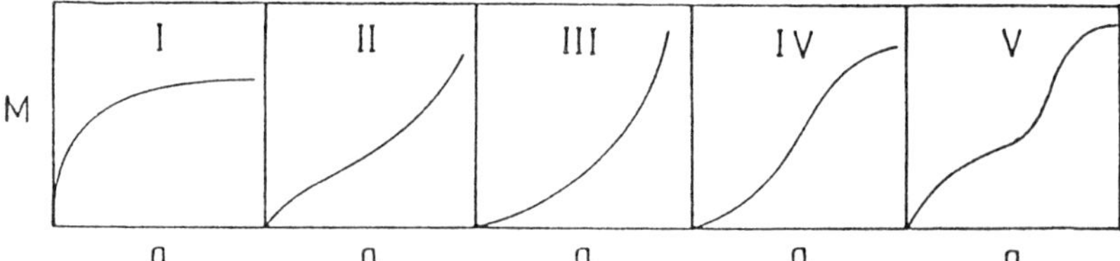

Figure 1 The five types of van der Waals adsorption isotherms proposed by Brunauer et al. (From Ref. 23.)

et al. [23] classified adsorption isotherms of materials into five general types (Fig. 1). If water-soluble crystalline components are present in foods, e.g., sugars or salts, the isotherm appears as concave shape type III. Most other foods result in sigmoid isotherm type II. The inflection point of the isotherm indicates the change of water-binding capacity or of the relative amounts of free and bound water. Type I is indicative of a nonswelling porous solid, such as silicate anticaking agents.

For most practical purposes, the isotherm is presented in an empirical or theoretical model equation. However, none of the isotherm models in the literature is valid over the entire water activity scale of 0–1. The GAB model is one of the most widely accepted models for foods over a wide range of water activities from 0.10 to 0.9. The details of the isotherm models with their parameters have been compiled by Rizvi [140], Okos et al. [131], Lomauro et al. [113,114], and Rahman [134].

3. Hysteresis

The difference in the equilibrium moisture content between the adsorption and desorption curves is called hysteresis and is shown in Figure 2. In region II of this figure, the water is less tightly held and is usually present in small capillaries, whereas in region III, water is held loosely in large capillaries or is free [59].

Hysteresis in sorption has important theoretical and practical implications in foods. The theoretical implications are evidence of irreversibility of the sorption process and the validity of the equations derived based on equilibrium thermodynamic functions. The practical implications deal with the effects of hysteresis on chemical and microbiological deterioration and with its importance in low- and intermediate-moisture foods [91]. Strasser [164] and Wolf et al. [178] maintain that changes in hysteresis could be used as an index of quality deterioration, since hystersis loops of foods change with storage time, but this is a poor method.

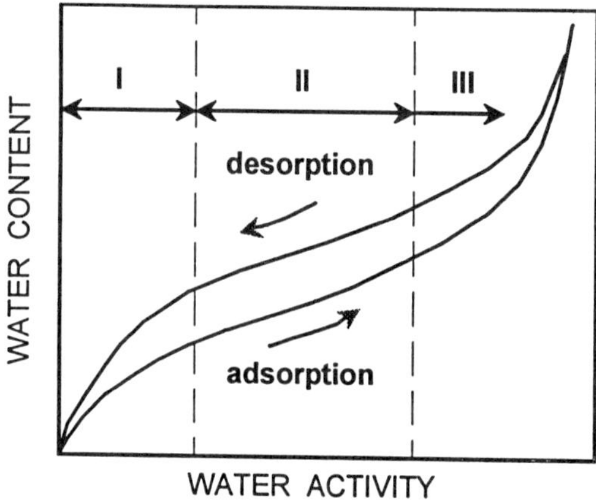

Figure 2 Sorption isotherm for typical food product showing hysteresis.

Factors Affecting Hysteresis

The desorption hysteresis loop usually ends at the monolayer, but in some cases it extends down to an activity of zero [99]. In foods, a variety of hysteresis loop shapes can be observed, depending on the type of food and the temperature [178]. The principal factors affecting hysteresis are composition of the product, isotherm temperature, storage time before isotherm measurement, pretreatments, drying temperature, and the number of successive adsorption and desorption cycles.

Types of foods affecting hysteresis: Variations in hysteresis can be grouped into three types of foods [91]:

1. Hysteresis in high-sugar foods: In high-sugar or high-pectin foods such as air-dried apple, hysteresis occurs mainly in the monomolecular layer water region, below the first inflection point of isotherm region I in Figure 2 [131]. Although the total hysteresis is large, there is no hysteresis above 0.65.
2. Hysteresis in high-protein foods: In pork, a moderate hysteresis begins at about 0.85 (i.e., in the capillary condensation region) [92].
3. Hysteresis in high-starch foods: In starchy foods a large hysteresis loop occurs, with a maximum at about water activity 0.70, which is within the capillary condensation region [131]. However, in kudzu starch hysteresis continues up to 0.90 [14].

Temperature effects on hysteresis: Total hysteresis decreases as sorption temperature increases [178]. Desorption isotherms usually give a higher water content than adsorption isotherms. In general, the types of changes encountered upon adsorption and desorption will depend on the initial state of the sorbent (amorphous vs. crystalline), the transitions taking place during adsorption, and the speed of desorption [92]. Chinachoti and Steinberg [30] found hysteresis in sugar-containing starch systems up to 0.60 and Bolin [16] in resin (with very high sugar content) up to 0.30. Tsami et al. [170] observed significant hysteresis below 0.5 or 0.6 and at temperatures about 30°C in fruits (raisin, currant, fig, prune, and apricot) and mentioned that absence of hysteresis at higher temperatures was due to the dissolution of sugars at high temperatures. The water activity below which a significant hysteresis effect was manifested was inversely proportional to the sugar content of the fruits [170]. The magnitude of the hysteresis loop decreased appreciably at higher temperatures. The effect of temperature was more pronounced on desorption isotherms than on adsorption isotherms [6].

Iglesias and Chirife [82] estimated and compared the isosteric heats of water adsorption and desorption for a number of foods and reported that the effect of temperature on the magnitude of hysteresis varied. There was no direct relationship between the observed differences in adsorption and desorption heats and the distribution of hysteresis along the isotherm. For some foods (thyme, winter savory, sweet marjoram, cooked trout, raw and cooked chicken, and tapioca), increasing temperature decreased or eliminated hysteresis, while for others the total hysteresis remained constant (ginger and nutmeg) or even increased (anise, cinnamon, chamomile, and coriander) [91].

Effects of the physicochemical nature of foods on hysteresis: The type of changes encountered upon adsorption and desorption will depend on (a) the initial state of the sorbent (amorphous versus crystalline), (b) the transitions taking place during adsorption, (c) the final water activity adsorption point, and (d) the rate of sorption. If the saturation point has been reached and the material has gone into sorption, rapid desorption may preserve the amorphous state due to supersaturation [91].

Some water remained after desorption at dry conditions even after prolonged storage due to hydrogen-bonded trapped water in amorphous sugar microregions as well as water of crystallinity [46,178].

Effects of the sorption cycle on hysteresis: In some cases hysteresis seems to be reproducible a second time [7,164], and for some cases the second sorption-desorption cycle resulted in the elimination of hysteresis [6].

Elimination of hysteresis upon the second or subsequent cycles may take place for a variety of reasons, including change in crystalline structure when a new crystalline form persists in subsequent cycles [8], swelling and increased elasticity of capillary walls resulting in a loss of power of trapping water [136,137], denaturation [42], surface-active agents [144], gelatinization [172], and even mechanical treatment, which may effect the capillary structure [91].

Theories of Sorption Hysteresis

Several theories have been formulated to explain the phenomenon of hysteresis. At present no theory has given complete insight into the several mechanisms responsible [170], and no quantitative prediction of hysteresis is available in the literature. Several theories have been proposed to explain hysteresis. The mechanisms proposed in the literature to cause hysteresis are discussed in the following sections.

Capillary condensation: This can mainly explain hysteresis in nonswelling porous solids. Capillary condensation can be explained using the Kelvin equation. Due to the presence of impurities (dissolved gases, etc.) the contact angle of the receding film upon desorption is smaller than that of the advancing film upon adsorption. Therefore, capillary condensation along the adsorption branch of the moisture sorption isotherm will be at a higher relative vapor pressure [92].

Ink bottle theory: Rao [138] assumed capillaries to be composed of narrow necks with a large pore, somewhat like an ink bottle. On adsorption, the capillary will not completely fill until the water activity corresponding to the larger radius of the pore is reached. During desorption, the smaller radius of the pore neck controls the emptying of the capillary, so that activity is lowered considerably [99]. This theory was confirmed by Labuza et al. [98] for a cellulose model system. Cohan [45] elaborated upon the open-pore theory by extending the bottleneck theory, including considerations of multilayer adsorption. This was based on the difference as affected by the shape of the meniscus.

Mechanisms of physicochemical changes in components: The physicochemical changes in food components also cause hysteresis, such as deformability and elastic stresses of the sorbent, a deformation of the polypeptide chains within the protein molecule [91], and the energy surplus of unfolding (swollen) protein phase transition [27].

Kapsalis [91] mentioned that adsorption from the dry state by biopolymers is due to (a) side chain amino groups, (b) end carboxylic and other groups, (c) peptide bonds, and (d) secondary structure. In general, below 0.50 water activity the main sites of sorption are the polar side chain groups. The contribution of the polypeptide chain becomes progressively more important at higher activities, for example, at 0.80 activity the peptide bonds account for almost half the adsorbed water in wool keratin. Deamination of methylation of side chain groups in wool and benzoylation in casein did not show any appreciable changes of hysteresis [123]. This suggests that it is the main chains of the biopolymers that are primary responsible [91].

Seehof et al. [152] supported the polar group interpretation of hysteresis, where binding mainly involves the free basic groups of the protein. Kapsalis [91] showed (from the data of Seehof et al. [152]) a correlation of the maximum amount of hysteresis with the sum of arginine, histidine, lysine, and cystine groupings. Besides the free basic groups of the protein molecule, sulfur linkages are also of prime importance in hysteresis [160]. In contrast to this work, hysteresis in casein was observed to be independent of the content of free amino groups [123]. Thus, a twofold nature of hysteresis was proposed: constant hysteresis, independent of the relative humidity desorption point, and hysteresis proportional to the amount adsorbed above the upper adsorption break of the isotherm [91].

Bettelheim and Ehrlich [9] suggested that in a swelling polymer, hysteresis depends on the mechanical constraints contributed by the elastic properties and cannot be interpreted by capillary condensation. Van Olphen [173] described retardation of adsorption due to the development of elastic stress in crystallites during the initial peripheral penetration of water between the unit layers. The shift toward higher relative vapor pressure during adsorption is caused by the activation energy required to open the unit layer stacks. The glass-rubber transition during adsorption and desorption may also cause hysteresis due to the nonequilibrium state of phase transition.

Structural collapse: With sorption, the capillary pores of the adsorbent become elastic and swell. Upon desorption, the removal of water causes shrinkage, and a general collapse of the capillary porous structure occurs. Alteration of structure causes elimination of hysteresis due to the absence of capillary condensation [91]. The collapse of capillaries during desorption also affects sorption hysteresis.

4. Water Activity Shift and Break in Isotherm

Water Activity Shift by Temperature

The isotherm shift due to temperature can usually be estimated by the well-known Clausius-Clapeyron equation [106]:

$$\ell n \frac{(a_w)_2}{(a_w)_1} = \frac{q + \lambda_w}{R} \left[\frac{1}{T_2} - \frac{1}{T_1} \right] \tag{1}$$

The slope of a plot of $\ell n a_w$ versus 1/T should give:

$$\text{slope} = \left[\frac{q + \lambda_w}{R} \right] \tag{2}$$

where q is the excess heat of sorption (kJ/kg) and λ_w is the latent heat of vaporization for water (kJ/kg). The excess heat of sorption was reviewed by Rahman [134] for a wide range of foods.

The typical water activity shift with temperature at constant moisture content is shown in Figure 3. The water activity shift by temperature is mainly due to the change in water binding, dissociation of water, physical state of water, or increase in solubility of solute in water.

It is widely accepted that an increase in temperature results in a decreased equilibrium moisture content in foods (Fig. 3A). Tsami et al. [170] found similar results for dried fruits up to a water activity of about 0.55–0.70. In that region the curves for several temperatures intersect. At water activity values higher than 0.7, there was an inversion of the effect of temperature (i.e., equilibrium moisture content increased with temperature) due to an increase in solubility of sugars in water. The intersection (or inversion) point depends on the composition of the food and the solubility of sugars [177]. For sultana raisins and currants the inversion point was about 0.55, likewise, 0.65 for fig, 0.70 for prune, 0.75 for apricot (possessing the lowest sugar content of fruit) [170], 0.55 for quince jam (Ferbar brand), and 0.65 for quince jam (Tapada nova) [148]. A similar intersection was also found by Saravacos et al. [147] for sultana raisins, by Weisser et al. [177] for sugar alcohol, and by Hellen and Gilbert [78] for biscuits. Apple does not show an intersection [143]. For products with protein or starch content, there is also no intersection point with an increase in temperature [6].

Water Activity Break Due to Physiochemical Change

In a pure component isotherm, the change of solute from the amorphous state to a crystal affects the isotherm. A break is observed in the isotherm, as shown in Figure 4A. In some foods one part of the solute (salts and sugar) is bound to a polymer (protein and starch) and the other part is crystalline or

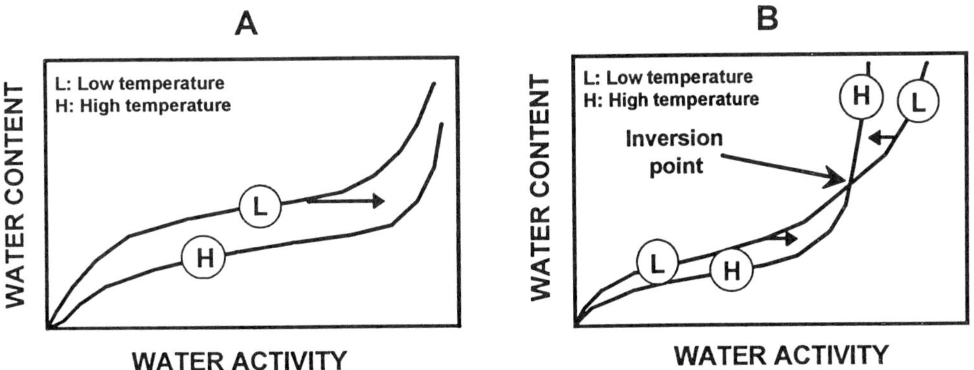

Figure 3 Water activity shift of food by temperature. (A) Shift without intersection. (B) Shift showing the point of intersection or inversion point.

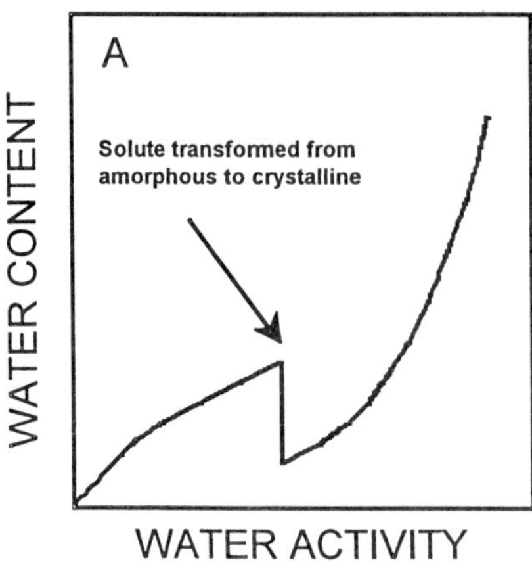

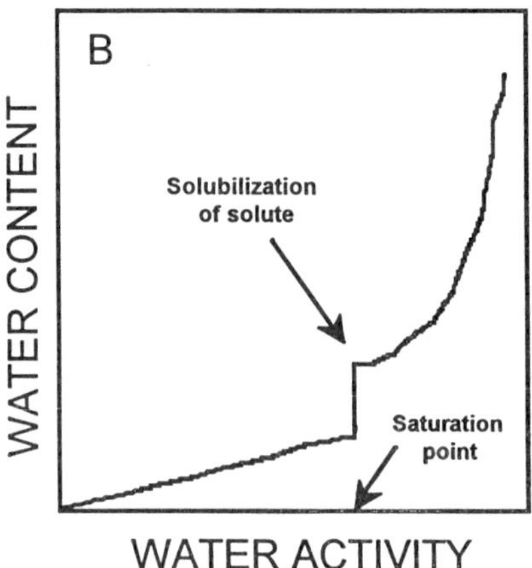

FIGURE 4 Water activity shift of food by physicochemical parameters. (A) Shift with solute transformation from amorphous to crystalline. (B) Shift with solubilization of solute.

amorphous. Bound and free forms of solutes are in equilibrium, which is strongly dependent on the actual water activity. If change in water activity take place slowly, this equilibrium may be maintained, whereas during rapid changes nonequilibrium conditions are likely to be attained.

Bound, crystalline, and amorphous solutes produce characteristic changes (i.e., break and shift) in the water sorption isotherm [65]. A typical curve showing break is shown in Figure 4B, where a break is also observed due to transformation of the solute from an amorphous to a crystalline state.

Concept of Local Isotherm

Rockland [141] proposed the concept of local isotherm to characterize the physical state or special type of water binding in foods. The local isotherm can be identified by graphical analysis of experimental sorption data according to Henderson's equation. The three localized isotherms may be distinguished by plotting experimental sorption data as $\ell n[-\ell n(1-a)]$ versus $\ell n M_w$. Three straight lines rather than a single straight line are observed, each being identified as a local isotherm. The three types of bound water that exist in foods are Type I (monolayer, unfrozen, localized hydrate, polar site, bound, oriented), Type II (multilayer, chemisorbed, intermediate), and Type III (mobile, free, capillary, solution).

Iglesias and Chirife [83] tested 235 isotherms of 71 different food products to confirm the Rockland local isotherm concept. They observed 174 isotherms having three local isotherms, whereas other showed isotherms having one to five local isotherms. The location of the point of intersection varied with the products type. The intersections of LI-1 and LI-2 should agree with the BET monolayer and LI-2 and LI-3 capillary water. Only 35% of isotherms tested showed the same first intersection as the BET monolayer. The second intercepts were found with a water activity range of 0.4–0.9. Karel [93] and Loncin [116] suggested that the capillary effects could only be expected above 0.90 water activity. Iglesias and Chirife [83] concluded that, although in a broad sense the local isotherms proposed by Rockland [141] may be related to the different modes of water binding, they cannot be used to give a precise and unequivocal definition of the physical state of water in foods. Moreover, the original Henderson equation should give only one curve—more curves present a poor fit. Thus, the Rockland concept of local isotherm cannot be used to characterize the physical state of water.

The interaction between a polymer and a solute may produce a break in the water sorption isotherm. The amount of sodium chloride bound to casein was highest at the saturation water activity ($a_w = 0.753$) [62,64,66] and decreased as water activity was increased. Similar results were found for paracasein and sodium chloride [74], for starch and sodium chloride [31], and for sucrose and starch

[32]. The isotherms of paracasein–sodium chloride [74], starch–sodium chloride [31], and starch-sucrose [32] mixtures were measured, and a shift at the saturation point of solute was observed. They plotted X_w versus log $(1 - a_w)$ as the Smith model for the systems, and two linear portions of the isotherm are observed with a break and shift at saturation. The bound, crystalline and amorphous solutes produce characteristic changes in the water sorption properties of polymer-solute-water [65]. The physical state of solutes also depends on the drying rate. Thus, a break in local isotherm can be used to estimate the polymer-solute-water interaction as well as any physicochemical change in the components of foods during processing and storage.

Isse et al. [86] described the range of moisture isotherm data of mango as two local isotherms for two different ranges of water activity. The net isosteric heat of sorption is used as a subdivision criterion for these two ranges. The first local isotherm is a water activity range where excess heat of sorption is positive and second local isotherm is when excess heat of sorption is negative. Thus, the second local isotherm should start at a value where net heat of sorption is zero.

6. Thermodynamic Properties Prediction

Thermodynamic properties such as the freezing point, boiling point and heat of sorption can also be predicted from water activity. For example the freezing point [Eq. (3)] and boiling point [Eq. (4)] can be estimated as follows [60]:

$$\ell n a_w = 9.6934 \times 10^{-3} \Delta + 4.761 \times 10^{-6} \Delta^2 \tag{3}$$

$$\ell n a_w = 1.1195 \times 10^{-4} \Delta^2 - 35.127 \times 10^{-3} \Delta \tag{4}$$

The kinetics of mass transport is known to depend on water activity. At low water content, diffusion is severely limited. Duckworth and Smith [52] measured diffusion of radiolabeled glucose and sulfate in pieces of dry vegetables. They found that the lowest level of moisture at which diffusion was detected was about 1.3 times the monolayer moisture value.

7. Porous Structure Investigation

Water sorption can be influenced by the surface area and porosity of the food material. The characteristics of a material (e.g., porous or nonporous) can be determined from sorption isotherms. The characteristic isotherms proposed by Aguerre et al. [1] are shown in Figure 5. It has been proposed in the literature that water activity could be used to calculate the food surface area as well as the pore size. However, Nagai and Yano [128] found that surface area and pore size calculated from the water sorption can be misleading. They suggested that water adsorped not only on the surface but mainly on the water-binding sites inside the structure, which does not increase with an increase in surface area. The active sites are due to the glucose residue. Water activity can also be used to determine the crystallinity of cellulose [63].

B. Factors Affecting Water Activity

1. Food Components

Proteins and starches adsorb much more water at low water activities than do fatty materials or crystalline substances like sugar. Water absorption in sugar also depends on the degree of grinding and the state (crystalline or amorphous). Pretreatment, such as heating, has little effect on proteins. On the other hand, such pretreatment increases the amount of water-impenetrable crystalline starch at the expense of amorphous starch. The smaller surface area available for adsorption means that less water will be adsorbed [99].

Sugars and salts present a difficult problem because the change from an amorphous to a crystalline state occurs fairly rapidly at normal temperatures [118]. This change releases water, which may be picked up by other materials if the sugar is present in a mixture such as dried milk. The material can then become sticky and lumpy, making it undesirable. The process of instantizing milk to pre-

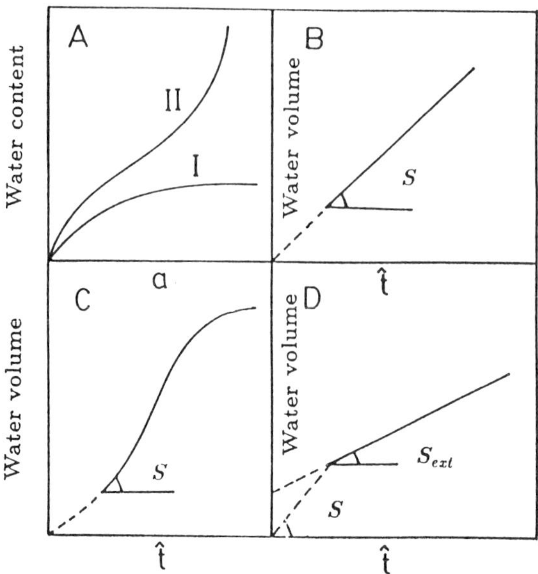

FIGURE 5 Characteristic isotherms of a nonporous solid (I) and microporous solid (II). (B) $\hat{t}$-plot (external surface sorption). (C) $\hat{t}$-plot (capillary condensation). (D) $\hat{t}$-plot (micropore and external surface sorption). (From Ref. 1.)

vent caking during storage was considered as early as 1930 by Troy et al. [169]. Salwin [146] observed that the equilibrium condition obtained is not an equal moisture content in all components in multicomponent mixture, but an equal activity.

2. Physicochemical State of Components

Many food components may be present in several states: crystalline solids, amorphous solids whether rubbery or glassy solids, aqueous solution, or bound to other components. Sorption in such systems is complex. Crystalline sugars sorb very little water, but amorphous sugars sorb substantially more water at the same conditions. The adsorption of water results in breaking of some hydrogen bond and an increase in mobility of sugars molecules, resulting eventually in the sugars transforming to the crystalline state. In this process the sugar loses water [95]. However, the sugar-polymer interaction and physical state (e.g., glassy) play an important role in separating out water from the system. Gelatinization followed by freeze-drying results in only minor differences in water-binding behavior up to water activity 0.94; above 0.95 the gelatinized samples adsorb considerably more water [172].

Saltmarch and Labuza [145] studied the effects of water activity and temperature on the transition of lactose from the amorphous to the crystalline state. Results from scanning electron microscopy indicated that lactose crystallized at 0.40, 0.33, and 0.33 water activity after one week at 25, 35, and 45°C, respectively. The crystallization of lactose in the hygroscopic whey powder appeared to be related to the water activity–related maximum for browning via the Maillard reaction.

Water activity also influences protein conformation. The schematic structures of starch at low and high water activities are given in Figure 6. The annealing effect of water, time, and temperature can alter the structure and functional properties of cereal starch, (i.e., amorphous to crystalline state) (Fig. 7).

3. Porous Structure of Foods

Structure or pore size and distribution of the material may also affect the sharply increasing region at higher water activity.

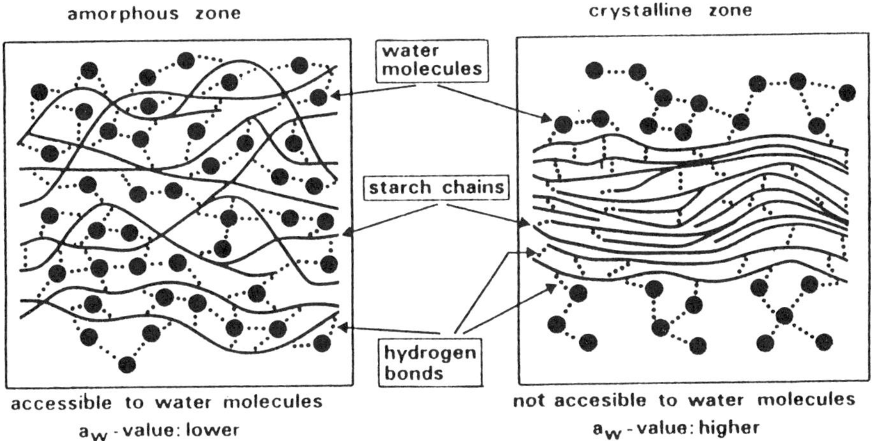

FIGURE 6 Schematic model of starch structure during amorphous and crystalline state. (From Ref. 124.)

4. Temperature

Above Freezing

The isotherm shift due to temperature can be estimated by the Clausius-Clapeyron equation as discussed earlier.

Below Freezing

Information in the literature on water activity of the unfrozen state below freezing is limited [91]. The vapor pressures of animal tissues over the temperature span of –26 to –1°C ranged from 13 to 20% lower than those of pure ice at the same temperature [54,80]. Other researchers demonstrated that the

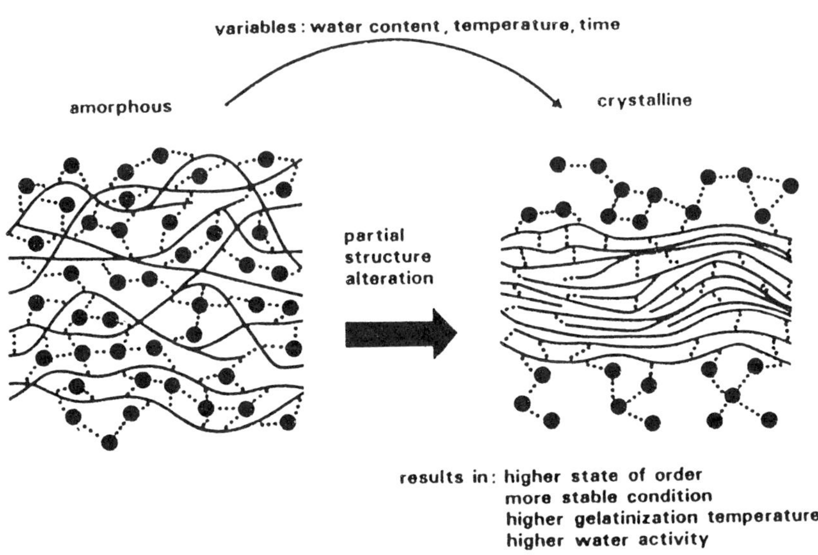

FIGURE 7 Annealing effects of starch induced by water, temperature, and time. (From Ref. 124.)

vapor pressures of frozen biological materials were equal to the vapor pressure of ice at the same temperatures [56,57,117,162]. Water activity values at subfreezing temperatures can be calculated (rather than measured) as [91]:

$$a_w = \frac{\text{Vapor pressure of solid water (ice)}}{\text{Vapor pressure of liquid supercooled water (not ice)}} = \frac{P_i^v}{(P_w^v)_{subcooled}} \quad (5)$$

The equation indicates that water activity does not depend on composition, but only on temperature. In a two-phase system (ice and solution) at equilibrium, the vapor pressure of solid water as ice crystals and the interstitial concentrated solution are identical, thus water activity depends only on the temperature, and not on the nature and initial concentration of solutes, present in the third or fourth phase (i.e., irrespective of the kind of food). This creates a basis to estimate the water activity of foods below freezing using Eq. (3). Thus, Fennema [57] concluded that changes in properties can occur below freezing without any change in water activity. These include changes in diffusional properties, addition of additives or preservatives, and disruption of cellular systems.

The water activity data of ice from 0 to –50°C is correlated with an exponential function as:

$$a_w = 8.727\left[\exp\left(-\frac{595.1}{T}\right)\right] \quad (6)$$

where T is in K. The maximum error in prediction is 0.012 unit water activity and the average is 0.0066, respectively. The data of Fennema [58] were used to develop the above correlation.

5. Pressure

The effect of pressure on the sorption isotherm is relatively small and negligible at reasonable pressure levels [131]. At constant moisture content, the variation of water activity with pressure can be derived thermodynamically as [131]:

$$\ell n\frac{a_2}{a_1} = \frac{\lambda_w}{\rho_w RT}[P_2 - P_1] \quad (7)$$

where a_1 and a_2 are the water activity at P_1 and P_2, R is the gas constant (82.05×10^{-3} m^3 atm/kg mole K or 8.314×10^3 m^3 Pa/kg mole K), T is the temperature (K), and P_1 and P_2 are total pressure (atm or Pa).

6. Surface Tension

The effect of capillary action on water activity can be estimated from the Kelvin equation as:

$$a_w = \exp\left(-\frac{\Delta PV_m}{RT}\right) \quad (8)$$

For spherical interface:

$$\Delta P = \gamma_s\left[\frac{1}{r_1} + \frac{1}{r_2}\right]\cos\theta \quad (9)$$

where ΔP is the pointing pressure (Pa), V_m is the liquid molar volume (18 m^3/kg mole), R is the gas constant (8.314 N m/kg mole K), T is the temperature (K), γ_s is the surface tension (N/m), cos θ is the contact angle, and r is the radius of curvature (m). If the droplet is spherical, then $r_1 = r_2$ and the above equation can be written as:

$$a_w = \exp\left[-\frac{2\gamma \cos \theta V_m}{rRT}\right] \tag{10}$$

If surface tension is reduced by a factor of 0.5, the ratio of water activities at the two conditions can be calculated using the above equation. The ratio is 0.995 at 20°C, thus the effect of surface tension cannot be measured. Chen and Karmas [29] reported that in intermediate food solutions, the water activity increased very little as the surface tension decreased. They suggested that ingredients that result in a reduction of surface tension should be avoided in order to attain low water activity. In contrast to Chen and Karmas [29], Alzamora et al. [2] found that surface tension did not appear to have any significant effect on water activity, at least within the range of apparatus error from 0.004 to 0.005 water activity unit.

II. WATER ACTIVITY IN FOOD PRESERVATION

A. Food Stability Diagram

The moisture sorption isotherm is an extremely valuable tool for food scientists because of its usefulness in predicting food stability. Most foods there have a critical moisture content below which the rate of quality loss is negligible. Quality is understood to include growth and toxin production by microorganisms as well as deterioration of sensory intensity such as crispness, hardness, caking, texture, color, flavor, or aroma [102]. A global food stability map was developed by Labuza and Karel [99,105] (Fig. 8).

B. Monolayer Stability

The monolayer value can be determined using the BET or GAB isotherm and is widely used to determine the stability of foods. The BET equation can be derived based on kinetic, statistical mechanics, or thermodynamic considerations. The equation can be written as:

$$\frac{a_w}{M_w(1 - a_w)} = \left[\frac{C - 1}{M_m}\right]a_w + \frac{1}{M_m C} \tag{11}$$

where a_w is the water activity, M_m is the BET monolayer, and C is the temperature dependence for sorption excess enthalpy, where:

$$C = \alpha\left[\exp\left(-\frac{Q_s}{RT}\right)\right] \tag{12}$$

where Q_s is the heat of sorption (kJ/kg) and α is the preexponent factor. The monolayer can be estimated from the slope of the linear line of the plot $a_w/M_w(1 - a_w)$ versus a_w. The BET equation is valid only within 0.05–0.50 water activity, thus values within that range should be used to estimate the monolayer value. The BET monolayer value of foods and food components are given in Tables 1–5. The monolayer value is generally at a water activity of 0.2–0.4 [102]. The maximum permissible water activity values of unpacked dry foods are given in Table 6.

In recent years the most widely accepted and represented model for sorption isotherms for foods has been the GAB. This is mainly due to its accuracy and its validity over a wide range of water activities from 0.1 to 0.9. The GAB isotherm was developed by Gugenheim, Anderson, and De Boer and can be written as:

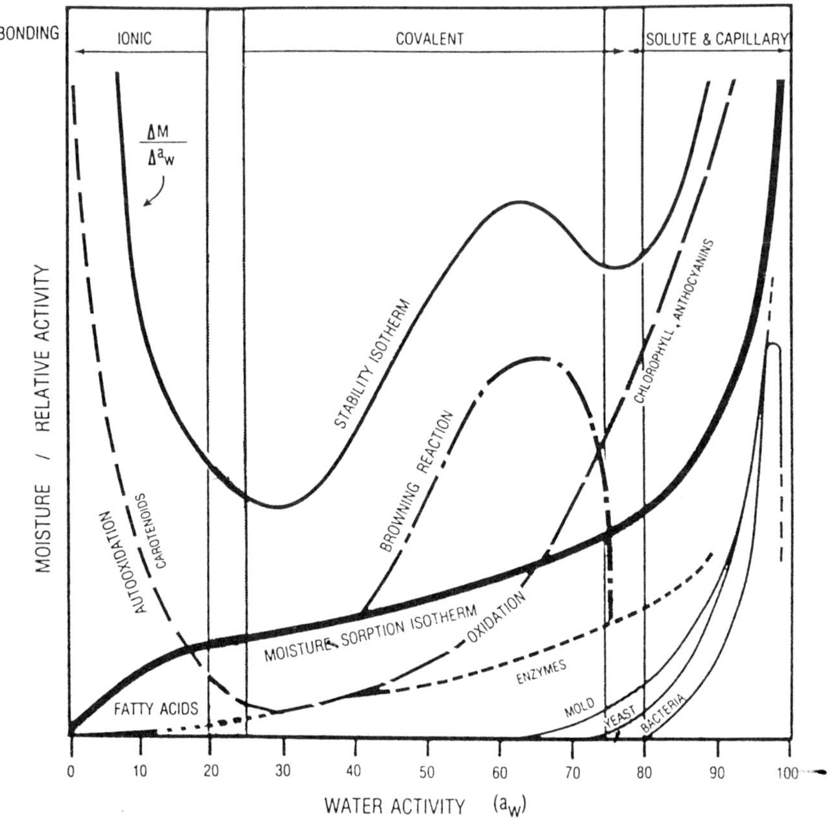

FIGURE 8 Food stability as a function of water activity. (From Ref. 142.)

$$M_w = \frac{M_m C k a_w}{(1 - k a_w)(1 - k a_w + C k a_w)} \tag{13}$$

where C and k are the model parameters and are related to the temperature. The GAB isotherm equation is an extension of the two-constant BET model and takes into account the modified properties of the sorbate in the multilayer region and bulk liquid properties through the introduction of a third constant k. The GAB model parameters and monolayer values have been compiled by Rahman [134] for a number of food products.

The quality of foods as received by the consumer depends not only on the initial composition but also on various quality changes occurring during processing, storage, and distribution. Many of these changes are affected by water content and the state of water in foods. The monolayer and nonsolvent water is particularly significant. The monolayer value should not be constructed as an amount of water uniformly covering the internal surfaces of foods, but rather as the amount of water strongly absorbed on specific polar groups in the foods [95].

As discussed by Labuza [102] the rate of quality loss begins to increase above water activity 0.20–0.3 for most chemical reactions in the aqueous phase (Fig. 8). At this water activity, the amount of water adsorbed on surfaces and in capillaries is enough to affect the overall dielectric properties such that the water can now behave as a solvent. Thus, chemical species can dissolve, become mobile, and are reactive. The higher the water activity, the faster the reaction rate because of the greater

Table 1 BET Monolayer Values of Food Components

Product	A/D	T (°C)	M_m	Ref.	Product	A/D	T (°C)	M_m	Ref.
Amylose	A	28.2	0.070	84	Gel[c]	A	30.0	0.111	84
Amylopectin	A	28.2	0.073	84		A	40.0	0.090	84
Agar	A	RT	0.135	84		A	50.0	0.101	84
CMC	A	25.0	0.095	84	Lactoglobulin[c]	A	25.0	0.059	84
Casein	A	30.0	0.076	84		A	40.0	0.056	84
CM	A	37.0	0.044	84	Lactose[c]	A	23.0	0.062	84
Collagen	A	25.0	0.096	84	Maltose[c]	A	23.0	0.050	84
	A	40.0	0.095	84	Pectic acid	A	29.0	0.073	84
Dextrin[a]	A	28.2	0.056	84	Pectin[d]	A	29.0	0.108	84
Lactose (amorphous)	—	—	0.060	94	Salmin	A	25.0	0.059	84
Sucrose (crystallize)	—	—	0.004	94		A	40.0	0.055	84
Dextrin	—	—	0.090	94	Serum[e]	A	20.0	0.061	84
Gelatin	—	—	0.087	102	Serum[f]	A	25.0	0.065	84
Gelatin	—	—	0.110	94		A	45.0	0.067	84
Starch (instant)	—	—	0.057	146		A	25.0	0.075	84
Starch (potato)	—	—	0.066	102	Starch gel[c]	A	20.0	0.092	84
Elastin	A	25.0	0.055	84		A	30.0	0.082	84
	A	40.0	0.056	84		A	40.0	0.078	84
Fish protein[b]	A	25.0	0.053	84		A	50.0	0.072	84
	A	35.0	0.051	84	Sucrose[c]	A	47.0	0.057	84
	A	42.0	0.045	84	Sugar[g]	D	35.0	0.069	84
Gelatin	A	20.0	0.133	84		D	47.0	0.064	84
Starch	—	—	0.110	94					

CMC: Carboxymethyl cellulose;
CM: cellulose microcrystal (7% mineral oil added).
[a]From corn starch.
[b]Protein concentrate (87% protein).
[c]Freeze-dried.
[d]Highly esterified.
[e]Bovine albumin.
[f]Horse albumin.
[g]Beet root.

TABLE 2 BET Monolayer Values of Animal Food Products

Product	A/D	T (°C)	M_m	Ref.	Product	A/D	T (°C)	M_m	Ref.
Animal flesh									
Beef (cooked)[a,b]	A	10.0	0.106	84	Chicken (raw)[c]	D	5.0	0.085	84
	A	22.2	0.075	84		D	45.0	0.052	84
	A	37.7	0.046	84		D	60.0	0.038	84
Beef (raw)[a,b]	A	10.0	0.069	84	Pork (cooked)[a,b]	A	4.5	0.046	84
	A	20.0	0.069	84	Pork (raw)[b]	D	19.5	0.068	84
	A	30.0	0.064	84	Salmon (raw)[a]	A	37.0	0.061	84
	A	40.0	0.055	84	Shrimp (cooked)		RT	0.056	84
	A	50.0	0.045	84	Trout (cooked)[a,d]	A	5.0	0.086	84
Cod (raw)	A	30.0	0.077	84		A	45.0	0.043	84
Chicken (cooked)[a,c]	A	5.0	0.073	84		A	60.0	0.033	84
	A	45.0	0.047	84	Trout (cooked)[d]	D	45.0	0.044	84
	A	60.0	0.033	84		D	60.0	0.044	84
Chicken (cooked)[c]	D	5.0	0.084	84	Trout (raw)[a,d]	A	5.0	0.079	84
	D	19.5	0.069	84		A	45.0	0.043	84
	D	45.0	0.050	84		A	60.0	0.035	84
	D	60.0	0.037	84	Trout (raw)[d]	D	5.0	0.088	84
Chicken (raw)[a,c]	A	5.0	0.083	84		D	45.0	0.088	84
	A	45.0	0.050	84		D	60.0	0.035	84
	A	60.0	0.037	84	Turkey[a,e]	A	22.0	0.056	84
Squid	D	5.0	0.111	25	Beef (ground)	—	—	0.061	146
	D	20.0	0.102	25	Beef (soup)	—	—	0.024	146
	D	40.0	0.073	25	Chicken	—	—	0.052	102
	A	5.0	0.115	25	Chicken (soup)	—	—	0.017	146
	A	20.0	0.103	25	Fish	—	—	0.049	102
	A	40.0	0.076	25	Beef (freeze-dried)	—	—	0.040	94

Shrimp	—	—	0.056	146					
Milk products									84
Cheese[a]	A	25.0	0.033	84	Yoghurt[a]	A	5.0	0.052	84
	D	25.0	0.035	84		A	25.0	0.041	84
Cheese[a]	A	25.0	0.033	84		A	45.0	0.030	84
	A	45.0	0.022	84		D	5.0	0.054	84
	D	25.0	0.037	84		D	25.0	0.042	84
Milk (nonfat)	A	1.7	0.101	84	Whey[f]	A	24.0	0.045	84
	A	15.6	0.075	84	Cheese (spray-dried)	—	—	0.022	146
	A	30.0	0.070	84	Milk (whole)	—	—	0.015	102
	A	37.7	0.067	84	Milk (instant)	—	—	0.057	102
	A	60.0	0.021	84	Milk (nonfat)	—	—	0.030	94
Eggs									
Egg (albumin)	A	25.0	0.063	84	Egg (whole)	A	17.1	0.037	84
	A	40.0	0.059	84		A	30.0	0.036	84
Egg (albumin)[a]	A	25.0	0.056	84		A	50.0	0.035	84
	A	26.5	0.056	84		A	70.0	0.033	84
	A	40.0	0.055	84		D	19.5	0.041	84
Egg[g]	A	25.0	0.052	84	Egg	—	—	0.068	102
	A	40.0	0.050	84	Egg (albumin)	—	—	0.050	102

[a]Freeze dried
[b]Fat-free basis
[c]White muscle
[d]Back muscle
[e]Outer breast muscle
[f]Protein concentrate (spray dried, 78% protein, 1.3% lactose)
[g]Heat-coagulated

Table 3 BET Monolayer Values of Nuts and Grains

Product	A/D	T (°C)	M_m	Ref.	Product	A/D	T (°C)	M_m	Ref.
Barley	A	25.0	0.086	84	Sorghum	A	21.1	0.066	84
Corn	D	4.5	0.083	84	Wheat	D	25.0	0.078	84
	D	15.5	0.077	84		D	50.0	0.060	84
	D	30.0	0.073	84	Wheat flour	A	20.2	0.059	84
	D	38.0	0.063	84		A	30.1	0.059	84
	D	50.0	0.059	84		A	40.8	0.056	84
	D	60.0	0.051	84		A	50.2	0.052	84
Rice (cooked)	D	19.5	0.081	84	Paranut	A	5.0	0.022	84
Rice (rough)	D	20.0	0.072	84		A	25.0	0.018	84
	D	23.3	0.071	84		A	60.0	0.011	84
	D	25.0	0.069	84		D	5.0	0.025	84
	D	30.0	0.070	84		D	60.0	0.012	84
Peanut	D	25.0	0.043	84	Soybean	A	30.0	0.026	84
Pekanut	A	5.0	0.019	84	Walnut (kernels)	A	22.5	0.007	84
	A	25.0	0.019	84	Orgeat	A	25.0	0.052	84
	A	45.0	0.016	84	Cocoa	—	—	0.039	102
	A	60.0	0.009	84	Coffee	—	—	0.083	102
	D	5.0	0.020	84	Macaroni (instant)	—	—	0.059	146
	D	45.0	0.017	84	Potato (dice)	—	—	0.060	94
	D	60.0	0.009	84					
Biscuit	S	30.0	0.040	5					
	S	70.0	0.029	5					
Fondant	S	30.0	0.006	5					
	S	70.0	0.008	5					
Biscuit[b]	S	30.0	0.022	5					
	S	70.0	0.021	5					

[a]Freeze-dried.
[b]Fondant-coated.

Table 4 BET Monolayer Values of Herbs and Spices

Product	A/D	T (°C)	M_m	Ref.	Product	A/D	T (°C)	M_m	Ref.
Anise	A	5.0	0.048	84	Hibiscus	A	25.0	0.022	84
	A	25.0	0.042	84		D	25.0	0.032	84
	A	45.0	0.036	84	Hops	A	25.0	0.040	84
	D	5.0	0.051	84		D	25.0	0.045	84
	D	25.0	0.046	84	Laurel	A	5.0	0.063	84
	D	45.0	0.045	84		A	25.0	0.045	84
Cardamon	A	5.0	0.073	84		A	45.0	0.031	84
	A	25.0	0.059	84		A	60.0	0.026	84
	A	45.0	0.047	84		D	45.0	0.045	84
	A	60.0	0.039	84	Nutmeg	A	5.0	0.054	84
	D	5.0	0.081	84		A	25.0	0.045	84
	D	45.0	0.051	84		A	45.0	0.037	84
Cinnamon	A	5.0	0.080	84		A	60.0	0.027	84
	A	25.0	0.061	84		D	5.0	0.065	84
	A	45.0	0.047	84		D	25.0	0.047	84
	D	5.0	0.103	84		D	45.0	0.039	84
	D	25.0	0.070	84	Peppermint	A	5.0	0.068	84
	D	45.0	0.054	84		A	25.0	0.068	84
Chamomile	A	5.0	0.062	84		A	45.0	0.047	84
	A	25.0	0.062	84		A	60.0	0.032	84
	A	45.0	0.041	84		D	5.0	0.079	84
	A	60.0	0.029	84		D	25.0	0.071	84
	D	5.0	0.072	84		D	45.0	0.045	84
	D	25.0	0.069	84	Sweet marjoram	A	5.0	0.063	84
	D	45.0	0.040	84		A	25.0	0.046	84

(continued)

Table 4 Continued

Product	A/D	T (°C)	M_m	Ref.	Product	A/D	T (°C)	M_m	Ref.
Clove	A	5.0	0.050	84		A	45.0	0.030	84
	A	25.0	0.041	84		A	60.0	0.021	84
	A	45.0	0.033	84		D	5.0	0.074	84
	A	60.0	0.016	84		D	25.0	0.052	84
	D	5.0	0.057	84		D	45.0	0.031	84
Coriander	A	5.0	0.054	84	Thyme	A	5.0	0.056	84
	A	25.0	0.054	84		A	25.0	0.047	84
	D	5.0	0.070	84		A	45.0	0.035	84
	D	25.0	0.061	84		A	60.0	0.031	84
	D	45.0	0.043	84		D	5.0	0.070	84
Fennel	A	5.0	0.028	84		D	25.0	0.049	84
	A	25.0	0.028	84		D	45.0	0.036	84
	A	45.0	0.025	84	Winter savory	A	5.0	0.073	84
	D	5.0	0.043	84		A	25.0	0.065	84
	D	25.0	0.034	84		A	45.0	0.037	84
	D	45.0	0.025	84		A	60.0	0.029	84
Ginger	A	5.0	0.074	84		D	5.0	0.091	84
	A	25.0	0.070	84		D	25.0	0.070	84
	A	45.0	0.047	84		D	45.0	0.035	84
	D	5.0	0.082	84					
	D	25.0	0.068	84					
	D	45.0	0.047	84					
Turmeric[a]	A	15.0	0.021	132					
	A	25.0	0.019	132					
	A	35.0	0.017	132					
	A	45.0	0.020	132					
Onion (powder)	—	—	0.036	146					

[a]Turmeric powder 150 μm ISS sieve (used a from 0.1 to 0.9).

Table 5 BET Monolayer Values of Fruits and Vegetables

Product	A/D	T (°C)	M_{bm}	Ref.	Product	A/D	T (°C)	M_{bm}	Ref.
Vegetables				84					84
Bean[a]	A	25.0	0.054	84	Mushroom[c]	A	25.0	0.049	84
Cabbage[a]	A	37.0	0.042	84		D	25.0	0.050	84
Carrot	D	19.5	0.045	84	Pea[a]	A	25.0	0.054	84
	A	37.0	0.046	84	Onion powder	—	RT	0.036	84
Celery[a]	A	5.0	0.063	84	Potato	D	19.5	0.075	84
	A	25.0	0.062	84		A	20.0	0.074	84
	A	45.0	0.034	84		A	25.0	0.052	84
	A	60.0	0.032	84		A	30.0	0.057	84
	D	45.0	0.036	84		A	40.0	0.061	84
Chives[a]	A	25.0	0.061	84		A	60.0	0.048	84
	A	60.0	0.014	84		A	80.0	0.037	84
Eggplant[a]	A	5.0	0.090	84	Radish[a]	A	5.0	0.060	84
	A	25.0	0.067	84		A	25.0	0.054	84
Green pea	D	19.5	0.050	84		A	45.0	0.035	84
Green pepper	A	22.2	0.028	84		A	60.0	0.021	84
Lentil	A	5.0	0.075	84		D	25.0	0.059	84
	A	25.0	0.069	84	Salsify[a]	A	45.0	0.052	84
	A	45.0	0.052	84		A	60.0	0.042	84
	D	5.0	0.093	84		D	45.0	0.057	84
	D	45.0	0.052	84	Spinach	A	37.0	0.043	84

(continued)

Table 5 Continued

Product	A/D	T (°C)	M_{bm}	Ref.	Product	A/D	T (°C)	M_{bm}	Ref.
Mushroom[a,b]	A	25.0	0.058	84	Beet root	D	20.0	0.055	84
Bean	—	—	0.042	102		D	35.0	0.049	84
Bean (lima)	—	—	0.054	102		D	47.0	0.050	84
Bean (navy)	—	—	0.052	146		D	65.0	0.039	84
Pea	—	—	0.036	102		A	25.0	0.058	84
Bean (small red)	—	—	0.045	146		A	45.0	0.064	84
Fruits									
Apple	D	19.5	0.042	84	Black currant	D	19.5	0.061	84
Avocado[a]	A	25.0	0.032	84	Grapefruit[a]	A	60.0	0.019	84
	A	45.0	0.027	84	Grapefruit[a]	D	5.0	0.058	84
	A	60.0	0.009	84	Raspberry	—	—	0.021	102
Avocado	D	25.0	0.033	84	Strawberries	—	—	0.048	102
Banana[a]	A	25.0	0.040	84	Orange juice	A	25.0	0.011	84
	A	45.0	0.031	84	Rhubarb	—	—	0.057	102

[a]Freeze-dried.
[b]Boletus.
[c]Pfifferling.

Table 6 Maximum Permissible Water Activity Values of Unpacked Dry Foods at 20°C

Food	Water activity	Food	Water activity
Baking soda	0.45	Sugars:	
Cracker	0.43	Crystaline fructose	0.63
Dried egg	0.30	Crystaline dextrose	0.89
Gelatin	0.43–0.45	Crystaline sucrose	0.85
Hard candy	0.25–0.30	Maltose	0.92
Chocolate, plain	0.73	Sorbitol	0.55–0.65
Chocolate, milk	0.68	Dehydrated meat	0.72
Potato flakes	0.11	Dehydrated vegetable:	
Flour	0.65	Pea	0.25–0.45
Oatmeal	0.12–0.25	Bean	0.80–0.12
Dried skim milk	0.30	Dried fruit:	
Dry milk	0.20–0.30	Apple	0.70
Beef-tea granule	0.35	Apricot	0.65
Dried soup	0.60	Date	0.60
Roast coffee	0.10–0.30	Peach	0.70
Soluble coffee	0.45	Pear	0.65
Starch	0.60	Plum	0.60
Macaroni, noodle, spaghetti, vermicelli	0.60	Orange powder	0.10

Source: Ref. 76.

solubility and increased mobility of the reactants. However, at some point no more species dissolve and therefore an increase in water activity decreases the concentration of the reacting species. Since the rate of a reaction is proportional to concentration on a molecular basis, the rate should reach a maximum and then fall as in Figure 8. Between this maximum and the monolayer, a semilog plot of rate versus water activity generally results in a straight line. For most dry foods, an increase in water activity by 0.1 unit in this region decreases the shelf life two- to threefold [102].

Figure 8 shows quality loss by oxidation below monolayer value. If a food is susceptible to oxidation of unsaturated fats, such as occurs in cereal grains, the rate increases as the water activity decreases below the monolayer. Oxidation and rancidification are aggravated by drying to very low moisture levels [146]. Attack by oxygen is also responsible for pigment instability and loss of vitamins and sometimes initiates nonenzymatic browning reactions [163]. The attachment of an oxygen molecule to a binding site of a protein would produce an incongruity in the aqueous covering sheath, which could disturb the hydration structure of neighboring sites [96]. Competition with oxygen is not the sole basis for explaining the protective effects of water. The bond energy of the adsorbed water would inhibit interactions between polar groups on adjacent carbohydate or protein molecules and thereby preserve rehydration ability, reconstitution ability, and texture of foods [146]. Moreover, with respect to fat oxidation, the catalytic effect of metallic compounds is reduced when they form coordination spheres with polar groups [171]. Water is important in lipid oxidation because it acts as a solvent, mobilizes reactants, and interacts chemically or by hydrogen bonding with other species. The basic protective function that water exhibited when the moisture content increased the absolute dry state could be accounted for by two factors: (a) water interacts with metal catalysts, making them less effective through changes in their coordination sphere, and (b) water hydrogen-bonds with hydroperoxides, tying them up so that they were no longer available for decomposition through initiation reactions. When moisture content is higher than the value at the monolayer, the solvent and mobilization properties of water become more important and the catalysts present are more easily mobilized and possible swelling of solid matrix exposes new catalytic sites, making oxidation rates even higher

[100]. Thus, foods having unsaturated fat should be kept at the critical water activity to maximize shelf life. The water activity at BET monolayer can be defined as the critical water activity.

In addition, the BET monolayer calculation is an effective method for estimating the amount of bound water to specific polar sites in dehydrated food systems [121]. Iglesias and Chirife [84] found that monolayer values decreased significantly with increasing temperature after studying 100 foods and food components. This may be due to the thermodynamics where higher temperatures increase the escaping tendencies of gas molecules. This is explained by Clausius-Clapeyron equation, where:

$$\ell n \frac{(a_w)_1}{(a_w)_2} = -\frac{Q_s}{R}\left[\frac{1}{T_2} - \frac{1}{T_2}\right] \tag{14}$$

C. Microbial Activity

1. Minimal Water Activity Limit

The minimal water activity is the limit below which a microorganism or group of microorganisms can no longer reproduce. Hypothetical curves showing effects of water activity are presented in Figure 9. The initial portion of this growth curve is composed of a lag phase during which the physiological machinery is being created for later growth. The lag period is increased with an increase in humectant content or a decrease in water activity. The growth or logarithmic phase also affected by water activity, as shown in Figure 9 [167].

Secondary metabolites are produced by some microorganisms that are highly toxic and/or carcinogenic to humans. The factors that affect sporulation can influence the formation of these metabolites. Beuchat [10] summarized the effects of water activity on spore formation and germination as well as toxin production by microorganisms commonly associated with foods and food spoilage. Minimal water activity values for growth and toxin production by microorganisms of public health significance are listed in Tables 7–11. The concern for food safety increases with increasing water activity. The water activity of some foods and their susceptibility to spoilage microorganisms are shown in Table 12 [12]. There is a critical water activity below which no microorganisms can grow. For most foods this is in the 0.6–0.7 water activity range. Pathogenic bacteria cannot grow below a water activity of 0.85–0.86, whereas yeast and molds are more tolerant of a reduced water activity of 0.80, but usually no growth occurs below a water activity of about 0.62 [36]. The critical limits of water activity may also be shifted to higher or lower levels by other factors, such as pH, salt, antimicrobial agents, heat treatment, and temperature to some extent.

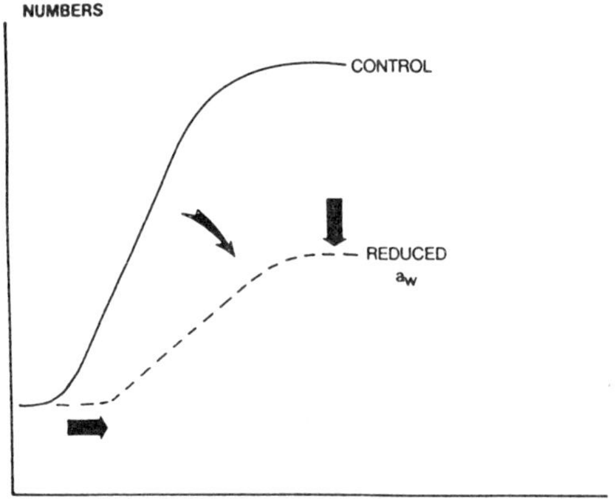

FIGURE 9 Hypothetical curves showing effects of water activity reduction on growth of bacteria. (From Ref. 167.)

TABLE 7 Minimal a_w for Growth and Toxin Production by Bacteria of Public Health Concern

Bacteria	Minimal water activity for		Toxin
	Growth	Toxin production	
Bacillus cereus	0.93–0.95	—	—
Clostridium botulinum	0.93–0.95	0.94–0.95	Type A
	0.93–0.94	0.94	Type B
	0.95–0.97	0.97	Type E
Clostridum perfringens	0.93–0.95	—	—
Salmonella spp.	0.92–0.95	—	—
Staphylococcus aureus	0.86–0.87	0.87–0.90	Enterotoxin A
	0.86–0.87	0.97	Enterotoxin B
Vibrio parahaemolyticus	0.94	—	—

Source: Ref. 12.

TABLE 8 Minimal a_w for Growth of Foodborne Pathogens in Laboratory Media at Optimum pH and Temperature

Pathogen	Minimal a_w	Pathogen	Minimal a_w
Campylobacter jejuni	0.990	*Salmonella* spp.	0.940
Aeromonas hydrophilia	0.970	*Escherichia coli*	0.935
Clostridium botulinum E	0.965	*Vibrio parahaemolyticus*	0.936
Clostridium botulinum G	0.965	*Bacillus cereus*	0.930
Shigella spp.	0.960	*Listeria monocytogenes*	0.920
Yersinia enterocolitica	0.960	*Staphylococcus aureus* (anaerobic)	0.910
Clostridium perfringens	0.945	*Staphylococcus aureus* (aerobic)	0.860
Clostridium botulinum A and B	0.940		

Water activity usually adjusted by the addition of sodium chloride.
Source: Ref. 33.

TABLE 9 Sodium Chloride Versus Glycerol in Minimal Water Activity Supporting Growth of Pathogenic Bacteria

Bacteria	a_w adjusted with	
	Sodium chloride	Glycerol
Clostridium botulinum E	0.966	0.943
Clostridium botulinum G	0.966	—
Escherichia coli	0.949	0.940
Clostridium perfringens	0.945	0.930
Salmonella spp.	0.941	—
Clostridium botulinum A and B	0.940	0.930
Vibrio parahaemolyticus	0.932	0.911
Bacillus cereus	0.930	0.920
Listeria monocytogenes	0.920	0.900
Staphylococcus aureus	0.860	0.890

Source: Ref. 36.

Table 10 Minimal Water Activity for Growth of Pathogenic Bacteria[a]

Bacteria	NaCl	KCl	Sucrose	Glucose
Listeria monocytogenes	0.920	—	0.920	—
Vibrio parahaemolyticus	0.936	0.936	0.940	—
Clostridium botulinum G	0.965	—	0.965	—
Clostridium botulinum E	0.972	0.972	0.972	0.975
Clostridium perfringens	0.945	—	—	0.946
Staphylococcus aureus	0.864	0.864	0.867	—

[a]In laboratory media, water activity adjusted with salts (NaCl and KCl) or sugars (sucrose and glucose).

Leistner and Rodel [108] found the rate of microbial death during frozen storage to be reduced by a decrease in temperature and no fluctuation. Thus, they suggested that the microbiological quality of frozen foods could be improved by initial storage of foods at –10°C (a_w = 0.90) to reduce the number of undesirable organisms followed by freezing at very low temperature (i.e., –30°C) [109].

2. Effects of Preservation on Microorganisms

Effect of Water Activity on Microorganisms

A decrease in water activity increases the osmotic stress to microbial cells because the cell always tries to maintain a slightly lower internal osmolality (i.e., water activity). This causes an influx of water into the cell to maintain surface integrity. It is the disruption of this process by solutes that leads to cell damage and death.

Effects of Solutes

In addition to osmotic stress, solutes may have other effects on microorganisms, including enzyme inhibition, cytoplasmic coagulation, and damage to the cell wall. Brown [20] showed that high lev-

Table 11 Minimal a_w for Growth of and Toxin Production by Molds of Public Health Concern

	Minimal water activity for		
Mold	Growth	Toxin production	Toxin
Alternaria alternata	—	<0.90	Altenuene, alternariol, alternariol monomethyl ether
Aspergillus flavus	0.78–0.80	0.83–0.87	Aflatoxin
Aspergillus parasiticus	0.82	0.87	Aflatoxin
Aspergillus orchraceus	0.77–0.83	0.83–0.87	Ochratoxin
Byssochlamys nivea	0.84	—	—
Penicillium cyclopium	0.81–0.85	0.87–0.90	Ochratoxin
Penicillium viridicatum	0.83	0.83–0.86	Ochratoxin
Penicillium ochraceus	0.76–0.81	0.80–0.88	Penicillic acid
Penicillium cyclopium	0.82–0.87	0.97	Penicillic acid
Penicillium martensii	0.79–0.83	0.99	Penicillic acid
Penicillium islandicum	0.83	—	—
Penicillium urticae	0.81–0.85	0.85–0.95	Patulin
Penicillium expansum	0.83–0.85	0.99	Patulin
Stachybotrys atra	0.94	0.94	Stachybotrym
Trichothecium roseum	0.90	—	Trichothecine

Source: Ref. 12.

Table 12 Water Activity of Some Foods and Susceptibility to Spoilage by Microorganisms

Range of a_w	Microorganisms generally inhibited by lowest a_w in this range	Examples of foods generally within this range of a_w
1.00–0.95	*Pseudomonas, Escherichia, Proteus, Shigella, Klebsiella, Bacillus, Clostridium perfringens,* some yeasts	Highly perishable foods (fresh and canned fruits, vegetables, meat fish) and milk; cooked sausages and breads
0.95–0.91	*Salmonella, Vibri parahaemolyticus, C. botulinum, Serratia, Lactobacillus, Pediococcus,* some molds, *Rhodotorula, Pichia*	Some cheeses (Cheddar, Swiss, Muenster, Provolone), cured meat, some fruit juice concentrates
0.91–0.87	Many yeasts (*Candida, Torulopis, Hansenula*), *Micrococcus*	Fermented sausage (salami), sponge cakes, dry cheeses, margarine
0.87–0.80	Most molds (*mycotoxigenic penicillia*), *Staphylococcus aureus,* most *Saccharomyces* (*baillii*) spp., *Debaryomyces*	Most fruit juice concentrates, sweetened condensed milk, chocolate syrup, maple and fruit syrups, flour, rice, pulses, fruit cake, country style ham, fondants, high sugar cakes
0.80–0.75	Most halophilic bacteria, mycotoxigenic aspergilli	Jam, marmalade, mazipan, glacé fruits, some marshmallows
0.75–0.65	Xerophilic molds (*Aspergillus chevalieri, A. candidus, Wallemia sebi*), *Saccharomyces bisporus*	Rolled oats, grained nougats, fudge, marshmallows, jelly, molasses, raw cane sugar, some dried fruits, nuts
0.65–0.60	Osmophilic yeasts (*Saccharomyces rouxii*), a few molds (*Aspergillus echinulatus, Monascus bisporus*)	Dried fruits, some toffees and caramels, honey
0.50	No microbial proliferation	Noodles, spaghetti, dried spices
0.40		Whole egg powder
0.30		Cookies, crackers, bread crusts
0.20		Whole milk powder, dried vegetables, corn flakes, dehydrated soups, some cookies, crackers

Source: Ref. 12.

els of salt with antibiotics like penicillin and cycloserine caused prokaryotic cell walls to become fragile while strong-walled eukaryotes survived. In that case, the controlling mode of action was damage to the cell wall. Water activity can explain only the osmotic stress. Genetic control is closely tied to the amino acid pool, especially betaine and proline, and the potassium level in the cell.

3. Adaptation with Changing Water Activity: Osmoregulation

A variety of mechanisms may avoid water loss or ensure water loss from microorganisms [69]. Such mechanisms are reviewed by Troller [167].

Sensing and Translation

While potassium may or may not be the trigger that initiates the process of osmoregulation, its transport into the cell is the primary modulatory event [75]. Helmer et al. [79], in a series of experiments, demonstrated that at least two and probably four K^+ transport systems exist in *Escherichia coli*. The first is accomplished, in part, by a series of three high-affinity genes. The first three genes in inner-membrane proteins of various molecular weights act as gatekeepers. The fourth gene alters its conformation in a manner that permits and intensifies transcription to maintain cytoplasmic K^+ content [55]. The second system is constitutive and requires ATP and a proton motive force to supply energy for net K^+ uptake. Helmer et al. [79] identified a proton motive force as supplying the primary energy to drive this reaction, whereas ATP supplies the energy to turn off K^+ transport. Another system, the K^+ export model, has only been postulated and is of some interest because of the potential existence of export-blocking proteins that might be synthesized by the cell in response to osmotic challenge. In this case, K^+ would not be pumped out of the cell but would be retained to trigger a metabolic response or to provide primary isoosmotic conditions across the membrane [167].

Accommodation

Christian [39] and Christian and Waltho [40] observed first that growth of *Salmonella oranienburg* at low water activity was stimulated by the addition of the amino acid proline. They observed a reversal of plasmolysis when exogenous proline was supplied to the bacteria growing at low water activity. Although uptake from the media may be one method of accumulating proline in response to water stress, most organisms appear to be able to synthesize proline. In fact, synthesis probably is the most common mechanism for accumulating proline in osmotically inhibited bacteria [167]. This is called a compatible solute. A number of osmoregulatory solutes protect proteins against denaturation by heat [69].

Measures [122] and Gould and Measures [73] showed that K^+ was required to maintain electrical neutrality or to balance the charges within cells exposed to environments with low water activity in which various amino acids, such as α-ketoglutarate and glutamic acid, accumulate intracellularly. The principal reaction involves conversion of α-ketoglutarate to glutamic acid by glutamate dehydogenase, an enzyme activated by K^+. Glutamic acid reduces the intracellular water activity to reverse plasmolysis by reducing relative amounts of K^+ and glutamate dehydrogenase. This leaves the cell at a balanced, osmotic null point by virtue of the increased glutamic acid pool. For some bacteria the process stops at this point, but for other organisms glutamic acid is converted to γ-aminobutyric acid or proline, neither of which are highly charged. Accumulation of high concentrations of glutamic acid would require concomitant acquisition of K^+ to keep the system at neutrality. This excessive amount of K^+ could be detrimental to the organism and, at the very least, costly in terms of energy expenditure. Both γ-aminobutyric acid and proline are remarkably efficient at reducing intracellular water activity without interfering in the cell's metabolism and for this reason have been termed compatible solutes [22].

Compatible protoplasmic solutes in bacteria include glycylbetaine, proline, glutamic acid, γ-aminobutyric acid, and glycerol. Polyols of various types are compatible protoplasmic solutes in many fungi (Table 13).

Exactly how these solutes are able to avoid interference is not fully understood [167]. Gould [69] suggested that specific binding between solutes and intracellular enzymes is not the mechanism. Wyn Jones and Pollard [180] suggested that because these solutes may be excluded from the hydra-

TABLE 13 Compatible Protoplasmic Solutes in Fungi

Solute	Genus	Solute	Genus
Mannitol	*Geotrichum*	d-Galactosyl-(1,1)-glycerol	*Ochromonas*
	Platymonas	Glycerol	*Chlamydomonas*
	Aspergillus		*Aspergillus*
	Dendryphiella		*Dunaliella*
	Penicillium		*Saccharomyces*
Cyclohexanetetrol	*Monochrysis*		*Debaromyces*
Arabitol	*Dendryphiella*	Erythritol	*Aspergillus*
	Saccharomyces		*Penicillium*
Sorbitol	*Stichococcus*		

Source: Ref. 21.

tion sphere of proteins, the term benign solutes might more accurately describe the nonparticipatory nature of these materials.

Genetic Adaptation

The genetic components controlling osmoregulation in microorganisms have been investigated for several years. *Escherichia coli* appears to have evolved a particularly advanced scheme for protection against osmotic stress through a proline-overproduced mutation, which confers osmotolerance. *Klebsiella pneumoniae* experiences an increase in intracellular free proline when it is exposed to high levels of NaCl. Thus, an enhanced level of osmoresistance in the organism results in its ability to fix nitrogen while under osmotic stress [167].

Changing Cell Metabolism

An important role of the membrane may be to exclude Na^+, which, if permitted to enter the cell, can quickly inactivate a number of vital enzymatic systems. Kanemasa et al. [89] attributed the barrier properties of *Staphylococcus aureus* membranes to Na^+. They found that Na^+ alters the types and amount of phospholipids within the membrane [167].

How a bacterial spore maintains such a low cytoplasmic water content or water activity even when suspended in pure water is not yet understood [69]. General rules for such mechanisms have yet to be developed.

4. Proposed Alternatives to the Water Activity Concept

The influence of water activity on the survival and death of microorganisms is complex. Problems with the water activity concept and proposed alternatives are discussed in the following sections.

Validity of Equilibrium Conditions

Water activity is defined at equilibrium, whereas foods with low and intermediate water content may not be in a state of equilibrium. Instead they may be in an amorphous metastable state, which is very sensitive to changes in moisture content and time.

In low-moisture and intermediate-moisture foods, the concept of water activity may be meaningless because the measured vapor pressure of water is no longer the equilibrium vapor pressure as defined in the literature. A stationary state may be reached under a given set of environmental conditions and mistaken for equilibrium. In moisture-sorption studies the situation is further complicated if the amorphous material undergoes a glass-rubber transition during the course of the measurement. Chirife and Buera [36] believe that an analysis of various literature data may throw some light on these aspects. An important comprehensive collaborative study within the framework of the European Corporation in the Field of Scientific and Technical Research (COST) was conducted to determine the precision of data (e.g., repeatability or reproducibility) in the determination of sorption isotherms.

In the water activity range of interest to microbial growth (0.6–0.9), the average standard deviation of all data from 24 laboratories was ±2.6% for the equilibrium moisture content of microcrystalline cellulose (MCC) and ±3.8% for that of potato starch. The repeatability was 2% for both MCC and potato starch. Chirife and Buera [36] also reported data on isotherms from different sources of the same material and found good reproducibility within a wide range of water activity. Lomauro et al. [115] concluded from a study of a large number of foods that a pseudoequilibrium was reached when the moisture content (dry basis) did not change by more than ±0.5% during three consecutive sample periods at intervals of no more than 7 days. This criterion for equilibrium moisture content was compared with the values obtained after 6 months of storage in closed mason jars, which were considered to be very close to the equilibrium moisture content. Lomauro et al. [115] concluded that the foods tested reached (or were very close to) equilibrium within 1 month, based on the above criterion. Various authors reported their equilibrium times for isotherm determinations of different food systems using the gravimetric static method over saturated salt solutions, and their equilibrium times ranged mostly between 1 and 4 weeks, depending on the temperature and relative humidity. Bizot et al. [13] utilized a practical equilibrium time of about 7 days (±0.02% water per 24 hr) for a 1-g sample, but they also stored their starch samples over saturated salt solutions for 2 years. They noted a slow drift in desorption pseudoequilibrium, but there was only a 1% difference in water content (dry basis) over this long time. Thus, water activities measured are likely to be close to equilibrium, and the differences should be within the uncertainties associated with the experimental determination of isotherm [36].

Chirife and Buera [36] reviewed sorption isotherms of fruits containing crystallizable sugars that constitute a nonequilibrium system. For example, in raisins the discontinuities in the isotherm are not noticed at water activity 0.30–0.90, suggesting that sugars remained amorphous even at very large $T - T_g$ (raisins contain 83% (dry basis) glucose and fructose). Sorption isotherms of other fruits reported in the literature do not show discontinuities [36]. Bolin [16], reporting water activity and equilibrium moisture values for raisins sealed in glass jars and held at 21 or 32°C up to 12 months, found little effect on the isotherm. This suggests that nonequilibrium effects are very slow, at least in these experiments. Chirife and Buera [36] also presented data on fruits, but they overlooked the crystallization of sugars in dairy products and formulated products having sugars or salts, as discussed in Chuy and Labuza [44] and Saltmarch and Labuza [145]. The consequence of recrystallization may be true for fruits, but cannot be generalized. There may be a real theoretical limitation of using water activity to measure food stability. This break in isotherm does have practical importance as far as detecting any change of structural components during storage.

Effects of Solutes

The microbial response may differ at a particular water activity when the latter is obtained with different solutes [35,36]. Thus, the proposed basis of a water activity limit for growth may not be universal.

Corry [48] reported that the survival of vegetative bacteria is influenced by nutrients in the food matrix. These influences show no consistent inhibitory pattern and are greatly affected by the matrix. Mugnier and Jung [125] studied the survival of bacteria and fungi in biopolymer gels with and without nutritive solutes. They observed that survival is increased at the point of mobilization of solute in the case of mannitol. While comparing a gram-positive bacterium and a gram-negative one, they concluded that low molecular weight compounds (C_3–C_5) had a deleterious effect on survival compared to higher molecular weight compounds (C_6–C_{12}), which had a protecting effect. The glass-rubber state of solutes may also play an important role, since higher molecular weight solutes have higher glass transition temperatures than low molecular weight solutes. The degree of protecting effect was in the order of mannitol > dextrin > ribose > glycerol. Thus, they [125] suggested that the concept of Duckworth and Kelly [53] and Seow [154] of discontinuity of properties of water could be used to explain the survival of microorganisms in polymer systems at various water activity levels. This concept is that above a certain amount of hydration (the mobilization point) there exists a second fraction of solute in the polymer system, which can serve as a true solvent for the microbial nutrient to maintain the organisms' metabolic activity [109]. Brown [19] stated that freeze-drying of microorganisms with a nonelectrolyte such as glycerol or sugar reduces mortality during dehydration, storage, and rehydration. This indicates that the nonelectrolyte functions directly as a solvent molecule for nutrients.

Slade and Levine [158] and Franks [61] proposed that water activity can serve as a useful, but not the sole indicator of microbial safety. Slade and Levine's [158] hypothesis stated that water dynamics or glass-rubber transition may be used instead of water activity to predict microbial stability. Slade and Levine [157] reported that for matched pairs of fructose and glucose at equal solute concentrations, fructose produced a much less stable system in which mold spores germinated much faster, i.e., the solute with the lower ratio of melting temperature to glass transition temperature ($\alpha = T_m/T_g$) allowed faster germination. Thus, the following order of antimicrobial stabilization was predicted by Slade and Levine [157]: glycerol ($\alpha = 1.62$) > glucose ($\alpha = 1.42$) > mannose ($\alpha = 1.36$) > fructose ($\alpha = 1.06$) and sucrose ($\alpha = 1.43$) > maltose ($\alpha = 1.27$). The germination of *Aspergillus flavus*, *Aspergillus niger*, and *Eurotium herbariorum* did not follow the above sequence [38]. For example, germination times for all three molds in fructose or glucose were always greater than in glycerol. In all cases, germination time increased when the water activity decreased; the relative effect, however, depended both on the solute type and the mold. None of the molds studied germinated in solution of propylene glycol (at $a_w = 0.85$ or 0.90; $\alpha = 1.27$) after 70 days of incubation at 28°C [38]. This is simply because the behavior of propylene glycol cannot be explained on the basis of mobility or water activity effects alone, since this molecule possesses specific antimicrobial effects already recognized in the literature [38]. However, further studies from the microbiology groups could clarify the above published results. Chirife and Buera [36] also showed evidence that above glass (i.e., rubbery) transition, microbial growth is inhibited in prunes and that below glass transition, microorganism growth is possible in wheat flour. This is most likely due to a region of high water activity right around the microbe.

Nelson [129] pointed out some potential reasons for the difference in microbial stability when different solutes are used to achieve the same water activity, but that do not mean that the concept of water activity is invalid, as suggested by Slade and Levine [159]. It has been shown repeatedly in the literature that each microorganism has a critical water activity below which growth cannot occur. For example, pathogenic bacteria cannot grow below a water activity of 0.85–0.86; yeasts and molds are more tolerant of reduced water activity, but usually no growth exists below a water activity of about 0.6 (Table 12).

Gould [71] acknowledged that in some instances solute effects may depend on the ability of the solute to permeate the cell membrane. Glycerol, for example, readily permeates the membrane of many bacteria and so does not initiate he same osmoregulatory response as nonpermanant solutes like sodium chloride and sucrose and therefore has a different, usually lower, inhibitory water activity.

Scott [151] noted that minimal water activity for the growth of microbial organisms was independent of the solutes employed to adjust the water activity of a medium. It was observed later that some solute were more inhibiting than others. Thus, the water activity of a medium is not the only determining factor regulating microbial response. The nature of the solute used also plays an important role [41,72]. This is referred to as specific solute effects by Chirife [34].

Figures 10 and 11 compare the minimal water activity supporting growth of various pathogenic bacteria when sodium chloride or glycerol are used to control the water activity. In all cases glycerol is less inhibitory than sodium chloride. Glycerol readily permeates the membrane of bacteria and does not initiate the same osmoregulatory response as the nonpermanent solute sodium chloride [34]. It is conclusive that the water activity limit varies with the type of solute used and the microorganisms in the medium. Thus, it is important to identify the range of variation and the possible causes of variation. Table 9 shows the effects of solute types on *Staphylococcus aureus*. The range of water activity varies from 0.860 to 0.966. Propylene glycol is more effective in *S. aureus*, which is explained by Chirife [34]. The points in Figure 11 are distributed on both sides of the line, indicating a high correlation of water activity with growth and fluctuation due to other factors in the microorganisms. The effect of solute type on different microorganisms should be clearly identified in order to recognize generalized trends or at least the limitation of validity.

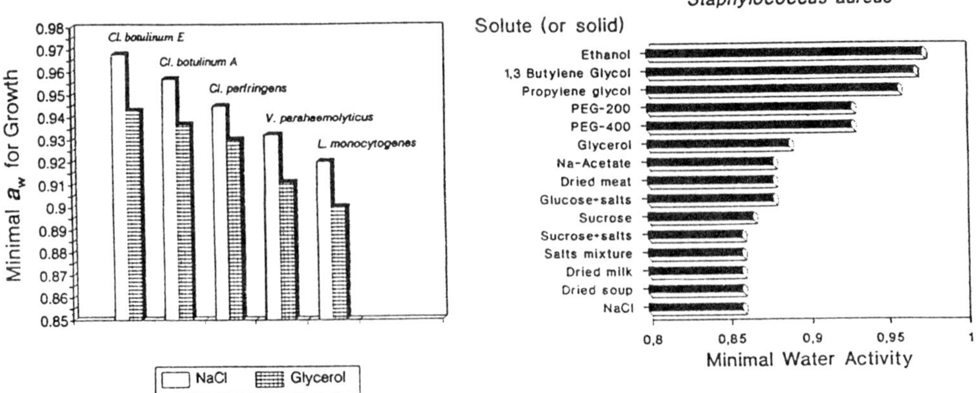

FIGURE 10 Effect of minimum water activity levels on growth of bacteria in different solutes. (From Ref. 36.)

Electron micrographs of *S. aureus* after growing in a medium containing different solutes were analyzed by Chirife [34]. Microbial cells subjected to sodium chloride and sucrose (a_w = 0.85) did not show any important morphological changes in the cells. Thus, the inhibitory effects of sucrose and sodium chloride against *S. aureus* cells was primarily related to their ability to lower water activity, specific solute effects are not significant.

Electron micrographs of *S. aureus* cells in solutions of propylene glycol (a_w = 0.92), 1,4-butylene glycol (a_w = 0.85), and polyethylene glycol 400 and 1000 (a_w = 0.85) showed that these solutes caused dramatic morphological modifications in the cells. These antibacterial effects may be attributed mainly to the effects of these molecules on the membrane enzymes responsible for peptidoglycan synthesis.

Anand and Brown [3] observed that polyethylene glycol was more inhibitory to yeast growth than were glucose and sucrose at a similar water activity. Marshall et al. [119] evaluated the inhibi-

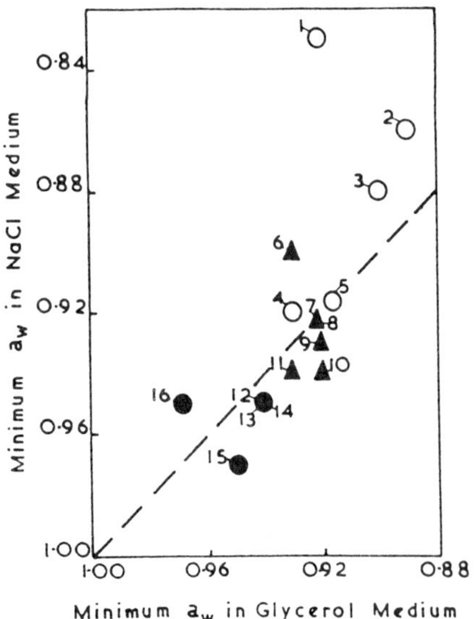

FIGURE 11 Relationship between the minimum water activity levels for growth of 16 species of bacteria in NaCl and in glycerol-adjusted medium. (From Ref. 41.)

tory effects of NaCl and glycerol at the same water activity on 16 species of bacteria. They found that glycerol was more inhibitory than NaCl to relatively salt-tolerant bacteria and less inhibitory than NaCl to salt-sensitive species. Lenovich et al. [111] showed that the type of solute influences resistance to sorbate in *Saccharomyces rouxii*, thus indicating an interactive effect of solute type and preservative.

D. Lipid Oxidation

Autooxidation of lipids occurs rapidly at low water activity levels, decreasing until a water activity range of 0.3–0.5 is reached [166]. At low water content especially in porous substrates in the complete absence of water, peroxidation of unsaturated lipids proceeds very rapidly. The addition of small quantities of water tends to produce a protective effect if the substrate is still free of oxidation products and reactive intermediates. However, reactions of oxidation products with proteins follow a more complex pattern [95].

In a model system consisting of methyl linoleate and lysozyme, the free radicals and other reactive species formed by the linoleate react with the protein, resulting in increased fluorescence, decreased enzyme activity, and decreased protein solubility. Water activity has an inhibitory effect on the initial oxidation of the lipid, but the secondary reactions of the lipid degradation products with the protein are accelerated by increasing water activity [90]. Schaich [149] studied free radical formation in proteins reacted with peroxidizing lipids and found that the amount and type of free radicals formed in the proteins were strongly affected by water activity. It appears that water facilitates recombination of free radicals and as a consequence the steady-state concentration of radicals, whereas various radical-initiated processes such as protein cross-linking increase at high water content.

E. Nonenzymic Activity

1. Consequences of Browning

Browning reactions in foods affect nutritional value as well as color and texture [103]. The induction period, defined as the time to visually detectable browning, is inversely proportional to the water activity [97,175]. Browning reactions are influenced by the type of reactant sugars and amines, pH, temperature, water activity, and the types of solutes or humectants used to adjust the water activity [166].

2. Types of Browning

There are three major pathways by which nonenzymatic browning can occur: high-temperature caramelization, ascorbic acid oxidation, and the Maillard reaction [103].

The browning reactions of sugars heated above their melting point in the absence of proteins or amino acids is called caramelization. This can be either beneficial or detrimental to the quality of a food product and can be prevented by avoiding high-temperature processing and low storage temperatures. It is enhanced in alkaline or acid conditions and is used to make commercial caramel colorings and flavors.

Ascorbic acid (vitamin C) oxidation, a second type of browning reaction, is catalyzed by low pH and elevated temperatures. The decomposition products resulting from the oxidation of ascorbic acid cause a brown discoloration as well as decreased nutritional value.

The Maillard reaction is a result of reducing compounds, primarily sugars, reacting with proteins or free amine groups. This changes both the chemical and the physiological properties of the protein. In general, the accumulation of brown pigments is the most obvious indication that Maillard browning has occurred in a food containing both carbohydrate and protein. It is used as an indicator of excessive thermal processing in the milk industry [103].

In the early stages of the Maillard reaction, the carbonyl group of the reducing sugar reacts with the free amino group of the amino acid to form a Schiff base and then the *N*-substituted glycosylamine as well as a molecule of water. Glycosylamines are converted to the 1-amino-1-deoxy-2-ketose by Amadori rearrangement (cyclization and isomerization). [112]. The Maillard reactions forming Amadori compounds do not cause browning but do reduce the nutritive value [120].

The advanced Maillard reaction has five pathways. The first two pathways start from the 1,2-enol or 2,3-enol forms of the Amadori product, yielding various flavor compounds. The third pathway is Strecker degradation, which involves oxidative degradation of amino acids by the dicarbonyls produced in the first two pathways. The fourth pathway involves transamination of the Schiff base. The fifty pathway starts with a second substitution of the amino-doxyketose. The final step of the advanced Maillard reaction is the formation of many heterocyclic compounds, such as pyrazines and pyroles [112]. Brown melanoidin pigments are produced in the final stage of the Maillard reaction. The pigments are formed by polymerization of the reactive compounds produced during the advanced Maillard reaction, such as unsaturated carbonyl compounds and furfural. The polymers have a molecular weight greater than 1000 and are relatively inert [120]. These pathways depend upon environmental conditions such as pH and temperature.

3. Factors Affecting Browning

The browning reaction rate increases from water activity at BET, sharply increases to a maximum, and then decreases (Fig. 8). Water can retard the rate of the initial glycosylamine reaction of which water is a product. This results in product inhibition by some of the intermediate reactions. A second factor is the dilution of reactive components with increasing water content. The mobility of the reactive species increases due to a decrease in viscosity with increasing water activity. However, the first two factors eventually overcompensate for the decreased viscosity at higher water activity, and thus the overall rate of browning decreases [103].

Wolf et al. [179] demonstrated that losses of free lysine and methionine were highly dependent on water activity, protein, and sugar. Thermal degradation of both amino acids followed first-order kinetics, and rates decreased at 65 and 115°C with increasing water activity. A more rapid decrease of lysine, tryptophan, and threonine at higher water activity in model systems is observed when heated at 95°C [107]. The retention of tryptophan was greater than lysine at water activity 0.75, but lysine retention was greater than that of tryptophan at water activity 0.22. At higher water activity, the Maillard reaction predominates and a rapid loss of lysine occurs. At lower water activity, browning proceeds at a slower rate and reactions involving the indole ring of tryptophan become significant [112].

Glucose utilization in a model system consisting of glucose, monosodium glutamate, corn starch, and lipid during nonenzymic browning was investigated by Kamman and Labuza [88]. The rates of glucose utilization at water activity 0.81 were higher than at 0.41. Lipid accelerated the reaction rates at 0.41 but had virtually no effect at 0.81 water activity. Liquid oil is more effective than shortening in increasing the degradation rate of glucose. These can be explained by the mobility of solutes in both water and oil [112].

Cerrutti et al. [28] studied browning in a model system consisting of lysine, glucose, sodium chloride, and phosphate buffer. They showed that water had little or no effect on the rate of glucose loss at water activity 0.90–0.95, but the rate was highly dependent on temperature and pH. Similar behavior was observed for accumulation of 5-hydroxymethylfurfural, fluorescent compounds, and brown pigments.

Seow and Cheah [155] found that nonenzymic browning decreased with an increase of water activity and temperature in a model system water-glycerol-sorbate-glycine at pH 4.

4. Maximum Browning Region

The region where the maximum browning occurs is usually near 0.65–0.80 water activity. In model freeze-dried foods the maximum browning rate is in the range of 0.40–0.67 [43], in whey powders at 0.44 [104], and in dehydrated foods in the range of 0.65–0.75 [174].

Petriella et al. [133] found that water had relatively less effect on the browning rates at water activity of 0.90–0.95. At this range pH and temperature are the determining factors. At very high water content, i.e., water activity greater than 0.95, moisture strongly inhibits the browning rate by diluting the reactive species [166].

Warmbier et al. [174] studied the influence of solutes on the maximal range of browning. For example, if glycerol is employed to reduce water activity, the range of maximal browning shifts from 0.65–0.75 to 0.40–0.50. They concluded that glycerol can influence the rate of browning at lower water activity values by acting as an aqueous solvent and thereby allowing reactant mobility at much lower moisture values than would be expected for water alone. The overall effect of glycerol or other liquid humectants on the maximum for nonenzymatic browning is to shift it to a lower water activity [103]. Obanu et al. [130], on the other hand, observing browning in glycerol–amino acid mixtures stored at 65°C, concluded that glycerol itself might participate in the browning reaction.

Moreover, Troller [166] mentioned that product quality relative to browning could be improved by reducing the water activity and, more importantly, the temperature during the final stage of drying. It is somewhat paradoxical that at water activity levels that minimize browning, autooxidation of lipids is maximized.

F. Enzymic Activity

Enzyme-catalyzed reactions can proceed in foods with relatively low water contents. Karel [95] summarized two features of results mentioned in the literature as follows:

1. The rate of hydrolysis increases with increasing water activity, with the reaction being extremely slow at very low activities.
2. At each water activity there appears to be a maximum extent of hydrolysis, which also increases with water content.

The apparent cessation of the reaction at low moisture cannot be because of irreversible inactivation of the enzyme, because upon humidification to a higher water activity, hydrolysis is resumed at a rate characteristic of the newly obtained water activity [95]. Silver [156] investigated a model system consisting of avicel, sucrose, and invertase and found that the reaction velocity increased with water activity. Complete conversion of the substrate was observed for water activities greater than or equal to 0.75. Below water activities of 0.75 the reaction continued to 100% hydrolysis. In solid media, water activity can affect reactions in two ways: lack of reactant mobility and alteration of active conformation of substrate and enzymatic protein [165]. Effects of varying the enzyme-to-substrate ratios on reaction velocity and the effect of water activity on the activation energy for the reaction could not be explained by a simple diffusional model, but required more complex postulates [95]:

1. The diffusional resistance is localized in a shell adjacent to the enzyme.
2. At low water activities, the reduced hydration produces conformational changes in the enzyme affecting its catalytic activity.

Tome et al. [165] tested the simple diffusional hypothesis on the basis of experiments in liquid systems in which water activity was reduced by the addition of glycerol, ethylene glycol, propylene glycol, diethylene glycol, sorbitol, methanol, or ethanol. In these solutions the effects of polyphenoloxidase on tyrosine were very similar to those obtained in solid systems. The optimum pH of activity is shifted slightly toward alkaline values. Three characteristics curves were observed: (a) for low water activity there was an almost total inhibition; (b) in the intermediate range, reaction rate was very dependent on water activity; and (c) for high water activity zones, activity was weakly affected by organic additives. In general, the rate increased rapidly with increasing water activity, and the reaction stopped at a certain level before all reactants were consumed; the higher the water content, the higher was the plateau. The authors were unable to find a correlation of enzyme activity with viscosity, solubility of oxygen and tyrosine, or dielectric constant. It also appeared that the more the mixture deviated from ideality, the more the enzymatic activity was inhibited, regardless of whether the deviation was positive or negative. Thus, solvent-water interaction is the main parameter in polyphenoloxidase inhibition.

The minimum water activities for enzymatic reactions in selected food systems are given in Table 14.

Table 14 Minimum Water Activity Values for Enzymatic Reactions in Selected Food Systems

Product/Substrate	Enzyme	T (°C)	Water activity threshold
Grains	Phytases	23	0.90
Wheat germ	Glycoside-hydrolases	20	0.20
Rye flour	Amylases	30	0.75
	Proteases	—	—
Macaroni	Phospholipases	25–30	0.45
Wheat flour dough	Proteases	35	0.96
Bread	Amylases	30	0.36
	Proteases	—	—
Casein	Trypsin	30	0.50
Starch	Amylases	37	0.40/0.75
Galactose	Galactosidase	30	0.40–0.60
Olive oil	Lipase	5–40	0.025
Triolein, trilaurin	Phospholipases	30	0.45
Glucose	Glucose oxidase	30	0.40
Lenoleic acid	Lipoxygenase	25	0.50/0.70

Source: Ref. 51.

G. Texture

In addition to the safety and nutrition aspects of food quality, texture is also important. Texture cannot be completely quantified by a single measurement as can percent moisture or pH. The affect of water activity on food texture is specific to the kind of food under consideration [17].

Rockland [141] defined food texture as a function of localized moisture sorption isotherms as follows: (a) region I (low water activity)—dry, hard, crisp, shrunken; (b) region II (intermediate water activity)—dry, firm, flexible; and (c) region III (high watery activity)—moist, juicy, soft, flacid, swollen, sticky (Fig. 12 and Table 15). A more detailed review is given in Bourne [17].

The effect of water activity on textural measurements for different types of foods is reviewed by Bourne [17]. At this time there are insufficient data to predict what the textural properties of a given type of food will be at a given water activity, and no sound theories exist to predict in advance the textural properties of a food at a given water activity [17].

Cenkowski et al. [26] studied the mechanical behavior of canola kernels by bringing them to equilibrium, by adsorption or desorption, at the same final moisture. The ratio of elasticity was 18–38% higher for kernels brought to equilibrium through adsorption than those through desorption for a moisture range of 9.5–7.0% (dry basis). At higher moisture contents, the differences in module of elasticity were not significant.

Table 15 Critical Values for Ingredients in Model Food Products

Moistness	Crispness	Chewiness	Toughness
Cereal		< 0.40	> 0.50
Fruit	> 0.30	< 0.50	< 0.30
Nuts		< 0.65	

Source: Ref. 15.

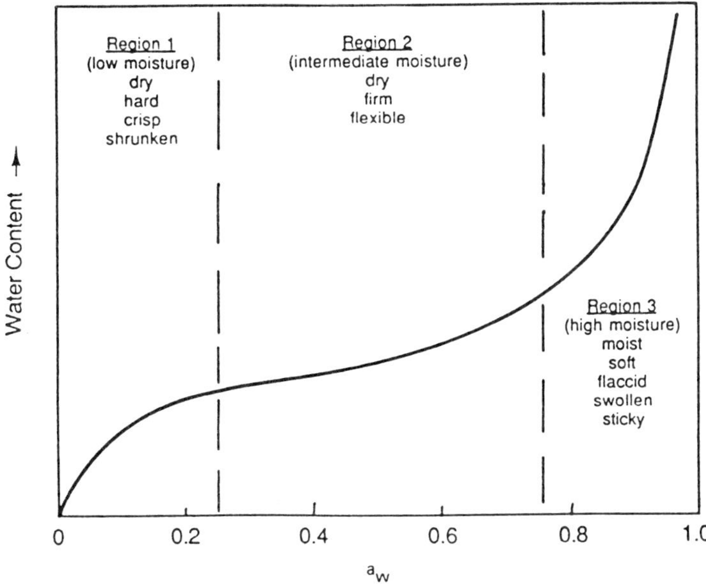

FIGURE 12 Food textures as a function of localized water sorption isotherms. (From Ref. 141.)

III. FURTHER APPLICATIONS OF WATER ACTIVITY

A. Process Design and Control

Water activity data are also important in food processing, such as by osmotic dehydration and air-drying. In drying operations, desorption isotherms at the process temperature are needed for design and control purposes. The endpoint of drying or osmotic dehydration processes can be determined by the equilibrium moisture content. In the drying process the foods equilibrate with air equilibrium relative humidity; in osmotic or salting processes foods equilibrate with the osmotic solution water activity. Thus, water activity plays an important role in designing, operating, and controlling drying processes and reverse osmosis.

B. Ingredient Selection

In many food products solutes such as salts, sugars, or polyols are added to depress water activity. The ability of a single solute to lower water activity enables an effective selection of solute in order to obtain satisfactory sensory quality and shelf life. Water activity prediction in foods with multiple solutes allows effective formulation of a final product.

C. Packaging Selection

When food materials are packed in semi-permeable membranes, the food will (a) collect moisture if its water activity is lower than the external relative humidity of the air or (b) lose moisture if its water activity is higher than the relative humidity. The sorption isotherm is necessary to predict the moisture-transfer rate through the packaging film or edible food coating so that shelf life can be predicted. The mathematical equations used to determine the isotherms for moisture transfer through packaging material are available in literature [49,135].

Heiss [77] described marking points on the sorption isotherm that can be used in packaging and storage of food materials. These marking points are shown in Figure 13. The initial point 1 and end-

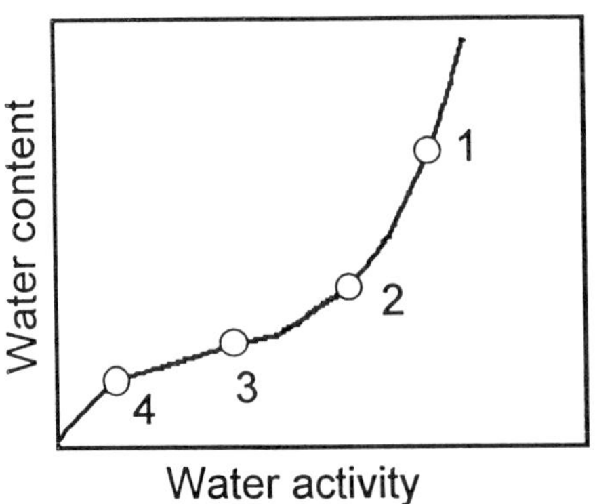

Figure 13 Schematic sorption isotherm with typical marking points. (From Ref. 77.)

point 4 can be easily determined from the processing and storage conditions, whereas the critical point must be determined for each product separately. Point 3 lies in the range of 0.15–0.25 for major foods; in this range foods are stable in respect to lipid oxidation, Maillard browning, and enzymic deterioration [101]. At the critical moisture content and at normal equilibrium relative humidity, some kinds of physical, chemical, or biological changes become so rapid that the food deteriorates before reaching the end of the desired storage time. The criteria are also very different depending on the type of food and the deterioration reaction that is first to appear organoleptically during storage.

D. Reducing Water Activity of Foods

It is important to control the water activity of foods to achieve the desired safety and quality. Water activity can be lowered or controlled by several methods.

1. Separating Out Water

Processes that can be used to remove water are drying, concentration, and dewatering by centrifuge. Drying is one of the oldest methods of food preservation. More details about drying are available in Chapters 9 and 10. Other processes are baking, freeze-drying, spray-drying, extrusion, frying, and membrane processes.

2. Adding Solute

Solutes can be added to foods to reduce water activity as well as to improve the functional and sensory properties of foods (e.g., salting of meat and fish and the sugaring of fruits). When only solute will be used to reduce water activity, then the specific antimicrobial effects and the cost of solutes or humectants should also be considered for food product formulation. The factors affecting the selection of humectants are summarized in Table 16. Water can be removed and solutes can be added

Table 16 Some Criteria for Humectants to Be Used in Foods

Safe
Approved by regulatory agencies
Effective at reasonable concentrations
Compatible with the nature of the food
Flavorless at use concentrations
Colorless and/or imparts no color changes to the food

Source: Ref. 166.

at the same time in a process of osmotic dehydration. More details of the process are described in Chapter 10.

REFERENCES

1. R. J. Aguerre, C. Suarez, and P. E. Viollaz, Swelling and pore structure in starchy materials, *J. Food Eng. 9*:71 (1989).
2. S. M. Alzamora, J. Chirife, and C. F. Fontan, Effect of surface active agents on water activity of IM food solutions, *J. Food Sci. 46*:1974 (1981).
3. J. C. Anand and A. D. Brown, Growth rate patterns of the so-called osmophilic and nonosmophilic yeasts in solutions of polyethylene glycol. *J. Gen. Microbiol. 52*:205 (1968).
4. C. Baird-Parker, M. Boothroyd, and E. Jones, The effect of water activity on the heat resistance of heat sensitive and heat resistant strains of salmonellae, *J. Appl. Bacteriol. 33*:515 (1970).
5. G. Balasubrahmanyam and A. K. Datta, Prevention of moisture migration in fondant coated biscuit, *J. Food Eng. 21*:235 (1994).
6. A. L. Benado and S. S. H. Rizvi, Thermodynamics properties of water on rice as calculated from reversible and irreversible isotherms, *J. Food Sci. 45*:1190 (1985).
7. S. W. Benson and R. L. Richardson, A study of hysteresis in the sorption of polar gases by native and denatured proteins, *J. Am. Chem. Soc. 77*:2585 (1955).
8. E. Berlin, B. A. Anderson, and M. J. Pallansch, Water vapor sorption properties of various dried milks and wheys, *J. Dairy Sci. 51*:1339 (1968).
9. F. A. Bettelheim and S. H. Ehrlich, Water vapor sorption of mucopolysaccharides, *J. Phys. Chem. 67*:1948 (1963).
10. L. R. Beuchat, Influence of water activity on sporulation, germination, outgrowth, and toxin production, *Water Activity: Theory and Applications of Food* (L. B. Rockland and L. R. Beuchat, eds.), Marcel Dekker, Inc., New York, 1987, p. 137.
11. L. R. Beuchat, Influence of water activity on growth, metabolic activities and survival of yeasts and molds, *J. Food Prot. 46*:135 (1983).
12. L. R. Beuchat, Microbial stability as affected by water activity, *Cereal Food World 26*:345 (1981).
13. H. Bizot, A. Buleon, N. Mouhous-Riou, and J. L. Multon, Some factors concerning water vapor sorption hysteresis in potato starch, *Properties of Water in Foods in Relation to Quality and Stability* (D. Simatos and J. L. Multon, eds.), Martinus Nijhoff, Dordrecht, 1985, p. 83.
14. K. Boki and S. Ohno, Moisture sorption hysteresis in kudzu starch and sweet potato starch, *J. Food Sci. 56*:125 (1991).
15. D. P. Bone, Practical applications of water activity and moisture relations in foods, *Water Activity: Theory and Applications to Foods* (L. B. Rockland and L. R. Beuchat, eds.), Marcel Dekker, New York, 1987, p. 369.
16. J. R. Bolin, Relation of moisture to water activity in prune and raisin, *J. Food Sci. 45*:1190 (1980).
17. M. C. Bourne, Effects of water activity on textural properties of food, *Water Activity: Theory and Applications to Food* (L. B. Rockland and L. R. Beuchat, eds.), Marcel Dekker, Inc., New York, 1987, p.
18. A. D. Brown, Microbial water stress, *Bacteriol. Rev. 40*:803 (1976).
19. A. D. Brown, Microbial water stress, *Bacteriol. Rev. 40*:803 (1976).
20. L. M. Brown, *Psychology 21*:408 (1982).
21. A. D. Brown, Compatible solutes and extreme water stress in eukaryotic microorganisms, *Adv. Microb. Physiol. 17*:181 (1978).
22. A. D. Brown and J. R. Simpson, The water relations of sugar-tolerant yeasts: the role of intracellular polyols. *J. Gen. Microbiol. 2*:589 (1972).
23. S. Brunauer, L. S. Deming, W. E. Deming, and E. Teller, On a theory of the van der Waals absorption of gases, *Am. Chem. Soc. J. 62*:1723 (1940).
24. H. Cakirlar and D. J. F. Bowling, *J. Exp. Bot. 32*:479 (1981).
25. C. A. Castanon and A. O. Barral, Sorption isotherms of raw squid (*Illex argentinus*) muscle, *Food Sci. Technol. 21*:212 (1988).

26. S. Cenkowski, Q. Zhang, and W. J. Crerar, Effect of sorption hysteresis on mechanical behavior of canola, *Trans. ASAE 38*(5):1455 (1995).
27. G. F. Cerofolini and C. Cerofolini, Heterogeneity, allostericity, and hysteresis in adsorption of water by proteins. *J. Colloid Interf. Sci. 78*:65 (1980).
28. P. Cerrutti, S. L. Resnik, A. Seldes, and C. Ferro-Fontan, Kinetics of deteriorative reactions in model food systems of high water activity: glucose loss, 5-hydroxymethyl furfural accumulation and fluorescence development due to nonenzymatic browning, *J. Food Sci. 50*:627 (1985).
29. C. C. Chen and E. Karmas, Effect of surface active agents on water activity in intermediate moisture foods, *Food Sci. Technol. 12*:68 (1979).
30. P. Chinachoti and M. P. Steinberg, Moisture hysteresis in due to amorphous sugar, *J. Food Sci. 51*:153 (1986).
31. P. Chinachoti and M. P. Stienberg, Interaction of sodium chloride with raw starch in freeze-dried mixtures as shown by water sorption, *J. Food Sci. 50*:825 (1985).
32. P. Chinachoti and M. P. Stienberg, Interaction of sucrose with starch during dehydration as shown by water sorption, *J. Food Sci. 49*:1604 (1984).
33. J. Chirife, Physicochemical aspects of food preservation by combined factors, *Food Control 4*:210 (1993).
34. J. Chirife, Specific solute effects with special reference to *Staphylococcus aureus*, *J. Food Eng. 22*:409 (1994).
35. J. Chirife and M. D. P. Buera, Water activity, glass transition and microbial stability in concentrated/semimoist food systems, *J. Food Sci. 59*:921 (1994).
36. J. Chirife and M. D. P. Buera, Water activity, water glass dynamics, and the control of microbiological growth in foods, *Crit. Rev. Food Sci. Nutr. 36*:465 (1996).
37. J. Chirife and C. F. Fontan, Water activity of fresh foods, *J. Food Sci. 47*:661 (1982).
38. J. Chirife, H. H. L. Gonzalez, and S. L. Resnik, On water dynamics and germination time of mold spores in concentrated sugar and polyol solutions, *Food Res. Int. 28*:531 (1996).
39. J. H. B. Christian, The water relations of growth and respiration of *Salmonella oranienburg* at 30°C. *Aust. J. Biol. Sci. 8*:490 (1955).
40. J. H. B. Christian and J. A. Waltho, The sodium and potassium content of nonhalophilic bacteria in relation to salt tolerance. *J. Gen. Microbiol. 25*:97 (1964).
41. J. H. B. Christian, Specific solute effects on microbial water relations, *Water Activity: Influences on Food Quality* (L. B. Rockland and G. F. Stewart, eds.), Academic Press, New York, 1981, p. 825.
42. D. S. Chung, Thermodynamic factors influencing moisture equilibrium of cereal grains and their products, Ph. D. thesis, Kansas University, Manhattan, KS.
43. C. Y. Chung and M. Toyomizu, Studies on the browning of dehydrated foods as a function of water activity. I. Effect of a_w on browning in amino acid-lipid systems. *Bull. Jap. Soc. Fish. 42*:697 (1976).
44. L. Chuy and T. P. Labuza, Caking and stickiness of dairy based food powders related to glass transition, *J. Food Sci. 59*:43 (1994).
45. L. Cohan, Hysteresis and the capillary theory of adsorption of vapors, *J. Am. Chem. Soc. 66*:98 (1944).
46. E. Cohen and I. Saguy, Effect of water activity and moisture content on the stability of beet powder pigments, *J. Food Sci. 48*:703 (1983).
47. D. L. Collins-Thompson, Food spoilage and food-borne infection hazards in the safety of foods, *The Safety of Foods*, 2nd ed. (H. D. Graham, ed.), Avi Publishing, Westport, CT, 1980.
48. J. E. L. Corry, The water relations and heat resistance of microorganisms. *Prog. Ind. Microbiol. 12*:73 (1973).
49. S. Desobry and J. Hardy, Modelling of the total water desorption rate from packaged moist food, *Int. J. Food Sci. Technol. 28*:347 (1993).
50. L. L. Diosady, S. S. H. Rizvi, W. Cai, and D. J. Jagdeo, Moisture sorption isotherms of canola meals, and applications to packaging, *J. Food Sci. 61*:204 (1996).
51. R. Drapron, Enzyme activity as a function of water activity, *Properties of Water in Foods* (D. Simato and J. L. Multon, eds.), Martinus Nijhoff Publishers, Dordrecht, 1985.

52. R. B. Duckworth and G. M. Smith, The environment for chemical change in dried and frozen foods. *Proc. Nutr. Soc. 22*:182 (1963).
53. R. B. Duckworth and C. E. Kelly, Studies of solution processes in hydrated starch and agar at low moisture levels using wide-line nuclear magnetic resonance. *J. Food Technol. 8*:105 (1973).
54. D. F. Dyer, D. K. Carpenter, and J. E. Sunderland, Equilibrium vapor pressure of frozen bovine muscle, *J. Food Sci. 34*:196 (1996).
55. W. Epstein and L. Lamins, Potassium transport in *Escherichia coli*: Diverse systems with common control by osmotic forces, *Curr. Trends Biochem. 5*:21 (1980).
56. O. Fennema, Enzyme kinetics at low temperature and reduced water activity, *Dry Biological Systems* (J. H. Crow, and J. S. Clegg, eds.), Academic Press, New York, 1978, p. 297.
57. O. Fennema, Water activity at subfreezing temperatures, *Water Activity: Influences on Food Quality* (L. B. Rockland, and G. F. Stewart, eds.), Academic Press, New York, 1981, p. 713.
58. O. R. Fennema, Water and ice, *Food Chemistry*, 2nd ed. (O. R. Fennema, ed.), Marcel Dekker, New York, 1985, p. 23.
59. M. Fortes and M. R. Okos, Drying theories: their bases and limitations as applied to foods, *Adv. Dry.*:119 (1980).
60. C. F. Fontan and J. Chirife, The evaluation of water activity in aqueous solutions from freezing point depression, *J. Food Technol. 16*:21 (1981).
61. F. Franks, Water activity: a credible measure of food safety and quality? *Trends Food Sci. Technol. 2*:68 (1991).
62. S. Gal, Solvent versus non-solvent water in casein-sodium chloride-water systems, *Water Relations in Foods*, Academic Press, New York, 1975, p. 183.
63. S. Gal, The need for and practical applications of sorption data, *Physical Properties of Foods*, (M. Pelig and E. B. Bagley, eds.), Avi Publishing, Westport, CT, 1983, p. 13.
64. S. Gal and R. Singer, Die Wasserdampfsorption von Casein Natrium-chloride Systemen bei hohen relativen Dampfdrucken, *Makromol. Chem. 87*:190 (1965).
65. S. Gal and D. Bankay, Hydration of sodium chloride bound by casein at medium water activities, *J. Food Sci. 36*:800 (1971).
66. S. Gal and M. Hunziker, Bindungsfähigkeit von Casein Für Natrium-chloride bei hohen Wasseractivitäten, *Makromol. Chem. 178*:1535 (1977).
67. J. M. Goepfert and R. A. Biggie, Heat resistance of *Salmonella typhimurium* and *Salmonella senftenberg* 775W in milk chocolate, *Appl. Microbiol. 16*:1939 (1968).
68. B. Gibson, The effect of high sugar concentrations on the heat resistance of vegetative microorganisms, *J. Appl. Bacteriol. 36*:365 (1973).
69. G. W. Gould, Present state of knowledge of a_w effects on microorganisms, *Properties of Water in Foods* (D. Simatos and J. L. Milton, eds.), Martinus Nijhoff Publishers, Dordrecht, 1985, p. 229.
70. G. W. Gould, Osmoregulation: Is the cell just a simple osmometer? The microbiological experience, *A Discussion Conference: Water Activity: A Credible Measure of Technological Performance and Physiological Viability?*, Faraday Division, Royal Society of Chemistry, Girton College, Cambridge, UK, July 1–3, 1985.
71. G. W. Gould, Osmoregulation: Is the cell just a simple osmometer? The microbiological experience, *A Discussion Conference: Water Activity: A Credible Measure of Technological Performance and Physiological Viability?*, Faraday Div., Royal Soc. Chem., Girton College, Cambridge, England, July 1–3, 1985.
72. G. W. Gould, Interference with homeostasis-food, *Homeostatic Mechanisms in Microorganisms* (J. G. Banks, R. G. Board, G. W. Gould, R. W. Mittenbury, eds.), Bath University Press, Bath, UK, 1988.
73. G. W. Gould and J. C. Measures, Water relations in single cells, *Phil. Trans. R. Soc. Lond. B. 278*:151 (1977).
74. J. J. Hardy and M. P. Steinberg, Interaction between sodium chloride and paracasein as determined by water sorption, *J. Food Sci. 49*:127 (1984).
75. F. M. Harold, Pumps and currents: a biological perspective. *Curr. Top. Membrane Transp. 16*:485 (1982).

76. R. Heiss, Shelf-life determinations, *Mod. Packag. 31*:119 (1958).
77. R. Heiss, *Haltbarkeit und Sorptionsverhalten wasserarmer Lebensmittel*, Springer-Verlag, Berlin, 1968.
78. H. J. Hellen and S. G. Gilbert, Moisture sorption of dry bakery products by inverse gas chromatography, *J. Food Sci. 50*:454 (1985).
79. G. L. Helmer, L. A. Laimins, and W. Epstein, Mechanisms of potassium transport in bacteria, *Membranes and Transport,* Vol. 2 (A. N. Martonosi, ed.), Plenum Press, New York, 1982.
80. J. E. Hill and J. E. Sunderland, Equilibrium vapor pressure and latent heat of sublimation for frozen meats, *Food Technol. 21*:1276 (1967).
81. B. P. Hills, C. E. Manning, Y. Ridge, and T. Brocklehurst, NMR water relaxation, water activity and bacterial survival in porous media, *J. Sci. Food Agric. 71*:185 (1996).
82. H. A. Iglesias and J. Chirife, Isoteric heats of water sorption on dehydrated foods. Part II. Hysteresis and heat of sorption comparison with BET theory, *Food Sci. Technol. 9*:107 (1976).
83. H. A. Iglesias and J. Chirife, On the local isotherm concept and modes of moisture binding in food products, *J. Agric. Food Chem. 24*:77 (1976).
84. H. A. Iglesias and J. Chirife, B.E.T. monolayer values in dehydrated foods and food components, *Food Sci. Technol. 9*:107 (1976).
85. H. A. Iglesias, P. Viollaz, and J. Chirife, Technical note: a technique for predicting moisture transfer in mixtures of packaged dehydrated foods, *J. Food Technol. 14*:89 (1979).
86. M. G. Isse, H. Schuchmann, and H. Schubert, Divided sorption isotherm concept: an alternative way to describe sorption isotherm data, *J. Food Eng. 16*:147 (1993).
87. B. J. Juven, J. Kanner, and H. Weisslowicz, Influence of orange juice composition on the thermal resistance of spoilage yeasts, *J. Food Sci. 43*:1074 (1978).
88. J. F. Kamman and T. P. Labuza, A comparison of the effect of oil versus plasticized vegetable shortening on rates of glucose utilization in nonenzymatic browning, *J. Food Proc. Pres. 9*:217 (1985).
89. Y. Kanemasa, T. Katayama, H. Hayashi, T. Takatsu, K. Tomochika, and A. Okabe, The barrier role of cytoplasmic membrane in salt tolerance mechanisms in *Staphylococcus aureus*, *Zentralbl. Bakteriol. Parasitenk. Infekt. Hyg. Abt. Suppl. 5*:189 (1976).
90. J. Kanner and M. Karel, Changes in lysozyme due to reactions with peroxidizing methyl linoleate in a dehydrated model system, *J. Agric. Food Chem. 24*:468 (1976).
91. J. G. Kapsalis, Influences of hysteresis and temperature on moisture sorption isotherms, *Water Activity: Theory and Applications to Food* (L. B. Rockland and L. R. Beuchat, eds.), Marcel Dekker, Inc., New York, 1987, p. 173.
92. J. G. Kapsalis, Moisture sorption hysteresis, *Water Activity: Influences on Food Quality* (L. B. Rockland and G. E. Stewart, eds.), Academic Press, New York, 1981, p. 143.
93. M. Karel, Recent research and development in the field of low-moisture and intermediate moisture foods. *Crit. Rev. Food Technol. 3*:329 (1973).
94. M. Karel, Physico-chemical modifications of the state of water in foods—a speculative survey, *Water Relations of Foods* (R. B. Duckwarth, ed.), Academic Press, London, 1975, p. 639.
95. M. Karel, The significance of moisture to food quality, *Developments in Food Science 2*, (H. Chiba, M. Fijimaki, K. Iwai, H. Mitsuda, and Y. Morita, eds.) Kodansha Ltd., Tokyo, 1979, p. 378.
96. I. M. Klotz and R. E. Heiney, Changes in protein topography upon oxygenation, *Proc. Natl. Acad. Sci. USA 43*:717 (1957).
97. I. J. Kopelman, S. Meydau, and S. Weinberg, *J. Food Sci. 42*:403 (1977).
98. T. P. Labuza and M. Rutman, The effect of surface active agents on sorption isotherms of model systems, Presented at Ann. Can. Chem. Eng. Conference, Oct. 18, 1967.
99. T. P. Labuza, Sorption phenomena in foods, *Food Technol. 22*:15 (1968).
100. T. P. Labuza, Kinetics of lipid oxidation in foods, *CRC Crit. Rev. Food Technol. (Oct.)*:355 (1971).

101. T. P. Labuza, The effect of water activity on reaction kinetics of food deterioration, *Food Technol. 34*:36 (1980).
102. T. P. Labuza, *Moisture Sorptions: Practical Aspects of Isotherm Measurement and Use*, Am Assoc. Cereal Chemists, St. Paul, MN, 1984.
103. T. P. Labuza and M. Saltmarch, The nonenyzmatic browning reaction as affected by water in foods, *Water Activity: Influences and Food Quality* (L. B. Rockland and G. F. Stewart, eds.), Academic Press, New York, 1981, p. 605.
104. T. P. Labuza and M. Saltmarch, Kinetics of browning and protein quality loss in whey powders under steady state and nonsteady state storage conditions, *J. Food Sci. 47*:92 (1981).
105. T. P. Labuza, S. R. Tannenbaum, and M. Karel, Water content and stability of low moisture and intermediate moisture foods, *Food Technol. 24*:543 (1970).
106. T. P. Labuza, A., A. Kaanane, and J. Y. Chen, Effect of temperature on the moisture sorption isotherms and water activity shift of two dehydrated foods, *J. Food Sci. 50*:385 (1985).
107. M. M. Leahy, and J. J. Warthesen, The influence of a Millard browning and other factors on the stability of free tryptophan, *J. Food Proc. Pres. 1*:25 (1983).
108. L. Leistner and W. Rodel, Microbiology of intermediate moisture foods, *Proceedings of the International Meeting on Food Microbiology and Technology* (B. Jarvis, J. H. B. Christian, and H. D. Michener, eds.), Medicina Viva Servizio Congress, Parma, Italy, 1979.
109. L. M. Lenovich, Survival and death of microorganisms as influenced by water activity, *Water Activity: Theory and Applications to Food* (L. B. Rockland and L. R. Beuchat, eds.), Marcel Dekker, Inc., New York, 1987, p. 119.
110. L. M. Lenovich, Mechanism of sorbate resistance in *Saccharomyces rouxii* at reduced water activity, Ph.D. thesis, Drexel University, PA, 1986.
111. L. M. Lenovich, R. L. Buchanan, and N. J. Worley, Effect of solute type on sorbate resistance in *Saccharomyces rouxii* at reduced water activity. Paper 134, 46th Annual Meeting of Inst. of Food Technologists, Dallas, TX, June 15–18, 1986.
112. H. K. Leung, Influence of water activity on chemical reactivity, *Water Activity: Theory and Applications to Food* (L. B. Rockland and L. R. Beuchat, eds.), Marcel Dekker, Inc., New York, 1987, p. 27.
113. C. J. Lomauro, A. Bakshi, and T. P. Labuza, Evaluation of food isotherm equations. Part I. Fruit, vegetable and meat products, *Food Sci. Technol. 18*:111 (1985).
114. C. J. Lomauro, A. Bakshi, and T. P. Labuza, Evaluation of food isotherm equations. Part II. Milk, coffee, tea, nuts, oilseeds, spices and starchy foods, *Food Sci. Technol. 18*:118 (1985).
115. C. J. Lomauro, A. S. Bakshi, and T. P. Labuza, Moisture transfer properties of dry and semimoist foods, *J. Food Sci. 50*:397 (1985).
116. M. Loncin, Basic principles of moisture equilibria, Report, Universität Karlsruhe, Lebensmittelverfahrenstechnik, 1973.
117. A. P. Mackenzie, The physico-chemical environment during freezing and thawing of biological materials, *Water Relations of Foods*, (R. B. Duckworth, ed.), Academic Press, New York, 1975.
118. B. Makower and W. Dye, Equilibrium moisture content and crystallization of amorphous sucrose and glucose, *J. Agric. Food Chem. 4*:72 (1956).
119. B. J. Marshall, D. F. Ohye, and J. H. B. Christian, Tolerance of bacteria to high concentrations of NaCl and glycerol in the growth medium, *Appl. Microbiol. 21*:363 (1971).
120. J. Mauron, The Maillard reaction in food: a critical review from the nutritional standpoint, *Progr. Food Nutr. Sci. 5*:5 (1981).
121. A. D. McLaren and J. W. Rowen, Sorption of water vapor by proteins and polymers: a review, *J. Polym. Sci. 7*:289 (1952).
122. J. C. Measures, Role of amino acids in osmoregulation of nonhalophilic bacteria, *Nature 257*:398 (1975).
123. E. F. Mellon, A. H. Korn, and S. R. Hoover, Water adsorption of protein. II. Lack of dependance of hysteresis in casein on free amino groups, *J. Am. Chem. Soc. 70*:1144 (1948).

124. K. Munzing, DSC studies of starch in cereal and cereal products, *Thermochim. Acta 139*:441 (1991).
125. J. Mugnier and G. Jung, Survival of bacteria and fungi in relation to water activity and solvent properties of water in biopolymer gels, *Appl. Environ. Microbiol. 50*:108 (1985).
126. D. I. Murdock and W. S. Hatcher, Effect of temperature on survival of yeast in 45° and 65°C Brix orange concentration, *J. Food Prot. 41*:689 (1978).
127. W. Murrell and W. J. Scott, The heat resistance of bacteria spores at various water activities, *J. Gen. Microbiol. 43*:411 (1966).
128. T. Nagai and T. Yano, Fractal structure of deformed potato starch and its sorption characteristics, *J. Food Sci. 55*:1334 (1990).
129. K. Nelson, Reactions kinetics of food stability: comparison of glass transition and classical models for temperature and moisture dependence, Ph.D. thesis, University of Minnesota, 1993.
130. Z. A. Obanu, D. A. Ledward, and R. A. Lawrir, Reactivity of glycerol in intermediate moisture meals. *Meat Sci. 1*:177 (1977).
131. M. R. Okos, G. Narsimhan, and R. K. Singh, Food dehydration, *Handbook of Food Engineering* (R. Heldman and D. B. Lund, eds.), Marcel Dekker, New York, 1992, p. 437.
132. V. S. Pawar, D. K. Dev, V. D. Paward, A. B. Rodge, V. D. Surve, and D. R. More, Moisture adsorption isotherms of ground turmeric at different temperatures, *J. Food Sci. Technol. 29*:170 (1992).
133. C. Petriella, S. L. Resnik, R. D. Lozano, and J. Chirife, Kinetics of deteriorative reactions in model food systems of high water activity: color changes due to nonenzymatic browning, *J. Food Sci. 50*:1358 (1985).
134. M. S. Rahman, *Food Properties Handbook*, CRC Press, Inc., Boca Raton, FL, 1995.
135. G. C. P. Rangarao, U. V. Chetana, and P. Veerraju, Mathematical model for computer simulation of moisture transfer in multiple package systems, *Food Sci. Technol. 28*:38 (1995).
136. K. S. Rao, Hysteresis in the sorption of water on rice, *Current. Sci. 8*:256 (1939).
137. K. S. Rao, Hysteresis loop in sorption, *Curr. Sci. 8*:468 (1939).
138. K. S. Rao, Hysteresis in sorption—V, *J. Phys. Chem. 45*:522 (1941).
139. L. Restaino, L., S. Bills, K. Tscherneff, and L. M. Lenovich, Growth characteristics of *Saccharomyces rouxii* isolated from chocolate syrup, *Appl. Environ. Microbiol. 45*:1614 (1983).
140. S. S. H. Rizvi, Thermodynamic properties of foods in dehydration, *Engineering Properties of Foods*, 2nd ed. (M. A. Rao, and S. Rizvi, eds.), Marcel Dekker, Inc., New York, 1995, p. 123.
141. L. B. Rockland, The practical approach to better low-moisture foods: water activity and storage stability, *Food Technol 23*:1241 (1969).
142. L. B. Rockland and L. R. Beuchat, Introduction, *Water Activity: Theory and Applications to Food* (L. B. Rockland and L. R. Beuchat, eds.), Marcel Dekker, Inc., New York, 1987, p. v.
143. G. N. Roman, M. J. Urbicain, and E. Rotstein, Moisture equilibrium in apples at several temperatures: Experimental data and theoretical consideration, *J. Food Sci. 47*:1484 (1982).
144. M. Rutman, The effect of surface active agents on sorption isotherms of model systems, M.S. thesis, Massachusetts Institute of Technology, Cambridge, MA, 1967.
145. M. Saltmarch and T. P. Labuza, SEM investigation of the effect of lactose crystallization on the storage properties of spray dried whey, *Studies of Food Microstructure* (D. N. Holcomb and M. Kalab), Scanning Electron Microscopy, Inc., Chicago, IL, 1981, p. 203.
146. H. Salwin, Defining minimum moisture contents for dehydrated foods, *Food Technol. 13*:594 (1960).
147. G. D. Saravacos, D. A. Tsiourvas, and E. Tsami, Effect of temperature on the water adsorption isotherms of sultana raisins, *J. Food Sci. 51*:381 (1986).
148. M. M. Sa and A. M. Sereno, Effect of temperature on sorption isotherms and heats of sorption of quince jam, *J. Food Sci. Technol. 28*:241 (1993).
149. K. M. Schaich, Free radical formation in proteins exposed to peroxidizing lipids, Sc.D. thesis, Massachusetts Institute of Technology, Cambridge, MA, 1974.
150. W. J. Scott, Water relations of food spoilage microorganisms, *Adv. Food Res. 7Z*:83 (1957).

151. W. J. Scott, Water relations of *Staphylococcus aureus* at 30°C, *Austr. J. Biol. Sci. 6*:549 (1953).
152. J. M. Seehof, B. Keilin, and S. W. Benson, The surface areas of proteins, V. The mechanism of water sorption, *J. Am. Chem. Soc. 75*:2427 (1953).
153. A. F. Senhaji, The protective effect of fat on the heat resistance of bacteria (II), *J. Food Technol. 12*:217 (1977).
154. C. C. Seow, Reactant mobility in relation to chemical reactivity in low and intermediate moisture systems, *J. Sci. Food Agric. 26*:535 (1975).
155. C. C. Seow and P. B. Cheah, Reactivity of sorbic acid and glycerol in nonenzymatic browning in liquid intermediate moisture model systems, *Food Chem. 18*:71 (1985).
156. M. E. Silver, The behavior of invertase in model systems at low moisture contents, Ph.D. Thesis, Massachusetts Institute of Technology, Cambridge, MA, 1976.
157. L. Slade and H. Levine, Non-equilibrium behavior of small carbohydrate-water systems, *Pure. Appl. Chem. 60*:1841 (1986).
158. L. Slade and H. Levine, Structural stability of intermediate moisture foods—a new understanding, *Food Structure—Its Creation and Evaluation*, (J. R. Mitchell and J. M. V. Blanshard, eds.), Butterworths, London, 1987, p. 115.
159. L. Slade and H. Levine, Beyond water activity: recent advances based on an alternative approach to the assessment of food quality and safety, *Crit. Rev. Food Sci. Nutr. 30*:115 (1991).
160. J. B. Speakman and C. J. Stott, The influence of drying conditions on the affinity of wool for water, *J. Text. Inst. 27*:T186 (1936).
161. W. H. Sperber, Influence of water activity on foodborne bacteria—a review, *J. Food Prot. 46*:142 (1983).
162. R. M. Storey and G. Stainsby, The equilibrium water vapor pressure of frozen cod, *J. Food Technol. 5*:157 (1970).
163. Stadtman and R. Earl, Nonenzymatic browning in fruit products, *Adv. Food Res. 1*:325 (1948).
164. J. Strasser, Detection of quality changes in freeze-dried beef by measurement of the sorption isobar hysteresis, *J. Food Sci. 34*:18 (1969).
165. D. Tome, J. Nicolas, and R. Drapron, Influence of water activity on the reaction catalyzed by polyphenoloxidase from mushrooms in organic liquid media, *Food Sci. Technol. 11*:38 (1978).
166. J. A. Troller, Water activity and food quality, *Water and Food Quality* (T. M. Hardman, ed.), Elsevier Applied Science, London, 1989, p. 1.
167. J. A. Troller, Adaptation and growth of microorganisms in environments with reduced water activity, *Water Activity: Theory and Applications to Food* (L. B. Rockland and L. R. Beuchat, eds.), Marcel Dekker, Inc., New York, 1987, p. 101.
168. J. A. Troller and J. H. B. Christian, Microbiol survival, *Water Activity and Food*, Academic Press, New York, 1978.
169. H. C. Troy and D. F. Sharp, Alpha and beta lactose in some milk products, *J. Dairy Sci. 13*:140 (1930).
170. E. Tsami, D. Marinos-Kouris, and Z. B. Maroulis, Water sorption isotherms of raisins, currants, figs, prunes and apricots, *J. Food Sci. 55*:1594 (1990).
171. N. Uri, Metal ion catalysis and polarity of environment in the aerobic oxidation of unsaturated fatty acids, *Nature, 177*:1177 (1956).
172. C. Van den Berg, F. S. Kaper, A. G. Weldring, and I. Wolters, Water binding by potato starch, *J. Food Technol. 10*:589 (1975).
173. H. Van Olphen, Thermodynamics of interlayer adsorption of water in clay. I. Sodium vermiculite, *J. Colloid Sci. 20*:822 (1965).
174. H. C. Warmbier, R. A. Schnickels, and T. P. Labuza, Nonenzymatic browning kinetics in an intermediate moisture model system: effect of glucose to lysine ratio, *J. Food Sci. 41*:981 (1976).
175. R. Warren and T. P. Labuza, Comparison of chemically measured availability by lysine with relative nutritive value measured by a tetrahymena bioassay during early stage of nonenzymic browning. *J. Food Sci. 42*:429 (1977).
176. H. Weisser, Influence of temperature on sorption equilibria, *Properties of Water in Foods* (D. Simatos and J. L. Multon, eds.), Martinus Nijhoff Publishers, Dordrecht, 1985.

177. H. Weisser, J. Weber and M. Loncin, Wasserdampfsorptionisotherms von Zuckeraustauschstoffen im Temperaturbereich von 25 bis 80°C, *Food Sci. Technol. 33*:89 (1982).
178. M. Wolf, J. E. Walker, and J. G. Kapsalis, Water vapor sorption hystersis in dehydrated foods, *J. Agric. Food Chem. 20*:1073 (1972).
179. J. C. Wolf, D. R. Thompson, J. J. Warthesen, and G. A. Reineccius, Relative importance of food composition in free lysine and methionine losses during elevated temperature processing, *J. Food Sci. 46*:1074 (1981).
180. R. G. Wyn Jones and A. Pollard, Towards a physical chemical characterization of compatible solutes, *Biophysics of Water* (F. Franks and S. F. Mathias, eds.), John Wiley & Sons, Chichester, 1985.

12

pH in Food Preservation

M. SHAFIUR RAHMAN
Horticulture and Food Research Institute of New Zealand, Auckland, New Zealand

I. EFFECTS OF pH ON MICROORGANISMS AND ENZYMES

A. pH Values of Foods

The importance of pH for food stability and food preservation are well documented. pH is the symbol for hydrogen ion concentration, which is a controlling factor in regulating many chemical, biochemical, and microbiological reactions. Hydrogen ion concentration is expressed in moles, and pH is the negative log-ion concentration. The pH scale ranges from 0 to 14. A neutral solution has a pH of 7.0. A lower scale reading indicates an acid solution, and a value above 7.0 indicates an alkaline solution. The pH scale is logarithmic rather than linear in character. Therefore, a pH of 3.0 is 10 times as acid as a pH of 4.0 [27].

The pH values of different food products are given in Tables 1 and 2. In general, fruits, soft drinks, vinegar, and wine have low pH values at which most bacteria will not grow, and these products have good keeping qualities. Most meats, seafoods, and raw milk have pH values greater than 5.6, which make them susceptible to bacterial spoilage and the possible growth of pathogens. Vegetables also have fairly high pH values and are more prone to bacterial spoilage [12]. pH may vary considerably, even within a given product. Some of the most important factors affecting the actual pH values of a product are variety of cultivar, maturity, seasonal variations due to growing conditions, geographical areas, handling and holding practices prior to processing, and processing variables [27].

B. Effects of pH on Microorganisms

1. pH Limits for Microbial Growth

Microorganisms require water, nutrients, and appropriate temperature and pH levels for growth. Table 3 lists the pH limits for growth of some important food spoilage and toxic microorganisms. Minimum pH values for toxin production by *Clostridium botulinum* types A and B in canned foods are given in Table 4. Minimum pH levels appear to be dependent on the type of foods. The effect of water activity together with pH on toxin production by *C. botulinum* type A and B in cooked, vacuum-packed potatoes is shown in Table 5. In general, heterptrophic bacteria tend to be the least acid-tolerant of common food microorganisms [12]. A pH value of 4.5 is important, because this is the pH below which *C. botulinum* is widely believed not to grow in foods. McClure et al.[35] reviewed the

Table 1 pH Values of Plant-Origin Foods

Food	pH	Ref.
Apples	2.90–3.30	12
Apples	3.33–3.84	40
Apple sauce	3.09–3.40	40
Apricots	4.18–4.67	40
Apricots (canned)	3.42–3.47	40
Asparagus (cooked)	6.03–6.10	40
Asparagus (canned)	5.20–5.32	40
Bananas	5.00–5.29	40
Bananas	4.50–4.70	12
Beans (cooked)	5.73–6.20	40
Beans (canned)	4.62–4.72	40
Beets (cooked)	5.23–5.90	40
Beets (canned)	4.92–4.98	40
Bread (white)	5.29–5.65	40
Bread (wholemeal)	5.47–5.61	40
Brussels sprouts	6.00–6.15	40
Butter (peanut)	6.28	40
Cabbage	5.40–6.00	12
Cabbage (green)	5.79–6.29	40
Cabbage (cooked)	6.38–6.82	40
Cantaloupe	6.17–7.13	40
Carrots	5.88–6.00	40
Carrots	4.90–6.00	12
Carrots (cooked)	5.58–5.88	40
Carrots (canned)	5.18–5.22	40
Cherries (red)	3.29–3.32	40
Cherries (black, canned)	3.82–3.93	40
Corn (canned)	5.90–6.44	40
Coconut	5.90–6.52	40
Cucumber	5.18–5.70	40
Figs (canned)	4.92–5.00	40
Grapes	2.80–3.80	40
Grapefruit	3.22–3.70	40
Grapefruit juice	3.00	12
Honey	3.70–3.78	40
Lemon juice	1.98–2.40	40
Lettuce	5.89–6.09	40
Lime juice	2.00–2.25	40
Melons	5.50–6.60	40
Olives (green)	3.38–4.00	40
Olives (ripe)	6.80	40
Onions	5.32–5.85	40
Oranges	3.30–4.30	40
Orange juice	3.60–4.30	12
Peaches	3.30–4.05	40
Peaches (canned)	3.70–3.82	40
Pears (Bartlett)	3.49–4.08	40
Pears (canned)	4.00–4.08	40
Peas (cooked)	6.22–6.88	40

TABLE 1 Continued

Food	pH	Ref.
Peas (canned)	5.71–6.00	40
Pineapple	3.20–3.64	40
Pineapple (canned)	3.39–3.50	40
Plums (Blue and Damson)	2.78–2.89	40
Plums (red and yellow)	3.62–4.95	40
Potatoes	5.40–5.90	12
Quince (stewed)	3.12–3.37	40
Raspberries	3.62–3.95	40
Rhubarb (stewed)	3.24–3.34	40
Spaghetti (cooked)	5.97–6.40	40
Spinach (cooked)	6.60–7.18	40
Strawberries	3.30–3.35	40
Tomatoes	3.99–4.75	40
Tomatoes (canned)	4.10–4.28	40
Vegetables soup (canned)	5.16	40
Vinegar	3.12	40
Watermelon	5.18–5.60	40
Worcestershire sauce	3.63–3.67	40
Yeast	5.65	40

factors affecting growth and toxin production of *C. botulinum*. They found that growth and toxin production may occur at pH values below 4.6 if there is strict anaerobiosis, a high concentration of protein, and various acidulants are used. Below about pH 4.2 most other food-poisoning microorganisms are well controlled, but microorganisms such as lactic acid bacteria and many species of yeast

TABLE 2 pH Values of Animal-Origin Foods

Foods	pH	Ref.
Beef (broth)	6.14–6.20	40
Beef (ground)	5.10–6.20	12
Beef (raw)	5.60	40
Buttermilk	4.41–4.83	40
Butter	6.10–6.40	12
Cheese (camembert)	7.44	40
Cheese (cheddar)	5.90	40
Cheese (cottage)	4.75–5.02	40
Cheese (roquefort)	5.41–6.10	40
Chicken	6.20–6.40	12
Cream	6.40–6.60	40
Eggs (whole)	6.58	40
Eggs (white)	7.96	40
Eggs (yolk)	6.10	40
Lobster	7.10–7.43	40
Milk	6.30–6.50	12
Oysters	5.68–6.17	40
Sardines	5.42–5.93	40
Shrimp	6.80–7.00	12
Soda crackers	5.65–7.32	40
Tuna (canned)	5.92–6.10	40

TABLE 3 Low pH Limits or Growth Range for Microbial Growth

Microorganism	Lowest or growth pH range value[a]	pH_{inside}	Ref.
Acetobacterium spp.	2.8–4.3	4.0–6.0	12
Bacillus acidocaldarius	2.0–5.0	5.9–6.1	12
Bacillus cereus	4.9		47, 66
Bacillus coagulans	3.7		7
Bacteria	4–9		12
Campylobacter jejuni	5.3		66
Clostridium botulinum	4.5		7
Clostridium botulinum (proteolytic)	4.7		47
Clostridium botulinum (proteolytic)	4.6		66
Clostridium botulinum (nonproteolytic)	4.7		47
Clostridium botulinum (nonproteolytic)	5.0		66
Clostridium perfringens	4.5		7
Clostridium perfringens	5.0		47
Clostridium thermoaceticum	5.0–8.0	5.7–7.3	12
Enterococcus faecalis	4.4–9.1	7.2–7.4	12
Escherichia coli	4.0		7
Escherichia coli	4.4–8.7	7.5–8.2	
Escherichia coli	4.4		47
Lactic acid bacteria	3.5		7
Listeria monocytogenes	4.3		47, 7
Lactobacillus species	3.0		66
Molds	1.5–11		5, 12
Most *Bacillus* spp.	4.0		7
Most yeasts and molds	<2.0–3.0		7
Pseudomonas spp.	5.0		66
Saccharomyces cerevisiae	2.35–8.6	6.0–7.3	12
Salmonella serovars	4.0		7
Salmonella	3.8		47
Salmonella spp.	4.0	66	
Staphylococcus aureus (toxin)	4.5		7
Staphylococcus aureus	4.0		47
Staphylococcus aureus (growth)	4.0		47
Vibrio parahaemolyticus	4.9		47
Yeasts	1.5–8.5		5
Yeasts	1.5–8		12
Yersinia enterocolitica	4.4		47
Yersinia enterocolitica	4.6		66

[a]Values may vary according to particular food substrate especially in the presence of organic acids.

and molds grow at pH values well below this. Many weak lipophilic organic acids act synergistically with low pH to inhibit microbial growth. Thus, propionic, sorbic, and benzoic acids are very useful food preservatives [28]. *Ichthyophonus hoferi*, an internal parasite of various fish species, was observed to grow at pH values of 3–7, from 0 to 25°C, and from 0 to 6% sodium chloride [57].

The efficacy of acids depends to a large extent on their ability to equilibrate, in their undissociated forms, across the microbial cell membrane and, in doing so, interfere with the pH gradient that is normally maintained between the inside (cytoplasm) of the cell and the food matrix surrounding

Table 4 Minimum pH Values for Toxin Production by *Clostridium botulinum* Types A and B in Canned Foods

Food	Minimum pH for toxin production
Prune pudding	5.44
Pears	5.42
Pimientos	5.25
Pineapple rice pudding	4.94
Pork and beans	4.93
Zucchini	4.86
Vegetable juice	4.84

Source: Ref. 63.

Table 5 Interaction of Water Activity and pH on Toxin Production by *Clostridium botulinum* Type A and B (Proteolytic) in Cooked Vacuum-Packed Potatoes

a_w	pH	Days to toxin detection
0.980	6.10	7
0.981	5.45	7
0.977	4.83	35
0.972	6.07	7
0.973	5.50	14
0.969	4.96	35
0.959	5.74	35
0.960	5.46	>35
0.964	4.95	>35

Source: Ref. 20.

it [15,21,28]. In addition to weak lipophilic acids, other preservatives widely used in foods include esters of benzoic acid, which are effective at higher pH values than organic acids. Inorganic acids, such as sulfate and nitrite, are most effective at reduced pH values, like organic acids. While these preservatives are employed at ppm levels of hundreds to thousands, the acids used principally as acidulants are often employed at percentage levels [28].

In many products, such as semi-dry sausages and cheeses, a combination of pH and water activity is used as a preservative. The combined inhibitory effects of pH and water activity on the survival of microorganisms are clearly additive [65]. In addition to pH, the type of acid also influences the extent of inhibition with water activity. Generally, citric and acetic acids tend to be more inhibitory in combination with water activity reduction than are hydrochloric or phosphoric acids [64]. The general effect of water activity and pH on growth of bacteria is shown in Figure 1 [65]. FDA's Good Manufacturing Practice Regulations (GMPR) governing processing requirements and the classification of foods are shown in Figure 2. Low-acid foods packaged in hermetically sealed containers must achieve commercially sterile conditions by either retorting, a combined treatment of pasteurization and water activity, or a combined treatment of pasteurization and acidification [31].

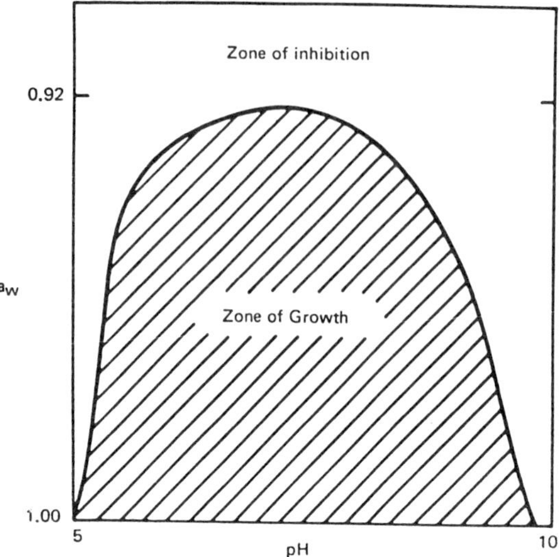

Figure 1 Effects of pH and water activity interaction on bacterial growth. (From Ref. 65.)

2. Mode of Action of pH

Booth and Kroll [12] and Corlett and Brown [17] summarized three regimes of action as follows:

1. Strong acids, which lower the external pH but do not themselves penetrate the cell membrane. These acids may exert their influence by the denaturing effect of low pH on enzymes present on the cell surface and by lowering of the cytoplasmic pH due to increased proton permeability when the pH gradient is very large.

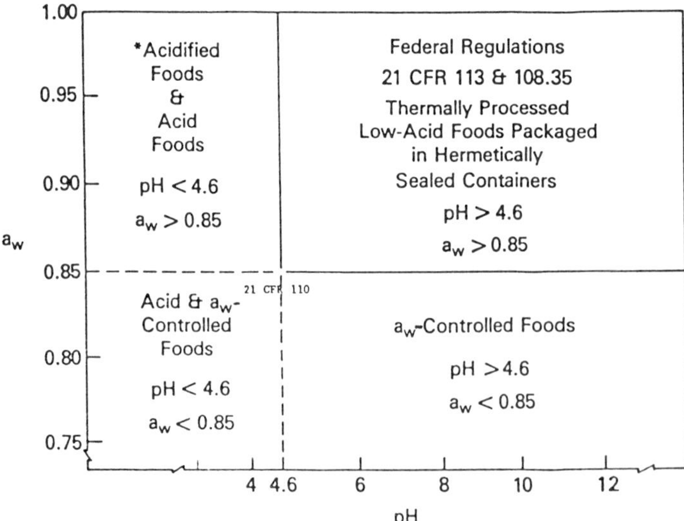

Figure 2 FDA Good Manufacturing Practice Regulations governing processing requirements and classification of foods. *Acidified foods—21 CFR 114 & 108.25. (From Ref. 31.)

2. Weak acids, which are lipophilic and penetrate the membrane. The primary effect of such acids is to lower cytoplasmic pH, and the undissociated acid may have specific effects on metabolism that amplify the effects of the weak acid.
3. Acid-potentiated ions, such as carbonate, sulfate, and nitrate, which are more potent inhibitors at low pH.

As long as the internal pH of a microorganism remains constant, the effect of external pH on growth rate must be due to (a) inactivation of one or more essential enzyme activities at the outer layers of cell, i.e., the outer membrane, the cell wall, the periplasm, and the inner membrane, and (b) reduction of transport systems for essential ions and nutrients [12].

At low external pH the passive influx of protons under the influence of the proton-motive force could be a major problem for cells attempting to regulate their cytoplasmic pH [52]. Most bacteria possess membrane-bound proton pumps, which exclude protons from the cytoplasm in order to generate a transmembrane electrochemical gradient of protons—the proton-motive force. Microorganisms that are tolerant of very low external pH, acidophilic bacteria and yeasts, have relatively low internal pH values. This may be a specific adaptation to the acidic environment [12]. External pH is a significant effector of metabolism, often changing the pattern of enzyme synthesis and the nature of end products of metabolism. Booth and Kroll [12] proposed the generalization that cells produce acidic products when growing at alkaline external pH and neutral or basic products at external acidic pH.

A further counter to acidification of the cytoplasm is the buffering capacity provided by the acidic and basic side chains of the proteins and by the phosphate groups of nucleic acids. In general, buffering capacities in bacteria are approximately 400 nmol H^+/pH unit/mg protein [11]. Booth and Kroll [12] found that the buffering capacity is finite and can be overcome, for example, in the presence of high concentrations of a weak acid at low external pH.

In most microorganisms a cytoplasmic pH close to neutrality is essential for growth. In some microorganisms, recovery of cells from sublethal acid injury did not require new macromolecular synthesis; extensive protein denaturation occurs in both bacterial and yeast cells and protein synthesis is required for recovery [9,41,46]. DNA damage has also been suggested to occur in cells incubated at low pH [56]. Two generalizations can be made about pH homeostasis in microorganisms. First, the optimum cytoplasmic pH is species dependent, i.e., acidophiles in the range of 4.5–6.0, neutrophiles at 7.5–8.0, and alkalophiles at 8.4–9.4. Generally, yeasts and fungi exhibit similar values to the acidophilic bacteria [13]. Second, microorganisms exhibit different capacities to regulate cystoplasmic pH and possibly also have different tolerances of pH perturbation. For the purpose of pH homeostasis, they can be classified as fermentative and respiratory microorganisms [12].

3. Effects of pH on Heat Stability of Microorganisms

When canning foods, one of the most important factors affecting sterilization times and temperatures is the actual pH of the food. The lower the pH, the lower the temperature required for sterilization. Consumer preference for acidic or nonacidic products also affect the selection of pH values. A pH of 4.6 is the usual dividing line between acid and nonacid foods. Foods classified according to their pH values are shown in Table 6 [27].

Bacterial spores are killed by heat more rapidly at low pH than at pH values near neutrality [2,10,16]. Anderson and Friesen [2] showed that the rate of destruction of *Bacillus stearothermophilus*

Table 6 Foods Classified According to Acidity

Group	Group name	pH range
I	Nonacid	7.0–5.3
II	Low or medium acid	5.3–4.6
III	Acid I	4.6–3.7
IV	Acid II	3.7–low

Source: Ref. 27.

spores suspended in acetate buffer was only slightly more rapid at pH 4 than at pH 7, but below pH 4 the rate of death was much more rapid and appeared to be proportional to the proton concentration. At pH values of 7.0 and 6.0, spores of *B. stearothermophilus* survived 60 minutes unharmed at 100°C in the presence of lactic acid and sodium phosphate buffer, whereas at pH 4.3 and 3.0 they died with D_{100} (decimal reduction times) values of 27 and 2.8 minutes, respectively. It was suggested that the enhanced death rate was due to toxic effects of undissociated lactic acid [49]. Watier et al. [68] measured the heat resistance of two strains of spoilage bacteria *Megasphaera cerevisiae* at temperatures of 50–60°C. The D_{50} values were lower at pH 5.2 and 6.0, while at pH 4 the heat resistance was 4.2 times higher; thus at low pH the destruction rate was much higher.

4. Enhancement of the Effects of Preservatives

The efficacy of any preservative depends on pH. Michener et al. [37] investigated 650 compounds for the ability to increase the susceptibility of spores to heat at pH 7. *Bacillus stearothermophilus* spores died much more rapidly at 100°C and mean pH 3.5 in the presence of lactic acid than in its absence. At pH 3.5, lactic acid ($pK_a = 3.87$) is approximately 70% undissociated, and it is possible that the enhanced death rate was due to the toxic effects of the undissociated acid [49].

C. pH Effects on Enzymes

Lipoxygenase catalyzes the oxidation of unsaturated fatty acids, resulting in an off-flavor. Asbi et al. [4] and Che Man et al. [14] studied the inactivation of soybean lipoxygenase by pH adjustment without employing heat treatment. Complete inactivation of lipoxygenase was irreversible when treated at pH 3.0 and below [4,14], and inactivation of urease was reduced to a commercially acceptable level at that pH. No effect on trypsin activity was observed up to pH 2. More than 70% protein dispersibility was retained in neutralized, full-fat soy flour after treatment with acid at about pH 3.0 [14].

Pectinesterase deesterifies pectin, which leads to cloud loss in citrus juice. The thermal inactivation rates of pectinesterase in citrus juices increased at lower pH values [6]. The effect of pH on the stability of thermolabile and thermostable pectinesterase was studied by Sun and Wicker [58]. Both isozymes showed stability over a wide pH range. Thermolabile pectinesterase was inactivated irreversibly at pH 2 and 12, whereas thermostable pectinesterase maintained almost the same activity as before pH treatment with slight inactivation at pH 12. Thus, stability and conformation of thermostable pectinesterase was less likely to be changed by low-pH treatment and the conformational change was nearly reversible.

The enzyme lysozyme has antimicrobial potential to prevent or delay microbial growth in a variety of foods, such as fresh fruits and vegetables, tofu, meats, seafoods, cheese, and wines. The relatively high thermal stability of lysozyme also makes it attractive for use in pasteurized and heat-sterilized food products, possibly allowing reduced thermal processes and minimizing nutritional and sensory quality loss [34]. Makki and Durance [34] studied the stability of lysozyme in aqueous buffer solutions at selected temperatures (73–100°C), pH values (4.2–9.0), and sucrose (0, 5, 15%) and sodium chloride (0, 0.1, 1 M) contents. Lysozyme was most stable at pH 5.2, and thermal stability decreased sharply as the pH increased to 9.0. At pH 7.2 and 9.0, sodium chloride had a clear stabilizing effect against the heat inactivation of lysozyme. Sucrose stabilized lysozyme against heat inactivation at 75°C but not at 91°C. Loss of activity followed first-order kinetics, and rate constant correlation was developed in the pH 5.2–7.2 range and temperatures between 73 and 100°C as follows:

$$\ln k = \frac{32.90 - 1.62 \times 10^4}{T + 1.19\,(\text{pH})} \tag{1}$$

where k is in min^{-1} and T in K. Ibrahim et al. [30] found that heat denaturation of lysozyme at increasing temperatures (80°C at pH 7.0 or over 90°C at pH 6.0) for 20 minutes resulted in progressive loss of enzyme activity, while greatly promoting bactericidal activity against gram-negative and gram-positive bacteria. They also observed that action is independent of catalytic function and kills bacteria through membrane damage mechanism as found from electron microscopy.

II. EFFECTS OF pH ON FOOD COMPONENTS

A. Effects of pH on Gel Formation

pH also affects many functional properties such as the color, flavor, and texture of foods, although the pH of a food is important for microbial growth. Acid fruit pulps form weak gels, which collapse under their own weight [43,69]. Texturization of these acid pulps below pH 3.5 involves an important dilution of the pulp, thus pulp should be neutralized by the addition of sodium hydroxide [32,39].

Egg albumen is an important food ingredient because of its ability to incorporate other ingredients through the formation of a three-dimensional gel matrix. Gel properties are dependent upon many variables, including pH and ionic strength as well as the salts present. Savoie and Arntfield [53] studied the gelation of ovalbumin in the presence of salts containing Ca^{2+} and Mg^{2+} at various pH values. The impact of this binding on gel structure was dependent on pH and the technique used to evaluate the structure. At pH 5 proteins tended to coagulate regardless of the type or amount of salt. At pH 7, the highest rigidity values from penetration measurements were obtained at salt levels of 0.005 M. At pH 9, the salt concentration for maximum rigidity varied with the type of salt, while the storage moduli from dynamic rheology were highest at 0.01 M.

A significant improvement in functional properties, including foaming, emulsifying, and gelling properties, were noted in dried egg white (7.5% water) having been heated in a dry state at 80°C for several days [38]. Mine [38] found that heating dried egg white proteins in the dry state at alkaline pH (under 9.5) was an effective method to obtain firm and elastic gels. The degree of unfolding of the proteins upon dry heating may play a crucial role in the gelling process of the proteins. The polymerization of the proteins was also enhanced by alkaline dry heating through sulfhydryl-disulfide interchange. Alkaline dry heating resulted in high molecular weight polymers of partially unfolded egg white proteins, which, in turn contribute in the formation of low molecular weight and narrow molecular distribution of the aggregate.

Sinapic acid and thomasidioic acid bind to canola protein and affect protein functionality, especially gel formation, upon heating of the protein. Rubino et al. [48] studied the influences of sinapic and thomasidioic acid on the rheological characteristics of canola protein gels. At pH 4.5, there was binding between sinapic acid and the canola protein through electrostatic interaction, while at pH 7 and 8.5, there appeared to be a hydrophobic association between thomasidioic acid and canola protein. The presence of either compound resulted in deterioration of the characteristics of heat-induced gels of the canola protein. Sunflower protein gelation is strongly pH dependent. Sanchez and Burgos [51] found that gelation was only possible in the pH range 7–11 for sunflower protein. The storage modulus reaches its maximum value at pH 8, and gels formed at pH 7 or above pH 9 are very weak. Gelation time increased with pH and decreased with protein concentration, and the storage modulus at pH 8 increased exponentially with protein concentration.

B. Effects of pH on Proteins

The behavior of proteins is pH dependent, and each protein has an isoelectric point where the contributions from positive and negative charges cancel out to give the molecule no net charge. At this pH, proteins tend to coagulate, and therefore it is to be expected that surface rheological parameters of proteins close to their isoelectric point will be maximal [18,29]. At the pI, charge-based contributions to repulsion will be minimal and steric stabilization will also be minimized since the proteins will be in their least expanded state [18]. The emulsion stability is greatest at the pI in some systems, such as gelatin [42], bovine serum albumin [8], pepsin [8], and soluble muscle protein [59]. This is probably due to the greater surface converge at the pI with compact protein structures. This, together with the tendency of the protein to coagulate at the pI, gives cohesive films with enhanced stabilizing action [18]. Proteins that have been shown to give less stable emulsions at the pI include low concentrations (0.004%) of either bovine serum albumin or lysozyme. This may be due to surface coverage of milk fat/whey proteins, which are highly unstable at pH 4.5–5.0, close to the isoelectric point [19]. Dalgleish [18] concluded that the stability of emulsions at the isoelectric point is dependent on protein concentration, the volume-surface area of the oil phase, and the pH.

At other pH values, proteins may show a distinct dependence on pH [18]. Bovine serum albumin showed increasing emulsifying activity as the pH increased from 4 to 9, and then decreased sharply as the protein changed conformation [3,45,67]. In the pH range 3–8, β-lactoglobulin did not change its emulsifying capacity, although it passed conformational transitions [67]. Shimizu et al. [55] found a strong dependence of the emulsifying power on pH, increasing from pH 3 to 9. The hydrophobocities of the different whey proteins vary with pH, but all of the proteins behave similarly, i.e., their surfaces become less hydrophobic as the pH is increased [54]. Increasing the ionic strength diminishes the charged-based interactions between proteins and will produce the same type of effects as changing the pH towards the isoelectric point [18].

Agboola and Dalgleish [1] studied the effects of pH and ethanol on the kinetics of destabilization of oil-in-water emulsions containing milk proteins. Under shear, emulsions containing caseinate were stable between pH 3 and 3.5 and also at pH ≥5.3, while those formed with β-lactoglobulin were stable below pH 4 as well as at pH ≥5.6. The kinetics of pH-induced aggregation in emulsions could be explained by orthokinetic flocculation, while ethanol-induced association in caseinate emulsions appeared to be a result of Ostwald ripening.

Caseins comprise approximately 80% of the total protein content in milk. Caseins are phosphoproteins precipitated from raw milk at pH 4.6 at 20°C. The α_{s1}-caseins contains more acidic amino acids than basic ones, with a negative net charge of 22 at pH 6.5 [70]. β-Lactoglobulin is remarkably acid-stable, resisting denaturation at pH 2.0. The protein assumes the shape of a prolate ellipsoid with an axial ratio of 2:1 and a hydration ratio of 35–40% [62]. It generally exists as a dimer resulting from the association of the monomer at the respective α-helical segments at the isoelectric pH of 5.2 and an alkaline pH range. β-Lactoglobulin A at low temperatures and high concentration, between pH 3.5 and 5.2, tends to form octamers as the predominant species. Below pH 3.5, β-lactoglobulin dissociates into monomers due to electrostatic repulsion between the subunits. Above pH 6.5, the dimers begin to dissociate. A transition in conformation occurs near pH 7.5, which is reversible and involves only a certain region of the molecule [36,60,61]. This reversible unfolding is followed by slow changes, which become increasingly irreversible with increasing pH [70].

Bovine α-lactalbumin is insoluble at the isolectric range between pH 4 and 5. In contrast, goat α-lactalbumin forms a clear solution over a wide pH range. In the physiological state, it exists in the calcium-bound form. The calcium is tightly bound and is not removed by isoelectric precipitation and dialysis against phosphate buffer [33,44].

C. Effects of pH on Vitamin Stability

The stability of vitamins also depends on the pH of the medium (Table 7). At higher water activity (>0.90), the rate of thiamine hydrochloride (vitamin B_1) degradation in a buffered solution of glycerol at 85–95°C is independent of water activity, but highly dependent on pH [24]. Thiamine hydrochloride was less stable in solutions of univalent ions than in glycerol or divalent ion solutions at the same water activity [50]. Riboflavin is also sensitive to pH.

III. METHODS OF CONTROLLING pH IN FOODS

The pH of foods can be altered by (a) adding acidulants, such as acetic, citric, ascorbic, and lactic acid and (b) the action of microorganisms in foods, such as cheese, yoghurt, meat, and alcoholic beverages. In the latter method, growth of spoilage microflora is controlled by the production of lactic or acetic acids.

In fermentation, carbohydrates and other reduced substrates are incompletely oxidized in the absence of an external electron acceptor. The fundamental principle is that all the electrons removed

TABLE 7 Stability of Vitamins Under Acid/Alkaline Conditions

Vitamin	Condition		
	pH = 7	Acid medium	Alkaline medium
Vitamin A	S	U	S
Vitamin D	S	—	U
Vitamin E	S	S	S
Vitamin K	S	U	U
Vitamin C	U	S	U
Vitamin B_1	U	S	U
Vitamin B_2	S	S	U
Vitamin B_6	S	S	S
Vitamin B_{12}	S	S	S
Niacin	S	S	S
Pantothenic acid	S	U	U
Biotin	S	S	S
Folic acid	U	U	S

S = stable; U = unstable.
Source: Ref. 40.

from a fermentable carbon source during oxidation to release energy must be consumed by the reduction of a carbon metabolite, resulting in the formation of a fermentation product [12]. The production of acid by fermentation plays a significant part in the preservation of food. In many dairy products, the production of lactic and acetic acids and hydrogen peroxide may also an important factor. In meat the reduction of pH and not the production of lactic acid is primarily responsible for the preserving action [12,25,26]. In dairy fermentations flavor production is very important. The principal flavor components include acetaldehyde and diacetyl, which are byproducts of the fermentation [12].

Fermentation has been applied to fish for many years and represents a low-level and affordable technology for use in tropical developing countries [22]. Fagbenro and Jauncey [23] studied fermented tilapia stored for 180 days. Fagbenro [22] developed a preservation method for raw shrimp heads by fermenting with 5% (w/w) *Lactobacillus plantarum* as inoculum at 30°C using 15% (w/w) cane molasses as a carbohydrate source. After incubation for 7 days, a desirable and stable pH of <4.5 was attained in the anaerobic treatments, which lasted 30 days after the start of fermentation. Addition of trona at 5% prior to fermentation restricted protein hydrolysis by inhibiting the activity of endogenous autolytic enzymes. Addition of onion extract at 5 ml/kg proved effective an antioxidant as the thiobarbituric acid–reactive substances remained low after 30 days of fermentation.

REFERENCES

1. O. Agboola and D. G. Dalgleish, Effects of pH and ethanol on the kinetics of destabilisation of oil-in-water emulsions containing milk proteins, *J. Sci. Food Agric. 72*:448 (1996).
2. R. A. Anderson and W. T. Friesen, The thermal resistance of *Bacillus stearothermophilus* spores, *Pharm. Acta Helv. 49*:295 (1974).
3. K. Aoki, M. Murata, and K. Hiramatus, *Analyt. Biochem. 59*:146 (1974).
4. B. A. Asbi, L. S. Wei, and M. P. Steinberg, Kinetics of inactivation of soybean lypoxygenase by acid, *Food Science and Technology in Industrial Development* (S. Maneepun and B. Phithakpol, eds.) Institute of Food Research and Product Development, Bangkok, 1988, p. 146.
5. ASHRAE Handbook, *Refrigeration Systems and Applications*, American Society of Heating, Refrigeration, and Air-Conditioning Engineers, Atlanta, GA, 1994.

6. C. D. Atkins and A. H. Rouse, Time-temperature relationships for heat inactivation of pectinesterase in citrus juices, *Food Technol. 8*:498 (1954).
7. A. C. Baird-Parker and G. W. Gould, Safety in the food industry: processing, *Food Safety in the Human Food Chain* (F. A. Miller, ed.), Centre for Agricultural Strategy, Reading, 1990, p. 45.
8. B. Biswas and D. A. Haydon, *Kolloid Z. 185*:31 (1962).
9. L. C. Blankenship, Some characteristics of acid injury and recovery of *Salmonella bareilly* in a model system, *J. Food Prot. 44*:73 (1981).
10. J. C. Blocher and F. F. Busta, Bacterial spore resistance to acid, *Food Technol. 37*:87 (1983).
11. I. R. Booth, Regulation of cytoplasmic pH in bacteria, *Microbiol. Rev. 49*:359 (1985).
12. I. R. Booth and R. G. Kroll, The preservation of foods by low pH, *Mechanisms of Action of Food Preservation Procedures* (G. W. Gould, ed.), Elsevier Applied Science, London, 1989, p. 119.
13. G. W. F. H. Borst-Pauwels, Ion transport in yeast, *Biochim. Biophys. Acta 650*:88 (1981).
14. Y. B. Che Man, L. Wei, and A. I. Nelson, Inactivation of lipoxygenase in whole soybeans by pH adjustment, *Food Science and Technology in Industrial Development* (S. Maneepun and B. Phithakpol, eds.) Institute of Food Research and Product Development, Bangkok, 1988, p. 137.
15. M. B. Cole and M. H. J. Keenan, Synergistic effects of weak-acid preservatives and pH on the growth of *Zygosaccharomyces baillii*, *Yeast 2*:93 (1986).
16. S. Condon and F. J. Sala, Heat resistance of *Bacillus subtilis* in buffer and foods different pH, *J. Food Prot. 55*:605 (1992).
17. D. A. Corlett and M. H. Brown, pH and acidity, *Microbial Ecology of Foods*, Vol. 1, *Factors Affecting the Life and Death of Microorganisms*, Academic Press, London, 1980.
18. D. G. Dalgleish, Protein-stabilized emulsions and their properties, *Water and Food Quality* (T. M. Hardman, ed.), Elsevier Applied Science, London, 1989, p. 211.
19. J. N. De Wit, G. Klarenbeek, and G. A. M. Swinkels, The emulsifying properties of whey protein concentrates, *Zuivelzicht 68*:442 (1976).
20. K. L. Dodds, Combined effect of water activity and pH on inhibition of toxin production by *Clostridium botulinum* in cooked, vacuum packed potatoes, *Appl. Environ. Microbiol. 55*:656 (1989).
21. T. Eklund, The effect of sorbic acid and esters of p-hydroxybenzoic acid on the protonmotive force in *Escherichia coli* membrane vesicles, *J. Gen. Microbiol. 131*:73 (1985).
22. O. A. Fagbenro, Preparation, properties and preservation of lactic acid fermented shrimp heads, *Food Res. Int. 29*:595 (1996).
23. O. A. Fagbenro and K. Jauncey, Chemical and nutritional quality of stored fermented fish (tilapia) silage, *Biores. Technol. 46*:207 (1993).
24. M. Fox, M. Loncin, and M. Weiss, Investigations into the influence of water activity, pH, and heat treatment on the breakdown of thiamine in foods, *J. Food Qual. 5*:161 (1982).
25. C. O. Gill and K. G. Newton, The development of aerobic spoilage flora on meat stored at chill temperatures, *J. Appl. Bacteriol. 43*:189 (1977).
26. S. E. Gilliland and M. L. Speck, Inhibition of psychrotrophic bacteria by lactobacilli and pediococci, *J. Food Sci. 40*:913 (1975).
27. W. A. Gould and R. W. Gould, *Total Quality Assurance for the Food Industries*, CTI Publications, Baltimore, 1988.
28. G. W. Gould, Biodeterioration of foods and an overview of preservation in the food and dairy industries, *Int. Biodeterio. Biodegrad. 36*:267 (1995).
29. P. J. Halling, Protein-stabilized foams and emulsions, *CRC Crit. Rev. Food Sci. Nutr. 15*:155 (1981).
30. H. R. Ibrahim, S. Higashiguchi, M. Koketsu, L. R. Juneja, M. Kim, T. Yamamoto, Y. Sugimoto, and T. Aoki, Partially unfolded lysozyme at neutral pH agglutinates and kills gram-negative and gram-positive bacteria through membrane damage mechanism, *J. Agric. Food Chem. 44*:3799 (1996).
31. M. R. Johnston and R. C. Lin, FDA views on the importance of a_w in good manufacturing practice, *Water Activity: Theory and Applications to Food* (L. B. Rockland and L. R. Beuchat, eds.), Marcel Dekker, New York, 1987, p. 287.
32. G. Kaletunc, A. Nussinovotch, and M. Peleg, Alginate texturization of highly acid fruit pulps and juices, *J. Food Sci. 55*:1759 (1990).

33. M. J. Kronman, J. Jeroszko, and G. W. Saga, Inter- and intramolecular interactions of α-lactalbumin. XII. Changes in the environment of aromatic residues in the goat protein, *Biochim. Biophys. Acta 285*:145 (1972).
34. F. Makki, and T. D. Durance, Thermal inactivation of lysozyme on influenced by pH, sucrose and sodium chloride and inactivation and preservative effect in beer. *Food Res. Int. 29*:639 (1996).
35. P. J. McClure, M. B. Cole, and J. P. P. M. Smelt, Effects of water activity and pH on growth of *Clostridium botulinum, J. Appl. Bacteriol. Symp. 76*(Suppl.):105S (1994).
36. H. A. McKenzie and W. H. Sawyer, Effect of pH on β-lactoglobulin, *Nature 214*:1101 (1967).
37. H. D. Michener, P. A. Tompson, and J. C. Lewis, Search for substances which reduce the heat resistance of bacterial spore, *Appl. Microbiol. 7*:166 (1959).
38. Y. Mine, Effect of pH during the dry heating on the gelling properties of egg white proteins, *Food Res. Int. 29*:155 (1996).
39. C. Mouquet and S. Guilbert, Texturization of pulps allows the fabrication of heat stable mango and passion fruit pieces, Food Ingredients Europe Conference Proceedings, Dusseldorf, 1992, p. 232.
40. A. Murray, From my flowsheet: roll out the kilderkin, *Food Ind. South Africa*, January:25 (1996).
41. A. L. Neal, J. O. Weinstock, and J. O. Lampen, Mechanisms of fatty acid toxicity for yeast, *J. Bacteriol. 90*:126 (1965).
42. G. E. Nielsen, A. Wall, and G. Adams, *J. Coll. Inter. Sci. 13*:441 (1958).
43. A. Nussinovitch and M. Peleg, Mechanical properties of a raspberry product texturized with alginate, *J. Food Proc. Pres. 14*:267 (1990).
44. N. Parris, J. M. Purcell, and S. M. Ptashkin, Thermal denaturation of whey proteins in skim milk, *J. Agric. Food Chem. 39*:2167 (1991).
45. K. N. Pearce and J. E. Kinsella, Emulsifying properties of proteins: evaluation of a turbidimetric technique, *J. Agric. Food Chem. 26*:716 (1978).
46. K. S. Przyblski and L. D. Witter, Injury and recovery of *Escherichia coli* after sublethal acidification, *Appl. Environ. Microbiol. 37*:261 (1979).
47. T. A. Roberts, Microbial growth and survival: developments in predictive modelling, *Int. Biodeterior. Biodegrad. 36*:297 (1995).
48. M. I. Rubino, S. D. Arntfield, C. A. Nadon, and A. Bernatsky, Phenolic protein interactions in relation to the gelation properties of canola protein, *Food Res. Int. 29*:653 (1996).
49. F. Ruiz-Teran and J. D. Owens, Sterilization of soybean cotyledons by boiling in lactic acid solution, *Lett. Appl. Microbiol. 22*:30 (1996).
50. J. Ryley, The effect of water activity on the stability of vitamins, *Water and Food Quality* (T. M. Hardman, ed.), Elsevier Applied Science, London, 1989, p. 325.
51. A. C. Sanchez and J. Burgos, Thermal gelation of trypsin hydrolysates of sunflower proteins: effect of pH, protein concentration, and hydrolysis degree, *J. Agric. Food Chem. 44*:3773 (1996).
52. D. Sanders and C. L. Slayman, Control of intracellular pH: predominant role of oxidative metabolism, not proton transport, in the eukaryotic microorganism *Neurospora, J. Gen. Physiol. 80*:377 (1982).
53. V. J. Savoie and S. D. Arntfield, Effect of pH and cations on the thermally induced gelation of ovalbumin, *J. Text. Stud. 27*:287 (1996).
54. M. Shimizu, T. Kamiya, and K. Yamauchi, The adsorption of whey proteins on the surface of emulsified fat, *Agric. Biol. Chem. 45*:2491 (1981).
55. M. Shimizu, M. Saito, and K. Yamauchi, *Agric. Biol. Chem. 49*:189 (1984).
56. R. P. Sinha, Toxicity of organic acids for repair-deficient strains of *Escherichia coli, Appl. Environ. Microbiol. 51*:1364 (1986).
57. B. Spanggaard and H. H. Huss, Growth of the fish parasite *Ichthyophonus hoferi* under food relevant conditions, *Int. J. Food Sci. Technol. 31*:427 (1996).
58. D. Sun and L. Wicker, PH affects marsh grapefruit pectinesterase stability and conformation, *J. Agric. Food Chem. 44*:3741 (1996).
59. C. E. Swift and W. L. Sulzbacher, *Food Technol. 17*:224 (1963).
60. C. Tanford, L. G. Bunville, and Y. Nozaki, The reversible transformation of β-lactoglobulin at pH 7.5, *J. Am. Chem. Soc. 81*:4032 (1959).

61. S. N. Timasheff, L. Mecanti, J. J. Basch, and R. Townend, Conformational transitions of bovine β-lactoglobulins A, B, and C, *J. Biol. Chem. 241*:2496 (1966).
62. S. N. Timasheff and R. Townend, Structural and genetic implications of the physical and chemical differences between β-lactoglobulins A and B, *J. Dairy Sci. 45*:259 (1962).
63. C. T. Townsend, L. Yee, and W. A. Mercer, Inhibition of the growth of *Clostridium botulinum* by acidification, *Food Res. 19*:536 (1954).
64. J. A. Troller, Effects of a_w and pH on growth and survival of *Staphylococcus aureus, Properties of Water in Foods* (D. Simatos, and J. L. Moulton, eds.), Martinus Nijhoff Publisher, Dordrecht, 1985.
65. J. A. Troller, Adaptation and growth of microorganisms in environments with reduced water activity, *Water Activity: Theory and Applications to Food* (L. B. Rockland and L. R. Beuchat, eds.), Marcel Dekker, New York, 1987, p. 101.
66. S. J. Walker, Chilled foods microbiology, *Chilled Food a Comprehensive Guide* (C. Dennis and M. Stringer, eds.), Ellis Horwood Ltd., West Sussex, 1992, p. 165.
67. R. Waniska, J. Shetty, and J. E. Kinsella, Protein-stabilized emulsions: effects of modification on the emulsifying activity of bovine serum in a model system, *J. Agric. Food Chem. 29*:826 (1981).
68. D. Watier, I. Chowdhury, I. Leguerinel, and J. P. Hornez, Survival of *Megasphaera cerevisiae* heated in laboratory media, wort and beer, *Food Microbiol. 13*:205 (1996).
69. G. Weiner and A. Nussinovitch, Succulent, hydrocolloid-based, texturized grapefruit products, *Food Sci. Technol. 27*:394 (1994).
70. D. W. S. Wong, W. M. Camirand, and A. E. Pavlath, Structures and functionalities of milk proteins, *Crit. Rev. Food Sci. Nutri. 36*:807 (1996).

13

Irradiation Preservation of Foods

M. SHAFIUR RAHMAN
Horticulture and Food Research Institute of New Zealand, Auckland, New Zealand

I. FOOD IRRADIATION PROCESS

A. Action of Ionization Irradiation

Ionization radiation interacts with an irradiated material by transferring energy to electrons and ionizing molecules by creating positive and negative ions [86]. The irradation process involves exposing the food, either prepackaged or in bulk, to a predetermined level of ionization radiation.

The radiation effects on biological materials are direct and indirect. In direct action, the chemical events occur as a result of energy deposition by the radiation in the target molecule, and the indirect effects occur as a consequence of reactive diffusible free radicals forms from the radiolysis of water, such as the hydroxyl radical (OH^-), a hydrated electron (e^-_{aq}), a hydrogen atom, hydrogen peroxide, and hydrogen [86]. Hydrogen peroxide is a strong oxidizing agent and a poison to biological systems, while the hydroxyl radical is a strong oxidizing agent and the hydrogen radical a strong reducing agent. These two radicals can cause several changes in the molecular structure of organic matter [33].

B. Sources of Ionization Irradiation

There are two classes of ionizing radiation: electromagnetic and particulate. They include γ-rays from radionuclides ^{60}Co or ^{137}Cs, x-rays generated from machine sources operated at or below 5 MeV, and electrons generated from machine sources operated at or below an energy level of 10 MeV [72,86]. The characteristics of different irradiation sources are summarized in Table 1. Mitchell [85] stated that although both isotopic and machine sources have identical impact on foods, consumers would react more favorably to machine sources than isotope sources because of the association of isotopes with the nuclear industry. All three sources require a large plant for economic viability. Much of the high cost of irradiation is associated with the need for heavy concrete shielding to protect the external environment when the source is in use. In addition, the plant must comply with hygiene and safety legislation relevant to such plants [56].

Table 1 Characteristics of Irradiation Sources

Radiation source	Characteristics
Cobalt-60	1. High penetrating power 2. Permanent radioactive source 3. High efficiency 4. Source replenishment needed 5. Low throughput
Electron breams	1. Low penetrating power 2. Switch on–switch off capability 3. High efficiency 4. High throughput 5. Power and cooling needed 6. Technically complex
X-rays	1. High penetrating power 2. Switch on–switch off capability 3. Low efficiency 4. High throughput 5. Power and cooling needed 6. Technically complex

Source: Ref. 56.

C. Dose and Dosimetry

The radiation dose (level of treatment) is defined as the quantity of energy absorbed during exposure [132]. Traditionally the dose of ionizing radiation absorbed by irradiated material has been measured in rads, which recently has been superseded by the gray (Gy), which is equal to 100 rad [86]. One gray represents one joule of energy absorbed per kilogram of irradiated product, and the energy absorbed depends on the mass, density, and thickness of the food [132].

Food irradiation doses are generally characterized as low (<1 kGy), medium (<1–10 kGy), and high (>10 kGy). The energy level used for food irradiation to achieve any technological purpose is normally extremely low, e.g., 0.1 or 1.0 kGy, which would be equivalent to a heat energy of 0.024°C or 0.24°C. The Codex Alimentarius Commission recommended 10 kGy as a maximum energy level or dose of ionizing radiation [15]. At 10 kGy the absorbed energy is equivalent to the heat energy needed to increase the temperature of water by 2.4°C. Loaharanu and Murrell [72] calculated on the basis that 10 kGy of ionizing energy is equivalent to a heat energy of 10 J/g and the heat capacity of water is 4.2 J/g °C, i.e., 10/4.2 = 2.4°C. Thus, it is a cold method of food preservation. The absence of noticeable change and small rise in temperature leads to difficulty in detecting whether food has been irradiated or not [56,85].

D. Scope of Irradiation

Moy [87] and Crawford and Ruff [17] summarized the potential applications of irradiation as disinfestation, shelf-life extension, decontamination, and product quality improvement.

1. Disinfestation

Disinfestation is an important postharvest treatment in food processing, and chemicals are usually used for this purpose. A low radiation dose of 0.15–0.50 kGy can damage insects at various stages of development that might be present on the food. Irradiation can damage an insect's sexual viability or its ability to reach adulthood [87].

2. Shelf-Life Extension

One type of shelf-life extension helps to inhibit sprouting of potatoes, yams, onions, and garlic at 0.2–0.15 kGy [78]. Another form delays the ripening and senescence of some tropical fruits such as bananas, litchies, avocados, papayas, and mangoes at 0.12–0.75 kGy [2]. Irradiation also extends the shelf life of perishable products like beef, poultry, and seafood by destroying spoiling microorganisms. Usually, fruits progressively lose their resistance to phytopathogens with ripening. When low dose radiation is used to delay ripening, a higher level of resistance is retained in fruits, and microbial development will also be delayed as an added benefit [132].

3. Decontamination

Disinfestation, the control of insects, in fruits can be achieved by radiation doses up to 3 kGy [132]. One form of decontamination uses a low dose (1.0–2.0 kGy) to pasteurize seafoods, poultry, and beef. Another form uses a higher dose (3.0–20 kGy) for sterilization of poultry, spices, and seasonings [87].

4. Product Quality Improvement

A higher juice yield could be obtained if fruits were first irradiated at a dose level of several kGy, thus improving product recovery. One study showed that the gas-producing factors in soybeans could be markedly decreased by a sequence of soaking, germination, irradiation, and subsequent drying of the beans. This required a dose of 7.5 kGy for maximum effect [42]. It also facilitates reduction of the need for chemicals used in food, such as nitrite, and salts. Moreover, irradiation does not leave any chemical residues in the foods [126]. The extent of doses and their purposes are summarized in Table 2.

E. Special Advantages of Irradiation

Hasegawa and Moy [42] identified at least three distinct benefits of using irradiation to preserve foods. Five advantages of irradiation are discussed below.

1. Minimal Food Losses

Radiation disinfestation and shelf-life extension can reduce the losses of fresh foods. A large portion of postharvest losses due to insect infestation can be controlled and minimized by irradiating foods such as grains, pulses, tubers, and fruits. The shelf life of tubers and some fruits can be extended through sprout inhibition or delayed ripening. Irradiation has great potential in the developing countries where food often spoils during the postharvest stage [42]. A potential added benefit of the application of irradiation of fruits is the increase in juice yield [132].

2. Improved Public Health

Many foods are contaminated with pathogenic microorganisms or parasites. The decontamination of these fresh foods by irradiation can greatly improve public health concerns. *Salmonella* in poultry is a prime source of foodborne illness. The use of irradiation up to 3.0 kGy to decontaminate poultry and up to 1.0 kGy to control *Trichinella spiralis* in pork carcasses is approved in the United States [87]. Irradiation can also be used to ensure the hygienic quality of solid food [70].

3. Increased International Trade

Many fresh foods are not candidates for international trade due to infestation by insects, infection by microorganisms, or limited shelf life. Irradiation can increase or improve the trade of fresh foods internationally by providing an effective quarantine procedure for infested or infected foods or prolonging shelf life [70,87].

Table 2 Extent of Dose and Purpose of Food Irradiation

Dose limit	Purpose	Dose limit (kGy)	Examples
Low dose (<1 kGy)	Sprouting inhibition	0.05–0.15	Potatoes, onions, garlic
	Insect and parasite disinfection	0.15–0.50	Cereals, pulses, dried fruit, pork
	Delay of ripening	0.50–1.00	Fresh fruits and vegetables
Medium dose (1–10 kGy)	Reduction of spoilage microorganisms	1.0–3.0	Fish, strawberries
	Reduction of nonspore pathogenic	2.0–7.0	Poultry, shellfish
	Microbial reduction in dry products	7.0–10.0	Herbs, spices
High dose (10–50 kGy)	Strerilization	25–50	Sterile diet meals
Very high dose (10–100 kGy)	Reduces or eliminates virus contamination	10–100	

Source: Refs. 59, 130.

4. Alternative to Fumigation

Various chemicals, such as ethylene dibromide, methyl bromide, and ethylene oxide, are used for fumigation of food and food ingredients. The use of chemical disinfestation treatments is rapidly diminishing due to their toxic nature and environmental impact, for example, the toxic nature of ethylene oxide and the ozone-depleting effect of ethylene dibromide [56,70]. Low-dose irradiation of 0.2–0.7 kGy can control insect infestation of grain and other stored products [70].

5. Increased Energy Savings

The energy used for irradiation of food is small compared to that used for canning, refrigeration, and/or frozen storage. The total energy used for refrigerated raw, cut-up chicken is 17,760 kJ/kg, for frozen chicken is 46,600 kJ/kg (3–5 weeks frozen storage), and for canned chicken meat is 20,180 kJ/kg. In comparison, refrigerated and irradiated raw cut-up chicken requires a total of 17,860 kJ/kg [12]. Moreover, a ban on CFC refrigerants could result in a higher cost of refrigerated food in the future, thus combined irradiation and chilling have great potential for saving energy during food processing [70]. The reduction in energy requirements can also contribute to an overall reduction of pollution caused by combustion products of traditional fuels [126].

II. EFFECTS OF IRRADIATION ON MICROORGANISMS AND FOOD COMPONENTS

A. Effects on Microorganisms

Ionization irradiation affects microorganisms, such as bacteria, yeasts, and molds, by causing lesions in the genetic material of the cell, effectively preventing it from carrying out the biological processes necessary for its continued existence [89]. The principal targets are nucleic acids and membrane lipids. Alteration in membrane lipids, particularly polyunsaturated lipids, leads to perturbation of membranes and to deleterious effects on various membrane functions, such as permeability. The activity of membrane-associated enzymes may be affected as a secondary effect of membrane lipid degradation [132]. Ionization radiation acts through changes induced in the DNA structure of the irradiated cells, which results in prevention of replication or function [72]. The energy levels used are sufficient to disrupt certain bonds in the molecules of DNA, thereby making cell reproduction impossible [85]. Nucleic acids, because of their large size, are the main targets of free radicals generated by irradiation [132]. Chromosomes of bacteria are intrinsically very sensitive, and lethal damage occurs as a result of exposure to irradiation. The ability of bacteria to repair a limited amount of such damage gives them considerably greater resistance to such radiation. The efficiency with which different bacteria repair the radiation-induced damage to their DNA varies considerably. The most sensitive vegetative bacterium is *Pseudomonas*, and the most resistant one is *Deinococcus* by a factor of about 100 [86].

Murano [89] reviewed the factors that affect the susceptibility of microorganisms to irradiation: dose level, temperature, atmosphere, medium, and type of organism.

In general, the higher the dose applied, the lower the number of survivors and the lower the temperature and the rate of reactions such as the formation of radicals from water molecules. These radicals can affect cells indirectly by interfering with normal cellular functions such as membrane transport. If the product is frozen, radical formation is practically inhibited [89]. The D value increased from 0.16 kGy at 5°C to 0.32 kGy at –30°C when *Campylobacter jejuni* was inoculated into ground beef [63]. In general, bacteria become more resistant to ionization radiation in the frozen state as well as in the dry state. In either state it is assumed that the contribution of indirect effects from the radiolysis of water is significantly reduced [86]. Off-flavor development in products irradiated in a dry state is less pronounced than that in moist products due to the low formation of free radicals at reduced moisture content [33]. Irradiation may have an effect, but this may only occur under specific conditions [89]. The composition of the irradiating medium will affect the survival of microorganisms. As a general rule, the simpler the life form, the more resistant it is to the effects of irradiation. For instance, viruses are more resistant than bacteria, which are more resistant than molds, which are more resistant than human beings. Also, some genera of bacteria are more resistant than others [89]. Bacterial spores are more resistant than their corresponding vegetative cells by a factor of about 5–15 [133].

The effectiveness of irradiation to control infectivity of foodborne parasites is summarized in Table 3 [72]. Low-dose irradiation (<1 kGy) offers a unique opportunity for controlling the infectivity of a number of foodborne parasites without changing the character of the food. Among the groups of foodborne parasites, trematodes appear to be the most sensitive to irradiation, followed by cestodes and protozoa [72]. The D values of various foodborne pathogens in fresh meat are given in Table 4.

Enzymes in foods must be inactivated prior to irradiation because they are more radiation resistant than microorganisms. Usually, enzyme inactivation is accomplished thermally. Generally, it may be said that complete inactivation of enzymes requires about 5–10 times the dose required for the destruction of microorganisms [33]. The D values of enzymes can be 50 kGy, and almost four D values would be required for complete destruction [22]. Thus, irradiated foods will be more susceptible to enzymatic attack than nonirradiated foods [33]. High resistance of enzymes to irradiation has been demonstrated with milk phosphatase, which was not destroyed by irradiation doses sufficient to sterilize milk [102]. Enzymes are affected by the indirect effects of the formation of free radicals in solvent phase, thus dilute solutions of enzymes are relatively more sensitive to irradiation than are concentrated solutes. Moreover, enzymes in their natural environments, as in foods, are relatively resistant [33]. The activity of enzymes is unaffected at normal doses, and thus it limits the achievable shelf life extension of fruits and vegetables [56].

B. Effects on Food Components

1. Effects on Proteins

The effects of irradiation on the nutritional qualities of foods are reviewed by Graham [33]. Low doses of irradiation may cause molecular uncoiling, coagulation, unfolding, even molecular cleavage and splitting of amino acids. Apparently peptide linkages were not attacked, and the main effects were concentrated around sulfur linkages and hydrogen bonds [33]. The sequence of protein bonds attacked by ionizing radiation is as follows: $-S-CH_3$, –SH, imidazol, indol, alpha-amino, peptide, and proline [22]. At 10 kGy radiation, overall increase in total free amino acids was observed mainly due to a rise in the levels of glycine, valine, methionine, lysine, isoleucine, leucine, tyrosine, and phenylalanine [33]. Irradiation is thought to bring about unfolding of the protein molecule, leading to the availability of more reaction sites [33].

Irradiation also affects the functional properties of proteins. In eggs, the doses required for effective *Salmonella* reduction have undesirable side effects, such as loss of viscosity in the white and off-flavors in the yolk [56]. Eggs irradiated with 6 kGy become thin and watery, possibly due to the destruction of alteration of ovomucin, the main thickening compound of egg albumin. Irradiation of milk resulted in an increase in rennet coagulation time and reduced heat stability [33].

Off-flavor development at high radiation doses is due to the presence of benzene, phenols, and sulfur compounds formed from phenylalanine, tyrosine, and methionine, respectively. Flavor changes and off-flavor resembling a burnt flavor were observed in irradiated milk [33]. Irradiation of cheese usually produces smoky off-flavors. Irradiation of soft cheeses at doses of 1–2 kGy is sufficient to reduce food pathogens and does not impair flavor quality, thus dose regulation is certainly the key for preventing off-flavor development [56].

2. Effects on Carbohydrates

Irradiation can break high molecular weight carbohydrates into smaller units, leading to depolymerization. This process is responsible for the softening of fruits and vegetables through breakdown of cell wall materials, such as pectin. Softening may be desirable, e.g., in reducing juice yield and in reducing the drying and cooking times of dehydrated products [56]. Sugars may be hydrolyzed or oxidized when subjected to gamma irradiation [132].

TABLE 3 Effect of Irradiation on Foodborne Parasites

Parasite	Minimum effective dose (kGy)	Effect of irradiation
Protozoa		
Toxoplasma gondii	0.09–0.7	Parasite killed or elimination of infectivity
Entamoeba histolytica	0.251	Killed cyst stage
Trematodes		
Fasciola hepatica	0.03	Inhibits maturation
Clonorchis sinensis	0.15–0.20	Inhibits maturation
Opisthorchis viverrini	0.10	Inhibits maturation
Paragonimus westermani	0.10	Inhibits maturation
Cestodes		
Taenia saginata	>3.00	Complete inactivation of larvae
	0.40	Prevents development in humans
	0.30	Elimination of infectivity
Taenia solium	0.20–0.70	Elimination of infectivity
Echinococcus granulosus	0.50	Elimination of infectivity
Nematodes		
Trichinella spiralis	0.10–0.66	Elimination of infectivity
	0.11	Sterilization of females
Angiostrongylus cantonesis	2.00–4.00	Decreased infectivity
Gnathostoma spinigirum	7.00	Reduced larval penetration
Anisakis species	6.00	Reduced larval penetration

Source: Ref. 72.

Table 4 Susceptibility of Various Foodborne Pathogens to Irradiation in Fresh Meat

Organism	D value (kGy)
Listeria	0.40–0.60
Salmonella	0.40–0.50
Escherichia coli O157:H7	0.25–0.35
Campylobacter	0.14–0.32
Yersinia	0.14–0.21
Aeromonas	0.14–0.19

Source: Ref. 103.

The irradiation of wheat at 0.2–10 kGy increased the initial levels of water-soluble reducing sugars by 5–92% compared to untreated samples [106]. Such overall increase in initial total reducing sugars resulted from the stepwise and random degradation of starch. These changes are highly advantageous in the generation of bread flavor and aroma by reducing sugar–amino acid reactions [33].

Irradiation of pure carbohydrates produced degradation products that have mutagenic and cytotoxic effects. However, these undesirable effects were produced using very high doses of irradiation [33].

3. Effects on Lipids

Irradiation initiates the normal process of autooxidation of fats which gives rise to rancid off-flavors. Highly unsaturated fats are more readily oxidized than less unsaturated fats. This process can be slowed by elimination of oxygen by vacuum or modified atmosphere [56]. In lipids, particularly unsaturated fatty acids, radiolytic decomposition occurs via a preferential break at the level of the carbonyl function of the double bond [39]. This decomposition induces the formation of some volatile compounds responsible for off-odors [80]. The formation of peroxides and volatile compounds and the development of rancidity and off-flavors have been reported [81,92]. The peroxide formed can also affect certain labile vitamins, such as vitamins E and K [33].

Lipids in cereals degraded only a high doses of irradiation, and no significant effects on iodine value, acidity, or color intensity of wheat flour lipids was observed. At 10 kGy, a 20% increase in total free lipids and a 46% decrease in bound lipids were observed [107]. Lipid-protein complexes, which are critical in baking, were not noticeably affected at low doses up to 2 kGy.

The volatile oil content of spices has a dose-dependent reduction effect in black pepper [121] and ginger [95] above 6 kGy. Similar reduction was also observed in Ashanti pepper berries when 47 essential oil compounds were analyzed individually at a dose of 10 kGy [96].

4. Effects on Vitamins

The extent of vitamin C, E, and K destruction by radiation depends on the dosage used. Thiamine is very labile to irradiation. Ascorbic acid in solution is quite labile to irradation but in fruits and vegetables seems quite stable at low-dose treatment [33]. Vitamins, particularly those with antioxidant activity, such as A, B_{12}, C, E, K, and thiamine, are degraded when irradiation is carried out in the presence of oxygen [132].

Irradiation can also partially damage vitamins C and B_1. Kilcast [56] stated that the literature referring to vitamin loss is misleading in many cases. Vitamin losses are often described at unrealistically high irradiation doses or under unrealistic conditions. In particular, vitamin C loss is often equated with ascorbic acid loss, ignoring the fact that irradiation converts ascorbic acid into dehydroascorbic acid, which is also active as a vitamin [56,132].

III. APPLICATIONS OF IRRADIATION IN FOODS

A. Foods of Plant Origin

Plant tissues showed a transient increase in respiration and ethylene production even at low radiation doses. The rate of respiration increased linearly with increasing dose of irradiation. The transient rate of respiration decreased back to preirradiated levels within 24 hours for 0.3 kGy, but more slowly at larger doses. Ethylene production also increased after irradiation, reaching a maximum at 1 kGy. It has been suggested that irradiation beyond 1 kGy caused membrane damage, since ethylene production is membrane associated [132]. Radiation caused a shift from glycolysis toward the pentose phosphate shunt in bananas and toward the glyoxylate cycle in bananas and mangoes [120]. At higher doses, climacteric fruits may not ripen normally and may develop uneven coloring and skin discoloration [132]. Fruits suffer physiological disorders when exposed beyond their limits of tolerance. These undesirable symptoms are mainly tissue softening and enzymatic browning [51]. Tissue softening is caused by partial depolymerization of cell wall polysaccharides, mainly cellulose and pectins [19], and by damage to cell membranes [128]. Enzymatic browning is an indication of cell decompartmentation due to damage of membranes, thus bringing phenolic substrates in contact with polyphenoloxidases [115]. Damage to the cell membrane may result in the loss of intracellular water, cell turgescence, and oxidative attack on polyunsaturated fatty acids of membrane lipids [132]. The oxidation can be minimized by irradiating in an atmosphere with reduced oxygen content, but then treatment efficiency is reduced. Voisine et al. [127] found that a high carbon dioxide atmosphere protected tissues from irradiation-induced loss of membrane proteins. Thus, low-dose irradiation combined with modified atmosphere is increasingly considered for control of microorganisms and delayed ripening. Couture and Willemot [16] showed the synergistic action of irradiation combined with high carbon dioxide to control mold development on strawberries. Table 5 shows the response of a number of fruits to irradiation. The applications of irradiation in specific plant materials are discussed below.

1. Spices

Irradiation is becoming increasingly important for the decontamination of spices. Spices imported into Western Europe are often heavily contaminated by pathogenic microorganisms as a consequence of open air-drying procedures [56].

Table 5 The Response of 23 Fruits to Irradiation

Effect	Results	Materials
Beneficial	Ripening delayed	Bananas, mangoes, papayas
	Senescence delayed	Sweet cherries, apricots, papayas
	Storage decay controlled	Tomatoes, strawberries, figs
Not beneficial	Lack of tolerance	Pears, avocados, lemons, grapefruit, oranges, tangerines, cucumbers, summer squash, bellpeppers, olives, plums, apples, table grapes, cantaloupes
	Ripening accelerated	Peaches, nectarines
	Irradiation tolerance only	Pineapples, lychees, honeydew melons

Source: Ref. 2.

The prevalent microorganisms of pepper are *Clostridium, Staphylococcus, Bacillus, Aspergillus*, and *Fusarium* species. An irradiation dose of 2.5 kGy reduced the fungal and bacterial load by 2 log cycles, and 7.5 kGy eliminated the fungal population of ground or whole pepper.

Yang et al. [136] found that irradiation of garlic bulbs with 0.15 kGy can inhibit sprouting and reduce weight losses during storage. Irradiation does affect the flavor compounds of garlic. Kwon and Yoon [60] reported that the content of diallyl disulfide in garlic was slightly reduced by irradiation of 0.05–0.10 kGy. There was no difference in the contents of diallyl disulfide between 0.05 kGy irradiated and unirradiated bulbs during storage [13]. The irradiated bulbs showed no difference in odor during storage when compared with untreated bulbs [18]. Wu et al. [135] evaluated the effects of 0.15 kGy on the content of volatile compounds in garlic bulbs during storage at room temperature. The content of diallyl disulfide decreased immediately after irradiation. At the end of 8 months of storage both irradiated and untreated samples showed a significant increase in diallyl disulfide.

Wu and Yang [134] studied the effect of 0.05 kGy irradiation on the volatile compounds of ginger rhizome. The quantities of some major volatiles were significantly decreased in irradiated rhizome after 3 months of storage.

2. Fruits and Vegetables

Berries

The postharvest shelf life of cherries, blueberries, and cranberries can be extended by low-dose irradiation [24]. Blueberries irradiated at 0.25, 0.5, 0.75, or 1.0 kGy can be stored at 1°C for 1, 3, 7 days, respectively, and 2 additional days at 15°C [82]. The firmness of Sharpblue berries was slightly affected, but the firmness of Climax berries was not affected by irradiation. Flavor and texture were negatively affected as the dose increased for berries of both cultivars. Weight loss, decay, peel color, total soluble solids, and titratable acidity were not affected by dosage. Irradiation at or below 0.75 kGy was determined not to be detrimental to postharvest quality. This treatment can be an effective alternative quarantine treatment to methyl bromide [82].

Mangoes

Mango preservation would be greatly benefited by irradiation treatment. The effects of irradation depend on the degree of maturity [132]. The optimal dose was 0.75 kGy for three-quarter ripe fruits at room temperature [2]. Combining irradiation with a mild heat treatment by hot water dip or vapor for 5 minutes at 50–55°C yielded even better results [6]. Surface scalding is a limiting factor, since at 0.25 kGy scalding occurred on mature green fruits and tolerance increased with maturity [2].

Carrots

Radiation doses up to about 0.1 kGy had little effect on the firmness of apples, carrots, and beets, but rapid softening occurred at higher doses [10,54]. The effective range for control of rotting and sprouting is 0.1–1 kGy, which means that substantial softening can occur [9]. Bourne [9] studied the kinetics of softening of carrots up to doses of 0–50 kGy. Two distinct regions were observed: a steep negative slope for doses up to 15 kGy and a shallow negative slope beyond 15 kGy. The two-stage softening rate curve is qualitatively similar to that for thermal softening of carrot.

Papaya

Papaya can tolerate up to 1 kGy γ-radiation before surface scald occurs. Surface color development is not disrupted up to 2 kGy, flavor and aroma up to 4 kGy, and tissue breakdown up to 5 kGy [2,3]. A dose of 0.75 kGy was considered optimal for retention of fruit firmness with only slight reduction of storage decay [3]. A hot water dip at 48.9°C for 20 minutes in combination with irradiation delayed ripening with an optimal dose of 0.75 kGy [2,6]. Hot water dip alone accelerated ripening, while irradiation alone provided only slight control of decay, and hot water dip was preferable to vapor heat treatment [132]. Respiratory activity was initially elevated immediately after irradiation and then returned to the level of untreated fruit within 24 hours [1]. Paull [99] studied the effect of 0.25 kGy on papaya and found the irradiated fruit softened more uniformly than nonirradiated fruit.

Zhao et al. [140] found that irradiation had no effect on the skin or flesh color development of ripening papaya. Irradiation induced immediate depolymerization and demethoxylation of papaya pectic substances, indicated by an increase in water-soluble pectin and a decrease in chelator-soluble and alkali-soluble pectin with a significant decline in methanol content. The linear decrease in firmness up to 1.5 kGy paralleled the change of pectin fractions. Pectin methylesterase was not affected, either immediately after irradiation or during ripening, by irradiation at doses up to 1.5 kGy.

Strawberries

Irradiation at doses 1, 2, and 3 kGy effectively prolonged the shelf life of strawberries stored at 4°C by 5, 13, and 16 days, respectively [122]. Maxie and Abdel-Kader [79] indicated that strawberries may tolerate an irradiation dose up to 2 kGy for reducing fungal infection without quality changes. The softening of strawberries occurred after irradiation [50,120,139]. The softer texture above 2 kGy may limit use of higher doses. The firmness of strawberries decreased as irradiation doses increased at 0.5, 1, and 2 kGy [137,138]. Success depends on cultivar: the firmer fruits of Tioga tolerated radiation better than the softer fruits of Brighton [6]. Several studies indicated that irradiation-induced texture change was associated with changes in pectic substances [45,54,114]. Water-soluble pectin increased and oxalate-soluble pectin decreased at 0 and 1 day after 1 and 2 kGy irradiation. Fruit firmness correlated with oxalate-soluble pectin content. Total pectin and nonextractable pectin were not affected by irradiation. The oxalate-soluble pectin content and firmness of irradiated strawberries increased slightly at the beginning of 2°C storage and then decreased as storage time increased [138]. Irradiation enhanced the sweetness of strawberries by reducing titratable acidity in comparison with untreated sample [73]. The depolymerization of carbohydrate polymers, such as starch and cellulose, may slightly increase the sugar content [19].

3. Cereals and Grains

Grains and cereals are treated with low doses of irradiation to eliminate fungi, since these organisms can produce mycotoxins [33]. Irradiation doses in the range of 0.2–1.0 kGy are effective in controlling insect infestation in cereals [48]. Increasing the dose to 5 kGy totally kills the spores of many fungi that survive lower doses [91]. In addition to its protective role from insects and microorganisms, irradiation also affects various quality criteria of cereal grains [58]. The amylograph peak viscosity and falling number values of flour decreased with increasing irradiation dose [75,93]. Rao et al. [105] also found that amylograph peak height and dough stability decreased with increasing dose. At 10 kGy, loaf volume and crumb grain were impaired. The overall bread quality of wheat was greatly reduced with medium-dose irradiation of 1–10 kGy [97]. Lai et al. [62] found that loaf volume and baking quality deteriorated above 5 kGy irradiation irrespective of the baking formula.

Increased cooking losses and inferior sensory scores were observed in Japanese noodles when wheat was irradiated in the range of 0.2–1.0 kGy [109,125]. Koksel et al. [58] studied the effects of irradiation up to 5.0 kGy on durum wheat and semolina properties and spaghetti cooking quality. The falling number and sedimentation values of wheat meals decreased with increasing dose levels. This indicated alterations in both starch and gluten components. Irradiation also caused important changes in spaghetti cooking properties. Koksel et al. [58] found that irradiation at 1 kGy may be useful for treating grain for insect control without adversely affecting quality. Above 1 kGy, irradiated samples exhibited lower scores for stickiness, firmness, and bulkiness compared to unirradiated samples due to deterioration in both starch and gluten.

Cowpeas can be preserved in polyethylene bags (100 μm) after ionizing treatment at doses less than 0.10 kGy without causing unfavorable nutritional consequences [23].

B. Foods of Animal Origin

Irradiation is effective in preventing or delaying the microbial spoilage of fresh meats and poultry. Mitchell [85] found, as did earlier studies [66], that irradiation at doses between 0.25 and 1 kGy under aerobic conditions increased microbiological shelf life but accelerated rancidity. A tallowy odor and flavor developed during storage. The fat was noticeably bleached, and peroxide accumulated more rapidly in the irradiated fat than in the control fat. Doses up to 2.5 kGy control *Salmonella, Campylobacter, Listeria monocytogenes, Streptococcus faecalis, Staphylococus aureus*, and *Escherichia coli* in poultry and other meats. Doses in excess of 2.5 kGy may cause changes in flavor, odor, and color, but these can be minimized by irradiating at low temperature or in the absence of oxygen [85].

Irradiation treatment is not effective in stopping the changes in meats that diversely affect consumer acceptance, such as oxidation of pigment to yield brown or gray discolorations, drip loss from the cut surface of lean tissue, and oxidation of meat lipids, which causes off-flavors, by atmospheric oxygen [85]. Thus, irradation coupled with vacuum packaging has the potential to extend the shelf life [85]. Table 6 shows the threshold dose for an identifiable irradiation flavor in meats.

Hydrogen generated during irradiation of frozen meats is a promising marker for distinguishing between irradiated and unirradiated frozen food [44]. Poultry meat in particular is known to be susceptible to color changes following irradiation [41]. A pink color is produced in fresh poultry when it is treated with irradiation [8]. Irradiated chicken breasts were found to exhibit increased greater redness (a values) when compared with unirradiated controls [83].

1. Poultry

At low-dose irradiation not all microorganisms are destroyed, and survivors such as *Moraxella, Acinetobacter, Lactobacillus*, and *Streptococcus* can cause spoilage [100]. A mixed microflora to a predominantly gram-positive microflora usually exists during the postirradiation stage [88]. Doses of 2–2.5 kGy are effective in controlling *Listeria* [46,69] and a dose of 1.0–2.5 kGy is adequate to eliminate *Pseudomonas aeruginosa*, and 2.5–5.0 kGy for elimination of *Serratia marcescens* [40]. A dose of 1.50 kGy was effective against *Staphylococcus aureus* in deboned chicken when irradiated in vacuum at 0°C and held at 35°C for 20 hours [117]. In deboned chicken meat a 90% decrease of viable *E. coli* cells can be achieved by doses of 0.27 kGy at 5°C and 0.42 kGy at –5°C [118].

In the Netherlands, a maximum dose of 3 kGy is permitted on an unconditional basis for poultry irradiation [53], and in Israel and South Africa a dose of 7 kGy is allowed to eliminate pathogenic bacteria [104]. Mulder [88] recommended a dose of 2.5–5 kGy to extend shelf life at chill temperatures for 6–14 days without insignificant organoleptic quality change. However, a dose as low as 0.5 kGy can induce a radiation odor, and 2.5 kGy can induce flavor changes (which may be removed on subsequent cooking) [40,88]. Table 7 shows the changes in odor of irradiated chicken carcasses during storage at 1.6°C.

The color of meat depends on three factors: the concentration of heme pigments, the chemical state of these pigments, and the physical light-scattering properties of the meat structure [76].

Patterson [100] studied the effects of irradiation under air, carbon dioxide, vacuum, and nitrogen on seven bacterial species inoculated onto sterile poultry meat. *Streptococcus faecalis* and *Staphylococcus aureus* were not sensitive to atmosphere; *Pseudomonas putida, Salmonella typhimurium, Escherichia coli, Moraxella phenylpyruvica*, and *Lactobacillus* species were more sensitive in atmospheres other than air. In general, a vacuum or carbon dioxide atmosphere during irradiation had the most lethal effect, and bacteria may be more resistant to irradiation if packaged under nitrogen than carbon dioxide or air [85,100].

TABLE 6 The Threshold Dose for an Identifiable Irradiated Flavor for Meats

Meats	Temperature (°C)	Threshold (kGy)	Ref.
Pork	5–10	1.75	116
Beef	5–10	2.50	116
Chicken	5–10	2.50	116
Lamb	5–10	6.25	116
Lamb	—	2.40	20

TABLE 7 Changes in Odor of Irradiated and Unirradiated Chicken Carcasses Stored at 1.6°C

Storage time (days)	Unirradiated	Irradiated 2.5 kGy	Irradiated 5.0 kGy
0	Fresh chicken	Slight irradiation odor	Irradiation odor
4	Fresh chicken	Fresh chicken odor	Slight irradiation odor
8	No odor	Fresh chicken odor	Fresh chicken odor
11	Slight off-odor	Chicken odor	Chicken odor
15	Putrid	Slight chicken odor	Slight chicken odor
18	Putrid	Stale chicken odor	Stale chicken odor
22	Putrid	Stale chicken odor	Stale chicken odor
31	Putrid	Stale chicken (sour)	Stale chicken odor

Source: Ref. 52.

2. Mutton Lamb

Irradiation of vacuum-packaged mutton backstraps at 4 kGy prevented the growth of bacteria for at least 8 weeks at 0–1°C [77], while *Brocothrix thermosphacta* and gram-negative bacteria grew on telescoped lamb carcasses irradiated at 2.4 kGy and stored at 5°C. The total population did not exceed 10^5 cfu/cm^2 during 16 weeks of storage [7]. However, at these high doses adverse effects on sensory attributes and increased volume of weep release were observed. Meat chunks irradiated at 1.0 kGy and 2.5 kGy were acceptable for 3 and 5 weeks, respectively, whereas minced meat was only good for 2 and 4 weeks, respectively [98]. In contrast, unirradiated meat, both chunks and minced, spoiled within 1 week of storage at 0–3°C [85].

3. Beef

Pseudomonas, Enterobacteriaceae, and *Brocothrix thermosphacta* were strongly inhibited in irradiated beef samples and sensory properties were not altered [85]. Rodriguez et al. [108] studied the effect of 2 kGy irradiation on fresh top round beef packed aerobically in polyethylene film. Psychrotroph counts on the untreated samples reached 10^7 cfu/cm^2 after storage for 8–11 days, while similar counts were not observed in irradiated samples until day 28 of storage. The shelf life of vacuum-packaged raw meat can be extended considerably by 1–5 kGy, which also yielded satisfactory sensory quality [20,25]. Grant et al. [36] studied the effect of 2 kGy irradiation on growth and toxin production by *Staphylococcus aureus* and *Bacillus cereus* on roast beef. Irradiation resulted in a 3–4 log reduction in numbers of both pathogens. Toxin production by both pathogens was also delayed by irradiation. Irradiated at 2 kGy, roast beef and gravy samples stored at 5°C and 10°C showed similar growth rates of *Listeria monocytogenes*, with a lag period of 6–9 and 2–4 days, respectively, compared to 1–2 days and less than 0.1 day in nonirradiated samples [37].

Postmortem aging of beef is typically done by holding carcasses or cuts between 1 and 4°C for up to 3 week. During that time, tenderness improves and a distinctive flavor develops. Snyder [113] suggested that carcass aging at high temperature followed by irradiation can reduce microbial numbers. Lee et al. [68] found that irradiation in conjunction with MAP containing 25% carbon dioxide and 75% nitrogen could be used for an accelerated aging process of beef at 30°C for 2 days. Their result was based on tenderness, chemical, visual, and microbiological effects. Moreover, if irradiated beef were chilled immediately after aging, this process could improve tenderness without excessive microbial growth and would be more efficient than aging carcasses at high temperatures followed by irradiation [68].

4. Pork

Sivinski and Switzer [112] found that a low-dose irradiation treatment between 0.30 and 1.0 kGy could be used to inactivate the parasite *Trichinella spiralis* in pork.

The effect of 1 kGy irradiation on vacuum-packaged pork stored up to 21 days at 4°C was studied. Radiation reduced the numbers of mesophiles, psychrotrophs, anaerobic bacteria, and staphylococci throughout storage. The effects of irradiation on the sensory characteristics of pork loin were minimal. Irradiation of pork strip loins (vacuum-packaged, pH 6.2–6.6) at 2.5 and 4.3 kGy reduced the number of viable bacteria present 3 and 5 log [84]. Significant organoleptic (color, odor, and flavor) changes occurred up to 1.0 kGy. Shelf life increased from 8 to 11.5 days for vacuum-packaged ground pork irradiated at 1 kGy when stored at 5°C [27,28] and from 4 to 6 weeks at 0°C [84]. The storage life of vacuum-packaged pork at 0°C and 5°C was doubled by 2.5 and 4.3 kGy treatment, but undesirable side effects included changes in the color and odor of uncooked meat [26]. The odor resulting from a dose of 2.5 kGy was sufficient to make the meat unacceptable. Treatment using a dose of 1.0 kGy was effective in extending storage life and produced only slight changes in color and odor [26]. After 12 days of refrigerated storage, *Lactobacillus* and coryneform bacteria predominated in the irradiated meat [27]. The microbial shelf life of vacuum-packaged pork loins stored at 2°C was extended from 41 days to 90 days when treated with 3 kGy [67]. Samples of vacuum-packaged ground fresh pork were irradiated at doses from 0.57 to 7.25 kGy and stored at 2°C for 35 days and analyzed. Surviving microflora were not detected in any sample that received an absorbed dose of 1.91 kGy or higher. *Staphylococcus, Micrococcus*, and yeast species predominated in samples that received a dose of 0.57 kGy [119]. There is not significant difference between irradiated samples and control for lipid oxidation of irradiated pork chops during storage [32].

After reviewing literature on irradiation in combination with modified atmosphere packaging, Mitchell [85] mentioned that substantial extensions in sensory shelf life can be achieved using doses from 0.5 to 1.75 kGy and modified atmospheres with 25–50% carbon dioxide (balance nitrogen). The product must be stored at 5°C or less to achieve extended shelf life. Irradiation in the presence of oxygen has a detrimental effect on chemical and sensory characteristics, resulting in rejection of the product [64]. The shelf life of irradiated and unirradiated pork loin in 100% nitrogen is shown in Table 8. The microflora of irradiated modified atmosphere packaging of pork consists almost exclusively of lactic acid bacteria [34]. Inoculation studies [35] showed the *Clostridium perfringens* was the most resistant and *Yersinia enterocolitica* the most sensitive of the pathogens.

5. Processed Meats

The amount of nitrite required in cured meats could be reduced by the use of lowering the chance of nitrosamine formation [20,21]. The nitrite levels in irradiated bacon can be reduced from normal levels of 120–150 to 20–40 mg/kg without a loss of organoleptic quality [110]. Wills et al. [131] treated vacuum-packaged sliced corn beef with radiation doses of 1, 2, and 4 kGy. The initial microbial load was reduced by 1, 2.5, and 5 log cycle, however, slight changes in aroma and flavor at 2 kGy were observed and storage life was doubled to about 5 weeks. Enterobacteriaceae were effectively inactivated by irradiation with doses of 1 or 2 kGy when sensory effects were minimal. The product can

TABLE 8 Shelf Life of Irradiated and Unirradiated Pork Loin in 100% Nitrogen

Irradiation dose (kGy)	Storage temperature (°C)	Shelf life (days)			
		Microbial	Color	Odor	Overall
0	5	14	9	16	9
1	5	21	35	26	21
0	25	2	<2	<2	<2
1	25	10	>14	2	2

Source: Ref. 65.

be stored up to 5–7 days when treated at 2 kGy. The ground beef patties irradiated at 2.0 kGy under vacuum remained unspoiled even after 60 days of refrigerated storage [90]. In minced meat, a slight reduction of pH to 5.2–5.3 was observed in vacuum-packed products at 2 kGy. Lactic acid bacteria were more resistant to radiation and became the dominant species during storage. The combination of pH reduction and irradiation prevented growth of Enterobacteriaceae even at 10°C incubation [31].

The sensory characteristics (flavor, texture, juiciness, and aftertaste) of ground beef patties irradiated at 2.0 kGy and stored under refrigerated conditions were studied by Murano et al. [90]. After one day, irradiated patties were significantly more juicy and tender than nonirradiated ones, but after 7 days no significant difference was observed.

6. Fish and Fish Products

Singh [111] and Nickerson et al. [94] reviewed the irradiation of meats and fish and their shelf life. Singh [111] found that control of pathogenic organisms and extension of shelf life of fresh fish can be achieved with relatively low doses of ≤2.5 kGy. However, *Clostridium botulinum* (A, B, E, and F) present in fish and fish products remained unaffected by the low doses of irradiation. Thus, storage at <3°C and oxygen availability need to be assured. In dried fish (moisture < 20%) a dose of 0.3 kGy is sufficient to control insects and their larvae [47], but at higher levels of moisture (20–40%) a dose of 0.5 kGy is required.

Mold growth can also contribute to spoilage, depending on the moisture level in the fish. Control of mold growth by irradiation alone requires doses of 3–5 kGy [47]. Bacterial spoilage is also a problem in semidry and fresh fish and fish products [111]. Typical doses up to 2.5 kGy are generally adequate to control the spoilage bacteria and extend the shelf life of fresh fish. The optimum irradiation doses and shelf life of freshwater and marine fish and shellfish are compiled by Singh [111]. In general, the shelf life extension using irradiation is dependent on the conditions of irradiation and storage and seem to vary according to species of fish.

Al-Kahtani et al. [4] studied the effects of gamma irradiation (1.5–10 kGy) and postirradiation storage up to 20 days at 2°C on tilapia and Spanish mackerel. They found that (a) total volatile basic nitrogen formation was lower in irradiated than in unirradiated fish, (b) in thiobarbituric acid values increased gradually during storage, (c) some fatty acids ($C_{14:0}$, $C_{16:0}$, and $C_{16:1}$) decreased upon irradiation, while others ($C_{18:0}$, $C_{18:1}$, and $C_{18:2}$) increased, (d) thiamine loss was more severe at higher doses (≥4.5), whereas riboflavin was not affected, and (e) doses higher than 3.0 kGy caused a decrease in α- and γ-tocopherols.

Chen et al. [14] studied the effects of low doses (2 kGy or less) of irradiation on pathogenic and spoilage microorganisms and on the sensory quality of carb products (white lump, claw, and fingers) throughout 14 days of ice storage. Irradiation effectively reduced spoilage bacteria, extending shelf life more than 3 days beyond control samples. Fresh crab odor and flavor were similar for treated and control samples, while off-flavors and odors developed more rapidly in controls during storage. Overall acceptability scores for irradiated crab samples were higher than for control samples for 14 days of ice storage.

IV. TECHNOLOGICAL PROBLEMS AND LIMITATIONS OF IRRADIATION

A. Major Problems of Irradiation

There are threshold radiation doses above which organoleptic changes and off-flavor development occur. But at low doses not all microorganisms and their toxins can be eliminated.

Successful implementation of a new technology depends on the availability of a proper infrastructure within a given country [70]. Irradiation has high capital costs and requires a critical minimum capacity and product volume for economic operation [124,126], although irradation has a low operating cost and requires low energy [70].

Willemot et al. [132] mentioned that variability in effects leads to difficulty in standardizing the irradiation treatment. The success of irradiation treatment depends on commodity and cultivar, radiation dose, degree of maturity, physiological status of fruits, temperature and atmosphere during and after treatment, pre- and postharvest treatments, and susceptibility of the microorganisms to be controlled [132]. Tolerance changes with the degree of maturity. The response of each individual batch of produce is difficult to predict, thus generalized dose levels and their effects on quality are difficult to determine.

The packaging materials used during irradiation should not produce or release undesired substances that may migrate into the food. Irradiation affects different packaging materials in different ways. At doses of 60 kGy and higher, some damage may occur to tin-coated steel and aluminum containers, but at sterilizing dose levels there should not be any affect. The enamels usually used must also be suitable. For example, for high-fat foods, oleoresin enamels are unsuitable, but they are suitable for enzyme-inactivated foods. Irradiation apparently depolymerizes butyl-rubber sealing compounds used in cans [33]. The shape of the container is also very important. A cubical form is most satisfactory for optimum dose distribution during irradiation [33].

El Makhzoumi [30] found that the effect of irradiation on plastic films depends on the nature of the packaging film, the number of layers of packaging film, the additives used in formulation, the temperature and oxygen content during treatment, the treatment dose, the dose actually absorbed, and contact with foods. At doses less than 20 kGy physical changes in flexible containers are negligible. Doses above 30 kGy cause brittleness in cellophanes, saran, and plioform, while 20 kGy or more can cause inconsequential physical changes in mylar, polyethane, vinyl, and polyethylene plastic films [33]. Physical damage can be considerably reduced by the addition of aromatic additives. At doses of 50 kGy the mechanical properties of polymers can be improved by cross-linking [49]. At doses from 3 to 25 kGy, water migration increases [74]. The properties of polyethylene terephthalate (PET) are well preserved during irradiation [57]. Table 9 shows the FDA approved levels of irradiation for different packaging materials.

An important problem in the irradiation of foods in plastic containers is the production of gas and volatile compounds, which may migrate to the food and cause off-flavors. At sterilizing doses nylon gives rise to little off-odor production, while with polyethylene, short fragmentations of the polymer are produced and enter the food [33]. Some food-packaging materials produce volatile compounds under certain conditions. Volatile compounds are formed in polyethylene, polyester terephthalate, and oriented polypropylene after irradiation doses from 5 to 50 kGy. Twenty-two compounds were identified for polyester terephthalate, 40 for oriented polypropylene, and only acetone was identified for polyethylene, which is a good candidate for irradiation of packaged food products. These

Table 9 Packaging Materials Approved by FDA for Use During Irradiation of Prepackaged Food

Type of material	Maximum dose (kGy)
Paper (kraft)	0.5
Paper (glassine)	10
Paperboard (wax-coated)	10
Cellophane (coated)	10
Polyolefin film	10
Polystyrene film	10
Rubber HCl film	10
Vinylidene chloride-vinyl chloride (copolymer film)	10
Nylon-6	10
Vegetable parchment	60
Polyethylene terephthalate film	60
Nylon-11	60
Vinyl chloride–vinyl acetate copolymer film	60
Acrylonitrile copolymers	60

Source: Ref. 132.

are hydrocarbons, ketones, and aromatic compounds [30]. Makhzoumi [30] mentioned that these compounds are able to migrate into a packaged food product and affect its quality. The kinetics of degradation show that some compounds remain trapped in the polymer and can be used as irradiation detectors.

Irradiated foods should be properly handled and stored after treatment to avoid deterioration, spoilage, and loss of nutritive value. Thus, handling, storage conditions, and packaging are also important postirradiation considerations [33].

B. Legal Aspects and Safety Issues

A joint FAO/IAEA/WHO Expert Committee on Food Irradiation (IJECFI) concluded that irradiation of food up to an overall average dose of 10 kGy causes no toxicological hazards and introduces no special nutritional or microbiological problems [129]. In 1993, the American Medical Association's Council on Scientific Affairs endorsed food irradiation as a safe and effective tool to increase food safety and reduce the incidence of foodborne illness, a view expressed earlier by the U.S. Department of Agriculture [72].

Currently 37 countries have approved the use of irradiation for processing or preservation of more than 40 food items or groups of food, either on an unconditional or on a restricted basis. Commercial applications of food irradiation are still limited and are presently carried out in 25 countries. The most common irradiated food products for commercial use are spices and dry vegetable seasonings [72]. Loaharanu and Murrell [72] proposed that the recent ban on the use of ethylene oxide for food by European Union could increase the quantity of spices and vegetable seasonings processed by irradiation in the near future. The fumigant ethylene oxide is reported to be carcinogenic, and methyl bromide may be harmful to the ozone layer [71].

The safety issues of irradiated foods can be classified as ([87]): (a) residual radioactivity, (b) free radicals and radiolytic products, (c) carcinogenic and mutagenic properties, (d) nutrient quality, (e) polyploidy, (f) toxicity, (g) microbiological safety, and (h) operator safety during processing.

The safety for human consumption of irradiated products has often been questioned [126]. The vast majority of toxicological studies and feeding trials have shown no evidence of toxic effects. However, some studies claimed to find evidence for polyploidy in irradiated wheat [56]. There is already much experience with the design, building, and operation of irradiation plants. Plants must be controlled and inspected by authorities in order to ensure both national and international radiological safety standards affecting the health and safety of workers and environmental pollution from radioisotopes [5].

Relatively small doses of irradiation can reduce the numbers of pathogenic organisms to a very low levels. This may cause a rise in the growth of secondary microflora, such as *Moraxella*, lactic acid bacteria, and yeasts. Thus, Mitchell [85] expressed concern that, in the absence of competing spoilage microorganisms postirradiation, toxin production by surviving pathogens may occur more quickly, making the food unsafe to eat before it is visibly spoiled [61]. Irradiation should not be an excuse for poor hygiene or used for reducing unacceptably high levels of microbial contamination [56].

C. Consumer Attitudes

The application and cost-effectiveness of irradiation as a method of controlling foodborne pathogens will depend on consumer attitude, regulatory actions, economics, and logistics associated with different situations. Roy [87] proposes four reasons for the slow commercialization of food irradiation: antinuclear activism, industry's hesitation, the time-consuming approval process, and insufficient consumer education. Griffith [38] believes that the major factors to determine the future of food irradiation are the development of a simple and reliable detection method, the harmonization of legislation, the commitment of the food industry, and consumer attitudes. Similar to genetic engineering techniques in food production, it is essential that consumer education and consultation with consumers be an integral part of future developments. Loaharanu [70] reviewed the results of consumer attitude surveys in different countries. He concluded that in advanced countries, consumers at large are still not knowledgeable about food irradiation and that they need accurate information about the safety, benefits, and limitations of food irradiation. Hashim et al. [43] also reviewed consumer attitudes toward irradiated poultry and recommended the following ways to increase acceptance of irradiated foods: (a) educational programs to increase consumer understanding of irradiation, (b) information about safety of irradiation via label or poster, (c) television shows, children interactions, and/or pamphlets or brochures, and (d) in-store sampling of cooked, irradiated poultry. Pohlman et al. [101] also showed that audiovisual presentations affect consumer knowledge about and attitude toward food irradiation.

REFERENCES

1. E. K. Akamine and T. Goo, Respiration of gamma irradiated fresh fruit, *J. Food Sci. 36*:1074 (1971).
2. E. K. Akamine and J. H. Moy, Delay in postharvest ripening and senescence of fruits, *Preservation of Food by Ionizing Radiation*, Vol. III (E. S. Josephson and M. S. Peterson, eds.), CRC Press, Boca Raton, FL, 1983.
3. E. K. Akamine and R. T. F. Wong, Extending shelf life of papayas with gamma irradation, *Hawaii Farm Sci. 15*:4 (1966).
4. H. A. Al-Kahtani, H. M. Abu-Tarboush, A. S. Bajaber, M. Atia, A. A. Bou-Arab, and M. A. El-Mojaddidi, Chemical changes after irradiation and post-irradiation storage in tilapia and Spanish mackerel, *J. Food Sci. 61*:729 (1996).
5. H. A. G. Ardeshir, Food irradiation: the U.K. perspective, *Food Science and Technology in Industrial Development* (S. Maneepun, P. Varangoon, and B. Phithakpol, eds.), Institute of Feed Research and Product Development, Bangkok, 1988, p. 499.
6. M. Baccaunaud, Effects de l'ionisation sur les fruits et legumes destines a la consommation en frais, *Ionisation des Produits Alimentaires* (J. P. Vasseur, ed.), Tec & Doc-Lavoisier, Paris, 1991, p. 329.
7. S. L. Beilken, B. Bill, F. H. Gran, I. Griffiths, J. J. Macfarlane, P. Vanderlinde, and P. A. Wills, Irradiation of vacuum packaged sheep carcases to extend chilled storage life, *Proceedings European Meeting Meat Research Works*, Helsinki, 1987, p. 177.
8. K. M. Blythe, The effect of irradiation on the organoleptic quality of fresh chicken carcases, M.Sc. thesis, The Queen's University of Belfast, Northern Ireland, 1990.
9. M. C. Bourne, Kinetic of softening of carrot by gamma radiation, *J. Text. Stud. 26*:553 (1995).
10. F. P. Boyle, Z. I. Kertesz, R. E. Glegg, and M. A. Connor, Effects of ionizing radiations on plant tissues. II. Softening of different varieties of apples and carrots by gamma rays, *Food Res. 22*:89 (1957).
11. J. K. Brecht, S. A. Sargent, J. A. Bartz, K. V. Chau, and J. P. Emond, Irradiation plus modified atmosphere for storage of strawberries, *Proc. Fla. State Hort. Soc. 105*:97 (1992).
12. A. Brynjolfsson, Energy and food irradiation, *Food Preservation by Irradiation*, Vol. II, STI/PUB/470, IAEA, Vienna, 1978.
13. L. N. Ceci, O. A. Curzio, and A. B. Pomilio, Effects of irradiation and storage on the flavor of garlic bulbs cv 'red', *J. Food Sci. 56*:44 (1991).
14. Y. P. Chen, L. S. Andrewa, and R. M. Grodner, Sensory and microbial quality of irradiated crab meat products, *J. Food Sci. 61*:1239 (1996).
15. Codex Alimentarius Commission, *Codex Alimentarius Vol. 1*, 2nd ed., Food and Agricultural Organization, Rome, Italy, 1992, p. 313.
16. R. Couture and C. Willemot, Combinaison d'une faible dose d'irradiation avec l'atmosphere controlee pour ralentir le murissement des fraises, *Proceedings of the International Conference on Technological Innovation in Freezing and Refrigeration of Fruits and Vegetables* (D. S. Reid, ed.), University of California, Davis, 1989, p. 40.
17. L. M. Crawford and E. H. Ruff, A review of the safety of cold pasteurization through irradiation, *Food Control 7*:87 (1996).

18. O. A. Curzio and A. M. Urioste, Sensory quality of irradiated onion and garlic bulbs, *J. Food Proc. Pres. 18*:149 (1994).
19. J. D'Amour, C. Gosselin, J. Arul, F. Castaigne, and C. Willemot, Gamma-radiation affects cell wall composition of strawberries, *J. Food Sci. 58*:182 (1993).
20. J. F. Dempster, Radiation preservation of meat and meat products: a review, *Meat Sci. 12*:61 (1985).
21. J. F. Dempster, L. McGuire, and N. A. Halls, Effect of gamma radiation on the quality of bacon, *Food Microbiol. 3*:13 (1986).
22. N. W. Desrosier and J. N. Desrosier, *Technology of Food Preservation*, 4th ed., Avi Publishing, Westport, CT, 1977.
23. Y. M. Diop, E. Marchioni, D. Ba, and C. Hasselmann, Radiation disinfestation of cowpea seeds contaminated by *Callosobruchus maculatus*, *J. Food Proc. Pres. 21*:69 (1997).
24. G. W. Eaton, C. Meehan, and N. Turner, Some physical effects of postharvest gamma radiation on the fruit of sweet cherry, blueberry, and cranberry, *J. Can. Inst. Food Technol. 3*:152 (1970).
25. A. F. Egan and P. A. Wills, The preservation of meats using irradiation, *CSIRO Food Res. Q. 45*:49 (1985).
26. A. F. Egan, D. Miller, B. J. Shay, and P. A. Wills, Preservation of vacuum-packaged pork using irradiation, *Food Science and Technology in Industrial Development* (S. Maneepun, P. Varangoon, and B. Phithakpol, eds.), Institute of Food Research and Product Development, Bangkok, 1988, p. 515.
27. R. M. Ehioba, A. A. Kraft, R. A. Molins, H. W. Walker, D. G. Olson, G. Subaraman, and R. P. Skowronski, Effect of low dose (100 krad) gamma radiation on the microflora of vacuum packaged ground pork with and without added sodium phosphates, *J. Food Sci. 52*:1477 (1987).
28. R. M. Ehioba, A. A. Kraft, R. A. Molins, H. W. Walker, D. G. Olson, G. Subaraman, and R. P. Skowronski, Identification of microbial isolates from vacuum packaged ground pork irradiated at 1 kGy, *J. Food Sci. 53*:278 (1988).
29. M. I. Eiss, Irradiation of spices and herbs, *Food Technol. Aust. 36*:362 (1984).
30. Z. El Makhzoumi, Effect of irradiation of polymeric packaging material on the formation of volatile compounds, *Food Packaging and Preservation* (M. Mathouthi, ed.), Blackie Academic & Professional, Glasgow, 1994, p. 88.
31. J. Farkas and E. Andrassy, Interaction of ionising radiation and acidulants on the growth of the microflora of a vacuum-packaged chilled meat product, *Int. J. Food Microbiol. 19*:145 (1993).
32. A. H. Fu, J. G. Sebranek, and E. A. Murano, Effect of irradiation treatment on selected pathogens and quality attributes of cooked pork chops and cured ham, *J. Food Sci. 59*306 (1994).
33. H. D. Graham, Safety and wholesomeness of irradiated foods, *The Safety of Foods* (H. D. Graham, ed.), Avi Publishing, Westport, CT, 1980, p. 546.
34. I. R. Grant, The microbiology of irradiated pork, *Dissert. Abstr. Int. C 53*:35 (1992).
35. I. R. Grant and M. F. Patterson, Effect of irradiation and modified atmosphere packaging on the microbiological safety of minced pork stored under temperature abuse conditions, *Int. J. Food Sci. Technol. 26*:521 (1991).
36. I. R. Grant, C. R. Nixon, and M. F. Patterson, Effect of low-dose irradiation on growth of and toxin production by *Staphylococcus aureus* and *Bacillus cereus* in roast beef and gravy, *Int. J. Food Microbiol. 18*:25 (1993).
37. I. R. Grant, C. R. Nixon, and M. F. Patterson, Comparison of the growth of Listeria monocytogenes in unirradiated and irradiated cook chill roast beef and gravy at refrigeration temperatures, *Lett. Appl. Microbiol. 17*:55 (1993).
38. G. Griffith, Irradiated foods: new technology, old debate, *Trends Food Sci. Technol. 3*:251 (1992).
39. K. Gruik and I. Kiss, *Acta Aliment. 16*:111 (1987).
40. T. Hanis, P. Jelen, P. Klir, J. Mnukova, B. Perez, and M. Pesek, Poultry meat irradiation—effect of temperature on chemical changes and inactivation of microorganisms, *J. Food Prot. 52*:26 (1989).
41. H. L. Hanson, M. J. Brushway, M. F. Pool, and H. Lineweaver, Factors causing color and texture differences in radiation-sterilized chicken, *Food Technol. 17*:108 (1963).

42. Y. Hasegawa and J. H. Moy, Reducing oligosaccharides in soybeans by gamma-radiation-controlled germination, *Joint FAO/IAEA Proc. Symp. Rad. Preserv. of Foods*, STI/PUB/317, 1973, p. 89.
43. I. B. Hashim, A. V. A. Resurreccion, and K. H. McWatters, Consumer attitudes toward irradiated poultry, *Food Technol. 50*:77 (1996).
44. C. H. S. Hitchcock, Determination of hydrogen in irradiated frozen chicken, *J. Sci. Food Agric. 68*:319 (1995).
45. L. R. Howard and R. W. Buescher, Cell wall characteristics of gamma irradiation refrigerated cucumber pickles, *J. Food Sci. 54*:1266 (1989).
46. C. N. Huhtanen, R. K. Jenkins, and D. W. Thayer, Gamma radiation sensitivity of *Listeria monocytogenes*, *J. Food Prot. 52*:610 (1989).
47. IAEA, *Radiation Preservation of Fish and Fishery Products* (STI/DOC/10/303), International Atomic Energy Agency, Vienna, 1989.
48. IAEA, International Atomic Energy Agency, *Analytical Detection Methods for Irradiated Foods. A Review of the Currant Literature*, IAEA-TECDOC-587, AIEA, Vienna, 1991.
49. P. Jacobs, *Polym. Plast. Technol. Eng. 17*:69 (1981).
50. C. F. Johnson, E. C. Maxie, and E. M. Elbert, Physical and sensory tests on fresh strawberries subjected to gamma irradiation, *Food Technol. 19*:119 (1965).
51. A. A. Kader, Potential application of ionizing radiation in postharvest handling of fresh fruit and vegetables, *Food Technol. 40*:117 (1986).
52. R. S. Kahan and J. J. Howker, Low-dose irradiation of fresh, non-frozen chicken and other preservation methods for shelf-life extension and for improving its public health quality, *Food Preservation by Irradiation*, Vol. II, Proceedings of Symposium, Wageningen, 1977, p. 221.
53. E. H. Kampelmacher, Irradiation for control of *Salmonella* and other pathogens in poultry and fresh meats, *Food Technol. 37*:117 (1983).
54. Z. I. Kertesz, R. E. Glegg, F. P. Boyle, G. F. Parsons, and I. M. Massey, Effect of ionizing radiation on plant tissues. III. Softening and changes in pectins and cellulose of apple, carrots beets, *J. Food Sci. 29*:40 (1964).
55. D. Kilcast, Irradiation of packaged food, *Food Irradiation and the Chemist* (D. E. Johnston and M. H. Stevenson, eds.), Royal Society of Chemistry, London, 1990, p. 140.
56. D. Kilcast, Food irradiation: current problems and future potential, *Int. Biodeterm. Biodegrad. 36*:279 (1995).
57. J. J. Killoran, Chemical and physical changes in food packaging materials exposed to ionizing radiation, *Rad. Res. Rev. 3*:369 (1972).
58. H. Koksel, S. Celik, and T. Tuncer, Effects of gamma irradiation on durum wheats and spaghetti quality, *Cereal Chem. 73*:507 (1996).
59. P. Kurstadt and F. Fraser, *Food Irradiation Technology Overview*, Nordion International, 1994.
60. J. H. Kwon and H. S. Yoon, Changes in flavor compounds of garlic resulting from gamma irradiation, *J. Food Sci. 50*:1193 (1985).
61. R. W. Lacy and S. F. Dealler, Food irradiation: unsatisfactory preservative, *Br. Food J. 92*:15 (1990).
62. S. P. Lai, K. F. Finney, and M. Milner, Treatment of wheat with ionizing radiations, IV. Oxidative, physical, and biochemical changes, *Cereal Chem. 36*:401 (1959).
63. J. D. Lambert and R. B. Maxcy, Effect of gamma radiation on *Campylobacter jejuni*, *J. Food Sci. 49*:665 (1984).
64. A. D. Lambert, J. P. Smith, and K. L. Dods, Physical, chemical and sensory changes in irradiated fresh pork packaged in modified atmosphere, *J. Food Sci. 57*:1294 (1992).
65. A. D. Lambert, J. P. Smith, K. L. Dods, and R. Charbonneau, Microbiological changes and shelf-life of MAP, irradiated fresh pork, *Food Microbiol. 9*:231 (1992).
66. C. H. Lea, J. J. Macfarlane, and L. J. Parr, Treatment of meat with ionizing radiations V. Radiation pasteurisation of beef for chilled storage, *J. Sci. Food Agric. 11*:690 (1960).
67. S. Lebepe, R. A. Molins, S. P. Charven, I. V. Farrar, and R. P. Skowronski, Changes in mi-

croflora and other characteristics of vacuum packaged pork loins irradiated at 3.0 kGy, *J. Food Sci. 55*:918 (1990).
68. M. Lee, J. Sebranek, and F. C. Parrish, Accelerated postmortem aging of beef utilizing electron beam irradiation and modified atmosphere packaging, *J. Food Sci. 61*:133 (1996).
69. S. J. Lewis and J. E. L. Corry, Survey of the incidence of *Listeria monocytogenes* and other *Listeria* spp. in experimentally irradiated and in matched unirradiated raw chickens, *Int. J. Food Microbiol. 12*:257 (1991).
70. P. Loaharanu, Food irradiation: current status and future prospects, *New Methods of Food Preservation* (G. W. Gould, ed.), Blackie Academic and Professional, Glasgow, 1995, p. 90.
71. P. Loaharanu, The status and prospects of food irradiation, *The IFTEC Symposium S-14 on Food Irradiation: Recent Developments and Future Prospects*, Hague, The Netherlands, 1992.
72. P. Loaharanu and D. Murrell, A role for irradiation in the control of foodborne parasites, *Trends Food Sci. Nutr. 5*:190 (1994).
73. R. T. Lovell and G. J. Flick, Irradiation of Gulf Coast area strawberries, *Food Technol. 29*:99 (1966).
74. F. Lox, *Proceedings 5th IAPRI Conf.*, Bristol, 1986, p. 1.
75. L. A. MacArthur and B. L. D'Appolonia, Gamma radiation of wheat. I. Effects on dough and baking properties, *Cereal Chem. 60*:456 (1983).
76. D. B. MacDougall, Instrument assessment of the appearance of foods, *Sensory Quality in Foods and Beverages. Definition, Measurement and Control* (A. A. Williams and K. K. Atkin, eds.), Ellis Horwood, Chichester, 1983, p. 121.
77. J. J. Macfarlane, I. J. Eustance, and F. H. Grau, Ionizing energy treatment of meat and meat products, *Proceedings of the National Symposium Ionization Energy Treatment of Foods*, Sydney, 1983.
78. A. Matsuyama and K. Umeda, Sprout inhibition in tubers and bulbs, *Preservation of Food by Ionizing Radiation*, Vol. III (E. S. Josephson and M. S. Peterson, eds.), CRC Press, Boca Raton, FL, 1983.
79. E. C. Maxie and A. Abdel-Kader, Food irradiation—physiology of fruits as related to feasibility of the technology, *Adv. Food Res. 15*:105 (1966).
80. C. Merrit, *Radiat. Res. Rev. 3*:353 (1972).
81. C. Merritt, P. Angelini, E. Wierbicki, and G. W. Shults, Chemical changes associated with flavor in irradiated meat, *J. Agric. Food Chem. 23*:1037 (1975).
82. W. R. Miller and R. E. McDonald, Low-dose electron beam irradiation: a methyl bromide alternative for quarantine treatment of floride blueberries, *Proc. Fla. State Hort. Soc. 108*:291 (1995).
83. S. J. Millar, B. W. Moss, D. B. Macdougall, and M. H. Stevenson, The effect of ionising radiation on the CIELAB colour co-ordinates of chicken breast meat as measures by different instruments, *Int. J. Food Sci. Technol. 30*:663 (1995).
84. D. E. Miller, Studies of techniques for extending the shelf life of vacuum packed pork, M. Phil. thesis, Griffith University, Brisbane, 1987.
85. G. E. Mitchell, Irradiation preservation of meats, *Food Aust. 46*:512 (1994).
86. B. E. B. Moseley, Ionizing radiation: action and repair, *Mechanisms of Action of Food Preservation Procedures* (G. W. Gould, ed.), Elsevier Applied Science, London, 1989, p. 43.
87. J. H. Moy, Food irradiation—lessons and prospects for world food preservation and trade, *Development of Food Science and Technology in South East Asia* (O. B. Liang, A. Buchanan, and D. Fardiaz, eds.), IPB Press, Bogor, 1993, p. 86.
88. R. W. A. W. Mulder, Ionising energy treatment of poultry, *Food Technol. Aust. 36*:418 (1984).
89. E. A. Murano, Irradiation of fresh meats, *Food Technol. 49*:52 (1995).
90. P. S. Murano, E. A. Murano, and D. G. Olson, Quality characteristics and sensory evaluation of ground beef irradiated under various packaging atmosphere, Presented at International Congress of Meat Science and Technology, San Antonio, 1995.
91. D. R. Murray, *Biology of Food Irradiation*, John Wiley & Sons, New York, 1990.
92. W. W. Nawar, Radiolytic changes in fats, *J. Rad. Res. Rev. 3*:327 (1972).
93. W. P. K. Ng, W. Bushuk, and J. Borsa, Effect of gamma ray and high-energy electron irradia-

tions on breadmaking quality of two Canadian wheat cultivars, *Can. Inst. Food Sci. Technol. J. 22*:173 (1989).

94. J. T. R. Nickerson, J. J. Licciardello, and J. Ronsivalli, Radurization and rancidation: fish and shellfish, *Preservation of Food by Ionizing Radiation*, Vol. III (E. S. Josephson and M. S. Peterson, eds.), CRC Press, Boca Raton, FL, 1983.
95. P. C. Onyenekwe and G. H. Ogbadu, Effect of gamma irradiation on the microbial population and essential oil of ginger, *Proc. 1st Natl. Conf. Nuclear Methods* (L. A. Dim, T. C. Akpa, M. C. Maiyaki, and S. P. Mallam, eds.), Zaria, Nigeria, 1992, p. 124.
96. P. C. Onyenekwe, G. H. Ogbadu, and S. Hashimoto, The effect of gamma radiation on the microflora and essential oil of Ashanti pepper (*Piper guineense*) berries, *Postharv. Biol. Technol. 10*:161 (1997).
97. O. Parades-Lopez and M. M. Covarrubias-Alvarez, Influence of gamma radiation on the rheological and functional properties of bread wheats, *J. Food Technol. 19*:225 (1984).
98. P. Paul, V. Venugopal, and P. M. Nair, Shelf life enhancement of lamb meat under refrigeration by gamma irradiation, *J. Food Sci. 55*:865 (1991).
99. R. E. Paull, Ripening behavior of papaya (*Carica papaya* L.) exposed to gamma irradiation, *Postharv. Biol. Technol. 7*:359 (1996).
100. M. Patterson, Sensitivity of bacteria to irradiated on poultry meat under various atmosphere, *Lett. Appl. Microbiol. 7*:55 (1988).
101. A. J. Pohlman, O. B. Wood, and A. C. Mason, Influence of audiovisuals and food samples on consumer acceptance of food irradiation, *Food Technol. 48*:46 (1994).
102. B. E. Procter and S. A. Goldblith, Food processing with ionizing radiations, *Food Technol. 5*:376 (1951).
103. T. Radomyski, E. A. Murano, and D. G. Olson, Elimination of pathogens of significance in food by low-dose irradiation: a review, *J. Food Prot. 57*:73 (1994).
104. A. H. Rady, R. J. Maxwell, E. Wierbicki, and J. G. Phillips, Effect of gamma irradiation at various temperatures and packaging conditions on chicken tissues, *Radiat. Phys. Chem. 31*:195 (1988).
105. S. R. Rao, R. C. Hoseney, K. F. Finney, and M. D. Shogren, Effects of gamma-irradiation of wheat on breadmaking properties, *Cereal Chem. 52*:506 (1975).
106. V. S. Rao, U. K. Vakil, C. Bandyopadhyay, and A. Sreenivasan, Effect of gamma irradiation on wheat on volatile flavor components of bread, *J. Food Sci. 43*:68 (1978).
107. V. S. Rao, U. K. Vakil, and Sreenivasan, Effects of gamma-irradiation of composition of wheat lipids and purothionines, *J. Food Sci. 43*:64 (1978).
108. H. R. Rodriguez, J. A. Lasta, R. A. Malbo, and N. Marchevsky, Low dose gamma irradiation and refrigeration to extend shelf life of aerobically packed fresh beef round, *J. Food Prot. 56*:505 (1993).
109. S. Shibata, T. Imai, H. Toyoshima, K. Umeda, and T. Ishima, Noodle-making quality of gamma-irradiated wheats, *J. Food Sci. Technol. (Tokyo) 21*:161 (1974).
110. H. Singh, radiation preservation of low nitrite bacon, *Radiat. Phys. Chem. 31*:165 (1988).
111. H. Singh, Extension of shelf-life of meats and fish by irradiation, *Shelf Life Studies of Foods and Beverages* (G. Charalamboue, ed.), Elsevier Science Publishers B. V., Amsterdam, 1993, p. 145.
112. J. Sivinski and K. Switzer, Low dose irradiation: a promising option for trichinasafe pork certification, Proceedings from the International Conference on Radiation Disinfection of Food and Agricultural Products, Honolulu, 1983.
113. C. F. Snyder, Method of tenderizing meat, U.S. patent 3,761,283 (1973).
114. L. P. Somogyi and R. J. Romani, Irradiation induced texture change in fruits and its relation to pectin metabolism, *J. Food Sci. 29*:366 (1964).
115. G. J. Strydom, J. Van Staden, and M. T. Smith, The effect of gamma radiation on the ultrastructure of the peel of banana fruits, *Environ. Exp. Bot. 31*:43 (1991).
116. S. Sudarmadji and W. M. Urbain, Flavor sensitivity of selected animal protein foods to gamma radiation, *J. Food Sci. 37*:671 (1972).

117. D. W. Thayer and G. Boyd, Gamma ray processing to destroy *Staphylococcus aureus* in mechanically deboned chicken meat, *J. Food Sci. 57*:848 (1992).
118. D. W. Thayer and G. Boyd, Elimination of *Escherichia coli* 0.157:H7 in meats by gamma irradiation, *Appl. Environ. Microbiol. 59*:1030 (1992).
119. D. W. Thayer, G. Boyd, and R. K. Jenkins, Low dose gamma irradiation and refrigerated storage in vacuo affect microbial flora of fresh pork, *J. Food Sci. 58*:717 (1993).
120. P. Thomas, Radiation preservation of foods of plant origin. Part V. Temperature fruits: pome fruits, stone fruit and berries, *CRC Crit. Rev. Food Sci. Nutr. 24*:357 (1986).
121. W. Uchman, W. Fiszer, I. Mroz, and A. Pawlik, The influence of radapertization upon some sensory properties of black pepper, *Nahrung 27*:461 (1983).
122. UFFVA, *Food Irradiation for the Produce Industry*, United Fresh Fruit and Vegetable Association, 1986, p. 1.
123. W. M. Urbain, Food irradiation, *Adv. Food Res. 24*:155 (1978).
124. W. M. Urbain, *Food Irradiation*, Academic Press, New York, 1982.
125. W. M. Urbain, *Food Irradiation* 2nd Edition, Academic Press, New York, 1986.
126. K. Vas, Food preservation, *Developments in Food Science 2* (H. Chiba, M. Fujimaki, K. Iwai, H. Mitsuda, and Y. Morita, eds.), Kodansha Ltd., Tokyo, 1979, p. 205.
127. R. Voisine, C. Hombourger, C. Willemot, F. Castaigne, and J. Makhlouf, Effect of high carbon dioxide storage and gamma irradiation on membrane deterioration in cauliflower florets, *Postharvest Biol. Technol. 2*:279 (1993).
128. R. Voisine, L. P. Vezina, and C. Willemot, Modification of phospholipid catabolism in microsomal membranes of γ-irradiated cauliflower (*Brassica oleracea* L.), *Plant Physiol. 102*:213 (1993).
129. WHO, Wholesomeness of Irradiated Food: Summaries of data considered by the Joint FAO/IAEA/WHO Expert Committee on the Wholesomeness of Irradiated Food, EHE/81.24, Geneva, Switzerland, 1980.
130. WHO, *Food Irradiation: A Technique for Preserving and Improving the Safety of Food*, WHO, Geneva, 1988.
131. P. A. Wills, J. J. Macfarlane, B. J. Shaw, and A. F. Egan, Radiation preservation of vacuum-packaged sliced corned beef, *Int. J. Food Microbiol. 4*:313 (1987).
132. C. Willemoti, M. Marcotte, and L. Deschenes, Ionizing radiation for preservation of fruits, *Processing Fruits: Science and Technology*, Vol. 1 (L. P. Somogyi, H. S. Ramaswamy, and Y. H. Hui, eds.), Technomic Publishing Company, Lancaster, PA, 1996, p.221.
133. C. Woese, Further studies on the ionizing radiation inactivation of bacterial spores, *J. Bacteriol. 77*:38 (1959).
134. J. J. Wu and J. S. Yang, Effects of gamma irradiation on the volatile compounds of ginger rhizome (*Zingiber officinale roscoe*), *J. Agric. Food Chem. 42*:2574 (1994).
135. J. Wu, J. Yang, and M. Liu, Effects of irradiation on the volatile compounds of garlic (*Allium sativum* L.), *J. Sci. Food Agric. 70*:506 (1996).
136. J. S. Yang, Y. H. Fu, and T. Y. Liu, *Effects of Irradiation Treatment on Nutritive Constituents of Garlics, Gingers, Onions and Potatoes* (Report No. 144), Food Industry Research and Development Institute, Hsinchu, Taiwan, 1979, p. 506.
137. L. Yu, C. A. Reitmeier, M. L. Gleason, G. R. Nonnecke, D. G. Olson, and R. J. Gladon, Quality of electron beam irradiated strawberries, *J. Food Sci. 60*:1084 (1995).
138. L. Yu, C. A. Reitmeier, and M. H. Love, Strawberries texture and pectin content as affected by electron beam irradiation, *J. Food Sci. 61*:844 (1996).
139. H. Zegota, Suitability of Dukat strawberries for studying effects of irradiation combined with cold storage, *Z. Lebensm. Unters. Forsch. 187*:111 (1988).
140. M. Zhao, J. Moy, and R. E. Paull, Effect of gamma-irradiation on ripening papaya pectin, *Postharv. Biol. Technol. 8*:209 (1996).

14

Nitrites in Food Preservation

M. SHAFIUR RAHMAN
Horticulture and Food Research Institute of New Zealand, Auckland, New Zealand

I. NITRITES

Preservatives are compounds used to delay or prevent the chemical and microbiological deterioration of foods. Nitrites and nitrates are used in many foods as preservatives and functional ingredients. It is a critical component used to cure meat, and it is known to be a multifunctional food additive. It is also a potent antioxidant.

Nitrites are white to pale yellow hygroscopic crystals. Sodium nitrite ($NaNO_2$) is markedly less hygroscopic than potassium nitrite (KNO_2). Nitrites are quite soluble in water and liquid ammonia but much less soluble in alcohol and other solvents. At room temperature, 1 part of water dissolves in 1 part of sodium nitrite or 3 parts of potassium nitrite [17].

II. ANTIMICROBIAL ASPECTS OF NITRITES

Sodium nitrite plays an important role in inhibiting the growth and toxin production of *Clostridium botulinum* in cured products [31]. Usually input concentrations in excess of 100 mg/kg are used to achieve protection against microflora [120]. Woods et al. [120] found that sodium nitrite at 200 mg/kg and pH 6.0 was capable of inhibiting strains of *Achromobacter*, *Aerobacter*, *Escherichia*, *Flavobacterium*, *Micrococcus*, and *Pseudomonas* species. *Salmonella*, *Lactobacillus*, and *Clostridium perfringens* are more resistant than other clostridia [120]. Tompkin et al. [108] found that times to first swell were 6.7, 29.8, 82.6, and 94.3 days when 0, 50, 100, and 156 μg/g of sodium nitrite was added to perishable canned cured meat. The primary effect of nitrite was in determining the length of the lag phase. Once swelling commenced, the rate at which the cans swelled was not significantly different at 50, 100, and 156 μg/g of sodium nitrite. At 50 μg/g nitrite a 75% probability of toxicity was predicted at 3 months. Hauschild et al. [39] concluded that (a) the degree of safety from *C. botulinum* toxin production can differ by several orders of magnitude, depending on the composition of formulation, and (b) large reductions in nitrite concentration could produce severe consequences.

A. Stages of Inhibition

Germination and outgrowth of bacterial spores involve five sequential steps: germination (becoming nonrefractile, stainable, and heat sensitive), swelling of the germinated spore, emergence of a new vegetative cell, elongation, and cell division [19]. The inhibitory effect of nitrite on bacterial spore formers is apparently due to inhibition of outgrowth during cell division. Duncan and Foster [19] identified two points of inhibition in the outgrowth process of anaerobic spores: up to 0.06% at pH 6.0 or 0.8–1% at pH 7.0 nitrite allowed emergence and elongation of vegetative cells but blocked cell division. Elongated cells that did not multiply eventually lysed, leaving the empty spore coats. With more than 0.06% nitrite at pH 6.0 or more than 0.8–1% at pH 7.0, the spores lost refractility and swelled, but no vegetative emerged. Even as much as 4% nitrite failed to prevent germination (complete loss of refractility) and swelling of the spores. Duncan and Foster [20] reported that sodium nitrite induced germination of *Clostridium sporogenes* spores. They found that the process of germination was accelerated by increased concentrations of sodium nitrite, low pH, and a high temperature of incubation. The increase in germination rate with increasing temperature and increasing nitrite concentration may be a result of the alteration of the tertiary structure of a spore protein, which in turn may be involved in the calcium–dipicolinic acid complex [20]. The stimulatory effect of nitrite on germination has a dual role in preservation: induction of spores to germinate, making them susceptible to a heating process, and inhibition of the outgrowth of any surviving spores [20]. Gould [32] also found that lower concentrations inhibited outgrowth of spores after germination, whereas higher concentrations inhibited germination itself.

Nitrite exerts a concentration-dependent antimicrobial effect on the outgrowth of spores from *C. botulinum* and other clostridia. The effectiveness of nitrite depends on several environmental factors in a very complex situation, such as foods. Thus, the concentration of nitrite required to prevent outgrowth of microorganisms varies with the types of media or food and environmental conditions.

B. Factors Affecting the Efficacy of Nitrites

1. Effects of pH

Nitrites have found to be most inhibitory to bacteria at an acidic pH [17]. Tarr [101–103] showed that the preservative action of nitrites in fish was greatly increased by acidification. In bacteriological medium, the inhibitory action was increased with decreasing pH, particularly at pH 6.0 and below. Grindley [35] suggested that the mode of preservation could be due to the formation of active nitrous acid. Jensen [48] suggested that the increased action of preservation at low pH was due to the undissociated active inhibitor nitrous acid. A 10-fold increase in the inhibitory effect of nitrite against *C. botulinum* was found when the pH was reduced from 7.0 to 6.0 [91]. A similar 10-fold increase per one unit decrease in pH was also observed for *Staphylococcus aureus* [14], *Bacillus* [21], and *C. sporogenes* [80].

The pH dependency of nitrite-induced bacterial inhibition also reflects the conversion of nitrite to nitrous acid [7]. Shank et al. [99] proposed a "nitrite cycle." The dynamics of nitrous acid production may be visualized in a cyclic reaction where nitrite undergoes a concomitant oxidation-reduction reaction resulting in the formation of nitrate, nitric oxide, and nitrogen dioxide. Nitrogen dioxide reacting with water would generate more nitrate with the nitrite reentering the cycle again (Figure 1). At low pH levels (pH 3–4), the cycle rapidly forms NO_3^- and NO. At intermediate pH levels (pH 4.5–5.5) the cycle rotates more slowly. The presence of HNO_2 is prolonged, thereby increasing its reaction potential. This is the level of maximum bactericidal activity. At higher pH levels (pH 6–7), the equilibrium shifts toward $NaNO_2$, the cycle is prevented from functioning, and no bactericidal effects are observed. Nitrous acid and nitric oxide have two fundamental areas of reaction: with the bacterial cell itself, and with various constituents of the medium, making them unavailable for subsequent metabolism. Either or both of these reactions could result in bacteriostasis. Further evidence for bound nitric oxide was presented by Frouin [25], who found that all measurable nitrite in various cured products could be volatilized under a high vacuum. This indicated that in meat systems nitrite is converted to nitric oxide and may produce complex equilibrium with other components.

At 20°C *C. perfringens* growth in laboratory medium was inhibited by 200 μg/ml nitrite and 3% salt or 50 μg/ml nitrite and 4% salt at pH 6.2. Fecal streptococci showed growth in the same medium with 400 μg/ml and 6% salt [30]. *Salmonella* showed visible growth within 1 week at 20°C in the presence of 400 μg/ml nitrite and 4% salt. Significant inhibition by salt and nitrite was achieved only at lower temperatures (10 or 15°C) and at pH 5.6 or 6.2. *Escherichia coli* was more resistant than

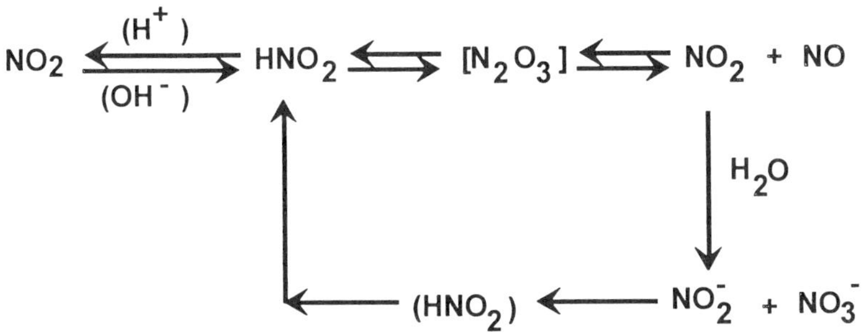

$$NO_2 \underset{(OH^-)}{\overset{(H^+)}{\rightleftharpoons}} HNO_2 \rightleftharpoons [N_2O_3] \rightleftharpoons NO_2 + NO$$

$$NO_2 \xrightarrow{H_2O} NO_2^- + NO_3^- \longrightarrow (HNO_2) \longrightarrow HNO_2$$

FIGURE 1 The dynamics of nitrous acid production in a cyclic reaction. (From Ref. 99.)

Salmonella. The inhibition was demonstrated only at the extremes of pH 5.6, salt 6%, nitrite 400 μg/ml, and temperature 10°C [29].

Survival of *Listeria monocytogenes* was detected after fermentation and drying, although their number was usually found to be reduced. Surveys of fermented meat products confirmed the presence of *L. monocytogenes* in finished products [117]. Junttila et al. [51] concluded that nitrite and nitrate additions to a meat product at officially approved levels did not eliminate *L. monocytogenes*. In broth cultures acidity and nitrite increased the inactivation rate of *L. monocytogenes* [8]. Whiting and Masana [117] studied the effect of nitrite (0–300 μg/ml) and pH on uncooked fermented meat products. The time to achieve a 4 log decline was greatly affected by pH, ranging from 21 days at pH 5.0 to less than 1.0 day at pH 4.0. Nitrite additions did not affect survival, suggesting that the effective component was the rapidly decreasing residual nitrite level. There is potential for production of bacteriocins by lactic acid bacteria in the starter cultures in fermented meats [117].

2. Effects of Oxygen

Nitrite is more inhibitory under anaerobic conditions [7,14,55]. Aerobically cultured *S. aureus* were able to grow in the presence of significantly higher concentrations of sodium nitrite than were cultures grown in an aerobic environment [14]. Buchanan and Solberg [7] studied the effect of pH and oxygen pressure on the bacteriostatic accumulation of sodium nitrite in *S. aureus*. They found that the magnitude of inhibition was dependent on the interaction of sodium nitrite concentration, initial pH, and partial pressure of oxygen. Aerobic cultures, after the initial pH decrease, showed a subsequent rise in pH to a level greater than the initial pH, whereas anaerobic cultures remain at the pH level of maximum pH decrease. Injury and cell destruction was most apparent at the lower pH level in the presence of nitrite concentration ≥500 ppm. However, 200 ppm sodium nitrite in cured meats offers significant protection against growth of *S. aureus*, particularly if the meat product is vacuum-packed. Buchanan and Solberg [7] suggested that nitrite may inhibit the growth of *S. aureus* by blocking the sulfhydryl sites of either coenzyme A or α-lipoic acid, thus blocking the normal metabolism of pyruvate.

3. Effects of Other Food Components

Temperature, salt concentration, and initial inoculum size also significantly influence the antimicrobial role of nitrite [28,85,87,92,93]. It has been reported that sodium chloride alone at concentrations of 9.0–10.5% can inhibit growth of and toxin production by *C. botulinum*. When nitrite was added in concentrations of 75 and 150 ppm, sodium chloride levels of 5.8 and 4.9% were required to inhibit toxin formation. The usual salt levels added to cured meat range from 2 to 3% of the weight of the product. This indicates that salt alone is not always a practical inhibitor of *C. botulinum* growth and toxin formation [81]. Pierson and Smoot [81] stated that the inhibitory effects of the interaction of sodium chloride and nitrite on various bacteria have been widely reported. Riemann [84] found significant inhibition of bacteria spores in a canned meat system due to interactions of sodium chloride with sodium nitrite or nitrate as well as pH interactions with sodium chloride. Others also found inhibitory effects resulting from the interaction of pH, sodium chloride, and sodium nitrite [88,90]. Baird-Parker and Baillie [5] found that most *C. botulinum* type A and proteolytic type B and F strains would grow in the presence of either 150–200 mg/kg of sodium nitrite or 6% sodium chloride at pH 6.0, but under the same conditions 200 mg/kg of sodium nitrite plus 3% sodium chloride inhibited almost all of the strains. The above results indicate that a combination of salt, nitrite, and pH can be synergistic with the inhibition [81].

Roberts et al. [94,95] studied the combined effect of the following factors on growth of *C. botulinum*: sodium chloride (0–4.5% w/v on water), sodium nitrite (100–300 μg/g), sodium nitrate (0–500 μg/g), sodium isoascorbate (0–1000 μg/g), polyphosphate (0–0.3% w/v), heat treatment (70–80°C), and storage temperature (15–35°C). Their findings can be summarized as follows:

1. Increasing nitrite, salt, or heat treatment, adding isoascorbate, polyphosphate or nitrate, or decreasing storage temperature significantly reduced toxin production.
2. The relative effect of increasing nitrite became less pronounced in the presence of isoascorbate or high salt levels.
3. Increasing salt or heat treatment, adding nitrate, or decreasing storage temperature had less effect if isoascorbate was present.
4. The addition of polyphosphate enhanced the effect of adding isoascorbate.

Roberts et al. [94] concluded that it served no purpose to assess which combinations give a guaranteed risk of toxin production, since minor changes in product formulation or production or in experimental conditions might significantly alter the ability to support toxin production and the variability of the system.

The combination of low nitrite (40 μg/g) plus sorbate or sorbic acid controlled the growth of *C. botulinum* as effectively as a higher level of nitrite alone (156 μg/g). The low level of nitrite (40 μg/g) alone had no significant effect on the growth of *C. botulinum*, but was included to ensure acceptable cured color and flavor [96]. Factors decreasing the toxin production of *C. botulinum* were the presence of potassium sorbate, increasing sodium chloride, decreasing pH, and decreasing storage temperature. Heat treatment interacted significantly with some other factors. The effect of sorbate (0.26% w/v) was greater at 3.5% sodium chloride than at 2.5%, at pH values below 6.0, and at low storage temperature [96].

Ethylenediaminetetraacetic acid (EDTA), isoascorbate and ascorbate enhance the antibotulinal efficacy of nitrite in canned meat. The degree of inhibition was inversely related to the level of iron and directly related to the level of EDTA. The use of isoascorbate and ascorbate has both positive and negative attributes, depending on their level in meat products. At moderate levels the synergistic effect is due to the sequestering action of isoascorbate or ascorbate on a cation, iron. On the other hand, excessive levels of ascorbate were shown to decrease the efficacy, because isoascorbate and ascorbate cause more rapid depletion of residual nitrite. EDTA more effectively sequesters iron, thus making iron less available for preventing nitrite inhibition [111]. Tompkin et al. [111] proposed using a minimum of isoascorbate to hasten the curing reaction and stabilize color and flavor and supplementing it with a low level of EDTA for improved botulinal protection.

When *Bacillus cereus* was inoculated into uncooked sausage in the presence of 500 mg/kg sodium isoascorbate and 200 mg/kg sodium nitrite and incubated for 48 hours at 20°C, no growth was demonstrated. Sodium isoascorbate alone had no inhibitory effects [82].

4. Effects of Heating

In a bacteriological medium, the inhibitory effect of nitrite is enhanced 10-fold after heating due to formation of an as-yet-unidentified substance. This is called the Perigo effect [80]. Perigo et al. [80] confirmed that the effect of unheated sodium nitrite was pH dependent and that 200–400 ppm of sodium nitrite was necessary to inhibit the growth of *C. sporogenes* at pH values around neutrality. They showed that as little as 3–5 ppm sodium nitrite heated in the medium for 20 minutes at 105–115°C could inhibit growth, and this inhibition was slightly dependent on the pH of the medium. The rate at which the inhibitory unknown substance is produced was maximal at a temperature of about

110°C. At temperatures exceeding 110°C the unknown substance appeared to break down or to react in such a way that its inhibitory activity declined. Perigo and Roberts [79] confirmed this effect in 30 clostridial strains including *C. botulinum* types A, B, E, and F (14 strains) and *Clostridium welchii* (8 strains). It was reported that a reducing agent such as and thioglycolate, ascorbate, or cysteine, and protein hydrolysate were the necessary components of the laboratory medium in order to produce the effect. Roberts and Gracia [89] showed that the inhibition was enhanced by the Perigo effect in 9 of 14 strains of *Bacillus* tested. *Streptococcus durans* (*faecium*) was also sensitive to this effect, whereas *Streptococcus faecalis* and *Salmonella* were more resistant.

It has been suggested that one or more new chemical species have been produced [80]. Evidence has been presented indicating that the media may contain substances such as Roussin black salt (iron thionitrosyl) [4] and nitrosothiols [45,65]. Involvement of sulfhydryl groups as well as a nitroso group is probably important [45,65]. Hansen and Levin [38] proposed that it could be that heat-induced Perigo inhibitors are distinct from these compounds. They suggested that a heat-induced inhibitor presumably of the Perigo type be compared with the nitrosothiols of thioglycolate and β-mercaptoethanol. Phase contrast microscopy revealed that inhibition of morphological events occurred either before germination or during early outgrowth, depending on inhibitor concentration The inhibitors derived from nitrites act at virtually every stage in the life cycle of *Bacillus*, suggesting that their mode of action is rather general and that inhibition may be the result of inactivation of several sensitive metabolic systems or steps. A synergistic inhibitory response could help to explain the elusive nature of the mode of action of nitrite curing salts as preservatives [38].

An inhibitor of *C. perfringens* is formed when low levels of nitrite are autoclaved with defined chemical medium. Only amino acids and mineral salts were involved in the production of this inhibitor. The toxic compound was formed at sublethal level from cysteine, ferrous sulfate, and sodium nitrite. S-Nitrosocysteine, unstable roussin red salt, and a complex of cysteine, iron, and nitric oxide were detected. Moran et al. [71] concluded that the observed inhibition may be due to the combined effects of sublethal concentrations of each compound.

The extended heat treatment of the meat may cause decomposition of proteins with liberation of amino acids, peptides, and, possibly, amines. Nitrite reacts with amines and amino acids forming N-nitroso compounds, either N-nitrosamines or N-nitrosamides, which are toxic and carcinogenic to animals and mutagenic to various species of microorganisms [114].

Johnston et al. [50] found the Perigo effect in minced pork. Johnston and Loynes [49] mentioned that the inhibitory effect of nitrite can be increased in the media by the addition of reducing agents, such as cysteine, thioglycolate, and ascorbate. These agents are known to aid in the reduction of nitrite and may affect the formation of nitroso-reductants. These intermediate carriers may transfer the nitroso group directly to components of the bacterial cells or release nitric oxide. Johnston and Loynes [49] found that the addition of reducing agents to meat suspensions decreased the redox potential and increased the inhibitory activity by Perigo factor formation. Ashworth and Spencer [3] studied the role of chemical additives in the formation of this inhibition in minced pork and found a similar effect. They added 0.1% of reducing agents in pork slurry containing nitrite and found that its inclusion increased the inhibitory effect of nitrite in sodium ascorbate, cysteine (free base), and thioglycollate, but with sodium formaldehyde sulfoxylate and sodium formaldehyde bisulfite, there was a marked decreased in inhibition.

Huhtanen and Wasserman [43] suggested that a potent anticlostridial inhibitor can be produced by addition of iron (ferrous or ferric) without autoclaving nitrite in the medium. They indicated that iron was a limiting factor and that sulfhydryl groups were probably necessary for its formation. Similarly, Custer and Hansen [16] found that lactoferrin (an iron-binding glycoprotein) and transferrin reacted with nitrite to form an inhibitor effective against spore outgrowth of *B. cereus*.

The Perigo inhibitor is formed at 105°C or higher, which exceeds the temperatures normally used in the processing of cured meats. Holley [42] mentioned that Perigo inhibitor is formed in a culture medium only when sulfhydryl groups and iron are present.

Nitrite reacts with various naturally occurring chemical components in a complex system of meat. The heating conditions normally used in the curing process speed up these reactions, and at the end of the process only about 10–20% of the originally added nitrite is analytically detectable. The residual nitrite level declines further during storage and distribution [12].

5. Effect of Irradiation

Pierson and Smoot [81] reviewed the effects of irradiation and found that a limited amount of added nitrite was required to successfully produce acceptable irradiated cured meat products.

C. Mode of Action with Microflora

A target can be selected biochemical knowledge of the undesired microorganisms, such as enzyme or a cellular component involved in a process essential to microbial survival or development [113]. The inhibitory action of sodium nitrite on *C. perfringens* is apparently at the cellular level, since microscopic examination of these organisms indicated no visible difference between inhibited and normal cells. Thus, damage was probably at a submicroscopic level [86]. Yarbrough et al. [123] showed that nitrite acts in three ways on bacterial cell metabolic processes as follows:

1. Nitrite interferes with energy conservation by inhibiting oxygen uptake, oxidation phosphorylation, and proton-dependent active transport.
2. Nitrite acts as an uncoupler, causing a collapse of the proton gradient.
3. Nitrite inhibits certain metabolic enzymes.

1. Inhibition of the Phosphoroclastic System

In the cell, oxidation of the substrate occurs with concomitant production of adenosine triphosphate (ATP). This can then be used subsequently as an energy source for the synthesis of new cellular material required for growth. In clostridia, an important source of ATP is the oxidation of pyruvate to acetate by phosphoroclastic system [120]. When nitrite is added to a suspension of cells of C. sporogenes incubated in medium containing glucose, there is a large and rapid decrease in the intracellular concentration of ATP and an excretion of pyruvate from the cells [119]. This increase in pyruvate indicated that the phosphoroclastic system is inhibited by nitrite [120]. Iron is a required nutrient for clostridial spore germination and outgrowth and botulinal toxin development. The growth of *C. sporogenes* and *C. botulinum* was inhibited by nitrite through interference with the phosphoroclastic system, resulting in an accumulation of pyruvic acid in the medium [118,119]. The inhibition was due to an interaction between nitrite and intracellular iron-bound protein, e.g., the reaction of nitric oxide with the nonheme iron of pyruvate:ferredoxin oxidoreductase [119]. Nitrite was also shown to inhibit the iron-sulfur enzyme, ferredoxin, in *C. botulinum* and *Clostridium pasteurianum* [10]. The addition of iron depleted residual nitrite levels in cured meats.

The phosphoroclastic system consists of two components: ferredoxin and pyruvate:ferredoxin oxidoreductase. Both of these contain nonheme iron moieties. Pyruvate:ferredoxin oxidoreductase consists of a single protein molecule containing thiamine pyrophosphate and a nonheme iron. Nitric oxide causes inhibition of the phosphoroclastic system by interacting with these components. Nitric oxide is a potent iron ligand that can form coordination complexes with nonheme iron. Pyruvate:ferredoxin oxidoreductase seemed to be more sensitive to nitric oxide [119,120]. Tompkin et al. [110] also suggested that nitric oxide reacted with iron-containing protein in *C. botulinum*. Reddy et al. [83] demonstrated the production of iron nitric oxide complexes using electron-spin resonance spectroscopy. Aliphatic and aromatic nitro compounds inhibit the ferredoxin possibly as a result of formation of S-nitrosothiols by reaction with cysteine residues [2]. Castellani and Niven [14] suggested that the bacteriostatic action of nitrite might be due to interference with the normal metabolism of a hypothetical pyruvate-sulfhydryl complex.

2. Inhibition of Enzyme Systems

At acid pH levels, sodium nitrite exists as nitrous acid, an extremely reactive molecule capable of interaction with a wide variety of substances including myoglobin, ascorbic acid, phenols, secondary amines, amino groups, and thiol groups [75]. Mirna and Hofmann [66] reported that although sodium nitrite reacts with both sulfhydryl (SH) groups and primary amino groups at pH 5.5, the reaction with SH groups is more rapid. Riha and Solberg [86] proposed that nitrite inhibition of *C. perfringens* may be due to a reaction of nitrous acid with SH-containing constituents of the bacterial cell. The nitrite could inhibit enzymes of glucose fermentation such as glyceraldehyde-3-phosphate dehydrogenase and aldolase in *C. perfringens* [75]. Nitrite also inhibited aldolase in *E. coli, P. aeruginosa*, and *S. faecalis* [123]. Nitrite inhibited the nitrogenase of *C. pasteurianum*, a system that comprises two nonheme iron–containing proteins. This inhibition was probably due to the reaction of nitric oxide with a component of the nitrogenase system [63].

McMindes and Sielder [62] reported that nitric oxide was the active antimicrobial principle of nitrite and that pyruvate decarboxylase may be an additional target for growth inhibition by nitrite. These observations are substantiated by the fact that the addition of iron to meats containing nitrite reduces the inhibitory effect of the compound [109]. Chelating agents like sodium ascorbate, EDTA, and polyphosphate enhance the antibotulinal action of nitrite. Muscle pigmentation is due to myoglobin and, to a lesser extent, hemoglobin remaining after carcass bleeding. Heart meat showed no inhibition of *C. botulinum* inoculum even with a 156 μg/g of sodium nitrite added to the product. Adding hemoglobin to the meat formulation reduced nitrite after processing and decreased botulinal inhibition. The degree of pigmentation of meat in descending order is as follows: heart meat, beef round and turkey thigh meat, pork ham, veal, and turkey breast. Tompkin et al. [110] offered the hypothesis that nitric oxide, which was formed from residual nitrite via nitrous acid, reacts with extracellular iron, thereby blocking some metabolic step essential for outgrowth. The reaction might involve the iron in ferredoxin or an enzyme in which iron plays an essential role [110]. The results of Miller and Menichillo [64] demonstrated that use of blood fractions that increased iron levels in beef above 30 μg/g interfered with the antibotulinal efficacy of sodium nitrite of 146 μg/g. Lucke [56] observed that blood sausages were associated with foodborne botulism in Germany. It is advisable to include additional microbial growth barriers when iron-containing compounds are added to cured meats [64].

Ingram [46] first postulated that nitrite inactivated enzymes associated with respiration. The active inhibitory agent outside the cell was closely correlated with nitrous acid, while the mechanism of action may vary for different physiological types of microorganism [123]. Nitrite was shown to inhibit active transport, oxygen uptake, and oxidative phosphorylation of *P. aeruginosa* by oxidizing the ferrous iron of an electron carrier, such as cytochrome oxidase, to the ferric form [97]. Since glucose transport in *S. faecalis* and *Streptococcus lactis* is not dependent on active transport or cytochromes, nitrite does not inhibit these organisms [97]. Nitrite inhibited the active transport of proline in *E. coli* but not group translocation by the phosphoenolpyruvate:phosphotransferase system [123]. Inhibition of other enzymes, particularly those containing sulfhydryl groups, can occur, but these effects usually occur at higher nitrite concentrations [120].

Damage to the cell wall or membrane was indicated by the graying or browning of *C. perfringens* cells incubated with inhibiting concentrations of sodium nitrite [75]. *Staphylococcus faecalis* or *S. lactis* was highly resistant to nitrite, although aldolase was sensitive to it. This suggested that these streptococci are impermeable to nitrite [123].

III. INTERACTIONS OF NITRITES WITH FOOD COMPONENTS

A water-soluble or low molecular weight compound is responsible in large part for nitrite depletion. Sebranek et al. [98] found nitrite bound to hot water–soluble and –insoluble meat residue. An amino acid or oligopeptide (probably with an SH group) could be involved in nitrite reduction. Fox and Nicholas [23] examined the effects of various compounds in meat slurries and found that histidine and reductants such as ascorbate and cysteine caused the nitrite depletion. Knowles et al. [52] investigated the interaction of nitrite with bovine serum albumin at gastric pH 2.5 and obtained 3-nitrotyrosine, 3,4-dihydroxyphenylalanine, and 6-hydroxynorleucine. Miwa et al. [69] mentioned that it is well known that primary amino acids react with nitrite to produce alcohol and nitrogen gas (van

Slyke reaction). Of the endogenous acidic substances tested, cysteic acid showed the greatest ability to decompose nitrite, accompanying the production of unidentified nitrogen compounds. Woolford et al. [121] studied the reaction of nitrite with isolated myosin and showed that a part of the lost nitrite was bound to the protein and identified 3-nitrotyrosine as a major product of the reaction. The nitric oxide formed from nitrite may partly bind to protein [66] or ferricytochrome [105]. Frouin [24] concluded that nitrite was rapidly broken down to nitric oxide in meat products and reacted with unsaturated carbon-carbon bonds [26]. If whole adipose tissue was treated with nitrite, it was bound to connective tissue, extracted lipid, and unsaturated carbon-carbon bonds. The experiments with various fatty acids and glycerides showed that binding was apparently related to the degree of unsaturation [33].

Fujimaki et al. [27] studied the fate of nitrite during curing and cooking in model solutions composed of myoglobin, sodium nitrite, and sodium ascorbate. After curing and cooking, nitrite was recovered as residual nitrite, nitrate, nitrosyl groups of denatured nitrosomyoglobin, and gaseous nitrogen compounds. Almost all of the nitrite was recovered as nitrate whenever greening occurred during the curing period. The gaseous nitrogen compounds were produced when both sodium nitrite and sodium ascorbate were abundant as compared with myoglobin; this reaction proceeded not during the curing period but at the cooking stage. The addition of sodium chloride to the model system increased residual nitrite and nitrosomyoblogin [27]. Emi-Miwa et al. [22] also studied the fate of nitrite added to whole meat, meat fractions, and model systems with added sodium ascorbate. They found residues such as nitrite, nitrate, nitrosothiol, denatured nitrosomyoglobin, and gaseous nitrogen compounds. Twenty percent of the total nitrite lost was changed to nitrosothiol-N during the specific interaction between nitrite and sulfhydryl groups of myosin [54]. Olsman [76] also stated that more than half of the free nitrite disappeared during the storage of canned cured meat as bound nitrite, probably as nitrosothiols formed with protein-bound thiol groups. The amount of bound nitrite increased with the addition of ferrous ions. This may be due to the the formation of ferric coordination complexes between cysteine residues and nitric oxide [77]. Cassens et al. [11] found typical distribution of nitrite to be in the following proportions: myoglobin 5–15%, nitrate 1–10%, nitrite 5–20%, gas 1–5%, sulfhydryl 5–15%, lipid 1–5%, and protein 20–30%.

Namiki and Kada [74] described formation of ethylnitrolic acid by the reaction of sorbic acid with sodium nitrite when heated at 90°C. The isolated compound ethylnitrolic acid showed a strong activity in comparison with the original materials. Ethylnitrolic acid, sorbic acid, and sodium nitrite were effective at concentrations of 0.025–0.05, 2–4, and 1.5–3 mg/ml, respectively. Namiki and Kada found that ethylnitrolic acid is necessarily formed in foodstuffs containing sorbic acid and sodium nitrite together [74].

Osawa et al. [78] concluded that the main mutagen formed by the nitrite or sorbic acid reaction was 1,4-dinitro-2-methyl pyrole. Piperine was also formed in the nitrite system. Food components, such as ascorbic acid, cysteine, and some phenolic compounds, were reported to react with nitrite, preventing the formation of nitrosamines in vitro as well as in vivo [34,67]. Ascorbic acid was reported to inhibit bacterial mutations induced by N-nitroso compounds [36]. The oxidative desmutagenic action of cabbage peroxidase [47] and mieroperoxidase against the mutagenic principles of tryptophan pyrolysate [122] have been reported. Ascorbic acid, cysteine, and other reducing substances were responsible for desmutagenic action against the mutagens of the sorbic acid or nitrite system.

IV. FUNCTIONAL AND SENSORY PROPERTY IMPROVEMENT

Taylor and Sumner [104] maintained that food additives play an important part in improving health, increasing food supply, enhancing that food's appeal, and improving convenience. Health benefits should be given the greatest consideration, while supply benefits are secondary in importance, and increased convenience and improved appeal are the least important.

Nitrite salts are used for curing meat, poultry, and fish products. Curing with nitrite results in development of a characteristic pink color and distinctive flavor [31]. The sequence of color changes during the curing of meat are as follows [120]: the initial purple-red color of myoglobin changes to the brown of metmyoglobin; in reducing conditions nitric oxide derived from nitrite converts metmyoglobin to the dark red nitrosylmyoglobin; if the meat is heated (e.g., in cooking) nitrosylmyoglobin is converted to the stable nitrosylhemochrome, which is pink. Ando et al. [1] found that 5'-inosinc acid, adenosine-5'-monophosphoric acid, reduced glutathione, glutamate, and Fe^{2+} influenced the cured color formation.

A nitrite content of 5 mg/kg can produce a satisfactory color for a short time, but it is generally believed that higher concentrations of about 20 mg/kg are necessary for commercial color stability. Nitrite concentrations of at least 50 mg/kg are thought to be necessary for correct flavor development [81,120]. The addition of nitrite at a minimum level of 50 ppm was necessary to achieve reasonably typical thuringer flavor and appearance characteristics in sliced and baked pizza topping products. At least 100 ppm added nitrite was necessary to produce these effects in fried thuringer. The effect of added nitrite above 100 ppm was negligible for further color development. Fresh, fried, or baked thuringer containing neither nitrite or nitrate was judged the most rancid and to have the poorest flavor and appearance quality. No nitrosamines were detected in thuringer regardless of initial nitrite (0–150 ppm) or nitrate (0–1500 ppm) concentration, storage conditions, or kitchen preparation method [18]. Pierson and Smoot [81] reviewed the minimum level of nitrite in different food products for color development. They were 20 ppm for cured meat and hams, 30 ppm for bacon, 25 ppm for wieners, 70 ppm for pork loins and country-style hams, 52 ppm for frankfurters, 26 ppm for franks, and 40 ppm for turkey frankfurters. Modification of fresh meat flavor is another change produced in meat by the addition of nitrite. A minimum level of 39–50 ppm nitrite was required to develop the appropriate flavor [81].

Nitrite added to meat has been associated with a delay in the development of oxidative rancidity [115]. When nitrite reacts with heme compounds to form cured meat pigments, the ferric iron (oxidized state, Fe^{3+}), which is active in lipid oxidation, is reduced to a ferrous ion (Fe^{2+}), which is an inactive catalyst [81]. The addition of nitrite to model lipid systems containing Fe^{2+} or Fe^{2+}-EDTA and aqueous beef extracts substantially reduced oxidation rates [57]. In bacon formulated without or with 15 ppm of nitrite, off-flavors were found to be high and to increase more rapidly [41]. A significant reduction in the formation of rancid off-flavors in pork during storage was observed when nitrite was added in the amount of 50 ppm or greater [58,59].

Nitrites and nitrates inhibit dairy cultures by their effect on the activity of a number of oxidoreduction enzymes, and as a consequence the natural ripening of milk is prevented and undesirable microflora is formed [6]. Lactic acid bacteria culture in yogurt has beneficial effects on health. Korenekova et al. [53] studied the effects of nitrites and nitrates on yogurt cultures up to a level of 100 mg/kg. They found that nitrites, depending on their concentration, were above to exert an inhibitory effect on a yogurt culture and that nitrates are not marked inhibitors of lactic bacteria. Thus, nitrates can be used in yogurt to preserve its quality without inhibiting lactic acid bacteria.

V. MEDICAL OR HEALTH ASPECTS

Two types of health benefits may be provided by food additives and food components: those that prevent or reduce the incidence of specific diseases and those that provide enhanced nutrition [104]. Many attempts have been made to develop agents that inhibit or inactivate undesired organisms but display little toxicity toward humans when ingested [113].

The National Academy of Sciences [72] concluded that 39% of dietary nitrite intake was from cured meat, 34% from baked goods and cereals, and 16% from vegetables. Cassens [13] found nitrate important in the total picture because it is found in substantial quantities in other foods, such as green leafy vegetables and root vegetables, and sometimes in drinking water. Cassens observed no detectable nitrate in cured meats.

Prolonged ingestion of sodium nitrite or sodium nitrate has been shown to cause methemoglobinemia, especially in infants. Methemoglobinemia causes production of abnormal hemoglobin [72]. The major adverse effect of nitrites is the possible induction of cancer. In rats nitrite increases the incidence of lymphoma when fed 250–2000 ppm nitrite in food or water [73]. Nitrite results in the formation of carcinogenic N-nitrosamines with secondary amines or substituted amides to form nitrosamides. More than 65 different nitrosamines detected in a variety of foods, including cheese, meats, mushrooms, and alcoholic beverages, have been found to be carcinogenic [60]. Epidemiological studies have indicated a possible link between exposure to high levels of nitrites and a high incidence of stomach and esophageal cancer [44,72]. Another well-known effect of nitrite is the lowering of oxygen transport in the bloodstream due to the oxidation of hemoglobin to methemoglobin [81]. Thus, nitrite levels should be reduced in cured products. Ascorbates or etythorbates are added to reduce nitosamine formation [31].

An oral challenge test with 30 mg of sodium nitrite may cause urticaria, intestinal disorders, or headache [40,70]. It may also cause cellular anoxia and inhibit the protective enzymatic activities of the intestinal mucosa, leading to increased permeability of the mucosa to other antigens. In addition, sodium nitrite may in some way enhance the effect of histamine present in many foods [37].

The lethal dose of nitrites in humans is 32 mg/kg body weight or 2 g [9] and 4–6 g [112]. In 1973, USDA established an expert panel on nitrates, nitrites, and nitrosamines. They concluded that (a) the use of sodium nitrate should be discontinued in all meat and poultry products, (b) the nitrite level permitted for curing of meat should be limited to 156 μg/g in canned, cured sterile products, (c) the permitted residual nitrite level should be reduced from 200 to 100 μg/g in cooked sausage products, 125 μg/g in canned and pickle-cured products, and 50 μg/g in canned cured sterile products [72], and (d) sodium nitrite (120 ppm) and potassium nitrite (140 ppm) should be added to bacon along with sodium ascorbate or erythorbate (550 ppm) to assist in the prevention of nitrosamine formation [44]. The regulations in 1986 for nitrite in bacon allow one of the following: (a) 120 ppm sodium nitrite or 148 ppm potassium nitrite plus 550 ppm sodium erythorbate or isoascorbate, (b) 100 ppm sodium nitrite or 123 ppm potassium nitrite plus 550 ppm sodium erythorbate or isoascorbate if a demonstration of adequate process control is met, or (c) 40–80 ppm sodium nitrite or 49–99 ppm potassium nitrite plus 550 ppm sodium erythorbate or isoascorbate plus 0.7% sucrose and a lactic acid bacterial culture (*Pediococcus*). The level of nitrites allowed is a maximum of 10 ppm in smoked cured tuna fish and 200 ppm (input not to exceed 500 ppm) in smoked cured stable fish, salmon, shad, cod roe, and home-curing mixtures. The level in smoked chub is fixed at 100–200 ppm. The use of nitrite in other products is limited to a maximum residual level of 200 ppm [17].

Product development efforts have resulted in an entire new generation of cured meat products that are low in fat and formulated with ingredients not previously used [61]. White [116] reported an average residual nitrite in cured meats of 52.5 ppm and a range of 0–195 ppm residual nitrite in wieners. Recently Cassens [13] found 5, 10, and 15 ppm residual nitrite on various cured meats in three trials of 164 samples. It is a reasonable conclusion that the current residual nitrite content of cured meats at retail in the United States is approximately 10 ppm. Cassens [13] stated that this change undoubtedly resulted from lowered ingoing nitrite, increased use of ascorbates, improved process control, and altered formulation. The mean value for the residual ascorbates was 209 ppm, nearly 40% of the maximum allowable addition of 550 ppm. The ascorbates routinely used are ascorbic acid, sodium ascorbate, erythorbic acid, and sodium erythorbate. Mitvish et al. [68] showed intragastric formation of N-nitrosamines in humans with higher doses of nitrate, but ascorbic acid inhibited their formation.

Nitric oxide is synthesized in the human body and is important to several physiological functions [13]. Cassens [13] maintained that nitrite and/or its reaction products are important in human physiology. It is known that nitric oxide is formed in the human body from nitrite. He reviewed the benefits of nitric oxide, which are: (a) it is a biological messenger important to the physiological functions of neurotransmission, blood clotting, blood pressure control, and immune system function, and (b) generation of salivary nitrite from dietary nitrate may also provide significant protection against gut pathogens in humans.

The risk of using nitrites and acquiring cancer from exposure to nitrosamines must be balanced against the risk of not using nitrites and acquiring botulism from cured meat. Such comparative risks may be obscure or difficult to quantitate [104].

REFERENCES

1. N. Ando, Y. Nagata, and T. Okayama, Proceedings of the 17th European Meeting of the Meat Research Workers, Bristol, England, 1971.
2. L. Angermeier and H. Simon, On the reduction of aliphatic and aromatic nitro-compounds by *Clostridia*; the role of ferredoxin and the stabilisation, *Hoppe-Seyeler's Z. Physiol. Chem. 364*:961 (1983).
3. J. Ashworth and R. Spencer, The perigo effect in pork, *J. Food Technol. 7*:111 (1972).
4. J. Ashworth, A. Didcock, L. A. Hargreaves, B. Jarvis, and C. L. Walters, Chemical and microbiological comparisons of inhibitors derived thermally from nitrite with an iron thionitrosyl (Roussin black salt), *J. Gen. Microbiol. 84*:403 (1974).
5. A, C. Baird-Parker and M. A. H. Baillie, The inhibition of *Clostridium botulinum* by nitrite and sodium chloride, Proc. Int. Symp. Nitrite in Meat Products, Zeist, The Netherlands, 1974, p. 77.
6. M. Baranova, P. Mal'a, and O. Burdova, Prestup dusicnanov do mlieka dojnic cestou traviaceho traktu, *Vet. Med. 38*:581 (1993).
7. R. L. Buchanan and M. Solberg, Interaction of sodium nitrite, oxygen, and pH on growth of *Staphylococcus aureus*, *J. Food Sci. 37*:81 (1972).
8. R. L. Buchanan, M. H. Golden, R. C. Whiting, and J. L. Smith, Nonthermal inactivation models for *Listeria monocytogenes*, *J. Food Sci. 59*:179 (1994).
9. E. H. W. J. Burden, The toxicology of nitrates and nitrites with particular reference to the potability of water supplies, *Analyst 86*:429 (1961).
10. C. E. Carpenter, D. S. A. Reddy, and D. P. Cornforth, Inactivation of clostridial ferredoxin and pyruvate-ferredoxin oxidoreductase by sodium nitrite, *Appl Environ. Microbiol. 53*:549 (1987).
11. R. G. Cassens, G. Woolford, S. H. Lee, and R. Goutefongea, Fate of nitrite in meat, Proc. 2nd Int. Symp. Nitrite Meat Prod., Centre for Agricultural Publishing and Documentation, Wageningen, Netherlands, 1977, p. 95.
12. R. G. Cassens, Use of sodium nitrite in cured meats today, *Food Technol. 49*:72 (1995).
13. R. G. Cassens, Residual nitrite in cured meat, *Food Technol. 51*:53 (1997).
14. A. G. Castellani and C. F. Niven, Factors affecting the bacteriostatic action of sodium nitrite, *Appl. Microbiol. 3*:154 (1955).
15. F. K. Cook and M. D. Pierson, Inhibition of bacterial spores by antimicrobials, *Food Technol. 37*:115 (1983).
16. M. C. Custer and J. N. Hansen, Lactoferrin and transferring fragments react with nitrite to form an inhibitor of *Bacillus cereus* spore outgrowth, *Appl. Environ. Microbiol. 45*:942 (1983).
17. P. M. Davidson and V. K. Juneja, Antimicrobial agents, *Food Additive* (A. L. Branen, P. M, Davidson, and S. Salminen, eds.), Marcel Dekker, New York, 1990, p. 83.
18. A. E. Dethmers, H. Rock, T. Fazio, and R. W. Johnston, Effect of added sodium nitrite and sodium nitrate on sensory quality and nitrosamine formation in thuringer sausage, *J. Food Sci. 40*:491 (1975).
19. C. L. Duncan and E. M. Foster, Effect of sodium nitrite, sodium chloride, and sodium nitrite on germination and outgrowth of anaerobic spores, *Appl. Microbiol. 16*:406 (1968).
20. C. L. Duncan and E. M. Foster, Nitrite-induced germination of putrefactive anaerobe 3679h spores, *Appl. Microbiol. 16*:412 (1968).
21. B. P. Eddy and M. Ingram, A slat-tolerant denitrifying *Bacillus* strain which 'blows' canned bacon, *J. Appl. Bacteriol. 19*:62 (1956).
22. M. Emi-Miwa, A. Okitani, and M. Fujimaki, Comparison of the fate of nitrite added to whole meat, meat fractions, and model systems, *Agric. Biol. Chem. 40*:1387 (1976).
23. J. B. Fox and R. A. Nicholas, 1974, Nitrite in meat. Effect of various compounds on loss of nitrite, *Agric. Food Chem. 22*:302 (1974).
24. A. Frouin, Nitrates and nitrites. The need to reconsider our conceptions and methods of analysis, 2nd Int. Symp. on Nitrite Meat Products, Pudoc, Wageningen, 1976.
25. A. Frouin, Nitrates and nitrites: reinterpretation of analytical data by means of bound nitrous oxide,

Proc. 2nd Int. Symp. Nitrite Meat Prod., Centre for Agricultural Publishing and Documentation, Wageningen, Netherlands, 1977, p. 115.
26. A. Frouin, D. Jondeau, and M. Thenot, Studies about the state and availability of nitrite in meat products for nitrosamine formation, Proc. 21st Eur. Meat Res. Workers, Berlin, 1975, p. 200.
27. M. Fujimaki, M. Emi, and A. Okitani, Fate of nitrite in meat-curing model systems composed of myoglobin, nitrite, and ascorbate, *Agric. Biol. Chem. 39*:371 (1975).
28. C. Genigeorgis and H. Riemann, Food processing and hygiene, *Food-Borne Infections and Intoxications* (H. Riemann and F. L. Bryan, eds.), Academic Press, New York, 1979, p. 613.
29. A. M. Gibson and T. A. Roberts, The effect of pH, water activity, sodium nitrite and storage temperature on the growth of enteropathogenic *Escherichia coli* and salmonellae in laboratory medium, *Int. J. Food Microbiol. 3*:183 (1986).
30. A. M. Gibson and T. A. Roberts, The effect of pH, sodium chloride, sodium nitrite and storage temperature on the growth of *Clostridium perfringens* and faecal streptococci in laboratory medium, *Int. J. Food Microbiol. 3*:195 (1986).
31. J. Giese, Antimicrobials: assuring food safety, *Food Technol. 48*:102 (1994).
32. G. W. Gould, Effect of food preservatives on the growth of bacteria from spores, 4th Int. Symp. Food Microbiology, *Microbial Inhibitors in Food* (N. M. Almqvist and Wiksell, eds.), Uppsala, 1964, p. 17.
33. R. Goutefongea, R. G. Cassens, and G. Woolford, Distribution of sodium nitrite in adipose tissue during curing, *J. Food Sci. 42*:1637 (1977).
34. J. I. Gray and C. J. Randall, The nitrite/N-nitrosamine problem in meats: an update. *J. Food Prot. 42*:168 (1979).
35. H. S. Grindley, The influence of potassium nitrate on the action of bacteria and enzymes, *Studies in Nutrition*, University of Illinois, Urbana, 1929, p. 359.
36. J. B. Guttenplan, *Nature 268*:368 (1977).
37. T. Haahtela and M. Hannuksela, Food additives and hypersensitivity, *Food Additives* (A. L. Branen, P. M. Davidson, and S. Salminen, eds.), Marcel Dekker, New York, 1990, p. 617.
38. J. N. Hansen and R. A. Levin, Effect of some inhibitors derived from nitrite on macromolecular synthesis in *Bacillus cereus*, *Appl. Microbiol. 30*:862 (1975).
39. A. H. W. Hauschild, R. Hilsheimer, R. Jarvis, D. P. Raymond, Contribution of nitrite to the control of *Clostridium botulinum* in liver sausage, *J. Food Prot. 45*:500 (1982).
40. W. R. Henderson and N. H. Raskin, "Hot-dog" headache: individual susceptibility to nitrite, *Lancet 2*:1162 (1972).
41. H. K. Herring, Effect of nitrite and other factors on the physicochemical characteristics and nitrosoamine formation in bacon, Proc. Meat Ind. Res. Conf., American Meat Institute, Chicago, 1973, p. 47.
42. R. A. Holley, Review of the potential hazard from botulism in cured meats, *Can. Inst. Food Sci. Technol. J. 14*:183 (1981).
43. C. N. Huhtanen and A. E. Wasserman, Effect of added iron on the formation of clostridial inhibitors, *Appl. Microbiol. 30*:768 (1975).
44. IFT, Expert panel on food safety and nutrition and committee on public information. Nitrate, nitrite and nitroso compounds in foods, *Food Technol. 41*:127 (1987).
45. K. Incze, J. Farkas, V. Mihalys, and E. Zuakl, Antibacterial effect of cysteine-nitrosothiol and possible precursors thereof, *Appl. Microbiol. 27*:202 (1974).
46. M. Ingram, The endogenous respiration of *Bacillus cereus*. II. The effect of salts on the rate of absorption of oxygen, *J. Bacteriol. 24*:489 (1939).
47. T. Inoue, K. Morita, and T. Kada, *Agric. Biol. Chem. 45*:345 (1981).
48. L. R. Jensen, *Microbiology of Meats*, 2nd ed., Garrard Press, Champaign, IL, 1945.
49. M. A. Johnston and R. Loynes, Inhibition of *Clostridium botulinum* by sodium nitrite as affected by bacteriological media and meat suspensions, *Can. Inst. Food Technol. J. 4*:179 (1971).
50. M. A. Johnston, H. Pivnick, and J. M. Samson, Inhibition of *Clostridium botulinum* by sodium nitrite in a bacteriological medium and in meat, *Can. Inst. Food Technol. J. 2*:52 (1969).

51. J. Junttila, J. Hirn, P. Hill, and E. Nurmi, Effect of different levels of nitrite and nitrate on the survival of *Listeria monocytogenes* during the manufacture of fermented sausage, *J. Food Prot. 52*:158 (1989).
52. M. E. Knowles et al., *Nature 247*:288 (1974).
53. B. Korenekova, J. Kottferova, and M. Korenek, Observation of the effects of nitrites and nitrates on yogurt culture, *Food Res. Int. 30*:55 (1997).
54. G. Kubberod, R. G. Cassens, and M. L. Creaser, Reaction of nitrite with sulfhydryl groups of myosin. *J. Food Sci. 39*:1228 (1974).
55. R. V. Lechowich, J. B. Evans, C. F. Niven, Effect of curing ingredients and procedures on the survival and growth of staphylococci in and on cured meats, *Appl. Microbiol. 4*:360 (1956).
56. F. K. Lucke, Heat inactivation and injury of *Clostridium botulinum* spores in sausage mixtures, *Fundamental and Applied Aspects of Bacterial Spores* (G. J. Dring, D. J. Ellar, and G. W. Gould, eds.), Academic Press, London, 1985, p. 409.
57. B. MacDonald, J. I. Gray, and L. N. Gibbins, Role of nitrite in cured meat flavor: antioxidant role of nitrite, *J. Food Sci. 45*:893 (1980).
58. B. MacDonald, J. I. Gray, Y. Kakuda, and M. L. Lee, Role of nitrite in cured meat flavor: chemical analysis, *J. Food Sci. 45*:889 (1980).
59. B. MacDonald, J. I. Gray, D. W. Stanley, and W. R. Usborne, Role of nitrite in cured meat flavor: sensory analysis, *J. Food Sci. 45*:885 (1980).
60. P. N. Magee and J. M. Barnes, Carcinogenic nitroso compounds, *Adv. Cancer Res. 10*:163 (1967).
61. R. W. Mandigo, Problems and solutions for low-fat meat products, Meat Ind. Res. Conf., Am. Meat Inst., Washington DC, 1991.
62. M. K. McMindes and A. J. Siedler, Nitrite mode of action: inhibition of yeast pyruvate decarboxylase (E.C. 4.1.1.1) and clostridial pyruvate:ferredoxin oxidoreductase (E.C. 1.2.7.1) by nitric oxide, *J. Food Sci. 53*:917 (1988).
63. J. Meyer, Comparison of carbon monoxide, nitric oxide, and nitrite as inhibitors of the nitrogenase from *Clostridium pasteurianum*, *Arch. Biochem. Biophys. 210*:246 (1981).
64. A. J. Miller and D. A. Menichillo, Blood fraction effects on the antibotulinal efficacy of nitrite in model beef sausages, *J. Food Sci. 56*:1158 (1991).
65. A. Mirna and K. Coretti, Uber den Verleib von Nitrite in Fleischwaren. II. Untersuchungen über chemische und bakteriostatische Eigenschaften verschiedener Reaktionsprodukte des Nitrites, *Fleischwirtschaft 54*:507 (1974).
66. A. Mirna and K. Hofmann, Uber der Verbleich von Nitrite in Fleischwaren. I. Umsetzung von Nitrit sulfhydryl Verbindungen, *Fleischwirtschaft 49*:1361 (1969).
67. M. Mirvish, L. Wallcave, M. Eagen, and P. Shubic, *Science 177*:65 (1972).
68. S. S. Mirvish, A. C. Grandjean, K. J. Reimers, B. J. Connelly, S. Chen, J. Gallagher, S. Rosinsky, G. Nie, H. Tuatoo, S. Payne, C. Hinman, and E. I. Ruby, Dosing time with ascorbic acid and nitrate, gum and tobacco chewing, fasting, and other factors affecting N-nitrosoproline formation in healthy subjects taking proline with a standard meal, *Cancer Epidem. Biomark. Prev. 4*:775 (1995).
69. M. Miwa, A. Okitani, H. Kato, M. Fujimaki, and S. Matsuura, Reaction between nitrite and low salt-soluble diffusable fraction of meat. Some compounds influencing nitrite depletion and producing unidentified-N compounds, *Agric. Biol. Chem. 44*:2179 (1980).
70. D. A. Moneret-Vautrin, C. Einhorn, J. Tisserand, Le role du nitrite de sodium dans les urticaires histaminiques d'origine alimentaire, *Ann. Nutr. Aliment. 34*:1125 (1980).
71. D. M. Moran, S. R. Tannenbaum, and M. C. Archer, Inhibitor of *Clostridium perfringens* formed by heating sodium nitrite in a chemically defined medium, *Appl. Microbiol. 30*:838 (1975).
72. NAS, *The Health Effects of Nitrate, Nitrite and N-Nitroso Compounds*, Committee on Nitrite and Alternative Curing Agents, National Research Council, National Academy Press, Washington, DC, 1981.
73. P. M. Newberne, Nitrite promotes lymphoma incidence in rats, *Science 204*:1079 (1979).
74. M. Namiki and T. Kada, Formation of ethylnitrolic acid by the reaction of sorbic acid with sodium nitrite, *Agric. Biol. Chem. 39*:1335 (1975).

75. V. O'Leary, and M, Solberg, Effect of sodium nitrite inhibition on intracellular thiol groups and on the activity of certain glycolytic enzymes in *Clostridium perfringens*, *Appl. Environ. Microbiol. 31*:208 (1976).
76. W. J. Olsman, Chemical behaviour of nitrite in meat products. I. The stability of protein-bound nitrite during storage, Proc. 2nd Int. Symp. Nitrite Meat Product, Centre for Agricultural Publishing and Documentation, Wageningen, Netherlands, 1977, p. 101.
77. W. J. Olsman, Chemical behaviour of nitrite in meat products. 2. Effect of iron and ethylenediaminetetraacetate on the stability of protein bound nitrite, Proc. 2nd Int. Symp. Nitrite Meat Prod., Centre for Agricultural Publishing and Documentation, Wageningen, Netherlands, 1977, p. 111.
78. T. Osawa, H. Ishibashi, M. Namiki, T. Kada, and K. Tsuji, Desmutagenic action of food components on mutagens formed by the sorbic acid/nitrite reaction, *Agric. Biol. Chem. 50*:1971 (1986).
79. J. A. Perigo and T. A. Roberts, Inhibition of clostridia by nitrite, *J. Food Technol. 3*:91 (1968).
80. J. A. Perigo, E. Whiting, and T. E. Bashford, Observations on the inhibition of vegetative cells of *Clostridium sporogenes* by nitrite which has been autoclaved in a laboratory medium, discussed in the context of sub-lethally processed cured meats, *J. Food Technol. 2*:377 (1967).
81. M. D. Pierson and L. A. Smoot, Nitrite, nitrite alternatives, and the control of *Clostridium botulinum* in cured meats, *CRC Crit. Rev. Food Sci. Nutr. 17*:141 (1982).
82. M. Raevuori, Effect of nitrite and erythrobate on growth of *Bacillus cereus* in cooked sausage and in laboratory media, *Zentralbl. Bakteriol. Hyg. I, Abt. Orig. B 161*:280 (1975).
83. D. Reddy, J. R. Lancaster, and D. P. Cornforth, Nitrite inhibition of *Clostridium botulinum*: electron spin resonance detection of iron-nitric oxide complexes, *Science 221*:769 (1983).
84. H. Riemann, Safe heat processing of canned cured meats with regard to bacterial spores, *Food Technol. 17*:39 (1963).
85. H. Riemann, W. H. Lee, and C. Genigeorgis, Control of *Clostridium botulinum* and *Staphylococcus aureus* in semipreserved meat products, *J. Milk Food Technol. 35*:514 (1972).
86. W. E. Riha and M. Solberg, *Clostridium perfringens* inhibited by sodium nitrite as a function of pH, inoculum size and heat, *J. Food Sci. 40*:439 (1975).
87. T. A. Roberts, The microbiological role of nitrite and nitrate, *J. Sci. Food Agric. 26*:1775 (1975).
88. T. A. Roberts, Inhibition of bacterial growth in model systems in relation to the stability and safety of cured meats, Proc. Int. Symp. Nitrite in Meat Products, Zeist, The Netherlands, 1974, p. 91.
89. T. A. Roberts and C. E. Gracia, A note on the resistance of *Bacillus* spp., faecal streptococci and *Salmonella typhimurium* to an inhibitor of *Clostridium* spp. formed by heating sodium nitrite, *J. Food Technol. 8*:463 (1973).
90. T. A. Roberts and M. Ingram, Inhibition of growth of *C. botulinum* at different pH values by sodium chloride and sodium nitrite, *J. Food Technol. 8*:467 (1973).
91. T. A. Roberts and M. Ingram, The effect of sodium chloride, potassium nitrate and sodium nitrite on the recovery of heated bacterial spores, *J. Food Technol. 1*:147 (1966).
92. T. A. Roberts, R. L. Gilbert, and M. Ingram, The effect of sodium chloride on heat resistance and recovery of heated spores of *C. sporogenes* (PA 3679/52), *J. Appl. Bacteriol. 29*:549 (1966).
93. T. A. Roberts, B. Jarvis, and A. C. Rhodes, Inhibition of *Clostridium botulinum* by curing salts in pasteurized pork slurry, *J. Food Technol. 11*:25 (1976).
94. T. A. Roberts, A. M. Gibson, and A. Robinson, Prediction of toxin production by *Clostridium botulinum* in pasteurized pork slurry, *J. Food Technol. 16*:337 (1981).
95. T. A. Roberts, A. M. Gibson, and A. Robinson, Factors controlling the growth of *Clostridium botulinum* types A and B pasteurized, cured meats. II. Growth in pork slurries prepared from 'high' pH meat (range 6.3–6.8), *J. Food Technol. 16*:267 (1981).
96. T. A. Roberts, A. M. Gibson, and A. Robinson, Factors controlling the growth of *Clostridium botulinum* types A and B in pasteurized, cured meats, *J. Food Technol. 17*:307 (1982).
97. J. J. Rowe, J. M. Yarbrough, J. B. Rake, R. G. Eagon, Nitrite inhibition of aerobic bacteria, *Curr. Microbiol. 2*:51 (1979).
98. J. G. Sebranek, R. G. Cassens, W. G. Hoekstra, and W. C. Winder, ^{15}N tracer studies of nitrite added to a comminuted meat product, *J. Food Sci. 38*:1220 (1973).

99. J. L. Shank, J. H. Silliker, and R. H. Harper, The effect of nitric oxide on bacteria, *Appl. Microbiol. 10*:185 (1962).
100. L. A. Shelef and J. A. Seiter, Indirect antimicrobials, *Antimicrobials in Foods* (P. M. Davidson and A. L. Branen, eds.), Marcel Dekker, New York, 1993, p. 539.
101. H. L. A. Tarr, Action of nitrites on bacteria, *J. Fish. Res. Board Can. 5*:265 (1941).
102. H. L. A. Tarr, Bacteriostatic action of nitrites, *Nature 147*:417 (1941).
103. H. L. A. Tarr, The action of nitrites on bacteria: further experiments, *J. Fish. Res. Board Can. 6*:74 (1942).
104. S. L. Taylor and S. S. Sumner, Risks and benefits of foods and food additives, *Food Additives* (A. L. Branen, P. M. Davidson, and S. Salminen, eds.), Marcel Dekker, New York, 1990, p. 663.
105. A. M. Taylor and C. L. Walters, *J. Food Sci. 32*:261 (1967).
106. B. J. Tinbergen, Low-molecular meat fraction active in nitrite reduction, Proc. Int. Symp. Nitrite in Meat Products, Wageningen, 1974.
107. R. B. Tompkin, The role and mechanism of the inhibition of *C. botulinum* by nitrite—is a replacement available?, Proc. 31st Ann. Reciprocal Meats Conference, Storrs, 1978.
108. R. B. Tompkin, L. N. Christiansen, and A. B. Shaparis, Variation in inhibition *C. botulinum* by nitrite in perishable canned comminuted cured meat, *J. Food Sci. 42*:1046 (1977).
109. R. B. Tompkin, L. N. Christiansen, and A. B. Shaparis, The effect of iron on botulinal inhibition in perishable canned cured meat, *J. Food Technol. 13*:521 (1978).
110. R. B. Tompkin, L. N. Christiansen, and A. B. Shaparis, Causes of variation in botulinal inhibition in perishable canned cured meat, *Appl. Environ. Microbiol. 35*:886 (1978).
111. R. B. Tompkin, L. N. Christiansen, and A. B. Shaparis, Iron and the antibotulinal efficacy of nitrite, *Appl. Environ. Microbiol. 37*:351 (1979).
112. H. J. Wagner, Vergiftung mit Pokelsalz, *Arch. Toxikol. 16*:100 (1956).
113. R. Widdus and F. F. Busta, Antibotulinal alternatives to the current use of nitrite in foods, *Food Technol. 36*:105 (1982).
114. A. E. Wasserman and C. N. Huhtanen, Nitrosamines and the inhibition of *Clostridia* in medium heated with sodium nitrite, *J. Food Sci. 37*:785 (1972).
115. B. M. Watts, Oxidative rancidity and discoloration in meat, *Adv. Food Res. 5*:1 (1954).
116. J. W. White, Relative significance of dietary sources of nitrate and nitrite, *J. Agric. Food Chem. 23*:886 (1975).
117. R. C. Whiting and M. O. Masana, *Listeria monocytogenes* survival model validated in simulated uncooked-fermented meat products for effects of nitrite and pH, *J. Food Sci. 59*:760 (1994).
118. L. F. J. Woods and J. M. Woods, The effect of nitrite inhibition on the metabolism of *Clostridium botulinum, J. Appl. Bacteriol. 52*:109 (1982).
119. L. F. J. Woods, J. M. Wood, and P. A. Gibbs, The involvement of nitric oxide in the inhibition of the phosphoroclastic system in *Clostridium sporogenes* by sodium nitrite, *J. Gen. Microbiol. 125*:399 (1981).
120. L. F. J. Woods, J. M. Wood, and P. A. Gibbs, Nitrite, *Mechanisms of Action of Food Preservation Procedures* (G. W. Gould, ed.), Elsevier Science Publishers, Essex, 1989, p. 225.
121. G. Woolford, R. G. Cassens, M. L. Greaser, and J. G. Sebranek, The fate of nitrite: reaction with proteins, *J. Food Sci. 41*:585 (1976).
122. M. Yamanaka, M. Tsuda, M. Nagao, M. Mori, and T. Sugimura, *Biochem. Biophys. Res. Commun. 90*:769 (1979).
123. J. M. Yarbrough, J. B. Rake, and R. G. Eagon, Bacterial inhibitory effects of nitrite: inhibition of active transport, but not of group translocation, and of intracellular enzymes, *Appl Environ. Microbiol. 39*:831 (1980).

15

Modified-Atmosphere Packaging of Produce

Leon G. M. Gorris and Herman W. Peppelenbos
Agrotechnological Research Institute, Wageningen, The Netherlands

I. MODIFIED-ATMOSPHERE PACKAGING—RATIONALE

Immediately after harvest the sensorial, nutritional, and organoleptic quality of fresh produce will start to decline as a result of altered plant metabolism and microbial growth. This quality deterioration is the result of produce transpiration, senescence, ripening-associated processes, wound-initiated reactions, and the development of postharvest disorders. In addition, microbial proliferation contributes markedly to postharvest quality loss. The relative importance of individual deterioration processes in determining the end of the shelf life will depend upon specific product characteristics as well as upon external factors. Low temperature and proper hygienic handling of the material are the prime factors that control these processes. In addition, modified-atmosphere packaging (MAP) is a preservation technique that may further minimize the physiological and microbial decay of perishable produce by keeping them in an atmosphere that is different from the normal composition of air [2,25,31,47,60,82].

MAP of respiring food products such as fresh and minimally processed produce requires a different approach than MAP of nonrespiring foods. With nonrespiring foods, modified atmospheres without oxygen are used to minimize oxidative deterioration reactions, such as brown discoloration of meat or rancidity of peanuts, or to reduce microbial proliferation, e.g., the growth of molds in cheese and bakery products. High gas barrier films or laminates are used to exclude the exchange of gases (especially O_2) through the package, which would result in a less beneficial in-package gas atmosphere. In contrast, respiring products stay metabolically active after harvest, and this activity is essential for keeping their quality.

Aiming at extension of the shelf life of respiring products through MAP, a prerequisite for a suitable packaging system will be that the composition of the gas atmosphere allows for a basic level of metabolism, which means that a certain amount of O_2 should be available. The required basic level of metabolism is highly at variance with different commodities (type, maturity) and heavily depends on the storage temperature and the degree of processing (trimming, cutting, slicing, etc.) applied. Due to the significant respiratory activity of the product, the gas atmosphere inside the package changes during the course of the storage period, and expert knowledge about these changes is necessary to tailor the package design of an individual product to optimize quality shelf life.

MAP of fresh and minimally processed fruits and vegetables is a preservation system that is nonsterile by design. Fruits and vegetables are characterized by an elaborate microflora, consisting of many different types of bacteria, molds, and yeasts, most of which are involved in the spoilage of the produce but are harmless to the human consumer. Microorganisms that are dangerous to humans (pathogens that are toxic or cause infectious diseases) normally cannot establish a dangerous population density because they have to compete with the spoilage microflora. However, packaging the produce will change the microenvironment perceived by the microorganisms and may well impair this safe balance. Consequently, evaluation of the impact of package design and use in the logistic chain is a mandatory exercise to assure consumer safety.

II. EARLY RESEARCH ON MODIFIED-ATMOSPHERE PACKAGING

Packaging techniques based on altered gas conditions have a long history. Ancient Chinese writings report the transport of fruits in sealed clay pots with fresh leaves and grass added. The respiratory activity of the various plant products generated a low-oxygen and high–carbon dioxide atmosphere, which retarded the ripening of the fruit [42,59]. In the beginning of the nineteenth century Berard demonstrated that fruit placed in closed containers did not ripen [12]. By the end of the nineteenth century, the first patent was granted covering the use of a CO_2/CO mixture to extend the shelf life of meat [31]. Extensive research on the use of altered gas conditions for fruits tarted early in the twentieth century, with the work of Kidd and West [66]. Commercial storage under altered gas conditions was undertaken in England in 1929, when apples were stored in 10% CO_2 and ambient O_2 [64]. Reduced O_2 concentrations and increased CO_2 concentrations also proved to be beneficial for harvested products other than apples. Products with a high potential for a successful commercial application in MAP include apple, banana, broccoli, cabbage, cherry, chicory, and Brussels sprouts. The first commercial application of MAP did not take place until 1974, when the technique was used for meat [31]. The use of modified atmospheres (MA) for storage and packaging has increased steadily over the years and contributed strongly to extending the postharvest life and maintaining the quality of fruits and vegetables [61]. In fact, the biggest growth in the use of MA has been for fresh fruits and vegetables, especially for minimally processed salads [31]. The technique of MA is now applied at a range of different sizes, i.e., for bulk storage packages (e.g., red currants), transport packages (e.g., bananas, strawberries), and consumer packages (e.g., apples, broccoli).

III. EFFECTS OF MODIFIED GAS ATMOSPHERES

The strategy of packaging produce under modified atmospheres is to slow down the metabolic activity of the product as well as of the growth of microorganisms (both spoilage and pathogenic) present by limiting O_2 supply and by application of an elevated level of CO_2. Because the same strategy underlies refrigerated storage, MAP of respiring produce is usually combined with this technique. Many commodities, for instance, avocados, mangoes, papayas, and cucumbers, are very sensitive to low-temperature injury and should not be stored below about 13°C. Commodities like apples, broccoli, and pears are not sensitive to chilling and can be stored near 0°C without ill effect [100].

A. Reduction of Oxidative Reactions

Plant parts such as seeds, fruits, leaves, or roots continue to live after harvest. The energy plant cells need to stay alive and/or to proceed with ripening is generated by aerobic respiratory processes. Respiration involves the consumption of O_2 and the production of CO_2. A reduction of respiration results in a lower energy supply and a reduced rate of changes within the product, like ripening [60]. To extend storage periods, conditions should be created that reduce respiration, for instance, by using low-temperature and low-O_2 concentrations. In general the reduction of the respiration rate is regarded to be the process that is most strongly affected by altered gas conditions [5,69]. For certain fruits, low O_2 levels inhibit the production and action of the plant hormone ethylene, which results in reduced ripening as well [25]. Because respiration has such an important central position in the overall metabolism of a plant (part), its measurement is often used as a general measure of metabolic rate. Specific metabolic changes, however, may occur without measurable changes in net respiration [64]. Nevertheless, good quantification of the effect of reduced O_2 on respiration rates is essential for MA, as this process helps to generate the modified atmosphere inside MA packages.

With both whole, fresh produce and minimally processed produce, oxidative reactions do not only relate to respiratory activity. In addition, oxygen also has an effect on the activity of certain enzymes present in bruised or wounded tissue. Such enzymes are involved in wound repair reactions and in the defense against intruding microorganisms. Their activity depends on the presence of oxygen and is driven by the metabolic activity of the produce. Most studied is polyphenol oxidase (PPO), an enzyme that causes browning of plant tissues. In the case of minimally processed product (i.e., chopped, cut, sliced, and peeled), the level of tissue injury is much higher than with whole produce. Consequently, the level of metabolic activity and thus the respiration rate of minimally processed produce is often orders of magnitude higher than that of the raw material. Also, enzymes such as PPO will be more active and may cause visible browning of cut surfaces. Such responses should be considered and overcome by choosing the correct MAP design. In the case that different types of minimally processed products are included in a MAP, which is often the case in mixed vegetable salads, conflicting levels of O_2 and CO_2 may be optimal for the individual components. A designer's solution for this problem needs to integrate all the different aspects that are important with regard to the quality features of the end product.

B. Fermentation Reactions

The most optimal MA condition for a product is often considered to be the O_2 concentration, which is as low as possible with regard to product respiration without initiating fermentative reactions [5,25]. Fermentative reactions lead to the production of compounds such as acetaldehyde, ethanol, lactic acid, and ethyl acetate. Alcoholic fermentation is always found in plant tissues exposed to an environment without O_2 [81]. An increased concentration of ethanol and/or ethyl acetate is often related to quality problems such as off-taste and off-odor [65,67]. A strong correlation was found between ethanol and ethyl acetate concentrations [65]. A relationship between other fermentative metabolites and off-flavors is less clear. With improved detection techniques, compounds like ethanol and acetaldehyde can even be detected at O_2 concentrations higher than those considered to be optimal for packaging of certain produce [80]. It seems that fermentation cannot be avoided completely and that it is not absolutely necessary to be avoided from the point of view of package design. Rather, it is important at what concentration of ethanol (or ethyl acetate) the consumer experiences off-odors or off-flavors. The package design should allow O_2 concentrations to be high enough to avoid an accumulation to that concentration. A complicating factor is that a relatively short period of too low O_2 concentrations can cause irreversible quality damage, because it has been found that strong off-flavors do not disappear once the favorable O_2 levels have been reestablished.

C. Selective Impact on Microbial Growth

For many minimally processed products, the main factor causing quality loss is not ripening or senescence, but microbial growth. The modified-atmosphere composition has a marked impact on the growth of spoilage microorganisms as well as on pathogens that occasionally occur in minimally processed produce [77]. The very low O_2 (typically 2–3%) and moderately high CO_2 (5–20%) levels prevailing inside a package slow down the proliferation of aerobic spoilage microorganisms [20,41,47,55,57–59,82]. The antimicrobial effect of CO_2 on microorganisms has been intensively documented [4,18,34–39,52,76]. However, it has been shown recently that only CO_2 levels well above 20–50% significantly affect the growth of psychrotrophic pathogens that are relevant to MA-packaged produce [10]. This contradicts the general belief that CO_2 has very pronounced antimicrobial properties. At levels of O_2 and CO_2 that are generally favorable for storage of produce, there is certainly no beneficial effect of CO_2 [10,29].

In in situ studies it was established that the specific conditions of MAP (reduced oxygen, increased carbon dioxide) can lead to marked changes in the epiphytic microflora, especially in chicory endive [11]. Thus, whereas there may be no direct antimicrobial effect of CO_2, there is an influence on the composition of the microflora and on the competition that pathogens may experience in this ecosystem. A specific safety hazard is that psychrotrophic, facultative aerobic pathogens such as *Listeria monocytogenes* are not suppressed under MA conditions that are optimal for respiring produce [10,15,28,29]. On the contrary, growth may be enhanced in certain cases [3,11], especially because the MA conditions diminish the growth of spoilage microorganisms that would be competitors of the pathogens.

IV. TYPES OF PACKAGES

With MAP, the gas composition surrounding the produce inside the package is different from the gas composition outside the package. Outside, the gas composition is always close to 78.1 kPa nitrogen, 20.95 kPa oxygen, 0.93 kPa argon, and 0.036 kPa carbon dioxide. Several different types of packages and packaging techniques have been developed to accommodate modified atmospheres around the produce, and these will be explained in detail below. The modification of the atmosphere generally implies a reduction of O_2 content and/or an increase of the CO_2 concentration, but in some cases changing the level of carbon monoxide (CO), ethylene, ethanol, or other compounds in the atmosphere can also contribute to shelf-life extension. Modified atmospheres can be created passively by the respiration activity of the product inside the package (product-modified MAP) or actively by introducing the desired gas mixture (gas packing). Other active ways of obtaining modified atmospheres are the use of gas generators and scrubbers (controlled-atmosphere packaging), evacuation of air (hypobaric storage, vacuum packaging), or addition of chemical systems that absorb or generate gases or volatile compounds (active packaging) in packages.

A. Modified-Atmosphere Packaging

In modified atmosphere packaging the gas composition within the package is not monitored or adjusted. Therefore, the term passive atmosphere packaging (PAM) is sometimes used in this respect. Depending on the oxygen sensitivity and metabolic activity of the product to be packaged, air or a predetermined gas mixture is used to flush packages before closing. The use of ambient air as the packaging gas obviously is most economic, but is an option mainly when the respiration activity under the prevailing storage conditions is high enough to reduce the in-pack O_2 level fast enough to lower levels that do not cause physiological or microbial deterioration. With produce highly sensitive to O_2 (e.g., many minimally processed fruits) or that have a low level of respiratory activity, flushing with a gas mixture composed of low oxygen and moderately high CO_2 is often used to shorten the time needed to reach the desired in-pack gas composition. After closing the package, the respiration of the product will cause a decrease in the oxygen content and an increase in the carbon dioxide content. These altered gas concentrations, however, cause a decrease in the respiration rate. Finally an equilibrium concentration inside the package is reached, which is the result of a balance between metabolic rates of the packed product and diffusion characteristics of the package materials. This explains the use of another term, equilibrium-modified atmosphere (EMA) packaging. The package is often designed in such a way that the equilibrium concentrations resemble the optimal gas concentrations found in experiments where products are stored under a range of stable gas conditions.

The course of the atmosphere modification is determined by three interacting processes: respiration of the commodity, gas diffusion through the commodity, and gas permeation through the film. Each of these processes is in turn strongly influenced by several commodity- and environment-generated factors. Respiration of a certain commodity depends, among others, on its physiological stage and temperature, O_2 and CO_2 partial pressures, relative humidity, and ethylene concentration. Gas diffusion is affected by temperature, gas gradient across the limiting barrier, and the commodity's mass, volume, respiration rate, membrane permeability, and gas diffusion path. Some of these variables may vary with the maturity stage of the product or even the degree of illumination. Some variables affecting gas permeation through the film are temperature, gas gradient across the film and film structure, water vapor gradient, thickness, and surface area. A change in product amount, free volume, or any of the variables listed above will affect the EMA and/or the time in which the steady-state conditions are established. Flushing a package with a premixed gas will influence the time needed to attain the EMA.

Strict temperature control in the distribution chain would be a prerequisite for optimal use of MAP in practice, but in most countries the cooling chain between production, distribution, retail, and the consumer has many uncontrolled links. The changes in the permeabilities of most packaging films to gases in response to changes in temperature are generally lower than changes in product respiration. Most of today's existing plastic films do not have the proper O_2:CO_2 permeability ratio to provide the ideal MA for many commodities at a given temperature. In view of all these variables and knowing that any change within or around the package will alter the dynamic equilibrium between the product and its environment, it is clear that knowledge about the limits of tolerance of a certain commodity is even more important for MAP than it is for controlled-atmosphere packaging.

B. Controlled-Atmosphere Packaging

In controlled-atmosphere packaging (CAP) the altered gas composition inside the package is monitored and maintained at a preset level by means of scrubbers and the inlet of gases. This method closely resembles the practices used in large controlled-atmosphere (CA) storage facilities where produce is stored essentially unpacked in bulk, except that CAP is used for storage or transport of smaller quantities of produce.

Additionally, new areas of attention in CA storage today are ultra low oxygen (ULO) storage and dynamic CA storage. Obviously these techniques can be used in CAP as well. ULO storage uses O_2 levels close to the minimum level required for maintenance of plant tissues; lower levels will induce disorders such as browning and tissue necrosis. Using ULO storage at 1–2°C with preset levels of 0.5–1% O_2 and 2–3% CO_2, for instance, Elstar apples can be stored for almost a whole year without unacceptable quality loss. In the case of dynamic CA storage, sometimes referred to as interactive CA storage, gas levels are not controlled at preset levels but are continuously adapted to the physiological response of the produce [98], for instance by monitoring fermentation products or cell degradation products. In this way, an optimal match is made between the physiological demand and tolerance of a product and the storage conditions. Although this concept is still in development for CA [87], packages comparable ideas have been described (see below).

C. Active Packaging

In some cases a package cannot be designed in such a way that optimal conditions will be reached passively. "Active packaging" can then provide a solution, by adding materials that absorb or release a specific compound in the gas phase. Compounds that can be absorbed are carbon dioxide, oxygen, water vapor, ethylene, or volatiles that influence taste and aroma. For some leafy vegetables carbon dioxide levels can induce browning of tissues, while for most fruits increased ethylene levels cause an acceleration of ripening. Even at rather low levels, depending on the type of produce, ethylene can induce senescence and maturation processes that reduce the fresh product quality. Inclusion of ethylene scrubbers like potassium permanganate counteracts the effect of ethylene, although the capacity of such scavengers is finite. In transport packages for grapes pouches are often added that slowly release sulfur-containing chemicals to reduce fungal growth. Recently research has been directed to replacing chemicals with compounds retrieved from plant tissues ("green chemicals"). How much of the active compounds needs to be added will depend on a range of interacting factors, such as production rates (carbon dioxide, ethylene), concentrations to be reached, how long the package should be functional, etc. Various possibilities exist, although precise control of O_2 in such packages is not possible [25].

Recently a number of new "intelligent" concepts have been introduced that involve more than only scrubbing or emitting compounds. These type of packages will only become "active" when a specific prerequisite has been met. Most of these packages focus on prevention of problems associated with anaerobic conditions. In one such system, holes are introduced in the package upon exposure to high temperatures for a certain time; originally, the holes are closed by solid hydrocarbons that have melting points between 10 and 30°C [25]. Because respiration of a product often increases at a faster rate than the diffusion of gases with a rise in temperatures, the hole in the package will prevent the depletion of O_2. Another idea is a sensor for ethanol mounted on a package that informs possible buyers of the history of the package in terms of possible mechanical damage or temperature abuse [25]. Yet another concept, which has seen some use in, for instance, France and the United States, is the "time-temperature indicator" or "time-temperature integrator (TTI). TTIs used now are in most cases small devices that, attached to the package, will indicate the combined time and temperature history of that product by a gradual change color [92,97]. TTIs integrate the time and temperature by specific enzymatic or chemical reactions that, ideally, have an identical rate constant to the quality or safety feature of the packed product. The consumer can compare the actual color at the time of intended purchase with the indicated sell-by limit color. A TTI is an elegant and user-friendly improvement that informs consumers of the expected shelf life at the point of sale. The concept could well be extended to the home situation.

D. Vacuum Packaging

Whereas MAP and CAP mostly operate at ambient pressure (101 kPa), storage at reduced atmospheric pressures has been experimented with and, in some cases, has been used for bulk storage (e.g., in the so-called hypobaric storage systems designed by Stanley Burg almost a quarter of a century ago [23,24]). In the Burg system, produce is stored under atmospheric pressure in the range of about 1–10 kPa at refrigerated temperatures. At this low pressure, a constant circulation of fresh air, substantially saturated with water (RH 80–100%), is maintained. Facilities to constantly scrub CO_2 and ethylene could be included as well. Although the system performed rather well, and shelf lives of different horticultural and floricultural products could be extended 3- to 10-fold, it was technically complex and for this reason was never used as widely as CA or MA storage. Vacuum packaging (VP) may be regarded as a special type of MAP, since part of the normal headspace is removed, leaving an altered initial atmosphere that is not controlled after packaging. VP puts quite a pressure strain on produce and is only suitable when the product is sufficiently durable.

Using a VP system—called a moderate VP system because it operates at 40 kPa—a significant prolongation of quality shelf life at 8°C was obtained with a range of minimally processed fruits and vegetables [48]. In this system, the initial gas composition is that of normal air, but because of the reduced partial gas pressure, the amount of O_2 available at the start of storage is about one third of the normal amount. As with MAP, the lower O_2 content stabilizes the postharvest product quality by slowing down the metabolism of the produce and the growth of spoilage microorganisms. Compared to refrigeration-only storage, refrigerated storage under moderate vacuum was found to improve microbial quality (e.g., red bell pepper, chicory endive, sliced apple, sliced tomato), sensory quality (e.g., apricot, cucumber) or both (e.g., mung bean sprouts and a mixture of cut vegetables). In some instances no beneficial effect (mushroom, green bell pepper, and a mixture of cut fruits) or an impeded decrease in sensory quality (strawberries, alfalfa) was noticed. With cut products (vegetables and fruits salad mixes, chicory endive, apple), VP strongly retarded enzymatic browning of the cut surfaces.

E. Modified-Humidity Packaging

MAP, CAP, and VP all focus on changing the metabolic gases oxygen and carbon dioxide. Modified humidity packaging (MHP), however, is designed for products where dehydration causes the most important quality losses, and therefore focuses on controlling water vapour levels. When products such as leafy vegetables or bell peppers are not packed, very soon quality losses can be observed (e.g., wilting and shriveling). In most "closed" packages such as MAP, CAP, and VP, the relative humidity is close to saturation due to the water exchange between the product and the headspace. This high humidity increases the probability of condensation and free water accumulating directly on the product, especially when the package is exposed to changing temperatures. Therefore, MHP systems are designed to control not only dehydration but also condensation.

The in-pack relative humidity (RH) is influenced by the rate of water loss (transpiration) of the product and the transmission rate for water vapor of the package, which are dependent on the prevailing water vapor pressure and temperature of storage. Temperature is one of the most important factors determining the in-pack RH. Weight loss relates more exactly to the vapor pressure deficit than to relative humidity, but at constant temperature weight loss has a linear relationship with relative humidities above 75–85% [50]. At higher temperatures the air can contain more water vapor, thereby decreasing the RH value. A package designed to have a high relative humidity at a high temperature will show condensation on the package surface or on the product if the temperature is decreased substantially. To counteract the effect of condensation, films have been developed which are coated with an antifog layer, due to which moisture forms a continuous layer rather than separate droplets on the surface of a film. This allows a clear view of the product and prevents water from forming a pool at the bottom of the package.

For many products transpiration must be reduced in order to maintain quality. Products with a large surface area such as lettuce and endive are very susceptible to wilting. Bell peppers and tomatoes also benefit from good control of relative humidity [9,88]. Reducing water loss is one of the main aspects related to packaging of minimally processed products, despite the usual emphasis on gas levels [25]. On the other hand, for products such as onions and flower bulbs, humidity should not be too high, as it results in increased sprouting. Like O_2 and CO_2, water vapor levels can be too high or too low, and an optimum level should be reached. For (Israeli) bell peppers this level was estimated to be 92% relative humidity at 8°C [83]. A lower relative humidity caused too much weight loss, while a higher relative humidity caused decay. Especially for products where water loss is the predominant cause of quality changes (e.g., bell pepper and tomato), MHP can be effectively used to minimize loss of quality. In such cases, the concentrations of oxygen and carbon dioxide in MHP are often close to that of ambient air.

Many commercially available packaging materials that have favorable gas-permeability characteristics for a certain commodity cannot be used because they have a rather low permeability for water vapor. When the in-pack RH is very high (≥95%), a small fluctuation in storage temperature results in condensation, which greatly enhances the proliferation and spread of spoilage microorganisms. Especially for fruits, the high RH conditions cause heavy losses due to microbial decay. Control of the in-package RH may be pursued through the use of packaging materials with high water vapor permeabilities, by inclusion of sachets containing water absorbers like $CaCl_2$, sorbitol, or xylitol in the package ("active packaging") or by use of packaging materials with suitable gas permeabilities onto which such desiccants are coated [6,88].

V. IMPORTANT PARAMETERS IN PACKAGE DESIGN

A. Product Characteristics

Before a package can be designed, detailed knowledge about the physiological characteristics of the product to be packed and the environmental conditions the package is exposed to after production is essential. Many specific parameters need to be known. Important are not only the optimal O_2, CO_2, and water vapor levels, but also the upper and lower limits of these components beyond which damage can be expected. When low O_2 and/or high CO_2 is beneficial, it becomes important to quantify the relationship between gas conditions and gas-exchange rates. For good quantification, O_2 uptake and CO_2 production should be measured under a range of O_2 and CO_2 concentrations. Such data sets, however, are still scarce.

Another important product aspect is the influence of light on color changes, for instance with chicory endive, which changes from the preferred yellow-white to the undesired green color under excess illumination. It is also important to know what mechanical properties of the package should have when delicate products such as berries are to be packaged without mechanical damage.

B. Package Characteristics

An important aspect of package design is the selection of the packaging material, and this can be a cumbersome exercise. Exama et al. [40] studied the possible application of 20 different types of polymer films and was still not able to find a suitable match with products with a high respiration rate. Using too high-barrier package film, O_2 will be fully depleted and fermentation will lead to off-odors and off-flavors. Also the right combination of low O_2 and high CO_2 is crucial. This highlights two decisive aspects in selecting films: (a) the permeability for O_2 and CO_2 at the temperature to be used, and (b) the ratio between O_2 and CO_2 permeability. A serious drawback is that gas permeability specifications given by film manufacturers are usually determined under conditions remote from the high-humidity refrigerated storage conditions of respiring produce. Thus, it is impossible to deduce only from the specifications provided by film manufacturers whether a specific film would provide for an in-package gas atmosphere with tolerable O_2 and CO_2 levels when applied in practice. Thus, the suitability of a film must be tested with the product under the correct practical conditions.

In addition to permeability of the metabolic gases, permeability for water vapor, ethylene, and volatiles can be important. A low permeability for water vapor can increase the risk for condensation. Condensation should always be avoided, since it generates an ideal climate for microbial growth. Also, discoloration of the product can result from condensation.

Currently, polyethylene (PE) and polyvinyl chloride (PVC) films are the most often used polymers. In the past decade a new type of film was introduced, with very small holes (microperforation) as the main pathway for diffusion [95]. The interesting aspect of these films is that diffusion of O_2 and CO_2 through the film is equal. This enables the creation of packages with both low O_2 and high CO_2 concentrations. Such atmospheres are especially suitable for minimally processed products but also for unprocessed products with extremely high respiration rates like asparagus, broccoli, mushrooms, or mung bean sprouts.

In addition to the selection of the type of film, other important aspects include thickness of the film, the surface area used, the package volume, and for films with microperforation, the number of holes per area. Film thickness, film area, and the number of holes influence the equilibrium gas composition inside the package. Varying package volume and the free volume inside the package influences the rate at which gas concentrations are changing. The final equilibrium concentrations will be equal, but the moment in time at which these concentrations are reached can differ by varying the volume.

C. Modeling

Since there are so many variables to take into account in package design, a trial-and-error type of approach can lead to numerous attempts to find the best package. The risk is that the best package will not be found. Sufficient control of the many different factors interacting in determining the atmosphere change in a MAP can only be achieved with the help of mathematical modeling [101]. Mathematical models may provide a means to determine and predict important packaging specifications. When optimal (equilibrium) gas conditions are known as well as the respiratory response to various O_2 and CO_2 concentrations, the suitable permeability characteristics of the package can be mathematically deduced. The most frequently used models that relate gas conditions to O_2 uptake and CO_2 production are based on Michaelis-Menten kinetics [5,25,68]. Although an inhibition of CO_2 on respiration is not found for all products, Peppelenbos and van 't Leven [78] examined which type of inhibition best described the influence of CO_2. The models of Banks et al. [5] or Peppelenbos et al. [79] can be applied best when not only respiratory CO_2 but also fermentative CO_2 production needs to be calculated. An example of gas-exchange modeling is given in Figure 1, where at low O_2 concentrations CO_2 production increases due to enhanced fermentation.

The description of gas diffusion through packaging materials is mostly done by applying Ficks law [25]. Although all models mentioned above are static, they can be incorporated into dynamic models to be used for the prediction of changing gas conditions inside a package [56]. Packages can be easily designed with such dynamic models by changing variables such as film type, surface area, and amount of product packed. A very useful extension of simulation models would be the incorpo-

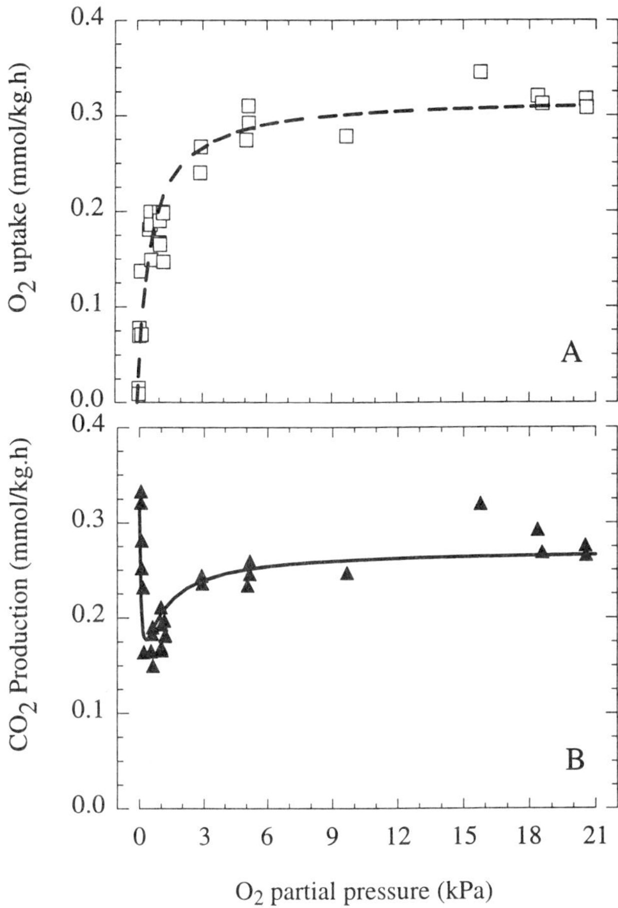

FIGURE 1 Gas exchange rates of strawberries (*Fragaria ananassa* cv. Elsanta) at 4°C. (A) O_2 uptake rates (□) with O_2 uptake model. (B) CO_2 production rates (▲) with CO_2 production model. (From Ref. 79.)

ration of expected variance of the achieved equilibrium conditions. Using expected variance, not only can optimal packages be designed, but so can (sub)optimal packages that are also safe. A survey of this variance has already been carried out for broccoli by Talasila et al. [91].

Since temperature in the distribution chain often cannot be strictly controlled, another interesting feature of dynamic modeling is the possibility of simulating products passing through the different links of the distribution chain. Using simulation, for instance, the dynamics of the gas composition inside the package can be evaluated in order to determine whether gas conditions will remain within the limits of tolerance of the commodity. When necessary, the use of different packaging films can be simulated in order to obtain the most optimal equilibrium modified atmosphere condition. The end result of the modeling exercise, however, could be that there is no suitable packaging film available commercially that would be suitable for use. Instead of not packing the product, this information could be used to further improve distribution chains or to give suggestions to the packaging industry, defining the requirements for new films in terms of their temperature sensitivity, CO_2:O_2 permeability ratio, etc.

Once proper models have been created and integrated in package design, they should be mandatory in the development of packages that achieve optimal gas conditions at dynamic temperatures encountered in practical situations and/or of packages that can overcome fluctuations in temperature, which temporarily cause gas conditions to exceed tolerance limits but do not affect product quality.

VI. MICROBIAL GROWTH UNDER MODIFIED ATMOSPHERES

A. Spoilage Microorganisms

Fresh fruits and vegetables normally have an elaborate spoilage microflora, due to intensive contact with various types of microorganisms during growth and postharvest handling. The high acidity of many fruits ($pH < 4.6$) limits spoilage to acid-tolerant molds, yeasts, and lactic acid bacteria. Vegetables generally have a pH around 6.0–7.0 and lack this intrinsic protection. Microbial spoilage of undamaged, healthy products can only be effectuated by microorganisms able to penetrate through the skin, which requires the presence of specific enzyme systems. In vegetables, pectinolytic gram-negative bacteria of the genera *Pseudomonas* and *Enterobacter* are often involved in spoilage. The effect of MAP, and in particular carbon dioxide, on spoilage organisms is distinctly selective, but it is possible to make some broad generalizations. Molds exhibit sensitivity, while yeasts are comparatively resistant. Different species of bacteria, on the other hand, vary greatly in sensitivity. For example, aerobic organisms such as *Pseudomonas, Micrococcus,* and *Bacillus* are inhibited by CO_2, while the *Lactobacillus* species are more resistant. On the other hand, facultative anaerobes such as *E. coli* are less affected by the level of CO_2 but more by the level of O_2. Most spoilage organisms that pose a quality problem in produce are aerobic, thus a limited supply of O_2 hampers their growth potential. Nitrogen has little inhibitory effect except in displacing oxygen.

Spoilage microorganisms usually pose no safety problem for consumers. The main concern is that application of modified atmospheres diminish the competition for oxygen, carbohydrates, other nutrients, and space between spoilage microorganisms and pathogens. This may allow the growth of certain pathogens to hazardous levels, especially during extended shelf life. In addition, the problem of product temperature abuse, either during manufacture, distribution, and retail or by the consumer, must also be considered. In those cases spoilage microorganisms may be important safety indicators, giving an organoleptic warning signal to consumers that the food product has been mishandled or kept beyond its shelf life and therefore may not be safe to eat. However, when technologies such as MAP are used to extend the product shelf life by suppressing spoilage organisms but not all hazardous pathogens, situations could occur in which the packaged food is organoleptically unspoiled but very unsafe to eat.

B. Pathogenic Microorganisms

Most fruit products have too low a pH to permit growth of pathogenic bacteria—only figs, peaches, and tomatoes, whose pH is potentially in the range 4.6–4.8, may permit pathogenic growth. In the early days of MAP, attention was focused primarily on anaerobic pathogens, especially proteolytic *Clostridium botulinum*, which produce a deadly toxin but do not grow below 10°C. Since nearly all MAP foods are refrigerated, focus has been mainly on the survival and outgrowth of cold-tolerant pathogens such as *Yersinia enterocolitica* and *Listeria monocytogenes* that can proliferate under low-oxygen conditions [15,82]. An important factor with respect to microbial safety is whether the MAP food is intended for direct consumption or requires heating before consumption. With MAP foods that are cooked before being eaten, vegetative pathogens should all be killed, providing the cooking instructions are properly followed. However, the majority of MAP produce is sold as "ready-to-eat." The main potential sources of pathogenic bacteria in fresh and minimally processed produce as the raw material, ingredients, plant workers, as well as the processing equipment and environment. Below, the main pathogens of possible concern to MAP are described.

1. *Clostridium botulinum*

Because of the potency of their toxin, the potential growth of *Clostridium* species in MAP foods has been of especially great concern [90]. The organisms can be present in soils and can thus come into contact with fruits and vegetables easily. *C. botulinum* is not markedly affected by the presence of CO_2, and growth is encouraged by the anaerobic conditions that may exist in MAP. Most strains of *C. botulinum* do not grow at temperatures below 10°C, although nonproteolytic *C. botulinum* types B, E, and F have been recorded as growing and producing toxins at temperatures as low as 3.3°C.

Botulism has been linked to coleslaw prepared from MA-packaged, shredded cabbage mixed with coleslaw dressing [63,89]. Shredded cabbage onto which spores of *C. botulinum* types A and B were inoculated and that subsequently was MA packaged and held at room temperature was found organoleptically acceptable after 6 days, yet type A toxin production was apparent on day 4. A pungent odor was produced and released on opening the bag, after which the cabbage smelt normal. A recent survey by Lilly et al. [72] on the incidence of *C. botulinum* in MAP and VP vegetables involving 1118 packages of a variety of precut produce (including cabbage, pepper, coleslaw, carrot, onion, broccoli, mixed vegetables, stir-fry vegetables, and various salad mixes) found that 1 package each of shredded cabbage, chopped green pepper, and Italian salad mix contained *C. botulinum* type spores, while an additional salad mix (main ingredient, escarole) contained by *C. botulinum* type A and type B spores. The overall incidence rate (0.36%) of *C. botulinum* spores thus may be quite low in commercially available precut vegetables.

2. *Listeria monocytogenes*

The widespread presence of this organism in the environment and its ability to grow at low temperatures makes it a pathogen of special concern. A serious outbreak of listeriosis was thought to be derived from cabbage fertilized with manure from infected sheep [86]. In many studies carried out since this outbreak of a variety of produce it was frequently established that *L. monocytogenes*, when the organism was inoculated onto vegetables, grew as well under MA or CA conditions as in air at 4°C or 15°C [3,13,16,17,29,43,44]. By now, growth of this pathogen under MA conditions has been reported for asparagus, broccoli, cauliflower, lettuce, and chicory endive. Studies by Carlin et al. [28] investigated the fate of *L. monocytogenes* in minimally processed foods in the presence of nonpathogens and at temperatures ranging from 3 to 20°C. It was shown that on unspoiled products *L. monocytogenes* would hardly grow more than 2 log units whatever the storage temperature, but that spoilage of the salad leaves would permit a rapid multiplication. Low storage temperatures reduced growth of *L. monocytogenes* more than that of the spoilage microflora and are therefore a factor improving safety. Carbon dioxide concentrations of 10–20% reduce spoilage development and growth of the spoilage microflora, whereas higher concentrations slightly increased growth of *L. monocytogenes*. On minimally processed green endive, it was found that that high inoculum concentrations overestimated the maximum growth of *L. monocytogenes*. Again, the epiphytic microflora of green endive leaves had a barrier effect against *L. monocytogenes*.

3. *Aeromonas hydrophila*

This psychrotrophic organism is also widespread in the environment and is mainly waterborne. It has been found in drinking water, fresh and saline water, and sewage water. Cytotoxic strains have been found on seafood, meats, and poultry, as well as on fresh produce, parsley, spinach, celery, and endive. The pathogen grows rapidly at refrigeration temperature [45]. Berrang et al. [14] observed that *Aeromonas* could grow to population densities exceeding 10^6 CFU/g within 2 weeks at 4°C on asparagus, broccoli, and cauliflower and that CA conditions did not markedly affect its growth potential.

4. *Yersinia enterocolitica*

Animals, specifically swine, are the predominant natural source of *Yersinia enterocolitica*. However, this cold-tolerant pathogen has also been isolated from raw vegetables. As with the former two pathogens, modified atmospheres optimal for produce do not hamper its growth at refrigeration temperature. With little oxygen (1.5%) present, carbon dioxide levels as high as 50% were found to be required before its growth was significantly reduced [10].

5. *Bacillus cereus*

This bacterium, a common contaminant of vegetables, does not usually grow below 10°C [53]. However, recent reports have shown that some enterotoxigenic strains can grow at temperatures as low as 4°C and produce toxin at 8°C. *B. cereus* is rather susceptible to the antimicrobial effects of CO_2 [10], and CO_2-rich environments severely reduce the ability of the spores to germinate.

6. *Salmonella* spp.

This organism is most commonly associated with animals and birds and is only present on vegetables through cross-contamination. Nevertheless, two large outbreaks of salmonellosis have been attributed to fresh produce, both involving tomatoes stored at ambient temperatures [54,99]. *Salmonella* species have also been implicated in smaller outbreaks in which raw bean sprouts and different types of melons were the vehicles. Although high levels of CO_2 retard the growth of *Salmonella*, generally the inhibitory effect on the organism is largely dependent on decreased temperature. Most *Salmonella* species are mesophilic bacteria, but many isolates survive well during storage at 5°C [46].

7. *Staphylococcus aureus*

S. aureus does not grow well under chill conditions or in the presence of competing microorganisms. The pathogen has been found on fresh produce and ready-to-eat vegetable salads. It is known to be carried by food handlers. Generally, CO_2 has an adequate inhibitory effect on growth of *S. aureus* when combined with low-temperature storage.

8. *Escherichia coli*

E. coli is a mesophilic bacteria often used as an indicator of fecal contamination. Enterotoxigenic *E. coli*, the common cause of travelers' diarrhea, is regularly detected on raw vegetables. Strains can grow at temperatures below 10°C, but not usually below 7°C. Some strains reportedly are able to grow and produce toxin at 5°C. Growth of this organism can be inhibited by high levels of CO_2. Enterohemorrhagic *E. coli* O157:H7 is recognized as an important emerging pathogen. Outbreaks of this pathogen have been associated with unpasteurized apple cider and cantaloupe. Also, broccoli is suspected to have carried this type of *E. coli*. MAP had no effect on pathogen growth on shredded lettuce and cucumber in experiments in which storage temperatures were 12°C and higher [1,33].

9. *Campylobacter jejuni*

This organism is still one of the major causes of bacterial enteritis. Poultry and other foods of animal origin are the main sources. The pathogen has been implicated in diseases caused by consumptions of fruits and vegetables. Cross-contamination of fresh produce with *C. jejuni* from poultry meats has been suspected. The pathogen has been found to survive sufficiently on sliced watermelon and papaya to be a risk for consumers. Total absence of oxygen was noted, whereas survival, without growth, was enhanced in an atmosphere of 100% N_2. Optimal growth has been documented to occur under atmospheres of reduced oxygen level and high temperatures (42–45°C). The minimum growth temperature reportedly is 32°C, so the risk with consumption of refrigerated MAP produce should be minimal.

10. Disinfectant Usage

The use of disinfectants to reduce the microbial load of minimally processed fresh salads is permitted and practiced in many countries around the world. It has been found that after disinfection, the surviving spoilage microorganisms have an increased growth rate and rapidly reach the same level found on nondisinfected products [11,28]. In the case of contamination with *L. monocytogenes* during processing, disinfection of salad leaves would reduce the antagonism from epiphytic bacteria and might increase growth of the foodborne pathogen. Therefore, the major role of disinfectants may be to prevent build-up of contamination in washing water during processing rather than to reduce microbial load of raw salad leaves. In the production of MAP produce, good manufacturing practices should be observed that avoid recontamination of disinfected produce with hazardous pathogens.

VII. RECOMMENDED MA CONDITIONS FOR PRODUCE

The benefits of CA and MA packaging vary greatly according to the plant product. Storing some products, like apples, under low O_2 conditions can increase the storage period by months. Also some products, like carrots [96], do not respond positively to low O_2 or high CO_2 concentrations. In generally, altered gas conditions are regarded as positive only within a certain range of concentrations, so-called optimum concentrations.

Much research effort has been devoted to the determination of the optimum gas concentrations for individual products [62,75,85]. Table 1 gives a recent update on recommended storage conditions for a range of fruits and vegetables. Traditionally, lists of recommended storage conditions have been developed by national research organizations conducting extensive laboratory research [59]. A common experimental procedure is to store products under a range of O_2 and CO_2 concentrations and to monitor quality changes. The lower O_2 limit for stored fruits is accomplished empirically by lowering the storage O_2 concentration until intolerable damage occurs. Each commodity and new cultivar requires a large investment in time, equipment, and materials [49,98]. By repeating the trials year after year, it is possible to sense the importance of climatic variation on product behavior [59].

Some caution is needed in the application of optimal concentrations. For the various apple cultivars, for instance, the advised optima differ according to country [75]. Growing conditions like climate and orchard factors influence crop growth and contribute to these differences [21,30,74]. Although this is probably also the case other plant products, this is never specified.

Another important aspect of optimal values for temperature, O_2, and CO_2 concentrations is that they are often established separately, although interactions between temperature, O_2, and CO_2 concentrations (and probably also humidity) are known. Optimal O_2 concentrations are found to shift to a higher value when CO_2 concentrations are increased [8,60,73,93], when products are more mature [19,60,71,94], when they are kept at a higher temperature [7,65,84,93], or when products sensitive to chilling injury are stored at too low temperatures [73]. For apples the CO_2 limit at low O_2 concentrations decreases when the temperature is decreased [66]. The effect of relative humidity is often an interacting factor as a cause for, or in the symptom expression of, a disorder. Relative humidity, however, is often not mentioned and frequently not (accurately) measured [87]. The conclusion is that it is very hard to recommend absolute values for optimum O_2 and CO_2 concentrations or O_2 and CO_2 limits for a product without knowledge of other factors. The understanding of actual physiological processes determining the potential of products for MA is developing steadily and will help the further development of optimal and safe packaging techniques.

VIII. FUTURE OUTLOOK

Fruit and vegetable consumers increasingly demand high-quality products. An important quality feature is freshness. No signs of senescence, decay, wilting, or shriveling are accepted. In general consumers are willing to pay more for products with better quality. Often consumers also have specific expectations as to ripening stage. For instance, consumers in Northern Europe want firm, not mealy tomatoes (i.e., the fruits should not be too ripe) but also tomatoes with taste and flavor (i.e., the fruits should not be harvested in a very unripe stage). These quality demands result in strict criteria for storage conditions, transport conditions, and shelf conditions. MA packaging is a tool of increasing importance in meeting these criteria.

MA packages are increasingly being used. The total European MAP market, handles on average 300 million package units of produce per year with a market value of about $1 billion [22,32]. In the United States, a market share of about $8 billion in MAP products has been predicted for the year 2000. Often successful applications are due to good control of the whole distribution chain, from

TABLE 1 Recommended Optimal Storage Conditions

Commodity	Temperature (°C)	RH (%)	$[O_2]$ (%)	$[CO_2]$ (%)	Potential
Fruits					
Apple	1–4	90–95	1–3	0–6	A
Avocado	5–13	90	2–5	3–10	B
Banana	12–14	85–95	2–3	8	A
Blackberry	0–2	90–95	5–10	15–20	A
Blueberry	0–2	90–95	2–5	12–20	A
Cherry	0–2	85–90	3–10	10–15	B
Kiwi	0–2	85–90	1–2	3–5	A
Mango	10–15	90	3–7	5–8	B
Melon	8–10	85–90	3–5	5–15	B
Nectarine	0–2	85–90	1–2	3–5	B
Peach	0–2	85–90	1–2	3–5	B
Pear	0–1	90–95	2–3	0–2	A
Persimmon	0–5	90	3–5	5–8	B
Plum	0–2	85–90	1–2	0–5	B
Raspberry	0–2	85–90	5–10	15–20	A
Red Currant	0–2	90	5–10	15–20	A
Strawberry	0–2	90	5–10	12–20	A
Sweet corn	0–2	90	2–4	5–10	B
Tomato	1–13	90	3–5	2–4	B
Vegetables					
Artichoke	0–2	90	3–5	0–2	B
Asparagus (green)	0–2	95	10–15	7–12	A
Broccoli	0–1	90–95	2–3	8–12	A
Brussels sprouts	0–1	90–95	2–4	4–6	A
Cabbage	0–1	95	2–3	3–6	A
Celery (stem)	0–2	90–95	3–5	1–4	B
Chicory (witloof)	0–2	90–95	2–3	5–10	A
Leek	0–2	90–95	3–5	3–6	B
Lettuce	0–2	90–95	2–3	2–5	B
Mung bean	0–2	90–95	1–2	1–3	A
Onion	0–2	70–80	1–4	2–5	B
Spinach	0–2	95	21	10–20	B

Value based on Proceedings 6th and 7th International Controlled Atmosphere Research Conferences (Ithaca 1993 and Davis 1997). Potential; A = excellent, B = fair. Products with a low potential or no potential are not listed.
Source: Adapted from Ref. 61.

the moment of packing the product until it is displayed on the retail shelf. When film permeabilities or respiration rates are not well characterized, package designers have to resort to empirical studies with MA packages that can be described as "pack and pray" [25]. A thorough understanding of principles and processes will lead to a more rational selection of packaging materials [60]. Nevertheless, even a package that has been well designed in a laboratory may not necessarily perform well in practice, when no information on the actual storage, transport, or shelf-life conditions were considered at the design stage.

Current state-of-the-art MAP systems for minimally processed produce have been optimized mainly for product quality. Safety and cost aspects have not yet been optimized. Also, quality deterioration still occurs, and further improvements can still be achieved through, for example:

A systematic approach to select appropriate gas conditions in MAP for specific products
Availability of more data on the interaction between produce and gas composition
Development of better computer software to aid in the selection of suitable packaging systems (gas compositions plus foils) for use under dynamic conditions (temperature, humidity)
Combating microbial hazards in MAP systems (i.e., psychrotrophic pathogens) using improved MAP systems or new hurdles to microbial growth
Use of "smart" films that compensate for temperature fluctuations by changing permeability properties
Studies on more environmentally friendly MAP systems (e.g., simple foils, biodegradable foils)
Minimizing packaging in MAP systems (including biocoatings as part of the packaging concept)

Table 2 lists these and other trends foreseen in modified atmosphere packaging.

A misconception of MAP is that it can overcome hygienic abuses in the production or handling of a product. MAP is not a panacea for the preservation of food products, but, if used correctly, it slows the natural deterioration of a product. There is no enhancement of product quality but, when starting with a good clean product, the initial fresh state of the product may be prolonged. Strict codes of practice should be enforced to ensure the maximum quality shelf life and safety of MA-packed foods.

With the increasing importance of MA worldwide, the role of temperature and of its control in the successful use of MAP should be considered. MA packages will not be a substitute for adequate temperature management. For all nontropical products only cooling is always better than only MA. But also when only MA can applied in cases where cooling is not possible, stable temperatures are necessary since no films are yet available that respond to temperature fluctuations of the packed product [40].

Modern consumer demand for convenient and fresh, wholesome produce together with the recent reorganization of the distribution chain will further stimulate the use and the broadening of the application area of MAP. Continued basic physiological and microbiological research on the action of MAs will minimize the risks of loss and safety associated with the use of MAs and will allow for

Table 2 New Developments in Modified Atmosphere Packaging

Packaging	Active packaging
Tailoring gas transmission/selectivity (new plastics, microperforation)	Absorbers (O_2, ethylene, water, off-flavors)
Water vapor transmission/selectivity (modified humidity storage)	Generators (CO_2, antifungals, flavors)
Biodegradability—environment (multi/mixed layered plastics fortified with biodegradable mass)	Controlled release (antimicrobials, antioxidants)
Composites (metals/cartons with plastics/liners)	Dynamic packaging
New concepts (high oxygen, noble gases, optimization)	Temperature dynamic films
Chain optimization (controlling basics: chilling, handling, logistics)	Humidity dynamic films
Minimal packaging	Biopackaging
Integrating functionalities	Biodegradable/Edible films
Simpler/Less films, better recyclabe films	Custom-made physical properties
	Biocoatings
	Edible, physical protection (invisible)
	Functional features (antimicrobials)

faster MAP optimization using models. New technological developments, especially in the area of more suitable packaging films, will contribute to the success of MA as a preservation technique. The main limiting factors for further expansion of MAP may arise out of environmental concerns. A solution for the huge waste problem, in the long run, may be found in the use of edible and biodegradable films that are able to create a modified atmosphere [51]. The integration of different preservative hurdles, such as refrigeration, MAP, active components, and/or green chemicals in accordance with the concept of combined processing [70], may not only minimize potential microbial problems but also contribute to optimized product quality. The range of powerful technologies we have at our disposal today will help to ensure a good supply of minimally processed, fresh, safe, and ready-to-eat products.

REFERENCES

1. U. M. Abdul-Raouf, L. R. Beuchat, and M. S. Ammar, Survival and growth of *Escherichia coli* O157:H7 on salad vegetables, *Appl. Environ. Microbiol. 59*:1999 (1993).
2. R. Ahvenainen, New approaches in improving the shelf life of minimally processed fruit and vegetables, *Trends Food Sci. Technol. 7*:179 (1996).
3. S. A. Aytaç and L. G. M Gorris, Survival of *Aeromonas hydrophila* and *Listeria monocytogenes* on fresh vegetables stored under moderate vacuum, *World J. Microbiol. Biotechnol. 10*:670 (1994).
4. R. C. Baker, R. A. Qureshim, and J. H. Hotchkiss, Effect of an elevated level of carbon dioxide containing atmosphere on the growth of spoilage and pathogenic bacteria at 2, 7, and 13-degrees-C, *Poultry Sci. 65*:729 (1986).
5. N. H. Banks, B. K. Dadzie, and D. J. Cleland, Reducing gas exchange of fruits with surface coatings, *Postharv. Biol Technol. 3*:269 (1993).
6. C. R. Barmore, Packaging technology for fresh and minimally processed fruits and vegetables, *J. Food Qual. 10*:207 (1987).
7. R. M. Beaudry, A. C. Cameron, A. Shirazi, and D. L. Dostal-Lange, Modified-atmosphere packaging of blueberry fruit: effect of temperature on package O_2 and CO_2, *J. Am. Soc. Hort. Sci. 117*:436 (1992).
8. R. M. Beaudry, Effect of carbon dioxide partial pressure on blueberry fruit respiration and respiratory quotient, *Postharv. Biol. Technol. 3*:249 (1993).
9. S. Ben-Yehoushua, B. Shapiro, Z. E. Chen, and S. Lurie, Mode of action of plastic film in extending life of lemon and bell pepper fruits by alleviation of water stress, *Plant Physiol. 73*:87 (1983).
10. M. H. J. Bennik, E. J. Smid, F. M. Rombouts, and L. G. M. Gorris, Growth of psychrotrophic foodborne pathogens in a solid surface model system under the influence of carbon dioxide and oxygen, *Food Microbiol. 12*:509 (1995).
11. M. H. J. Bennik, H. W. Peppelenbos, C. Nguyen-the, F. Carlin, E. J. Smid and L. G. M. Gorris, Microbiology of minimally processed, modified atmosphere packaged chicory endive, *Postharv. Biol. Technol. 9*:209 (1996).
12. J. E. Berard, Memoire sur la maturation des fruits, *Ann. Chim. Phys. 16*:152 (1819).
13. M. E. Berrang, R. E. Brackett, and L. R. Beuchat, Growth of *Listeria monocytogenes* on fresh vegetables stored under controlled atmosphere, *J. Food Prot. 52*:702 (1989).
14. M. E. Berrang, R. E. Brackett, and L. R. Beuchat, Growth of *Aeromonas hydrophila* on fresh vegetables stored under a controlled atmosphere. *Appl. Environ. Microbiol. 55*:2167 (1989).
15. L. B. Beuchat, Pathogenic bacteria associated with fresh produce, *J. Food Prot. 59*:204 (1995).
16. L. R. Beuchat and R. E. Brackett, Survival and growth of *Listeria monocytogenes* on lettuce as influenced by shredding, chlorine treatment, modified atmosphere packaging and temperature, *J. Food Sci. 55*:755,870 (1990).
17. L. R. Beuchat and R. E. Brackett, Behavior of *Listeria monocytogenes* inoculated into raw tomatoes and processed tomato products, *Appl. Environ. Microbiol. 57*:1367 (1991).
18. E. Blickstad, S. O. Enfors, and G. Molin, Effect of high concentrations of CO_2 on the microbial flora of pork stored at 4-degrees-C and 14-degrees-C, *Psychrotrophic Microorganisms in Spoil-*

age and Pathogenicity (T. A. Roberts, G. Hobbs, and J. H. B. Christian, eds.), Academic Press, 1981.
19. M. R. Boersig, A. A. Kader, and R. J. Romani., Aerobic-anaerobic respiratory transition in pear fruit and cultured pear fruit cells, *J. Am. Soc. Hort. Sci. 113*:869 (1988).
20. R. E. Brackett, Influence of modified atmosphere packaging on the microflora and quality of fresh bell peppers, *J. Food Prot. 53*:255 (1990).
21. W. J. Bramlage, M. Drake, and W. J. Lord, The influence of mineral nutrition on the quality and storage performance of pome fruits grown in North America, *Acta Hort. 92*:29 (1980).
22. A. L. Brody, A perspective on MAP products in North America and Western Europe, *Principles of Modified-Atmosphere and Sous Vide Product Packaging* (J. M. Farber and K. L. Dodds, eds.), 1995, p. 13.
23. S. P. Burg, and E. A. Burg, Fruit storage at subatmospheric pressures, *Science 153*:314 (1966).
24. S. P. Burg, Hypobaric storage and transportation of fresh fruits and vegetables, *Postharvest Biology and Handling of Fruits and Vegetables* (N. F. Haard and D. K. Salunkhe, eds.), Avi Publ. Co., Westport, CT, 1975, p. 172.
25. A. C. Cameron, P. C. Talasila, and D. W. Joles, Predicting film permeability needs for modified atmosphere packaging of lightly processed fruits and vegetables, *HortScience 30*:25 (1995).
26. A. C. Cameron, B. D. Patterson, P. C. Talasila, and D. W. Joles, Modeling the risk in modified-atmosphere packaging: a case for sense and respond packaging, Proc. 6th Nat. Contr. Atm. Res. Conf., Ithaca, NY, June 15-17, 1993, p. 95.
27. F. Carlin and C. Nguyen-the, Fate of *Listeria monocytogenes* on four types of minimally processed green salads, *Lett. Appl. Microbiol. 18*:222 (1994).
28. F. Carlin, C. Nguyen-the, and A. Abreu da Silva, Factors affecting the fate of *Listeria monocytogenes* on minimally processed fresh endive, *J. Appl. Bacteriol. 78*:636 (1995).
29. F. Carlin, C. Nguyen-the, A. Abreu da Silva, and C. Cochet, Effect of carbon dioxide on the fate of *Listeria monocytogenes*, aerobic bacteria and on the development of spoilage in minimally processed fresh endive, *Int. J. Food Microbiol. 32*:159 (1996).
30. P. M. Chen, D. M. Borgic, D. Sugar, and W. M. Mellenthin, Influence of fruits maturity and growing district on brown core disorder of Bartlett pears, *HortScience 21*:1172 (1986).
31. N. Church, Developments in modified-atmosphere packaging and related technologies, *Trends Food Sci. Technol. 5*(11):345 (1994).
32. B. P. F. Day and L. G. M. Gorris, Modified atmosphere packaging of fresh produce on the West-European market, *Int. J. Food Techn. Mark. Pack. Anal. 44*:32 (1993).
33. C. Diaz and J. H. Hotchkiss, Comparative growth of *Escherichia coli* O157:H7, spoilage organisms and shelf-life of shredded iceberg lettuce stored under modified atmospheres, *J. Sci. Food Agric. 70*:433 (1996).
34. T. Eklund, The effect of CO_2 on microbial growth and on uptake processes in bacterial membrane vesicles, *Int. J. Food Microbiol. 1*:179 (1984).
35. T. Eklund and J. Jarmund, Microculture model studies on the effect of various gas atmospheres on microbial growth at different temperatures, *J. Appl. Bacteriol. 55*:119 (1983).
36. S. O. Enfors and G. Molin, The influence of high concentrations of CO_2 on the germination of bacterial spores, *J. Appl. Bacteriol. 45*:279 (1978).
37. S. O. Enfors and G. Molin, Effect of high concentrations of CO_2 on growth rate of *Pseudomonas fragi, Bacillus cereus*, and *Streptococcus cremoris, J. Appl. Bacteriol. 48*:409 (1980).
38. S. O. Enfors and G. Molin, The influence of temperature on the growth inhibitory effect of CO_2 on *Pseudomonas fragi* and *Bacillus cereus, Can. J. Microbiol. 27*:15 (1981).
39. S. O. Enfors and G. Molin, The effect of different gases on the activity of microorganisms, *Psychrotrophic Microorganisms in Spoilage and Pathogenicity* (T. A. Roberts, G. Hobbs, and J. H. B. Christian, eds.), Academic Press, 1981,
40. A. Exama, J. Arul, R. W. Lencki, L. Z. Lee, and C. Toupin, Suitability of plastic films for modified atmosphere packaging of fruits and vegetables, *J. Food Sci. 58*:1365 (1993).
41. J. M. Farber, Microbiological aspects of modified-atmosphere packaging technology - a review. *J. Food Prot. 54*:58 (1991).

42. J. D. Floros, Controlled and modified atmospheres in food packaging and storage, *Chem. Eng. Progr. 6*:25 (1990).
43. G. A. Francis and D. O'Beirne, Effect of gas atmosphere, antimicrobial dip and temperature on the fate of *Listeria innocua* and *Listeria monocytogenes* on minimally processed lettuce, *Int. J. Food Sci. Technol. 32*:141 (1997).
44. R. M. Garcia-Gimeno, G. Zurera-Cosano, and G. Amaro-Lopez, Incidence, survival and growth of *Listeria monocytogenes* in ready-to-use mixed vegetable salads in Spain, *J. Food Safety 16*:75 (1996).
45. R. M. Garcia-Gimeno, M. D. Sanchez-Pozo, M. A. Amaro-Lopez, and G. Zurera-Cosano, Behaviour of *Aeromonas hydrophila* in vegetable salads stored under modified atmosphere at 4 and 15 degree C, *Food Microbiol. 13*:369 (1996).
46. D. A. Golden, E. J. Rhodehamel, and D. A. Kautter, Growth of *Salmonella* spp. in cantaloupe, watermelon and honeydew melons, *J. Food Prot. 56*:194 (1993).
47. L. G. M. Gorris and H. W. Peppelenbos, Modified atmosphere and vacuum packaging to extend the shelf life of respiring food products, *HortTechnology 2*:303 (1992).
48. L. G. M. Gorris, Y. de Witte, and E. J. Smid, Storage under moderate vacuum to prolong the keepability of fresh vegetables and fruits, *Acta Hort. 368*:479 (1994).
49. C. D. Gran and R. M. Beaudry, Determination of the low oxygen limit for several commercial apple cultivars by respiratory quotient breakpoint, *Postharv. Biol. Technol. 3*:259 (1993).
50. W. Grierson and W. F. Wardowski, Relative humidity effects on the postharvest lifes of fruits and vegetables, *HortScience 13*:22 (1978).
51. S. N. Guilbert, N. Gontard, and L. G. M. Gorris, Prolongation of the shelf-life of perishable food products using biodegradable films and coatings, *Lebensm. Wiss. Technol. 29*:10 (1996).
52. Y. Y. Hao and R. E. Brackett, influence of modified atmosphere on growth of vegetable spoilage bacteria in media, *J. Food Prot. 56*:223 (1993).
53. S. M. Harmon and D. A. Kautter, Incidence and growth potential of *Bacillus cereus* in ready-to-serve foods, *J. Food Prot. 54*:372 (1991).
54. C. W. Hedberg, K. L. MacDonald, and M. T. Osterholm, Changing epidemiology of foodborne disease: a Minnesota perspective, *Clin. Infect. Dis. 18*:671 (1994).
55. Y. S. Henig, Storage stability and quality of produce packaged in polymeric films, *Symposium: Postharvest Biology and Handling of Fruits and Vegetables* (N. F. Haard and D. K. Salunkhe, eds.), Avi Publ Co., 1975, p. 144.
56. M. L. A. T. M. Hertog, H. W. Peppelenbos, L. M. M. Tijskens, and R. G. Evelo, Modified atmosphere packaging: optimisation through simulation, Proc. 7th Int. Contr. Atm. Res. Conf., 13-18 July 1997, Davis, CA.
57. C. B. Hintlian and J. H. Hotchkiss, The safety of modified atmosphere packaging: a review, *Food Technol. 40*(12):70 (1986).
58. J. H. Hotchkiss and M. J. Banco, influence of new packaging technologies on the growth of microorganisms in produce, *J. Food Prot. 55*:815 (1992).
59. J. Jameson, CA storage technology—recent developments and future potential, Proc. of COST94 workshop, 22-23 April 1993, Milan 1995, p. 1.
60. A. A. Kader, D. Zagory, and E. L. Kerbel, Modified atmosphere packaging of fruits and vegetables, *Crit. Rev. Food Sci. Nutr. 28*:1 (1989).
61. A. A. Kader, Modified atmospheres during transport and storage, *Postharvest Technology of Horticultural Crops* (A. A. Kader ed.), publ. Nr. 3311, Univ. Calif., 1992, p. 85.
62. A. A. Kader, A. summary of CA and MA requirements and recommendations for fruits other than pome fruits, Proc. 6th Int. Contr. Atm. Res. Conf., Ithaca, NY, June 15-17, 1993, p. 859.
63. D. A. Kautter, T. Lilly Jr., and R. Lynt, Evaluation of the botulism hazard in fresh mushrooms wrapped in commercial polyvinylchloride film, *J. Food Prot. 55*:372 (1991).
64. S. J. Kays, *Postharvest Physiology of Perishable Plant Products*, AVI, Van Nostrand Reinhold, New York, 1991.
65. D. Ke, L. Goldstein, M. O'Mahony, and A. A. Kader. Effects of short-term exposure to low O_2 and CO_2 atmospheres on quality attributes of strawberries, *J. Food Sci. 56*:50 (1991).

66. F. Kidd and C. West, Brown heart, a functional disease of apples and pears, Special report no. 12, Food Inv. Board, Dep. Sci. Ind. Res., 1923, p. 1.
67. M. Larsen and C. B. Watkins, Firmness and concentrations of acetaldehyde, ethyl acetate and ethanol in strawberries stored in controlled and modified atmospheres, *Postharv. Biol. Technol. 5*:39 (1995).
68. D. S. Lee, P. E. Haggar, J. Lee, and K. L. Yam, Model for fresh produce respiration in modified atmospheres based on principles of enzyme kinetics, *J. Food Sci. 56*:1580 (1991).
69. L. Lee, J. Arul, R. Lencki, and F. Castaigne, A review on modified atmosphere packaging and preservation of fresh fruits and vegetables: physiological basis and practical aspects—Part I, *Pack. Technol. Sci. 8*:315 (1995).
70. L. Leistner and L. G. M. Gorris, Food preservation by hurdle technology, *Trends Food Sci. Technol. 6*:41 (1995).
71. P. D. Lidster, G. D. Blanpied, and E. C. Lougheed, Factors affecting the progressive development of low-oxygen injury in apples, Proc. 4th Nat. Contr. Atm. Res. Conf., Raleigh, NC, 1985, p. 57.
72. T. Lilly Jr., H. M. Solomon, and E. J. Rhodehamel, Incidence of *Clostridium botulinum* in vegetables packaged under vacuum or modified atmosphere, *J. Food Prot. 59*:59 (1996).
73. E. C. Lougheed, Interactions of oxygen, carbon dioxide, temperature and ethylene that may induce injuries in vegetables, *HortScience 22*:791 (1987).
74. M. T. Luton and D. A. Holland, The effects of preharvest factors on the quality of stored conference pears. I. Effects of orchard factors, *J. Hort. Sci. 61*:23 (1986).
75. M. Meheriuk, CA storage conditions for apples, pears and nashi, Proc. 6th Int. Contr. Atm. Res. Conf., June 15-17 1993, Ithaca, NY, p. 819.
76. G. Molin, The resistance to CO_2 of some food related bacteria, *Eur. J. Appl. Microbiol. Biotechnol. 18*:214 (1983).
77. C. Nguyen-the and F. Carlin, The microbiology of minimally processed fresh fruits and vegetables, *Crit. Rev. Food Sci. Nutr. 34*:371 (1994).
78. H. W. Peppelenbos and J. van 't Leven, Evaluation of four types of inhibition for modelling the influence of carbon dioxide on oxygen consumption of fruits and vegetables, *Postharv. Biol. Technol. 7*:27 (1996).
79. H. W. Peppelenbos, L. M. M. Tijskens, J. van 't Leven, and E. C. Wilkinson, Modelling oxidative and fermentative carbon dioxide production of fruits and vegetables, *Postharv. Biol. Technol. 9*:283 (1996).
80. H. W. Peppelenbos, H. Zuckermann, and S. Robat, Alcoholic fermentation of apple fruits at various oxygen concentrations. Model prediction and photoacoustic detection, Proc. 7th Int. Contr. Atm. Res. Conf., Davis, CA, 1997,
81. P. Perata and A. Alpi, Plant responses to anaerobiosis, *Plant Sic. 93*:1 (1993).
82. C. A. Phillips, Review: modified atmosphere packaging and its effects on the microbiological quality and safety of produce, *Int. J. Food Sci. Technol. 31*:463 (1996).
83. V. Rodov, S. Ben-Yehoshua, T. Fierman, and D. Fang, Modified-humidity packaging reduces decay of harvested bell pepper fruit, *HortScience 30*:299 (1995).
84. M. E. Saltveit and W. E. Ballinger, Effects of anaerobic nitrogen and carbon dioxide atmospheres on ethanol production and postharvest quality of 'Carlos' grapes, *J. Am. Soc. Hort. Sci. 108*:462 (1983).
85. M. E. Saltveit, A summary of CA and MA requirements and recommendations for the storage of harvested vegetables, Proc. 6th Int. Contr. Atm. Res. Conf., June 15-17, 1993, Ithaca, NY, p. 800.
86. W. F. Schlech III, P. M. Lavigne, R. A. Bortolussi, A. C. Allen, E. V. Haldane, A. J. Wort, A. W. Hightower, S. E. Johnson, S. H. King, E. S. Nicholls, and C. V. Broome, Epidemic listeriosis-evidence for transmission by food, *N. Engl. J. Med. 308*:203 (1983).
87. S. P. Schouten, R. K. Prange, J. Verschoor, T. R. Lammers, and J. Oosterhaven, Improvement of quality of Elstar apples by dynamic control of ULO conditions, Proc. 7th Int. Contr. Atm. Res. Conf., 13-18 July, 1997, Davis, CA,

88. A. Shirazi and A. C. Cameron, Controlling relative humidity in modified atmosphere packages of tomato fruit, *HortScience 27*:336 (1992).
89. H. M. Solomon, D. A. Kautter, T. Lilly, and E. J. Rhodehamel, Outgrowth of *Clostridium botulinum* in shredded cabbage at room temperature under a modified atmosphere, *J. Food Prot. 53*:831 (1990).
90. H. Sugiyama and K. H. Yang, Growth potential of *Clostridium botulinum* in fresh mushrooms packaged with semipermeable plastic film, *Appl. Microbiol. 30*:964 (1975).
91. P. C. Talasila, A. C. Cameron, and D. W. Joles, Frequency distribution of steady-state oxygen partial pressures in modified-atmosphere packages of cut broccoli, *J. Am. Soc. Hort. Sci. 119*:556 (1994).
92. P. S. Taoukis and T. P. Labuza, Applicability of time-temperature indicators as shelf life monitors of food products, *J. Food Sci. 54*:783 (1989).
93. M. Thomas, A quantative study of the production of ethyl alcohol and acetaldehyde by cells of the higher plants in relation to concentration of oxygen and carbon dioxide, *Biochem. J. 19*:927 (1925).
94. M. Thomas and J. C. Fidler, Zymasis by apples in relation to oxygen concentration, *Biochem. J. 27*:1629 (1933).
95. P. Varoquaux, G. Albagnac, C. Nguyen-the, and F. Varoquaux, Modified atmosphere packaging of fresh bean sprouts, *J. Sci. Food Agric. 70*:224 (1996).
96. J. Weichmann, Physiological response of root crops to controlled atmospheres. Proc. 2nd Nat. Contr. Atm. Res. Conf., April 5-7, 1977, East Lansing, MI, 1977, p. 667.
97. J. H. Wells and R. P. Singh, Application of time-temperature indicators in monitoring changes in quality attributes of perishable and semiperishable foods, *J. Food Sci. 53*:148 (1988).
98. A. S. Wollin, C. R. Little, and J. S. Packer, Dynamic control of storage atmospheres, Proc. 4th Nat. Contr. Atm. Res. Conf., 1985, Raleigh, NC, p. 308.
99. R. C. Wood, C. Hedberg, and K. White, A multistate outbreak of *Salmonella javiana* infections associated with raw tomatoes, CDC Epidemic Intelligence Service, 40th Ann. Conf. Atlanta, U.S. Dept. of Health and Human Services, Public Health Service, 1991, p. 69.
100. Zagory and A. A. Kader, Modified atmosphere packaging of fresh produce, *Food Technol. 42*:70 (1988).
101. D. Zagory, J. D. Mannapperuma, A. A. Kader, and R. P. Singh, Use of a computer model in the design of modified atmosphere packages for fresh fruits and vegetables, Proc. 5th Int. Contr. Atm. Res. Conf., Wenatchee, June 14-16, 1989, p. 479.

16

Combined Methods for Food Preservation

LOTHAR LEISTNER
International Food Consultant, Kulmbach, Germany

I. INTRODUCTION

The microbial stability and safety of most traditional and novel foods is based on a combination of several preservative factors (called *hurdles*), which microorganisms present in the food are unable to overcome. This is illustrated by the so-called hurdle effect, first introduced by Leistner [1]. The hurdle effect is of fundamental importance for the preservation of foods, since the hurdles in a stable product control microbial spoilage, food-poisoning, as well as desired fermentation processes [1,2]. Leistner and coworkers acknowledged that the hurdle concept illustrates only the well-known fact that complex interactions of temperature, water activity, pH, redox potential, etc. are significant for the microbial stability of foods. From an understanding of the hurdle effect, hurdle technology [3] was derived, which allows improvements in the safety and quality of foods using deliberate and intelligent combinations of hurdles. Over the years insight into the hurdle effect has been broadened and the application of hurdle technology extended. In industrialized countries hurdle technology is currently of particular interest for minimally processed foods, whereas in developing countries foods storable without refrigeration, due to stabilization by hurdle technology, are at present of paramount importance. The application of deliberate and intelligent hurdle technology is increasing rapidly worldwide. This concept is also referred to as food preservation by combined methods, combined processes, combination preservation, or combination techniques. At present, the term hurdle technology is most often used.

In Europe, a 3-year research project on food preservation by combined processes, supported by the European Commission, to which scientists from 11 European countries have contributed, fostered the application of hurdle technology [4,5]. The hurdle technology concept proved successful, since an intelligent combination of hurdles secures microbial stability and safety as well as the sensory quality of foods [6–8], it provides convenient and fresh foods to the consumers, and it is cost-efficient for producers since it demands less energy during production and storage.

II. PRINCIPLES OF COMBINED PRESERVATION METHODS

Many preservation methods are used for making foods stable and safe, e.g., heating, chilling, freezing, freeze-drying, drying, curing, salting, sugar addition, acidification, fermentation, smoking, oxygen removal. However, these processes are based on relatively few parameters or hurdles, i.e., high temperature (F value), low temperature (t value), water activity (a_w), acidification (pH), redox potential (Eh), preservatives, and competitive flora. In some of the preservation methods mentioned, these parameters are of major importance, in others they are only secondary hurdles [2,6].

The critical values of these parameters for the death, survival or growth of microorganisms in foods have been determined in recent decades and are now the basis of food preservation. However, it must be kept in mind that the critical value of a particular parameter changes if other preservative factors are present in the food. For instance, the heat resistance of bacteria increases at low a_w and decreases in the presence of some preservatives, whereas a low Eh increases the inhibition of microorganisms caused by a reduced a_w. The simultaneous effect of different preservative factors could be additive or even synergistic. Furthermore, as mentioned earlier, the microbial stability and safety of many foods is based on the combined effects of hurdles. For instance, mildly heated canned foods ("half-preserved" or "three-quarter-preserved") need refrigeration during storage, and fermented sausages are only stable and safe if both the a_w and the pH are in an appropriate range.

Therefore, in food preservation the combined effect of preservative factors must be taken into account, which is illustrated by the hurdle effect.

A. Hurdle Effect

For each stable and safe food a certain set of hurdles is inherent, which differs in quality and intensity depending on the particular product. In any case the hurdles must keep the "normal" population of microorganisms in this food under control. The microorganisms present (at the start) in a food should not be able to overcome (leap over) the hurdles present, otherwise the food will spoil or even cause food-poisoning. In previous publications [6,7] some examples were given to illustrate the hurdle effect. Since these examples are quite helpful for an understanding of the hurdle effect, as well as for the applications of combined methods in food preservation, they are briefly repeated here.

Figure 1 shows eight examples of the hurdle effect. Example No. 1 represents a food that contains six hurdles: high temperature during processing (F value), low temperature during storage (t value), water activity (a_w), acidity (pH), redox potential (Eh), and preservatives (pres.). The microorganisms present cannot overcome these hurdles, and thus the food is microbiologically stable and safe. However, No. 1 is only a theoretical case, because all hurdles are of the same height, i.e., have the same intensity, and this rarely occurs. A more likely situation is presented in No. 2, since the microbial stability of this product is based on hurdles of different intensity. In this particular product the main hurdles are a_w and preservatives, whereas other less important hurdles are storage temperature, pH, and redox potential. These five hurdles are sufficient to inhibit the usual types and numbers of microorganisms associated with such a product. If there are only a few microorganisms present at the start (No. 3), then a few or only low hurdles are sufficient for the microbial stability of the product. The ultra-clean or aseptic processing of perishable foods is based on this principle. The same proves true if the initial microbial load of a food (e.g., high-moisture fruits or carcass meat) is substantially reduced (e.g., by application of steam), because after such a reduction fewer microorganisms are present at the start, which are more easy to inhibit. On the other hand, if due to bad hygienic conditions too many undesirable microorganisms are initially present (No. 4), even the usual inherent hurdles in a product may be unable to prevent spoilage or food-poisoning. Example No. 5 is a food rich in nutrients (N) and vitamins (V), which foster the growth of microorganisms (called the booster or trampoline effect), and thus the hurdles in such a product must be enhanced, otherwise they will be overcome. Example No. 6 illustrates the behavior of sublethally damaged microorganisms in food. If, for instance, bacterial spores in meat products are damaged sublethally by heat (see Sec. IV.B), then the vegetative cells derived from such spores lack "vitality," and are therefore inhibited by fewer or lower hurdles. In some foods stability is achieved during processing by a sequence of hurdles that are important in different stages of the ripening process and lead to a stable final prod-

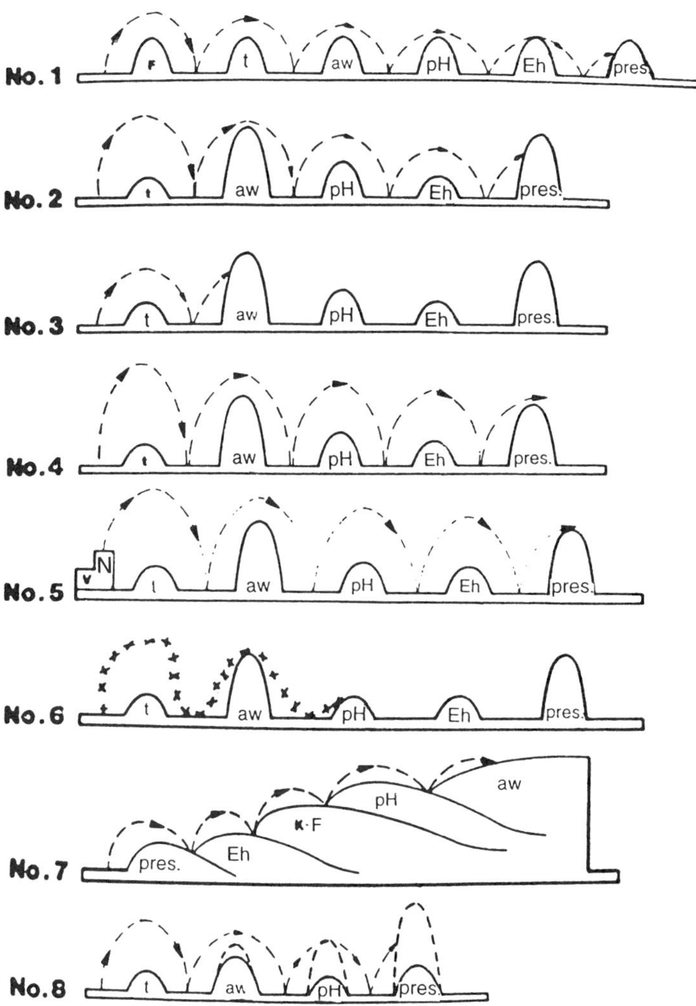

FIGURE 1 Illustration of the hurdle effect, using eight examples. See text for details. (From Refs. 6,19.)

uct. Example No. 7 illustrates the sequence of hurdles in fermented sausages (see Sec. IV.B). Finally, No. 8 illustrates the possible synergistic effect of hurdles, which probably relates to a multitarget disturbance of the homeostasis of microorganisms in foods (discussed in Sec. III.D).

B. Hurdle Technology

A better understanding of the occurrence and interaction of different preservative factors (hurdles) in foods is the basis for improvements in food preservation. If the hurdles for a food are known, the microbial stability and safety of this food might be optimized by changing the intensity or quality of these hurdles. From an understanding of the hurdle effect, hurdle technology has been derived [3], which means that hurdles are deliberately combined in the preservation of traditional and novel foods. Using an intelligent mix of hurdles it is possible to improve not only the microbial stability and safety but also the sensory and nutritive quality as well as the economic aspects of a food. It is important that the water content in the product be compatible with its microbial stability, and if an increased a_w is compensated by other hurdles (pH, Eh, etc.) this food becomes more economical. Even the pet food industry now employs this principle. Stable pet food was formerly produced with an a_w of 0.85, which required addition of excessive amounts of propylene glycol, which might have adversely affected the pet's health. But now, due to the application of hurdle technology, pet foods are microbiologically stable at ambient temperatures with an a_w of 0.94, and they are more healthy, tasty, and economical [6]. Hurdle technology is increasingly used for food design in industrialized and developing countries for optimizing traditional foods and for making new products according to needs. For instance, if energy preservation is the goal, then energy-consuming hurdles such as refrigeration are replaced by other hurdles (a_w, pH, or Eh) that do not demand energy but still ensure a stable and safe food [1]. Furthermore, if we want to reduce or replace preservatives (e.g., nitrite) in meats, we could emphasize other hurdles in the food (e.g., a_w, pH, refrigeration, or competitive flora), which would stabilize the product [9]. More recent examples related to the application of hurdle technology will be give in subsequent sections of this chapter.

C. Total Quality

Stanley [10] proposed that the hurdle technology approach could be applicable to a wider concept of food preservation than just microbial stability, but that, in order for it to work, a precise knowledge of the effectiveness of each hurdle for a given commodity would be required. Furthermore, he suggested distinguishing between positive and negative hurdles for the quality of foods. Certainly hurdle technology is applicable not only to safety, but also to quality aspects of foods, although this area of knowledge has been much less explored than the safety aspect. McKenna [11] emphasized that while hurdle technology is appropriate for securing the microbial stability and safety of foods, the total quality of foods is a much broader field and encompasses a wide range of physical, biological, and chemical attributes. The concept of combined processes should work towards the total quality of foods rather than the narrow but important aspects of microbial stability and safety. But at present the tools for applying hurdle technology to total food quality are still not adequate, and this is equally true for predicting food quality by modeling. However, researchers should appreciate the wider power of the hurdle technology concept, and the food industry should use the available tools of combined processes for as many quality enhancements as possible [11].

Some hurdles, e.g., Maillard reaction products, influence the safety as well as the quality of foods, because they have antimicrobial properties and at the same time improve the flavor of the products; this also applies to nitrite used in the curing of meat. The possible hurdles in foods can influence the stability and safety, as well as the sensory, nutritive, technological, and economic properties of a product, and they may be negative or positive in securing the desired total quality of a food. Moreover, the same hurdle could have a positive or a negative effect on foods, depending on its intensity. For instance, chilling to an unsuitable low temperature is detrimental to fruit quality (chilling injury), whereas moderate chilling is beneficial. Another example is the pH of fermented sausages, which should be low enough to inhibit pathogenic bacteria, but not so low as to impair taste. If the intensity of a particular hurdle in a food is too small it should be strengthened; on the other hand, if it is detrimental to the total food quality it should be lowered. By this adjustment, the hurdles in foods should be kept in the optimal range, considering safety as well as quality [7,12].

D. Potential Hurdles

For the advanced application of hurdle technology a continually increasing number of preservative factors (hurdles) has become available (Table 1). The most important hurdles in common use for the preservation of foods, applied as either process or additive hurdles, are temperature (high or low), water activity (a_w), acidity (pH), low redox potential (Eh), preservatives (e.g., nitrite, sorbate, sulfite), and competitive microorganisms (e.g., lactic acid bacteria). In addition, more than 50 hurdles of potential use for foods of animal or plant origin, which improve the stability and/or the quality of these products, have been identified and described [12,13], and the list is by no means complete. At present, physical, nonthermal processes (high hydrostatic pressure, mano-thermo-sonication, oscillating magnetic fields, pulsed electric fields, light pulses, etc.) are receiving considerable attention, since

TABLE 1 Potential Hurdles for Foods of Animal or Plant Origin to Improve the Stability and/or the Quality of These Products

Temperature (low or high)
pH (low or high)
a_w (low or high)
Eh (low or high)
Modified atmosphere (nitrogen, carbon dioxide, oxygen, etc.)
Packaging (aseptic packaging, vacuum or modified-atmosphere or active packaging, edible coatings, etc.)
Pressure (high)
Radiation (microwaves, UV, irradiation, etc.)
Other physical processes (mano-thermo-sonication, high electric field pulses, oscillating magnetic field pulses, radiofrequency energy, photodynamic inactivation, etc.)
Microstructure (emulsions, fermented sausage, ripened cheese, etc.)
Competitive flora (lactic acid bacteria, etc.)
Preservatives (organic acids, lactate, acetate, sorbate, ascorbate, glucono-delta-lactone, phosphates, propylene glycol, diphenyl, parabens, free fatty acids and their esters, phenols, monolaurin, chelators, Maillard reaction products, ethanol, spices and their extracts, nitrite, nitrate, sulfite, carbon dioxide, oxygen, ozone, chlorine, smoke, antioxidants, pimaricin and other antibiotics, lysozyme, chitosan, lactoperoxidase, nisin and other bacteriocins, pectine hydrolysate, protamin, hop extracts, etc.)

Source: Adapted from Refs. 12,13.

in combination with conventional hurdles they are of potential use for the microbial stabilization of fresh-like food products with little induced degradation of sensory and nutritional properties. These novel processes are used to achieve a reduction of the microbial load, while the growth of any remaining microorganisms is inhibited by additional, conventional hurdles. Another group of hurdles of special interest at present in industrialized as well as in developing countries are "natural preservatives" (spice extracts, lysozyme, chitosan, pectine hydrolysate, protamine, paprika glycoprotein, hop extracts, etc.). Moreover, the microstructure of some foods (e.g., emulsions, fermented sausages, ripened cheese), which might be a considerable hurdle to microbial growth, is the subject of research (see Sec. IV.B). However, not all of the potential hurdles for food preservation will be commonly applied, and certainly not to the same food product.

III. BASIC ASPECTS OF FOOD PRESERVATION

Food preservation implies exposing microorganisms to a hostile environment in order to inhibit their growth, shorten their survival, or cause their death. The feasible responses of the microorganisms to such a hostile environment determine whether they grow or die. More basic research is needed in this area, because a better understanding of the physiological basis for the growth, survival, and death of microorganisms in food products could open new dimensions for food preservation [7]. Furthermore, such an understanding would be the scientific basis for an efficient application of hurdle technology in the preservation of foods. Recent advances have been made by considering the homeostasis, metabolic exhaustion, and stress reactions of microorganisms, as well as by introducing the concept of multitarget preservation for gentle yet effective preservation of foods [8,14].

A. Homeostasis

A key phenomenon that deserves more attention in food preservation is the interference by the food with the homeostasis of microorganisms [15]. Homeostasis is the tendency to uniformity or stability in the normal status (internal environment) of organisms. For instance, the maintenance of a defined pH within narrow limits is a feature and prerequisite of living organisms [16]; this applies to higher organisms as well as to microorganisms. Much is already known about homeostasis in higher organisms at the molecular, subcellular, cellular, and systematic levels in the fields of molecular biology, biochemistry, physiology, pharmacology, and medicine [16]. This knowledge should now be transferred to microorganisms important in the toxicity and spoilage of foods. If the homeostasis of microorganisms, i.e., their internal equilibrium, is disturbed by preservative factors (hurdles) in foods, they will not multiply but will remain in the lag-phase or even die before their homeostasis is reestablished. Thus, food preservation is achieved by disturbing the homeostasis of microorganisms in a food temporarily or permanently [7].

Gould [15] has pointed out that during evolution a wide range of more or less rapidly acting mechanisms (e.g., osmoregulation to counterbalance a hostile water activity in food) have developed in microorganisms that act to keep important physiological systems operating, in balance, and unperturbed even when the environment around them is greatly perturbed [17]. In most foods microorganisms are operating homeostatically in order to react to environmental stresses imposed by the preservation procedures applied. The most useful procedures employed to preserve foods are effective in overcoming the various homeostatic mechanisms the microorganisms have evolved in order to survive extreme environmental stresses [17]. The repair of a disturbed homeostasis demands much energy, and thus the restriction of energy supply inhibits repair mechanisms in microbial cells and leads to a synergistic effect of preservative factors (hurdles). Energy restrictions for microorganisms are, for example, caused by anaerobic conditions, such as in vacuum or modified-atmosphere packaging of foods. Therefore, low a_w (and/or low pH) and low redox potential act synergistically [17]. Such interference with the homeostasis of microorganisms or entire microbial populations provides an attractive and logical focus for improvements in food-preservation techniques [17].

B. Metabolic Exhaustion

Another phenomenon of practical importance is the metabolic exhaustion of microorganisms, which could lead to "autosterilization" of foods. This was first observed by us many years ago [18]. Mildly heated (95°C core temperature) liver sausage was adjusted to different water activities by the addition of salt and fat, and the product was inoculated with *Clostridium sporogenes* PA 3679 and stored at 37°C. Clostridial spores that survived the heat treatment vanished in the product during storage if the products were stable. Later this type of behavior for both *Clostridium* und *Bacillus* spores was regularly observed during storage of shelf-stable meat products (SSP), especially F-SSP [19] (see Sec. IV.B). The most likely explanation is that bacterial spores that survive the heat treatment are able to germinate in these foods under less favorable conditions than those under which vegetative bacteria are able to multiply [6]. Therefore, during storage of these products some viable spores germinate, but the germinated spores or vegetative cells derived from these spores die. Thus, the spore counts in stable hurdle-technology foods actually decrease during storage, especially in unrefrigerated foods. Also during studies in our laboratory of Chinese dried meat products, we observed the same behavior of microorganisms [20]. If these meats were contaminated after processing with staphylococci, salmonellae, or yeasts, the counts of these microorganisms on stable products decreased quite fast during unrefrigerated storage, especially on meats with a water activity close to the threshold for microbial growth. The same phenomenon was observed by Latin American researchers [21–24] in studies of high-moisture fruit products (HMFP) (see Sec. IV.C) because the counts of a variety of bacteria, yeasts, and molds that survived the mild heat treatment decreased quite fast in the products during unrefrigerated storage because the hurdles applied (pH, a_w, sorbate, sulfite) did not allow growth.

A general explanation for this behavior might be that vegetative microorganisms that cannot grow will die, and they die more quickly if the stability is close to the threshold for growth, storage temperature is elevated, antimicrobial substances are present, and the organisms are sublethally injured (e.g., by heat) [7]. Apparently, microorganisms in stable hurdle-technology foods strain every possible repair mechanism for their homeostasis in order to overcome the hostile environment. By doing this they completely use up their energy and die. This leads eventually to autosterilization of such foods [14].

Thus, due to autosterilization hurdle-technology foods, which are microbiologically stable, become even more safe during storage, especially at ambient temperatures. So, for example, salmonellae that survive the ripening process in fermented sausages will vanish more quickly if the product is stored at ambient temperature, and they will survive longer and possibly cause foodborne illness if the products are stored under refrigeration [7]. It is also well known that salmonellae survive in mayonnaise at chill temperatures much better than at ambient temperature. Unilever laboratories in Vlaardingen have confirmed metabolic exhaustion in water-in-oil emulsions (resembling margarine) inoculated with *Listeria innocua*. In these products listeria vanished faster at ambient (25°C) than at chill (7°C) temperature, at pH 4.25 > pH 4.3 > pH 6.0, in fine emulsions more quickly than in coarse emulsions, and under anaerobic conditions more quickly than under aerobic conditions. From these experiments it was concluded that metabolic exhaustion is accelerated if more hurdles are present, and this might be caused by increasing energy demands to maintain internal homeostasis under stress conditions (P. F. ter Steeg, personal communication).

C. Stress Reactions

A limitation to the success of hurdle-technology foods could be stress reactions of microorganisms. Some bacteria become more resistant (e.g., toward heat) or even more virulent under stress, when they generate stress shock proteins. The synthesis of protective stress shock proteins is induced by heat, pH, a_w, ethanol, etc. as well as by starvation. These responses of microorganisms under stress might hamper food preservation and could turn out to be problematic for the application of hurdle technology. On the other hand, the activation of genes for the synthesis of stress shock proteins, which help organisms to cope with stress situations, should be more difficult if different stresses are received at the same time. Simultaneous exposure to different stresses will require energy-consuming synthesis of several or at least much more protective stress shock proteins, which in turn may cause the microorganisms to become metabolically exhausted [8]. Therefore, multitarget preservation of foods could be the way to avoid synthesis of such stress shock proteins, which otherwise could jeopardize the microbial stability and safety of hurdle-technology foods [14]. Further research on stress shock proteins and the different mechanisms that govern their formation seems warranted in relation to hurdle-technology foods.

D. Multitarget Preservation

The multitarget preservation of foods should be the ultimate goal for the gentle but most effective preservation of foods [8]. It has been suspected for some time that different hurdles in a food might not have just an additive effect on microbial stability, but might act synergistically [1]. Example No. 8 in Figure 1 illustrates this. A synergistic effect could be achieved if the hurdles in a food hit, at the same time, different targets (e.g., cell membrane, DNA, enzyme systems, pH, a_w, Eh) within the microbial cells and thus disturb the homeostasis of the microorganisms present. If so, the repair of homeostasis as well as the activation of stress shock proteins would become more difficult [7]. Therefore, employing different hurdles simultaneously in the preservation of a particular food should achieve optimal microbial stability. In practical terms, this could mean that it is more effective to use different preservatives in small amounts than one preservative in larger amounts, because they might act synergistically [12].

It is anticipated that the targets in microorganisms of different preservative factors (hurdles) for foods will be elucidated, and that hurdles could then be grouped in classes according to their targets. A mild and effective preservation of foods, i.e., a synergistic effect of hurdles, is likely if the preservation measures are based on intelligent selection and combination of hurdles taken from different target classes [7]. This approach seems not only valid for traditional food-preservation procedures, but for modern processes (e.g., food irradiation, ultra-high pressure, mano-thermo-sonication, etc.) as well. An example of a multitarget novel process is the application of nisin, which damages the cell membrane, in combination with lysozyme and citrate, which then are able to penetrate easily into the cell and disturb the homeostasis with different targets [14].

Food microbiologists could learn in this respect from pharmacologists, because the mechanisms of action of biocides have been studied extensively in the medical field. At least 12 classes of biocides are already known to have more than one target within the microbial cell. Often the cell membrane is the primary target, becoming leaky and disrupting the organism, but biocides also impair the synthesis of enzymes, proteins, and DNA [25]. Multidrug attack has proved successful in the medical field to fight bacterial infections (e.g., tuberculosis) as well as viral infections (e.g., AIDS), and thus a multitarget attack of microorganisms should be a promising approach in food microbiology [14].

IV. APPLICATION OF HURDLE TECHNOLOGY

Foods based on combined preservation methods (hurdle technology) are prevalent in industrialized as well as in developing countries. In the past and often still today hurdle technology was applied empirically without knowing the governing principles in the preservation of a particular food. But with a better understanding of these principles and improved monitoring devices, the deliberate application of hurdle technology has advanced.

Some general aspects of hurdle technology will be briefly discussed here, and examples of the application of combined preservation methods in industrialized and in developing countries will be given.

A. General Aspects

With regard to the application of hurdle technology, one can differentiate between intermediate-moisture foods, high-moisture foods, and integer foods, because they differ somewhat in the types of hurdles used or the mode of hurdle application. Therefore, these three groups will be discussed separately.

1. Intermediate-Moisture Foods

Intermediate-moisture foods (IMF) have a a_w range of 0.90–0.60, and thus water activity is their primary hurdle for securing microbial stability and safety. However, IMF are often stabilized by additional hurdles, such as heating, preservatives, pH, redox potential, and competitive microflora [26]. These foods are easy to prepare and storable without refrigeration, thus they are cost and energy efficient. Traditional IMF based on meat, fish, fruits, and vegetables are common and much liked in different parts of the world, because they are tasty, nutritious, and, in general, safe. On the other hand, newly developed, tailor-made IMF (except for certain candy bars) have not achieved the expected breakthrough in human nutrition. Some reasons for this disappointing performance are the poor palatability of most novel IMF due to the high concentration of humectants and the need to introduce high amounts of antimicrobial additives (chemical overloading), which may cause health concerns and pose legal problems [7].

Traditional IMF are today the prevalent food items in developing countries. Few food manufacturers in these countries have the ability to measure a_w, and few recognize the relevance and significance of water activity to the preservation of their foods. Thus, the application of hurdle technology in the processing of IMF in developing countries is done empirically. Only recently have changes occurred. An outstanding example of such recent developments is the CYTED-D program (Science and Technology for Development) of Latin America, which was sponsored by Spain and to which Argentina, Brazil, Chile, Costa Rica, Cuba, Mexico, Nicaragua, Puerto Rico, Uruguay, and Venezuela have contributed. A project in this program, entitled Development of Intermediate Moisture Foods (IMF) Important to Ibero-America, had the objective to identify and evaluate foods of Latin America that are storable without refrigeration. This study comprised fruits, vegetables, and bakery products as well as foods derived from fish, milk, meat, and other miscellaneous products. About 260 food items of the region were approved as microbiologically stable and safe at ambient temperatures. The properties (a_w, pH, preservatives, food composition, etc.) and production technology for these products were measured and described [27,28], and it was concluded that most of the approved products were intermediate-moisture foods. However, some had higher a_w values, sometimes as high as 0.97–0.98, and nevertheless were stable and safe without refrigeration, and it turned out that the stability of these high-moisture foods was caused by a combination of several empirically applied hurdles. This observation was the starting point for Latin America to apply intentional hurdle technology [3] to high-moisture foods, especially to tropical and subtropical fruits storable without refrigeration (see Sec. IV.C). In the opinion of Latin American scientists [28] the technological achievements of their region deserve a closer look, in particular by developing countries where refrigeration is scarce. Since IMF are often not satisfactory from the sensory point of view and contain high levels of additives, the application of hurdle technology to stabilize high-moisture foods that also will need no refrigeration seems to have great potential [28].

The Latin American CYTED-D study demonstrated a promising approach for improving the stability of foods in developing countries, which could be applied in other regions as well. Following this concept, the properties of already available, microbiologically stable and safe food items should first be thoroughly studied and described. Then the preservative factors (hurdles) effective in these foods and the principles behind their microbial stability and safety must be elucidated. Third, if feasible, the preservation and quality of these foods should be improved by the intentional application of hurdle technology [7]. Using this concept, Tapia et al. [29] identified the hurdles in food items studied within the CYTED-D program, and by critical evaluation of the hurdles traditionally applied to certain foods, they assessed the microbial stability and safety of these products. They demonstrated that similar hurdles are active in the same type of foods in different countries of Latin America. However, there were also surprising differences indicating over- or underprocessing of the same food in different countries of the same region [29]. This insight could lead to the avoidance of some nonessential preservatives or, on the other hand, to improved stability, safety, and quality of some food items by fortification of certain hurdles.

Latin American researchers [28,29] found that reduced water activity was the main hurdle in the IMF they described within the CYTED-D project, however, in practically all of these products additional hurdles were present that contributed considerably to stability and safety. This observation has confirmed the opinion expressed by Leistner and Rödel [26] that several hurdles are inherent in most IMF. Tapia et al. [29] listed many traditional IMF of Latin America, derived from fruits, vegetables, meat, milk, fish, etc., in which several identified hurdles contributed to microbial stability and safety. Very often the reduced a_w was combined with a reduced pH. However, in some salted fish and shrimp the pH was >8.0, and thus in these foods the elevated pH might contribute to preservation. Many meat, fish, fruit, and dairy products contained in addition preservatives (nitrite, smoke, benzoate, sorbate, sulfites, spices, Maillard products, etc.) and sometimes competitive microorganisms. Maillard reaction products, which are generated during caramelization of sweetened condensed milk (dulce de leche), are probably an important hurdle for this Latin American food item. A special case concerns the candied fruits common in this region. Their heavy sugar coating acts as a physical barrier (hurdle) against microbial contamination after the heat process, therefore these foods are stable during storage in spite of the absence of preservatives and sometimes rather high pH. Quite often a thermal treatment is used to inactivate heat-sensitive microorganisms during the manufacturing process and to improve microbial stability, and a vacuum in sealed containers is achieved by hot filling of the products. Once the container is opened, the redox potential increases and then the microbial stability of these IMF against mold and yeasts must be secured by the preservatives present. Additional hurdles are employed in particular in those IMF that have rather high a_w and/or pH values. In the IMF studied it was exceptional if the a_w was apparently the only hurdle present, but in most IMF products three to five hurdles have been identified [29]. How many hurdles, besides a_w, are active in IMF depends on the type of product, however, it may be concluded that the microbial stability and safety of IMF in general is based on a combination of several preservative factors.

It is obvious that a thorough study of traditional IMF using up-to-date methodology would be of benefit to developing countries. However, such studies would also be useful to industrialized countries, because traditional products are an abundant source of innovative ideas that can be used in food design. For instance, we learned from traditional Chinese sausage (*la chang*) that a sausage could be preserved in the raw state even without fermentation, and we found that in traditional *charqui* of Brazil fermentation takes place even at an $a_w < 0.90$ if halophilic pediococci are involved. Heat inactivation of most pathogenic bacteria, including staphylococci, is achieved in some Chinese IMF meats by just applying 50°C for several hours. Another interesting aspect of traditional IMF meats is the bactericidal effect of Maillard reaction products towards toxic bacteria (see Sec. III.B). Apparently the growth inhibition of xerotolerant molds on unpackaged Chinese IMF meats with $a_w < 0.69$ is also supported by Maillard reaction products, which, therefore, are important hurdles for traditional IMF [6].

2. High-Moisture Foods

Intermediate-moisture foods are prevalent in developing countries, whereas in industrialized countries high-moisture foods (HMF) are more common, where they are often only minimally processed and, due to their fresh-like properties and convenience, appeal greatly to the consumer. However, since in HMF the watery activity is above 0.90, this hurdle is less prominent and other hurdles are needed to secure microbial stability and safety of foods during storage. Therefore, HMF are often chilled or frozen, and low-temperature storage is widely used. But refrigeration is energy consuming and thus costly, and in case of temperature abuse the stability and safety of the foods might be jeopardized. Therefore, besides the low-temperature hurdle for HMF, additional hurdles (e.g., heating, pH, Eh, a_w, preservatives, competitive flora), often applied in combination by means of hurdle technology, are significant. In the past the hurdles were applied emperical for HMF, but today often intentionally.

An example of a product involving the empirical use of hurdles in HMF is Italian mortadella. This meat product is an emulsion-type sausage, which is traditional and very common in Italy and can be stored without refrigeration as long as the sausage is uncut. Italian mortadella contains, due to a mild heat process (78°C core temperature), viable bacterial spores. However, the growth of bacilli and clostridia is inhibited in genuine mortadella by a slightly decreased a_w (<0.95). This effect, achieved using salt, sugar, milk powder, and drying, was accomplished in the past without knowledge of the mechanisms involved. Traditional recipes continued to achieve the desired a_w level of <0.95 [9,19,30]. In Germany meat products preserved using a similar principle (pasteurization and a_w adjustment to <0.95 by drying the sausage) are popular too. These a_w-SSP are stable at ambient temperature (see Sec. IV.B).

Another example of a shelf-stable emulsion-type European sausage is the Gelderse rookworst, typical of The Netherlands, in which intentional hurdle technology has been used for a number of years. A major hurdle of this product is reduced pH, which is adjusted to 5.4–5.6 by the addition of 0.5% glucono-delta-lactone, and thus Rookworst belongs to the group of pH-SSP products. This sausage is stable for several weeks at ambient temperature if vacuum-packaged and reheated at 80°C for about one hour in the pouch. The heat treatment eliminates vegetative organisms, and bacterial spores are apparently not of much concern, as their numbers decrease during the heat process and surviving spores are inhibited by the pH and other hurdles present (e.g., nitrite). Gelderse rookworst is exported from The Netherlands in large quantities to Britain, and with pH > 5.4 is acceptable from the sensory point of view. The binding of water and fat in rookworst is not a problem, in spite of the relatively low pH, if pork rinds and/or phosphates are added to the product [3,19]. More examples of the application of intentional hurdle technology to HMF in industrialized countries will be presented in Section IV.B.

As mentioned before, IMF are prevalent in developing countries. However, there is a trend in developing countries to move gradually away from IMF towards HMF [31]. IMF contain high amounts of humectants (e.g., sugar and/or salt) as well as fungistatic preservatives (e.g., sorbate, benzoate, sulfite), which are undesirable from a sensory or nutritional point of view, and they have a less appealing texture and appearance than HMF. Therefore, efforts are being made to improve the quality of IMF by decreasing the sugar and salt contents, as well as by increasing the moisture content and a_w without sacrificing the microbial stability and safety of the products if stored without refrigeration [19]. It might be expected that high-moisture, fresh-like foods that nevertheless are storable at ambient temperature, because they have been stabilized by intentional hurdle technology, will be on the increase in developing countries as soon as the application of advanced hurdle technology has been mastered [31]. Recent examples that support this trend are high-moisture fruit products of Latin America and dried meat products of China (see Sec. IV.C).

3. Integer Foods

Whole or integer foods are not comminuted and consist of large pieces of plant or animal tissue, nevertheless, their microbial stability and safety might be improved by application of hurdle technology. Two approaches are common: the use of coatings that contain and maintain inhibitory substances in order to protect the surface of the foods against microbial deterioration or a dewatering and impregnation process, which consists of soaking foods in highly concentrated solutions of humectants or other food additions [12,19].

An example of a food to which a surface layer is applied is pastirma, a traditional beef product of Moslem countries, which can be stored for several months at ambient temperatures and is eaten in the raw state [32]. The stability and safety of pastirma is based on reduced a_w (0.90–0.85) in combination with several additional preservative factors [31,33]. The interior of this product is stabilized by dry-curing of meat stripes (~5 cm thick) with salt and nitrate, which is reduced by bacteria to nitrite, and the removal of water by drying and pressing of the meat as well as by the growth of lactic acid bacteria, which decrease the pH to about 5.5. These hurdles secure the inhibition of spoilage and pathogenic bacteria, including salmonellae, within the meat. The surface of pastirma is covered with an edible paste (3–5 mm thick) containing 35% freshly ground garlic and other spices (paprika, kammon, mustard, as well as fenugreek as a binder). This surface coating prevents the growth of molds on the product during storage, even at elevated humidities and temperatures. Thus, at least five hurdles (a_w, nitrite, pH, competitive flora, and garlic) are used in preservation of pastirma.

Torres [34] studied the surface microbial stability of model foods using coatings that maintain preservatives and the desired low pH. He confirmed that a low pH in the surface layer greatly improved the effectiveness of sorbic acid in this coating. Guilbert [35] used superficial edible layers for easily perishable tropical fruits and achieved preservation without affecting the integrity of the food pieces. He pointed out that the formulation of edible films and coatings must include a component that can form an adequately cohesive and continuous matrix as well as the addition of a plasticizing agent to overcome brittleness. Specific agents (antimicrobials, antioxidants, organic acids, nutritional additives, flavors, coloring, etc.) can be incorporated into edible films to achieve functional effects localized on the surface [36].

Edible coatings as well as osmotic dehydration represent two ways to apply hurdle technology to solid foods without affecting their structural integrity [36]. The so-called osmotic dehydration is a dewatering and impregnation process, which consists of soaking foods (fruits, vegetables, meat, cheese, fish, etc.) in concentrated solutions of humectants (sucrose, sodium chloride, etc.) and which is employed for solute transfer from a solution into the product [37]. With this process, also called direct formulation, not only water activity–lowering agents are inserted into the food, but also preservatives and nutrients, as well as substances that control the pH, texture, and flavor. This would build up positive hurdles to improve the stability as well as the quality of food products [36,38]. In contrast to traditional soaking techniques (e.g., salting as used in cheese making, fish or meat curing, sugar addition as used in candying and semi-candying), which further solute impregnation and limit water loss, osmotic dehydration generally involves significant water removal (40–70 g of water is lost per 100 g of initial product) with limited and controlled solute incorporation (5–25 g of solute is gained per 100 g of initial product). Under typical operating conditions used for fruit and vegetables, mass transfer mainly occurs during the first 2 hours, thereafter, mass transfer rates become progressively slower until water loss stops, whereas solute gain continues to increase steadily. A soaking process does not generally produce stable products, but soaking must be used as a preprocessing step before complementary processing steps, such as drying, freezing, pasteurization, canning, frying, and/or the addition of preservative agents [39].

B. Applications in Industrialized Countries

The deliberate and intelligent application of hurdle technology started in the mid-1970s in Germany [1] and was first used for the preservation of meat products [3]. Soon this concept was applied to a variety of food items in industrialized as well as in developing countries. Here we will discuss some examples of advanced hurdle technology employed in industrialized countries for fermented, heated, or chilled foods, and in the design of healthful foods as well as in relation to the trend toward less packaging of foods.

1. Fermented Foods

In fermented foods (e.g., fermented sausages, raw hams, ripened cheeses, and pickled vegetables) a sequence of hurdles leads to a stable and safe product. Using the sequence of hurdles shown in Figure 1 (No. 7), toxic and spoilage organisms are inhibited in salami and the desired competitive flora (K-F) (lactic acid bacteria) is selected. Because the hurdles inherent to the salami process were studied in our laboratory [7,19,33,40] and confirmed by others, they will be briefly discussed here and are again illustrated in Figure 2.

Important hurdles in the early stages of the ripening process of salami are nitrite and salt (pres.), which inhibit many of the bacteria in the initial product. However, other bacteria are able to multiply, use up the oxygen, and thus cause the redox potential of the product to decrease. This in turn enhances the Eh hurdle, which inhibits aerobic organisms and favors the selection of lactic acid bacteria. These are the competitive flora (c.f.), which flourish by metabolizing the added sugars, causing a decrease in pH value (i.e. an increase of the pH hurdle). In long-ripened salami nitrite is depleted and the lactic acid bacteria vanish, while the Eh and pH increase. Only the water activity hurdle (a_w) is strengthened with time, and it is then largely responsible for the stability of long-ripened raw sausage [6,33]. Since this sequence of hurdles was revealed, the production of fermented sausages became less empiric and more advanced, and the knowledge has been used to achieve the required inhibition of *Clostridium botulinum, Listeria monocytogenes*, and *Staphylococcus aureus* as well as the inactivation of *Salmonella* spp. and verotoxins producing *Escherichia coli* in salami during fermentation and ripening. The sequence of hurdles that secures the microbial stability and safety of raw hams is also well known [41]. The sequence of hurdles for the proper fermentation process of other foods (e.g., ripened cheeses and pickled vegetables) is also important, and it would be challenging to elucidate them.

In Germany fermented sausages are categorized in two groups: quick- and slow-ripened products (Table 2). Quick-ripened products amount to about 80% of sausage production and slow-ripened products only to 20%. In quick-ripened products the a_w is rather high, they still contain much water, and are therefore less expensive. To compensate for this high a_w a low pH is essential in such products for microbial stability. In contrast, slow-ripened products, which are more expensive due to the long drying period, have a low a_w and can have a rather high pH, which makes them much more flavorful. These differences illustrate nicely the interchangeability of hurdles in a food. Different hurdles that achieve microbial stability can be chosen in order to achieve different features related to sensory properties and price.

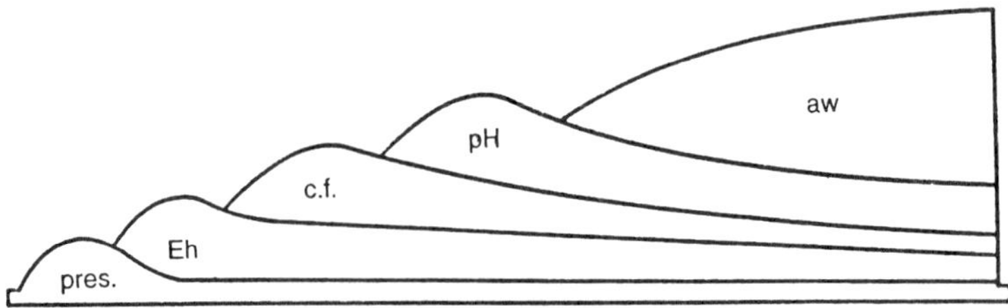

FIGURE 2 Sequence of hurdles occurring during the ripening and drying of fermented sausages (salami): pres., addition of nitrite-curing salt; Eh, decrease of redox potential; c.f., growth of competitive flora; pH, acidification; a_w, decrease of water activity during the drying process. (Adapted from Refs. 19,33.)

A feature peculiar to fermented sausages (probably also for ripened cheeses) is their microstructure, which influences the desired ripening process as well as the survival of pathogenic bacteria in the product. Thus, the microstructure is an important hurdle for the stability of salami [7,19]. Electron microscopy studies [42] have revealed that the natural flora as well as added starter cultures are not evenly distributed in fermented sausages, but accumulate in little cavities, i.e., the ripening flora can only grow in small nests. These nests are 100-5000 μm apart, and thus large areas of the sausage must be influenced by metabolites (e.g., nitrate reductase, catalase, organic acids, bacteriocins) accumulated in such cavities. Thus, from small nests of desirable bacteria the entire fermentation of the product must be accomplished, and the pathogenic bacteria (e.g., salmonellae or listeria) must be inactivated even though they might be located in distant areas of the food matrix. Within each nest the bacteria, either in pure or mixed cultures, are in keen competition for nutrients and impair each other by their metabolic products. In nests of mixed cultures, generally the lactic acid bacteria prevail due to their tolerance of low Eh, pH, and a_w. At the beginning of sausage fermentation in these nests the lactobacilli appear vigorous and metabolically active, whereas at the end of the ripening process the lactobacilli have degenerated and possibly died [19,42]. Small and equal distances between nests of desirable bacteria in the sausage matrix should be advantageous, since this would foster the proper ripening process and the inactivation of pathogenic bacteria. The thorough mixing of meat and fat particles in the sausage batter before stuffing it into casings would bring about a more even distribution of bacteria in the sausage matrix. If starter cultures are used, they should be added in a fashion that favors an even distribution, which could be better achieved by using liquid starter cultures rather than powders [7,19].

The microstructure is not important only for salami (and cheese), but for other foods too. In concentrated oil-in-water emulsions the bacteria form small colonies, and in water-in-oil emulsions bacteria growth is confined to the water droplets, which can lose their integrity due to coalescence [43]. The impact of microstructure on microbial growth, survival, and death in foods has theoretical and practical implications. Certainly, under these circumstances predictive modeling of the behavior of microorganisms is difficult. On the other hand, it is possible to influence the number, size, and distance of microbial nests in such foods, and thus their safety, stability, and quality, by the recipes for the product and the technology applied [7,19]. The microstructure is definitely an important hurdle for certain foods, and therefore is listed in Table 1. Further studies on the behavior of submerged bacterial colonies in food matrices seem warranted [44].

2. Heated Foods

Heat-processed high-moisture foods based on hurdle technology which are stable at ambient temperature have been named shelf-stable products (SSP) [3]. They offer the following advantages: the mild heat treatment (70–110°C) improves the sensory and nutritional properties of the food, and the lack of refrigeration simplifies distribution and saves energy during storage. SSP are heated in sealed containers (casings, pouches, or cans), which avoid recontamination after processing. However, because of the mild heat treatment these foods still contain viable spores of bacilli and clostridia, which are inhibited by adjusting of a_w, pH, and Eh, and, in the case of autoclaved sausages, by sublethal injury of the spores. At present four different types of SSP foods are distinguished—F-SSP, a_w-SSP,

Table 2 Criteria of Quick- and Slow-ripened Fermented German Sausages

Criterion	Quick-ripened products	Slow-ripened products
a_w	0.95–0.90	0.90–0.65
pH	4.8–5.2	5.4–6.0
Time (weeks)	1–2	4–8
Production share	80%	20%

Table 3 Different Shelf-Stable Products (SSP) and Their Primary Hurdles[a]

Product	Primary hurdles
F-SSP	Sublethal damage of bacterial spores
a_w-SSP	Slightly reduced water activity
pH-SSP	Slightly increased acidity
Combi-SSP	Combination of equal hurdles

[a]All are mildly heated foods, stable at ambient temperature, nevertheless having fresh product characteristics.

pH-SSP, and Combi-SSP—depending on their primary hurdles (Table 3), though additional hurdles can foster the safety and stability of these products [6,7,19].

F-SSP

In F-SSP [3,6,33], the sublethal damage of the spores is the primary hurdle, which is achieved by a mild heat treatment. Such sausages with an adjusted a_w (Bologna-type sausage < 0.97; liver and blood sausages < 0.96) in PVDC casings (impermeable to water vapor and air) are heated in counterpressure autoclaves to $F_0 > 0.4$. They are stored unrefrigerated for several weeks and have caused no problems with regard to food poisoning or spoilage because guidelines for their processing have been suggested and followed [45]. F-SSP, due to metabolic exhaustion of the microorganisms, even autosterilize during storage (see Sec. III.B). Casings are more advisable than cans for F-SSP, because during chilling of cans after autoclaving, some water condensation may occur inside the lid, and if drops of water fall back on the surface of the sausage mix, the critical a_w increases and clostridia may start to grow. If autoclaved sausages fill the casings tightly, water condensation cannot occur. Therefore F-SSP in casings are more stable than in cans with headspace [46].

a_w-SSP

The stability of a_w-SSP [3,6,19,30,33] is primarily caused by a reduction of water activity below 0.95, and guidelines for such products have been suggested [33]. Examples of traditional a_w-SSP meats are Italian mortadella and German Brühdauerwurst. A large variety of such meat products is now on the market, most of which are snack items. The shelf life of a_w-SSP at ambient temperature is even better than for fermented sausages, since in a_w-SSP due to the heat treatment (internal temperature >75°C) lipases are inactivated and these products are less prone to become rancid.

pH-SSP

In pH-SSP [3,6,33] increased acidity is the primary hurdle. This is the principle applied in Gelderse rookworst, the shelf-stable meat product discussed in Section IV.A. Other traditional meat products of the pH-SSP type are brawns (jelly sausages), which are adjusted to an appropriate pH by the addition of acetic acid. Such products are composed of a brine (pH < 4.8) made of water, gelatine, salt, sugar, agar-agar (2%), and spice and a solid phase made of Bologna-type sausage in cubes with a_w of <0.98. Both components are mixed (two parts brine to three parts meat), filled in casings, and heated to an internal temperature of >72°C but no higher than 80°C. If the product is in equilibrium, it should have a final pH < 5.2, and then it is storable for several days at ambient temperatures. Outside the meat industry, pH-SSP are common as heat-pasteurized fruit and vegetable preserves with pH < 4.5, which are bacteriologically stable and safe in spite of a mild heat treatment. In such products vegetative microorganisms are inactivated by heat and the multiplication of surviving bacilli and clostridia is inhibited by the low pH. Since bacterial spores are able to germinate at lower pH levels than those at which vegetative bacilli and clostridia are able to multiply, in pH-SSP, as in F-SSP and a_w-SSP, the number of spores tends to decrease during storage due to metabolic exhaustion of microorganisms. On the other hand, while the heat resistance of bacteria and their spores is enhanced with decreasing a_w, it is diminished with decreasing pH. Thus, pH-SSP need less heat treatment for inactivation of microorganisms than do a_w-SSP.

Combi-SSP

In Combi-SSP [6,19,47] a combination of rather equal hurdles is applied [6]. Our experimental work suggests that even small enhancements of individual hurdles in a food have a distinct effect on the microbial stability of a product. For instance, for the stability and safety of a food it is of significance whether the F_0 is 0.3 or 0.4, the a_w is 0.975 or 0.970, the pH is 6.5 or 6.3, and the Eh value is somewhat higher or lower. Every small improvement or reinforcement of a hurdle brings some weight to the balance, and the sum of these weights determines whether a food is microbiologically unstable, uncertain, or stable (Fig. 3). In other words, many small steps in the direction of stability can swing the balance from an unstable into a stable state [6]. We followed this procedure in our product design of Bologna-type sausages as Combi-SSP. Different types of Brühwurst (wieners, Bockwurst, Fleischwurst, Fleischkäse, etc.) have been developed which proved microbiologically stable and safe for at least one week at 30°C. The initial spore load of the sausage mix is low, because spice extracts instead of natural spices are used, nitrite (100 ppm) with curing salt must be added, and these products are heated to a core temperature of >72°C and are adjusted to an a_w and pH of <0.965 and <5.7, respectively. These products are repasteurized after vacuum-packaging for 45–60 minutes (depending on the diameter of the products) at 82–85°C [48]. Combi-SSP require strict rules for food design and process control (see Sec. V.) The Combi-SSP concept is applicable not only to meat products, but to other foods too. For instance, an Italian past product (tortellini) has been stabilized using as hurdles a water activity reduction and a mild heat treatment, as well as modified atmosphere or ethanol vapor during storage combined with moderate chilling temperatures [49,50]. Another example is paneer, a dairy product of India, which was developed as a Combi-SSP [51,52] and will be discussed later (Sec. IV.C). In both cases the thesis work of young scientists was ground breaking.

The Federal Centre for Meat Research in Germany demonstrated the efficiency of the application of hurdle technology in an extensive study (supported by the Medical Corps of the German Army) on 75 meat products with fresh product characteristics that nevertheless were storable without refrigeration [19,47,48]. In this study eight categories of meat products were selected and optimized; one represented fermented sausages and most of the others F-SSP, a_w-SSP, and Combi-SSP (Table 4). Since these meats must be suitable for army provisions during military exercises, they had to be stable and safe for at least 6 days at 30°C as well as flavorful and nutritious. It was necessary to define the manufacturing of these meats in detail, and for this purpose we introduced a linkage between hurdle technology and HACCP [19,53]. In the manufacturing plants processing the recommended meats for the army, no microbiological tests must be carried out, but time, temperature, pH, and a_w are strictly controlled. These measurements should be done on-line or at least close to the line. Fortunately, a precise instrument for this purpose became available [54], which allows a_w determi-

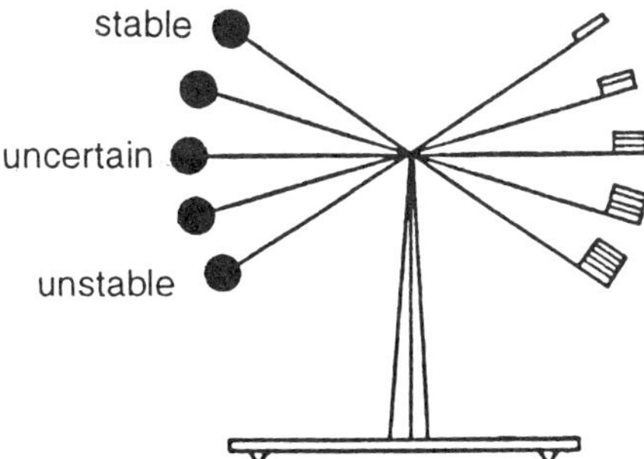

FIGURE 3 The balance should illustrate that even small enhancements of different hurdles can result in a substantial improvement in the microbial stability of a food. (Adapted from Refs. 6,19.)

Table 4 Categories of German Meat Products Fresh-like and Storable Without Refrigeration Due to Hurdle Technology

1.	Quick-ripened fermented sausage
2.	Mini-salami (two different technologies)
3.	Brühwurst and liver sausage as F-SSP
4.	Dried Brühwurst as a_w-SSP
5.	Repasteurized Brühwurst as a_w-SSP
6.	Brawns and Brühwurst as pH-SSP
7.	Items of Brühwurst as Combi-SSP
8.	Brühwurst in autoclaved flat pouches

Sources: Adapted from Refs. 19,47,48.

nation for meat products within 10–20 minutes. This army study might be used for other situations where hurdle technology and HACCP should be linked [47].

3. Chilled Foods

The results of hurdle technology are most obvious in high-moisture foods that become shelf stable at ambient temperature due to an intelligent application of combined methods for preservation. However, the use of hurdle technology is appropriate for chilled foods too, because in the case of temperature abuse, which can easily happen during food distribution, the stability and safety of chilled foods could break down, especially if low-temperature storage is the only hurdle. Therefore, it is reasonable to incorporate into chilled foods (e.g., sous vide dishes) some additional hurdles that will act as a back-up in case of temperature abuse. This type of safety precaution for chilled foods can be called "invisible technology" [8,19], implying that additional hurdle act as safeguards in chilled foods ensuring that they remain microbiologically stable and safe during storage in retail outlets as well as in the home.

For many chilled foods additional hurdles are already routinely used. This is in particular true for refrigerated foods for which modified-atmosphere packaging (MAP) is employed. Raw and minimally processed vegetables are among the most popular ready-to-eat foods that are stored chilled in MAP [55,56]. In Europe MAP is mainly applied to salads, potatoes, carrots, and cabbage, whereas in the United States the fresh-cut produce market is served mainly with salads, paprika, onions, cabbage, mushrooms, endive, and spinach. MAP is intended to suppress microbial growth, to retard the respiration, ripening, and aging of vegetables, and to inhibit oxidative reactions requiring oxygen. In combination with appropriate chilling, the shelf life of foods using MAP can be up to 5–7 days. Immediately before sealing, a gas of defined composition is introduced into the packages, which are generally stored at 4–7°C. Typically, the gas composition in MAP for fresh and minimally processed vegetables is 2–5% oxygen and 5–10% carbon dioxide. The modified composition of the gas atmosphere in MAP systems has a marked impact on the growth of spoilage microorganisms as well as pathogens [57–59]. Minimally processed vegetables are not sterile, and numerous genera of spoilage bacteria, yeasts, and molds are frequently encountered, as are some cold-tolerant pathogens (*E. coli* 0157:H7, *Bacillus cereus, Yersinia enterocolitica, Aeromonas hyrophila, Listeria monocytogenes*) [60]. The CO_2 in MAP does not inhibit pathogens directly; the competitive flora that grows in MAP at reduced oxygen and increased carbon dioxide levels might act as a hurdle and thus suppress the pathogens mentioned [58]. However, occasionally *L. monocytogenes* under certain conditions in MAP systems will proliferate to hazardous levels. Additional preservative factors (hurdles), e.g., *Enterococcus mundtii* as a protective culture, may be integrated into the MAP system to assure safety better than is currently possible [61].

4. Healthful Foods

Consumers demand healthful foods that are low in fat, salt (sodium), and cholesterol. Because of this trend muscle foods derived from red meat, poultry, or fish are modified. However, reductions in salt or fat as well as their substitutes and replacers might diminish the microbial stability and safety of foods, since important hurdles change. Compensation for these deficits could be achieved by application of intelligent hurdle technology [62,63].

Several substitutes and/or enhancers for sodium chloride (i.e., various other salts and hydrolysates) and more than 100 fat replacers (i.e., fat mimetic systems or synthetic fat replacers) are available today and of potential use for modified muscle foods. However, the microbial consequences of their use are generally unknown. Of major importance for the stability and safety of low-fat and/or low-salt muscle foods is an increased water activity (a_w). The acidity (pH) might also be unfavorable, since low-salt muscle foods have a slightly higher pH than normal products. If fat is replaced by proteins, the pH and/or the buffering capacity could increase, decreasing the stability and safety of these products. Added preservatives might be diluted and thus rendered less effective, since low-fat products contain more water due to an increased water-holding capacity caused by fat replacers. Furthermore, the redox potential (Eh) might change, because proteinaceous fat replacers might increase the Eh-buffering capacity of a food, whereas foods with a high water content should have a low Eh capacity. The microstructure of some foods (e.g., emulsions, fermented products) is significant for their microbial stability and safety. This microstructure could be changed by the addition of some fat replacers, but this effect has not yet been investigated. More data are needed on the impact of different fat replacers and salt substitutes (sodium), as well as of relevant combinations, on the important preservative factors (hurdles) of modified foods. The obtained quantitative data would serve as the basis for the microbial stability management of novel low-fat and/or low-salt muscle foods and would make the design by intelligent hurdle technology feasible.

To keep low-fat and/or low-salt muscle foods microbiologically stable and safe, refrigeration during storage of these products must be perfect. Since this cannot always be guaranteed, the weak hurdles in modified products should be backed up by alternative preservation factors. Microbiologists should take an active part in the design, processing, and marketing of low-fat and/or low-salt muscle foods, and their close cooperation with technologists will prove fruitful. Hitherto in the design and production of low-fat and/or low-salt muscle foods, the nutritional aspects have been much more emphasized than the microbial aspects. However, the latter should not be neglected, because these foods will only be continuously accepted by the consumer if they have no toxic effects and do not spoil easily [62,63].

5. Packaging of Foods

Packaging is an important hurdle for most foods, since it supports the microbial stability and safety as well as the sensory quality of the product. Little research has been done on packaging of hurdle-technology foods specifically. The appropriate packaging materials and procedures must be selected and applied with respect to the individual properties of the food items in question [8,64].

Industrialized countries have the tendency to overpackage their foods, whereas in developing countries there is a lack of knowledge and of simple packaging materials for foods. Therefore, further exploration into the application of easy-to-use, cheap, environmentally friendly, and efficient packaging for hurdle-technology foods of developing countries should be a challenging and even a lucrative prospect [64].

Packaging in industrialized countries is sophisticated, because absorbers, scavengers, scrubbers, getters or emitters for gas components as well as desiccants, antimicrobial packaging materials, tuned infrared films, edible food coatings and films, time-temperature integrator tags, microwave doneness indicators, etc., are employed for food items [65,66]. These procedures and devices can be summarized under the term "active packaging." Japan was in many respect and for several decades the world champion of food packaging, and many developments in active or smart packaging originated in that country. However, since the mid-1980s several Japanese packaging experts have cast doubts on these developments and questioned whether these strategies in food packaging should be pursued in the future. According to the opinions of Japanese packaging experts [67], five generations of food packaging are distinguishable: First came the function of containing—since prehistorical times people have used vessels, pots, baskets, etc. as containers for foods. The second generation of packaging had the function of transporting—wrappings, bottles, barrels, etc. used for food transport and distribution. The third generation extended the shelf life of foods by hermetically sealing the packages, which made

Table 5 Generations of Food Packaging

First—Containing: People used packaging (pots, baskets, etc.) since prehistorical days for containing food stuffs.
Second—Transporting: Packaging is used to transport food stuffs in bottles, barrels, containers, cartons, etc.
Third—Sealing: This made canning of foods, vacuum, modified-atmosphere, and aseptic packaging possible.
Fourth—Create Quality: Active packaging improves the food through scavengers, absorbers, getters, emitters, etc.
Fifth—Less Packaging: Just-in-time distribution of foods or advanced hurdle-technology foods provide the shelf life.

Source: Adapted from Refs. 67,68.

canning, vacuum- and modified-atmosphere as well as aseptic packaging possible. In the fourth generation, the packaging enhanced quality during distribution—the packaging itself had a processing function in addition to the common processing of foods. Such systems are called smart or active or functional packaging, and they are sophisticated but costly and wasteful. In addition they create environmental problems, and in a densely populated country like Japan already have had severe consequences. Therefore, recycling and/or reduced packaging have become necessary goals. Japan has led the way in the fifth generation of packaging—less packaging of foods [67]. In this fifth generation, the shelf life and quality of foods will not depend on the packaging but on the food itself or the mode of distribution (Table 5).

In the opinion of Japanese experts, future packaging should provide only information (name and picture of product, nutritional and shelf life data, etc.) and some convenience to the consumer. The required shelf life of the food will be secured either by aseptic packaging combined with just-in-time delivery systems or by development of stable and safe hurdle-technology foods.

Aseptic (ultra-clean) packaging of food is perfectly executed in Japan because factory workers carry out orders with discipline and diligence. If due to ultra-clean packaging only a few microorganism are at the start in the food, few hurdles of low intensity are needed to inhibit microbial growth, and thus ultra-clean packaging is an important component of hurdle technology [6,7]. Ultra-clean packaged foods need only simple packaging if a quick delivery at low temperature is guaranteed. In Japan such foods are transported at close to the freezing point, and just-in-time distribution systems have been widely introduced, because they provide the needed shelf life at low costs due to reduced inventory at the point of sale [67]. Deliveries are often made three times a day, and this just-in-time system has proved successful and is expected to be increasingly used. However, the frequency of the transportation needed will be a burden on the environment.

Another approach that could eliminate elaborate packaging is to design foods that are stable, safe, and of superior sensory quality in spite of minimal packaging. In this respect Japanese authors initially discussed the promotion of intermediate-moisture foods [67]. However, after learning about hurdle technology applied to high-moisture foods, the latter concept became a promising option for achieving the fifth generation of food packaging (K. Ono, personal communication) [68].

C. Applications in Developing Countries

There exists an abundant variety of preserved foods in developing countries, because the gap between harvest peaks has to be bridged and the taste as well as the nutritional properties of foods can be enhanced by preservation. Most of the preserved foods in developing countries, which must be storable without refrigeration, are based on the empiric use of hurdle technology. However, several foods have already been optimized by the intentional application of hurdles. The state-of-the art in the use of hurdle technology in some countries of Latin America, Asia, and Africa has recently been presented in a comprehensive review [31]. Therefore, only a few examples will be discussed here.

1. Fruits in Latin America

During the last decade a novel process for the preservation of high-moisture fruit products (HMFP; $a_w > 0.93$) has been developed in seven Latin American countries (Argentina, Costa Rica, Cuba, Mexico, Nicaragua, Puerto Rico, and Venezuela) and has been applied to peach halves, pineapple slices, mango slices and purée, papaya slices, chicozapote slices, purée of banana, plum, passion fruit, and tamarind, as well as whole figs, strawberries, and pomalanca [22,23]. The new technologies were based on the combination of a mild heat treatment (blanching for 1–3 minutes with saturated steam), slight reduction in water activity (to 0.98–0.93 by addition of glucose or sucrose), lowering of pH (to 4.1–3.0 by addition of citric or phosphoric acid), and the addition of antimicrobials (potassium sorbate or sodium benzoate plus sodium sulfite or sodium bisulfite) in moderate amounts to the syrup of the products (Table 6). During storage of HMFP the sorbate and in particular the sulfite levels decreased, and the a_w fell (i.e., the a_w hurdle increased) due to the hydrolysis of sucrose [23].

Combined methods technology (hurdle technology) was applied in these novel processes [21–23,69,70]. The minimal processes proved inexpensive, energy efficient, simple to carry out (little capital investment), and satisfactory for preserving fruits in situ. The resulting fresh-like products were still scored highly by a consumer panel after 3 months of storage at 35°C for taste, flavor, color, and especially texture, which is often problematic for canned fruits. Thus, according to Latin American researchers [22,23,69,71,72] combined methods allow the storage of fruits, without losses between seasonal harvest peaks, for direct domestic consumption or for further processing to confectionery, bakery goods, and dairy products, or for preserves, jams, and jellies. Fruit pieces can also be utilized as ingredients in salads, pizzas, yoghurt, and fruit drink formulations [23,71]. Moreover, these novel HMFP will open new possibilities for export markets.

The high-moisture fruit products stabilized by hurdle technology proved shelf stable and safe for at least 3–8 months of storage at 25–35°C. Due to the blanching process the initial microbial counts were substantially reduced, and during the storage of the stabilized HMFP the number of surviving bacteria, yeasts, and molds decreased further, often below the detection limits [21–24,69–72]. Banana purée challenged with yeasts, molds, clostridia, and bacilli (known to spoil fruits) and stored at ambient temperatures for 120 days remained stable if proper hurdles were applied (mild heat treatment, adjustment of a_w to 0.97 and pH to 3.4, addition of 100 ppm potassium sorbate, 400 ppm sodium bisulfite, and 250 ppm ascorbic acid). The inoculated microorganisms declined and often vanished below the detection limit [70]. These favorable microbiological results obtained with HMFP are probably due to metabolic exhaustion of the microorganisms present in the stabilized products (see Sec. III.B). Since HMFP during storage at ambient temperatures become apparently sterile, pathogenic and toxigenic microorganisms are not likely to be a hazard for these foods [31].

Alzamora and coworkers expressed the opinion that HMFP technologies as developed in Latin America will attract much attention in many developing countries, because they are easy to implement and will improve considerably the quality of stored fruits [22]. They even believe that the use-

Table 6 Typical Process Used for Preservation of High-Moisture Fruit Products (HMFP)

Hurdles	Intensity
1. Heat treatment[a]	Saturated steam for 1–3 min
2. Water activity[b]	a_w reduction to 0.98–0.93
3. Acidification[c]	pH adjustment to 4.1–3.0
4. Preservative (I)	1000 ppm sorbate or benzoate
5. Preservative (II)	150 ppm sulfite or bisulfite

[a]Dependent on size and type of fruit.
[b]Adjusted with sucrose, glucose, maltodextrin, etc.
[c]Adjusted, if necessary, with phosphoric or citric acid.
Source: Adapted from Refs. 22,23,71,72.

fulness of combined methods preservation (hurdle technology) for HMFP may give rise to an explosion of research on minimally processed fruits and the application of this innovative process by the food industry [23]. The advances in Latin America in fruit preservation are impressive and recently have been confirmed by Indian researchers, who concluded that "hurdle technology is seen as a promising technique for the preservation of fresh fruits and vegetables" [73]. However, the preservation of HMFP must certainly be based on guidelines for good manufacturing practice (GMP) in order to be successful under industrial or even artisan conditions [74]. For instance, the reuse of syrup may become a risk for build-up of spoilage flora (e.g., *Zygosaccharomyces bailii*, which could be sorbate-resistant), and therefore the reuse of syrup in HMFP processes should only occur after pasteurization.

2. Meats in China

Even though the per capita consumption of meat, compared to western countries, is rather low in China, nevertheless, due to the large population, this country has the largest meat consumption worldwide. Pork is the preferred meat in China, however, beef, water buffalo, sheep, poultry, and rabbit are also used as raw materials [75]. Only about 15% of the available meat of China is processed [76], but this is still a large amount and meat products are an important and precious part of the diet. Two categories of meat products exist in China side by side: Chinese meats and western meats. Of the western meat products, autoclaved emulsion-type sausages have recently gained ground in China. Their technology has been derived from the German F-SSP (see Sec. IV.B), but the a_w of the Chinese products is not adjusted and therefore they require severe heat treatment ($F_0 > 5.0$), which is not beneficial for the taste but does result in a shelf-stable product. On the other hand, the recipes for traditional Chinese meat products date back centuries, and their microbial stability and safety are based on combined preservative factors empirically applied. Studies are in progress to identify the hurdles inherent to these traditional meats [77]. The concept of hurdle technology was introduced to China by the author of this chapter. Work related to application of hurdle technology to meat products is being carried out at present in several Chinese research institutes, and recent publications in Chinese (Refs. 78–82) have made the concept popular in the Peoples Republic of China.

Traditional Chinese meat products are quite simple to prepare without expensive equipment, have a typical flavor, are ready-to-eat, and are storable for an extended time without refrigeration. The traditional meats of China listed in Table 7 are all intermediate-moisture foods, which implies that they might be too salty or too sweet, too tough, and too dark due to the formation of Maillard reaction products. The water activity is the primary hurdle in these meats, but if additional hurdles are strengthened, then the a_w can be raised, which often improves the sensory quality of the products. However, the microbial stability and safety of the meats must not be jeopardized by an increased a_w. Therefore, intentional and intelligent hurdle technology is increasingly applied. Some examples of such endeavors will be cited.

Table 7 Traditional Chinese Meat Products[a]

Cured meats (Yan La)	Chinese bacon—La Rou Pressed duck—Ban Ya Silk rabbit—Cha Si Tu Cured chicken—Yuan Bao Ji
Dried meats (Rou Gan)	Dried meat—Rou Gan Sweet meat—Rou Pu Meat floss—Rou Song
Sausage (La Chang)	Guangdong La Chang, Sichuan La Chang, etc.
Raw ham (Ho Tui)	Yü Nam Ho Tui, Jin Hua Ho Tui, etc.

[a]All are intermediate-moisture foods and are storable without refrigeration.
Source: Adapted from Refs. 31,77.

Rou gan is a typical dried meat product of China prepared mainly from beef using a technology that has not changed for hundreds of years, but improvements are possible and desirable. Consumers now prefer products with a softer texture, lighter color, and less sugar addition. Shafu is a modified dried meat that fulfills these expectations. The a_w of rou gan is in the range of 0.60–0.69, whereas shafu has an a_w of 0.74–0.76, because its moisture is higher and the sugar content lower. Additional hurdles in shafu are nitrite and ascorbic acid, which improve the color and delay rancidity, whereas rou gan is cured with nitrate only. Furthermore, the microbial stability of shafu is improved by the selection of raw material with low microbial load, low temperature during curing but relative high temperature and shorter times during heating, minimizing of recontamination after the heat process, as well as vacuum-packaging of the final product in order to inhibit mold growth and to delay rancidity. Therefore, shafu is microbiologically stable and safe for several months at ambient temperatures, and thus has the same shelf life as the traditional rou gan [76]. Both products have low residual levels of nitrite and nitrate, contain few microorganisms including in general no pathogenic or toxigenic bacteria, and thus are safe meats.

A similar approach was chosen to improve the quality of Islamic dried beef in dices, which is also storable for several months without refrigeration. The traditional product is simply salted and dried, and is therefore very salty and easily becomes rancid. The modified product contains less salt, is cured with nitrate to improve color, and some ascorbic acid is added to delay rancidity. Furthermore, vacuum-packaging is applied to reduce oxidation and to avoid mold growth [83]. More examples of the application of hurdle technology to Chinese meat products have been given in a recent review article [31].

Finally, the stabilization of the Chinese sausage, which was achieved in cooperation with Taiwan [84], will be mentioned. Chinese sausage is highly esteemed by Asian consumers and differs from the fermented sausages common in western countries: it is also processed raw, but with little fermentation. Several types of la chang (also called lup cheong) are known; the Sichuan type is more spicy and the Cantonese type more sweet, whereas Taiwanese la chang is softer. In mainland China finished products of la chang have the following properties: a_w 0.85–0.70, pH 5.9–5.7, NaCl 3–5%, sugar 4–20%, total count $< 10^6$/g, shelf life 2–3 months without refrigeration and if vacuum-packaged 4–5 months. To the coarsely ground meat and fat from pork, sugar, salt, soy sauce, liquor, spices, nitrate (or nitrite), and sometimes sodium ascorbate are added. The mix is stuffed into casings of small diameter, and the sausages are dried quickly over charcoal at 45–60°C to $a_w < 0.92$, and then further at ambient temperature to $a_w < 0.80$, in order to avoid an increase of lactic acid bacteria counts and the resultant sour taste. Thus, la chang is a typical meat product preserved by combined factors. In general, before consumption la chang is sliced and heated in rice or vegetable dishes [33,77,78,84].

The Taiwanese variety of Chinese sausage contains more moisture and therefore can have an a_w as high as 0.94. This improves the sensory properties since the products are softer, but it decreases their microbial stability and safety, because in such products lactic acid bacteria still grow and lead to a sour taste, and *Staphylococcus aureus* might cause food poisoning. After addition of 3.5% sodium lactate and 1.0% sodium acetate, the Taiwanese sausage remains tasty but is rendered microbiologically stable and safe, even if stored for several weeks without refrigeration. These additives reduce the a_w of the product but also have some antimicrobial effects [84]. Challenge tests using inocula of *S. aureus, Listeria monocytogenes*, or *Salmonella* spp. confirmed that Taiwanese sausage, modified by intentional hurdle technology, was stable and safe [84]. Furthermore, they demonstrated that in this product the number of pathogens decreased faster during storage at 25°C than at 10°C. This was probably due to a more rapid metabolic exhaustion of the pathogens at the higher storage temperature (see Sec. III.B).

3. Dairy Products in India

Lectures in India by the author of this chapter on food preservation using combined preservative factors, as well as reports on this subject published by Indian scientists (e.g., Refs. 52, 73, 85, 86), have stimulated research about the application of hurdle technology to traditional and modified Indian food products. Work related to food preservation using hurdle technology is at present carried out at several research institutes in India. In India vegetarian food is common, but dairy products are also much appreciated. Some recent examples of the application of hurdle technology to Indian dairy products will be discussed.

Paneer is a traditional, cottage cheese–type product fried in cubes with oil and onions, to which a sauce containing salt, spices, and often tomatoes is added. This food is frequently consumed and much liked in the northern provinces of India because of its nutritive value and characteristic taste. However, paneer spoils bacteriologically within 1–2 days at room temperature (which in Indian can reach 35°C), and this is a strong drawback for its industrial production. Sterilized paneer in cans has severe sensory limitations with regard to flavor, texture, and color. Therefore, together with a visiting scientist from India, Dr. K. Jayaraj Rao, at the German Meat Research Institute, Kulmbach, a mildly heated paneer in hermetically sealed containers with the desired flavor (like fresh), color (little browning), and texture (not too hard) was developed [51]. This product was stabilized by hurdle technology, and thus is stable and safe for several weeks without refrigeration. The following combinations of hurdles proved effective with this product: $a_w = 0.97$, heating to F_0 of 0.8, pH = 5.0 or alternatively $a_w = 0.96$, $F_0 = 0.4$, pH = 5.0 [51]. After his return to India, Rao conducted a thorough study of the application of hurdle technology to fried paneer in cubes made from buffalo milk. The product with gravy was packed either in tins or flexible retort pouches, and a set of hurdles ($a_w = 0.95$, $F_0 = 0.8$, pH 5.0, and 0.1% potassium sorbate) was chosen, which had maximum lethal and inhibitory effects on microorganisms and minimal effects on textural and chemical characteristics [52]. The water activity of paneer and gravy was lowered using humectants, such as dahi, skim milk powder, salt, and glycerol. The pH was adjusted by changing the dahi:skim milk powder ratio. The resulting product had a keeping quality of 2 weeks at 45°C, 1 month at 30°C, or >3 months at 15°C, which was limited by textural changes (hardness, cohesiveness, gumminess, springiness, and chewiness) as well as by chemical changes (browning, oxidation, lipolysis, and loss of available lysin), but not by microbial spoilage. The stabilized product was compared with fresh samples from restaurants and was found to be equally acceptable. In the opinion of Rao [52], this method of paneer preservation has large scope for alterations in product formulations depending on regional taste preferences, without affecting the keeping quality of the product. Via paneer hurdle technology has been introduced in India, and its application to other indigenous foods is now anticipated.

A recent example is dudh churpi. This is a popular dairy product of the Himalayan region of India (Bhutan, Sikkim, Darjeeling) made of milk from yaks or cows and is stable for several months without refrigeration. Most important for dudh churpi is the texture (elasticity), since people living at high altitudes chew it as an "energy tablet." The sensory quality and microbial stability of dudh churpi was optimized by Hossain [86] using combined methods (hurdle technology) involving heating, acid coagulation, addition of sugar and sorbate, smoking, drying, and packaging in closed containers. In this detailed and diligent study dudh churpi was scientifically explored and then a feasible optimization of the product was suggested. Thus, hurdle technology was applied to improve the quality of a traditional food of a remote region, and at the same time the scientific basis of this study opened new avenues for food science and industrial food production in India.

V. DESIGN OF HURDLE-TECHNOLOGY FOODS

The application of hurdle technology is useful for the optimization of traditional foods as well as in the development of novel products. There are similarities to the concepts of predictive microbiology and hazard analysis critical control point (HACCP). The three concepts have related but different goals: hurdle technology is primarily used in food design, predictive microbiology for process refinement, and HACCP for process control. In product development these three concepts should be combined. Therefore, we have suggested for the design of foods a 10-step procedure including all three concepts (Table 8), and this approach proved suitable when solving real product development tasks in the food industry [19,47,87]. However, these 10 steps should still be considered tentative until further practical experience with the application of the suggested user guide for food design has accumulated in the food industry.

In food design different types of researchers, including microbiologists and technologists, must work together. The microbiologist should determine which types and intensity of hurdles are needed

Table 8 Steps for Food Design Using an Integrated Concept Comprising Hurdle Technology, Predictive Microbiology, and HACCP or GMP

1. For the modified or novel food product the desired sensory properties and the desired shelf life are tentatively defined.
2. A feasible technology for the production of this food must be outlined.
3. The food is manufactured according to this technology, and the resulting product is analyzed for pH, a_w, preservatives, or other inhibitory factors. Temperatures for heating (if intended) and storage as well as the expected shelf life are defined.
4. For preliminary microbial stability testing of the food product, predictive microbiology might be employed.
5. The product is challenged with toxic and spoilage microorganisms, using somewhat higher inocula and storage temperatures than would be "normal" for this food.
6. If necessary, the hurdles in the product are modified, taking multitarget preservation and the sensory quality of the food (i.e., total quality) into consideration.
7. The food is again challenged with relevant microorganisms, and if necessary the hurdles in the food are modified again. Predictive microbiology for assessing the safety of the food might be helpful at this stage too.
8. After the established hurdles of the modified or novel foods are exactly defined, including tolerances, the methods for monitoring the process are agreed on. Preferably physical methods for monitoring should be used.
9. The designed food should now be produced under industrial conditions, because the possibilities for a scale-up of the proposed manufacturing process must be validated.
10. For the industrial process the critical control points (CCPs) and their monitoring must be established, and therefore the manufacturing process should be controlled by HACCP. If HACCP is not appropriate, guidelines for the application of manufacturing control by GMP must be defined.

Source: Adapted from Refs. 19,47,87.

for the necessary safety and stability of a particular food product, and the technologist should determine which ingredients or processes are proper for establishing these hurdles in a food, taking into account the legal, technological, sensory, and nutritive limitations. Because the engineering, economic, and marketing aspects must also be considered food design is indeed a multidisciplinary endeavor [7,19].

Predictive microbiology [88,89] is a promising concept that involves computer-based and quantitative predictions of microbial growth, survival, and death in foods, and thus should be an integral part of advanced food design (see Table 8, steps 4 and 7). However, the predictive models constructed so far handle only up to four different factors (hurdles) simultaneously. There are numerous hurdles to be considered which are important for the stability, safety, and quality of a particular food (see Sec. II.D). It is unlikely that all or even a majority of these hurdles could be covered by predictive modeling. Thus, predictive microbiology cannot be a quantitative approach to the totality of hurdle technology. However, it does allow quite reliable predictions of the fate of microorganisms in food systems, while considering few but the most important hurdles. Because several hurdles are not taken into account, the predicted results are fortunately often on the safe side, i.e., the limits indicated for growth of pathogens in foods by the models available are in general more prudent than the limits in real foods. Nevertheless, predictive microbiology will be an important tool for advanced food design, because it can narrow down considerably the range over which challenge tests with relevant microorganisms need to be performed. Although predictive microbiology will never render challenge testing obsolete, it may greatly reduce both the time for and costs of product development [7,12,19].

After the food has been properly designed, its manufacturing process must be effectively controlled, for which purpose the application of HACCP might be suitable. However, in a strict sense the HACCP concept only controls the hazards of foods and not their stability or quality [90]. Even in commercial practice safety and quality issues will often overlap if HACCP is applied [19]. Since for hurdle-technology foods microbial safety and stability as well as sensory quality, i.e., the total quality of the food (see Sec. II.C), is essential, the HACCP concept might be too narrow for this purpose if it relates only to biological, chemical, and physical hazards. Therefore, the HACCP concept should be broadened in order to cover the microbial safety (food poisoning) and stability (spoilage) of foods as well as their sensory quality. If this is not acceptable, the production process should be controlled by good manufacturing practice (GMP), and rules or guidelines for the production of each food item must be defined [33]. For hurdle-technology foods of developing countries GMP guidelines are often more acceptable because the application of HACCP poses practical difficulties where many small producers prevail.

VI. CONCLUSIONS

The stability, safety, and quality of most preserved foods are based on empirical application of combined methods for preservation and more recently on knowingly employed hurdle technology. The deliberate and intelligent application of hurdle technology allows gentle but efficient preservation of safe, stable, nutritious, and tasty foods and is advancing worldwide. Moreover, knowledge of the physiological mechanisms of growth, survival, and death of pathogenic and spoilage microorganisms in foods is increasing, since homeostasis, metabolic exhaustion, and stress reactions of microorganisms in relation to the hurdle effect are now better understood. Therefore the multitarget preservation as the ultimate goal for the effective preservation of foods is becoming more likely to be achieved soon.

In industrialized countries, the hurdle-technology approach is at present of most interest for minimally processed, fresh-like foods which are mildly heated or fermented, and for underpinning the microbial stability and safety of foods coming from future lines, such as healthful foods with less fat and/or salt or advanced hurdle-technology foods that require less packaging. For refrigerated foods chill temperatures are the major and sometimes the only hurdle. But if exposed to temperature abuse during distribution this hurdle breaks down, and spoilage or toxin formation could happen. Therefore, additional hurdles should be incorporated as safeguards into chilled foods using an approach called "invisible technology."

In developing countries the intentional application of hurdle technology for foods that remain stable, safe, and flavorful even if stored without refrigeration has made impressive strides (especially in Latin America with the development of novel high-moisture fruit products). However, much interest in intentional hurdle technology is also emerging for meat products in China as well as for dairy products of India. There is a general trend in developing countries to move gradually away from intermediate-moisture foods because they are often too salty or too sweet and have a less appealing texture and appearance than high-moisture foods. However, deliberate hurdle technology should be applied to high-moisture foods without sacrificing microbial stability and safety, especially of foods stored without refrigeration. Therefore, if hurdle-technology foods become more sophisticated, they will require a thorough understanding of the principles involved as well as more back-up of their production by guidelines based on good manufacturing practice (GMP) and, where appropriate, by application of the HACCP concept.

Hurdle-technology foods are in general less robust than traditional food products, which are often overprocessed and thus possess a large margin of safety. Therefore, if modified hurdle-technology foods are produced the applied processes must be exactly defined and controlled. For the design of hurdle technology foods a 10-step procedure has been suggested, which comprises hurdle technology, predictive microbiology, and HACCP (or GMP guidelines). This procedure has proved suitable for solving real product development tasks in the food industry, but it is open to further improvements. Hurdle technology should not lead to the addition of too much additives, but actually should reduce the amount of additives used even if their number might increase. It is of paramount importance that additional hurdles be introduced into a food product only after careful consideration of the necessity and in essential amounts, otherwise an undesirable chemical overloading of the food might result.

Combined methods used for tissue preservation are by no means a new development, as pointed out by Chirife et al. [91] in their study on mummification in ancient (>3000 years ago) Egypt. In the opinion of these authors, embalmed mummies contained at least three hurdles, namely, reduced a_w (0.72), increased pH (10.6), and preservatives (spices, aromatic plants). Today the action of combined preservative factors is much better understood, and intentional and intelligent application of such factors is progressing. Further applications of hurdle technology for the optimization of traditional foods as well as in the design of new foods are anticipated.

REFERENCES

1. L. Leistner, Hurdle effect and energy saving, *Food Quality and Nutrition* (W. K. Downey, ed.), Applied Science Publishers, London, 1978, p. 553.
2. L. Leistner, W. Rödel, and K. Krispien, Microbiology of meat products in high- and intermediate-moisture range, *Water Activity: Influences on Food Quality* (L. B. Rockland and G. F. Stewart, eds.), Academic Press, New York, 1981, p. 855.
3. L. Leistner, Hurdle technology applied to meat products of the shelf stable product and intermediate moisture food types, *Properties of Water in Foods in Relation to Quality and Stability* (D. Simatos and J. L. Multon, eds.), Martinus Nijhoff Publishers, Dordrecht, 1985, p. 309.
4. L. Leistner and L. G. M. Gorris, *Food Preservation by Combined Processes*, FLAIR Final Report, EUR 15776 EN, European Commission, Brussels, 1994.
5. L. Leistner and L. G. M. Gorris, Food preservation by hurdle technology, *Trends Food Sci. Technol. 6*:41 (1995).
6. L. Leistner, Food preservation by combined methods, *Food Res. Int. 25*:151 (1992).
7. L. Leistner, Principles and applications of hurdle technology, *New Methods of Food Preservation* (G. W. Gould, ed.), Blackie Academic & Professional, London, 1995, p. 1.
8. L. Leistner, Food protection by hurdle technology, *Bull. Jpn. Soc. Res. Food Prot. 2*:2 (1996).
9. L. Leistner, I. Vuković, and J. Dresel, SSP: meat products with minimal nitrite addition, storable without refrigeration, Proceedings of 26th Eur. Meeting Meat Res. Workers, Vol. II, Colorado Springs, 1980, p. 230.
10. D. W. Stanley, Biological membrane deterioration and associated quality losses in food tissues, *Crit. Rev. Food. Sci. Nutr. 30*:487 (1991).
11. B. M. McKenna, Combined processes and total quality management, *Food Preservation by Combined Processes* (L. Leistner and L. G. M. Gorris, eds.), FLAIR Final Report, EUR 15776 EN, European Commission, Brussels, 1994, p. 99.
12. L. Leistner, Further developments in the utilization of hurdle technology for food preservation, *J. Food Eng. 22*:421 (1994).
13. L. Bøgh-Sørensen, Description of hurdles, *Food Preservation by Combined Processes* (L. Leistner and L. G. M. Gorris, eds.), FLAIR Final Report, EUR 15776 EN, European Commission, Brussels, 1994, p. 7.
14. L. Leistner, Emerging concepts for food safety, Proceedings of the 41st ICoMST, San Antonio, TX, 1995, p. 321.
15. G. W. Gould, Interference with homeostasis—food, *Homeostatic Mechanisms in Micro-organisms* (R. Whittenbury, G. W. Gould, J. G. Banks, and R. G. Board., eds.), Bath University Press, Bath, 1988, p. 220.
16. D. Häussinger, ed., *pH Homeostasis—Mechanisms and Control*, Academic Press, London, 1988.
17. G. W. Gould, Homeostatic mechanisms during food preservation by combined methods, *Food Preservation by Moisture Control, Fundamentals and Applications* (G. V. Barbosa-Cánovas and J. Welti-Chanes, eds.), Technomic, Lancaster, PA, 1995, p. 397.
18. L. Leistner and S. Karan-Djurdjić, Beeinflussung der Stabilität von Fleischkonserven durch Steuerung der Wasseraktivität, *Fleischwirtschaft 50*:1547 (1970).
19. L. Leistner, *Food Design by Hurdle Technology and HACCP*, Adalbert-Raps-Foundation, Kulmbach, 1994.

20. H.-K. Shin, *Energiesparende Konservierungsmethoden für Fleischerzeugnisse, abgeleitet von traditionellen Intermediate Moisture Foods*, Ph.D. thesis, Universität Hohenheim, Stuttgart-Hohenheim, 1984.
21. S. Sajur, *Preconservación de Duraznos por Métodos Combinados*, M.S. thesis, Universidad Nacional de Mar del Plata, Argentina, 1985.
22. S. M. Alzamora, M. S. Tapia, A. Argaiz, and J. Welti, Application of combined methods technology in minimally processed fruits, *Food Res. Int. 26*:125 (1993).
23. S. M. Alzamora, P. Cerrutti, S. Guerrero, and A. López-Malo, Minimally processed fruits by combined methods, *Food Preservation by Moisture Control, Fundamentals and Applications* (G. V. Barbosa-Cánovas and J. Welti-Chanes, eds.), Technomic, Lancaster, PA, 1995, p. 463.
24. M. S. Tapia de Daza, A. Argaiz, A. López-Malo, and R. V. Díaz, Microbial stability assessment in high and intermediate moisture foods: special emphasis on fruit products, *Food Preservation by Moisture Control, Fundamentals and Applications* (G. V. Barbosa-Cánovas and J. Welti-Chanes, eds.), Technomic, Lancaster, PA, 1995, p. 575.
25. S. P. Denyer and W. B. Hugo, eds., *Mechanisms of Action of Chemical Biocides: Their Study and Exploitation*, Blackwell Scientific Publications, London, 1991.
26. L. Leistner and W. Rödel, The stability of intermediate moisture foods with respect to micro-organisms, *Intermediate Moisture Foods* (R. Davies, G. G. Birch, and K. J. Parker, eds.), Applied Science Publishers, London, 1976, p. 120.
27. J. M. Aguilera, J. Chirife, M. S. Tapia de Daza, J. Welti-Chanes, and E. Parada Arias, *Inventario de Alimentos de Humedad Intermedia Tradionales de Iberoamérica*, Unidad Profesional Interdisciplinaria de Biotecnología, Instituto Politécnico National, México, 1990.
28. J. Welti, M. S. Tapia de Daza, J. M. Aguilera, J. Chirife, E. Parada, A. López Malo, L. C. López, and P. Corte, Classification of intermediate moisture foods consumed in Ibero-America, *Rev. Esp. Cienc. Tecnol. Aliment. 34*:53 (1994).
29. M. S. Tapia de Daza, J. M. Aguilera, J. Chirife, E. Parada, and J. Welti, Identification of microbial stability factors in traditional foods from Iberoamerica, *Rev. Esp. Cienc. Tecnol. Aliment. 34*:145 (1994).
30. L. Leistner, F. Wirth, and I. Vuković, SSP (Shelf Stable Products)—Fleischerzeugnisse mit Zukunft, *Fleischwirtschaft 59*:1313 (1979).
31. L. Leistner, Use of combined preservative factors in foods of developing countries, *The Microbiology of Foods* (B. M. Lund, A. C. Baird-Parker, and G. W. Gould, eds.), Chapman and Hall, London (in press).
32. T. El-Khateib, U. Schmidt, and L. Leistner, Mikrobiologische Stabilität von türkischer Pastirma, *Fleischwirtschaft 67*:101 (1987).
33. L. Leistner, Shelf-stable products and intermediate moisture foods based on meat, *Water Activity: Theory and Applications to Food* (L. B. Rockland and L. R. Beuchat, eds.), Marcel Dekker, Inc., New York, 1987, p. 295.
34. J. A. Torres, Microbial stabilization of intermediate moisture food surfaces, *Water Activity: Theory and Applications to Food* (L. B. Rockland and L. R. Beuchat, eds.), Marcel Dekker, Inc., New York, 1987, p. 329.
35. S. Guilbert, Technology and application of edible protective film, *Food Packaging and Preservation* (M. Matathlouthi, ed.), Elsevier Applied Science Publishers, New York, 1986, p. 371.
36. S. Guilbert, Edible coatings and osmotic dehydration, *Food Preservation by Combined Processes* (L. Leistner and L. G. M. Gorris, eds.), FLAIR Final Report, EUR 15776 EN, European Commission, Brussels, 1994, p. 65.
37. C. R. Lerici, D. Mastrocola, A. Sensidoni, and M. Dalla Rosa, Osmotic concentration in food processing, *Preconcentration and Drying of Food Materials* (S. Bruin, ed.), Elsevier Applied Science Publishers, Amsterdam, 1988, p. 123.
38. A. L. Raoult-Wack, S. Guilbert, and A. Lenart, Recent advances in drying through immersion in concentrated solutions, *Drying of Solids* (A. S. Mujumdar, ed.), International Science Publishers, New York, 1992, p. 21.

39. A. L. Raoult-Wack, Recent advances in the osmotic dehydration of foods, *Trends Food Sci. Technol. 5*:255 (1994).
40. L. Leistner, Stable and safe fermented sausages world-wide, *Fermented Meats* (G. Campbell-Platt and P. E. Cook, eds.), Blackie Academic & Professional, London, 1995, p. 160.
41. L. Leistner, Allgemeines über Rohschinken, *Fleischwirtschaft 66*:496 (1986).
42. K. Katsaras and L. Leistner, Distribution and development of bacterial colonies in fermented sausages, *Biofouling 5*:115 (1991).
43. M. Robins, T. Brocklehurst, and P. Wilson, Food structure and the growth of pathogenic bacteria, *Food Technol. Int. Eur.*:31 (1994)
44. J. W. T. Wimpenny, L. Leistner, L. V. Thomas, A. J. Mitchell, K. Katsaras, and P. Peetz, Submerged bacterial colonies within food and model systems: their growth, distribution and interactions, *Int. J. Food Microbiol. 28*:299 (1995).
45. H. Hechelmann and L. Leistner, Mikrobiologische Stabilität autoklavierter Darmware, *Mitteilungsbl. Bundesanst. Fleischforsch. Kulmbach* (84):5894 (1984).
46. H. Hechelmann, L. Leistner, and R. Albertz, Ungleichmäßiger a_w-Wert als Ursache für mangelhafte Stabilität von F-SSP, *Jahresber. Bundesanst. Fleischforsch. Kulmbach 1985*:C 27 (1985).
47. L. Leistner and H. Hechelmann, Food preservation by hurdle-technology, Proceedings of Food Preservation 2000, Vol. II, U.S. Army Natick, Research, Development and Engineering Center, Natick, MA, 1993, p. 511.
48. H. Hechelmann, R. Kasprowiak, S. Reil, A. Bergmann, and L. Leistner, *Stabile Fleischerzeugnisse mit Frischprodukt-Charakter für die Truppe*, BMVg FBWM 91-11, Dokumentations- und Fachinformationszentrum der Bundeswehr, Bonn, Germany, 1991.
49. P. Giavedoni, *Azioni Combinate nella Stabilizzazione degli Alimenti*, Ph.D. thesis, Università degli Studi di Udine, Udine, Italy, 1994.
50. P. Giavedoni, W. Rödel, and J. Dresel, Beitrag zur Sicherheit und Haltbarketi von frischen gefüllten Teigwaren, abgepackt in modifizierter und in einer Äthanol-Gas-Atmosphäre, *Fleischwirtschaft 74*:639 (1994).
51. K. J. Rao, J. Dresel, and L. Leistner, Anwendung der Hürden-Technologie in Entwicklungsländern, zum Beispiel für Paneer, *Mitteilungsbl. Bundesanst. Fleischforsch. Kulmbach 31*:293 (1992).
52. K. J. Rao, *Application of Hurdle Technology in the Development of Long Life Paneer-Based Convenience Food*, Ph.D. thesis, National Dairy Research Institute, Karnal, India, 1993.
53. L. Leistner, Linkage of hurdle-technology with HACCP, Proceedings of 45th Reciprocal Meat Conference, Chicago, 1992, p. 1.
54. W. Rödel, R. Scheuer, and H. Wagner, A new method of determining water activity in meat products, *Fleischwirtschaft Int. 1990*:22 (1990).
55. L. G. M. Gorris and H. W. Peppelenbos, Modified atmosphere and vacuum packaging to extend the shelf-life of respiring products, *Hort. Technol. 2*:303 (1992).
56. I. J. Church and A. L. Parsons, Modified atmosphere packaging technology: a review, *J. Sci. Food Agr. 67*:143 (1994).
57. C. Nguyen-the and F. Carlin, The microbiology of minimally processed fresh fruits and vegetables, *Crit. Ref. Food Sci. Nutr. 34*:371 (1994).
58. M. H. J. Bennik, H. W. Peppelenbos, C. Nguyen-the, F. Carlin, E. J. Smid, and L. G. M. Gorris, Microbiology of minimally processed, modified-atmosphere packaged chicory endive, *Postharv. Biol. Technol. 9*:209 (1996).
59. C. A. Phillips, Modified atmosphere and its effects on the microbiological quality and safety of produce, *Int. J. Food Sci. Technol. 31*:463 (1996).
60. L. R. Beuchat, Pathogenic microorganisms associated with fresh produce, *J. Food Prot. 59*:204 (1995).
61. M. H. J. Bennik, W. van Overbeek, E. J. Smid, and L. G. M. Gorris, Biopreservation for the control of *Listeria monocytogenes* on minimally processed, modified atmosphere stored vegetables, *Lett. Appl. Microbiol.* (in press).
62. L. Leistner, Microbial stability of low fat and/or low salt meat products, Proceedings of 43rd ICOMST, Auckland, New Zealand, 1997, p. 414.

63. L. Leistner, Microbial stability and safety of healthy meat, poultry and fish products, *Production and Processing of Healthy Meat, Poultry and Fish Products* (A. M. Pearson and T. R. Dutson, eds.), Blackie Academic and Professional, London, 1997, p. 347.
64. L. Leistner, Stable hurdle technology foods and packaging—worldwide, *J. Pack. Sci. Technol. Jpn. 6*:4 (1997).
65. A. L. Brody, Active packaging 2001, *Meat Int. 2*(9/10):42 (1991).
66. T. P. Labuza, An introduction to active packaging for foods, *Food Technol. 50*(4):68 (1996).
67. K. Ono, *Packaging Design and Innovation*, material for a Third Country Training Programme in the field of food packaging, conducted February 20–March 5, 1994, Singapore.
68. Kentaro Ono, Snow Brand Tokyo, Japan, personal communication.
69. A. López-Malo, E. Palou, J. Welti, P. Corte, and A. Argaiz, Shelf-stable high moisture papaya minimally processed by combined methods, *Food Res. Int. 27*:545 (1994).
70. S. Guerrero, S. M. Alzamora, and L. N. Gerschenson, Development of a shelf-stable banana purée by combined factors: microbial stability, *J. Food Prot. 57*:902 (1994).
71. A. Argaiz, A. López-Malo, and J. Welti-Chanes, Considerations for the development and stability of high moisture fruit products during storage, *Food Preservation by Moisture Control, Fundamentals and Applications* (G. V. Barbosa-Cánovas and J. Welti-Chanes, eds.), Technomic, Lancaster, PA, 1995, p. 729.
72. M. S. Tapia de Daza, S. M. Alzamora, and J. Welti Chanes, Combination of preservation factors applied to minimal processing of foods, *Crit. Rev. Food Sci. Nutr. 36*:629 (1996).
73. N. K. Rastogi, J. S. Sandhi, P. Viswanath, and S. Saroja, Application of hurdle/combined method technology in minimally processed long-term non-refrigerated preservation of banana and coconut, Abstracts for ICFoST '95, Mysore, India, 1995, p. 109.
74. L. Leistner, Use of hurdle technology in food processing: recent advances, *Food Preservation by Moisture Control, Fundamentals and Applications* (G. V. Barbosa-Cánovas and J. Welti-Chanes, eds.), Technomic, Lancaster, PA, 1995, p. 377.
75. L. Leistner, Fermented and intermediate moisture products, Proceedings of 36th ICOMST, vol. III, Havana, Cuba, 1990, p. 842.
76. W. Wang and L. Leistner, Shafu: a novel dried meat product of China based on hurdle-technology, *Fleischwirtschaft 73*:854 (1993).
77. W. Wang und L. Leistner, Traditionelle Fleischerzeugnisse von China und deren Optimierung durch Hürden-Technologie, *Fleischwirtschaft 74*:1135 (1994).
78. W. Wang and L. Leistner, Hurdle technology applied to traditional meat products, *Meat Res. 1995*(3):8 (1995), in Chinese.
79. W. Wang and L. Leistner, Application of hurdle technology in the development of food products. Part I, *Meat Res. 1996*(1):42 (1996), in Chinese.
80. W. Wang and L. Leistner, Application of hurdle technology in the development of food products. Part II, *Meat Res. 1996*(2):42 (1996), in Chinese.
81. X. Q. Zhu, Developments in the theory of food preservation and its applications in foreign countries, *Meat Res. 1996*(2):39 (1996), in Chinese.
82. X. Q. Zhu and L. Leistner, Water activity and food preservation, *Meat Res. 1996*(3):48 (1996), in Chinese.
83. R. Xia and Q. N. Hsu, Processing method for Islamic dried beef in dices, *Meat Res. 1996*(3):32 (1996), in Chinese.
84. J. C. Kuo, J. Dresel, and L. Leistner, Effects of sodium lactate and storage temperature on growth and survival of *Staphylococcus aureus, Listeria monocytogenes* and *Salmonella* in Chinese sausage, *Chin. Food Sci. 21*:182 (1994).
85. J. S. Berwal, Hurdle technology for shelf-stable food products, *Ind. Food Industry 13*:40 (1994).
86. S. A. Hossain, *Technological innovation in manufacturing dudh churpi*, Ph.D. thesis, University of North Bengal, Siliguri, India, 1994.
87. L. Leistner, User guide to food design, *Food Preservation by Combined Processes* (L. Leistner and L. G. M. Gorris, eds.), FLAIR Final Report, EUR 15776 EN, European Commission, Brussels, 1994, p. 25.

88. T. A. McMeekin, J. N. Olley, T. Ross, and D. A. Ratkowsky, *Predictive Microbiology: Theory and Application*, Research Studies Press Ltd., Taunton, Somerset, 1993.
89. P. J. McClure, C. de W. Blackburn, M. B. Cole, P. S. Curtis, J. E. Jones, J. D. Legan, I. D. Ogden, M. W. Peck, T. A. Roberts, J. P. Sutherland, and S. J. Walker, Modelling the growth, survival and death of microorganisms in foods: the UK Food Micromodel approach, *Int. J. Food Microbiol. 23*:265 (1994).
90. M. D. Pierson and D. Corlett jr., *HACCP: Principles and Applications*, Van Nostrand Reinhold, A Division of Wadsworth Inc., 1992.
91. J. Chirife, G. Favetto, S. Ballesteros, and D. Kitic, Mummification in ancient Egypt: an old example of tissue preservation by hurdle technology, *Lebensm. Wiss. Technol. 24*:9 (1991).

17

Nonthermal Preservation of Liquid Foods Using Pulsed Electric Fields

HUMBERTO VEGA-MERCADO, M. MARCELA GÓNGORA-NIETO, GUSTAVO V. BARBOSA-CÁNOVAS, AND BARRY G. SWANSON
Washington State University, Pullman, Washington

I. INTRODUCTION

There are many different forms in which to apply electric energy for food pasteurization. These include ohmic heating [1–3], microwave heating [4–6], low electric field stimulation [7,8], high-voltage arc discharge [9-12], and high-intensity pulsed electric field (PEF) application [13–15].

Ohmic heating is one of the earliest forms of electricity applied to food pasteurization [1]. This method relies on the heat generated in food products when an electric current is passed through them. Getchell [2] described the ohmic heating method in milk pasteurization. A 220 V, 15 kW alternating current supply was applied to milk through carbon electrodes in an electrical heating chamber. The milk was heated to and held at 70°C for about 15 seconds. It has been reported that ohmic heating is suitable for viscous products and foods containing particles, and this method is considered to be a promising technique for the aseptic processing of foods [3].

Microwave heating has been extensively applied in everyday households and the food industry [4]. Many food materials possess very low values of static conductivity. However, when they are subjected to microwave fields, they exhibit very high values of alternating field conductivity and consume considerable energy [5]. The heat generated by microwaves is used for heating processes. Studies on microbial inactivation using microwave energy have concluded that microbial death is caused solely by thermal mechanisms [6].

Low electric field stimulation has been explored as a method of bacterial control of meat. In electrical stimulation of meat an electric field of 5–10 V/cm is applied as alternate current (ac) pulses to the sample through electrodes fixed at opposite ends of the long axis of the muscle [7]. Recently, a very low field (0.4 V/cm) has been applied in a 6 liter treatment medium in search of an easy, safe, and practical method to eliminate bacteria for food-processing purposes. Several species of bacteria in saline solution were inactivated [8]. Salt solutions and their concentrations play a very important role in this method.

Inactivation of microorganisms and enzymes contained in food products by electric discharge began in the 1920s with the Electropure process for milk [16], which consisted of passing an electric current through carbon electrodes and heating milk to 70°C to inactivate *Mycobacterium tuber-*

culosis and *Escherichia coli*. Beattie and Lewis [17] demonstrated a lethal effect of electrical discharges on microorganisms when the applied voltage used to treat food was increased 3000–4000 volts. The electrohydraulic treatment was introduced in the 1950s to inactivate microorganisms suspended in liquid foods. The inactivation of microorganisms was attributed to a shock wave generated by an electric arc that prompted the formation of highly reactive free radicals from chemical species in food [14]. Gilliland and Speck [18] applied pulsed electric discharges at different energy levels for the inactivation of *E. coli, Streptococcus faecalis, Bacillus subtilis, Streptococcus cremoris,* and *Micrococcus radiodurans* suspended in sterile distilled water as well as for trypsin and a protease from *B. subtilis*.

Sale and Hamilton [19] demonstrated the nonthermal lethal effect of homogeneous electric fields on bacteria such as *E. coli, Staphylococcus aureus, Micrococcus lysodeikticus, Sarcina lutea, B. subtilis, Bacillus cereus, Bacillus megaterium, Clostridium welchii*, and yeasts such as *Saccharomyces cerevisiae* and *Candida utilis*. In general, an increase in the electric field intensity and number of pulses was found to lead to an increase in the inactivation of microorganisms (Fig. 1, Table 1). Other factors that influence microbial inactivation by pulsed electric fields are the treatment temperature, pH, ionic strength, and conductivity of the medium containing the microorganisms [9,20–26].

Formation of pores on cell membranes by high-intensity pulsed electric fields (HIPEFs) is not entirely understood. Zimmermann et al. [27], applying the dielectric rupture theory, concluded that membrane rupture is caused by an induced transmembrane potential approximately 1V larger than the natural potential of the cell membrane.

The reversible or irreversible rupture (or electroporation) of a cell membrane depends on factors such as intensity of the electric field, number of pulses, and duration of the pulses [28–31]. The plasma membranes of cells becomes permeable to small molecules after being exposed to an electric field; permeation then causes swelling and the eventual rupture of the cell membrane (Fig. 2).

In September of 1996, the U.S. Food and Drug Administration (FDA), based in Washington, D.C., released a "letter of no objection" for the use of pulsed electric fields to treat liquid eggs. To meet the FDA requirements [32] in filing a new and a novel process, it is necessary to (a) establish an active and continuous dialog with the FDA during process development, (b) meet with the FDA to describe the process, (c) invite the FDA to a site visit (pilot and production facility), and (d) draft and provide the FDA with an outline of the proposed filing.

The objective of the FDA is to conduct a scientific evaluation of the process to determine if the aseptically produced product poses a potential public health hazard and if all of the critical factors necessary to render the product commercially sterile are monitored and controlled. The filing information of the new process must contain:

1. Equipment design: a description of the system, control mechanisms used, and fail-safe procedures

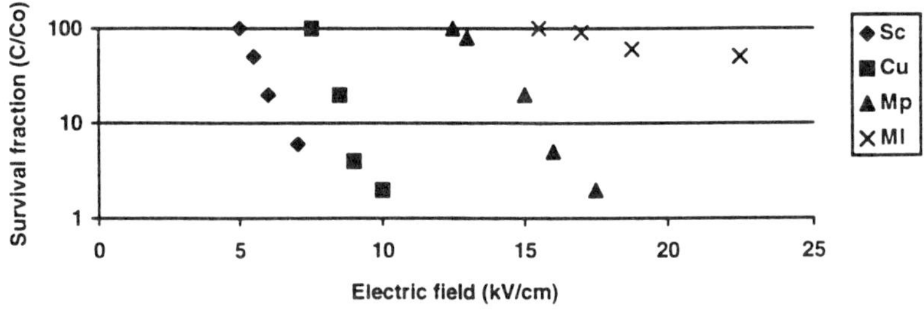

Figure 1 Relationship between survival fraction and electric field strength (10 20-μs pulses). (**Sc** = Saccharomyces cerevisiae; **Cu** = Candida utilis; **Mp** = Motile pseudomonad; **Ml** = Micrococcus lysodeikticus. (Adapted from Ref. 59.)

TABLE 1 Activity of *Staphylococcus aureus* After Pulsed Electric Field Treatment

Electric field (kV/cm)	Survivors (%)	Protoplasts not lysed (%)
0.00	100	100
9.25	100	100
14.25	35	43
19.50	0.9	16
24.00	0.3	3
27.50	0.6	2

Source: Adapted from Ref. 59.

2. Product specifications: a full description of the product, including physical/chemical aspects, critical factors, and influence of processing on the critical factors
3. Process design: a complete description of the critical/processing conditions used in the manufacture of the product
4. Validation: a physical demonstration of the accuracy, reliability, and safety of the process

In the area of pulsed electric fields, there are many possible project-development designs related to (a) unknown destruction kinetics of microbial pathogens (e.g., *Clostridium botulinum*), (b) identification of proper indicator organisms, (c) uniformly delivered treatment, (d) impact of processing conditions (e.g., temperature, pH, moisture, and lipid content), (e) identification/monitoring of critical factors (e.g., surface, intensity), and (f) food additives.

II. ENGINEERING ASPECTS OF PULSED ELECTRIC FIELDS

The concept of pulsed power is simple: electric energy at low power levels is collected over an extended period and stored in a capacitor. That same energy can then be discharged almost instantaneously at very high levels of power. The generation of pulsed electric fields requires two major devices: a pulsed power supply and a treatment chamber, which converts the pulsed voltage into pulsed electric fields.

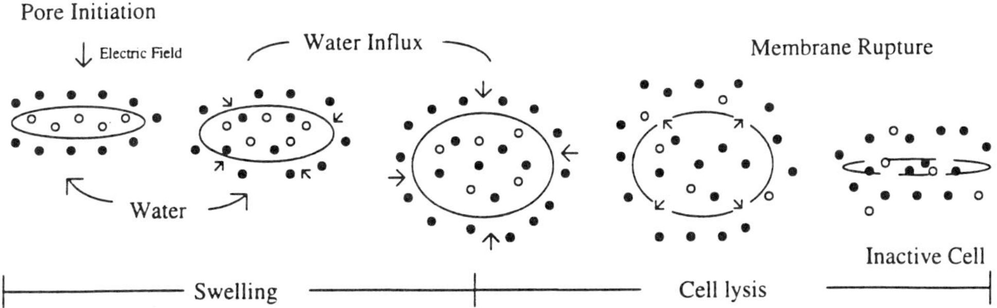

FIGURE 2 Mechanism of cell inactivation. (Adapted from Ref. 30.)

A. Bench-Top Unit

A commercial electroporator (i.e., GeneZapper, IBI-Kodak Company, Rochester, NY) may be used as a bench-top pulsed power supply. This unit provides a maximum of 2.5 kV pulses. The instrument consists of a capacitor (7 μF), charge and discharge switches, and a wave controller. The wave controller may be connected to the electroporator to improve the discharge pattern. Treatment cuvettes with a 0.1-cm electrode gap and 100-μl volume may be used for PEF treatments, which give a maximum field intensity of approximately 25 kV/cm. Appropriate voltage and current monitors should be attached to the GeneZapper to measure the pulsed electric field treatments. Figure 3 illustrates the major components of the GeneZapper. This bench-top unit provides a convenient method for determining the inactivation kinetics for selected microorganisms.

B. Lab Scale Pulser

Exponential decay electric pulses could be generated by discharging a capacitor into a chamber containing the food (Figs. 4–6). Current designs for power supplies are able to provide up to 40 kV. Capacitors of 5 μF are used to store the electric energy that is discharged across metal electrodes, creating the electric field used to inactivate microorganisms and enzymes. A mercury ignitron spark gap may be used as the discharge switch. This type of unit may be employed for inactivation studies in a continuous mode.

Pulsed voltage across the treatment chamber may be monitored by a resistance voltage divider. Electric current may be monitored by a Rogowski coil connected to a passive integrator. Both voltage and current waveforms may be monitored using a digital oscilloscope.

C. Treatment Chambers

A static PEF treatment chamber consists of two electrodes held in position by insulating materials that also forms an enclosure containing food materials. Uniform electric fields can be achieved by parallel plate electrodes with a gap sufficiently smaller than the electrode surface dimension. Disk-shaped, round-edged electrodes can minimize electric field enhancement and reduce the possibility of dielectric breakdown of fluid foods. A continuous flow-through treatment chamber (Fig. 7) was developed at Washington State University (WSU) to test the flow-through concept using low flow rates. The chamber consisted of two electrodes, a spacer, and two lids. Each electrode was made of stainless steel, whereas the spacer and lids were made of polysulfone. A flow channel was provided between the two electrodes to eliminate dead corners as well as to ensure uniform treatment.

The operating conditions for the parallel plate continuous chamber were as follows: chamber volume 20 or 8 cm^3; electrode gap 0.95 or 0.51 cm; PEF intensity 35 or 70 kV/cm; pulse width 2–15 μs; pulse rate 1 Hz; and food flow rate 1200 or 600 cm^3/min. Cooling of the chamber was accomplished by circulating water at a selected temperature through jackets built into the two stainless steel electrodes.

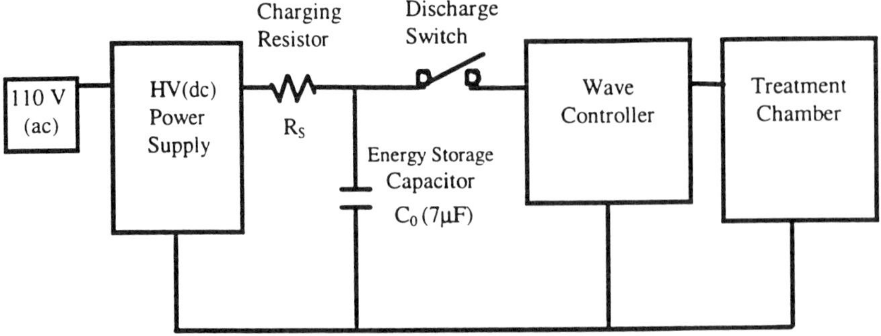

Figure 3 Major components of commercial electroporator GeneZapper.

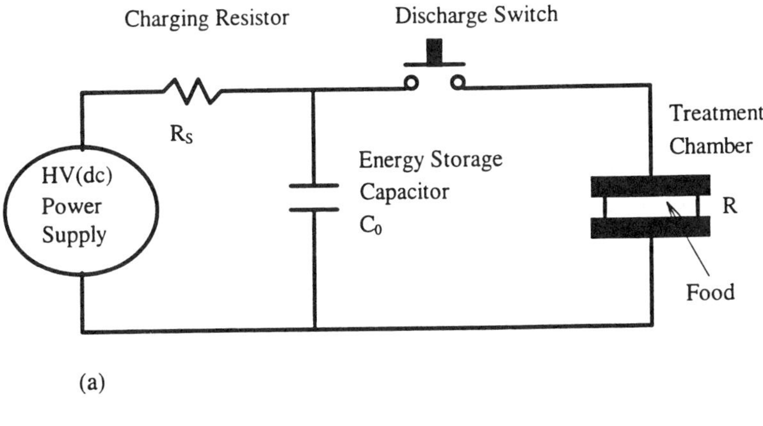

(a)

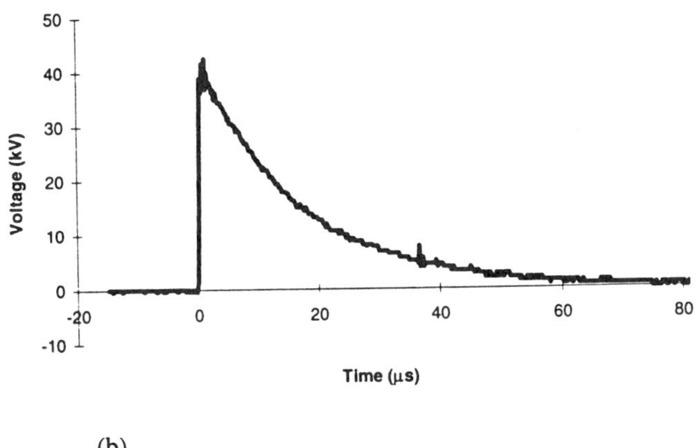

(b)

FIGURE 4 (a) A simplified circuit for producing exponential decay pulses and (b) a voltage trace across the treatment chamber.

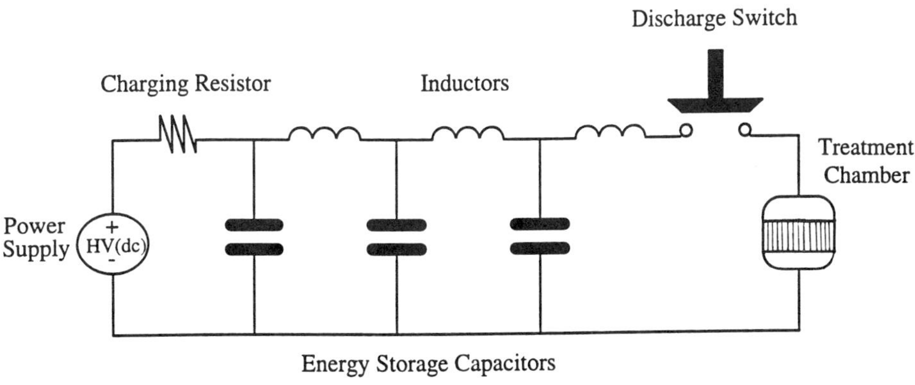

FIGURE 5 Typical pulser configuration for high-intensity pulsed electric fields.

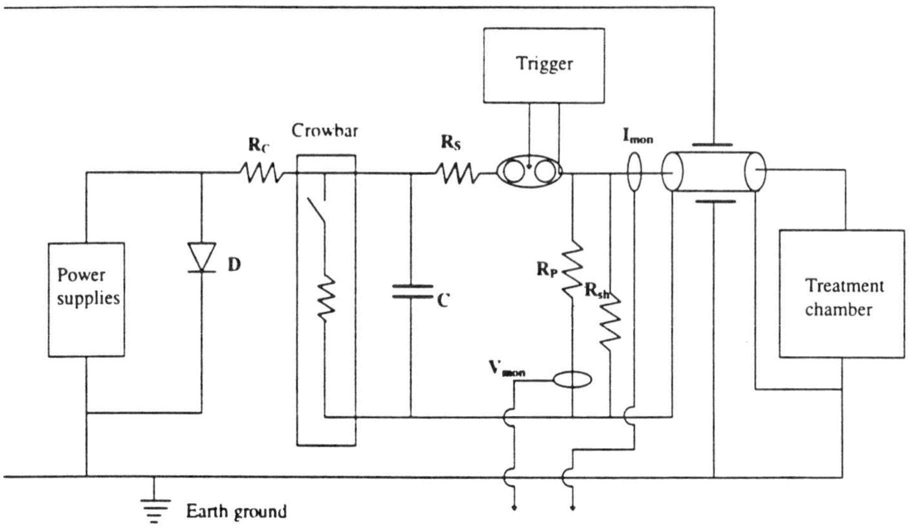

FIGURE 6 Current set-up of the PEF facility at Washington State University. The pulser has a 16 kJ/s charging power supply, 40 kV peak charging voltage, and 10 Hz pulse repetition rate. C = Storage capacitor; D = power supply protection diode; R_c = charging resistor; R_s = series resistor; R_{sh} = shunt resistor; R_p = voltage-measuring resistor; I_{mon} = current monitor; V_{mon} = voltage monitor.

It should also be pointed out that a completely sealed treatment chamber is dangerous. When the test fluid experiences a spark, high pressure develops rapidly and the chamber may break apart. A pressure-released device must be included in the treatment chamber design to ensure safety of the operation.

A coaxial treatment chamber (Fig. 8) with a uniform field distribution along the fluid path was designed at WSU. The fluid is fed into the chamber through the bottom region and treated product exits at the top of the chamber. The protruded surface, located at the outer grounded electrode, en-

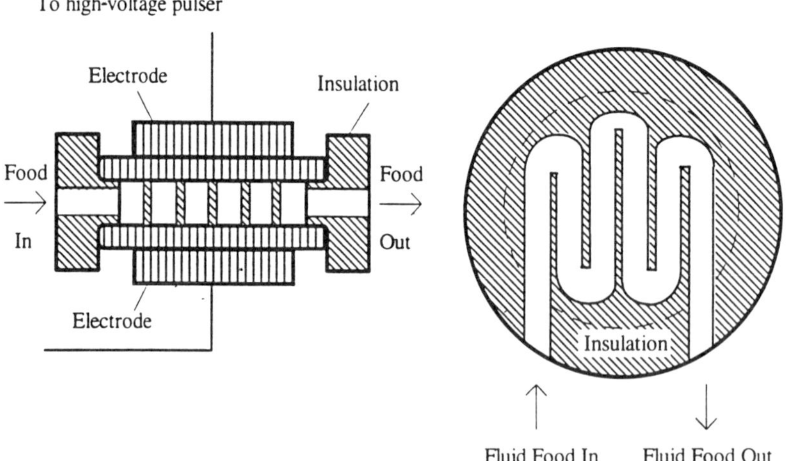

FIGURE 7 Schematic drawing of a flow-through treatment chamber. Fluid inside the chamber is baffled to avoid dead spots.

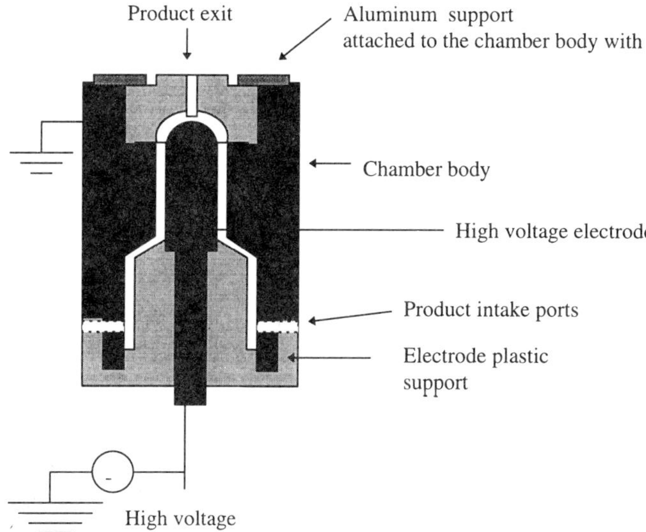

FIGURE 8 Schematic of the PEF continuous treatment chamber.

hances and makes uniform the electric field within the treatment region while it reduces the field intensity in other regions of the fluid path. Cooling fluid is circulated to control the temperature between the inner high-voltage electrode and the outer grounded electrode. The gap in the coaxial electrode or the liquid food thickness along the direction of the electric field can be selected by changing the diameter of the inner electrode.

D. Pulsed Electric Field Process Design

1. HACCP Principles and PEF Technology

The PEF process is summarized in Figure 9. The key operations are the receiving of raw materials, PEF treatment, aseptic packaging operation, and finished product storage and distribution. The following analysis [33] is based upon the seven principles of hazard analysis and critical control points (HACCP).

Hazard Assessment

Microbial hazards are the main concern throughout the PEF operation. Raw materials contain spoilage microbes and pathogens that may spoil the ingredient or raw material or may be harmful to the consumer. Storage facilities for raw materials may increase the risk of microbial contamination from soil and water deposits. The cleanliness of processing equipment plays a key role in preventing microbial contamination, thus the multiple assembly parts must always be properly sanitized. Inappropriate aseptic packaging operations and storage conditions may result in spoilage of the product.

Chemical hazards to consider are the presence of antibiotic and pesticide residuals on raw materials, electrically induced chemical reactions, and excessive detergent-sanitizer residues from processing and packaging equipment. Physical hazards include foreign matter in raw materials (e.g., stones, rubber, plastic, metal, eggshells), metal particles from the treatment chamber after a spark, and plastic or rubber pieces from seals.

The final risk classification may be defined in terms of the product (milk, apple juice, eggs, soups, etc.). Six microbiological hazard characteristics, as well as chemical and physical hazard characteristics, are defined by the National Advisory Committee on the Microbiological Criteria for Foods (NACMCF) and will be used to classify PEF products. In general, the final hazard classification should occur between risk categories IV and VI as defined by the NACMCF.

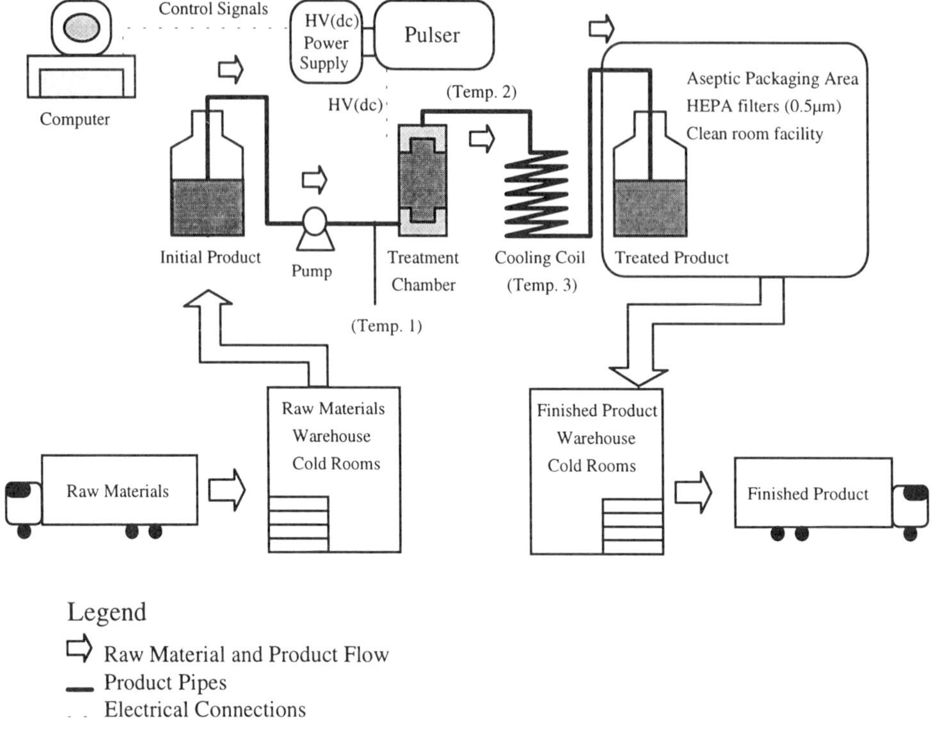

Figure 9 Pulsed electric field unit operations layout.

Critical Control Points: Determination, Limits, Procedures, and Corrective Actions

The following critical control points (CCPs) should be selected to ensure the safety of PEF products: receiving and storage section, PEF treatment section, and aseptic packaging section.

The main factors considered and monitored for each CCP are handling and processing time, temperature of material, and cleanliness of equipment and utensils. The treatment conditions (electric field intensity, pulsing rate, input voltage, input current, and chamber temperature) should be monitored and recorded on a continuous basis. Uniform PEF treatment requires the design and construction of a pulser that accomplishes variable pulsing rates, charging rates, voltage settings, pulse widths, and pulse shapes. Pulser components such as power source, computerized controls, triggering mechanism, overloads, dummy loads, and treatment chamber should comply with defined specifications and characteristics such as maximum operating temperature, maximum voltage and current outputs, and reliability (mean time between failures, yields, etc.). The reliability of the pulser may be measured in terms of number of pulses with correct energy level per unit of time as well as total pulses per unit of time. Monitoring devices may include oscilloscopes for voltage and current measures, and pulse counters.

Standard operating procedures (SOPs) should be in place to define aspects such as reception, storage, and preparation of raw materials, to ensure proper handling, and to reduce the risk of contamination. The pulsing and packaging units must have procedures to specify the assembly and disassembly of the machinery. Cleaning specifications such as frequency and type of detergents and sanitizers to use should be established to prevent contamination between products. The operational parameters for PEF treatments must be specified for each food product based upon its microbial risk, initial microbial counts, physical and chemical characteristics (e.g., pH, ionic strength, composition), and the maximum time to complete the processing of each food (i.e., time from initial discharge of raw materials to the end of the packaging operation). Alternative procedures must define the corrective actions associated with deviations from process specifications or CCP limits. Quality assurance procedures must be developed for the approval or rejection of PEF-treated products based on the CCP limits and corrective actions.

Record Keeping

Record keeping is a key aspect not only in a PEF operation, but in any successful manufacturing operation. The status of raw materials, process and packaging sequence, as well as storage and shipping procedures must be reflected in the batch or lot documents. Proper design of the documents is an important and difficult task because the documents must provide enough space for critical measurements without confusing the operator.

2. Hazard and Operability Study (HAZOP) Principles and PEF Technology

The main concern of individuals working in a PEF facility is the voltage intensity, which reaches the kilovolt range. A typical pulser configuration is presented in Figure 10. A high-voltage power supply is selected to charge the capacitor (eventually more than one) and a discharge switch releases the stored electric energy from the capacitor through the product in the form of an electric field. The power supply, capacitor, and treatment chamber must be confined in a restricted access area with interlocked gates. The gates will turn off the pulser if they are opened while the power supply is on. Emergency switches must be accessible in case of a process failure. Also, discharging bars must be provided to discharge the elements in the circuit before maintenance or inspection of the unit occurs. To prevent the leakage of high voltage through any fluid (food or refrigerant) in contact with the treatment chamber, all connections to the chamber will be isolated and the pipes carrying materials to or from the chamber connected to ground.

Electrical and mechanical devices such as pumps, computers, and packaging machines must be protected using safeguards. Proper warning signs must be in place regarding the safety hazards (high voltage, high-intensity electric field) in the processing area. The information related to the operation and maintenance procedures must be contained in standard operating procedures (SOPs). The personnel involved in the PEF operation must be trained and instructed in these SOPs.

The selection of appropriate detergents and sanitizers must comply with the FDA and USDA/FSIS regulations or these of equivalent organizations in other countries. Proper protection devices such

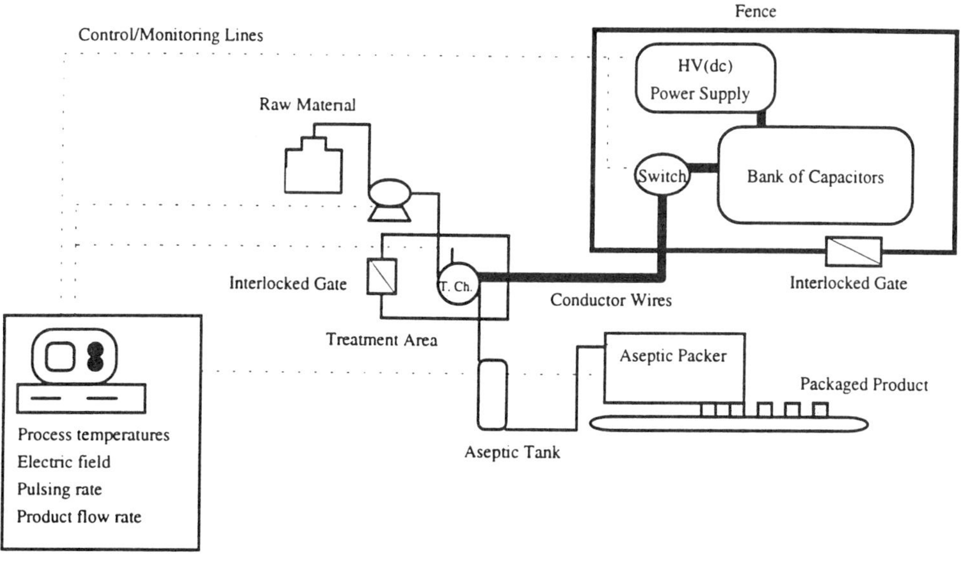

FIGURE 10 Schematic diagram of a PEF equipment configuration.

as face masks or goggles, aprons, boots, and gloves must be used by employees while applying and removing the cleaning solutions. A complete procedure must be in place to define what kind, when, where, and how to use the cleaning and sanitizing solutions. Proper record keeping is required to avoid contamination of the products with detergent or sanitizer solutions.

A complete layout of the facility including details about location of utilities, location of equipment, and emergency exits must be available. Changes in the configuration of the facility must be reflected in the layout.

E. Currently Used PEF Technology

PurePulse Technologies Co., a subsidiary of Maxwell Laboratories in San Diego, California, owns three U.S. patents to preserve fluid foods such as dairy products, fruit juices, and fluid eggs by treatment with high-intensity electric discharges from about 5 to 100 kV/cm with flat-topped exponentially decaying pulse shapes. Pulse duration is controlled to prevent electrical breakdown of the food product; the typical duration is between 1 and 100 μs with repetition rates between 0.1 and 100 Hz [11,34]. The patents describe both a batch and continuous processing system and recommend that HIPEF treatments be applied to preheated liquid foods, which enhance microbial inactivation and shelf-life stability.

Dunn and Pearlman [11] reported more than five logarithmic cycles of microbial count reduction (5D reduction) of naturally occurring microorganisms in orange juice after 35 pulses of 100 μs at a voltage intensity of 33.6–35.7 kV/cm and a process temperature of 42–65°C. The shelf life of orange juice was increased from 3 days to one week with no significant change in odor or taste. A 3D reduction of *E. coli* (ATCC-10536) inoculated in homogenized and pasteurized milk exposed to 23 pulses of 100 μs at 28.6–42.8 kV/cm was also reported. When a similar test run was carried out using milk seeded with *Salmonella dublin* prior to treatment with 36.7 kV/cm and 40 pulses of 100 μs at 63°C, no *Salmonella* and only 20 cfu/ml of milk bacteria was found. These results may suggest that deactivation from the PEF treatment process is selective and that *S. dublin* are preferentially deactivated over the milk bacteria. Yogurt inoculated with *Streptococcus thermophilus, Lactobacillus bulgaris*, and *Saccharomyces cerevisiae* was treated with 20 100-μs pulses at 23–38 kV/cm at a process temperature of 63°C, resulting in a 2D reduction of the lactic acid bacteria and *S. cerevisiae* [11].

The *ELSTERIL* process, developed by Krupp Maschinentechnik GmbH (Hamburg, Germany) in the late 1980s and early 1990s, is used for the sterilization and pasteurization of liquid and electrically conductive media [13,35,36]. Krupp Maschinentechnik GmbH, in association with the University of Hamburg, reported microbial inactivation when PEF was applied to fluid foods such as orange juice and milk [36]. A microbial inactivation exceeding 4D has been found for *Lactobacillus brevis* inoculated in milk and treated with 20 pulses of 20 μs at 20 kV/cm, *S. cerevisiae* inoculated in orange juice and treated with 5 pulses of 20-μs at 4.7 kV/cm, and *E. coli* inoculated in sodium alginate and treated with 5 pulses of 20 μs at 14 kV/cm [35,36]. However, no inactivation of the endospores of *B. cereus* or the ascospores of *Bacillus nivea* was reported [36]. A substantial reduction in ascorbic acid and lipase activity was observed in milk treated with the *ESTERIL* process [36]. The taste of milk and orange juice did not significantly change after the electric field treatments [36].

The disruption of cell membranes to release fat from animal cells was conducted using a process called *ELCRACK* (Krupp Maschinentechnik GmbH, Hamburg, Germany). The *ELCRACK* process consists of the exposure of a slurry of comminuted fish or slaughterhouse offal to high-intensity electric pulses that break down cells, leading to increased fat recovery during the separation step after it is pumped through one or more treatment chambers [35].

Washington State University has a patent for the design and development of a static PEF chamber and has filed another for the design and development of a continuous PEF chamber intended for processing liquid foods with PEF treatments [37–40].

III. APPLICATIONS OF PEF IN FOOD PROCESSING

The application of PEF as a food-processing tool is gaining popularity, since it represents a nonthermal alternative to conventional pasteurization and sterilization methods. The PEF approach, which does not involve the use of added preservatives, is expected to be more appealing to consumers who are skeptical about the use of chemicals in foods. Furthermore, the PEF treatment, being a nonthermal process, may also have no significant detrimental effect on heat-labile components present in foods such as vitamins.

The major disadvantage of PEF operation is the initial investment. A pilot plant–size pulser may cost around \$250,000. Other units for industrial use are available at prices that range from \$450,000 to \$2,000,000.

A. Inactivation of Microorganisms

Raw and reconstituted apple juice, peach juice, skim milk, beaten eggs, and pea soup exposed to PEFs of 25–45 kV/cm were treated using the chamber designed at Washington State University. *E. coli* inoculated in skim milk and exposed to 60 pulses of 2-μs width at 45 kV/cm and 35°C was reduced by 2D [25]. A reduction of 6D was observed in liquid egg inoculated with *E. coli* and treated with an electric field of 25.8 kV/cm and 100 pulses of 4 μs at 37°C [41]. *E. coli* and *B. subtilis* inoculated in pea soup and exposed to PEFs of 25–33 kV/cm (10–30 pulses of 2 μs) provided a limited inactivation (<1.5D) when the process temperature of pea soup was below 53°C, while microbial inactivation was 4.4D with process temperatures between 53 and 55°C [26].

1. Simulated Milk Ultrafiltrate (SMUF)

The inactivation of *E. coli* varied as a function of the electric field intensity, number of pulses, and pH. Low field intensity (20 kV/cm) resulted in insignificant inactivation of microorganisms independent of temperature and pH ($p > 0.05$).

Meanwhile, inactivation of *E. coli* increased with an increase in the number of pulses and an increase in the electric field from 40 to 55 kV/cm. The inactivation was more significant at pH 5.69 than at pH 6.82 ($p < 0.05$) (Figs. 11 and 12). The temperature effect (10 or 15°C) on the inactivation for these experiments was not statistically significant ($p > 0.05$). Table 2 summarizes the inactivation results after eight pulses for each of the experimental conditions.

The role of pH in the survival of microorganisms is related to the ability of the organisms to maintain the cytoplasm pH near neutrality [42]. Membrane permeability increase due to formation of pores in the cell wall during PEF treatment [30] and the rate of transport of hydrogen ions may also increase due to the osmotic imbalance around the cell. Thus, a reduction in cytoplasm pH may

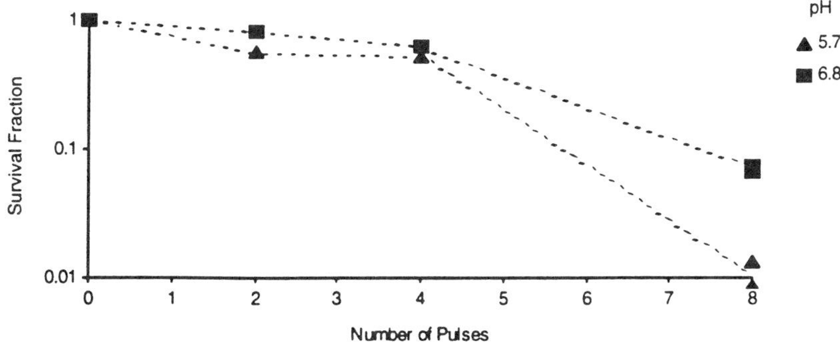

Figure 11 Inactivation of *E. coli* suspended in SMUF, using 40 kV/cm at 10°C, two samples per each experimental condition. (From Ref. 26.)

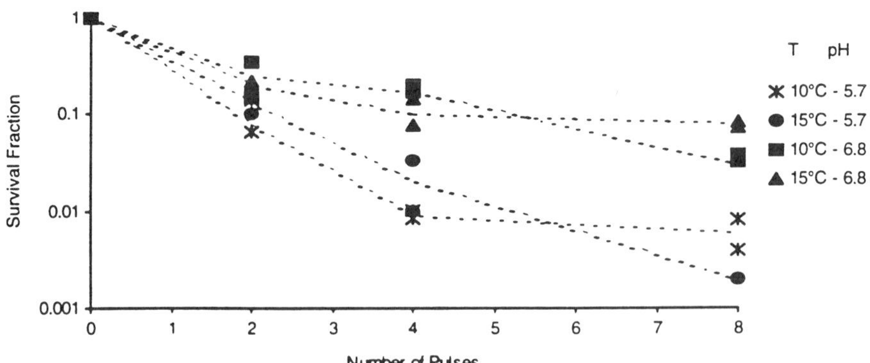

FIGURE 12 Inactivation of *E. coli* suspended in SMUF, using 55 kV/cm, two samples per each experimental condition. (From Ref. 26.)

be observed because a higher number of hydrogen ions are available than at a neutral pH. The change in pH within the cell may induce chemical modifications in fundamental compounds such as DNA or ATP, as discussed by Wiggins [43] and Dolowy [44]. Also, oxidation and reduction reactions such as those proposed by Gilliland and Speck [18] may occur within the cell structure induced by the PEF treatment.

The ionic strength of the solution also plays an important role in the inactivation of *E. coli*. An increase in the ionic strength increases the electron mobility through the solution, resulting in a decrease in the inactivation rate. The reduced inactivation rate in high–ionic strength solutions can be explained by the stability of the cell membrane when exposed to a medium with several ions [30]. The effect of ionic strength can be observed in Figure 13, where a difference of 2.5 log cycles was obtained between the 0.168 and 0.028M solutions.

The growth stage of *E. coli* affected the effectiveness of PEF treatments (36 kV/cm at 7°C, two and four pulses). Cells in the logarithmic phase were most sensitive to the electric field treatments compared to cells in the stationary and lag phase (Fig. 14) as reported by Pothakamury et al. [45]. Figures 15 and 16 present the effect of temperature on the log-cycle reduction of *E. coli* using exponentially decaying pulses and square wave pulses of 35 kV/cm. The rate of inactivation increases with an increase in the temperature. Coster and Zimmermann [46] suggested synergistic effects of high-intensity electric fields with moderate temperatures. The rate of inactivation increased when square wave pulses were used compared to exponentially decaying pulses. Similar results were re-

TABLE 2 Effect of Processing Parameters on the Inactivation of *E. coli* Suspended in SMUF after Eight Pulses

		Number of log-cycle reduction	
Description		10°C	15°C
20 kV/cm	pH 5.7	0.00[a]	0.20[a]
	pH 6.8	0.00[a]	0.06[a]
40 kV/cm	pH 5.7	1.95[b]	1.85[b]
	pH 6.8	1.16[c]	1.00[c]
55 kV/cm	pH 5.7	2.22[d]	2.56[d]
	pH 6.8	1.45[e]	1.10[c]

Log cycle reduction data with similar superscripts are not significantly different at $\alpha = 0.05$, two samples per each experimental condition.
Source: Ref. 26.

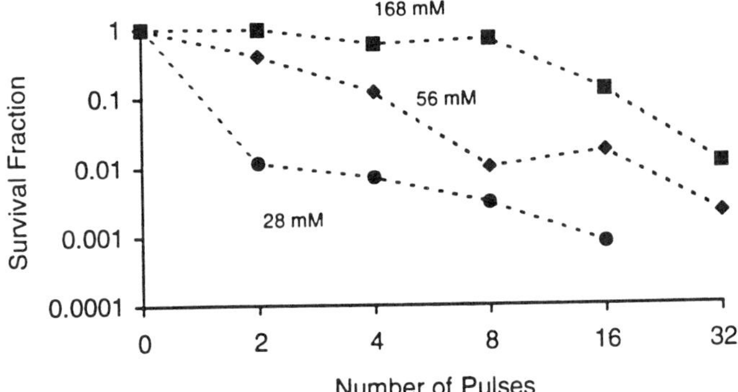

Figure 13 Effect of ionic strength on the inactivation of *E. coli* suspended in SMUF, at 40 kV/cm and 10°C, two samples per each experimental condition. (From Ref. 26.)

ported for *S. aureus* when exposed to PEF at 9 and 16 kV/cm and *L. delbrueckii* and *B. subtilis* when exposed to 9, 12, and 16 kV/cm. Figures 17, 18, and 19 present the reported results by Pothakamury et al. for *S. aureus, L. delbrueckii*, and *B. subtilis* suspended in SMUF.

2. Pea Soup

PEF inactivation of *E. coli* and *B. subtilis* suspended in pea soup depends on the electric field intensity, number of pulses, pulsing rate, and flow rate [49] (Table 3). The maximum bulk temperature of the peak soup achieved during the PEF treatment was 55°C and is a function of both flow rate and pulsing rate. PEF treatments with a bulk temperature below 53°C resulted in limited microbial inactivation (<1.64D). Microbial inactivation dependence on process temperature may be explained by changes in the sensitivity of the microorganisms to PEF when the temperature exceeds 53°C. Thermal inactivation of microorganisms was avoided by cooling treated pea soup to 20°C. Thermal inactivation of *E. coli* requires up to 10 minutes at 61°C when suspended in bouillon [50].

PEF inactivation of *B. subtilis* and *E. coli* decreased almost 2D when the microorganisms were mixed together in pea soup. Figures 20, 21, and 22 summarize the inactivation of *E. coli, B. subtilis*, and the mixture of organisms suspended in pea soup and exposed to selected treatment conditions [49]. There is a significant difference in the inactivation levels ($p < 0.05$) between *E. coli* alone and *E. coli* mixed with *B. subtilis*. PEF inactivation of *E. coli* alone reached 6.5D after 30 pulses at 30 kV/cm and flow rate of 0.5 liters/min, while an inactivation of 4.0D was observed when *E. coli* was mixed with *B. subtilis*. *B. subtilis* alone had 5.0D when exposed to 33 kV/cm at 4.3 Hz and 0.5 liters/min, while only 2.0D were observed when mixed with *E. coli* and exposed to 20 pulses at 30 kV/cm, 4.3 Hz and 0.75 liters/min or 3.5D after 30 pulses.

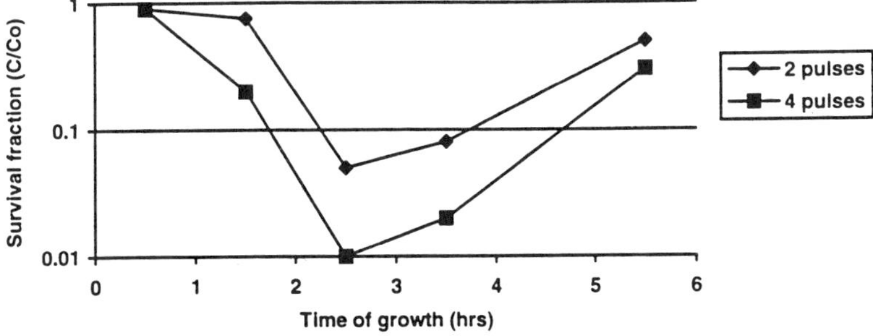

Figure 14 Effect of growth stage on the pulsed electric field inactivation of *E. coli* suspended in SMUF. (From Ref. 45.)

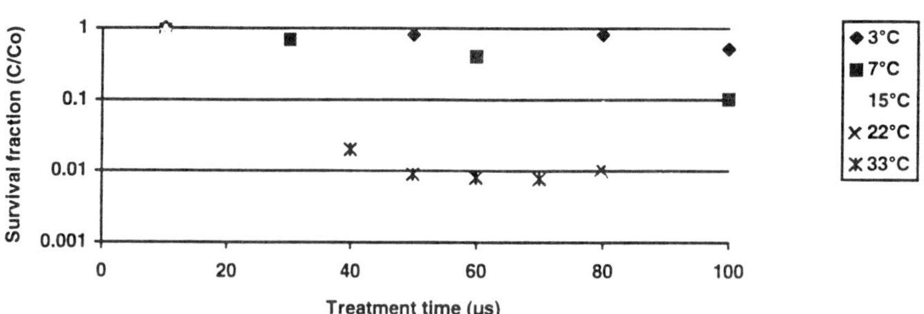

Figure 15 Effect of temperature on PEF inactivation of *E. coli* suspended in SMUF, using exponential decay pulses. (From Ref. 45.)

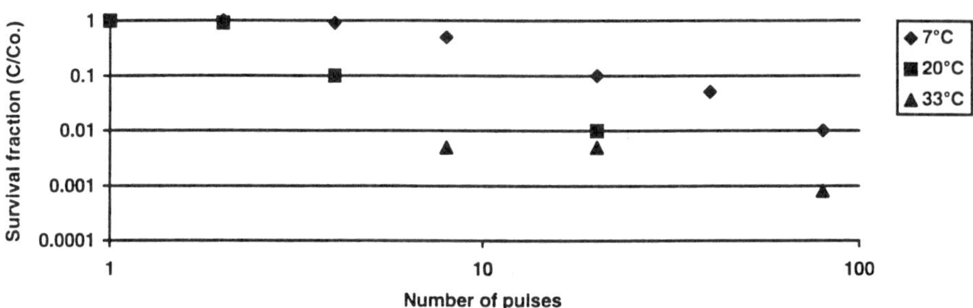

Figure 16 Effect of temperature on PEF inactivation of *E. coli* suspended in SMUF, using square wave pulses. (From Ref. 45.)

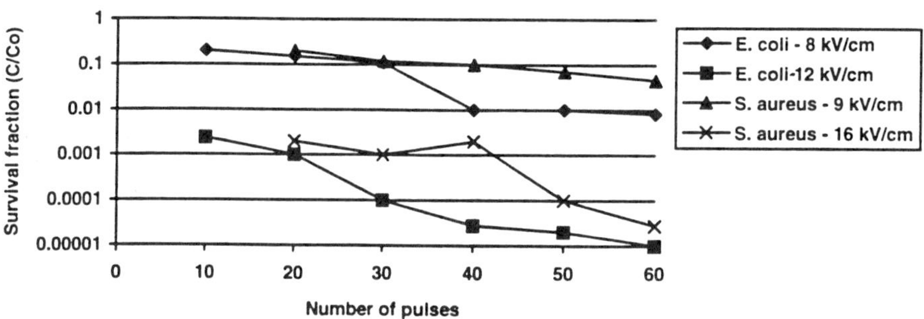

Figure 17 Inactivation of *E. coli* and *S. aureus* in SMUF by PEF. (a) Simplified circuit for exponential decay pulse generation. (b) Voltage trace of an exponential decay pulse. (From Ref. 47.)

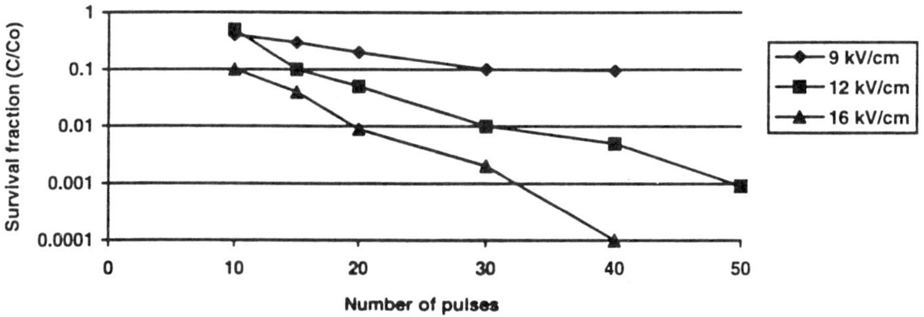

Figure 18 Inactivation of *L. delbrueckii* suspended in SMUF. (From Ref. 47.)

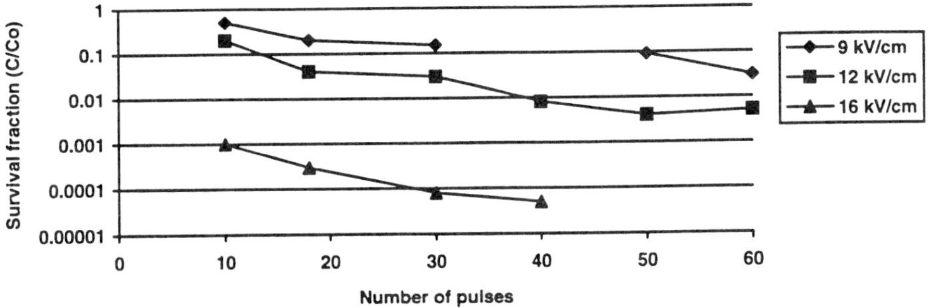

FIGURE 19 Inactivation of *B. subtilis* suspended in SMUF. (From Ref. 47.)

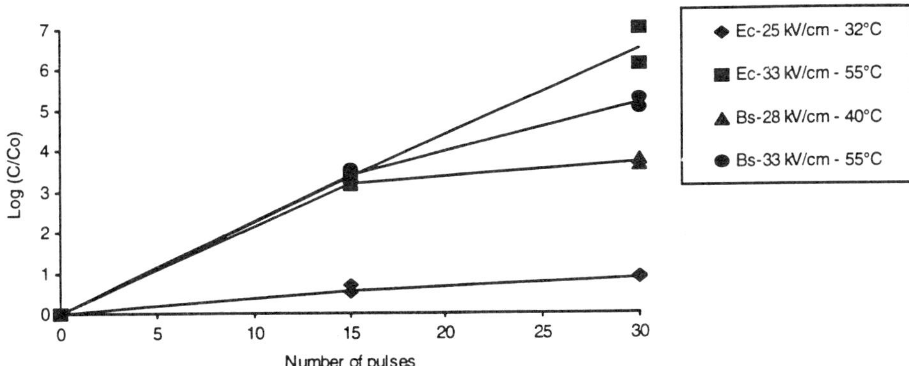

FIGURE 20 Inactivation of microorganisms suspended in pea soup using PEF at 0.5 liter/min and 4.3 Hz (Ec is *E. coli*; Bs is *B. subtilis*). (From Ref. 49.)

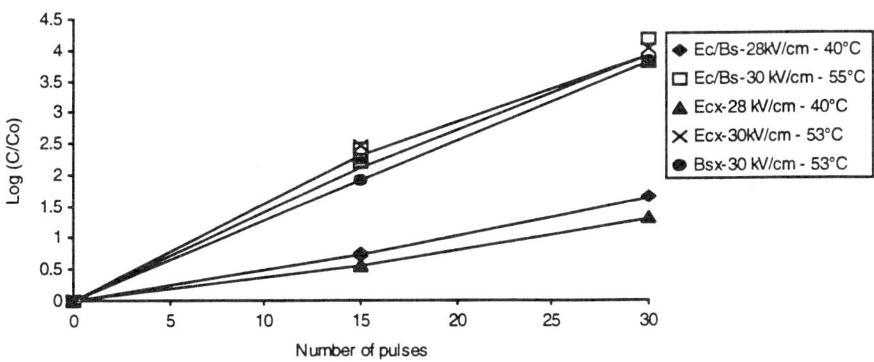

FIGURE 21 Inactivation of mixture of microorganisms suspended in pea soup using PEF at 0.5 liter/min and 4.3 Hz (Ec is *E. coli*; Bs is *B. subtilis*; Ec/Bs is the overall inactivation for the mixture of microorganisms; Ecx is the inactivation of *E. coli* in the mixture; Bsx is the inactivation of *B. subtilis* in the mixture). (From Ref. 49.)

Table 3 Inactivation of an *E. coli–B. subtilis* Mixture Suspended in Pea Soup Using PEF

Flow rate frequency	Number of pulses	28 kV/cm		30 kV/cm	
		Process temperature (°C)	Log reduction (D)	Process temperature (°C)	Log reduction (D)
0.5 liter/min 4.3 Hz	15	43	0.7	55	2.3
	30	39	1.6	55	4.0
0.7 liter/min 6.7 Hz	15	41	0.7	53	4.4
	30	41	0.7	55	4.8
0.75 liter/min 4.3 Hz	10	32	0.8	41	1.1
	20	31	1.0	42	1.0

Source: Ref. 49.

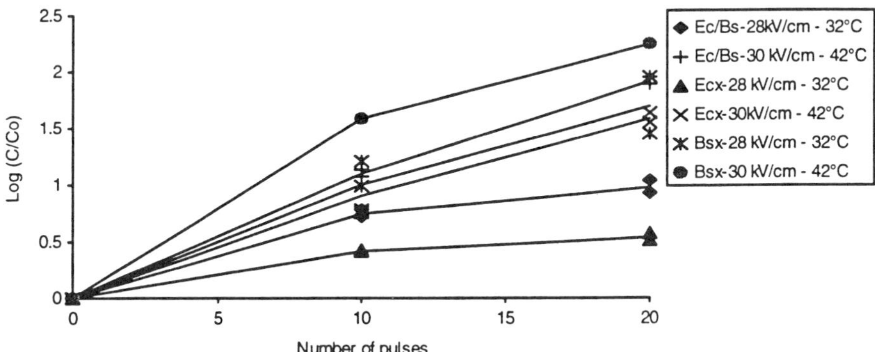

FIGURE 22 Inactivation of mixture of microorganisms suspended in pea soup using PEF at 0.75 liter/min and 4.3 Hz (Ec is *E. coli*; Bs is *B. subtilis*; Ec/Bs is the overall inactivation for the mixture of microorganisms; Ecx is the inactivation of *E. coli* in the mixture; Bsx is the inactivation of *B. subtilis* in the mixture). (From Ref. 49.)

The results for the inactivation of *E. coli* and *B. subtilis* using PEF demonstrate the feasibility of the technology for preservation of foods containing suspended particles and gelatinized starch.

3. Liquid Eggs

High-intensity PEF (26 kV/cm) treatment in continuous flow systems (continuous recirculation and simple pass) inactivates *E. coli* inoculated in liquid egg 6D with a peak processing temperature of 37.2 ± 1.5°C (Table 4, Figs. 23 and 24). PEF treatments with 4-μs pulses were more effective than 2-μs pulses (Figs. 25 and 26), which may be explained by the amount of energy applied to the liquid egg [41]. Figure 27 illustrates the effect of energy input in the inactivation of *E. coli*, with energy input (in Joules) calculated as follows:

$$\text{Energy/pulse} = 0.5\ C\ V^2$$

where C is the capacitance, 0.5 μF for 2-μs pulses and 1.0 μF for 4-μs pulses, and V is the measured potential across the treatment chamber (15.6 kV). The total energy input (in Joules) after n pulses is calculated by:

$$\text{Total energy} = n * \text{Energy/pulse}$$

TABLE 4 Treatment Conditions for Liquid Egg Exposed to PEF

Description	Operating conditions	
	Treatment 1	Treatment 2
Pulse duration (μs)	2	4
Capacitance (μF)	0.5	1
Input voltage (kV)	40	30
Input flow rate (liter/min)	0.5	0.5
Input pulse rate (Hz)	1.25, 2.5	1.25, 2.5
Peak voltage (kV)	15.5	15.5
Peak current (kA)	8.0	8.0
Electric field intensity (kV/cm)	26	26
Pulse energy (J)	60	120
Maximum temperature (°C)	37	37

Source: Ref. 41.

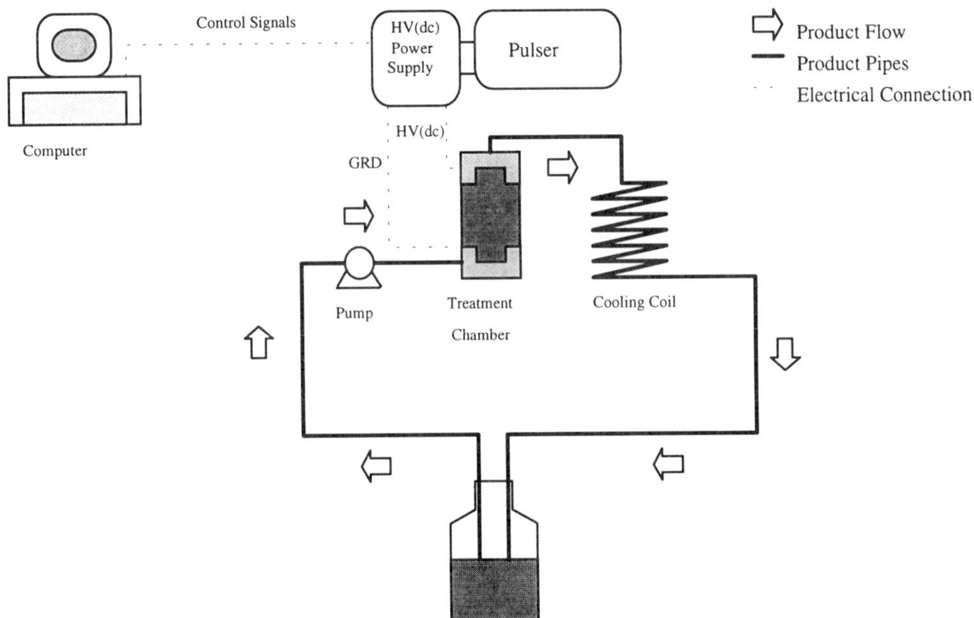

FIGURE 23 Continuous recirculation PEF operation.

The survival fraction of *E. coli* in liquid egg is reduced almost 6D with 12,000 J applied in pulses of 4 μs (Fig. 26). Grahl et al. [36] nearly reached 5D by exposing *E. coli* suspended in sodium alginate to an electric field of 14 kV/cm with five pulses of 20 μs. Zhang et al. [51] observed a 6D reduction in *E. coli* suspended in potato dextrose agar and exposed to 64 pulses of 40 kV/cm at 15°C and a 9D reduction using 70 kV/cm and *E. coli* suspended in simulated milk ultrafiltrate (SMUF) [52].

Proteins, an important nutrient for microbial growth, diminished the effectiveness of the PEF treatment [18,53]. The inactivation of microorganisms using PEF is more difficult in food materials than in buffer solutions [51]. In general the bactericidal effect of PEF is inversely proportional to the

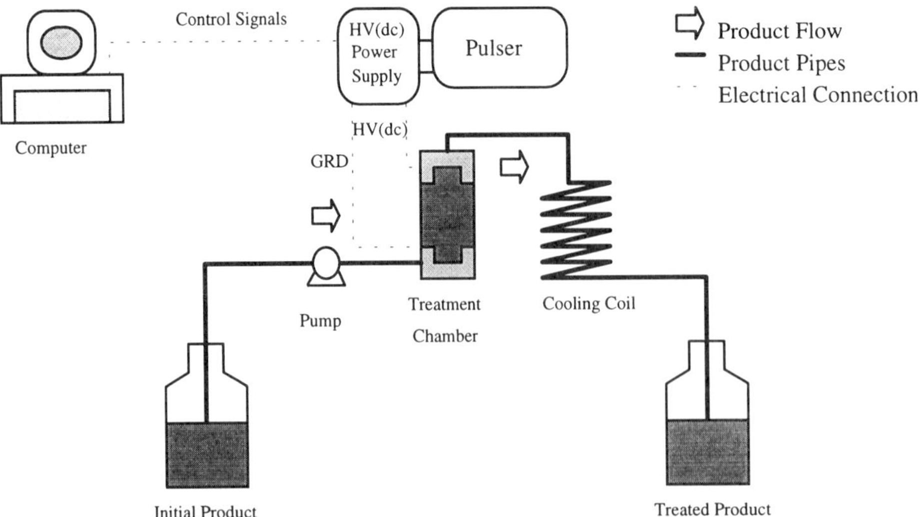

FIGURE 24 Single-pass PEF operation.

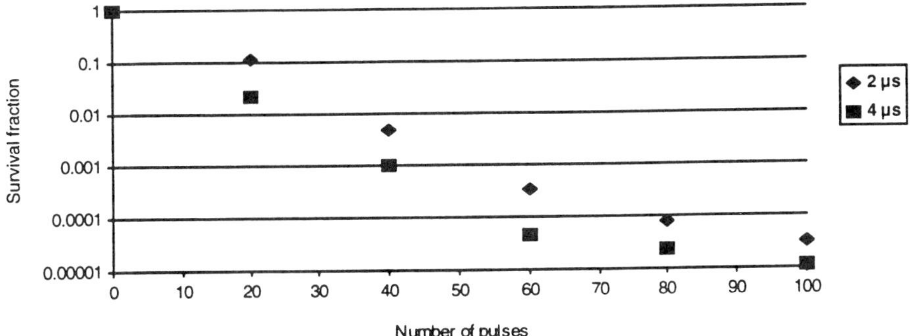

Figure 25 *E. coli* in liquid egg after PEF treatment at 26 kV/cm and 37°C in a continuous recirculation system. (From Ref. 41.)

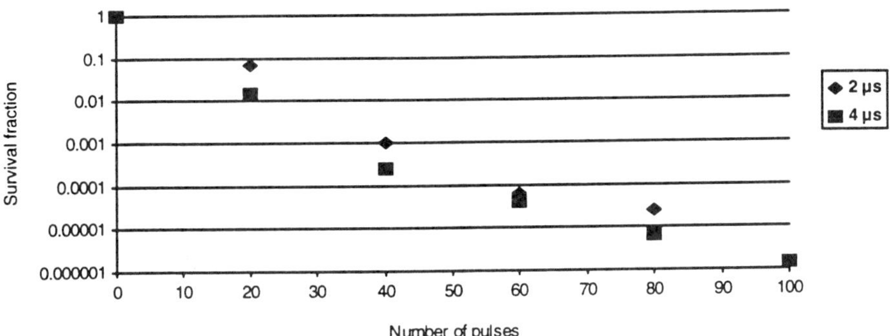

Figure 26 *E. coli* in liquid egg after PEF treatment at 26 kV/cm and 37°C in a stepwise system. (From Ref. 41.)

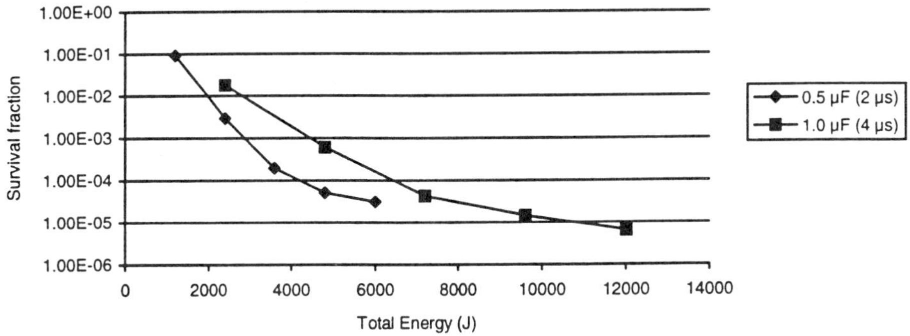

Figure 27 *E. coli* in liquid egg after PEF treatment at 26 kV/cm and 37°C as a function of input energy. (From Ref. 41.)

ionic strength and increases with electric resistivity [26,54]. The electric resistance of liquid egg (1.9 Ω) is low compared to other foods and makes necessary the exposure of liquid egg to a large number (100) of pulses.

There was no significant difference ($p > 0.05$) in the effectiveness of PEF treatment when the pulse rate varied from 1.25 to 2.50 Hz, as the inactivation of *E. coli* in liquid egg was at least 4D if the number of pulses and pulse width remained constant. There was also no significant difference ($p > 0.05$) between the inactivation of *E. coli* using continuous recirculation or stepwise treatments.

4. Apple Juice

Commercial apple juice ultrafiltrated and exposed to different PEF treatments showed no changes in pH, acidity, vitamin C, glucose, fructose, and sucrose content [55] as summarized in Table 5.

The inactivation of *S. cerevisiae* suspended in apple juice is affected by the intensity of the electric field, treatment time, and number of pulses [56,57]. Figure 28 illustrates the microbial count of *S. cerevisiae* as a function of peak field intensity when two pulses were used and the selected field intensities were 13, 22, 35, and 50 kV/cm. The rate of inactivation increases with an increase in field intensity [57]. Microbial inactivation is a function of the number of pulses, as illustrated in Figure 29. An inactivation of 6D is reported after 10 pulses of 35 kV/cm at 22–34°C.

The shelf life of PEF-treated apple juice increases over 3 weeks when stored at either 4 or 25°C, as illustrated in Figure 30.

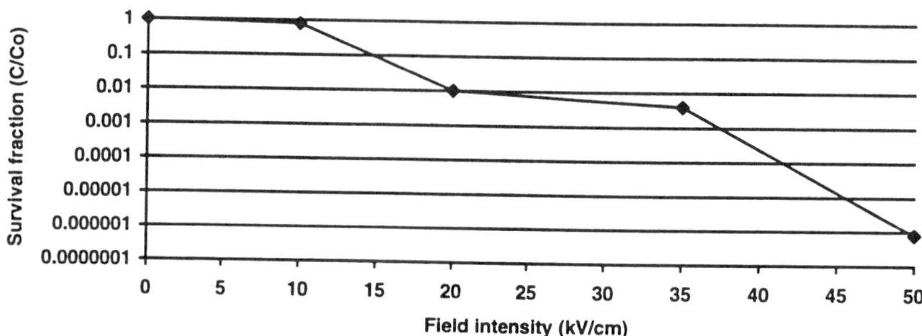

Figure 28 Survival fraction of *S. cerevisiae* as a function of peak field intensity when two 2.5-μs pulses were applied. (From Ref. 57.)

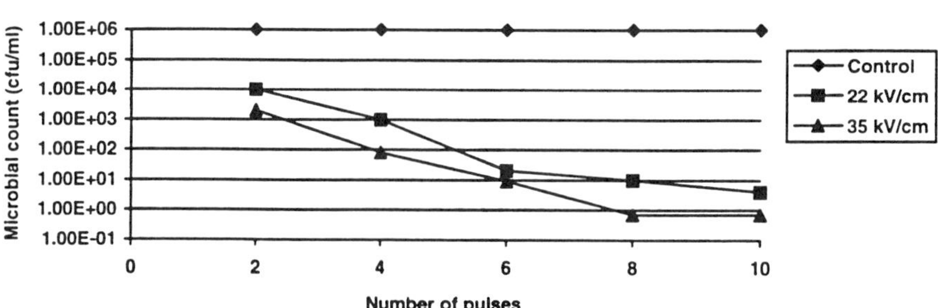

Figure 29 Microbiological count of *S. cerevisiae* in apple juice as a function of the number of 2.5-μs pulses. (From Ref. 57.)

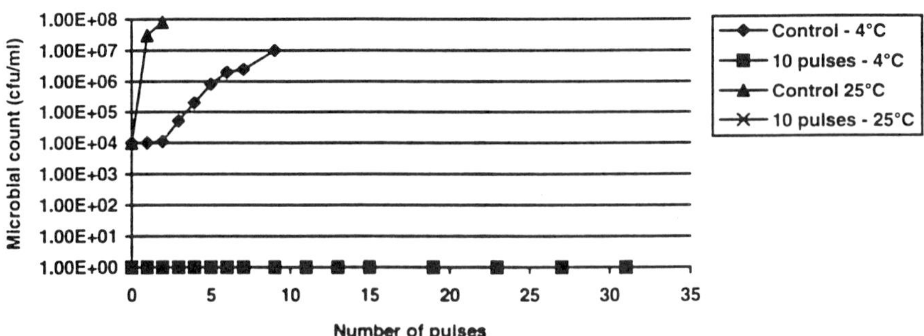

Figure 30 Shelf life of apple juice after PEF treatment of 10 2.5-μs pulses at 36 kV/cm. (From Ref. 57.)

TABLE 5 Apple Juice Chemical Properties Before and After PEF

Sample	pH	Acidity (mallic acid)	Vitamin C (mg/100 g)	Glucose	Fructose	Sucrose
Control	4.10 ± 0.02	2.63 ± 0.02	1.15 ± 0.01	2.91 ± 0.33	4.95 ± 0.64	2.18 ± 0.25
PEF-T1	4.36 ± 0.03	2.67 ± 0.02	1.02 ± 0.02	2.87 ± 0.06	4.96 ± 0.11	2.25 ± 0.06
PEF-T2	4.18 ± 0.01	2.75 ± 0.07	1.12 ± 0.00	3.01 ± 0.34	5.08 ± 0.67	2.21 ± 0.31
PEF-T3	4.09 ± 0.01	2.63 ± 0.02	1.02 ± 0.00	2.90 ± 0.09	4.89 ± 0.13	2.13 ± 0.06
PEF-T4	4.23 ± 0.01	2.61 ± 0.00	1.15 ± 0.24	2.57 ± 0.25	4.33 ± 0.47	2.43 ± 0.13

The data presented are average values of two experiments each carried out in duplicate.
Source: Ref. 55.

5. Skim Milk

Treatment in a Static Chamber System

PEF treatment inactivates *E. coli* in skim milk at 15°C. The principal parameters influencing the microbial inactivation are the applied electric field intensity and treatment time, which can be expressed by the number of pulses (n) when the width of each pulse is fixed [15].

The *E. coli* survival fraction decreases when milk is treated with an increasing number of pulses at a constant field intensity (Fig. 31). The rate of inactivation of *E. coli* increases with an increase in the electric field intensity at a constant number of pulses (Fig. 32). Less than one log reduction in *E. coli* population was observed for PEF treatments of 20, 25, and 30 kV/cm and 64 pulses at 15°C. However, PEF treatments at 45 kV/cm, 64 pulses, and 15°C lead to a nearly 3 log cycle reduction [58]. The reported results are consistent with those of Dunn and Pearlman [11], but these authors mentioned that the treatment temperature increased up to 43°C.

Similar *E. coli* inactivation was obtained with 20 kV/cm PEF in saline solution [59]. Hülsheger et al. [21] reduced the population 4 log cycles by applying 20 kV/cm PEF for *E. coli* inoculated in phosphate buffer, and Grahl et al. [36] reached a nearly 5 log cycle reduction by treating *E. coli* suspended in sodium alginate solution with 26 kV/cm PEF. The inactivation of *E. coli* in potato dextrose agar by applying 64 pulses of 40 kV/cm at 15°C resulted in a 6 log cycle reduction. Notice that PEF inactivation kinetics in semisolid products are different from the PEF inactivation kinetics in fluids because *E. coli* cells are fixed in a gel matrix, which increases uniformity of inactivation [51].

Inactivation of *E. coli* in skim milk by PEF treatment in a static chamber satisfied Hülsheger's model (Table 6) because the destruction of this microorganism in skim milk followed a first-order kinetic for both the electric field intensity and number of pulses.

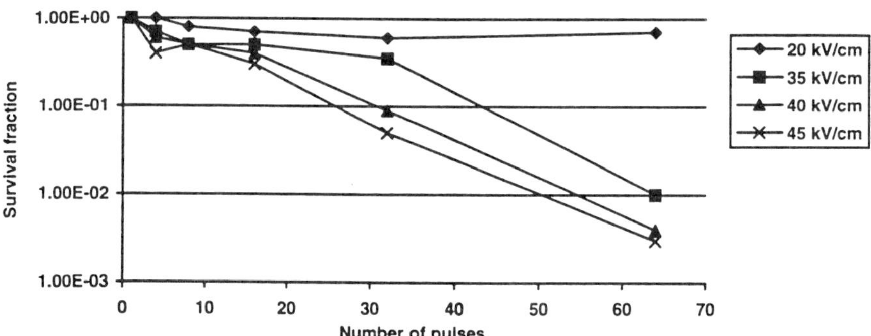

FIGURE 31 Inactivation of *E. coli* in skim milk at 15°C in a static chamber at several field intensities. (From Ref. 58.)

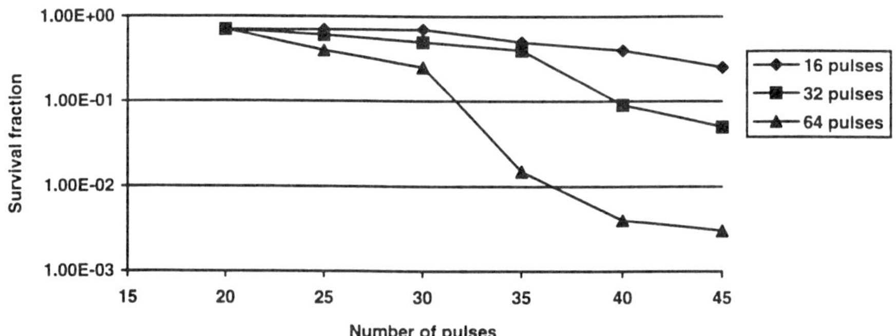

FIGURE 32 Inactivation of *E. coli* in skim milk at 15°C in a static chamber with different number of pulses. (From Ref. 58.)

Table 6 Kinetics Constant of Hülsheger's Model for *E. coli* Inactivation in Skim Milk by PEF[a]

Electric field intensity (kV/cm)	Number of pulses (n)	n_{min}	E_c (kV/cm)	k (kV/cm)	R^2
35	<64	15.2	—	5.6	82.9
40	<64	13.0	—	6.1	95.8
45	<64	11.0	—	8.0	98.5
<45	16	—	18.7	2.9	83.3
<45	32	—	20.4	3.9	86.1
<45	64	—	19.9	2.7	92.4

R^2 = correlation coefficient for regression analysis ($p = 0.05$).
[a]Treatment in a static chamber.
Source: Ref. 58.

Martin et al. [58] reported that the minimum number of pulses (n_{min}) necessary to inactivate the microorganisms in skim milk at 45 kV/cm using a static chamber is 11 and 15 pulses at 35 kV/cm, respectively. The critical electric field (E_c) is 19.9 kV/cm with 64 pulses at 45 kV/cm, which is higher than the value reported by Grahl et al [36] for *E. coli* suspended in sodium alginate solution (14 kV/cm). Zhang et al. [51] calculated 17.5 kV/cm E_c for *E. coli* in semisolid model foods.

It is more difficult to reduce the survival fraction of microorganisms present in skim milk than in buffer solutions and model foods because the composition of skim milk is complex (i.e., high protein content 33–40 g/liter) [60]. These substances diminish the lethal effect of PEF in microorganisms because they absorb free radicals and ions, which are active in the breakdown of cells [18,53]. Moreover, the inactivation of bacteria by PEF is a function of solution resistance, which is inversely proportional to ionic strength. Survival fractions decrease when medium resistance increases and ionic strength decreases [26,54]. The measured resistivity of skim milk is 310 Ω cm and that of buffer solutions is even higher. Since dilution of milk increases the resistivity and decreases protein concentration, the effectiveness of PEF treatment is improved. The inactivation rate of *E. coli* suspended in skim milk:water (1:2.3) and exposed to 40 kV/cm in a static chamber at 15°C is higher than when less diluted skim milk (1:1) is used (Fig. 33).

Treatment in a Continuous System

PEF treatment in a continuous-flow chamber also inactivates *E. coli* inoculated in skim milk. An increase in field intensity or number of pulses produces greater bacterial inactivation (Figs. 34 and 35) and microorganism death follows first-order kinetics with both field intensity and number of pulses (Table 7). The E_c when PEF treatment was carried out in a continuous system at 30 kV/cm

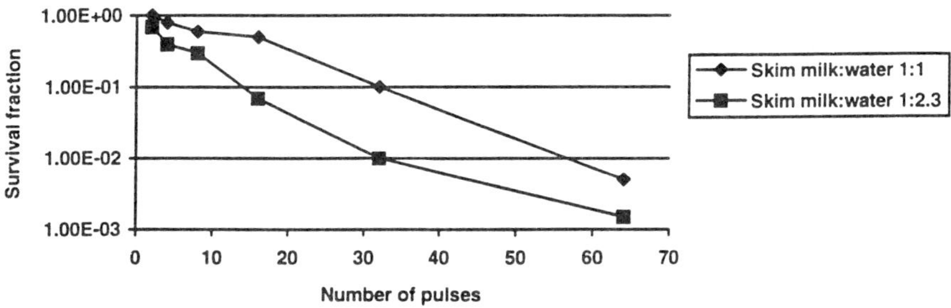

Figure 33 Effect of skim milk dilution in the inactivation of *E. coli* by 35 kV/cm PEF treatment in a static chamber at 15°C. (From Ref. 58.)

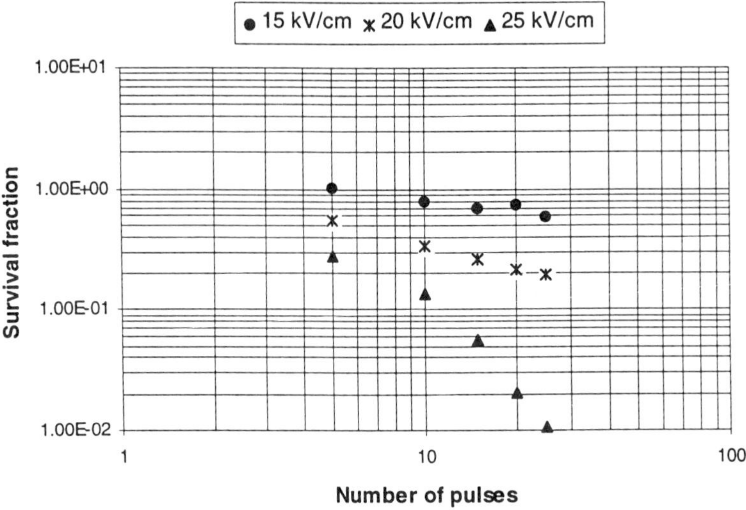

FIGURE 34 Inactivation of *E. coli* in skim milk at 15°C in a continuous chamber at different field intensities. (From Ref. 58.)

maximum electric field intensity was between 12.34 and 14.62 kV/cm, and n_{min} ranged from 1.9 to 5.4 pulses. These values were lower than those obtained in the same treated product using the static system.

In general, PEF treatment in continuous systems is more effective in terms of microorganism inactivation than in static systems due to the treatment uniformity being greater. Moreover, in this study even though both chambers are of the parallel plate type, the treatment volume in a static chamber is higher (14.5 ml) than the continuous-flow chamber (8 ml). Therefore, the energy density (defined as energy divided by volume) is higher in continuous systems.

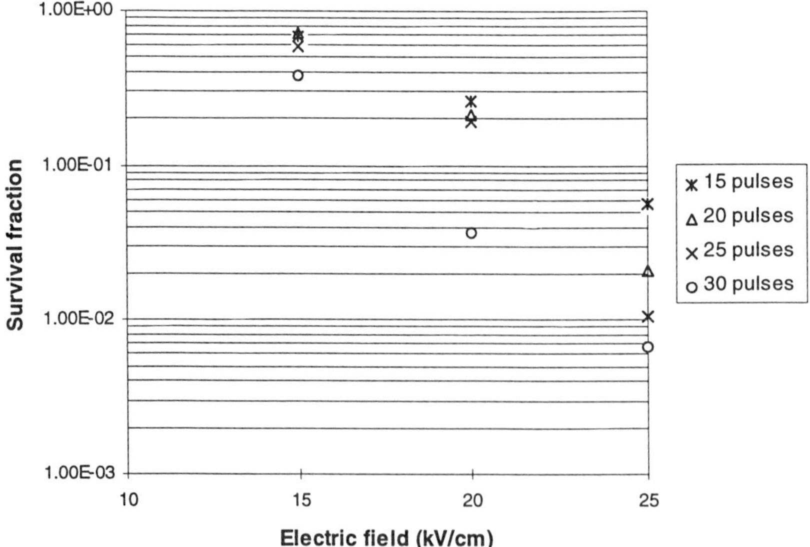

FIGURE 35 Inactivation of *E. coli* in skim milk at 15°C in a continuous chamber with different number of 1.8-ms pulses. (From Ref. 58.)

TABLE 7 Kinetics Constant of Hülsheger's Model for *E. coli* Inactivation in Skim Milk by PEF[a]

Electric field intensity (kV/cm)	Number of pulses (n)	n_{min}	E_c (kV/cm)	k (kV/cm)	R^2
15	<30	5.4	—	3.9	91.8
20	<30	1.9	—	9.5	99.7
25	<30	2.7	—	5.8	95.5
<30	15	—	13.82	4.3	98.5
<30	20	—	14.62	2.2	96.8
<30	25	—	14.44	2.2	93.8
<30	30	—	12.34	3.5	99.2

R^2 = correlation coefficient for regression analysis ($p = 0.05$).
[a]Treatment in a continuous-flow chamber.
Source: Ref. 58.

The effectiveness of PEF treatment also depends on pulse duration, which increases the *E. coli* inactivation because the energy applied in each pulse is higher. Applying 25 pulses of 0.7 μs each at 25 kV/cm in a continuous-flow chamber reduces the survival fraction of *E. coli* inoculated in skim milk less than one log cycle, but a treatment in the same chamber with the same number of pulses and field intensity and a 1.8-μs duration pulse reduces the survival fraction by more than two log cycles (Fig. 36).

B. Denaturation of Proteins

1. Alkaline Phosphatase

The activity of alkaline phosphatase (ALP) in pasteurized milk products has public health significance, since the presence of active ALP indicates inadequate pasteurization or cross-contamination with raw milk [61]. In fresh raw milk, ALP is present in association with the membrane of fat globules; in skim milk it is in the form of lipoprotein particles.

The inactivation of ALP by PEF is a function of the field intensity, the fat content of the milk, and the concentration of ALP. The activity of ALP decreases with an increase in field intensity [61].

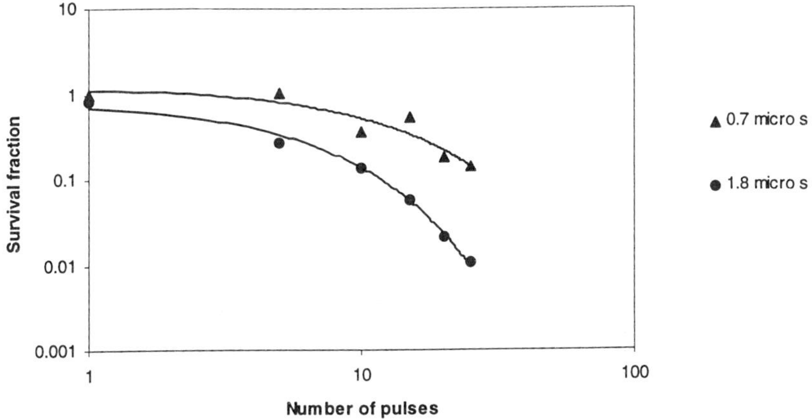

FIGURE 36 Effect of pulse duration in the inactivation of *E. coli* in skim milk at 15°C by 25 kV/cm PEF treatment in a continuous chamber. (From Ref. 58.)

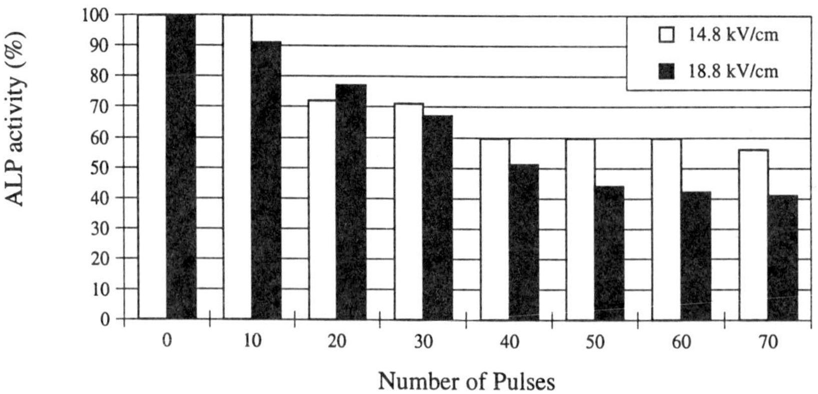

FIGURE 37 PEF inactivation of ALP diluted in UHT pasteurized 2% milk. (From Ref. 61.)

A reduction of 43–59% in ALP activity is reported when the enzyme is suspended in 2% milk and exposed to 70 pulses of 0.40–0.45 ms at 14.8–18.8 kV/cm (Fig. 37). Seventy pulses of 0.74 msec of a field strength of 22 kV/cm applied to 2 mg/ml ALP in SMUF reduced the ALP activity by 65% (Fig. 38). The activity of ALP dissolved in UHT-pasteurized 2% and 4% milk was reduced 59% when exposed to 70 pulses of 0.40 msec at 18.8 kV/cm, while a 65% reduction was observed in nonfat milk as illustrated in Figure 39. ALP suspended in milk (1 ml raw milk in 100 ml 2% milk) using 13.2 kV/cm and 43.9°C after 70 pulses showed a reduction of 96% in activity, whereas heat treatment at 43.9°C for 17.5 minutes showed only a 30% reduction (Fig. 40). Castro [61] demonstrated a reduction in initial velocity of fluoroyellow production of ALP as a function of number of pulses, as illustrated in Figure 41. Castro also found that PEF-treated ALP is more susceptible to trypsin proteolysis (70 pulses of 0.78 msec at 22.3 kV/cm), as illustrated in Figure 42. The inactivation of ALP is attributed to conformational changes induced by PEF [31,61].

2. Plasmin and a Protease from *Pseudomonas fluorescens* M3/6

The proteolytic enzyme plasmin and a protease from *Pseudomonoas fluorescens* M3/6 were also inactivated using pulsed electric fields. A 90% inactivation of plasmin activity was observed during 30 and 45 kV/cm, 10–50 pulses of 2 μs duration, and a process temperature of 10 and 15°C [61] as presented in Figures 43 and 44. Meanwhile, 80% inactivation was found for a protease extracted from *P. fluorescens* when dispersed in Triptych Soy Broth and exposed to 20 pulses of 2 μs at 11–18 kV/

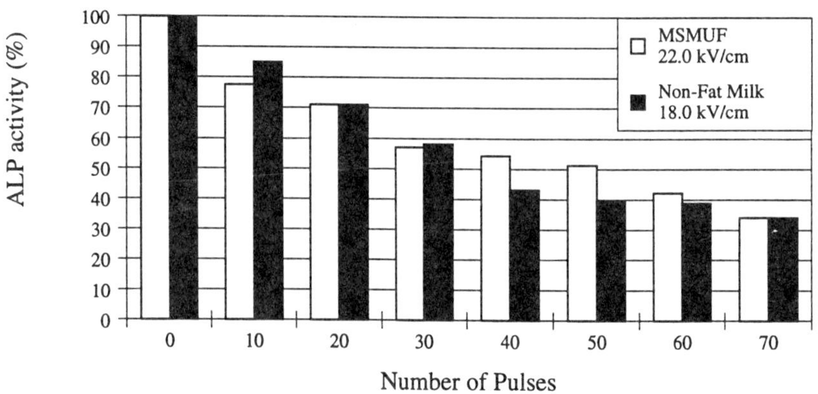

FIGURE 38 PEF inactivation of ALP diluted in MSMUF or nonfat milk. (From Ref. 61.)

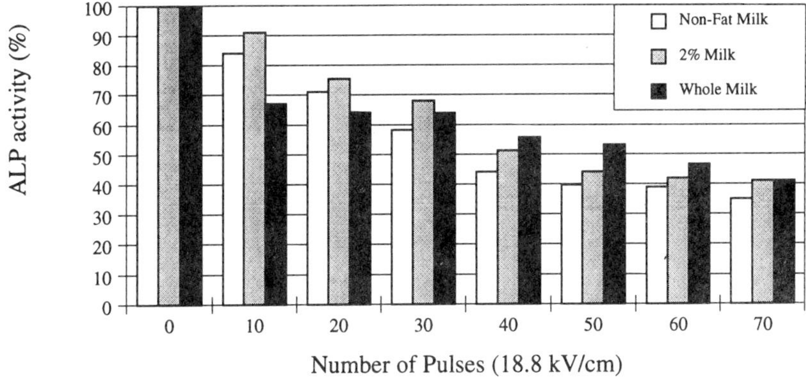

FIGURE 39 PEF inactivation of ALP diluted in UHT pasteurized nonfat, 2%, and whole milk. (From Ref. 61.)

cm and 20–24°C. A 60% inactivation was detected when inoculated in sterilized skim milk and exposed to 98 pulses of 2 μs at 15 kV/cm and 50°C (Fig. 45); no inactivation was detected when inoculated in a sterilized casein–Tris buffer and exposed to a PEF treatment similar to that for skim milk. The decreased effectiveness of PEF in the inactivation of the protease in skim milk and the casein–Tris buffer may be attributed to a protective role of the substrate (i.e., casein) against conformational changes of the enzyme induced by the electric fields [63].

The susceptibility of casein to proteolysis varies as a function of treatment conditions [63]; a HIPEF treatment of 25 kV/cm at 0.6 Hz and 30°C was found to increase the proteolytic activity in skim milk inoculated with a protease from *P. fluorescens* M3/6. However, 14 or 15 kV/cm at 1 or 2 Hz and 30°C had no significant effect on the susceptibility of casein in skim milk proteolysis, and no significant change was observed in the susceptibility of casein suspended in a casein–Tris buffer when exposed to treatment conditions similar to those for skim milk [63].

The inactivation of the protease from *P. fluorescens* M3/6 when exposed to PEF does not depend on the presence of calcium in the media containing the protease (Fig. 46). The inactivation is the same for the three solutions containing 0, 10, or 15 mM calcium. The proteolytic activity of the protease was reduced 30% after exposure to 20 pulses of 700 μs at 6.2 kV/cm and 15–20°C [64].

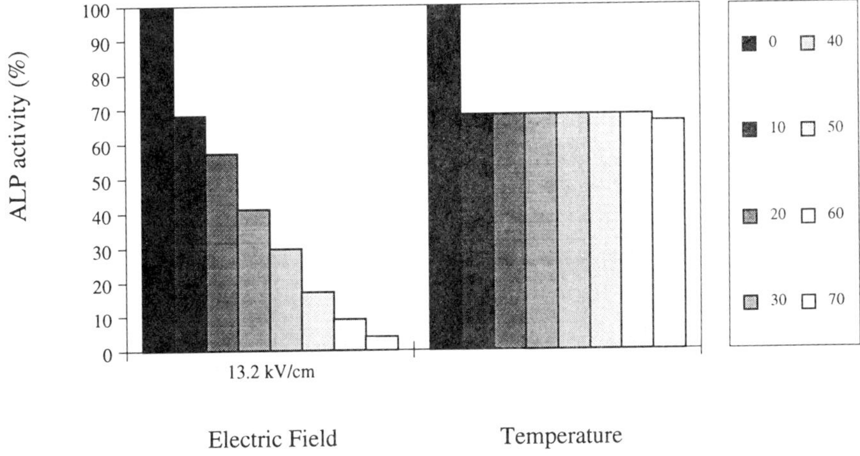

FIGURE 40 Inactivation of alkaline phosphatase by PEF or heating at 44°C for 17.5 minutes. (From Ref. 61.)

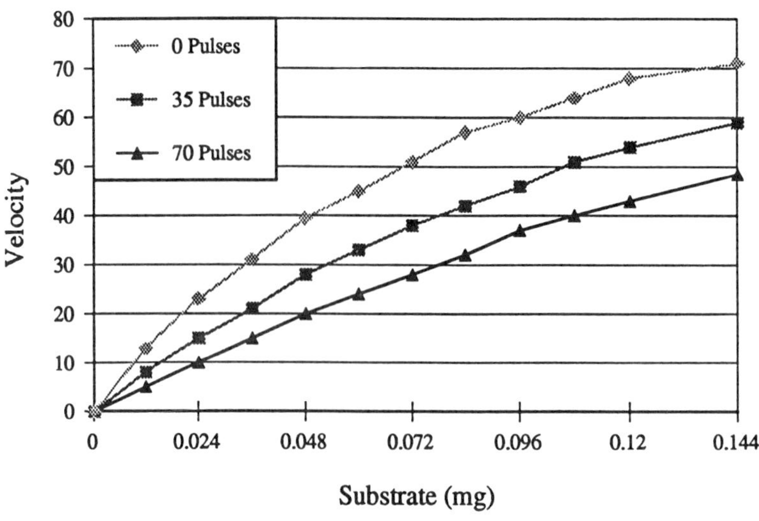

FIGURE 41 Initial velocity of fluoroyellow (FY)-producing reaction of ALP in MSMUF treated with 0.78-ms pulses of 22.3 kV/cm of PEF versus the substrate concentration. (From Ref. 61.)

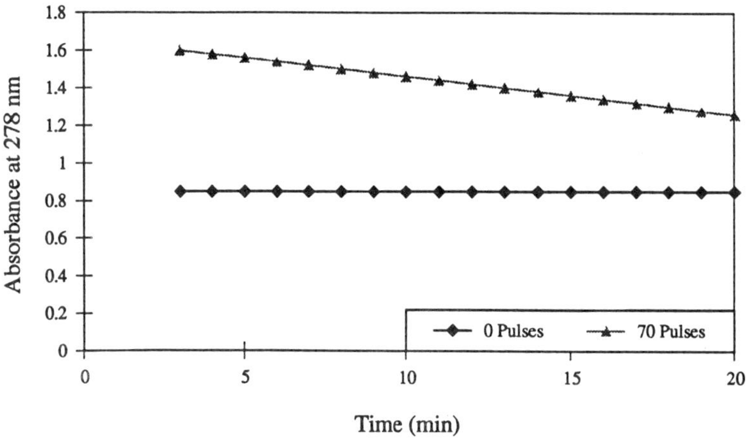

FIGURE 42 Trypsin digestion of native and PEF-treated alkaline phosphatase. (From Ref. 61.)

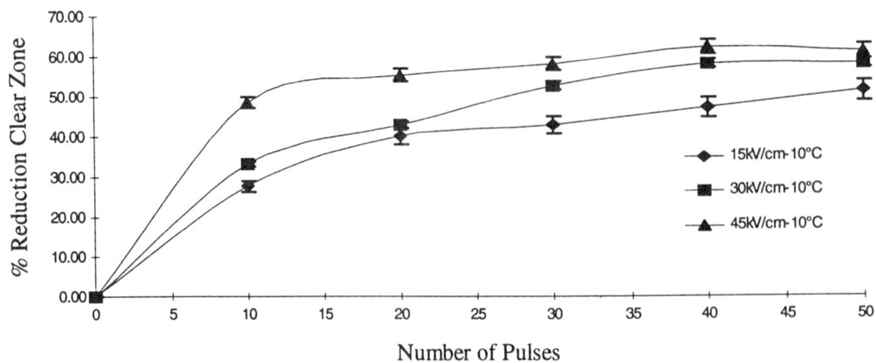

FIGURE 43 PEF inactivation of plasmin at 10°C. (From Ref. 62.)

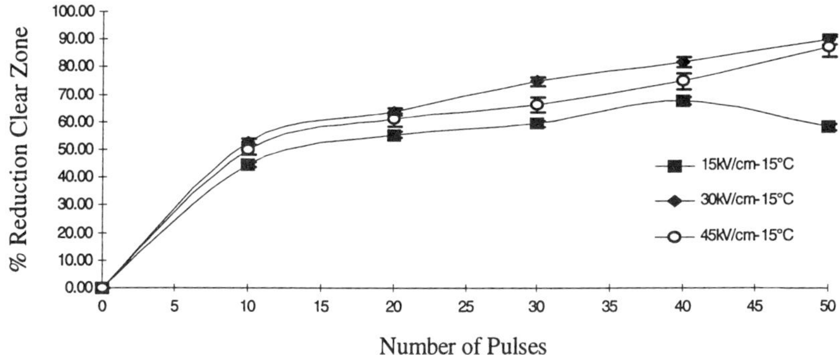

FIGURE 44 PEF inactivation of plasmin at 15°C. (From Ref. 62.)

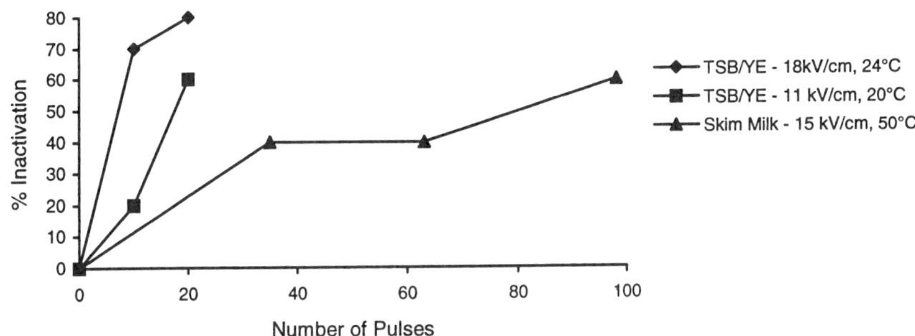

FIGURE 45 Inactivation of a protease from *P. fluorescens* M3/6 in Triptych soy broth enriched with yeast extract (TSB/YE, pulsing rate of 0.25 Hz) and skim milk (pulsing rate 2 Hz) using 2-μs pulses. (From Ref. 63.)

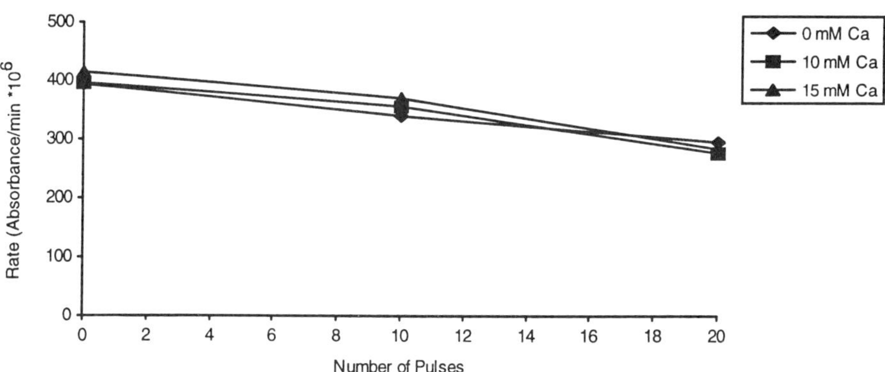

FIGURE 46 PEF inactivation of protease from *P. fluorescens* M3/6 at 6.2 kV/cm. (From Ref. 64.)

In contrast to PEF, thermal inactivation of the protease suspended in SMUF does vary with calcium content. Heated samples containing either 10 or 15 mM calcium retained 71% of the original activity compared to a 12% retention on samples without calcium after 5 minutes of heating, followed by a steady decrease in activity as a function of the heating time (Fig. 47).

The analysis by HPLC using the hydrophobic interaction column (HIC) of PEF (20 pulses, 15 mM Ca^{2+}) and heat-treated (5 minutes, 15 mM Ca^{2+}) samples showed differences in the retention time and peak high of the eluted protein when compared to nontreated samples (Table 8).

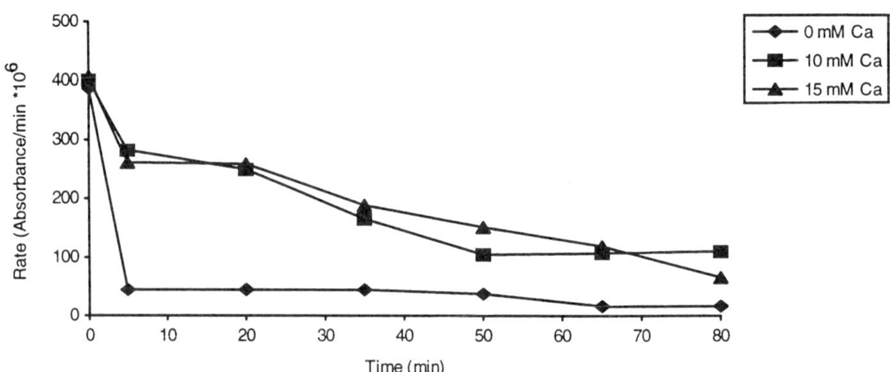

FIGURE 47 Thermal inactivation of protease from *P. fluorescens* M3/6. (From Ref. 64.)

TABLE 8 Hydrophobic Changes of Protease Suspended in SMUF Induced by PEF and Thermal Treatments

Sample high	Retention time (min)	Peak (mm)
Control	6.01	22.9
20 pulses[a]	5.96	25.4
Heat-treated[a]	5.93	20.6

[a]15 M Ca^{2+}.
Source: Ref. 64.

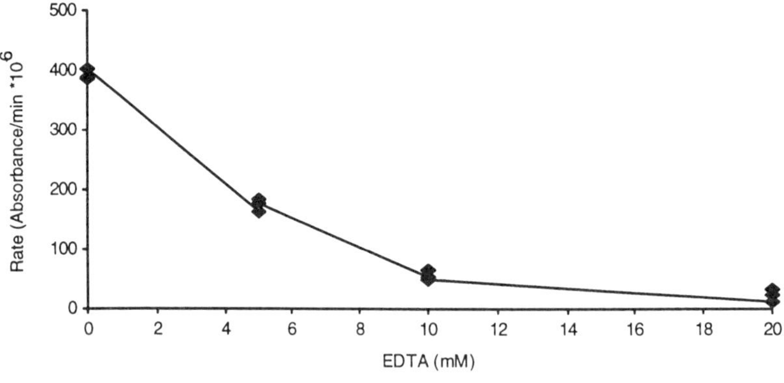

FIGURE 48 Inhibitory effect of EDTA on a protease from *P. fluorescens* M3/6. (From Ref. 64.)

EDTA has a significant inhibitory effect on the proteolytic activity of the protease (Fig. 48). This result is similar to reported data for the protease from *P. fluorescens*. PEF treatment of samples containing EDTA enhanced the inactivation of the protease in SMUF (Fig. 49).

IV. FINAL REMARKS

The research on pulsed electric fields as a nonthermal process needs to include not only the inactivation of microorganisms, but the inactivation of enzymes, retention of vitamins, and the effect of

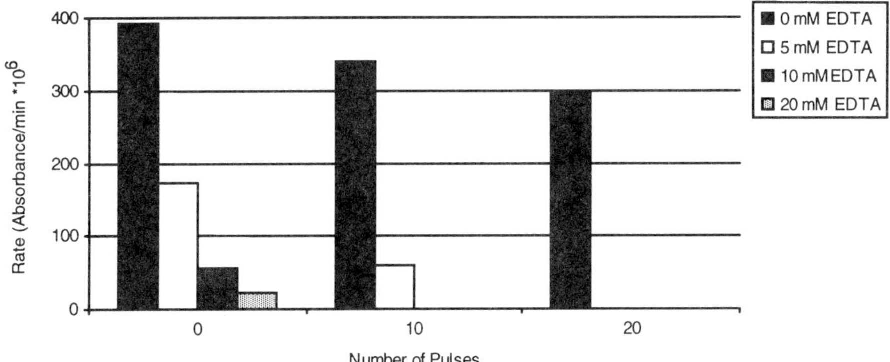

FIGURE 49 PEF inactivation of a protease form *P. fluorescens* M3/6 in SMUF with EDTA. (From Ref. 64.)

PEF treatments on other food components. The reported inactivation of enzymes, as well as the increased proteolysis of casein following exposure to PEF, suggests that detailed research is needed in areas other than preservation. Pulsed electric fields could be utilized as an effective hurdle when used in combination with other preservation factors such as pH and water activity or as a complementary step with mild thermal processes.

REFERENCES

1. A. K. Anderson and R. Finkelstein. A study of the Electropure process of treating milk, *J. Dairy Sci. 2*:374 (1919).
2. B. E. Getchell, Electric pasteurization of milk, *Agric. Eng. 16* (10):408 (1935).
3. S. Palaniappan and S. K. Sastry, Electrical conductivity of selected juices: influences of temperature, solids content, applied voltage, and particle size, *J. Food Proc. Eng. 14*:247 (1991).
4. N. E. Bengtsson and T. Ohlsson, Microwave heating in food industry, *Proc. IEEE 62* (1):44 (1974).
5. C. A. Balanis, *Advanced Engineering Electromagetics*, John Wiley & Sons, New York, 1989.
6. IFT, Microwave food processing, *Food Technol. 43* (1):117 (1989).
7. R. A. Lawrie, *Meat Science,* 4th ed., Pergamon Press, New York, 1985.
8. Y. Li, M. F. Slavik, C. L. Griffis, J. T. Walker, J. W. Kim, and R. E. Wolfe, Destruction of salmonella in poultry chiller water using electrical stimulation, *Trans. ASAE 37* (1):211 (1994).
9. A. Mizuno and Y. Hori, Destruction of living cells by pulsed high-voltage application, *IEEE Trans. Ind. Appl. 24*:387 (1988).
10. S. Palaniappan and S. K. Sastry, Effects of electricity on microorganisms: a review, *J. Food Proc. Pres. 14*:393 (1990).
11. J. E. Dunn and J. S. Pearlman, Methods and apparatus for extending the shelf life of fluid food products, *U. S. patent* 4,695,472 (1987).
12. S. Jayaram, G. S. P. Castle, and A. Margaritis, Effects of high electric field pulses on *L. brevis* at elevated temperatures, Proceedings of IEEE Industry Applications Society Annual Meeting, 1991, p. 647.
13. B. Mertens and D. Knorr, Developments of nonthermal processes for food preservation, *Food Technol. 46* (5):124 (1992).
14. W. Sitzmann, High voltage pulse techniques for food preservation, *New Methods of Food Preservation* (G. W. Gould, ed.), Backie Academic & Professional, Chapman & Hall, New York, 1995, p. 236.
15. B. L. Qin, Q. Zhang, G. V. Barbosa-Cánovas, and B. G. Swanson, Inactivation of microorganisms by pulsed electric fields of different voltage waveforms, *IEEE Trans. Dielectr. Electr. Insul. 1* (6):1047 (1994).

16. J. C. Fetterman, The electrical conductivity method of processing milk, *Agric. Eng 9*:107 (1928).
17. J. M. Beattie and F. C. Lewis, The electric current (apart from the heat generated). A bacteriological agent in the sterilization of milk and other fluids, *J. Hyg. 24*:123 (1925).
18. S. E. Gilliland and M. L. Speck, Mechanism of the bactericidal action produced by electrohydraulic shock, *App. Microbiol. 15*:1038 (1967).
19. A. J. H. Sale and W. A. Hamilton, Effect of high electric fields on microorganisms. I. Killing of bacteria and yeast, *Biochim. Biophys. Acta. 148*:781 (1967).
20. H. E. Jacob, W. Foster, and H. Berg, Microbial implication of electric field effects. II. Inactivation of yeast cells and repair of their cell envelope. *Z. Allg. Mikrobiol. 21*:225 (1981).
21. H. Hülsheger, J. Potel, and E. G. Niemann. Electric field effects on bacteria and yeast cells, *Radiat. Environ. Biophys. 22*:149 (1983).
22. M. Sato, K. Tokita, M. Sadakata, and T. Sakai, Sterilization of microorganisms by high-voltage pulsed discharge under water, *Kagaku Hogaku Ronbunshu 4*:556 (1988).
23. U. R. Pothakamury, H. Vega-Mercado, Q. Zhang, G. V. Barbosa-Cánovas, and B. G. Swanson, Effect of growth stage and temperature on the inactivation of *E. coli* by pulsed electric fields, *J. Food Prot. 59* (11):1167 (1996).
24. Q. Zhang, A. Monsalve-González, G. V. Barbosa-Cánovas, and B. G. Swanson, Inactivation of *E. coli* and *S. cerevisiae* by pulsed electric fields under controlled temperature conditions, *Trans. ASAE 37*:581 (1994).
25. Q. Zhang, B. L. Qin, G. V. Barbosa-Cánovas, and B. G. Swanson, Inactivation of E. coli for food pasteurization by high intensity short duration pulsed electric fields, *J. Food Proc. Pres. 19*:103 (1995).
26. H. Vega-Mercado, U. R. Pothakamury, F. J. Chang, G. V. Barbosa-Cánovas, and B. G. Swanson, Inactivation of *E. coli* by combining pH, ionic strength and pulsed electric field hurdles, *Food Res. Int. 29* (2):117 (1996).
27. U. Zimmermann, G. Pilwat, and F. Riemann, Dieletric breakdown on cell membranes, *Biophys. J. 14*:881 (1974).
28. R. Benz and U. Zimmermann, Pulse-length dependence of the electrical breakdown in lipid bilayer membranes, *Biochim. Biophys. Acta 597*:637 (1980).
29. D. Knorr, M. Geulen, T. Grahl, and W. Sitzmann, Food application of high electric field pulses, *Trends Food Sci. Technol. 5*:71 (1994).
30. T. Y. Tsong, Review on electroporation of cell membranes and some related phenomena, *Biochim. Bioeng. 24*:271 (1990).
31. T. Y. Tsong, Electroporation of cell membranes, *Biophys. J. 60*:297 (1991).
32. J. W. Larkin and S. H. Spinak, Regulatory aspects of new/noval technologies, *New Processing Technologies Yearbook* (D. I. Chandarana, ed.), National Food Processors Association (NFPA), Washington, D.C., 1996, p. 86.
33. H. Vega-Mercado, L. O. Luedecke, G. M. Hyde, G. V. Barbosa-Cánovas, and B. G. Swanson, HACCP and HAZOP for a pulsed electric field processing operation, *Dairy, Food Environ. Sanit. 16* (9):554 (1996).
34. A. H. Bushnell, J. E. Dunn, R. W. Clark, and J. S. Pearlman, High pulsed voltage systems for extending the shelf-life of pumpable food products, U. S. patent 5,235,905 (1993).
35. W. Sitzmann, Keimabtotung mit Hilfe elecktrischer Hochspannungsimpulse in pumpfähigen Nährungsmitteln. Vortrag anläblich des Seminars. Mittelsanforderung in der Biotechnologie. Ergebnisse des indirekt-spezifischen Programma des BMFT, Germany, 1990, p. 1986.
36. T. Grahl, W. Sitzmann, and H. Märkl, Killing of microorganisms in fluid media by high-voltage pulses, Proceedings of the 10th Dechema Biotechnol. Conference Series 5B, Verlagsgesellsellschaft, Hamburg, Germany, 1992, p. 675.
37. O. Martin, Q. Zhang, A. J. Castro, G. V. Barbosa-Cánovas, and B. G. Swanson, Pulse electric fields of high voltage to preserve foods. Microbiological and engineering aspects of the process, *Span. J. Food Sci. Technol. 34*:1 (1994).
38. B. L. Qin, Q. Zhang, G. V. Barbosa-Cánovas, B. G. Swanson, and P. D. Pedrow, Continuous

flow electrical treatment of flowable food products, Washington State University, Pullman, WA (1977).
39. Q. Zhang, F. J. Chang, G. V. Barbosa-Cánovas, and B. G. Swanson, Engineering aspects of pulsed electric field pasteurization, *J. Food Eng. 25*:261 (1995).
40. Q. Zhang, B. L. Qin, G. V. Barbosa-Cánovas, B. G. Swanson, and P. D. Pedrow, Batch Mode Food Treatment Using Pulsed Electric Fields, U.S. Patent No. 5,549,041 (1995).
41. O. Martin, H. Vega-Mercado, B. L. Qin, F. J. Chang, G. V. Barbosa-Cánovas, and B. G. Swanson, Inactivation of *E. coli* suspended in liquid eggs using pulsed electric fields, *J. Food Proc. Pres. 21*:193 (1997).
42. D. A. Corlett and M. H. Brown, pH and acidity, *Factors Affecting the Life and Death of Microorganisms*, Academic Press, New York, 1980.
43. P. M. Wiggins, Cellular functions of a cell in a metastable equilibrium state, *J. Theor. Biol. 52*:99 (1975).
44. K. Dolowy, Uniform hypothesis of cell behavior—movement, contact inhibition of movement, adhesion, chemotaxis, phagocytosis, pinocytosis, division, contact inhibition of division, fusion, *J. Theor. Biol. 52*:83 (1975).
45. U. R. Pothakamury, A. Monsalve-González, G. V. Barbosa-Cánovas, and B. G. Swanson, Inactivation of *E. coli* and *S. aureus* in model foods by pulsed electric field technology, *Food. Res. Int. 28* (2):167 (1995).
46. H. G. L. Coster and U. Zimmermann, The mechanisms of electrical breakdown in the membrane of *Valonia utricularis*, *J. Membrane Biol. 22*:73 (1975).
47. U. R. Pothakamury, A. Monsalve-González, G. V. Barbosa-Cánovas, and B. G. Swanson, High voltage pulsed electric field inactivation of *B. subtilis* and *L. delbrueckii*, *Span. J. Food Sci. Technol. 35* (1):101 (1995).
48. U. R. Pothakamury, Preservation of foods by nonthermal processes, Ph.D. thesis, Washington State University, Pullman, WA 1995.
49. H. Vega-Mercado, O. Martín, F. J. Chang, G. V. Barbosa-Cánovas, and B. G. Swanson, Inactivation of *E. coli* and *B. subtilis* suspended in pea soup using pulsed electric fields, *J. Food Proc. Pres. 20* (6):501 (1996).
50. J. M. Jay, High temperature food preservation and characteristics of thermophilic microorganisms, *Modern Food Microbiology*, 4th ed., Van Nostrand Reinhold, New York, 1992, p. 335.
51. Q. Zhang, F. J. Chang, G. V. Barbosa-Cánovas, and B. G. Swanson, Inactivation of microorganisms in a semisolid model food using high voltage pulsed electric fields, *Food Sci. Technol. (LWT). 27* (6):538 (1994).
52. Q. Zhang, F. J. Chang, G. V. Barbosa-Cánovas, and B. G. Swanson, Inactivation of *E. coli* for food pasteurization by high intensity short duration pulsed electric fields, *J. Food Proc. Pres. 17*:469 (1994).
53. M. Allen and K. Soike, Sterilization by electrohydraulic treatment, *Science 10*:155 (1966).
54. H. Hülsheger, J. Potel, and E. G. Niemann, Killing of bacteria with electric pulses of high field strength, *Radiat. Environ. Biophys. 20*:53 (1981).
55. M. V. Simpson, G. V. Barbosa-Cánovas, and B. G. Swanson, Influence of PEF on the composition of apple juice, Internal Report, Washington State University, Pullman, WA, 1995.
56. B. L. Qin, Q. Zhang, G. V. Barbosa-Cánovas, B. G. Swanson, and P. D. Pedrow, Pulsed electric field treatment chamber design for liquid food pasteurization using a finite element method, *Trans. ASAE. 38* (2):557 (1995).
57. B. L. Qin, Q. Zhang, G. V. Barbosa-Cánovas, B. G. Swanson, P. D. Pedrow and R. G. Olsen, A continuous treatment system for inactivating microorganisms with pulsed electric fields, Proceedings of IEEE/IAS meeting, Orlando, FL, October 1995.
58. O. Martin, B. L. Qin, F. J. Chang, G. V. Barbosa-Cánovas, and B. G. Swanson, Inactivation of *E. coli* in skim milk by high intensity pulsed electric fields, *J. Food Eng.* (in press).
59. W. A. Hamilton and A. J. H. Sale, Effect of high electric fields on microorganisms I. Mechanism of action of the lethal effect, *Biochem. Biophys. Acta 148*:789 (1967).

60. H. D. Goff and A. R. Hill, Chemistry and physics, *Dairy Science and Technology Handbook*, Vol. 1: *Principles and Properties* (Y. H. Hui, ed.), VCH Pub. Inc., New York, p. 1.
61. A. J. Castro, Pulsed electric field modification of activity and denaturation of alkaline phosphatase, Ph.D. thesis, Washington State University, Pullman, WA, 1994.
62. H. Vega-Mercado, J. R. Powers, G. V. Barbosa-Cánovas, and B. G. Swanson, Plasmin inactivation with pulsed electric fields, *J. Food Sci. 60*:1143 (1995).
63. H. Vega-Mercado, J. R. Powers, O. Martín-Belloso, O. L. Luedecke, G. V. Barbosa-Cánovas, and B. G. Swanson, Effect of pulsed electric fields on the susceptibility of proteins to proteolysis and inactivation of an extracellular protease from *P. fluorescens* M 3/6, Proceedings of ICEF7, Seventh International Congress on Engineering and Food, The Brighton Center, Brighton, UK, 13–17 April, 1997, p. C73.
64. H. Vega-Mercado, Inactivation of proteolytic enzymes and selected microorganisms in foods using pulsed electric fields, Ph.D. thesis, Washington State University. Pullman, WA, 1996.

18

Preserving Foods with Electricity: Ohmic Heating

M. SHAFIUR RAHMAN
Horticulture and Food Research Institute of New Zealand, Auckland, New Zealand

I. THE OHMIC HEATING PROCESS

A. What is Ohmic Heating?

Ohmic heating operates by direct passage of electric current through the food product, with heat generated as a result of electrical resistance. A schematic diagram of an ohmic heating system is shown in Figure 1. In conventional heating methods, heating travels from heated surface to the product interior by means of both convection and conduction, which is time consuming, especially with longer conduction or convection paths. Electro-resistive or ohmic heating is volumetric by nature, thus has potential to reduce overprocessing.

There are many advantages of ohmic heating. It provides rapid and even or uniform heating, resulting in less thermal damage to the product. The operating costs are low, and 90% of its energy is converted to heat compared with 70% for microwaves [37]. The low penetration depth of microwaves into solid food also causes thermal nonuniformity. Ohmic heating should not exhibit this problem. The absence of a hot surface with ohmic heating reduces fouling problems and thermal damage to the product, which sometimes limit the operation of conventional ultra heat treatment (UHT) systems [9].

Factors important to the successful use of ohmic heating include type of the product, flow rate, temperature rise, heating rate, holding time [31].

B. The Economics of the Ohmic Heating Process

Allen [1] compared the cost of installation and operation of two ohmic food-processing systems to that of conventional retorting, freezing, and heating in a conventional tubular heat exchanger. Tables 1 and 2 show cost comparisons for low-acid and high-acid foods. Acidity is one of the important factors in predicting the desired level of sterilization. The components included in the cost analyses were labor, energy, packaging, equipment maintenance and repairs, plant supplies, and interest and depreciation on the processing and filling equipment. Ohmic operational costs were found to be comparable to those for freezing and retort processing of low-acid food products (Table 1) [1]. Table 2

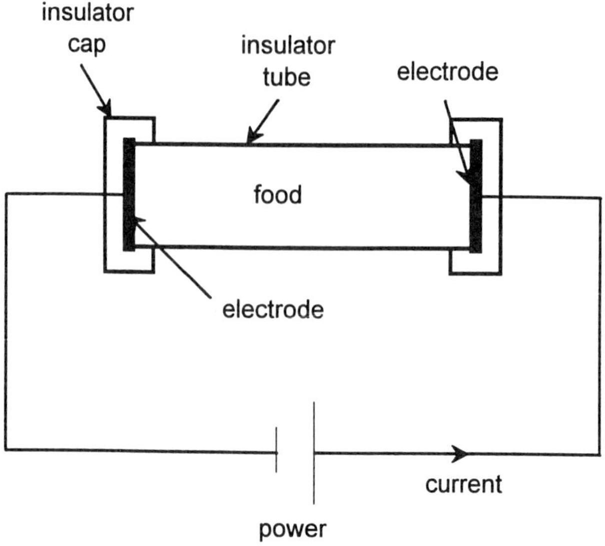

FIGURE 1 Schematic diagram of an ohmic heating process.

shows more cost in case of ohmic heating for high-acid foods. If quality considerations are important and ohmic heating can produce products of clearly superior quality to those produced using conventional technologies, this process should be considered. There are also risks associated with start-up costs and unknown market potential, although the economics and technology appear to be favorable [47]. Thus, these factors should also be considered when selecting ohmic heating process for foods.

The following process and safety criteria should be maintained for a successful unit [8,9]:

1. Efficient particle pumping machinery
2. An electrical design that avoids electrolysis due to either electrode breakdown or localized product scorching
3. Noncontaminating electrodes, which have good contact with the food material
4. Control of heating rate and flow rate of food
5. An effective aseptic packaging process for sterilization
6. Overall cost-effectiveness

TABLE 1 Installation and Processing Costs of Low-Acid Products Processed in Retorting, Freezing, and Ohmic Heating

			Ohmic heating	
Depreciation schedule	Retorting	Freezing[a]	OH75 system	OH300 system
Installation[b]	$2,137,771	$1,543,991	$8,543,600	$9,176,400
5-year	$0.33/lb	$0.26/lb	$0.49/lb	$0.25/lb
10-year	$0.32/lb	$0.24/lb	$0.36/lb	$0.21/lb

[a]Additional distribution cost for frozen low-acid product is 0.53¢/lb.
[b]Installation costs include purchase, delivery, and site installation of processing and filling equipment.
Source: Ref. 1.

TABLE 2 Installation and Processing Costs of High-Acid Products Processing in Retorting, Freezing, and Ohmic Heating

Depreciation schedule	Commercial tubular heat exchanger	Freezing[a]	Ohmic heating			
			Hot fill		Aseptic fill	
			OH75	OH300	OH75	OH300
Installation[b]	$524,200	$1,890,675	$1,122,000	$1,302,800	$3,943,600	$4,576,400
5-year	$0.08/lb	$0.08/lb	$0.17/lb	$0.11/lb	$0.39/lb	$0.17/lb
10-year	$0.08/lb	$0.08/lb	$0.14/lb	$0.10/lb	$0.28/lb	$0.14/lb

[a] Installation costs include purchase, delivery, and site installation of processing and filling equipment.
[b] Additional distribution cost for frozen low-acid product is 2.53¢/lb.
Source: Ref. 1.

C. Heat Generation

The most important questions in ohmic heating process design are [21]: Where is the cold spot in the medium? What is the lethal treatment delivered to the cold spot? How is the lethal treatment ensured? If the particle heats faster than the liquid portion, the particle's coldest spot is at its surface. Thus, prediction equations to estimate heat generation in the medium is an important issue in ohmic heating system design.

The key to the ohmic process is the rate of heat generation, the electrical conductivity of food material, and the way the food flows through heater [8,10]. The heat generation by electrical energy due to electrical resistance can be written as:

$$Q = I^2R = \sigma V^2 \tag{1}$$

where Q is the heat generated (W), I is the current (A), V is the voltage gradient (volt), σ is the electrical conductivity (S/m), and R is the electrical resistance (ohm). The electrical resistance can be written from Ohm's law as:

$$R = \frac{V}{I} \tag{2}$$

The electrical conductivity can be defined as:

$$\sigma = \left(\frac{1}{R}\right)\left(\frac{L}{A}\right) \tag{3}$$

where L is the length (m), A is the cross-sectional area (m^2), R is the resistance (ohm), and σ is the electrical conductivity (S/m). The specific electrical resistance is $1/\sigma$ (ohm m).

For liquids like orange and tomato juice electrical conductivity increased linearly with temperature regardless of the mode of heating used, and it decreased with increasing solid content [28]. Palaniappan and Sastry [28] found that electrical conductivity tended to increase as particle size decreased, but a general conclusion cannot be reached without accounting for particle shapes and orientations.

The electrical conductivity of vegetable tissue undergoes a sharp increase at about 60°C during conventional heating due to breakdown of the cell walls [27]. When cellular tissue is heated ohmically, the electrical conductivity temperature curve becomes more and more linear as the voltage gradient is increased. A possible cause may be electroosmosis during ohmic heating [32].

Some fondants, syrups, and fats are natural electrical insulators, while strong brines and pickles are too acidic and have high electrical conductivity. The addition of salt to an emulsion is known to decrease the electrical resistance. Nonconducting materials will not be heated by electrical resistance and heating is mainly by thermal conduction. Examples of these are fat, oil, nuts, air, alcohol, bone, and ice [47].

The rate of heat generation depends on the local electric field strength. When a two-phase mixture heat electrically, the two phase will generate heat at the same rate if the liquid and solid have the same electrical resistance [8]. When the fluid is of a higher electrical conductivity than the particles, nonuniform heating can occur in the fluid surrounding the particles if heating takes place under static conditions [33].

The ratio of heat generation in solid to liquid is [8]:

$$\frac{Q_s}{Q_L} = \frac{\rho\sigma_s\sigma_L}{(\sigma_s + 2\sigma_L)^2} \tag{4}$$

where ρ is the density (kg/m^3), σ_s and σ_L are the electrical conductivity of solid and liquid, respectively, and Q_s and Q_L are the heat generation in solid and liquid phases, respectively. The above equation, derived for sphere neglecting conduction effect, indicates that relative heating rate is a function of both solid and liquid electrical conductivities. Fryer [8] showed that for Ψ $(=\sigma_s/\sigma_L)$

between 1 and 4 the rate of heat generation in the solid is greater than that in the surrounding liquid. Outside of this region, less heat, is generated in the solid than in the liquid. Moreover, the shape of the particle and its orientation can cause over- or underheating. Fryer [8] showed that in the case of elongated particles, the particle will overheat the fluid if placed parallel to the electric field or underheat it if it is placed at right angles to the field.

The voltage gradient can be longitudinal or transverse to the flow, depending on ohmic heater design. If voltage is applied longitudinally along the axis of flow, the first moving regions at the center receive a lower current density because of lower temperature and greater resistance, thus nonuniformity can become more pronounced under this condition. In the case of transversely applied voltage, the center element would also receive a lower thermal treatment [32].

Particle shape and orientation also play an important role in the effective electrical conductivity of medium containing particles [32]. If the mixture consists of long thin particles, their orientation relative to the electrical field has a significant influence on electrical properties as well as the relative heating rates of phases [6]. If the solids have an aspect ratio close to unity, the effect of orientation on overall electrical conductivity is small. Spheres, cylinders with length equal to diameter, and cubes have an aspect ratio close to unity [35].

Around materials of high or low electrical conductivity, the electric field can be distorted [9]. Factors affecting the heating rate in ohmic heating of foods with high solids content are (a) electrical conductivities of fluid and the particles, (b) particle size, shape, concentration, and specific heat, (c) viscosity of the fluid, and (d) orientation relative to the electrodes and to other particles [9,19].

A particle does not heat uniformly during ohmic heating because of the nonuniform nature of the electric field within the system [21]. Zones of high and low heat generation in a fluid around a particle of low electrical conductivity are shown in Figure 2. The current density is high on the sides of the particle and is low in the front and rear regions, since current attempts to bypass the solid. This nonuniformity of heating can be reduced by increasing fluid mixing, reducing the viscosity of fluids, improving the rotation of particles, and the presence of other particles in the medium [32].

A temperature gradient exists within the particulate if it was heated faster than the fluid. This is due to the low electrical resistance of solid particulates. A sharp temperature gradient within a solid was predicted by De Alwis and Fryer [6]. Finite element analysis showed temperatures of 120°C at the center and 80°C near the surface. Electrical conductivity variations of about ±5% may produce temperature differences of up to 10°C in parts of the fluid, which may be acceptable as a uniform heating rate [7]. Fryer et al. [9] demonstrated that differential heating effects were not confined to solids if the viscosity of the solution is significant. The chemical marker used also indicated a higher temperature at the center than near the surface of food particulates ohmically heated using a 5 kW system. Higher lethality both microbiologically and chemically was observed at the center of meatballs than near the surface [19]. Kim et al. [19] concluded that ohmically processed low-conductiv-

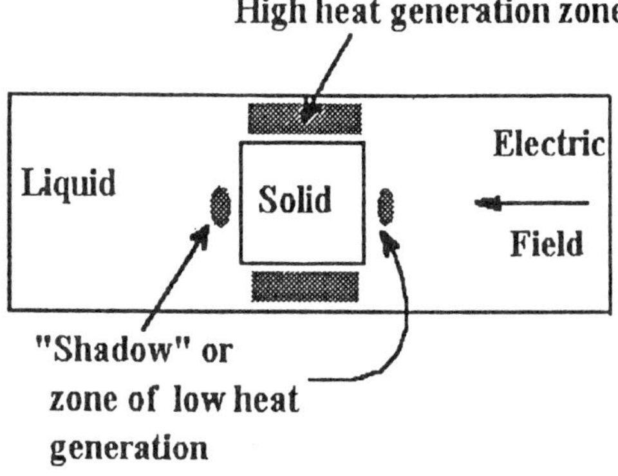

FIGURE 2 Zones of high and low heat generation in the fluid around a particle of low electrical conductivity. (From Ref. 32.)

ity particulates heat faster than the fluid. The microbiological safety of the entire food product can be ensured on the basis of the measured fluid temperature and the conservative laminar flow residence time limit [18,19].

Sastry and Palaniappan [34] showed that for a static ohmic heater, changes in the solid:liquid ratio resulted in changes in the rate of heating for each phase. The heating rate of the particles was initially slower than that of the liquid but then exceeded that of the liquid as the product temperature increased. This was observed when a low-conductivity particle concentration increased. Low electrical conductivity particles in a high electrical conductivity fluid could heat faster or slower than the fluid, depending on their concentration [33]. Higher solids concentration is important in ensuring more rapid heating of the solid phase. If a low electrical conductivity particle is placed within a high-conductivity fluid, the fluid tends to heat faster than the particle due to the bypass of current from the particle, and thus some local field variations occur. When the solids concentration is increased, the fluid fraction decreases and a greater proportion of overall current flows through the solid phase and solids tend to heat faster than liquid [32]. Sastry [32] stated that the use of high solids concentrations is important to ensure relatively rapid heating of the solid phase. Moreover, particle-to-particle interaction may also affect heating.

Larkin and Spinak [21] concluded that the limited published information about the many critical factors inherent in ohmic heating of food requires careful analysis of the system to ensure that the product receives adequate thermal treatment.

D. Electrolysis

In metallic conductors the current is carried by electrons. When electrical potential is applied, the electrons are forced to flow in one direction without appreciable movement of matter. In electrolytes, the current is carried by ions. An electrical potential causes these charged particles of the medium to move—the positive ions in the direction of the current and the negative ions in a positive direction—thus increasing the mobility of the charges particles [11].

The major electrolytic effect is the dissolution of the metallic electrodes, which may contaminate the product. Alternating current at low frequency (e.g., 50 and 60 Hz) has an electrolytic effect similar to that of direct current [31]. Reznick [31] stated that one way to overcome this problem is by utilizing high-frequency current. There is no apparent metal dissolution if an alternating current above 100 kHz is used.

Stainless steel and specially treated pure carbon electrodes can be used with success. Reznick [31] reported that stainless steel electrodes had been used in the industry for more than 3 years, and no marks of metal dissolution were observed, thus replacement was not necessary.

Every product has a specific critical current density above which arcing is likely to occur. Current density can be estimated from the current divided by area of the electrode. When the limiting current density is known, total current call be derived from power and voltage, and then minimum area of the electrode is determined [31].

Although ohmic heating is useful in reducing fouling on the surface, heat is generated within electrodes, which may foul during ohmic heating. Thus, it is usually necessary to cool electrodes by some external means [32].

II. THE EFFECTS OF OHMIC HEATING ON FOODS AND FOOD COMPONENTS

A. Effects on Microorganisms and Enzymes

Extrathermal lethal effects exist in the product due to the presence of electricity [32]. Palaniappan et al. [26] reviewed the nonthermal effects of electricity on microorganisms and concluded that data available in the literature are limited and inconclusive. Some studies showed that there was no additional effect of electricity on bacterial kill other than heat generation by electric current [14].

TABLE 3 D and z Values of Yeast Cells While Maintaining Identical Temperature Histories in Conventional and Ohmic Heating

	D (s) values at different temperatures				
Treatment[a]	49.75°C	52.3°C	55.8°C	58.8°C	z values (°C)
Conventional	294.6	149.7	47.2	16.9	7.19
Ohmic	274.0	113.0	43.1	17.8	7.68

[a]Treatments are not significantly different at 95% confidence level.
Source: Ref. 29.

High voltage can destroy microorganisms by causing pores (i.e., electroporation) to develop in microbial cell walls. Low-voltage treatments decreased microbial counts over long periods of time without heating to lethal temperatures. At low voltage the bactericidal effect on *Escherichia coli* depended on the current passing through the suspension, the presence of chloride-containing compounds, and the amount of time the cells were left standing in the medium after treatment [30]. Low-voltage alternating current at 50 Hz, 10–200 mA, and stainless steel electrodes were used in the above study. The temperature measured after the treatment was below 40°C in all experiments. In the case of growth of *E. coli* it was concluded that there was not stimulative effect of low-voltage electric current [40]. Sastry [32] found that under ohmic heating conditions, neither are the voltages high enough to cause electroporation nor are residence times enough long to result in low-voltage effects. Thus, only thermal effects and rapidity of ohmic heating cause microbial death. The D and z values of yeast cells *zygo Saccharomyces bailii* obtained by conventional and ohmic heating are given in Table 3. The results show no significant difference between treatments while maintaining identical temperature histories in each case. If mild thermal sublethal electrical treatments were applied before heat treatment, D values of *E. coli* were significantly reduced in only some cases (Table 4) [29]. The reasons for this behavior are not entirely clear. Palaniappan et al. [29] suggested that an optimum combination of electrical pretreatment and thermal treatment may reduce the inactivation requirement. At temperatures above or below this level, the effect of electrical pretreatment seems diminished. Thus, combinations of sublethal electrical pretreatment appear to offer potential for increased bacterial inactivation in certain cases.

Mizrahi et al. [25] studied the inactivation of enzymes by electrconductive heating of corn on the cob immersed in an isoelectrical conductivity aqueous system. They reported complete peroxidase inactivation by ohmic heating in less than 3 minutes as compared to 17 minutes in the case of boiling water blanching. Palaniappan et al. [26] attributed the enzyme inactivation to heat only, and no other mechanism was considered in their study.

B. Electroosmosis

Electric fields are known to enhance diffusion across membranes [3,42], and applied electric fields can influence the mobility of color in beetroot [13]. Schreier et al. [36] found the betanin diffusion

TABLE 4 D and z Values of *E. coli* in Conventional and Electrical Pretreatment

	D (s) values at different temperatures				
Treatment	60.0°C	64.5°C	68.5°C	71.0°C	z values (°C)
Conventional	164.79	48.19	22.97[a]	11.83	9.87
Electrical pretreatment	180.41	51.17	21.19[a]	11.79	9.36

[a]Significantly different at 95% confidence level.
Source: Ref. 29.

from beetroot to be greater during 50 Hz electrical heating than during conventional heating. The enhanced diffusion in electro-osmosis may be due to the result of increased transport through the cell membrane and the surrounding bulk solution, and the ions in solution may behave similarly to that of direct current. The dye efflux was proportional to the surface area of a given sample, suggesting that the enhancement followed the rules for mass transfer coefficient. The beetroot dye efflux (kg/m^2s) was linearly increased with the increasing applied field (V/cm).

The applied voltage on plant tissue induces electroporation of the cell membrane caused by the induced membrane potential [20,41]. When induced membrane potential reaches a critical value, rupture results in the formation of pores in the cell membrane and increases permeability. This phenomenon is called electroporation.

Imai et al. [17] studied the effects of frequency on the heat generation in ohmic heating (50 Hz to 10 kHz; 40 V/cm) of white radish. They found that 50 Hz gave the sharpest initial rise in temperature and the shortest time to raise the temperature at the midpart of radish to 80°C. Heating rates above 60°C were found to be almost the same and linear for all frequencies. NMR imaging suggested that the initial rapid heating at low frequency was caused by the electroporation of radish tissue membrane, resulting in the reduction of its impedance. The pressurization (400 MPa, 25°C for 10 min) of radish eliminated the sharp initial increase in temperature.

In many cases, loss of soluble solids during water blanching of vegetables adversely affects the quality of the product. Solute loss may also have an environmental impact, since wastewater contains solutes that cannot be discarded without treatments. Conventional hot water blanching is a slow process. Thus, large vegetables are usually diced before water blanching to increase the surface area of the product that is exposed to the hot water. An increase in surface area also increases solute loss. The need for size reduction can be avoided using ohmic heating [25]. This method is capable of heating a large product (with a relatively small surface-to-volume ratio) very quickly and uniformly regardless of its size or shape, thus enabling a considerable shortening of blanching time with no need for size reduction [24]. Mizrahi [24] mentioned that a reduction of one order of magnitude in solute losses may be achieved by blanching with ohmic heating. This is due to a favorable combination of low surface-to-volume ratio and short process time in the case of whole vegetable blanching.

C. Effects on Functional Properties

A typically slow heating rate activates protease to degrade myofibrillar proteins before the protease can be thermally inactivated [44]. Protease could be inactivated by rapid heating methods to minimize proteolytic activity without the use of enzyme inhibitors [12,45]. The gel strength of surimi made from walleye pollock, white croaker, threadfin bream, and sardine was improved when samples were ohmically heated instead of in a 90°C conventional water bath [38,39].

Yongsawatdigul et al. [45] found an increase in gel functionality of Pacific whiting surimi when processed ohmically. Surimi (78% moisture and 2% sodium chloride) without enzyme inhibitors was heated conventionally to 90°C after holding at 55°C. Gels heated slowly in a water bath exhibited poor gel quality, while the ohmically heated gels without holding at 55°C showed a more than twofold increase in shear stress and strain over conventionally heated gels. Degradation of myosin and actin was minimized by ohmic heating, resulting in a continuous network structure as shown in scanning electronic micrographs. Thus, ohmic heating with rapid heating rate was an effective method for maximizing gel functionality without the addition of enzyme inhibitors.

D. Effects on Sensory Quality

Ohmic heating is currently used for sterilization and pasteurization of a number of products. Recently, a wide variety of low- and high-acid products as well as refrigerated extended shelf-life products have been developed. These were found to have texture, color, flavor, and nutrient retention comparable to or better than those of traditional processing methods such as freezing, retorting, and aseptic processing [47]. Various combinations of meats, vegetables, pasta, and fruits can be successfully processed when accompanied by appropriate carrier medium and suitable process controls [47]. Zoltai and Swearingen [47] found that particulate size is typically limited to 1 in^3 for the following reasons: (a) to ensure that sufficient clearance past the electrodes is maintained, (b) because aseptic particulate fillers are capable of filling particulates up to 1 in^3 without damage to the product, and (c) because particulates larger than 1 in^3 would require cutting prior to consumption and thereby reduce convenience. Larger particulates, such as pasta or bamboo shoots, may occasionally be processed, since they are elongated, flexible, and relatively flat [47]. The particulate concentration in most ohmic formulations ranges from 20 to 70%. Extremely low or high concentrations require special consideration of size, shape, and texture [47]. Particle density, carrier-medium viscosity, and electrical conductivity of the medium should also considered in formulation. Extremely dense particulates in a low-viscosity carrier medium will tend to sink at various points in the system, which will result in overprocessing. Conversely, very light particulates have the capacity to float in the medium. For product formulation and ohmic treatment design, control of ionic concentrations can be used to enhance product heating rates [32].

Yang et al. [43] found that the average overall sensory score of six products was 7.3 on a 1–9 hedonic scale, where 7 is "good" and 8 is "very good" [19]. Imai et al. [17] studied the change of hardness in white radish when ohmic heating was used. One sample was heated in hot water at 95°C and another sample was ohmically heated at 50 Hz and 40 V/cm until the centers of both samples reached 90°C. They found that the sample cooked in hot water was wholly softened, especially in outer areas, while the ohmic heated sample had almost the same outer area breaking strength as the original raw sample, but was softened in the inner region. Thus, ohmic heating at low frequency is very effective technology for rapid heating of agricultural products to improve texture.

Some foods processed by ohmic heating will never taste the same as those processed using conventional heating. Thus, fine-tuning in product formulation should be considered. Moreover, products with improved sensory properties could also be developed. In ohmic heating the heating rate of the food product is dependent on the product formulation [21]. Thus, an understanding of the effect of formulation changes on process design with adequate control and on sensory quality is necessary. A processor needs to use ingredients with adjusted electrical conductivities. When an ingredient is added to the product during the batching phase, the length of the holding time for the batch may be critical. This may be due to transference of the electrolyte from the ingredient to the rest of formulation. The presence of nonelectrical conducting particles, such as fat in formulation, may also affect the particle heating.

The physical and chemical properties of the product play an important role in determining how much lethal treatment is required. The critical factors are size and shape, moisture content, viscosity, electrolytes, pH, specific heat, thermal conductivity, solid:liquid ratio, and electrical conductivity [21].

III. APPLICATIONS IN THE FOOD INDUSTRY

In the early twentieth century, pasteurization of milk was achieved by pumping fluid between parallel plates having different voltages [2]. More details of the electric pasteurization of milk have been reviewed by Palaniappan et al. [26]. The technology of ohmic heating gradually disappeared due to the lack of suitable electrode materials and adequate control systems. As discussed by Sastry [32], recently interest in ohmic heating has been revived at least partly due to the availability of improved electrode materials. Ohmic heating has shown significant promise in a number of food processes, including sterilization and pasteurization [33]. Ohmic heating is potentially a very important development in aseptic high-temperature, short-time (HTST) processing. A number of applications have, been found in areas such as thawing and baking [15]. The applications of ohmic heating in blanching and in enhanced diffusion processes have been discussed in an earlier section. Nineteen plants using ohmic heating are operating worldwide, and the technique has won awards for innovation [5].

Conventional thawing by water immersion heating can be classified as surface heating, where heat is supplied to the surface by convection and is conducted into the block. The thaw water cannot be discarded in a sewer system due to adverse environmental impact. Moreover, long thawing time may also increase microbial growth. It has been suggested that for fast thawing of fish, electrical resistance heating, dielectric heating, or microwave heating should be used [4]. The thawing time could be reduced two to three times compared to air or water immersion thawing [22]. Ohmic thawing has lower operational costs and physical and sensory properties with this process were similar to the conventional thawing [16]. Runaway heating may be a problem, since all current tends to flow through the thawed portion due to lower resistance. Luzuriaga and Balaban [22] stated that this problem could be avoided by proper control of the ohmic thawing process.

In rapid heating and coagulation of sausages, ohmic heating in a plastic (PTFE) tube and ring-formed graphite electrodes was used. An even heating throughout the material was achieved, and arcing and skin formation on the electrodes were avoided by optimization of the material flow and current density [23].

Boiling may occur because there is no upper temperature limit for ohmic heating and the system may be pressurized. It is possible to achieve temperatures above the boiling point even under these elevated pressure conditions [46].

No ohmic heating systems have been developed for use in the home. This is due to the problem of safety, as the food is "live" when it is cooked. Moreover, scaling to a smaller size for domestic applications is also a problem [37].

REFERENCES

1. K. Allen, V. Eidman, and J. Kinsey, An economic engineering study of ohmic food processing, *Food Technol. 50*:269 (1996).
2. A. K. Anderson and R. Finkelstein, A study of the electropure process of treating milk, *J. Dairy Sci. 2*:374 (1919).
3. P. H. Barry and A. B. Hope, Electroosmosis in membranes: effect of unstirred layers and transport numbers, *Biophys. J. 9*:700 (1969).
4. G. H. O. Burgess, C. L. Cutting, J. A. Lovern, and J. J. Waterman, *Fish Handling and Processing*, Chemical Publishing, New York, 1967, p. 189.
5. Events to come, *Food Aust. 49*:224 (1997).
6. A. A. P. DeAlwis and P. J. Fryer, A finite element analysis of heat generation and transfer during ohmic heating of food, *Chem. Eng. Sci. 45*:1547 (1990).
7. A. A. P. De Alwis and P. J. Fryer, Operability of the ohmic heating process: electrical conductivity effects, *J. Food Eng. 15*:21 (1992).
8. P. Fryer, Electrical resistance heating of foods, *New Methods of Food Preservation* (G. W. Gould, ed.), Blackie Academic and Professional, Glasgow, 1995, p. 205.
9. P. J. Fryer, A. A. De Alwis, E. Koury, A. G. F. Stapley, and L. Zhang, Ohmic processing of solid-liquid mixtures: heat generation and convection effects, *J. Food Eng. 18*:101 (1993).
10. P. Fryer and Z. Li, Electrical resistance heating of foods, *Trends Food Sci. Technol. 4*:364 (1993).
11. S. Glasstone and D. Lewis, *Elements of Physical Chemistry*, Macmilland & Co. Ltd., London, 1960.
12. D. H. Greene and J. Babbitt, Control of muscle softening and protease parasite interactions in arrowtooth flounder (*Athererthes stomias*), *J. Food Sci. 55*:579 (1990).
13. K. Halden, A. A. P. De Alwis, and P. J. Fryer, Changes in electrical conductivity of foods during ohmic heating, *Int. J. Food Sci. Technol. 25*:9 (1990).
14. C. W. Hall and G. M. Trout, *Milk Pasteurization*, Van Nostrand Reinhold, New York, 1968.
15. H. He and R. C. Hoseney, A critical look at the electric resistance oven. *Cereal Chem. 68*:151 (1991).
16. T. Henderson, M. O. Balaban, and A. Teixeira, Ohmic thawing of frozen shrimp: preliminary technical and economic feasibility, Proceedings of the Workshop on Process Optimization and Minimal Processing of Foods, Superior de Biotechnologica, Porto, Portugal, September 23, 1993.
17. T. Imai, K. Uemura, N. Ishida, S. Yoshizaki, and A. Noguchi, Ohmic heating of Japanese white raddish *Rhaphanus sativus* L., *Int. J. Food Sci. Technol. 30*:461 (1995).
18. H. Kim, Y. Choi, T. C. S. Yang, I. A. Taub, P. Tempest, P. Skudder, G. Tucker, and D. L. Parrott, Validation of ohmic heating for quality enhancement of food products, *Food Technol. 50*:253 (1996).

19. H. J. Kim, Y. M. Choi, A. Yang, T. Yang, I. A. Taub, J. Giles, C. Ditusa, S. Chall, and P. Zoltai, Microbiological and chemical investigation of ohmic heating of particulate foods using a 5 kW ohmic system, *J. Food Proc. Pres. 20*:41 (1996).
20. K. Kinoshita, M. Hibino, and H. Itoh, Events of membrane electroporation visualized on a time scale from microsecond to seconds, *Guide to Electroporation and Electrofusion* (D. C. Chang, B. M. Chassy, J. A. Saunders, A. E. Sowers, eds.), Academic Press, New York, 1992, p. 24.
21. J. W. Larkin and S. H. Spinak, Safety considerations for ohmically heated, aseptically processed, multiphase low-acid food products, *Food Technol. 50*:242 (1996).
22. D. A. Luzuriaga and M. O. Balaban, Electrical conductivity of frozen shrimp and flounder at different temperatures and voltage levels, *J. Aquatic Food Prod. Technol. 5*:41 (1996).
23. Y. Malkki, and A. Jussila, Electrical resistance heating of food emulsions, *Engineering and Food*, Vol. 1, (B. M. McKenna, ed.), Elsevier Applied Science Publishers, Essex, 1984.
24. S. Mizrahi, Leaching of soluble solids during blanching of vegetables by ohmic heating, *J. Food Eng. 29*:153 (1996).
25. S. Mizrahi, I. J. Kopelman, and J. Perlman, Blanching by electroconductive heating, *Int. J. Food Sci. Technol. 10*:281 (1975).
26. S. Palaniappan, S. K. Sastry, and E. R. Richter, Effects of electricity on microorganisms: a review, *J. Food Proc. Pres. 14*:393 (1990).
27. S. Palaniappan and S. K. Sastry, Electrical conductivities of selected solid foods during ohmic heating, *J. Food Proc. Eng. 14*:221 (1991).
28. S. Palaniappan and S. K. Sastry, Electrical conductivity of selected juices: influences of temperature, solids content, applied voltage and particle size, *J. Food Proc. Eng. 14*:247 (1991).
29. S. Palaniappan, S. K. Sastry, and E. R. Richter, Effects of electroconductive heat treatment and electrical pretreatment on thermal death kinetics of selected microorganisms, *Biotechnol. Bioeng. 39*:225 (1992).
30. A. Pareilleux and N. Sicard, Lethal effects of electric current on *Escherichia coli*, *Appl. Microbiol. 19*:421 (1970).
31. D. Reznick, Ohmic heating of fluid foods, *Food Technol. 50*:250 (1996).
32. S. K. Sastry, Ohmic heating, *Minimal Processing of Foods and Process Optimization: An Interface* (R. P. Singh and F. A. R. Oliveira, eds.), CRC Press, Boca Raton, FL, 1994, p. 17.
33. S. K. Sastry and Q. Li, Modeling the ohmic heating of foods, *Food Technol. 50*:246 (1996).
34. S. K. Sastry and S. Palaniappan, Ohmic heating of liquid particle mixtures, *Food Technol. 46*:64 (1992).
35. S. K. Sastry and S. Palaniappan, Influence of particle orientation on the effective electrical resistance and ohmic heating rate of a liquid particle mixture, *J. Food Proc. Eng. 15*:213 (1992).
36. P. J. R. Schreier, D. G. Reid, and P. J. Fryer, Enhanced diffusion during electrical heating of foods, *Int. J. Food Sci. Technol. 28*:249 (1993).
37. E. Scott, Ohmic heating hits commercial scale, *Food Technol. NZ30*(7):8 (1995).
38. M. Shiba, Properties of kamaboko gels prepared by using a new heating apparatus, *Nippon Suisan Gakkaishi 58*:895 (1992).
39. M. Shiba and T. Numakura, Quality of heated gel from walleye a pollack surimi by applying joule heat, *Nippon Suisan Gakkaishi 58*:903 (1992).
40. K. Shimada and K. Shimahara, Effect of alternating current on growth lag in *Escherichia coli* B, *Gen. Appl. Microbiol. 23*:127 (1977).
41. A. E. Sowers, Mechanisms of electroporation and electrofusion, *Guide to Electroporation and Electrofusion* (D. C. Chang, B. M. Chassy, J. A. Saunders and A. E. Sowers, eds.), Academic Press, New York, 1992, p. 119.
42. J. A. Wesselingh and R. Krishna, *Mass Transfer*, Ellis Horwood, Chichester, 1988.
43. T. C. S. Yang, J. S. Cohen, R. A. Kluter, and M. G. Driver, Feasibility of applying ohmic heating and split-phase aseptic processing for ration entree preservation, Tech. Rept. TR-94/021, U.S. Army Natick RD&E Center, Natick, 1994.
44. J. Yongsawatdigul, J. W. Park, and E. Kolbe, Electrical conductivity of Pacific whiting surimi paste during ohmic heating, *J. Food Sci. 60*:922 (1995).

45. J. Yongsawatdigul, J. W. Park, E. Kolbe, Y. A. Dagga, and M. T. Morrissey, Ohmic heating maximizes gel functionality of Pacific whiting surimi, *J. Food Sci. 60*:10 (1995).
46. L. Zhang and P. J. Fryer, Modelling heat generation and transfer in laminar flow of food materials, Presented at the Sixth International Congress on Engineering and Food, Makuhari Messe, Chiba, Japan, May 23-27, 1993.
47. P. Zoltai and P. Swearingen, Product development considerations for ohmic processing, *Food Technol. 50*:263 (1996).

19

High-Pressure Treatment in Food Preservation

Enrique Palou*, Aurelio López-Malo*, Gustavo V. Barbosa-Cánovas, and Barry G. Swanson
Washington State University, Pullman, Washington

I. INTRODUCTION

Consumer trends and therefore food markets are changing and will change more in the future [1]. Foods with high quality and more fresh-like attributes are in demand, consequently less extreme treatments and/or fewer additives are required. Gould [2] identified some food characteristics that must be attained in response to modem consumer demands: less heat and chill damage, more freshness, less acid, and less salt, sugar, and fat. To satisfy these demands, some changes in the traditionally used preservation techniques must be achieved. From a microbiological point of view, these changes have important and significant implications. Moreover, food safety is an aspect of increasing importance, and improvements in microbial control must be attained. Therefore, in order to satisfy market requirements, the safety and quality of foods will be based on substantial improvements in traditional preservation methods, or the use of "emerging technologies." On the other hand, the use of Good Manufacturing Practices in the food industry and the maintenance of hygiene standards in food service establishments and homes are essential to the control of foodborne diseases. However, these procedures alone may be insufficient to ensure microbial safety, making food-preservation methods necessary [3].

One "new" or emerging technology receiving a great deal of attention is high hydrostatic pressure. Studies examining the effects of high pressure on foods date back to the end of the nineteenth century, but renewed research and commercialization efforts worldwide could place high-pressure–treated foods on several markets soon [4–7]. In April 1990, the first high-pressure product, a high-acid jam, was introduced to the Japanese retail market. In 1991, yogurts, fruit jellies, salad dressings, and fruit sauces were also introduced, and two Japanese fruit juice processors installed semi-continuous high-pressure equipment for citrus juice bulk processing [8].

The basis of high hydrostatic pressure is the Le Chatelier principle, according to which any reaction, conformational change, or phase transition that is accompanied by a decrease in volume will

**Current affilliation*: Universidad de las Américas-Puebla, Cholula, Mexico.

be favored at high pressures, while reactions involving an increase in volume will be inhibited [7,9]. However, due to the complexity of foods and the possibility of changes and reactions that can occur under pressure, predictions of the effects of high-pressure treatments are difficult, as are generalizations about any particular type of food. However, a tremendous amount of information has been generated in the past decade, and evidence has been recorded about the effects of high pressure on food systems, including microbial inactivation, chemical and enzymatic reactions, and structure and functionality of biopolymers [4,5,7]. The study of chemical and microbiological changes of foods processed by high hydrostatic pressure will determine their safety and quality, but commercial feasibility must include research on the design and construction of plant and equipment for the high-pressure processing of foods. Integration of the large amount of available information to design an efficient process is necessary. This chapter examines the current situation of high-hydrostatic-pressure technology with a view of its feasible use in the food industry.

II. HYDROSTATIC PRESSURE TREATMENT OF FOOD

High-pressure technology was originally used in the production of ceramics, steels, and superalloys. In the past decade, high-pressure technology was expanded to include the food industry. High pressure presents unique advantages over conventional thermal treatments [10,11], including application at low temperatures, which improves the retention of food quality [12]. High-pressure treatments are independent of product size and geometry, and their effect is uniform and instantaneous [10,13–15]. The principle of isostatic processing is presented in Figure 1; the food product is compressed by uniform pressure from every direction and then returns to its original shape when the pressure is released [16].

A. Terminology

The term "high pressure" is meaningless unless it is related to pressures experienced on earth. At the deepest point in the oceans the pressure is about 100 MPa and at the center of the earth it is 360 GPa [17]. In commercial applications the highest pressure used is around 5–6 GPa; which is applied for diamond grit production [17]. High-isostatic-pressure technology is the application of pressure uniformly throughout a product, and is essentially applied for isostatic pressing, quartz growing, chemical reactors, and simulators [18]. Quartz crystals are grown from a strong alkaline solution of sodium hydroxide at a pressure of up to 200 MPa and a temperature of up to 420°C. Some chemical reactions are carried out at high pressure to increase the yield of the reaction. For example, low-density polyethylene is synthesized at a pressure of 200 MPa and a temperature of 350°C. High-pressure vessels are also used as simulators to test equipment that would be used in a high-pressure environment, e.g., deep in the ocean [19]. The food industry employs the technique of isostatic pressing for applying high pressures to foods.

A high-pressure system consists of a high-pressure vessel and its closure, pressure-generation system, temperature-control device, and material-handling system [20]. Once loaded and closed, the vessel is filled with a pressure-transmitting medium. Air is removed from the vessel by means of a low-pressure fast-fill-and-drain pump, in combination with an automatic deaeration valve, and high hydrostatic pressure is then generated. High-pressures can be generated by direct or indirect compression or by heating the pressure medium [18].

Direct compression is generated by pressurizing a medium with the small diameter end of a piston (Fig. 2). The large diameter end of the piston is driven by a low-pressure pump. This direct compression method allows very fast compression, but the limitations of the high-pressure dynamic seal between the piston and the vessel internal surface restricts the use of this method to small-diameter laboratory or pilot plant systems.

Indirect compression uses a high-pressure intensifier to pump a pressure medium from a reservoir into a closed high-pressure vessel until the desired pressure is reached (Fig. 2). Most industrial isostatic pressing (cold, warm, or hot) systems utilize the indirect compression method.

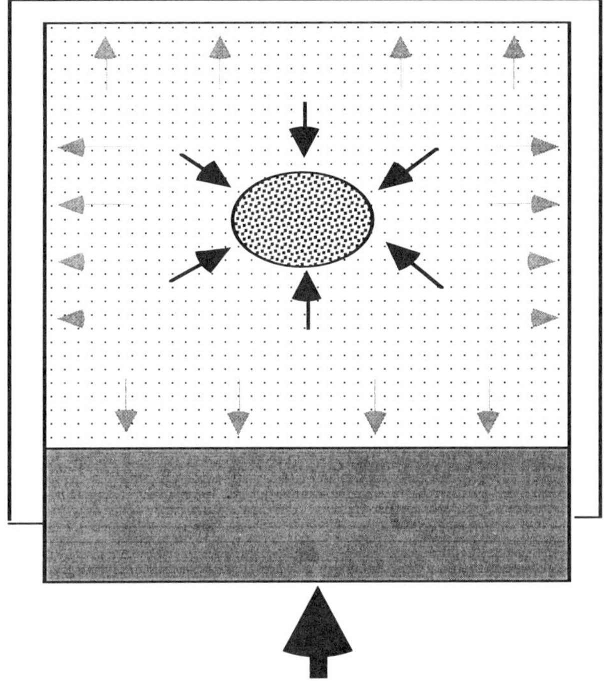

Figure 1 The principle of isostatic processing. (Adapted from Ref. 16.)

Heating of the pressure medium utilizes expansion of the pressure medium with increasing temperature to generate high pressure. Heating of the pressure medium is therefore used when high pressure is applied in combination with high temperature and requires very accurate temperature control within the entire internal volume of the pressure vessel [19].

The isostatic pressing systems may be operated as cold isostatic, warm isostatic, or hot isostatic systems [21], depending on the application.

Cold isostatic pressing is essentially a forming technique used in the metal, ceramics, carbon, graphite, and plastic industries. Powdered materials are filled in an elastomer mold and subjected to high pressure. High-pressure machines work at ambient temperature and use as pressurization fluid a liquid such as water, emulsified water, or oil. Applied pressure is in the range of 50–600 MPa. The cold isostatic pressure process uses "wet bag" or "dry bag" configurations. In the wet bag method, the mold is filled outside the pressure vessel. The mold is then placed in the pressure vessel, which is filled with the pressure medium. In the dry bag method, the mold is fixed in the pressure vessel and separated from the pressure medium by an elastomer tool [19]. The cycle time in a wet bag method is a few minutes, while the cycle time in a dry bag method varies between 20 and 60 seconds. Cold and warm isostatic pressure systems are most similar to future food applications. Both the dry bag (in bulk) and wet bag (in-container) process options are of interest for food processing [20].

Warm isostatic pressing is also a forming technique. Isostatic pressure is applied in combination with temperatures between ambient and 250°C. A warm isostatic pressure system is used in situations where a chemical reaction develops during pressurization.

Hot isostatic pressing is used primarily in the metallic and ceramic industries. The material is uniformly heated and pressurized. The temperature employed is as high as 2000–2200°C, while pressure is 100–400 MPa. The pressure medium used is a gas such as argon, nitrogen, helium, or air. The cycle time typically varies between 6 and 12 hours [18].

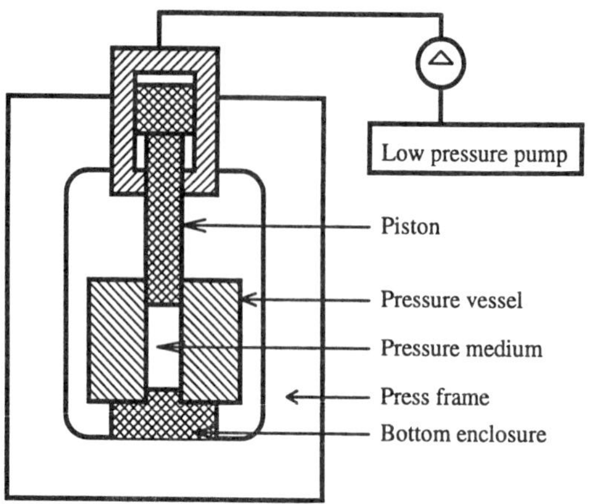

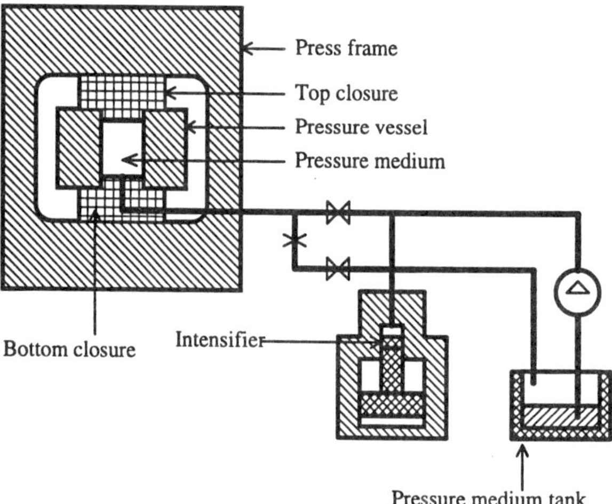

Figure 2 Generation of high pressure by direct (top) and indirect (bottom) compression of the pressure-transmitting medium. (Adapted from Ref. 19.)

The cold isostatic pressing equipment originally developed for ceramic application was modified to meet additional requirements for food processing. While the pressure medium used in the vessel is water containing an anti-rusting agent or synthetic oil to protect the pressure vessel against corrosion, food processing requires use of potable water or emulsified potable water [19].

The pressure vessel is the most important component of high-hydrostatic-pressure equipment. Crossland [17] mentioned several considerations that must be taken into account in vessel design: it is necessary to design the high-pressure vessel to be dimensionally stable in a safe-fail way. The vessel does not yield in service; if it fails it should fail with leak before fracture. There is also the problem of fatigue, which at the highest pressure cannot be avoided, but an acceptable economic lifetime must be achieved. It is also necessary to establish the minimum number of cycles to failure in order to determine the desired frequency of inspection.

B. Commercial Equipment

The Japanese are the leading manufacturers of high-pressure vessels. The major Japanese companies manufacturing high-pressure vessels are Mitsubishi Heavy Industries Ltd., Kobe Steel Ltd., and Nippon Steel Ltd. Other manufacturers of high-pressure equipment include Engineered Pressure Systems, ABB Autoclave Systems Inc., ACB, NKK Corp., Flow International, and Autoclave Engineers [6,19,22]. Table 1 provides specifications of some commercially available high-pressure systems and some details on pressure vessels and maximum operating pressures. As indicated in the table, the larger the vessel, the lower the maximum operating pressure [6]. The first high-pressure food-processing vessel was manufactured by Mitsubishi Heavy Industries (Tokyo, Japan). The pressure vessels manufactured by Mitsubishi Heavy Industries varied in capacity from 0.6 to 210 liters and the maximum working pressures from 400 to 700 MPa [23]. The increase in the high-pressure unit capacity generally reduces the maximum pressure that can be achieved. To minimize the reduction of equipment life due to repeated use of the pressure vessel, the Mitsubishi pressure vessel is made up of double cylinders. The inner surface of the pressure vessel is preloaded with a high compression stress. The parts of the vessel that come in contact with the pressurizing medium are made of stainless steel. The pressurizing and decompressing cycle is fast. Maximum pressure is attained in 90 seconds. High pressure and a long holding time imply that a great load is applied to the seal of the vessel cover. A self-seal packing with high durability and reliability is used. The seals can withstand repeated opening and closing of the pressure vessel and application of high pressure without leakage. A piston driven by a hydraulic cylinder is used to generate the required pressure [24].

Kobe Steel Ltd. developed a small test pressure vessel and one of the largest pressure vessels available today, with an internal volume capacity of 9400 liters and a maximum working pressure of ~200 MPa. The small test pressure vessel uses a piston for pressurization, and the oil hydraulic sys-

TABLE 1 Specifications of Seleted Commercially Available High-Pressure Vessels

Vendor/Model	Diameter (m)	Length (m)	Volume (L)	Maximum operating pressure (MPa)
Mitsubishi Heavy Industries				
MFP 700	0.06	0.2	0.6	700
MCT 150	0.15	0.3	6.0	420
FP-30V	3.00	7.0	50	420
FP-40L	4.00	17.0	210	400
Kobe Steel	0.06	0.2	NA	700[a]
	2.00	3.0	9400	196[a]
ABB Autoclave Systems				
Quintus	0.09	0.225	NA	900[a]
Quintus	0.30	1.250	100	900[a]
Quintus	0.50	2.500	500	900[a]
Engineered Pressure Systems	0.09	0.55	3.5	1,380
	0.1	1.0	8.5	1,030
	0.1	1.0	37	690
	0.6	2.5	700	550
	0.6	4.5	1,250	410
	1.0	4.0	3,150	200
	1.7	4.0	9,000	100

NA = Not available.
[a] Maximum temperature ≈80°C.
Source: Ref. 6.

tem and operational panel are compactly packaged as part of the equipment. Operation is fully automated, and the temperature inside the pressure vessel can be recorded. The equipment also allows the use of a pressure-control program [25].

The research high-pressure food-processing system developed by ABB Autoclave Systems, Inc (Vasteras, Sweden) consists of two components: the process module and the control module. The process module consists of a cabinet, which contains the Quintus prestressed wire-wound pressure vessel, the electrohydraulic pumping system, and a hot water circulation system. The system can reach 900 MPa within 4 minutes. Temperature is maintained by circulating water in channels between the wire winding and the cylinder wall of the pressure vessel. The programmable control module in the ABB high-pressure research vessel monitors and controls the process time, pressure, and temperature. A microprocessor is used to control food loading into the press, press cycling and downloading to a conveyor [26]. The cost of treating foods in 100- and 500-liter systems is approximately $0.25 and $0.07 dollars per batch, respectively [27]. The high-pressure systems Quintus models from ABB Autoclave Systems vary in their dimensions from 0.09 to 0.5 m (internal diameter) and from 0.225 to 2.5 m (internal length) with an internal capacity up to 500 liters and a maximum operating pressure of 900 MPa. Pressure vessels with other dimensions available from ABB Autoclave Systems include: 0.045 m diameter × 0.3 m length with a maximum pressure of 1200 MPa; 0.11 m diameter × 0.26 m length with a maximum pressure of 830 MPa; 0.32 m diameter × 1.25 m length (100 liters) with a maximum pressure of 900 MPa; and 0.50 m diameter × 2.50 m length (500 liters) with a maximum pressure of 900 MPa. The fatigue value can be maintained infinitely by replacing a shrunk wear-liner every 30,000 cycles. Changing the liner is convenient and inexpensive. ABB Autoclave Systems is designing and constructing a high-pressure vessel to work in batch mode with a maximum operating pressure of 1700 MPa. The internal diameter of the pressure vessel will be 0.076 m, and the height will be 0.18 m [28].

A warm isostatic pressing system is available from Engineered Pressure Systems Inc. (EPSI), a subsidiary of National Forge Co. (Andover, MA). The system consists of a double-ended, lined pressure vessel with plug closures. The design parameters of the high-pressure systems constructed by EPSI are listed in Table 1. Recently EPSI developed a laboratory-scale pressure vessel with the following specifications: 0.1 m (internal diameter) × 2.5 m (internal height) with maximum operating pressure of 680 MPa and maximum operating temperature of 90°C. While the design pressure is 750 MPa, the maximum operating pressure is 680 MPa. An electrohydraulic intensifier pump with a motor pressurizes the vessel to the operating pressure in 5 minutes or less [29]. An advanced laboratory-scale pressure vessel with a useful diameter of 0.024 m, length of 0.04 m, and maximum pressure of 800 MPa is also available from EPSI.

The processing cost as a function of time and pressure is presented in Figure 3. Processing at 400 MPa with a holding time of 10 minutes is twice as expensive as processing at 800 MPa with no hold time [16]. The combination of pressure, time, and temperature at which the product is processed must therefore be evaluated carefully. A low maximum operating pressure can cause drastic reductions in the fabrication costs. High-pressure processing may be combined with moderately high temperatures, so the operating pressures required are not extremely high [30].

C. Processing Operation

A sterile container filled with food is sealed and placed in the pressure chamber for pressurizing. Ethylene-vinyl alcohol copolymer (EVOH) and polyvinyl alcohol (PVOH) films are recommended for packaging food for high-pressure treatment [31]. Also, the existing multilayer plastic and some aluminum packages may be used for high-pressure processing. No deformation of the package occurs because the pressure is uniform [27]. The shape of the package needs to be designed to fill the vessel volume as far as possible to increase the economical feasibility of the process. The basis for applying high pressure to foods is to compress the water surrounding the food. At room temperature the volume of water decreases, as presented in Figure 4 [5,32]. Because liquid compression results in a small volume change, high-pressure vessels using water do not present the same operating hazards as vessels using compressed gases [5].

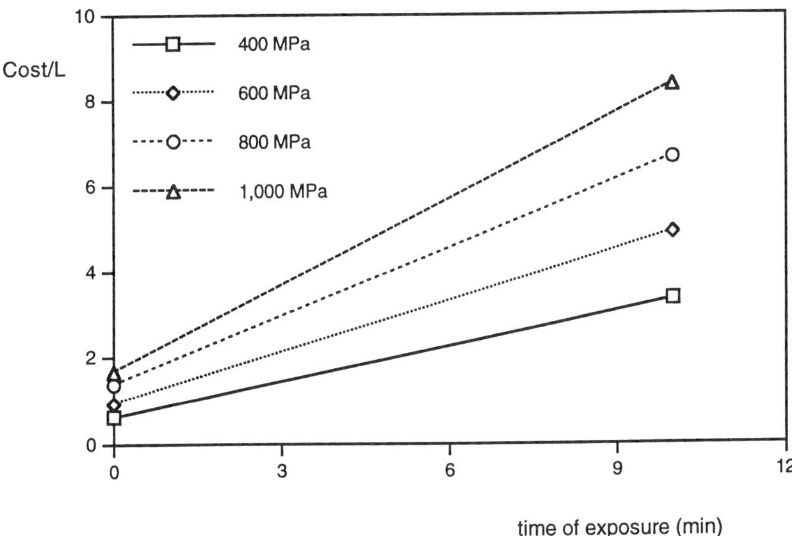

Figure 3 Process cost as a function of exposure time at different pressures. (Adapted from Ref. 16.)

The capacity of a high-pressure plant depends on three factors: the number of cycles that can be done in a given time, the volume of the product, and the number of high-pressure vessels available. The cycle time is determined by the time needed to handle the food product, including loading, unloading, opening and closing the high-pressure vessel, the pressure holding time, and the pressurization and decompression rates. The productivity of the batch system is increased by a reduction in the pressurizing-decompressing cycle. The pressurizing time is reduced by increasing the delivery rate of the pump [19]. When the required operating pressure is attained, the pumping rate is reduced. At the end of the specified holding time, the pressure vessel is decompressed in two stages to avoid sudden release of pressurized water [24]. Batch processing reduces the risk of large quantities of food

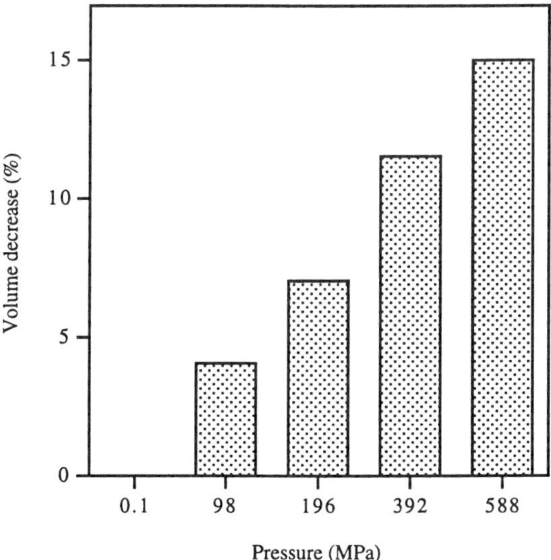

Figure 4 Decrease in water volume with pressure. (Adapted from Ref. 5.)

becoming contaminated by the lubricants or wear particles from the machinery. Different types of food can be processed in a batch system, without the danger of cross-contamination or the need to clean the equipment after each run [19]. The technical advantage of the batch-type pressure vessel is the simplicity of fabrication when compared to a continuous flow pressure vessel operating at pressures as high as 400–900 MPa [6,19,21]. A batch system pressure vessel with a processing capacity of 600 liters/hr of liquid food at a maximum operating pressure of 420 MPa was used to commercially produce grapefruit juice in Japan [24]. The production rate of the batch process can be increased by operating pressure vessels in sequence with no lag in the processing times so the system operates sequentially.

Food is subjected to high pressure for a specified time period. The holding time in the pressure vessel depends on the type of food and process temperature. At the end of the processing time the chamber is decompressed to remove the treated batch. A new batch of food is placed in the pressure vessel and the cycle begins again [13]. As in heat sterilization of packed products, the water in contact with the product should have drinking quality and the lubricating and rust protecting products should be allowed to be used in food processing. For bulk processing, the high-pressure equipment should be part of an aseptic line. Parts in contact with the food product should be clean and sterile. The engineering challenges of the application of high pressure in the food industry are primarily the construction of pressure vessels to handle large volumes of food and withstand the high pressures; the pressure vessel should have a short cycle time, be easy to clean, and be safe to operate with accurate process controls. It is desirable to develop a continuous process of pressurization for industrial purposes at reasonably low capital and operating costs [6,18,19]. Most of the challenges are being met to some extent, but research is still needed to further develop high-pressure technology and the necessary equipment.

D. Commercial Applications

Current industrial applications of high hydrostatic pressure are presented in Table 2. Hayashi [33] presents a list of pressure-processed foods in the Japanese market. The first pressure-processed foods in the human history were strawberry, kiwi, and apple jams (Meidi-ya Food Co.). The jams were produced using high-pressure treatment without application of heat. The jams were vivid and natu-

TABLE 2 Some Commercial High-Pressure–Treated Foods

Product	Processing conditions	Package	Company
Jams Fruit dressings Fruit sauces Fruit jellies Yogurts	400 MPa, 20°C 10–30 min	Plastic cup (100–125 g)	Meidi-ya
Grapefruit juice	120–400 MPa, 20°C 2–20 min	Glass bottle (200–800 g)	Pokka
Mandarin juice	300–400 MPa, 20°C 2–3 min	Glass bottle (500 g)	Wakayama
Sugar-impregnated tropical fruits for sherbets and ice creams	50–200 MPa	Paper cups (130 g)	Nisshin
Beef tenderization	100–250 MPa, 20°C 30 min–3 hr		Fuji Chiku & Mutterham
Rice cake	400 MPa, 45–70°C 10 min		Echigo Seika

Source: Refs. 9–11.

ral in color and taste. The list includes fruit sauces and desserts (Meidi-ya Food Co.), mandarin (Wakayama Co.) and grapefruit juices (Pokka Co.), and unrefined rice wine called "nigori-sake." The new sake has a white color and fresh flavor, instead of a brown color and cooked smell. Cheftel [9] mentioned that many of the products presented in Table 2 are acidic foods, hence they have an intrinsic safety factor; also some of the products are stored and sold refrigerated, consequently oxidative reactions are retarded. Hayashi [33] reported that at least seven food companies now sell high-pressure–processed foods, including salted raw squid and fish sausages. Knorr [10] mentioned that progress in various additional products is underway, but information is not available due to its confidential nature. In the United States and Europe, developments are being made in fruit products, ready meals, dairy products, meats, fish, and others. Avocado paste (guacamole) is now produced in Mexico using high pressure. Several new industrial applications are expected soon.

III. EFFECTS OF HIGH PRESSURE ON BIOLOGICAL MATERIALS

Pressure is an important thermodynamic variable and can affect a wide range of biological structures, reactions, and processes [34]. Pressure primarily affects the volume of a system [35]. The influence of pressure on the reaction rate may be described by the transition state theory: the rate constant of a reaction in a liquid phase is proportional to the quasi equilibrium constant for the formation of the active reactants [36,37]. Based on this assumption, Heremans [35], van Eldik et al. [36], and Tauscher [37] reported that at constant temperature, the pressure dependence of the reaction velocity constant (k) is due entirely to the activation volume of the reaction (ΔV^*):

$$\left(\frac{\delta \ln k}{\delta P}\right)_T = -\frac{\Delta V^*}{RT} \tag{1}$$

where P is the pressure, R is the gas constant ($8.314\ cm^3 \cdot MPa \cdot K^{-1} \cdot mol^{-1}$), and T is temperature (K).

Water is the most important food ingredient in many food products, thus, its characteristics under pressure are very important. Compared to gases, water is nearly incompressible; adiabatic compression of water increases the temperature by about 3°C per 100 MPa [37]. Self-ionization of water is also promoted by high pressure. The water-freezing characteristics can be changed by the application of pressure [38]. At approximately 1000 MPa, water freezes at room temperature, while the freezing point decreases to –22°C at 207.5 MPa. This event promotes opportunities for subzero storage of foods without ice crystal formation, fast thawing of frozen foods by pressurization, and increasing food freezing by decompression of pressurized foods held below 0°C [34].

In an aqueous system, water molecules surrounding an ionized group align themselves according to the influence of the electrostatic charge, giving a more compact arrangement. Ionization of the acidic or basic groups found in many biomolecules, such as proteins, involves a volume decrease and, therefore, will be enhanced by increased pressure [4,39]. Microorganisms, chemical, biochemical, and enzymatic reactions, as well as some functional properties of biomolecules are affected, to some extent, by high pressure.

A. Microorganisms

Food preservation is based primarily on the inactivation, growth delay, or prevention of spoilage and pathogenic microorganisms and works through factors that influence microbial growth and survival. Gould [1] classified the major food-preservation technologies as those that act by preventing or slowing down the growth of microorganisms (low temperature, reduced water activity, less oxygen, acidification, fermentation, modified-atmosphere packaging, addition of preservatives, compartmentalization in water-in-oil emulsions) and those that act by inactivating microorganisms (heat pasteurization and sterilization, ionizing radiation, high hydrostatic pressure, pulsed electric fields). In addition, there is a strong emphasis on the use of techniques in combination, applying the hurdle technology concept [40], which could act by inhibiting or inactivating microorganisms, depending on the combination of hurdles applied.

Few techniques act primarily by inactivation, of which heat is by far the most commonly used. However, there is much interest in alternative, nonthermal preservation processes such as high pressure. From a food-safety point of view, techniques that inactivate pathogenic and spoilage flora are preferred over those based on preventing or slowing down growth of microorganisms. This fact stimulates the interest in high pressure as a preservation technique to inactivate microorganisms in foods.

In heat sterilization and pasteurization, the applied treatment inactivates and considerably reduces the number of microorganisms initially present. However, the food sensory and nutritional characteristics are also strongly affected. High hydrostatic pressure can be applied without temperature elevation, thus other food characteristics could be maintained. Nonthermal process applied to food preservation without the collateral effects of heat treatments are being seriously studied and tested. The pressure sensitivity of microorganisms varies with the type of microorganism. Gould [1] reports that as high-pressure targets, microorganisms may be divided into those that cause food poisoning and those that cause food spoilage. Microorganisms can be further divided into those that are relatively pressure sensitive and those that are pressure resistant. Regarding the pressure sensitivity, the most important categories are the vegetative and spore forms of microorganisms; in general the vegetative forms are inactivated by pressures between 400 and 600 MPa, while spores of some species may resist pressures higher than 1000 MPa at ambient temperatures. Gram-positive bacteria are more pressure resistant than gram-negative ones [3,41,42]. Among the gram-positive bacteria, Earnshaw [42] reported that *Staphylococcus* is one of the most resistant and can survive treatment at 500 MPa for more than 60 minutes.

1. Vegetative Cells

The relative pressure sensitivity of the vegetative forms of microorganisms has made them the obvious first targets for the preservation of foods by high pressure, and particularly for low pH foods and other foods in which the intrinsic preservation systems already operating ensure that the pressure-resistant food poisoning or spoilage spore-formers that may survive are unable to grow [1]. Increased opportunities must originate when combinations of pressure with some of the other already well-established inhibitory food-preservation techniques are applied using the hurdle technology approach. High-hydrostatic-pressure treatments can be considered as a new hurdle that can be used in combination with other traditional microbial stress factors such as pH, water activity (a_w), and preservatives [40,43]. However, if high pressure is to be used instead of other stress factors, the kinetics of microbial pressure inactivation must be known as well as the spore resistance of toxigenic bacteria [1].

The extent of microbial inactivation achieved at a particular pressure treatment depends on a number of interacting factors, including type and number of microorganisms, magnitude and duration of high-pressure treatment, temperature, and composition of the suspension media or food [3,44,45]. Other experimental variables that must be taken into account include the compression and decompression rates [42].

Type and Number of Microorganisms

Zobell [46] reported that most bacteria are capable of growth at pressures around 20–30 MPa; barophiles are organisms that can grow at pressures higher than 40–50 MPa, and those that survive for prolonged periods at pressures >200 MPa are named baroduric or barotolerant. The pressure effects on several pathogenic and spoilage microorganisms inoculated in a pork slurry can be used to illustrate the microbial response to high hydrostatic pressure and the differences among species and microbial forms [41]. Figure 5 presents the response to high-hydrostatic-pressure treatments of vegetative forms of *Escherichia coli* and *Saccharomyces cerevisiae* and spores of *Bacillus cereus* inoculated in a pork slurry. *E. coli* counts were almost unaffected at pressures lower than 203 MPa, but treatments at 304 MPa or higher pressures drastically reduced the initial inocula (10^6–10^7 cfu/g). More than 6 log cycles of *E. coli* were reduced at pressures higher than 405 MPa for 10 minutes. For *S. cerevisiae*, less than 2 log cycle reductions were observed at pressures lower than 304 MPa and more than 6 log cycles at pressures higher than 405 MPa. The *B. cereus* spore counts were not reduced considerably (less than 1 log cycle) even in treatments at 608 MPa for 10 minutes.

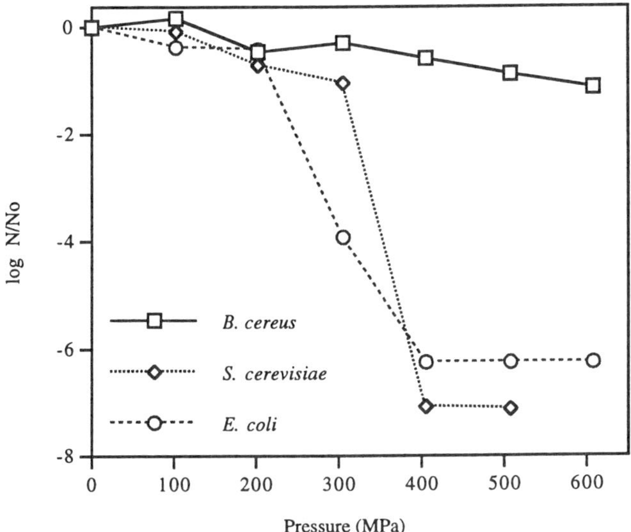

Figure 5 High hydrostatic pressure effects on *Bacillus cereus* spores and *Saccharomyces cerevisiae* and *Escherichia coli* vegetative cells inoculated into pork slurries. The inoculated samples were subjected to pressure treatments for 10 minutes at 25°C. N and N_0 represent the number of survivors and initially inoculated microorganisms, respectively. (Adapted from Ref. 41.)

Patterson et al. [3,47] reported important information about the effect of high pressure on foodborne vegetative pathogens. Styles et al. [48] and Takahashi et al. [49] reported that gram-negative bacteria are pressure sensitive. Among these *Vibrio parahaemolyticus* is one of the most sensible—a 6 log reduction in the initial population can be attained in treatments at 200 MPa for 20 minutes. On the other hand, gram-positive pathogens such as *Listeria monocytogenes* and *Staphylococcus aureus* required for a 6 log reduction in the initial inoculum treatments for 20 minutes at 340 and 400 MPa, respectively [3]. Another pressure-resistant pathogenic bacteria is *E. coli* O157:H7; Takahashi et al. [49] reported that for a 6 log cycle reduction, the treatment must be at 700 MPa for 13 minutes. Shigehisa et al. [41] postulated that the more complex membrane structure of gram-negative bacteria is more susceptible to environmental changes like those caused by high-pressure treatments.

Earnshaw [42] reported that little difference was observed in the overall rates of *S. carnosus* inactivation for treatments at 650 MPa with different initial populations ranging from 10^7 to 10^3 cfu/ml. When 10^2 cfu/ml were initially inoculated, a faster inactivation rate was observed. It is obvious that with a great number of initially inoculated microorganisms the effects of specific high-pressure treatments can be observed, but studies with realistic numbers of microorganisms are also needed. The pressure sensitivity of microorganisms may vary with the species and probably with the strain of the same species and with the stage of the growth cycle at which the organisms are subjected to the high-hydrostatic-pressure treatment. In general, cells in the exponential phase are more sensitive to pressure treatments than cells in the log or stationary phases of growth [3,42,46,50]. Isaacs et al. [51] reported that freshly inoculated *E. coli* cultures growing rapidly were more sensitive than *E. coli* cultures that had reached the growth stationary phase. No effect after 7 minutes at 200 MPa with cells in the stationary phase and around 5 log cycle reductions with young cells from the log phase were observed. Also, it has been established that the cell age distribution in inoculation studies might be an important factor in the result obtained after high-pressure treatments. Bacteria in the stationary phase are smaller and more spherical than in the log phase, when they are rapidly growing, rod-shaped, and exhibit an accelerated metabolism. Isaacs et al. [51] stated that the greater resistance to pressure when the cell metabolism slows down may reflect the accumulation of cell components, which can reduce the effects of high pressure.

Considerably variation in pressure sensitivity between strains isolated from different foods or culture collections has been reported for the same microorganisms. Patterson et al. [3] reported a 3 log reduction for *L. monocytogenes* from a culture collection after 30 minutes at 375 MPa. The same pressure treatment reduced the initial population $>10^4$- and 10^7-fold, for strain Scott A and a chicken isolate, respectively. Cheftel [9] reported that when the pressure resistance of various microorganisms is compared, the survival fractions determined by different investigators vary by a factor of 1 to >8 for different species of the same genus (*Salmonella*) or by a factor of 1.5–3.5 for different strains of the same microorganism (*L. monocytogenes*).

Extent and Duration of High-Hydrostatic-Pressure Treatments

Generally, an increase in pressure increases microbial inactivation. However, increasing the duration of the treatment does not necessarily increase the lethal effect. Above 200–300 MPa, the inactivation ratio of vegetative cells increases with pressure or process time [9].

Table 3 presents some results about the effects of high pressure level and exposure time on several microorganisms. As mentioned before, the microbial response to high-pressure treatments depends on the type of microorganism. For each microorganism, there is a pressure-level threshold at which no effects of increasing the exposure time are detected. There also exists a pressure level at which increasing treatment time causes significant reductions in the initially inoculated microbial counts. The intrinsic conditions of the suspension media such as pH, a_w, and nutrients may influence the pressure threshold, which can increase or decrease depending on the microorganism and variation of intrinsic, extrinsic, and processing factors. Kinetic studies at pressures over the pressure threshold are needed. With reliable kinetic data, the pressure sterilization or pasteurization of foods can be predicted and achieved.

Temperature

The temperature during pressurization can have a significant effect on the inactivation of microbial cells. Several authors [9,52–55] observed that the resistance to pressure of an endogenous or inoculated microbial strain is maximal at normal temperatures (15–30°C) and decreases significantly at higher or lower temperatures. Freezing temperatures (–20°C) in treatments ranging from 100 to 400 MPa for 20 minutes enhance microbial inactivation when compared with high-pressure treatments at 20°C [49]. Hashizume et al. [56] reported that *S. cerevisiae* cells were more effectively inactivated by high-pressure treatments at elevated (40°C) or subzero (–10 and –20°C) temperatures.

The decrease in pressure resistance of vegetative cells at low temperatures (<5°C) may be due to changes in the membrane structure and fluidity, weakening of hydrophobic interactions, and crystallization of phospholipids [9]. On the other hand, moderate heating (40–60°C) may also enhance the pressure microbial inactivation, resulting in some cases in a lower minimal inactivation pressure [9,52]. Ogawa et al. [57,58] reported an enhanced inactivation of natural flora and inoculated microorganisms in mandarin juice treated at 40°C in combination with pressures in the range 400–450 MPa. Figure 6 presents the effect of temperature on the high-pressure inactivation of two psychrotrophic bacteria, *Pseudomonas fluorescens* and *Listeria innocua*, and one thermotolerant, *Citrobacter freundii*, inoculated in minced beef muscle [52]. The psychrotrophic bacteria were more sensitive to the effects of pressure at low temperatures, and *C. freundii* was more sensitive at 35 and 50°C. For the three bacteria, the greatest inactivation was obtained when combining the pressure treatment with 50°C [52].

Zobell [46] observed that the maximum microbial growth temperature at increased pressure is generally a few degrees higher than the optimum growth temperature at atmospheric pressure. Sale et al. [59] reported an increasing sensitivity to pressure of *Bacillus coagulans* spores with temperature. Roberts and Hoover [60] evaluated the effect of combinations of pressure at 400 MPa with heat against spores of *B. coagulans* and at 25°C observed less than 1 log cycle reduction in the initial inocula. As the temperature during pressurization increased, the effectiveness of the high-pressure treatment increased; 2 and 4 log cycle reductions were observed at 45°C and 70°C, respectively.

Composition of Suspension Media or Food

The effect of pressure on inhibition or inactivation of microorganisms in combination with major environmental variables should be analyzed for a better understanding of the mode of action of this preservation technique. The extent of microbial reduction achieved by a high-hydrostatic-pressure treatment depends on a number of interacting factors including the composition of the media or food [44,45]. Hoover et al. [4], Oxen and Knorr [61], Patterson et al. [3], Pothakamury et al. [61], and Palou et al. [44,45] stated that many food constituents appear to protect microorganisms from the effects of high pressure. Therefore, it is important to evaluate high-hydrostatic-pressure treatments for each individual case.

Dring [62] observed that nonnutritive solutions reduce the microorganism's barotolerance; Marquis [63] reported that an enriched media protects microorganisms against pressure and suggested that free amino acids and vitamins are available and therefore the cells are protected. Hashizume et al. [56], Ogawa et al. [57], and Oxen and Knorr [61] observed that the pressure resistance of fungi increases as the sugar (sucrose, fructose, glucose) concentration on the media increases.

A baroprotective effect of reduced water activity for organisms that can grow at reduced a_w has been reported [10,11,14,44,45,64]. Oxen and Knorr [61] observed that high-hydrostatic-pressure treatment at room temperature and 400 MPa for 15 minutes inactivated the yeast *Rhodotorula rubra* when the a_w of the suspension media was higher than 0.96, while the number of survivors was higher when the a_w was depressed. Similar treatment at 30°C achieved 7 log cycle reductions in the initial inocula at a_w 0.96; 2 log cycles were reduced at a_w 0.94, and no reductions were observed when the a_w of the suspension media was 0.91 [14]. At higher temperatures (45°C) the yeast was inactivated even at low a_w values. Oxen and Knorr [14] demonstrated that pressure-temperature combination treatments result in faster and greater yeast inactivation in comparison with treatments where only temperature or pressure is applied. For heat inactivation treatments for 15 minutes at 70–80°C and atmospheric pressure were required to inhibit *R. rubra*.

Palou et al. [44] reported the effect of reduced a_w (or increasing soluble solids concentration) on *Zygosaccharomyces bailii* inhibition suspended in laboratory model systems adjusted to pH 3.5. Figure 7 presents the effect of the model system soluble solids concentration on the viability of *Z. bailii* without (0.1 MPa) and with a high-pressure treatment of 5 minutes at 345 MPa. *Z. bailii* grew well without the pressure treatment in the range of sugar concentration studied (2–59% w/w), rapidly reaching high counts ($\approx 10^7$ cfu/ml). This yeast is osmotolerant and its growth occurs in media containing up to 60% (w/w) glucose, with a_w 0.85. The soluble solids concentration in the model systems considerably affects the recovery counts after high-pressure treatment, being higher as the soluble solids concentration increased. Greater counts were observed for the experiments with sugar concentrations >40%. In experiments with less than 20% soluble solids, complete inhibition (<10 cfu/ml) of *Z. bailii* was observed. Ogawa et al. [57] reported that for *S. cerevisiae* inoculated in concentrated fruit juices, the number of surviving microorganisms depends on the juice-soluble solids concentration and observed that the inactivation effect at pressure ≤ 200 MPa decreased as juice concentration increased. Hashizume et al. [56] also reported an increase in the number of surviving *S. cerevisiae* cells with increasing concentrations of sucrose (0–30% w/w) when pressurized at 260 MPa for 20 minutes at 25°C.

Palou et al. [44] reported a linear relationship between the water activity depression factor ($1\text{-}a_w$) and the log of *Z. bailii* survival fraction (N/N_0) after a high-pressure treatment at 345 MPa for 5 minutes:

$$\log(N/N_0) = -4.599 + 45.538\,(1\text{-}a_w) \tag{2}$$

This relationship indicates that as the a_w of the model system decreased, the number of surviving *Z. bailii* increased. The results obtained by Palou et al. [44] without pressure treatment demonstrate that *Z. bailii* was adapted to grow at the selected a_w values, therefore the observed high-pressure effects can be attributed to this variable and not to the effects of reduced a_w on the extension of the lag phase

Table 3 Effect of High-Hydrostatic-Pressure Treatments on Selected Microorganisms

Microorganism	Substrate or suspension media	Treatment conditions		Decimal reductions	Ref.
Saccharomyces cerevisiae	Satsuma mandarin juice	250 MPa	5 min	≈2	57
			10	≈4	
			30	6	
		300 MPa	5 min	≈5	
			10	6	
Aspergillus awamori	Satsuma mandarin juice	250 MPa	5 min	≈3	57
			10	≈4	
			30	>4	
		300 MPa	5 min	5	
Listeria innocua	Minced beef muscle	330 MPa	10 min	≈2	52
			20	≈3	
			30	≈5	
		360 MPa	5 min	≈1	
			10	≈2	
			20	≈4	
			30	≈6	
Listeria monocytogenes	10 mM phosphate buffer saline pH 7.0	300 MPa	10 min	<1	47
			20	≈1	
			30	≈2.5	
		350 MPa	10 min	≈4	
			20	≈5.5	
			30	≈6.5	
		400 MPa	10 min	≈6	
			20	≈8	
			30	≈8	

Vibrio parahaemolyticus	100 mM phosphate buffer pH 7.0 with 3% NaCl	103 MPa	20 min	<1	48
			40	≈1	
		138 MPa	10 min	≈1.5	
			20	≈2	
			30	≈4	
		172 MPa	10 min	≈2.5	
			20	≈4.5	
			30	≈6	
Salmonella typhimurium	63 mM phosphate buffer pH 7.0	241 MPa	30 min	<1	67
		276 MPa	10 min	<1	
			20	≈1	
			30	≈1.5	
		345 MPa	10 min	≈1.8	
			20	≈2.5	
			30	≈3	
Total plate count	Fresh-cut pineapple	200 MPa	5 min	0.6	77
		270 MPa	5 min	1.8	
		≈4°C	15	1.6	
		340 MPa	5 min	1.9	
		≈4°C	15	3.0	
			40	2.1–2.9	

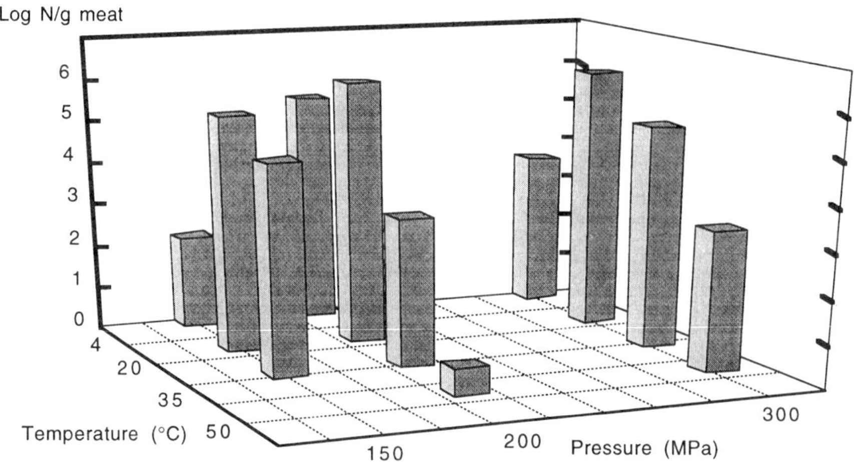

Figure 6 Effect of temperature on the pressure inactivation of *Pseudomonas fluorescens* (150 MPa, 20 min), *Citrobacter freundii* (200 MPa, 20 min) and *Listeria innocua* (300 MPa, 20 min) inoculated in minced beef muscle. (Adapted from Ref. 52.)

or on the reduction of the growth rate. At $(1-a_w) > 0.07$, reductions smaller than 1 log cycle were observed [44]. In comparison, Pandya et al. [65] reported for *Z. bailii* suspended in buffer solutions ($a_w \approx 1.0$) with pH 4.0, 5.0, and 6.0 that 7 log cycles were reduced in treatments at 304 MPa for 10 minutes, resulting in the total inhibition of the initial inocula.

The resistance to inhibition at reduced a_w values may be attributed to cell shrinkage, which probably causes a thickening in the cell membrane that reduces membrane permeability and fluidity [44]. The increased baroresistance of microorganisms at low a_w may also be attributed to a partial cell

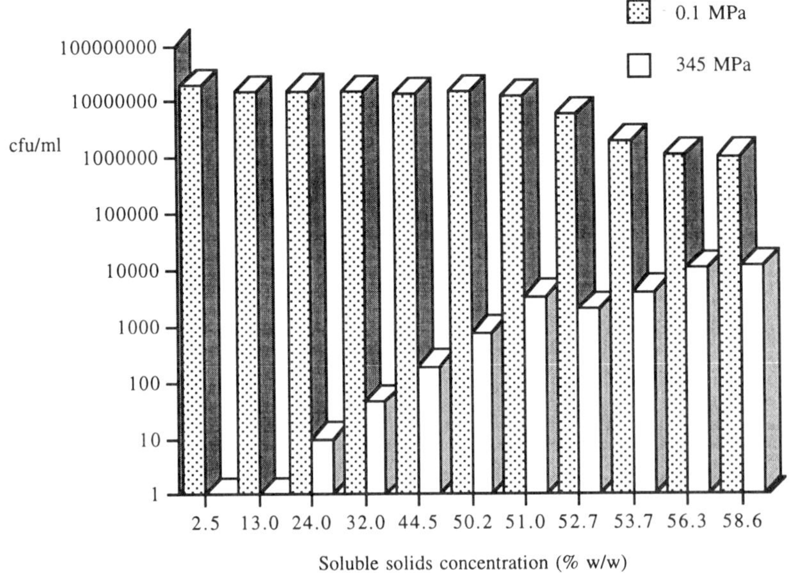

Figure 7 Effect of soluble solids concentration and high pressure (345 MPa, 5 min) on the viability of *Zygosaccharomyces bailii*. (Adapted from Ref. 44.)

dehydration due to the osmotic pressure gradients between the internal and external fluids, which may result in small cells and thicker membranes and an increased pressure resistance [14]. The baroprotective effect of reduced a_w reveals that inhibition of microorganisms by high pressure depends not only on pressure level and extent of the treatment, but also on the interactions with other intrinsic and extrinsic variables that influence the microbial response [44]. The design of effective pressure treatments that assure microbial stability of foods will depend on an understanding of the relationships between microorganisms and food components.

Kinetics of Microbial Inactivation

The patterns of high-hydrostatic-pressure inactivation kinetics observed with different microorganisms are quite variable. Some investigators indicate first-order kinetics in the case of several bacteria and yeast [52,55,56,66]. Other authors observed a change in the slope and a two-phase inactivation phenomenon, the first fraction of the population being quickly inactivated, whereas the second fraction appears to be much more resistant [9]. The pattern of inactivation kinetics is also influenced by pressure, temperature, and composition of the medium [54]. For a broader use of high pressure in food processing, it is of special interest to determine the process conditions for pressure pasteurization in view of industrial applications [9]. To increase microbial safety and assure microbial stability of foods processed by high pressure, the pressure treatment must ensure a satisfactory reduction in the initial microbial counts, thus kinetic analysis and the pressure dependence of microbial-inactivation rates are needed.

Several scientific reports [41,48,67] demonstrate the efficacy of high-hydrostatic-pressure treatments against different microbial species. However, few of them reported kinetic data, which would be necessary for product and process design. For low-acid foods in particular, the kinetic information for pathogenic bacteria and spores will be indispensable in terms of food safety. The kinetic nature of high-pressure inhibition and inactivation may be different from that detected in heat treatment and other food-processing methods. Deviations from first-order kinetics, occurrence of survivor "tails" in death kinetics, and the possibility of cell recovery after pressurization have been observed [3,42].

Earnshaw [42] stated that there is clear evidence from his and other laboratories that pressure-mediating death is not first order, and inactivation curves often present pronounced survivor tails. Thus the D and z concepts commonly used in thermal processing cannot be usefully applied to describe pressure processes. However, there is a difference with heat-treatment kinetic evaluation and sampling, and therefore the survivor enumeration is not continuous. Determination of microbial pressure-inactivation kinetics depends on pressure increase and decrease rates since it includes a decompression step to perform sampling. The pressure increase and decrease are not always reported, and the come-up time to reach the pressure is therefore not always taken into account in the logarithmic representation of the number of survivors. Cheftel [9] also mentioned that some publications do not indicate precisely and fully the experimental variables such as pressure, temperature, and time conditions. Moreover, Earnshaw [42] notes that in experiments that evaluate high-pressure-inactivation kinetics, the starting population must always be given. Cheftel [9] suggested that only inoculation with high numbers of a specific strain permits precise determination of the extent of inactivation as a function of processing conditions.

In many reported survival curves, it is unclear if time zero experiments are growth controls without pressure treatment or are pressure treatments that only take into account the come-up time to reach the pressure level. Figure 8 presents the come-up times for pressure treatments ranging from 241 to 517 MPa [68]; the time to release the pressure was less than 15 seconds. The pressure come-up times exert an important effect on the yeast survival fraction. *Z bailii* counts decreased as pressure increased, and at pressures greater than 414 MPa, 2 log cycle reductions were observed. Cheftel [9] reported that the rate of pressure increase and decrease is often neglected as an experimental variable in high-pressure microbial-inactivation studies, and the initial population (N_i) can be notably reduced during the come-up time (Fig. 8). Palou et al. [45] reported an important effect of the come-up time at pressures of 345 and 517 MPa on *Z. bailii* log reductions in a food model system (a_w 0.98 and 0.95; pH 3.5), furthermore, total inhibition of the initially inoculated cells was achieved when only the time to reach 689 MPa was applied. Therefore, the effects of the rate of pressure increase and decrease need to be evaluated and reported.

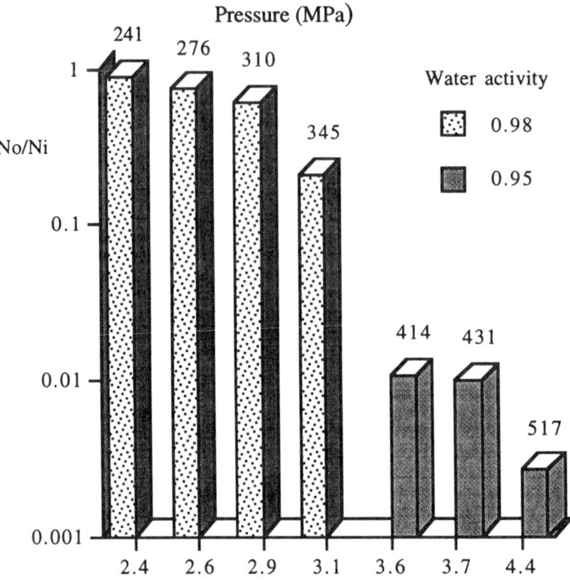

FIGURE 8 Effect of initial water activity (a_w), pressure, and pressure come-up time on *Zygosaccharomyces bailii* initial population reduction (N_0/N_i). N_0 = Yeast count (cfu/ml) after the come up time for the working pressure; N_i = yeast initial population (cfu/ml). (From Ref. 68.)

Palou et al. [68] observed first-order kinetics for *Z. bailii* inoculated in food model systems with pH 3.5 and a_w 0.98 and 0.95. The logarithm of the survival fraction decreased linearly with time, and yeast cells were inactivated more rapidly with increasing pressure treatments (Fig. 9). The experimental points with a pressure treatment duration of "0 minutes" express the effect of the come-up time to reach the working pressure and correspond to the initial population (N_0) for the kinetic analysis. Hashizume et al. [56] observed that pressure inactivation kinetics for *S. cerevisiae* at 25°C and $a_w \approx 0.99$ follows a first-order kinetic model. The death velocity constants or inactivation rates (k) can be calculated from the reciprocal of the slope of the survival curves following a traditional kinetic analysis. Table 4 presents the velocity constants reported by Palou et al. [68] for *Z. bailii* and those calculated from the data reported by Hashizume et al. [56]. As can be seen, the k values suggest that *S. cerevisiae* is more pressure resistant than *Z. bailii.* To compare the effectiveness of pressure treatments and to optimize process conditions, the calculation of D values can be used to compare the resistance of microorganisms. By analogy with the analysis of thermal destruction or inactivation of microorganisms, a decimal reduction time (D value) can be defined as the time needed to reduce 90% the initial population and can be calculated as $D = 2.303/k$.

Calculated D values for *Z. bailii* [68] and *S. cerevisiae* [56] are presented in Table 4. Other reports detected first-order inactivation rates and decimal reduction times for bacterial inactivation with high hydrostatic pressure. Carlez et al. [52] observed that the number of surviving *Pseudomonas fluorescens*, *Citrobacter freundii*, and *Listeria innocua* inoculated in a minced meat product decreased exponentially with process time. Carlez et al. [52] reported decimal reduction times at 20°C of $D_{230\ MPa}$ = 14.7 minutes for *C. freundii*, $D_{250\ MPa}$ = 23.8 minutes for *P. fluorescens*, and $D_{230\ MPa}$ = 6.5 minutes and $D_{360\ MPa}$ = 5 minutes for *L. innocua.* Smelt and Rijke [55] reported first-order kinetics for high-pressure inactivation of *E. coli* in a physiological saline solution at 20°C, with D values of 25.9, 8.0, 2.5, and 0.8 minutes for treatments at 200, 250, 300, and 350 MPa, respectively. The pressure resistance of microorganisms depends on the type of microorganism and composition of the

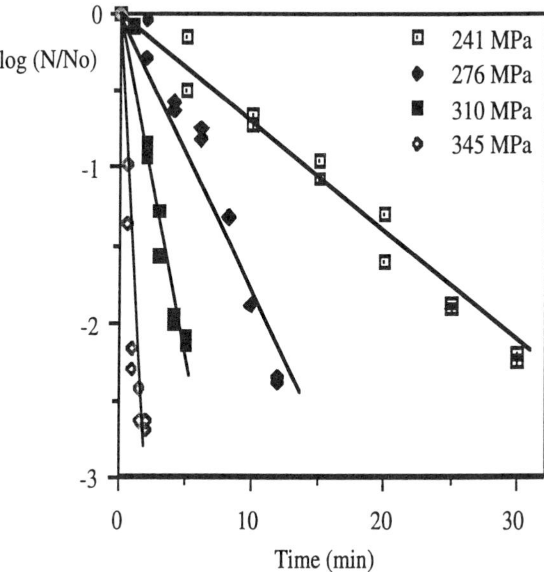

FIGURE 9 First-order pressure-inactivation kinetics of *Zygosaccharomyces bailii* in a laboratory model system with water activity 0.98 and pH 3.5. (Adapted from Ref. 68.)

suspension media. Many factors that could affect the microbial response under high-pressure treatments, such as temperature, gas solubility, ionic strength, pH, and cavitation, are also modified by pressure. Thus, the microbial inactivation curves that can be obtained during a high hydrostatic pressure treatment also depend on these factors.

Figure 10 presents the fit of the experimental data obtained from Palou et al. [68] and Hashizume et al. [56] to Eq. (1). The positive slope obtained from the plots of pressure inhibition rates (ln k)

TABLE 4 Effect of Initial Water Activity (a_w) and Pressure on Inactivation Rates (k) and Decimal Reduction Times (D) of *Zygosaccharomyces bailii* at 21°C and *Saccharomyces cerevisiae* at 25°C

a_w	Pressure (MPa)	k (min^{-1})	D (min)
Z. bailii[a]			
0.98	241	0.176	13.12
	276	0.478	4.82
	310	1.128	2.04
	345	2.833	0.81
0.95	414	0.902	2.55
	431	1.099	2.09
	517	2.645	0.87
S. cerevisiae[b]			
≈0.99	210	0.025	94.00
	240	0.067	34.63
	250	0.094	24.60
	270	0.187	12.30

[a]From Ref. 68.
[b]From Ref. 56.

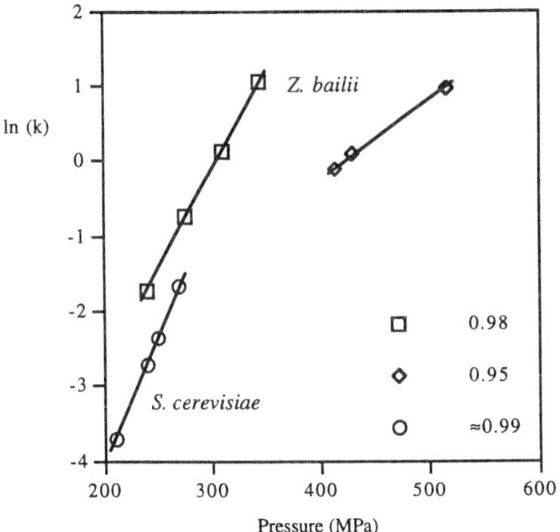

FIGURE 10 Effect of pressure and initial water activity on the inactivate rate (k) of *Zygosaccharomyces bailii* at 21°C [68] and *Saccharomyces cerevisiae* at 25°C [56].

versus pressure, and therefore a negative activation volume, indicates that a decrease in volume is related to the yeast-inhibition process. The activation volumes reported by Palou et al. [68] are presented in Table 5. The values obtained are quite different between system compositions and between yeast species. The apparent ΔV*, calculated considering the activation process of microbial inhibition as one step, indicates the volume variation between the "activated comples" and initial states of the yeast pressure inhibition "reaction" [68]. A negative activation volume represents a reaction favored by increased pressure, so a reaction with a greater absolute ΔV* value indicates that increments in pressure can accelerate the response—in this case the yeast inactivation rate.

A "pressure z value" can be defined as the pressure increment needed to reduce or increase the D value by a factor of 10 and is calculated as the reciprocal of the slope in a plot of log D versus pressure. Table 5 also presents z values for *Z. bailii* in food model systems at a_w 0.95 and 0.98 [68]. The z value for yeast inactivation in media with a_w 0.98 was 2.6 times smaller than that obtained for media with a_w 0.95, showing that *Z. bailii* was more pressure sensitive in the former condition. These results indicate that the composition of the media influences the microbial response under high-hydrostatic-pressure treatments and demonstrates the baroprotective effect of high sugar concentration or reduced a_w reported by several authors [44,45,56,57,61,69]. The z value (68 MPa) calculated from the data reported by Hashizume et al. [56] indicates that *S. cerevisiae* is more sensitive to changes in pressure than *Z. bailii*, which can be attributed to response differences between microorganisms to high-pressure treatments and composition of the suspension medium.

In biological systems, the volume changes associated with ionization can be involved in the mechanism of microbial inactivation [45]. Enhanced ionization under high-pressure treatments is

TABLE 5 Activation volume (ΔV*) and z Values of *Zygosaccharomyces bailii* at 21°C and *Saccharomyces cerevisiae* at 25°C

a_w	Pressure range (MPa)	ΔV* ($cm^3 \cdot mol^{-1}$)	z(MPa)
Z. bailii[a]			
0.98	241–345	−65.2	85.8
0.95	414–517	−25.3	222.7
S. cerevisiae[b]			
≈0.99	210–270	−83.9	68.0

[a]From Ref. 68.
[b]From Ref. 56.

reported for water and acid molecules [70]. Palou et al. [68] mentioned that in the conditions of their work, and knowing that during pressurization a decrease in the pKa of the acids and pH reduction is expected, a temporary reduction in pH and an increase in the dissociated form of the acid can be present during pressurization. The pH changes could enhance the effects of high-pressure treatments on microorganisms and favor the first-order kinetics observed for pressure inactivation of *Z bailii*.

Earnshaw [42], Isaacs et al. [51], Metrick et al. [67], and Patterson et al. [3] reported that for several process conditions and different microorganisms, the inactivation curves do not follow a first-order kinetics pattern. Figure 11 presents a comparison of high-pressure treatments at 20°C for *Yersinia enterocolitica* suspended in a pH 7.0 phosphate buffer [3]. Similar exponential decay curves have been reported for *E. coli*, *L. monocytogenes*, *S. typhimurium*, and *S. enteritidis* by the same authors at relatively high pressures. As can be seen for treatments at 275 and 300 MPa, the survival curves cannot be described by linear relations; various authors have proposed the logistic model to describe this pattern. However, the reliability of the logistic model for other organisms under different conditions and process variables needs to be investigated [71]. If the logistic model describes the experimental data, the time required to reduce the initial population by a factor of 10 (D value) can be calculated and used as a comparison parameter. Figure 12 presents the process time required to achieve a 6D reduction in the initial inocula for three pathogens reported by Patterson et al. [3].

There are a number of possible theories to explain the tail effect [42]: tailing is a normal characteristic associated with the inactivation or resistance mechanisms, is independent of the mechanisms of survival and/or inactivation, and is the result of microbial population heterogeneity or is the result of experimental errors. Another possible reason for tailing is microbial adaptation and recovery during and after pressure treatment.

Microbial Cell Recovery After Pressurization

The high-hydrostatic-pressure effects on microorganisms are often determined by plate counts, and usually dilution and plating are made just after treatment. Survival, as measured immediately after pressure release, may differ from that determined after a repair period in the food or in an enriched medium [9]. Recovery after pressure treatment is a very important consideration for process efficacy and death kinetics assessment [3]. Metrick et al. [67] suggested that the lack of nutrients in phosphate

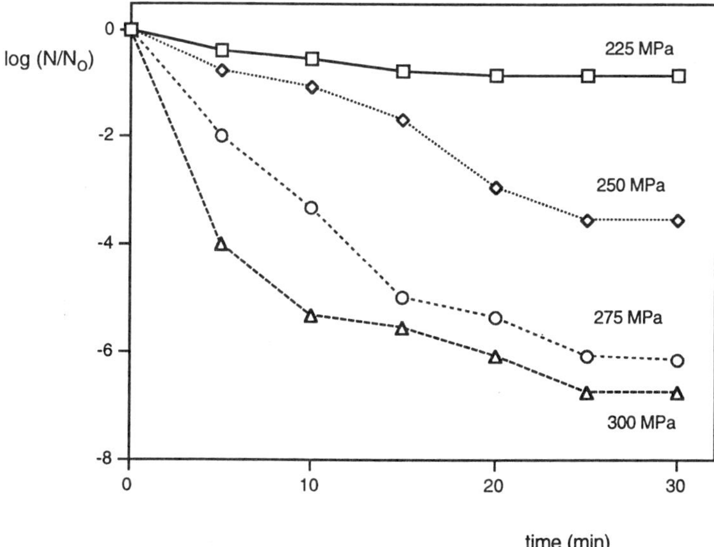

FIGURE 11 Effect of pressure on the inactivation of *Yersinia enterocolitica* in 10 mM buffered saline solution (pH 7.0) at 20°C. N = Number of survivors; N_0 = initial number of microorganisms. (Adapted from Ref. 3.)

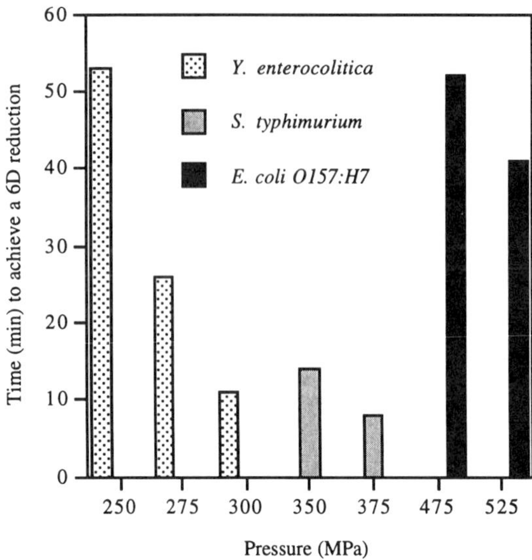

Figure 12 Predicted time to achieve a 6 D reduction in the initial population of pathogenic bacteria suspended in a pH 7.0 phosphate buffer and subjected to high-hydrostatic-pressure treatments. (Adapted from Ref. 3.)

buffer prevents the recovery of the pressure-damaged cells. The ultimate fate of the injured cells will depend on the conditions after pressure treatment. However, the fact that pressure can cause injury may be advantageous when high hydrostatic pressure is combined with other preservation methods [3].

The possibility of cell recovery exists, and in many cases pressure-treated microorganisms may not be detected in plate count methods because of their failure to initiate growth when they are plated immediately after treatment. However, if the repair mechanism remains intact, the microorganisms may be capable of regeneration and growth. Isaacs et al. [51] observed that for *E. coli* suspensions treated at 200 MPa for 0–6 minutes and plated on selective (Mac-Conkey and Eosin Methylene Blue agars) and nonselective (Tryptone Soya agar) media, the survival fraction was greater for bacteria plated on the nonselective agar. This was attributed to the inhibitory ingredients contained in the selective media, indicating that there is a proportion of microorganisms which, after pressurization, can repair and reproduce, whereas the added stress caused by culturing on selective media inhibits the repair process.

Figure 13 indicates that for pressure treatments that lead to significant reductions in the initial population of *Z. bailii*, but not enough to inactivate the initially inoculated cells, further incubation reveals that the survivors can grow [45]. Yeast cells subjected to physical or chemical hurdles may become injured or sublethally stressed, the recovery of these cells requires generally more incubation time and the injured survivors are capable of growth even in systems with 1000 ppm potassium sorbate and reduced a_w. In treatments rendering complete inhibition of 10^5 *Z. bailii* cfu/ml after 48 hours, the same result was obtained with a longer (120 hr) incubation period [45]. Carlez et al. [53] demonstrated that minced meat inoculated with *Pseudomonas* strains treated for 20 minutes at 300 or 450 MPa results in no microbial growth detected after 2 or 6 days of storage, respectively, when stored at 3°C. However, the *Pseudomonas* counts increased with a longer storage. The lag time before reappearance of microbial growth was related to the intensity of the pressure treatment. Carlez et al. [53] defined two different pressure levels: a lower one that causes microbial injury and delays growth and a higher one that induces complete inactivation of vegetative microorganisms.

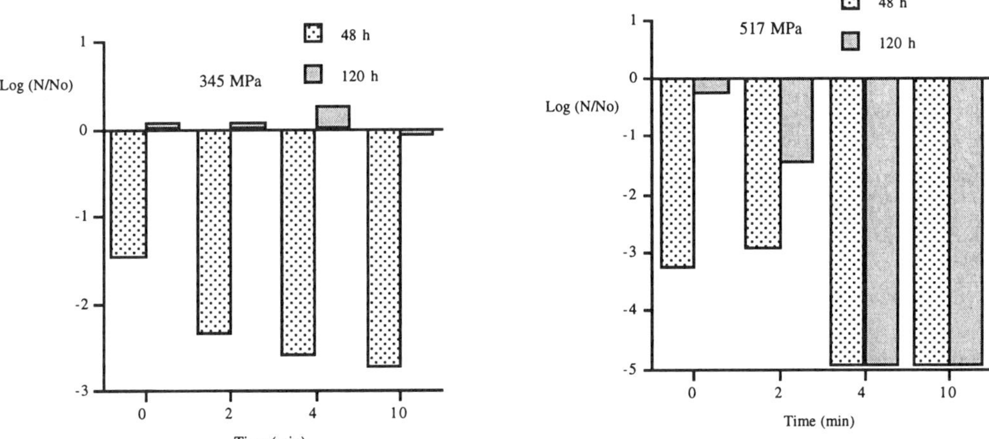

FIGURE 13 Effect of high-pressure treatments (345 and 517 MPa) on the log of the *Zygosaccharomyces bailii* survival fraction (N/N_0) after a recuperation period of 48 and 120 hours at 25°C. (From Ref. 45.)

2. Microbial Spores

Bacterial spores have demonstrated pressure resistance [59,72], and it has been suggested that spore proteins are protected against solvation and ionization. The structure and thickness of the bacterial spore coats are believed to account for this high resistance. Microbial spores suspended in foods and laboratory model systems could be inactivated by high-pressure treatments, but compared with the requirements for vegetative cells the treatment conditions must be extreme: higher pressures and long exposure times at elevated temperatures [73]. Hydrostatic pressure can cause spore germination. Some authors suggested a high-pressure treatment to induce spore germination and a subsequent treatment to inactivate the germinated microorganisms [10,11]. Gould and Sale [72] studied the germination of *Bacillus* spores and demonstrated that treatments at 25 MPa and 50°C for 30 minutes cause the germination of 50–64% of the initially inoculated spores.

Crawford et al. [74] reported a decrease in viable *Clostridium sporogenes* spores with increasing pressure up to 690 MPa; higher pressures were ineffective in further reducing the spore counts. Crawford et al. [74] postulated that spore germination occurred at pressures lower than 690 MPa and the germinated cells were inactivated by the pressure treatment, leading to a maximum reduction in the initially inoculated spore counts. At higher pressures, a small fraction of the spore population may have remained highly resistant and capable of resisting pressures of 830 MPa. When pressure-temperature were combined at 690 MPa and 80°C for 20 minutes, the treatment was more effective, with a significant reduction in the *C. sporogenes* spore count [74]. Mallidis and Drizou [75] confirmed the extraordinary pressure resistance of *Bacillus stearothermophilus* spores and concluded that the simultaneous application of moderate pressure (up to 30 MPa) and heat cannot be used as a preservation method, since the effect of pressure on the reduction of heat resistance is low at high temperatures. Nakayama et al. [76] observed no appreciable reductions in the spore viability for six *Bacillus* strains, including *B. stearothermophilus*, *B. coagulans*, *B. subtilis*, *B. licheniformis*, and *B. megaterium*, in pressure treatments at 5–10°C for 40 minutes up to pressures of 981 MPa or at 588 MPa for 120 minutes. Therefore, there is no possibility of pressure sterilization of low-acid foods at low temperatures. Hayakawa et al. [78] reported successful treatments for *B. stearothermophilus* spores when pressure treatments were combined with moderate temperature (70°C). Pressure treatments were four to six compression-decompression cycles at 600 MPa with 5 minutes of holding time. These treatments reduced by 6 to 7 log cycles the initial inocula. Lechowich [79] mentioned that there are no published reports on the high-pressure resistance of *C. botulinum* spores, and their ability to support high pressure at low or high temperatures is unknown. There is a need for data concerning the high-pressure resistance of *Bacillus* and *Clostridium* spores. With this information the safety aspects of high hydrostatic pressure treatment of low-acid foods can be documented [79].

Spores from yeast and molds are easily inactivated at pressures of 300 (*Aspergillus oryzae*) or 400 MPa (*Rhyzopus javanicus*) at ambient temperatures [9]. However, Butz et al. [80] demonstrated that ascospores of heat-resistant molds such as *Byssochlamys nivea* are extremely pressure resistant. For the inactivation of *B. nivea* ascospores, pressures above 600 MPa and temperatures above 60°C were needed. No effects on spore viability after treatment at 70°C and 500 MPa for 60 minutes were observed; around 3 log cycle reduction was observed at 600 MPa after 60 minutes; approximately 5 log cycles were reduced at 700 MPa; and total and rapid inactivation of *B. nivea* within a few minutes (<10) were observed at 800 MPa.

Bacterial spores represent a challenge for high-pressure technology, and more information about their resistance is required. Data are needed on the destruction of *C. botulinum* spores and the most resistant spore-forming species, and studies of the high-pressure process need to include inoculation and challenge testing.

3. High-Pressure Mechanism of Action

In the inactivation of microorganisms by high pressure, the membrane is the most probable key site of disruption [4]. Inactivation of key enzymes, including those involved in DNA replication and transcription, is also mentioned as a possible inactivating mechanism [4]. The lethal effect of high pressure on vegetative microorganisms is thought to be the result of a number of possible changes taking place simultaneously in the microbial cell [3]. Shimada et al. [81] suggested that the structural impact of the high hydrostatic pressure on yeast cells occurred directly in the membrane system, particularly in the nuclear membrane. Besides membrane damage, a decrease in pH due to the enhancement of the ionic dissociation resulting from electrostriction (pressure causes the separation of electrical charges because the electrical charges "organize" water molecules around them, with a resulting decrease in the total volume of the system) during high pressure treatments was reported by Cheftel [9]. Smelt [82] observed that the intracellular pH decreased under pressurization and associated the pH drop with the loss of ATPase activity and the reduction of the proton efflux from the cell interior. Knorr [10] reported that the reduced Na/K ATPase activity during and after pressurization can be related to a decrease in the bilayer membrane fluidity. Smelt [82] postulated that to maintain the internal pH homeostasis, membrane-bounded ATPase acts as an ion pump. High pressure can denature the enzyme or cause a dislocation in the membrane, thus microbial cells could die by internal acidification.

Microbial death is attributed to permeabilization of the cell membrane after a high-pressure treatment [5]. Pressure affects several biochemical reactions, and this kind of disturbance may be attributed to volume changes during compression, thus any biological process would be affected when high pressure is applied [4]. Farr [5] established that protein denaturalization can be attributed to changes in the chain conformational arrangement. Water pH is reduced from 7.00 (0.1 MPa and 25°C) to 6.27 when 101 MPa is applied [63], and water volume is also reduced as shown in Figure 4. These effects can also contribute to microbial inactivation by high pressure.

High-hydrostatic-pressure treatments can alter the membrane functionality such as active transport or passive permeability, and therefore perturb the physicochemical balance of the cell [83]. The physical state of the lipids that surround membrane proteins plays a crucial role in the activity of membrane-bound enzymes, and there is considerable evidence that pressure tends to loosen the contact between attached enzymes and membrane surfaces as a consequence of the changes in the physical state of lipids that control enzyme activity [35]. Jaenicke [84] reported that pressures in the range of 101–304 MPa denature several enzymes and treatments at 304 MPa make the phenomena irreversible. The activity of succinate, formate, and malate dehydrogenases in *E. coli* decreases with an increase in pressure. The dehydrogenases are completely inactivated when subjected to a pressure of 100 MPa for 15 minutes at 27°C [85]. Thus, the microbial inactivation mechanism by high pressure can be attributed, at least partially, to enzyme inactivation [4].

Perrier-Comet et al. [83] observed that yeast cell volume variations during a pressure treatment at 250 MPa for 15 minutes can be divided into three phases. A first phase of volume decrease occurs during the come-up time to reach the pressure; a second phase occurs during the holding time when the cell volume still decreases although pressure remains constant. Volume decrease is attributed to mass transfer between external and cellular media. A third phase of volume variation is attributed to membrane compression. The initial cell volume was not recovered during the decompression or after returning to atmospheric pressure. An irreversible mass transfer (mainly water) occurs during the holding time of a pressure treatment.

High-pressure treatment also induces morphological changes in microbial cells. Separation of the cell wall and disruption in the homogeneity of the intermediate layer between the cell wall and the cytoplasmic membrane occur. Isaacs et al. [51] demonstrated with electron microscopy studies that ribosomal destruction in cells of *E. coli* and *L. monocytogenes* result in metabolic malfunctions that can cause cell death. Mackey et al. [86] also observed by electron microscopy that the nuclear material appearance changes considerably in *L. monocytogenes* and *Salmonella thompson* after being treated at 500 MPa for 10 minutes. Hayakawa et al. [78] reported morphological changes in the *B. stearothermophilus* spore surface after pressurization for six cycles of 5 minutes each under 600 MPa at 70°C and observed that every spore was completely ruptured after this process. These observations were attributed to a weakening of the physical strength of the spore coat and rupture of the coat as a result of the pulsed pressure treatment. For *Schizosaccharomyces pombe*, after a treatment at 100 MPa the nuclear membrane was damaged and fragmented [87]. In the same study, a pressure treatment above 250 MPa dramatically changed the cytoplasmic substance, the cellular organelles could hardly be detected, and the fragmented nuclear membrane was barely visible. The outer cell shape, observed by scanning electron microscopy (SEM), of *S. cerevisiae* was almost unaffected by high-pressure treatments up to 300 MPa, but at pressures higher than 500 MPa there was disruption and damage to the cell wall [81]. Transmission electron microscopy (TEM) revealed that the inner structures were damaged, especially the nuclear membrane, even at 100 MPa [81]. The damage profile of high-pressure treatments revealed that *S. pombe* cells were more affected at low pressure stresses than were *S. cerevisiae* cells [87].

Isaacs et al. [51] observed that pressure treatments at 200–400 MPa cause leakage of UV-absorbing materials from *E. coli* cells. Shimada et al. [81] reported that leakage of intracellular UV-absorbing substances from *S. cerevisiae* began to be released at relatively low pressures (100 MPa). When pressure was 200 MPa for 10 minutes, the leakage gradually increased. Leakage of internal substances was related to cell viability, and at 300 MPa most of the *S. cerevisiae* cells were inactivated, corresponding to the great concentration of UV-absorbing substances. This can be attributed to increased permeability and fluidity and might also provide evidence about the mechanism of high-pressure inactivation [70].

To explain the response of microorganisms to different pressures, high-pressure effects on several biological molecules have been studied; protein denaturation, lipid phase change and enzyme inactivation can perturb the cell morphology, genetic mechanisms, and biochemical reactions. However, the mechanisms that damage the cells are still not fully understood [83].

B. Chemical and Biochemical Reactions

The application of pressure influences biochemical reactions since most of these reactions involve a change in volume. Hoover et al. [4] reported that pressure affects reaction systems in two apparent ways: by reducing the available molecular space and by increasing interchain reactions. Thus, reactions involved with the formation of hydrogen bonds are favored by high pressure since bonding results in a decrease in volume [4]. However, Masson [88] reported that hydrogen bonds are insensitive to pressure. Cheftel [9] mentioned that various biochemical studies indicate that pressures above 100–200 MPa often cause: (a) the dissociation of oligomeric structures into their subunits, (b) partial unfolding and denaturation of monomeric structures, (c) protein aggregation, probably as a consequence of unfolding, and (d) protein gelation if protein concentration and pressure are high enough.

High pressure can denature protein molecules. Pressure denaturation of proteins is a complex phenomenon depending on the protein structure, pressure range, temperature, pH, and solvent composition. Oligomeric proteins are dissociated by relatively low pressures (200 MPa), whereas single-chain protein denaturation occurs at pressures greater than 300 MPa. Pressure-induced denaturation is sometimes reversible, but renaturation after pressure release may take a long time. Protein denaturation becomes irreversible beyond a given pressure threshold, which depends on the protein, or at high protein concentrations that enhance aggregation [9]. Figure 14 presents a schematic diagram illustrating the effect of temperature on the denaturation of proteins. This kind of diagram delimits regions where the protein is active or denatured; at high temperature pressure stabilizes the protein against temperature denaturation [9,35]. Heremans [35] mentioned that the fact that one can "cook" an egg with pressure is the result of the unique phase diagram of proteins. There is evidence that a similar phenomenon occurs with other biomolecules such as polysaccharides [35,89], microorganisms and bacteriophages [90], and phospholipids [35].

Pressure may affect the secondary, tertiary, and quaternary structure of proteins. The main targets of pressure are the electrostatic and hydrophobic bonds in protein molecules. High pressure causes deprotonation of charged groups and disruption of salt bridges and hydrophobic bonds, thereby resulting in conformational and structural changes of proteins. Structural transitions are accompanied by large hydration changes. Hydration changes are the major source of volume decreases associated with dissociation and unfolding of proteins [88]. Hydrophobic interactions in proteins can be either disrupted or stabilized according to the magnitude of the applied pressure [39]. The disruptive effect of high hydrostatic pressure on the ionic and hydrophobic interactions and hydrogen bonds of milk casein micelles would allow independent movement of micelle fragments along with caseins and calcium phosphate [39] and cause a conformational change in proteins. In fresh milk, casein micelles are large complex structures consisting of many molecules of different individual caseins and calcium phosphate, maintaining the structural integrity. The structural and spatial distribution changes to the micelle components during pressurization would prevent recovery of the micelle original structures after high-pressure treatment [39].

Other food components that can be affected by high-pressure treatments are lipids. Ohshima et al. [91] observed that in cod muscles exposed to high-pressure treatments in the range of 202 to 608 MPa for 15 and 30 minutes, the peroxide value of the extracted oils increased with increasing pressure and processing time. The presence of fish muscle accelerates the lipid oxidation after high-

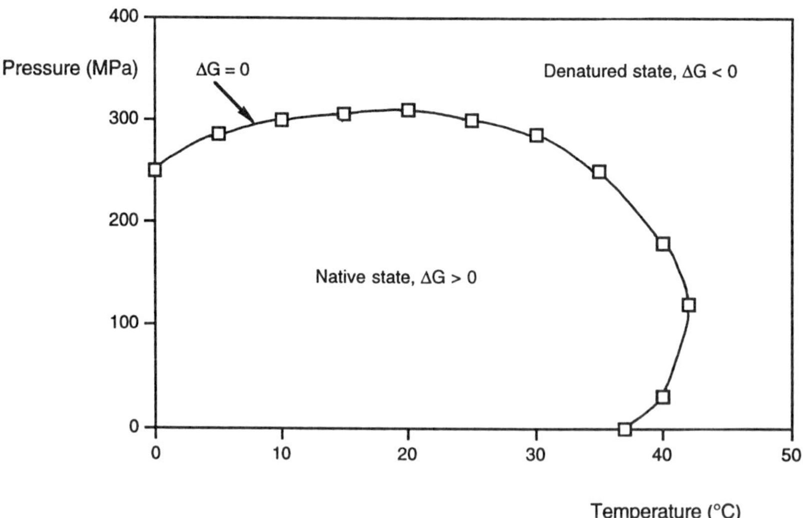

FIGURE 14 Schematic diagram illustrating the effect of temperature and pressure on the denaturation of proteins. (Adapted from Ref. 35.)

pressure treatments, while the isolated oil extract was relatively stable against autoxidation in treatments up to 608 MPa [92]. Cheah and Ledward [93] observed the effects of high-pressure treatments on lipid oxidation in rendered pork fat. Pork samples treated at 800 MPa for 20 minutes and stored at 50°C present a shorter induction time for lipid oxidation and greater peroxide and TBA (2-thiobarbituric acid) values than untreated pork samples. When pork was stored at 25°C the induction periods were longer than at 50°C, and at 4°C after 8 months of storage the peroxide value was higher than in the untreated samples. The effects of a_w were also evaluated by Cheah and Ledward [93], who reported that after 4, 6, and 8 days of storage at 50°C, the lipid oxidation in terms of peroxide and TBA values of high-pressure–treated (800 MPa, 20 min) pork fat was inhibited to some extent at a_w values higher or lower than 0.44. In another study, Cheah and Ledward [94] reported that pressure treatments had little effect on lipid oxidation of minced pork, in terms of TBA value, below 300 MPa but increased proportionally at higher pressures. Pressure treatments in the range of 300–400 MPa appear to be critical for inducing marked changes in pork meat; many structural changes are induced at these pressures. These results may restrict the application of high-pressure technology for meat-based products due to induced oxidation, at least in the mentioned pressure range.

The application of high-hydrostatic-pressure treatments, in combination with moderate or elevated temperatures, may influence chemical reactions inherent to food systems such as the Maillard reaction. This reaction, which is manifested by the development of brown color in many processed foods, is known to be highly pH and temperature dependent. However, few studies have dealt with the effects of high pressure on the Maillard reaction. Tamaoka et al. [95] reported that brown color development was inhibited at pressures in the range of 200–400 MPa in xylose-lysine systems (pH 8.2) when heated at 50°C. They also pointed out that Maillard reactions involving xylose-lysine, xylose–β-alanine, or glutaraldehyde–β-alanine are inhibited by high-pressure treatments. Hill et al. [96] compared the rate of browning of glucose-lysine systems at 50°C, in the pH range of 5.1–10.1, with and without the application of high pressure at 600 MPa. Hill et al. [96] reported that at initial pH of 8.0 or 10.1, pressure enhances browning, while at pH 6.5 and 5.1 the effect is the opposite. At 600 MPa the rate of browning was reduced significantly. These observations were attributed to the pH decrease in the systems during pressurization. At pH 6.5 and 5.1 the system buffer capacity is due to the carboxylic acid group of the amino acid, and decreases of about 1.2 pH units occurred. Hill et al. [96] also demonstrated by HPLC and UV spectra that the composition of the reaction products is similar in samples with the same intensity of browning, whether the samples were treated with high pressure or not. Acid hydrolysis of proteins is enhanced under high-pressure treatment, whereas hydrolysis of corn starch and locust bean gum is unaffected [12,97]. Pressure treatments at 392–490 MPa at temperatures of 45–50°C enhanced the susceptibility of wheat, corn, and potato starches to α-amylase action [32].

C. Enzymatic Reactions

Exposure to high pressure may activate or inactivate enzymes. Pressure inactivation of enzymes is influenced by pH, substrate concentration, the subunit structure of the enzyme, and temperature during pressurization [4]. Pressure effects on enzyme activity are expected to occur with the substrate-enzyme interaction. If the substrate is a macromolecule, then the effects may be on the conformation of the macromolecule, which can make the enzymic action easier or more difficult [35]. Pressure enzyme inactivation can also be attributed to an alteration of intermolecular structures or conformational changes at the active site. Inactivation of some enzymes pressurized to 100–300 MPa is reversible. Reactivation after decompression depends on the degree of distortion of the molecule. The chances of reactivation decrease with an increase in pressure beyond 300 MPa [84,98].

Earnshaw [34] mentioned that of particular significance is the apparent lack of pressure effect on some food enzymes, including those that affect food quality, such as proteases, lipases, esterases, and oxidases. Some enzymes, such as phosphatase, are relatively pressure sensitive and can be inactivated by pressures in the range of 400–800 MPa. The enzymes alkaline phosphatase and lactoperoxidase have been successfully used as process markers in milk heat treatments. Quality control markers such as these enzymes will be needed for high-pressure processing of dairy products. Fig-

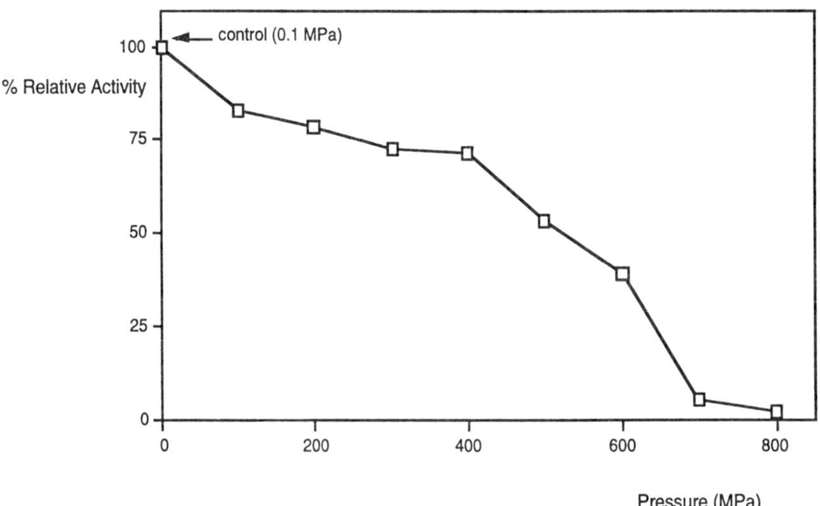

Figure 15 Relative activity of enzyme alkaline phosphatase in raw milk after 20 minutes of pressure treatment. (Adapted from Ref. 99.)

ure 15 presents the effect of 20-minute pressure treatments on alkaline phosphatase in raw milk; the enzyme activity decreases as pressure level increases. However, there is a need to establish microbial-quality marker relations to ensure, for example, the destruction of *Mycobacterium tuberculosis* [99].

Heremans and Heremans [100] reported that chymotrypsin is an enzyme that exhibits a pressure-induced conformational change. Its optimal activity is at neutral pH, and the activity disappears at pH 10, which is attributed to rupture of the salt bridge in the vicinity of the active site. Raman spectroscopy studies indicated that pressure inactivates the enzyme due to the destabilizing effects of pressure on the salt bridge and revealed that below 400 MPa the conformational change is reversible [100].

Homma et al. [101] mentioned that high hydrostatic pressure is one of the new technologies that can be used for tenderizing meat or accelerating meat conditioning. Pressure induces changes in the muscle that could be derived from the physical force and/or increase in the proteolytic activity of meat enzymes. The proteolytic activity of enzymes in meat is enhanced by the application of high pressures [101]. The total activities of cathepsin B, D, L, and acid phosphatase in the muscle increase when subjected to pressures ranging between 100 and 500 MPa for 5 minutes at 2°C. Cathepsin H and aminopeptidase B are resistant to pressure treatment. An increase in the activity of cathepsin B1 may account, in part, for the tenderization of meat by pressure-heat treatment [102]. Ashie et al. [103] reported a reduction in the proteolytic activity of fish muscle enzymes with increasing pressure in the range of 100–300 MPa (30 min) and increasing crude inhibitor (α_2-macroglobulin) concentration (0.1–0.3%) at constant pH (5.5, 6.0, or 6.5). The combination of treatments could enhance the texture of fish muscle, since it favors the inactivation of proteolytic enzymes. This kind of result could have a widespread use in surimi and other minced fish products, which have the problem of undesirable protease activity resulting in gel softening.

Enzymes are generally inactivated in vegetables by hot water blanching. Disadvantages of blanching include thermal damage, leaching of nutrients, and possible environmental pollution due to the production of high biochemical oxygen demand effluent. High-pressure treatment can fulfill the requirements of hot water blanching while avoiding mineral leaching and accumulation of wastewater. High-pressure treatment produces less effluent because less water is required than in hot water blanching [104]. Quaglia et al. [105] reported that pressure treatment at 900 MPa for 10 minutes reduces the peroxidase activity 88% in green peas—comparable to traditional water blanching. However, the pressurization treatment resulted in greater ascorbic acid and firmness retention. Lower pressure levels decreased the enzyme activity less than 50%, even when pressure was combined with moderate temperatures (39–60°C). Anese et al. [106] also observed in peroxidase from a carrot cell-free extract that a complete loss of enzyme activity was achieved only when the pressure treatment was applied at 900 MPa for 1 minute. Enzyme activation was observed for treatments in the range of 300–500 MPa. For polyphenoloxidase (apple cell–free extract) it was observed that at pH 7.0, 5.4, and 4.5 a significant reduction in enzyme activity occurred in pressure treatments at 900 MPa for 1 minute. For both enzymes a pH dependence on residual activity after the pressure treatment was observed. Eshtiaghi and Knorr [104] reported that addition of citric acid could lead to increased polyphenoloxidase inactivation since pH reduction enhances the pressure effects on enzyme inactivation. Denaturation and inactivation of enzymes occur only when very high-pressure treatments are applied; the activation effects that could be presented at relatively low pressures could be attributed to reversible configuration and/or conformation changes on enzyme and/or substrate molecules [106,107]. Seyderhelm et al. [108] evaluated the effects of high-hydrostatic-pressure treatments on selected enzymes including catalase, phosphatase, lipase, pectinesterase, lypoxygenase, peroxidase, polyphenoloxidase, and lactoperoxidase and reported that peroxidase was the most barostable enzyme with 90% residual activity after 30 minute treatment at 60°C and 600 MPa. Therefore peroxidase could be used as an enzyme indicator for high-pressure treatments.

Pressurization at 100 and 200 MPa causes hardly any inactivation of pectinesterase [58]. Pectinesterase in juices such as Satsuma mandarin juice is inactivated when pressurized to 300–400 MPa. Purified pectinesterase is also inactivated at pressures of 300 MPa or higher. The inactivation is irreversible, and the pectinesterase is not reactivated during storage at 0°C or transportation. The activity of pectinesterase from mandarin juice remains at low levels during 90 days of storage at 0°C after pressure treatments at 400–600 MPa. Soluble solids such as sugars, proteins, and lipids exert a protective action against pectinesterase inactivation by high pressure or heat [58]. Polyphenoloxidase is often described as a soluble enzyme, localized mainly in the cytosol of plant cells, and is also associated with particulate cell fractions [109]. It is well established that polyphenoloxidases from different sources may have different molecular sizes and conformations. Thus, it is expected that the polyphenoloxidases may respond differently during and following high-pressure treatments. It is also anticipated that important differences will occur when the enzyme activity is analyzed in whole foods, extracts, or commercial enzymes. In untreated onion cells, phenolic compounds are confined to vacuoles and spatially separated from the polyphenoloxidase by the tonoplast; after pressurization (>100 MPa) the cell and the tonoplast are disrupted and phenolic oxidation products are formed. Polyphenoloxidase is no longer separated from the substrate, and enzymatic browning begins [109]. The activity of polyphenoloxidase increases five times when slices of Bartlett pears are pressurized at 400 MPa and 25°C for 10 minutes (Fig. 16). Further increase in pressure does not increase the enzyme activity. On the other hand, pressurization of homogenates of apples, bananas, or sweet potatoes did not result in activation of polyphenoloxidase [110]. Gomes and Ledward [111] reported a reduction in polyphenoloxidase activity from a crude potato extract with increasing pressure (400–800 MPa for 10 min). In contrast, when the crude extract of mushroom was treated at 400 MPa for 10 minutes, an enhancement in the activity was observed. Pressures above 300 MPa inactivated polyphenoloxidase in apple slices [112]. Cano et al. [113] studied the combination of high pressure and temperature on peroxidase, polyphenoloxidase, and pectin methylesterase activities of fruit-derived products. Optimal inactivation of peroxidase in strawberry puree was achieved using 230 MPa and 43°C. Pressurization-depressurization treatments caused a significant loss of strawberry polyphenoloxidase up to 230 MPa. Combinations of high pressure and 35°C effectively reduced peroxidase in orange juice. The effects of pressure and temperature on pectin methylesterase activity in orange juice were similar to those for peroxidase.

There is some evidence of changes in the enzyme-substrate interactions during pressurization and, therefore, changes in enzyme reaction kinetics. Some beneficial aspects of enzyme activation or reduced enzyme activity by high pressure can be used to retain or increase food quality.

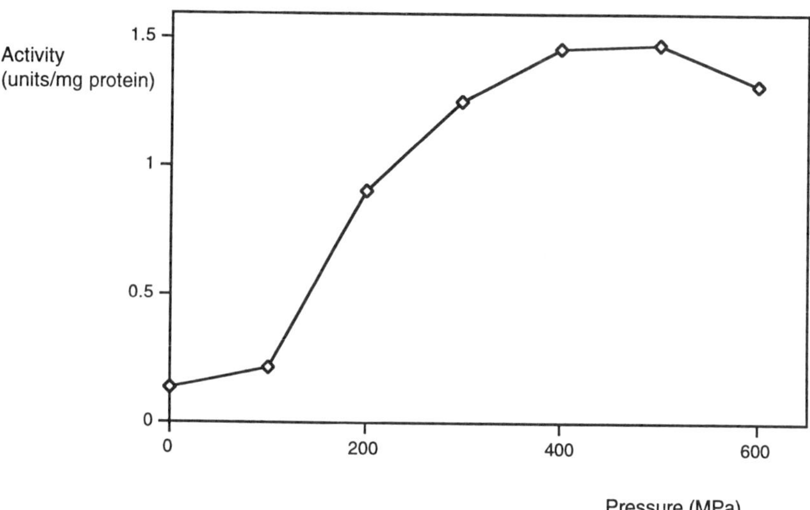

FIGURE 16 Activation of polyphenoloxidase from Bartlett pears by high pressure. (Adapted from Ref. 110.)

D. Functional Properties

The functional properties of biological molecules are usually dependent on conformation and conformational changes. The interactions between solvent and solute molecules and inter- and intramolecular interactions of the solute are influenced when subjected to pressure. Therefore, either beneficial or detrimental changes can be produced as a result of a high-pressure treatment [99]. Hydrogen bonding, which stabilizes protein structures (α-helix and β-pleated sheets) is influenced by pressure, but to a lesser extent than ionic or hydrophobic interactions. Hydrogen bond formation results in the shortening of interatomic distances with the corresponding volume decrease and is, therefore, enhanced by high pressure [99].

Phase changes in proteins and lipids are accompanied by the application of high pressure; these modifications offer opportunities to develop new products with unique rheological properties [34]. The structure of food proteins and polysaccharides can be changed by high-hydrostatic-pressure treatments and confer different rheological properties and mouthfeel. Earnshaw [34] explained that some Japanese researchers claim that these changes are desirable and that the gel quality of surimi can be improved with high-pressure treatments.

Pressure has an influence on meat ultrastructure with similar changes to those observed during the age-conditioning of meat, therefore the juiciness and tenderness are affected [114]. Plant structures containing entrapped air will be affected by high pressure when the air volume is compressed. However, in some cases the vacuoles and pores can be filled with the surrounding fluid, after which the material can maintain its structural integrity with increased density [34]. Adiabatic heating occurs in most food materials subjected to high pressure, and this is proportional to the compressibility of the food:air ratio. Entrapped air in the food matrix and cell vacuoles is very compressible and will increase the food or system temperature [34]. Some vegetable structures are resistant to pressure, while others can exhibit significant softening and severe color changes after pressurization. The effect of high-pressure treatments will depend on the kind of vegetable or fruit, physical characteristics, and maturity. Dumay et al. [115] evaluated the effects of high-hydrostatic-pressure treatments on the physicochemical characteristics of dairy creams and oil/water model system emulsions. Pressure treatments at 450 MPa did not affect the structure of the emulsions. Coalescence or emulsion breakdown was not observed. In the case of model emulsions, important rheological changes were observed, depending on the surface-active protein present and the effects of high pressure on unfolding and/or aggregation. Emulsions containing sodium casinate remained unchanged after treatment, while in those containing β-lactoglobulin the pressure treatment induced changes in their rheological behavior.

Fat crystallization by high-hydrostatic-pressure treatments is another interesting aspect to be considered. Bechheim and Abou El Nour [116] observed that fat crystallization increased with the extent of the pressure treatment; maximum changes were found in the pressure range of 300–350 MPa. Dumay et al. [115] stated that this behavior can be used for aging ice cream mixes and physical ripening of dairy cream in butter making. These can be considered as potential applications of high-pressure technology.

Exposure to high pressure unfolds protein molecules. Unfolding results in alterations of the functional properties of the protein [39]. Foaming, emulsifying, gelling, and water-binding capacity of proteins may be influenced. Proteins treated with high pressures may lead to the development of a range of functional food ingredients prepared from food proteins by controlled unfolding [39]. It is well known that pH and ionic strength influence protein aggregation and gelation under heating conditions; this is also observed when protein solutions are subjected to high hydrostatic pressure. Cheftel [9] mentioned that it is likely, yet not fully demonstrated, that protein aggregation and gelation occur under pressurization as well as after pressure release.

E. Sensory Properties

The principal advantage of high-pressure technology is its relatively small effect on food composition and, hence, on sensory and nutritional attributes. Generally, pressure has little effect on food nutritional characteristics. However, more research is needed before solid judgments can be made [34]. Alemán et al. [77] studied the effect of high hydrostatic pressure on the natural flora present in fresh-cut pineapple. Treatments of 340 MPa for 15 minutes at −4, 21, or 38°C considerably reduced the initial counts of mesophilic bacteria, yeast, and molds, suggesting an increased shelf life in comparison with untreated pineapple samples. Alteration of the structure of starch and protein by high pressure can be utilized so that rice can be cooked in a few minutes [4]. Grapefruit juice manufactured by high-pressure technology does not possess the bitter taste of limonene present in conventional thermal-processed grapefruit juice [117]. Peaches and pears processed at 410 MPa for 30 minutes remained commercially sterile for 5 years [4]. Pressure treatment of nonpasteurized citrus juices provides a fresh-like flavor with no loss of vitamin C and a shelf life of approximately 17 months [5]. The internal structure of tomatoes becomes tough, tissues of chicken and fish fillets become opaque, and prerigor beef is tenderized [4].

The jams obtained by high-pressure processing retain the taste and color of fresh fruit, unlike the conventional jams produced by heat. In Japan, high-pressure processing is utilized for the manufacture of jams, marmalades, and sauces from strawberry, orange, and other fruits. The desired plastic container is filled with a mixture of raw materials consisting of fruits, fruit juice, sugar, and acidulants. The container is sealed and subjected to a pressure of 400–600 MPa for 1–30 minutes. Strawberry jam can be obtained by pressurization at 400 MPa for 15 minutes and strawberry puree by pressurization at 400 MPa for 10 minutes. Pressurization allows the permeation of sugar solution into the fruits as well as commercial preservation of the jam [118].

El Moueffak et al. [119] reported that duck foie gras pressurized at 50°C and 400 MPa for 10–30 minutes or at 300 MPa for 30 minutes present attractive sensory characteristics, with less fat melting, softer texture, less cooked flavor, and a microbiological quality close to that obtained with traditional heat treatment. Pressure treatments up to 150 MPa did not change the sensory and instrumental color of minced beef muscle [53]. However, treatments for 10 minutes at pressures higher than 350 MPa turned the meat surface to a grayish tone, which corresponds to a decrease in the instrumental a* value. The grayish tone was even more noticeable when 450 MPa were applied for 10 minutes.

Eshtiaghi et al. [120] observed that the color of dried green beans, carrots, and potatoes dried without pretreatment was dark due to enzymatic browning, while pressure or water-blanched pretreated vegetables retained an acceptable color. The pretreatments applied were water blanching in boiling water (carrot and green beans for 7 min and potatoes for 4 min) and pressure treatment at 600 MPa for 15 minutes at 70°C. The texture of dried and rehydrated pressure-pretreated green beans, carrots, and potatoes resulted in textures near that of raw vegetables.

Butz et al. [109] reported that diced onions subjected to high-hydrostatic-pressure treatments lose their typical pungency and characteristic odor due to an intense decrease in dipropyldisulfide content and an increase in 2-methyl-pent-2-enal. Dipropyldisulfide is the compound associated with the odor of fresh onion. Diced onions presented no major changes in appearance immediately after a 30-minute pressure treatment at 300 MPa and 25°C [109]. However, a slight glassy appearance, typical of steamed onions, was observed. In the same study, onions treated at 350 MPa and stored for 24 hours at 20°C exhibited an intense brown color. Onions treated at 300 MPa started to brown, whereas samples pressurized at 100 MPa remained unchanged. Any pressure treatment above 100 MPa induces browning of diced onions, and the rate of browning increases with increasing pressure. Microscopic evaluation of high-pressure–treated (300 MPa) onions revealed a severe damage in the vacuoles of the epidermis cells, with the liberation of substrates for polyphenoloxidase activity [109].

Takahashi et al. [121] evaluated the effect of high-hydrostatic-pressure treatments on Satsuma mandarin juice and reported that juice pressurized up to 600 MPa for 5 or 10 minutes at 20–22°C did not change in chemical composition including soluble solids, acidity, amino nitrogen, vitamin C, and essential oil contents. The pressure-treated juices presented no off-flavor, and dimethyl sulfide, the characteristic compound found in off-flavor juices, was not detected. High-pressure–treated juices had very high scores in the sensory evaluation. Hayashi [33] reported some characteristics of food proteins treated by high-pressure technology: (a) beef muscle pressurized at 400 MPa for 10 minutes looked like raw ham and the taste of pressurized beef was intact, even when the surface seemed slightly baked; (b) in shrimp treated at 400 MPa for 10 minutes, no apparent changes in color or shape were observed. However, shrimp meat was coagulated as in boiled shrimp.

There are few studies regarding high-pressure effects on nutritional characteristics of pressure-treated foods. Elgasim and Kennick [122] reported that a pressure treatment at 103 MPa for 2 minutes improved the apparent digestibility of meat protein and had no adverse effect on the apparent biological value, net protein utilization, or protein efficiency ratio.

A wide variety of effects and changes in food flavor, texture, physical appearance, and structure could result after the application of pressure, and these changes will depend on the type of food and its composition and structure.

F. Gelation and Gelatinization Processes

The process of gel formation is the macroscopic consequence of the denaturation, on a molecular level, of proteins and other biomacromolecules such as polysaccharides. The denatured state forms a gel or a precipitate, depending on the physical and chemical environmental characteristics.

Egg yolk subjected to a pressure of 400 MPa for 30 minutes at 25°C forms a gel. While a pressure of 500 MPa renders egg white partially coagulated and opaque, a pressure of 600 MPa causes complete gelation. Pressure-induced gels of egg white possess a natural flavor, displaying no destruction of vitamins and amino acids, and are more easily digested when compared to heat-induced gels. The gels retain the original color of the yolk or the white and are soft, lustrous, and adhesive when compared to heat-induced gels. While the strength of the gels increases, the adhesiveness decreases with an increase in the applied pressure. However, the hardest gel formed by high-pressure (500 MPa) treatment exhibits one-sixth the strength of heat-induced gels. Gumminess of pressure-induced gels is considerably less than gumminess of heat-induced gels. Gels of egg white produced at 600 and 700 MPa deform readily without fracture. Cohesiveness of pressure-induced gels increases with increases in applied pressure. The force deformation curves of pressure and heat-induced gels of egg yolk and egg white are presented in Figure 17 [123]. Ibarz et al. [124] studied the viscoelastic characteristics of egg gels formed under several high-hydrostatic-pressure conditions. For egg yolk samples, gels were formed at pressures above 500 MPa, while for whole egg and egg white samples, gels were formed at pressures over 600 MPa. During amplitude sweep and frequency sweep tests storage modu-

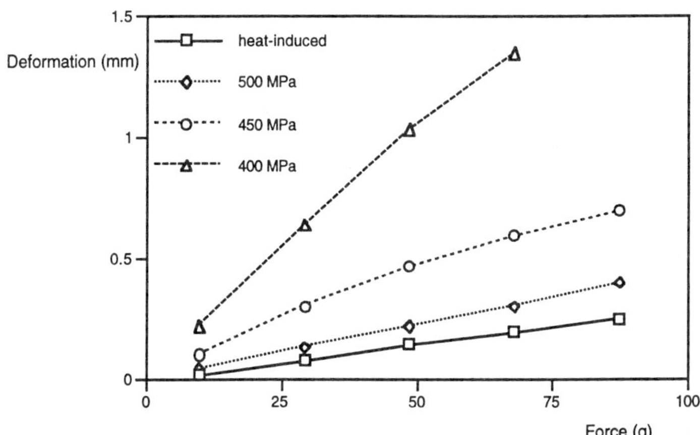

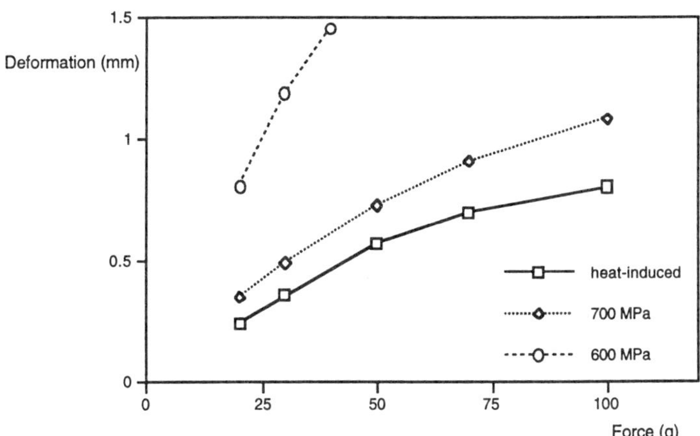

FIGURE 17 Force deformation curves of pressure- and heat-induced gels of egg yolk (top) and egg white (bottom). (Adapted from Ref. 123.)

lus (G′) and loss modulus (G″) of gels increased when the treatment pressure increased. G′ was greater than G″ in every case studied. G′ values for boiled egg gels were greater than for pressurized gels.

In Japan, a hydrostatic pressure of 400 MPa is used to induce gelation of pollack, sardine, skip jack, and tuna-based surimi. Squid-based surimi is obtained by pressurization of extracted muscle protein at 600 MPa. Pressure-induced surimi gels are organoleptically superior to heat-induced surimi gels [5]. Gelation can be used for adhesion-binding of small size muscles or fish fillets, restructuring of minced fish or deboned meat, and molding of surimi or pieces of gelled surimi into seafood analogs. The possibility of obtaining acceptable gels simultaneously with commercial sterilization at a temperature as low as 0°C is of tremendous practical interest to the surimi industry [22].

The mechanism of high-pressure–induced gelation is different from heat-induced gelation. Gelation by high-pressure treatment is attributed to a decrease in the volume of the protein solution. On the other hand, the application of heat results in violent movement of protein molecules leading to the destruction of noncovalent bonds, denaturation, and formation of a random network. The rearrangement of water molecules around amino acid residues in pressure-induced gels produces glossy and transparent gels compared to opaque gels obtained by high temperatures [123].

Heremans [35] mentioned that important differences in the mechanical properties for temperature- and pressure-induced gels are expected. Hayashi [33] reported that pressure-coagulated food proteins, i.e., egg, soy protein, beef, pork, and fish meat, are more glossy, transparent, dense, smooth, and soft compared to boiled ones. These unique textural properties obtained by the pressurization offer ways to create new food materials.

Gelatinization is the transition of starch granules from the birefringent crystalline state to a nonbirefringent, swollen state. Starch can be gelatinized using pressure or heat. The pressure at which starch gelatinizes depends on the source of starch. Gelatinization may be stimulated by increased temperatures of pressurization [125]. High-pressure may also produce an upward shift of gelatinization temperature of about 3–5°C per 100 MPa. Pressures higher than 150 MPa do not further enhance the gelatinization temperature. The effect of high pressure on gelatinization is due to the stabilization of hydrogen bonds, which maintains the starch granule in the original state [89]. High-pressure treatments on starches produce unique properties, which are different from those formed by heat-gelatinization. Pressure-treated starches keep the granular structure intact, while heat treatment destroys starch granules and dissolves the starches to give transparent solutions [33].

IV. COMBINATION TREATMENTS

The increasing demand for foods with reduced amounts of chemical additives and less physical damage is opening new opportunities for the hurdle-technology concept of food preservation [40,70]. The commercial challenge of minimally processed foods provides a strong motivation to study food-preservation systems that combine traditional microbial stress factors or hurdles, while introducing "new" variables for microbial control, such as high pressure. High pressure presents unique advantages over conventional thermal treatments [11]. However, many reported data indicate that commercial pasteurization or sterilization of low-acid foods using high pressure is very difficult without using some additional factors to enhance the inactivation rate. Factors such as heat, antimicrobials, ultrasound, and ionizing radiation can be used in combination with high pressure. These approaches will not only help to accelerate the rate of inactivation, but can also be useful to reduce the pressure level and hence the cost of the process while eliminating the commercial problems associated with sublethal injury and survivor tails.

High pressure can be used to reduce the severity of the factors traditionally used to preserve foods. The use of high pressure in combination with mild heating has considerable potential [3]. Studies have shown that the antimicrobial effect of high pressure can be increased with heat, low pH, carbon dioxide, organic acids, and bacteriocins such as nisin [45,60,75,126,127]. Mackey et al. [50] observed that *L. monocytogenes* cells were sensitized to pressure by butylated hydroxyanisole, potassium sorbate, and acid conditions. Therefore, if the hurdle concept could be applied to the optimization of high hydrostatic pressure for low-acid foods, a combination of moderate treatments including pressure can lead to a food-preservation method effective against bacterial spores [60].

Knorr [10,11], Papineau et al. [127], and Popper and Knorr [126] reported enhanced pressure inactivation of microorganisms when combining pressure treatments with additives such as acetic, benzoic, or sorbic acids, sulfites, some polyphenols, and chitosan. These combination treatments allow lower processing pressure, temperature, and/or time of exposure. Roberts and Hoover [60] evaluated the effect of combinations of pressure at 400 MPa, heat, time of exposure, acidity, and nisin concentration against *B. coagulans* spores. Sublethally injured spores by the pressurization in combination with heat and acidity caused *B. coagulans* spores to become more sensitive to nisin. Acidic foods could be protected from spore outgrowth with the combined treatment. Hauben et al. [128] studied the lethal inactivation and sublethal injury of *E. coli* by high pressure and combinations of high-pressure treatments with lysozyme, nisin, and/or EDTA. High-pressure treatments from 180 to 320 MPa disrupted the outer membrane of bacterial cells, causing periplasmic leakage and sensitization to lysozyme, nisin, and EDTA, demonstrating that sublethal injury can be usefully applied in a hurdle-technology approach as an effective food-preservation method.

Crawford et al. [74] evaluated the combination of high hydrostatic pressure, heat, and irradiation to eliminate *C. sporogenes* spores in chicken breast. These authors reported no significant differences in the number of surviving spores between samples that were first irradiated and then pressurized or vice versa. However, there was a significant difference between samples exposed to combined treatments and those that were only irradiated, with the combined processes being more effective. No survivors of the initial inoculated spores were observed with a 6 kGy irradiation dose followed by pressurization at 690 MPa and 80°C for 20 minutes. Crawford et al. [74] concluded that a combination of lower doses of irradiation and high pressure is more useful in eliminating *C. sporogenes* spores than the application of either process alone.

Earnshaw et al. [70] mentioned that there is no synergistic antimicrobial action between sorbic acid and pressure up to 400 MPa when applied to *Z. bailii* and attributed this lack of synergy to the modification of sorbic acid dissociation constant under pressurization. Tauscher [37] mentioned that the carboxylic acids commonly used as food preservatives show enhanced ionization when subjected to high pressure. However, Palou et al. [45] demonstrated that increased antimicrobial effects can be obtained when combining high pressure and potassium sorbate to inactivate *Z. bailii* in laboratory model systems with reduced a_w and pH (Fig. 18). The initial inoculum (10^5 *Z. bailii* cfu/ml) was completely inactivated in systems with a_w 0.98 in the presence of potassium sorbate with pressures ≥345 MPa for more than 2 minutes; without potassium sorbate the pressure had to be applied at 517 MPa for 4 minutes. In laboratory systems with a_w 0.95 and without potassium sorbate the pressure must be ≥517 MPa for 10 minutes, and with potassium sorbate the treatment time could be reduced to 4 minutes [45].

Citric and sorbic acids were included in the laboratory model systems used by Palou et al. [45]. Thus, a temporary reduction in pH and an increase in the dissociated form of the acids could be present, which will depend on the pressure level. This effect could decrease the antimicrobial effectiveness of potassium sorbate during the time of exposure, since the major antimicrobial action is attributed to the undissociated form of the acid. However, the result of high-hydrostatic-pressure treatments would depend not only on the previously mentioned effects but on the consequences of high pressure in the biological systems involved. The results presented in Figure 18 reveal a synergistic antimicrobial action between potassium sorbate and high pressure at both a_w. For the same pressure level, the holding time required to inhibit *Z. bailii* is shorter in the presence of the preservative. High-pressure damage to *Z. bailii* renders cells more susceptible to other antimicrobial agents (low pH, potassium sorbate), probably due to the exposure of critical cell surface targets.

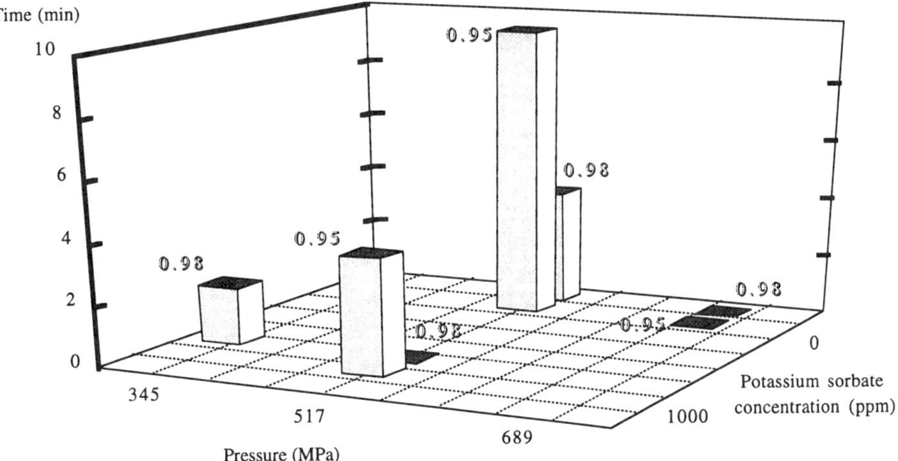

FIGURE 18 Combinations of high-pressure treatments (pressure and exposure time), potassium sorbate concentration, and water activity to inactivate *Zygosaccharomyces bailii* in model systems. (Adapted from Ref. 45.)

There is theoretical evidence about the pH shift during pressurization, but accurate models relating to microorganisms in food systems and direct measurement of pH are not available [70]. A detailed understanding of pH modification in foods might allow the design of food formulation that maximizes effective reversible antimicrobial pH shifts during pressurization [70]. The volume changes associated with ionization can be involved in the action of high hydrostatic pressure in biological systems. Earnshaw et al. [70] mentioned that water and acid molecules show increased ionization under high pressure. Thus, during the pressurization and holding time of a high-pressure treatment, an increase of proton concentration and a pH reduction are expected. High pressure decreases the pKa of certain acids that correspond to a decrease in the pH of solutions or buffers containing these acids [129]. Decreases in the pKa are more important for phosphoric acid than for citric acid. The pH of water or phosphate buffers decreases reversibly by 0.2–0.3 pH units per 100 MPa of applied pressure [130]. These effects can contribute, in a cooperative manner, to enhance the pressure effects on microorganisms, and the knowledge and understanding of their effects may aid to design effective pressure-combined processes. For most of the possible combined processes, the primary goal consists of identifying the factors or treatments that could sensitize microorganisms to pressure [9] or recognize the factors or treatments that could cause the microbial death in sublethal pressure-injured microbial cells. However, the protective effects that could exert food components made necessary the assessment of each combination process in each particular food product.

V. POTENTIAL APPLICATIONS

Cheftel [9] mentioned that potential applications of high pressure include decontamination of raw milk and some curds and cheeses made from raw milk, reduction of the intensity of thermal processing for prepared chilled meals containing thermosensitive food constituents, and the sanitation and increase of the refrigerated shelf life of spreads, emulsified sauces, essential oils, aromatic extracts, and herbs.

The observed tails in some survivor curves made small probabilities of survival difficult to obtain and call into question the efficiency of the high-pressure process [1]. Many reported data show that commercial pasteurization or sterilization of nonacidic foods is very difficult or impossible without using some additional factors or methods to enhance the inactivation rate. In practice, processing should be carried out at a relatively high pressure or in combination with moderate temperatures in order to inactivate the microbial cells. To optimize high-hydrostatic-pressure preservation techniques, the combined effects of stress factors on the growth of key microorganisms, the storage-dependent changes in food systems after high pressure treatments, and their shelf-life–limiting factors must be understood [68].

The effects of pressure in food products and foodborne microorganisms were first reported in 1899 by Hite. These early studies include results on microbial inactivation in milk and meat. In 1914, the results were expanded to fruit products [131]. The success of these pioneer works was attributed to the combined effect of low pH and high hydrostatic pressure; in some products such as berries the presence of natural organic acids (sorbic, benzoic) also contributed. However, for low-acid foods the results were not the same. Papineau and Schmerseal studied the effect of high pressure on *S. cerevisiae* inoculated in fruit juices. Figure 19 presents some of their results [4]. The survival fraction was reduced sixfold with treatments at 284 MPa for 30 minutes in apple, orange, and cranberry juices. Based on these kind of studies, Hoover et al. [4] mentioned that the antimicrobial effect of high hydrostatic pressure depends on the microbial flora initially present in the food and food composition.

Since high pressure has less effect on microbial spores than on vegetative cells, products with a pH lower than 4.0 can theoretically be processed successfully with this technology. Pressure inactivation of yeast and molds has been reported in citrus juices [57,58]. Juices pressurized at 400 MPa for 10 minutes at 40°C did not spoil during 2–3 months of storage. Capellas et al. [132] reported that high-pressure treatments effectively reduced the bacterial flora of fresh goat milk cheese and significantly extended the refrigerated storage life. No surviving *E. coli* were detected in the cheese after 60 days of storage (2–4°C) in inoculation studies after treatments at 400–500 MPa for 5–10 minutes. Potential pressurized juices and fruit salads are promising in Europe [9]. Other potential pressurized

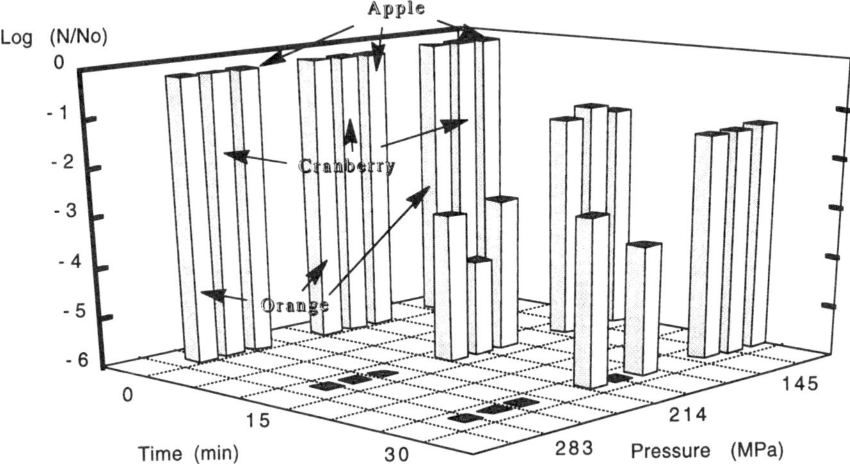

FIGURE 19 Effect of high-pressure treatments on *Saccharomyces cerevisiae* inoculated in apple, cranberry, and orange juices. (Adapted from Ref. 4.)

acid foods are sauces, vegetables, acidified seafoods, and possibly some wines (to replace SO_2).

High-pressure technology offers a unique opportunity to develop new foods of high nutritional and sensory quality, novel texture, more convenience, and with an increased shelf life [20]. High-pressure treatment has the potential to improve the microbiological safety and quality of foods, including meat, milk, and dairy products [3]. Carlez et al. [53] reported that high-pressure processing of minced meat contains microbial growth during storage at 3°C, and that growth delay was around 2–3 days after a high–pressure treatment at 200 MPa for 20 minutes, approximately 6–9 days at 300 MPa for 20 minutes, 10–12 days at 400 MPa for 20 minutes, and about 13–15 days if the treatment was applied at 450 MPa for 20 minutes.

Pressure depresses the freezing point of water and the melting point of ice, making it possible for various high-density forms of ice to be obtained [38]. These effects of pressure on the solid-liquid phase diagram of water have several potential applications in food technology, including pressure-assisted freezing, pressure-assisted thawing, and nonfrozen storage at low temperature under pressure. The greatest potential appears to be in high-pressure–assisted freezing and thawing [38]. High-pressure processing thaws frozen foods much faster than conventional thawing methods. Two kg of frozen beef are thawed in 80 minutes when pressurized to 200 MPa, whereas thawing the same amount of beef at atmospheric pressure takes 7 hours [53]. The flavor and juiciness of the pressure-thawed beef are equivalent to the flavor and juiciness obtained by thawing beef under low humidity at 5°C. The surface of the pressure-thawed beef is slightly discolored. Meat stored at 3°C after pressurization for 20 minutes at 200 MPa and 20°C begins to spoil after 2 days. The spoilage of meat begins after 6 days when pressurized at 300 MPa and after 6–13 days when pressurized at 400–450 MPa [53]. However, meat remains unaffected by microbial spoilage under refrigerated temperatures after a pressure treatment for 1 hour at 540 MPa and 52°C.

The perspectives for high-pressure food applications depend on further research, both at academic and industrial levels [9]. Research and development work should include shelf-life testing and appropriate microbial challenge testing, as well as incorporation of Good Manufacturing Practices in high-hydrostatic food processing [35]. Understanding of the pressure-inactivation mechanism may help in the design of successful bactericidal treatments by combining pressure with mild heat treatments and/or other traditional microbial stress factors, since it appears that the effects of high pressure on microorganisms are more complicated than previously thought [53]. More studies dealing with kinetic data of microbial inactivation, interactions between pressure and food constituents, and storage-dependent changes are needed.

VI. FINAL REMARKS

No specific official requirements are needed in some countries that relate to the control of new food-processing technologies [34]. However, the regulations of food safety are relevant, and there are still several important questions about safety and nutritional value that need to be answered.

The unique physical and sensory properties of food processed by high-pressure technology offer new chances for food product development such as minimally processed or raw meat and fish, long shelf-life convenience foods with fresh and natural colors, new types of food gels, and frozen foods with improved quality [34]. Identification of commercially feasible applications is probably the most difficult of the challenges for high-pressure technology [20]. Products processed by high pressure need to have inherent added value or an increased profitability due to the expensive process cost. Two important questions need to be answered: will the consumers accept a high-pressure-processed product, and are they prepared to pay an extra cost for a high-pressure-processed food? The commercial application of high-pressure technology in the food industry will depend largely on the economical feasibility of the process. The capital cost associated with the equipment purchase and installation is an important obstacle for its commercial implementation [20]. The cost of the high-pressure vessel represents the main fraction of the total cost of an industrial high-pressure processing plant and will depend on the maximum working pressure and the vessel dimensions and processing capacity.

Earnshaw [34] mentioned that it is unlikely that pressure processing will replace canning or freezing, nevertheless it could find applications for expensive foods with short shelf lives and high-value ingredients such as flavors, vitamins, and functional biopolymers that are heat-sensitive. In Europe, Japan, and the United States there is significant commercial interest in the development of high-pressure food processing, and millions of dollars have been invested in research and development.

High hydrostatic pressure is not a cheap technology, and a systematic approach must be taken to search for processing options to ensure that high-pressure treatments can be successfully and economically applied to a wide range of products [133]. The feasibility studies must include effective equipment design solutions and precisely defined minimum required pressures and time cycles. Continuous operation is also a major task.

REFERENCES

1. G. W. Gould, The microbe as a high pressure target, *High Presure Processing of Foods* (D. A. Ledward, D. E. Johnston, R. G. Earnshaw, and A. P. M. Hasting, eds.), Nottingham University Press, Nottingham, 1995, p. 27.
2. G. W. Gould, Ecosystem approaches to food preservation, *J. Appl. Bacteriol. 73*:58S (1992).
3. M. F. Patterson, M. Quinn, R. Simpson, and A. Gilmour, Effects of high pressure on vegetative pathogens, *High Pressure Processing of Foods* (D. A. Ledward, D. E. Johnston, R. G. Earnshaw, and A. P. M. Hasting, eds.), Nottingham University Press, Nottingham, 1995, p. 47.
4. D. G. Hoover, C. Metrick, A. M. Papineau, D. F. Farkas, and D. Knorr, Biological effects of high hydrostatic pressure on food microorganisms, *Food Technol. 43*(3):99 (1989).
5. D. Farr, High pressure technology in the food industry, *Trends Food Sci. Technol. 1*:14 (1990).
6. U. R. Pothakamury, G. V. Barbosa-Cánovas, B. G. Swanson, and R. S. Meyer, The pressure builds for better food processing, *Chem. Eng. Prog. (March)*:45 (1995).
7. D. A. Ledward, High pressure processing—the potential, *High Pressure Processing of Foods* (D. A. Ledward, D. E. Johnston, R. G. Earnshaw, and A. P. M. Hasting, eds.), Nottingham University Press, Nottingham, 1995, p.1.
8. A. Williams, New technologies in food preservation and processing: part II, *Nutr. Food Sci.* (1): 16 (1994).
9. J. C. Cheftel, High-pressure, microbial inactivation and food preservation, *Food Sci. Technol. Int. 1*:75 (1995).
10. D. Knorr, Hydrostatic pressure treatment of food: microbiology, *New Methods of Food Preservation* (G. W. Gould, ed.), Blackie Academic and Professional, New York, 1995, p. 159.

11. D. Knorr, High pressure effects on plant derived foods, *High Pressure Processing of Foods* (D. A. Ledward, D. E. Johnston, R. P. Earnshaw, and A. P. M. Hasting, eds.), Nottingham University Press, Nottingham, 1995, p. 123.
12. J. C. Cheftel, Effects of high hydrostatic pressure on food constituents: an overview, *High Pressure and Biotechnology* (C. Balny, R. Hayashi, K. Heremans, and P. Masson, eds.), Colloque INSERM, Vol. 224, John Libbey Eurotext, Montrouge, France, 1992, p. 195.
13. F. Zimmerman and C. Bergman, Isostatic pressure equipment for food preservation, *Food Technol. 47*(6):162 (1993).
14. D. Knorr, Effects of high-hydrostatic pressure process on food safety and quality, *Food Technol. 47*(6):156 (1993).
15. G. D. Alemán, E. Y. Ting, S. C. Mordre, A. C. O. Hawes, M. Walker, D.F. Farkas, and J. A. Torres, Pulsed ultra high pressure treatments for pasteurization of pineapple juice, *J. Food Sci. 61*:388 (1996).
16. S. Olsson, Production equipment for commercial use, *High Pressure Processing of Foods* (D. A. Ledward, D. E. Johnston, R. G. Earnshaw, and A. P. M. Hasting, eds.), Nottingham University Press, Nottingham, 1995, p. 167.
17. B. Crossland, The development of high pressure processing, *High Pressure Processing of Foods* (D. A. Ledward, D. E. Johnston, R. G. Earnshaw, and A. P. M. Hasting, eds.), Nottingham University Press, Nottingham, 1995, p. 7.
18. G. Deplace and B. Mertens, The commercial application of high pressure technology in the food processing industry, *High Pressure and Biotechnology* (C. Balny, R. Hayashi, K. Heremans, and P. Masson, eds.), Colloque INSERM, Vol. 224, John Libbey Eurotext, Montrouge, France, 1992, p. 469.
19. G. V. Barbosa-Cánovas, B. P. Swanson, U. R. Pothakamury, and E. Palou, *Nonthermal Preservation of Foods*, Marcel Dekker, New York, 1997.
20. B. Mertens, Hydrostatic pressure treatment of food: equipment and processing, *New Methods of Food Preservation* (G. W. Gould, ed.), Blackie Academic and Professional, New York, 1995, p. 135.
21. B. Mertens and G. Deplace, Engineering aspects of high pressure technology in the food industry, *Food Technol. 47*(6):164 (1993).
22. J. C. Cheftel, Applications des hautes pressions en technologie alimentaire, *Ind. Aliment. Agric. 108*:141 (1991).
23. Mitsubishi Heavy Industries, Technical data 1992.
24. K. Hori, Y. Manabe, M. Kaneko, T. Sekimoto, Y. Sugimoto and T. Yamane, The development of high pressure processor for food industries, *High Pressure and Biotechnology* (C. Balny, R. Hayashi, K. Heremans, and P. Masson, eds.), Colloque INSERM, Vol. 224, John Libbey Eurotext, Montrouge, France, 1992, p. 499.
25. T. Kanda, T. Yamauchi, T. Naoi, and Y. Inoue, Present status and future prospects of high pressure food processing equipment, *High Pressure and Biotechnology* (C. Balny, R. Hayashi, K. Heremans, and P. Masson, eds.), Colloque INSERM, Vol. 224, John Libbey Eurotext, Montrouge, France, 1992, p. 521.
26. ABB Autoclave Systems, Technical data, 1993.
27. R. J. Swientek, High hydrostatic pressure for food preservation, *Food Proc.* (November):90 (1992).
28. ABB Autoclave Systems, Technical data, 1994.
29. Engineered Pressure Systems, Technical data, 1994.
30. C. E. Morris, High pressure builds up, *Food Eng.* (Oct):113 (1993).
31. M. Masuda, Y. Saito, T. Iwanami, and Y. Hirai, Effects of hydrostatic pressure on packaging materials for food, *High Pressure and Biotechnology* (C. Balny, R. Hayashi, K. Heremans, and P. Masson, eds.), Colloque INSERM, Vol. 224, John Libbey Eurotext , Montrouge, France, 1992, p. 545.
32. R. Hayashi, Application of high pressure to food processing and preservation: philosophy and development, *Engineering and Food*, Vol. 2 (W. E. L. Spiess and H. Schubert, eds.), Elsevier, London, 1989, p. 815.

33. R. Hayashi, Advances in high pressure processing technology in Japan, *Food Processing: Recent Developments* (A. G. Gaonkar, ed.), Elsevier, London, 1995, p. 85.
34. R.P. Earnshaw, High pressure food processing, *Nutr. Food Sci. 2*:8 (1996).
35. K. Heremans, High pressure effects on biomolecules, *High Pressure Processing of Foods* (D. A. Ledward, D. E. Johnston, R. G. Earnshaw, and A. P. M. Hasting, eds.), Nottingham University Press, Nottingham, 1995, p. 81.
36. R. van Eldik, T. Asano, and W. J. Le Noble, Activation and reaction volumes in solution. 2, *Chem. Rev. 89*:549 (1989)
37. B. Tauscher, Pasteurization of food by hydrostatic pressure: chemical aspects, *Z. Lebensm. Unters. Forsch. 200*:3 (1995).
38. M. T. Kalichevsky, D. Knorr, and P. J. Lillford, Potential food applications of high pressure effects on ice-water interactions, *Trends Food Sci. Technol. 6*:253 (1995).
39. D. E. Johnston, B. A. Austin, and R. I. Murphy, Effects of high hydrostatic pressure on milk, *Milchwissenchaft 47*(12):760 (1992).
40. L. Leistner, Principles and applications of hurdle technology, *New Methods of Food Preservation*, (G. W. Gould, ed.), Blackie Academic and Professional, New York, 1995, p. 1.
41. T. Shigehisa, T. Ohmori, A. Saito, S. Taji, and R. Hayashi, Effects of high hydrostatic pressure on characteristics of pork slurries and inactivation of microorganisms associated with meat and meat products, *Int. J. Food Microbiol. 12*:207 (1991).
42. R. G. Earnshaw, Kinetics of high pressure inactivation of microorganisms, *High Pressure Processing of Foods* (D. A. Ledward, D. E. Johnston, R. G. Earnshaw, and A. P. M. Hasting, eds.), Nottingham University Press, Nottingham, 1995, p. 37.
43. G. W. Gould and M. V. Jones, Combination and synergistic effects, *Mechanisms of Action of Food Preservation Procedures* (G. W. Gould, ed.), Elsevier, London, 1989, p. 401.
44. E. Palou, A. López-Malo, G. V. Barbosa-Cánovas, J. Welti-Chanes, and B. G. Swanson, Combined effect of high hydrostatic pressure and water activity on *Zygosaccharomyces bailii* inhibition, *Lett. Appl. Microbiol. 24*:417 (1997).
45. E. Palou, A. López-Malo, G. V. Barbosa-Cánovas, J. Welti-Chanes, and B. G. Swanson, High hydrostatic pressure as a hurdle for *Zygosaccharomyces bailii* inactivation, *J. Food Sci. 62*:855 (1997).
46. C. E. Zobell, Pressure effects on morphology and life processes of bacteria, *High Pressure Effects on Cellular Processes* (A. M. Zimmerman, ed.), Academic Press, New York, 1970, p. 85.
47. M. F. Patterson, M. Quinn, R. Simpson, and A. Gilmour, Sensitivity of vegetative pathogens to high hydrostatic pressure treatment in phosphate-buffered saline and foods, *J. Food Prot. 58*:524 (1995).
48. M. F. Styles, D. G. Hoover, and D. F. Farkas, Response of *Listeria monocytogenes* and *Vibrio parahaemolyticus* to high hydrostatic pressure, *J. Food Sci. 56*:1404 (1991).
49. Y. Takahashi, H. Ohta, H. Yonei, and Y. Ifuku, Microbicidal effect of hydrostatic pressure on satsuma mandarin juice, *Int. J. Food Sci. Technol. 28*:95 (1993).
50. B. M. Mackey, K. Forestiere, and N. S. Isaacs, Factors affecting the resistance of *Listeria monocytogenes* to high hydrostatic pressure, *Food Biotechnol. 9*(1&2):1 (1995).
51. N. S. Isaacs, P. Chilton, and B. Mackey, Studies on the inactivation by high pressure of microorganisms, *High Pressure Processing of Foods* (D. A. Ledward, D. E. Johnston, R. G. Earnshaw, and A. P. M. Hasting, eds.), Nottingham University Press, Nottingham, 1995, p. 65.
52. A. Carlez, J. P. Rosec, N. Richard, and J. C. Cheftel, High pressure inactivation of *Citrobacter freundii*, *Pseudomonas fluorescens* and *Listeria innocua* in inoculated minced beef muscle, *Lebensm. Wiss. Technol. 26*:357 (1993).
53. A. Carlez, J. P. Rosec, N. Richard, and J. C. Cheftel, Bacterial growth during chilled storage of pressure-treated minced meat, *Lebensm. Wiss.Technol. 27*:48 (1994).
54. H. Ludwing, C. Bieller, K. Hallbauer, and W. Scigalla, Inactivation of microorganisms by hydrostatic pressure, *High Pressure and Biotechnology* (C. Balny, R. Hayashi, K. Heremans, and P. Masson, eds.), Colloque INSERM, Vol. 224, John Libbey Eurotext, Montrouge, France, 1992, p. 25.

55. J. Smelt and G. Rijke, High pressure treatment as a tool for pasteurization of foods, *High Pressure and Biotechnology* (C. Balny, R. Hayashi, K. Heremans, and P. Masson, eds.), Colloque INSERM, Vol. 224, John Libbey Eurotext, Montrouge, France, 1992, p. 361.
56. C. Hashizume, K. Kimura, and R. Hayashi, Kinetic analysis of yeast inactivation by high pressure treatment at low temperatures, *Biosci. Biotechnol. Biochem. 59*:1455 (1995).
57. H. Ogawa, K. Fukuhisa, and H. Fukumoto, Effect of hydrostatic pressure on sterilization and preservation of citrus juice, *High Pressure and Biotechnology* (C. Balny, R. Hayashi, K. Heremans, and P. Masson, eds.), Colloque INSERM, Vol. 224, John Libbey Eurotext, Montrouge, France, 1992, p. 269.
58. H. Ogawa, K. Fukuhisa, Y. Kubo, H. Fukumoto, Inactivation effect of pressure does not depend on the pH of the juice, *Agric. Biol. Chem. 54*(5):1219 (1990).
59. A. J. H. Sale, G. W. Gould, and W. A. Hamilton, Inactivation of bacterial spores by hydrostatic pressure, *J. Gen. Microbiol. 60*:323 (1970).
60. C. M. Roberts and D. G. Hoover, Sensitivity of *Bacillus coagulans* spores to combinations of high hydrostatic pressure, heat, acidity and nisin, *J. Appl. Bacteriol. 81*:363 (1996).
61. P. Oxen and D. Knorr, Baroprotective effects of high solute concentrations against inactivation of *Rhodotorula rubra*, *Lebensm. Wiss. Technol. 26*:220 (1993).
62. G. J. Dring, Some aspects of the effects of hydrostatic pressure on microorganisms, *Inhibition and Inactivation of Vegetative Microbes* (S. A. Skinner and V. Hugo, eds.), Academic Press, New York, 1976, p. 257.
63. R. E. Marquis, High pressure microbial physiology, *Adv. Microbial Physiol. 14*(1):159 (1976).
64. P. Rovere and A. Maggi, Approccio alle alte pressioni: una nuova tecnologia a disposizione dell'industria alimentare, *Indust. Conserve 70*(1):45 (1995).
65. Y. Pandya, F. F. Jewett, and D. G. Hoover, Concurrent effects of high hydrostatic pressure, acidity and heat on the destruction and injury of yeasts, *J. Food Prot. 58*:301 (1995).
66. P. Butz and Ludwig, Pressure inactivation of microorganisms at moderate temperatures, *Physica 139 & 140B*:875 (1986).
67. C. Metrick, D. G. Hoover, and D. F. Farkas, Effects of high hydrostatic pressure on heat-resistant and heat-sensitive strains of *Salmonella*, *J. Food Sci. 54*:1547 (1989).
68. E. Palou, A. López-Malo, G. V. Barbosa-Cánovas, J. Welti-Chanes and B. G. Swanson, Kinetic analysis of *Zygosaccharomyces bailii* by high hydrostatic pressure, *Lebensm. Wiss. Technol. 30*:703 (1997).
69. S. Fujii, K. Obuchi, H. Iwahashi, T. Fujii, and Y. Komatsu, Saccharides that protect yeast against hydrostatic pressure stress correlated to the mean number of equatorial OH groups, *Biosci. Biotechnol. Biochem. 60*:476 (1996).
70. R. G. Earnshaw, J. Appleyard, and R. M. Hurst, Understanding physical inactivation processes: combined preservation opportunities using heat, ultrasound and pressure, *Int. J. Food Microbiol. 28*:197 (1995).
71. M. Peleg, A model of microbial gowth and decay in a closed habitat based on combined Fermi's and the logistic equations, *J. Sci. Food Agric. 71*:225 (1996).
72. G. W. Gould and A. J. H. Sale, Initiation of germination of bacterial spores by hydrostatic pressure, *J. Gen. Microbiol. 60*:335 (1970).
73. D. G. Hoover, Pressure effects on biological systems, *Food Technol. 47*(6):150 (1993).
74. Y. J. Crawford, E. A. Murano, D. G. Olson, and K. Shenoy, Use of high hydrostatic pressure and irradiation to eliminate *Clostridium sporogenes* spores in chicken breast, *J. Food Prot. 59*:711 (1996).
75. C. G. Mallidis and D. Drizou, Effect of simultaneous application of heat and pressure on the survival of bacterial spores, *J. Appl. Bacteriol. 71*:285 (1991).
76. A. Nakayama, Y. Yano, S. Kobayashi, M. Ishikawa, and K. Sakai, Comparison of pressure resistance of spores of six *Bacillus* strains with their heat resistances, *Appl. Environ. Microbiol. 62*:3897 (1996).
77. G. D. Alemán, D. F. Farkas, J. A. Torres, E. Wilhelmsen, and S. Mcintyre, Ultra-high pressure pasteurization of fresh cut pineapple, *J. Food Prot. 57*:931 (1994).

78. I. Hayakawa, T. Kanno, K. Yoshiyama, and Y. Fujio, Oscillatory compared with continuous high pressure sterilization on *Bacillus stearothermophilus* spores, *J. Food Sci. 59*:164 (1994).
79. R. V. Lechowich, Food safety implications of high hydrostatic pressure as a food processing method, *Food Technol. 47*(6):170 (1993).
80. P. Butz, S. Funtenberger, T. Haberdtzl, and B. Tauscher, High pressure inactivation of *Byssochlamys nivea* ascospores and other heat resistant moulds, *Lebensm. Wiss. Technol. 29*:404 (1996).
81. S. Shimada, M. Andou, N. Naitto, N. Yamada, M. Osumi, and R. Hayashi, Effects of hydrostatic pressure on the ultrastructure and leakage of internal substances in the yeast *Saccharomyces cerevisiae*, *Appl. Microbiol. Biotechnol. 40*:123 (1993).
82. J. P. Smelt, Some mechanistic aspects of inactivation of bacteria by high pressure, Proceedings of European Symposium—Effects of High Pressure on Foods, University of Montpellier, 1995.
83. J. P. Perrier-Cornet, P. A. Marénchal, and P. Gervais, A new design intended to relate high pressure treatment to yeast cell mass transfer, *J. Biotechnol. 41*:49 (1995).
84. R. Jaenicke, Enzymes under extreme conditions, *Ann. Rev. Biophys. Bioeng. 10*:1 (1981).
85. R. Y. Morita, Effect of hydrostatic pressure on succinic, malic and formic dehydrogenases in *Escherichia coli*, *J. Bacteriol. 74*:251 (1957).
86. B. M. Mackey, K. Forestiere, N. S. Isaacs, R. Stenning, and B. Brooker, The effect of high hydrostatic pressure on *Salmonella thompson* and *Listeria monocytogenes* examined by electron microscopy, *Lett. Appl. Microbiol. 19*:429 (1994).
87. M. Sato, H. Kobori, S. A. Ishijima, Z. H. Feng, K. Hamada, S. Shimada, and M. Osumi, *Schizosaccharomyces pombe* is more sensitive to pressure stress than *Saccharomyces cerevisiae*, *Cell Struct. Func. 21*:167 (1996).
88. P. Masson, Pressure denaturation of proteins, *High Pressure and Biotechnology* (C. Balny, R. Hayashi, K. Heremans, and P. Masson, eds.), Colloque INSERM, Vol. 224, John Libbey Eurotext, Montrouge, France, 1992, p. 89.
89. J. M. Thevelein, J. A. Van Assche, K. Heremans, and S. Y. Gerlsma, Gelatinization temperature of starch, as influenced by high pressure, *Carbohydrate Res. 93*:304 (1981).
90. P. Groβ and H. Ludwig, Pressure-temperature phase diagram for the stability of bacteriophage T4, *High Pressure and Biotechnology* (C. Balny, R. Hayashi, K. Heremans, and P. Masson, eds.), Colloque INSERM,Vol. 224, John Libbey Eurotext, Montrouge, France, 1992, p. 57.
91. T. Ohshima, T. Nakagawa, and C. Koizumi, *Seafood Science and Technology* (E. G. Bligh, ed.), Fishing News Books, Oxford, United Kingdom, 1992, p. 64.
92. T. Ohshima, H. Ushio, and C. Koizumi, High pressure processing of fish and fish products. *Trends Food Sci. Technol. 4*:370 (1993).
93. P. B. Cheah and D. A. Ledward, High pressure effects on lipid oxidation, *JAOCS 72*:1059 (1995).
94. P. B. Cheah and D. A. Ledward, High pressure effects on lipid oxidation in minced pork, *Meat Sci. 43*:123 (1996).
95. T. Tamaoka, N. Itoh, and R. Hayashi, High pressure effect on Maillard reaction, *Agric. Biol. Chem. 55*:2071 (1991).
96. V. M. Hill, D. A. Ledward, and J. M. Ames, Influence of high hydrostatic pressure and pH on the rate of Maillard browning in a glucose-lysine system, *J. Agric. Food Chem. 44*:494 (1996).
97. K. Hayashi, S. Takahashi, H. Asano, and R. Hayashi, *Pressure-Processed Food* (R. Hayashi, ed.), San-Ei Shuppan Co., Kyoto, 1990, p. 277.
98. C. Suzuki and K. Suzuki, The gelation of ovalbumin solutions by high pressure, *Arch. Biochem. Biophys. 102*(3):367 (1963).
99. D. E. Johnston, High pressure effects on milk and meat, *High Pressure Processing of Foods* (D. A. Ledward, D. E. Johnston, R. G. Earnshaw, and A. P. M. Hasting, eds.), Nottingham University Press, Nottingham, 1995, p. 99.
100. L. Heremans and K. Heremans, Raman spectroscopic study of the changes in secondary structure of chymotrypsin: effect of pH and pressure on the salt bridge, *Biochem. Biophys. Acta 999*:192 (1989).
101. N. Homma, Y. Ikeuchi and A. Suzuki, Effect of high pressure treatment on the proteolytic enzymes in meat, *Meat Sci. 38*:219 (1994).

102. L. B. Kurth, Effect of pressure-heat treatments on cathepsin B1 activity, *J. Food Sci. 51*:663 (1986).
103. I. N. A. Ashie, B. K. Simpson, and H. S. Ramaswamy, Control of endogenous enzyme activity in fish muscle by inhibitors and hydrostatic pressure using RSM, *J. Food Sci. 61*: 350 (1996).
104. M. N. Eshtiaghi and D. Knorr, Potato cubes response to water blanching and high hydrostatic pressure, *J. Food Sci. 58*:1371 (1993).
105. G. B. Quaglia, R. Gravina, R. Paperi, and F. Paoletti, Effect of high pressure treatments on peroxidase activity, ascorbic acid content and texture in green peas, *Lebensm. Wiss. Technol. 29*:552 (1996).
106. M. Anese, M. C. Nicoli, G. Dall'Aglio, and C. R. Lerici, Effect of high pressure treatments on peroxidase and polyphenoloxidase activities, *J. Food Biochem. 18*:285 (1995).
107. C. Balny and P. Masson, Effects of high pressure on proteins, *Food Rev. Int. 9*:611 (1993).
108. I. Seyderhelm, S. Bouguslawski, G. Michaelis, and D. Knorr, Pressure induced inactivation of selected enzymes, *J. Food Sci. 61*:308 (1996).
109. P. Butz, W. D. Koller, B. Tauscher, and S. Wolf, Ultra-high pressure processing of onions: chemical and sensory changes, *Lebensm. Wiss. Technol. 27*:463 (1994).
110. M. Asaka and R. Hayashi, Activation of polyphenol oxidase in pear fruits by high pressure treatment, *Agric. Biol. Chem. 55*:2439 (1991).
111. M. R. A. Gomes and D. A. Ledward, Effect of high-pressure treatment on the activity of some polyphenoloxidases, *Food Chem. 56*:1 (1996).
112. A. Ibarz, E. Sangronis, G. Barbosa-Cánovas, and B. G. Swanson, Inhibition of polyphenoloxidase in apple slices during high hydrostatic pressure treatments, 1996 IFT Annual Book of Abstracts, New Orleans, 1996, p. 100.
113. M. P. Cano, A. Hernández, and B. De Ancos, High pressure and temperature effects on enzyme inactivation in strawberry and orange products, *J. Food Sci. 62*:85 (1997).
114. J. J. Macfarlane, Pre-rigor pressurization of muscle: effects on pH, shear value and taste panel assessment, *J. Food Sci. 38*:294 (1973).
115. E. Dumay, C. Lambert, S. Funtenberger, and J. C. Cheftel, Effects of high pressure on the physico-chemical characteristic of dairy creams and model oil/water emulsions, *Lebensm. Wiss. Technol. 29*:606 (1996).
116. W. Buchheim and A. M. Abou El Nour, Induction of milk fat crystallization in the emulsified state by high hydrostatic pressure, *Fat Sci. Technol. 10*:369 (1992).
117. S. Nagatsuji, The fat of the land under pressure, *Look Japan* (Oct):28 (1992).
118. Y. N. Horie, K. I. Kimura, and M. S. Ida, Jams treated at high pressure, U.S. patent 5,075,124 (1991).
119. A. El Moueffak, C. Cruz, M. Antoine, M. Montury, G. Demazeau, A. Largeteau, B. Roy, and F. Zuber, High pressure and pasteurization effect on duck foie gras, *Int. J. Food Sci. Technol. 30*:737 (1995).
120. M. N. Estiaghi, R. Sute, and D. Knorr, High-pressure and freezing pretreatment effects on drying, rehydration, texture and color of green beans, carrots and potatoes, *Food Sci. 59*:1168 (1994).
121. K. Takahashi, Sterilisation of microorganisms by hydrostatic pressure at low temperatures, *High Pressure and Biotechnology* (C. Balny, R. Hayashi, K. Heremans, and P. Masson, ed.), Colloque INSERM, Vol. 224, John Libbey Eurotext, Montrouge, France, 1992, p. 297.
122. E. A. Elgasim and W. H. Kennick, Effect of pressurization of pre-rigor beef muscles on protein quality, *J. Food Sci. 45*:1122 (1980).
123. M. Okamoto, Y. Kawamura, and R. Hayashi, Application of high pressure to food processing: textural comparison of pressure- and heat-induced gels of food proteins, *Agric. Biol. Chem. 54*(1):183 (1990).
124. A. Ibarz, E. Sangronis, L. Ma, G. V. Barbosa-Cánovas, and B. G. Swanson, Viscoelastic properties of egg gels formed under high hydrostatic pressure, 1996 IFT Annual Book of Abstracts, New Orleans, 1996, p. 180.
125. R. Hayashi and A. Hayashida, Increased amylase digestibility of pressure-treated starch, *Agric. Biol. Chem. 53*:2543 (1989).

126. L. Popper and D. Knorr, Applications of high-pressure homogenization for food preservation, *Food Technol. 44*(7):84 (1990).
127. A. M. Papineau, D. G. Hoover, D. Knorr, and D. F. Farkas, Antimicrobial effect of water-soluble chitosana with high hydrostatic pressure, *Food Biotechnol. 5*:45 (1991).
128. K. J. A. Hauben, E. Y. Wuytack, C. C. F. Soontjens, and C. W. Michiels, High-pressure transient sensitization of *Escherichia coli* to lysozyme and nisin by disruption of outer-membrane permeability, *J. Food Prot. 59*:350 (1996).
129. H. Stapelfeldt, P. H. Petersen, K. R. Kristiansen, K. B. Qvist, and L. H. Skibsted, Effect of high hydrostatic pressure on the enzymic hydrolysis of β-lactoglobulin B by trypsin, thermolysin and pepsin, *J. Dairy Res. 63*:111 (1996).
130. S. Funtenberger, E. Dumay, and J. C. Cheftel, Pressure-induced aggregation of β-lactoglobulin in pH 7.0 buffers, *Lebensm. Wiss. Technol. 28*:410 (1995).
131. B. H. Hite, N. J. Giddings, and C. E. Weakly, The effects of pressure on certain microorganisms encountered in the preservation of fruits and vegetables, Bull. 146 W. Va. Univ. Agric. Exp. Sta., Morgantown, 1914, p. 1.
132. M. Capellas, M. Mor-Mur, E. Sendra, R. Pla, and B. Guamis, Populations of aerobic mesophils and inoculated *E. coli* during storage of fresh goat's milk cheese treated with high pressure, *J. Food Prot. 59*:582 (1996).
133. G. Deplace, Vessel design, *High Pressure Processing of Foods* (D. A. Ledward, D. E. Johnston, R. G. Earnshaw, and A. P. M. Hasting, eds.), Nottingham University Press, Nottingham, 1995, p. 137.

20

Surface Treatments and Edible Coatings in Food Preservation

ELIZABETH A. BALDWIN
Citrus and Subtropical Products Laboratory, Agricultural Research Service, U.S. Department of Agriculture, Winter Haven, Florida

Consumer interest in health, nutrition, and food safety combined with environmental concerns has renewed efforts in edible coating research. Renewable and abundant resources are available for use as film-forming agents that could potentially reduce the need for synthetic packaging films that add to waste-disposal problems. Alternatives to petroleum-based packaging include naturally occurring lipid, resin, protein, and carbohydrate film formers and their derivatives. In fact, coating techniques had been in use for decades, and even centuries, before the development of plastic polymers. For example, beeswax was used to coat citrus fruit to retard water loss in China during the twelfth and thirteenth centuries [1], and "larding" (coating food with fat) to prevent desiccation was practiced in sixteenth-century England [2]. The use of synthetic and natural waxes and resins to coat fresh fruits and vegetables has been researched and practiced in the United States, the United Kingdom, and Australia since the 1930s [3–9]. Development of edible coatings for use on meat products was first reported in the late 1950s [10–14].

Currently, edible coatings and films are commonly used on many commodities, such as candies, fresh fruits and vegetables, and processed meats. New research seeks to expand and improve coating technologies and materials to further enhance food stability and quality. Other surface treatments for foods include application of antioxidants, acidulants (or other pH-control agents), fungicides, preservatives, and mineral salts, some of which are more extensively covered in other chapters of this volume.

I. RATIONAL FOR USING EDIBLE COATING AND SURFACE TREATMENTS

Edible coatings serve many purposes in food systems. Coatings are used to improve appearance or texture and reduce water loss. Examples include the "waxing"of apples and oranges to add gloss and reduce shrinkage due to water loss or the coating of candies to reduce stickiness [15,16]. Use of antioxidants and sulfites to preserve fresh appearance in minimally processed fruits and vegetables or processed foods is well reviewed by Sapers [17] and Sherwin [18]. Antioxidants are used to reduce browning of cut apple and pear [19–22], potato [23,24], mushrooms [25,26] and shellfish [17,27,28] and to preserve the color of fish [29], which is more extensively reviewed in another chapter. Fungicides are used to reduce decay of whole fruits [30]; and salts, such as calcium, are reported to delay ripening, increase firmness, improve appearance, and enhance disease resistance of fruits [31–35]. Preservatives, acidulants, and to some extent antioxidants and sulfites reduce surface microbial populations on fresh-cut produce and processed foods [36,37]. Many of these treatments are covered in more detail in other chapters.

This review will concentrate on the use of edible coatings alone and as carriers of antioxidants [38–41], preservatives [41–46], acidulants [40,41,47], salts [34], and fungicides [48–55]. Coatings have been shown to increase the efficiency of preservatives [41,44], but fungicides sometimes have reduced activity in a coating system [51,52]. Edible coatings can also encapsulate flavor for preservation, storage, or controlled release in food systems [56]. Another advantage is the retention of flavor volatiles in coated fruit [57]. Coatings and films can slow deteriorative changes in coated products by reducing desiccation and oxidation and, in some cases, by creating a modified atmosphere around coated products [15,16]. Coatings can reduce the migration of lipids (fats and oils) in confectionary products [58] and fried foods [16,58,59] or separate components of different water activity [60,61]. Formulations can be designed to carry desired additives (including antioxidants, acidulants, chelators, preservatives, and fungicides) that help to extend product stability and, therefore, shelf life [62]. Edible films and coatings can also improve mechanical handling properties and structural integrity of various food products by helping fruit slip over packing lines with less injury [63] or by holding toppings in place during product distribution [64].

II. MECHANISM OF ACTION

A. Permeability Properties of Coatings

Permeability of films and coatings to water vapor, gas, solute, or lipids is an important property to consider when selecting film materials or for tailoring coatings for specific commodities. Permeability of a barrier is calculated from a combination of Ficks first law of diffusion and Henry's law of solubility. This is used to determine flux of a permeate through a nonporous barrier, assuming that the barrier has no imperfections. Permeance is a measure of flux without accounting for barrier thickness and used for performance evaluation of a film rather than describing its property. Transmission rate describes permeance without accounting for film thickness or the partial pressure gradient of the permeate. Resistance describes the ability of a material to serve as a barrier to the diffusion of a permeate. A detailed discussion of terms, equations, and theories of permeability is presented by Donhowe and Fennema [64] and McHugh and Krochta [65].

Permeability properties of edible films are often unpredictable due to the absence of a homogeneous structure and the often hydrophilic nature of most formulations [65]. The chemical composition and structure of the film-forming polymer affects film permeability in general. Highly polar materials with a high degree of hydrogen bonding exhibit low gas permeability, especially under conditions of low humidity, but are poor barriers to moisture. Nonpolar materials, such as lipids, provide good moisture barriers, but are permeable to gases such as oxygen. The type of functional group on a polymer can also have an effect, depending on the resulting chain interaction and motion and whether or not the functional group is hydrophilic or hydrophobic. Ionic functional groups create strong polymer chain interactions, which restrict chain motion. This usually results in good oxygen barriers, but also hydrogen bonding with water and subsequent water absorption at high relative humidities (RH), which, in turn, results in high rates of water vapor permeation. In addition, absorption of water disrupts intermolecular chain interaction, which increases permeability in general. This is the reason that films are often more permeable at high RHs [66]. Nonpolar groups result in a much less effective oxygen barrier film when present as the side chain but improve water permeability slightly.

Addition of low molecular weight components, or plasticizers, can affect film permeability and flexibility, often increasing both (especially water vapor permeability) by disruption of polymer chain hydrogen bonding [65,67]. These components are generally added to decrease film brittleness by increasing elasticity/flexibility, resulting in less cracking and flaking of coatings.

The structure of the film-forming polymer is also important in terms of influencing permeability properties of a film. Polymer chain packing, whether it is tight or loose due to bulky side chains, results in increased or decreased permeability properties, respectively [65]. Molecular weight and crystalline structure of a polymer can have an effect [67]. Lipids can exist in different crystalline states, which result in different barrier properties, with the higher degree of crystallinity resulting in lower permeability. Temperature affects polymer mobility [68] and thus permeability. Higher temperatures (above the "glass" transition state) result in polymers that are more mobile (plastic amorphous state) and have relatively increased permeability properties compared to when they exist as "glasses" or in brittle form at lower temperatures [65]. Even without going through a structural transition, oxygen transmission through protein films was affected by temperature [68]. Orientation of polymers to the flow of permeate can affect permeability properties. For example, the packed arrangement of wax crystals perpendicular to the direction of gas flow presents a better barrier [69] than when parallel to the direction of flow.

Cross-linking of polymer chains with ions or enzymes can lower permeability values as well as change the pH (depending on the isoelectric point in the case of protein films) [65,70]. The addition of hydrophobic materials (lipids) to a hydrophilic film former to make a composite coating can sometimes improve the moisture-barrier properties of the hydrophilic film former. This was demonstrated for a matrix of methylcellulose and hydroxypropyl methylcellulose combined with saturated C_{16} and C_{18} fatty acids laminated with beeswax and with a chitosan film containing lauric acid [71–73]. This can also be achieved by forming bilayer films from hydrophilic and hydrophobic materials. An example of this was reported for hydroxypropyl methylcellulose and a blend of stearic and palmitic acids [60,74].

1. Effect on Water Loss

Water loss usually occurs in the vapor phase. Water vapor permeability describes the movement of water vapor through a film or coating per unit area and thickness, and determines the vapor pressure difference across the film at a specific temperature and humidity [75]. If pores, cracks, or pinholes form on the film surface, then water vapor flows through these areas directly, which is different from the dissolving and diffusion of water vapor through a film barrier [65]. Water vapor transfer through films is dependent on environmental conditions such as temperature and humidity, and thus should be tested under the conditions expected to be encountered by a specific product. Generally, the more hydrophilic the film-forming material, the more permeable the film will be to water vapor.

2. Effect on Gas Exchange of Fresh Fruits and Vegetables

Creation of a Modified Atmosphere for Coated Fresh Produce and Effect on Ripening

Cells of plant tissues, such as harvested fruits and vegetables, are physiologically active in that they consume oxygen (O_2) and produce carbon dioxide (CO_2) as they respire. When fruits or vegetables are sealed in semi-permeable plastic packaging or coating, a modified atmosphere (MA) is created within the packaging—or in the internal atmosphere of the fruit, in the case of edible coatings—depending on the permeability of the film or coating. During storage, fruit respiration continues to consume O_2 and release CO_2 [15]. If O_2 levels fall too low (below 1–3%, depending on the produce and storage temperature), anaerobic reactions can occur, which result in off-flavors, abnormal ripening, and spoilage [76,77]. Climacteric-type fruit are often harvested immature and ripen off the mother plant with an accelerated respiration pattern and ethylene production [76]. The high rates of respiration and ethylene production, which turn on genes regulating ripening and senescence, contribute to a relatively short shelf life for this type of produce. Ethylene production, like respiration, is a process that requires O_2. Low O_2 (below 8%) and high CO_2 (above 5%) concentrations slow down respiration and retard ethylene production and, therefore, ripening [77]. High storage temperatures increase fruit or vegetable respiration [76] and exacerbate the effect of a coating or other packaging on the internal atmosphere of the coated produce. Low temperature, on the other hand, slows down fruit ethylene production and respiration, thus minimizing the effect of a film or coating in terms of modifying the atmosphere inside a fruit.

Retardation of Weight Loss and Surface Desiccation

Fruits and vegetables also lose water to the surrounding air in the form of water vapor in a process called transpiration. This entails the movement of water from fruit cells to the surrounding atmosphere following a gradient of high water concentration (~100% RH in fruit intercellular spaces or internal atmosphere) to low water concentration (% humidity of the storage environment). For this reason fresh produce is often stored under conditions of high RH (90–98%) to minimize water loss, subsequent weight loss, and shriveling [78]. Edible coatings can help retard this movement of water vapor, but become more permeable to water vapor and gases under conditions of high RH as explained above.

3. Effect on Stability of Lightly Processed Fruits and Vegetables

Light or minimal processing of fresh produce indicates cutting, slicing, coring, peeling, trimming, or sectioning of fruits and vegetables. Since fresh-cut produce, like intact products, have an active metabolism, the processing operations result in a series of chemical and biochemical reactions that can lead to deteriorative changes. These reactions include increased respiration, ethylene production, rapid senescence, undesirable color changes, flavor changes, synthesis of secondary metabolites, and increased microbial growth [79–81]. Many of these reactions are plant wound responses [82–84] due to injury incurred by the minimal processing. Methods used to extend the storage life of lightly processed products are application of antioxidants, acidulants, preservatives, mineral salts, and osmotic agents. Some applications can be made using an edible coating as a carrier of these active compounds that retard browning or discoloration, microbial growth, and softening, etc. In addition, coatings with the appropriate permeability properties and under certain conditions can create a modified atmosphere around the product and retard respiration, oxidation, and desiccation. Several reviews have been published on this subject [79–81,85]. Reduction of surface water activity can increase product stability. This can be achieved by infiltration of fruit slices or pieces with fruit juices, sucrose syrups, or glycerol with or without a coating made up of a suitable water-soluble polymeric material [79].

B. Structural Integrity and Appearance of Coated Products

Coatings on fruits and vegetables can act as lubricants to reduce surface injury, scarring, and chafing [1,86]. With less wounding of fruit, decay due to opportunist wound pathogens is lessened. In addition, the act of applying certain types of coatings reduces surface microbial populations [87]. For these reasons, waxed citrus fruit experience less decay compared to unwaxed fruit [88]. For food consisting of multiple components, a film can be used to secure the components to the product during marketing [64]. Waxes are also used to encase cheeses to prevent surface molding during the ripening and/or aging process [89]. Resins, zein protein, and microemulsions of waxes can impart a high gloss to the coated product [90,91]. Shellac, polyethylene, and carnauba wax microemulsions are used on fruits [15], and carnauba, shellac, and zein have been applied to candies and confectioneries as well [16,38,92]. Zein has been tested on tomato fruit, but thus far has not been used commercially on fruit [93]. Candelilla wax microemulsions impart a glossy appearance, especially when combined with gelatin protein [94]. Carbohydrate coatings, such as cellulose or pectin, result in an attractive nonsticky sheen when applied to products when dry, but often give an undesirable slippery texture when products become wet with condensation, as is often the case after removal from chilled storage. The polysaccharide film formers, however, do not result in the high gloss finish obtained with shellac, carnauba wax, or zein coatings.

III. MATERIALS USED IN EDIBLE COATING AND FILM FORMULATIONS

Many and diverse materials are used in coating or film formulations. Descriptions of the most common main ingredients or film formers are given below. The United States is generally considered the leader in worldwide food regulation, thus, when possible, approval rating based on the U.S. Food and Drug Administration Federal Code of Regulations [95] is also given. Generally recognized as safe (GRAS) status covers direct food additives for their intended use at a quantity not to exceed the amount reasonably required to accomplish the intended physical, nutritional, or other technical effect in food; that are of appropriate food grade; and used with good manufacturing practices (GMP) (FDA, 21 CFR, 1996). The GRAS status is shown without differentiating between initial GRAS status (FDA 21 CFR, Part 182) and reaffirmed as safe with minor restrictions and within specified ranges and uses or purposes and under conditions prescribed (Part 184). Part 180 refers to food additives permitted in food on an interim basis or in contact with food pending additional study; part 172 contains other direct food additives that are not GRAS including food preservatives, coatings, and films, and special dietary and nutritional additives; Part 173 contains secondary direct food additives including polymer substances and adjuvants for food treatment, enzyme preparations, and specific usage additives; and Part 175 contains indirect food additives including adhesives and components of coatings (FDA 21 CFR, 1996).

A. Lipids

Lipids include a group of hydrophobic compounds, which are neutral esters of glycerol and fatty acids. They also include "waxes," which are esters of long-chain monohydric alcohols and fatty acids [69]. A list of lipid components commonly used in coatings, along with their status according to U.S. FDA 21 CFR [95], is given in Table 1. Fatty acids and alcohols lack structural integrity and durability in their free form to be good film formers. Due to the fragile nature of these compounds, lipids are often incorporated into a structural matrix of some other compound such as a polysaccharide [64]. Lipid components are, therefore, incorporated in composite coatings made up of a least two materials. The supporting matrix, if made up of hydrophilic polymers, may affect film resistance to water vapor transmission [96]. Generally, oils are not as resistant to gases and water vapor transfer as are the solid-state waxes [60,74]. Stearyl alcohol was the most resistant to O_2 transmission, probably due to its ability to crystallize as overlapping platelets with an orientation perpendicular to the direction of gas flow [97,98]. Generally, coatings that include lipid solids up to 75% can be used to improve coating performance without diminishing moisture-barrier properties, but below 25% solids, permeability increase was observed under test conditions [99].

1. Oils

Paraffin oil, mineral oil, castor oil, rapeseed oil, acetylated monoglycerides, and vegetable oils (peanut, corn, soy) have been used alone or in combination with other ingredients to coat food products. White mineral oil is a petroleum-based product, being a mixture of liquid paraffinic and naphthenic hydrocarbons. It is approved for use as a food-release agent and as a protective coating for fresh fruits and vegetables [69]. Fatty acids and polyglycerides are derived from vegetable or talo oils and are considered GRAS [95]. They are commonly used with glycerides as emulsifiers. Among several oils tested, paraffin oil had the greatest resistance to water followed by vegetable oil and light mineral oil [15]. Acetoglycerides are synthetic fats where acetic acid is substituted for a portion of a naturally occurring fat or oil resulting in mono- or diacetotriglycerides or combinations of these compounds. Although acetic acid is a fatty acid, it does not occur as a glyceride in natural fats. Its water vapor permeability was found to decrease with increasing acetylation [100,101]. Acetoglycerides are highly flexible in polymorphic form and stable in crystalline form. These modified fats have use as protective coatings and as plasticizers [102]. Most oils used in coatings are considered direct food additives with varying restrictions according to the U.S. FDA [95]. Paraffin and mineral oil are allowed as release agents and lubricants, defoamers, or components of coating, and castor oil is approved as a release agent and component of coatings. Vegetable oils are generally considered GRAS, while acetylated monoglycerides are considered food additives with few restrictions other than that they be made from edible fat [95].

2. Waxes

Paraffin, carnauba, beeswax, candelilla, and polyethylene waxes have been used to coat food products, alone or in combination with other ingredients. Paraffin wax is a distillate fraction of crude petroleum [103]. Synthetic paraffin wax is formed from the catalytic polymerization of ethylene and

Table 1 Common Lipid Components of Coatings

Lipid	Classification	U.S. FDA 21 CFR #
Oils		
Acetylated monoglyceride	Removable hot-melt strippable coating	172.828, 175.230
Castor oil	Component of coatings—candy, tablets	172.876
Fatty acids from edible sources: capric, lauric, myristic, oleic, pamitic, stearic;	Release agent; lubricant; protective coating for raw fruits and vegetables	172.860
Lard oil	GRAS, edible oil	182.70
Mineral oil	Removable hot-melt strippable coating	175.230
Peanut oil	GRAS, edible oil	182.70
Rapeseed oil	GRAS, emulsifier, stabilizer	184.1555
Salts of fatty acids	Binder, emulsifier, anticaking agent	172.863
Soy oil	GRAS, edible oil	182.70
Synthetic isoparaffinic petroleum hydrocarbons	Component of coatings on fruits and vegetables	172.884
Tallow	GRAS, edible oil	182.70
White mineral oil	Component hot melt coating for frozen meat acetylated monoglyceride	172.878
Waxes		
Beeswax	GRAS, surface finishing agent	184.1973
Candelilla	GRAS, lubricant, surface finishing agent	184.1976
Carnauba	GRAS, lubricant, surface finishing agent	184.1978
Paraffin wax	Component of coating	175.250, 175.300
Petroleum wax	Component of microcapsules for flavorings, defoamer, protective coating for cheese, and raw fruits and vegetables	172.886

GRAS = Generally recognized as safe by U.S. Food and Drug Administration.

is allowed for food use in the United States. It is used as a protective coating for raw fruits, vegetables, and cheese, as a chewing gum base and defoamer, and as a component in the microencapsulation of flavorings [69]. Carnauba wax is the exudate of palm tree leaves from the tree of life (*Copernica cerfera*), found in Brazil. Beeswax or "white wax" is secreted by honeybees, and candelilla wax is an exudate of the candelilla plant (*Euphorbia antisphilitica*), which is a reedlike plant that grows in Mexico and southern Texas. These natural waxes are considered GRAS [95] in the United States and are used in chewing gum, hard candy, and edible coatings. Polyethylene wax is oxidized polyethylene or the basic resin produced by the mild oxidation of polyethylene, a petroleum-based product. This substance is allowed for use in edible coatings for fruits and nuts where the peel or shell is not normally ingested [69,95,105]. Coatings made with this wax are more permeable to gases than shellac and most other polymers [105].

3. Emulsions

Emulsion coatings are oil or wax dispersed in water or some other hydrophilic solution. A macroemulsion has dispersed wax particle sizes ranging from 2×10^3 to 1×10^5 Å, and microemulsions from 1×10^3 to 2×10^3 Å. Melted wax disperses in water in a manner similar to oil [106]. Carnauba wax and beeswax form stable microemulsions with the appropriate emulsifiers, forming a glossy coating, while macroemulsions generally impart little shine to the coated product [69].

B. Resins

Resins are a group of acidic substances that are produced and secreted as a wound response by specialized plant cells of trees and shrubs. Synthetic resins are petroleum-based products [69]. A list or resins commonly used in coatings, along with their status according to 4% U.S. FDA 21 CFR [95], is shown in Table 2.

1. Shellac

Shellac resin is secreted by the insect *Laccifer lacca* found in India. Shellac is composed of aleuritic and shelloic acids [107] is compatible with waxes, and gives coated products a high gloss appearance. This compound is permitted as an indirect food additive (FDA 21 CFR 175.300: resinous and polymeric coatings for food contact surfaces), but is nevertheless commonly used in coatings for fresh fruits and candies where the coated surface is consumed. Apparently this is allowed because a petition has been submitted for GRAS status [69]. Shellac and other resins have relatively low permeability to gases and moderate permeability to water vapor [105,108].

2. Wood Rosin

Wood rosin is manufactured from oleoresins of pine trees. The rosin is the residue left after distillation of volatiles from the crude resin [69]. It is less expensive than shellac, and specific esters of maleic anhydride–modified wood rosin are approved for use in coatings for citrus. Similarly, coumarone-

TABLE 2 Common Resin Components of Coatings

Resins	Classification	U.S. FDA 21 CFR #
Copal	Resinous and polymeric coatings	175.300
Coumerone indene	Resinous and polymeric coatings	172.215
Damar	Resinous and polymeric coatings	175.300
Elemi	Resinous and polymeric coatings	175.300
Shellac	Resinous and polymeric coatings	175.300
Terpene	Moisture barrier on soft gelatin capsules or powders of ascorbic acid	172.280
Wood rosin	Coatings on fresh citrus fruit	175.300

indene resin is approved for use specifically on citrus fruit. Coumarone-indene resin is a petroleum and/or coal tar by-product that is available in several grades. It is most often used as a component of the "solvent waxes" for citrus [69,109]. These so-called waxes consist mostly of resins with little or no actual wax component and small amounts of petroleum solvent (U.S. FDA 21 CFR 172.250), ethanol (U.S. FDA 21 CFR 184.1293), or isopropanol (U.S. FDA 21 CFR 173.240, 173.340) [3,69,95]. They have low viscosity and rapid drying rates. Resin solution waxes contain a small amount of lipid (usually oleic acid), morpholine (U.S. FDA 21 CFR 172.235) or potassium or ammonium hydroxide (U.S. FDA 21CFR 184.1631 and 184.1139, respectively), and other ingredients [95,110,111].

3. Other Resins

Terpene resin is obtained from polymerization of terpene hydrocarbons derived from wood and is approved as a direct food additive [95]. It is allowed for use as a moisture barrier in soft gelatin capsules. Other resins allowed only for food contact include copal, damar, and elemi, which are used in the pharmaceutical industry [69,95].

C. Proteins

Proteins have been used for their film-forming abilities for nonfood applications since ancient times as a component of glue, paint leather finishes, paper coatings, and inks. More recently protein materials, such as the milk protein casein and corn protein zein, have been used as edible coatings for extruded meats as well as nuts and confectionery items, respectively [112]. A list of proteins commonly used in coatings and films, along with their status according to the U.S. FDA 21 CFR [95], is given in Table 3. Film-forming proteins of plant origin include corn zein, wheat gluten, soy protein, peanut protein, and cottonseed protein, of which all but the latter are considered GRAS [95]. Keratin, collagen, gelatin, casein, and milk whey proteins are film formers derived from animal sources, of which casein and whey proteins are GRAS [95]. Adjustment of protein film pH can alter film formation and permeability properties, as was demonstrated for soy protein, wheat gluten [113,114],

Table 3 Commonly Used Protein Materials in Coatings

Protein Materials	Classification	U.S. FDA 21 CFR #
Casein/Sodium caseinate	GRAS, GMP	182.90, 182.1748
Collagen		
Cottonseed (modified products)	Food additive	172.894
Gelatin	Microapsules for flavorings (succinylated gelatin	172.230
Fish protein concentrate	Food suppliment	172.385
Fish protein isolate	Food suppliment	172.340
Keratin		
Peptones	GRAS, Nutrient suppliments	184.1553
Soy protein isolate	Migrating to food from paper products	182.90
Wheat gluten	GRAS, stabilizer, thickener, surface finishing agent	184.1322
Whey	GRAS, GMP	184.1979
Zein	GRAS, surface finishing agent	184.1984

GRAS = Generally recognized as safe by the U.S. Food and Drug Administration; GMP = good manufacturing practices.

and casein [70]. Most protein films are hydrophilic and, therefore, do not present good barriers to moisture. However, dry protein films such as zein, wheat gluten, and soy present relatively low permeabilities to O_2 [113].

1. Milk Proteins

Milk protein products include casein (80% of total milk protein) and whey (20% of total milk protein) and combinations of both [65,115]. They can result in films of different properties depending on the commercial source and method of extraction [70,115].

Casein

Caseins are soluble in aqueous solutions and form flexible colorless films. The addition of lipid compounds and adjustment of pH reduced the water vapor permeability of casein films [116], while the addition of whole milk, sodium caseinate, and nonfat dry milk or whey into polysaccharide films decreased the water vapor permeability of these hydrophilic films [117]. Lactic acid–treated casein films retained more sorbic acid preservative, improving the microbial stability of dehydrated apricot and papaya in intermediate-moisture food test systems [118].

Whey

Whey proteins produce films similar to those produced from caseinates. Heating is required to form intermolecular disulfide bonds, which produces films that are water insoluble and brittle, requiring plasticizers [112,115].

2. Collagen and Gelatin

Collagen is the major component of skin, tendon, and connective tissues in animals [112]. This material is partially digested with acid or enzymes to produce edible collagen casings. Collagen casings for meat products was one of the first examples of edible film application in modern times. Gelatin is formed from the partial hydrolysis of collagen [112] and is also allowed as a component of microcapsules for flavorings and for soft capsules in the pharmaceutical industry [95]. It is soluble in aqueous solutions, forming a flexible, clear, oxygen permeable film.

3. Wheat Gluten

The gluten complex is a combination of gliadin and glutenin polypeptides with some lipid and carbohydrate components [112,119]. It is soluble in aqueous alcohol, but alkaline or acidic conditions are required for the formation of homogeneous film-forming solutions [113]. This material also requires plasticizers to increase flexibility, for the films are excessively brittle [112,119]. These films have high water permeability, but are good barriers to O_2 and CO_2 [119].

4. Corn Zein

Zein is a prolamine derived from corn gluten and is insoluble in water except at very low or high pH. This is due to its high content of nonpolar amino acids. It is soluble in aqueous alcohol and dries to a glossy grease-resistant surface. The film is, however, brittle, and plasticizers are required to add plasticity [112,119]. It has been used as a substitute for shellac because of its high-gloss appearance, faster drying rate, and increased stability during storage [119].

5. Soy Protein

Soy protein is available as concentrate (70% protein) or isolate (90% protein). Film formation is enhanced by heating, which partially denatures the protein, allowing formation of disulfide bonds. This was shown to lower water vapor permeability. Enzymatic digestion can increase cross-linking [120]. The pH must be adjusted away from the isoelectric point of the soy protein (~4.6) in order to form films. In Asia, films are formed from heated soy milk and are used for wrapping food products [112].

6. Peanut Protein

Peanut protein films can be formed by two methods. The first is surface film formation, using protein/lipid solutions derived from roasted peanut, partially defatted peanut flour, and protein concentrate. This produces films with rough surface texture and poor mechanical properties. Films can also be produced by deposition method at pH 9.5 from peanut protein concentrate. This method showed promise for development as an edible film [121].

Zein, casein, and soy proteins have been used on confectionery, fruit, and vegetable products as well as on eggs. Self-supporting sheets of edible proteins have been developed that dissolve in water for microencapsulation of flavorings. Casein, soy [123], peanut, corn protein, and collagen have been used to form free-standing films or sheets for wrapping of foodstuffs [112]. Films made of proteins can add a nutritional component to coated foods especially if formulated to include diet-limiting amino acids [112,119]. Allergies to food proteins can be a concern when developing coatings and films from these materials. Gluten intolerance and allergic reactions to milk proteins (casein and whey) and associated lactose is common and may require labeling.

D. Carbohydrates

Polysaccharides are used in food systems as thickeners, stabilizers, gelling agents, and emulsifiers [125]. They comprise an abundant and renewable resource of hydrophilic film-forming agents with a wide range of viscosities, relatively low permeability to gases, but little resistance to water vapor transfer. A list of commonly used polysaccharides in coatings and films, along with their status according to the U.S. FDA 21 CFR [95], is given in Table 4.

1. Cellulose

Cellulose is the most abundant polysaccharide on the planet, being a major component of plant cell walls. Cellulose is made up of repeating glucose units in β-1,4 linkage. In its natural form, cellulose is not soluble in water, but derivatized forms such as sodium carboxymethyl-, methyl-, hydroxypropyl- and hydroxypropyl methylcellulose (CMC, MC, HPC and HPMC, respectively) are more soluble [102,125]. These derivatives have different permeabilities to water vapor and gases and are good film formers [102,106]. The degree of substitution and type of substitution (ionic and nonionic) for functional groups and polymer chain length affect permeability, solubility, and viscometric properties [124,125]. The derivatives CMC and MC are GRAS, while HPC and HPMC are approved as direct food additives for the purpose of film former, stabilizer, thickener, and suspending agent [95]. The latter two derivatives are not permitted for food use in all countries. However, several commercial coatings were developed from cellulose polymers including TAL Pro-long (Coutaulds Group, London), Semperfresh (United Agriproducts, Greeley, CO), and Nature Seal (EcoScience Corp., Orlando, FL). Another cellulose product is called cellulon fiber, which is bacterial cellulose produced by aerobic fermentation of glucose by a strain of *Acetobacter*. It has a fine fiber structure physically but is not chemically different from plant cellulose. It has been applied to surimi to aid in the binding of water [127], but has no reported uses as a coating or surface agent.

2. Pectin

Pectins are a complex mixture of polysaccharides that are also components of plant cell walls [129]. They are commercially obtained from citrus peel and apple pomace [124]. These polymers are mainly long chains of α-1,4-linked galacturonic acid units with varying degrees of esterification with methyl groups. The degree of esterification (DE) affects solubility and gelation properties; pectins with DE above 50% are labeled high-methoxyl and below 50% DE, low-methoxyl pectins [102,124]. As with cellulose polymers, the chain length also affects solubility and viscosity. When used as a film former in coatings, this polymer gives a somewhat glossy, nonsticky surface and LM pectins can be cross-linked with calcium ions to form gels [102]. Coatings made with pectin materials generally have high water vapor transmission rates [130] due to their hydrophilic nature [131], which can be improved by the addition of paraffin or beeswax. The tensile strength of pectinic acid films increases

TABLE 4 Commonly Used Polysaccharides in Coatings

Polysaccharides	Classification	U.S. FDA 21 CFR #
Agar	GRAS, drying and flavoring agent, stabilizer, thickener, surface finisher	184.1115
Alginate	GRAS, emulsifier, stabilizer, thickener	184.1011
Carageenan	Emulsifier, stabilizer, thickener, gelling agent	172.620
Salts of carrageenan	Emulsifier, stabilizer, thickener	172.626
Chitosan	Approved in Canada	
Dextrin	GRAS, formulation aid, processing aid, stabilizer, thickener	184.1277
Ethyl cellulose	Binder, filler, component of protective coatings for vitamin and mineral tablets	172.868
Furcelleran	Emulsifier, stabilizer, thickener	172.655
Salts of furcelleran	Emulsifier, stabilizer, thickener	172.660
Gellan gum	Stabilizer, thickener	172.665
Gum arabic (acacia gum)	GRAS, emulsifier, formulation aid	184.1330
Gum ghatti	GRAS, emulsifier	184.1333
Gum karaya	GRAS, formulation aid, stabilizer, thickener	184.1349
Gum tragacanth	GRAS, emulsifier, formulation aid	184.1351
Locust bean gum	GRAS, stabilizer, thickener	184.1343
Guar gum	GRAS, emulsifier, formulation aid, firming agent	184.1339
Hydroxypropyl cellulose	Emulsifier, film former, protective colloid, thickener	172.870
Hydroxypropyl methylcellulose	Emulsifier, film former, protective colloid, thickener	172.874
Methylcellulose	GRAS, GMP	182.1480
Methyl ethyl cellulose	Aerating, emulsifying, or foaming agent	172.872
Modified starch	Food additive	172.892
Pectins	GRAS, GMP	184.1588
Sodium carboxymethyl cellulose	GRAS, GMP	182.1745
Xanthan gum	Stabilizer, emulsifier, thickener, suspending agent	172.695

GRAS = Generally recognizerd as safe by the U.S. Food and Drug Administration;
GMP = good manufacturing practices.

with a decrease in methoxyl content [132] because the removal of ester groups leads to increased cross-bonding between residual carboxyl groups. Miers et al. [132] reported that pectin coatings were of acceptable strength with methoxyl contents of 4% or less and intrinsic viscosities of 3.5 or above. Pectins are generally considered GRAS [95].

3. Chitin

Next to cellulose, chitin is the second-most abundant polysaccharide on the planet, being a component of fungal and green algae cell walls and the skeletal substance of invertebrates [133]. It is a β-1,4-linked polymer of 2-acetamido-2-deoxy-D-glucan [129]. Partial deacetylation of chitin results in chitosan, which has been shown to induce plant-defense responses and inhibit growth of fungi [134,135]. Use of this polymer as film former and natural preservative resulted in the commercial coating Nutri-Save (Nova Chem, Halifax, NS, Canada). Methylation of the polymer resulted in a twofold resistance to CO_2 [15,136]. This allows it to retard ripening of climacteric fruits [137]. It, however, has relatively low resistance to water vapor transfer compared to lipid materials. Chitosan has not yet received approval for food use in the United States, but it is approved in Canada.

4. Starch

Amylaceous materials (amylose, amylopectin, and derivatives) have also been used to make coatings. These films have been reported to be semipermeable to CO_2 but highly resistant to O_2 [138]. Most starch consists of 25% amylose and 75% amylopectin, with one notable exception being hybrid corn, which contains 50–80% amylose. Amylose is a polymer of α-1,4-linked glucose, and amylopectin consists of an amylose backbone with side chains of α-1,6-linked glucose. Of the two polymers, amylose is a better film former and amylopectin is more useful as a thickening agent. Some derivatives, such as an hydroxypropyl amylose, showed low permeability to O_2 and improved water solubility [102] increased elongation properties, but no resistance to water vapor. Dextrins (partially hydrolyzed starch molecules, i.e., reduced in size as measured by dextrin equivalent or DE) are used as film formers, encapsulating agents, and flavor carriers. Coatings made from such polymers have lower permeability to water vapor compared to starch films [102,139] and may have resistance to O_2 [140]. Carboxylated dextrins are used as encapsulating agents [129]. Raw starch and dextrin products are considered GRAS, while modified starch products (modified by acid, bleach, esterification, or oxidized using chlorine) are approved as direct food additives [95].

5. Seaweed and Gum Polymers

Seaweed Products

Seaweed products such as carrageenan, alginates, and agar make good film formers or gels. Carrageenan consists of sulfate esters of 3,6-anhydro-α-*D*-galactopyranosyl units [124,141], alginates are salts of alginic acid (1,4-polyuronic acid with poly-β-*D*-mannopyranosylurinic and poly-α-*L*-gulopyranosyluronic acid blocks) [102,124,129,142], and agar is made up of β-1,4-*D*-galactopyranosyl linked to 3,6-anhydro-α-*L*-galactopyranosyl, partially esterified with sulfuric acid. Of these three, agar is more for formation of gels (currently used as a culture medium) and the other two polymers as gel and film formers. Alginate gels are relatively heat stable [124]. Carrageenan is approved as a direct food additive as an emulsifier, stabilizer, and thickener, while agar is GRAS [95]. In Japan there is a commercial carrageenan-based coating called Soageena (Mitsubishi International Corp., Tokyo).

Gum Products

Certain gum products are exudates from plants found mostly in Africa and Asia produced in response to injury [124,139,143], some of which are seed and fermentation products. Gums are complex heteropolysaccharides including gum arabic or acacia gum from the tree *Acacia senegal* and related species (D-galactopyranosyl, L-rhamnopyranosyl, L-arabinopyranosyl, L-arabinofuranosyl, and D-glucopyranosyluronic acid units with calcium, magnesium, and potassium ions) [144,145]. Gum arabic forms an aqueous solution of low viscosity and can form stable emulsions with most oils [129]. It is used in the confectionery industry as a stabilizer, adhesive, and flavor fixative. It has also been used to coat pecans [146] and is a good emulsifier [124].

Other less-used gums include gum tragacanth, whose film-forming properties are more useful in nonfood products such as hair and hand lotions and creams [129]; gum karaya, which forms smooth films that require plasticizer; locust bean gum, whose film-forming properties have been used in the textile industry as a finishing agent and as a common component of cosmetics, sauces, and salad dressings; guar gum, which is also used as a film former in the textile industry (both locust bean and guar gum are galactomannans) [124]; xanthan gum, a fermentation product with a cellulosic backbone [124], used as a thickener in sauces, gravies, frostings, fruit gels, and coatings to prevent moisture migration during frying [129] and in salad dressings with propylene glycol alginate [147]; and gellan gum, another fermentation product developed and patented by Kelco [148], which is also used in glazes, icings, and jams/jellies [129]. Of these gums, gum arabic (acacia gum), gum ghatti, guar gum, and locust bean gum (carob gum) are considered GRAS, while xanthan and gellan gum are approved as direct food additives for use in glazes, icings, frostings, jams, jellies and as stabilizer, emulsifier, thickener, etc. [95].

IV. ADDITIVES AS INDIVIDUAL TREATMENTS OR IN COATING FORMULATIONS

Materials other than film formers are added to edible coatings for basically two reasons. One is to improve the structural, mechanical, or handling properties of a coating. The other is to improve the quality, flavor, color, or nutritional properties of the coated product [62]. In the latter case, the coating acts as a carrier of useful compounds that have a desired effect on the coated item.

A. Plasticizers, Emulsifiers, and Surfactants

1. Plasticizers

Plasticizers are usually low molecular weight compounds that impart increased strength and flexibility to coatings, but also increase coating permeability to water vapor and gases [64,102]. Commonly used plasticizers, along with their status according to the U.S. FDA 21 CFR [95], are listed in Table 5. Common plasticizers include polyols such as glycerol, sorbitol, mannitol, propylene glycol, and polyethylene glycol (molecular weight 200–9500). Sucrose, sucrose fatty acid esters, and acetylated monoglycerides also can be used as plasticizers. Of these, glycerol, sorbitol, and propylene glycol are considered GRAS [95].

2. Emulsifiers and Surfactants

Emulsifiers can be classified as surface-active agents or as macromolecular stabilizers. Macromolecular stabilizers are proteins, gums, and starches that stabilize emulsions [149]. Commonly used emulsifiers and surfactants, along with their status according to the U.S. FDA 21 CFR [95], are listed in Table 5. Surface-active agents reduce surface water activity and can effect the rate of moisture loss from a food when used as a coating. This was shown with glycerol monopalmitate and glycerol monostearate and other 16- to 18-carbon fatty alcohols [102]. Reduction of surface water activity at the water/oil interface helps to both form and stabilize emulsions, which is important for shelf life properties of emulsion coatings. The hydrophilic-lipophilic balance (HLB) of surfactants ranks these compounds according to their hydrophobic and hydrophilic portions, which has an effect on their performance as emulsifiers. For example, sodium lauryl sulfate is a very hydrophilic surfactant with a HLB value of 40. Generally, surfactants with low HLB values are effective for water-in-oil emulsions, and those with high HLB values are more useful for oil-in-water emulsions [69]. Some common emulsifiers are acetylated monoglyceride, lecithin (GRAS) and lecithin derivatives, ethylene glycol monostearate, glycerol monostearate, sorbitan fatty acid esters (TWEENS), and palm and corn

TABLE 5 Commonly Use Plasticizers, Emulsifiers, and Surfactants

Compounds	Classification	U.S. FDA 21 CFR #
Acetylated monoglycerides	Emulsifiers, component of coating	172.828
Corn oil	Edible oil	
Ethoxylated mono-, diglycerides	Emulsifiers	172.834
Glycerol	GRAS, GMP	182.1320
Glycerol monopalmitate (Ethoxylated mono- and diglycerides)	Emulsifier	172.834
Glycerol monostearate (Ethoxylated mono- and diglycerides)	Emulsifier	172.834
Hydroxylated lecithin	Emulsifier	172.814
Lecithin	GRAS, GMP	184.1400
Manitol	Permitted food additive, component of resinous and polymeric coatings	180.25, 175.300
Oleic acid	Lubriant, binder, defoaming agent	172.862
Palm oil	GRAS, cocoa butter substitute	184.1259
Polyethylene glycol	MW 2200-9500; coating, binder, plasticizer, lubricant	172.820

Polysorbate 60	Emulsifier, foaming agent	172.836
Polysorbate 65	Emulsifier	172.838
Polysorbate 80	Emulsifer, dispersing agent, surfactant, wetting agent	172.840
Propylene glycol	GRAS, solvent, thickener, component of resinous and polymeric coatings	184.1666, 175.300
Propylene glycol alginate	Component of coatings for citrus	172.212
Sodium lauryl sulfate	Emulsifier, whipping agent, surfactant, wetting agent	172.822
Sodium stearoyl lactylate	Surfactant, emulsifier, stabilizer	172.846
Sorbitan monooleate	Emulsifier for clarification of cane or beet sugar juice	173.110
Sorbitan monostearate	Emulsifier	172.842
Sorbitol	Component of resinous and polymeric coatings	175.300
Sucrose	GRAS, GMP	184.1854
Sucrose fatty acid esters	Emulsifiers, texturizers, components of protective coatings for fresh fruit	172.859

GRAS = Generally recognized as safe by the U.S. Food and Drug Administration;
GMP = good manufacturing practices.

oil (GRAS) [95]. Surfactants help coatings adhere to coated surfaces. Most natural waxes also have emulsifying properties since they are comprised of long-chain alcohols and esters [69].

B. Fungicides and Biocontrol Agents

1. Fungicides

Fresh fruits and vegetables are susceptible to a variety of postharvest decay types that can be reduced by treatment with fungicides with and without a coating or "wax" [150]. About 20 compounds have been developed and tested for use as postharvest pesticides over the last 30 years [62], but many have been banned or not reregistered in the United States and other nations. Use of fungicides applied in fruit coatings has been reported for citrus including benomyl, imazalil, and thiabendazole (TBZ), of which only the latter two are currently registered for use on citrus in the United States. These fungicides are applied in solvent or water waxes, but this results in reduced ability to inhibit mold growth compared to application as an aqueous suspension. It is thought that encapsulation of the fungicide in the wax is the reason for its reduced efficiency [53]. Use of fungicides in fruit coatings has been reported for stone fruits (methyl-1-(butylcarbamoyl)-2-benzimidazolecarbamate or benomyl) [49], papayas (TBZ) [54], strawberries 3-(3,5-dichlorophenyl)-*N*-(1-methylethyl)-2,4-dioxo-1-imidazolidinecarboxamide, iprodione or Roveral® [55], tomatoes (*N*-[(trichloromethyl)thio]-4-cyclohexene-1,2-dicarboximide or captan) [151], and apples (Roverol) [152], as well as captan and benomyl on raspberries [153].

2. Biological Control Agents

Antagonistic yeasts and bacteria have been shown to inhibit mold growth and thus prolong the shelf life of fresh fruits and vegetables [154,155]. The mechanisms of action are reported to be the production of an antibiotic compound, competition for nutrients at wound sites on fresh produce, direct interaction with the pathogen, and induction of host defense responses [155]. These compounds have been successfully applied in fruit coatings and were shown to delay spoilage of citrus fruit by this method [156,157]. Two commercial products approved and available on the U.S. market are Biosave® (EcoScience Corp., Orlando, FL), which contains an antagonist bacteria (*Pseudomonas syringae*), and Aspire® (Ecogen Corp. Langhorne, PA), which contains an antagonist yeast (*Candida oleophila*) for control of decay on apples and citrus fruits.

C. Preservatives

Chemical preservatives such as salt, nitrites, and sulfites have long been used to prolong the shelf life of food products [36]. Coatings can also act as carriers of antimicrobial agents for lightly processed and processed food products [62]. Commonly used preservatives, along with their U.S. FDA 21 CFR [95], are listed in Table 6.

1. Benzoates, Sorbates, and Other Short-Chain Organic Acids

Preservatives such as benzoic acid and benzoates are most effective at pH 2.5–4.0 with the undissociated form of benzoic acid being most effective (pK_a 4.2), rendering this preservative ineffective above pH 4.5 [36]. This preservative controls yeasts and molds more effectively than bacteria and is considered GRAS to a maximum of 0.1% in the United States and up to 0.15–0.25% in other countries [95]. Sorbic acid and sorbates are also most effective in the undissociated state against fungi and certain bacteria [36]. This preservative is also considered GRAS for most products in accordance with good manufacturing practices up to a level of 0.1% the United States [95]. Sorbates are permitted in all countries of the world for preservation of various food products in the range of 0.15–0.25%. Acetic, lactic, propionic, fumaric, and citric acids also can be used in coatings and contribute to antimicrobial activity. Use of coatings as carriers of preservatives such as benzoates and sorbates improved their performance when applied to cut fruit or cheese analogs. This may be due to prevention of diffusion of preservatives into the food tissue or the fact that more preservative is present on the cut surface

TABLE 6 Commonly Used Preservatives

Preservatives	Classification	U.S. FDA 21 CFR #
Acetic acid	GRAS, curing or pickling agent, or in food at levels not to exceed GMP	184.1005
Benzoic acid	GRAS, antimicrobial agent	184.1021
Calcium disodium EDTA	Preservative, color retention	172.120
Citric acid	GRAS, GMP	184.1033
Dehydroacetic acid	Preservative for cut or peeled squash	172.130
Disodium EDTA	Preservative, color retention	172.135
Fumaric acid	Nutritional additive	172.350
Lactic acid	GRAS, antimicrobial agent	184.1061
Methylparaben	GRAS, antimicrobial agent	184.1490
Natamycin	Mold inhibitor for sliced cheeses	172.155
Potassium sorbate	GRAS, GMP	182.3640
Propionic acid	GRAS, antimicrobial agent	184.1061
Propylparaben	GRAS, antimicrobial agent	184.1670
Sodium benzoate	GRAS, antimicrobial agent	184.1733
Sodium nitrate	Preservative, color fixative for fish and meat	172.170
Sodium nitrite	Preservative, color fixative for fish and meat	172.175
Sorbic acid	GRAS, GMP	182.3089

GRAS = Generally recognized as safe by the U.S. Food and Drug Administration; GMP = good manufacturing practices.

due to the thickness of the coating. Use of coatings to establish a surface pH that favors the active form of sorbic acid and other preservatives is also a possibility. This was demonstrated with a zein coating on a cheese analog for sorbate [158]; with MC/fatty acid films with sorbate in a test system [46,159]; with chitosan, MC, and HPMC films with sorbate in a test system; and with a CMC/soy protein coating on cut apple with sorbate and benzoate [41].

2. Parabens

Alkyl esters of *p*-hydroxybenzoic acid, or parabens, are effective antimicrobial agents, especially against yeasts and molds. In the United States methyl and propyl parabens are considered GRAS up to 0.1% [95]. In the United Kingdom methyl, ethyl, and propyl parabens are permitted in food, while other countries allow butyl ester parabens as well [36].

3. Sulfites

Sulfites or sulfur dioxide and its various salts are effective antimicrobials for the control of yeasts, molds, and especially bacteria, and they prevent enzymatic browning in foods. The effectiveness of this preservative is greatest when the acid is undissociated at pH <4 (pK_a of sulfur dioxide = 1.76 and 7.20). Although sulfur dioxide and various sulfite salts are considered GRAS in the United States, they cannot be used on meats, in food products that are sources of thiamine, or on raw fruits and vegetables [95] due to the elicitation of allergenic responses in a certain segment of the population [36].

4. Sucrose Esters and Chitosan

Sucrose esters are approved as emulsifiers and are an ingredient in edible coatings along with cellulose in Tal Pro-long (Courtaulds Group) and Semperfresh (United Agr. Products), or with guar gum in Nu-coat Flo (Surface Systems International Ltd.) Sucrose esterified with palmitic and stearic acids showed some antimicrobial activity against certain molds at levels of 1%. Chitosan, the film former of the edible coating Nutri Save (Nova Chem. Ltd.), has been shown to inhibit growth of fungi on plants by inducing plant defense responses [134,135].

5. Other Natural Antifungal Compounds

Unripe fruit are often more resistant to decay possibly due to the presence of some antifungal compounds. Several of these compounds were found in unripe mango including 5,12-*cis*-heptadecenyl resorcinol and 5-pentadecenyl resorcinol. A similar situation was discovered in unripe avocado where an antifungal compound, 1-acetoxy-2-hydroxy-4-oxoheneicosa-12,15-diene and 1-acetoxy-2,4-dihydroxy-*n*-heptadeca-16-ene. High levels of CO_2 enhanced concentrations of the antifungal avocado diene in treated fruits [160], presenting a possible explanation for the antimicrobial action of this compound, at least in the case of avocado fruit.

An antifungal essential oil and a long-chain alcohol, thought to be 17-pentatriacontanol, showed antifungal activity by inhibiting the mycelial growth of *Aspergillus carneus*. This compound was isolated from *Achyranthes aspera*, an herb with reported medicinal properties [161]. Isothiocyanates derived from mustard and horseradish have also been shown to have antimicrobial activity [162]. Another compound, pyrrolnitrin, was isolated from the bacteria *Pseudomonas cepacia,* which in turn was isolated from apple leaves. This microbe was found to have antagonistic activity toward *Penicillium expansum*, *Botrytis cinerea*, and *Mucor* species due to the production of the secondary metabolite pyrrolnitrin. This compound was applied to harvested strawberries and was found to retard various storage rots [163]. None of these natural fungicides have been approved for human consumption, but they offer promising alternatives to synthetic fungicides and preservatives.

D. Antioxidants

Antioxidants are compounds that inhibit or prevent the oxidation reaction caused by free radicals, with or without oxidation enzymes, that cause discoloration or browning of certain fruit and vegetable tissues and rancidity of fats [17,18]. This can effect the color or flavor of meat, fish, mushrooms, fruit, and vegetable products. Commonly used antioxidants, along with their U.S. FDA 21 CFR [95] are listed in Table 7.

1. Phenolic Antioxidants

The phenolic structure of certain compounds suppresses free radical formation, which delays the autooxidative process in fat or oil by acting as a proton donor [18]. Approved phenolic antioxidants include butylated hydroxyanisole (BHA), butylated hydroxytoluene (BHT), and esters of gallic acid such as propyl gallate and tertiary butyl hydroquinone (TBHQ). Natural antioxidants are also effective, such as the tocopherols and lecithin. The antioxidants BHA, BHT, tocopherol, and lecithin are GRAS, while TBHQ is approved as a direct food additive and propyl gallate as an indirect food additive component of coatings [95]. These antioxidants are also approved for food use in many countries, especially BHA, BHT, and tocopherol [18]. Coatings have been used as carriers of antioxidants to retard rancidity of meat and nut products and discoloration of lightly processed fruits and vegetables [16,39–41]. For whole apples, aqueous dips in the antioxidant diphenylamine (DPA) (300–3000 ppm) or ethoxyquin help to reduce surface discoloration known as scald [152,164].

2. Other Antioxidants and Antibrowning Agents

Some agents such as cinnamic and benzoic acids (both GRAS) [95] are effective browning inhibitors in combination with ascorbic acid since, like sulfites, they inhibit polyphenol oxidase (PPO) activity [21]. This enzyme is responsible for the browning that occurs when monophenolic compounds of plants or shellfish are hydroxylated to *o*-diphenols and subsequently to *o*-quinones in the presence of oxygen and PPO in plants and shellfish. The PPO enzyme requires copper, and thus complexing and chelating agents such as ethylenediamine tetraacetic acid (EDTA) and citric acid can inhibit enzymatic browning [17]. Ascorbic acid and its derivatives, erythorbic acid, ascorbic acid-2-phosphate, and -triphosphate, are effective inhibitors of enzymatic browning for cut apple [21,165].

TABLE 7 Commonly Used Antioxidants

Antioxidants	Classification	U.S. FDA 21 CFR #
Anoxomer	Antioxidant	172.105
Ascorbic acid	GRAS, GMP	182.8013, 182.5013
Ascorbic acid-2-phosphate		
Ascorbic acid-3-phosphate		
Ascorbyl palmitate	GRAS, GMP	182.3149, 172.110
BHA	GRAS, GMP	182.3169, 172.115
BHT	GRAS, GMP	182.3173
L-Cysteine	GRAS, improve biological quality of total protein in a food	184.1271, 172.320
Diphenylamine (DPA)	Surface treatment of apples for scald disorder	
Erythorbic acid	GRAS, GMP	182.3041
Ethoxyquin	Antioxidant	172.140
4-Hydrocymethyl-2-6-di-*tert*-butylphenol	Antioxidant	172.150
Lecithin	GRAS, GMP	184.1400
Potassium bisulfite	GRAS, GMP, raw fruits and vegetables	182.3616
Potassium metabisulfite	GRAS, GMP, raw fruits and vegetables	182.3637
Propyl gallate	GRAS, antioxidant	184.1660
Rosemary	GRAS, flavoring	182.10, 182.20
Sodium bisulfite	GRAS, GMP, raw fruits and vegetables	182.3739
Sodium metabisulfite	GRAS, GMP, raw fruits and vegetables	182.3766
Sodium sulfite	GRAS, GMP, raw fruits and vegetables	182.3798
TBHP	Antioxidant	172.190
TBHQ	Antioxidant	172.185
Tocopherols	GRAS, GMP	182.8890
α-Tocopherols	GRAS, inhibitors of nitrosamine formation	184.1890
α-Tocopherol acetate	GRAS, GMP	182.8892

GRAS = Generally recognized as safe by the U.S. Food and Drug Administration;
GMP = good manufacturing practices.

Ascorbyl palmitate, cinnamic acid, benzoic acid, and β-cyclodextrin were reported to be effective browning inhibitors in juice [21]. Ascorbic acid, erythorbic acid, and ascorbyl palmitate are GRAS [95], but other ascorbic acid derivatives are, so far, not approved. The amino acid cysteine is also an effective inhibitor of PPO [166]. Rosemary extract (and its constituents carnosol, carnosic acid, and rosmarinic acid) is a source of natural antioxidants [167,168]. Citric acid and EDTA have been incorporated into coatings as browning inhibitors for cut apples, potatoes, and mushrooms [41,169]. The amino acid cysteine inhibits PPO activity by reacting with quinone intermediates as well as reduced glutathione. Inorganic halides such as sodium or calcium chloride ($CaCl_2$) also inhibit PPO activity [17]. Resorcinol and its derivatives, such as 4-hexylresorcinol, inhibit browning tyrosinase isozymes in mushrooms and may inhibit PPO by serving as a substrate for this enzyme. These compounds have antimicrobial activity as well [22]. They are not yet approved for food use in the United States.

E. Mineral and Growth Regulator Treatments

1. Calcium

The mineral calcium has many postharvest uses [31]. Postharvest dips of calcium chloride ($CaCl_2$) can reduce symptoms of bitter pit or small brown lesions that occur on apples as well as scald [152,170]. Calcium or $CaCl_2$ dips or infiltrations on whole or cut fruit has been reported to increase fruit firmness for apple [32,171], peach, [172], blueberry [33], and strawberry [173], and to delay ripening and decay of avocado [174], mango [175,176], apple [177], pear [178], peach [179], strawberry [173], and potato [34]. The reasons for these benefits range from alleviating disorders resulting from calcium deficiency, to the effect of calcium on cell walls that makes them more resistant to decay [177] and firmer in texture, to adversely affecting conidial germination and germ-tube elongation [180]. The mineral salt $CaCl_2$ is considered GRAS [95].

2. Growth Regulators

The polyamines putrescine and spermidine altered texture when infiltrated into apples [32] and spermine and spermidine increased firmness in sliced strawberries [181]. Growth regulators such as 2,4-dichlorophenoxyacetic acid (2,4-D) and 2,4,5-trichlorophenoxyacetic acid (2,4,5-T) were added to fruit wax as antisenescent compounds to extend the shelf life of mandarin oranges [182]. Maleic hydrazide (250 ppm) and 2,4,-D were added to wax emulsions to delay ripening of mango fruits [183]. Gibberellic acid (150 ppm) suppressed sprouting of yam tubers for one month [184]. The only approved postharvest growth regulator treatment for fruits in the United States is for lemons destined for long-term storage. These fruit are sometimes treated with 2,4,-D to delay senescence of the button (calyx and residual stem) to reduce infection from *Alternaria* [185].

V. FUMIGATION AND GAS TREATMENTS

Hydrogen peroxide is an antibacterial agent based on its oxidative properties [36], effective at concentration ranges of 0.01–0.1% (U.S. FDA 21 CFR 184.1366) [95]. It is especially effective against gram-negative bacteria such as coliforms. There is little information as to its effect on fungi. It is mostly used to extend the shelf life of dairy products and is considered GRAS as an antimicrobial agent for treatment of milk for the making of cheese, whey, and starch [95]. It is reportedly effective as a vapor phase treatment fumigant for fresh table grapes. Previously fumigation with sulfur dioxide was used, but there are concerns about adverse effects on some sensitive individuals (U.S. FDA 21 CFR 182.3862) [95,186]. Acetaldehyde vapor (0.25–0.5%) significantly reduced decay of harvested table grapes [187], while acetic acid vapor fumigation reduced fungal decay of grapes, apples, oranges, tomatoes, and strawberries [188,189]. Other natural fruit and plant volatiles have been found to exhibit fungistatic activity. Of these volatiles, benzaldehyde, methyl salicylate and ethyl benzoate, 1-hexanol, *trans*-2-hexenal 2-nonanone, and furan compounds were found to be particularly effective [190–193]. The antimicrobial activity of CO_2 is greatest against molds and gram-negative psychrotrophic bacteria in the concentration range of 10–100% (U.S. FDA 21 CFR 184.1240) [36,95]. The mechanism of action is not known, but may be related to lack of O_2, acidification of intracellular contents, or effect on enzymes. Various concentrations of CO_2 used in modified-atmosphere packaging (MAP) are often within the microbistatic range [194,195]. It has been shown to reduce brown rot of package cherry fruit [196].

VI. FRUIT QUARANTINE TREATMENTS

Fruit flies are major pests worldwide, and their fruit hosts must be treated to kill 100% of the immatures inside the fruits prior to export to uninfested areas of the world. One of the main treatments currently used is methyl bromide fumigation, which is scheduled to be phased out over the next few years since it is a suspected stratospheric ozone depleter [197]. Other treatments include cold storage, hot air, vapor heat, and hot water treatments [198]. Unfortunately, most of these treatments impart surface or internal quality damage to many fruits. Recently, use of CA and edible coatings has been investigated as alternative treatment alone or in combination with currently used methods [199–201]. Preliminary findings suggest that lowered O_2 and elevated CO_2 may contribute to fruit fly mortality [200,201]. Fruit coatings are already approved as a disinfestation treatment for surface mites on cherimoyas and limes from South America [202].

VII. SURFACE PREPARATION AND COATING TECHNIQUES

Edible coatings can be applied by dipping products in coating materials and then allowing excess coating to drain as it dries and solidifies [64]. This was first reported for the Florida citrus industry [8] where the fruit were submerged into a tank of emulsion coating. Commodities are then generally conveyed to a drier where water or solvent is removed or coated items can be allowed to air-dry under ambient conditions [203]. Sometime a porous basket can be used to drain excess coating. Some emulsion coatings are applied with a foam applicator where a foaming agent is added to the coating or compressed air is blown into the applicator tank. The agitated foam is applied to commodities as they move by on rollers and cloth flaps or brushes distribute the emulsion over the surface of the commodity [203]. Excess coating is then removed by squeegees and sometimes recirculated. This type of emulsion contains little water and, therefore, dries quite quickly, but inadequate coverage is often a problem. Edible coating can be sprayed, which is especially useful to obtain a thinner and more uniform coat or if only one part or side of a product is to be coated [64]. This is the most popular method for coating whole fruits and vegetables, especially with the development of high-pressure spray applicators and air-atomizing systems. Overhead drip emitters can also deliver coating to fruit and brushes below. The fruit tumble over rotating brush beds that become saturated with coating from overhead spray or drip applicators. As the fruit tumble and rotate over the saturated brushes they become uniformly covered with coating [203]. Controlled drop applicator allows drops of coating to be shattered in microdrops that coating are delivered through a spray. Free-standing films can be cast such that the thickness can be controlled by spreading or pouring. A spreader with adjustable height can be used to control the thickness of the film, which is then allowed to dry [64]. Pan coating of tablets and candies involves a pan, which is enclosed and perforated along the side panels. The coating is delivered by a pump to a spray gun(s) mounted in various parts of the pan. The coating is atomized by the spray guns [203]. Solutions of antioxidant, preservatives, or other aqueous materials can be applied by dip or spray. In most cases fruit and vegetable products are sanitized with chlorine and/or *o*-phenylphenate (SOPP) and dried on brushes or with fans as much as possible prior to coating. Sanitizing solutions are covered in U.S. FDA 21 CFR 178.1010 [95].

Coating of breakfast cereal products usually involves sugar-based materials, which presents a difficult situation since sugar is hydrophilic. With minimal water as solvent, sugar is heated to a hard candy condition and sprayed on cereal, which is further heated to fuse the sugar. Otherwise sugar is spun into a blanket on which cereal pieces are placed prior to application of a second blanket of sugar. The whole sugar/cereal sandwich is then compressed and dried [16].

VIII. REPORTED APPLICATIONS IN FOODS

A. Fresh Intact Fruits and Vegetables

Edible coatings have been applied to a diverse array of whole fruits and vegetables since the 1930s and 1940s [15]. Coatings of vegetable or mineral oil have been applied to tropical fruits (mango, pineapple, banana, papaya, guava, and avocado) with varying degrees of benefit in terms of extending shelf life [204–207]. Sometimes oils, such as mineral oil, are used alone to coat fruits such as tomatoes or limes, but in such cases they remain in a liquid state as a thin film on the surface of the fruit. The purpose of coating these fruits with oils is to help them slip over equipment (lubricant), add a slight sheen, delay ripening, delay weight (water) loss, and delay yellowing in the case of limes [15,86,208,209]. Use of a cellulose based film reduced the number of viable *Salmonella montevideo* cells on the surface of tomatoes in addition to retarding the ripening process [210]. Casein-lipid coatings reduced moisture loss from citrus, apple, and zucchini [65]. Shellac, carnauba wax, and polyethylene wax retard moisture loss and add shine to apples, citrus, and other fruit [15]. Waxing of potatoes with paraffin wax did not adversely affect respiration, but it did reduce sprouting and synthesis of chlorophyll (green pigment) and solanine (toxic glycoalkaloid) [211]. The Tweens' lecithin and hydroxylated lecithin surfactants, and applied films were also useful in inhibiting chlorophyll and solanine synthesis in the peel of potato tubers [212,213]. Creation of a modified atmosphere within the coated potatoes may have inhibited greening due to an affect on the synthesis of these undesirable compounds. In contrast, lipid and hydrocolloid coatings inhibited degreening of lemons and limes [209] probably by inhibiting chlorophyll breakdown due to a modified atmosphere. The modified atmosphere induced by coatings can also be useful in retarding ripening as demonstrated with zein or chitosan on tomato [93,137] and cellulose on mango and tomato [169].

B. Lightly Processed Fruits and Vegetables, Dried Fruit, and Nut Products

Coating of lightly processed fruits and vegetables is a new field [79,85,214], while coatings have been reported on nut and dried fruit products since the 1940s and 1950s [215]. Casein-lipid coatings reduced moisture loss [65], and a cellulose coating inhibited surface drying and the resulting color change (whitening) of peeled carrots [47,216,217]. There have been several reports of coatings applied to cut apple. Dextrin coatings prevented oxidative browning of apple slices [140], while a cellulose coating with antioxidants reduced cut apple discoloration more effectively than a solution of antioxidants alone [41]. A chitosan/lauric acid coating inhibited browning of cut apple [218], while caseinate/lipid coatings reduced moisture loss [70] and an alginic acid/casein/lipid coating reduced water loss and browning of cut apple [79]. A bilayer coating of polysaccharide and lipid decreased water loss, respiration, and ethylene production in cut apple [219]. Pectin coatings were used to coat almonds and reportedly held salt and antioxidants on the surface while providing a nonoily texture [39]. Coating nuts with hydrogenated oils or acetylated monoglyceride containing an antioxidant increased their shelf life by retarding development of oxidative rancidity [40,220]. Gum arabic has also been used to coat pecans [146]. Starch (amylose) and whey protein isolate coatings were used to reduce oxidative rancidity during storage of certain products such as nuts, cereal, beans [220–223], candies, and dried fruits [220]. This can also be controlled with antioxidants such as tertiary butylhydroquinone (TBHQ) alone or in a coating, as was shown for peanuts [40]. For coated candied fruit and dates, pectin coatings reduced stickiness [39]. Acetylated monoglycerides, hydrogenated coconut oil, or confectioner's butter stabilized dried fruit pieces in cake or bread mix [224] and raisins in cereals [225,226]. Mineral oil, beeswax, vegetable oils, and acetylated monoglycerides have also been used to reduce clumping and stickiness of raisin products [225–229]. Vegetable oil blends as coatings for dried fruit have poor flavor stablility [230]. Two commercial products, Spraygum® and Sealgum® (Colloides Naturels, Inc., Bridgewater, NJ), are based on gum acacia and gelatin. These products have been used to coat chocolates, nuts, cheese, and pharmaceutical products. In addition, they reduced darkening of potatoes in combination with $CaCl_2$ [34].

C. Processed Food and Animal Products

Edible coatings are also applied to processed foods to restrict the movement of moisture and gases, especially O_2. Edible coatings can be used to prevent moisture loss and absorption or transfer between components of differing water activity [16]. Edible coatings protect some commodities, such as meat and nut products, from oxidative rancidity, fat absorption, breading loss during frying [231,232], and oil migration from such products as chocolate using hydrocolloids [233]. Amylose coatings provided a nonsticky surface at RH $< 80\%$, prevented fat migration from cheese and chocolate products, and retained volatile flavors [220]. Nonedible wax coatings are used to encase some cheese products [89,234]. Meat fats, vegetable oils, and acetylated monoglycerides, mono-, di-, and triglycerides, waxes, and mixtures of these components have been used to coat meats and meat products including frozen chicken pieces and pork chops [231,235]. These coatings protected the meat products from dehydration [16]. Cornstarch-alginate coatings reduced moisture loss and improved juiciness of coated meat products [139], while HPC films reduced moisture loss from chicken [236]. Gelatin coatings have been used to coat or wrap meats [112] and form soft capsules in the pharmaceutical industry. Casein-lipid coatings reduced moisture loss from chicken eggs [65]. Emulsions of oil, water, and sugar have been used to coat cereal products to limit the entrance of water and therefore the amount of drying required [237]. Candies are sometimes coated with chocolate for flavor and reduction of moisture loss [238] or carnauba wax to add shine and reduce stickiness [16]. Corn zein and the derivatives MC and HPMC are oil repellent and reduced oil absorption for potato balls or meat when included in batters or when applied as coatings [124,236,239]. These compounds also improved adhesion of batters or coatings to food products [125]. Coatings can delay the uptake of liquid by dry cereal products so that they last longer in milk before becoming soggy [221]. Such coatings often have a sugar base to impart a sweet flavor [16].

Heterogeneous products of differing water activities, such as ice cream cones with ice cream, or frozen filled pies where a dry material is in contact with a moist filling, have shown benefit from a barrier film between the two components. Such films had a polysaccharide component in a bilayer system with palmitic/stearic acid or saturated C_{16} and C_{18} fatty acids and beeswax [72,74]. For example, a lipid-cellulose coating showed promise as a barrier to internal moisture migration between two components of a bread/tomato sauce product [71,72].

IX. LEGAL ASPECTS

Discussion of the various coating components discussed thus far in this review reflects rulings by the U.S. FDA. The United States allows coating of fresh produce with restrictions as to what can be used as a coating material or ingredient and, in some cases, the type of produce (usually dependent on whether the coated fruit or vegetable peel is normally consumed or not, i.e., (apple versus avocado). Other countries, however, do not allow coating of fruits and vegetables at all. For example, a ban has been reported in Norway on imports of waxed fruit. The Norwegian policy is that foods making significant nutritional contributions to the diet should be as free from additives as possible [240]. The German government tried to ban waxed apples, but the European Commission overruled them with a directive that allows apples, pears, and some other products to be imported from countries where waxing or coating of fruit is legal as long as the formulations contain beeswax, candelilla wax, carnauba wax, and/or shellac. Japan accepts shellac- and carnauba wax–coated citrus from the United States, but will not accept fruit coated with petroleum-based waxes that are legal in the United States for certain fruit, including citrus. The use of the cellulose derivative HPC is allowed in coating formulations in the United States, but not in New Zealand. Other countries such as Thailand and Australia are evaluating the applicability and acceptability of various coating film formers and additives [241,242].

The U.S. FDA ruled in 1993 that all waxed fruit must be subjected to ingredient labeling regulations at the retail level [243]. Retailers are permitted to use collective names for coating ingredients, however, which should be prominently visible in the retail area where the commodity is displayed. The label or sign should indicate that the commodity is waxed or coated with food-grade animal-, vegetable-, petroleum-, beeswax-, and or shellac-based wax or resin to maintain freshness [242]. The United Kingdom proposed a similar labeling scheme requiring labels indicating all postharvest treatments applied to fresh produce [243]. Processed foods in the United States are required to contain a label on the package listing all additives. Coatings, preservatives, and antioxidants are all considered to be additives and, therefore, must be listed as individual ingredients. Postharvest use of fungicides, however, is not required to be labeled in the United States.

X. CONSUMER ATTITUDES

Consumers are becoming increasingly educated about health and nutrition and are concerned about what goes on or into their food. Some consumer groups are concerned with the waxes and coatings themselves, and others fear that, in the case of fresh produce, harmful pesticide residues may be sealed inside the fruits by the coatings. There is also the issue of the use of animal products in coatings. Although there is a federal law in the United States about labeling coated produce, it is not always enforced. This presents a problem for Moslems, vegetarians, and Orthodox Jews, who have concerns about animal-derived products, and for those with certain protein allergies and intolerances. Enforcement of the regulations described in the previous section requiring a conspicuous sign would allow consumers to make an informed choice. Nevertheless, the main reason for coating produce such as apples and citrus is cosmetic. The fact remains that coated (shiny) fruit sell better than uncoated ones [244]. Some consumer groups want postharvest fungicide treatments to be listed as well as coating types [245].

In conclusion, edible coatings alone or as carriers of useful additives serve many functions for all types of food products. They improve the external and internal quality characteristics of diverse commodities. Coatings can reduce dehydration and oxidation as well as the resulting undesirable changes in color, flavor, and texture. Waxes and other coatings delay ripening and senescence of fresh produce and can increase the microbial stability of lightly processed fruits, vegetables, and some processed products. Coatings show promise as environmentally friendly quarantine treatments. Most coating materials are produced from renewable, edible resources and can even be manufactured from waste products that represent disposal problems for other industries.

REFERENCES

1. R. E. Hardenburg, Wax and related coatings of horticultural products. A bibliography, *Agr Res. Bull. No. 965*, Cornell Univ., Ithica, NY, 1967, p.1.
2. T. Labuza and R. Contrereas-Medellin, Prediction of moisture protection requirements for foods, *Cereal Food World 26*:335 (1981).
3. H. J. Kaplan, Washing, waxing and color adding, *Fresh Citrus Fruits* (W. F. Wardowski, S. Nagy, and W. Grierson, eds.), AVI Publishing Co., Westport, CT, 1986, p. 379.
4. S. A. Trout, E. G. Hall, and S. M. Sykes, Effects of skin coatings on the behavior of apples in storage, *Aust. J. Agr. Res. 4*:57 (1953).
5. C. W. Hitz and I. C. Haut, Effects of waxing and pre-storage treatments upon prolonging the edible storage qualities of apples, *Univ. of MD Agr. Exp. Sta. Tech. Bull.* No. A14, (1942).
6. L.L, Claypool, The waxing of deciduous fruits, *Am. Soc. Hort. Sci. Proc. 37*:443 (1940).
7. D. V. Fisher and J. E. Britton, Apple waxing experiments, *Sci. Agri. 21*:70 (1940).
8. H. Platenius, Wax emulsions for vegetables, *Cornell Univ. Agri. Exp. St . Bull.* No. 723, (1939).
9. R. M. Smock, Certain effects of wax treatments on various varieties of apples and pears, *Proc. Am. Soc. Hort. Sci. 33*:284 (1935).
10. S. K. Williams, J. L. Oblinger, and R. L. West, Evaluation of a calcium alginate film for use on beef cuts, *J. Food Sci. 43*:292 (1978).
11. C. R. Lazarus, R. L. West, J. L. Oblinger, and A. Z. Palmer, Evaluation of a calcium alginate coating and a protective plastic wrapping for the control of lamb carcass shrinkage, *J. Food Sci. 41*:639 (1976).
12. R. D. Earl, Method of preserving foods by coating same, U.S. patent 3,395,024 (1968).
13. M. E. Zabik and L. E. Dawson, The acceptability of cooked poultry protected by an edible actylated monoglyceride coating during fresh and frozen storage, *Food Technol.17*:87 (1963).
14. R. C. Meyer, A. R. Winter, and H. H. Weiser, Edible protective coatings for extending the shelf life of poultry, *Food Technol. 12*:146 (1959).
15. E. A. Baldwin, Edible coatings for fresh fruits and vegetables: past, present, and future, *Edible Coatings and Films to Improve Food Ouality* (J. M. Krochta, E. A. Baldwin, and M. O. Nisperos-Carriedo, eds.), Technomic Publishing Company, Lancaster, PA, 1994, p. 25.
16. R. A. Baker, E. A. Baldwin, and M. O. Nisperos-Carriedo, Edible coatings and films for processed foods, *Edible Coatings and Films to Improve Food Quality* (J. M. Krochta, E. A. Baldwin, and M. O. Nisperos-Carriedo, eds.), Technomic Publishing Company, Lancaster, PA, 1994, p. 89.
17. G. M. Sapers, Browning of foods: control by sulfates, antioxidants, and other means, *Food Technol. 47*:75 (1993).
18. E. R. Sherwin, Antioxidants, Food Additives (A. L. Branen, P. M. Davidson, and S. Salminen, eds.), Marcel Dekker, Inc., New York, 1990, p. 139.
19. J. D. Ponting, R. Jackson, and G. Watters, Refrigerated apple slices: preservative effects of ascorbic acid, calcium and sulfites, *J. Food Sci., 37*:434 (1972).

20. G. M. Sapers and F. W. Douglas, Measurement of enzymatic browning at cut surfaces and in juice of raw apple and pear fruits, *J. Food Sci. 52*:1258 (1987).
21. G. M. Sapers, K. B. Hicks, J. G. Phillips, L. Garzarella, D. L. Pondish, R. M. Matulaitis, T. H. McCormack, S. M. Sondey, P. A. Seib, and Y. S. El-Atawy, Control of enzymatic browning in apple with ascorbic acid derivatives, polyphenol oxidase inhibitors, and complexing agents, *J. Food Sci. 54*:997 (1989).
22. A. Monsalve-Gonzalez, G. V. Barbosa-Canovas, A. J. McEvily, and R. Iyengar, Inhibition of enzymatic browning in apple products by 4-hexylresorcinol, *Food Technol. 49*:110–118 (1995).
23. L. Giannuzzi, A. M. Lambardi, and N. E. Zaritzky, Diffusion of citric and ascorbic acids in pre-peeled potatoes and their influence on microbial growth during refrigerated storage, *J. Sci. Food Agri. 68*:311 (1995).
24. G. M. Sapers and R. L. Miller, Heated ascorbic/citric acid solution as browning inhibitor for pre-peeled potatoes, *J. Food Sci. 60*:762 (1995).
25. R. Noble and K. S. Burton, Postharvest storage and handling of mushrooms: physiology and technology, *Postharvest News Info. 4*:125 (1993).
26. G. M. Sapers, R. L. Miller, F. C. Miller, P. H. Cooke, and S. Choi, Enzymatic browning control in minimally processed mushrooms, *J. Food Sci. 59*:1042 (1994).
27. P. S. Taoukis, T. P. Labuza, J. H. Lillemo, and S. W. Lin, Inhibition of shrimp melanosis (black spot) by ficin, *Lebensm. Wiss. Technol. 23*:52 (1990).
28. A. J. McEvily, R. Iyengar, and S. Otwell, Sulfite alternative prevents shrimp melanosis, *Food Technol. 45*:80 (1991).
29. D. H. Wasson, K. D. Reppond, and T. M. Kandianis, Antioxidants to preserve rockfish color, *J. Food Sci. 56*:1564 (1991).
30. J. W. Eckert and J. M. Ogawa, The chemical control of postharvest diseases: subtropical and tropical fruits, *Ann. Rev. Phytopathol. 23*:421 (1985).
31. B.W. Poovaiah, Role of calcium in prolonging storage life of fruits and vegetables, *Food Technol. 40*:84 (1986).
32. C.Y. Wang, W. S. Conway, J. A. Abbott, G. F. Kramer, and C. E. Sams, Postharvest infiltration of polyamines and calcium influences ethylene production and texture changes in 'Golden Delicious' apples, *J. Am. Soc. Hort. Sci. 118*:801 (1993).
33. E. J. Hanson, J. L. Beggs, and R. M. Beaudry, Applying calcium chloride postharvest to improve highbush blueberry firmness, *HortScience 28*:1033 (1993).
34. G. Mazza and H. Qi, Control of after-cooking darkening in potatoes with edible film-forming products and calcium chloride, *J. Agric. Food Chem. 39*:2163 (1991).
35. W. S. Conway, C. E. Sams, and A. Kelman, Enhancing the natural resistance of plant tissues to postharvest diseases through calcium applications, *HortScience 29*:751 (1994).
36. P. M. Davidson and V. K. Juneja, Antimicrobial agents, *Food Additives* (A. L. Branen, P. M. Davidson, and S. Salminen, eds.), Marcel Dekker, Inc., New York, 1990, p. 83.
37. S. Doores, pH control agents and acidulants, Food Additives (A. L. Branen, P. M. Davidson, and S. Salminen, eds.), Marcel Dekker, Inc., New York, 1990, p. 477.
38. C. Andres, Natural edible coating has excellent moisture and grease barrier properties, *Food Proc.* (Dec):48 (1984).
39. H. A. Swenson, J. C. Miers, T. H. Schultz, and H. S. Owens, Pectinase and pectate coatings. II. Application to nut and fruit products, *Food Technol. 7*:232 (1953).
40. M. W. Hoover and P. J. Nathan, Influence of tertiary butylhydroquinone and certain other surface coatings on the formation of carbonyl compounds in granulated roasted peanuts, *J. Food Sci. 47*:246 (1981).
41. E. A. Baldwin, M. O. Nisperos, X. Chen, and R. D. Hagenmaier, Improving storage life of cut apple and potato with edible coating, *Postharvest Biol. Technol.* 151–163 (1996).
42. S. Guilbert, Use of superficial edible layer to protect intermediate moisture foods: application to the protection of tropical fruit dehydrated by osmosis, *Properties of Water in Foods* (C.C. Seow, ed.), Elsevier Applied Science Publishers, Ltd., London, 1988, p. 119.

43. J. A. Torres, Sorbic acid stability during processing of an intermediate moisture cheese analog, *J. Food Proc. Pres. 13*:409 (1989).
44. J. A. Torres and M. Karel, Microbial stabilization of intermediate moisture food surfaces. III. Effects of surface preservative concentration and surface pH control on microbial stability of an intermediate moisture cheese analog, *J. Food Proc. Pres. 9*:107 (1985).
45. F. Vojdani and J. A. Torres, Potassium sorbate permeability of polysaccharide films, chitosan, methyl cellulose and hydroxy propyl methyl cellulose, *J. Food Proc. 12*:33 (1989).
46. F. Vojdani and J. A. Torres, Potassium sorbate permeability of methyl cellulose and hydroxy propyl methyl cellulose coatings: effects of fatty acids, *J. Food Sci. 55*:941 (1990).
47. S. A. Sargent, J. K. Brecht, J. J. Zoellner, E. A. Baldwin, and C. A. Campbell, Edible films reduce surface drying of peeled carrots, *Proc. Fla. State Hort. Soc. 107*:245 (1994).
48. J. M. Wells, Heated wax-emulsions with benomyl and 2,6-dichloro-4-nitroaniline for control of postharvest decay of peaches and nectarines. *Phytopathology 62*:129 (1971).
49. J. M. Wells and L. G. Reaver, Hydrocooling peaches after waxing: effects on fungicide residues, decay development and moisture loss, *HortScience 11*:107 (1976).
50. G. E. Brown, Benomyl residues in Valencia oranges from postharvest applications containing emulsified oil, *Phytopathology 64*:539 (1974).
51. J. W. Eckert and M. J. Kolbezen, Influence of formulation and application method on the effectiveness of benzimidazole fungicides for controlling postharvest diseases of citrus fruits, *Neth. J. Plant Path. 83 (supp)*:343 (1977).
52. G. E. Brown, S. Nagy, and M. Maraulia, Residues from postharvest nonrecovert spray applications of imazalil to oranges and effects on green mold caused by *Penicillium digitatum*, Plant Dis. 67:954 (1983).
53. G. E. Brown, Efficacy of citrus postharvest fungicides applied in water or resin solution water wax, *Plant Dis. 68*:415 (1984).
54. H. M. Couey and G. Farias, Control of postharvest decay of papaya, *HortScience 14*:719 (1979).
55. A. E. Ghaouth, J. Arul, R. Ponnampalam, and M. Boulet, Chitosan coating effect on storability and quality of fresh strawberries, *J. Food Sci. 56*:1618 (1991).
56. G. A. Reineccius, Flavor encapsulation, *Edible Coatings and Films to Improve Food Quality* (J. M. Krochta, E. A. Baldwin, and M. O. Nisperos-Carriedo, eds.), Technomic Publishing Company, Lancaster, PA, 1994, p. 105.
57. M. O. Nisperos-Carriedo, P. E. Shaw, and E.A. Baldwin, Changes in volatile flavor components of Pineapple orange juice as influenced by the application of lipid and composite film, *J. Agric. Food Chem. 38*:1382 (1990).
58. K. L. Nelson and O. W. Fennema, Methyl cellulose films to prevent lipid migration in confectionery products, *J. Food Sci. 56*:504 (1991).
59. I. S. Saguy and E. J. Pinthus, Oil uptake dairying deep-fat frying: factors and mechanism, *Food Technol. 49*:142 (1995).
60. S. L. Kamper and O. Fennema, Water vapor permeability of edible bilayer films, *J. Food Sci. 49*:1478 (1984).
61. S. L. Kamper and O. Fennema, Use of an edible film to maintain water vapor gradients in foods, *J. Food Sci. 50*:382 (1985).
62. S. L. Cuppett, Edible coatings as carriers of food additives, fungicides and natural antagonists, *Edible Coatings and Films to Improve Food Quality* (J. M. Krochta, E. A. Baldwin, and M. O. Nisperos-Carriedo, eds.), Technomic Publishing Company, Lancaster, PA, 1994, p. 121.
63. W. M. Mellenthin, P. M. Chen, and D. M. Borgic, In-line application of porous wax coating materials to reduce friction discoloration of 'Bartlett' and 'd'Anjou' pears, *HortScience 17*:215 217 (1982).
64. G. Donhowe and O. Fennema, Edible films and coatings: characteristics, formation, definitions and testing methods, *Edible Coatings and Films to Improve Food Quality* (J. M. Krochta, E. A. Baldwin, and M. O. Nisperos-Carriedo, eds.), Technomic Publishing Company, Lancaster, PA, 1994, p. 1.

65. T. H. McHugh and J. M. Krochta, Milk-protein-based edible films and coatings, *Food Technol. 48*:97 (1994).
66. I. G. Donhowe and O. Fennema, The effect of relative humidity gradient on water vapor, permeance of lipid and lipid-hydrocolloid bilayer films, *JAOCS 69*:1081 (1992).
67. H. J. Park, C. L. Weller, P. J. Vergano, and R. F. Testin, Permeability and mechanical properties of cellulose-based edible films, *J. Food Sci. 58*:1361 (1993).
68. A. Gennadios, C. L. Weller, and R. F. Testin, Temperature effect on oxygen permeability of edible protein-based films, *J. Food Sci. 58*:212 (1993).
69. E. Hernandez, Edible coatings from lipids and resins, *Edible Coatings and Films to Improve Food Quality* (J. M. Krochta, E. A. Baldwin, and M. O. Nisperos-Carriedo, eds), Technomic Publishing Company, Lancaster, PA, 1994, p. 279.
70. R. J. Avena-Bustillos and J. M. Krochta, Water vapor permeability of caseinate-based edible films as affected by pH, calcium cross linking and lipid content, *J. Food Sci. 58*:904 (1993).
71. J. J. Kester and O. Fennema, An edible film of lipids and cellulose ethers: barrier properties to moisture vapor transmission and structural evaluation, *J. Food Sci. 54*:1384 (1989).
72. J. J. Kester and O. Fennema, An edible film of lipids and cellulose ethers: performance in a model frozen-food system, *J. Food Sci. 54*:1390 (1989).
73. D. W. S. Wong, F. A. Gastineau, K. S. Gregorski, S. L. Tillin, and A. E. Pavlath, Chitosan-lipid films: microstructure and surface energy, *J. Agric. Food Chem 40*:540 (1992).
74. S. L. Kamper and O. Fennema, Water vapor permeability of an edible, fatty acid, bilayer film, *J. Food Sci. 49*:1482 (1984).
75. ASTM, Standard test methods for water vapor transmission of materials, ASTM *Book of Standards*, E96-80, Philadelphia, PA, 1980.
76. R. H. Wills, T. H. Lee, D. Graham, W. B. McGlasson, and E. G. Hall, *Postharvest, an Introduction to the Physiology and Handling of Fruits and Vegetables*, AVI Publishing Co., Inc., Westport, CT, 1981.
77. A. A. Kader, Biochemical and physiological basis for effects of controlled and modified atmospheres on fruits and vegetables, *Food Technol. 40*:99 (1986).
78. J. L. Woods, Moisture loss from fruits and vegetables, *Postharvest News Info. 1*:195 (1990).
79. D. W. S. Wong, M. W. Camarind, and A. E. Pavlath, Development of edible coatings for minimally processed fruits and vegetables, *Edible Coatings and Films to Improve Food Quality* (J. M. Krochta, E. A. Baldwin, and M. O. Nisperos-Carriedo, eds.), Technomic Publishing Company, Lancaster, PA, 1994, p. 65.
80. E. A. Baldwin, M. O. Nisperos-Carriedo, and R. A. Baker, Edible coatings for lightly processed fruits and vegetables, *HortScience 30*:35 (1995).
81. E. A. Baldwin, M. O. Nisperos-Carriedo, and R. A. Baker, Use of edible coatings to preserve quality of lightly (and slightly) processed products, *Crit. Rev. Food Sci. Nutr. 35*:50 (1995).
82. R. F. MacLeod, R. A. Kader, and L. L, Morris, Stimulation of ethylene and CO_2 production of mature-green tomatoes by impact bruising, *HortScience 11*:604 (1976).
83. T. Boller and H. Kende, Regulation of wound ethylene synthesis in plants, *Nature* 286:259 (1980).
84. N. E. Hoffman and S. F. Yang, Enhancement of wound-induced ethylene synthesis by ethylene in preclimacteric cantaloupe, *Plant Physiol. 69*:317 (1982).
85. S. Guilbert, N. Gontard, and L. G. M. Gorris, Prolongation of the shelf-life of perishable food products using biodegradable films and coatings, *Lebensm. Wiss. Technol. 29*:10 (1996).
86. J. Hartman and F. M. Isenberg, Waxing vegetables, *New York Agri. Extent. Ser. Bull. No. 965*, 1956, p. 3.
87. R. G. McGuire and E. A. Baldwin, Compositions of cellulose coatings affect populations of yeasts in the liquid formulation and on coated grapefruits, *Proc. Fla State Hort. Soc., 107*:293 (1994).
88. J. Waks, M. Schiffmann-Nadel, E. Lomaniec, and E. Chalutz, Relation between fruit waxing and development of rots in citrus fruit during storage, *Plant Dis. 62*:869 (1985).
89. N. N. Potter, *Food Science*, AVI Publishing Co., Inc., Westport, CT, 1986, p. 372.
90. R. D. Hagenmaier and R. A. Baker, Internal gases, ethanol content and gloss of citrus fruit coated

with polyethylene wax, carnauba wax, shellac or resin at different application levels, *Proc. Fla. State Hort. Soc. 107*:261 (1994).
91. R. D. Hagenmaier and R. A. Baker, Lavered coatings to control weight loss and preserve gloss of citrus fruit, *HortScience 30*:296 (1995).
92. H. B. Cosler, Method of producing zein-coated confectionery, U.S. patent 2,791,509 (1957).
93. J. J. Park, M. S. Chinnan, and R. L. Shewfelt, Edible coating effects on storage life and quality of tomatoes, *J. Food Sci. 59*:568 (1994).
94. R. D. Hagenmaier and R. A. Baker, Edible coatings from candelilla wax microemulsions, *J. Food Sci. 61*:562 (1996).
95. U.S. FDA, *Code of Federal Regulations 21*:parts 1-99 (1966).
96. J. J. Kester and O. Fennema, Tempering influence on oxygen and water vapor transmission through a stearyl alcohol film, *JAOCS 66*:1154 (1989).
97. J. J. Kester and O. Fennema, Resistance of lipid films to oxygen transmission, *JAOCS 66*:1130 (1989).
98. J. J. Kester and O. Fennema, Resistance of lipid films to water vapor transmission, *JAOCS 8*:1139 (1989).
99. M. Martin-Polo, A. Voilley, G. Blond, B. Colas, M. Mesnier, and N. Floquet, Hydrophobic films and their efficiency against moisture transfer. 2. Influence of the physical state, *J. Agric. Food Chem. 40*:413 (1992).
100. R. O. Feuge, Acetoglycerides—new fat products of potential value to the food industry, *Food Technol. 9*:314 (1955).
101. N. V. Lovegren and R.O. Feuge, Permeability of acetostearin products to water vapor, *Agric. Food. Chem. 2*:558 (1954).
102. J. J. Kester and O. R. Fennema, Edible films and coatings: a review, *Food Technol. 40*:47 (1986).
103. H. Bennett, *Industrial Waxes*, Vol. 1, Chemical Pub. Co., New York, 1975.
104. Eastman Chemicals, Water-emulsion fruit and vegetable coatings based on EPOLENE® waxes, Pub. No. F-257B, August 1986.
105. R. D. Hagenmaier and P. E. Shaw, Permeability of shellac coatings to gases and water vapor, *J. Agric. Food Chem. 39*:825 (1991).
106. E. Hernandez and R. A. Baker, Candelilla wax emulsion, preparation and stability, *J. Food Sci. 56*:1382 (1991).
107. W. C. Griffin, Emulsions, *Kirk-Othmer Encyclopedia of Chemical Technology*, 3rd ed., Vol. 8, 1979, p. 913.
108. R. D. Hagenmaier and P. E. Shaw, Gas permeability of fruit coating waxes, *J. Am. Soc. Hort. Sci. 117*:105 (1992).
109. Nevill Chemical Co., Production information, resins, plasticizer, nonstaining antioxidants and chlorinated paraffins, Brochure No. NCCFA188, 1988, p. 1.
110. D. J. Hall, Innovations in citrus waxing—an overview, *Proc. Fla State Hort. Soc. 94*:258 (1981).
111. S. R. Drake, J. K. Fellman, and J. W. Nelson, Postharvest use of sucrose polymers for extending the shelf-life of stored golden delicious apples, *J. Food Sci. 52*:1283 (1987).
112. A. Gennadios, T. H. McHugh, C. L. Welter, and J. M. Krochta, Edible coatings and films based on proteins, *Edible Coatings and Films to Improve Food Quality* (J. M. Krochta, E. A. Baldwin, and M. O. Nisperos-Carriedo, eds.), Technomic Publishing Company, Lancaster, PA, 1994, p. 201.
113. A. Gennadios, A. H. Brandenburg, C. L. Weller, and R. F. Testin, Effect of pH on properties of wheat gluten and soy protein isolate films, *J. Agric. Food Chem. 41*:1835 (1993).
114. A. H. Brandenburg, C. L. Weller, and R. G. Testin, Edible films and coatings from soy protein, *J. Food Sci. 58*:1086 (1993).
115. J. R. Maynes and J. M. Krochta, Properties of edible films from total milk protein, *J. Food Sci. 59*:909 (1994).
116. J. M. Krochta, A. E. Pavlath, and N. Goodman, Edible films from casein-lipid emulsions for lightly-processed fruits and vegetables, *Engineering and Food*, Vol. 2, *Preservation Processes and*

Related Techniques (W. E. Spiess and H. Schuberts, eds.), Elsevier Science Pub., New York, 1990, p. 329.

117. N. Parris, D. R. Coffin, R. F. Joubran, and H. Pessen, Composition factors affecting the water vapor permeability and tensil properties of hydrophilic films, *J. Agric. Food Chem. 43*:1432 (1995).
118. S. Guilbert, Use of superficial edible layer to protect intermediate moisture foods: Application to protection of tropical fruits dehydrated by osmosis, *Food Preservation and Moisture Control* (C. C. Seow, ed.), Elsevier Applied Science Pub., Let., Essex, UK, 1988, p. 199.
119. A. Gennadios and C. L. Weller, Edible films and coatings from wheat and corn proteins, *Food Technol. 44*:63 (1990).
120. Y. M. Stuchell and J. M. Krochta, Enzymatic treatments and thermal effects on edible soy protein films, *J. Food Sci. 59*:1332 (1994).
121. A. Jangchud and M. S. Chinnan, Development of surface and depsition films from peanut protein, *CoFE 1995 - Proc. Conf. Food Eng.* (M. Okos, ed.), 1995, p. 1.
122. S. Okamoto, Factors affecting protein film formation, *Cereal Sci. Today 25*:256 (1978).
123. A. Gennadios and C. L. Weller, Edible films and coatings from soymilk and soy protein, *Cereal Foods World 36*:1004 (1991).
124. G. R. Sanderson, Polysaccharides in foods, *Food Technol. 35*:50, 83 (1981).
125. A. J. Ganz, Cellulose hydrocolloids, Food Colloids (H. P. Graham, ed.), AVI Pub. Co., Westport, CT, 1977, p. 383.
126. M. S. Chinnan and H. J. Park, Effect of plasticizer level and temperature on water vapor transmission of cellulose-based edible films, *J. Food Proc. Eng. 18*:417 (1995).
127. R. A. Kent, R. S. Stephens, and J. A. Westland, Bacterial cellulose fiber provides an alternative for thickening and coating, *Food Technol. 45*:108 (1991). .
128. G. O. Aspinall, *Polysaccharides*, Pergamon Press., Oxford, UK, 1970.
129. M. O. Nisperos-Carriedo, Edible coatings and films based on polysaccharides, *Edible Coatings and Films to Improve Food Quality* (J. M. Krochta, E. A. Baldwin, and M. O. Nisperos-Carriedo, eds.), Technomic Publishing Company, Lancaster, PA, 1994, p. 305.
130. T. H. Schultz, J. C. Miers. H. S. Owens, and W. D. Maclay, Permeability of pectinase films to water vapor, *J. Phys. Colloid. Chem. 53*:1320 (1949).
131. T. H. Schultz, H. S. Owens, and W. D. Maclay, Pectinase films, *J. Colloid. Sci. 53*:53 (1948).
132. J. C. Miers, H. A. Swenson, T. H. Schultz, and H. S. Owens, Pectinase and pectate coatings. I. General requirements and procedures, *Food Technol. 7*:229 (1953).
133. General news: Chitosan-derivative keeps apples fresh, *Postharvest News Info. 2*:75 (1991).
134. M. Walker-Simmons, D. Jin, C. A. West, L. Hadwiger, and C. A. Rayan, Comparison of proteinase inhibitor-inducing activities and phytoalexin elicitor activities of a pure fungal endopolygalacturonase, pectic fragments, and chitosans, *Plant Physiol. 76*:833 (1984).
135. P. Stossel and J. L. Leuba, Effect of chitosan, chitin and some aminosugars on growth of various soilborne phytopathogenic fungi, *Phytopathol. Z. 111*:82 (1984).
136. C. M. Elson and E. R. Hayes, Development of the differentially permeable fruit coating "Nutri-Save®" for modified atmosphere storage of fruit, *Controlled Atmosphere for Storage and Transport of Perishable Agricultural Commodities* (S. M. Blankenship, ed.), North Carolina State Univ., Raleigh, NC, 1985, p. 248.
137. A. El Gaouth, R. Pannampalam, F. Castaigne, and J. Arul, Chitosan coating to extend the storage life of tomatoes, *HortScience 27*:1016 (1992).
138. J. C. Rankin, I. A. Wolff, H. A. Davis, and C. E. Rist, Permeability of amylose film to moisture vapor, selected organic vapors, and the common gases, *Indus. Eng. Chem. 3*:120 (1958).
139. L. Allen, A. I. Nelson, M. P. Steinberg, and J. N. McGill, Edible corn-carbohydrate food coatings. I. Development and physical testing of a starch-algin coating, *Food Technol. 17*:1437 (1963).
140. D. G. Murray and L. R. Luft, Low-D. E. corn starch hydrolysates, *Food Technol. 27*:32 (1973).
141. E. R. Morris, D. A. Rees, and G. Robinson, Cation-specific aggregation of carrageenan helices: domain model of polymer gel structure, *J. Mol. Biol. 138*:349 (1980).

142. R. L. Whistler and J. R. Daniel, Functions of polysaccharides in foods, *Food Additives*, (A. L. Branen, P. M. Davidson, and S. Salminen, eds.), Marcel Decker, Inc., New York, 1990, p. 395.
143. R. L.Whistler and J. N. BeMiller, Industrial Gums, Polysaccharides and Their Derivatives, 2nd ed., Academic Press, New York, 1973.
144. A. M. Prakash, M. Joseph, and M. E. Mangino, The effects of added proteins on the functionality of gum arabic in soft drink emulsion systems, *Food Hydrocoll. 4*:177 (1990).
145. Osman, M. E., P. A. Williams, A. R. Menzies, and G. O. Phillips, Characterization of commercial samples of gum arabic, *J. Agric. Food Chem. 41*:71 (1993).
146. F. W. Arnold, U.S. patent 3,383,220 (1963).
147. D. J. Pettitt, J. E. B. Wayne, J. J. Renner Nantz and C. F. Showmaker, Rheological properties of solutions and emulsions stabilized with xanthan gum and propylene glycol alginate, *J. Food Sci. 60*:528 (1995).
148. Kelco, Gellan gum—multifunctional gelling agent, Technical Bull., Merck and Co., Rahway, NJ, 1990.
149. E. Art, Emulsifiers, *Food Additives* (A.L. Branen, P. M. Davidson, and S. Salminen, eds), Marcel Dekker, Inc., New York, 1990, p. 347.
150. C. L. Wilson and P. L. Pusey, Potential for biological control of postharvest plant diseases, *Plant Dis. 69*:375 (1985).
151. J. A. Domenico, A. R. Rahman, and D. E. Wescott, Effects of fungicides in combination with hot water and wax on the shelf life of tomato fruit, *J. Food Sci. 37*:957 (1972).
152. C. R. Little, H. J. Taylor, and F. McFarlane, Postharvest and storage factors affecting superficial scald and core flush of 'Granny smith' apples, *HortScience 20*:1080 (1985).
153. B. L. Goulart, P. E. Hammer, K. B. Evensen, W. Janisiewicz, and F. Takeda, Pyrrolnitrin, captan + benomyl, and high CO_2 enhance raspberry shelf life at 0 or 18C., *J. Amer. Soc. Hort. Sci. 117*:265 (1992).
154. S. Droby, E. Chalutz, and C. L. Wilson, Antagonistic microorganisms as biological control agents of postharvest diseases of fruits and vegetables, *Postharvest News Info. 2*:169 (1991).
155. M. E. Wisniewski and C. L. Wilson, Biological control of postharvest diseases of fruits and vegetables: recent advances, *HortScience 27*:94 (1992).
156. R. Potjewijd, M. O. Nisperos, J. K. Bums, M. Parish, and E. A. Baldwin, Cellulose-based coatings as carriers for *Candida guillermondii* and *Debaryomyces* sp. in reducing decay of oranges, *HortScience 30*:1417 (1995).
157. R. G. McGuire and R. D. Hagenmaier, Shellac coatings for grapefruits that favor biological control of *Penicillium digitatum* by *Candida oleophila*, *Biol.. Cont. 7*:100 (1996).
158. J. A. Torres, J. O. Bouzas, and M. Karel, Microbial stabilization of intermediate moisture food surfaces. II. Control of surface pH, *J. Food Proc. Pres. 9*:93 (1985).
159. D. C. Rico-Pena and J. A. Torres, Sorbic acid and potassium sorbate permeability of an edible methyl cellulose-palmitic acid film: water activity and pH effects, *J. Food Sci., 56*:497 (1991).
160. D. Prusky and N. T. Keen, Involvement of preformed antifungal compounds in the resistance of subtropical fruits to fungal decay, *Plant Dis. 77*:114 (1993).
161. T. N. Misra, R. S, Singh, H. S. Pandey, C. Prasad, and F. P. Singh, Antifungal essential oil and a long chain alcohol from *Achyranthes aspera*, *Phytochemistry 31*:1811 (1992).
162. P. J. Delaquis and G. Mazza, Antimicrobial properties of isothiocyanates in food preservation, *Food Technol. 49*:73 (1995).
163. F. Takeda, W. J. Janisiewicz, J. Toitman, N. Mahoney, and F. B. Abeles, Pyrolnitrin delays postharves fruit rot in strawberries, *HortScience 25*:320 (1990).
164. A. L. Snowden, Pome fruits, apples and pears, *A Color Atlas of Post-Harvest Diseases and Disorders of Fruits and Vegetables*, Vol. 1: *General Introduction and Fruits*, CRC Press, Inc., Boca Raton, FL, 1990, p. 170.
165. G. M. Sapers, R. L. Miller, F. W. Douglas Jr., and K. B. Hicks, Uptake and fate of ascorbic acid-2-phosphate in infiltrated fruit and vegetable tissue, *J. Food Sci. 56*:419 (1991).
166. R. C. Richard, P. M. Goupy, J. J. Nicolas, J. Lacombe, and A. A. Pavia, Cysteine as an in-

hibitor of enzymatic browning. I. Isolation and characterization of addition compounds formed during oxidation of phenolics by apple polyphenol oxidase, *J. Agric. Food Chem. 39*:841 (1991).
167. S. Lai, J. I. Gray, D. M. Smith, A. M. Booren, R. L. Crackel, and D. J. Buckley, Effects of oleoresin rosemary, tertiary butylhydroquione, and sodium tripolyphosphate on the development of oxidative rancidity in restructured chicken nuggets, *J. Food Sci. 56*:616 (1991).
168. E. N. Frankel, S. Huang, R. Aeschbach, and E. Prior, Antioxidant activity of a rosemary extract and its constituents, carnosic acid carnosol, and rosmarinic acid, in bulk oil and oil-in-water emulsion, *J. Agric Food Chem. 44*:131 (1996).
169. M. O. Nisperos-Carriedo, E. A. Baldwin, and P. E. Shaw, Development of an edible coating for extending postharvest life of selected fruits and vegetables, *Proc. Fla. State Hort.. Soc., 104*:122 (1991).
170. B. Mitcham, H. Andris, and C. H. Crisosto, Apple disorders, *Perish. Handl. Newslett. 86*:2 (1996).
171. J. Stow, Effect of calcium ions on apple fruit softening during storage and ripening, *Postharvest Biol. Technol. 3*:1 (1993).
172. H. Javeri, R. Toledo and L. Wicker, Vacuum infusion of citrus pectinmethylesterase and calcium effects on firmness of peaches, *J. Food Sci. 56*:739 (1991).
173. J. M. Garcia, S. Herrera, and A. Morilla, Effects of postharvest dips in calcium chloride on strawberry, *J. Agric. Food Chem. 44*:30 (1996).
174. R. B. H. Wills, M. C. C. Yuen, M. Utami, and D. Utami, Effect of calcium infiltration on delayed ripening of Minyak avocado, *Asian Food J. 4*:43 (1988).
175. A. Mootoo, Effect of post-harvest calcium chloride dips on ripening changes in 'Julie' mangoes, *Trop. Sci. 31*:243 (1991).
176. B. P. Singh, D. K. Tandon, and S. K. Kalra, Changes in postharvest quality of mangoes affected by preharvest application of calcium salts, *Sci. Hort. 54*:211 (1993).
177. W. S. Conway and C. E. Sams, Possible mechanisms by which postharvest calcium treatment reduces decay in apples, *Phytopathology 74*:208 (1984).
178. D. Sugar, Enhanced resistance to postharvest decay in Bosc pears treated with calcium chloride, *Acta Hort. 256*:201 (1989).
179. R. B. H. Wills and M. S. Mahendra, Effect of postharvest application of calcium on ripening of peach, *Aust. J. Exp. Agric. 29*:751 (1989).
180. R. J. McLaughlin, A review and current status of research on enhancement of biological control of postharvest diseases of fruits by use of calcium salts with yeasts, Biological Control of Postharvest Diseases of Fruits and Vegetables, Workshop Proc., U.S. Dept of Agri., Agri. Res. Ser. ARS-92 (C. L. Wilson and E. Chalutiz, eds.), 1990, p. 184.
181. T. Ponappa, J. C. Scheerens, and A. R. Miller, Vacuum infiltration of polyamines increases firmness of strawberry slices under various storage conditions, *J. Food Sci. 58*:361 (1993).
182. B. B. Lodh, S. De, S. K. Mukherjee, and A. W. Bose, Storage of mandarin oranges. I. Effects of hormones and wax coatings, *J. Food Sci. 28*:519 (1963).
183. H. Subramanyam, N. V. Moorthy, V. B. Dalal, and H. C. Srivastava, Effect of a fungicidal wax coating with or without growth regulator on the storage behavior of mangoes, *Food Sci., (Mysores) 11*:236 (1962).
184. E. C. Nnodu and S. O. Alozie, Using gibberellic acid to control sprouting of yam tubers, *Trop. Agric. 68*:329 (1992).
185. D. K. Saulukhe and B. B. Desai, Citrus, *Postharvest Biology of Fruits and Vegetables*, Vol. I, CRC Press, Boca Raton, FL, 1986, p. 59.
186. C. F. Forney, R. E. Rij, R. Denis-Arrue, and J. L. Smilanick, Vapor phase hydrogen peroxide inhibits postharvest decay of table grapes, *HortScience 26*:1512 (1991).
187. I. Avissar and E. Pesis, the control of postharvest decay in table grapes using acetaldehyde vapors, *Ann. Appl. Biol. 118*:229 (1991).
188. D. L. Sholberg and A. P. Guance, Fumigation of fruit with acetic acid to prevent postharvest decay, *HortScience 30*:1271 (1995).

189. A. L. Moyls, P. L. Sholberg, and A. P. Guance, Modified-atmosphere packaging of grapes and strawberries fumigated with acetic acid, *HortScience 31*:414 (1996).
190. C. Wilson, J. D. Franklin, B. I. Otto, Fruit volatles inhibitory to *Monilinia fructicola* and *Botrytis cinera, Plant Dis. 71*:316 (1987).
191. B. Hardin, Natural compounds inhibit decay fungi, *Agric. Res.* (Sept):21 (1993).
192. S. F. Vaughn, G. F. Spencer, and B. S. Shasha, Volatile compounds from raspberry and strawberry fruit inhibit postharvest decay fungi, *J. Food Sci. 58*:793 (1993).
193. S. Vaughn and M. K. Ehlenfeldt, Natural volatile furan compounds inhibit blueberry and strawberry decay fungi, *Proc. 6th Intro. Cont. Atmos. Res. Conf.*, Cornell Univ., Ithica, NY, June 15–17, 1993, p. 393.
194. I. J. Church and A. L. Parsons, Modified atmosphere packaging technology: a review, *J. Sci. Food Agric. 67*:143 (1995).
195. A. L. Brody, Integrating aseptic and modified atmosphere packaging to fulfill a vision of tomorrow, *Food Technol. 50*:56 (1996).
196. R. M De Vries-Paterson, A. L. Jones, and A. C. Cameron, Fungistatic effects of carbon dioxide in a package environment on the decay of Michigan sweet cherries by *Monilinia fructicola, Plant Dis.75*:943 (1991).
197. Addition of methyl bromide to list of Class I substances and phaseout schedule, *Fed. Reg. 58*:65028 (1993).
198. J. L. Sharp and G. J. Hallman, *Quarantine Treatments for Pests of Food Plants*, Westview Press, Boulder, CO, 1994.
199. G. J. Hallman, Controlled atmospheres, *Insect Pests and Fresh Horticultural Products: Treatments and Responses* (R. E. Paull and J. W. Armstrong, eds.), CAB International, Wallingford, UK, 1994, p. 121.
200. G. H. Hallman, M. O. Nisperos-Carriedo, E. A. Baldwin, and C. A. Campbell, Mortality of Caribbean fruit fly immatures in coated fruits, *J. Econ. Entomol. 87*:752 (1994).
201. G. J. Hallman, R. G. McGuire, E. A. Baldwin, and C. A. Campbell, Mortality of feral Caribbean fruit fly (Diptera: Tephritidae) immatures in coated guavas, *J. Econ. Entomol. 88*:1353 (1995).
202. Treatment schedule TIO2b: soapy water and wax. *U.S. Dept. Agric., Animal Plant Health Inspec. Serv. Treatment Manual*, 1993, p.
203. L. Grant and J. K. Burns, Application of coatings, *Edible Coatings and Films to Improve Food Quality* (J. M Krochta, E. A. Baldwin, and M. O. Nisperos-Carriedo, eds.), Technomic Publishing Company, Lancaster, PA, 1994, p. 189.
204. V. B. Dalal, P. Thomas, N. Nagaraja, G. R. Shah, and B. C. Amla, Effect of wax coating on bananas of varying maturity, *Indian Food Packer 24*:36 (1970).
205. V. B. Dalal, W. E. Eipeson, and N. S. Singh, Wax emulsion for fresh fruits and vegetables to extend their storage life, *Indian Food Packer 25*:9 (1971).
206. P. B. Mathur and H. C. Srivastava, Effect of skin coatings on the storage behavior of mangoes, *Food Res. 20*:559 (1955).
207. A. N. Bose and G. Basu, Studies on the use of coating for extension of storage life of fresh Fajli mango, *Food Res. 19*:424 (1954).
208. J. C. Ayres, A. A. Krafy, and L. C. Peirce, Delaying spoilage of tomatoes, *Food Technol. 9*:100 (1964).
209. E. A. Baldwin, M. O. Nisperos-Carriedo, R. D. Hagenmaier, and R. A. Baker, Use of lipids in edible coatings for food products, *Food Technol.* 56–62 (1996).
210. R. Zhuang, L. R. Beuchat, M. S. Chinnan, R. L. Shewfelt, and Y. W. Huang, Inactivation of *Salmonella montevideo* on tomatoes by applying cellulose-based edible films, *J. Food Prot. 59*:808 (1996).
211. M. T. Wu and D. K. Salunkhe, Control of chlorophyll and solanine synthesis and sprouting of potato tubers by hot paraffin wax, *J. Food Sci. 37*:629 (1972).
212. P. A. Poapst, I. Price, and F. R. Forsyth, Prevention of post storage greening in table stock potato tubers by application of surfanctants and adjutants, *J. Food Sci. 43*:900 (1978).

213. M. T. Wu and D. K. Salunkhe, Responses of lecithin- and hydroxylated lecithin-coated potato tubers to light, *J. Agric. Food Chem. 26*:513 (1978).
214. E. A. Baldwin, M. O. Nisperos-Carriedo, and R. A. Baker, Edible coatings for lightly processed fruits and vegetables, *HortScience 30*:35 (1995).
215. E. A. Baldwin, M. O. Nisperos-Carriedo, and R. A. Baker, Use of edible coatings to preserve quality of lightly (and slightly) processed products, *Crit. Rev. Food Sci. Nutr. 35*:509 (1995).
216. L. R. Howard and T. Dewi, Sensory, microbiological and chemical quality of mini-peeled carrots as affected by edible coating treatment, *J. Food Sci. 60*:142 (1994).
217. R. J. Avena-Bustillos, L. A. Cisneros-Zevallos, J. M. Krochta, and M. E. Saltveit, Optimization of edible coatings on minimally processed carrots using response surface methodology, *ASAE 36*:801 (1993).
218. E. Pennisi, Sealed in edible film, *Sci. News 141*:12 (1992).
219. D. W. S. Wong, S. J. Tillin, J. S. Hudson, and A. E. Pavlath, Gas exchange in cut apples with bilayer coatings, *J. Agric. Food Chem. 42*:2278 (1994).
220. L. Jokay, G. E. Nelson and E. L. Powell, Development of edible amylaceous coatings for foods, *Food Technol. 21*:1064 (1967).
221. R. Danials, Coatings for cereal-type products, *Edible Coatings and Soluble Packaging*, Noyes Data Corp., Park Ridge NJ, 1973, p. 229.
222. M. S. Cole, U.S. patent 3,479,191 (1969).
223. J. I. Mate, E. N. Frankel, and J. M. Krochta, Whey protein isolate edible coatings: effect on the rancidity process of dry roasted peanuts, *J. Agric. Food Chem. 44*:1736 (1996).
224. R. A. Shea, June 23, 1970, U.S. patent 3,516,836 (1970).
225. E. Lowe, E. L. Durkee, and W. E. Hamilton, U.S. patent 3,046,143 (1962).
226. G. G. Watters and J. E. Brekke, U.S. patent 2,909,435 (1959).
227. S. P. Kochhar and J. B. Rossell, A vegetable oiling agent for dried fruits, *J. Food Technol. 17*:661 (1982).
228. E. Lowe, E. L. Durkee, W. E. Hamiliton, G. G Watters, and A. I. Morgan, Jr., Continuous raisin coater, *Food Technol. 17*:109 (1963).
229. G. G. Watters and J. E. Brekke, Stabilized raisins for dry cereal products, *Food Technol. 15*:236 (1961).
230. N. Goldenberg, The oiling of dried sultanas, *Chem. Ind. 21*:956 (1976).
231. R. Daniels, Edible Coatings and Soluble Packaging, Noyes Data Corp., Park Ridge, NJ, 1973.
232. Food gums stick it out in dry mix glazes, *Prep. Foods 1*:53 (1993).
233. N. C. Brake and O. R. Fennema, Edible coatings to inhibit lipid migration in a confectionery product, *J. Food Sci. 58*:1422 (1993).
234. A. H. Johnson and M. S. Pererson. eds., *Encyclopedia of Food Technology*, AVI Publishing Co., Inc., Westport, CT, 1974, p. 178.
235. C. D. Bauer, G. L. Neuser, and H. A. Pinkalla, U.S. patent 3,406,081 (1968).
236. V. M. Balasurbramaniam, M. S. Chinnan, P. Mallikarjunan, and R. D. Phillips, The effect of edible film on oil uptake and moisture retention of a deep-fat fried poultry product, *J. Proc. Eng. 20*:17 (1997).
237. A. A. Lyall and R. J. Johnston. U.S. patent 3,959,670 (1976).
238. B. Biquet and T. P. Labuza, Evaluation of the moisture permeability characteristics of chocolate films as an edible moisture barrier, *J. Food Sci. 53*:989 (1988).
239. P. Mallikarjunan, M. S. Chinnan, V. M. Balasubramaniam, and R. D. Phillips, Edible coatings for deep-fat frying of starchy products, *Lebensm. Wiss. Technol. 30*:709–714 (1997).
240. Waxed fruit banned in Norway, *Postharvest News Info. 1*:433 (1990).
241. New ACIAR project to investigate edible coatings, *Postharvest News Info. 3*:103 (1992).
242. Waxing fruit—the debate continues, *Citrograph* (1993).
243. UK government considers food labeling scheme, *Postharvest News Info. 3*:3 (1992).
244. C. Lecos, How to shine an apple, *Food Drug Admin. 16*:8 (1992).
245. J. D. Aylsworth, Debate over waxing heats up, *Fruit Grower* (May):6 (1992).

21

Encapsulation and Controlled Release in Food Preservation

RONALD B. PEGG AND FEREIDOON SHAHIDI
Memorial University of Newfoundland, and PA Pure Additions, Inc., St. John's, Newfoundland, Canada

I. INTRODUCTION

Additives are incorporated into foods for a variety of reasons. For example, antimicrobial agents are added to various foodstuffs in an effort to ward off the early onset of microbial growth, antioxidants are used to prolong the shelf life of lipid-containing foods by protecting triacylglycerols and phospholipids against oxidative degradation, flavoring and coloring agents are added for the purpose of enhancing the sensory characteristics of the food, while various carbohydrate-based additives are employed to improve the rheological and textural properties of the product in question.

In recent years, there has been a growing trend toward reducing permitted levels of many food additives and, where possible, replacing those chemically derived with alternatives perceived to be of natural origin [1,2]. The use of synthetic antioxidants in foods, such as butylated hydroxyanisole (BHA) and butylated hydroxytoluene (BHT), is being reevaluated because of a growing concern over their possible carcinogenic effects [3]. Thus, strategies have been developed for the isolation, purification, and identification of antioxidants from natural sources for use in food systems. Unfortunately, many natural ingredients are less potent at equivalent addition levels or are more restricted in their applicability than their synthetic counterparts. Thus, a novel strategy to increase the effectiveness and range of application of many types of natural functional ingredients is to make use of microcapsular delivery systems. Because of the wide availability of encapsulated ingredients, many food products whose development was thought to be technically unfeasible are now possible. Such ingredients are products of a process in which the active ingredient has been enveloped in a coating or "capsule," thereby conferring many useful properties to or eliminating undesirable properties from the original ingredient.

A. Basis of Encapsulation

Encapsulation has been used by the food industry for more than 60 years. In a broad sense, encapsulation technology in food processing includes the coating of minute particles of ingredients (e.g., acidulants, fats, and flavors) as well as whole ingredients (e.g., raisins, nuts, and confectionery products), which may be accomplished by microencapsulation and macro-coating techniques, respectively. The science of encapsulation deals with the manufacture, analytical evaluation, and application of encapsulated products. Despite its long history, the technology that has been developed for the food industry remains relatively unsophisticated compared to many other fields of application. This is a consequence of the limitations imposed on the food industry for the use of edible, low-cost ingredients and processing.

King [4] notes that it is important for the food scientist to distinguish between encapsulation versus entrapment of food ingredients. He states that encapsulation may be defined as a process of forming a continuous thin coating around encapsulants (i.e., solid particles, droplets of liquids, or gas cells), which are wholly contained within the capsule wall as a core of encapsulated material. On the other hand, entrapment refers to the trapping of encapsulants within or throughout a matrix (e.g., gel, crystal), but a small percentage of the entrapped ingredients will normally be exposed at the particle surface, whereas this would not be so for the encapsulated product. The material that is entrapped is generally a liquid but could be a solid particle or gas and is referred to by various names, such as core material, payload, actives, fill, or internal phase (5). The material that forms the coating is referred to as the wall material, shell, or coating.

The food industry applies encapsulation for a number of reasons [6–8]:

1. Encapsulation/entrapment can protect the core material from degradation by reducing its reactivity to its outside environment (e.g., heat, moisture, air, and light).
2. Evaporation or transfer rate of the core material to the outside environment is decreased/retarded.
3. The physical characteristics of the original material can be modified and made easier to handle. For example, a liquid component can be converted to solid particles; lumping can be prevented; the core material can be distributed more uniformly throughout a mix by giving it a size and outside surface; hygroscopicity can be reduced; flowability and compression properties can be improved; dustiness can be reduced; and density can be modified.
4. The product can be tailor-designed to either release slowly over time or release at a certain point (i.e., to control the release of the core material so as to achieve the property delay until the right stimulus.
5. The flavor of the core material can be masked.
6. The core material can be diluted when only very small amounts are required, yet still achieve a uniform dispersion in the host material.
7. It can be employed to separate components within a mixture that would otherwise react with one another.

B. Benefits and Types of Microcapsules

Microencapsulation is defined as the technology of packaging solids, liquids, or gaseous materials in miniature, sealed capsules that can release their contents at controlled rates under specific conditions [9,10]. The miniature packages, called microcapsules, may range from submicrometer to several millimeters in size and have a multitude of different shapes, depending on the materials and methods used to prepare them. Generally speaking, the microcapsule has the capability of modifying and improving the apparent shape and properties of a substance. More specifically, the microcapsule has the ability to preserve a substance in the finely divided state and to release it as occasion demands.

Microcapsules offer the food processor a means with which to protect sensitive food components, ensure against nutritional loss, utilize otherwise sensitive ingredients, incorporate unusual or time-release mechanisms into the formulation, mask or preserve flavors and aromas, and transform liquids into easily handleable solid ingredients [11]. The unusual properties afforded by encapsulated ingredients provide the food technologist with greater flexibility and control in developing foods that are more flavorful and nutritious to meet the expectations of today's consumers.

Various properties of microcapsules that may be changed to suit specific ingredient applications include composition, mechanism of release, particle size, final physical form, and cost. Before considering the properties desired in encapsulated products, the purpose of encapsulation must be clear. In designing the encapsulation process, the following questions should be taken into consideration:

1. What functionality should the encapsulated ingredients provide to the final product?
2. What kind of coating material should be selected?
3. What processing conditions must the encapsulated ingredient survive before releasing its content?
4. What is the optimum concentration of the active material in the microcapsule?
5. By what mechanism will the ingredient be released from the microcapsule?
6. What are the particle size, density, and stability requirements for the encapsulated ingredient?
7. What are the cost constraints of the encapsulated ingredient?

The architecture of microcapsules is generally divided into several arbitrary and overlapping classifications (Fig. 1). One such classification is known as matrix encapsulation. This is the simplest structure in which a sphere is surrounded by a wall or membrane of uniform thickness, resembling that of a hen's egg. In this design, the core material is buried to varying depths inside the shell. This microcapsule has been termed a single-particle structure (Fig. 1A). It is also possible to design microcapsules that have several distinct cores within the same capsule or, more commonly, numerous core particles embedded in a continuous matrix of wall material. This type of design is termed the aggregate structure (Fig. 1B). The particles in the aggregate structure need not be all of the same material, and if one wishes, control of the particle size can be achieved. This technique has been accomplished with numerous materials to improve size-distribution properties [13]. Another well-known design for a microcapsule is a multiwalled structure in which the different concentric wall layers can have the same, or quite different compositions. In this case, the multiple walls are placed around a core in order to achieve multiple purposes related to the manufacture of the capsules, their subsequent storage, and controlled release.

The theory and application of microcapsular delivery systems encompasses a variety of engineering techniques and scientific disciplines, thus making it difficult to present a systematic view of the total effort being made in this field or to acquire a total picture of it. This chapter summarizes the art of microencapsulation as it relates to the food industry and presents current information on the process of encapsulation. To accomplish this, a comprehensive examination of the various encapsulating matrices currently used by the food industry is included. In addition to their general description, the advantages and disadvantages they offer as encapsulating agents when forming microcapsules are discussed. An in-depth examination of the various microencapsulation techniques follows. This includes the processes of spray-drying, spray-cooling and spray-chilling, fluidized bed coating, extrusion, centrifugal extrusion, lyophilization, coacervation, centrifugal suspension separation, cocrystallization, liposome entrapment, interfacial polymerization, and inclusion complexation. Afterwards, encapsulated ingredients and their application to various food systems are considered with reference to some of their common uses. Finally, what is meant by controlled release and some of the mechanisms surrounding it is discussed.

II. THE ENCAPSULATION MATRIX

In order to encapsulate a food ingredient, the first requirement is the selection of an appropriate coating material, referred to as the encapsulating matrix. In the literature, many researchers have referred to the coating material as the shell, wall material, or encapsulating agent [1].

Coating substances, which are basically film-forming materials, can be selected from a wide variety of natural or synthetic polymers, depending on the material to be coated and the characteristics desired in the final microcapsules. The composition of the coating material is the main determinant of the functional properties of the microcapsule and of how it may be used to improve the per-

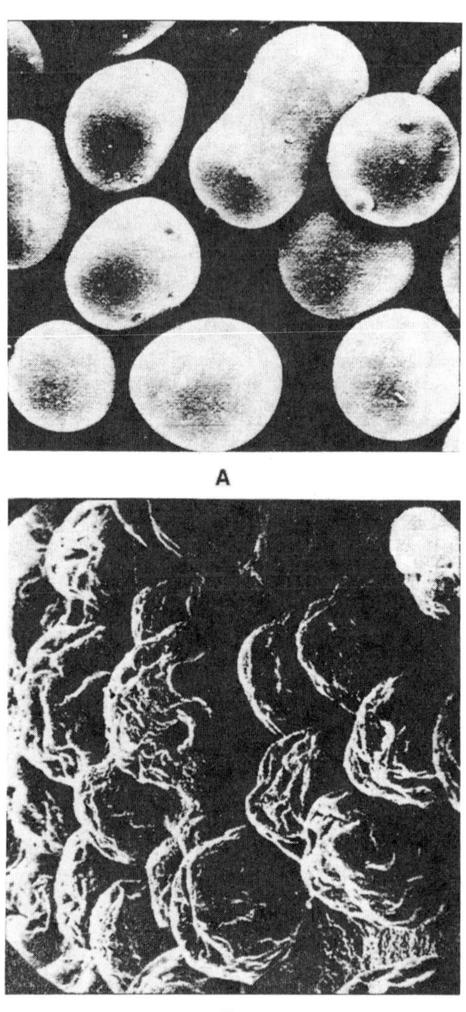

FIGURE 1 Photomicrographs of different food ingredients: (A) microencapsulated potassium chloride; (B) vitamin A capsules in ethyl cellulose. (From Ref. 12.)

formance of a particular ingredient. An ideal coating material should exhibit the following characteristics:

1. Good rheological properties at high concentration and easy workability during encapsulation
2. The ability to disperse or emulsify the active material and stabilize the emulsion produced
3. Nonreactivity with the material to be encapsulated both during processing and on prolonged storage
4. The ability to seal and hold the active material within its structure during processing or storage
5. The ability to completely release the solvent or other materials used during the process of encapsulation under drying or other desolventization conditions

TABLE 1 Coating Materials for Encapsulation of Food Ingredients

Carbohydrate	Starch, maltodextrins, corn syrup solids, dextran, modified starch, sucrose, cyclodextrins
Cellulose	Carboxymethylcellulose, methylcellulose, ethycellulose, nitrocellulose, acetylcellulose, cellulose acetate-phthalate, cellulose acetate-butylate-phthalate
Gum	Gum acacia, agar, sodium alginate, carrageenan
Lipid	Wax, paraffin, beeswax, tristearic acid, diacylglycerols, monoacylglycerols, oils, fats, hardened oils
Protein	Gluten, casein, gelatin, albumin, hemoglobin, peptides

Source: Ref. 12.

6. The ability to provide maximum protection to the active material against environmental conditions (e.g., oxygen, heat, light, humidity)
7. Solubility in solvents acceptable in the food industry (e.g., water, ethanol)
8. Chemically nonreactivity with the active material
9. Possession of specified or desired solubility properties of the capsules and release properties of the active material from the capsule
10. Inexpensive, food-grade status

Because no single coating material can meet all of the criteria listed above, in practice either coating materials are employed in combinations or modifiers such as oxygen scavengers, antioxidants, chelating agents, and surfactants are added. Some commonly used coating materials presented in Table 1 are discussed in detail below.

A. Carbohydrates

The ability of carbohydrates to absorb and adsorb volatiles from the environment or to retain them tenaciously during the drying process has important implications and applications for flavor encapsulation. In fact, carbohydrates are the most commonly used coating material in flavor-encapsulation processes.

The mechanisms by which carbohydrates retain volatiles during processing such as freeze- and spray-drying as well as extrusion are not fully understood but most probably involve physical interactions [14]. It has been postulated that the formation of microregions during freeze-drying, which contain highly concentrated solutions of carbohydrate and volatiles, results in molecular association of the carbohydrate through hydrogen bonding. This in turn creates a stable network and traps the volatiles [15]. For example, it has been reported that loss of volatiles from lactose during freeze-drying increased when the material changed from an amorphous solid to a crystalline one [16]. Formation of cracks in the microregion structure might have accounted for this [14].

The two major processes used for encapsulating food flavorings are spray-drying and extrusion [17]. Both of these depend primarily upon carbohydrates as the encapsulation matrix [18]. Although one can find examples of encapsulation using fats (e.g., spray chilling), proteins (gelatin), and inorganics (fused silica) as wall materials, carbohydrates constitute the majority of encapsulation matrices. While many compounds are classified as carbohydrate, discussion here does not include all such compounds. Some are discussed under different headings and classifications.

1. Maltodextrins and Corn Syrup Solids

Starch is one of the most naturally abundant polymers found on earth. It has been extracted from numerous food sources including corn, tapioca, potato, wheat, rice, and waxy maize.

Starch comprises polymers of glucose units linked together primarily by α-(1$\rightarrow$4) bonds and secondarily by α-(1$\rightarrow$6) bonds. The two polymer types found in starch are amylose, a straight-chain polymer, and amylopectin, a branched-chain polymer. With its long, straight chains, amylose is known for forming strong, flexible films. On the other hand, due to its extensive branching, amylopectin is not a strong film former, but is noted for clarity and stability when forming gels and may show a slightly greater tendency towards absorption or binding of flavors. The content of amylose and amylopectin in starch granules varies depending on the source. When mixed with water and provided with enough heat, starch granules swell sufficiently to form pastes that can produce strong films; however, the viscosity of native starch is too high for most encapsulation processes.

Maltodextrins, $(C_6H_{12}O_5)_nH_2O$, are nonsweet nutritive polysaccharides consisting of α-(1→4) linked D-glucose units. However, in order to be termed maltodextrin, they must possess a reducing sugar content or "dextrose equivalence" (DE) of less than 20. Maltodextrins are prepared as white powders or concentrated solutions by partial hydrolysis of corn starch with safe and suitable acids or enzymes. If the DE equals or exceeds 20, they are referred to as corn syrup solids. DE, expressed as a percentage, is a measure of the reducing power of a sample compared to an equal weight of dextrose. Common designations of maltodextrins are 5, 10, 15, and 18 DE, while commercial corn syrup solids have 20, 25, 36, and 42 DE [19]. Products with a DE greater than 42 cannot be easily dried and hence are sold only as concentrates/syrups. Because maltodextrins and corn syrup solids are so closely related to one another in terms of their physical and chemical properties as well as their applicability to food ingredient encapsulation, they will be discussed jointly. A flow diagram for the production of maltodextrins and corn syrup solids from corn starch is presented in Figure 2.

In the production of maltodextrins and corn syrup solids, starch is only partially hydrolyzed by acid or enzymes; thus, the resulting products are heterogeneous mixtures of various chain length glucose polymers. The higher the DE, the higher the concentration of product that can be put into

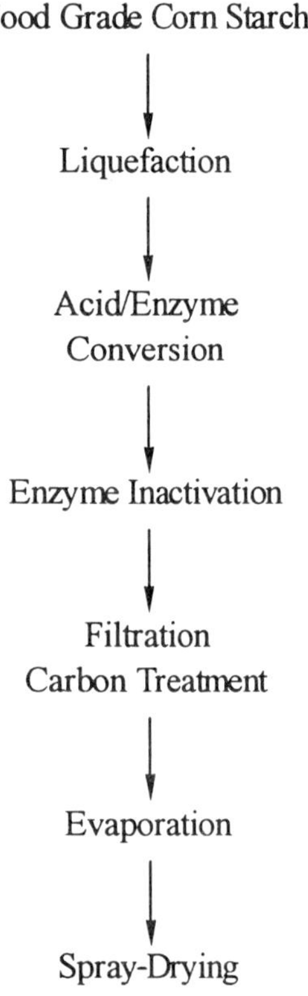

FIGURE 2 Flow diagram for the production of maltodextrin and corn syrup solids from corn starch. (From Ref. 19.)

solution. In spray-dried encapsulations, increased levels of soluble solids at a low viscosity is a major factor in the efficiency of production. In spray-dried encapsulation of citrus oils, Anandaraman and Reineccius [20] reported that the higher the DE of the corn syrup solids used, the longer the stability of the encapsulated oil. Bangs and Reineccius [21] found intermediate- or lower-DE products to be more efficient for spray-dried encapsulation of volatile artificial flavor compounds. It was postulated that a balanced polymer length might aid in trapping the volatiles as the surface of the droplet dries.

These hydrolyzed starches offer the advantages of being relatively inexpensive (approximately one third that of modified starches), bland in flavor, and low in viscosity at high solids content. However, the major problem with these products is the lack of emulsification properties [22]. Since most active materials (especially flavors) are insoluble in aqueous solutions, they must be present as emulsions. Thus, emulsion stability is viewed as an important consideration when selecting a coating material. Maltodextrins and corn syrup solids lack lipophilic characteristics and have virtually no emulsion-stabilizing effect on water-insoluble components [18]. It is also found that maltodextrins and corn syrup solids do not retain volatile compounds well during spray-drying; corn syrup solids typically perform more poorly, and retention often ranges between 65 and 80% [18]. The retention capacity changes significantly as DE values change. Raja et al. [23] investigated the use of maltodextrins with varying DE values for encapsulating cardamon oil. The reason for the poor retention of volatiles by maltodextrins and corn syrup solids was believed to be their poor film-forming capabilities (which is why they are sometimes referred to as carriers and not encapsulators). The wet encapsulation matrix must form a film around the droplets of active material in order to effectively retain them during the drying process and water removal. It is considered that since maltodextrins and corn syrup solids have no emulsification properties, they produce coarse emulsions and therefore result in poor flavor retention during drying [24].

Maltodextrins and corn syrup solids vary greatly in protecting encapsulated ingredients from oxidation. There is a strong dependence of associative stability on the DE of the hydrolyzed starch. The encapsulated product with the highest DE is extremely stable and would have a shelf life of years without use of an antioxidant [20]. Several factors have been attributed to the outstanding protection afforded by high-DE coating materials. It has been considered that the higher-DE systems are less permeable to oxygen and therefore offer better protection to encapsulated ingredients [18]. One should also keep in mind that the presence of glucose in the encapsulation system has a considerable effect on the antioxidative properties.

2. Modified Starch

Starch presents an interesting situation with regard to flavor binding. Because the amylose fraction forms helical structures, starch can entrap flavor molecules, thereby producing very stable complexes [25]. However, starch is hydrophilic and hydrolysates derived from it afford virtually no emulsification properties to the compound being encapsulated.

In its natural state, starch is cold water insoluble. One method used to modify its viscosity and cold water solubility is pyroconversion or dextrinization. In dextrinization, starch is heated in dry granular form, generally in the presence of acid or alkali. Partial hydrolysis of starch granules ensues as well as repolymerization to form more highly branched polymers. The extent of this process can be varied to yield products with different solubility and viscosity characteristics. Dextrins have increased cold water solubility and lower solution viscosity than gelatinized native starch. However, if heated too long, the products become darker and stronger reaction flavors are noted. Unfortunately, these strong color and flavor characteristics and a lack of lipophilic emulsifying qualities make dextrins less than ideal for encapsulation, especially of oil-based products.

The lack of emulsification properties of native starch creates two significant problems. The first is poor flavor retention. The fineness of the infeed emulsion has a strong influence on determining the extent of flavor retention during drying. The second problem relates to the stability of the flavor emulsion once reconstituted in the final product. If the carrier provides no emulsification to the flavor, then the flavor rapidly separates from the product and forms a ring at the top. Thus, for a compound to function as an emulsifier, it must contain both lipophilic and hydrophilic groups. To overcome this problem, starches can be modified chemically to change their functional characteristics. For example, the U.S. Food and Drug Administration (FDA) has approved the reaction of starch with 1-octenylsuccinic anhydride to form a modified starch containing both hydrophobic and hydrophilic groups. This level of substitution, usually in the range of 0.02%, results in a product that is vastly different from that of the native starch. The addition of lipophilic moieties along the starch polymer permits the formation of emulsions with tight alignment of the polymer around an oil droplet. This stabilization is extremely important for encapsulation of lipid products. Modified starch provides excellent retention of volatiles during spray-drying and can be used at a higher infeed solids level than gum acacia (also known as gum arabic). While gum acacia is generally limited to use at about 35% infeed solids level, modified starch can typically be used at levels approaching 50% [18]. The high solids levels helps to reduce the loss of encapsulated ingredients and increases spray-dryer throughout.

The emulsification properties of lipophilic starches as well as the oil retention in the spray-dried powders are reported to be equal to or greater than that of gum acacia [26,27]. Modified starch also excels in promoting emulsion stability. One means of doing so is to produce small particle-size droplets. Solutions of gum acacia produced an average emulsion droplet size of about 3 μm, and modified starch gave droplets of less than 2 μm. The emulsions made with modified starch were physically more stable than those made with the standard gum acacia [17]. Reineccius [18] pointed out that modified starches do have some disadvantages. For example, they are not considered natural for labeling purposes, they often have an undesirable off-flavor, and they do not afford good protection to oxidizable flavorings.

3. Cyclodextrins

Cyclodextrins are chemically and physically stable molecules formed by the enzymatic modification of starch. They have an ability to form complexes with a wide variety of organic compounds within their ringed structure. The ability of these unusual molecules to form inclusion complexes, which can change the physical and chemical properties of guest molecules, offers a variety of potential uses to the food industry. Although cyclodextrins have been studied for a century and their ability to form inclusion complexes has been recognized for at least 40 years, they were not utilized for food applications until the 1970s when Japan and Hungary began producing them commercially.

Cyclodextrins were discovered in 1891 when Villiers reported their appearance in rotting potatoes. In 1904, Schardinger characterized them as cyclic oligosaccharides and identified *Bacillus macerans* as the bacterium that produced cyclodextrin glycosyltransferase (CGTase), the enzyme responsible for the generation of cyclodextrins from starch. Because of Schardinger's studies, cyclodextrins were initially referred to as Schardinger dextrins. Of more significance was the fact that his work set the direction for future research, pointing it toward a study of the structure of cyclodextrins and their commercial production. French [28] has provided a detailed history of the development of cyclodextrins up to 1956.

Today, cyclodextrins are produced from starch by selected microorganisms such as *B. macerans* and *Bacillus circulans*, which have CGTase activity. After cleavage of starch by the enzyme, the ends are joined to form circular entities with β-(1→4) linkages. Because cyclodextrins are closed circular molecules, glucoamylases and β-amylases cannot hydrolyze them as there is no reducing end group, which is necessary to initiate hydrolysis. The cyclic dextrins formed contain six-, seven-, or eight-glucose monomers; these are referred to as α-, β-, and γ-cyclodextrin, respectively. The glucose monomers are joined to one another in a doublenut-shaped ring, giving the cyclodextrins a molecular structure that is relatively rigid and has a hollow cavity of specific diameter and volume. Depending upon the enzyme used and the conditions under which the reaction is performed, the ratio of cyclodextrins can vary from various mixtures to a single cyclodextrin being formed.

Figure 3 shows the chemical structure of β-cyclodextrin, the predominant cyclodextrin produced by CGTase enzymes. Polar hydroxyl group of the glucose monomers are located on the rim of the molecule and are directed away from the cavity. These groups interact with water, giving cyclodextrins their aqueous solubility properties, and will interact with polar groups of some molecules to form hydrogen bonds. While the outer surfaces (top and bottom) are hydrophilic, the internal

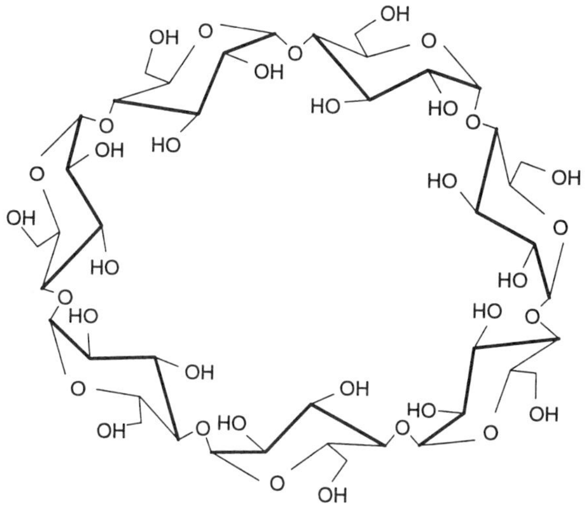

FIGURE 3 Chemical structure of β-cyclodextrin.

cavity has a relatively high electron density and is hydrophobic in nature due to the hydrogen and glycosidic oxygen atoms being oriented to the interior of the cavity.

Organic molecules of suitable size, shape, and hydrophobicity are able to interact noncovalently with cyclodextrins to form stable complexes. Several forces, such as van der Waals forces, hydrophobic interaction, and dipole-dipole interaction, are involved in the binding of guest molecules to the cyclodextrin cavity. These forces are sufficiently efficacious to form a stable complex but are not so secure that the guest molecule can be released from the complex to become available for its intended effect [29].

The dimensions of the cyclodextrin's cavity allow some selectivity for complexation of guest molecules. Strong binding results if more interaction occurs between the walls of the cyclodextrin and the guest molecule. If the molecule to be encapsulated is small compared to the cavity, only part of its surface is in contact with the walls and the full potential of the guest molecule to interact with the cyclodextrin is not realized. For molecules containing five or fewer carbon atoms, the smaller cavity of α-cyclodextrin affords more interaction between the molecule and the cavity walls. Better complexation results than if β- or γ-cyclodextrin were used. On the other hand, large bulky molecules, such as anthracene, fit into the cavity of the γ-cyclodextrin better than that of α- or β-cyclodextrin. In fact sometimes molecules are too large to fit into the cavity of one or more of these. Thus, the guest molecule might be totally excluded from the cavity or only a portion of it would fit. The more of the molecule that can fit into the cavity, the stronger the binding. Some of the physical properties of cyclodextrins are summarized in Table 2 [30].

β-Cyclodextrin deserves special attention, as it is the most readily available cyclodextrin. In preliminary studies it is generally used and is known to be able to form inclusion complexes with flavor ingredients of molecular masses ranging between 80 and 250 daltons. Lindner [31] reported that the molecules of nearly all natural spices and flavors fit into this range. Much research has focused on the ability of cyclodextrins to prevent the volatilization of flavors and essences from spices, flavor extracts, and lipids. Nagatomo [32] reported that cyclodextrins improve the stability of spices for use in sausages and other meat products. Spices that have been included in cyclodextrins have demonstrated controlled flavor release. In addition, thermal stability is improved when fats are added to them. Nagatomo [32] also noted that cyclodextrins preserved the flavor of cookies, vegetable pastes, biscuits, citrus fruits, Japanese onions, garlic, celery, and a variety of other products. Pagington [33] reported that the strong odor of onion oil, garlic oil, and pyrazines was restricted by cyclodextrin use, but complexing with cyclodextrins prevented their flavor from being lessened in processing and released their flavor directly into the mouth.

TABLE 2 Physical Properties of Cyclodextrins

Type of cyclodextrin	Physical properties						
	Number of glucose units	molecular weight	Molecular dimensions (Å)			Solubility at 25°C (g/100 ml H_2O)	$[\alpha]_D^{20}$ (H_2O, 1%)
			Inside diameter	Outside diameter	Height		
α	6	973	5.7	13.7	7.0	14.50	150.5°
β	7	1135	7.8	15.3	7.0	1.85	162.5°
γ	8	1297	9.5	16.9	7.0	23.20	117.4°

Source: Ref. 12.

Natural pigments, such as carotenoids and anthocyanins, can also be stabilized by a cyclodextrin complex [32]. Pigments can be masked or color tones intensified. It has been reported that the colors can be changed through the inclusion complexation process. Cyclodextrin complexes can protect ingredients from oxidation, light-induced reactions, thermal decomposition, and evaporation loss. Crystalline complexes are stable and improve processing conditions, handling, and storage of food ingredients.

4. Modified Cyclodextrins

Although β-cyclodextrin forms a stable microcapsular structure, water solubility of β-cyclodextrin complexes is generally a problem. The solubility of α- and γ-cyclodextrin at room temperature is 12.8 and 25.6 g/100 ml water, respectively, whereas, for β-cyclodextrin it is only 1.8 g/100 ml. As temperature increases, the solubility of cyclodextrins also increases, but their solubility can change when a guest is complexed. If the guest molecule is highly soluble in water, then the inclusion complex is more soluble than the cyclodextrin itself. The polar or ionic moiety of the guest molecule projects out of the cavity and contributes to the solubility of the complex along with the interaction of the hydroxyl groups of the cyclodextrin. On the other hand, complexation of the cyclodextrin with a guest that is not soluble or only partially soluble in water generally results in a decrease in the solubility of the cyclodextrin. Although solubility of the complex is generally less than that of the cyclodextrin, it is greater than that of the guest molecule.

The solubility of cyclodextrins can be improved by substituting hydroxyl groups along the rim of the cyclodextrin molecule [34]. By chemical modification, cyclodextrins can attain characteristics very different from those of the original material. Moreover, cyclodextrins can be incorporated into polymer structures. One such polymer can be produced by linking cyclodextrin rings with suitable agents such as epichlorohydrin in order to obtain insoluble copolymers in the form of water-swelling beads. Some of these polymers retain the ability of the cyclodextrin to form complexes with various compounds, especially those with hydrophobic groups.

It has been reported that if cyclodextrins are linked to polyethers, water-soluble polymers are produced [35]. Initial studies have shown that heptamino-β-cyclodextrin can be cross-linked by hexamethylene diisocyanate. If the degree of polymerization is high enough, the cyclodextrins bound within the matrix become insoluble. The chemical structure of an experimental polymer produced by Amaizo is shown in Figure 4.

5. Sucrose

As the most commonly used ingredient in the food industry, sucrose (β-D-fructofuranosyl-α-D-glucopyranoside) provides sweetness and is used as a bulking agent, texture modifier, preserving agent, and fermentation substrate in food applications [36]. Sucrose is also useful as a carrier for food ingredient encapsulation because of the following properties: (a) quick dissolution in water producing a clear solution; (b) heat stability; (c) nonhygroscopicity; (d) indefinite shelf life under ambient conditions; and (e) inexpensive nature [37].

In extrusion processing, sucrose and other mono- and disaccharides provide flavor, sweetness, energy, texture, stabilization, water activity control, and color control [38]. Sucrose and maltodextrin mixtures are the most commonly used coatings for extrusion encapsulation [39,40]. Flink and Karel [15] reported that retention of volatiles by carbohydrates during lyophilization was roughly in the order of sucrose > maltose ≥ lactose > glucose ≫ dextran T-10. In the case of lactose, its crystal form as well as the structure of the volatile influenced the amount of absorption [41].

Sucrose is used for encapsulating food flavors by a process known as cocrystallization [37,42–44]. However, before it can be used, its chemical structure needs to be modified from a single perfect crystal to that of a microsized, irregular, agglomerated form, before cocrystallization occurs. This modified structure has an increased void space and surface area, which provide a porous bed or base for the incorporation of active ingredients.

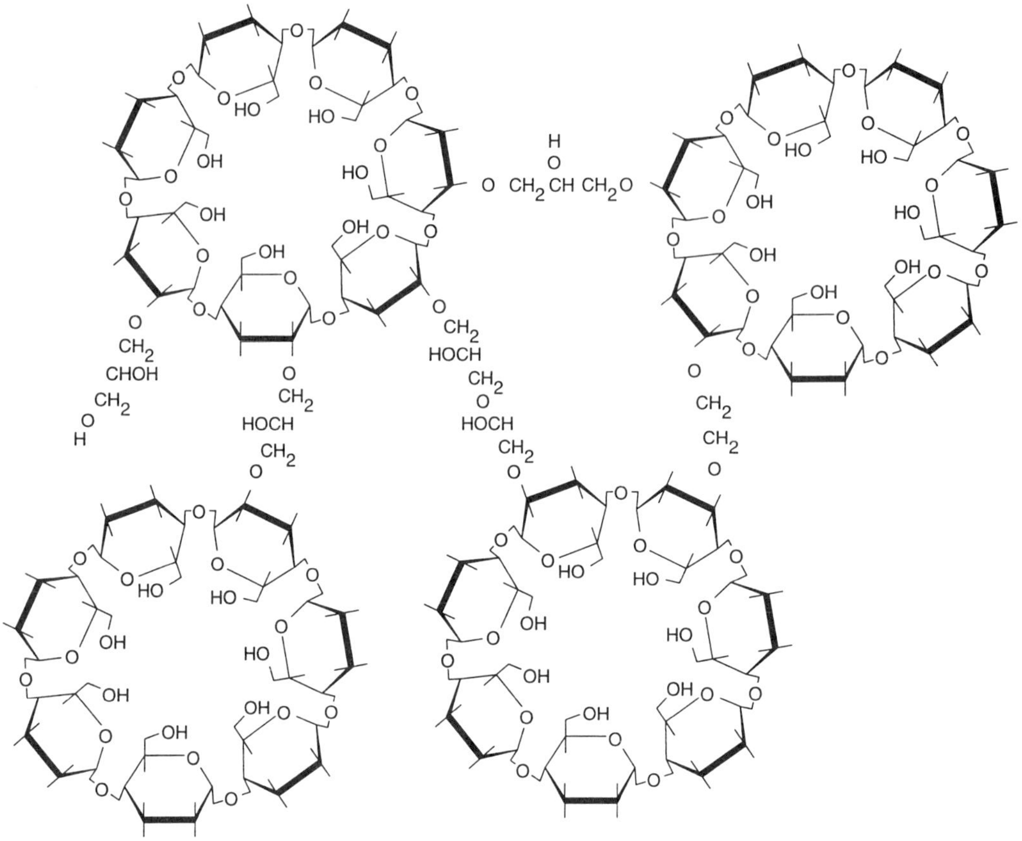

FIGURE 4 Structure of a polymeric, modified β-cyclodextrin. (From Ref. 30.)

6. Chitosan

Chitosan is the principal product from the alkaline hydrolysis of chitin, a main constituent of the exoskeleton of crustaceans such as crab. It consists of 2-deoxy-2-aminoglucopyranosyl residues joined by β-(1→4) linkages. Complex coacervate capsule formation can occur between chitosan, a cationic polyglucosamine, and carrageenan or alginic acid, which are anionic in nature.

Gel bead formation can be achieved by interaction of chitosans with low molecular counterions such as polyphosphates. The gelling properties of chitosans allow for a wide range of applications, the most attractive being coating of foods and pharmaceuticals and gel entrapment of biochemicals, plant embryos and whole cells, microorganisms, or algae [45,46]. Such entrapment offers diverse uses including microencapsulation and controlled release of flavors, nutrients, or drugs. Because chitosan has been shown to be an effective agent, concurrent cell permeabilization and immobilization using chitosan-containing complexes in coacervate capsules have been explored [45,46].

Polycationic chitosan molecules can be incorporated with oppositely charged polymers to form coacervate capsules of good mechanical strength. The permeability of these coacervate capsules can be controlled by altering either the type of chitosan and/or the counterion [47].

7. Cellulose

Cellulose is the main constituent of plant cell walls. It consists of glucopyranosyl residues joined by β-(1→4) linkages. Together with some other inert polysaccharides, cellulose constitutes the indigestible carbohydrate fraction of plant foods, referred to as dietary fiber. The importance of dietary fiber in human nutrition appears to be mainly the maintenance of intestinal mobility (peristalsis).

Cellulose as an edible film for food preservation and other functional ingredients in food processing has attracted much research interest [48–50]. As an edible film for food coatings, the permeability of cellulose coatings can be modified by combining them with other coating materials [51]. It was found that methyl- and hydroxypropyl methylcellulose mixed with lauric, palmitic, stearic, and arachidic acids significantly lowered the permeation rate relative to cellulose ether films containing no fatty acids [52]. Cellulose has always been used in encapsulation of water-soluble food ingredients such as sweeteners and acids. Furthermore, it can be used to encapsulate enzymes and cells [53].

B. Gums

One class of material often exploited for its encapsulating capabilities is hydrocolloids or, more commonly, gums. These compounds are long-chain polymers that dissolve or disperse in water to give a thickening or viscosity-building effect [54]. Gums are generally used as texturing ingredients, but their secondary effects include encapsulation [55], stabilization of emulsions, suspension of particulates, control of crystallization, and inhibition of syneresis (i.e., the release of water from fabricated foods) [56,57]. Additionally, several gums are capable of forming gels.

Food gums are obtained from a variety of sources. Although most gums are obtained from plant materials such as seaweed, seeds, and tree exudates, others are products of microbial biosynthesis, and still others are produced by chemical modification of natural polysaccharides. Some gums commonly used as coating materials for food ingredient encapsulation are discussed below.

1. Seaweed Extracts

Alginates, agar, and carrageenan are extracts from red (*Rhodophyceae*) and brown (*Phaeophyceae*) algae, collectively referred to as seaweeds [58]. Their use in encapsulation processes is well documented. The major source of alginates used for industrial production is the giant kelp (*Macrocystis pyrifera*). Algae are extracted from alkali from seaweed, and the polysaccharide is usually precipitated from the extract by addition of acids or calcium salts.

Alginates include a variety of products made up of β-D-mannuronic and α-L-guluronic acids joined by α-(1→4) linkages. They are arranged either in regions composed solely of one unit or the other, referred to as M-blocks and G-blocks, or in regions where the two units alternate [58]. Both the ratio of mannuronic to guluronic acid and the structure of the polymer determine the solution properties of the alginate. Alginates are powerful thickening, stabilizing, and gel-forming agents and are utilized in a variety of foods. At a level of 0.25–0.5%, they improve and stabilize the consistency of fillings for baked products, salad dressings, and milk chocolate and prevent the formation of large ice crystals in ice cream during storage. They are also used as an encapsulating agent. It has been reported that water-soluble alginate was capable of forming encapsulated liquid capsules [59]. Viscous high-fat food can also be encapsulated with calcium alginate [60].

Agar is a heterogeneous complex mixture of related polysaccharides having the same backbone chain structure. Its main components are β-D-galactopyranose (galactose) and 3,6-anhydro-α-L-galactose, which alternate through 1-4 and 1-3 linkages. The chains are esterified to a low extent with sulfuric acid. Deemed as one of the most potent gel-forming agents, agar produces perceptible gelation at concentrations as low as 0.04%. The gelling properties of the gum, the heat resistance of its gels, and the differential between the gel-forming and melting temperatures are the primary reasons for selecting agar. Chlorella gar has been used for the encapsulation of flavors [61].

Carrageenan is composed of β-D-galactose and 3,6-anhydro-D-galactose, which is partially sulfated as 2-, 4-, and 6-sulfates and 2,6-disulfates. The galactose residues are alternatively linked by 1-3 and 1-4 linkages. Carrageenan utilization in food processing is based on the ability of the polymer to gel, to increase solution viscosity, and to stabilize emulsions and various dispersions. Gels from carrageenan are thermoreversible. Because of its reactivity with certain proteins, the gum has found use at low concentrations (typically 0.01–0.03%) in a number of food products. A process for producing capsules containing meat soup or juice with agar-agar, carrageenan, or pectin coatings has been developed Hoashi [62].

2. Exudate Gums

Gum arabic (acacia), gum ghatti, gum karaya, and gum tragacanth are referred to as exudate gums. Among these, gum arabic, which is a natural vegetable colloid obtained by exudation from the trunk and branches of leguminous plants of the Acacia family, primarily *Acacia senegal*, is the most commonly used encapsulation coating material [63,64]. Although there are several hundred species of Acacia, only a few are gum products, and these are located in the subdesert region of Africa.

Gum acacia is a mixture of closely related polysaccharides, with an average molecular weight range of 260–1160 kDa. Gum acacia primarily consists of D-glucuronic acid, L-rhamnose, D-galactose, and L-arabinose, with about 5% protein. This protein fraction is responsible for the emulsification properties of the gum. The gum also exists as a mixed salt of sodium, calcium, magnesium, and potassium. Owing to the complex character of this polymer, the stereochemical organization of the molecule is not completely understood, even though the qualitative and quantitative analysis of the sugars is. A hypothesis of the structure of gum acacia is presented in Figure 5.

Gum acacia is the traditional gum of choice for flavor encapsulation via spray-drying. It is an outstanding natural emulsifier and rates well based on criteria used in evaluating a flavor carrier. Because beverage applications account for a large proportion of dry flavorings used, emulsion stability in the finished product is one of the most important criteria in carrier selection. It has the advantage of being considered natural in virtually all countries. An interesting and unique property of gum acacia is its low viscosity in aqueous solutions. Although solutions containing up to 50% gum can be prepared, the solution viscosity starts to rise steeply at concentrations of greater than 35%. Most other gums yield solutions with a high viscosity at concentrations as low as 1%. It is impossible to effectively atomize these very viscous emulsions, and thus these other gums are not especially useful as flavor encapsulants.

Gum acacia is also applied as a flavor fixative in the production of powdered aroma concentrates. While modified food starches are superior to traditional gum acacia in emulsion stability, gum acacia produces quite stable emulsions. The emulsions are then spray-dried. In this process, the polysaccharide forms a film surrounding the oil droplet, which then protects the oil against oxidative degradation. Compared to maltodextrins, gum acacia gives superior aroma retention during drying, and very little aroma is lost during storage at humidities below the water monolayer level [65]. New-generation gums (blends of West African gums) have been shown to be superior even to modified starches for stabilizing flavor emulsions [18]. Protection of oxidizable flavorings by gum acacia varies with the source of the gum. The traditional gum acacia is not quite as good as the modified food starch/corn syrup solids blend and quite inferior to the blends of West African gums [18]. Blends of gum acacia with maltodextrins and the new West African gum acacia can be used to encapsulate flavors and offer excellent stability to oxidation [66].

C. Lipids

1. Wax

Waxes are important derivatives of higher alcohols, such as C_{12}–C_{28}, which are esterified to long-chain fatty acids. Traditionally, wax coatings have been applied to fresh fruits and vegetables to extend their postharvest storage life. Edible waxes are significantly more resistant to moisture transport than most other lipid or nonlipid coatings. It has been reported that waxes are most effective in blocking moisture migration, paraffin wax being the most resistant followed by beeswax [67–69]. For this reason, waxes are commonly used as lipid coatings for encapsulation of food ingredients, particularly for the encapsulation of water-soluble ingredients. In 1980, petroleum wax was permitted for use by the FDA in formulating microcapsules for encapsulation of spice-flavoring substances in frozen pizza [70].

The great resistance of paraffin and beeswax coatings to diffusion of water is related to their molecular compositions. Paraffin wax consists of a mixture of long-chain, saturated hydrocarbons,

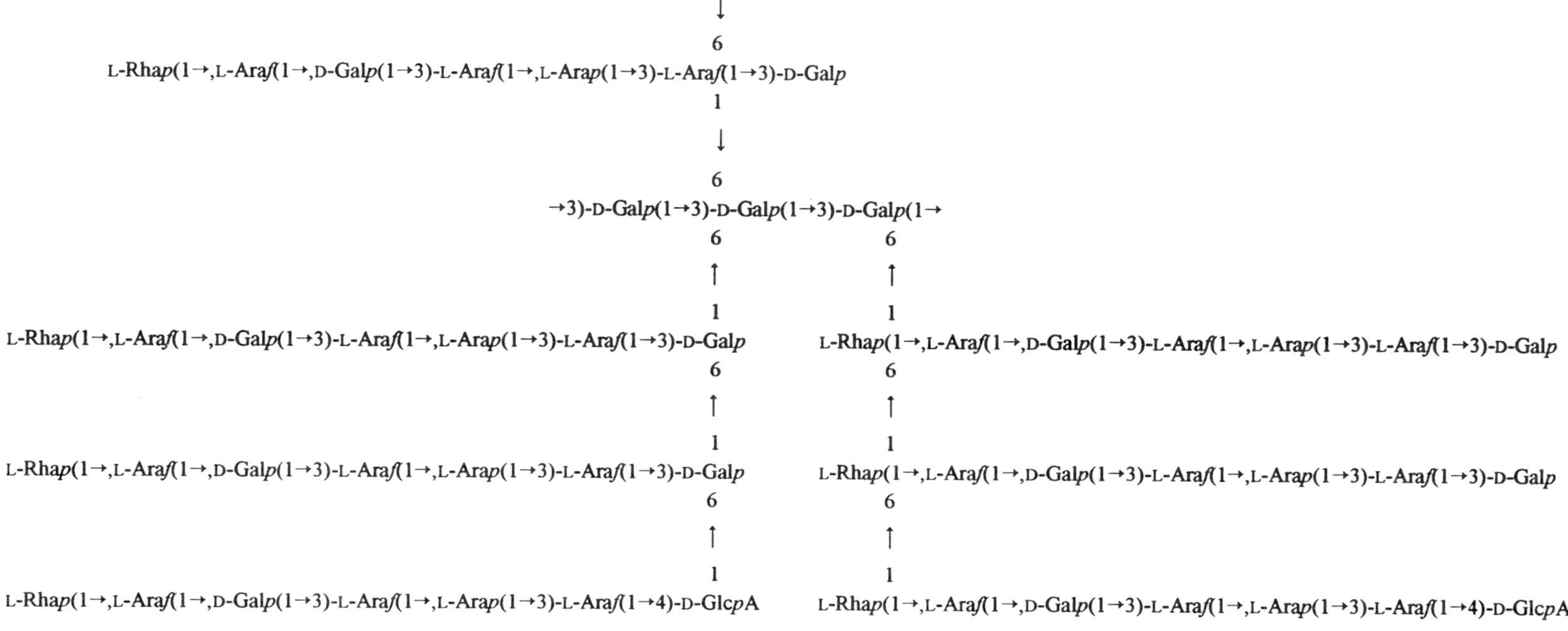

Figure 5 Proposed structure of gum acacia (*Acacia senegal*), where L-Rha*p* = β-L-rhamnopyranosyl; L-Ara*f* = β-L-arabinofuranosyl; D-Gal*p* = β-D-galactopyranosyl; L-Ara*p* = β-L-arabinopyranosyl; and D-Glc*p*A = β-D-glucuronopyranosyl acid. (From Ref. 64.)

whereas beeswax consists of 71% hydrophobic, long-chain ester compounds, 15% long-chain hydrocarbons, 8% long-chain fatty acids, and 6% other compounds [71,72]. The absence of polar groups in paraffin and the relatively low level in beeswax account for their significant resistance to moisture transport.

2. Acetoacylglycerols

Acetylation of glycerol monostearate by reaction with acetic anhydride yields 1-stearodiacetin. This acetylated monoacylglycerol displays unique characteristics of solidifying from the molten state into a flexible, waxlike solid.

It is found that the barrier properties of acetoacylglycerol improve as the degree of acetylation increases. This is due to removal of free hydroxyl groups, which would otherwise interact directly with migrating water molecules or other small polar molecules. The lower permeability through the acetoacylglycerol film prepared from technical grade monoacylglycerols might be a consequence of difference in crystal packaging or the number of free hydroxyl groups [68]. Although the water vapor permeability of acetylated monoacylglycerol films is considerably less than that of most polysaccharide films, it is greater than the permeability values of ethyl- and methylcellulose [73].

3. Lecithins

Lecithin plays a significant role as a surface-active substance in the production of emulsions. Pure lecithin is a water-in-oil (W/O) emulsifier with a hydrophile-lipophile balance (HLB) value of about 3. Because commercially used lecithins are complex mixtures of lipids, their HLB values vary considerably.

Major phospholipids of raw soya lecithin are listed in Table 3 [74]. The ethanol-insoluble fraction is suitable for stabilization of W/O emulsions and the ethanol-soluble fraction for oil-in-water (O/W) emulsions. To increase the HLB value, "hydroxylated lecithins" are prepared by controlled partial oxidation of unsaturated acyl residues with hydrogen peroxide or benzoyl peroxide [74].

Lecithin vesicles have recently been used for encapsulation of food enzymes since the formation of lecithin capsules can be achieved under relatively low temperatures. Using lecithin vesicles to encapsulate lysozyme and pepsin, it was found that the encapsulating efficiency was best when the pH was close to the isoelectric point of each enzyme [75].

Blended with other coating materials, lecithin will change the structure of microcapsules formed. Studies on the encapsulation of β-galactosidase in lecithin-cholesterol liposomes prepared by dehydration-rehydration (DR) and reverse-phase evaporation (RE) by Matsuzki et al. [76] revealed that encapsulation efficiency decreased as cholesterol content increased. A mixture of lecithin and polyethylene has been used for encapsulating other active ingredients, such as sweeteners and flavor compounds [77]. As a nutrient, lecithin has also been encapsulated as a dietary supplement [78].

4. Liposomes

A liposome (or lipid vesicle) is defined as a structure compound of lipid bilayers that encloses a number of aqueous or liquid compartments [79]. Prepared by a variety of techniques, liposomes consists of one, a few, or many concentric bilayer membranes whose size varies from about 25 nm

Table 3 Percentage of Phosphatidyl Compounds in Unfractionated and Fractionated Soy Lecithin

Type	Unfractionated	Ethanol-soluble fraction	Ethanol-insoluble fraction
Phosphatidylethanolamine	32.6	32.5	32.6
Phosphatidylcholine	32.6	65.1	4.6
Phosphatidylinositol	34.8	2.4	62.8

Source: Ref. 74

to several μm in diameter (Fig. 6).

Over the past 20 years, liposomes have been studied extensively in the medical and pharmaceutical areas because of their potential use as targetable carriers of drugs and bioactive macromolecules [80]. Liposome microencapsulation technologies have been developed almost to the point where they can be employed in a variety of commercial applications. Recently, there has been interest in the use of liposomes in the food industry for development of new food products with improved characteristics, especially for encapsulation or immobilization of enzymes.

Liposomes are prepared from phospholipids such as those from egg yolk or soybean lecithins. Semi-synthetic phospholipids with fatty acid chains of defined length and saturation as well as cholesterol are also used for specific purposes. The choice of the type of phospholipid and the amount of cholesterol play important roles in determining liposomal stability during storage and their fate in injected animals [80]. Virtually any substance, regardless of solubility, electrical charge, molecular size, or other structural characteristics, can be incorporated into liposomes, provided that the substance does not interfere with liposome formation [80]. Water-soluble materials may be entrapped in the aqueous phase of liposomes, whereas lipid-soluble materials will be incorporated into the lipid phase.

Liposome structure is determined by its method of preparation. Although various techniques exist for preparing liposomes [81,82], they are generally divided into three classes, namely multilamellar vesicles (MLV), small unilamellar vesicles (SUV), and large unilamellar vesicles (LUV).

Multilamellar vesicles were first prepared by Bangham et al. [83]. In a typical preparation, a solution of phospholipids in chloroform is evaporated producing a thin film, which is then hydrated with an aqueous solution. The main advantage of MLV is that the lipid and the aqueous solution to be encapsulated are not subjected to harsh treatments such as exposure to organic solvents or to high-intensity ultrasound. However, a major disadvantage of MLV is their heterogeneous size distribution (diameters 0.2–2.0 μm) and their low encapsulation efficiency (5–14%) [84].

Small unilamellar vesicles were first prepared from MLV by sonication. High-intensity ultrasound results in MLV of a much smaller size (25–50 nm in diameter). A second method for producing SUV involves injection of lipid dissolved in ethanol into the desired aqueous phase. The resulting vesicles had diameters in the range of 30–110 nm, while a third technique involves pumping of MLV through a French pressure cell to produce liposomes with diameters in the range of 30–50 nm [82]. The main disadvantage of SUV is their small diameter and consequently their low capture volume.

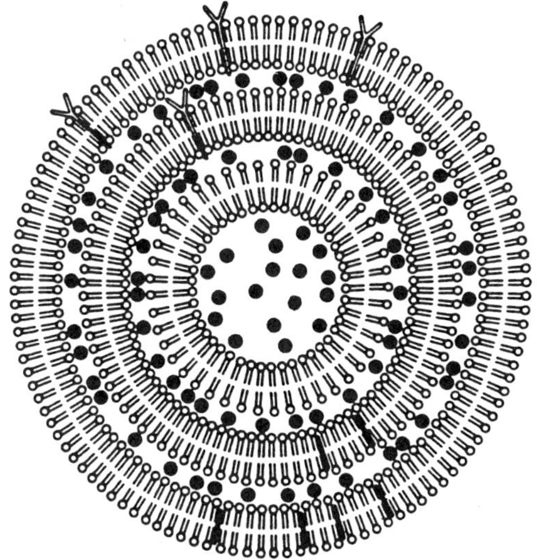

FIGURE 6 Molecular organization of a liposome. (From Ref. 12.)

Several methods are available for production of LUV whose size ranges from 100 to 500 nm; these are often the most useful liposomes. The three common methods of preparing LUV are infusion, reverse-phase evaporation, and detergent dilution. In general, LUV are more homogeneous than MLV and have a higher encapsulation efficiency than SUV.

A serious drawback of the liposome preparations listed above for their application in foods has been the use of organic solvents. Liposome microencapsulation using a microfluidizer eliminates this problem because the method does not utilize any organic solvent or detergent. The two most common microencapsulation techniques, spray-drying and extrusion, encounter major problems with flavor encapsulation, the occurrence of oxidative reactions, and inability to implement procedures for intermediate-moisture foods [79]. A limitation of the use of liposomes in some food applications may be their lack of stability in the presence of moderate levels of oils or hydrophobic proteins.

D. Proteins

As an important nutrient in food, proteins possess many desirable functional properties. These properties allow them to be good candidates for coating materials for the encapsulation of food ingredients. The most commonly used protein for this purpose is gelatin, even though other proteins are equally useful.

Gelatin is a water-soluble protein derived from collagen and is a valuable coating material partially because it is nontoxic, inexpensive, and commercially available. In addition to a good film-forming properties, gelatin has other ideal chemical and physicochemical characteristics that lend themselves to microencapsulation. For example, gelatin forms thermally reversible gels when warm aqueous suspensions of polypeptides are cooled. With an aqueous solution of gelatin, the change between the gel and solid state is quite definite. However, when the gelatin concentration in the aqueous solution is lower than about 1%, definite gelation cannot be observed even by cooling. These characteristic properties are effectively used for formation of capsules.

The isoelectric point of gelatin and its derivatives can be changed depending upon the method of preparation [85]. By changing the pH of the aqueous solution, either polycationic or polyanionic effects are exhibited by gelatin. This property is used for coacervation formation.

Gelatin is often used in combination with gum acacia to form coating films. Gum acacia, a hydrocolloid derived from plant sources, consists mainly of carboxylic acid functional groups. When the pH is lower than its isoelectric point, gelatin becomes polycationic, and hence there is an interaction between polycationic gelatin and polyanionic gum acacia resulting in the formation of a coacervation. As an example, if pigskin gelatin (isoelectric point pH 8–8.5) in aqueous solution is mixed with gum acacia at pH 4.0–4.5, a complex coacervation will form because of ionic attraction between the negatively charged acacia gum and the positively charged gelatin [85]. Fixing (insolubilization) of this structure can be achieved by the use of cross-linking agents such as ionized calcium. The type of gelatin and gum acacia selected and the formation and fixing procedures employed ultimately influence coating permeability [85]. Coating formation can also be achieved by a solvent-evaporation technique.

Protein-encapsulated tallow and vegetable oils have been applied to produce animal feeds [86]. Proteins can also be used, together with other coating materials, to form microcapsules. A mixture of protein and carbohydrate has been applied to an encapsulation process of oily substances [87,88].

III. MICROENCAPSULATION TECHNIQUES

A. Spray-Drying

Spray-drying is the most widely utilized encapsulation method in the food industry and is typically used for the preparation of dry, stable food additives and flavors. The process is economical, flexible in that it offers substantial variation in encapsulation matrix, adaptable to commonly used processing equipment, and produces particles of good quality [89–91]. In fact, spray-drying production costs are lower than those associated with most other methods of encapsulation. It is also one of the

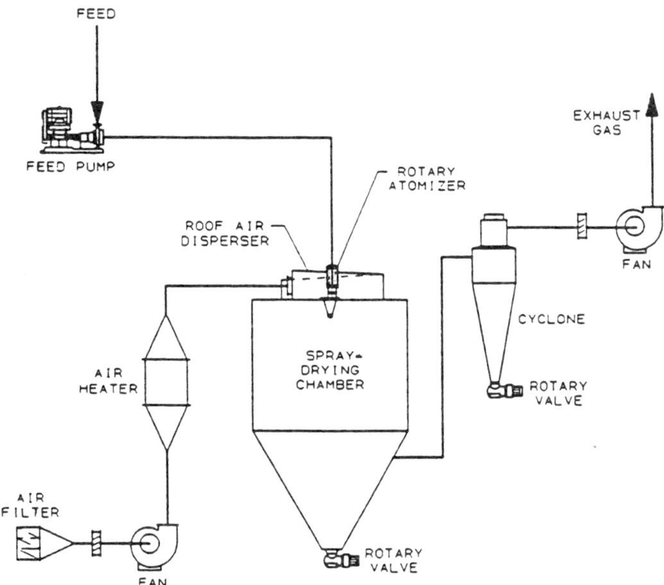

Figure 7 Typical spray-drying operation consisting of a drying chamber fitted with an atomizer, air heater, fan or blower, and cyclone for product recovery. (From Ref. 123.)

oldest encapsulation techniques, having been employed in the 1930s to prepare the first encapsulated flavors using gum acacia as the coating [92].

Although spray-drying is most often considered a dehydration process and is used in the preparation of dried materials such as powdered milk, it can be used as an encapsulation process when it entraps "active" materials within a protective matrix formed from a polymer or melt. The process is conducted in a spray dryer such as the one shown in Figure 7 and involves the following basic steps.

1. Preparation of the Dispersion or Emulsion

The initial step in spray-drying an encapsulated food ingredient is the selection of a suitable wall material or encapsulating agent. The ideal choice should have adequate emulsifying properties; be a good film former; have low viscosity at high solids levels (<500 cps at ≥45% solids levels); exhibit low hygroscopicity; release the coated ingredients when reconstituted in a finished food product; be low in cost, bland in taste, and stable in supply; and afford good protection to the encapsulated ingredients [22,93]. A food-grade hydrocolloid such as a gelatin, vegetable gum, modified starch, dextrin, or nongelling protein [11] is generally used as an encapsulating agent.

Once a wall material or combination has been selected, it must be hydrated. It is desirable to use a particular infeed solids level that is optimum for each encapsulating agent or the combination chosen. Research has shown that infeed solids level is the most important determinant of flavor retention during the spray-drying process [18]. Increasing the solids level up to the point that the additional solids are no longer soluble benefits flavor retention by decreasing the required drying time to form a high solids surface film around the drying droplets. Once the droplet surface reaches about 10% moisture, flavor molecules cannot diffuse through this surface film while the relatively smaller water molecules continue to do so and are lost to the drying air [94–97].

A high infeed solids level means that this semipermeable membrane forms quickly and thus assists flavor retention. It is possible to pump and atomize infeed materials that contain encapsulating agent solids in excess of the solubility limits. Insoluble solids offer no barrier to the diffusion of flavor molecules and therefore do not improve flavor retention during drying. It has been found that there is an optimum infeed solids level that is unique to each wall material [93,98,99].

Once the encapsulating agent or mixture has been solubilized (with or without heating), the flavor or ingredient to be encapsulated is added to the mixture and then thoroughly dispersed into the system. A typical ratio of encapsulating agent to core material is 4:1, but in some applications higher flavor loads are used. Brenner et al. [100] have obtained a patent for a process that produces high-load spray-dried flavorings. They claim that high surface oils and poor flavor retention during drying are largely due to particle shrinkage and cracking during the dehydration process. A cracked particle surface results in substantial flavor loss during drying. Brenner et al. [100] used a combination of polysaccharides (e.g., gum arabic, starch derivatives, and dextrinized and hydrolyzed starches) and polyhydroxy compounds (e.g., sugar alcohols, lactones, monoethers, and acetals) to form an encapsulating mixture that remained plastic during spray-drying. Using this plastic encapsulating agent, Brenner et al. [100] reported to have spray dried infeed materials with a flavor load of up to 75% (based on dry solids). Mass balance data showed oil recoveries of 80% at this high loading. However, higher flavor loads typically result in an unacceptable loss of flavors in the dryer. For example, Emberger [101] has shown that compared to a 10% loading, only 33–50% of the flavor was retained during drying when a 25% flavor load was used.

2. Homogenization of the Dispersion

Prior to spray-drying, the mixture is homogenized in order to create small droplets of flavor or ingredient within the encapsulating solution. The creation of a finer emulsion increases the retention of flavor during the drying process [6]. Sometimes addition of an emulsifier is required and the dispersion is then homogenized prior to spray-drying. However, considerable process variation exists within the industry in this respect. Risch and Reineccius [102] reported a direct relationship between the degree of homogenization and the retention of orange peel oil during spray-drying. Therefore, it appears advantageous to efficiently homogenize the dryer infeed material. Water-soluble materials may also be encapsulated by the treatment of homogenization. Instead of having a clearly defined core and coating, the product consists of a homogeneously blended matrix of the polymer entrapping the core. These products are sometimes described as matrix particles or entrapped ingredients. They are also said to be covered with a very fine film of coating.

3. Atomization of the Infeed Emulsion

The core/wall material mixture is fed into a spray-dryer where it is atomized through a nozzle or spinning wheel. The single-fluid, high-pressure spray nozzle and the centrifugal wheel are two types of widely used atomizers; the industry is nearly equally divided between their use. While each type of atomizer has its advantages and disadvantages, nothing in the literature suggests that one type is superior to the other.

Atomization parameters have a significant effect upon the particle size distribution of the resultant powders. Several researchers have reported that larger particles result in improved flavor retention, but Reineccius and Coulter [16] found that particle size had no effect on flavor retention. On the other hand, studies by Chang et al. [103] indicated that there is an optimum particle size for flavor retention. Part of the controversy is cleared up by Bomben et al. [97], who showed that particle size is insignificant if high infeed solids were used. This might explain why some authors found a relationship between particle size and flavor retention while others have not. Although particle size may have a minimal influence on flavor retention during drying, it is often desirable to produce large particles to aid in dispersion upon reconstitution. Small particles are often difficult to disperse and tend to float on liquid surfaces. Larger particles can be obtained by using a large orifice, low atomization pressure (pressure nozzle only), high infeed solids, high infeed viscosity, low wheel speed (centrifugal wheel atomization only), or some type of agglomeration technique [104].

4. Dehydration of the Atomized Particles

When hot air flowing in either a co-current or countercurrent direction contacts the atomized particles, water is evaporated and a dried product consisting of starch or encapsulating matrix containing small droplets of flavor or core is formed. As the atomized particles fall through the gaseous medium, they assume a spherical shape with the oil encased in the aqueous phase. This explains why most spray-dried particles are water soluble. The rapid evaporation of water from the coating during its solidification keeps the core temperature below 100°C in spite of the high temperatures used in the process [105]. The particles' exposure to heat is in the range of a few seconds at most [11]. Thus, the main advantage to this method is its ability to handle many heat-labile materials. However, since a flavor may contain as many as 20–30 different components (alcohols, aldehydes, esters, ketones, etc.) with boiling points ranging from 38 to 180°C, it is possible to lose certain low-boiling point aromatics during the drying process [89]. The dried particles fall to the bottom of the dryer and are collected, or they may be separated by a gas-solid separation unit such as a dust cyclone. Spray-dried ingredients typically have a very small particle size (generally <100 μm), which makes them highly soluble but may present separation problems in dry blends. Separation can be prevented and fluidity improved by a separate agglomeration step in which the encapsulated particles are treated with steam to induce their cohesion and form large particles. Factors such as coating structure may also effect the solubility of the spray-dried microcapsules [106].

B. Spray-Cooling and Spray-Chilling

Spray-cooling and spray-chilling are two encapsulation processes that are similar to spray-drying in that both involve dispersing the core material into a liquefied coating material and spraying through heated nozzles into a controlled environment [107]. However, unlike spray-drying, there is no water to be evaporated. Other principal differences between these processes and spray-drying lie in the temperature of the air used in the drying chamber and in the type of coating applied. Spray-drying employs hot air to volatilize the solvent from a coating dispersion; in contrast, spray-cooling and spray-chilling use air cooled to ambient or refrigerated temperatures. The core and wall mixture are atomized into the chilled air, which causes the wall to solidify around the core.

Microcapsules produced by spray-chilling and spray-cooling are insoluble in water due to the lipid coating. Consequently, these techniques tend to be utilized for encapsulating water-soluble core materials such as minerals, water-soluble vitamins, enzymes, acidulants, and some flavors.

In spray-cooling, the coating substance is typically some form of vegetable oil or its derivatives. However, a wide variety of other encapsulating materials may also be employed. These include fat and stearin with melting points of 45–122°C as well as hard mono- and diacylglycerols with melting points of 45–65°C. Taylor [89] indicated that mono- and diacylglycerols facilitate dispersion of the encapsulate in the finished, reconstituted food products and may also be considered as part of the overall emulsification system.

In spray-chilling, the coating is typically a fractionated or hydrogenated vegetable oil with a melting point in the range of 32–42°C. Coating materials with even lower melting points can be used, but their end products may require specialized handling and storage conditions [89]. Furthermore, in spray-chilling there is no mass transfer (i.e., evaporation from the atomized droplets), and therefore these solidify into almost perfect spheres to give free-flowing powders. Through atomization, it gives an enormous surface area and an immediate as well as intimate mixing of these droplets with the cooling medium.

Spray-chilling is used primarily for the encapsulation of solid food additives such as ferrous sulfate, acidulants, vitamins, and solid flavors, as well as for heat-sensitive materials or those that are not soluble in typical solvents [89]. Liquids may also be encapsulated following their conversion to a solid form, perhaps by freezing. The end product of the process, resembling fine beadlets of a large particle size, are water soluble but release their contents at or around the melting point of the wall material. With the ability to select the melting point of the wall, this method of encapsulation can be used for controlled release. The process is therefore suitable for protecting many water-soluble materials, such as spray-dried flavors, which may otherwise be volatilized from a product during thermal processing. Spray-chilled products have applications in bakery products, dry soup mixes, and foods containing high levels of fat [92].

Lamb [108] pointed out the importance of maintaining optimum temperatures during processing, as this can affect the fat's polymorphism, a phenomenon that describes the ability of a substance to exist in more than one crystalline form. He also noted that if a fat, for example, a powdered triacylglycerol, is permitted to exit from a chiller at too high a temperature, heat generated by polymorphism tended to reverse the encapsulating process and return the powder to a melt or perhaps a pasty mass.

C. Fluidized Bed Coating

Fluidized bed coating, also referred to as air suspension coating or the Wurster process, is a common technique used for commercial production of encapsulated ingredients for the food industry. In general, it has been found that dense particles with a narrow particle size distribution and good flowability are most suitable for encapsulation by fluid bed. Ideally a particle size distribution between 50 and 500 μm is best, although it is possible to encapsulate particles ranging from 35 to 5000 μm [8].

Solid particles to be sprayed are suspended in an upward-moving column of air in a fluidized bed chamber at a controlled temperature and humidity. Depending upon the specific application, the air flow may be heated or cooled [107]. Once the moving fluid bed of particles has reached the prescribed temperature, the encapsulation coating material is introduced to the system. Great variations in available wall materials exist. Cellulose derivatives, dextrins, emulsifiers, lipids, protein derivatives, and starch derivatives are examples of typical coating systems, and they may be used in a molten state or dissolved in an evaporable solvent. The coating is atomized through spray nozzles at the top of the chamber, whose droplets are of smaller size than the substrate being coated. The atomized particles travel down into the particle stream and deposit as a thin layer on the surface of suspended core material. The turbulence of the air column is sufficient to keep the coated particles suspended, allowing them to tumble and become uniformly coated. Upon reaching the top of the air stream, the particles move into the outer, downward-moving column of air, which returns them to the fluidized bed with their coating nearly dried (Fig. 8). The particles pass through the coating cycle many times per minute [10]. With each successive pass, the random orientation of the particles further ensures their uniform coating. In the case of hot melts, the coating is hardened by solidification in cool air. In the case of solvent-based coatings, the coating is hardened by evaporation of the solvent in hot air. The amount of coating applied can be regulated by controlling the length of time (i.e., residence time)

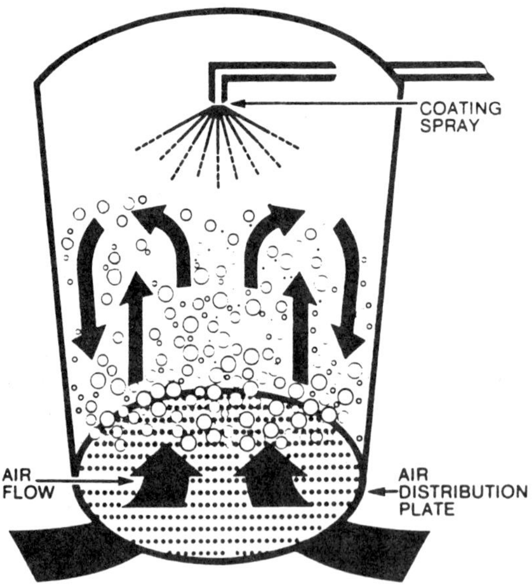

FIGURE 8 Schematic representation of a conventional air suspension system. (From Ref. 123.)

that the particles are in the chamber. In order to achieve a good degree of coating, the process lasts anywhere from 2–12 hours to complete. After this period, only 0.2–1.5% of the particles remain uncoated.

D. Extrusion

Encapsulation of food ingredients by extrusion is a relatively new process compared to spray-drying. Extrusion used in this context is not the same as extrusion used for cooking and texturizing of cereal-based products. Actually, extrusion, as applied to flavor encapsulation, is a relatively low-temperature entrapping method, which involves forcing a core material dispersed in a molten carbohydrate mass through a series of dies into a bath of dehydrating liquid. The pressures and temperatures employed are typically <100 psi and seldom exceed 115°C, respectively [22]. Upon contact with the liquid, the coating material, which forms the encapsulating matrix, hardens and entraps the core material. Isopropyl alcohol is the most common liquid used for the dehydration and hardening process. The extruded filaments or strands are broken into small pieces, dried to mitigate hygroscopicity (an anticaking agent such as calcium triphosphate can facilitate this), and sized.

Schultz et al. [109] were pioneers in the extrusion/encapsulation process. They emulsified orange peel oil in a molten dextrose mass, poured it on stainless steel sheets, and let it cool. The pulverized product exhibited good stability and flavor retention over a 6-month period. Combining the basic formulation of Schultz et al. [109] with extrusion, Swisher [110] created a novel encapsulating process that is similar to the one currently used today in the flavor industry. The primary benefit claimed in his patent [111] was the maintenance of fresh flavor in encapsulated citrus oils, which otherwise would readily oxidize and yield objectionable off-flavors during storage. He conducted an accelerated shelf-life test on encapsulated orange peel oil that contained an antioxidant and found that its shelf life was about one year. Figure 9 shows the key steps for flavor encapsulation by extrusion.

Swisher [111] added an essential oil such as orange peel oil, containing an antioxidant and a dispersing agent, to an aqueous melt of core syrup solids (42 DE) and glycerine. The core syrup melt contained from 3 to 8.5% moisture and was held at a temperature ranging from 85 to 125°C—typically 120°C. The flavor/core syrup mixture was agitated vigorously while blanketed under nitrogen to form an oxygen-free emulsion. This emulsion was forced through a die into a hot immiscible liquid (e.g., vegetable or mineral oil), which was then rapidly cooled or extruded into pellets and allowed to solidify. The hardened pellets or solid globules were ground to a desired particle size, washed with isopropanol to remove surface oil, and then dried under vacuum to yield a free-flowing granular material containing 8–10% flavoring.

The extrusion process of encapsulation has remained largely unchanged since Swisher's patent [111]. Most research developments to date concern the composition of the material that forms the encapsulating matrix. For example, Beck [112] replaced the high-DE corn syrup solids with a combination of sucrose and maltodextrin: a melt consisting of about 55% sucrose and 41% maltodextrin (10–13 DE). Even though the low-DE maltodextrin/sucrose matrix was considerably less hygroscopic than that used by Swisher [110,111], Beck continued to employ an anticaking agent and even recommended pyrogenic silica rather than tricalcium phosphate. The flavor load obtained by Beck [112] ranged from 8 to 10%, with 12% considered as a practical maximum.

Barnes and Steinke [113] were awarded a patent for developing a modified food starch in place of sucrose in a similar process. Because chemically modified starches can possess good emulsification properties, the authors hypothesized that an emulsifying starch, with its lipophilic characteristics, would absorb the flavor oils into the matrix. The maltodextrin was, therefore, used primarily to provide bulk and some viscosity control. Barnes and Steinke [113] claimed that the use of emulsifying starches in the encapsulation matrix would permit increasing the loading capacity up to 40% flavoring.

Another benefit cited by the authors was that the total replacement of sucrose with emulsifying starches resulted in a product that was "sugar-free." This might have some advantages in marketing of a final food product. Sucrose substituted with modified starches also provided greater flexibility to manufacturers. Because sucrose will invert to glucose and fructose at low pH and high

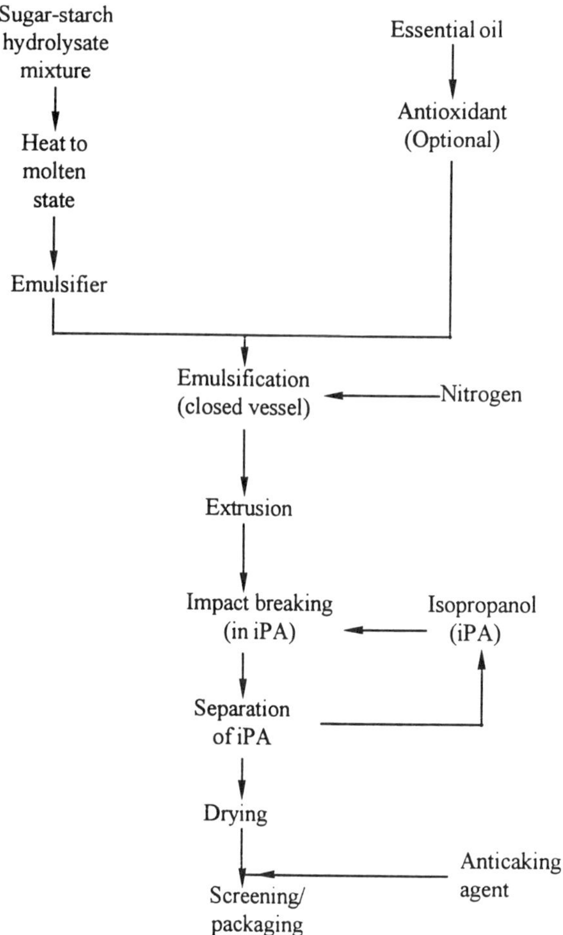

Figure 9 Flow diagram of encapsulation of food flavors via extrusion processing. (From Ref. 17.)

temperature, the resulting product would be more hygroscopic and readily participate in nonenzymatic browning reactions. Therefore, the replacement of sucrose permitted longer cooking times, larger bath sizes, and higher cooking temperatures. Barnes and Steinke [113] also claimed that fruit juices, fruit essences, volatile substances, and propylene glycol could be encapsulated in this way using their encapsulation matrix. In order to successfully encapsulate fruit essences, it was first necessary to remove water and low molecular weight alcohols from the essence. The essence was then incorporated into an edible oil so that it would form an emulsion with the encapsulation matrix. For example, orange juice concentrate (42% water) could be encapsulated at 10–15% loading levels with their process. This was a substantial improvement considering that prior formulations using sucrose were limited to 5–6% juice solids loading and could only be used with concentrates containing <20% water.

Miller and Mutka [114,115] were awarded two patents for flavor encapsulation via extrusion. The first patent [114] involved a process for the encapsulation of orange juice solids, while the second dealt primarily with optimization of the extrusion process. It was their intent to improve the flavor load and encapsulation efficiency. A study of the effect of cooking temperature on flavor load and encapsulation efficiency indicated that high-load products (>22%) had an optimum cooking temperature of about 123°C. As shown in Table 4, temperatures above or below this value resulted in poorer encapsulation efficiencies. Since the cooking temperature is basically determined by moisture con-

TABLE 4 Influence of Cooking Temperature on Encapsulation Efficiency

Oil encapsulated (%)	Encapsulation efficiency (%)	Cooking temperature (°C)
20.5	63.5	118
22.9	70.9	122
21.1	65.3	126
19.3	59.8	130
19.2	59.4	134

Source: Ref. 17.

tent, Miller and Mutka [115] postulated that too little moisture reduced emulsification effectiveness, while too much moisture hindered encapsulation. A cooking temperature of 123°C corresponded to about 5% moisture.

From the work of Miller and Mutka [115], optimization of cooking temperature, emulsifier concentration, and pressurization of the cooking vessel resulted in an improved encapsulation efficiency at high flavor loadings. Although their patent claims that loadings of up to 35% could be used, only one example with loading as high as 27.6% was cited. The majority of examples demonstrated feasibility at flavor loadings from 15 to 20%, but still such levels are well over the traditional 8–10% flavor loadings achieved in commercial applications.

The extrusion process is particularly useful for heat-labile substances and has been used to encapsulate flavors, vitamin C, and colorants. According to Risch [5], extrusion provides true encapsulation in that the core material is completely surrounded by the wall material. When the material contacts the isopropanol and the wall is hardened, all residual oil or core material is removed from the surface. The absence of residual surface oil and the complete encapsulation gives products manufactured in this manner an excellent shelf life. This technique produces larger particles, which can be used when visible flavor pieces are desirable. The primary advantage of extrusion is unquestionably its outstanding protection of the flavor against oxidation. For example, an accelerated shelf-life test on encapsulated orange peel oil containing no antioxidants was reported to be in excess of 4 years [110]. In terms of its weaknesses, extrusion is considerably more expensive than spray-drying. In fact, process costs are estimated to be nearly double those of spray-drying. Twenty percent flavor loading is standard for spray-drying, while extrusion delivers less flavor per unit weight because its loading is currently running in the 8–12% range. Finally, one must realize that extrusion is a high-temperature batch process. The flavorings must be able to tolerate 110–120°C temperatures for a substantial period of time without deterioration.

E. Centrifugal Extrusion

Centrifugal extrusion is another encapsulation technique that has been investigated and used by some manufacturers. A number of food-approved coating systems have been formulated to encapsulate products such as flavorings, seasonings, and vitamins. These shell materials include gelatin, sodium alginate, carrageenan, starches, cellulose derivatives, gum acacia, fats/fatty acids, waxes, and polyethylene glycol.

Developed by scientists in the United States, centrifugal extrusion is a liquid coextrusion process utilizing nozzles consisting of concentric orifices located on the outer circumference of a rotating cylinder (i.e., head) [116]. The encapsulating cylinder or head consists of a concentric feed tube through which coating and core materials are pumped separately to the many nozzles mounted on the outer surface of the device. While the core material passes through the center tube, coating material flows through the outer tube. The entire device is attached to a rotating shaft such that the head rotates around its vertical axis. As the head rotates, the core and coating materials are co-extruded through the concentric orifices of the nozzles as a fluid rod of core sheathed in coating material. Centrifugal force impels the rod outward, causing it to break into tiny particles. By the action of surface tension, the coating material envelopes the core material, thus accomplishing encapsulation. The capsules are collected on a moving bed of fine-grained starch, which cushions their impact and absorbs unwanted coating moisture. Particles produced by this method have diameters ranging from 150 to 2000 μm [117].

Another extrusion-based development is a process for encapsulating water-soluble lipids as particles of 1–15 mm. In this process, a core material is fed down a vertical tube while the coating material, a viscous solution of sodium alginate, simultaneously flows through a ring-shaped opening around the base of the tube, forming a membrane across the bottom of the device. The extruding core material pushes against the membrane until it eventually breaks off and carries a portion of the membrane with it. Upon spinning, the particles assume a spherical shape and become encapsulated. Passage through a bath of aqueous calcium acetate, calcium glutamate, or calcium lactate finishes this film-forming process by converting the coating to a water-insoluble calcium salt.

F. Lyophilization

Lyophilization or freeze-drying is a process used for the dehydration of almost all heat-sensitive materials and aromas. It has been used to encapsulate water-soluble essences and natural aromas [118,119] as well as drugs [120]. Except for the long dehydration period required (commonly 20 hours), freeze-drying is a simple technique, which is particularly suitable for the encapsulation of aromatic materials.

Because the entire dehydration process is carried out at low temperature and low pressure, it is believed that the process should have a high retention of volatile compounds. Model system investigations by Thijssen and coworkers [96,121] and Flink and Karel [15,122] indicated that the retention of volatile compounds during lyophilization was dependent upon the chemical nature of the system; flavor retention increased when the molecular weight of the carbohydrate wall materials decreased and the level of total soluble solids increased (up to about 20%).

For the production of citrus aroma powders to be used as natural flavor ingredients in soft drink dry mix formulations, Kopelman et al. [118] proposed the use of a freeze-drying method. By simply dissolving various blends of corn syrup solids and sugars (mono- and disaccharides) in an aroma solution at a 25% (w/w) level followed by lyophilization, these authors claimed that approximately 75% of the initial aroma volatiles could be retained in the optimal maltodextrin-sucrose mixture [118].

Freeze-drying methods can also be used for other encapsulation processes. For example, Kirby and Gregoriadis [120] used freeze-drying in the development of a technique known as DRV (dehydration-rehydration vesicles) for liposome entrapment. Upon the controlled addition of water, up to 70% of the water-soluble drugs present were entrapped in the formed liposomes. It has been reported that preparation of coatings only entrapped drugs that could be freeze-dried again and the liposomal structural integrity was apparently preserved. Intact liposomes with most of their contents still entrapped were obtained upon rehydration [80].

G. Coacervation

Coacervation, also called phase separation, was developed and patented in the 1950s by the National Cash Register Company in the United States and was used as a means of producing a two-component ink system for carbonless copy papers. Because of the very small particle size attainable with this process (ranging from a few submicrometers to 6 mm), coacervation is regarded by many as the original and true microencapsulation technique [123].

Coacervation involves the separation of a liquid phase of coating material from a polymeric solution followed by the coating of that phase as a uniform layer around suspended core particles. The coating is then solidified. In general, the batch-type coacervation processes consist of three steps, as summarized below, and are carried out under continuous agitation [9].

1. Formation of a Three-Immiscible-Chemical Phase

In the first step, a three-phase system consisting of a liquid manufacturing vehicle phase, a core material phase, and a coating material phase is formed by either a direct addition or in situ separation technique. In the direct-addition approach, the coating-insoluble waxes, immiscible polymer solutions, and insoluble liquid polymers are added directly to the liquid-manufacturing vehicle, provided that it is immiscible with the other two phases and is capable of being liquefied. In the in situ separation technique, a monomer is dissolved in the liquid vehicle and then subsequently polymerized at the interface.

2. Deposition of the Coating

Deposition of the liquid polymer coating around the core material is accomplished by controlled physical mixing of the coating material (while liquid) and the core material in the manufacturing vehicle. Deposition of the liquid polymer coating around the core material occurs if the polymer is sorbed at the interface formed between the core material and the liquid vehicle phase; this sorption phenomenon is a prerequisite to effective coating. Continued deposition of the coating is promoted by a reduction in the total free interfacial energy of the system brought about by a decrease of the coating material surface area during coalescence of the liquid polymer droplets.

3. Solidification of the Coating

Solidification of the coating is achieved by thermal, cross-linking, or desolventization techniques and forms a self-sustaining microcapsule entity. The microcapsules are usually collected by filtration or centrifugation, washed with an appropriate solvent, and subsequently dried by standard techniques such as spray or fluidized bed drying to yield free-flowing, discrete particles.

Simple coacervation deals with systems containing only one colloidal solute (e.g., gelatin), while complex coacervation deals with systems containing more than one solute (e.g., gelatin and gum acacia [124] or gelatin and polysaccharide [125]). Coacervation may also be subdivided into nonaqueous phase separation and aqueous phase separation techniques.

Aqueous phase separation has been used to encapsulate citrus oils, vegetable oils, and vitamin A. It requires a hydrophilic coating, such as gelatin or gelatin–gum acacia, and water-insoluble core particles. The resulting microcapsules may contain payloads of 85–90% and can release their contents by pressure, hot water, or chemical reaction. For nonaqueous phase separation, the coating is usually hydrophobic and the core may be water soluble or water immiscible. This process has been investigated for the encapsulation of solid food additives such as ferrous sulfate [11].

Coacervation is a very efficient but expensive process. It has found limited used in flavor encapsulation [7,126] because of the high costs associated with the technology and difficulties encountered with the level of flavor that can be incorporated into the microcapsules [89]. Another reason cited by various industries for the limited use of coacervation is problems associated with finding of suitable encapsulating materials that are food approved. According to Blenford [92], the technology is limited primarily to encapsulated ink systems used in carbonless office forms and to encapsulated fragrances that are applied in the form of "scratch-and-sniff" strips in promotional literature. However, work is currently in progress on this technology, and we may see its commercial application in the future [127,128].

H. Centrifugal Suspension Separation

Centrifugal suspension separation is a more recent microencapsulation process. The process has been patented [129,130] and was first applied commercially in February 1987 to a chemical produced in Europe. The process in principle involves suspending core particles in a pure, liquefied coating material, and then pouring the suspension over a rotating disk apparatus under such conditions that excess liquid between the core particles spreads into a film thinner than the core's particle diameter. The excess liquid is atomized into tiny droplets, separated from the coated product, and recycled. The core

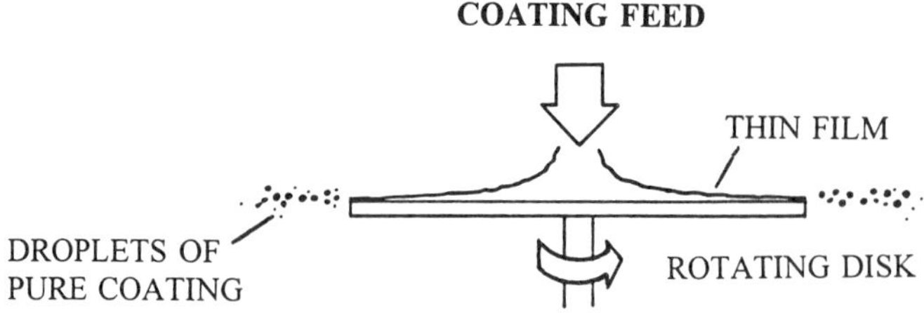

A Establishing Particle Size for Pure Coating

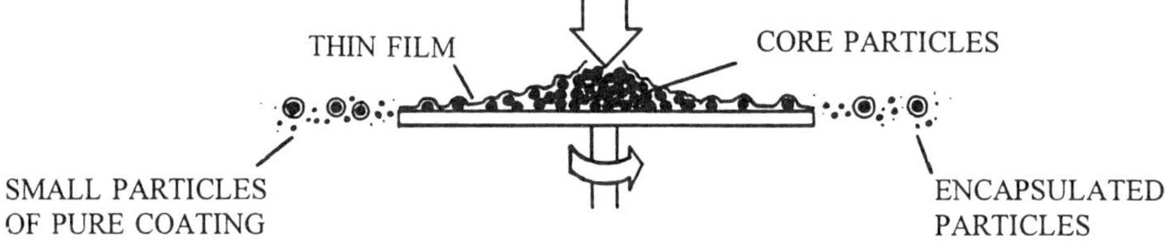

B Encapsulation by Suspension Separation

Figure 10 Representation of rotational suspension separation system. (From Ref. 129.)

particles leave the disk with residual liquid still around them, which forms the coating. The particles are hardened by chilling and drying [131]. The principle behind this process is illustrated in Figure 10.

Centrifugal suspension separation is a continuous, high-capacity process that takes seconds to minutes to coat core particles. The process can handle a wide variety of core materials, including those that are temperature sensitive, and coating materials in solid, liquid, or suspension states without presenting aggregation problems. Furthermore, the process handles each particle only once and under most conditions produces no uncoated particles. The process has been used successfully to coat particles ranging from 30 μm to 2 mm. Coatings have been produced with thicknesses ranging from 1 to 200 μm. Microcapsules have been prepared with payloads ranging from 1 to 97%, depending on the diameter size of the particle. Another advantage associated with centrifugal suspension separation is that the size distribution of the encapsulated particles resembles that of the uncoated particles.

I. Cocrystallization

Cocrystallization is a new encapsulation process utilizing sucrose as a matrix for the incorporation of core materials. Although granulated sugar is composed of solid, dense, monoclinic spherical crystals with a limited surface area, it is not suitable as an encapsulating agent for flavor encapsulation. In order for flavors to be incorporated into the matrix, the structure of sucrose must be modified from a single perfect crystal to a microsized, irregular, agglomerated form to increase void space and surface area [37,132]. It involves spontaneous crystallization, which produces aggregates of micro- or fondant-size crystals ranging from 3 to 30 μm while causing the inclusion of entrapment of all nonsucrose materials within or between sucrose crystals [133]. Use of the cocrystallization process allows many types of food ingredients—either single ingredients or combinations of ingredients—to be incorporated permanently into a crystalline sucrose aggregate, thus providing interesting and useful characteristics.

Sucrose syrup is concentrated to the supersaturated state and maintained at a temperature high enough to prevent crystallization. A predetermined amount of core material is then added to the concentrated syrup with vigorous mechanical agitation, thus providing nucleation for the sucrose/ingredient mixture to crystallize. As the syrup reaches the temperature at which transformation and crystallization begin, a substantial amount of heat is emitted. Agitation is continued in order to promote and extend transformation/crystallization until the agglomerates are discharged from the vessel. The encapsulated products are then dried to the desired moisture (if necessary) and screened to a uniform size [43,44]. It is very important to properly control the rates of nucleation and crystallization as well as the thermal balance during the various phases. The essential steps for the preparation of cocrystallized flavor are presented in Figure 11.

The agglomerates form a loose network, bonded together by point contacts. The encapsulated materials are located primarily in the interstices between crystals. Due to the porosity of the agglomerates, it is easy for an aqueous solution to rapidly penetrate the agglomerate and release the core materials for dispersion and/or dissolution.

The cocrystallization process offers several advantages; for example, it can be employed to achieve particle drying. In the highly saturated solution, nucleation and crystallization proceed at a rapid rate and the resulting heat of crystallization can be used to affect particle dehydration by evaporation. By means of the cocrystallization process, core materials in a liquid form can be converted to a dry powdered form without additional drying. Because the flavor or core material is well entrenched in the modified sucrose matrix, there is no tendency for flavor material to separate from or

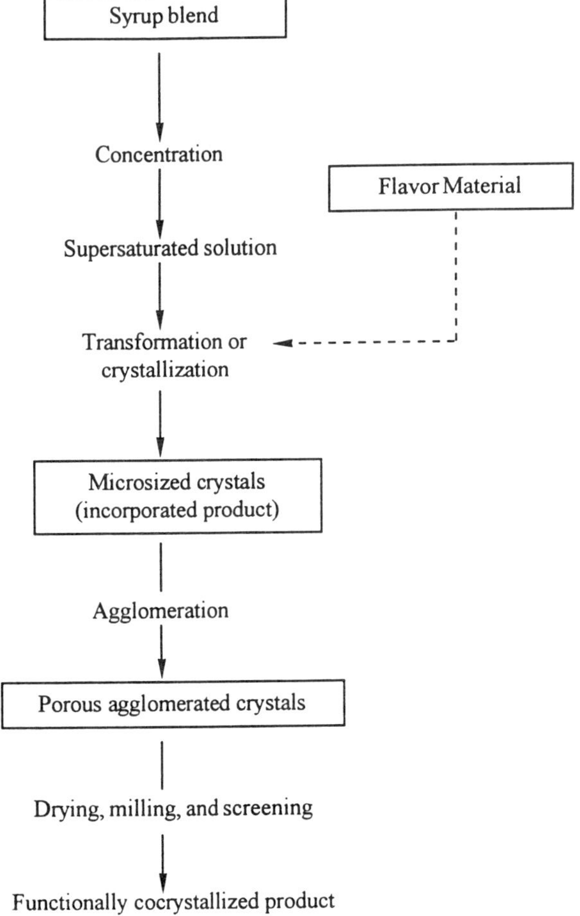

FIGURE 11 Essential steps for the preparation of a cocrystallized flavor. (From Ref. 37.)

settle out during handling, packaging, or storage. Additionally, all cocrystallized sugar/flavor products offer direct tableting characteristics because of their agglomerated structure and thus offer significant advantages to the candy and pharmaceutical industries [134].

J. Liposome Entrapment

Numerous methods of liposome entrapment have been developed [79,80,135]. Preparations obtained vary widely in vesicle size distribution, number of bilayersper vesicle, and encapsulation efficiency.

Liposomes consist of an aqueous phase that is completely surrounded by a phospholipid-based membrane. When phospholipids, such as lecithin, are dispersed in an aqueous phase, the liposomes form spontaneously. One can have either aqueous or lipid-soluble material enclosed in the liposome. However, liposome entrapment for many flavor compounds is not possible because liposomes will not form for materials that are soluble in both the aqueous and lipid phases [5]. From a physicochemical point of view, the formation of liposome structures may be illustrated by phase diagrams. A simplified phase diagram of the 1,2-dipalmitoyl phosphatidylcholine–water system is shown in Figure 12 [136]. Addition of water decreases the transition temperature of the phospholipid to a limiting value (T_c), which is the minimum temperature required for water to penetrate between the layers of lipid molecules. When the system is cooled below T_c, the hydrocarbon chains adopt an ordered packing. The structure of this phase, known as the gel, is lamellar and the hydrocarbon chains extended [136]. Each type of phospholipid molecule is characterized by a phase-transition temperature. Below T_c, its fatty acyl chains are in a quasicrystalline array, while above T_c, the chains are in a fluidlike state.

There are two principal requirements for liposome microencapsulation. First, the lipid of choice must have a negative Gibb's free energy value (ΔG) for bilayer structure formation, because a negative ΔG value between two states of system indicates a favorable reaction. Second, sufficient energy must be put into the system to overcome the energy barrier. Close to room temperature, the value of ΔG for the formation of liposomes is always negative and, therefore, favorable. Even though thermodynamics are favorable, this does not mean that the reaction will proceed automatically; it is usually necessary to overcome an energy barrier in order to initiate a reaction. Different lipids and types of

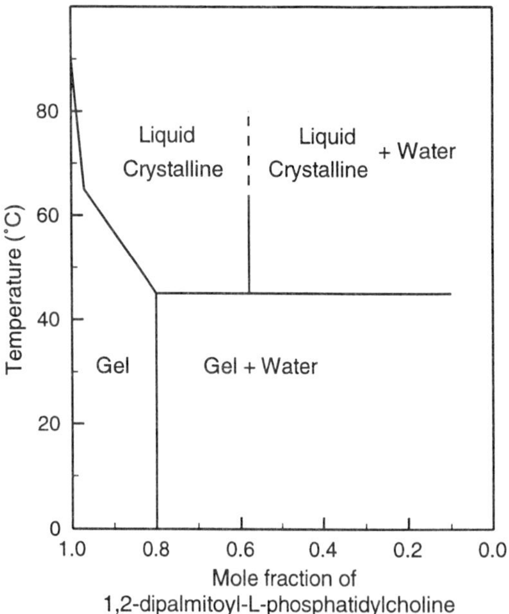

FIGURE 12 Phase diagram of the 1,2-dipalmitoyl-L-phosphatidylcholine-water system. (From Ref. 12.)

energy input may be used to produce different varieties of liposomes for specific purposes. Some methods commonly employed are described below.

1. Microfluidization

The microfluidization technique is based on the dynamic in specially designed microchannels. The resulting momentum and turbulence allows the lipid emulsion to overcome the energy barrier ($\Delta G^{\ddagger}$). An air-driven microfluidizer operates at pressures of up to 10,000 psi. A pump driven by compressed air is used to pump the aqueous emulsion of lipids, and the single-feed stream is split into two fluidized ones. The two flows interact with one another at ultrahigh velocities and in precisely defined microchannels.

Mayhew and Lazo [137,138] found that small (0.1 μm in diameter) liposomes with high solute-capture efficiency could be easily formed by microfluidization technology. At an initial lipid concentration of 300 mM, up to 75% of cytosine arabinoside was captured in the aqueous space of these liposomes. Advantages of microfluidization include: (a) a large volume of liposomes can be formed in a continuous and reproducible manner; (b) the average size of the liposomes can be adjusted; (c) very high capture efficiencies (>75%) can be obtained; (d) the solutes to be encapsulated are not exposed to sonication, detergents, or organic solvents; and (e) the resulting liposomes appears to be stable and do not aggregate or fuse.

2. Ultrasonication

Ultrasonic dispersion is often used for the preparation of SUV; the lipid emulsion overcomes the energy barrier through ultrasound absorption. In one approach, phospholipids are sonicated by immersing a metal probe directly into a suspension of large liposomes. In a second method, the lipid dispersion is sealed in a glass vial, which is then suspended in an ultrasonic cleaning bath. Bath sonication requires longer periods (up to 2 hours) than probe sonication (only a few minutes), but it has the advantage that it can be carried out in a closed container under nitrogen or argon and does not contaminate the lipid with metal from the probe tip [82].

3. Reverse-Phase Evaporation

This technique has been developed for the preparation of LUV in which lipids in mixed aqueous-nonpolar solvents form inverted micelles (i.e., the lipid tails are inserted into the nonpolar phase and the head groups surround water droplets). When the nonpolar solvent is removed by rotary evaporation under vacuum, the gel-like intermediate phase changes into large unilamellar and oligolamellar vesicles. This procedure produces liposomes of quite uniform size, ranging from 0.1 to 1.0 μm in diameter, with high encapsulation efficiency of up to 65% in low ionic strength media. However, its disadvantage is that components are exposed to both organic solvents and sonication. This may result in the denaturation of proteins and other molecules of similar stability [81].

K. Interfacial Polymerization

Interfacial polymerization happens when two different polymeric solutions are brought together. These two reactive polymeric species, each solubilized in a different liquid, react with one another when one liquid is dispersed in the other. The polymerization reaction takes place at the interface between the two polymeric liquids.

The interfacial polymerization process can be used to encapsulate solutions or dispersions of hydrophobic materials. It can also be used to encapsulate aqueous solutions or dispersions of hydrophilic substances. In the interfacial polymerization microencapsulation process, both the dispersed and continuous phases serve as a source of reactive polymeric species. In general, an interfacial polymerization reaction proceeds at a rapid rate that results in the formation of a very thin film having physical property characteristics of a semipermeable membrane. Properties of the film are markedly influenced by the reaction time [139].

The ultimate capsule size of interfacial polymerization is determined by the size of the first monomer. In general, the capsule size ranges from about 1 μm to several millimeters. This capsule size is a direct function of the agitation rate [139]. It is found that an increase in the concentration of the emulsifier yields a narrow size distribution range and a reduction of the average particle size. The patent application for the microencapsulation process utilizing the principle of interfacial polymerization was filed by IBM (serial No. 813,425) in 1959 [139]. However, use of interfacial polymerization for food systems is limited since most coatings are not food grade.

L. Inclusion Complexation–Molecular Inclusion

Molecular inclusion is another means of achieving encapsulation. Unlike other processes discussed to this point, this technique takes place at a molecular level, and β-cyclodextrin is typically used as the encapsulating medium [24]. As previously noted, β-cyclodextrin is a cyclic glucose oligomer, consisting of seven β-D-glucopyranose units linked by α-(1→4) bonds. Due to its molecular structure, β-cyclodextrin has limited solubility, a hydrophobic center, and a relatively hydrophilic outer surface, all of which affect the compound's formation of complexes.

The β-cyclodextrin molecule forms inclusion complexes with compounds that can fit dimensionally into its central cavity. These complexes are formed in a reaction that takes place only in the presence of water. Molecules that are less polar than water (i.e., most flavor substances) and have suitable molecular dimensions to fit inside the cyclodextrin interior can be incorporated into the molecule. In aqueous solution, the slightly nonpolar cyclodextrin interior is occupied by water molecules. This situation is energetically unfavorable, and therefore the sites occupied by water are readily substituted by the less polar guest molecules. Cyclodextrin complexes are relatively stable and their solubility in aqueous solutions is reduced compared to the uncomplexed cyclodextrin. Therefore, the complexed cyclodextrins readily precipitate out of solution and can be recovered simply by filtration.

The complexing of a cyclodextrin with a guest compound can be accomplished by three methods [33]:

1. Stirring or shaking the cyclodextrin and guest molecules to form a complex, which could then be easily filtered and dried. In some cases, complexation of an insoluble guest can only be accomplished through dissolution of the guest in a water-soluble solvent.
2. Blending of solid β-cyclodextrin and guest with water to form a paste. Solvent should not be used. This method is particularly applicable for oleoresins.
3. Forcing a gas through the solution for complexation to occur. This method is seldom used.

It should be emphasized that there are several variations to these basic techniques, but in all methods both the cyclodextrin and the guest molecules must be solubilized. If the guest material is insoluble in water, it is necessary to dissolve it in another solvent such as alcohol.

The composition of the cyclodextrin complex formed depends greatly upon the molecular weight of the guest molecule in question. Because one molecule of cyclodextrin will normally include only one guest molecule, the loading depends upon the compounds included. It should be noted that the theoretical maximum loading is not always obtained. For example, Pagington [140] stated that dimethyl sulfide should be complexed at 5.5%, but only 2% loading has been observed.

It has been reported that cyclodextrins have a variable affinity for different guest compounds. This may be used to advantage or it can be disadvantageous. Some researchers have made use of the variable binding properties offered by β-cyclodextrin to selectively remove bitter compounds from orange and grapefruit juices [141]. Variable binding properties can also be a disadvantage when it comes to the encapsulation of flavor compounds. Reineccius and Risch [24] formed zero (isoeugenol) to 100% inclusions (ethyl hexamoate and linalool) when they added a model flavor system to β-cyclodextrin in an ethanol-water mixture. The losses of flavor compounds were due to the lack of inclusion rather than a loss during the subsequent complex recovery and/or drying steps. Once the complex was formed, it was quite stable to evaporation.

The variable inclusion properties of cyclodextrins would result in a dry flavor quite different from that of the original flavor when the flavor is comprised of a broad range of flavor molecules (e.g., an artificial flavor that contains short-chain esters and longer-chain character impact compounds). However, flavors such as orange, which have been included in β-cyclodextrin, may not be distinguishable from fresh orange even by trained taste panels [142].

There are substantial data in the literature that document excellent protection for substances treated with cyclodextrins [7,142–144]. As previously mentioned, the cyclodextrin-guest complex formed is very stable to evaporation. Szente and Szejtli [144] reported only about a 5% loss of included volatiles after 2 years of storage at room temperature. However, more important is the oxidative stability of the included guest compounds. Many reports have demonstrated that inclusion complexes are quite stable to oxidation [142,144].

As with all processes, there are limits to the application of cyclodextrin complexation in the formation of flavors [145]:

1. There is a limited amount of flavor content in formulations (average 9–14% by weight).
2. The size and polarity of flavors to be complexed limits the usefulness of the process.
3. Cyclodextrin can act as an artificial enzyme, sometimes enhancing the rate of hydrolysis of some ester-type flavor components. This can result in undesirable adulteration of the flavor.
4. The water-solubility of β-cyclodextrin flavor complexes is generally much lower than that of spray-dried and other microencapsulated samples.

IV. ENCAPSULATED INGREDIENTS AND THEIR APPLICATION

Microencapsulation can potentially offer numerous benefits to the materials being encapsulated. Various properties of active materials may be changed by encapsulation. For example, handling and flow properties can be improved by converting a liquid to a solid encapsulated form. Hygroscopic materials can be protected from moisture, and the stability of ingredients volatile or sensitive to heat, light, or oxidation can be protected, thereby extending their shelf life. Materials that are otherwise incompatible can be mixed and used safely together. Currently there are several hundred types of microcapsules being utilized as food additives in North America [7], some of which are described below.

A. Acidulants

Acidulants are added to foods for a variety of reasons. They can be used as flavor modifiers, preservation aids, and processing acids. In addition, they facilitate the development of a wide variety of textural effects in foods because of their interaction with other macro- and micromolecules such as proteins, starches, pectins, and gums [146].

Unencapsulated food acids can react with food ingredients to produce many undesirable effects. These include decreased shelf life of citrus-flavored and starch-containing foods (e.g., pudding and pie fillings in which the acid hydrolyzes the starch), loss of flavor, degradation of color, and separation of ingredients. Encapsulated food acids overcome these problems and others because they preclude oxidation and provide controlled release under specific conditions. Moreover, encapsulated acids reduce hygroscopicity, reduce dusting, and provide a high degree of flowability without clumping.

Encapsulation of acids in a time-release matrix is suggested as a means of avoiding undesirable reactions of acidulants with other food ingredients. The matrix used for forming the encapsulating coat in the acid products is generally a partially hydrogenated vegetable oil, although maltodextrin and emulsifiers are also available for this purpose. The encapsulated acids can be released at the appropriate time in the processing operation either by heating to the melting point of the coating material or by contact with water or a combination of these methods. Several applications of encapsulated acidulants are given below.

1. Meat-Processing Aids

In the meat industry, encapsulated acids, such as lactic, citric, and glucono-δ-lactone (GDL), are used to assist in the development of color and flavor in meat emulsions, dry sausage products, uncooked processed meats, and meat-containing products, such as pasta meals. Fat encapsulation allows the acid to survive the blending process, giving a uniform dispersion within the meat. Later, the encapsulated acid controls the drop in pH and prevents the meat from prematurely setting [8].

Cured meat products, especially dry and semi-dry sausages (e.g., summer sausages, pepperoni, hard salami), have historically been prepared using lactic acid–producing bacterial cultures to develop flavor and lower the pH. Bacteria is added to the meat emulsions and allowed to proliferate until a sufficient amount of lactic acid is generated. Upon its production, the pH drops, binding occurs, and flavor develops. However, such products often tend to have inconsistent flavor, color, and textural characteristics from batch to batch. Uncoated lactic acid and citric acid cannot be added to meat during curing because they react almost instantly with the meat, rendering it unsuitable for further processing. Contamination is especially troublesome where the meat processor may use fermented raw meat as the source of bacteria rather than frozen cultures. However, an encapsulated acid, which is formulated for delayed release under smokehouse temperatures, can be used as an alternative to the cultures. Acidification by encapsulated acids can improve emulsification and protein binding of emulsified meat and poultry products and impart the "tangy" flavor found in fermented sausages without the complicated use of lactic acid starter cultures. Encapsulation permits addition of the acidulants prior to stuffing without premature denaturation/binding of meat.

About 25 years ago, encapsulated acids in a heat-rupturable inert vehicle such as ethyl cellulose were developed [147]. The encapsulated acids were mixed with nitrite-treated ground meats, and upon thermal processing the acid was released bringing about a lowering in the pH of the meat and giving rise to rapid development and stabilization of cured meat color. The more acidic conditions of the meat assisted the production of nitrous acid or dinitrogen trioxide from the exogenous sodium nitrite. Both nitrous acid and dinitrogen trioxide are nitrosating species, which interact with the prosthetic heme group of myoglobin to form the cooked cured-meat pigment.

The effect of encapsulated food acids on restructured pork from prerigor sow meat was studied by Cordray and Huffman [148]. Results from sensory panels showed that sodium acid pyrophosphate (SAP) and encapsulated GDL treatments yielded products with a more intense flavor than that of the control sample; objective analysis revealed no difference in shear value, tensile strength, water-holding capacity, cooked yield, or chilled yield. Significantly more of the total meat pigment was converted to nitrosohemochromogen in the GLD treatment than in the control sample. Lactic acid can also be encapsulated by plating it onto a particle calcium lactate carrier and then encapsulating the carrier and acid with a molten edible lipid [149].

2. Dough Conditioners

The baking industry has long been aware of the need for stable acids and baking soda for use in wet and dry mixes to control the release of carbon dioxide during processing and subsequent baking. Products commonly encapsulated for bakery applications include a variety of leavening system ingredients, as well as vitamin C, acetic acid, lactic acid, potassium sorbate, sorbic acid, calcium propionate, and sodium chloride.

Use of ascorbic acid (vitamin C) for the strengthening and conditioning of bread and roll doughs provides many positive effects to the finished products. Examples of these are stronger sidewalls, uniform crust color, and improved slicing, in addition to a stronger structure, which support the addition of other protein-rich ingredients (such as soybean flour, nonfat milk powder, and wheat germ). However, because ascorbic acid degrades rapidly in the presence of water and oxygen, most of the acid is destroyed before it is needed. Encapsulated in an edible coating, ascorbic acid imparts some of the effect of an oxidizing agent when used alone in natural breads. In combination with bromate, it enables greater amounts of protein-rich ingredients to be utilized without disturbing the grain of the bread to any great extent [150].

For yeast-raised doughs, encapsulated salt, potassium sorbate, and sorbic acid are employed because they do not allow the pH to drop too early in the baking process, allowing the yeast to grow. Once baked, however, the mold-inhibiting properties of these ingredients are released into the dough [8].

3. Other Encapsulated Acidulants

Acids are frequently used as liquids but would be easier to handle if they could be supplied in solid forms. Seighman [151] developed a method for encapsulation of food-grade phosphoric acid in a dispersion containing a film-forming agent (hydrogen octenylbuane-dioate-amylodextrin) and a matrix-forming ingredient (modified and hydrolyzed starches). The dispersion is thermal processed and then extruded into cold aqueous alcohol to solidify the matrix-forming ingredients and to allow the film-forming agent to harden to a vitreous structure.

B. Flavoring Agents

The development and production of artificial or natural flavors and spices is an ever-expanding field in the food industry. The vast majority of flavor compounds used are a liquid at room temperature, and constituents of the flavors tend to show sensitivity towards air, light, irradiation, and elevated temperatures. Moreover, these flavor concentrates are oily and lipophilic materials, which can be difficult to work with. Therefore, it is necessary to employ a process to convert these flavor compounds to a more useable form. One of the purposes behind encapsulation in the food industry is the conversion of liquid flavors to dry powders. Microencapsulated flavors provide the convenience of a solid form over a liquid one, with reduced volatility and less oxidation [17,90,140]. Microencapsulation has become an attractive option to transform liquid food flavorings into stable and free-flowing powders, which are easier to handle and incorporate into a dry food system.

The flavor industry depends heavily on encapsulation as a means of providing solid flavor compounds that offer them protection until consumption. Flavoring agents and spices are encapsulated by a variety of processes and provides numerous advantages to food processors. Processes for flavor encapsulation and encapsulated flavorings prepared during the least 35 years are summarized in Table 5.

Examples of commonly used encapsulated flavors are citrus oils, mint oils, onion and garlic oils, spice oleoresins, and whole spices. Citrus oils are very susceptible to oxidation due to sites of unsaturation in their mono- and sesquiterpenoid structure. Oxidative deterioration results in the development of off-flavors described as painty or turpentinelike. Encapsulated citrus oil, prepared by spray-drying in a maltodextrin matrix, have a greater stability than unprotected oil [20].

Because flavors are often volatile materials, the stability of the dry microcapsules is an important consideration. Microcapsules must be stable for an extended period of time. Many volatile liquids can be encapsulated and subsequently dried to form free-flowing powders with minimal loss of activity during storage. Table 6 illustrates the stability of encapsulated flavors as a function of storage time in microcapsules of various particle sizes under ambient conditions [13].

Flavors encapsulated by inclusion complexation in β-cyclodextrin were protected against volatilization and attack by oxidation [140,144]. Storage stability of flavors encapsulated in β-cyclodextrin under "nonstress" conditions at room temperature showed that molecular encapsulation, in most cases, provides an almost perfect preservation of flavors for up to 10 years [144] (Table 7).

There has been a great expansion in the development of techniques to encapsulate flavors. A spray-dried composition comprising a volatile and/or a liable component in a carrier can be further encapsulated in an extruded glassy matrix. Such a procedure of double-encapsulation has recently been developed by Levine et al. [191]. Excellent reviews of microencapsulation technology as it applies to food flavors have been written [9,17,89,105,152,154,157]. However, it should be noted that details about these techniques are difficult to obtain because they are often trade secrets.

TABLE 5 Literature on Flavor Encapsulation

Subjects	Ref.
Overall reviews	7,9,11,89,152–156
Spray drying	90,98,99,105,152,157–159
Coacervation systems synthetic film formers	153,160
Cheese flavor technology	161–166
Flavor oils	167–171
Lemon and citrus oils	172,173
Safflower oil	174
Essential oil for bakery mixes	175
Volatile flavorings (aroma)	42,126,176–180
Use of cyclodextrins	181
Use of extrusion coating	182
Use of fluidized bed by spraying	182
Use of sorbitol and other ingredients	183
Use of water-insoluble coatings	55
Flavor food ingredients encapsulation	184–186
Coffee and tea flavor encapsulation	145,187,188
Seasonings	61,189
Spray-dried spice oils	106
Artificial flavors	24
Flavors from microorganisms	190

Source: Ref. 12.

C. Sweeteners

Sweeteners are often subjected to the effects of moisture and/or temperature. Encapsulation of sweeteners, namely sugars and other nutritive or artificial sweeteners, reduces their hygroscopicity, improves their flowability, and prolongs their sweetness perception. Sugar that has been encapsulated with fat and incorporated in a chewing gum requires more shear and higher temperatures to release its sweetness than uncoated sugar, which dissolves more rapidly in the mouth.

TABLE 6 Stability of Microencapsulated Flavors

			Flavor content in microcapsules	
Encapsulated flavor	Average capsule size (μm)	Storage period (days)	(%) Initial	(%) Final
Cassia	750	730	87.8	86.1
	20	730	63.1	59.2
	600	400	90.2	89.9
Lemon	250	500	70.5	76.3
	40	730	74.0	67.9
	20	730	60.1	59.9
Lime	1,000	409	92.5	89.6
Peppermint	500	732	75.3	74.6
	20	730	58.5	56.3

Source: Ref. 12.

TABLE 7 Changes in the Flavor Content of Cyclodextrin-Spice Complexes after 10 Years Under Normal Storage Conditions

Sample	Flavor content of the samples (%)	
	In 1977	In 1987
Garlic oil	10.2–10.4	10.0–10.3
Onion oil	10.4–10.6	10.2–10.4
Caraway oil	10.5	9.9–10.2
Thyme oil	9.4–9.8	9.0–9.2
Lemon oil	8.9–9.1	8.6–8.8
Anise oil	9.0–9.2	9.0–9.3
Peppermint	9.4–9.7	9.0–9.2
Marjoram	8.8–9.0	8.0–8.2
Orange	9.0–9.5	6.0–7.0
Tarragon	10.0–10.3	8.8–9.0
Mustard	10.8–11.0	11.0–11.2

Source: Ref. 12.

Patents for the encapsulation of sweeteners were awarded mainly in the 1980s, as the technical development of encapsulation allowed their commercial manufacture. Among these, aspartame is the most widely studied. Aspartame is the methyl ester of a dipeptide made from two amino acids, phenylalanine and aspartic acid (aspartate). Although this white, odorless, and crystalline powder has a very intense sweetness (approximately 180–220 times sweeter than sucrose), potential for its use in food has, in the past, been limited. At high temperatures, aspartame degrades into the amino acids aspartic acid and phenylalanine, accompanied by a loss of sweetness. This internationally marketed sweetener has now been encapsulated by many methods.

Patents awarded to Cea et al. [192,193] mainly involve the encapsulation of APM (L-aspartyl-L-phenylalanine methyl ester) as a chewing gum composition. It has been claimed that the encapsulated APM overcomes difficulties experienced in the use of APM with respect to its stability in the presence of water or elevated temperature [192,193]. Yang and coworkers developed a process for encapsulating aspartame in a film composed of high molecular weight polyvinyl acetate and a hydrophobic plasticizer (mono- or diacylglycerol with fatty acid chains of 16–22 carbon atoms) [186,194,195]. In this process, active ingredients, including soluble dietary fibers, flavoring agents, and drugs, can also be encapsulated. The product can be used to give chewing gum an extended shelf life, with highly controlled release of active ingredients [186].

A process developed by Cherukuri and coworkers can be used to produce a stable delivery system. It comprises a dipeptide or amino acid sweetener or flavorant or mixture thereof encapsulated in a mixture of fat and high-melting-point polyethylene wax [196–198].

Gas chromatographic analyses were used to measure the retention of orange, synthetic peppermint, and natural lemon flavors, which had been cocrystallized, and then storage in polyethylene bags under ambient conditions. Data indicated no significant change in flavor retention for up to 15 weeks of storage. Results from oxidation studies [199] showed that peanut butter–flavored products had a very good shelf life, even after storage for an appreciable period of time. Chen et al. [43,44,199,200] have published a number of patents in this area. Some typical examples of products encapsulated by cocrystallization are listed in Table 8.

D. Colorants

Natural colors such as annatto, β-carotene, and turmeric present solubility problems during their use and may create dust clouds. Encapsulated colors are easier to handle and offer improved solubility,

TABLE 8 Examples of Products Encapsulated by Cocrystallization

Flavored sugar crystals	Brown sugar, chocolate, honey, molasses, and peanut butter granules
Fruit juice crystals	Cranberry, grape, orange, raspberry, and strawberry juices
Essential oil powders	Cinnamon, lemon, orange, and peppermint oils
Dry flavors	Barbecue, beef fat, butterscotch, chocolate, maple, and smoke flavors
Volatile substances	Acetaldehyde and diacetyl

stability to oxidation, and control over stratification from dry blends. Synthetic colors, together with other food ingredients, can also be encapsulated for improving their stabilities [201].

A technique for solubilizing oily substances in micellar solutions of protein and carbohydrates was applied by Ono [202] in order to achieve encapsulation of two oil-soluble pigments—paprika oleoresin and β-carotene. The pigment in oil was solubilized in an aqueous solution containing 60% (w/w) corn syrup solids and 1% (w/w) polypeptone. The solubilized mixture obtained was solidified by vacuum-drying at 60°C and formed into granules by crusting and sieving. These granules containing approximately 12% pigment-containing oil underwent virtually no discoloration during storage for 20 days at 60°C or when subjected to irradiation from a fluorescent lamp. Dispersibility of the pigments in water was improved by their encapsulation in a protein-carbohydrate matrix [202].

Ciliberto and Kramer [203] developed an encapsulation process for producing granular water-soluble food ingredients, which otherwise deteriorated on exposure to the atmosphere (such as coloring agent). It was claimed that the resulting coated particles had a long shelf life and were still substantially instantaneously soluble in water.

Studies on encapsulation of preformed cooked cured-meat pigment (CCMP) showed that the CCMP may be stabilized effectively by its encapsulation in food-grade starch-based wall materials. The color stability of the treated meat products was found to be similar to their nitrite-cured analog [204].

E. Lipids

Lipids contribute to more than 30% of the dietary energy of North Americans, and similar figures apply to many other affluent societies. Use of lipids/fats is commonplace in food-processing practices, but the susceptibility of lipids to oxidative degradation during processing and storage is always a concern; particular attention must be paid to foodstuffs containing higher proportions of polyunsaturated fatty acids (PUFA). One possible way to protect lipid moieties against oxidative deterioration is via encapsulation. Early research in this area was mainly focused on production of encapsulated lipids for animal feed [174,205–207], but more recently, encapsulated high-fat powders or shortenings have been available in food formulations for human consumption [208].

Because of the pro-health benefits of fish oils, encapsulated oils have been available in health food stores, pharmacies, and supermarkets for a number of years. These fish oils contain long-chain omega-3 PUFA, such as eicosapentaenoic acid (EPA), docosahexaenoic acid (DHA), and docosapentaenoic acid (DHA), whose beneficial effects have been ascribed to their ability to lower blood serum triacylglycerol and cholesterol levels [209,210]. While DHA is essential for proper functioning of the eye and may have a structural role in the brain, EPA serves as a precursor to eicosanoid compounds [211] and has therapeutic benefits in human cardiovascular diseases [212,213]. It should be noted that fish oils are exceptionally susceptible to autoxidation and can form complex mixtures of high molecular weight oxidation products. Shukla and Perkins [214] reported that because of the unknown health effects of the oxidative polymeric materials and their high level in some encapsulated oils, caution should be exercised when ingesting fish oil capsules on a regular basis. However, encapsulation can enhance the oxidative stability of these oils.

Gejl-Hansen and Flink [215] freeze-dried an aqueous emulsion of linoleic acid in a maltodextrin coating in the presence of detergents. The microencapsulated linoleic acid was not susceptible to oxidative deterioration even though more effective encapsulating wall materials could have been used. Ono and Aoyama [88] reported that vacuum-dried rice brain oil embedded in granules containing corn syrup solids and pork polypeptone did not undergo much oxidation upon exposure to air at a high temperature for a few weeks. Taguchi et al. [216] reported the oxidative stability of sardine oil embedded in spray-dried egg white powder and use of the product as a source of omega-3 PUFA for fortification of cookies. These authors reported that use of microencapsulated sardine oil fortified cookies did not affect their sensory quality.

The antioxidative effects of spray-dried powders at various water activities prepared from alcoholic solutions of gliadin, linoleic acid, and palmitic acid were compared with powders prepared by simple mixing of these components in the same portions and against gelatin or starch powders substituted for gliadin by Iwami et al. [217]. It is reported that the microcapsules obtained from the experiment were highly resistant to oxidative deterioration during long-term storage at different a_w [217]. Shahidi and Wanasundara [218] spray-dried an emulsion of seal blubber oil, containing 21–26% long-chain omega-3 fatty acids, with either β-cyclodextrin, corn syrup solids, or maltodextrins. They found that β-cyclodextrin was the most effective entrapping agent and prevented oxidative deterioration of seal blubber oil.

F. Vitamins and Minerals

Most vitamins cannot be synthesized by the body and must be supplied by the diet [219]. Because vitamins are such important nutritional and dietary factors, processed foods are often enriched or fortified with vitamins. Table 9 presents the recommended daily allowances for vitamins A, D, E, K, C, B_6, B_{12}, folic acid, thiamine, riboflavin, and niacin as compiled by the National Academy of Sciences' Food Nutrition Board [220]. Vitamins and minerals are often added to dry mixes to fortify a variety of foods.

Encapsulation of vitamins and minerals offers many advantages as it reduces off-flavors contributed by certain vitamins and minerals, permits time release of the nutrients, enhances stability of vitamins to extremes in temperature and moisture, and reduces each nutrient's reaction with other ingredients. Encapsulation also improves flow properties and reduces dusting when nutrients are added to dry mixes. Both fat- and water-soluble vitamins may be encapsulated with a variety of coatings to provide many advantages. Hall and Pondell [221] developed a process to encapsulate vitamin or mineral particles. The coating matrix for this process is chiefly ethyl cellulose together with propy-

Table 9 Recommended Dietary Allowances

Vitamin	Men	Women	Children to age 11
Fat-soluble			
Vitamin A (retinol, μg)	1000	800	400–700
Vitamin D (cholecalciferol, μg)	5–10	5–10	10
Vitamin E (α-tocopherol, mg)	10	8	6–7
Vitamin K (μg)	45–80	45–65	15–30
Water-soluble			
Vitamin C (mg)	60	60	40–45
Vitamin B_1 (thiamine, mg)	1.5	1.1	0.7–1.0
Vitamin B_2 (riboflavin, mg)	1.7	1.3	0.8–1.2
Niacin (mg)	19	15	9–13
Vitamin B_6 (pyridoxine, mg)	2.0	1.6	1.0–1.4
Vitamin B_{12} (μg)	2.0	2.0	0.7–1.4
Folic acid (μg)	200	180	50–100

lene glycol monoester and acetylated monoglycerol. Vitamins and minerals can also be encapsulated in fat [222] or in starch matrices [223].

For encapsulation of water-soluble vitamins, ethyl cellulose is useful because it is water insoluble and coatings with increased thickness reduce the water permeability of the prepared capsules. Thiamine enrichment of some bakery products such as devil's food cake, ginger snaps, and soda crackers, has always been unsuccessful due to vitamin destruction in the neutral or alkaline pH. A procedure for microencapsulating thiamine in an ethyl cellulose coating to protect it from alkaline conditions experienced in bakery products and to mask its undesirable bitter taste has been developed [224].

Riboflavin, thiamine, and niacin are partially destroyed during the processing and cooking of pasta products. Studies on unprotected versus encapsulated thiamine, riboflavin, and niacin in cooked enriched spaghetti showed that concentrations of the three B vitamins tested were higher in cooked pasta that contained encapsulated vitamins [225].

Lipid-soluble vitamins lose their activity due to isomerism, anhydro-vitamin formation, oxidation, and photochemical reactions [140]. Losses of vitamins in fortified foods can be minimized if they are added as cyclodextrin complexes [140] or gelatin-encapsulated beadlets [226]. It was found that the stability of vitamin A in skim milk was substantially increased by encapsulation in gelatin. Loss of the vitamin in fortified milk powder was minimal even when heated at 100°C for 9 minutes or stored at 28°C for 40 weeks [226]. Table 10 presents the stability data of vitamin A palmitate, of 325,000 units per gram potency, encapsulated in a modified gelatin film [13]. The data indicate that the rate of vitamin A degradation under the test conditions is significantly reduced by microencapsulation.

A well-designed phase-separation technique for encapsulation of vitamin A has been developed by Markus and Peleh [227]. The matrix components used consisted of substituted cellulosic materials, fatty acids, or a variety of proteins. Antioxidants such as butylated hydroxytoluene and ethoxyquin were incorporated in the formulations. It has been claimed that the capsules prepared with substituted cellulosic materials protected vitamin A best from degradation [227].

Iron compounds have been encapsulated to improve the color, odor, and shelf life of fortified products. Encapsulation reduced the ability of iron to react with other food ingredients and also lightened the color of an unspecified type of electrolytic iron [228]. The process for encapsulation of ferrous sulfate was developed by Jackel and Belshaw [229] in the 1970s. It is reported that encapsulated $FeSO_4$, a fine, white, free-flowing powder, can withstand 6-month storage without any deteriorative change. Harrison et al. [230] examined the effect of iron in various forms on the oxidation of lipids in white flour. When subjected to an accelerated stability test (stored at 50°C), flours enriched with ferrous sulfate, fat enriched with ferrous sulfate, electrolytic iron powder, and carbonyl iron powder developed an unacceptable oxidized flavor after 8 days. However, oxidation was not detected in flour stored at room temperature for 2 years [230].

Soy milk beverages have gained attention as possible alternatives to cow's milk. However, soy milk is nutritionally inferior to cow's milk with respect to its calcium content. Attempts to fortify soy milk with calcium have been unsuccessful since soy protein was coagulated and precipitated by cal-

Table 10 Stability of Vitamin A Palmitate at 45°C and 75% Relative Humidity

Time (days)	Percentage of potency retained	
	Raw oil	Microcapsulated
5	86.1	98.3
15	84.2	97.8
42	76.2	94.2
56	69.9	94.1

Source: Ref. 12.

cium [231,232]. Hirotsuka et al. [232] found that calcium coated with lecithin to form liposomes could be added to soy milk without undesirable calcium-protein interactions. The technology was successful in fortifying 100 g of soy milk with an additional 120 mg of calcium.

G. Enzymes

Enzymes are being used increasingly in the food industry for a wide variety of applications. Encapsulation of enzymes could enhance their properties in a number of very different ways. The first and foremost of these concerns is stability. The complex biochemical structure of the enzyme can make it highly vulnerable to inactivation by other components or conditions within the food system. By segregating it inside a microcapsule, it can be maintained in conditions that could otherwise be very harmful to it. A variety of other stabilizing materials can be encapsulated alongside the enzyme to protect them from different antagonistic effects. Inhibitory agents and harmful ions will be excluded from the capsule. Penetrating ions can be removed by buffers or chelating agents, and oxidative damage may be prevented by the use of antioxidants. Thermostabilizers such as sugars will protect against extreme processing conditions such as dehydration or freezing. Further stabilization may be achieved by simply maintaining the enzyme in a concentrated form rather than allowing it to become diluted into the bulk-food phase.

As long as it remains encapsulated, the enzyme will be isolated from its substrate and therefore latent and passive within the food matrix. By selecting a capsule with appropriate properties, we can choose when, where, and how it will interact with its intended substrate. By altering the surface properties of the microcapsules, they can often be made to accumulate at a particular microscopic location within the food. When they eventually break down, the enzyme activity will be concentrated at the intended target site rather than nonspecifically dispersed throughout the food. In this way, enzymes can be used much more selectively and with far greater efficiency than their normal usage would allow.

The timing of enzyme release can be controlled by selecting a microcapsule according to its stability properties within a particular food system. A low stability will lead to early release in the food process, whereas a more stable one will allow postponement of its release. This is very useful where early release is undesirable and enzyme action is not needed until a later step of a multistage process.

Considerable progress in research for the control of cheese ripening using encapsulated enzymes has been achieved [161,162,164,233–238]. Principles involved in this application provide a good illustration of how encapsulation can be applied generally in the food industry, as has been reviewed by Kirby and Law [165]. Other enzymes such as lipase [239,240] and invertase [241,242] have also been encapsulated for applications in food processing.

H. Microorganisms

Encapsulation of viable bacterial cells has several advantages over encapsulation of isolated cheese-ripening enzymes. The stability of enzymes in intact cells is greater than in extracts. Furthermore, production achieved by cells is easily manipulated by controlling substrate concentration in microcapsules [237].

Cells of *Brevibacterium linens* were successfully entrapped in milk fat–coated microcapsules by Kim and Olson [238]. It is believed that the bacteria, using methionine to produce methanethiol and other sulfur compounds, makes a major contribution to the Cheddar cheese flavor of low-fat cheese products. Microencapsulated microorganisms may be useful in reducing the ripening time of blue cheese or in imparting blue cheese flavor to other foods. Spores of *Penicillium roqueforti* have been encapsulated in a milk fat coating matrix [156]. The microenvironment provided by the microcapsules enhanced methyl ketone production by spore enzymes. However, it should be noted that there are fewer examples of encapsulated microorganisms, especially for food use, than of enzymes.

I. Gases

Some hard candies can be made with entrapped carbon dioxide gas [239]. The confections made with encapsulated carbon dioxide produce a sizzling effect on the tongue as the candy melts in the mouth. The candy is produced by incorporating gas at a pressure of 50–1000 psi into the molten sugar. Concentrations of carbon dioxide in the candy range from 0.5 to 15 ml/g of sugar [239]. Gas can also be injected into the encapsulation system and be coated together with the foaming and aromatic core mixtures [179].

J. Other Food Additives

Almost all food additives can theoretically and technically be encapsulated. However, only some encapsulated additives are commercially available because many factors have to be taken into consideration before the process leads to commercial manufacture. Research has been done to encapsulate food preservatives such as monocapric acid [243] and oleic acid [244]. A process for preparing a coated-particle salt substitute composition was described by Meyer [245]. Recent studies suggested that encapsulated antioxidants could be beneficial to food preservation [246]. It is expected that many new encapsulated food ingredients will be produced, which could contribute greatly to further development of food processing and preservation.

V. CONTROLLED RELEASE MECHANISM AND EFFECTS

Encapsulation allows reactive ingredients to be separated from their environment until their release is desired. Although separation is indeed the objective of encapsulation, release mechanisms of the core material must be considered as well. In fact, when designing a custom encapsulated ingredient, one must determine the desired release mechanism and a method for quality measurement. A well-controlled release of core material is a very important property of microcapsules. For example, a substance in formulated food may be released upon consumption but prevented from diffusing throughout the product during processing operations (e.g., flavors, nutrients). Similarly an additive may be released in a specific processing step but protected in preceding operations (e.g., acids, leavening agents, cross-linking agents) (247).

Because the physical and chemical properties of volatile compounds are governed by their structures and cannot be changed, one has to manipulate the choice of the encapsulation matrix as well as the formulation of the flavor itself if the flavor is a compound one. By picking a capsule matrix with limited selectivity, which may in fact be chosen to discriminate against vapor pressure differences and the desired flux rate (to release slowly or quickly but uniformly), flavor imbalances can be minimized. Additionally, if the flavor is a formulated one, there may be some opportunity to choose flavor compounds that will have similar release rates. Such well-controlled release-delivery systems present the food technologist with exciting opportunities for improving the performance of existing food processes, as well as for the development of entirely new ones [166,247]. However, in order to address the issue of controlled release, one needs to examine the basic principles of controlling the release of encapsulated materials and then consider which technologies can be applied in the food industry. The various mechanisms of release from controlled release-delivery systems in consumer products are provided in Table 11 [248].

A. Release Rate

Release rates that are achievable from a single microcapsule are generally zero, half, or first order. Zero order occurs when the core is a pure material that may be released through the wall of a microcapsule as a pure material. Half-order release generally occurs with matrix particles, while first-order release occurs when the core material is actually a solution trapped within a solid matrix [247]. As the solute material releases from the capsule, a desired concentration of solute is reached.

Table 11 Mechanisms of Release from Controlled-Release Delivery Systems in Consumer Products

Diffusion-controlled release	Membrane-controlled release
Pressure-activated release	Tearing or peeling release
Solvent-activated release	Osmotically controlled release
pH-sensitive release	Temperature-sensitive release
Melting-activated release	Hybrid release

Source: Ref. 248.

A mixture of microcapsules will include a distribution of capsules varying in size and wall thickness. The effect, therefore, is to produce a release rate different from zero, half, or first order because of the ensemble of microcapsules. Thus, it is desirable to carefully examine the experimental basis of the release rate from an ensemble of microcapsules and to recognize the deviation from theory due to the distribution in size and wall thickness [7]. Numerous factors affecting the release rate of core materials are summarized in Table 12.

B. Release Mechanisms

The coating not only protects the core material from moisture, light, oxygen, other food ingredients, and additional external agents [123], but it allows/assists in controlling the release of core materials. Thus, release of the core material is dependent upon the type and geometry of the particle and the wall material used to form the microcapsule. These factors dictate the mechanism of release for the capsule, which may be based on solvent effects, diffusion, degradation or particle fracture [249]. A variety of release mechanisms that have been proposed for microcapsules are summarized below.

1. Fracturation or Pressure-Activated Release

A number of controlled release systems prepared primarily by coacervation technology depend on pressure for release of the active core [250]. The coating can be fractured or broken open by external forces, such as pressure, shearing, and ultrasonics, or by internal forces, as would occur in a microcapsule having a permeation-selective coating. Both fracturation and diffusion involve the controlled release of volatile materials, however, a slow release of core material from the capsule in the case of fracturation is a detriment rather than an attribute. A completely impermeable capsule is needed that releases only on rupture. For example, capsules made from hardened fats or waxes are insoluble in water but can be made to release their contents by mechanical breakage, e.g., shear, or by increasing the temperature to the melting point of the fat (see Sec. V.B). The act of chewing is the most commonly used mechanical release means. It is also possible to get release of the core substance by incorporation of a swelling agent into the core substance or by an electromagnetic method using discharge or magnetic force. The force-fractured release is accomplished in a relatively shorter time beginning at certain controlled conditions compared to the other release mechanisms.

Table 12 Parameters Affecting the Release Rate of Core Materials

Coating properties	Density, crystallinity, orientation, solubility, plasticizer level, cross-linking, pretreatments
Capsule properties	Size, wall thickness, configuration, conformity, coating layers, posttreatment
Experimental parameters	Temperature, pH, moisture, solvent, mechanical action, partial pressure differential (inside and outside of coating)

Source: Ref. 12.

2. Diffusion

This mechanism acts to limit the release of core material from within the capsule to the surface of the particle by controlling the rate of diffusion of the active compound. The bulk of the capsule material itself may control release (i.e., matrix-controlled release) or a membrane may be added to the capsule for controlling release (i.e., membrane-controlled release). Most microcapsules have thin walls, which can function as a semipermeable membrane. Furthermore, because microcapsules are very small, they have a very large surface area per unit weight. Hence, controlled release is frequently accomplished through a diffusion-controlled process [251].

Diffusion release depends upon the kinetic relationship between the core and wall materials and the rate at which the core material is able to pass through the outer wall. It is strictly governed by the chemical properties of the microcapsule and by the physical properties of the wall material such as the matrix structure and pore sizes [249]. Diffusion is a permeation process driven by a concentration gradient or interchain attractive forces [252]. In other words, it is controlled by the solubility of a component in the matrix (this establishes a concentration gradient in the matrix for driving diffusion) and the permeability of the component through the matrix. In the absence of cracks, pinholes, or other flaws, the primary mechanism for core materials to flow through a wall or coating is by activated diffusion, i.e., the penetrant dissolves in the film matrix at the high concentration side, diffuses through the film driven by a concentration gradient (i.e., Fick's law, $I_A = -D_{AB}\, dC_A/dy$, where I_A is the flux of the core material in the y direction, D_{AB} is the diffusivity, and dC_A/dy is the concentration gradient), and evaporates from the other surface. It should be noted that if the food component were not soluble in the matrix, it would not enter the matrix to diffuse through, irrespective of the matrix's pore size.

Diffusion also depends upon the size, shape, vapor pressures, and polarity of the penetrating molecules as well as the segmental motion of polymer chains [252,253]. This also includes interchain attractive forces such as hydrogen bonding and van der Waals interactions, degree of cross-linking, and the amount of crystallinity [254]. In general, cross-linking of a matrix has little meaning in most food applications. Very few situations exist where the matrix can be cross-linked considering the limitations imposed by requiring food-approved materials [251]. However, cross-linking of proteins as a consequence of Maillard reactions can occur and possibly influence the diffusion of solutes in heated protein-based encapsulation matrices (e.g., gelatin). Thus, the greater the degree of cross-linking, the less the rate of diffusion through the matrix (hence, a readily controllable process of making a controlled-release capsule).

The problem of uniform releasing of the aroma of an encapsulated flavor into food should be noted. Because a flavor consists of aroma compounds with a range of volatility, their release, for example, into the head space of a food package, will not be uniform and therefore a balanced characteristic food aroma may not be achieved [255]. The volatility or vapor pressures of these different compounds and their resistances to diffusion will affect their rate. Thus, aromas could become imbalanced as the constituents diffuse through the capsule.

For most physical methods, it is known that the success of encapsulation depends on the formation of a metastable amorphous structure, a glass, with a very low permeability to organic compounds encapsulated within it. In drying processes, the presence of sugar and/or polymers in the encapsulation system reduces the water content. Reduction of water content lowers the glass transition temperature and the resulting amorphous matrix is impermeable to organic compounds as well as to oxygen. However, permeability to water remains finite. This phenomenon, also known as the selective diffusion theory of Thijssen and Rulken [256], is the basis for encapsulation using spray-drying and freeze-drying [247]. In spray-drying, upon droplet formation, rapid evaporation from the surface produces a surface layer in which the selective diffusion mechanism operates. In freeze-drying, upon water crystallization, the nonfrozen solution is viscous and the diffusion of core materials is retarded. At the beginning of freeze-drying, the surface of this solution becomes an amorphous solid in which selective diffusion comes into play.

The permeability of the coating structure can be changed by controlled conditions. The physical state of the food polymer has a considerable role in influencing diffusion and thus release of the core

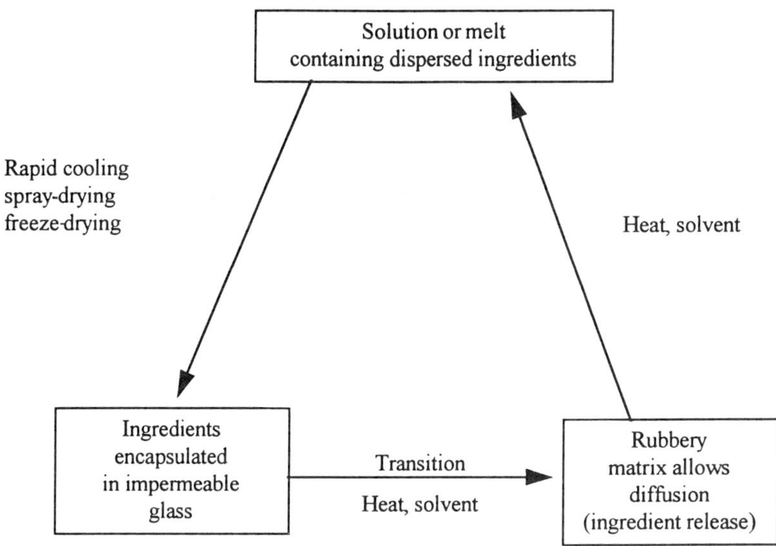

FIGURE 13 Preparation and release of core ingredients from microcapsules. (From Ref. 247.)

material. The physicochemical principles governing the softening or glass transition of the encapsulating materials have been studied by several researchers [257–260]. These investigations have shown that the release occurs when the glassy, impermeable structure undergoes a transition to a more mobile rubbery state (Fig. 13). Thus, the glass/rubber transition of a matrix material is a relevant consideration when evaluating release properties. The relation of transition temperature to the composition of encapsulating formulations has been studied by To and Flink [261] and by Levine and Slade [258] in starch-derived encapsulating agents. It must be noted, however, that even after the critical moisture content or the critical temperature is exceeded, the rate of release is also a function of water content, temperature, and time [262]. This fact allows the generation of controlled-release systems. The maltodextrins and similar materials with controlled collapse temperatures are not only important as encapsulating agents, but are also extremely useful in protecting enzymes and other sensitive biological materials during dehydration and subsequent storage. The principles are similar in that the sensitive materials are placed in a medium in which their mobility is restricted.

3. Solvent-Activated Release

Solvent-activated release is the most common controlled-release mechanism used in the food industry. Since most encapsulating matrices are water soluble, the water in the food product dissolves away the microcapsule, thereby liberating its content to the food, or it causes the capsule to swell to either begin or enhance the release of the core material. However, water-insoluble coatings can also be dissolved by selecting an appropriate solvent. Encapsulated agents are often added to dry food products such as dry beverages and cake and soup mixes. The encapsulated flavors in these products are released upon rehydration [251]. Their release may be a sudden burst or a continued or delayed delivery regulated by controlling the rate of wall solubility, the swelling of the wall material, pH effects, or changes in the ionic strength of the surrounding medium [249].

Although most traditional wall materials will rapidly release the core material once they are rehydrated, microcapsule matrices may be modified to release the active material at a desired point in time. Osmotically controlled release is similar to solvent-activated release in that the core of the particle adsorbs a solvent (usually water) over time and swells until the capsule bursts [248]. For any food ingredient that is first encapsulated in a hydrophilic matrix and then coated with a lipophilic one, osmotically controlled release functions to a limited extent. The encapsulated product will eventually swell and either expand the surface coating, causing cracks or fractures, or rupture entirely.

4. Melting-Activated Release

The integrity of the coating can be destroyed by thermal means. This mechanism of release involves the melting of the capsule wall (or a protective coating that has been placed on the capsule wall) to release the active material. Because numerous meltable materials are approved for food use (e.g., lipids, waxes, and modified lipids), this method of release is easily accomplished. Yet the applications are limited. In general, salts, nutrients, leavening agents, and some water-soluble flavoring agents have been protected by hydrophobic coatings to curtail release of the active ingredient into the food until the baking process. The hydrophobic coating and core material must be immiscible with one another in order to avoid migration of the active ingredient through the wall material. This limits the usefulness of the technique for many flavor applications. On the other hand, an already encapsulated flavor prepared by spray-drying can be coated with a hydrophobic matrix via centrifugal coating or the fluidized bed technique. In this manner, the secondary coating on the flavor provides melt-release properties [263]. The major problem with this approach, however, is the dilution of the flavoring by additional wall material and the extra cost involved.

5. Biodegradation and pH-Sensitive Release

Release from microcapsules can be accomplished by biodegradation processes if the coatings lend themselves to such degradative mechanisms. Lipid coatings may be degraded by the action of lipases [264]. Karel and Langer [247] released enzymes from liposomes using pH as a stimulant to initiate release. They postulated that pH changes destabilized the phospholipid-based liposomal structure, thereby releasing the enzymes from the liposome core.

VI. CONCLUSIONS

This chapter has focused on the art of microencapsulation and has presented an up-to-date account of the process as it relates to the food industry. Although microencapsulation has been extensively used by the pharmaceutical and chemical industries for many years, its applications to the food industry have lagged and require further improvement. Food ingredients are encapsulated for a variety of reasons including protection from volatilization during storage, protection from undesirable interactions with other food components, minimization of flavor interactions or light-induced deteriorative reactions, and protection against oxidation. Other benefits include ease of handling and mixing, uniform dispersion, and improved product consistency during and after processing. Yet compared with single living cells, the capsules prepared to date are too simplistic, and more development is needed before this technology can be widely applied to the food industry. Because the art of microencapsulation encompasses numerous research fields (e.g., chemistry, engineering, processing, and microbiology), innovative strategies by the food scientist, particularly in the area of controlled release of encapsulated food ingredients, and ever-changing technology offer the industry new and exciting areas for research and development.

REFERENCES

1. J. A. Bakan, Microencapsulation, *Encyclopedia of Food Science* (M. S. Peterson and R. Johnson, eds.), AVI Pub. Co., Inc., Westport, CT, 1978, p. 499.
2. C. J. Kirby, Microencapsulation and controlled delivery of food ingredients, *Food Sci. Technol. Today* 5(2):74 (1991).
3. S. M. Barlow, Toxicological aspects of antioxidants used as food additives, *Food Antioxidants* (B. J. F. Hudson, ed.), Elsevier Applied Science, London, 1990, p. 253.
4. A. H. King, Encapsulation of food ingredients: A review of available technology, focusing on hydrocolloids, *Encapsulation and Controlled Release of Food Ingredients* (S. J. Risch and G. A. Reineccius, eds.), ACS Symposium Series No. 590, American Chemical Society, Washington, DC, 1995, p. 26.

5. S. J. Risch, Encapsulation: Overview of uses and techniques, *Encapsulation and Controlled Release of Food Ingredients* (S. J. Risch and G. A. Reineccius, eds.), ACS Symposium Series No. 590, American Chemical Society, Washington, DC, 1995, p. 2.
6. C. Andres, Encapsulation ingredients: I, *Food Proc. 38*(12):44 (1977).
7. R. J. Versic, Flavor encapsulation: An overview, *Flavor Encapsulation* (G. A. Reineccius and S.J. Risch, eds.), ACS Symposium Series No. 370, American Chemical Society, Washington, DC, 1988, p. 1.
8. T. J. DeZarn, Food ingredient encapsulation: An overview, *Encapsulation and Controlled Release of Food Ingredients* (S. J. Risch and G. A. Reineccius, eds.), ACS Symposium Series, No. 590, American Chemical Society, Washington, DC, 1995, p. 74.
9. R. D. Todd, Microencapsulation and the flavor industry, *Flav. Ind. 1*:768 (1970).
10. R. E. Sparks, Microencapsulation, *Kirk-Othmer Encyclopedia of Chemical Technology,* 3rd ed., Vol. 15 (M. Grayson and E. David, eds.), Wiley and Sons Inc., New York, 1981, p. 470.
11. L. L. Balssa and G. O. Fanger, Microencapsulation in the food industry, *Crit. Rev. Food Technol. 2*:245 (1971).
12. F. Shahidi and X.-Q. Han, Encapsulation of food ingredients, *Crit. Rev. Food Technol. 33*:501 (1993).
13. J. A. Bakan, Microencapsulation of foods and related products, *Food Technol. 27*(11):34 (1973).
14. M. A. Godshall, The role of carbohydrates in flavor development, *Food Technol. 42*(11):71 (1988).
15. J. Flink and M. Karel, Effects of process variables on retention of volatiles in freeze-drying, *J. Food Sci. 35*:444 (1970).
16. G. A. Reineccius and S. T. Coulter, Flavor retention during drying, *J. Dairy Sci. 52*:1219 (1989).
17. G. A. Reineccius, Flavor encapsulation, *Food Rev. Int. 5*:147 (1989).
18. G. A. Reineccius, Carbohydrates for flavor encapsulation, *Food Technol. 46*(3):144 (1991).
19. M. M. Kenyon, Modified starch, maltodextrin, and corn syrup solids as wall materials for food encapsulation, *Encapsulation and Controlled Release of food Ingredients* (S. J. Risch and G. A. Reineccius, eds.), ACS Symposium Series No. 590, American Chemical Society, Washington, DC, 1995, p. 42.
20. S. Anandaraman and G. A. Reineccius, Stability of encapsulated orange peel oil, *Food Technol. 40*(11):88 (1986).
21. W. E. Bangs and G. A. Reineccius, Influence of dryer infeed matrices on the retention of volatile flavor compounds during spray drying, *J. Food Sci. 47*:254 (1982).
22. G. A. Reineccius, Part II—Flavor encapsulation, *Source Book of Flavors*, 2nd ed. (G. A. Reineccius, ed.), Chapman & Hall, New York, 1994, p. 605.
23. K. C. M. Raja, B. Sankarikutty, M. Sreekumar, A. Jayalekshmy, and C. S. Narayanan, Material characterization studies of maltodextrin samples for the use of wall material, *Stärke 41*:289 (1989).
24. G. A. Reineccius and S. J. Risch, Encapsulation of artificial flavors by β-cyclodextrin, *Perf. Flav. 11*(4):1 (1986).
25. J. Solms, Interaction of non-volatile and volatile substances in foods, *Interactions of Food Components* (G. G. Birch and M. G. Lindley, eds.), Elsevier, London, 1986, p. 189.
26. W. King, P. Trubiano, and P. Perry, Modified starch encapsulating agents offer superior emulsification, film forming and low surface oil, *Food Prod. Dev. 10*(10):54 (1976).
27. P. C. Trubiano and N. L. Lacourse, Emulsion-stabilizing starches, *Flavor Encapsulation* (S. J. Risch and G. A. Reineccius, eds.), ACS Symposium Series No. 370, American Chemical Society, Washington, DC, 1988, p. 45.
28. D. French, The Schardinger dextrins, *Advances in Carbohydrate Chemistry*, Vol. 12 (M. L. Wolfram, ed.), Academic Press, New York, 1957, p. 189.
29. A. R. Hedges, W. J. Shieh, and C. T. Sikorski, Use of cyclodextrins for encapsulation in the use and treatment of food products, *Encapsulation and Controlled Release of Food Ingredients* (S. J. Risch and G. A. Reineccius, eds.), ACS Symposium Series No. 590, American Chemical Society, Washington, DC, 1995, p. 60.

30. D. E. Pszczola, Production and potential food applications of cyclodextrins, *Food Technol. 42*(1):96 (1988).
31. K. Lindner, Using cyclodextrin aroma complexes in catering, *Nahrung 26*:675 (1982).
32. S. Nagatomo, Cyclodextrins: Expanding the development of their functions and applications, *Chem. Econ. Eng. Rev. 17*:28 (1985).
33. J. S. Pagington, β-Cyclodextrin and its uses in the flavor industry, *Developments in Food Flavors* (G. G. Birch and M. G. Lindley, eds.), Elsevier Applied Science Publishers Ltd., London, 1986, p. 131.
34. Vast range of potential uses for cyclodextrins, *Food Eng. 59*:36 (1987).
35. W. Saenger, Cyclodextrin inclusion compounds in research and industry, *Angew. Chemie Int. Ed. 19*:344 (1980).
36. M. W. Nicol, Sucrose in food systems, *Carbohydrate Sweeteners in Food and Nutrition* (P. Koivistoinen and L. Hyvonen, eds.), Academic Press, New York, 1980, p. 151.
37. A. C. Chen, M. F. Veiga, and A. B. Rizzuto, Cocrystallization: An encapsulation process, *Food Technol. 42*(11):87 (1988).
38. G. R. Huber, Carbohydrates in extrusion processing, *Food Technol. 45*(3):160 (1991).
39. Encapsulated spice, seasoning extracts add flavor, color to dry-mix products, *Food Eng. 53*(8):67 (1981).
40. Greater stability for encapsulated flavors, *Food Eng. 54*(11):59 (1981).
41. J. W. Marvin, R. A. Berhard, and T. A. Nickerson, Interaction of low molecular weight adsorbents on lactose, *J. Dairy Sci. 62*:1546 (1979).
42. J. A. Kitson and H. S. Sugisawa, Sugar captured flavor granules, *Can. Inst. Food Sci. Technol. J. 7*:15 (1974).
43. A. C. Chen, C. E. Lang, C. P. Graham, and A. B. Rizzuto, Crystallized, readily water-dispersible sugar product, U.S. patent 4,338,350 (1982).
44. A. C. Chen, C. E. Lang, C. P. Graham, and A. B. Rizzuto, Crystallized, readily water-dispersible sugar product containing heat sensitive, acidic or high invert sugar substances, U.S. patent 4,362,757 (1982).
45. D. Knorr and R. A. Teutonico, Chitosan immobilization and permeabilization of *Amaranthus* tricolor cells, *J. Agric. Food Chem. 34*:96 (1986).
46. D. Knorr and J. Berlin, Effects of immobilization and permeabilization procedures on growth of *Chenopodium rubrum* cells and amaranthine concentration, *J. Food Sci. 52*:1397 (1987).
47. Y. Pandya and D. Knorr, Diffusion characteristics and properties of chitosan coacervate capsules, *Proc. Biochem. 26*(2):75 (1991).
48. I. K. Greener and O. Fennema, Barrier properties and surface characteristics of edible, bilayer films, *J. Food Sci. 54*:1393 (1989).
49. I. K. Greener and O. Fennema, Evaluation of edible, bilayer films for use as moisture barriers for food, *J. Food Sci. 54*:1400 (1989).
50. J. J. Kester and O. Fennema, An edible film of lipids and cellulose ethers: Barrier properties to moisture vapor transmission and structural evaluation, *J. Food Sci. 54*:1383 (1989).
51. F. Vojdani and J. A. Torres, Potassium sorbate permeability of polysaccharide films: Chitosan, methylcellulose and hydroxypropyl methylcellulose, *J. Food Proc. Eng. 12*:33 (1990).
52. F. Vojdani and J. A. Torres, Potassium sorbate permeability of methylcellulose and hydroxypropyl methylcellulose coatings: Effect of fatty acids, *J. Food Sci. 55*:841 (1990).
53. D. S. B. Poncelet, D. Poncelet, and R. J. Neufeld, Control of mean diameter and size distribution during formulation of microcapsules with cellulose nitrate membranes, *Enzyme Microb. Atechnol. 11*:29 (1989).
54. M. Glicksman, Background and classification, *Food Hydrocolloids*, Vol. 1 (M. Glicksman, ed.), CRC Press, Boca Raton, FL, 1982, p. 3.
55. T. J. Carroll, D. Feinerman, R. J. Huzinec, and D. J. Piccolo, Gum composition with plural time releasing flavors and method of preparation, U.S. patent 4,485,118 (1984).
56. M. Glicksman, The hydrocolloids industry in the '80s—Problems and opportunities, *Gums and*

Stabilisers for the Food Industry. Interaction of Hydrocolloids (G. O. Phillips, G. J. Wedlock, and P. A. Williams, eds.), Pergmon Press, Oxford, UK, 1982, p. 299.
57. M. Glicksman, Food applications of gums, *Food Carbohydrates* (D. R. Lineback and G. E. Inglett, eds.), AVI Publishing Co., Inc., Westport, CT, 1982, p. 270.
58. J. D. Dziezak, Focus on gums, *Food Technol. 45*(3):116 (1991).
59. Meiji Seika Kaisha Ltd, Encapsulated liquid foods, Japanese patent 17,941/71 (1971).
60. I. A. Veliky and M. Kalab, Encapsulation of viscous high-fat foods in calcium alginate gel tubes at ambient temperature, *Food Struct. 9*:151 (1990).
61. Y. Tanaka, Encapsulated food stuffs and process for the production of same, British patent 1,489,539 (1977).
62. C. Hoashi, Food product with capsules containing meat soup or juice, U.S. patent 4,844,918 (1989).
63. F. Thevenet, Acacia gums: Stabilizers for flavor encapsulation, *Flavor Encapsulation* (S. J. Risch and G. A. Reineccius, eds.), ACS Symposium Series No. 370, American Chemical Society, Washington, DC, 1988, p. 37.
64. F. Thevenet, Acacia gums: Natural encapsulation agent for food ingredients, *Encapsulation and Controlled Release of Food Ingredients* (S. J. Risch and G. A. Reineccius, eds.), ACS Symposium Series No. 590, American Chemical Society, Washington, DC, 1995, p. 51.
65. M. Rosenberg and I. J. Kopelman, Microencapsulation of food ingredients—Process, application and potential, Proceedings of the 6th International Congress of Food Science and Technology, 1983, p. 142.
66. B. R. Bhandari, E. D. Dumoulin, H. M. J. Richard, I. Noleau, and A. M. Lebert, Flavor encapsulation by spray drying: Application to citral and linalyl acetate, *J. Food Sci. 57*:217 (1992).
67. T. H. Schultz, J. C. Miers, and W. D. Maclay, Permeability of pectinate films to water vapor, *J. Phys. Colloid Chem. 53*:1320 (1949).
68. S. L. Kamper and O. Fennema, Water vapor permeability of edible bilayer films, *J. Food Sci. 49*:1478 (1984).
69. S. L. Kamper and O. Fennema, Water vapor permeability of an edible, fatty acid, bilayer film, *J. Food Sci. 49*:1482 (1984).
70. Food and Drug Administration, United States of America, *Fed. Reg. 45* (1980).
71. A. P. Tulloch, The composition of beeswax and other waxes secreted by insects, *Lipids 5*:247 (1970).
72. A. P. Tulloch and L. L. Hoffman, Canadian beeswax: Analytical values and composition of hydrocarbons, free acids, and long-chain esters, *J. Am. Oil Chem. Soc. 49*:696 (1972).
73. N. V. Lovergren and R. O. Feuge, Permeability of acetostearin products to water vapor, *J. Agric. Food Chem. 2*:558 (1954).
74. H.-D. Belitz and W. Grosch, Lipids, *Food Chemistry*, Springer-Verlag, Berlin, 1987, p. 128.
75. K. Koide and M. Karel, Encapsulation and stimulated release of enzymes using lecithin vesicles, *Int. J. Food Sci. Technol. 22*:707 (1987).
76. M. Matsuzaki, F. McCafferty, and M. Karel, The effect of cholesterol content of phospholipid vesicles on the encapsulation and acid resistance of β-galactosidase from *E. coli*, *Int. J. Food Sci. Technol. 24*:451 (1989).
77. W. F. Hopkins and T. J. Carroll, Encapsulated active ingredients, process for preparing them and their use in ingested products, European patent application EP 0,252,374 A1 (1988).
78. H. H. Hatanaka, Egg lecithin process, U.S. patent 4,844,926 (1989).
79. H-H. Y. Kim and I. C. Baianu, Novel liposome microencapsulation techniques for food applications, *Trends Food Sci. Technol. 2*:55 (1991).
80. G. Gregoriadis, Encapsulation of enzymes and other agents in liposomes, *Chemical Aspects in Food Enzymes* (A. J. Andrews, ed.), Royal Society of Chemistry, London, 1987, p. 94.
81. F. Szoka and D. Papahadjopoulos, Comparative properties and methods of preparation of lipid vesicles (liposomes), *Annu. Rev. Biophys. Bioeng. 9*:467 (1980).
82. D. W. Deamer and P. S. Uster, Liposome preparation: Methods and mechanisms, *Liposomes* (M. J. Ostro, ed.), Marcel Dekker, New York, 1983, p. 27.

83. A. D. Bangham, M. W. Hill, and N. G. Miller, Preparation and use of liposomes as models of biological membranes, *Methods Memb. Biol. 1*:1 (1974).
84. M. R. Niesman, Liposome encapsulated manganese chloride as liver specific contrast agent for magnetic resonance imaging, *Diss. Abstr. Int. B 49*:3555 (1989).
85. A. Kondo, Microencapsulation utilizing phase separation from an aqueous solution system, *Microcapsule Processing and Technology* (J. Wade van Valkenburg, ed.), Marcel Dekker, Inc., New York, 1979, p. 70.
86. W. N. Garrett, Y. T. Yang, W. L. Dunkley, and L. M. Smith, Energy utilization, feed-lot performance and fatty acid composition of beef steers fed protein encapsulated tallow or vegetable oils, *J. Animal Sci. 42*:1522 (1976).
87. F. Ono, New encapsulation technique with protein-carbohydrate matrix, *J. Jpn. Food Sci. Technol. 27*:529 (1980).
88. F. Ono and Y. Aoyama, Encapsulation and stabilization of oily substances by protein and carbohydrate, *J. Jpn. Food Sci. Technol. 26*:13 (1979).
89. A. H. Taylor, Encapsulation systems and their applications in the flavor industry, *Food Flav. Ingred. Packg. Proc. 5*(9):48 (1983).
90. H. B. Heath, The flavor trap, *Food Flav. Ingred. Packg. Proc. 7*(2):21 (1985).
91. M. Meyers, High performance encapsulation (HPE). Applications in meat processing technology, *Agro-Food-Industry Hi-Tech* (September/October):23 (1995).
92. D. Blenford, Fully protected, *Food Flav. Ingred. Packg. Proc. 8*(7):43 (1986).
93. W. E. Bangs, Development and characterization of wall materials for spray dried flavorings production, *Diss. Abstr. Int. B 46*(4):1011 (1985).
94. P. J. A. M. Kerkhoff and H. A. C. Thijssen, Retention of aroma components in extractive drying of aqueous carbohydrate solutions, *J. Food Technol. 9*:415 (1974).
95. P. J. A. M. Kerkhoff and H. A. C. Thijssen, The effect of process conditions on aroma retention in drying liquid foods, *Aroma Research*, Proceedings of the International Symposium, Wageningen Centre for Agricultural Publishing and Documentation, Wageningen, The Netherlands, 1975, p. 26.
96. L. C. Menting, B. Hoogstad, and H. A. C. Thijssen, Aroma retention during the drying of liquid foods, *J. Food Technol. 5*:127 (1970).
97. J. L. Bomben, B. Bruin, H. A. C. Thijssen, and R. L. Merson, Aroma recovery and retention in concentration and drying of foods, *Adv. Food Res. 20*:1 (1973).
98. G. A. Reineccius and W. E. Bangs, Spray drying of food flavors: III. Optimum infeed concentrations for the retention of artificial flavors, *Perf. Flav. 10*(1):27 (1985).
99. M. M. Leahy, S. Anandaraman, W. E. Bangs, and G. A. Reineccius, Spray drying of food flavors: II. A comparison of encapsulating agents for the drying of artificial flavor, *Perf. Flav. 8*(5):49 (1983).
100. J. Brenner, G. H. Henderson, and R. W. Bergensten, Process of encapsulating an oil and product produced thereby, U.S. patent 3,971,852 (1976).
101. R. Emberger, Aspects of the development of industrial flavor materials, *Flavour '81, Weurman Symposium* (P. de Gruyter Schreier, ed.), Berlin, 1981, p. 620.
102. S. J. Risch and G. A. Reineccius, Spray-dried orange oil: Effect of emulsion size on flavor retention and shelf stability, *Flavor Encapsulation* (S. J. Risch and G. A. Reineccius, eds.), ACS Symposium Series No. 370, American Chemical Society, Washington, DC, 1988, p. 67.
103. Y. I. Chang, J. Scire, and B. Jacobs, Effect of particle size and microstructure properties on encapsulated orange oil, *Flavor Encapsulation* (S. J. Risch and G. A. Reineccius, eds.), ACS Symposium Series No. 370, American Chemical Society, Washington, DC, 1988, p. 87.
104. D. M. Jones, Controlling particle size and release properties: Secondary processing techniques, *Flavor Encapsulation* (S. J. Risch and G. A. Reineccius, eds.), ACS Symposium Series No. 370, American Chemical Society, Washington, DC, 1988, p. 158.
105. J. Brenner, The essence of spray-dried flavors: The state of the art, *Perf. Flav. 8*(20):40 (1983).
106. L. Y. Sheen and S. T. Tsai, Studies on spray-dried microcapsules of ginger, basil and garlic essential oils, *J. Chinese Agric. Chem. Soc. 29*:226 (1991).

107. J. A. Bakan and J. L. Anderson, Microencapsulation, *The Theory and Practice of Industrial Pharmacy* (L. Lachman, H. A. Lieberman, and J. L. Kang, eds.), Lea and Febiger, Philadelphia, 1970, p. 384.
108. R. Lamb, Spray chilling, *Food Flav. Ingred. Packg. Proc. 9*(12):39 (1987).
109. T. H. Schultz, K. P. Dimick, and B. Makower, Incorporation of natural fruit flavors into fruit juice powders. I. Locking of citrus oils in sucrose and dextrose, *Food Technol. 10*(1):57 (1956).
110. H. E. Swisher, Solid essential oil-containing components, U.S. patent 2,809,895 (1957).
111. H. E. Swisher, Solid essential oil-flavoring components, U.S. patent 3,041,180 (1962).
112. E. E. Beck, Essential oils, U.S. patent 3,704,137 (1972).
113. J. M. Barnes and J. A. Steinke, Encapsulation matrix composition and encapsulate containing same, U.S. patent 4,689,235 (1987).
114. D. H. Miller and J. R. Mutka, Preparation of solid essential oil flavor composition, U.S. patent 4,499,122 (1985).
115. D. H. Miller and J. R. Mutka, Solid essential oil composition, U.S. patent 4,610,890 (1986).
116. W. Schlameus, Centrifugal extrusion encapsulation, *Encapsulation and Controlled Release of Food Ingredients* (S. J. Risch and G. A. Reineccius, eds.), ACS Symposium Series No. 590, American Chemical Society, Washington, DC, 1995, p. 96.
117. E. L. Anderson, W. W. Harlowe, L. M. Adams, and M. C. Marshall, Process and apparatus for the production and collection of microcapsules, European patent application EP 152,285 (1985).
118. I. J. Kopelman, S. Meydav, and P. Wilmersdorf, Freeze drying encapsulation of water-soluble citrus aroma, *J. Food Technol. 12*:65 (1977).
119. I. J. Kopelman, D. J. Mangold, and S. Weinberg, Storage studies of freeze-dried lemon crystals, *J. Food Technol. 12*:403 (1977).
120. C. J. Kirby and G. Gregoriadis, A simple procedure for preparing liposomes capable of high encapsulation efficiency under mild conditions, *Liposome Technology*, Vol. I (G. Gregoriadis, ed.), CRC Press, Boca Raton, FL 1984, p. 19.
121. W. H. Rulkens and H. A. C. Thijssen, Retention volatile compounds in freeze-drying slabs of malto-dextrin, *J. Food Technol. 7*:79 (1972).
122. J. Flink and M. Karel, Retention of organic volatiles in freeze-dried solution of carbohydrates, *J. Agric. Food Chem. 18*:295 (1970).
123. J. D. Dziezak, Microencapsulation and encapsulated ingredients, *Food Technol. 42*(4):136 (1988).
124. L. A. Luzzi and R. J. Gerraughty, Effects of selected variables on the extractability of oils from coacervate capsules, *J. Phar. Sci. 53*:429 (1964).
125. G. R. Chilvers, A. P. Gunning, and V. J. Morris, Coacervation of gelatin-XM6 mixtures and their use in microencapsulation, *Carbohydrate Polym. 8*:55 (1988).
126. L. L. Balssa, Flavor encapsulation, U.S. patent 3,495,988 (1970).
127. C. Arneodo, A. Baszkin, J.-P. Benoit, and C. Thies, Interfacial tension behavior of citrus oils against phases formed by complex coacervation of gelatin, *Flavor Encapsulation* (S. J. Risch and G. A. Reineccius, eds.), ACS Symposium Series No. 370, American Chemical Society, Washington, DC, 1988, p. 132.
128. A. H. King, Flavor encapsulation with alginates, *Flavor Encapsulation* (S. J. Risch and G. A. Reineccius, eds.), ACS Symposium Series No. 370, American Chemical Society, Washington, DC, 1988, p. 122.
129. R. E. Sparks and N. S. Mason, U.S. patent 4,675,140 (1987).
130. R. E. Sparks, N. S. Mason, P. Autant, A. Cartillier, and R. Pigeon, U.S. patent 5,186,937 (1993).
131. R. E. Sparks, I. C. Jacobs, and N. S. Mason, Centrifugal suspension-separation for coating food ingredients, *Encapsulation and Controlled Release of Food Ingredients* (S. J. Risch and G. A. Reineccius, eds.), ACS Symposium Series No. 590, American Chemical Society, Washington, DC, 1995, p. 87.
132. A. Awad and A. C. Chen, A new generation of sucrose products made by cocrystallization, *Food Technol. 47*(1):146 (1993).
133. J. W. Mullin, Crystallization kinetics, *Crystallization* (J. W. Mullin, ed.), CRC Press, Butterworths, London, 1972, p. 174.

134. A. B. Rizzuto, A. C. Chen, and M. F. Veiga, Modification of the sucrose crystal structure to enhance pharmaceutical properties of excipient and drug substances, *Pharm. Technol. 8*(9):32 (1984).
135. G. A. Reineccius, Liposomes for controlled release in the food industry, *Encapsulation and Controlled Release of Food Ingredients* (S. J. Risch and G. A. Reineccius, eds.), ACS Symposium Series No. 590, American Chemical Society, Washington, DC, 1995, p. 113.
136. D. Chapman, R. M. Williams, and B. D. Ladbrook, Physical studies of phospholipids: VI. Thermotropic and lyotropic mesomorphism of some 1,2-diacylphosphatidylcholines (lecithin), *Chem. Phys. Lipids 1*:445 (1967).
137. E. Mayhew, S. Conroy, J. King, R. Lazo, G. Nikolopoulus, A. Siciliano, and W. J. Vail, High-pressure continuous-flow system for drug entrapment in liposomes, *Drug and Enzyme Targeting*, Part B (R. Green and K. J. Widder, eds.), Academic Press, Inc., New York, 1987, p. 64.
138. E. Mayhew, R. Lazo, W. J. Vail, J. King, and A. M. Green, Characterization of liposomes prepared using a microemulsifier, *Biochem. Biophys. Acta 755*:169 (1984).
139. A. Kondo, Microencapsulation by interfacial polymerization, *Microcapsule Processing and Technology* (J. Wade van Valkenburg, ed.), Marcel Dekker, Inc., New York, 1979, p. 35.
140. J. S. Pagington, Molecular encapsulation with β-cyclodextrin, *Food Flav. Ingred. Packg. Proc. 7*(9):51 (1985).
141. P. E. Shaw, J. H. Tatum, and C. W. Wilson, Improved flavor of Navel orange and grapefruit juices by removal of bitter components with β-cyclodextrin polymer, *J. Agric. Food Chem. 32*:832 (1984).
142. L. L. Westing, G. A. Reineccius, and F. Caporaso, Shelf life of orange oil: Effects of encapsulation by spray-drying, extrusion, and molecular inclusion, *Flavor Encapsulation* (S. J. Risch and G. A. Reineccius, eds.), ACS Symposium Series No. 370, American Chemical Society, Washington, DC, 1988, p. 110.
143. K. Lindner, L. Szente, and J. Szejtli, Food flavoring with β-cyclodextrin-complexed flavor substances, *Acta Aliment. 10*:175 (1981).
144. L. Szente and J. Szejtli, Stabilization of flavors by cyclodextrins, *Flavor Encapsulation* (S. J. Risch and G. A. Reineccius, eds.), ACS Symposium Series No. 370, American Chemical Society, Washington, DC, 1988, p. 148.
145. L. Szente, M. Gal-Fuzy, and J. Szejtli, Tea aromatization with β-cyclodextrin complexed flavors, *Acta Aliment. 17*:193 (1988).
146. L. E. Werner, Encapsulated food acids, *Cereal Foods World 25*(3):102 (1980).
147. W. E. Delaney, Meat color stabilization, U.S. patent 3,560,222 (1971).
148. J. C. Cordray and D. L. Huffman, Restructured pork from hot processed sow meat: Effect of encapsulated food acids, *J. Food Prot. 48*:965 (1985).
149. P. J. Percel and D. W. Perkins, Process of preparing a particulate food acidulant, U.S. patent 4,537,784 (1985).
150. Specially coated natural ingredient improves bread, *Food Proc. 34*(1):10 (1973).
151. J. T. Seighman, Process for encapsulating liquid acids and product, U.S. patent 4,713,251 (1987).
152. W.M. McKernan, Microencapsulation in the flavor industry. I, *Flav. Industry 3*:596 (1972).
153. W.M. McKernan, Microencapsulation in the flavor industry. II, *Flav. Industry 4*:70 (1973).
154. S. Anandaraman and G. A. Reineccius, Microencapsulation of flavor, *Food Flav. Ingred. Packg. Proc. 1*(9):14 (1980).
155. U. Marquardt, Spices, Flavors and aromas, *Eur. Food Drink Rev.* (Autumn):101 (1990).
156. L. S. Jackson and K. Lee, Microencapsulation and the food industry, *Lebens. Wiss. Technol. 24*:289 (1991).
157. J. R. Bedford and D. R. Ashworth, Encapsulated flavors—Their applications and development, *Food Flav. Ingred. Packg. Proc. 5*(2):13 (1983).
158. R. A. Youngs, Spray drying encapsulation—Today's review, *Food Flav. Ingred. Packg. Proc. 8*(10):31 (1986).
159. W. E. Bangs and G. A. Reineccius, Characterization of selected materials for lemon oil encapsulation by spray drying, *J. Food Sci. 55*:1356 (1990).

160. C. I. Beristain, A. Vazquez, H. S. Garcia, and E. J. Vernon-Carter, Encapsulation of orange peel oil by co-crystallization, *Lebens, Wiss. Technol. 29*:645 (1996).
161. E. L. Magee, Microencapsulation of cheese ripening systems in milkfat, *Diss. Abstr. Int. B 40*:2103 (1979).
162. E. L. Magee and N. F. Olson, Microencapsulation of cheese ripening systems: Production of diacetyl and acetol in cheese by encapsulated bacterial cell-free extract, *J. Dairy Sci. 64*:616 (1981).
163. S. D. Braun, N. F. Olson, and R. C. Lindsay, Microencapsulation of bacterial cell-free extract to produce acetic acid for enhancement of cheese flavor, *J. Food Sci. 47*:1803 (1982).
164. S. D. Braun and N. F. Olson, Encapsulation of proteins and peptides in milkfat: Encapsulation efficiency and temperature and freezing stabilities, *J. Microencap. 3*(2):115 (1986).
165. C. J. Kirby and B. A. Law, Development in the microencapsulation of enzymes in food technology, *Chemical Aspects of Food Enzymes* (A. T. Andrews, ed.), Royal Society of Chemistry, 1987, p. 106.
166. C. J. Kirby, and B. A. Law, Recent development in cheese flavor technology: Application of enzyme microencapsulation, *Biotechnology in the Food Industry* Proceedings of the Conference, Online International Ltd., London, UK, 1986, p. 17.
167. Food additives: Microcapsules for flavoring oils, *Fed. Reg. 33* (232, Nov. 28):17752 (1968).
168. Food additives: Microcapsules for flavoring oils, *Fed. Reg. 33* (242, Dec. 13):18488 (1968).
169. L. L. Balssa and J. Brody, Microencapsulation—The blachem way, *Food Eng. 40*(11):88 (1968).
170. E. R. Jensen, Encapsulated flavors, Canadian patent 866,713 (1971).
171. Microencapsulation extends shelf life of marginally stable ingredients, *Food Proc. 42*(7):40 (1981).
172. Lemon flavor for fish survives frying and baking, *Food Proc. 31*(5):24 (1970).
173. E. Palmer, Method for encapsulating materials, U.S. patent 3,989,852 (1976).
174. W. Abe, Y. Yamamoto, R. Uehara, K. Ogiwara, and T. Satoh, Studies on feeding encapsulated safflower oil to milking cows and fattening steers, *Jpn. J. Zootechnol. Sci. 47*:639 (1976).
175. Bush Boake Allen Ltd., Encapsulated flavors, British patent 1,327,761 (1973).
176. R. T. Darragh and J. L. Stone, Fats with encapsulated flavors, U.S. patent 3,867,556 (1975).
177. C. Andres, Encapsulated concentrates retain full-flavor profile balance, *Food Proc. 42*(12):57 (1981).
178. How the flavor is sealed, *Food Proc. Ind. 50*(600):36 (1981).
179. J. Tuot, Foamed capsules having edible shells enclosing aromatic components, UK patent application GB 2,144,701 A (1985).
180. R. T. Liu, W. R. Nickerson, and C. H. Anderson, Process for the preparation of flavorant capsules (for various food uses), U.S. patent 4,576,826 (1986).
181. J. Szejtli, L. Szente, and E. Banky-Elod, Molecular encapsulation of volatile, easily oxidizable labile flavor substances by cyclodextrins, *Acta Chim. Acad. Sci. Hung. 101*(1/2):27 (1979).
182. R. S. Johnson, Encapsulation of volatile liquids, European patent 0.070,719 B1 (1985).
183. M. Glass, Sorbitol containing mixture encapsulated flavor, U.S. patent 4,388,328 (1983).
184. C. Andres, Unique encapsulation flavors maximize flavor/aromas—Instantly release oils of mustard, onion, spices, *Food Proc. 33*(5):23 (1972).
185. Y. C. Wei, S. R. Cherukuri, F. Hriscisce, and D. J. Piccolo, Elastomer encapsulation of flavors and sweeteners, long lasting flavored chewing gum compositions based there on and process of preparation, U.S. patent 4,590,075 (1986).
186. R. K. Yang, Encapsulation composition for use with chewing gum and edible products, European patent application EP 0,229,000 A2 (1986).
187. J. Szejtli, L. Szente, and L. Szenta, Method for aromatizing tea, and the aromatizing product, Swiss patent CH 656,778 A (1986).
188. L. Szente and J. Szejtli, Molecular encapsulation of natural and synthetic coffee flavor with β-cyclodextrin, *J. Food Sci. 51*:1024 (1986).
189. Heat glaze coating encapsulates flavor, *Food Proc. 30*(6):32 (1969).
190. S. D. Braun, N. F. Olson, and R. C. Lindsay, Production of flavor compounds: Aldehydes and alcohols from leucine by microencapsulated cell-free extracts of *Streptococcus lactis* var. multigenes, *J. Food Biochem. 7*:23 (1983).

191. H. Levine, L. Slade, B. V. Lengerich, and J. G. Pickup, Double-encapsulated compositions containing volatile and/or labile components, and processes for preparation and use thereof, U.S. patent 5,087,461 (1992).
192. T. Cea, J. D. Posta, and M. Glass, Encapsulated APM and method of preparation, U.S. patent 4,384,004 (1983).
193. T. Cea, J. D. Posta, and M. Glass, A chewing gum composition incorporating encapsulated L-aspartyl-L-phenylalanine methyl ester, European patent EP 0,067,595 B1 (1987).
194. A. M. Schobel and R. K. Yang, Encapsulated sweeteners composition for use with chewing gum and edible products, U.S. patent 4,824,681 (1989).
195. R. K. Yang and A. M. Schobel, Chewing gum composition with encapsulated sweetener having extended flavor release, U.S. patent 4,911,934 (1990).
196. S. R. Cherukuri and G. Mansukhani, Sweetener delivery systems containing polyvinyl acetate, U.S. patent 4,816,265 (1989).
197. S. R. Cherukuri and G. Mansukhani, Multiple encapsulated sweetener delivery system, U.S. patent 4,933,190 (1990).
198. S. R. Cherukuri, G. Mansukhani, and K. C. Jacob, Stable sweetener delivery system for use with cinnamon flavors, U.S. patent 4,839,184 (1989).
199. A. C. Chen, A. B. Rizzuto, and M. F. Veiga, Cocrystallized sugar-nut product, U.S. patent 4,423,085 (1983).
200. A. C. Chen, S. J. Drescher, and C. P. Graham, Maple sugar product and method of preparing and using same, U.S. patent 4,159,210 (1979).
201. J. W. Kinnison and R. S. Chapman, Extrusion effects on colors and flavors, *Snack Food 61*(10):40 (1972).
202. F. Ono, Solubilization of fats and oils by solid components of soy sauce: X. Encapsulation and stabilization of oil-soluble pigment in a protein-carbohydrate matrix, *J. Jpn. Food Sci. Technol. 26*:346 (1979).
203. P. Ciliberto and S. Kramer, Non-caking, water-soluble, granular coated food ingredient, U.S. patent 4,288,460 (1981).
204. F. Shahidi and R. B. Pegg, Encapsulation of the preformed cooked cured-meat pigment, *J. Food Sci. 56*:1500 (1981).
205. J. Bitman, T. R. Wrenn, L. P. Dryden, and L. F. Edmondson, Feeding encapsulated vegetable fats to increase the polyunsaturation of milk, cheese, and meat, XIX International Dairy Congress, 1E, 1974, p. 107.
206. L. F. Demondson, R. A. Yoncoskie, N. H. Rainey, F. W. Douglas, and J. Bitman, Feeding encapsulated oils to increase the polyunsaturation in milk and meat fat, *J. Am. Oil Chem. Soc. 51*(3):72 (1974).
207. V. F. Kristensen, Influence on the fatty acid composition of milkfat on feeding dairy cows increasing amounts of encapsulated soybean oil, XIX International Dairy Congress, 1E, 1974, p. 109.
208. Encapsulated ingredients, *Food Technol. 42*(4):158 (1988).
209. A. P. Simopoulos, Omega-3 fatty acids in health and disease and growth and development. A review, *Am. J. Clin. Nutr. 54*:438 (1991).
210. A. P. Simopoulos, Fatty acids, *Functional Foods, Designer Foods, Pharmafoods, Nutraceuticals* (I. Goldberg, ed.), Chapman & Hall, Glasgow, 1994, p. 355.
211. L. M. Branden and K. K. Carroll, Dietary polyunsaturated fats in relation to mammary carcinogenesis in rats, *Lipids* 21:285 (1986).
212. J. Dyberg, Linolenate-derived polyunsaturated fatty acids and prevention of atherosclerosis, *J. Nutr. Rev. 44*:125 (1986).
213. J. Mehta, L. M. Lopez, D. Lowton, and T. Wargovich, Dietary supplementation with omega-3 polyunsaturated fatty acids in patients with stable coronary disease. Effects of indices of platelet and neutrophil function and exercise performance, *Am. J. Med. 84*:45 (1988).
214. V. K. S. Shukla and E. G. Perkins, The presence of oxidative polymeric materials in encapsulated fish oils, *Lipids 26*:23 (1991).

215. F. Gejl-Hansen and J. M. Flink, Freeze-dried carbohydrate containing oil-in-water emulsions: Microstructure and fat distribution, *J. Food Sci. 42*:1049 (1977).
216. K. Taguchi, K. Iwami, F. Ibuki, and M. Kawabata, Oxidative stability of sardine oil embedded in spray-dried egg white powder and its use in n-3 unsaturated fatty acid fortification of cookies, *Biosci. Biotech. Biochem. 56*:560 (1992).
217. K. Iwami, M. Hattori, S. Nakatani, and F. Ibuki, Spray-dried gliadin powders inclusive of linoleic acid (microcapsules): Their preservability, digestibility and application to bread making, *Agric. Biol. Chem. 51*:3301 (1987).
218. F. Shahidci and U. N. Wanasundara, Oxidative stability of encapsulated seal blubber oil, *Flavor Technology—Physical Chemistry, Modification, and Process* (C.-T Ho, C.-T. Tan, and C.-H. Tong, eds.), ACS Symposium Series 610, American Chemical Society, Washington, DC, 1995, p. 139.
219. Hoffmann-La Roche, Inc., Vitamins, Part II: General considerations, *Encyclopedia of Food Science and Technology* (Y. H. Hui, ed.), John Wiley and Sons, Inc., New York, 1992, p. 2687.
220. J. Giese, Vitamin and mineral fortification of foods, *Food Technol. 49*(5):110 (1995).
221. H. S. Hall and R. E. Pondell, Encapsulated nutrients, U.S. patent 4,182,778 (1980).
222. Fat matrix encapsulation controls ingredients release-reactions are temperature-specific, *Food Proc. 37*(5):72 (1976).
223. J. Eden, R. Trksak, and R. Williams, Starch-based particulate encapsulation process, U.S. patent 4,755,397 (1988).
224. L. D. Morse, P. A. Hammes and W. A. Boyd, Devil's food cake and other alkaline bakery goods, U.S. Patent 3,821,422 (1974).
225. P. T. Berglund, J. W. Dick, and M. L. Dreher, Effect of form of enrichment and iron on thiamin, riboflavin and niacinamide, and cooking parameters of enriched spaghetti, *J. Food Sci. 52*:1376 (1987).
226. J. M. deMan, L. deMan, and T. Wygerde, Stability of vitamin A beadlets in nonfat dry milk, *Milchwissenschaft 41*:468 (1986).
227. A. Markus and Z. Pelah, Encapsulation of vitamin A, *J. Microencap. 6*:389 (1989).
228. C. Andres, Fat matrix encapsulation controls ingredients release-reactions are temperature-specific, *Food Proc. 37*(1):72 (1976).
229. S. S. Jackel and F. Belshaw, Encapsulated ferrous sulphate protects baking mixes, flour from rancid flavors, *Food Proc. 32*(5):28 (1971).
230. B. N. Harrison, G. W. Pla, G. A. Clark, and J. C. Fritz, Selection of iron sources for cereal enrichment, *Cereal Chem. 53*(1):78 (1976).
231. K. Weingartner, A. Nelson, and J. Erdman, Effect of calcium addition on stability and sensory properties of soy beverages, *J. Food Sci. 48*:256 (1983).
232. M. Hirotsuka, H. Taniguchi, H. Narita, and M. Kito, Calcium fortification of soymilk with calcium-lecithin liposome system, *J. Food Sci. 49*:111 (1984).
233. H. W. Schafer, Increasing the recovery of milk proteins in cheese made by direct acidification and encapsulation of enzymes for acceleration of cheese ripening, *Diss. Abst. Int. B 36*:1127 (1975).
234. S. D. Braun, Microencapsulated multi-enzymes to produce flavors and recycle cofactors, *Diss. Abstr. Int. B 46*:366 (1985).
235. C. J. Kirby, B. E. Brooker, and B. A. Law, Accelerated ripening of cheese using liposome-encapsulated enzyme, *Int. J. Food Sci. Technol. 22*:355 (1987).
236. B. Law and C. Kirby, Microencapsulated enzymes for cheese technology, *North Eur. Food Dairy J. 53*(6):194 (1987).
237. M. El-Soda, L. Pannell, and N. Olson, Microencapsulated enzyme systems for the acceleration of cheese ripening, *J. Microencap. 6*:319 (1989).
238. S. C. Kim and N. F. Olson, Production of methanethiol in milk fat-coated microcapsules containing *Brevibacterium linens* and methionine, *J. Dairy Res. 56*:799 (1989).
239. F. P. Colten, J. J. Halik, R. J. Ravallo, and J. L. Hegadorn, Gasified candy enrobed with oleaginous material, U.S. patent 4,275,083 (1981).

240. M. Iso, T. Shirahase, S. Hanamura, S. Urushiyama, and S. Omi, Immobilization of enzyme by microencapsulation and application of the encapsulated enzyme in the catalysis, *J. Microencap. 6*:165 (1989).
241. I. Garcia, R. B. Aisina, O. Ancheta, and C. Pascual, Action of gastric juice on microencapsulated invertase, *Enzyme Microb. Technol. 11*:247 (1989).
242. J. Mansfeld, A. Foerster, A. Schellenberger, and H. Dautzenberg, Immobilization of invertase by encapsulation in polyelectrolyte complexes, *Enzyme Microb. Technol. 13*:240 (1991).
243. Taiyo Chemical Industries Co., Ltd., Encapsulated preservative, Japanese patent 5,136,334 (1976).
244. J. Rozenblat, S. Magdassi, and N. Garti, Effect of electrolytes, stirring and surfactants in the coacervation and microencapsulation (of oleic acid) process in presence of gelatin, *J. Microencap. 6*:515 (1989).
245. D. R. Meyer, Process for preparing a coated-particle salt substitute composition, U.S. patent 4,734,290 (1989).
246. D. J. Deeble, B. J. Parsons, and G. O. Phillips, Studies on food antioxidants encapsulated in β-cyclodextrin, *Food and Agriculture. Agricultural Research and Development in Wales*, 6th Conference, Cardiff, UK, 1989, p. 26.
247. M. Karel and R. Langer, Controlled release of food additives, *Flavor Encapsulation* (G.A. Reineccius and S. J. Risch, eds.), ACS Symposium Series No. 370, American Chemical Society, Washington, DC, 1988, p. 177.
248. L. Brannon-Peppas, Controlled release in the food and cosmetic industries, *Polymeric Delivery Systems* (M. A. El-Nokay, D. M. Piatt, and B. A. Charpentier, eds.), ACS Symposium Series No. 520, American Chemical Society, Washington, DC, 1993, p. 42.
249. C. Whorton, Factors influencing volatile release from encapsulation matrices, *Encapsulation and Controlled Release of Food Ingredients* (S. J. Risch and G. A. Reineccius, eds.), ACS Symposium Series No. 590, American Chemical Society, Washington, DC, 1995, p. 134.
250. D. Pendergrass, The use of controlled delivery in print materials, Second Workshop on the Controlled Delivery in Consumer Products, Controlled Release Society, Secaucus, NJ, May 13-15, 1992.
251. G. A. Reineccius, Controlled release techniques in the food industry, *Encapsulation and Controlled Release of Food Ingredients* (S. J. Risch and G. A. Reineccius, eds.), ACS Symposium Series No. 590, American Chemical Society, Washington, DC, 1995, p. 8.
252. C. Mellenheim and N. Passy, Choice of packages for foods with specific considerations of water activity, *Properties of Water in Foods* (D. Simatos and J. L. Multon, eds.), Martinus Nijhoff Publishing, Dordrecht, The Netherlands, 1985, p. 375.
253. B. Pascat, Study of some factors affecting permeability, *Food Packaging and Preservation: Theory and Practice* (M. Mathlouthi, ed.), Elsevier Applied Science Publishers, London, England, 1986, p. 7.
254. C. A. Kumins, Transport through polymer films, *J. Polym. Sci. Part C 10*:1 (1965).
255. P. I. Lee, Controlled release of volatile multicomponent active agents: Physical considerations, Second Workshop on the Controlled Delivery in Consumer Products, Controlled Release Society, May 13–15, Secaucus, NJ, 1992.
256. H. A. C. Thijssen and W. H. Rulkens, Retention of aromas in drying food liquids, *De Ingenieur 80*(47):45 (1968).
257. M. Karel and J. Flink, Some recent developments in food dehydration research, *Adv. Drying 2*:103 (1983).
258. H. Levine and L. Slade, A polymer physico-chemical approach to the study of commercial starch hydrolysis products (SHPs), *Carbohydrate Polym. 6*(3):213 (1986).
259. N. Peppas, Diffusional release from polymeric carriers, Second Workshop on the Controlled Delivery in Consumer Products, Controlled Release Society, May 13–15, Secaucus, NJ, 1992.
260. L. Slade and H. Levine, Glass transitions and water-food structure interactions, *Advances in Food and Nutrition Research*, Vol. 38 (J. Kinsella, ed.), Academic Press, San Diego, 1994.
261. E. C. To and J. M. Flink, 'Collapse,' a structural transition in freeze dried carbohydrates: I. Evaluation of analytical methods, *J. Food Technol. 13*:551 (1978).

262. M. Karel, Effects of water activity and water content on mobility of food components, and their effects on phase transitions in food systems, *Properties of Water in Foods* (D. Simatos and J. L. Multon, eds.), Martinus Nijhoff Publishing, Dordrecht, Netherlands, 1985, p. 153.
263. C.-T. Tan, Y. C. Kang, M. A. Sudol, C. K. King, and M. Schulman, Method of making controlled release flavors, U.S. patent 5,064,669 (1991).
264. K. Yazawa, R. Arai, M. Kitajima, and A. Kondo, Method of producing oil and fat encapsulated amino acids, U.S. patent 3,804,776 (1974).

22

Light and Sound in Food Preservation

M. SHAFIUR RAHMAN
Horticulture and Food Research Institute of New Zealand, Auckland, New Zealand

I. LIGHT IN FOOD PRESERVATION

A. Ultraviolet Radiation

Ultraviolet (UV) radiation has long been known to be the major factor in the bactericidal action of sunlight. It is mainly used in sterilizing air and thin liquid films due to low penetration depth. When used at high dosage there is a marked tendency toward flavor and odor deterioration before satisfactory sterilization is achieved. But low-level radiation at carefully applied doses can often usually extend the shelf life of foods without damaging quality [10].

The technique of using UV radiation to kill off bacteria in water is well known. UV irradiation is safe, environmentally friendly, and more cost effective to install and operate than conventional chlorination. It does not affect the taste of the water as does chlorine. High-intensity UV-C lamps have become available, which can increase the potential of destroying surface bacteria on food [22]. UV radiation has been used in dairy plants for many years. It is also being used in the ice cream industry and in meat and vegetable processing plants [52].

1. Ultraviolet Radiation in Food Preservation and Deterioration

Food-Preservation Enhancement by UV Radiation

Ultraviolet irradiation is being applied commercially in the use of bactericidal ultraviolet lamps in various food-processing applications: tenderizing or aging of meat, curing and wrapping of cheeses, prevention of surface mold growth on bakery products, and air purification in bottling and food-processing establishments and over pickle vats [10].

The lethal effect of ultraviolet light on microorganisms has been well documented. The practical application of this has been controversial because of the type and intensity of radiation, methods of estimating lethality, and other factors. A study of the germicidal powers of ultraviolet light shows that 3–83% of the yeast and 33–72% of the molds were killed in apple cider through layers varying from 2 to 25 mm in thickness [10]. A greater part of the light was absorbed by coloring agents. Incident energy levels of 253.7 nm inhibited 90% of *Bacillus megatherium* at 1100 mWs/cm^2 and 90% of *Sarcina lutea* at 19,800 mWs/cm^2 [2]. There was a 90% reduction in the microbial count of apple juice. Coupled with effective refrigeration, this could be of commercial significance.

It is generally agreed that the wavelength for maximum germicidal effect is 2600 Å. Low-pressure mercury-vapor lamps have a maximum output at 2537 Å, a value close to the peak wavelength for bactericidal effectiveness. The lethal action varies with the time of exposure and intensity of light. Other influential factors include temperature, hydrogen ion concentration, and the number of organisms per unit area exposed. The relative humidity affects the death rate of bacteria suspended in air, this being most noticeable at relative humidities greater than 0.50, at which point an increase in relative humidity results in a decreased death rate [10].

Spores of bacteria are generally more resistant to ultraviolet light than vegetative bacteria; *B. subtilis* is reported to be 5–10 times more resistant than *Escherichia coli*. Molds are more resistant than vegetative bacteria, while yeasts differ less from bacteria in this respect. It has been suggested that some mold species may be protected by fatty or waxy secretions on the cell surface, which shield them from the rays. Pigments apparently also afford some protection; dark-pigmented spores are more resistant to UV irradiation than nonpigmented types [10].

Short exposures, even long enough to cover one or more life cycles of the organism, are more efficient than higher radiation intensities for brief periods. This presumably is due to the fact that during certain stages of the life cycle the susceptibility to ultraviolet radiation is increased.

The effect of UV irradiation on bacteria and fungi such as *Penicillium* and *Aspergillus* has been reported by Kleczkowski [62]. UV radiation has been reported to inhibit fungal development in grapes [23], kumquats, and oranges [99]. Moy et al. [87] combined UV and gamma radiations for the preservation of papaya. Combined methods can avoid high doses of gamma and UV radiations.

Lu et al. [75] studied the efficacy of gamma rays (0.1–3 kGy), electron beams (0.1–5.0 kGy), or UV radiation (4.4–73.3 kerg/mm^2) to preserve Walla Walla onions up to 4 weeks at 20–25°C. UV-radiated onions exhibited the greatest percentage of marketable product and reduction in postharvest rot. Sprouting was observed with control, UV, and electron beam–irradiated onions but not with gamma-irradiated onions. No significant total sugar, pH, moisture, ascorbic acid, color, texture, or sensory quality changes were observed in the onions irradiated with UV. The optimum UV doses were in the range of 35.8–73.3 kerg/mm^2 for Walla Walla onions. In addition, UV irradiation is much more economical and safer to use than gamma or electron beam irradiation.

Ranganna et al. [97] studied the efficacy of ultraviolet radiation treatment in the control of both soft rot and dry rot diseases of potato tubers for short-term storage of 3 months. They used four UV radiation dose levels (75, 100, 125, and 150 kerg/mm^2) and three incubation levels each for *Fusarium solani* fungi (0, 1, and 2 days) and the bacteria *Erwinia carotovora* var. *carotovora* (0, 6, and 12 hr). The highest UV dose level was found to be more effective than the other three in controlling infection by the above fungi and bacteria. Visual observations of potato quality found no significant changes in the tuber qualities such as firmness and color.

Maple sap is susceptible to microbial infection, which lowers the quality of the syrup. Schneider et al. [109] studied the reduction of living cells of bacteria (*Pseudomonas-25* and *Pseudomonas–11*) and yeast (*Cryptococcus albidus*) strains suspended in maple sap when exposed to UV radiation of different intensities and for different lengths of time. The two bacterial strains were equally as sensitive as the yeast. Increase in exposure time had the same effect regardless of the method of UV irradiation employed.

Freshly pressed apple juice or fresh cider contains many microorganisms that cause deterioration within 2 days at room temperature unless they are inhibited or destroyed. The microbial population of fresh cider was greatly reduced and storage life prolonged without affecting the flavor by specially designed UV lamps [46]. Harrington and Hills [46] found that percentage reduction of microbial populations was affected by the clarity of the cider, the length of UV exposure, and the presence of potassium sorbate. This is very suitable where the initial microbial count is high and refrigeration is not adequate.

Meat becomes tender upon storage as a result of enzymic activity. This process is speeded up at relatively high temperatures, which favor the growth of surface microorganisms. By controlling such growth with ultraviolet light, the advantages of high-storage temperature can be better utilized and loss of meat is controlled. In this particular case, irradiation alone is the less likely active factor. The lamps employed emit rays not only in the germicidal 2537 Å range but also in the 1850 Å range. These shorter waves convert atmospheric oxygen to ozone; irregular and shaded areas of an irradiated surface are sterilized by the ozone. UV radiation is also used in storage vats and other tanks over both conveyers and for final treatment of both caps and stoppers [10].

Putrefaction of fresh meat can occur in a few hours as a result of the action of spoilage bacteria. UV radiation at a wavelength of 253.7 nm was effective in destroying surface bacteria on fresh meat by 2 log cycles (99% reduction) decrease on smooth surface beef after a radiation dose of 150 mWs/cm^2. A further increase in dose level to 500 mWs/cm^2 reduced bacteria 3 log cycles. Since UV radiation does not penetrate most opaque materials, it was less effective on rough-surface cuts of meat, such as round steak, because bacteria were partly shielded form the radiation. No deleterious effects on color (redness) or general appearance were observed, and UV irradiation of meat carcasses could also effectively increase the lag phase of bacterial growth until adequate cooling of the surface has occurred [115].

The physical appearance of a cut of meat in the display case is the most important factor determining consumer selection of a beef product. Reagan et al. [98] mentioned that significant increases in shelf life may be obtained by exposure of beef muscle and fat surface to UV light (maximum wavelengths 3660 Å for 2 min). Decreases in initial count and/or attenuation of the bacteria present on retail cuts via the use of UV light resulted in increased consumer acceptability, higher muscle color ratings, and increased shelf life beef [98].

Kaess and Weidemann [60] found that continuous UV (0.2–24 $\mu W/cm^2$) irradiation of psychrophilic microorganisms growing on muscle slices at 0°C and 0.993 equilibrium relative humidity resulted in an extension of the lag phase of *Pseudomonas* and of the molds *Thamnidium* and *Penicillium*, but not of the yeast *Candida scottii*. A minimum intensity of 2 $\mu W/cm^2$ at the meat surface is necessary to prolong storage life substantially. Lower-equilibrium relative humidity did not substantially increase UV effects. The relative extension of storage life at 10°C was comparable to that obtained at 0°C. Simultaneous use of UV radiation (0.2 $\mu W/cm^2$) and ozone (0.5 mg/m^3) produced synergistic effects with molds, but not with bacteria [60].

The use of UV radiation is effective in inhibiting the action of spoilage bacteria on fish and seafood [22]. UV radiation at 254 nm and doses of 300 mWs/cm^2 from a photochemical reactor of 4.8 Ws/cm^2 from a high-intensity UV-C lamp (40 sec at 120–180 mW/cm^2) reduced surface microbial counts on mackerel by 2 or 3 log cycles [52]. Huang and Toledo [52] found that Spanish fresh mackerel kept 7 days longer than untreated sample when the skin surface was treated with high-intensity UV light and stored in ice at −1°C. When UV irradiated and packed in 0°C ice, surface microbial counts on vacuum-packaged mackerel lagged 4 days behind those on mackerel wrapped in 1 mil polyethylene [52].

The use of UV radiation has some disadvantages, for example, it does not penetrate most opaque materials and it is less effective on rough surfaces [22]. Huang and Toledo [52] found that rough-surface fish such as croaker and mullet had little surface bacterial count reduction with a UV-C-13 lamp at doses 120–180 mW/cm^2 for up to 50 seconds. They found that spray-washing with water containing 10 ppm chlorine by itself or in combination with UV radiation was necessary to reduce surface bacterial counts on rough-surface fish to the same extent as on smooth-surface fish.

Quality Defects Caused by UV

Fat oxidation by photochemical action results in off-flavors such as rancidity, tallowiness, fishiness, cardboard flavor, and oxidized flavor [32]. Coe and Le Clerc [21] attributed rancidity to the ultraviolet light range of the spectrum. Packaging materials having the ability to screen ultraviolet light have been developed for food products. Ellickson and Hasenzahl [32] observed that processed cheese in normal cellophane wax-coated wrappers becomes oxidized within 12 hours, and within 48 hours the top slice became inedible. This process can be retarded by incorporating a substituted benzophenone within the wax coating normally applied to certain types of cheese wrappers. Hirsch [50] stated that the meaty portion of bacon is subject to fading when exposed to UV light. Fading can be reduced appreciably through vacuum-packaging and a UV barrier on the packaging material. A good vacuum-packaging operation will deliver a bacon package with no less than 28 inches of vacuum. By incorporating polyvinylidine chloride (PVDC) into the packaging material, both oxygen and UV light are screened and the product survives a considerably longer period of time without fade. Generally, retinoids are very susceptible to oxidation because of their alkyl chains with highly conjugated double bonds [110]. Shimoyamada et al. [110] found that retinol and retinoic acid bound to β-lactoglobulin were less susceptible to light-induced oxidation by UV light irradiation than those that were free or bound to bovin serum albumin. They found different mechanisms of protection against light-induced oxidation compared to enzymatic oxidation.

Iwanami et al. [58] studied effect of UV radiation on a lemon flavor composed of lemon oil, water (pH 6 phosphate beffer), and ethanol. Three compounds of aldehyde were newly identified as photoreaction products of citral. Limonene, terpinolene, and nonanal decreased, while *p*-cymene increased after UV radiation. Other components, such as sesquiterpene hydrocarbons, citronellal, linalool, and terpineols, were slightly changed. These results suggested that citral is a UV-unstable component in lemon flavor and that the photolysis of citral could affect other components in lemon flavor during UV radiation [58].

2. Disinfection Effects of UV Radiation

UV rays, which are nonionizing radiation, have been used extensively in the industrial disinfection of equipment, glassware, and air for many years [39]. The bactericidal effect of ultraviolet light is widely used for sanitation purposes. It is particularly effective in destroying airborne organisms and consequently becomes an important sanitary aid to in-plant installations. It may eliminate detrimental contamination and keep away objectionable invaders. Cerny [17] found that high-intensity UV irradiation may be used in the sterilization of packing materials for aseptic packaging.

The penetrating power of ultraviolet rays is very low, so that lethal action is confined to organisms on or near the surface of irradiated materials. Aerial disinfection is severely limited by the presence of dust particles in the atmosphere. Several different UV lamps are available commercially for food industry applications for processing or disinfection.

3. UV Mode of Action

A number of conflicting theories have been proposed with regard to the mode of action of ultraviolet light. These include indirect lethal action resulting from the production of hydrogen peroxide, and various chemical and physiochemical changes in the constituents of the cell. The production of hydrogen peroxide is not generally considered to be the mechanism by which ultraviolet light induces its effect, although organic peroxides may be involved. It has been suggested that substances in the cell nucleus are involved in the destructive action by ultraviolet light. UV wavelengths of 200–290 nm penetrate cell membranes to disrupt DNA molecules, preventing cell replication [10]. Also, the degradation of the cell walls destroys bacteria and causes a germicidal effect [2,120].

B. Visible Light Radiation

The germicidal effect of sunlight is due largely to the ultraviolet radiation received at the earth's surface. The wavelength is 290–300 μm. Altitude and latitude and clarity of the atmosphere affect its effectiveness. Visible light having electromagnetic radiation of wavelength 400–750 μm is absorbed by relatively few of the compounds present in nonphotosynthetic organisms. Light that is not absorbed has little or no effect. This is also true for the longer ultraviolet wavelengths of 300-400 μm. Ultraviolet radiation with a wavelength of less than 300 μm, on the other hand, is strongly absorbed by proteins and nucleic acids. Relatively small doses of such radiation will cause chromosome breakage, genetic mutation, inactivation of enzymes, or death [10].

Cool white fluorescent illumination (14.5 W/m^2 for 72 hr) of apples at 2°C enhances red color without hampering fruit quality and storability potential [102]. Saks et al. [101] studied the Dort and Ofra cultivars of strawberry illuminated at 14.5 and 17.5 W/m^2 in white fluorescent light at 2°C. A 2-hour treatment was sufficient to overcome the genetic limitation of white shoulders in Dorit and poor red color in Ofra. Illumination enhanced both external and internal fruit color with no effect on quality attributes (e.g., freshness of calyx, fruit firmness, and fruit decay) when treated and kept in storage simulating air or sea transport followed by shelf life (18°C). The treatment reduced fruit rot in both cultivars. In fruit inoculated with *Botrytis cinerea*, the most common storage pathogen of strawberry, the appearance of disease symptoms was delayed.

C. Photoreactivation

If microorganisms are treated with dyes (e.g., erythrosin), they may become sensitive to damage by visible light. This effect is known as *photoreactivation*. Some food ingredients coud induce the same reaction. Such dyes are said to possess photodynamic action [10].

Spores may occasionally fail to show photoreactivation when inactivated with ultraviolet light, whereas the corresponding vegetative cells sometimes show photoreactivation. The simplest explanation for this is that radiation damages genetic material in the spore and that certain bacteria may produce diploid spores as a result of specific disruptions. Such bacterial spores exhibit two types of radiation inactivation curves: *B. subtilis, B. brevis*, and *B. mesentericus* are inactivated in a single-hit fashion, whereas *B. megaterium, B. cereus*, and *B. mycoides* are affected by a multiple hit. In all cases there is no effect on spore survival if the postirradiation medium is changed from yeast extract to a purely chemically defined medium [10].

When surface microbial contamination is the major cause of spoilage in certain seafood, the application of intense, short pulses of incoherent, continuous, broad-spectrum light can be used to increase the shelf life. The extension is achieved through two processes: by the destruction of spoilage-causing microorganisms and by the inactivation of enzymes. These effects are obtained through a complex photothermal and photochemical mechanism mediated by the use of wavelengths of less than 300 nm. The pulsed light waves transfer thermal energy to a thin surface layer without raising the interior temperature of the product [22]. Colby and Flick [22] concluded that increased efficiency is possible by the use of dyes or other chemical compounds that selectively bind to either microorganisms or enzymes, thereby increasing their susceptibility to the pulsed electromagnetic waves.

D. Pulsed Light

Pulsed light with an intensity 20,000 times that of normal sunlight is applied at rates up to 20 flashes per second. The flashes of this high-intensity light are a very quick way to transfer large amounts of thermal energy to the surface of a material, raising a thin surface layer to a high enough temperature to affect vegetative cells on that surface. This method inactivates microorganisms through a combination of photothermal and photochemical reactions. The UV light needs to be filtered out from the pulsed light when it is used to treat UV-sensitive foods, thus most of the remaining energy is in the visual and IR spectrum and the inactivation mode is photothermal [77].

II. SOUND IN FOOD PRESERVATION

Ultrasound is sound energy with a frequency range that covers the region from the upper limit of human hearing, which is generally considered to be 20 kHz, to beyond hundreds of MHz [122]. Ultrasonic technology was first developed as a means of submarine detection in World War I, and developments in this area have continued to the present in different fields of technology [65]. The two applications of ultrasound in foods are (a) characterizing a food material or process, such as estimation of chemical composition, measurements of physical properties, nondestructive testing of quality attributes, and monitoring food processing; and (b) directly affecting food preservation or processing. Low-power ultrasound is used in a variety of capacities as a sensing medium. Only preservation and processing aspects will be considered in this chapter. Applications of sound to directly improve processes and products is less popular in food manufacturing [40]. High-intensity sound is mainly used for such applications with a frequency in either the sonic ($<$18 kHz) or the ultrasonic ($\geq$18 kHz) range, depending on the application [40].

A. Generation and Propagation of Sound

1. Generators

The first ultrasonic apparatuses were piezoelectric generators of quartz submerged in oil that generated ultrasonic waves of a very high frequency but low intensity (10 W/cm^2). Modern ultrasonic instruments consist essentially of a piezoelectric generator having a crystal of zirconate titanate that changes its shape under the effect of an electric field of 100 V and a frequency of 20 kHz, supplied by a standard 50–60 Hz, 120–220 V converter. The high-intensity equipments are characterized by relatively low frequencies, up to about 100 kHz, by continuous (as distinct from pulsed) operation, and by power levels from 10 kW upward [45]. This electric energy, transformed into mechanical energy of the same frequency, is transmitted to a titanium alloy disruptor horn. The horn transmits and amplifies this energy onto its tip, which is submerged into the menstruum being ultrasonicated. Amplification depends on the volume and shape of the disruptor [103]. Berlan and Mason [8] reviewed the available generators. The geometry of the chamber and sonicating horn is also important, and the effect of any given ultrasonication treatment is inversely related to the volume of the chamber [25].

2. Propagation and Attenuation into the Medium

Sound involves the propagation and transmission of vibrational energy above the upper limits of audible sound. Ultrasonic vibrations pass through a body as a system of pulsating energy waves propagated by alternating compression-expansion zones. As a wave propagates through a relaxing medium, its amplitude decreases or attenuates and sound energy is lost. Ultrasonic attenuation is a measure of the relative amplitudes of a wave at two locations in space [45]. When moving through a liquid they cause the phenomenon known as *cavitation*. This involves formation of tiny vapor bubbles or voids within the liquid. The collapse of these cavities is responsible for the creation of pressures up to several hundred atmospheres. This cavitation occurs at high frequency or at very low amplitude [53,80,108].

Three causes of attenuation are diffraction, scattering, and absorption. Diffraction and scattering are properties of the shape and macroscopic structure of the material. The diffraction effects are constant if the sound source is inside the Fresnel zone, i.e., near the field. Scattering contributes less than 1% to absorption, thus is negligible [49]. Ultrasonic waves absorbed in liquid depend on viscosity, thermal conductivity, and thermal relaxation. Absorption due to viscosity and thermal conduction takes place in liquids as well as in gases, but the thermal conduction effects are usually negligible compared with those due to viscosity [9,45].

As a result of the specific absorption of acoustic energy by materials, particularly at their interfaces, a selective temperature increase may take place. Arkhangel'skii and Statnikov [1] theoretically found that the temperature change due to absorption at a solid wall, under given conditions, was 0.1°C for water and about 1°C for air. These results were also verified experimentally [40]. Floros and Liang [40] indicated that a significant temperature increase (5–60°C) may occur in sugar solutions as a result of ultrasonication, particularly at high sugar concentrations. High-intensity acoustic energy passing through a solid medium generates a rapid series of alternating constructions and expansions, much like when a sponge is squeezed and released repeatedly. This mechanism is known as *rectified diffusion* [40]. Sound travels by sine waves that have a node and an antinode. At the node position, the velocity is zero and the pressure is maximum. At the antinode position, the velocity is maximal and the pressure is zero [40].

The cavitation threshold (minimum oscillation of pressure for cavitation to occur: amplitude of pressure) depends on (a) dissolved gas content (liquids saturated with gases have a very low threshold, which increases linearly with the vapor pressure of the liquid); (a) hydrostatic pressure; (c) specific heat of the gas bubble; (d) tensile strength of the liquid, and (e) temperature [103].

Sala et al. [103] summarized many different parameters that influence the efficacy of ultrasound, all of which should be at their optimum for achieving maximum cavitation. Thus, ultrasonication conditions should be carefully chosen and controlled for maximum benefit [8].

B. Consequences of Ultrasound

The beneficial or deteriorative use of sound depends on its chemical, mechanical, or physical effects on the process or products [40]. In this section, the applications of sound in food processing are summarized from selected reviewed papers.

1. Effect on Microorganisms

Process Applications

Harvey and Loomis [47] first reported the lethal effect of ultrasound in microorganisms [103]. Most investigations in this field refer to the degree of killing attained in microbial cultures when they are submitted to defined ultrasonic treatment [10]. Bacteria, especially spores, are very resistant and require hours of ultrasonication [80,107].

Ultrasonics has been reported to pasteurize milk [18,47] and destroy cells at high frequencies [112]. An increase in total counts at low frequencies was reported by Huhtanen [54,55] and Stone and Fryer [116]. This is probably due to the breaking up of clumps and bacteria, which normally occur in milk. A decrease in *Salmonella typhimurium* counts were time dependent, with the greatest decrease occurring after a 30-minute ultrasonication treatment at 50°C. Stone and Fryer [116] also showed 30-minute ultrasonic treatments to be more effective than shorter treatments. Lee et al. [66] showed a 4 log reduction in *Salmonellae* with a 10-minute ultrasonic treatment in peptone water and a 0.78 log reduction in chocolate milk treated for 30 minutes. Thus, treatment duration needed for microbial reduction depends on the substrate or medium.

Ultrasound is used to free bacteria adhered to surfaces to facilitate the removal of contaminated flora [26]. *Salmonella* and *Campylobacter* have long been associated with poultry and poultry products. Lillard [70] showed that bacteria are firmly attached to poultry skin, and although bactericides are lethal to samonella in processing water, they do not seem to access bacteria that are firmly attached to or entrapped in poultry skin [59,71]. Daufin and Saincliviert [24] reported that most bacteria from milk films on metal surfaces can be affected by ultrasonic waves of 80 kHz. Sams and Feria [104] studied ultrasonic treatment (47 kHz) of broiler drumsticks by dipping in deionized water at 25 or 40°C for 15 or 30 minutes in the presence of lactic acid with pH adjusted to 2 or 4. They found no significant effect of treatment and offered the explanation that the irregular broiler skin surface may provide some level of physical protection for bacteria against cavitation. Stumpf et al. [117] and Miller [85] observed that ultrasound waves are transmitted most efficiently over flat surfaces, whereas irregular surfaces reflected or refracted the waves, creating stationary waves, which greatly reduced cavitation. *Salmonellae* that were attached to broiler skin were reduced 1–1.5 log by sonication in peptone at 20 kHz for 30 minutes, by less than 1 log by chlorine alone, but by 2.5–4 log by sonicating skin in chlorine solution [72]. The difference in the results from Sams and Feria [104] may be due to the use of skin pieces rather than whole drumsticks, the use of peptone, or the greater synergistic effect of chlorine and sonication compared to sonication and heat [73].

Mode of Action

The mechanical disruption of cells by the very intense currents generated by ultrasound is the main lethal effect on microorganisms. Dorothy [29] concluded from electron micrographs that ultrasonic energy destroyed microorganisms by physical forces rather than chemical ones. Later Elliott and Winder [33] attributed bacterial destruction to thermal effects due to high-temperature hot spots. Since the volume of the high-temperature region is small, only a very small number of cells were affected [103]. Suslick [113] found that highly reactive chemical radicals and reaction products, such as hydrogen peroxide, are also lethal. Now most authors agree that, among other factors, cavitation is the mechanical effect due to extreme variations in pressure, which are responsible for the destruction of bacterial cells [73,103,108,111]. It has been mentioned that microorganisms can withstand high pressure but are incapable of withstanding the rapid alternating pressures produced during cavitation [103].

2. Effect on Enzymes

El'Piner [34] reviewed the effect of ultrasound on enzymes and other food components. The effect of ultrasound on enzymes depends on [103] (a) the ultrasonic field, (b) the molecular structure of enzymes, and (c) the nature of the sonicating medium, especially the nature of the dissolved gas. Inactivation effects generally require long irradiation periods and the presence of oxygen and are reduced if hydrogen replaces oxygen in the medium or when antioxidants are present [20,43,105]. Dunn and Macleod [31] found that the formation of free radicals by cavitation affects enzyme inactivation, which is related to the dissolved gas.

At low temperature, catalases are resistant to ultrasound [103] and yeast invertase is fairly resistant at low concentration [1,93], while pepsin is resistant at low concentration [90]. Ribonuclease is not inactivated in the presence of either oxygen or hydrogen, and in some cases serum aminopeptidase shows a similar nature [30]. However, lysozyme, alcohol dehydrogenase, hyalurodinase, lactate dehydrogenase, malate dehydrogenase, polyphenoloxidase, and other oxidases are much more sensitive [20,43,61,83,105].

3. Process and Quality Enhancement

Ultrasound is used in processes such as liquid degasification, homogenization, mixing, emulsification, crystallization, and the aging of meat, liquors, and wines. Ultrasonic-assisted cutters are also used in the food industry to enhance cutting performance and achieve a faster, cleaner, sharper cut with minimal waste due to self-cleaning blades.

Microwave Ultrasonics

Shukla [111] stated that the active force in microwave ultrasonics (above 16 kHz) is mechanical in nature due to cavitation, unlike heat in microwave heating.

Emulsification

Processes and apparatus utilizing the principle of ultrasound have found extensive use in emulsification and are potentially useful for cleaning operations. Ultrasonic treatment of wine induces cavitation. Dissolved bases such as carbon dioxide and sulfur dioxide are driven away in a rather drastic manner. Ultrasonic preservation of wine is therefore not advisable [10]. Ultrasonic homogenization can produce uniform solutions with reduced particle size. Martinez et al. [78] found ultrasonic homogenization of expressed human milk prevented fat loss during tube feeding.

Drying and Acoustically Assisted Diffusion

Mechanisms: A number of mechanisms may enhance water removal during drying or other mass transfer unit operations. These mechanisms include temperature increase at the interfaces, pressure fluctuation due to cavitation, developing microchannels by fracture, turbulence at the boundary, and structure change of the medium. The heat effect was assumed to be responsible for the significant increase in diffusion of sodium ions through living frog skin under ultrasound [67].

The contraction caused by acoustic energy releases a minute quantitity of water, and enhanced migration takes place during acoustic drying and dewatering [36]. In more dense materials, the alternating acoustic stress facilitates dewatering by either maintaining existing channels for water movement or creating new ones. Microscopic channels are created in directions normal to wave propagation during rarefaction or parallel to wave propagation during compression [89]. This mechanism also affects pressure gradient at gas/liquid interfaces, which enhance evaporation. Although the pressure variation introduced by the sound wave is very low, its effect is strong because of the rapid rate of pressure oscillation [12].

Acoustic waves cause extreme turbulence known as *acoustic streaming* or *microstreaming* at the interfaces [92]. This increases convective mass transfer by reducing the diffusion boundary layer. Furthermore, this has significance where ordinary mixing is not possible [1].

The mechanism that prevents binding or bridge formation may also affect the process, such as filtration and cleaning surfaces [40]. In the drying of foods, for example, sound may reduce the water-binding energy [123]. This was supported by the results of Huxsoll and Hall [56] for sonic drying of wheat and corn.

Acoustic drying may solve some problems faced when drying heat-sensitive materials, and it may be useful in removing the bound water from certain foods [40]. The effect of acoustics on the drying rate and permeation through membranes are shown in Figure 1 [40]. Acoustic radiation increases the efficiency of drying by extending the constant rate period [11,44]. Floros and Liang [40] reviewed the estimated acceleration due to sound and found that it varied from 1.00 to 4.75 factors. This indicates that in several cases the improvement is not large enough to justify commercialization, but in gelatin, yeast, and orange powder the rates are doubled or tripled, thus acoustic drying is beneficial.

Diffusion of a number of substances through membranes by acoustic is also reviewed by Floros and Liang [40], and the increment varied from 1.2 to 6.0 fold. Thus, ultrasound may result in significant improvements in the processes involving membranes, such as filtration, ultrafiltration, reverse osmosos, and dialysis [40].

Factors affecting acoustically enhanced diffusion: The factors affecting acoustically enhanced diffusion, reviewed by Floros and Liang [40], are as follows:

1. Temperature: Huxsoll and Hall [56] found an approximate 20% net increase in drying rate for whole wheat dried at low temperature (21°C). At higher temperature

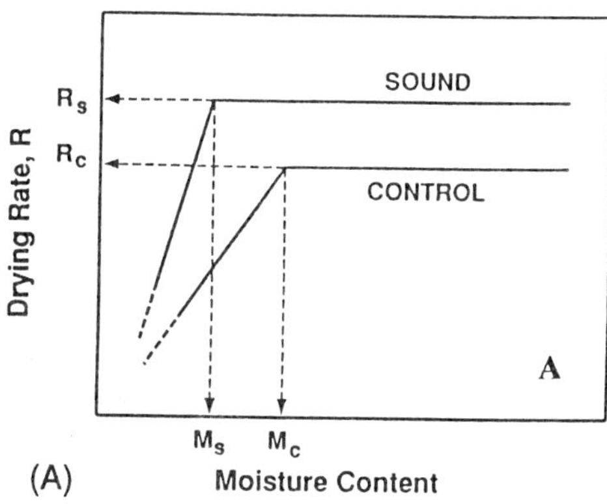

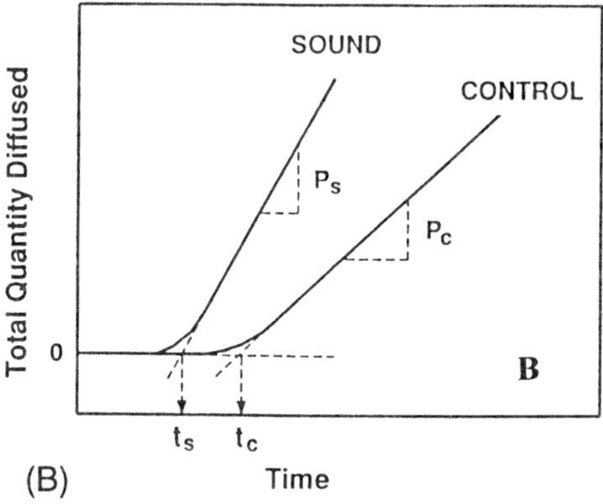

FIGURE 1 Rate acceleration by acoustic: (A) drying and (B) membrane separation. (From Ref. 40.)

(94°C), the positive effect of sonication diminished to about 6%. Similarly, for green rice at 20.5°C, sound caused a 15% net increase, but no improvement at 40°C [88].

Conflicting results were found for drying yeast cake, when moisture migration increased by 80% at 25°C and 200% at 37°C [13]. Thus, Floros and Liang [40] concluded that temperature can make the effect of sound either more or less pronounced, depending on the specific products or processes.

2. Acoustic Intensity: Intensity is a measure of the amount of acoustic energy traveling through a given area [40]. The effect of acoustic intensity on diffusion is shown in Figure 2 [40]. The experimental results are summarized by Floros and Liang [40] as follows: acceleration of diffusion by sound is a function of intensity [86], the function is nonlinear [37,68], and cavitation produced by high intensity sound negatively affects diffusion through membranes [37,51]. As shown in Figure 2, there is a threshold intensity value below which the effect of acoustic on diffusion cannot be observed. For acoustic drying the threshold value is about 130 dB [38] or 145 dB [1,13]. Above the threshold, an optimum intensity is observed where the effect of acoustic energy on diffusion is maximal [37]. Above the optimum, diffusion may be retarded due to the extreme turbulence at interfaces or vapor locks in porous media by violent cavitation.
3. Acoustic Frequency: Theoretically the diffusion coefficient is a function of sound frequency [3]. Howkins [51] observed a slight increase in diffusion as the frequency increased from 20 kHz to 1 MHz. In acoustic drying mostly frequencies in the audible (sonic) range are used [40]. An optimum value of 8.1 kHz was found for potatoes [6] and of 6–10 kHz for other materials [12]. Ultrasonic energy (≤18 kHz) is used in case of liquid-liquid extraction and membrane separation, such as osmosis. Floros and Liang [40] concluded that the effect of frequency is not clearly identified in the literature.
4. Direction of Acoustic Wave: The effect of sound on diffusion is maximal when it propagates in the same direction as the diffusion flow, and minimal when propagation is opposite to the direction of diffusion flow. When the direction is to diffusion flow, then the results are between the two extremes [68]. The theoretical analysis and discussion of Arkhangel'skii and Statnikov [1] support the above observations [40].

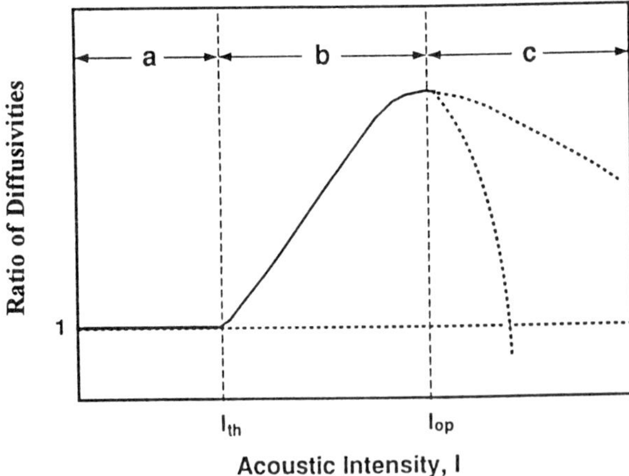

Figure 2 Possible effects of acoustic intensity on diffusion. (From Ref. 40.)

5. Pulsation of Acoustic Wave: Sound waves may be applied in a continuous or a pulsed (on-off) mode [40]. Lehmann and Krusen [67] and Mortimer et al. [86] found that both continuous and pulsed acoustic energy of the same average intensity resulted the same effect on diffusion, whereas Dinno et al. [27] stated that pulsation causes a significantly larger effect.
6. Medium Properties: Concentration, viscosity, and porosity may alter the effect of sound on diffusion [40]. At a low sugar concentration of 10°Brix, sound accelerated the diffusion of water into the apple tissue 5- to 10-fold after 6 hours. At a medium concentration of 30°Brix, sound had no effect, while at 60°Brix some acceleration of water migration from the tissue was observed [69]. Thus, effect of sonication on osmosis is concentration dependent.
7. Other Factors: The positioning of sample in a sound field at the node or antinode may affect the outcome, but conclusive evidence for this does not exist. Other factors, such as gravitational forces, pressure, and acoustic impedance compared to the medium impedance may also affect diffusion by acoustic forces [40].

Biotechnological Processes

Ultrasound has also been used to break cell walls to investigate cellular components [84]. The application of ultrasound to biotechnological processes has recently gained attention because treatment caused by ultrasounic waves may release many useful components from living cells, such as intracellular enzymes. For example, glucose oxidase was continuously released from *Aspergillus* sp. under mild ultrasound [57] and vacuole-located pigment from *Beta vulgaris* cells were enhanced for repeated harvesting by the use of 1.02 MHz ultrasound [64,64]. This is due to the change in cell membranes and walls of microorganisms or plants that permitted intracellular substances such as enzymes or metabolites to be released [100].

Lactose-hydrolyzed milk products are thought to have therapeutic value for people who cannot tolerate the lactose that remains in normal milk products. Usually lactose-hydrolyzed yoghurt produced by fermentation of lactose-hydrolyzed milk or by the simultaneous addition of β-glactosidase and lactic acid bacteria [119]. It is well known that lactic acid bacteria have β-galactosidase activity, which can be used to hydrolyze lactose in fermented milk. Sakakibara et al. [100] studied milk fermentation with *Lactobacillus delbrueckii* under ultrasonic irradiation in a 450 cm^3 bioreactor with a polyethylene film bottom. Ultrasonic treatment increased the hydrolysis of lactose in milk but decreased cell viability. However, the viable cell count increased again when the ultrasound was stopped, because ultrasound did not destroy the cell propagation ability of surving cells. When the sonication power was 17.2 kW/m^2 and the sonication period 3 hours, 4.9×10^8 cfu/cm^3 of the viable cell count and 55% lactose hydrolysis were attained. In contrast, the viable cell count was 2×10^9 cfu/cm^3 and 35.6% of lactose was hydrolyzed during control fermentation. Sakakibara et al. [100] found that both aspects could be enhanced if sonication were carried out under optimum conditions for β-glactosidase activity and lactic acid bacteria viability, e.g., at suitable pH and temperature. Matsuura et al. [82] found that fermentation of wine, beer, and sake can be accelerated 50–65% by ultrasonic treatment.

Functional Properties

Sala et al. [103] observed the following effects of ultrasound: (a) the reversible reduction of viscosity of aqueous solutions of starch, gum arabic, gelatin, and other macromolecules, (b) the depolymerization of starch and polymerization of dextrans to high molecular weight, and (c) the breakdown of DNA to fragments retaining the native configuration.

High-intensity sound affects the structural properties of fluids, particularly their viscosity [40]. Usually Newtonian fluids maintain their Newtonian characteristics, but dilatent and thixotropic fluids tend to either stiffen or become less viscous [36,89]. High intensity sound also permits protein breakdown and hydrolysis, simple cell lysis, and protein particulation and may help retain vitamins and other heat-sensitive ingredients [111].

Ultrasound showed potential for improving the mechanical strength of milk protein films. Banerjee et al. [4] studied the effects of ultrasound frequency (168 kHz and 520 kHz), acoustic power (low, medium, and high), and exposure time (0.5 and 1 hr) on the functional properties of whey protein concentrate and sodium caseinate films. The average tensile strength of the ultrasound-treated caseinate films was 224% higher than that of controls. The treatment was more effective on sodium caseinate than on whey protein concentrate film. Resistance to puncture was improved for both types of film treated at an acoustic power of 5.22 W. Stronger films can be formed by increasing exposure time. The improvement by ultrasound could be due to the reduction of particle size in a film-forming solution, which results in increased molecular interaction and products a film with greater rigidity and compactness. Elongation at break, water vapor permeability, and moisture content of films were not affected by the ultrasound treatment [4].

The application of ultrasound treatment can produce a coating with improved physical and mechanical properties, e.g., brightness, hardness, compactness, and adhesion strength [124]. Thus, food preservation by edible coating can be enhanced by forming improved functional coating on the surface. Sonication has been used to alter the resistance of proteins from cow's milk and hen's eggs to proteolysis [91].

Chen et al. [19] studied the effects of ultrasonic conditions and storage in acidic solution on changes in molecular weight and polydispersity of treated chitosan. The results showed that chitosan was degraded faster in dilute solutions and in lower-temperature solutions. The degradation increased with prolonged ultrasonic time, and chitosan was degraded during storage in an acidic solution at ambient temperatures [19].

Synthesis of Active Functional Organic Compounds

Low [74] clearly identified the value of ultrasound to organic synthesis as its ability not only to accelerate known reactions, particularly those that are heterogeneous, in solvent systems, but also to generate new chemistry that is not available using existing methodologies. He discussed two distinct areas: generation of a number of synthetically useful reagents and the effects of adding electron-transfer agents to sonochemical reactions, and the chemistry of π-allyltricarbonyliron lactone and lactam complexes, which have been shown to be useful intermediates for the synthesis of a variety of biologically active or functionalized compounds. Bremner [14] also reviewed recent applications of ultrasound in organic synthesis. Thus, ultrasound is a tool that is valuable in synthetic development.

C. Thermo-Sonication

Sonication in conjunction with higher temperatures was shown to inhibit lipolytic activity and completely eliminate bacterial contaminants in human milk [79]. Ordonez et al. [94,95] reported that *thermoduric streptococci* as well as a strain of *Staphylococcus aureus* were rendered much more susceptible to damage by the combined effect of ultrasonic (20 kHz) and heat treatment than by either treatment separately. Sanz et al. [106] found a marked decrease in heat resistance (without killing) of *Bacillus stearothermophilus* spores by ultrasound at 20 kHz regardless of the heating temperature and the storage time between both treatments. Its heat resistance was reduced from one half to one third of original value. The heat resistance at 105°C of *Bacillus cereus* can *Bacillus licheniformis* decreased after a previous ultrasound treatment at 20 kHz [15]. Garcia et al. [41] showed that ultrasonic treatment (20 kHz, 150 W) of *Bacillus subtilis* spores in distilled water or milk resulted in little or no decrease in heat resistance, but simultaneous ultrasonic and heat treatments were effective. This effect became smaller at the higher-temperature treatment. This may be due to the increased vapor pressure, decreased viscosity and, as a consequence, reduction in the intensity of cavitation [103].

D. Mano-Thermo-Sonication

Mano-thermo-sonication decreased heat resistance of *Bacillus subtilis* to about one-tenth that of a simply heated control in the range 100–112°C. It was also effective with other microorganisms, such as spore-formers, vegetative cells, and yeast. The lethality of mano-thermo-sonication was 6–30 times greater than that of the corresponding heat treatment at the same temperature, depending on the microorganism, when Sala et al. [103] studied the survival curves of *Aeromonas hydrophila, S. cerevisiae, B. coagulans*, and *B. stearothermophilus*. Sala et al. [103] found that with combined heat-ultrasound under pressure, the sensitizing effect could be retained even at temperatures above the boiling point of medium. The efficacy of mano-thermo-sonication depended on the intensity of the ultrasound (sonication time, amplitude, and instrument output) and the extent of the pressure [103]. An ultrasonication treatment at 45°C under pressure failed to show any lethal effect on spores of *B. subtilis* var. *niger*, but heat resistance was reduced from 0.1 to 0.015 minutes at 112°C under pressure (20 kHz, 117 μm, 300 kPa) [103]. Thus, the effect is not additive, but synergistic.

1. Mode of Action in Mano-Thermo Sonication

The physical effects of ultrasound could sensitize target molecules and/or structures. Thus, heat plus pressure is more effective [103]. Scherba et al.[108] mentioned that extremely high-temperature hot spots developed by cavitation cause heat inactivation. Extreme pressure changes and shock waves may cause physical damage to the cell. Ultrasonication disrupted the spore *exosporium* and released dipicolinic acid and low molecular weight polypeptides from the cortex of some bacterial spores [7,96]. This cortex degradation would lead to rehydration of the protoplast, resulting in a loss of heat resistance [42].

2. Effects of Mano-Thermo-Sonication on Enzymes

The effects of mano-thermo-sonication on lipoxygenase, peroxidase, polyphenoloxidase, and *Pseudomonas fluorescens* extracellular protease and lipase are summarized by Sala et al. [103]. The enzyme-inactivation efficacy of heat is increased by a factor that depends on the nature of the enzyme and the treatment conditions. The role of the different bonds and interactions involved in protein structure stabilization are not equally important in maintaining the native structure of the catalytic center of each of the enzymes, and they are not equally affected by heat and ultrasound.

Enzyme inactivation by a combined treatment of heat and ultrasound under pressure is not additive but a synergistic. The magnitude of the synergistic effect depends on the bonds and interactions of the active sites for catalysis, the nature of the groups participating in the catalysis, and the molecular weight of the protein [103].

The synergistic effect at constant pressure decreased with an increase in temperature in lipoxygenase and *P. fluorescens* protease [103]. This may be due to less effective collapse by an increase of the water vapor pressure in the cavitation bubble and the lower number of cavitation collapses available to release energy [103].

Lu and Whitaker [76] suggested a mechanism involving the release of the heme moiety of the enzyme previous to the denaturation of the liberated apoprotein during heat inactivation of peroxidase. A similar mechanism is also involved in mano-thermo-sonication [103]. Sala et al. [103] mentioned that it must be clarified whether the potentiating effects of ultrasound on peroxidase destruction are related to an increase in the dissociation rate or to the hindering of heme binding to the apoprotein. This could be due to an increase in denaturation rate or in the rate of heme destruction [103].

Sala et al. [103] summarized other factors effecting mano-thermo-sonication efficiency for enzyme inactivation: the effect (a) is almost independent of the ionic strength in the range 0–1, (b) increases as pH changes from 5 to 8, and the level of increment also depends on temperature and types of enzymes, (c) diminishes with increasing enzyme concentration, and (d) increases with soluble solids concentration [103]. The concentration effect is due to the increasing intensity of cavitation [28,35].

Sala et al. [103] concluded, based on the data in the literature, that the resistance of most microorganisms and enzymes to ultrasound is so high that it would probably produce extensive undesirable quality changes. Thus, a combination of ultrasound with other treatments, such as heat and pressure, should have a much better chance for practical use in the future.

REFERENCES

1. M. E. Arkhangel'skii and Y. G. Statnikov, Diffusion in heterogeneous systems, *Physical Principles of Ultrasounic Technology*, Vol. 2 (L. D. Rozenberg, ed.), (translated from Russian by J. S. Wood), Plenum Press, New York, 1973.
2. R. Bachman, Sterilization by intense UV radiation, *Brown Boveri Rev. 62*:206 (1975).
3. A. S. Bakai and N. P. Lazarev, Effect of acoustic waves on the diffusion of interstitial impurity atoms in a solid, *Sov. Phys. Solid State 28*:1373 (1986).
4. R. Banerjee, H. Chen, and J. Wu, Milk protein-based edible film mechanical strength changes due to ultrasound process, *J. Food Sci. 61*:824 (1996).
5. H. Barkworth, Taints and off-flavors of milk, *Dairy Ind. 3*:367 (1938).
6. L. G. Bartolome, J. E. Hoff, and K. R. Purdy, Effect of resonant acoustic vibration on drying rates of potato cylinders, *Food Technol. 23*:321 (1969).
7. J. A. Berger and A. G. Marr, Sonic disruption of spores of Bacillus cereus, *J. Gen. Microbiol. 22*:147 (1960).
8. J. Berlan, and T. J. Mason, Sonochemistry: from research laboratories to industrial plants, *Ultrasonics 30*:203 (1992).
9. J. Blitz, *Fundamentals of Ultrasonics*, 2nd ed., Butterworths, London, 1967.
10. G. Borgstrom, *Principles of Food Science*, The Macmillan Co., London, 1968.
11. Y. Y. Borisov and N. M. Gynkina, Acoustic drying, *Physical Principles of Ultrasound Technology*, Vol. 2 (L. D. Rozenberg, ed.) (translated from Russian by J. S. Wood), Plenum Press, New York, 1973.
12. R. M. G. Boucher, Drying by airborne ultrasonics, *Ultrason. News 3*:8 (1959).
13. R. M. G. Boucher, Ultrasonics in processing, *Chem. Eng. 68*:83 (1961).
14. D. H. Bremner, Recent advances in organic synthesis utilizing ultrasound, *Ultrason. Sonochem. 1*:S119 (1994).
15. J. Burgos, J. A. Ordonez, and F. J. Sala, Effect of ultrasonics waves on the heat resistance of *Bacillus cereus* and *Bacillus licheniformis* spores, *Appl. Microbiol. 24*:497 (1972).
16. L. Buruiana, Action of sunlight on milk, *Biochem. J. 31*:1452 (1937).
17. G. Cerny, Sterilization of packaging materials for aseptic packagings. 2. Investigation of the germicidal effects of UV-C rays, *Verpackungs-Rundschau 28*:77 (1977).
18. L. S. Chambers and N. J. Gaines, Some effect of intense audible sound on living organisms and cells, *J. Cell. Comp. Physiol. 1*:451 (1932).
19. R. H. Chen, J. R. Chang, and J. S. Shyur, Effects of ultrasound conditions and storage in acidic solutions on changes in molecular weight and polydispersity of treated chitosan, *Carbohyd. Res. 299*:287 (1997).
20. W. T. Coarkley, R. C. Brown, C. J. James, and R. K. Gould, The inactivation of enzymes by ultrasounic cavitation, *Arch. Biochem. Phys. 159*:722 (1973).
21. M. R. Coe and Le Clerc, Photochemical action, a cause of rancidity, *Cereal Chem. 9*:519 (1932).
22. J. Colby and G. J. Flick, Shelf life of a fish and shellfish, *Shelf Life Studies of Foods and Beverages* (G. Charalambous, ed.), Elsevier Science Publishers B. V., Amsterdam, 1993, p. 85.
23. L. L. Creasy, and M. Coffee, Phytoalexin production potential of grape berries, *J. Am. Soc. Hort Sci. 113*:230 (1988).
24. G. Daufin and M. Saincliviert, Use of ultrasounic vibrations for removal of bacteria contained in film milk fixed on metal surfaces, *Ann. Technol. Agric. 16*:195 (1967).
25. R. Davies, Observations on the use of ultrasound waves for the disruption of microorganisms, *Biochim. Biophys. Acta 33*:491 (1959).
26. E. Dewhurst, D. M. Rawson, and G. C. Steele, The use of a model system to compare the efficiency of ultrasound and agitation in the recovery of *Bacillus subtilis* spores from polymer surfaces, *J. Appl. Bacteriol. 61*:357 (1986).
27. M. A. Dinno, L. A. Crum, and J. Wu, The effect of therapeutic ultrasound on electrophysiological parameters of frog skin, *Ultrasound Med. Biol. 15*:461 (1989).

28. A. Dognon and Y. Simonot, Actions des ultrasounds sur les suspensions. Influence de la concentration des particles. *C. R. Acad. Sci. 227*:1234 (1948).
29. H. Dorothy, The effect of ultrasounic waves upon *Klebsiella pneumoniae, Saccharomyces cervisiae*, and influenza virus A. Am, *J. Bacteriol. 57*:279 (1949).
30. C. A. Dubs, Ultrasounic effects on isoenzymes, *Clin. Chem. 12*:181 (1966).
31. F. Dunn and R. M. Macleod, *J. Acoustic Soc. Am. 40*:932 (1968).
32. B. E. Ellickson and V. Hasenzahl, Use of light-screening agent for retarding oxidation of process cheese, *Food Technol. 12*:577 (1958).
33. T. A. Elliott and M. C. Winder, Effect of ultrasounic wave on the bacterial flora of milk, *J. Dairy Sci. 38*:598 (1955).
34. I. E. El'Piner, *Ultrasound: Physical Chemical and Biological Effects*, Consultants Bureau, New York, 1964, p. 149.
35. I. E. El'Piner and M. D. Surova, Acceleration of protein degradation process in an ultrasonic field, *Doklady Akad Nauk SSSR 94*:243 (1954).
36. D. Ensminger, Acoustic and electroacoustic methods of dewatering and drying, *Drying Technol. 6*:473 (1988).
37. H. V. Fairbanks, Use of ultrasound to increase filtration rate, Ultrasonics Intl. Conf. Proc., IPC Science and Technology Press, Guildford, England, 1973, p. 11.
38. H. V. Fairbanks, Ultrasonically assisted drying of fine particles, *Ultrasonics 12*:260 (1974).
39. M. L. Fields, *Fundamentals of Food Microbiology*, AVI Publishing, Westport, CT, 1978.
40. J. D. Floros and H. Liang, Acoustically assisted diffusion through membranes and biomaterials, *Food Technol. 48*:79 (1994).
41. M. S. Garcia, J. Burgos, B. Sanz, and J. A. Ordonez, Effect of heat and ultrasonic waves on the survival of two strains of *Bacillus subtilis*, *J. Appl. Bacteriol. 67*:619 (1989).
42. G. W. Gould and G. J. Dring, Heat resistance of bacterial endospores and concept of an expanded osmoregulatory corex, *Nature 258*:402 (1975).
43. P. Grabar, Voinovitch, and R. O. Prudhome, Action des ultrasonides sur une oxidase, *Biochem. Biophys. Acta 3*:412 (1949).
44. P. Greguss, The mechanism and possible applications of drying by ultrasonic irradiation, *Ultrasonics 1*:83 (1963).
45. S. Gunasekaran and C. Ay, Evaluating milk coagulation with ultrasonics, *Food Technol. 48*:74 (1994).
46. W. O. Harrington and C. H. Hills, Reduction of the microbial population of apple cider by ultraviolet irradiation, *Food Technol. 22*:117 (1968).
47. E. Harvey and A. Loomis, The destruction of luminous bacteria by high frequency sound waves, *J. Bacteriol. 17*:373 (1929).
48. E. Harvey and A. Loomis, High speed photomicrography on living cell subjected to supersonic vibrations, *J. Gen. Physiol. 15*:147 (1932).
49. F. K. Herzfeld and T. A. Litovitz, *Absorption and Dispersion of Ultrasonic Waves*, Academic Press, New York, 1959.
50. A. Hirch, *Flexible Food Packaging*, Van Nostrand Reinhold, New York, 1991.
51. S. D. Howkins, Diffusion rates and effect of ultrasound, *Ultrasonics 8*:129 (1969).
52. Y. W. Huang and R. Toledo, Effect of high doses of high and low intensity UV irradiation on surface microbiological counts and storage life of fish, *J. Food Sci. 47*:1667 (1982).
53. D. E. Hughes and W. L. Nyborg, Cell disruption by ultrasound, *Science 138*:108 (1962).
54. C. N. Huhtanen, Effect of ultrasound on disaggregation of milk bacteria, *J. Dairy Sci. 49*:1008 (1966).
55. C. N. Huhtanen, Effect of low frequency ultrasound and elevated temperatures on isolation of bacteria from raw milk, *Appl. Microbiol. 16*:470 (1968).
56. C. C. Huxsoll and C. W. Hall, Effects of sonic irradiation on drying rates of wheat and shelled corn, *Trans. ASAE 13*:21 (1970).
57. Y. Ishimori, I. Karube, and S. Suzuki, *Enzyme Microb. Technol. 4*:85 (1982).

58. Y. Iwanami, H. Tateba, N. Kodama, and K. Kishino, Changes of lemon flavor components in an aqueous solution during UV irradiation, *J. Agric. Food Chem. 45*:463 (1997).
59. W. O. James, R. L. Brewer, J. C. Prucha, W. O. Williams, and D. R. Parham, Effects of chlorination of chill water on the bacteriologic profile of raw chicken carcasses and giblets, *J. Am. Vet. Assn. 200*:60 (1992).
60. G. Kaess and J. F. Weidemann, Effects of ultraviolet irradiation on the growth of micro-organisms on chilled beef slices, *J. Food Technol. 8*:59 (1973).
61. H. Kashkoolo, J. Roony, and R. Rooxby, Effects of ultrasound on catalase and malate-dehydrogenase, *J. Acoust. Soc. Am. 67*:1798 (1980).
62. A. Klezkowski, Methods of inactivation by ultraviolet radiation, *Methods Virol. 4*:93 (1968).
63. N. J. Kilby and C. S., Hunter, *Appl. Microbiol. Biotechnol. 33*:448 (1990).
64. N. J. Kilby and C. S. Hunter, *Appl. Microbiol. Biotechnol. 34*:478 (1991).
65. L. E. Kinsler, A. R. Frey, A. B. Copens, and J. V. Sanders, *Fundamentals of Acoustics*, 3rd ed., John Wiley & Sons, New York, 1982.
66. B. H. Lee, S. Kermasha, and B. E. Baker, Thermal, ultrasonic and ultraviolet inactivation of *Salmonellae* in thin films of aqueous media and chocolate, *Food Microbiol. 6*:143 (1989).
67. J. F. Lehmann and F. H. Krusen, Effect of pulsed and continuous application of ultrasound on transport of ions through biologic membranes, *Arch. Phys. Med. Rehab. 35*:20 (1954).
68. I. Lenart and D. Auslander, The effect of ultrasound on diffusion through membranes, *Ultrasonics 18*:216 (1980).
69. H. Liang, Modeling of ultrasound assisted and osmotically induced diffusion in plant tissue, Ph.D. thesis, Purdue University, West Lafayette, IN, 1993.
70. H. S. Lillard, Incidence and recovery of salmonellae and other bacteria from commercially processed poultry carcasses at selected pre- and post-evisceration steps, *J. Food Prot. 52*:88 (1989).
71. H. S. Lillard, Factors affecting the presistence of *Salmonella* during the processing of poultry, *J. Food Prot. 52*:829 (1989).
72. H. S. Lillard, Bactericidal effect of chlorine on attached salmonellae with and without sonication, *J. Food Prot. 56*:716 (1993).
73. H. S. Lillard, Decontamination of poultry skin by sonication, *Food Technol. 48*:72 (1994).
74. C. M. R. Low, Ultrasound in synthesis: natural products and supersonic reactions?, *Ultrason. Sonochem. 2*:S153 (1995).
75. J. Y. Lu, C. Stevens, P. Yakubu, and P. A. Loretan, Gamma, electron beam and ultraviolet radiation on control of storage rots and quality of Walla Walla onions, *J. Food Proc. Pres. 12*:53 (1987).
76. A. T. Lu and J. R. Whitaker, Some factors affecting rates of heat inactivation and reactivation of horseradish peroxidase, *J. Food Sci. 39*:1173 (1974).
77. C. Manvell, Minimal process of food, *Food Sci. Technol. Today 11*:107 (1997).
78. F. E. Martinez, I. D. Desai, A. G. F. Davidson, S. Nakai, and A. Radcliffe, Ultrasonic homogenization of expressed human milk to prevent fat loss during tube feeding, *J. Pediatr. Gastroenterol. 6*:593 (1987).
79. F. E. Martinez, A. G. F. Davidson, J. D. Anderson, S. Nakai, I. D. Desai, and A. Radcliffe, Effects of ultrasounic homogenization of human milk on lipolysis, IgA, IgG, lactoferrin and bacterial content, *Nutr. Res. 12*:561 (1992).
80. T. J. Mason, Sonochemistry: A technology for tomorrow, *Chem. Ind.* 47 (1993).
81. M. Matsudaira and A. Sato, Effect of supersonic ray on enzymes, *Tohoku J. Exp. Med. 22*:412 (1933).
82. K. Matsuura, M. Hirotsune, Y. Nunokawa, M. Satoh, and K. Honda, Acceleration of cell growth and ester formation by ultrasonic wave irradiation, *J. Ferment. Bioeng. 77*:36 (1994).
83. R. M. McCleod and F. Dunn, Effects of ultrasounic cavitation on trypsin, chymotrypsin and lactate dehydrogenase solutions, *J. Acoustic Soc. Am. 42*:527 (1967).
84. H. Mett, B. Schacher, and L. Wegman, Ultrasonic disintegration of bacteria may lead to irreversible inactivation of lactamase, *J. Antimicrob. Chemother. 22*:293 (1988).
85. F. Miller, *College Physics,* 5th ed., Harcourt Brace Jovanovich, New York, 1982.

86. A. J. Mortimer, B. J. Trollope, E. J. Villeneuve, and O. Z. Roy, Ultrasound enhanced diffusion through isolated frog skin, *Ultrasonics 26*:348 (1988).
87. J. H. Moy, T. McElhandy, and C. Matsuzaki, Combined treatment of UV and gamma radiation of papaya for decay control. Food Preservation by Irradiation Proceedings at an IAEA, FAO, Who Symposium, Wageningen, 1977.
88. H. S. Muralidhara and D. Ensminger, Acoustic drying of green rice, *Drying Technol. 4*:137 (1986).
89. H. S. Muralidhara, D. Ensminger, and A. Putnam, Acoustic dewatering and drying (low and high frequency): State of the art review, *Drying Technol. 3*:529 (1985).
90. G. M. Naimark and W. A. Mosher, Effects of sonic on the proteolytic activity of pepsin, *J. Acoustic Soc. Am. 25*:289 (1953).
91. S. K. Nikolov, B. S. Mikhalenko, and L. N. Selikhova, Proteolysis of the proteins of cow's milk and hen's eggs after ultrasonic treatment, *Prikl. Biok. Mikrobiol. 6*:196 (1970).
92 W. L. Nyborg, Acoustic streaming, *Physical Acoustics*, Vol. 2B (W. P. Mason and R. N. Thurston, eds.), Academic Press, New York, 1965, p. 265.
93. A. I. Oparin, M. S. Bardinskaya, and El'Piner, Action of ultrasonic waves on yeast invertase, *Dok. Akad. Nauk. SSSR 99*:423 (1954).
94. J. A. Ordonez, B. Sanz, P. E. Hernandez, and P. Lopez-Lorenzo, A note on the effect of combined ultrasonic and heat treatments on the survival of thermoduric streptococci, *J. Appl. Bacteriol. 56*:175 (1984).
95. J. A. Ordonez, M. A. Aguilera, M. L. Garcia, and B. Sanz, Effect of combined ultrasonic and heat treatment (thermo-ultrasonication) on the survival of a strain of *Staphylococcus aureus*, *J. Dairy Res. 54*:61 (1987).
96. P. Palacios, J. Burgos, L. Hoz, B. Sanz, and J. A. Ordonez, Study of substances released to ultrasonic treatment from *Bacillus stearothermophilus* spores, *J. Appl. Bacteriol. 71*:445 (1991).
97 B. Ranganna, G. S. V. Raghavan, and A. C. Kushalappa, effect of ultraviolet radiation on control of diseases for short-term storage of potatoes (*Solanum tuberosum* L.), *Harvest and Postharvest Technologies for fresh fruits and vegetables* (L, Kushwaha, R. Serwatowski, and R. Brook, eds.), American Society of Agricultural Engineers, St. Joseph, Michigan, 1995, p. 293.
98. J. O. Reagan, G. C. Smith, and Z. L. Carpenter, Use of ultraviolet light for extending the retail caselife of beef, *J. Food Sci. 38*:929 (1973).
99. V. Rodov, S. Ben-Yehoshua, J. J. Kim, B. Shapiro, and Y. Ittah, Ultraviolet illumination induces scoparone production in kumquant and orange fruit and improves decay resistance, *J. Am. Soc. Hort. Sci. 117*:788 (1992).
100. M. Sakakibara, D. Wang, K. Ikeda, and K. Suzuki, Effect of ultrasonic irradiation on production of fermented milk with *Lactobacillus delbrueckii, Ultrason. Sonochem. 1*:S107 (1994).
101. Y. Saks, A. Copel, and R. Barkai-Golan, Improvement of harvested strawberry quality by illumination: colour and *Botrytis* infection, *Postharv. Biol. Technol. 8*:19 (1996).
102. Y. Saks, L. Sonego, and R. Ben-Arie, Artificial light enhances red pigmentation, but not ripening, of harvest 'Anna' apples, *HortScience 25*:547 (1990).
103. F. J. Sala, J. Burgos, S. Condon, P. Lopez, and J. Raso, Effect of heat and ultrasound on micro-organisms and enzymes, *New Methods of Food Preservation* (G. W. Gould, ed.), Blackie Academic and Professional, Glasgow, 1995, p. 176.
104. A. R. Sams and R. Feria, Microbial effects of ultrasonication of broiler drumstick skin, *J. Food Sci. 56*:247 (1991).
105. L. Santamaria, A. Castellani, and F. A. Levi, Hyalurodinase inactivation by ultrasonic waves and its mechanisms, *Enzymologia 15*:285 (1952).
106. B. Sanz, P. Paklacios, P. Lopez, and J. A. Ordonez, Effect of ultrasonic waves on the heat resistance of *Bacillus stearothermophilus* spores, *Fed. Eur. Microbiol. Soc. 18*:251 (1985).
107. B. Sanz, P. Palacios, P. Lopez, and J. A. Ordonez, Effect of ultrasonic waves on the heat resistance of *Bacillus stearothermophilus* spores, *Fundamental and Applied Aspects of Bacterial Spores* (G. J. Dring, D. J. Ellar, and G. W. Gould, eds.), Academic Press, New York, 1985, p. 251.

108. G. Scherba, R. M. Weigel, and J. R. O'Brien, Quantitative assessment of the germicidal efficacy of ultrasonic energy, *Appl. Environ. Microbiol. 57*:2079 (1991).
109. I. S. Schneider, H. A. Frank, and C. O. Willits, Maple syrup. XIV. Ultraviolet irradiation effects on the growth of some bacteria and yeasts, *Food Res. 25*:654 (1960).
110. M. Shimoyamada, H. Yoshimura, K. Tomida, and K. Watanabe, Stability of bovine β-lactoglobulin/retinol or retinoic acid complexes against tryptic hydrolysis, heating and light-induced oxidation, *Food Sci. Technol. 29*:763 (1996).
111. T. P. Shukla, Microwave ultrasonics in food processing, *Cereal Foods World 37*:332 (1992).
112. R. Springer and G. Koehl, Observations and experiments on the destruction of microorganisms by ultrasonic waves, *Lebensmitt. Rdscn. 47*:489 (1951).
113. K. S. Suslick, Homogeneous sonochemistry, *Ultrasounds. Its Chemical, Physical and Biological Effects* (I. S. Suslick, ed.), VCH Publishers, New York, 1988.
114. V. C. Stebnitz and H. H. Sommer, The oxidation of butterfat. I. The catalytic effect of light, *J. Dairy Sci. 20*:181 (1937).
115. R. A. Stermer, M. Lasater-Smith, and C. F. Brasington, Ultraviolet radiation—an effective bactericide for fresh meat, *J. Food Protect. 50*:108 (1987).
116. D. L. Stone and T. F. Fryer, Disruption of bacterial clumps in refrigerated raw milk using an ultrasonic cleaning unit, *NZ J. Dairy Sci. Technol. 19*:221 (1984).
117. P. K. Stumpf, D. E. Green, and F. W. Smith, Ultrasonic disintegration as a method of extracting bacterial enzymes, *J. Bacteriol. 51*:487 (1946).
118. H. L. Templeton and H. H. Sommer, Wrappers for processed cheese, *J. Dairy Sci. 20*:231 (1937).
119. T. Toba, K. Arihara, and S. Adachi, *Jpn. J. Zootech. Sci. 56*:835 (1985).
120. S. Varga, R. A. Keith, P. Michalik, G. G. Sims, and L. W. Regier, Stability of lean and fatty fish fillets in hypobaric storage, *J. Food Sci. 6*:1487 (1980).
121. A. R. Williams, D. A. Stafford, A. G. Callely, and D. E. Hughes. Ultrasonic dispersal of activated sludge flocs, *J. Appl. Bacteriol. 33*:656 (1970).
122. P. M. Withers, Ultrasonic, acoustic and optical techniques for the non-invasive detection of fouling in food processing equipment, *Trends Food Sci. Technol. 7*:293 (1996).
123. X. Xiong, G. Narsimhan, and M. R. Okos, Effect of composition and pore structure on binding energy and effective diffusivity of moisture in porous food, *J. Food Eng. 15*:187 (1991).
124. Y. Zhao, C. Bao, R. Feng, and Z. Chen, Electroless coating of copper on ceramic in an ultrasonic field, *Ultrason. Sonochem. 2*:S99 (1995).

23

Packaging and Food Preservation

ROBERT H. DRISCOLL AND JANET L. PATERSON
The University of New South Wales, New South Wales, Australia

I. THE NEED FOR PACKAGING

A. Why Package Foods?

For most of recorded history humans have consumed food close to the point of harvest. Natural containers (gourds, leaves, etc.) helped with short transport, Mesolithic humans used baskets and Neolithic humans used metal containers and discovered pottery. Four thousand years ago sealed pottery jars were used to protect against rodents, and in 1550 B.C. glass making was an important industry in Egypt. The Phoenicians developed the blow-pipe, but glass remained expensive until the eighteenth and nineteenth centuries. Tin-plating iron became possible in A.D. 1200 and as steel replaced iron, this method became useful after A.D. 1600. In 1825 Oersted first extracted aluminum: it cost \$545/lb in 1852, \$11/lb in 1855, \$0.57/lb in 1892, and \$0.14/lb in 1942 [5].

One hundred years ago there was little use for packaging in the food industries. Products were sold at small local stores or markets, often from bulk containers with the consumer providing a container for his or her purchase. Butter, for example, was sold from large blocks, and wrapped in pieces of paper for the consumer to take home. Liquids were kept in kegs and scooped out into the customers' containers. Then near the turn of the century, the need to keep biscuits fresh led to a development of a folding carton with a wax lining and a printed label wrapped around the container. Packaging evolved rapidly, so that now practically all products are sold in packages. In the industrialized world less than 2% of food spoils between production and consumption, whereas in developing countries, 30–50% of all food is wasted, largely due to inadequate packaging. Worldwide about 150 billion containers are used annually by the food and beverage packaging industry.

Packaging dates back to when people first started moving from place to place. Originally skins, leaves, and bark were used for food transport. The Greek and Roman times saw the rise of pottery and the start of glass. In the 1400s timber chests were first used, and after 1850 paper and glass started to be used substantially as processes were developed for mass production. Napoleon Bonaparte was involved in the invention of canning. More recently plastics were developed, particularly the first commercial plastics in the United States around 1935–42.

The primary purposes of packaging are containment and protection. Containment refers to holding goods in a form suitable for transport, whereas protection refers to safekeeping goods in a way that prevents significant quality deterioration. The term packaging covers packaging materials and ancillary materials such as glues, closures, packing, inks, etc. Some secondary reasons for packaging are as follows:

1. Nonuniform food provision: food production varies from year to year (droughts, etc.) as well as within a year (harvest times). This affects product quality and price. Packaging is one means of spreading the product availability over time. It also has affected international trade by making shipping of food products possible, allowing seasonal products to be more accessible out of season.
2. Maintaining quality: some factors that affect the quality of food are oxidation, microorganisms, color loss, dehydration, wilting, moisture uptake, crystallization, nonenzymatic browning, and enzyme activity. The common methods of food preservation include canning, salting, drying, preservation, and freezing. One method of preserving quality is to package.
3. Marketing: a package can help sell a product. Nowadays companies mainly use trade names for products instead of generic names to enhance customer recognition of the package.
4. The first commercially successful films (e.g., cellulose nitrate) were developed and have revolutionized packaging possibilities.
5. Eating styles (see below, under trends in packaging).
6. Countrywide economic unions such as the European Economic Union, transport, and trade are all factors that have enhanced the development of packaging, so that over the last three decades packaging has grown in volume and importance into one of the most significant areas of food production.

B. Trends in Packaging

A major change in eating habits in Australia has been the trend away from meat, vegetable, and potato meals to "snacking" and "grazing" on as many as six meals a day. Two thirds of the nation changed in this manner from 1983 to 1993. The microwave oven was the right equipment at the right time to facilitate this change. The first "grazing" meals were microwaveable. "TV dinners," which were conventional aluminium tray meals covered with paperboard cartons, were followed by the introduction of leak-proof paperboard trays covered in a heat-proof polyester. More recently the polyester CPET has proved a successful material by itself. It is expensive and still requires an outer carton, but it is recyclable. Restorable pouches are also now being used for microwaveable meals. Another new development is microwave susceptor packs, in which the packaging contains a metallized patch or thin layer to channel microwave energy into heat. Finally, there is the trend to sterilized bowl or "lunch bucket" containers, where the product is cooked in a retort at (typically) 121°C in a sealed laminate structure, opened by the consumer (FAEZO lid) and covered with a special lid or cling wrap, then placed in the microwave. There is a large range of convenient foods, the preparation of which varies from adding to (instant noodles) or throwing into (pouches) boiling water to eating cold from the package.

Other trends include fresh food, nutritious meals, and healthful products, e.g., low fat, low salt, low calorie, low cholesterol, high fiber. The growing export market has led to packaging styles suitable for the eating styles of Southeast Asia. In America the "health craze" has all but finished, with producers targeting teenagers and the convenience food markets.

Large companies may have their own packaging departments, with the objective of developing new forms of packaging or upgrading existing forms of packaging. These departments must find technical solutions to the conflicting demands of low cost, product preservation, attractive presentation, and container utility.

A code of practice has been designed in Australia to help the industry apply legislation in order to prevent misleading or deceptive packaging. A package that meets the requirements of the code is not considered deceptive by state authorities, but if it does not comply with the code, the onus is on the manufacturer to prove it is not deceptive. The code covers both imported and Australian goods. Its main emphasis is on volume. A few examples of the code are discussed below.

The following free space proportions for food products are permitted:

1. Cakes, confectionery, biscuits—40%
2. Chocolate thins—40%
3. Aerosols—40%
4. Other articles—25%

The code specifies package sizes for bubble packs and lists exceptions to the above rules, e.g., chips and transparent packages. Total free cavity size must be less than 15%, except for aerosols.

The Weights and Measures Act of 1915 is currently being modernized. This requires a quantity statement, a packer identification, standard sizes, and restricted free space in packages.

The General Foodstuffs Act (Pure Food Act 1908) requires packaged food to be labeled with the name of the food, ingredients, name and address of manufacturer, packer or vendor, and country of manufacture. The Consumer Protection Regulation (Date Stamping in 1978) requires packaged food with a durable life of less than 2 years to be stamped with a "use by" or "packed" date.

The Plant Diseases Act of 1924 contains requirements for labeling and packaging of fresh fruit and certain vegetables. The Bread Act of 1969 provides for nominal weights of prepackaged breads. There are similar Acts for Fish and Oysters (1935), Poultry (1969), and Butter and Cheese (1979). All contain regulations as to what is contained on the label and what is to be packaged.

C. The Ideal Package?

There is no such thing as the ideal package. Packaging materials continue to improve so that we may come closer to the ideal, but there are a number of conflicting requirements that a package must satisfy, e.g., high visibility yet low contact with the environment. Compare, for example, the packaging requirements of ice cream, breakfast cereals, and vegetables. The requirements of packaging will be affected by the climate, predicted duration of storage, and product (Table 1).

An ideal package would meet the following requirements:
Zero toxicity
High product visibility
Strong marketing appeal
Moisture and gas control
Stable performance over a large temperature range
Low cost and availability
Suitable strength in compression, wear, and puncture characteristics
Easy machine handling and suitable friction coefficient for film
Closure characteristics such as opening, sealing and resealing, pouring
Proper labeling
Protection from loss of flavor and odor, leaching, and migration from package materials
Controlled transmission of required gases (e.g., oxygen for meat color)

The choice of package is always a compromise between the different objectives outlined above.

1. Displaying the Product

For a package to be effective, it must present the product well. In this section factors that contribute to presentation effect in the store are considered.

1. Type of store: the protective packaging may have flaps that can be opened to give a ready-made display for the product, whereas some stores may remove the protective packaging to display the product directly on the shelves, leading to a preference for

TABLE 1 Average Climates

Location	Temp (°C)	%RH
Adelaide, Australia	30	38
Algeria	25	63
Alice Springs	36	31
Alice Springs	32	36
Ankara, Turkey	20	60
Ankara, Turkey	11	81
Argentina	29	81
Astraham, Russia	25	60
Athens, Greece	27	47
Athens, Greece	16	71
Baghdad, Iraq	22	73
Bangkok, Thailand	34	92
Barrow, Alaska	4.5	85
Belem, Brazil	31	98
Boston	2	72
Brazil	31	89
Canberra	28	56
Capetown, South Africa	25	85
Chad	29	47
Chile	24	77
Darwin	32	78
Darwin	33	78
Darwin, Australia	25.5	60
Dublin, Ireland	13.5	80
Egypt	18	71
Eismitte, Greenland	−10.5	85
France	6	88
Germany	2	89
Ghat, Libya	50	5–15 (10)
Great Britain	5	89
Greenland	−7	85
Hakodate, Japan	4	80
Hobart, Australia	22	58
Hobart, Australia	20	66
Hong Kong	18	77
Hong Kong	19	84
Indonesia	29	95
Jakarta, Indonesia	30	94
Japan	8	73
Kenya	25	74
Khartoum, Sudan	38	21
Kuala Lumpur, Malaysia	33	97
London	10	81
Los Angeles	18	67
Madagascar	30	89
Madras, India	33	80
Manila, Philippines	23	82
Manila, Philippines	33	85
Melbourne, Australia	26	58

TABLE 1 Continued

Location	Temp (°C)	%RH
Melbourne, Australia	24	64
Merida, Mexico	37	84
Mexico	31	74
Miami	23	81
Moose Factory, Canada	16.5	80
Moscow, Russia	19.5	70
New Zealand (North)	23	71
New Zealand (South)	21	65
Nigeria	30	40
North Korea	−3	74
Ottawa	−6	83
Peking	1	50
Peoria, United States	23.5	72.5
Perth, Australia	29	51
Perth, Australia	27	58
Peru	22	85
Pusan, South Korea	12	65
Pyogyang, North Korea	7	66
Rome, Italy	15	83
Santiago, Argentina	32	80
Shanghai, China	13	89
Singapore	30	82
Singapore	31	76
South Africa	26	72
South Korea	0	78
Spain	9	86
Sydney, Australia	26	68
Sydney, Australia	13	70
Taipei, Taiwan	21	90
Tehran, Iran	15	61
Tokyo, Japan	12	75
Toronto, Canada	3	76
Tucumam, Argentina	15	40-50 (45)
Ulan Bator, Mongolia	−4	78
USSR	−9	82
Washington	12	72

Data taken from a range of sources and combined.
Data represents yearly averages.

rectangular containers (high product volume-to-shelf space ratio). So, for example, end-of-aisle displays may require different cartoning than shelf displays, where the outer carton is discarded.

2. Type of lighting: direct sunlight (shop windows, open air displays) will fade many inks and cause many plastics to become brittle. Also some plastics (e.g., cellulose) reflect light so well that the contents are obscured, so less glossy polypropylene may be used. Customers tend to prefer more gentle, indirect lighting.

3. The item should be displayed in a way that gives all the required information about the product quickly. For example, the product should be visible through the packaging; if not, a picture of the product on the wrapping will help it sell. A short brand name is more effective than a long one, as it can be displayed in larger letters and takes less time for the consumer to identify.
4. The choice of colors, print size, organizing material on the label, brand name, key words, and many other factors will affect the success of the package in grabbing the consumers' attention.

Two particular properties of the packaging material are relevant here:

1. Clarity (haze): a measure of the ability of the material to permit light to pass through it. The optical properties of a clear packaging material are a critical part of the display of the package and can enhance the product. A film for overwrapping, for example, must have high clarity. Haze (the opposite of clarity) is technically defined as the percentage of light which in passing through the film is deviated by more than 25 degrees by forward scattering. This scattering is caused by small particles, either on the surface (surface haze) or in the bulk (bulk haze).
2. Gloss (specular reflectance): the ability of the material to reflect light in the same way that a mirror does. The opposite of gloss is matte, caused by diffuse reflectance of light. Film with high gloss has a shiny appearance and so has aesthetic appeal to the consumer.

2. Marketing Versus Technical Requirements

The main places where decisions are made about the type of packaging are the manufacturing and marketing departments. The packaging department must be a master of many disciplines, such as graphics design, printing, creating models (mock-ups), packaging machinery, suppliers, legal requirements, marketing requirements, and product requirements.

From the point of view of marketing, the following factors are important:

1. Visibility
2. Information
3. Function (how well the package works, e.g., ease of opening, protection, microwaveability, portion size, product size, etc.)
4. Ease of handling (transport, store, shelf)
5. Consumer appeal
6. Cost of packaging relative to product value

In practice these objectives may not totally agree with those of the manufacturers. They will be more concerned about the following product characteristics:

1. Dimensions, volumes, densities
2. Availability of equipment and ability of packaging materials to be used with existing equipment
3. How to get product into the package
4. Speeds of packaging lines and other such aspects

Resolution of these differences is an important aspect in developing a suitable product. An R&D department may combine several specialties. For example, a company might bring together product developers, packaging developers, basic science experts (storage stability, microbiology, etc.), and process developers (chemical engineers) to develop a suitable product/package combination.

3. The Environment

Environmentalists are concerned about the consequences of current methods of disposal of packaging. In June 1990, Germany passed a law requiring all plastic packaging to be made from materials that can be either recycled or burned without producing noxious fumes (many halogenated hydrocarbons produce poisonous fumes). It will also become illegal to use printers' inks that contain heavy metals or otherwise restrict recyclability, and every package will have to state its composition. By 1995 the EEC will require 95%of all plastics to be recycled, leading to lower-cost packaging and increased pressure to use recycled plastic in contact with food. In Australia pressure from environmentalists has led to all sorts of "green" claims—enviro-paks, biodegradable products, and especially recycling.

Recent clean-ups around Australia (Clean Up Australia Day, Mr. Ian Kiernan) have highlighted the problem of plastic, with the majority of the 20,000 tons cleaned up annually being plastic. The trend is towards making more plastics recyclable. Roadside litter consists of (by number, not weight) 48% paper, 17% plastic, 4% bottles, 4% cans, and 26% miscellaneous (auto parts, etc.).

The question is whose responsibility it should be to address the packaging-disposal problem. Should the food industry be self-regulating with regards to packaging? Who should pay for collection and sorting? Are products packaged excessively?

Not all materials are easily recycled. Many materials are not economic to collect and reuse. For example, 62% of aluminum cans sold are recovered, by only 26% of glass.

Aluminum used in other applications (e.g., foil, containers) has a poor recovery rate. The ability to recycle depends on factors such as market size, ease of recovery, and energy saving in reuse.

The packaging industry usually has in-house recycling, for example, collecting skeletal webs and other waste plastic, regrinding and remelting in plastics extruders. An alternative method of recovering value from waste is by incineration, where a component of its energy can be used.

Recycling is not the only solution. Biodegradability refers to the breakdown of a packaging material when exposed to the environment. Cellulose (paper) breaks down extremely slowly by bacterial action after being buried, but some modern plastics can break down in 1–2 months. For example Hi-cone rings (6-packs) are made from photodegradable polyethylene. Another solution is to reduce the amount of material used. In other words, recycle, reuse, and reduce.

One of the largest changes affecting the food industry has been the recent advances in aseptic packaging. Fruit juices and milk are aseptically packaged on a large scale, and recent research has led to the development of particulate packaging techniques. The main research difficulty has been in guaranteeing F_0 values for solid particles in a diphase mixture.

In defense of packaging, note that:

1. Packaging accounts for less than 10% of the total waste in Australia.
2. Packaging reduces food losses from 30–50% (in developing countries with limited or no packaging) to 2–3% (countries with developed packaging infrastructure),
3. When comparing oranges with orange juice, the packaged concentrate and the ability to use its byproducts (especially the peel) in large quantities close to the point of processing mean that packaging *reduces* wastage and transport fuel in a way that would not be possible without packaging. Household waste is reduced as only the portion of the food required is distributed.
4. Only a small percentage of resources is required for package.
5. Our ability to recycle is growing.
6. There is a trend to use less material (e.g., glass, aluminum, plastic).

To consider point 5 in more detail, there is a growing pressure to recycle materials and for industry to use materials which can be recycled. One of many difficulties with recycling is that there is no control over how the consumer uses the plastic container (e.g., for pesticides, chemicals), and the container may become contaminated. The ink on labels is also difficult to remove during recycling. Thus some countries have regulated that only industrial plastic waste can be recycled for food containers.

Regrind material can be sandwiched between other materials, provided that the virgin plastic layer has sufficient barrier properties. The main difficulty then with reused plastics is the migration of contaminants into the food.

In the United States the Plastics Recycling Task Force has submitted guidelines for the safe use of recycled plastics to the U.S. FDA. Existing U.S. regulations require that all components of food packaging must be of a suitable purity for food contact and that the packaging will not contaminate or adulterate the food. In Australia existing legislation is inadequate. The SICRO Food Research Lab recommends that total quality control is required for both recycled materials and for virgin material. They have suggested the following guidelines:

1. the source of the recycled material must be known and controlled.
2. Regular testing of recycled materials be initiated, testing for known contaminants,
3. virgin barrier materials must be used to sandwich recycled materials.
4. The CSIRO must work with industry in testing packaging and testing food for contaminants.

An example of reducing the amount of material is glass bottles. Coca-Cola-Amotil's original coke bottle used large amounts of glass. Recently they have reduced the amount of packaging even further by switching to PET, using up to 30% recycled PET in a laminate structure. Note that 1-liter refillable bottles must have a deposit (20) in South Australia (South Australia Beverages Act, 1978), the result of which is a 97% recycle rate, with each bottle being used an average of 30 times. Beer cans provide a more extreme example. The switch from steel cans to aluminum has reduced the weight of the can from 91 to 17g. In addition, the aluminum can is successfully recycled.

In an American study comparing aluminum, glass, and PET soft drink containers, a cradle-to-grave analysis was made that showed that PET bottles were 25% more energy efficient than glass and 65% more efficient than aluminum and had less impact on resources at all levels of possible recycling (1987 NAPCOR study commissioned to Franklin Associates). Compare this result with the Clean Up Australia Day result above [4].

D. Packaging Needs for Specific Foods

Different foods will require different packaging solutions. In this section, packaging requirements of the major food groups are considered.

1. Bakery Goods

Foods in this group include breads, cakes, croissants, biscuits, and pastries. They are characterized by being based on milled cereal grain (flour), with baking being the means of gelatinizing the starch granules into a consumable form.

Deterioration in product quality for this group is due to:
Crystallization of starch granules, causing staling
Moisture uptake, causing a reduction in crispness
Mold growth, as bakery products typically have a high moisture content
Rancidity from exposure of lipids to oxygen
Drying out of the interior (crumb) of a loaf of bread

The main forms of packaging for bakery products have distribution rather than protection as their primary objective. Product quality is ensured by specifying short shelf lives. For example, bread should be consumed as quickly as possible. It is difficult to specify a packaging material for bread that could meet the competing requirements of keeping the crust dry and the crumb moist. Furthermore, moisture movement from starch to protein within the crumb leads to texture changes as the bread ages. Since the shelf life is short, the goal of bread packaging is partially moisture control, but its main purpose is to allow the product to be distributed safely and hygienically. The material used should be inexpensive, as bread is a staple. Materials such as waxed paper, cellophane, and polyethylene are commonly used in bread packaging.

For many bakery products, even this level of packaging is excessive. Cakes and pastries are often distributed in cardboard boxes that are not air-tight, made possible with a waxed paperboard support. Packages containing pies must have airholes, so that the package can "breathe" moisture during heating or cooling, preventing crust uptake of moisture.

2. Breakfast Cereals

This food groups covers ready-to-eat (RTE) breakfast cereals such as muesli, corn flakes, shredded wheat, rice puffs, and nutrigrain. In many cases the product name is a strong link to the cereal grain used in its production.

The main factor causing quality deterioration is moisture uptake, therefore moisture barrier properties are required. Since moisture uptake is slow, the degree of barrier required is not high. Typically cardboard boxes with waxed paper liners are used. For shredded wheat, the package must be able to breathe to prevent the development of rancid odors. Polyethylene bags are increasingly being used for liners, and for high-sugar products foil laminates may be used to provide a higher degree of moisture barrier.

3. Chocolates and Confections

This group of food products is characterized by high sugar (and often high fat) content. Confections have been manufactured at least since the time of the ancient Egyptians, and probably as fruit or honey concoctions even earlier.

Factors causing deterioration include uptake of moisture in crystalline sweets (leading to stickiness), loss of moisture in noncrystalline sugar compounds, bloom in chocolates (separation of sugar appearing as a fine grey coating on the chocolate surface), and oxidation of fats in high-fat compounds (although chocolate fat is not subject to rancidification). Syrups need to be contained and may be affected by loss of aromas. Temperature can affect product texture (especially melting of chocolate, but more generally viscoelastic and elastic properties of solid sugar-water mixes) and lead to component separation (e.g., oil from water phases).

Packaging requirements for these products are generally moderate. Until recently most sweets received minimal packaging, and then mainly as required for transport and distribution of the product rather than for product protection. Sweets were distributed in cardboard boxes, which were designed as display items, so that they could be opened in sweet counters, the contents exposed directly to ambient air.

Cellulose wraps are a distinctive form of packaging for some sweets, using cellophane's twist-retention properties. Aluminum foil, glassine paper, polyethylene bags, laminates, and metallized polyesters are also widely used. To prevent "show-through" with chocolate (unsightly "wet" spots on plastics), opaque packaging or pearlized polypropylene is commonly used.

4. Dairy Products

There is a wide range of products in this particular food group, with the number of products having increased rapidly over the past two decades. Examples are:

Milk products, including whole milk, evaporated milk, skim milk, low-fat milk, milk powder
Yogurt and other fermented milk products
Cheeses, divided into cream and hard, processed cheeses
Ice cream
Cream, including sour cream and whipped cream
Butter

Dairy products are high in fat (raw milk contains about 4% fat) and are susceptible to a range of bacteria, including *Mycobacterium tuberculosis, Staphylococcus, Lactobacillus* such as *Bacillus cereus, Listeria monocytogenes*, and *Salmonella*. They are also generally high in protein (protein denaturation at high temperatures, such as in ultra-high-temperature (UHT) processing) and water content. Because of the high water activity, milk products must be sterilized or pasteurized, the resulting safe storage time depending strongly on storage temperature. In addition, the milk may be homogenized, giving a stable dispersion of fat globules through the water phase. A specific problem for dairy products such as milk is susceptibility to light, with vitamin B_2 being destroyed and generating oxidized flavors. Thus, high-barrier packaging and light protection are the main packaging requirements.

Milk products are packaged in bottles, cartons, plastic bottles (HDPE), small laminate packs for single serves, and tins or helical foil-laminate containers for powders. Cartons give better light protection than bottles (nonwaxed better than waxed). Yogurts are packaged in tubs, for example, of polystyrene. Cheeses are packaged in wax (outer layer protection), foils, nylon, and laminate plastics. Ice cream is packaged in tubs, glassine paper, waxed paper, and paper wraps. Cream is packaged in cartons, glass jars, or plastic (HDPE) tubs, and butter in glassine or plastic tubs. UHT processed dairy products are packaged in polyethylene-line paperboard, increasingly with aluminum foil as an added barrier protection.

The major problem with dairy products is their short shelf life. For milk, two innovations have increased the lifetime: sterile processing (UHT milk) and aseptic packaging. These allow milk to be sold unrefrigerated.

5. Meats

Meats are divided into fresh and processed meats. Processed meats can be further categorized as fermented and nonfermented meats. The ability to process and package meats has had an impact on the meat industry. Processing techniques such as brining and fermenting have developed slowly over millennia, but refrigeration was the key to the development of packaging as a preservation method for meat products over the last century.

Fresh meat deteriorates by three main mechanisms: color change (due to changes in the valence states of iron in blood), microbial spoilage, and dehydration. Thus, in addition to preventing product contamination, fresh meat packaging needs to prevent moisture loss, allow some oxygen transmission (to maintain color), and prevent spoilage. Some color degradation occurs due to light.

Meat is often packaged in meal-size portions in plastic trays covered with shrink-wrapped PVC, cellulose, or pliofilm, but where longer storage is required, it is packaged with PVDC or a copolymer of vinylidene chloride (VDC) and vinyl chloride (VC), which is relatively impervious to moisture and oxygen. The blocks are then vacuum-packed or freeze-packed. Color changes due to freezing may also affect the perceived meat quality (slow freezing causes translucency, fast freezing causes a grey and opaque surface appearance).

A typical supermarket pack consists of a tray of polystyrene or paperboard, a drip tray to collect blood and fluid seepage, and an oriented plastic shrink-wrap, which allows oxygen migration but resists moisture movement (e.g., PVC, EVA, cellophane, or pliofilm).

6. Savory Snack Foods

Savory snack foods include extruded and fried high-starch products such as potato chips and pretzels, as well as nuts, cheese spreads, and some crackers. Breakage, staling, and rancidity are problems. Packaging consists of polyethylene/polypropylene bags, and helical twist laminates of foil and paperboard are used (composite cans). Glassine bags are used for potato chips.

7. Fresh Fruits and Vegetables

Fresh fruit and vegetables are affected by a large number of deterioration factors, such as vermin, insects, bruising, condensation, certain gases [such as carbon dioxide (CO_2) and ethylene (C_2H_4)], temperature fluctuation, diseases that affect appearance, anaerobic fermentation, wilting, rotting, and ripening.

Packaging requirements are largely determined by the fact that fruits and vegetables respire during storage. The package may need to breathe to allow product ripening. Ethylene ingress may be prevented to avoid senescence before transport and then enhanced to complete ripening just prior to sale. Four compounds need to be controlled: CO_2, O_2, H_2O, and C_2H_4. In some cases, other low molecular weight gases may be introduced to control physiological disorders.

Examples of materials used in vegetable film shrink wraps are polyethylene, cellulose acetate, and pliofilm.

8. Other Foods

The following foods deserve for various reasons special consideration:

Coffee requires high-barrier packaging to contain aromas and flavors.

Small packs of sauces, spices, jams, etc. used by airlines and restaurants are generally aseptically filled and require high barrier properties.

Dried foods have few packaging requirements except for preventing the product from reabsorbing moisture and being contaminated.

Frozen foods require packaging that can cope with low temperatures without becoming too brittle (polypropylene, for example, becomes brittle).

wines require both high O_2 and SO_2 barriers. Wine bags may be complex laminates. Bottles and corks are an effective packaging container for this purpose, although taints produced by corks have spurred interest in plastic cork alternatives. Some protection from light is required to maintain flavor.

E. Summary

This section has given a brief cross section of the food industry and packaging of specific food groups. In the next sections we will tackle packaging from the opposite point of view, looking at the properties of different packaging materials.

II. TYPES OF PACKAGING MATERIALS

A. Plastics

1. Background

Introduction

The first type of packaging material to be discussed is plastic; technologically this is a most complex class of materials. In general packaging, materials may be grouped into rigid and flexible structures. Plastic film, foil, paper, and textiles are flexible materials, whereas wood, glass, metals, and hard plastics are examples of rigid materials.

The volume of plastics produced each year now exceeds the amount of steel consumed, and practically none of it is recycled. Plastics now account for about 25% of household waste, although less than 20% of the containers we use are plastic. Half of these are used for milk and various carbonated drinks. However, using other forms of packaging would double packaging costs, quadruple the amount of waste products, would take more energy to produce, and reduce the number of new jobs per year created by the growing packaging industry (e.g., 2500 in Australia). There is increasing recognition of the need to recycle. The main difficulties are separation of plastics (manual labor) and purity of the final product (energy problems).

Because of their lower unit cost and lower energy consumption during manufacture, plastics have tended to replace the traditional packaging materials, glass, paper and metals, in situations where high barrier properties are not required by the product. Although no ideal barrier plastic exists, developments in laminates and copolymers are slowly reducing the competitive edge of glass and metal containers.

History

The following dates are important in the development of plastics:

1843: Malayan *gutta percha*, a shellac molding material, was the first semi-natural plastic.
1870: Searching for a substitute for ivory for constructing billiard balls, Hyatt made pyroxylin from cotton and nitric acid and then reacted this with camphor, producing celluloid.
1909: Leo Baekeland reacted phenol with formaldehyde with a catalyst, hexemethyleneteramine, under pressure (to stop foaming), producing the first synthetic resin, called bakelite.
1919: Casein was developed as a film.
1927: Cellulose acetate and polyvinyl chloride were developed.
1935: ICI reacted ethylene under high pressure with trace O_2, giving LDPE, which suited the newly developed technique of blow-molding.
1953: Karl Zeigler produced HDPE from ethylene using the catalysts titanium tetrachloride and triethyl aluminium.

Since 1953, a range of other important plastics have been developed.

Definitions of Terms

mer: a unit, e.g., a basic unit in a chemical structure
polymer: many units
plastic: deforms, can be shaped by a shear force in excess of a plastic yield minimum.
thermoplastic: shaped by reducing the yield stress by heat softening
thermosetting: initially shaped by heat, but shape is then set permanently
crystalline: regular, periodic shape
amorphous: opposite to crystalline, random shape
isotactic: possessing a preferred orientation of molecules (same feel)
atactic: molecules in structure are randomly oriented (no feel)

2. Manufacturing

Manufacture of Plastic

Polyethylene is a plastic that is produced from oil or coal through extraction of ethylene gas. Ethylene (C_2H_4) is a monomer, a molecule that can combine with itself through breaking the double bond to produce long-chain macromolecules that are highly inert to the environment. Polyethylene is a polymer. Many other monomers are derived from ethylene.

The chain of mers can twist and kink around the C–C bonds so that the resulting macromolecule can vary from a straight line to a sphere. The degree of linearity has a large effect on the resulting plastic properties. As the polymer is cooled after melting or manufacture, the chains will link to each other, either by van der Waals forces, by ionic attraction, or by cross-link bonding. If the macromolecules are straight and regular they will fit well together, giving higher density and crystallinity.

The kinked and twisted molecules are stretchable, sometimes to many times their unstretched length, so the plastic can be oriented. Branching can occur along the chains but is relatively rare, requiring reaction to occur at a C–H bond. Higher temperatures will permit more of these random branching reactions to occur. Branching reduces crystallinity by preventing ordered molecular arrangements.

Most polymers are from difunctional monomers; a few are tri- or multifunctional, leading to three-dimensional configurations instead of long chains. These structures are more rigid at high temperatures.

The plastics are divided into two groups: thermoplastic and thermoset. Thermoplastic means that the plastic may be heated and cooled without losing its structure, while thermoset plastics, once cooled, cannot be reheated without breakdown of the macromolecule. When a thermoplastic material is heated, it becomes pliable and can be shaped as required. Linear polymers tend to be thermoplastic, whereas cross-linked polymers are thermosetting. Cross-linking occurs when atoms join across polymers, e.g., sulfur, which is used to cross-link isoprene to give us the rubber used in car tires, or oxygen, which causes aging in the same rubber by cross-linking and so reducing the rubber flexibility.

The structure of natural rubber (isoprene) is (leaving out irrelevant hydrogen atoms):

$$\begin{array}{c} \qquad\quad CH_3 \\ \qquad\quad | \\ \cdots C - - C = C - C \cdots \\ \qquad\qquad\quad | \\ \qquad\qquad\quad H \end{array}$$

The C after the double bond can point up or down, leading to isomers (isoprene and "gutta percha") with different properties. By analogy with isomers, the basic units can also combine in different orientations, leading to isotactic polymers with ordered structures, which can more easily be crystalline, and atactic structures, where the orientation continuously varies along the chains, preventing crystallinity. For example, polyethylene (PE) varies from atactic PE with a density of 900 kg/m^3 to crystalline PE with a density of 1000 kg/m^3. Common forms of PE are low-density PE (LDPE, 920 kg/m^3), linear low-density PE (LLDPE), medium-density PE (MDPE), and high-density PE (HDPE, 960 kg/m^3).

Bonds can be broken in a number of ways (degradation of the polymer). A C–C bond has an energy of 6.1×10^{-19} J. UV light at 300 nm has an energy of 6.1×10^{-19} J/photon, so is close to the same energy. Visible light at around 600 nm has about half as much energy. If thermal or strain energy are also present, then visible light will also degrade plastic (but more slowly than UV).

Glass temperature refers to the temperature at which individual molecular energy is too low to allow the macromolecules to slide past each other. The melt temperature is the transition temperature from a solid to a free liquid and is higher than the glass temperature. Below the glass temperature, the plastic is brittle, so some plastics cannot be used for freezing, but between the two temperatures the plastic can be shaped and bent more easily.

A plasticizer acts by reducing the average molecular size, acting as a lubricant between the large macromolecules so that they slide over each other.

Methods of polymerizing include:

1. Addition: unsaturated compounds open up their double bonds under heat and pressure. The macromolecule forms in three stages: initiation, propagation, and termination. Initiation is the initial rupture of the C=C bond to produce energetic combination sites. In propagation, further monomers react with a growing chain. In termination, two growing chains meet each other and combine, preventing any further reaction. For high molecular weights, this should be delayed as long as possible.
2. Condensation: after initiation, molecules combine, releasing low molecular weight compounds such as water.
3. Ionic (either anionic or cationic polymerization): initiation occurs through the introduction of ionic compounds, which become integrated into the polymer, producing a class of plastics called ionomers (surlyn).

Ethylene is the main monomer used in the production of plastics ($CH_2=CH_2$). The hydrogen atoms can be replaced by halides (chlorine, fluorine, bromine), phenyl, or other such groups to produce further monomers, all containing the central ethylene double bond. These polymers are generally thermoplastic and are called polyolefins.

Large chemical companies usually manufacture plastic in the form of small plastic beads, or resin. These granules are transported in bulk to factories for manufacture into either rigid or flexible materials. In either case, the resin is fed into the hopper of an extruder, which brings the resin temperature to above its plastic melting point but below its degradation temperature. It is then ready to be shaped by passing through a die plate, containing die inserts that shape the product as required.

Construction of Rigid Plastics

Blow molding: Many plastics are shaped using the process called blow molding. This is a process for producing hollow parts in one operation, and is primarily used to make plastic bottles. A hot plastic tube, called a parison, is inflated against a cold mold. The final shape will have a thin neck, ideal for many packaging applications. The parison is produced by extruding the plastic at high temperature and pressure.

Variations on blow molding include coextrusion blow molding, injection blow molding, and injection stretch molding. Labeling can also be done in the mold.

Injection molding: Injection molding involves direct injection of molten plastic into a shaped space between two dies. The dies separate after the plastic cools to allow the final shape to fall out. This is used for making solid plastic objects such as caps, closures, and plastic toys.

Thermoforming: Many plastic containers are formed by thermoforming. This involves heating a plastic sheet until it softens and then shaping it by stamping it between two cooled molds. A growing requirement of molded plastics is that they be microwaveable, i.e., have a softening point well above 100°C. They must also be shelf-stable. Plastics are generally slow to degrade under ambient conditions, but they may discolor or become brittle.

Construction of Plastic Sheets and Films

Approximately 10 different polymers account for the majority of packaging films. In general, a film is defined as a layer of thickness of less than 0.010 inch (0.25 mm). Thicker plastic is referred to as a sheet.

Films may be used as liners, wraps, or overwraps, depending on how they are applied to the product. Liners are plastic films applied to the inside of the container, often to protect the container from the product, such as plastic linings sprayed into the inside of metal drums or cans. Wraps are applied directly to the product, usually by making a tube from the film, sealing one end, filling the tube with product, then sealing the top end. An overwrap is applied around the carton, often heat shrunk onto the carton to provide barrier protection.

Films may also be used for bags, envelopes, and pouches. A common method of closing plastic bags is with plastic clips that can also double as price tags, as in bread packaging. Envelopes are bags sealed on three sides, with the last side folded over and sealed after filling with product, used mainly for flat items. Pouches are constructed and filled on the processing line automatically. A typical pouch is a complex laminate, possibly of oriented polypropylene (outside), LDPE, aluminum foil, adhesive, and a copolymer film on the inside.

Films are manufactured using an extruder. The plastic resin granules and other ingredients are fed into the extruder and compressed and heated until the plastic flows out of the die. The two main die shapes are slit circular and slit rectangular. The extruded plastic may then be stretched while still hot and flexible:

1. Monoaxially: the die is designed to give a wide sheet of plastic film. This can then be stretched in the machine direction by up to 80 times its original length, which has the effect of thinning the sheet and orienting the molecular structure. The result is that the plastic can be stretched easily in one direction after cooling, but not in the other.

The thickness of the film is checked, the film cooled, and then would onto the mill roll.

2. Biaxially: jaws and rollers are used to stretch the film apart in both directions at once (Tenter process). The molten film is poured onto a casting wheel and then through a series of rollers, which stretch the film in the machine direction. The film then passes through the Tenter section oven, where jaws are used to grip the softened plastic and pull it apart in the transverse direction. The result is a stiff plastic. A second method is called the bubble method, where the plastic is extruded as a continuous cylinder, and a bubble of air is blown up the center of the cylinder, forcing biaxial expansion. As the plastic heat-sets it will shrink slightly before assuming a fixed shape.

The plastic may then be coated prior to the mill roll in several ways, to enhance its properties. A common example is coating a plastic that is difficult to print with a printable plastic on one or both sides. Coating will be discussed in more detail below.

Another common coating produces metallized plastic. For this technique, the plastic passes quickly through an evacuated chamber in which aluminum is vaporized. This deposits a metal layer several molecules thick (~200 Å), resulting in a metallized appearance. This material is often used to replace foils as it is cheaper but does not have the same barrier properties as foil (although there will be some improvement over untreated film). The metal coat also substantially reduces UV light reaching the product. Metallized films are a common base for eye-catching graphics. However, the metal does not attach firmly to the plastic and may be scraped off (a surface lacquer presents this), and the plastic is not sealable on the metallized side. Metallized plastics also suffer from metal oxidation, affecting print adhesion, giving a maximum effective lifetime of about 3 months.

The film may also be pearlized, a technique that distributes fine air bubbles throughout the plastic (or cavitating the surface) resulting in a pearly white opaque appearance. This is useful for preventing "show-through," the wet appearance where fats directly contact the plastic, and a particular problem with chocolate wrapping. Pearlized plastics also tear more easily.

The ideal film should have good barrier properties, be chemically stable, temperature resistant, and heat-sealable, resist grease, be strong, not allow migration, have good transparency and gloss, be printable, and be inexpensive to make. Since no plastic will satisfy all of these requirements, the actual plastic chosen for a particular function will be a compromise.

3. Properties

Plastics

Most of the properties discussed in the following section are explained where first used. One key property relevant to packaging equipment, slip, is not covered in detail, so will be explained here. From the perspective of machining, plastic film must have the right friction coefficient to pass smoothly through the packaging equipment. If the slip is too high, the result is different package sizes in tube-formed packaging due to the film sliding past the forming rollers as well as a loss of registration in printing (see below). The product slides easily inside the package, interfering with sealing the pack. On the other hand, too low a slip can cause the film to stick to hot surfaces or folding box surfaces and can cause the product not to slide down to the bottom of the pack, also affecting pack sealing and presentation. The product should be able to slide back and forwards in the pack relatively easily for tight sealing at both ends.

This section also serves to introduce and describe the primary packaging plastics.

Polyolefins

Polyethylene: Polyethylene (PE) is the result of polymerization of ethylene gas and has the formula $(CH_2)_n$. It was invented by ICI in the 1930s. Two main manufacturing processes result in different polyethylene products. The first is called low-density polyethylene (LDPE), and the second is high-density polyethylene (HDPE).

LDPE is formed at high pressures (1000–3000 atm). This results in long branched chains, weakly linked to each other by van der Waals forces (but strong overall force due to length). The branching is random, and so LDPE is an atactic polymer. Thus neighbouring chains can slip past each other, allowing the material to bend easily (flexible). As a result, the printability of LDPE is poor. However, many plastics with poor printability can be made printable by corona treatment, in which an ionic discharge is used to sensitize one side of the plastic.

LDPE has low density, as the long chains exhibit branching, so that the molecules are not able to fit closely together. This irregularity in structure also results in a lower melting point and a less crystalline (ordered chain) structure.

LDPE is tough, semi-transparent (poor clarity), flexible, and has a waxy feel. It resists most chemicals below 60°C, and it resists water moderately but not gases (poor O_2 barrier).

LDPE is usually used as thin sheets or laminated to other packaging materials. It is used for bag manufacture (bread, diapers), for low-temperature storage (due to its low barrier properties), and for packaging rice. The melting point of LDPE is 105°C.

HDPE is produced at low temperatures and pressures of about 10 atm. This gives rise to an ordered molecular structure, which is called an isotactic polymer. The Ziegler process is used, employing a catalyst.

HDPE is stiffer, harder, less flexible, and waxy. Higher temperatures are required to produce thermoplasticity (melting point = 134°C).

HDPE is used for making containers, e.g., crates, bottles, bags, tubs, plastic knives and forks, etc. It can be steam-sterilized, where as LDPE cannot. HDPE bottles are opaque and can be used to contain detergent and milk.

HDPE resists fats and oils better than LDPE. However, it does not seal easily.

The above simplifies the total PE picture. Available PEs include low-, medium-, and high-density PE, linear low (LLDPE) and medium densities (with twice the hot-tack strength), copolymers of PEs, copolymers with vinyl acetate (EVA) and ionomer film (surlyn). PE may be coextruded with nylon, saran, and EVOH sandwiched inside. Such films have high strength, flexibility, clarity, and especially barrier properties, and are used for bag-in-the-box, pouches, cups and lids, etc. Usually the form of PE used will be a trade-off between barrier properties required and cost (e.g., for 50-μm films, LDPE = 11.3/m^2, LLDPE = 12.2/m^2, 7.5% EVA = 14.8/m^2, LDPE/surlyn CO-EX = 22.3/m^2).

Polypropylene: This monomer has the formula CH_2=CH–CH_3. Polypropylene (PP) was developed using polymerization catalyst technology by Guilio Natta in 1954.

Polypropylene forms a regular, highly ordered polymer at low pressures in the presence of certain catalysts called isotactic polypropylene. It has high crystallinity (high clarity and gloss), is hard, heat resistant (higher softening point, 150°C), exhibits good memory, flex crack resistance, puncture resistance, and stiffness.

It is resistant to chemicals (except aromatic and chlorinated hydrocarbons).

PP has excellent moisture and average gas barrier properties. It can be printed on, and is ideal for reverse or surface printing. Cast polypropylene has excellent heat sealability.

PP is used for injection-molded containers and blister packs, laminations, carton overwraps, snack food bags, and confectionery bags. It may be coated (e.g., with PVDC or acrylic) and may contain additives. PP has poor heat stability, so precise heat control is required in the packaging equipment.

PP film may be stretched during production yielding an oriented (OPP) or biaxially-oriented film (BOPP) of high clarity, strength, and resistance to water vapor and gases (e.g., for wrapping snack foods), and coated with saran or acrylic for better barrier and heat sealability.

PP causes contact transparency ("show-through"), and is available from 15 to 75 μm. It may be pearlized ($CaCO_3$ + heat generates CO_2 bubbles, resulting in reduced barrier properties and strength, but no show-through) or white (pigmented, better print density, saves undercoat, reduced show-through). It is used for confectionery, especially chocolate.

PP may be coextruded with PE (1.5 μm) for heat sealability.

Polyvinyl chloride: This monomer has the formula CH_2=CH–Cl. The term vinyl means that a halogen has been substituted for a hydrogen atom. Polyvinyl chloride (PVC) has low crystallinity (so has good transparency when pure), but higher interchain bonding than PE due to the Cl^- halogen, so is harder and stiffer. For this reason plasticizers may be added during manufacture. The Australian Standard 3010—Plastics for Food Contact: PVC—allows four plasticizers for food involving flexible sheets and films, e.g., vitafilm (for meat) with particular oxygen diffusion characteristics.

PVC has good feel and printability. It is highly inert. It is glossy and resistant to moisture, fats, and gases.

There is great variety in PVC compositions, e.g., stabilizers, impact aids, lubricants, and other additives are present in large proportions. The stabilizer is necessary because the decomposition point for PVC is close to the melting point. For food products, the extraction of the stabilizer from the PVC must be less than 1 ppm of the stabilizer, and the stabilizer must be a calcium-tin or dioctyl-tin system (not butyl-tin, used for nonfood products), containing tinuven for UV protection.

PVC is used in the biaxial-stressed form, e.g., for shrink-wrapping of cheese and meat. It is also used for thermoformed containers, e.g., for chocolates, as well as for plastic pipes and toys. It heat-shrinks after stretching and can be thermoformed.

Polyvinylidene chloride: This polymer is similar to PVC, except that there is a double chlorine substitution, giving $CH_2{=}CCl_2$. Polyvinylidene chloride (PVDC) has a more ordered structure, with high crystallinity and softness. It has excellent barrier properties (especially to O_2) and is commonly used as a copolymer with PVC.

PVDC possesses reasonable clarity, good feel, good printability, and strength. It is difficult to cut as it lacks stiffness, so is hard to machine by itself and is too expensive for use as a pure monofilm (except for household use). It resists chemicals, has low water, gas, aroma, and flavor permeability (due to chlorine ions), and high strength. It has high chemical stability and is hygrophobic.

PVDC is used with PVC as a copolymer to coat other packaging materials to provide good barrier properties. It is used in laminates and is an important shrink film, with excellent cling properties.

PVDC cannot be reprocessed because it degrades (melt point 162°C). This makes coextrusion lamination difficult as well, although it can still easily be used in coating from solution. When used in coextrusion it must be copolymerised first (for example with vinyl chloride) to give better temperature stability.

Polytetrafluoroethylene: The formula for this monomer is $CF_2{=}CF_2$. Polymerisation produces long, straight unbranched chains, with ionic intermolecular bonding, so polytetrafluoroethylene (PTFE) is strong and crystalline.

Since the bonds C–C and C–F are strong, this material is extremely inert and has a high softening point (340°C).

PTFE possesses a high gloss and is waxy in feel. It is used in coating cookware (nonstick surfaces) and forming and packaging where easy-to-clean nonstick surfaces are required.

Polystyrene: This polymer results when an ethylene hydrogen is replaced by a phenyl radical ($CH_2{=}CH{-}C_6H_5$). It is a synthetic rubber, which does not degrade over time.

Polystyrene (PS) is amorphous (random packing), low in density, and brittle. In the pure form it has good clarity and printability. It is often used for loose bulk packaging, especially for packing fragile materials. Originally this involved aeration of the liquid plastic with fluorocarbons, but environmental aspects have encouraged CFC-free production and reusability of loose packaging materials (also called void packaging).

PS is usually copolymerized, especially with butadiene, to give high-impact polystyrene (HIPS, see below), which has a less brittle structure. Yogurt and ice cream tubs are common examples. Margarine tubs may be made from ABS (acrylonitrile butastyrene).

PS is also used for disposable plates and cups. PS can be foamed with hexane to form expanded polystyrene (EPS), a low-cost, low-density material that is easily formed into holding trays, cups, etc.

Polystyrene is the plastic of choice for thermoforming because it has strength, low cost, formability, and sealability. PE and PP are only used where specifically required; PP containers can be filled at high temperatures.

Although PS provides a good barrier to gases, it is permeable to water vapor. It resists grease, acids, alcohols, and alkalis (i.e., chemically stable). Oriented polystyrene is a most useful packaging film and is heat shrinkable. The melting point of HIPS is 100°C.

Other Thermoplastics

Polyesters: Polyesters are plastics formed by the polymerization of esters. In general they have reasonable clarity and poor feel and printability, but are strong, versatile, with good heat resistance. This is useful for boil-in-the-bag type applications, where the plastic must sustain temperatures of 100°C without deforming or softening. When metallized polyesters are also used for snack and coffee pouches.

Ethylene glycol and terephthalic acid yield polyethylene terephthalate (PET). PET has high strength and chemical stability. It is used for blow-molded bottles and some films, increasingly for thermoformed trays, for shrink wraps, and for boil-in-the-bag products. It has excellent clarity, comparable to that of glass. It can be printed, metallized, and laminated. It has high barrier properties for a plastic film. Due to efficient recycling, PET is a cheap plastic. It is only used in 12- and 15-μm gauges for film and is commonly used as the outer layer of a laminate structure due to its gloss and temperature stability (melt point 254°C). It is used for microwave pie wraps (micro-perforated).

Forms of PET include APET (atactic PET), CPET (crystalline PET), OPET (oriented PET), and PETG (copolymer PET/cyclohexane dimethanol; melting point 265°C). Polyethylene naphthalene (PEN) shows promise due to its low permeability to gases. There is growing interest in recycling polyesters due to their initially high cost.

The advent of the PET range has revolutionized the packaging industry, allowing plastic to compete directly with glass bottles. Blow-molded PET bottles have less weight, lower production and energy costs, and are droppable, while still have excellent clarity and gloss. The bottle is molded as follows: the closure is first injection molded, producing a little bottle with the screw top already formed, then transferred hot to a blow-mold die. The wall thickness is computer-controlled. The production equipment is specific to PET. Current production rates are about 1/sec. They are not suitable for beer, which requires a light barrier (opacity).

Cellulose: This was the first transparent film to be used (invented by Du Pont) in packaging and was widely used until the advent of polypropylene. It is biodegradable. A common name is cellophane, a trademark name of British Cellophane Ltd.

Cellulose is clear, stiff, printable, and glossy, but has poor feel and moisture resistance. It has good heat-dimensional tolerance. It is not heat sealable. It is naturally antistatic, so is good for powders (e.g., milk powder).

Cellulose is still widely used today, due to the addition of coatings that have given cellophane great adaptability. Examples of coatings are nitrocellulose (making cellophane flexible and durable), PVC, PE, and PVDC (oxygen-barrier properties).

Cellulose can be laminated to foil, paper, and some plastic films.

It is able to retains folds (dead fold), so that if it is twisted, it retains that twist, making it ideal for individual candy wrappings. It is also tearable. It is used for cookies, confectionery, and pastries in situations where vapors may need to "breathe" to prevent surface molding.

Cellulose is sold on the basis of weight per unit area, not thickness, due to the variation in thickness that occurs with moisture absorption. It is available in weights of 22–60 g/m^2. Some typical cellulose codes are given in Table 2.

Cellulose acetate: This product is made from cellulose and acetic anhydride. Cellulose acetate has good clarity, is printable in sheet form, but has poor feel and barrier properties. It is becoming obsolete, although it is still used in laminates and thermoformed blister packs.

Other thermoplastics: Other biodegradable biopolymers are being developed but have not yet become competitive. The raw material for the new thermoplastics is usually starch, which is blended with conventional oil-based polymers.

Polyamides (*nylons*): these are made from condensation of a diacid (e.g., adipic acid) and a diamine (e.g., hexamethylene diamine). polyamides have high crystallinity, strength, impact strength, puncture and stress-crack resistance, flexibility, and melting (255°C) and softening points. They also have good chemical resistance.

Polyamides are used for boil-in-the-bag type products, frozen foods, fish, meat, vegetables, processed meat and cheese, always in lamination. They have low water barrier and high gas permeability.

TABLE 2 Some Typical Cellulose Codes

Code	Explanation
A	Anchored (describes lacquer coating)
/A	Copolymer coated from dispersion
B	Opaque
C	Colored
D	Coated one-side only
F	For twist wrapping
M	Moisture-proof
P	Plain (non–moisture-proof)
Q	Semi–moisture-proof
S	heat sealable
/S	Copolymer coated from a solvent
T	Transparent
U	For adhesive tape manufacture
X	Copolymer coated on one side
XX	Copolymer coated on both sides

A large range of different nylons exists. Polyvinylidene-coated or metallized nylon is about twice the price of pure nylon film.

Nylon is available as cast (18–100 μm) and biaxially oriented (BON) nylon (12, 15, and 20 μm). It can be metallized or PVDC-coated for better barrier properties and is usually laminated. As a film it must be biaxially oriented to give printing and machining stability.

Polycarbonates: These are formed from condensation of carbonic acid in the presence of aliphatic or aromatic dihydroxy compounds. They are amorphous.

Most polycarbonates are tough, stiff, hard, and transparent (high clarity), with high softening points, so can be cook in oven or sterilized. they have poor barrier properties, and cost three times as much as polypropylene. They are used for plastic tableware and fruit juice containers.

Acrylonitrile: This is an excellent gas barrier (like EVOH and PVDC). It possesses good chemical resistance and has a melting point of 170°C. It is easy to thermoform because of its high stiffness.

Acrylomitrile (AN) forms copolymers such as Lopac and Borex, which have potential as high-barrier plastics. Acrylonitrile buta-styrene (ABS) is still used, but polyacrylonitrile (PAN) has been replaced by PET.

Pliofilm: This is a rubber hydrochloride formed by combining polyisoprene (natural rubber) with hydrochloric acid. It is a printable, good-feel, opaque film with good heat sealing characteristics and grease resistance.

Pliofilm is no longer used much as it is not easy to machine and is not very durable. However it was one of the first thermoplastics. It is chemically stable, requires a plasticizer, and has poor barrier properties.

Ethylene vinyl alcohol: This film has high oxygen-barrier properties, but hydroxyl groups make it hydrophilic, which increases its permeability. Thus it must be sandwiched between materials with good water-barrier properties, such as polypropylene of low-density polyethylene, to be effective. However, its oxygen barrier properties make it a highly desirable film, competing with PVDC for this role.

Ethylene vinyl alcohol (EVOH) is more expensive than PVDC, but it is easier to process and is recyclable. The melting point of EVOH is 185°C.

Ionomers: Surlyn is the brand name of a range of Du Pont ionomer resins (invented by Rees in 1961). An ionomer resin has both ionic and covalent bonds. The ionic bonds are due to sodium or zinc cations.

Surlyn has low-temperature heat sealability (about 90°C), good hot tack (four times better than LDPE, LLDPE, and EVA), formability, toughness, and chemical inertness. Surlyn has some haze. To achieve a coefficient of friction of about 0.2 (as required by modern packaging machinery), slip agents must be added.

Surlyn is used for shrink-wrapped meat, cheese blocks, fish, individual candy wrapping, pet food bags, potato chips, snack foods, drink tetra pak cartons, margarine tubs, cookies, frozen foods, nuts, etc. as part of a laminate structure, especially as the inner heat-sealing layer.

Relative Costs

A list of relative costs per unit mass on the basis that PET costs 1 unit (based on U.S. prices in 1986) is given in Table 3.

4. Processes

Copolymers

Condensation of combinations of certain homopolymers with each other can produce complex copolymers with properties different from the individual constituents. Depending on how the individual monomers combine, a great variety of properties can result.

Let A and B be two monomers: They can combine (a) in strict sequence, e.g., A–B–A–B–A– . . . , (b) in blocks, e.g., A–A–A–A–B–B–B–B– . . . , or (c) in branching chains, with a central chain of one monomer. Some examples are ethylene-vinyl acetate (EVA, a good heat sealant but with poor barrier properties), ethyl-vinyl alcohol (EVOH, a hygrophilic plastic), vinyl chloride copolymers, and polystyrene copolymers.

Lamination

Since no single film can satisfy all packaging requirements, plastic films may be combined by lamination or coextrusion. Lamination is a technique for bonding films together to give a film with the properties of both constituents. By combining the qualities of choice from the raw material films, a laminate can be tailor-made for its particular application. Each layer in the resulting laminate may exhibit different properties from its free state, such as mutual layer reinforcement in which cracks in a brittle layer are prevented from propagating by a high elongation (elastic) layer. This effect depends on good adhesion between the layers.

Three factors affect the adhesion between layers:

1. Viscosity/shear rate match during melding. To be coextruded the melt flow viscosities should be similar (a ratio of within 3:1), otherwise one of the plastics will flow with respect to the other, preventing bonding.
2. Temperature, pressure and period of contact, to build the bond.

Table 3 Relative Costs per Unit Mass on the Basis of PET Costs

PET:1.00
HIPS:0.82
EVOH+:5.83
Nylon:2.00
HDPE:0.75
EVOH:4.00
PP:0.85
LDPE:0.70
PAN:2.17

3. Functionality of adjacent resin layers, i.e., that they are sufficiently similar in structure to mix at the contact surfaces.

If these factors are not all present, then an adhesive layer is necessary and the plastics may be cold-bonded with a tie-layer of resin adhesive. Adhesives are discussed below.

A typical triple-layer film would be composed as follows:
Properties of outside layer: high gloss, printable, good lamination, possibly metallized, high slip
Properties of middle layer: strength, stiffness, barrier properties, possibly opaque
Properties of inner layer: easy to seal (hot seal, good hot tack properties, or good cold seal properties), low migration rates, barrier properties

A laminating machine has the following components:

Continuous feed roll with a feeder "on-the-fly" splicer, which can cut off the old roll and join on the new (There may be several rolls feeding film into the machine at once.)
Tensioning rollers to give exact control over the tension in the plastic
Lamination stage where the primary and secondary web are combined. (A web refers to the film as it passes through the machinery.)
Compression rollers to push the layers together
A takeup (rewind) roller to collect the final laminate

Note that plastics can be laminated with papers and foils as well as plastics. Laminations with paper will tend to use water-based adhesives, since the solvent (water) will absorb into the paper base away from the adhesion zone, allowing the glue to set quickly. Lamination also allows reverse printing, where one plastic layer is printed (in mirror image) before lamination in order to sandwich the printing inside the laminate for greater protection.

Lamination of plastic films is achieved by one of four processes:

1. Adhesive lamination, where a continuous glue source is fed as a flat film between consecutive layers of the laminate. This may be a hot glue (a plastic resin fed from an extruder, for example) or a pressure-sensitive glue applied as a high-viscosity liquid to one web, then dried in a tunnel oven to remove the solvent, and then wrapped with the secondary web around a chill roller. The glue may physically bond (e.g., to paper or board) or chemically bond. This technique is also used for labels.
2. Extrusion lamination, where extruders replace the feel roll, and the layers are coextruded.
3. Thermal lamination, where heat is used to join the webs by partially melting.
4. Wax (or hot melt) lamination, where a thermoplastic glue or wax is used instead of a pressure-sensitive glue. The rollers must have the capacity to be heated to a suitable temperature.

Typical speeds for lamination are 100–200 m/min.

Coatings

To enhance plastic film properties such as printability, coatings are often used. Aqueous and solvent coatings are applied to the substrate through water dispersions or emulsions, solvent solutions, or waxing.

Aqueous and solvent coatings are applied to the substrate using direct roll, direct gravure, or reverse gravure methods. The coating liquid is then dried and cured by infrared heat or a hot-air oven. The coated film is then cooled before being wound onto a roll core. Modern machines allow simultaneous coating on both sides.

Examples of coating are:

1. Foil lidding: a lacquer coating is applied to the foil, which can be heat-sealed to the container but is peelable for convenient opening. On the outer side a clear nitrocellulose lacquer is applied.
2. PVDC may be coated onto paper or PP by reverse gravure. PVDC does not bond well to other plastics, so a prime coat may be applied first. The prime coat may be nitrocellulose, vinyl, acrylic, or shellac.
3. Cohesives (cold seal, latex-based) coatings are often gravure-coated onto films during printing (on the opposite face). Cold seals are pressure-sensitive, so can be sealed more quickly than hot seals, which require adequate dwell-time.
4. Colored vinyl lacquers may be applied to outer packaging surfaces, e.g., to give colored aluminum foil.
5. Paraffin waxes are used as coating bases for papers, especially for inner wraps or liners for biscuits or cereals. The wax may be blended with resins, synthetic rubbers, and polymers to meet specific requirements.

Adhesives

Adhesive layers are used to bond film layers together to construct a laminate and are also polymers. These tie layers will affect the mechanical properties of the final construction. They are usually expensive and must have lower melt temperatures than the layers they are bonding. The adhesives are mostly olefin copolymers, polyurethane, or polyester dissolved in solvents. If no solvent is used, the glue may be extruded hot between the films. Examples of a hot-melt adhesive are copolymers of ethylene and acetic acid and surlyn.

Adhesives must "wet" the film surfaces to provide laminate strength, or the laminate will tend to separate into its individual components when used. They may be categorized as liquid, solid, solution, or emulsion [1]:

1. Liquids are monomers that react with trace water to form polymers, such as cyanoacrylate ("superglue"), reactive liquids that combine chemically such as epoxy resins, and pressure-sensitive adhesives (congealed liquid resin in a rubber matrix, with the property that they can be peeled off again).
2. Solids are powder adhesives activated by the addition of moisture, solvent, or heat. Hot melts consist of a polymer, resin, wax, and stabilizers. The hot melt is supplied in solid form, applied in the hot liquid form, and then resolidifies as it cools. No solvent is required and they set very quickly.
3. The two main classes of solutions are a water-based (traditional glues with natural polymers such as starch, flour, casein, animal glue, dextrin, gum arabic, etc., and newer synthetic resins in solution such as PVA, cellulose ethers, and vinyl pyrrolidones) and (b) other solvents (natural, e.g., resin, shellac, bitumen, rubber, and gums, and synthetic, e.g., nitrocellulose, urethanes, nitrile rubbers, epoxies, and cellulose acetate).
4. Emulsions dry faster than solutions and have a greater range of properties. Synthetic resin emulsions now dominate the packaging industry. They include PVAc, acrylic resins, polychloroprenes, etc.

Heat Seals

A major requirement of plastic film is heat sealability. A heat seal is the fusion of two surfaces under the influence of heat, pressure, and time. The two surfaces are partially melted by the heat applied by a pair of heated jaws. For plastics such as polypropylene, which has a higher melting point, a copolymer coating may be applied on one or both sides to give heat sealability. Some important terms used in heat sealing are:

Dwell time—time for heat to penetrate outer films to reach the two layers being bonded, i.e., time the plastic must remain under heat and pressure.
Heat seal threshold—minimum temperature at which a heat seal threshold of 200 *g* force per 25 mm is obtained. This temperature should be about 80–120° C, i.e., a balance between too high (wrong layers melt) and too near ambient (no melting). The lower the temperature, the faster the sealing, since less heat must be lost before resolidification of the plastic.
Pressure—about 1 atmosphere pressure is applied.
Useful heat seal range—range of temperatures over which the films may be sealed. The upper limit is the point where distortion of the plastics starts to occur.
Maximum linear film speed—the maximum speed at which film can be passed through the sealing machine.
Hot tack—the strength of the initial partially molten seal is called the hot tack. If the hot tack is not good then stresses on the film before cold tack is achieved can reduce the integrity of the seal.
Cold tack—the final "cold" strength of the bond.

5. Package Types

Plastic Bags

Bags are formed from sheet or film plastic by folding and heat-sealing as required. Some bags have folds in the base so that when packed, they expand to a rectangular shape. Handles may be inserted during folding and heat-sealed into the folds. The bags must be strong enough to resist breakage under the design load, but also must not break when being loaded.

Bags may be preformed, in which case they may be "wicketed," or formed from the source plastic sheet during packaging (usually by forming tubes from the plastic). Wicketing is the process of punching small carry holds at one end of the bag with which to hold the bag during loading. The holes must be carefully designed to carry the load of the product entering the bag, yet must tear off easily so that the next bag becomes available.

Plastic Closures

A closure must perform five functions:

1. Contain, to the same level as the remainder of the package.
2. Allow access, so that the consumer can retrieve the product in a convenient way. The ability of the package to be functional in this regard is an important marketing consideration.
3. Restrict access, e.g., tamper-evident and child-resistant caps.
4. Protect the product, keeping out dirt, moisture, etc.
5. Be economic.

The closure may also be used for advertising or barcoding.

A plastic screw cap lid has three main components:

1. Thc cap itself.
2. A linear (HDPE wad), adhesively attached to the cap in most cases.
3. The screw, which interlocks with connecting lugs in the finish of the container but does not provide a good barrier seal.

The closure must be applied with the correct torque. Insufficient torque leads to leakage, whereas too much torque makes removal difficult. Tamper-evident attachments to the screw cap are commonly used with plastic beverage bottles, consisting of a ratchet ring under the cap which becomes detached when the customer removes the lid.

A dispensing closure is one that allows the product to be dispensed without removing the closure. Examples of dispensing closures are lids such as flip-tops (e.g., gable-top cartons), pump action, aerosols, and opening pourers, which allow small amounts of the product to be removed easily.

Aerosols are cans containing a liquid product layer and compressed gas propellant. The propellant may also be present as a liquid, which boils as product is used to replenish the driving pressure. The product may be dispensed as a fine spray, mist, dust, or foam. An example of an aerosol application for the food industry is in instant whipped cream. The advantages of aerosols are simple dispensing and complete exclusion of air. They are, however, expensive and an explosion risk, and since 1974 the use of CFCs has been a suspected cause of damage to the ozone layer. There is now an almost total ban on CFC propellants. In Australia CFCs have been totally replaced by trigger pump packs and inert gas pressure packs.

The advent of flexible packages (replacing rigid containers) has led to a need to develop closures for such packages. Methods include plastic zippers, pressure-sensitive tape, metal bands, plastic clips, and twist ties (e.g., breads).

Child-resistant lids have locking devices that either prevent turning unless squeezed or require pressure to open. Tamper-evident seals are discussed in more detail elsewhere.

Oven-Safe Containers

Ovenable plastic-based food trays have become an essential component of convenience foods. The three main plastics used are polypropylene (PP), polystyrene (PS), and crystallized PET (CPET). PP is suitable for microwaves but not for the higher conventional oven temperatures. Foam PS has a still lower melt temperature, but now PS low-density blends have been developed with heat deflection temperatures (HDTs) of 190°C, which is suitable for microwaves. These are cheaper than CPET. CPET is stable from –40 to 220°C, is clear, resists fats, oils, water, and O_2 on the shelf, but is more expensive. Some instant snack meals have complex laminate structures with metal FAEZO (full aperture easy opening) lids and snap on plastic overlids. Aluminum foil trays have good appearance, stack well, have good barrier properties, and can be dual-ovenable. Paperboard trays are ovenable in a range of sizes.

B. Metals

1. Canning

The most common use of metals for packaging is in tin and aluminum cans. The metal provides a highly effective barrier between the food product and the environment. Thus, the critical concepts of canning are to ensure that the product in the can is biologically stable and that the seal provided by the metal is complete. Food stability for nonpowders is usually achieved by thermal processing.

Canning was invented by Appert in the nineteenth century in response to the need to supply Napoleon's army with good-quality food. He used glass bottles, but an Englishman named Durand used metal and pottery at about the same time. The two ideas together gave us tin cans.

Tin cans are made of sheet steel coated with 0.5-mm tin. The coating is applied by tinning, which is electrolytic deposition of the tin at about 10 g/m^2, and hot-dip, which uses about 30 g/m^2. The steel is rolled and ribbed (for added strength) and either sealed with solder (usually 95% lead, 5% tin), or, more commonly, welded. The resulting tube ends are flanged, and the lids at both ends are attached by a double seam without solder. Since steel corrodes rapidly in the presence of acidic substances, the tin acts as a barrier. Some cans are lacquered internally for high-acid products (pH < 3) or for products that change color in the presence of tin. Foods that contain sulfur produce a blackening of the tin. The steel can provides almost perfect barrier protection and, due to its structural strength and ability to handle pressure, can be retorted (cooked under pressure) after sealing.

Lacquering is an important part of can manufacture. The lacquer is a resin, such as an acrylic (which resists high temperatures), alkyd, epoxy, phenolic, polybutadiene, or vinyl resin. The lacquering must be complete. Small gaps in the coating can lead to the iron being exposed. As iron corrodes it produces hydrogen gas, which can blow the can. The development of lacquers has meant that tin-free cans are possible.

In competition with traditional tin-plated steel cans, modern cans may be made from tin-free steel, aluminum, or laminates. Laminates or composite cans are often fiber-foil containers, such as helical-wound tubes, with metal ends. The fiber may be paperboard. The first layer would on will be the liner, followed by other layers until finally the printing layer is wound on. The layers are sealed together with adhesives, which generally contribute the majority of the structural strength.

The closure is traditionally the seamed lid, to be opened by the consumer with a can opener. A major development in canning has been the customers' preference for convenience over cost, so that pull tabs (now out of favor because of pollution), zip tops, pop tops, and ring pulls (where the ring remains attached to the can) have been adopted. Other examples are sardine cans (peel-back lids using a key to lever open the lid) and the recently developed full aperture easy opening (FAEZO) cans, where a ring tab is used to peel back the entire top lid. The lid is scored during manufacture to a precise metal thickness—too thick and the can is difficult to open, too thin and the package integrity is endangered.

Not all cans are retorted. The term "general line" is used for containers that are not hermetically sealed for heat processing, which account for about 16% of the tinplate market worldwide. Advantages of tinplate here are strength (impact, puncture), barrier, formability, printability, and product compatibility. They tend to be used for higher-value products, as the painted tin can look very effective.

The processes of manufacturing steel cans are described well in many textbooks and will not be covered here. Aluminum can manufacture is described below.

2. Aluminum

Introduction

Aluminum is used increasingly for canning, due to its lightness, low cost, corrosion resistance, availability, and recyclability. Aluminum is also used extensively in many noncanning applications, such as:

1. Foil packaging, e.g., chocolate, household or industrial foil
2. Bottle closures and overwraps, e.g., caps, wine bottles
3. Convenience food containers and lids, e.g., frozen-stored/oven-heated, single portion sizes, yogurt tub lids
4. Kitchenware, e.g., saucepans, cutlery
5. Special applications, e.g., shrimp freeze blocks
6. Laminates

Properties

Aluminum makes up 7.9% of the earth's crust and is attacked by acidic solutions (especially food acids—pH < 4). Special inks had to be developed to work with aluminum, due to its smooth metal surface and high reflectance. A common solution with formed containers is to put all of the print on the lid. The main properties of aluminum are lightness (three times lighter than steel), strength (alloys are as strong as steel), corrosion resistance, electrical conductivity (twice that of copper), appearance, and ease of recycling. It has the barrier properties of steel, but without the corrosion problem. It is highly attractive in appearance, as it reflects about 85% of the incident light, so stands out from other products. It can be bonded with paper (e.g., chewing gum and cigarette wrappers), allowing easier printing. It has excellent strength, so that thin films can be made. It can be extruded into complex shapes, such as roof guttering.

Manufacture

Aluminum foil became common after the electrolytic method of extracting aluminum metal from bauxite was developed independently by Hall in the United States and Heroult in France in 1883. Bauxite contains 75% hydrated alumina (Al_2O_3) in mono- and trihydrate forms, as well as oxides of iron, silicon, clay, etc.

Aluminum Foil

Foil may be used for formed or semi-rigid containers. Many instant meals are packed in cooking and eating trays made of aluminum, with different compartments commonly formed in the tray to separate the meal components, especially with frozen foods.

Aluminum foil is made of solid sheet aluminum rolled to a thickness of less than 0.15 mm and sold on cardboard rolls. For food applications it is sold at high purity (99.8%) except for formed containers, where it is strengthened (like steel) by the addition of 1–1.5% manganese. After rolling, the aluminum is work-hardened, so it is brittle and crinkly. The aluminum can be softened by slow heating and slow cooling (annealing, about 24-hr heating at 300°C, which also cooks off the processing lubricating oils, then 24-hr cooling is typical), giving a soft metal of low strength and high flexibility (ductility) suitable for household use and most other food applications—this is called zero temper. By rolling and strain hardening ("quenching" when hot), a more brittle, stronger material is produced. This is called H temper, with a number added according to the degree of hardening. Hard tempers must be used whenever a high degree of forming is required. The metal may also be normalized, a process of air-cooling heated metal that is intermediate between quenching and annealing.

Foil Laminates

Aluminum foil is difficult to use on modern fast packaging equipment because of creases, tearing, and marking effects. Thus, additional treatments are common. Lamination can be difficult for the same reasons, but once laminated the resulting plies have excellent machining and visual properties. For example, sodium silicate may be used to glue foil to vegetable parchment (cigarette foil). The foil may be printed, coated, seal applied, and laminated in a step called converting. The web may also be embossed (embossing roller), giving a textured matte appearance, which reduces glare and makes separation in refrigerated storage easier.

Rigid Foil Containers

Rigid aluminum foil containers have varying strengths but offer the ultimate in convenience for all food processing, packaging, display, and consumer requirements. They are produced as follows. The foilstock (after hot-rolling) is cold-rolled to 7 μm, the final rolling being two-ply. The plies are separated and annealed (cold-rolling causes high degrees of work hardening). The foil may then be coated. The web is then slitted to size and then die-formed using complex dies, which control the degree of drawing of the aluminum to retain uniform strength. Lids may be cardboard laminated with aluminum. They are not affected by heat, and can be heated, immersed, and frozen.

Pinhole Defects

In practice, aluminum foil of fine gauge has minute pinhole defects due to the tolerances of the rollers, crystal size, and lubricants used, which allow transmission of air and water. The aluminum foil is graded according to the number of defects as: grade 0, <200 holes/m^2; grade 1, 200–900 holes/m^2; grade 2, 900–3000 holes/m^2; and grade 3, >5000 holes/m^2. The size of the holes is also significant. For this reason aluminum is commonly bonded with polyethylene for commercial food-packaging applications, so that thicknesses of the order of 10 μm can be used. The resulting barrier properties are far superior to those of plastics and plastic laminates.

Aluminum Tubing

Aluminum is also used for squeezable tubes (e.g., toothpaste or tomato paste tubes). However, this use is becoming less common due to the following problems:

1. Neck finishing is expensive.
2. Inks tend to crack and peel off after squeezing.
3. Some products are affected by aluminum, so a lining may be necessary.
4. Plastic laminates cost about 20% less.
5. Aluminum tubes are more subject to contamination.

Plastic laminates (usually including a foil layer) are increasingly being substituted for products such as sauces, peanut butter, cheese, etc. The main problem with aluminum tubing has been the integrity of the side seam.

Aluminum Cans

Cans are made by cutting a blank (a disk of aluminum) from coiled sheet, the skeletal web being recycled, and then drawing the blank into a cup. The walls may then be ironed by forcing the cup through a series of annular rings (dies) until the cup has the required height [drawn and wall-ironed (DWI) process], with a bottoming die forming a raised dome in the base. Alternatively, the cup may be redrawn, giving a thicker wall [drawn and redrawn (DRD) process]. The thinner walled (DWI) cans are suitable for carbonated beverages, while DRD cans are suitable for steam sterilization and retorting. The body shell is trimmed to length, chemically washed, and then given a chemical etch primer (chromate phosphate) so that later coatings will stick. The lid (two-piece design) is then added.

The can may have a reduced neck diameter for improved appearance, better stacking, and saving metal. Ring-pull tabs or FAEZO openings may be used. Printing may be done before drawing the can (i.e., on the blank), the ink design stretching with the can, or may be applied by offset printing to the final can. Lacquers (e.g., vinyl or epoxy) may be applied internally for acid products to prevent interaction between the product and the can and externally for protecting the ink and providing the right slip properties on the base. The final can has a base thickness of about 0.020 inch and a wall thickness of 0.0065 inch.

Current Research

Some areas of current research are dent recovery (using laminates), foil machinability, pinhole defects, closure opening force, and recycling (Australia has the highest recovery rate in the world). Apart from cans, aluminum recycling is difficult due to lamination with other materials, food and moisture contaminants, and the low value and volumes of material, so the trend is to reduce the amount of aluminum required.

Foil food containers are difficult to microwave due to arcing and heat energy reflectance. However, there is some interest in developing aluminum food trays for microwave use. All packaging materials either transmit (glass, plastic, paperboard), absorb (susceptors such as metallized polyester laminated to paper, useful for browning and crisping), or reflect (metals). Thus, the tray must be open to the microwave energy at the top, so that the food cooks more slowly but more evenly.

C. Glass

1. History

Glass was first manufactured by humans thousands of years ago, possibly as an offshoot of pottery as glazes, and dates to 12,000 B.C. Pressing glass in molds to form cups or bowls dates from 1200 B.C. and blowpiping was invented by the Phoenicians in 300 B.C. In the third century A.D. clear glass was discovered, for example, cast glass using flat stones, used for church windows. Until the industrial revolution, glass was mainly used for high-quality tableware.

With mass production, however, glass started to become ubiquitous, first through the cork-sealed narrow-necked bottle, then from about 1850 on the wide-necked jar and from 1920 the screw top jar. The glass bottle is an almost ideal form of packaging for a large variety of products. It is inert to most substances, the product is visible, the cylindrical shape is good for loading, stacking, and holding, and it is cheap to manufacture and versatile in design.

The glass bottle is used for milk, jams, soft drinks, wines, beer, and spirits and for many food products. It is highly inert, shows the product well, is available, easy to mold, and cheap, has almost perfect barrier properties (including barriers to odors), and is recyclable. However, it is brittle, and some product loss will occur through breakage. Because it is fragile, high weights are required per product unit, and for a while research was directed at reducing the high weight ratios by coatings (e.g., surlyn), which allow the glass to be handled at much higher packaging speeds. The coating reduced breakage, but this research ended when the PET bottle became available. More recently, environmental considerations have revived the idea. Over 75 billion glass containers are used annually by the food industries.

2. Glass Manufacture

Glass is the result of heating silica, soda ash, and limestone to over 1500°C, with the small addition of minerals for color or strength. As the mixture melts, the compounds fuse and become easy to shape. This may be done by sucking the melt into heat-resistant molds or by blowing semi-molten glass into rough shape in a mold and then pressing this into a second mold where a jet of compressed air forces the glass into the final shape. Crystallization is prevented by cooling the final product quickly, so that the final product is amorphous and thus transparent. Annealing is a process of reheating the glass, and then gradually cooling to remove stresses (also used for metals). Safety glass is laminated and toughened.

Special glasses include Pyrex, produced by the addition of borosilicate and having resistance to thermal shock, amber glass, used to inhibit ultraviolet radiation for beer bottles, and crystal with added lead (Note that wine can leach lead from glass!). Glass may be corroded by application of hot concentrated alkali. Leaching tests for lead, cadmium, arsenic, and zinc are conducted on glass with high contents of these minerals, but in most cases the fusion process of glass production prevents traceable amounts of these elements from escaping. In general, glasses are not retortable due to thermal shock and the expense of Pyrex glasses, but if retorted must be cooled under pressure to prevent thermal shock.

3. Glass Containers

The main components of the container are the cylindrical main part, the bottom, the neck (called the finish), the closure (the screw cap), and the label. The cylinder shape is chosen as maximizing strength for a given volume (the sphere is a better shape, but not convenient for packaging). Glass is not well suited for sharp corners.

The main components of the cap are a lacquer, wad, liner, and cover. Caps may be plastic or metal, and the type of closure might be thread, lug, friction, snap-cap, roll-on, cork, crown, twist-off, etc.

The various types of glass containers have a range of names:

1. Bottles (most used—round, narrow neck to facilitate pouring and closure, for liquids and powders.
2. Jars (wide-mouthed bottles)—neckless, allowing fingers or utensils to be easily inserted. Used for liquids, solids, nonpourable liquids such as sauces, jellies, and pastes.
3. Tumblers (open-ended jars)—shaped like drinking glasses. Used for jams, condiments, jellies.
4. Jugs (bottles with carrying handles)—short, narrow necks designed for pouring.
5. Carboys (shipping containers), shaped like short-necked bottles, usually used with a wooden crate holder.
6. Vials and ampoules (small glass containers), occasionally used for spices etc., but mainly used by the pharmaceutical industry.

The main uses of glass for packaging are in milk bottles, condiments, baby foods, instant coffee, and drinks. Glass is not used for frozen products or for ground or roasted coffee because of breakage costs and the difficulty of vacuum flushing.

D. Timber, Cardboard, and Papers

1. Timber

Wood is commonly used in box construction, but the use of wood for individual packaging (such as cigars) has decreased since the advent of plastics. Examples of timber for packaging are cases, boxes, and casks for long-distance transport.

2. Cardboard

The next choice of packaging material to be considered is cardboard. This may be a protective package (see previous section) or a presentation package. Folding carton construction consists of taking a two-dimensional flat piece of board (excellent for storage) and cutting, scoring, folding, and then gluing (or locking) it into a three-dimensional rigid box. The cardboard will usually be laminated to paper to allow printing and presentation.

For transport purposes, the fiberboard must resist relative humidity and temperature effects. Relative humidity affects the moisture content of the board (which is hygroscopic), raising it from 6–7% safe moisture (at manufacture) to 14%, which point the board becomes like a piece of rag. Temperature strongly influences the rates of diffusion of gases and moisture into the package. Thus the transport requirements will depend on ambient conditions. The board should be designed for the optimum lifetime of the projected job.

The various elements of cuts, tucks, locks, flaps, and folds may be assembled in an endless variety of ways, although the most common is rectangular for packing and storage convenience. Since cardboard is obviously highly versatile, the details for construction vary widely. The actual specifications may include (a) grade of board, ink type, glue type (e.g., hot melt glue or water-based), etc., (b) performance criteria, such as handling strength and crushing strength, and (c) size.

One of the major uses of cardboard is as corrugated cardboard, a concept developed by Albert Jones in 1871, adopting the method for making ruffles from collars and paper for sweatbands in tall hats. Jones hand-cranked paper, and in 1874 Oliver Long patented gluing paper to both sides of fluted paper. Corrugated cardboard consists of two liner boards covering a central corrugated sheet. The liner boards are made of kraft, test liner, or low-grade pulp covered with kraft. The corrugated paper is made of straw paper or kraft paper. The main factor determining the board properties are the corrugations. A useful modification is the double- and triple-wall container. The board is usually printed at the time of manufacture.

Choosing a carton for a specific job depends on the capacity of the carton to meet the requirements for that job. There is a trend to replace subjective tests (e.g., cartons must run on certain packaging machines) with more scientific objective tests (e.g., compression strength). The choice of carton for a specific job will depend on:

Carton load, both the internal weight of the product and the external load applied.
Warehousing conditions, such as stack heights, ambient warehouse conditions.
Storage life.
Type of handling, e.g., fork, manual, palletizer.
In-use conditions (especially RH).
Item size, determining critical dimensions of the carton.
Maximizing pallet efficiency by using the available space.
Size and style: accuracy of the dimensions, which affects packaging machine performance (e.g., 1/64 inch), prefolds, print areas, etc.
Protective properties: different cartons protect against different agents, e.g., moisture or odor. These vapors can enter/leave through the cardboard itself, through the creases, through the glue seals or through the gaps between folds. Moisture vapor protection (MVP) is generally achieved through waxing the cardboard and is measured in terms of the resulting water vapor transfer rate (WVTR) measured in $g/m^2/day$.

A waxed board is difficult to print on, so the board may be laminated with white paper.

Glue

The glue must be chosen that will seal within the time the box is in the packaging machine. A common glue is dextrin, which is water-based, the water being absorbed into the cardboard as the glue sets. Hot glues may also be used.

Inks

The inks used must be of the specified hue (and reproducible), resist fading, and resist rubbing.

Opening Cartons

For ease of use, some quick reliable means of opening the carton is usually built in. Examples are perforated thumbnail openings, fold-and-tear openings designed to assist pouring, or designated areas for cutting.

Board Strength

The board must be strong enough for packaging, handling, storage, and intended use. It creates a bad impression with the user if the pack bulges, so bulge strength is important. The board must also resist compression and a degree of impact. The strength of the cardboard chosen for the protective package is related to the strength of the product package, so that if a weak carton is used, a strong external box is necessary for delivery to the point of sale.

Note that the air moisture content (relative humidity) has a large effect on the strength of cardboard. Other problems with cartons include pallet integrity, ropes denting boxes, weak cartons (bottoms fall out), pallet stack collapse (compression), overweight cartons, and forklift damage.

3. Papers

History and Manufacture

Paper bags were used in the seventeenth century. A bag-making machine was developed in 1852 by Wolle in the United States. The gusseted bag (1873) and multiwalled bag (1925) were later important developments.

Paper is defined as sheets of material thinner than 0.23 mm and lighter than 220 g/m^2. Paper and board are produced from wood pulp (treated with calcium bisulfite or caustic soda to break down the lignin structure), rags, and other waste. Paper is decomposed by bacterial action over a period of time. Thus, paper is ultimately environmentally friendly. However, paper has had tough competition as a packaging material over recent years due to extensive use of plastics. Treatments of paper to make it more competitive include paraffin and waxes (waterproofing) and plastic coating (added strength, water and gas resistance. Paper can also be laminated with aluminum.

Paper is produced from wood pulp, treated by the addition of soda, calcium sulfite, or calcium sulfate (depending on the end-use). The pulp is milled into a continuous sheet and bleached (chlorine, caustic soda, and sodium hypochlorite). After drying it may be treated with various chemical coats to enhance its performance.

Paper is designated by the weight of a ream of paper of given size. Usually a ream is 500 pages, but variation in size makes it difficult to directly compare two papers. For printing, the standard size is 24 × 36 inches (6 ft^2 or 0.55742 m^2). Some of the standard papers used are:

1. Bond papers (17″ × 22″): soda pulp, uncoated bleached, finished to give wet strength and a good printing surface.
2. Tissues (24″ × 36″): lightweight semi-bleached or bleached, finished to give wet strength, with open or closed fiber formation.
3. Litho papers (25″ × 38″): smooth printing surface but not as strong as bond paper, used in magazines.
4. Kraft papers (24″ × 36″): unbleached equivalent to bond paper, but of heavier basis weight (and hence greater strength), and cheaper.

Other special types of paper include glassine, greaseproof paper (glassine paper that has not been calendared and is free of wood pulp, water resistant, and heavily milled), vegetable parchment (boil-proof and fat impervious, due to treatment with sulfuric acid), and waxed papers. This paper is not moisture-proof, so may be waxed or laminated.

Glassine is used extensively because of its inherent resistance to grease, oils, and fats and is the densest paper made. It is made from straw, which is pulped and purified, then hydrated at high temperature until it partly gelatinizes. The resulting sheet is fed into a calendar, where it is rolled under high-pressure steam to give a transparent paper. It can be laminated. It is used for dry products such as cereals and biscuits after being waxed (paraffin).

Paper laminations commonly used include paper/aluminum (for strength and excellent resistance to moisture and air) and paper/plastic (good for heat sealing as the plastic can bond across the seal, but also good for writing on).

Regenerated cellulose is cellulose precipitated out of solution. Cellophane is clear cellulose regenerated from a viscose solution.

Paperboard

Waste paper can be used to produce board for cartons. Some types are:

1. Chip board: waste paper blended with wood pulp to give a flexible grey board (not suited for printing).
2. Manila-lined board: a top liner of ground wood pulp covering newspaper or other waste paper pulp. This can be printed on.
3. Clay-coated board: same as manila-lined, except that the top-liner is coated with white mineral powder bonded to the surface. This important innovation, coupled with fast electrostatic printing, has allowed the carton to be an attractive way of presenting goods.

Waste paper may also be turned into paper pulp for molding, e.g., in egg cartons, molded trays, and vegetable holders. Recycled paper is weak and discolored, making it poor for packaging. It must be used in conjunction with virgin paper. Recycled cardboard does not crease accurately, so boxes cannot be erected as accurately or quickly.

E. Ceramics

The term ceramic describes any nonmetal nonorganic material produced by high temperatures, such as glass and pottery. The raw material is molded into the required shape and then fired. Once fired the material cannot be easily modified, as it is brittle and inert. If the material has been applied in a thin coat to another substance before firing, it is called a glaze. The most common use of ceramics in the food industry is, of course, pottery.

The chemical composition of most ceramics is silica (SiO_2), alumina (Al_2O_3), and water. Glass is almost pure silica, whereas clays have large amounts of alumina present. The main chemical structure of the fired product is the tetrahedral SiO_4 complex, although other stable structures like this may be present. Most clays are reddish brown, due to the presence of iron, the exception being kaolin (China clay).

During firing the clay shrinks as water is removed. Thus control of the amount of water in the clay is important, and an initial drying stage is necessary to remove unbound moisture before firing. The firing temperature is also critical. For pottery it should be above 1000°C. This is well below the melting point of the clay, but is high enough to cause the clay structure to break down into Al_2O_3 and SiO_2 molecules, which then react exothermally (sintering). A glass results from complete melting and then cooling.

Vitreous enamel is a finish applied to metals. This is done by coating the metal in powdered glass, and then heating above the melting point of the glass. Legislation nowadays prevents the use of cadmium or lead in vitreous glazes.

III. FOOD - PACKAGING INTERACTION

A. Monomers and Health

Several monomers have been linked with health problems, the most significant of these being vinyl chloride. Table 4 reviews the history of growing concern about monomers.

An epidemiological link was shown between exposure to vinyl chloride monomer (VCM) and angiosarcoma, although this was not proven (see history records above). Eighteen months after establishing this link, the U.S. Food and Drug Administration (FDA) published a proposal to regulate VCM in food packaging and food-related plastics. FDA believes that film, caps, gaskets, can coatings, and adhesives are safe, but rigid and semi-rigid PVC (bottles and sheets) may allow migration.

The FDA proposal is still not law, but residual levels of VCM have been reduced to 1–2 ppb (Ethyl Corporation). Analytical techniques for detecting VCM have improved. In 1982, the maximum safe level for VCM was set at 1 ppm. Leaching of PVC into beer led to a ban on PVC in beer. Detection of VCM is performed by gas chromatography using a flame ionization detector, infrared absorption, and/or mass spectrometer. The National Health and Medical Research Council (NHMRC) standard in Australia is <50 ppb VCM for utensils, <10 ppb in film, and 0 ppb in foods.

Symptoms of VCM poisoning are now well documented. If VCM is present in the air at a greater concentration than 500 ppm, then poisoning occurs. The following symptoms may occur:

Sensitivity to cold
Reduced circulation to extremities
Raynaud's disease
Paresthesia (pins and needles)
Mutagenic effects
Visual disturbance
Bone resorption, digits
Scleroderma
Liver fibrosis
Death (if >120,000 ppm)

Other cancers apart from liver have been reported (brain, lung, and skin).

The mechanism involved is believed to be as follows: VCM is metabolized in the cytochrome P450 component of cells. If excess VCM is inhaled, the capacity of the liver is overwhelmed, and metabolism to carcinogenic intermediates occurs, causing tumors.

Table 4 History of Growing Concern About Monomers

Date	Item
1949	Russian study: 30% of PVC workers had liver disease.
1965	Cleaners of PVC kettles (used for PVC polymerization) developed Acro-osteolysis and Raynaud's disease. Exposure levels estimated at 500 ppm
1971	Studies in Italy: animals developed tumors at 30,000 ppm
1972	Angiosarcomas at 250 ppm were observed in rats
1974	At the B.F. Goodrich Chemical Co., PVC plant, Louisville, Kentucky, a doctor (Dr. Creech) noted that 4 workers died of a rare form of liver cancer after 20 years of exposure to levels of 1–21 ppm
	Other reports were then uncovered from other factors; with 17 cases from 1 plant
1982	Maximum safe levels for VCM residues set to 1 ppm

VCM = Vinyl chloride monomer (sometimes simply abbreviated as VC).

For these reasons, tight safety precautions have been implemented in industry, especially during PVC manufacture, PVC is inherently unstable, and ventilation is required. Air filtration is required to remove dust from the processing environment. Safety gear (which covers the whole body) is required when cleaning or servicing. Gas masks are required for operators, no welding torches are allowed in PVC kettles, and CO_2 extinguishers must be available. Portable detectors can be worn, sensitive to 1.40 ppm. Monitoring for atmospheric VCM is required on site and over a 2 km radius by air samples. Strippers for air and water emissions are required at all plants.

PVC also has been linked to other problems. Combustion of PVC (e.g., shrink film) will produce HCl, CO, CO_2, and 75 other known decomposition products, some of which are toxic and others of which are carcinogenic. These come from breakdown and reaction of the inert PVC macromolecules.

Other monomers may also be carcinogenic. Acrylonitrile monomer (AN) may also prove to be carcinogenic. Vinylidene chloride monomer (VdCM) is currently being checked (tumors in rats are observed at high concentrations). The Food and Agriculture Organization of the United Nations (FAO) considers VC, AN, styrene, and di(2-ethylhexyl) philiatate to likely carcinogens.

B. Migration

1. Purity of Plastics

Since polymers are inert, they do not react or combine with a food. Even if swallowed they will not react in the body. But in many cases other compounds are produced by polymerization reaction or are deliberately added to the plastic to add certain properties. Examples of such substances are given in Table 5.

2. Migration

Migration in plastics packaging refers to the transfer of compounds from the plastic to the food product. This might be by leaching or diffusion. Remember that plastics are never 100% pure, and the

Table 5 Other Ingredients of Plastics

A. Polymerization by-products	Source
Monomers	Unreacted residues
Low molecular weight polymers	From incomplete reaction
Catalysts	To increase polymerization rate
Solvents, emulsifiers, and wetting agents	Used in facilitating polymerization
Impurities	From source materials
B. Deliberate additives	Purpose
Antioxidants	To prevent plastic deterioration during processing
Antiblock	To roughen plastic surface
Heat and light stabilizers	Especially for UV protection
Plasticizers	To make polymer more flexible, e.g., butadiene for polystyrene
Pigments	
Lubricants	So that plastic slides across processing equipment
Impact improvers	To make plastic less brittle
Flame retardant	To slow plastic combustion
Brighteners	To improve hue
Bacteriocidal agents	To prevent bacterial damage
Antistatic agents	To prevent static electricity effects during processing
Surface-deposited metals	For greater barrier properties

nonplastic components are of great variety. The main components that have caused problems are amides (slip agents), heat-degradation products from the polymer base, and ink components. Particularly slip, block, and antifogging agents are originally chosen for their ability to migrate to the surface of the plastic, because that is where they are effective. Therefore, they must be biologically inert.

Migration may also occur from the food to the plastic, in some cases resulting in plasticizing of the package if the vapors are water or certain solvents. This can result in loss of mechanical strength. The food may lose valuable volatiles, such as odors, CO_2, water, or flavors. For example, a fruit juice in polyethylene will lose limonene to the plastic (scalping) and increase ascorbic acid degradation. Thus aseptically packaged orange juice has a shelf life of about 6 months in plastic [2].

The rate of migration for a particular food product/packaging combination is affected by the concentration of mobile impurities, the surface area of the package, and the temperature. Much legislation now exists specifying tolerances for the allowable amounts of migration. Therefore it is necessary to test the extractibility of compounds from plastics. Some of the problems with testing are:

1. The compounds may not be known.
2. Only small amounts are transferred, so analytical measurement is difficult.
3. Transfer is slow, thus long transfer periods are required so that a measurable amount is transferred. However, many food products are only stable for short periods of time.
4. The range of different possible test conditions is very large, with factors such as sunlight, temperature, and ambient humidity all potentially affecting extractibility.

3. Leaching

Direct contact between plastic and a food product can result in components of the packaging being leached out into the product, changing the flavors of the food. The flavors imparted by the plastic are affected by (a) residual volatiles from the polymer manufacture or from coextrusion, (b) processing and heat sealing temperatures, (c) solvents (ink, plasticizers, lubricants, etc.), and (d) adhesives. Examples of reported off-flavors include (a) styrene flavor in sour cream, (b) catty flavor in pork from a ketone solvent in can lacquer, (c) "unpleasant flavor" in a fruit soft drink from the layer adhesive in a laminate, (d) turpentine-like off-flavor in cola from EVA closure liner, (e) waxy and burnt odors from overheated PE, and (f) residual catalysts and antioxidant additives. These affect the quality as perceived by the consumer. Food processors, plastics converters, and consumers perceive such effects differently. But standard tests and nomenclature are being steadily introduced. Some methods used to determine if there is a problem are discrimination tests (triangle, paired samples), intensity of off-odors/flavors, and descriptions of the types of off-flavors. Various chemicals and names have been tested for their effectiveness in characterizing the ranges of odors, such as "glue," "solvent," "turpentine," "bitter," "musty," "astringent," "musty," "sanitizer," "vinyl-like," "fruity," and "chlorine." As recycled material is increasingly added to film manufacture, this problem is increasing [6].

4. Global Migration

Instead of testing for each possible migratory compound, a test for the total migration of all compounds from a packaging material is usually used. A sample of the material is placed in contact with the food product under controlled conditions. At the end of the test, the total weight of substance leached from the plastic is measured (by evaporation of the simulant). For fats, since extraction of the leached compounds from the simulant is more difficult, the weight of the plastic before and after the experiment is determined. The difficulty with this method is that some of the oil is absorbed by the plastic, and the amount of oil absorbed has to be estimated first. The result is expressed in weight per unit surface area (e.g., mg/m^2).

5. Mechanisms for Migration

Flexible forms of packaging generally allow some degree of diffusion of gas or moisture from the surrounding air. In papers, diffusion occurs through pores in the matrix paper structure. In metal foils, diffusion is entirely through pinholes in the foil. For plastics the diffusion mechanism is called activated diffusion or solution diffusion and occurs in three stages: the gas dissolves into the plastic on one side, diffuses through the film, and then evaporates from the other side. This process is very dependent on temperature and gas concentration. Diffusion is modeled by Fick's laws. This concentration may be measured on a mass or molar basis. Other driving forces include thermal and pressure gradients. Within a product, capillary flow, liquid diffusion, or vapor diffusion are the main transport mechanisms.

The transport mechanism for plastics is called activated vapor diffusion. In activated vapor diffusion there are three stages: adsorption, diffusion, and desorption. Adsorption occurs at the surfaces, diffusion describes its transport through the plastic structure, and desorption is release from the plastic. Pore effects tend to apply mainly to thin films, and porous flow decreases rapidly as film thickness increases.

An assumption behind activated vapor diffusion is that the substance permeating does not react in any way with the plastic. Exceptions to which diffusion does not apply include water in cellophane, solvents, and liquids—these cases are called anomalous diffusion.

The rate of diffusion is especially affected by the cavity size within the plastic. For example, increased crystallinity and increased cross-links reduce pore size, whereas long-chain nonionic plastics have increased pore size. Highly polar polymers (hydroxyl, halogens) have excellent gas barriers but poor water barrier properties. The permeances for different materials are given in Table 6.

A common unit for expressing permeability is the barrer (adopted by ASTM): 1 barrer = 10^{-11} (ml at STP) cm / (cm^2 sec cm Hg), where 1 ml = 1 cm^3. Film thickness is often measured in gauge, defined as 1 gauge = 10^{-5} inch, so 100 gauge = 10^{-3} inch = one thousandth of an inch = one mil.

Permeability is affected by the following factors:

- The degree of intermolecular bonding (as above).
- Free volume between chains (void volume, orientation, and chain stiffness).
- The solubility of the permeating gas in the film matrix (dependent on moisture and temperature).
- Temperature, which tends to increase the mean distance between chains, also increasing gas permeability in most cases.
- The type of gas, for which rate of permeation is dependent on molecular size. For example, oxygen diffuses 5 times faster than nitrogen, and carbon dioxide 25 times faster than nitrogen, on average. For most materials permeability will decrease with decreasing temperature.
- Damage to the film, for example, creasing the plastic can also increase permeability.
- Presence of water for polar compounds. Water has a large effect because it is polar, so polymers with hydrogen bonding will attract water molecules at the expense of other bonds. As the forces holding the chains are weakened, the permeability will again increase. Water is in effect acting as a plasticizer. Hydrophobic polymers are not affected by the amount of water.
- Chemical inertness, the previous point about water being a special case.
- Glass transition temperature.

TABLE 6 Permeances (kg $H_2O/m^2 \cdot hr \cdot mmHg \times 10^{-7}$)

Paperboard	3333
Polypropylene	137
Cellophane and polyethylene	102
Polyethylene terephthalate	50
PET/PE laminate	19
Polyester/Foil/PE	1

Therefore, barrier properties can be increased by increasing the density and crystallinity of the material. The cohesive energy density is a measure of the amount of energy forcing the chains together, due to the polarity of the chains. The greater the cohesive energy, the less will be the void space and the lower will be the gas permeability.

Calculating permeation rates for laminates is based on steady-state conditions where constant amounts of material must flow through each film. This is analogous to heat flow or electrical current. The individual resistances are x/P_m, where x = film thickness.

C. Measurement Methods

1. Food Simulants

The food in contact with the plastic has some effect on the rate of extraction, and therefore the composition of the food itself must be considered. Because of the enormous range of constituents of food, an effective way of grouping foods for migration studies was chosen. The five groups are dry solids, neutral liquids, acidic liquids, alcoholic liquids, and fatty foods. To study the different groups, a representative food for each group was chosen, called a food simulant. Extraction data for dry powders are not required, but the simulants for the remaining groups are distilled water (or sodium bicarbonate solution for high pH foods), dilute acetic acid, ethanol/water mixture, and fatty food simulants. The choice of fatty food simulant (the group most affected by migration) is difficult, and even small traces of oil (e.g., 0.1–0.2% in orange juice), can have large effects.

For fats, several studies were conducted of the migration of radioactive additives from plastics into various oil-rich compounds (sunflower oil, tricaprylin, paraffin, and *n*-heptane). None of the proposed simulants were universally effective for the different plastics (PVC and HDPE) and different additives tested, and in fact there was a great range of migration results for each. For example, tricaprylin caused swelling of PVC and high extraction rates. In general pure simple organic solvents tended to dissolve the plastic, so were not suitable simulants, and edible oils showed too great a range in results. As a result, Figge and his team developed a synthetic triglyceride as a fat simulant, which gave similar results to fatty foods.

Temperature has a marked effect on migration rates. By running tests at elevated temperatures, it is possible to study migration effects at ambient temperatures in a shorter period of time. Standardized temperatures and times are specified in many tests to allow accelerated migration testing.

2. Measurement of Permeabilities

There are several standard methods for measuring permeability:

1. Pressure increase (ASTM D 1434): the film, supported on filter paper, is subject to a vacuum and the pressure in the cell monitored by manometer.
2. Concentration increase (ASTM D 3985—isostatic method): the test cell contains inert gas, into which the test gas diffuses. The concentration is measured by chemical analysis, GC, special probes, etc. This allows different RHs to be tested. An example is the Ox-Tran.
3. Volume increase: change in volume due to gas permeation is measured. This method is not often used.
4. Detector film: a plastic detector film impregnated with a chemical that reacts with the permeating gas by changing color is layered onto the test film and its color monitored spectrophotometrically.

An interesting example of the last method is one developed by Rooney and Holland for oxygen measurement. Certain organic dyes (e.g., methylene blue, endoperoxide, dimethylanthracene) react with intense light to become "activated." In turn these dyes react with oxygen diffusing through the film to raise oxygen atoms from their ground state to a more energetic singlet state, where they remain for microseconds before decaying. The singlet state is highly reactive and reacts with a suitable acceptor material such as rubrene. Activated dyes remain energetic for months if stored in the dark, so the detector film (cellulose acetate) can be prepared beforehand and laminated to the test film as required. The laminate is mounted in a test cell with a limited headspace, and after an initial period of scavenging the O_2 in the head space, the film reacts only with O_2 traveling through the test film.

3. Measurement of Water Vapor Transmission Rates

The standard method (the gravimetric method) is to place a desiccant into an aluminum dish, seal it with test film and wax, and place it into a controlled environment, either temperature (25°C and 75% RH) or tropical (38°C and 90% RH) [see ASTM D 895 and 1251 (cycling)]. Modern methods include IR detection of trace amounts. Film detectors ($CoCl_2$) are also used—pink when saturated and blue when dry.

4. Shelf Life

Shelf life may be measured in order to (a) determine the shelf life of existing products, (b) study specific factors (materials, temperature), or (c) determine the shelf life of a new product. The main methods used are (a) literature (similar products), (b) shelf-turnover time (gives estimate of required shelf-life), (c) randomly purchasing samples and testing quality, and (d) accelerated shelf-life testing (ASLT). The criteria may be organoleptic or microbial.

5. High-Barrier Plastics

There is still not great demand for high-barrier plastics because of the high raw material costs, reduced production speeds due to lamination, relatively small market size, uncertain shelf life, and reliability of the seal. The barrier properties of any plastic can be increased simply by increasing the thickness, but then processing speed and cost are affected. To overcome the high material costs, thin films of gas-resistant plastics are usually sandwiched between thick films of moisture-resistant plastic.

For barrier packaging, the major plastic films are EVOH, PVDC, MXD6, AN copolymer, amorphous nylon, PET, and PVC. All have good oxygen barrier properties compared with PE, PP, and PS, which have better moisture barrier properties. Thus thin layers of high-barrier materials are frequently coextruded with thick layers of PE or PS. After all of this, though, a weak seal, poor join, or absorbent coating can ruin the efforts to provide barrier protection.

IV. PACKAGING EQUIPMENT

A. Standard Package Tests

The main hazards for a package involve transportation, storage, and seal failure. Several tests have been devised to standardize testing of package materials. The different levels of package for a single product from product package to outer protective layers (transport package) may be tested individually or in conjunction. From an economics point of view, the objective is to just achieve the required strength or barrier properties and then to maintain this optimum throughout the lifetime of the package. Some Australian and American tests are described below.

1. Water Vapor Permeability Test

The American standard D1251 refers to the water vapor permeability of packages by a cyclic method. Each cycle consists of exposure of the packaged sample to an atmosphere at −17.8°C for 1 day, then at 37.8°C and 90% relative humidity for 6 days. A desiccant is sealed within the package, and the package sealed in the normal manner with the normal packaging equipment. The mass of the package is measured at the start of each cycle, and the procedure is continued until a constant rate of moisture uptake is achieved.

The Australian Standard D895-79 refers to water vapor permeability of finished packages when closed and sealed, provided the outer body wall is not hygroscopic (like cardboard). A desiccant is sealed into the package and the package placed in a temperature (37.8°C) and relative humidity (90%)

controlled environment and weighed at intervals from 3 days to bimonthly. Plots of weight against time for the constant period of uptake are used to calculate the WVTR.

The two standards test different characteristics. The first tests migration under varying conditions so that any leakage of the package will cause the package to "breathe," exchanging air with the outside environment as the air density changes. The second test, for static conditions, is designed to test the steady-state flow of moisture through the package.

2. Drop Tests

Protective cartons have their sides numbered in a specific way, based on the position of the manufacturer's join. When the box is cut prior to folding, all sides will be joined except one pair, which is called the manufacturer's join. Holding a carton with this join on your right-hand side, the six faces are labeled as:

Face 1—Top
Face 2—Right side
Face 3—Bottom
Face 4—Left side
Face 5—Front (facing you)
Face 6—Back (facing away from you)

The edges are numbered according to the adjoining faces (e.g., 12, 36), and the corners similarly (corner 125, 346). Drop testers are devices that hold the carton in a known orientation and height before dropping onto concrete or steel plate. A typical test would include drops on the bottom corner, bottom edge, flat bottom, flat end, and vertical edge for a filled and empty carton (10 tests). Damage is specified in terms of the six faces as distances from edges and corners.

3. Vibration Testers

A vibration machine is a table containing the carton fitted to oscillating cams that vibrate the table at a known frequency. This simulates the effect of transporting by truck. Damage to the carton is assessed in the same manner as in the previous test.

4. Incline-Impact Tester

In this test the package is placed in tandem on a dolly (small-wheeled table) and slid down a 10° incline of varying distance to hit a solid wall.

5. Compression Tests

The most important compression test is the vertical stacking test, where a weight equivalent to several layers of the filled box is placed on a carton for 30 days. If the box compresses more than 2 inches or the side walls bulge more than 3/4 inch, the package is failed. An alternative form is to add increasing weight (using an Instron, for example) until the package fails.

6. Summary

In general packaging tests should be chosen that are appropriate to the transport and storage requirements of the carton. The tests must be well designed so that it is not possible to criticize the technique used at a later stage, for example, once the packages start collapsing in the warehouse, spilling product all over the floor.

B. Printing and Design

Printing design and appearance of the final package are essentials to package development. The product itself will partially define the form of package, for example, the package barrier requirements; this is called the basic package. The steps in developing a good design after developing the basic package are to:

1. Define the product: obtain information on the product being sold; size, weight, intended market, ingredients, etc.
2. Define the market: determine who the expected or intended consumer is, for example, demographic and cultural detail. This will affect the graphics, which is probably the main product advertisement, but it may also affect other aspects of design.
3. Check the basic package design: Is it compatible with the intended market? If not, redesign the basic package.
4. Study the competition: obtain samples of the competition product and study the packaging, looking especially at the product image they convey.
5. Consider engineering aspects, such as detailed construction drawings, packaging machine limitations.
6. Design the printing. This is a large and diverse area, which will be considered in more detail in the following sections.

Note that very few new products succeed. Eighty percent of new product ideas are scrapped, 14% more fail the initial consumer tests, and more fail in test marketing. For the product to succeed, the packaging must be effective at presenting the product to consumers and challenging them to change their buying patterns. Although consumers may have heard of the product through media advertising, the package itself is the most important factor in their decision to buy. On average, a consumer spends about 4 seconds per item deciding between brand names. The supermarket manager will give preference to established products that sell well, so the new item must fight for shelf space. More than 200 new products are offered to a supermarket each week on America. The lifetime of a new product is also short. A new package or product remains innovative for less than a year.

1. Printing

There are three main forms of type: Roman, Gothic, and Italic. All other forms are basically variations on these three types. There are about 100 commonly used different types.

Roman type is characterized by serifs that are fine horizontal lines attached to the letters (for example, what you are reading now). Gothic type is sans-serif (without serifs) and is simple and bold. Italic type is used for emphasis and is inclined.

Type size is specified either in points or by some common name. A point is about 1/72 of an inch (0.05448 mm). A type width of 12 points thus gives about 6 letters to the inch. Some common type sizes are:

6 point (nonpareil)—very small type
8 points (brevier)—small type
12 points (pica)—normal letter type size

The sequence of point sizes commonly available is: 3, 4, 5, 6, 7, 8, 9, 10, 11, 12, 14, 16, 18, 24, 30, 36, . . . 144. The weight of a typeface is defined as lightface, standard, boldface, and extrabold.

The design may be produced by skilled draftsmen, but increasingly is produced by CAD (computer-aided design).

2. Types of Printing Equipment

After the final design has been approved, a high-resolution negative photographic image is produced (typically of the order of 1000 lines per inch). For multicolor labels, a negative image of each principal color used is produced. Light shining through the negative activates a photosensitive material, softening it so that it is preferentially etched away by dilute acid solution, leaving a raised positive image. The trend is toward using plastic as the photosensitive material. This positive imaged sheet is then wrapped around a steel roll, the circumference of the roll being chosen to match an integer multiple of the repeat length. One roll represents one color.

There are three main methods for printing. The first method is flexography. The plastic to be printed (called a web) is loaded on a continuous-feed roller (allowing a used plastic roll to be interchanged with a replacement roll without stopping the machine). The plastic sheet is then fed through a series of tensioning rollers. It then comes into close contact with the first printing roll (with the raised positive image), which rotates in contact with rollers in an ink bath at a precisely controlled viscosity. The plastic sheet may then be dried (for high-speed or high-quality requirements) by hot air tunnels before passing to the next press (stack press system). The plastic is lined up for the second and subsequent color applications using registration marks before passing across the next color roll. After all the colors have been applied, the plastic is dried and cooled (to prevent shrinking during cooling once on the final roll, which can crush the roll or cause star stress patterns on roll ends, preventing smooth flow in the packaging equipment) and loaded on the take-up roll. The ink solution is picked up by the fountain roll (rubber) and spread onto an anilox roll, which controls the amount of ink. This then coats the plate, which pushes at 4 atm pressure onto the impress roll.

The second form of printing is by rotogravure press. Negative images are placed onto stainless steel rolls directly by etching, then chroming. The image can be seen on the roll but is too fine to be felt. The roll rotates in an ink bath as before, but this time the ink is scraped off the roll by a razor-edged high tensile steel blade called a doctor blade, leaving a fine coating of ink in the roll image. This allows a precisely defined amount of ink to be applied, leading to higher-quality, sharper images. The cost of the image rolls (the plate) makes this method more expensive than flexogravure, and changeover times between jobs are much longer.

Third, there are several ways of printing letters. Some printers (letterpress) use individual metal hammers with the letter engraved in bold relief, striking an inked ribbon, to place the letter on the package label, like a typewriter. The package label is then applied to the package. This is used particularly for metallic printing. However, for the additional lettering such as due by dates, where the printing must be done at the point of packaging, ink-jet printers are increasingly being used. These work by sending a thin jet of ink through a fine tube directly onto the package, while heating the tube. Modern ink-jet printers have a keyboard for supplying essential information such as fonts, input code, bar codes, graphics, security, time and dates, etc. Thermochromic inks change color during processing, confirming that the product has passed through the equipment correctly.

3. Bar Codes

Improvements in industrial automation have led to a need for a method of reporting what is happening in the factory. Increasingly product packaging is coded, and the main methods for doing this are by ink-jet and laser coding (etching the top of the pack through a stencil).

Statistical product evaluation may include production line efficiency, inventory control, sorting, management, and marketing data, so the most convenient form of code is one that can be read by computer. Manual methods have problems with mistakes, time required, illegible entries, and deliberate falsification. A common form of modern package coding is bar coding, where the product information is coded as alternating black and white lines, which can be scanned by a bar code reader (as is commonly used in supermarkets). In this way reporting is automated.

Bar coding originated with a Harvard Business School honors thesis in 1932 on grocery store item processing automation. But it was not until 1973 that the Universal Product code became standardized as Code 3 of 9. The information in the bar code is scanned in (by contact scanners or noncontact scanners) and passed to a computer for processing. The bar code itself is put onto the package by an encoder (printer), which may be a mechanical dot matrix, a thermal, an ink-jet, a laser, or an electrostatic printer or a printing press as part of the label design.

The system of bar coding used in Australia is called the International Article Numbering (EAN) system. For products there is a 13-digit and an 8-digit (small-sized products) bar code. Most retailers will refuse a product without a bar code. EAN Australia supplies a manufacturer with a 5-digit manufacturer number. The manufacturer allocates a 5-digit product number. The country issuing the manufacturer number is coded as 2digits, and the final digit is a check digit. The bar code, to be useful, must be clear and scan easily.

Optical Character Recognition (OCR) might be the system of the future for factory floor reporting. OCR is the technology of getting a computer to scan ordinary words and to interpret them correctly. At present there is little standardization, and available packages have not proved reliable at recognizing characters. But the systems are improving at great speed, so will probably eventually replace bar codes. In fact OCR predates bar coding. The major drawbacks of OCR at present are that the reading wand must touch the code to read it and the equipment is expensive.

4. What Are Labels Made From?

The label no longer needs to be made of paper. Plastic labels are increasingly used, for example, shrink-sleeve labeling where the label is first printed, then supplied to the applicator as a tube on a reel. The tube is opened and slipped over the product, which is the passed through a hot air shrink tunnel. Plastics used include PVC, oriented polystyrene, polyester film (polyethylene terephthalate, PET), and oriented polypropylene, all monoaxial films with transverse shrink. Certain sleeves that resist light (metallized) and UV may prolong product life.

Plastic labels may also be applied in the mold (e.g., HDPE labels by ACI), as on some household cleaning products at present (Jif, Domestos) because the label is robust against the product, labeling is faster and cheaper, and the final package is free from paper and glue, so is more fit for recycling.

5. Inks

An ink is a combination of a coloring medium (pigment or dye), a resin (binder), a solvent, and other additives. The objective is to get the pigment onto the right position on the package and keep it there.

The coloring medium is usually a pigment (insoluble powders dispersed through the resin), rather than a dye, which dissolves in the solvent. Pigments can be inorganic (e.g., white titanium dioxide) or organic (e.g., cyan blue, benzidine yellow). The color intensity depend son the refractive index and particle size of the pigment. Dyes should be used on metallized and pearlized surfaces because they are transparent and allow the background to partially shine through, creating good graphic effects. But dyes are degraded by light over long periods of time.

The resin binder acts as the carrier for the pigment and binds it to the surface of the film. It adds a finish to the surface, either matte or gloss, and protects the pigment.

The solvent may be a hydrocarbon (inexpensive), alcohol (good solvency and fast evaporation), glycol ether (in a mix to reduce evaporation rate), ketone, or ester. It dissolves the binder. The amount of solvent controls the viscosity, which has a large effect on the ink thickness for both rotogravure and flexo presses.

Other additives include wetting agents, waxes (to give slip and rub resistance), plasticizers (to give the base resin flexibility), and adhesion promoters.

The ideal is to have a good-quality presentation. The perceived color depends on the nature of the incident light, since color is caused by absorption of specific wavelengths preferentially, so that what we observe is the sum total of the visible wavelengths reflected. Thus the light source is important. Sunlight (5400K at noon), due to the high temperature of the sun and the relative evenness of its spectrum, gives white light. A candle (about 1900K) looks yellow, a 100 W tungsten lamp (2800K) is still a bit yellow, and a fluorescent day tube (4800K) is better. Certain types of light only radiate certain frequencies. Thus, for color matching, color booths often need to be set up to ensure that the colors do in fact match. A color-rendering index (CRI) of over 90% is required for these booths (100% is perfect daylight). A normal fluorescent tube has a CRI of about 50%.

6. Product Launches

An area relevant to design is the product launch, or more generally, promotional packaging, a growing trend resulting from the high costs of advertising. This includes sample distribution through extra wraps (e.g., on magazines), samples in existing products, and miniature samples as giveaways. Attention may be grabbed by the use of new labeling techniques such as holographics. The sample pack will cost about 20 cents.

C. Packaging Equipment

Packaging equipment refers to the equipment at the end of the processing line which applies the packaging to the product, whether this be the inner wrap, product container, outer container, or packaging container. Most packaging equipment, especially that involved with folding flaps and boxes, looks complicated. This is due to the replacement of human labor by automatic equipment to do complex tasks such as folding. There is tremendous variety of packaging equipment.

1. History of Packaging

The first packaging equipment simply replaced mundane jobs and were treadle-operated or steam-powered. The individual operations tended to be combined with time to produce more powerful machines. Laminators grew out of the paper industry. Extruders came from pasta products. The screw extruder was developed in 1879 but not adapted to plastics until the 1950s. In 1910 a machine was developed to form, cut, and twist-wrap candy kisses, and about 1913 equipment was invented to fold cookie cartons with paper liners. After World War II most equipment was still custom-designed. After the war efforts to integrate the whole packaging line and generalize operations were made. Pouchmaking was developed in 1945–1952, and vacuum-packed pouches in the mid-1950s. The current trend is to automatic high-speed packaging lines.

2. Cup-Making

A plastic cup filling line is described below as an example of a packaging line. We trace this system from the resin granules through to cartons. The steps involved are as follows:

1. The extruder takes the granules and converts them to film, as described earlier.
2. The mill roll is fed through a preheating stage to a thermoformer, where radiant heat is used to heat the plastic film to its softening point. The required shape is then stamp-pressed by molds into the film. The molds may be arranged on a roll cylinder, stamping into a reverse roll cylinder, or simply be two mold impressions being stamped together. This is also called pressure forming. The molds must be kept at precise temperatures, so are often water-cooled using electronic sensors.
3. The waste plastic after the cups have been cut out is recirculated to the extruder. The cups are stacked and counted.
4. The cups are unstacked as required by the product line and printed. The cup is rimmed and filled. In a continuous system the cups may be filled before being punched from the plastic film.
5. The cup lid is applied using cold-seal pressure-sensitive glue. The composition of the lid was discussed earlier in the course.
6. The containers are counted.
7. The cups are then layer-packed, shrink-wrapped by layer, and case-packed.
8. The cartons are palletized, and the pallet is shrink-wrapped for transport.
9. The pallets are fork-lifted to storage.

This simple example introduces the ideas of the required balance between the various packaging components such as cup production, raw packaging material supply, lids, and product. The next example looks at a pressure-forming machine in more detail.

3. Automated Pressure-Forming Machine

These machines are generally suited for cup and lid production, so that the complete unit package can be constructed by the one machine by interchanging dies. The plastic lids are designed to snap onto the base. Film drawn off the material roll is transported through a heating zone (not under tension or it will stretch, so a chain transport system must be used) (radiant heat, e.g., infrared ceramics) to the tool. A system of cams pulls the required length of plastic into the forming zone. The tool is a series of molds fixed to a platen. One platen is stamped under pressure into the opposite fixed platen by a cam. The platen lifts away from the thermoformed sheet, which is then transported to a unit for punching out the shapes from the base sheet (called the skeletal web).

Because of the speed and power of these machines, safety requirements are very stringent, and all components must be shielded from the operator during operation. This gives thermoforming equipment its typical cabinet shape, with operator windows to observe that it is functioning correctly.

The cams and chain gear may be motor driven, pneumatically operated, or a combination of both. Some components need water-cooling, others need heating (such as the sheet feed zone).

The formed cups or lids are then nested into columns and counted, then pushed out into stacking chutes where they can be sleeve-packed and cartoned or sent for filling.

By interfacing the product line with the package production line, it is possible to form, fill, and seal continuously. The web from the material roll is heated, thermoformed, filled, cooled, and sealed (under vacuum or gas flushed as necessary), and then the filled cup is punched from the web. The label is then adhesive-applied to the cup. For small cups, piston dosing machines are used for filling. These extrude a measured amount of product into many trays simultaneously. The product is fed to the dosing unit by a pump from a buffer tank, which is kept at the correct temperature (cool for high-viscosity products, warm or hot for low-viscosity products).

The base is usually PVC or PET for clear packs and ABS or HIPS for opaque. PP is too difficult to form, fill, and seal. ABS is used for butter and margarines as it is more resistant to oils and fats. The lid may be aluminum foil lacquered on one side for sealing and printed/coated on the other, or it may be some combination of metallized polyester, paper, etc.

For solid products, blister packaging may be used. Blister packs consist of a card (the cardboard printed backing), the product, and a clear plastic "blister" sealed to the card. The packaging machine puts this package together in reverse order. Web from the material roll is heated to its softening point, then a mold fitted with vacuum equipment sucks the plastic into a blister shape.

The final packaged product must be checked before being palletted for storage. Scanning might consist of a person checking, a mechanical system such as a gate pushed by the product, a weight check (load cell), an infrared scanner, or laser or optical scanning. Optical scanning is increasingly used. This system is able to read bar codes and interpret the data, for example, sending information to the stock control database or to control downstream packaging equipment such as palletizers. It may also check the shape of the product so that rejects can be eliminated.

4. Portion and Single-Serve Packaging

Portion packs are the small butter, milk, jam, etc. containers used in airline flights, restaurants, and hotels. Single-serve packs refer to fruit juice packs, yogurt tubs, etc. These are usually form, fill, and seal (FFS), but about 40% of the single-serve market consists of preformed containers. The advantages of preformed containers are preprinting, embossing, reduced capital investment, and obviating the difficulty of recycling after lidding failure in a FFS machine. Preformed tubs may be produced by injection molding, and as well as HIPS and ABS, polypropylene may be used for its higher water resistance, laminability to EVOH, shelf life and cold-resistance. Two contrary properties are required of the lid: reliable seal and easy opening.

5. Retorting and Aseptic Packaging

Retorting is the traditional method of processing low- to medium-acid foods. Plastic pouches are increasingly being used. The choice of plastic is narrow—it must have a melting point of 121–130°C—and because it is inexpensive polypropylene is usually used. The problems with this method are that processing times are 1.5–2 hours, so generally the PP-based pouch softens and needs support, the containers tend to lose shape, and food texture and taste may be altered.

Aseptic processing is based on ultra-high-temperature (UHT) or high-temperature, short-time (HTST) processing, which takes seconds. The cooled product is then poured into the containers and sealed. The carton containers used are commonly sevenfold laminations, for example (from the product side):

Polyethylene: hermetic heat seal and inert layer for product.
Ionomer: adhesion layer for foil
Aluminum foil: moisture, gas, and light barrier.
Polyethylene: adhesion coating to bind the paperboard to the foil.
Paperboard: shape, stiffness, and strength (about 70% of weight)
Print: reverse gravure, decoration, labeling requirements, etc.
Polyethylene: waterproofs paperboard, protects ink, gives glossy appearance

Aseptic retort packages have been designed using flexible aluminum foil–PE films, which have sophisticated sterile locks allowing the product to be poured into the bag, resealing the bag after withdrawal of the insertion spout. Various forms of locks exist for this purpose. Recycling of these boxes is difficult.

D. Protective Packaging

1. What Is Protective Packaging?

Protective packaging is a term applied to packaging primarily designed to protect the goods, rather than for appearance or presentation, so generally is used to apply to the outer containers used for transporting goods from the manufacturer to the point of sale, and filling materials inside the outer container, e.g., nylon barrier-sealed bubble packaging (computer parts), urethane expanding foam, PE foam package "cushions," and PS loose-fill packaging.

The most widely used protective package is the outer carton. All packaging is protective as one of its primary functions, so it is more accurate to call this transport packaging or tertiary packaging (on the basis of the primary packaging being in contact with the product, secondary for grouping units together for single purchase, and tertiary being for grouping secondary packaging for convenience of distribution).

A pallet is the frame base for carrying the transport packs and typically measures 1165 × 1165 × 150 mm (national standards vary, but typical is 1100 mm^2). The transport packages are loaded onto the pallet and secured to it by shrink-wrapping, straps, string and/or glue, or a similar method. This complete unit is called the unit load. About 50 transport packages may be loaded per pallet in four or five layers. They are loaded onto the pallet in a way that either maximizes the unit load stability (by overlapping layers, called locking) or the compressive load strength (by vertical columns of cartons) and maximizes the number of transport packages to the load. These transport packages may double as display cartons for large items, or, if the product is small, the transport package will contain a number of display cartons. The product itself is packaged in the primary package.

Environmentally the best way of securing the pallet load is to use the minimum shrink-wrap and use a hot-melt adhesive between cartons. The adhesive can be applied in blobs (conventional) or as an atomized spray (more economical, less carton tearing and defacing, increased secured area, and less danger of operator burns).

2. Carton Conveying

Downstream from the packaging equipment, after the primary packages have been put into cartons and the cartons sealed, a large factory will have specialized conveyors for transporting and sorting the cartons, allowing products from different processing lines to be cartoned with the same equipment, then sorted onto individual product pallets. The speed of the sorter is limited by:

The carton length (more 200 mm cartons can be processed than 400 mm cartons)
The sorter type and gap—the gap between cartons is essential, allowing diverting mechanisms to push across the conveyor then pull back before the next carton arrives. A pop-up device requires minimal gap. The gap is created by either gradual increases in con-

veyor speed, "pulling" a gap, or sort controllers, which set a constant optimum gap (e.g., 300 mm) irrespective of carton length.

Conveyor speed—the maximum speed is usually set by noise, maintenance, and carton-tracking requirements.

Tracking the carton—if a carton moves outside its window, missorts will occur. If friction is used to move the cartons, there will be large degrees of randomness in carton position. In practice, this limits speeds to about 30 m/min. For direct drive conveyors, this may be raised to 103 m/min.

Diversion speeds—at low carton rates, right-angle takeoffs work, but at high speeds, the take-off must be angled (e.g., at 80 cartons per minute (cpm) a 45° deflection is required).

Infeed and outfeed—the flow of cartons on and off the conveyor must be controlled, usually by means of buffer feed conveyors.

A general principle for selecting a protective container is to choose the one that does the job at the least cost. Possible material choices are cardboard, fiberboard, steel containers, rail cars (in bulk), possibly strengthened with plastic or steel ties. High-value, fragile products may be additionally packaged in foam plastic.

External shrink wrapping may be used for high-quality, air-sensitive products, with nitrogen flushing. Products may be given a second protective packaging cover before being transported. Once the product has been boxed and palletized, the whole pallet can be covered in plastic and heated, causing the plastic to shrink and cling tightly to the cartons.

3. Tamper-Evident Packaging

Another aspect of protective packaging involves primary packaging designed to prevent anyone from opening the package before purchase. Two cases have highlighted this problem:

1. In England in 1985, glass chips were found in bottles of Heinz baby food. Was it an accident (e.g., mold damage) or deliberate?
2. In Chicago in 1982, 7 people died from cyanide injected into supermarket packs of Tylenol. The immediate result was that tamper-evident packaging became a requirement for over-the counter pharmaceuticals.

Other cases of extortion or sabotage have also been reported. Less serious problems are shop pilfering, tampering and sampling from packs, resulting in product contamination, loss of freshness, and stock loss.

In the mid-1970s, child-resistant packaging became an issue, leading to the development of child-proof lids for poisonous products. Tamper-resistant refers to the ability of the packaging to resist tampering (or opening), for example, for child protection, whereas tamper-evident refers to the ability of the packaging to reveal that it has been opened. Examples are:

1. Clear plastic sealed wraps over the whole primary package
2. Plastic ring seals on bottle closures (e.g., milk)
3. Pop-lids, which give a distinctive noise when first opened (e.g., baby foods) as vacuum is released
4. The aluminum roll-on closure for bottles, where twisting the cap tears a perforated strip that is retained on the neck
5. Shrink-wrap labels with a perforated neck, torn when opened
6. A separate lid used as a tool to open the neck of the container (e.g., toothpastes).
7. Inherently tamper-evident packs, such as steel cans, paperboard cartons, sealed sachets, pouches, and bags

A tamper-evident closure on a PET bottle adds about 2 cents to the product cost.

E. Computers in Packaging

1. Introduction

Computer technology has affected most aspects of the food industry, from raw material inventory through to distribution of the finished product. Some applications of computers in packaging include:

1. Automation of packaging equipment, with modern equipment able to sense when the next item is ready for packaging. Labeling and counting are automated, and productivity reports generated automatically. Packaging fillers are able to detect and respond to problems such as damaged filling heads and compensate automatically. Increasingly computer intelligence is used to automate packaging operations, resulting in increased packaging speeds and less downtime.
2. Automatic palletizing, where cartons are handled by robot palletizers and stacked ready for the forklift without human intervention.
3. The computer may draw the labels. The specifications for the package can be entered into the computer, the typeset and layout chosen, colors tested on the display unit, and high-quality thermal prints produced for evaluation. By having an electronic "picture" of the package available to study, a lot of false starts can be avoided in developing the new design.
4. The packaging itself may be designed by a computer. For example, the artwork might be computer-drawn, a CAD program may be used to design the food package, and the computer that draws the design may also control the equipment used to manufacture the package (computer-aided manufacture, or CAM), providing greater versatility in developing new designs quickly.

2. Palletizers

Modern PLC control systems have changed the emphasis in plant operation from unit operations (such as packaging) to overall integrated operations, and as a result palletization has now become an integral part of the food-production system. Manual palletizing is not a pleasant job, and in conjunction

TABLE 7 Water Vapor Transmission Rates

Material	WVTR, temperate ($g/m^2/day$)	WVTR, tropical ($g/m^2/day$)
Glassine	600	2000
Veg. Parch.	500	1400
Nylon	200	800
Kraft	10	33
Polyester	10	—
PT paper	8	24
PVC	5	16
Polypropylene	3	9
Pliofilm	1.5	7
Waxed paper	1.5	7
Polyethylene	1	7
Cellulose	1	6
PVDC (saran)	0.5	—
Aluminum foil	0.1	0.2

with other difficulties in uniformity, etc., modern plants tend towards automatic stacking of boxes on the pallet.

Initially palletizers were designed for a single layer only. A complete pallet layer would be held and arranged on a feed line, requiring factory space. Even multilayer palletizers required a hold-up of boxes as they arrived off the line to organize each layer. The resulting automatic palletizer was usually much faster then the line speed. The robot palletizer requires no hold-up space. Each item is individually and accurately transferred to the pallet by the robot arm as it arrives.

The robot arm is programmed with the required pallet configuration and carton dimensions. As each carton arrives, it is first detected (by pressure switch, for example), picked up (by suction or a gripper head on the arm) from the conveyor, and swung onto the pallet. Modern robots are capable of moving about 10 items per minute. They can be rapidly reallocated to other jobs.

The main criteria for a palletizer are (a) sturdy design (for 24 hr operation), (b) few mechanical movements, (c) high availability, (d) quick carton size changes, (e) speed, (f) minimum space requirements, (g) ease of operation, and (h) hygienic, gentle, safe, and quiet. Robot palletizers are defined in terms of the number of degrees of freedom in the horizontal and vertical directions. Software is available to design the stacking of the cartons on the pallet in order to maximize the available space, for example, tipping cases sideways to fill the center gap. The palletizer is able to learn these designs and implement them. Multiple product lines can feed to the same centralized palletizer that reads the bar codes before directing the product to the correct pallet.

In practice, commercial robots are slow, expensive, and not food safe, and their applications in the food industry so far have been few. However, as robots improve, they have potential for use in dangerous environments and with boring jobs (such as meat processing). Other applications of robots are for automated guided vehicles (AGV), for use in automatic storage and retrieval systems, and for vision systems.

REFERENCES

1. Bolton, CMC Chemicals, *AIP National News October*, 1992.
2. H. Mannheim, J. Miltz, and A. Letzter, Interaction between polyethylene laminated cartons and aseptically packaged citrus juices, *J. Food Sci. 52*(3):737 (1987).
3. Neales, *Sydney Morning Herald*, August 23, 1993.
4. Puplick and B. Nicholls, *Completely Wrapped: Packaging*, Report of Waste Management and the Australian Environment, Packaging Environment Foundation of Australia, 1992.
5. Sacharow and R. C. Griffin, *Principles of Food Packaging*, 2nd ed., AVI, 1980.
6. Thompson et al., Method for evaluating package-related flavors, *Food Technol. 47*:90 (1994).

24

Hazard Analysis and Critical Control Point (HACCP)

ANNE PERERA
Food and Nutrition Consultancy Service, Auckland, New Zealand

TITUS DE SILVA
Montana Wines Ltd., Auckland, New Zealand

I. BACKGROUND

A. The Need for an Effective Food Safety Assurance Method

Food safety has been of concern since the Middle Ages, and regulatory measures have been enforced to prevent the sale of adulterated or contaminated food. Many rules and recommendations advocated in religions or historical texts are evidence of a concern to protect people against foodborne diseases and food adulteration. Augsburg in 1276 ordered meat that was not freshly slaughtered to be sold at a specially designated stand, and Florentines forbade the sale on Monday of meat that had been on sale the previous day [1]. Motarjemi et al. [2] summarized the present concern and emphasized the need for an effective food safety program [2].

Since the time of Robert Malthus, the ever-increasing world population has challenged economists to address the issue of food supply worldwide. Technological progress in food processing has brought about greater awareness of problems associated with food preservation. International and national organizations have enforced laws and regulations to achieve quality and safety in food preparation and preservation in order to safeguard the consumer from foodborne infections and intoxication.

B. The Development of HACCP

The principles of the Hazard Analysis and Critical Control Point (HACCP) technique were applied to the chemical processing industry, particularly in Great Britain, more than 40 years ago [3]. The Pillsbury Company, together with NASA and the U.S. Army's Research, Development and Engineering Center at Natick, first developed this system to ensure the safety of astronauts' food [2,4,5]. Since then the HACCP system has grown to become a universally recognized and accepted method for food-safety assurance. The World Health Organization (WHO) has recognized the importance of the HACCP system for prevention of foodborne diseases for over 20 years and has played an important role in its development and promotion. One of the highlights in the history of the HACCP system was when in 1993 the Codex Guidelines for the Application of the HACCP system were adopted by the FAO/WHO Codex Alimentarius Commission, requiring them for international trade [2].

In the United States, large chemical industries have also adopted hazard control programs. The U.S. Occupational Safety and Health Administration (OSHA) introduced the HACCP technique to reduce accidents.

C. The Benefits of HACCP

Lack of food-safety systems costs the food industry million of dollars annually through waste, reprocessing, recalls and resulting loss of sales. It is now recognized internationally that the most cost-effective approach to food safety is through the application of the HACCP technique. By adopting an effective food safety system based on HACCP, the industry can minimize the potential for things to go wrong and ensure the safety of food products. HACCP is entirely complementary and adds essential safety elements to existing processing systems such as Good Manufacturing Practices and ISO 9000 standards.

Food producers as well as retailers and consumers will be benefited as a result of an effective HACCP program. The benefits of the HACCP system are summarized in Table 1 [2].

II. TERMINOLOGY

The following terms are important to the HACCP concept:

Hazard: a biological, chemical, physical, or other property in a food product that has the potential to harm the consumer or cause illness. It can occur in the ingredients or at any stage in the life of a product. The term can thus be applied to foreign material, chemical residues, and/or microbiological contamination.

Hazard Analysis: the identification of biological, chemical, or physical and/or other hazards associated with ingredients, production practices, processing, storage, distribution, retailing, and use.

Sensitive Ingredient: an ingredient known to have been associated with a hazard and about which there is a concern.

Control Point: an operational step in a manufacturing and distribution process that may be controlled in order to maintain quality and meet regulatory requirements.

Critical Control Point (CCP): an operational step in a manufacturing process that results in injury or risk to consumer if not controlled. At critical control points, controlling measures can be exercised to eliminate or minimize any form of hazard.

Critical Limit: one or more prescribed tolerances that must be met to ensure that a CCP effectively controls a health hazard.

Table 1 Benefits of the HACCP System

Ensures safety of food products through preventive measures rather than through final inspection and testing
Capable of identifying all potential hazards
Easy to introduce technological advances in equipment design and processing procedures related to food products
Directs resources to the most critical part of the food-processing system
Encourages confidence in food products by improving the relationship among regulatory bodies, food processors, and the consumer
Promotes continuous improvement of the system through regular audits
Focuses on safety issues in the whole chain from raw materials to consumption
Complements the quality management system (e.g., ISO 9000)

Hazard Analysis and Critical Control Point (HACCP): a scientific, rational, and systematic approach to identification, assessment, and control of hazards during production, processing, manufacturing, preparation, and use of food to ensure that it is safe when consumed. The HACCP system provides a preventive and thus a cost-effective approach to food safety.
HACCP Plan: a document that sets out the procedures based on the principles of HACCP to be followed to assure food safety.
HACCP System: the organizational structure, procedures, processes, and resources needed to implement the HACCP plan.
Validation: a review of the HACCP plan to ensure that all elements of the plan are accurate and correct.
Verification: the use of methods, procedures, and/or tests to ensure that the requirements of the HACCP system have been fulfilled.
Hazard vs. Risk: these terms are not identical. The risk is the probability of occurrence of the hazard in the future, whereas hazard is the cause of injury. The risk can be minimized by a proper or adequate control of hazard. The severity is related to the level of the hazard. It can be life threatening in some circumstances. A hazard may be severe in one product but moderate in another product.
ISO 9000: the generic standard that specifies minimum requirements to be fulfilled by organizations in order to meet a customer's needs. It does not specifically address the issue of food safety, but it addresses the need to identify and comply with regulatory requirements that are applicable to the product and/or process. Enforcement of regulatory requirements is an attempt to protect the consumer from harmful food products. ISO 9000 standard and HACCP techniques are complementary. A quality management system does not guarantee food safety unless the hazards are identified and controlled. HACCP techniques should therefore be used as a tool to support the quality management system ISO 90000.
Good Manufacturing Practices (GMP): applied to food processing, a code of practice for controlling and optimizing the process that recognizes the need to have an HACCP system in place to produce safe and cost-effective foods. In order to meet the requirements of GMP, regulatory bodies provide well-defined guidelines for food-processing operations.
Total Quality Management (TQM): the management philosophy that seeks continuous improvement in the quality of performance of all processes, products, and services of an organization.

III. STEPS IN THE HACCP PROGRAM

HACCP study begins with the selection of a team consisting of members drawn from various disciplines in the food-processing operation. The success of the HACCP program depends upon the constitution of the HACCP team. Table 2 shows the responsibilities of the various members of the HACCP team [6]. It is a good idea to carry out an HACCP study for each new product in each plant. Figure 1 shows the steps in the implementation of the HACCP program. Hazard analysis critically examines the quality of all ingredients, processing steps, and the product itself. The CCPs can be identified by analyzing the hazards in each processing step. HACCP plan is managed by regular monitoring and reviewing of the system through implementation of corrective action when necessary.

A. Types of Hazards

Hazards in food processing can be classified into three types: biological, chemical, and physical.

1. Biological Hazards

Biological hazards are associated with microorganisms that cause foodborne infections and intoxication. A proper food-safety program must take into consideration all potential safety hazards inherent in the handling of food.

TABLE 2 Responsibilities of HACCP Team Members

Team member	Responsibility
Project leader	Convenes and chairs meeting Ensures that HACCP principles are correctly applied
Production manager	Constructs flow charts Advises on production issues and process capability
Technical expert	Advises on technical issues Identifies hazards and recommend solutions
Engineer	Supplies information on performance of equipment and machinery Makes recommendations on new machinery, equipment, or processes that may be required
Others (as required)	Provides information on specialist areas
Secretary	Records proceedings of meetings

2. Chemical Hazards

Chemicals are used in food products, such as pesticides during the growing stage or as food additives during formulation and processing stages. Types and concentration levels of chemicals are important for safety aspects. Some examples of hazardous chemicals are heavy metals, such as lead, tin, cadmium, copper, and mercury; food additives, such as certain preservatives, colorings, and conditioners; and others, such as solvents, cleaning agents, paints, and adhesives.

3. Physical Hazards

Physical hazards are caused by foreign matter that can enter into a food product at any stage from the processing of raw materials to consumption of the finished product. Foreign matter may be visible to the naked eye or may be dissolved or dispersed in the food product. The physical form of the foreign matter can vary from powder to particulate matter, depending upon its type and origin. The detection of foreign matter in a food product is not easy because of the variety and infrequent occurrence.

Some of the common types of foreign matter associated with physical hazards in food are insects, spiders, worms, etc. (although these organisms are nonhazardous in themselves, they may carry pathogenic microorganisms), parts of animals, birds, metals, machine parts, glass pieces, plastic objects, sand, stones, dirt, cigarette butts, and plastic dressing strips.

B. Sources of Hazards

Contamination of food can occur under a wide variety of conditions. Prior knowledge of the potential hazards and their origin can be useful in monitoring and controlling such hazards. Hazards may originate from five main sources: raw materials, processing steps, machinery, handling of foods or ingredients, and environmental conditions.

1. Raw Materials

Raw materials are the primary source of contamination. Failure to follow basic quality assurance procedures on raw materials may lead to food products that are unsafe for consumption. The common quality assurance procedures carried out on raw materials are related to:

identification and labeling
storage conditions
handling requirements

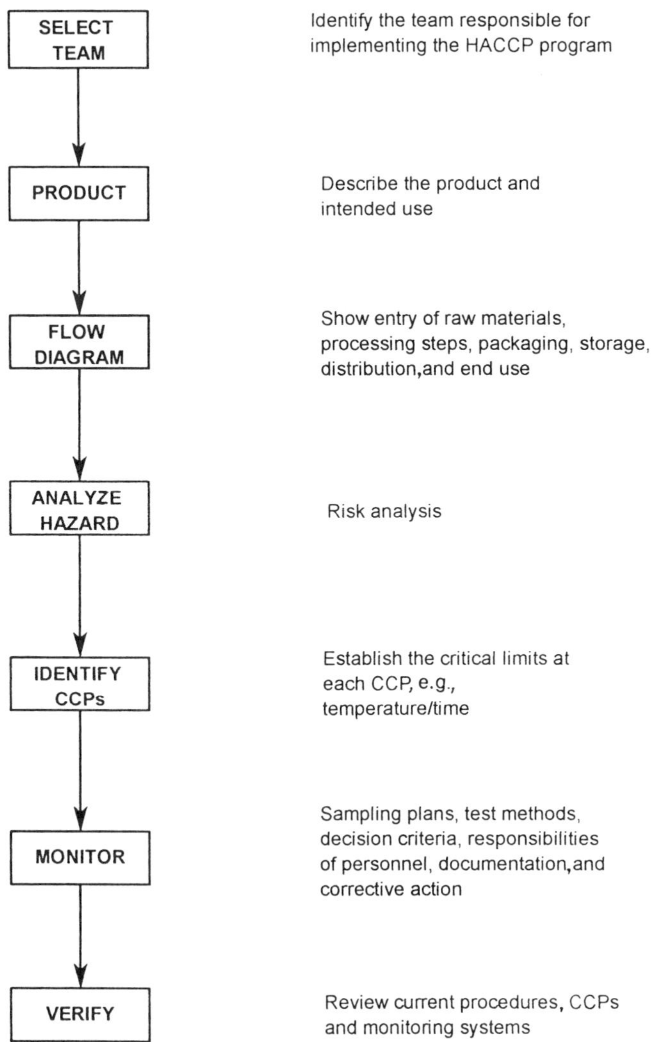

Figure 1 Implementation of the HACCP program.

preparation and processing
isolation of unsuitable raw materials

The raw materials that are most likely to cause microbial hazards are meat, poultry, fish, and dairy products. The level of microbial contamination depends upon [7]:

the source
the refining and handling process
the packaging material
the storage conditions

The HACCP technique has been closely associated with the protection of the consumer from microbiological hazards. Such narrow focus has been criticized on the grounds that microbiological hazards account for only one type of hazard [8].

Raw materials may carry traces of chemicals and foreign matter. Heavy metals such as lead, arsenic, mercury, tin, and cadmium are thought to be of greatest health concern. They occur in vegetables grown in contaminated soils. Packaging material such as lead capsules in wine bottles and solder form the side seams of cans are potential sources of contamination. Adhesives, coatings, and resins used in packages may cause health hazards unless they comply with health and safety requirements [9].

Cleaning chemicals, solvents, lubricants, and dirty or incompletely washed bottles may contaminate food products. Excessive amounts of sulfur dioxide used in the sterilization process prior to filling of bottles result in high levels of sulfur dioxide in the food product (e.g., high sulfur dioxide levels in wine).

Preservatives, colorings, flavors and conditioners are common additives used in food processing, and excessive amounts of these can be harmful. The permissible levels of food additives are governed by the food regulations in each country.

Fruits and vegetables sprayed with pesticides can retain high levels of pesticide residues unless the applications are carefully monitored. Foreign objects such as stones and insect parts may also be found in fruits and vegetables. An unhygienic working environment can promote contamination with rodent bait, insecticides, insects, etc.

2. Processing Steps

Uncontrolled processing operations can lead to hazardous situations. Failure to maintain processing conditions such as temperature/time, delays in processing, using incorrect formulations and procedures, and following unauthorized processing techniques may all result in contamination and/or microbiological growth. Mercury thermometers in the processing area can be a potential hazard, and most industries prohibit the use of such thermometers in their factories.

3. Machinery

Unclean and unhygienic equipment can easily promote the growth of microorganisms or other hazards. Failure to maintain the sterility of equipment when it is required results in microbial contamination. Proper setting up of equipment should also be followed. Glass bottles can get chipped at the filler or capping machine if the machines are not properly set up. Imperfectly made containers can also contaminate the food with the material of the container. Metal pieces from meshes or metal parts and nuts and bolts can easily get mixed with the food product if the machines are not regularly maintained.

Preventive maintenance of machinery is an important aspect in a safety-management program. If safety requirements are ignored, the layout of machinery and equipment can be a potential hazard. The machinery should be examined at intervals that ensure safe operation. Any change in engineering should be such that it is not hazardous.

4. Handling of Foods

With the introduction of highly automated, high-speed machinery, vast volumes of food products are processed, stored, and transported to distribution centers and retail chains. Therefore, food safety depends upon processing characteristics as well as handling during transport, storage, and customer use. Hazards may develop due to inadequate temperature control during storage, transportation, retail handling, and home storage. Products such as chilled/frozen entrees and meal components are preserved by refrigeration. Hazards could be developed if these products are stored at higher temperatures or used beyond their recommended shelf life [10].

Failure to rotate stocks of dated products can result in outdated products reaching the customer. Abuse by the customer is possible in the absence of clear storage or preparation instructions. Lack of knowledge about handling, cooking, and storage of foods increases the risk of hazard occurrence.

Personal hygiene is extremely important in any food-serving establishment. If adequate precautions are not taken, food handlers can transmit pathogenic bacteria. Personal articles such as jewelry can get mixed with foods during preparation.

5. Environmental Conditions

Hazards due to environmental conditions may affect raw materials, processing, and machinery. Pollution of water and soil can have alarming results through the food chain. Through regulatory requirements, most countries monitor and control the disposal of domestic and industrial wastes to prevent the entry of hazardous material into food products. Environmental contamination may be due to foreign matter, chemicals such as sprays, and contaminants in water.

C. Some Measures for Controlling Hazards

1. Measures at the Processing and Packaging Stage

Raw Materials

In a food-processing environment, raw materials constitute one of the most important areas that must be carefully controlled. The food producer has no direct control over the quality of incoming raw materials. Until adequate control can be established over the entire range of raw materials to prevent the entry of or eliminate harmful or potentially harmful organisms and residues, constant vigilance must be maintained. This is particularly true of "sensitive ingredients"—those ingredients that have been historically associated with a known hazard, such as eggs, fish, milk, cheese, shellfish, etc. The processes used to remove or destroy microorganisms in raw materials and packaging are shown in Table 3[7]. Some of the controls that can be established in order to ensure that incoming raw materials do not cause a health hazard are summarized in Table 4.

Packaging materials can also be a source of health hazards since most users are unaware of the materials used. Controls can be established by specifying the recommended types of packaging material. Bulk containers that are used to transport food products should have a cleaning program in place, which should be audited. Only permissible products should be transported in bulk containers that are used to carry food products. Other controls include tamper-proof seals, inspection of samples on delivery, and maintaining appropriate storage conditions.

Processing Steps

A wide variety of methods can be employed to control the hazards that may be encountered during the processing operations. The type of control mechanism depends upon the processing method or methods employed in the plant.

Temperature and pressure recorders are common in most food-processing systems. With advances in electronic technology, thermometers are not often used, and if they are employed, mercury thermometers should be avoided. Control charts, log sheets, and other records can be employed to monitor the temperature and pressure in a processing environment.

Batch records should clearly state the type and quantities of ingredients used in production. Products that require a use-by date should be controlled at the source. All products including the ingredients used should have a batch number or a lot number to enable traceability. Finished products should be maintained at the specified temperature, and products held under quarantine should be clearly labeled to prevent them from being dispatched.

Table 3 Processing Methods to Control Microorganisms in Raw Materials and Packaging

Method	Control parameter
Heat treatment	Time, temperature, humidity
Filtration	Pore size, filter integrity
Irradiation	Dosage and density of load
Chemical	Concentration, pH, temperature

TABLE 4 Control of Incoming Raw Materials to Ensure Safety

1. Be highly selective of sources and suppliers of materials and their ability to produce and deliver a safe product consistently by implementing an approved supplier policy.
2. Establish specifications for raw materials taking into consideration those characteristics that are critical to quality and safety.
3. Avoid using the cheapest price as the sole criterion for purchase. Relate the price to risk assessment.
4. Review any new ingredients introduced into the system. Instruct the supplier to inform you of any changed characteristics of the raw material, since even minor changes may affect the final quality.
5. Carry out periodic audits at the supplier's premises.
6. Instruct the supplier to have an HACCP and a QA program in place. Provide encouragement and support if necessary. Developing a partnership can be mutually beneficial.
7. Inform the supplier to label the raw materials accurately and provide assurance in the form of compliance certificate when they are delivered.
8. Carry out periodic tests on raw materials on a random basis on delivery.
9. Monitor storage conditions of raw materials both at the supplier and the producer.
10. Request representative samples for inspection prior to delivery.
11. Encourage the raw material suppliers to develop safe packaging of ingredients.

Operating storage tanks at positive pressure can create problems of cross-contamination between liquid and gas lines. Such cross-contamination can be avoided by using nonreturn valves at appropriate locations. Changes made in a process should be controlled through a change control procedure, which should include a reassessment of the hazards and CCPs.

Plant and Machinery

Hazards due to plant and machinery can be controlled by the development and maintenance of physical equipment and accessories used to manufacture a food product. It is necessary to thoroughly clean and sterilize all equipment and utensils before and after processing. It is also important to recognize the importance of HACCP principles in planning the layout of engineering equipment. A hazard control program requires that every production line in the plant be correctly laid out showing the operation and interrelation of all machinery and equipment. A preventive maintenance program should be in place that indicates the frequency at which equipment should be checked. When a change in machinery or machine settings occurs, the hazards must be reassessed. Critical measuring equipment such as thermometers, weighing scales, etc. should be calibrated by organizations authorized by the national bodies to do so, so that the measurements can be traced to a national standard.

Line lubricants, grease, and chemicals used for cleaning of equipment should be recognized as safe and should be purchased from an approved supplier. The operators are the closest to the machinery, and they should be adequately trained to identify potential hazards. Unusual observations should be immediately investigated.

Storage and Distribution

Hazards due to storage, dispatch, and distribution are associated with storage conditions, stock rotation, and physical location. Specific storage conditions can be monitored by the use of temperature/time records, while the physical location can be observed for cleanliness and freedom from vermin and dirt. Scheduled cleaning and vermin control programs should be in place and monitored in the storage area.

Products released for dispatch should be physically located away from the quarantine area. The use of status stickers such as HOLD, QUARANTINE, REJECT, and PASSED will prevent substandard products from being dispatched. The design of food storage areas should take into account access to goods, personnel, forklifts, ease of cleaning, drainage, lighting and ventilation. Refrigerated foods such as sous vide and other chilled foods have gained popularity worldwide. These foods are more susceptible to mishandling than frozen or shelf-stable products and therefore need to be managed carefully. Training of personnel in the safe handling of foods during transport, monitoring temperature/time records in refrigerated trucks, maintaining cleanliness and hygiene, and correct delivery procedures are some of the controls that can be exercised to reduce or eliminate the hazards due to transport and storage. Routine inspections and audits can be used to monitor the effectiveness of the storage, dispatch, and distribution system.

Premises

Control methods that can be used to prevent the occurrence of hazards and for safe operation within the premises of the food-production area depend upon the proper design and layout of processing areas. Several control measures can be used to prevent potential hazards:

- Control of pests in storage and manufacturing areas
- A scheduled maintenance program
- Temperature/time records in manufacturing and storage areas
- Cleaning program for walls, floors, and ceiling
- Monitoring the temperature of water used for sterilizing/sanitizing of equipment
- Regular disposal of rubbish bins in order to prevent the entry of rodents and pests

Health and safety regulations also provide some measurers of controls over the hazards that originate in a food-processing facility. Regular audits should be carried out to ensure that health and safety regulations are observed.

Personnel

An HACCP program should take into account the hazards due to poor handling of food in a production facility and at food serving stations. In catering facilities and chilled and frozen food premises, the physical health and cleanliness of staff represent a major risk [11]. Food handlers can be a major source of pathogenic bacteria. Incidences have been noted where personal articles such as pens, paper, jewelry, metal items, cigarette butts, and chewing gum have been incorporated into food products. Entry of these hazards can be controlled by introducing a policy covering prohibition of smoking, chewing gum, and wearing jewelry, maintaining personal hygiene, use of clean uniforms, regular medical care, and regular audits to monitor the effectiveness of documented procedures for handling food.

All garments should be clean and free of soil. Freshly laundered garments should be provided for food handlers on a daily basis. Wherever appropriate, head covers should be worn. Besides being unpleasant, hair is also a source of microorganisms [12]. Workers who handle food should not have cuts or infectious diseases. Such workers should be prevented from handling food. Touching prepared food with bare hands should be avoided. Suitable hand washing and drying facilities should be provided near work stations.

2. Measures of Postprocessing and Packaging Stages

Contamination of food products can also occur at postprocessing and packaging stages. Food manufacturers and retailers should aware of the need to handle food in a safe and hygienic manner. Manufacturers and retailers have the responsibility to ensure that food products are not abused by the consumer after purchase.

Retail

Before food reaches the consumer, the retailer is responsible for maintaining the safety of all food products in his care. The retailer must store the food at the recommended temperature, and adequate care has to be taken when food is handled. The control measures therefore relate to monitoring temperature/time records during storage, inspection of equipment and facilities, auditing, training of staff, and use of tamper-proof and tamper-evident packaging.

Food Service

Food is presented to the consumer in a variety of ways. Some food service systems are prone to microbiological and other hazards. The techniques of HACCP that can be applied to food-production systems are also valid in food-service systems. The presence in food of certain microorganisms or their metabolic products in amounts sufficiently large to cause illness when consumed is a major concern. The population is at greater risk in this respect from the food-service establishments [13].

Hazards that need to be controlled are linked with several factors, including the composition of menus and individual food items, particularly raw materials not subjected to further processing, storage, preparation, handling and holding procedures. The control methods include:

Selection of suppliers
Inspection of raw materials on arrival
Temperature/time control in storage and food-handling areas
Monitoring personal hygiene and food-handling practices
Sanitation of utensils and handling equipment
Provision of adequate covers to protect from insects
Control of the entry of insects

The Consumer

Food preparation: Outbreaks of food poisoning due to poor handling of food in the home are not uncommon. Food spoilage due to pathogenic microorganisms as well as hazards from foreign objects can occur during the preparation of food. Consumer awareness of the potential hazards of handling foods in the household is important in order to ensure the safety of foods prepared at home. The hazards can be controlled by checking containers prior to purchase, handling the product correctly on the way home, properly storing ingredients and food, keeping kitchen equipment clean, preparing food correctly, and managing the pantry appropriately [14].

Food usage: Prepared food products such as ham, cheese, and sauces may be consumed directly or may be incorporated into other foods. Hazards can occur due to consumer abuse. Limited control methods are possible in the hands of the consumer. Controls can be exercised through the provision of consumer information as to how the food product should be handled, used, and stored. Warning labels such as use-by dates and storage conditions, use of temperature/time indicators on sensitive and high-risk food items, and packaging design that minimizes abuse by the consumer are also ways in which the food processor can help the consumer minimize hazards.

D. Food Recalls: Purpose and Consequences

It is essential to have an effective product recall procedure in place to safeguard the consumer. Product recalls are not uncommon; some examples of products recalled and the reasons include pickles and relishes (glass fragments), peanut butter (foam rubber particles), dinner kits (mold), super trim milk (cleaning spirits), fruit bars (wire fragments), canned meat (can damage), canned asparagus (can seam damage), canned tomatoes (faulty cans), natural peanuts (stones), and peanut butter (*Salmonella*).

Food manufacturers are required by the regulatory authorities to inform the public of the harmful characteristics of the food product, the circumstances in which the use of the food product is harmful, and safe procedures for its disposal. If the food product creates an imminent risk of death, serious illness, or serious injury, an immediate recall may be ordered by the regulatory authorities. Penalties may be imposed for a contravention of a compulsory product recall order.

IV. CRITICAL CONTROL POINTS (CCPs)

A control point in a food processing system leads to an unacceptable health risk and can be effectively controlled to prevent or eliminate health hazards to an acceptable level of low risk.

A. Classification

Critical control points can be generally classified as CCP1 and CCP2 [15]. CCP1 is defined as a step or location in a food-processing system that on its own effectively eliminates a hazard, e.g., metal detection in food products and sterilization. CCP2 is defined as a step or location in a food-processing system that contributes to the control of a hazard but does not guarantee elimination, e.g., inspection and pasteurization.

It is important to distinguish between CCPs and control points that are less critical in ensuring food safety. Several key points to note in determining the CCPs [6]. are:

1. CCPs should not be restricted to a minimum or a maximum number.
2. CCP is specific to a product and process.
3. CCPs should not be duplicated.
4. CCPs should only be introduced when it is necessary to eliminate or reduce a health hazard.
5. CCPs should always be developed by consulting an expert when there is doubt about a product or a process.
6. Development of CCPs requires the use of common sense.

The presence of a control downstream should not be considered a reason to neglect controls in preceding steps. For example, wines are tested for pesticide residues prior to bottling, and even then the grape grower has the responsibility to control the spray program. Any opportunity to eliminate or minimize the occurrence of a hazard should not be overlooked.

B. Location of CCPs

HACCP techniques enable the food processor to identify hazards and risks, focus on where they pose a threat to the safety of food, and develop the means to control them. The actual location of a CCP depends upon the type of hazard, the ingredients, packaging, processing procedures, storage and handling. Emphasis should be placed on prevention of entry rather than detection after they have been introduced. CCPs should be introduced as early as possible in the food processing system and close to the origin of the hazard. All precautions must be taken to prevent the entry of new ones [15].

Hazards associated with raw materials should be controlled at the source, i.e., the supplier. This minimizes the risk of entry of hazards and avoids unnecessary inspection of raw ingredients on receipt. Therefore, preprocessing techniques such as washing and sorting will be more effective in controlling hazards.

Food-production processors are often associated with more than one CCP. For example, in the production of entrees in conventional, cook/chill, and cook/freeze food service systems, time-temperature is a CCP throughout production in each of the models. Equipment and personal sanitation are also CCPs that should be regularly monitored using standards and criteria established by the food-processing system [16]. The inspection of the finished product usually verifies the effectiveness of controls placed so far.

C. Identification of CCPs

True CCPs have often been confused with control points and as a result a large number of CCPs are identified, making the HACCP system unworkable. For example, in a commercial process of smoked fish, it is possible to identify many individual steps, but only three can be considered critical: salt penetration, smoking, and storage [8]. A useful tool for identifying true CCPs is the CCP decision tree developed by the Codex Alimentarius Committee on Food Hygiene (Fig. 2) [17,18].

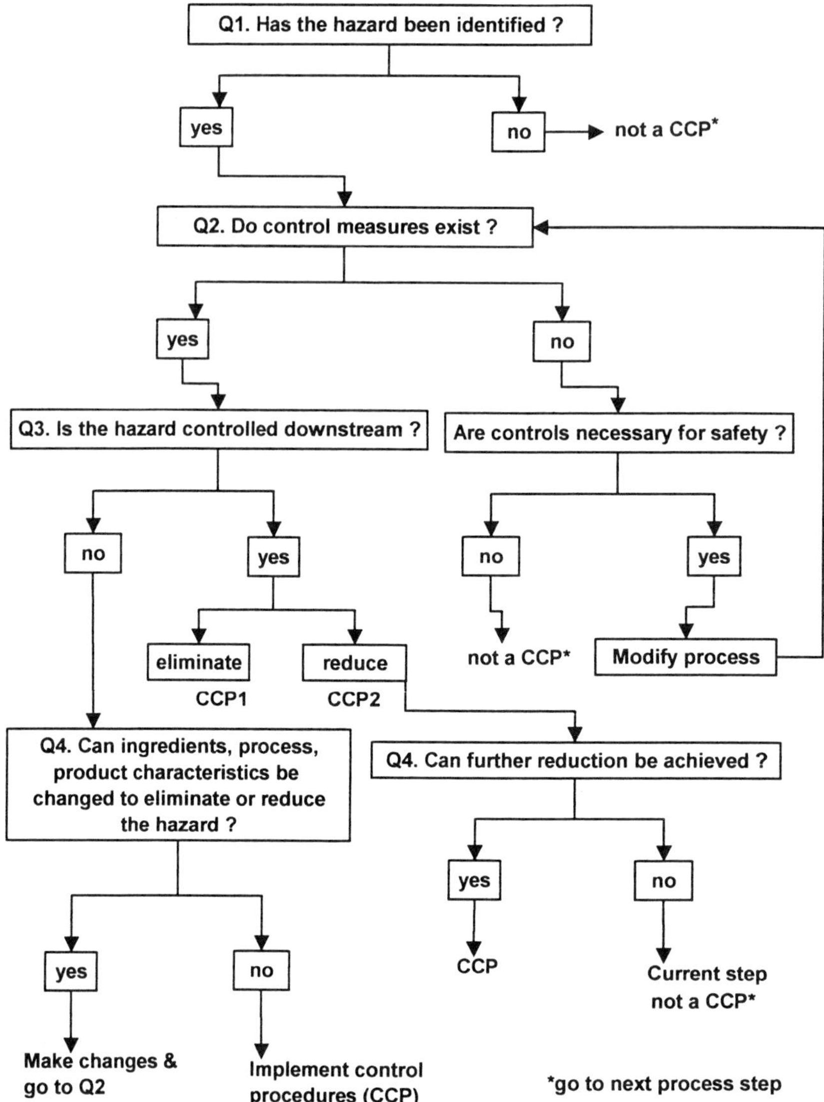

FIGURE 2 Decision tree used to locate CCPs in a process flow.

V. IMPLEMENTATION OF HACCP

A. The Use of Flow Charts

1. Symbols

Flow charts are used to show the various steps of the food-production operation. These include the entry of ingredients, all processing steps, packaging, storage, distribution, and handling by the consumer. These operations can be shown using five standard symbols as shown in Table 5.

2. Block Diagram

A block diagram gives an overview of the food-manufacturing process. The process steps are shown inside boxes and the entry of ingredients are indicated by arrows. Figure 3 shows a block diagram for the production of chicken and vegetable salad.

TABLE 5 Symbols Used in Flowcharts

Activity	Symbol	Description
Operation	○	Change in physical (e.g., cutting of meat), chemical (e.g., pH), or microbiological (e.g., sterilization) property of a material, mixing ingredients or separation of components (e.g., separation of bone from meat).
Inspection	□	Control step to check the product or process.
Transportation	➡	Change in location without any change in the product itself.
Delay	D	Temporary stoppage of the process until the arrival of the next processing step. The delay associated with a process itself (e.g., sterilization) is not represented by this symbol.
Permanent storage	▽	Keeping the product under conditions suited to the product in order to minimize deterioration (e.g., warehouse storage prior to dispatch).
Combined operation	⧈	Combination of operation and inspection.

3. Process Flow Diagram

A process flow diagram shows in detail the various steps of the food-processing operation. Figure 4 shows a process flow diagram for the production of chicken and vegetable salad.

B. Assessing the Hazard Potential

An important step in implementing an HACCP program is the assessment of potential hazards. Hazard analysis requires a knowledge of the pathogenic organisms or any agent that could cause spoilage of the product and be harmful to consumer. A broad understanding of how these hazards could arise is also essential for a complete assessment. An assessment of potential hazards involves a detailed examination of the following: raw materials, process, product, and end usage. Different systems have been used to assess the hazards associated with food products [19,20]. Suitable layouts for the assessment of the hazards are described in the sections that follow. The absence of a hazard is indicated by a minus sign and the efficacy of a hazard removal process or the extent to which the hazard can occur is shown by the number of plus signs (e.g., +++ for high, ++ for moderate, + for low, and – for none).

1. Assessment of Raw Materials

The hazards related to raw materials can be grouped under microbial, foreign matter, and those associated with transportation and storage. At this stage of assessment consideration is not given to any controls in the process that may be introduced to eliminate or reduce the hazard. A suitable layout for assessing raw materials is shown in Table 6.

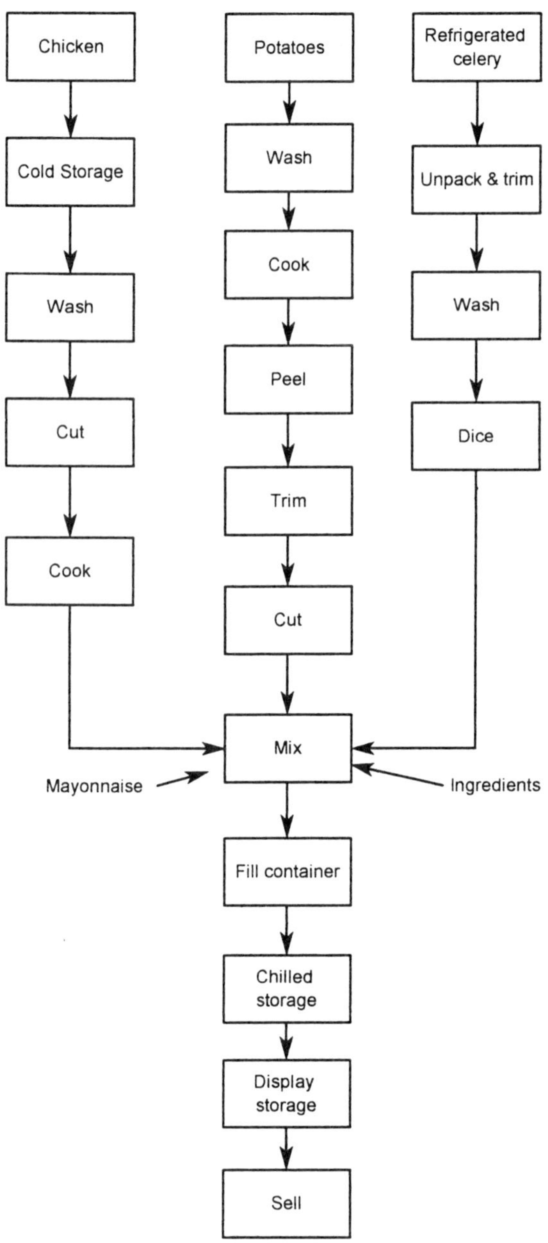

FIGURE 3 Block diagram for the production of chicken and vegetable salad.

Microbial Contamination

Some food products are more prone to microbial contamination than others, e.g., food products such as fish and meat are more likely to be contaminated with microorganisms than are fruits and vegetables. Chlorinated water and food ingredients such as salt generally do not carry microorganisms.

Microbial Growth

Several factors are important for the growth of microorganisms. These factors are discussed in other chapters and must be taken into consideration in assessing the risk potential.

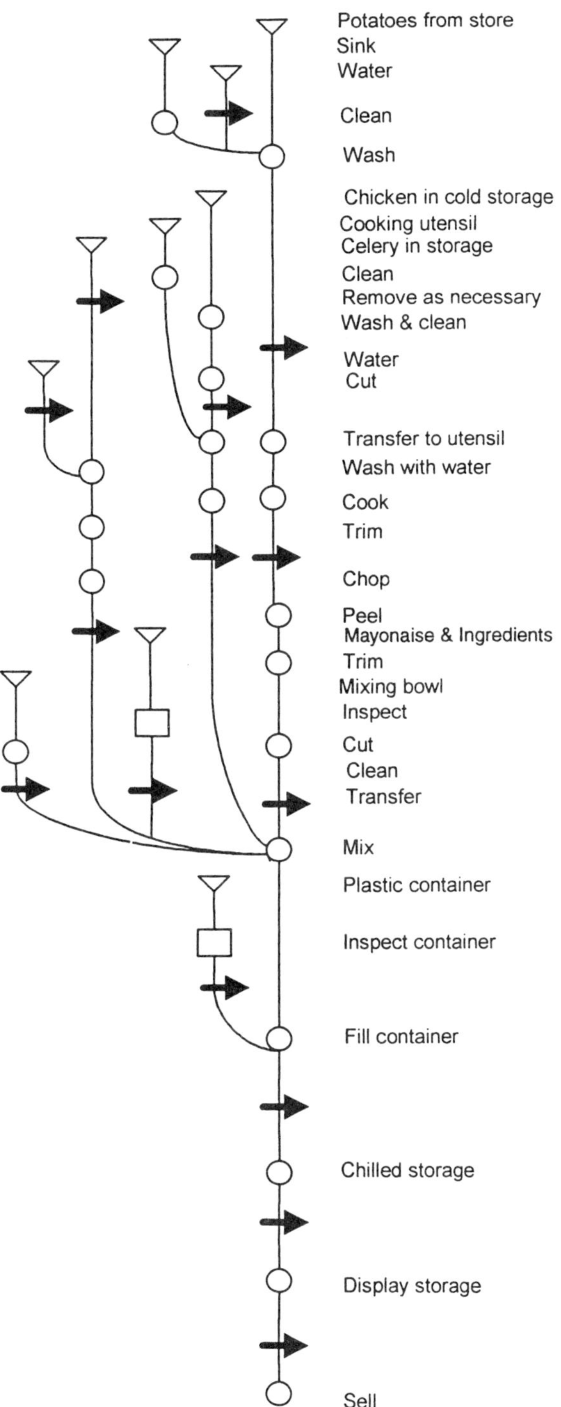

FIGURE 4 Process flow diagram for the production of chicken and vegetable salad.

TABLE 6 Assessment of Raw Materials

Raw Material	Contamination	Growth	Foreign Matter	Transport	Storage	Risk
Potatoes	+++	+	+++	–	–	High
Chicken	++++	+++	–	++	++	High
Celery	+	+	++	–	–	Low
Mayonnaise	–	–	++	–	–	Moderate
Ingredients	–	–	+	–	–	Low
Trays	–	–	+	–	–	Moderate

Production of ....Chicken and Vegetable Salad........

Foreign Matter

The common types of foreign matter that are of concern in a food-processing environment are soil, chemicals, pesticide residues, metal objects, personal articles, etc., and the possibility of such contamination should also be assessed.

Transportation

Some food products may be damaged or undergo deterioration during transport. The raw materials are assessed on the extent of damage or deterioration that can occur under uncontrolled conditions.

Storage

The assessment of raw materials on the basis of storage conditions takes into consideration the hazards that can occur under uncontrolled conditions during storage in the food-production facility. Products such as salt and sugar do not require special storage conditions, whereas maintaining a prescribed temperature/time is important for chilled foods.

Assessing the Risk

Risk can be high or low, for example, cooked food such as fish, meat, and eggs have a low risk in contrast to uncooked food. Even if the hazard is eliminated at a later stage of the process, any risk associated with raw materials should not be ignored.

2. Assessment of the Process

Assessment of the process involves analysis of each step. A decision tree can be used to identify hazards (Figure 2). A suitable layout for analyzing the process steps is shown in Table 7. At each step of the process, consideration has to be given to:

Efficacy of microbial destruction
Microbial contamination and growth (e.g., during handling)
Foreign matter removal/destruction
Foreign matter introduction
Equipment
Degree of control

Microbial Destruction

Microbial destruction is a critical factor in food processing and is described as high (e.g., sterilization), low (e.g., sanitation), or none.

Microbial Contamination/Growth

During the handling of food items, microbial contamination can occur. However, microbial growth will take place only if the material is a suitable substrate and stored at a condition suitable for growth.

Foreign Matter Removal/Destruction

Processes such as sieving, washing, inspection, and metal detection are designed to remove or reduce foreign matter present in a food product. The assessment is based on the efficacy of the process.

Foreign Matter Introduction

New foreign matter can be introduced during certain steps in the process. For example, plastic pieces can become embedded in the food product during a packaging operation, and metal pieces or shavings can get into the product via faulty machinery.

Table 7 Assessment of the Process

Process	Microbial Destruction	Microbial Contamination/ Growth	Foreign Matter Removal/ Reduction	Foreign Matter Introduction	Equipment	Degree of Control
Potatoes						
Washing	–	+	+++	–	+	Moderate
Cooking	++	–	–	–	–	Moderate
Peeling/Trimming Cutting	–	+	+++	+	+	High
Chicken						
Washing	–	+	++	+	+	Low
Cutting	–	–	–	+	+	Low
Cooking	+++	–	–	–	–	High
Celery						
Washing	–	+	+++	+	+	Moderate
Trimming/Chopping	–	–	++	–	+	High
Mixing	–	+++	+	++	++	High
Filling	–	–	++	+	+	Low
Labelling	–	–	–	–	–	None
Chilling/Storage	–	+++	–	–	–	High

Production of ......Chicken and Vegetable Salad......

Equipment

Any equipment that comes into direct contact with food items can contaminate the food if it has not been thoroughly cleaned. On the other hand, carton closure equipment that does not come in contact with food directly cannot be considered a possible hazard source.

Degree of Control

The degree of control necessary to eliminate or reduce a hazard can be classified as high, low, or moderate. If the hazard is not eliminated downstream, a high degree of control is necessary. If vegetables are washed for the purpose of removing foreign material and intended to be consumed uncooked, the washing step is critical and needs a high degree of control because of possible microbial contamination. However, if the vegetables are intended to be cooked, the washing step is for the purpose of removing soil and dirt and therefore needs a low degree of control.

Displaying raw chilled seafood, poultry, and meat needs a high degree of control because it is important to keep the display temperature below 4°C in order to limit bacterial growth [21]. The consequences of failing to do so can be serious. Process steps such as filling trays and handling of food require a moderate degree of control.

3. Assessment of the Product

The methods by which food should be stored and delivered depend on the characteristics of the food product. The product is evaluated on the basis of hazards associated with its stability (Table 8). Consideration has to be given to storage conditions, packaging requirements, and delivery instructions necessary to prevent the product from undergoing deterioration or spoilage.

Storage

Products such as bottled wine, canned foods, and jam do not require special storage conditions, and hence there are no hazards associated with storage. For perishable products such as meat, fish, and ice cream, the storage conditions are critical and hence the hazards are classified as high. Vegetables require special storage conditions, but if abused, the hazards that can occur are less critical. The hazards associated with such products are classified as low/moderate compared with those associated with meat and fish.

Delivery

Food is now transported over long distances and is subject to a wide variety of storage conditions and handling techniques. Food ingredients are also transported to food producers, and in some instances contaminated substances may be transported in the same vehicle [22]. The hazards that can occur during the transport of food substances are assessed as to the extent of damage or deterioration.

Few or no hazards are associated with food products that do not require special storage conditions or handling techniques. Some food products packed in glass or plastic containers require special storage conditions during transport to prevent damage and subsequent spoilage. The hazards associated with such products are classified as low. Chilled and frozen foods require special storage conditions (temperature/time), and they may be subject to unfavorable conditions of temperature over long periods during transport. These food products require special packaging. The hazards associated with such products are classified as high.

Table 8 Assessment of the Product

Product	Storage	Delivery	Degree Of Control
Chicken and Vegetable Salad	+++	++	High

Degree of Control

The degree of control indicates the extent to which controls should be exercised to eliminate or reduce the occurrence of the hazard and is classified as high or low. Controls are not necessary for food products such as canned foods, sugar, and salt. When chilled and frozen food products are transported, a high degree of control is necessary. Some products that are transported over short distances in insulated packaging may not require special storage conditions, and failure to maintain control over temperature/time may not have serious consequences. The degree of control necessary for this category of food substance is classified as low.

4. Assessment of End Use

During the last few years food manufacturers have been aware of the increase in the cases of consumer dissatisfaction and complaints about food products [14]. Mishandling of products that leads to deterioration of quality has been cited as one of the causes of consumer dissatisfaction. Hazards can sometimes occur in the hands of the consumer as a result of inappropriate usage and abuse by the consumer.

Inappropriate Usage

Food products that can be safely consumed by the general population cause no risk. Some food products that can be safely consumed by a section of the population may not be tolerated by others, although the effects of such a hazard may not be significant. The hazards associated with the usage of such foods are classified as low. The hazards that can occur, for example, as a result of incorrect labeling or wrong formulation of food products made especially for certain groups of people, such as the elderly, children with allergies, and diabetic patients, are critical and are therefore classified as high.

Abuse by the Consumer

Food products that do not require special storage conditions or handling cause no hazards in the hands of the consumer. However, some food products have a low risk of being abused and require moderate care in handling, for example, bread left outside and open to air soon develops mold. Some foods such as cooked meat require special handling by the consumer to prevent spoilage that may not be obvious, and the consumption of such food may cause serious illness (e.g., *Salmonella* poisoning). These food products have a high risk of abuse.

Degree of Control

Controls are not necessary for food products that cannot be abused. Foods that have a low risk of abuse require a moderate degree of control, and other food products that require special handling need a high degree of control. A suitable layout to assess the end use of a food product is presented in Table 9.

C. Overall Evaluation

It is possible now to focus on preventive and control measures for identified hazards using the HACCP block diagram and evaluation schedule.

Table 9 Assessment of End Use

Product	Improper Usage	Consumer Abuse	Degree Of Control
Chicken and Vegetable Salad	–	++	High

1. HACCP Block Diagram

The purpose of the HACCP block diagram is to show the information that relates to the potential hazards and highlight the CCPs and criteria for control. HACCP block diagrams for the production of chicken and vegetable salad are shown in Figures 5 through 8.

2. Evaluation Schedule

The information from the HACCP block diagram is transferred to a table that shows the process/item, parameters to be controlled, control limits, the frequency with which the parameters will be monitored, the appropriate action in case of failure of control, and the person responsible for the task [17].

The parameters should be monitored and recorded by the person responsible for the task at stated intervals. The recording instruments are checked to ensure that they are working properly. It may also be necessary to record measurements such as temperature and time. When samples are taken for tests such as microbiological analysis, the sampling time, log number, and other relevant data must also be recorded. Regular audits by the supervisor will ensure that monitoring is continued according to the prescribed schedule. The verification of the evaluation schedule is an important part of the HACCP system. Table 10 shows an evaluation schedule for the production of chicken and vegetable salad.

3. Overall Risk Assessment and Reduction

At this stage the food producer should assess the risks associated with the raw materials, production process, and handling of the finished product in order to minimize the occurrence of hazards. Processes such as washing of raw materials and heating and handling of products and containers should be critically examined to improve the effectiveness of the HACCP system. As the result of a review, if changes are made to the ingredients, formulation, production process, procedures, equipment, or handling of the finished product, the HACCP program itself must be reviewed accordingly.

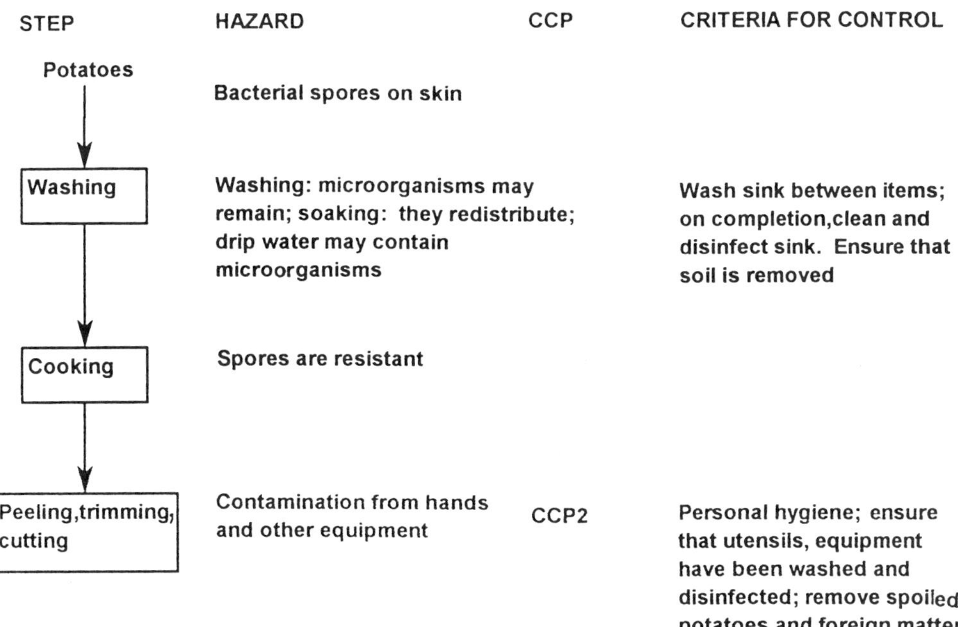

FIGURE 5 HACCP block diagram for the processing of potatoes.

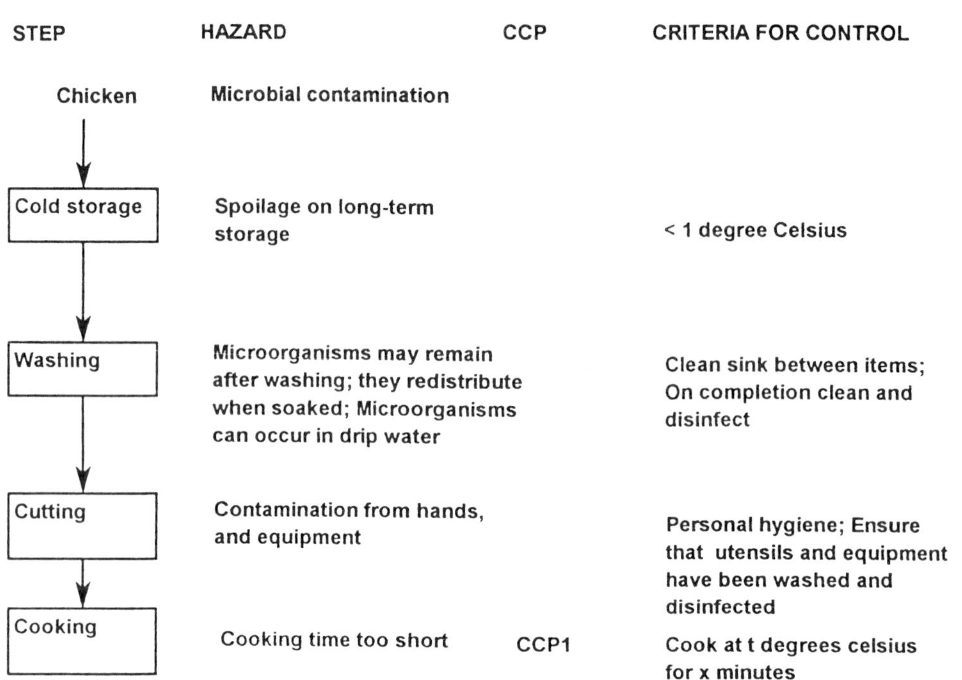

FIGURE 6 HACCP block diagram for the processing of chicken.

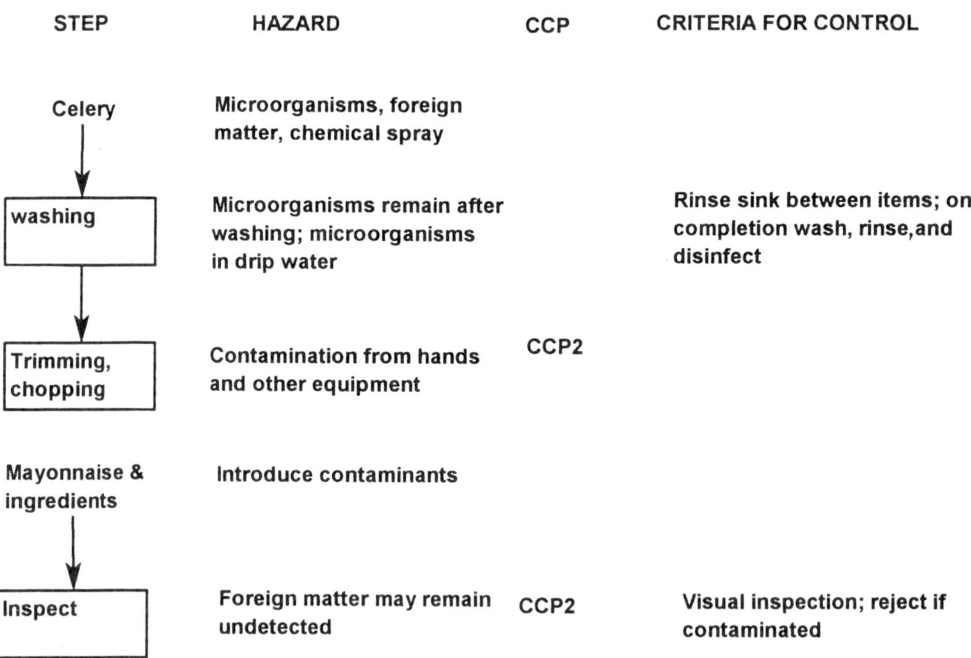

FIGURE 7 HACCP block diagram for the processing of celery.

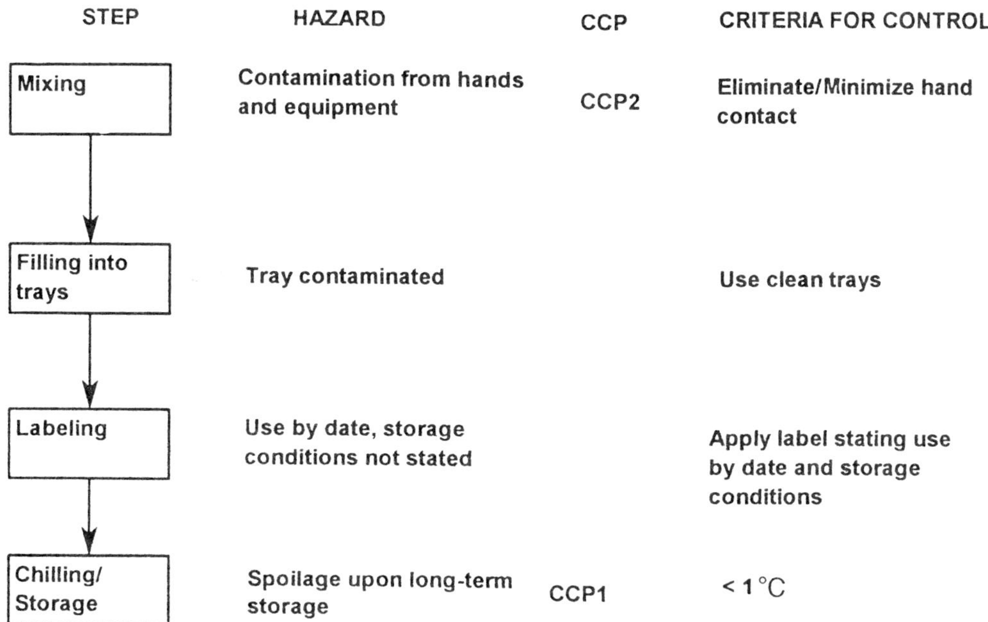

FIGURE 8 HACCP block diagram (processing steps).

4. Consumer Protection

The primary objective of an HACCP system is to protect the consumer from harm caused by hazards associated with food products. A survey carried out by a task force has revealed four causes of consumer dissatisfaction with food products [14]: unfamiliarity and expectations, price, defects, and mishandling by the consumer.

The food producers have a responsibility to ensure that all food products are adequately labeled. Information such as product description (particularly for new products), ingredients, shelf life, storage conditions, and special precautions, if any, should be provided to create awareness of the nature of the product and to prevent mishandling. Retailers as well as manufacturers can go a step further by providing information in the form of leaflets that explain handling of food products on the way home, storing of ingredients and prepared foods, and the potential hazards involved with handling foods in the household. The food producers as well as the retailers have an important role to play in educating the consumer as to food safety.

D. Management of the HACCP Program

1. Review, Audit, and Recall Processes

Review

The HACCP program, just like the quality management system, is dynamic. With advances in food technology, food producers are constantly looking at new, rapid, and safe ways of processing food. The HACCP program should be flexible enough to adapt itself to changing circumstances. New knowledge gained about microorganisms also presents a challenge to already developed and implemented HACCP programs. An HACCP program should also take into account the variability and adaptability of agents responsible for hazards to human health.

It is easy for organizations that have a quality management system in place to incorporate an HACCP program into the already existing system. Procedures such as management responsibility,

Table 10 Evaluation Schedule for Chicken and Vegetable Salad

Step/Item	Item to be controlled	Control limit	Frequency of test	Control method	Action	Person responsible
Peeling, trimming, cutting potato CCP2	Equipment cleanliness, foreign matter	Clean equipment, no foreign matter	Continuous	Visual	Reclean, remove foreign matter	Operator
Cooking chicken CCP1	Temperature, time	*t* degrees Celsius *x* minutes	Continuous	Dial thermometer clock	Cook longer	Operator
Trimming, chopping celery CCP 2	Equipment cleanliness, foreign matter	Clean equipment, no foreign matter	Continuous	Visual	Reclean, remove foreign matter	Operator
Inspect mayonnaise and ingredients CCP2	Foreign matter	No foreign matter	Every batch	Visual	Reject batch	Supervisor
Mixing equipment CCP2	Equipment cleanliness, handling	Visually clean, no dirt/residue, no hand contact	Every production	Visual	Reclean	
Mixing operation CCP2			Every production	Monitor procedure	Hold batch for evaluation, train operator	Supervisor
Chilling/Display storage CCP1	Temperature	<0°C	Every batch	Dial thermometer	Quarantine product for tests, investigate cause	Supervisor

management review, document approval and issue can be applied to the HACCP system. The function of the management representative then is to maintain the system through regular audits and reviews. The management representative responsible for the program should ensure that all new and current product specifications, standards of practice, changes to procedures and equipment, engineering and microbiological data, safety controls, and monitoring systems are reviewed regularly by the HACCP team. It is the responsibility of the team to determine, in relation to current practices or new procedures, (a) the potential hazards in ingredients, products, and risks, (b) whether the hazard can be eliminated or minimized, (c) the effectiveness of a terminal heat treatment, (d) the possibility of recontamination, and (e) the hazards associated with handling, storage, distribution, and product usage.

A hazard analysis form (Fig. 9) can be used to report the results found by the team. All accidents, misuse of ingredients, unsafe environment, and safety issues must be recorded and reported to the appropriate authorities. Authority must also be given to the operators to stop the process if in their opinion it is unsafe to operate. All safety issues must be dealt with immediately, and timely action must be taken to eliminate unsafe practices and equipment.

Audit

An HACCP audit can be defined as a systematic and independent examination to determine whether (a) HACCP activities and related results comply with planned arrangements, (b) these arrangements are implemented effectively, and (c) the arrangements are suitable to achieve the objectives. A schedule of audits must be prepared and carried out as planned. There is no standard yet for the HACCP system equivalent to the ISO 9000 standard series. However, the HACCP system can be audited against the specified requirements of the system.

HACCP audits should provide (a) an assessment of the adequacy of the existing system, (b) a benchmark against which improvements can be made and evaluated, (c) evidence that contractual and legal requirements have been met, and (d) feedback on safety issues. HACCP audits are carried out in a manner similar to quality system audits and typically apply to, but are not limited to, records and activities associated with control points and critical control points, training, and reviews. All noncompliances must be dealt with at the earliest opportunity, and products related to these noncompliances should be kept under quarantine for thorough investigation.

The Product Recall Process

A reliable and well-tested method of recall should be in place to deal with a food item which has been established to be contaminated with a harmful ingredient. Government regulations place a legal responsibility on food producers who recall food products for safety-related reasons to notify the authorities in writing within a specific period of initiating a recall (Fig.10). The traceability information will enable the affected product in the warehouse, in retail outlets, and in the hands of the consumer to be isolated. A suggested plan for a product recall is shown in Figure 11.

The text of the advertisement placed in the daily print media should comply with the statutory requirements and include:

The name of the product and the producer
The pack size and a description of the packaging
Any other details necessary for identification
The reason for recall
The necessity to identify and quarantine the stock
The manner of disposal
If the hazard to the consumer is serious, indications of clinical symptoms and advice to consult a medical practitioner
A toll-free telephone number to provide assistance to consumers

A sample recall notice is shown in Figure 12.

HAZARD ANALYSIS FORM

A. HAZARD

Identified hazard (include details such as product name, code, pack size etc.)

Identified by:..Date:..................................Time:................
Location:..

Potential effect on health/safety...
....
....
Risk..

B. ANALYSIS

Contributing factors...
....
....
....

C. SOLUTION

Recommended method for reduction/elimination of hazard..
....
....
....

D. IMPLEMENTATION

Responsibility...By when..

E. MONITORING

Results of audit...
....
....
Approved by: **Date:**

FIGURE 9 Sample hazard analysis form.

When the recall has been terminated, the recall team should review the effectiveness of the recall procedure and recommend changes, if necessary. The recall team should document the following information:

The name of the product and pack size
The reason for recall
The cause of the problem
The chronological history of the recall events with action taken
The effectiveness of the recall
The total cost of the recall
The corrective action taken
The effectiveness of the corrective action

OUR FACTORY LIMITED

PRODUCT RECALL FORM

REFERENCE:	**NAME OF OUTLET:**
CONTACT PERSON:	**DATE:**
TELEPHONE:	**FAX:**

1. PLEASE REMOVE THE FOLLOWING PRODUCTS FROM SALE TO CUSTOMERS IMMEDIATELY

PRODUCT CODE SIZE

----

2. REASON FOR RECALL

----

3. PLEASE RETURN YOUR STOCK TO:

NOTES:

1. Please return the form to..
2. If you are not holding stock send a NIL return
3. All stock will be replaced at Our Factory expense

I have returned today..(units)
to..

-- Date:
Manager

Our Phone: Our Fax:

FIGURE 10 Recall form.

In case of a recall, the accuracy of information and the speed with which action is taken is important.

2. Training

HACCP has now been accepted as the most cost-effective means of controlling hazards related to microbiological, physical, and chemical contamination of foods. The implementation of the HACCP

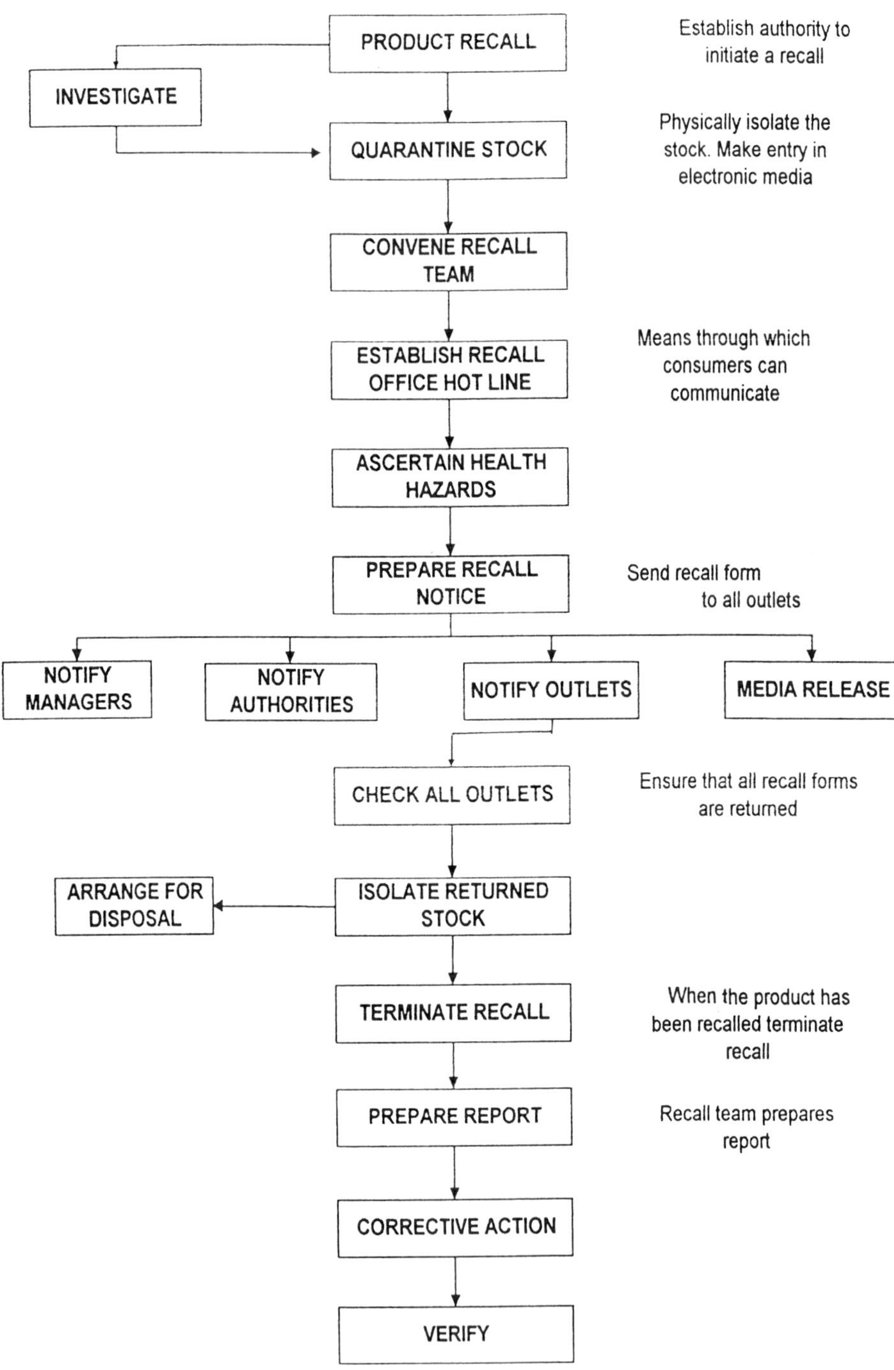

Figure 11 Recall plan.

FOOD RECALL NOTICE

FOOD PRODUCTS LIMITED STRAWBERRY JAM

FOOD PRODUCTS LIMITED IS RECALLING 375G JARS OF STRAWBERRY JAM, BECAUSE OF THE DISCOVERY OF GLASS IN TWO JARS OF FOOD PRODUCTS' STRAWBERRY JAM. THE RECALL APPLIES TO THE PRODUCT HAVING THE CODE L23A6.

THE PRODUCT HAVING THE ABOVE CODE SHOULD NOT BE CONSUMED.

AS A SAFETY MEASURE WE ARE RECALLING ALL SUPPLIES OF THIS PRODUCT WITH THE ABOVE IDENTIFICATION ON THE LABEL.

THERE HAVE BEEN NO REPORTS OF INJURY OR ILLNESS. HOWEVER, ANY PERSON CONCERNED ABOUT THEIR HEALTH AS A RESULT OF CONSUMING THE PRODUCT SHOULD SEEK MEDICAL ADVICE.

PLEASE RETURN THE PRODUCT TO THE POINT OF PURCHASE FOR A REFUND OR PHONE TOLL FREE.............

THE RECALL DOES NOT APPLY TO ANY OTHER FOOD PRODUCTS LIMITED STRAWBERRY JAM HAVING A DIFFERENT CODE OR ANY OTHER FOOD PRODUCTS LIMITED PRODUCT.

WE SINCERELY APOLOGIZE FOR ANY INCONVENIENCE CAUSED BY THIS RECALL.

FOOD PRODUCTS LIMITED
(ADDRESS)
FAX

FIGURE 12 Recall notice.

plan is a team exercise. Training and education are essential if full benefits are to be achieved [23]. Food producers have the responsibility to produce safe food, and the regulatory bodies need to be competent in monitoring the HACCP programs.

HACCP training should provide (a) knowledge of concepts, principles, and benefits of the HACCP program, (b) practical skills and knowledge necessary for the implementation of the HACCP program, and (c) the skills needed for further development of the HACCP program. Regulatory authorities, senior managers, shop floor personnel, and technical managers should be involved in the HACCP training program. The training program itself should target the needs of each of the above groups. For example, regulatory authorities need to have knowledge of the concepts, principles, and benefits of the HACCP programs, and shop floor personnel need to have practical skills.

Scientific data should be used to support the training progress. Software on HACCP programs is now available, and the material presented in such programs can be an essential tool for training.

3. HACCP in the Overall Quality System

The survival of a business depends upon its ability to satisfy the customer's needs and expectations at an affordable cost, which is achieved through the establishment of an effective quality management system. It provides a basis for continuous improvement. The effectiveness of a quality management system can be assessed against standards such as ISO 9000.

Table 11 The Seven Principles of HACCP

Principle	Step	Activity
1	Hazard analysis	Construct a flow diagram of the process, identify and list all potential hazards
2	Identification of CCPs	Identify CCPs using the decision tree
3	Establishing critical limits	Establish critical limits for all preventive measures associated with each CCP
4	Monitoring	Establish monitoring requirements and monitor at stated intervals
5	Corrective action	Decide on corrective action to be taken when the process deviates from specified limits
6	Documentation	Establish effective documentation necessary to implement the HACCP system
7	Verification	Establish procedures for verification to ensure that HACCP system is working correctly

The quality management system should be designed to meet the individual needs of the organization. Food manufacturers, while being aware of the quality of the foods, should also realize that food safety is an absolute requirement. Regulatory authorities all over the world enforce laws aimed at protecting the consumer from harmful food. HACCP is a management tool that focuses attention on food safety and complements the quality management system.

Implementation of a quality management system and an HACCP program both require teamwork. The requirements of an HACCP program are embodied in the seven principles as defined by the National Advisory Committee on Microbiological Criteria for Food [24]. These are summarized in Table 11.

The procedures relating to the HACCP system can be developed using the seven principles as the basis. The clauses in the ISO 9000 standard, such as responsibility, management review, and organization, can be applied while keeping the focus on safety rather than on quality. An HACCP system can be incorporated into the quality management system by making reference to individual clauses such as process control, corrective and preventive actions, internal quality audits, document control, and others common to both systems. The HACCP system should not be limited to the seven principles and should also include procedures relating to HACCP planning, customer complaints, product recall, and control of HACCP records.

4. Benefits of Implementing an HACCP System

The HACCP system is a preventive and cost-effective approach to food safety and is more effective in preventing foodborne illnesses than traditional approaches. Application of an effective HACCP system leads to both economic as well as certain intangible benefits, summarized in Table 12.

VI. CASE STUDY: PRODUCTION OF CHICKEN AND VEGETABLE SALAD

A. Product Description

Chicken and vegetable salad is a ready-to-eat chilled product prepared with cooked chicken, freshly sliced celery, and cooked, peeled, and diced potato in a dressing mixture. Chicken breasts (skinless, boneless) and other ingredients are bought at the market. The dressing mixture consists of a pasteur-

Table 12 Benefits of Implementing the HACCP System

Economic benefits
1. Maintain shelf life of the product as stated on the package
2. Fewer customer returns and claims
3. Efficient operation of machinery
4. Minimal or no reprocessing and corrective action
5. Lower wastage
6. A consistent product
7. Better control of the process allowing for remedial action before a problem occurs

Intangible benefits
1. Satisfied customers and staff
2. Enhanced reputation of the organization and better image
3. Greater consumer confidence in the product
4. Possibility of predicting potential hazards
5. Greater involvement of staff in day-to-day operations
6. Clean and hygienic working environment

ized mayonnaise base blended with spices. The product is put into plastic trays, which are covered with a peelable foil membrane and packaged in a cardboard outer package. The product is refrigerated and brought to the refrigerator case as required.

B. Assessing the Hazard Potential

A block diagram and a process flow chart for the production of chicken and vegetable salad have already been described (Figs. 3 and 4). The next step in the HACCP technique is to assess the hazards associated with the raw materials, processing steps, product, and end usage. It will then be possible to determine the raw materials and the processes that are critical. The whole process is then examined in order to establish the critical control points.

1. Raw Materials

Of the raw materials, the most hazardous item is chicken. Chicken may be contaminated with pathogenic bacteria. The storage conditions after collection and during transport are also critical. Potatoes may carry bacterial spores on the skin. Celery may be contaminated with bacteria and spray residues. Although chicken, potato, celery, and other ingredients may be contaminated with foreign matter, washing and/or inspection prior to use will reduce the risk of contamination. Mayonnaise, being acidic, may be contaminated with acid-tolerant bacteria.

2. Process

The process assessment of chicken and vegetable salad is given in Table 7. Potatoes are washed for the purpose of removing foreign matter from the skin. Cooking of chicken is the only lethal step in the whole process and requires careful control of temperature and cooking time. Potatoes are also cooked, but the bacterial spores residing on the skin are not destroyed. After cooking, there are several steps (peeling, trimming, and chopping) that involve manual handling. All such steps are hazardous, and there are no lethal steps downstream. Washing of celery may reduce the level of vegetative pathogens but will not eliminate them. Microbial contamination can occur during the mixing operation. The salad is hand-filled into trays and exposed to the environment. The frequent manual handling during the filling operation is a health hazard. The temperature of chilled storage and display cabinets must be controlled to prevent the growth of microorganisms. In the last column (degree of control) in Table 6, the steps recognized as high and moderate should be considered for the identification of CCPs in the final analysis.

3. Product

Microbial growth can occur if the storage conditions of the salad are not controlled at the food producer's premises and during transport and distribution (Table 8).

4. End Use

The chicken and vegetable salad is a food product that is prepared for the general population. However, after purchase, abuse such as leaving the container open or keeping it at warm temperature can cause the product to deteriorate and be a health hazard if consumed (Table 9).

C. HACCP Block Diagram

The decision tree (Figure 2) can be used as a tool to determine the CCPs. The block diagrams are shown in Figures 5 through 8. For example, the potato skins could be contaminated with bacterial spores. The decision tree can be applied in order to determine whether the step of washing the potatoes is a CCP:

Q1. Hazard: bacterial spores
Q2. Control measures exist.
Q3. A subsequent step (peeling) will reduce the hazard downstream. The level of control achieved: reduction
Q4. Further reduction at this step is not possible.
The washing step is therefore not a CCP.

Similarly, it can be shown that the step of cooking potatoes is not a CCP because the bacterial spores are not destroyed by cooking. The peeling, trimming, and cutting steps are classified as hazards because of manual handling. Again, the decision tree can be applied to determine the category of CCP:

Q1. Hazard: microbiological
Q2. Control measures exist.
Q3. Subsequent steps will not eliminate the hazard or reduce the hazard to an acceptable level.
Q4. The raw materials, process, or product cannot be altered to eliminate or reduce the hazard to an acceptable level.

Therefore, the peeling, trimming, and cutting steps are CCPs.

The decision tree is applied to all the steps to determine the CCPs. The chicken can undergo spoilage during storage. However, the storage step is not classified as a CCP because there is ample opportunity to inspect for spoilage during the subsequent steps of washing, cutting, and cooking. Cooking the chicken for a sufficiently long period can destroy microorganisms. Hence it is a lethal step and is classified as CCP1.

The steps of peeling and cutting after cooking can introduce contaminants from hands, knives, and other equipment. However, bacteria cannot multiply in the presence of high acid ingredients such as mayonnaise, vinegar, pickle, etc. Therefore, the mixing step can give absolute control if sufficiently high-acid ingredients are added and thoroughly blended. Thus, the peeling and cutting steps are not CCPs.

Celery may be contaminated with foreign matter and microorganisms. The washing step will remove foreign matter but may not reduce the risk of vegetative pathogens to an acceptable level. While the reduction of foreign matter is an important control point, the trimming and chopping operation after the washing step will provide ample opportunity to ensure that foreign matter is removed. Therefore, the washing step is not classified as a CCP. However, if there is sufficient evidence to show that the washing step does reduce the level of microbial contamination to an acceptable level, the washing step could be considered as a CCP.

The plastic trays into which the salad is filled are new. Although the trays may have foreign matter, they are inspected prior to use and hence contamination with foreign matter can be eliminated. The process of chilling and storage of the finished product are CCPs (CCP1), and storage under correct conditions will prevent the growth of microorganisms.

D. Evaluation Schedule

The steps that have been identified as CCPs are controlled as specified in the evaluation schedule (Table 10).

E. Overall Risk Assessment and Reduction

It is now possible to consider how risks can be reduced by making some changes to the process.

Vegetables are a major source of contamination. Potato and celery can be purchased from an approved grower. Celery may be blanched or washed with chlorinated water to reduce the bacterial load. Potato can be washed, peeled, trimmed, cooked and diced, or diced and cooked.

Contamination can occur during the manual handling operations, particularly after the cooking step. The use of disposable gloves during manual handling will minimize contamination.

Pathogenic bacteria do not grow in acid media and contamination during the mixing step can be controlled by adjusting the pH to 4.5 or lower. If this is not practicable, the formulation can be tested and the recipe then followed accurately.

Tamper-proof or tamper-evident packs can be used as a precaution against tampering and possible contamination. The risk of abuse by the consumer can be minimized by providing warning labels that give instructions on storage after purchase.

Metal detectors can be installed on line to detect metal objects in the food.

The finished product can be subjected to microbiological tests to detect the presence of specific microorganisms.

F. Good Manufacturing Practices

In preparing the salad, there are several operations that are control points but not Critical Control Points. These steps can be controlled by implementing good manufacturing practices (GMP).

Inspection and storage of raw materials are essential for controlling the quality of incoming ingredients. The cleaning of equipment and the food-preparation area should be closely monitored. Food-grade detergents must be used to clean equipment and utensils that come into contact with food. Use of disposable gloves during manual handling operations should be considered to avoid any contamination of food from hands.

Training in personal hygiene and correct operation of machinery is essential for those who are working in a food-production or food-service facility. Training programs in food safety and sanitation that include short courses, workshops, and training manuals should be developed for the continuous education of all employees in food-production and food-service establishments.

A preventive maintenance program for machinery including chillers must be carried out to prevent breakdown of machine parts and ensure continuous operation. During maintenance, all food products and food containers should be removed from the vicinity, and production should commence only when the machines are cleaned/sanitized and restored to proper working condition.

All food products must bear a batch number or a lot number to enable traceability. Information such as shelf life and storage conditions should be provided on the label to prevent abuse by the consumer. The use of status stickers such as HOLD, REJECT, PASSED, and “QUARANTINE”, having separate storage areas for each category, will prevent unintentional dispatch of goods other than those released for sale.

The GMP program should also include regular audits of production facilities, procedures, and test methods. Audit reports should be used as a basis for quality improvement in the plant.

REFERENCES

1. *The New Encyclopaedia Britannica 8*:695 (1983).
2. Y. Motarjemi, F. Kaferstein, G. Moy, S. Miyagawa and K. Miyagishima, Importance of HACCP for public health and development: The role of the World Health Organization, *Food Control 7*:77 (1996).
3. O. P. Snyder, HACCP—an industry food safety self-control program—Part I. *Dairy, Food Environ. Sanit. 12*:26 (1992).
4. R. Vail, Fundamentals of HACCP, *Cereal Foods World 39*:393 (1994).
5. B. Nordmark, HACCP, *Food Technol. N.Z. 30*:18 (1995).
6. R. Kirby, HACCP in practice, *Food Control 5*:231 (1994).
7. E. Underwood, Good manufacturing practices—a means of controlling biodeterioration, *Int. Biodeter. Biodegrad. 36*:449 (1995).
8. E. S. Garrett and M. Hudak-Roos, Use of HACCP for seafood surveillance and certification, *Food Technol. 44*:159 (1990).
9. H. Bauman, HACCP: concept, development and application, *Food Technol. 44*:156 (1990).
10. H. E. Baumann, The origin of the HACCP system and subsequent evolution, *Food Sci. Technol. 8*:66 (1994).
11. J. Brooks and M. Reeves, Methods of controlling hazards, *Managing Food Safety*, Food Technology Research Centre, Department of Food Technology, Massey University, Palmerston North, 1995, p. 12.
12. J. S. Avens, Safety of food service delivery systems in schools, *Safety of Foods* (H. D. Graham, ed.), AVI Publishing Company Inc., Connecticut, 1980, p. 758.
13. B. A. Munce, Hazard analysis critical control points and the food service industry, *Food Technol. Aust. 36*:214 (1984).
14. T. D. Beard, HACCP and the home: the need for consumer education, *Food Technol. 45*:123 (1991).
15. J. Brooks and M. Reeves, "Critical control points," *Managing Food Safety*, Food Technology Research Centre, Department of Food Technology, Massey University, Palmerston North, 1995, p. 2.
16. B. J. Bobeng and B. D. David, HACCP Models for quality control of entree production in food service systems, *J. Food Prot. 40*:632 (1977).
17. Microbiology and Food Safety Committee of the National Food Processors Association, HACCP implementation: a generic model for chilled foods, *J. Food Prot. 56*:1077 (1993).
18. J. Brooks and M. Reeves, "Evaluating control points," *Managing Food Safety*, Food Technology Research Centre, Department of Food Technology, Massey University, Palmerston North, 1995, p.5.
19. R. A. Savage, Hazard analysis critical control point: a review, *Food Rev. Int. 11*:575 (1995).
20. J. Brooks and M. Reeves, Methods of controlling hazards, *Managing Food Safety*, Food Technology Research Centre, Department of Food Technology, Massey University, Palmerston North, 1995, p. 1. (section 7).
21. F. L. Bryan, Application of HACCP to ready-to-eat chilled foods, *Food Technol. 44*:70 (1990).
22. D. L. Archer, The need for flexibility in HACCP, *Food Technol. 44*:174 (1990).
23. T. Mayes, HACCP training, *Food Control 5*:190 (1994).
24. Microbiology and Food Safety Committee of the National Food Processors Association, Implementation of HACCP in a food processing plant, *J. Food Prot 56*:548 (1993).

25

Commercial Considerations: Managing Profit and Quality

Anne Perera
Food and Nutrition Consultancy Service, Auckland, New Zealand

Gerard La Rooy
Business Process Improvement, Auckland, New Zealand

We were delighted to be invited by the editor to write this last, but in our view by no means the least important chapter of this handbook, a book that we are confident will make a significant contribution to the international food-preservation profession and one we expect to be used by many in the field. While the handbook is essentially a technical publication written for technical people, it is appropriate to keep in mind the books' fundamental purpose, which can be stated in its simplest form as "caring for the customers." Putting caring for the customers in a practical context, we can say that organizations must have the intent and appropriate capability to develop and produce wanted and risk-free products and services. When we say risk-free, we mean not only risk-free for the customer—however important that requirement is—it is also very much applicable to the organization's objectives. Indeed, the proper management of risk is particularly about not endangering the continuity of existence of the business enterprise.

In determining the content of this chapter, we had to make a choice between breadth and depth. We felt it was more appropriate to go for breadth in order to provide the reader with a reasonable range of topics so that anyone interested can undertake further reading as they see fit.

When working on the structure of this chapter, we were mindful of the thrust of this part of the handbook—enhancing food preservation by indirect approach. For us this means that the food-preservation profession must be prepared to go beyond their traditional fields of expertise and see their contribution in a much wider context. It also reinforces our view that irrespective of the method of preservation employed, the selection of the method must not only be based on sound technical grounds, but also on appropriate business considerations. Please note that while the chapter as a whole was compiled by us jointly, the specialized sections were contributed by either Anne or Gerard individually.

I. MANAGING PROFIT

A. The Business Environment

For a business to be successful it must have sound processes, up-to-date and profitable products, and well-managed services. Successful companies, besides being highly competent in their respective technical fields, also need to be very skilled in business management. While this need for business skill is generally accepted, it should be appreciated that we are witnessing a fundamental change in what business management ought to be about. Business structures are changing from "tall" to ones with few levels and with fewer "functional silos." This means, for example, that food preservation is becoming more fully integrated with marketing, production, and customer service. To illustrate the points made so far, we will look at the case of "flat earth thinking" versus "round earth thinking."

1. Management Structures and Practices

Peter Scholtes of Joiner Associates has likened the traditional style of management with the belief that the earth was flat. He explains that people who believed that the earth was flat would ask questions like: "What happens when I sail my ship until I reach the edge?" Once science proved that the earth was round, a sudden shift in thinking occurred, and questions about falling off the edge became irrelevant.

What this means is that organizations that are flat earth based (and we believe many organizations are) will need to undergo some revolutionary rather than evolutionary change. The revolutionary changes needed will make many current management practices irrelevant, redundant, and in many cases quite wrong. Once the big change in thinking has been accepted and the practices implemented, we can employ the process of evolutionary change for further development and improvement.

To illustrate how current structures and practices are becoming irrelevant or inappropriate, we can construct a simple comparative table (Table 1). Looking at the table, we can observe that some of the items are either one way or the other, e.g., directive driven or direction led. Other items are more continuous (tall vs. flat). If we consider the complete table, however, there can be little doubt that we are dealing with a dichotomous situation—the traditional versus the new. Expecting "Traditional" organizations to gradually change into "New Age" ones is a bit like expecting an ocean liner to start sprouting wings and gradually change into a 747 aircraft. Putting it another way, we sometimes need revolution before there can be evolution. These changes in business have and are having a profound effect on the food-preservation profession, and practitioners need to be fully aware of these changes lest they be left behind with their flat-earth thinking.

2. The Changing Role of the Food Profession

The science of food preservation too is undergoing much change. It should be appreciated, however, that we are not just concerned with technological change. On the contrary, we believe that the major change and challenges facing the profession will come form the changes to the business environment referred to earlier.

Consequently there is a need for the profession to become more integrated with the totality of the business. What this means in practice is that professionals have to assume a wider role and accept greater responsibility for the success of the companies that employ them. To identify the specific change in emphasis we believe is necessary to ensure the profession's further effectiveness and relevancy, it is useful to employ another comparative table (Table 2). While most items in the table are fairly self-explanatory, the question of functional (personal) objectives versus company objectives warrants some examination. For many technical personnel the main work focus tends to be on objectives that are very close to the person in question and on the quality of the actual process. "Doing things right," it could be called. For example, a product development professional will have expectations about: the quality of the development process itself and what is to be ready by when. He or she will also be concerned with the robustness of data and with the appropriateness and quality of any tests and experiments. It is our view that this almost exclusive focus on functional objectives is not sufficient in today's commercial environment, let alone in tomorrow's. What is required of the New

Table 1 The Changing Business Environment

Traditional (flat earth)	New Age (round earth)
1. Directive driven	1. Direction (vision) led
2. Management through power and position	2. Management through earned influence
3. Controlled by rules	3. Guided by values
4. Compliance	4. Commitment
5. Corporate/center/CEO focused Executives near CEO Tall (many levels)	5. Customer focused: Executive near staff Flat (fewer levels)
6. Vertical functional silos	6. Integrated value chain
7. Functional specialists	7. Integrated resources
8. Internal customer is the next level	8. Internal customer is the next process in the value chain
9. Emphasis on control: Centralized decision making Risk aversion Reliance on experience Imposed results Low rate of improvement	9. Emphasis on improvement: Dispersed decision making Experimentation Learn to learn Agreed achievements Fast rate of improvement
10. Industrial relations	10. People development
11. Huge differences in reward levels	11. Smaller differences in reward levels
12. Staff are data gatherers (information for the center)	12. Staff are information users (information for the coalface)
13. Information belongs to the selected few	13. Information sharing and openness

TABLE 2 Conventional and Business-Aligned Approaches

Conventional	Business Aligned
Narrow view	Wider role
Advisory	Accountable
Product emphasis	Customer focus (external *and* internal)
Production driven	Market led
Cost unawareness	Profit appreciation
Quality control	Quality management
Risk avoidance	Quantified assessment of exposure
Functional skills	Business knowledge
Functional/Personal objectives	Company/Organization objectives

Age professional is, in addition to technical competence and focus, a marked increased in appreciation of company objectives, overall results, and the commercial "levers" that drive them. We can term this demand for additional understanding and emphasis ensuring that we do the right things. The important point is that unless we do the right things as well as doing things right, our efforts may be ill-directed and as a consequence largely wasted.

B. Commercial Requirements

Most food-preservation technologists are bound to be employed in a commercial enterprise of one type or another at some time during their working lives. Even for those with entirely academic careers, there is still the issue of increasing commercialism in the management of academic institutions. Consequently it is important for professionals in any technical field to have a reasonable appreciation of business fundamentals as well as the commercial and other expectations of the enterprise for which they work. Understanding these fundamentals and expectations will enable the technologist to contribute to the organization's success to the fullest extent possible.

The work of technical staff can profoundly affect, both positively and negatively, the financial performance of most organizations. Sound technical developments can open up new business opportunities, lead to greater efficiency, and secure a stronger position in the marketplace. Misdirected efforts, on the other hand, are likely to result in increased and unnecessary complexity, higher costs, poorer asset utilization, and lower profitability. It is worth noting that profit is not a dirty word but is in fact a vital prerequisite to growth and long-term success. Profitability is a must if funds are to be available for investment in new products, processes, and technologies.

1. Revenue, Cost, and Assets

Of the many factors that impact on a business's financial performance, the following are the most critical:

Revenue
Costs
Assets employed

Food professionals can and often do have a very significant impact on all three. Before looking at how this may come about, it is important to understand the way the factors affect overall financial performance and their interrelationship.

Revenue

Assume there are two firms, X and Y, with annual sales revenues of $50 million and $40 million, respectively (Fig. 1). Which is the better performing company? In terms of sales revenue the answer is firm X, but more information is needed to determine which is the sounder firm, e.g., costs.

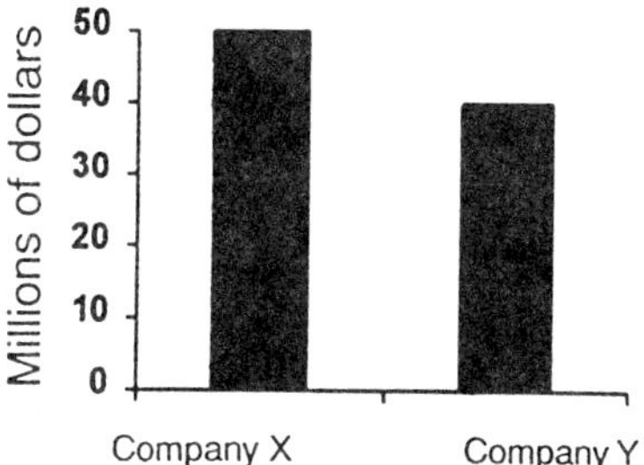

FIGURE 1 Revenue companies X and Y.

Costs

Assume the costs are \$40 million for company X and \$20 million for company Y (Fig. 2). With a margin of \$10 million for firm X and \$20 million for firm Y, which is the better company? Looking at just the margin, it is clearly firm Y, but before we award them the annual prize for performance we need to look at the resources employed to generate the \$20 million.

Assets Employed

When assessing financial performance the resources are considered to comprise the assets employed by the organization, such as buildings, land and plant (fixed assets), and funds tied up in inventories and debtors (current assets—these can be turned into cash reasonably quickly).
If the total assets used by the companies are \$20 million for X and \$50 million for Y, which now shows the better performance?

$$\text{Performance} = \frac{\text{Margin}}{\text{Assets}} \times 100\%$$

$$\text{Performance for company X} = \frac{\$10}{\$20} \times 100\% = 50\%$$

$$\text{Performance for company Y} = \frac{\$20}{\$50} \times 100\% = 40\%$$

Company X shows the better return on the money invested in it.

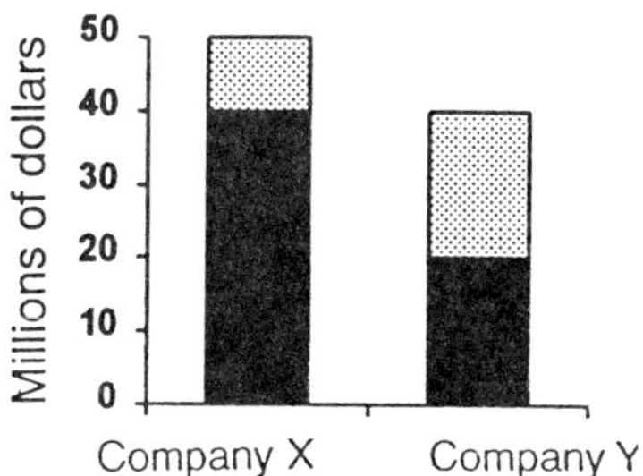

FIGURE 2 Costs companies X and Y.

Observations

For company X:

The lower margin may mean the business deals in low-value items.
Lower use of assets could mean quick turnover of stocks, calculated as follows:

$$\frac{\text{Annual sales}}{\text{Average inventory value}}$$

If, for example, half of the \$20 million were tied up in inventories, the stock turnover per year would be 50/10 = 5 times per annum.
The low asset figure may also mean low plant book values (plant could be obsolete).
The old plant may be costly to operate and hence be the reason for the low margin.

For Company Y:

Company Y appears to deal in higher-value items.
The higher margin may attract competitors to move in on the market.
Stock turnover (again assuming half of the assets are tied up in inventory) would be \$40/\$25 = 1.6 times per annum.
The company may have a more modern plant, which may mean a lower cost structure and hence the higher margin.

2. Fixed and Variable Costs

Many actions and decisions made by technical staff have a significant effect on costs. It is therefore important for the decision makers to appreciate the difference between fixed and variable costs. Let us consider the basic definitions first.

Fixed Costs

In brief, fixed costs are those costs not affected by changes in output. An example of a fixed cost would be the cost of monthly machine rental which we would incur no matter what the output. Note, however, that any production beyond machine capacity would mean a step in our fixed costs, i.e., cost of renting another machine. Also, the passage of time can lead to increases or decreases in fixed costs. Consequently, the earlier statement on fixed costs needs some qualification: Fixed costs are not affected by changes in output within a specified output range, and fixed costs remain constant within a specified time period.

Variable Costs

Variable costs vary with changes in output, and the unit cost is generally deemed to be constant, although that is by no means always the case in reality. For example, bulk purchase may bring the cost per unit down, or in other cases the cost of additional units may be higher than the average. When comparatively small quantity ranges are considered, however, variable costs may be taken as truly variable, i.e., the cost per unit stays constant and total variable cost can be found by multiplying the unit costs by the quantity used.

Fixed Versus Variable Costs

In many business situations, including the technical field, there are often trade-offs to be considered between fixed and variable costs. It is important to have a clear understanding of the difference between fixed and variable costs as well as the relationship between the two.

Assume a company wishes to launch a new product and has been given two different manufacturing proposals.

TABLE 3 Comparing Fixed and Variable Costs

	Proposal 1	Proposal 2
Fixed cost pa	$100,000	$200,000
Variable cost per unit	$10	$7
Selling price per unit	$15	$15
Sales volume units pa	40,000	40,000

1. Proposal 1: Low fixed cost, high variable cost
   Fixed cost (equipment, etc.) = $100,000 pa
   Variable (material, labor) = $10 per unit
   Selling price = $15 per unit
   Sales volume = 40,000 units pa
2. Proposal 2: Higher fixed cost, lower variable costs: The second proposal is more mechanized and hence requires less expensive material and less labor. Fixed costs are consequently $100,000 higher, but variable costs are lower at $7 per unit.
3. Initial comparison: The two proposals can be readily compared by means of a table (Table 3).
4. Comparing profit and break-even points: To determine which of the two proposals is better in terms of the data supplied, we need to calculate (a) total profit for the year and (b) the break-even point, i.e., the volume at which total revenue equals total cost. Again the two proposals can be compared (Table 4).
5. Observations: While proposal 2 is more profitable, proposal 1 is less risky. If sales are evenly spread throughout the year, it would take the first proposal 6 months to break even, while for proposal number 2 the break-even point could occur at 7½ months. The break-even analysis, while not a sophisticated tool, is quite useful in illustrating the differences between proposals. Take care, however, as in reality fixed costs have a habit of going up (or sometimes down), for example, when needing to rent an extra piece of equipment. Also, variable costs are rarely as linear as assumed here, although they can usually be taken as linear between reasonably small quantity intervals.

3. Price, Margins, and Costs

Before discussing the impact of technical staff on costs, it is necessary to introduce a number of cost-related concepts. The approach to the management of costs in an organization is very much depen-

TABLE 4 Comparing Profits and Break-Even Points

	Proposal 1	Proposal 2
Sales revenue = quantity × selling price	600,000	600,000
Fixed cost	(100,000)	(200,000)
Variable cost = quantity × variable cost per unit	(400,000)	(280,000)
Profit	100,000	120,000
Contribution margin = selling price – variable cost	$5	$8
Break-even point = fixed cost ÷ contribution margin	20,000 units	25,000 units

dent on how costs are viewed—as a starting point or as the end point. Figure 3 illustrates different ways of considering costs.

Traditional Cost Plus

The traditional method of determining a selling price was to add up all direct and indirect costs (e.g., overheads) and add a suitable margin. This approach works when there is little or no competition (e.g., due to import control). Unfortunately, some companies will persist with this approach even when, because of changed conditions, it is no longer appropriate (Fig. 3, method 1).

Market Price

In a free market the price is set by that market and has nothing to do with costs or margins. The change from "tradition" to "market price" method took place in some countries during the 1950s and 1960s but in many other countries during the 1970s and early 1980s. The principal effect of the change to a market (and as a rule lower) price was reduced margins, since costs were generally considered as "fixed" (Fig. 3, method 2).

Required Margins

The squeeze on margins meant that many organizations experienced unsatisfactory returns on their invested capital. However, to attract new capital, returns and hence margins must be adequate. Since prices are largely beyond the control of the organization and margins must be achieved, the real situation is that:

PRICE — MARGIN → COST

This means that cost is the important element and one that management needs to focus on constantly.

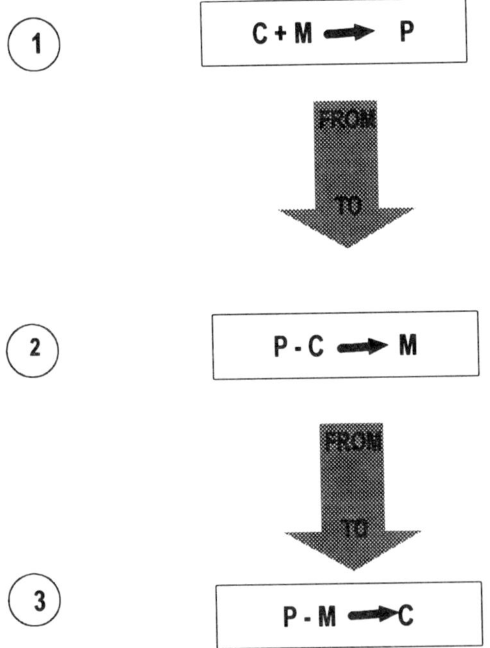

Figure 3 C, M, P, relationships, where C = cost, M = margin, and P = price.

C. Technical Impact on Business

Most managers do not fully appreciate the impact that technical resources have (or could have) on the business in general and on revenue generation in particular. What is more, it is quite likely that the majority of technical staff are also not completely aware of their contribution. We will look briefly at four types of impact:

Impact on revenue
Impact on costs
Impact on assets
Impact on cash

1. Impact on Revenue

In many cases the technical staff provide the lifeblood for the organization's future, as all products and processes require constant renewal and replacement if there is to be security of future revenue.

Types of Impact

When considering the impact that technical professions have on an organization's revenue generation, it is necessary to appreciate that there are shorter-term and longer-term impacts.

1. Shorter-term impacts (tactical or current revenue): Projects coming under this heading are generally concerned with existing products and processes or new products within existing competencies (e.g., product line extensions, product improvements, ingredient substitutions, etc.).
2. Longer-term impacts (strategic or future revenue): These are concerned with new technology and products outside current competencies.

Conceptual and Analytical Soundness

An important point to note is that a project can be analytically correct but conceptually wrong at the same time. In other words, the arithmetic is fine but the underlying premises are flawed. Technical staff should seek advice if required and take particular care in (a) setting out all underlying assumptions clearly, (b) defining the scope of the project accurately (including in some cases any matters which, while outside the project, may still be relevant, (c) listing all agreed expectations in terms of both magnitude and probability, and (d) assessing the economies of the proposal soundly and presenting the justification in a logical manner.

Reliable Tangible Cost and Benefit Information

Proposals need details about the various types of costs and benefits, that is: one-off costs and benefits (e.g., purchase of new packaging printing dies, the sale of some piece of equipment made possible by the project) and ongoing costs and benefits (e.g., increased production energy costs, reduced labor costs).

Intangible Cost and Benefit Information

It is important to included all intangibles, even if it is difficult to estimate dollar figures. It may in fact be better not to try to show dollar figures, as it often leads to much argument and debate. Make sure, however, that all intangibles are clearly identified and list them separately, preferably at the end of the presentation. Examples of intangibles include:

Enhanced brand image
Improved competitive position
Increased customer satisfaction
Improved staff morale

Adverse impact on the quality of life of people in the neighborhood
Waste generated by proposed process not suitable for recycling

The point to remember is that intangibles should not be dismissed as of little importance because of our inability to put precise dollar figures on them. On the contrary, with many project the intangibles (the "soft" issues) are often more important than the "hard" (dollar) issues.

Technical Revenue Contribution Reporting

When it comes to ongoing management, technical function professionals may assist their cause by ensuring that the revenue contribution is duly reported on and acknowledged (Table 5).

2. Impact on Costs

Technical staff can and often do have a very significant impact on costs. Frequently the technical professionals are the members of an organization who make the decisions that affect not only the total costs of a development but which more particularly determine the balance between fixed and variable costs. Understanding that balance is important especially when dealing with new proposals. Some examples are outlined below.

1. Increased R&D costs: Deciding to spend more on R&D (higher fixed cost) in order to gain lower product costs (variable costs).
2. Higher specification levels: Insisting on a higher specification than may be necessary is likely to push up the costs of production (higher variable costs). The decision can also result in higher fixed cost if, for example, the close tolerances specified require a more expensive piece of equipment than would be the case with a less demanding specification.
3. Product line extension: Every time a new product is added to a company's product portfolio, it is likely that production run sizes and frequencies will be affected (usually adversely). Consequently, the set-up cost (a fixed cost) has to be recovered over a smaller number of units, thus driving up the total cost per unit.

3. Impact on Assets

We have already considered briefly the importance of the assets employed when assessing a company's performance. As with costs, the actions and decisions of the technical professional can have considerable impact. Again some examples can illustrate the position.

Current Assets

When launching a new product, a company invariably needs to increase its working capital (current assets), for example, (a) extra funds tied up not only in finished goods but also in new ingredients and components and (b) additional funds required to finance the increased debtors resulting from the sales of the new product.

Fixed Assets

Technical developments often lead to companies investing in new processes and equipment. It is important that this new investment does not dilute the company's financial performance, that is, the return on the new investment should not be less than the norm for that particular industry and preferably be better than the company's current average return. Exceptions may occur when companies invest in new processes and facilities for strategic reasons, provided of course that this fact is understood and accepted by the decision makers.

4. Impact on Cash

Companies may be profitable and still fail if they run out of cash and cannot pay their bills. While a proper discussion on cash management is beyond the scope of this chapter, it should be noted that

Table 5 Sample of Technical Revenue Contribution Report

	Year to date	Budget	Variance	Last year	Estimate for remainder of year
Net sales $ 'old' revenue					
Net sales $ 'new' revenue					
Total net sales					
% New revenue of total net sales					
Technical costs					
% Technical costs of new revenue					

profit and cash are not the same things. It is important that the technical professional has some understanding of that difference, since again technical decisions can have considerable impact. (If you wish to pursue this topic we suggest you talk to the financial staff in your organization about how organizations absorb and release cash. You may also like to find out about cash flow evaluations.)

D. Technical Responsibilities

Companies employ food professionals to carry out specific roles in various technical areas, such as:

Research and development
Quality assurance
Technical services

In these roles, food professionals have the opportunity to put into practice what they have learned during academic training. Professional development, recognition, and experience gained are part and parcel of the rewards.

1. Research and Development

Research and development, commonly referred to as R&D, is a vital activity for any progressive company. Failure by companies to develop new products and processes will not only hamper growth, it may lead to the decline and possible demise of the organization. We believe that this is especially so in the case of food companies.

Food companies depend on their technically qualified professionals to carry out the functions of new product/process developments to move their companies forward. This is achieved by identifying products/processes that have the potential to increase the profitability of the company. The research and development function of a food company must thus be closely associated with both marketing and production functions. New product ideas are generated by companies through:

Study of market trends
Brainstorming with the company
Requests from customers
Competitive pressures

The ones that are selected are passed on to the R&D department for development.

Product Development Processes and Procedures

The product development process will depend on the type of company, the nature of the business, the company philosophy, and management style. Some companies have a formal procedure in place to ensure that product/process development is carried out systematically, while others may develop new products in an informal manner.

As a rule, a more formal approach gives better and more dependable results, especially in larger companies. The main steps comprising a product development procedure are shown in Figure 4. Developing new products can be quite expensive. Companies therefore must subject all new ideas to rigorous screening and prioritizing to ensure that only those ideas with a high probability of success are approved for development. Also, if a partially developed product is no longer likely to meet expectations it is better to abandon the idea. The costs incurred must be considered as "sunk costs", and it would be incorrect to spend any more resources. Another point to note is that products developed at considerable expense principally as technical projects, without a proper input of marketing requirements or with insufficient understanding of business objectives, are unlikely to see the light of day. It is possible to prevent such situations by managing the product-development process.

Product Development Considerations

Food professionals working on an R&D team need to have access to information on new product launches in their specific product categories, preferably from around the world, and to set up a prod-

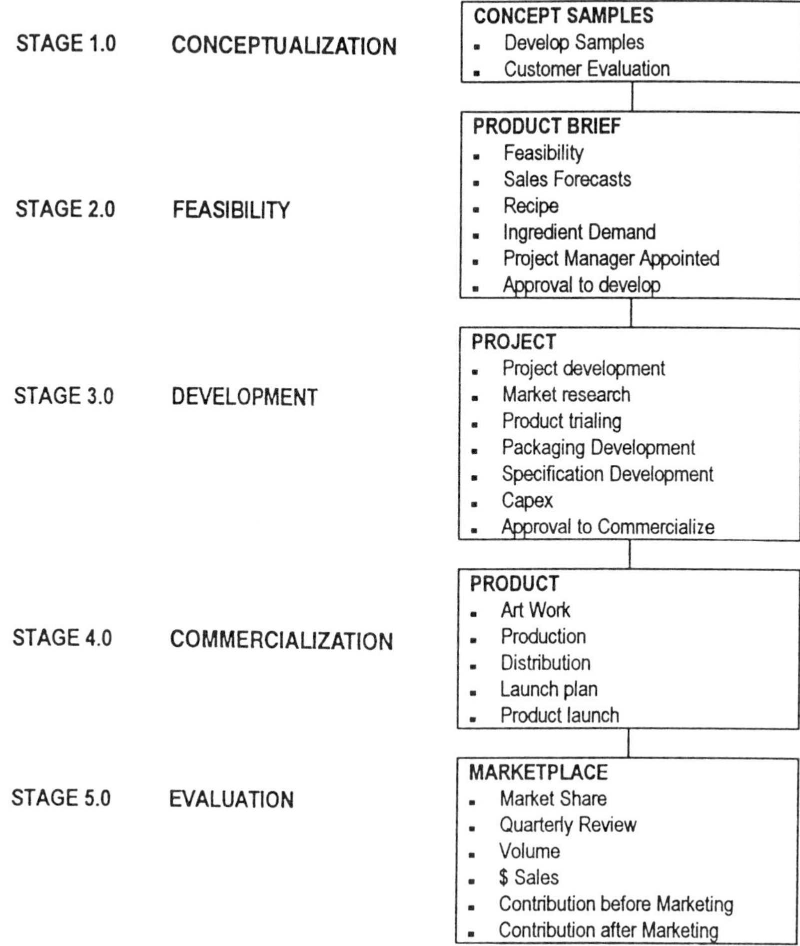

FIGURE 4 Main steps in product development procedure.

uct-awareness library. They also need to build their own networks with other professionals in the areas of ingredients, packaging, and analytical laboratories related to product quality food composition, and nutrition. Some large food companies have their own technical centers and in-house support services, while others have to depend on outside facilities to provide such services. In dealing with organizations and personnel outside their own company, food professionals involved especially in new product development need to be aware of the importance of maintaining confidentiality about commercially sensitive information.

Professionalism in R&D requires practitioners to be familiar with rules and regulations pertinent to their area of work and to abide by them at all times. There are no universal food regulations, and it is the responsibility of the food professionals to know the food regulations of their own country as well as the relevant information on regulations of countries importing their products. In this regard it is important to understand that different countries have different labeling requirements. While the artwork on packages is normally handled by the marketing department, the R&D personnel have to take an active role in providing and/or approving information that is to be shown on the labels of new products.

In addition to new product development, most food companies carry out continuous product improvement through ingredient substitution and/or process improvement. No matter which R&D activity food professionals are involved in, they need to be aware of the cost implications and try to avoid unnecessary expense. Developing product specifications is a function that R&D staff must carry out with essential input from the marketing and production departments. Such teamwork is necessary to ensure that the marketing expectations are matched by the production capabilities of the company. Once the specifications are developed and approved, they become the basis for quality assurance, which is another area in which food professionals are employed.

2. Quality Assurance

The primary role of the quality assurance (QA) function is to assure that the quality of the products processed and marketed by the company is of the standard required, in other words, that the products are made in accordance with specifications. It is the responsibility of the QA function to ensure that the raw materials received comply with the standards that have been mutually agreed to by the company and the supplier. If the incoming material is not of the agreed quality, the QA department should intervene and take the necessary action. In most cases this means rejection of the incoming material. Only in exceptional circumstances should reworking be considered. In any case, reworking is as a rule a costly exercise, which should be avoided if possible.

During processing samples are generally taken for routine checks of quality. Specific quality parameters of the product in question are recorded on QA forms, which are held in a file. This enables the product history to be traced back if required. Such records also provide valuable process performance information and may serve as evidence when responding to consumer complaints and/or in the event of a product recall. Staff involved in QA must take all precautions and all necessary steps to prevent any problems from occurring. Such problems may be related to the contamination of a product with undesirable substances, be they of physical, chemical, or microbiological origin. Programs such as Hazard Analysis Critical Control Point (HACCP; see chapter 24) are becoming increasingly important to food companies in monitoring and preventing potential problems. It should be noted that hazards are not confined to the food products. On the contrary, manufacturing hazards can and do impact significantly on the environment.

3. Other Technical Services

Research and development and quality assurance are not the only technical functions of food professionals. For example, those with interests in sales and marketing can become technical sales personnel or product managers. Their skills are valuable in providing technical background on the products, ingredients, and equipment/machinery that they sell or market. In addition, they are in a good position to relate to a company's R&D and QA staff, who generally are the customers for such goods and services.

Although cost is often a considerable factor in decision making, in many instances the decision of a client company to purchase from one supplier and not the other is determined by the type of technical service provided. Leading ingredient companies employ food professionals to work closely with their clients and often tailor-make ingredients, products, or equipment so suit their specific needs. Some food professionals from supplier companies work in close association with the product development staff of client companies in developing new products. In such cases the client company may negotiate exclusive rights for the ingredient and or the system for a given period of time. In situations where new ingredients are not yet permitted by the regulatory authorities of some countries, the supplier companies or their agents in the countries concerned may intervene and often succeed in obtaining approval. Food professionals play an important role in preparing submissions for approval, which often require substantial technical input by way of research and supporting evidence on the safety of the ingredient in question.

II. MANAGING QUALITY

A. The Quality Scene

Over the last decade or so there has been a considerable increase in the understanding and acceptance of the importance of quality. The technical professional needs not only to be in tune with this change in thinking, he or she must also take a pro-active stance because it often is the technical area in which one finds the origins of both quality successes and failures.

1. Importance of Quality

To assess the nature (or the customer's perception) of quality it is useful to employ Dr. Kano's model of quality, which identifies three different aspects of quality: basics, satisfiers, and delighters.

The Basics

The basics are the fundamental aspects and features of a product (or service) that are considered to form part of its inherent functionality. Basics are all those ordinary things that the customer expects to be up to the mark (e.g., size, flavor, functionality). Failure to meet the basic requirements usually makes the customer angry.

The Satisfiers

Satisfiers are aspects that make the customer feel happy about his or her purchase. They are important to the customer but are only effective if the basic requirements have been met.

The Delighters

Delighters are extra features or services that clearly exceed the customer's expectations. They can be counterproductive if the basics are not met. A faulty product coupled with a delighter will make the customer cynical.

Points to Note

The following are worth noting:

1. Most customer complaints are likely to be concerned with the basics.
2. Remember, failure to meet the basics tends to make the customer angry.
3. The quality mode is very relevant in all areas of management, be it marketing, product development, or production.

2. Origins, Sources, and Causes of Quality Problems

Quality problems usually manifest themselves when a customer complains or when quality assurance staff detect an out-of-specification product. To find the root causes of complaints we need to dig deep. It is a bit like peeling an onion—each layer needs to removed before the next one is revealed.

Origins

First we need to identify the three fundamental origins of quality failures. These are:

1. Faulty design (design mistakes, specification shortcomings)
2. Faulty execution (faults in manufacture or assembly, shortcomings in delivery or service)
3. Incorrect use (intentional or unintentional).

It should be noted that many, if not most, of the quality failures experienced by firms are the result of faulty design, even though they appear to have their origins in faulty execution. For example:

1. In many cases firms settle on design features and specifications without having a proper understanding of what the operational systems are capable of delivering in a consistent manner, e.g., unrealistic mix tolerances.
2. Quite often designs will carry within them the seeds of faulty execution: parts that can be assembled incorrectly, awkward design that will encourage assembly shortcuts, unwarranted production complexity, lack of standardization.

Sources

Once we have identified the fundamental origin of failure, we can look for the source by identifying each specific process, operation, or procedure involved with the failure. In most instances we will need a concise description of the process or operation concerned. When faulty execution is the origin of failure, two types of sources are possible: those within the organization and hence under our direct control (internal) and those outside of the organization, usually because of faulty materials or services (external).

Even where the source is external, there will always be some contributing internal factor. For example:

1. When a supplier is responsible for poor-quality components (external source), we also need to identify the appropriate internal contributing factors (e.g., inadequate or unrealistic purchasing specification, price-only purchasing policy, or undue emphasis on credit).
2. Management of the inward goods acceptance process is inadequate.
3. The line staff is unable to react quickly when faulty materials are encountered.

Causes

While having an understanding of the origins and sources of quality failures is a necessary first step, there is little we can do about prevention and improvement unless we also understand the cause of our problems. All quality problems arise from a failure to meet customer expectations. Most of the time a failure is the result of variability in one or more processes. While the question of process variability is discussed in more depth in Section II.B, it is worth noting here that understanding the variability of processes is an essential prerequisite to reducing quality problems. To reduce the variability in processes we need to appreciate what causes it. There are two different types of causes of variability: common causes and special causes.

Common-cause (statistical) variability includes the random variation in results or performance which is due to the system itself rather than any specific action. This variation is considered "normal" for the particular system. As common-cause variability is an inherent feature of the process, improvement in process performance is dependent on making changes to the system. Pep talks to staff, exhortations, or slogans are quite useless in this situation. Unless the actual system is changed for the better, variability will not decrease.

Special causes are those extraordinary and often one-time events that cause a temporary increase (or decrease) in variability. There is no need to change the system of operation to avoid variability due to a special cause.

3. Quality Culture and Processes

It is important to note that in the end improved quality can only come about if there exits in the organization an all-pervading quality culture. This means not quality one day and quantity the next. It means that the entire organization should be quality aligned with regard to management practices and systems. In addition, there needs to be an organization-wide agreement to regard quality issues principally as opportunities for improvement rather than as problems to blame on someone. In practice it means that all quality issues need to be viewed in terms of origin, source, and cause (as outlined earlier) and that there are in place sound, standard, and accepted process-improvement techniques in the company. It is important to appreciate that it is only when the underlying processes are improved in a permanent fashion that one can expect a sustainable increase in quality performance.

B. Understanding and Reducing Variability

It is very important for technical staff to thoroughly appreciate the effect process variability can have on business performance. There are two principal impact areas:

1. At the time of product and process development the R&D specialist has a major role in ensuring that potential process variability is understood and minimized.

2. During production the quality assurance professional can do much in introducing the right monitoring systems and training so that operational staff can learn about process performance and subsequently work to reduce its variability.

1. Costs Associated with Variability

To introduce how variability can reduce quality and performance and drive up costs, we will use as an example the costs involved in maintaining a painted house. The lasting quality of paint on a wooden house provides a classic demonstration of the effects of variability in performance and the costs associated with it.

Variability in Paint Performance

Have you ever had the unenviable task of preparing a clapboard house for repainting? You must have wondered as you toiled away why some of the paint had flaked away while other bits stubbornly remained in spite of vigorous wire brushing and scraping. You may also have seen houses that obviously haven't been repainted for several decades yet still have some paint showing. While some of the remaining paint would be easy to peel off, other remnants would still be difficult to remove.

Costs of Paint System Performance

Consider the case of a house with painting costs over the first 35 years of its life as shown in Table 6. If there was no variability in the performance of the paint, all the paint would last exactly the same length of time. Remember that some small parts of the paint system remain intact for several decades, so if all the paint performed as well as the best parts a 35-year life would be possible. If there was absolutely no variability, we could predict the life of the paint system with total confidence. We coud plan the job of painting (Job A) with precision, as on a given day (known in advance) all of the paint would fall off the house. In addition, there would be no need for scraping and sanding. The table shows that the difference in per m^2 cost between the two scenarios is \$32 (\$48–\$16), which is the cost due to the variability in performance of the paint system. Even if we take a more modest paint life of 21 years, the cost would still be considerably less. What the example shows very clearly is that there are costs associated with variability and that high variability means high costs.

TABLE 6 Costs Associated with Variability in Paint Performance

	\$ per m^2 (labor and material)
A. When new: Three-coat finish	8
B. After 7 years: Making good spot failures, spot prime, and two coats	7
C. After 14 years: Making good spot failures, spot prime, and two coats	12
D. After 21 years: Complete repaint including sand back, refill prime, coat	12

Paint life costs:

Year	Job	\$	Accumulated
0	A	8	8
7	B	7	15
14	C	7	22
21	D	12	34
28	B	7	41
35	C	7	48

Other Costs

While the dollar costs associated with high variabilities are very important, they are in the majority of cases reasonably self-evident and quantifiable. But there can be other important costs.

1. The costs of a poor reputation: Without any doubt the most important of all costs is the detrimental effect variability has on one's reputation in the marketplace. If there is one thing that annoys customers more than anything else, it is a lack of consistency. Nothing will damage a reputation more quickly and more permanently than a high level of variability in the quality of products and services.
2. Staff-related costs: These costs are associated with pride of workmanship and staff morale. High variability in our processes is likely to cause many internal problems, such as scheduling difficulties, quality-of-fit problems, and reworking—all of which are bound to lead to reduced job satisfaction and, as a consequence, to poorer performance.

2. Managing and Reducing Variability

In many organizations process variability is given insufficient weight, resulting in quality problems and subsequent loss of customer confidence. It should also be appreciated that variability reduction is not something that happens by itself. On the contrary, it requires focused management attention, sound systems, and properly trained staff.

Reasons for Variability in Performance

To understand why we often experience quite large variability in performance, we need to identify the various sources. In the case of the painted house, they can be divided into four categories:

1. Timber surface: The original timber surface is likely to be a source of considerable variability in the performance of the paint system. Variation in surface finish, moisture content, and resin will all contribute to that variability.
2. Materials: If you buy paint supplies from a reputable source you can be fairly certain that the variability will be low. However, in spite of a manufacturer's best efforts to keep paints uniform, there will always be some variability. This is likely to be greater if a particular paint comes from more than one batch. The upshot is that some of the paint will perform better than average and some worse.
3. Application: The way the paint is applied provides an enormous potential for variability in performance. Using a brush, for example, results in large differences in coat thickness. Even employing different painters for different parts of the job is likely to lead to an increase in variability. In addition, one must take into account the varying painting conditions, such as air temperature and humidity, which will increase variability in performance.
4. Position: This is a critical factor in how long a paint system is going to last. Areas of paint exposed to wind, rain, and full sun are likely to fail earlier than those protected by porches or eaves.

Understanding Variability

When advocating the reduction of variability, we should remember that most if not all regular company reporting is about reporting averages. For example, most standard costing systems are designed to report average monthly performance against some predetermined (average) standard. Averages by their very natures will mask variability, so while we may aim to meet a particular better average performance, we could be unwittingly steering our ship in the wrong direction.

Take, for instance, a case involving three teams with the average daily outputs over a month as shown in Tables 7 and 8: on the face of it, Team B is the best, and if we did not look further we

TABLE 7 Case Study of Three Teams: Average Output

	Team A	Team B	Team C
Average units/day	1000	1100	1000

could conclude that Teams A and C should pull their socks up and work like Team B. If we were to dig a little deeper, we would learn that the highest and lowest daily figures were as shown in Table 8. If we also learned that production in excess of 1200 units per day was considered to be too hard on the equipment and that such high rates of throughput tended to lead to quality problems, which is now the better performing team? In terms of output variability, it is Team A, because it turns in the most consistent performance. Perhaps Teams B and C could learn something from Team A about how to avoid very low throughput days, while Team A could be encouraged to talk to Team C about what they do to reach 1200. As for Team B, the first aim needs to be "no production beyond 1200 units per day" with the subsequent focus on reducing throughput variability further by improving low-day output.

Best, Worst, or Average Performance

In the case of the painted house, it is quite obvious that the worst performing parts of the paint system determine the painting costs over the lifetime of the house. The best performing parts are those difficult-to-remove bits that we strike when trying to get the house back to bare wood. While the average performance of the total paint system could be 20 years, it this useful information? After all, would a paint system with an average life of say 25 years necessarily be a better proposition? What about if in the latter case the performance range was 5–45 years (rather than 7–33 years)? Averages can be highly dangerous statistics, and while politicians and other public figures can be masters at using and misusing averages, in business we see much misunderstanding as well.

Accumulation of Effects

Because customers experience the final result of the various processes, one needs to be aware of the potential accumulation of the variability effects of all processes. Considering the case of the painted house again, the accumulation of effects can be looked upon as a lottery. Some areas are unlucky, in that they have a combination of all of the worst aspects, while others have a combination of the best with the best. Between the two extremes, we have a multitude of combinations, which provide either better- or worse-than-average performance. As a consequence, we see some early failures occurring in quite unexpected places (e.g., away from the weather), while other areas of paint continue to perform well in spite of being much more exposed.

If we apply this concept of accumulation of effects to several business processes, we can see why some customers can be "unlucky" when it comes to customer service. Take, for instance, the following situation. We supply goods to customers and have the following performance statistics: stock availability 90%, order accuracy 90%, invoicing accuracy 90%. Table 9 shows the average effect on our customers (assuming the three statistics are independent of each other). We can observe that we have on average one "unlucky" customer per 1000 orders who experiences the "worst with worst

TABLE 8 Case Study of Three Teams: Highest and Lowest Output

	Team A	Team B	Team C
Highest daily	1100	1500	1200
Lowest daily	900	700	800

TABLE 9 Customer Performance Statistics

Problems	Percentage
None	72.9
One	24.3
Two	2.7
Three	0.1
Total	100.0

with worst" combination. They are the ones who are likely to tell their friends "Don't deal with that firm, they cannot do anything right." Using this example, it is clear why some people end up with a "lemon" of a car, while other buyers of the same model are very satisfied. In the business of food preservation, the problems are very much the same.

3. Improvement Opportunities

The important point to note is that process variability will, generally speaking, not reduce by itself. On the contrary, it requires special effort and attention to detail. There are many techniques available, but any detailed discussion is beyond the scope of this chapter. We will therefore confine ourselves to identifying opportunities to reduce variability and to the principles underlying improved process performance.

Minimizing Potential Variability

It is by seeking to minimize variability in the first place that the technical professional can make a significant contribution to the ultimate performance of the product. During the development phase there are many opportunities for this, but unfortunately the opportunities are seldom taken advantage of to their full extent. Some examples include:

1. Understanding the capability of the processes and developing the product and specifications accordingly. There is little point in specifying a process to ±1°C when the actual variability is ±2°C.
2. Looking for opportunities to standardize materials and processes wherever appropriate. Undue and unnecessary diversity does invariably lead to higher variability of output.
3. Making the definition of the process control methods and information required part of the product or process design and specification.

Minimizing Operational Variability

Assuming we have taken all important steps to minimise potential variability, we must now put in place the systems and procedures necessary to understand and minimize operational variability. The main point is that processes must be monitored and properly understood before changes and adjustments are made, as shown in Figure 5.

Benefits of Reducing Variability

In spite of what intuition may sometimes tell you, reducing variability through systematic improvement of the process will invariably lead to reductions in cost, increased customer satisfaction, and other benefits.

1. Lower costs: The ability to set a lower target weight when packing a valuable product and being sure that that customer is not shortchanged.
2. Happier customers: Lack of consistency in products and services will harm a company's reputation. Customers will be happy when their expectations are under-

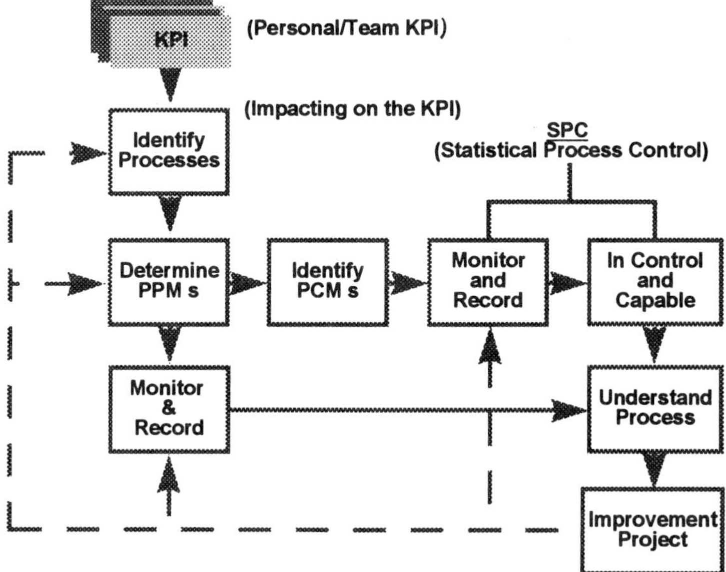

FIGURE 5 Systematically reducing variability. KPI = key performance indicator; PPM = process performance measure; and PCM = process control measure.

stood and satisfied. If you aim to exceed their expectations, be confident that you can exceed them in a consistent manner. Make sure the statement "exceeding customer expectations" is not just a slogan. Raising customers' expectations without the possibility of consistent delivery is bound to have an adverse affect on your business.

3. Inventory savings: Reduced product variability also makes possible savings in inventory. Improved consistency is a prerequisite for just-in-time (JIT) management. It should be remembered that "out of spec" products as a rule become a liability to the company, requiring "write downs" or additional expense or both.
4. Increased capacity: Processes that are under proper control as a rule operate better with resulting increases in output.
5. Happier staff: reduced process variability means fewer production problems, less panic, and fewer complaints, all of which help staff to feel better about their jobs and themselves.

C. Managing Customer Complaints

It is no secret that many companies do not like receiving customer complaints. They are usually seen as embarrassing, troublesome, and possibly frivolous. But we need to manage them well if we are to maintain good customer relations and turn the complaints into opportunities for improvement. To do this we need to (a) ensure we have the correct culture, (b) design the right system and operate it properly, and (c) carry out proper analysis, report, and follow-up.

1. Assessing the Situation

The Correct Culture

Before one designs and implements any system it cannot be overemphasized how important it is to have the correct culture in place. This means treating a customer who complains as a person doing us a considerable favor. The customer is telling us something about our product or service that we probably did not know but ought to. We should accept that many other customers may have similar feelings about our product or service except but do not bother to complain.

Points to remember include:

Are customer complaints principally seen as a nuisance?
Who is responsible for managing complaints, and who actually handles them?
Are complaints dealt with by some junior and possibly untrained person?
Is it possible for customers to get "shoved around"?
Are we sure that complaints are handled professionally every time?
Are complaints seen as important?
A good check on the perceived importance of customer complaints is to look at how they are reported. Are they included in the monthly report? If yes, are they hidden somewhere at the back, or do they form part of the important summarized results?

Do We Have the Right System?

Once an organization has accepted that customer complaints are extremely important and introduced the necessary managerial and cultural changes, it should ensure that it also implements a sound customer complaint–management system and related procedures. The most important requirement is confidence that the system and procedures will ensure the appropriate response in every situation. In addition, the procedures should ensure a good standard of recording to provide the organization with a proper source of information for any required follow-up and subsequent improvement of the processes. If they are not up to scratch, redesign them.

Points to consider include:

Is there in place an up-to-date written procedure for dealing with complaints?
Does the procedure ensure an appropriate response in every case?
Can we be confident that all (or at least most) of our customers are happy with the way the organization deals with complaints?
Does the system ensure proper recording, monitoring, and reporting?
Do the complaints lead to improvements?

2. Developing a New Approach

If an organization considers customer complaints to be important, it must be prepared to develop, implement, and maintain a sound management system.

Defining the Rules

A crucial part of any new customer complaint–management system is defining the rules that are to govern its operation. It is important to identify all likely incidents and link each one of them to an appropriate response. A useful tool for doing this is a decision table, which will enable the organization to define precise rules for action in each case. To construct a decision table it is necessary to identify all situations (conditions). These would include product or service faults (e.g., wrong weight, minor contaminant, serious contaminant, life-threatening fault, rude salesperson, poor after-sales service) and disposition of the customer (e.g., happy, neutral, angry, threatening action). It is also necessary to agree what is the appropriate response (e.g., days to respond, call on the customer, etc.).

Complaint Classification

Once all conditions have been identified, a complaint classification table can be constructed. The classifications are to identify a predetermined response for every situation. Table 10 shows five different responses denoted by the letters A, B, C, D, and E. If all of the conditions, actions, and appropriate response rules have been defined, there will be a basis for a sound and reliable procedure.

The Appropriate Responses and Operation

Once there is agreement on the complaint classifications, the appropriate responses need to be linked to them. Table 11 shows the general idea: by determining the complaint classification the person

TABLE 10 Complaint Classifications

Complaint	Customer			
Wrong weight	A	A	B	C
Minor contaminant	A	A	C	D
Serious contaminant	B	C	D	E
Life threatening	E	E	E	E
Rude salesperson	A	B	C	D
Poor service	A	B	C	D

dealing with the complaint can select the correct response every time. If required, another column ("else") can be added to meet any situation that does not fit A, B, C, D, or E.

The Responsibility Matrix

To help define who is responsible for what, one can employ a responsibility matrix. The idea is to relate people and required actions in a single table (see Table 12).

3. Recording, Monitoring, and Reporting

To turn quality problems into improvement opportunities, it is essential to record and monitor all relevant information. Depending on the size of the organization and the number of complaints, anything from a simple manual system to a sophisticated computer system may be required. For most organizations a modest PC-based system will prove the best solution. However, irrespective of whether the system is computer-based or not, the general principles are essentially the same: the system must provide for ease of monitoring, analysis, and subsequent reporting.

Origins, Sources, and Causes

In an earlier section we looked at the origin, sources, and causes of quality problems. It is important that the customer-complaint system and procedures and the staff operating it are capable of differentiating between origins, sources, and causes. If there is no such distinction, it is unlikely that the management of customer complaints will lead to process improvement and a subsequent reduction in quality failures. Of particular concern is that many organizations to not appear to appreciate the difference between common-cause and special-cause quality problems.

In organizations where the difference between common and special causes is not understood, one often finds major changes to systems and procedures because of some out-of-the-ordinary fail-

TABLE 11 The Response Rules

	Complaint Classification				
Action	A	B	C	D	E
Days to respond	3	1	1	I	I
Standard letter	Y	N	N	N	N
Special letter	N	Y	O	N	N
Vouchers	2	3	N	N	N
Call on customer	N	O	Y	Y	Y
Retrieve product	N	O	Y	Y	Y
Recall procedures	N	N	N	O	Y
Advise Q/A Manager	N	O	Y	Y	Y
Advise GM	N	N	O	Y	Y

I = Immediately; Y = yes; N = no; O = optional.

Table 12 Responsibilities for Customer Complaints

Functions or actions	Staff responsible: Customer services clerk	Customer manager	Area sales representative	Dispatch clerk	Etc.
Complaint receipt	R				
Fill out form	R	A			
Determine classification	A	R			
Initiate response	A	R			
Call on customer			R		
Send replacement				R	
Etc.					

R = Response; A = assist.

ure. Also, failures due to common-cause variability may be explained to customers as one-time problems that are unlikely to recur.

Recording Details

For a customer complaint system to be effective, the proper recording of all necessary details is essential. Figure 6 shows an example of monthly recording of the details of origins, sources, and causes. Similarly, other information such as complaint classification and the nature of the complaint (basic, satisfier, or delighter) should be recorded. An action and improvement log should also be maintained.

Analysis and Reporting

If a good recording and monitoring system is in place, subsequent analysis and reporting can be tailored to fulfill particular needs. A company may settle for a simple form of analysis and reporting, in which case the task is fairly straightforward. On the other hand, it may decide to be quite sophisticated and opt to employ advanced analytical techniques and reporting methods. Whatever approach is adopted, it is crucial that the analysis and reporting have some real meaning leading to improvements in performance. The following points should be kept in mind:

No.	Origin				Source	Cause	
	Design	Execution		Use	Description of Source	Common	Special
		Int	Ext				
001	✓				Specification limits beyond machine quality.	✓	
002		✓			New operator.		✓
003	✓				Scraps of packing material can occasionally enter the product.	✓	
004			✓		Outside contractor working on machine failed to clean grease off weighing bucket.		✓
031							
032							
Total	10	19	2	1		22	10

Figure 6 Origins, sources, and causes of complaints.

1. More sophisticated reports: The information collected should allow one to identify possible correlations, for example, between (a) complaint classifications and origins; (b) particular types of complaints such as contamination and origins; (c) origins and causes; (d) reductions in complaints and improvement projects. (A word of caution: When embarking on a program of more sophisticated analysis, be sure to employ a suitable numerate employee. Basic conceptual errors can make any analysis at best useless, at worst dangerous.)
2. Action/Improvement reporting: The main source of information for this will be the action and improvement log, although information may also come from other sources such as customers and operations. Control of dates is important (e.g., date items first raised, original agreed completion date, latest agreed completion date).
3. Place in company reporting: Once the analysis to be done and reports to be produced have been determined, it is necessary to decide on the content of the regular monthly report. It is also necessary to decide who will be responsible for the reporting and where the report should be included.

4. Improving the Process

Ensuring the correct customer response in every case is a crucial feature of any complaint system. This part of the system is "outward focused," which is very important to maintaining good customer relations. However, to improve performance and hence the reputation as a quality company, it is necessary to have an inward focus as well.

Customer complaints can provide valuable information about underlying weaknesses. However the important point to note here is that what shows up as customer complaints are effects or results and that to prevent or lessen complaints we must identify and improve the underlying processes. It may be said that this statement is rather obvious, however it is a fact that many organisations do not bother to search out the processes involved with the object of affecting permanent improvements. It should be remembered that most management reporting (not just customer complaints) is about results with little or no appreciation of what are the relevant processes.

Managing Improvements

A system of recording and monitoring will identify many areas where improvements are both possible and necessary. Improvements should be managed in a structured and somewhat formal manner. Some points to remember include:

The customer complaint system, if managed correctly, will provide factual information as a basis for improvement rather than emotive assertions such as: We always have the wrong weights; We never deliver on time; or Our product is full of contaminants.

The emphasis needs to be on prevention and improvement rather than culprit-finding.

It must be quite clear whether problems are common cause (systems) or special cause types.

Generally speaking, improvements in results or a reduction in complaints, comes about by focusing on and improving the underlying processes.

It is necessary to exercise patience. Improvement of the process will take time, but if done properly the improvement will be permanent and hence very worthwhile.

Identifying Processes

If it is working correctly the system will identify the process that needs to be improved, e.g., staff induction/training, design procedures, specification development, production tolerances, or process monitoring. Of particular interest are complaints that have their origin in product design and development. Design is the start of the sequence of operation, and hence any failings in design will affect all subsequent processes.

As mentioned earlier, a large number of a firm's quality failures have their origin in faulty design, although they may, at first sight, appear to have their origin in faulty execution. If products are designed and developed with problem prevention in mind, many subsequent quality failures can be avoided.

Important areas to consider include:

Standardization of ingredients where appropriate
Simplifying processes where possible
Designing intermediate tests into the process, including statistical process control

Once a reasonable quantity of data has been collected, it should be possible to identify the specific processes giving rise to particular complaints, such as:

Ingredient quantity control
Mixing
Cooking

Note, however, that to find the real source of the problem, considerably more investigation is usually needed. For example, complaints may be traced to the cooking process, but the real source may be insufficient operator training, out-of-date manuals, or a lack of maintenance of the temperature-control equipment.

5. Management Commitment

As should by now be clear, the proper management of customer complaints requires much thought and effort as well as attention to detail. In addition, considerable management support and involvement is required, especially if improvement is to be made. It really comes down to the following questions:

Is the organization serious about the customer's expectations and the company's reputation?
Is it prepared to make the effort to turn problems into opportunities?

Even if not all problems can be solved, we can still benefit from some of the concepts and techniques discussed, provided the organization knows where it wants to go and ensures a consistency of signals.

6. Consistency Is All-Important

Consistency of signals to both customers and staff is vital if one is to succeed in reducing complaints. For example, do not allow quality policies and procedures to be overridden arbitrarily by edicts from above. Be sure to provide ongoing support for the complaint system's operation and resulting improvement projects.

In order to ensure that output from the system is given due recognition, it is important to (a) check the company's current culture and practices, (b) decide what the current shortcomings are and what changes are warranted, (c) obtain agreement from affected managers and staff to develop a plan of action, and (d) implement and enjoy the improvements in performance.

The process of improving an organization's performance and enhancing its reputation as a quality company is all about doing many things better. Focusing on customer complaints is a good starting point.

REFERENCES

The following articles, together with a number of others, have been republished in book form titled "Systematic Business Success" by Profile Books Ltd., Auckland, New Zealand, February 1998.

1. G. La Rooy, How to handle customer complaints, NZ Business Magazine, April (1993).
2. G. La Rooy, Variability in business—hidden impediments or latent opportunities, NZ Business Magazine, February (1995).
3. G. La Rooy, Opportunities in waiting, NZ Business Magazine, March (1995).
4. G. La Rooy, Coalface information, NZ Business Magazine, October (1996).
5. G. La Rooy, Breakeven revisited, NZ Business Magazine, August (1997).

Index